Europe

Areas of China included in Chapter 10 (Central Asia) and Chapter 11 (East Asia)

*The annexation of Crimea by Russia in March of 2014 was declared invalid in a resolution passed by the United Nations General Assembly.

The Sights, Sounds, and Tastes of World Regions

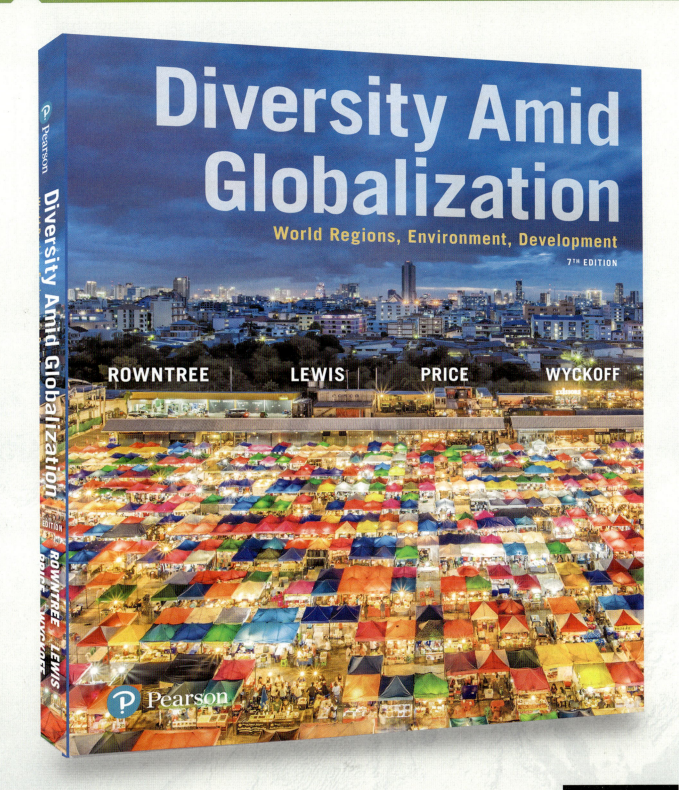

Diversity Amid Globalization

World Regions, Environment, Development

7TH EDITION

ROWNTREE | LEWIS | PRICE | WYCKOFF

Providing Pathways for Students to Explore...

Everyday Globalization

Our Plastic Bag World

Looking for something to celebrate? How about International Plastic Bag–Free Day— every July 3, people all over the world shop without using plastic grocery bags, pick up trash from beaches and roadsides, even commit to banning plastic bags in their communities.

One ban-the-bag organization estimates that one million plastic bags are used each minute. Besides being a pollution problem, manufacturing new bags takes millions of barrels of oil. In the United States alone, 12 million barrels of oil are used each year to make plastic bags.

True, some plastic bags can be recycled, but most are not, creating a global problem. Millions of plastic bags litter the countryside in China, where it's called "white pollution." Kenyans and South Africans refer to discarded white bags wafting about on the winds as their national flower (Figure 2.4.1). Even if these wayward bags were collected and buried in a landfill, they'd take several hundred years to decompose, releasing climate-changing methane gas.

Banning bags, or making them more expensive, seems to reduce usage. Some 20 U.S. states and over 200 cities have plastic bag ordinances. China claims to have an outright ban on plastic bags, but enforcement seems lax. Paris may be the biggest city with a full ban, and flood-prone Bangladesh may have the most compelling reason for a ban: engineers say that loose plastic bags clog drains, worsening flooding.

1. Does your community restrict or discourage the use of plastic shopping bags? If not, do you know of one? If so, what do people use instead?

2. Does your community recycle plastic bags? Where do the bags eventually end up?

▲ Figure 2.4.1 Waste Plastic Bags Litter the World's Landscapes

Everyday Globalization

Soccer Balls from Sialkot

Sports equipment manufacturing is a highly globalized business. All of the balls used in Major League Baseball, for example, are made by the Rawlings company in Costa Rica. Soccer balls, on the other hand, are closely associated with the city of Sialkot in northern Pakistan. Sialkot produces roughly 40 million soccer balls a year, or about 40 percent of the world's total supply (Figure 12.4.1). Its firms specialize in hand-stitched balls, used in World Cup competition and generally regarded as the best. Around 8 percent of Sialkot's soccer balls are exported to the United States.

Sialkot's soccer-ball industry faces a number of challenges. The city produces an array of other sporting goods, but its businesses have been losing customers to companies based in lower-cost countries, particularly China. As a result, Pakistani officials are keen to find new techniques that would allow them to maintain a competitive edge. In 2014, a team of researchers from Yale and Columbia Universities and Pakistan's Lahore School of Economics developed new procedures that would significantly reduce waste in the raw materials used to make high-quality soccer balls. They were disappointed, however, to learn that the factory workers resisted these new techniques, mostly because they did nothing to increase worker pay. In 2015, Pakistan's government announced its support for a new facility in Sialkot devoted to producing high-quality soccer balls using mechanized methods, hoping to eliminate the need for hand stitching.

1. How might Sialkot's sports-equipment firms respond to the challenge posed by Chinese exporters? What risks might they face in doing so?

2. How many of your own sporting goods are imported? What countries have supplied them?

▲ Figure 12.4.1 Soccer Balls from Sialkot The Pakistani city of Sialkot is noted for its export-oriented sporting-goods industry. Here see workers sewing panels for an Adidas AG "Brazuca Replica Glider" soccer ball.

Geographers at Work

Exploring High-Latitude Siberia

Kelsey Nyland's fascination for polar regions—and geography—blossomed when she was an undergraduate at George Washington University. Initially a geology major, Nyland "didn't know what geography was," but an arctic environments course and subsequent fieldwork in Alaska hooked her. She found that in geography, you "get to see places, not just learn about them in classrooms abstractly," and eventually became an undergraduate research assistant for Dr. Nikolay Shiklomanov, who leads a program that monitors permafrost change throughout the Arctic and Antarctic. As permafrost warms and thaws, it can undermine roads, damage infrastructure, and release additional greenhouse gasses. "What's interesting is how [the thawing] permafrost actually affects infrastructure, how it affects people," Nyland notes. Her interest in both physical and human environments made geography an obvious fit—"everything I was doing was lining up with geography"—and she found that this is true for many students.

Siberian Research Nyland continued in the Master's program in Geography at George Washington and her interests returned her to Siberia (Figure 9.3.1). By 2015, Nyland had traveled to Siberia for five different field courses and learned all about using Landsat images to remotely monitor changing permafrost conditions. She also developed an additional research project (using historical photographs) examining how depopulating Siberian cities look decades after people had left.

▲ Figure 9.3.1. Geographer Kelsey Nyland observes material transport by Romantic Glacier in the Polar Ural Mountains, Russia during an expedition after the Tenth International Conference on Permafrost (July 2012).

Russian economic and urban policies impact the natural environments surrounding Siberian cities. Nyland's diverse interests in physical and human geography demonstrate how good research benefits from a thorough background in both parts of the discipline.

...the Sights, Sounds, and Tastes of World Regions

◀ **Figure 12.22 Hindu Temple** Hinduism's religious architecture often entails lavish sculpture and bright colors, as can be seen on this temple.

Explore the **Sights** of Sacred Pushkar Lake

http://goo.gl/EORY1w

Exploring **Global Connections**

India's Emerging Computer Game Industry

India is well known for its information technology. Most Indian IT firms got their start by subcontracting for large western firms, often by providing specified forms software as well as a large array of back-office services. Companies such as Infosys, with revenue of US$8.7 billion in 2015, have been moving up the technology ladder, emerging as global leaders in certain niches of software engineering. But until recently, India lacked the "start-up" business culture that has been vital in developing cutting-edge technologies in Silicon Valley and other tech hubs. This is beginning to change, however, as a new generation of Indian entrepreneurs launch their own firms. Some of these companies are following in the steps of established firms, such as Amazon and Uber. Many of the smaller Indian start-ups, however, are focusing on developing video games for both Indian and global customers (Figure 12.5.1).

Developing an Indian Gaming Industry Video game-making in India got its start in the same way that many other tech industries did: by taking on certain highly specialized tasks, such as modeling the movements of racing cars, for American or European firms. Now companies like Mumbai-based Yellow Monkey Studios make their own games designed for a global audience, such as "Socioball," described by the company as a "stylish new isometric puzzle game." The rise of the Indian gaming industry was evident in the 2015 and 2016 meetings of Pocket Gamer Connect, one of the world's leading mobile gaming events, in Bengaluru. Firms as large as Intel, Amazon, and Google sent representatives to the event, eager to establish a presence in this fast-growing sector.

A number of Indian gaming firms are connected with the film and television industries. Mumbai-based Reliance Games, for example, has racked up more than 70 million downloads of smartphone games linked to such movies as *Real Steel*, *Catching Fire*, and *Pacific Rim*. India's own massive film industry plays an increasingly important role as well. Ubisoft Pune, for example, has had marked success in smartphone music games that are endorsed by Bollywood stars. Game developers hope that such Indian-themed content will win over a global audience.

Obstacles and Opportunities India's gaming industry has been aided by the spectacular rise of smartphones in the country, now roughly as numerous in India as they are in the United States. But it has also been held back by India's poorly developed mobile communications infrastructure, which basically operates at 2G speeds. An equally serious problem is the lack of credit cards, held by only about 8 percent of Indians, which hinders purchase through online app stores. But, not surprisingly, Indian entrepreneurs are working on these problems through such measures as third-party wallet companies.

▲ **Figure 12.5.1 Indian Video Game Development** The Indian information technology industry has recently begun to take on video game development. Pictured here is Jaspreet Binda, head of the Entertainment Division of Microsoft India, with a game console at his office in Delhi, India.

However it is examined, the fast-growing Indian gaming industry is a highly globalized phenomenon. A prime example is UTV Software Communications, which aims to be the first Indian company to become a fully global player in the video game market—and which was acquired by Disney Enterprises in 2011. But globalization cuts both ways. A small Indian studio called On The Couch Entertainment recently created a game called Rooftop Mischief, but did not have the expertise to generate the accompanying music. For this feature the company turned to outsourcing, getting the necessary music from a firm in the United Kingdom.

Google Earth Virtual Tour Video

http://goo.gl/Pqu0xp

1. What cultural advantages might help India in its quest to develop video games for the global market?

2. What other obstacles might India face in trying to develop a start-up business culture?

Structured Learning Path...

Physical Geography and Environmental Issues
Stretching from Texas to the Yukon, the North American region is home to an enormously varied natural setting and to an environment that has been extensively modified by human settlement and economic development.

Population and Settlement
Settlement patterns in North American cities reflect the diverse needs of an affluent, highly mobile population. The region's sprawling suburbs are designed around automobile travel and mass consumption, whereas many traditional city centers struggle to redefine their role within the decentralized metropolis.

Cultural Coherence and Diversity
Cultural pluralism remains strong in North America. Currently, more than 49 million immigrants live in the region, more than double the total in 1990. The tremendous growth in the numbers of Hispanic and Asian immigrants since 1970 has fundamentally reshaped the region's cultural geography.

Geopolitical Framework
Cultural pluralism continues to shape political geographies in the region. Immigration policy remains hotly contested in the United States, and Canadians confront persistent regional and native peoples' rights issues.

Economic and Social Development
North America's economy recovered in many settings after the harsh economic downturn between 2007 and 2010. Still, persisting poverty and many social issues related to gender equity, aging, and health care challenge the region today.

The Critical Themes of Geography present a consistent thematic structure covering five themes in each regional chapter, making navigation and cross-regional comparisons easy for students and instructors. Themes include *Physical Geography and the Environment, Population and Settlement, Cultural Coherence and Diversity, Geopolitical Framework,* and *Economic and Social Development.*

Regional-specific Learning Objectives at the beginning of each chapter connect to end-of-chapter material, review questions, and MasteringGeography activities, forming a cohesive student experience.

Review Questions at the end of each thematic section help students check their comprehension as they read, emphasizing the most important points for clarity.

Port Metro Vancouver is now North America's third largest port (and Canada's largest). The sprawling dock facilities, cargo and cruise terminals, and shipyards symbolize the close connections between this dynamic North American city and the global economy. Vancouver (2.5 million people) is also one of North America's most culturally diverse cities and its healthy economy has attracted a sizable foreign-born population—45 percent of its residents. The city's Chinese population is among North America's largest, along with sizable immigrant communities from South Asia and the Philippines. Future trade with Asia promises further growth and even stronger connections across the Pacific. Signed in 2016, the **Trans-Pacific Partnership (TPP)**, a comprehensive agreement among 12 countries including Canada and the United States, is designed to stimulate Pacific trade and lower tariff barriers between member states. If ratified by all participants, the TPP guarantees that Vancouver's bustling port and its regional economy will continue to grow in the 21st century.

Similar processes of globalization have fundamentally refashioned many portions of North America. Large foreign-born populations are found in many North American settings. Tourism brings in millions of additional foreign visitors and billions of dollars, which are spent everywhere from Las Vegas to Disney World. North Americans engage globally in more subtle ways: eating ethnic foods, enjoying the sounds of salsa and Senegalese music, and surfing the Internet from one continent to the next. Globalization is also a two-way street, and North American capital, popular culture, and power are ubiquitous. By any measure of multinational corporate investment and global trade, the region plays a role that far outweighs its population of 360 million residents.

> North America, one of the world's wealthiest regions with two highly urbanized, mobile populations, helps drive the processes of globalization.

Defining North America

North America is a culturally diverse and resource-rich region that has seen tremendous, sometimes destructive, modification of its landscape and extraordinary economic development over the past two centuries (Figure 3.1). As a result, North America is one of the world's wealthiest regions, with two highly urbanized, mobile populations that help drive the processes of globalization and have the highest rates of resource consumption on Earth. Indeed, the region exemplifies a **postindustrial economy** shaped by modern technology, innovative information services, and a popular culture that dominates both North America and the world beyond (Figure 3.2).

Politically, North America is home to the United States, the last remaining global superpower. In addition, North America's largest metropolitan area, New York City (20 million people), is home to the United Nations and other global political and financial institutions. North of the United States, Canada is the region's other political unit. Although slightly larger in area than the United States (3.83 million square miles [9.97 million square kilometers] versus 3.68 million square miles [9.36 million square kilometers]), Canada's population is only about 11 percent that of the United States.

The United States and Canada are commonly referred to as "North America," but that regional terminology can be confusing. As a physical feature, the North American continent commonly includes Mexico, Central America, and often the Caribbean. Culturally, however, the U.S.–Mexico border seems a better dividing line, although the growing Hispanic presence in the southwestern United States, as well as ever-closer economic links across the border, makes even that regional division problematic. In addition, while Hawaii is a part of the United States (and included in this chapter), it is also considered a part of Oceania (and discussed in Chapter 14). Finally, Greenland (population 56,000), which often appears on the North American map, is actually an autonomous territory within the Kingdom of Denmark and is mainly known for its valuable, but diminishing, ice cap.

LEARNING Objectives

3.1 Describe North America's major landform and climate regions.

3.2 Identify key environmental issues facing North Americans and connect these to the region's resource base and economic development.

3.3 Analyze map data to identify and describe major migration flows in North American history.

3.4 Explain the processes that shape contemporary urban and rural settlement patterns.

3.5 List the five phases of immigration shaping North America and describe the recent importance of Hispanic and Asian immigration.

3.6 Provide examples of major cultural homelands (rural) and ethnic neighborhoods (urban) within North America.

3.7 Describe how the United States and Canada developed distinctive federal political systems and identify each nation's current political challenges.

3.8 Discuss the role of key location factors that explain why economic activities are located where they are in North America.

3.9 List and explain contemporary social issues that challenge North Americans in the 21st century.

REVIEW

3.1 Describe North America's major landform regions and climates, and suggest ways in which the region's physical setting has shaped patterns of human settlement.

3.2 Identify the key ways in which humans have transformed the North American environment since 1600.

3.3 Describe four key environmental problems that North Americans face in the early 21st century.

■ **KEY TERMS** Trans-Pacific Partnership (TPP), postindustrial economy, boreal forest, tundra, prairie, urban heat island, acid rain, renewable energy sources, fracking

...for Active Learning

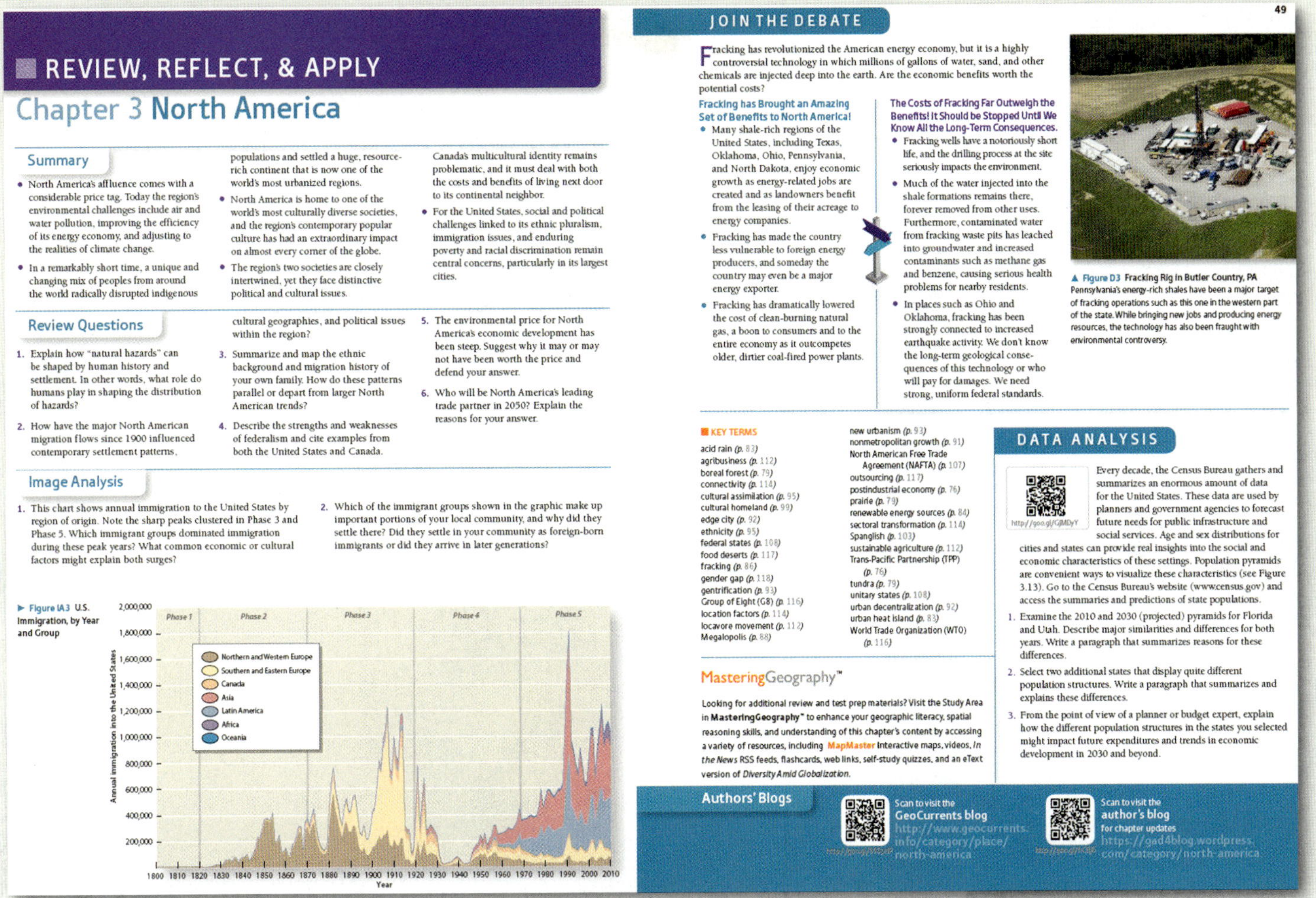

■ REVIEW, REFLECT, & APPLY
Chapter 3 North America

Summary

- North America's affluence comes with a considerable price tag. Today the region's environmental challenges include air and water pollution, improving the efficiency of its energy economy, and adjusting to the realities of climate change.
- In a remarkably short time, a unique and changing mix of peoples from around the world radically disrupted indigenous populations and settled a huge, resource-rich continent that is now one of the world's most urbanized regions.
- North America is home to one of the world's most culturally diverse societies, and the region's contemporary popular culture has had an extraordinary impact on almost every corner of the globe.
- The region's two societies are closely intertwined, yet they face distinctive political and cultural issues.
- Canada's multicultural identity remains problematic, and it must deal with both the costs and benefits of living next door to its continental neighbor.
- For the United States, social and political challenges linked to its ethnic pluralism, immigration issues, and enduring poverty and racial discrimination remain central concerns, particularly in its largest cities.

Review Questions

1. Explain how "natural hazards" can be shaped by human history and settlement. In other words, what role do humans play in shaping the distribution of hazards?

2. How have the major North American migration flows since 1900 influenced contemporary settlement patterns.

3. Summarize and map the ethnic background and migration history of your own family. How do these patterns parallel or depart from larger North American trends?

4. Describe the strengths and weaknesses of federalism and cite examples from both the United States and Canada.

5. The environmental price for North America's economic development has been steep. Suggest why it may or may not have been worth the price and defend your answer.

6. Who will be North America's leading trade partner in 2050? Explain the reasons for your answer.

Image Analysis

1. This chart shows annual immigration to the United States by region of origin. Note the sharp peaks clustered in Phase 3 and Phase 5. Which immigrant groups dominated immigration during these peak years? What common economic or cultural factors might explain both surges?

2. Which of the immigrant groups shown in the graphic make up important portions of your local community, and why did they settle there? Did they settle in your community as foreign-born immigrants or did they arrive in later generations?

▶ Figure IA3 U.S. Immigration, by Year and Group

Legend:
- Northern and Western Europe
- Southern and Eastern Europe
- Canada
- Asia
- Latin America
- Africa
- Oceania

JOIN THE DEBATE

Fracking has revolutionized the American energy economy, but it is a highly controversial technology in which millions of gallons of water, sand, and other chemicals are injected deep into the earth. Are the economic benefits worth the potential costs?

Fracking has Brought an Amazing Set of Benefits to North America!
- Many shale-rich regions of the United States, including Texas, Oklahoma, Ohio, Pennsylvania, and North Dakota, enjoy economic growth as energy-related jobs are created and as landowners benefit from the leasing of their acreage to energy companies.
- Fracking has made the country less vulnerable to foreign energy producers, and someday the country may even be a major energy exporter.
- Fracking has dramatically lowered the cost of clean-burning natural gas, a boon to consumers and to the entire economy as it outcompetes older, dirtier coal-fired power plants.

The Costs of Fracking Far Outweigh the Benefits! It Should be Stopped Until We Know All the Long-Term Consequences.
- Fracking wells have a notoriously short life, and the drilling process at the site seriously impacts the environment.
- Much of the water injected into the shale formations remains there, forever removed from other uses. Furthermore, contaminated water from fracking waste pits has leached into groundwater and increased contaminants such as methane gas and benzene, causing serious health problems for nearby residents.
- In places such as Ohio and Oklahoma, fracking has been strongly connected to increased earthquake activity. We don't know the long-term geological consequences of this technology or who will pay for damages. We need strong, uniform federal standards.

▲ Figure D3 Fracking Rig in Butler Country, PA Pennsylvania's energy-rich shales have been a major target of fracking operations such as this one in the western part of the state. While bringing new jobs and producing energy resources, the technology has also been fraught with environmental controversy.

■ KEY TERMS

acid rain (p. 83)
agribusiness (p. 112)
boreal forest (p. 79)
connectivity (p. 114)
cultural assimilation (p. 95)
cultural homeland (p. 99)
edge city (p. 92)
ethnicity (p. 95)
federal states (p. 108)
food deserts (p. 117)
fracking (p. 86)
gender gap (p. 118)
gentrification (p. 93)
Group of Eight (G8) (p. 116)
location factors (p. 114)
locavore movement (p. 112)
Megalopolis (p. 88)

new urbanism (p. 93)
nonmetropolitan growth (p. 91)
North American Free Trade Agreement (NAFTA) (p. 107)
outsourcing (p. 117)
postindustrial economy (p. 76)
prairie (p. 79)
renewable energy sources (p. 84)
sectoral transformation (p. 114)
Spanglish (p. 103)
sustainable agriculture (p. 112)
Trans-Pacific Partnership (TPP) (p. 76)
tundra (p. 79)
unitary states (p. 108)
urban decentralization (p. 92)
urban heat island (p. 83)
World Trade Organization (WTO) (p. 116)

MasteringGeography™

Looking for additional review and test prep materials? Visit the Study Area in **MasteringGeography™** to enhance your geographic literacy, spatial reasoning skills, and understanding of this chapter's content by accessing a variety of resources, including **MapMaster** interactive maps, videos, *In the News* RSS feeds, flashcards, web links, self-study quizzes, and an eText version of *Diversity Amid Globalization*.

DATA ANALYSIS

http://goo.gl/GjMGyY

Every decade, the Census Bureau gathers and summarizes an enormous amount of data for the United States. These data are used by planners and government agencies to forecast future needs for public infrastructure and social services. Age and sex distributions for cities and states can provide real insights into the social and economic characteristics of these settings. Population pyramids are convenient ways to visualize these characteristics (see Figure 3.13) and access the summaries and predictions of state populations.

1. Examine the 2010 and 2030 (projected) pyramids for Florida and Utah. Describe major similarities and differences for both years. Write a paragraph that summarizes reasons for these differences.

2. Select two additional states that display quite different population structures. Write a paragraph that summarizes and explains these differences.

3. From the point of view of a planner or budget expert, explain how the different population structures in the states you selected might impact future expenditures and trends in economic development in 2030 and beyond.

Authors' Blogs

Scan to visit the GeoCurrents blog
http://www.geocurrents.info/category/place/north-america

Scan to visit the author's blog for chapter updates
https://gad4blog.wordpress.com/category/north-america

NEW! Two-page *Review, Reflect, & Apply* sections at the end of each chapter provide a robust interactive review experience, including: a concise chapter summary, *Review Questions* that bridge multiple chapter themes, *Image Analysis* questions, *Join the Debate* activities that encourage students to explore both sides of a complex issue affecting the region, *Data Analysis* activities, as well as QR links to author blogs.

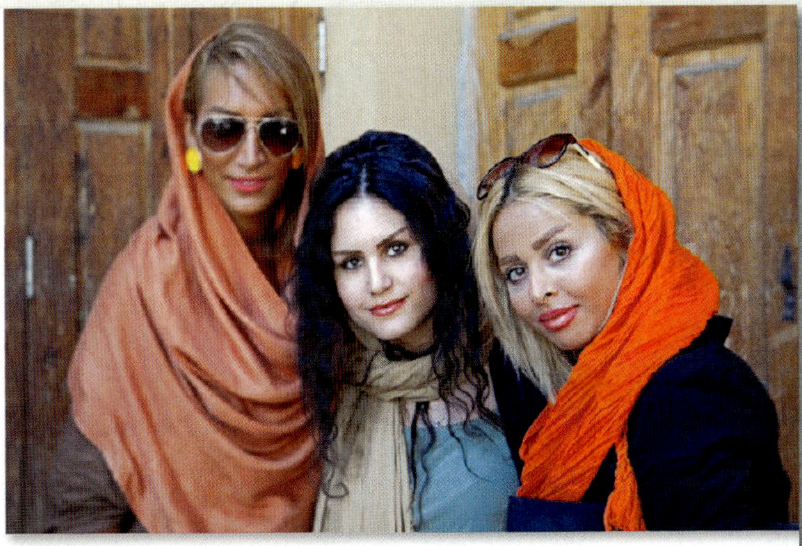

UPDATED! Visual Questions within each chapter section give students more opportunity to stop and practice visual analysis and critical thinking.

▲ Figure 7.45 **Iranian Women** These fashionably dressed young women in Tehran suggest how Iran's more urban and affluent residents have embraced many elements of Western culture. **Q: What unique challenges confront educated women in Iran? Unique opportunities?**

Continuous Learning
Before, During, and After Class...

BEFORE CLASS

Mobile Media & Reading Assignments Ensure Students Come to Class Prepared

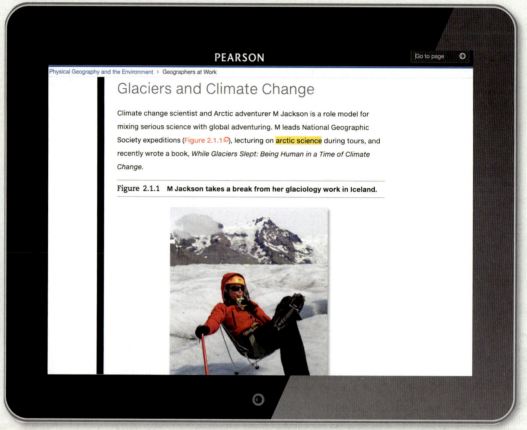

Glaciers and Climate Change

Climate change scientist and Arctic adventurer M Jackson is a role model for mixing serious science with global adventuring. M leads National Geographic Society expeditions (Figure 2.1.1), lecturing on arctic science during tours, and recently wrote a book, *While Glaciers Slept: Being Human in a Time of Climate Change.*

Figure 2.1.1   M Jackson takes a break from her glaciology work in Iceland.

UPDATED! Dynamic Study Modules help students study effectively by continuously assessing student performance and providing practice in areas where students struggle the most. Each Dynamic Study Module, accessed by computer, smartphone, or tablet, promotes fast learning and long-term retention.

NEW! Interactive eText 2.0 gives students access to the text whenever they can access the Internet. eText features include:

- Now available on smartphones and tablets
- Seamlessly integrated videos and other rich media
- Accessible (screen-reader ready)
- Configurable reading settings, including resizable type & night reading mode
- Instructor and student note-taking, highlighting, bookmarking, and search

Pre-Lecture Reading Quizzes are easy to customize and assign

UPDATED! Reading Questions ensure that students complete the assigned reading before class and stay on track with reading assignments. Reading Questions are 100% mobile ready and can be completed by students on mobile devices.

...with MasteringGeography™

DURING CLASS

Engage students with Learning Catalytics

What has Teachers and Students excited? Learning Catalytics, a "bring your own device" student engagement, assessment, and classroom intelligence system, allows students to use their smartphone, tablet, or laptop to respond to questions in class. With Learning Catalytics, you can:

- Assess students in real-time using open ended question formats, including sketching, ranking, image upload, and word clouds, to uncover student misconceptions and adjust a lecture accordingly.

- Automatically create groups for peer instruction based on student response patterns, to optimize discussion productivity.

> *"My students are so busy and engaged answering Learning Catalytics questions during the lecture that they don't have time for Facebook."*
>
> Declan De Paor, *Old Dominion University*

Continuous Learning Before, During, and After Class...

AFTER CLASS

Easy to Assign, Customizable, Media-Rich, and Automatically Graded Assignments

NEW! Geography Videos from sources such as the BBC, *Financial Times*, and Television for the Environment's *Life* and *Earth Report* series are included in MasteringGeography. These videos provide students with a sense of place with applied real-world examples of geography in action, and allow students to explore a range of locations and topics.

NEW! *MapMaster 2.0 Interactive Map Activities* are inspired by GIS, allowing students to layer various thematic maps to analyze spatial patterns and data at regional and global scales. Now fully mobile, with enhanced analysis tools, such as split screen, the ability for students to geolocate themselves in the data, and the ability for students to upload their own data for advanced map making. This tool includes zoom, and annotation functionality, with hundreds of map layers leveraging recent data from sources such as the PRB, the World Bank, NOAA, NASA, USGS, United Nations, the CIA, and more.

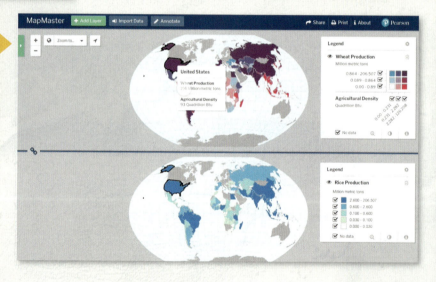

NEW! *Mobile Field Trip Videos* have students accompany photographer and pilot Michael Collier in the air and on the ground to explore the processes and stories of iconic landscapes in North America and beyond.

...with MasteringGeography™

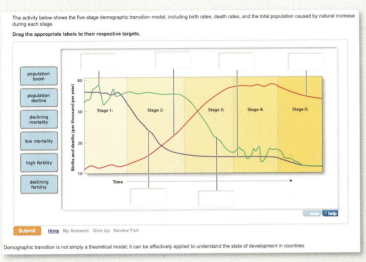

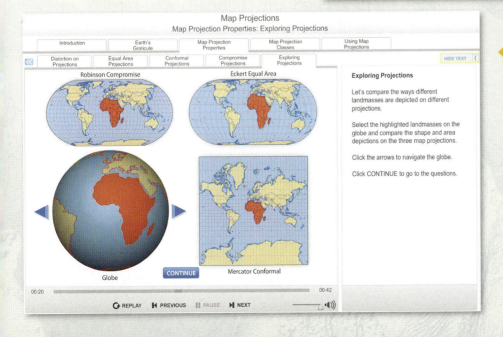

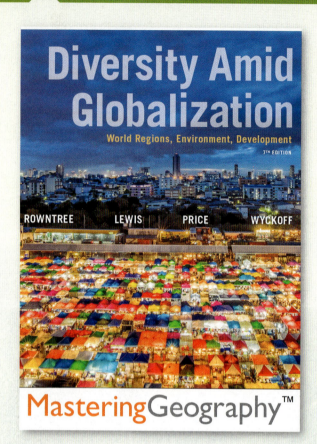

MasteringGeography™ provides you with everything you need to prep for your course and deliver a dynamic lecture, in one convenient place. Resources include:

LECTURE PRESENTATION ASSETS FOR EACH CHAPTER

- PowerPoint Lecture Outlines
- PowerPoint clicker questions
- Files for all illustrations, tables, and photos from the text

Measuring Student Learning Outcomes

All of the MasteringGeography assignable content is tagged to learning outcomes from the book, the National Geography Standards, and Bloom's Taxonomy. You also have the ability to add your own learning outcomes, helping you track student performance against your specific course goals. You can view class performance against the specified learning outcomes and share those results quickly and easily by exporting to a spreadsheet.

TEST BANK

- The *Test Bank* in Word format
- TestGen Computerized *Test Bank*, which includes all the questions from the printed *Test Bank* in a format that allows you to easily and intuitively build exams and quizzes

TEACHING RESOURCES

- *Instructor Resource Manual* in Microsoft Word and PDF formats
- Pearson Community Website (https://communities.pearson.com/ northamerica/s/)
- *Goode's World Atlas, 23rd Edition*

Diversity Amid Globalization

World Regions, Environment, Development

7TH EDITION

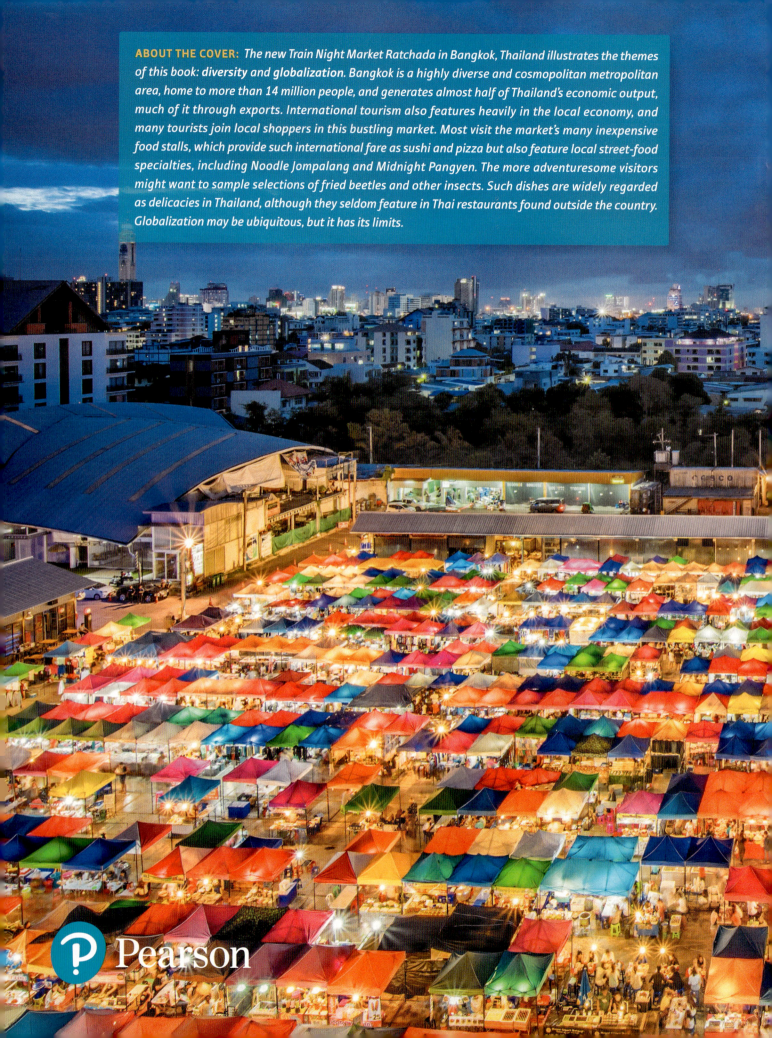

P Pearson

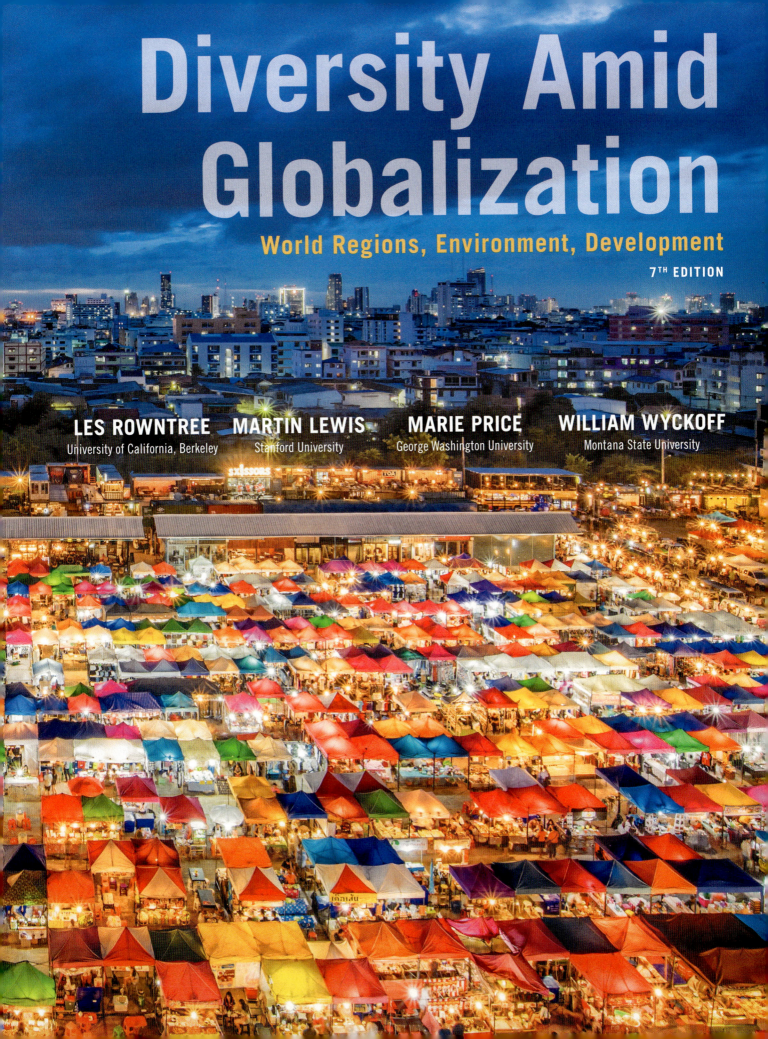

Diversity Amid Globalization

World Regions, Environment, Development

7TH EDITION

LES ROWNTREE
University of California, Berkeley

MARTIN LEWIS
Stanford University

MARIE PRICE
George Washington University

WILLIAM WYCKOFF
Montana State University

Executive Editor, Geosciences Courseware: Christian Botting
Director, Courseware Portfolio Management: Beth Wilbur
Content Producer: Brett Coker
Managing Producer: Michael Early
Courseware Director, Content Development: Ginnie Simione Jutson
Development Editor: Veronica Jurgena
Geosciences Courseware Editorial Assistant: Emily Bornhop
Senior Content Producer: Tim Hainley
Rich Media Content Producer: Mia Sullivan
Full-Service Compositor: SPi Global
Full-Service Project Manager: Michelle Gardner
Copyeditor: Melanie Brown
Cartography and Illustrations: International Mapping

Art Coordinator: Kevin Lear, International Mapping
Design Manager: Mark Ong
Cover and Interior Designer: Jeff Puda
Rights & Permissions Project Manager: Laura Perry, Cenveo Publishing Services
Rights & Permissions Management: Ben Ferrini
Photo Researcher: Kristin Piljay
Manufacturing Buyer: Maura Zaldivar-Garcia, LSC Communications
Executive Product Marketing Manager: Neena Bali
Senior Field Marketing Manager: Mary Salzman
Marketing Assistant: Ami Sampat
Cover Photo Credit: Thapakorn Karnosod / Getty Images

Library of Congress Cataloging-in-Publication Data
Names: Rowntree, Lester, 1938- author.
Title: Diversity amid globalization : world regions, environment, development / Les Rowntree, University of California, Berkeley, Martin Lewis, Stanford University, Marie Price, George Washington University, William Wyckoff, Montana State University.
Description: Seventh edition. | San Francisco : Pearson, [2017]
Identifiers: LCCN 2016047494 | ISBN 9780134539423
Subjects: LCSH: Geography. | Globalization.
Classification: LCC G128 .R68 2017 | DDC 910--dc23
LC record available at https://lccn.loc.gov/2016047494

ISBN 10: 0-134-53942-7; ISBN 13: 978-0-134-53942-3 (Student Edition)
ISBN 10: 0-134-61025-3; ISBN 13: 978-0-134-61025-2 (Books à la Carte Edition)
2 17

www.pearsonhighered.com

BRIEF CONTENTS

CONTENTS

1 Concepts of World Geography   2

2 Physical Geography and the Environment   46

13 Southeast Asia 538

14 Australia and Oceania 586

*D*iversity Amid Globalization, Seventh Edition, is an issues-oriented textbook for college and university world regional geography classes that explicitly recognizes the vast geographic changes taking place because of globalization. With this focus, we join the many scholars who see globalization as the most fundamental reorganization of the world's socioeconomic, cultural, and geopolitical structures since the Industrial Revolution. That premise provides the point of departure and underlying assumptions of this book.

As geographers, we think it essential for our readers to understand and critically appraise the two interactive themes that are reflected in this book's title, *Diversity Amid Globalization*. First, the convergence of environmental, cultural, political, and economic systems through the processes of globalization has numerous consequences, both obvious and obscure, at every scale of analysis and in every part of the world. Second, many forms of diversity persist—and sometimes even expand— despite the leveling tendencies of globalization. It is also increasingly apparent that globalization generates its own resistance, which can range from celebrations of local products and customs, to grassroots opposition to global trade deals, to hard-edged nationalism and hostility toward immigration. Clearly, globalization is a ubiquitous, politically charged, and complex phenomenon that demands sustained geographical analysis.

New to the Seventh Edition

- New chapter opener vignettes and photos have been added that highlight recent events and global linkages, with accompanying maps that pinpoint vignette locations. This edition also features more focused and consistent introductions in Chapters 3–14, placed under the heading "Defining the Region."

- New *Geographers at Work* sidebars profile a working geographer's research, intellectual development, and views on the discipline. Many of these sidebars include maps or remote sensing images.

- New *Sights of the Region* features provide mobile-ready Quick Response (QR) links from photos to online Google Maps, enabling students to browse web maps and community-contributed photos of the diverse geographies featured in the print book. Students use mobile devices to scan Quick Response (QR) codes to get immediate online access and connect print images with dynamic online web maps and photos.

- New *Sounds of the Region* features provide QR links to sound clips that help give students a sense of culture and natural environments around the world, highlighting language, music, and the soundscapes of both natural and urban environments.

- New *Tastes of the Region* features in each regional chapter explore culinary traditions and innovations associated with different parts of the world. These QR links to websites provide recipes and other pertinent information on food production and consumption, as well as material on cultural aspects of regional cuisines.

- The new end-of-the-chapter format—*Review, Reflect, & Apply*—asks students to answer broad-based questions spanning concepts and regions. Two of the three components of this feature, *Image Analysis* and *Data Analysis*, provide concrete exercises based on the analysis of graphic images and demographic or socioeconomic data. The third, *Join the Debate*, frames two opposing viewpoints on controversial issues and asks students to assess their claims and weigh in on their own.

- New *Mobile Field Trip Videos* have students accompany renowned geoscience photographer Michael Collier in the air and on the ground to explore iconic landscapes that have shaped North America and beyond. Students scan QR codes in the print book to get instant access to these media, which are also available for assignment with quizzes in MasteringGeography.

Revised Features in the Seventh Edition

- Although the sidebar categories from the previous edition have been largely retained, most individual sidebars have been replaced. The *Everyday Globalization* feature has been sharpened and framed around new topics. The *Working Toward Sustainability* and *Exploring Global Connections* sidebars have been revamped, many with QR links to new Google Earth Virtual Tour videos. Most *People on the Move* sidebars are new as well.

- Extensive current events updates have been provided in each chapter, many maps have been substantially revised, and new photos are found throughout the book. All climate maps now have enlarged climographs for improved viewing, and most religion maps have been redrafted to reflect the more accurate information that has recently become available. The two data tables in each regional chapter have been completely updated with the most recently available data.

- The learning-path feature has been sharpened and enhanced in each chapter. *Learning Objectives* are now numbered for easy reference and have been revised; key terms are now repeated at the end of each section and in the end-of-chapter materials; concept check questions have been added to all four sidebars; tabular data on population and development measures have been integrated with MapMaster and now include questions related to mapping activities. This edition also features a more consistent use of questions that are linked to selected figures.

New and Updated Features in Chapter 1:
Concepts of World Geography

- This chapter has been tightened and is now focused more sharply on geography. The order of the introductory sections has been rearranged so that *Geography Matters* (which introduces the basic concepts of the discipline) now comes before *Converging Currents of Globalization*. A new U.S. Rust Belt map illustrates the functional region concept.

- The *Globalization* section of Chapter 1 has itself been reordered to follow the main themes of the regional chapters (environment, population, culture, geopolitics, and economic and social development). The "Thinking Critically about Globalization" discussion has been revised and shortened, and new information has been provided on digital information flows.

- The thematic sections of the introductory chapter have also been updated, revised, and provided with new examples and photos.

- Several key terms have been added, including *space*, *place*, and *territory*. The existing sidebars in this chapter have been revised, and a new sidebar on conflict mapping has been added.

New and Updated Features in Chapter 2:
Physical Geography and the Environment

- The discussion of plate tectonics in the *Geology* section has been substantially revised, with new diagrams and photos illustrating plate movement and natural hazards. New Mobile Field Trips QR link to concise videos on earthquake faults, climate change, volcanic activity, cloud dynamics, cloud forests, oil sands, and forest fires.

- The *Global Climates* section has been revised and expanded, with enhanced discussions of both climate change itself and international mitigation efforts. The 2015 Paris Agreement goals are introduced, as are various national climate plans. Weather and climate diagrams, images, and data have been revised and updated.

- The *Bioregions and Biodiversity* section is now highlighted by earlier placement in the chapter and through a new bioregions photo spread with detailed captions. Additional new features here include discussions of nature and the global economy, climate change and the natural world, and the current extinction crisis.

Thematic Organization

Diversity Amid Globalization is organized around the conventional world regions of Sub-Saharan Africa, Europe, Latin America, East Asia, South Asia, and so on. We have, however, added two distinctive regions that are often excluded from the standard world regional scheme: Central Asia and the Caribbean. Our 12 regional chapters further depart from the treatment found in traditional world regional textbooks by employing a thematic framework that avoids extensive descriptions of each individual country. Instead, we have placed such "country by country" material online in the MasteringGeography Study Area.

In *Diversity Amid Globalization,* each regional chapter is organized around five key geographic themes. First is *Physical Geography and Environmental Issues,* in which we not only describe the physical geography of each region but also outline major environmental issues, including climate change and energy use. Next comes *Population and Settlement,* where we examine each region's demographic trends, migration patterns, land use systems, and settlement configurations, including urbanization. Our third theme, *Cultural Coherence and Diversity,* covers the traditional topics of language and religions and examines the ethnic and cultural tensions resulting from globalization. Topics of popular culture, including sports, music, and food, are also included in this section. We turn next to the *Geopolitical Framework,* examining the political geography of each region. Here we take on such contentious issues as ethnic conflicts, border disputes, separatism, regionalism, systems of alliance, and global terrorism. Each regional chapter concludes with an extended discussion of *Economic and Social Development.* In this section we explore changing economic frameworks at the local, national, regional, and global scales, and examine as well such social issues as health, education, and gender inequalities.

These 12 regional chapters follow two substantive introductory chapters that provide the fundamentals of both human and physical geography. The first chapter, "Concepts of World Geography," begins with an introduction to the discipline of geography, highlighting its major concepts. It then turns to the geographic dimensions of globalization, including a discussion of the costs and benefits of globalization, as outlined by both proponents and opponents. In the next section, "The Geographer's Toolbox," students are informed about such matters as map-reading, cartography, aerial photos, remote sensing, and GIS. This introductory chapter concludes with an analysis of the key themes that are used to structure the regional chapters, as well as a description of the data tables that are found throughout the text.

Chapter 2, "Physical Geography and the Environment," provides an overview of the principles of physical and environmental geography that are discussed extensively in the first section of each regional chapter. Here we emphasize geomorphology and basic geological processes; environmental hazards and degradation; weather, climate, and global warming; global bioregions and biodiversity; hydrology and water stress; and changing energy profiles.

Chapter Features

- **Structured learning path.** Every chapter begins with an explicit set of learning objectives to provide students with the larger context of each chapter. Review questions and key terms after each section allow students to refresh and test their learning. Each chapter ends with an innovative, graphically rich "Review, Reflect, & Apply" section, where students are asked to employ what they have absorbed from the chapter in an active-learning framework.

- **Comparable regional maps.** Of the various maps found in each regional chapter, many are constructed on the same themes and with similar data so that readers can easily draw comparisons between different regions. All regional chapters have comparable maps of physical geography, climate, environmental issues, population density, language, and geopolitical issues. Most of these chapters also have similar maps on migration and religion.

- **Additional maps pertinent to each region.** The regional chapters also contain many additional maps illustrating important geographic topics such as global economic issues, social development, and ethnic tensions.

- **Comparable regional data sets.** Two thematic tables in each regional chapter facilitate comparisons between regions and provide important information on the characteristics of the area under consideration. The first table provides essential population data for each country within the region, including fertility rates, the proportions of the population under 15 and over 65 years of age, and net migration rates. The second table presents economic and social development data for each country, including gross national income per capita, gross domestic product growth, life expectancy for men and women, percentage of the population living on less than $3.10 per day, child mortality rates, and the United Nations gender inequality index.

- Consistent **Sidebar essays, many with Google Earth Virtual Tour Videos.** Each of the regional chapters has five sidebars focused on particular geographic themes. To facilitate visual geographic understanding, several sidebars in each chapter contain QR "hot links" to Google Earth Virtual Tour Videos. Critical thinking questions at the end of each sidebar ask students to reflect on how these topics apply to their own lives. The sidebar themes are as follows:

  - **WORKING TOWARD SUSTAINABILITY,** which features case studies of sustainable environmental projects throughout the world, emphasizing positive ecological and social initiatives and their results.

  - **EXPLORING GLOBAL CONNECTIONS,** which investigates the many ways in which activities in different parts of the world are linked together, showing students that in our globalized world, regions are neither isolated nor discrete.

  - **PEOPLE ON THE MOVE,** which seeks to capture the human geography behind contemporary migration, exploring how people relocate, legally and not so legally, as they respond to the varied currents and expressions of globalization.

  - **EVERYDAY GLOBALIZATION,** a shorter sidebar feature that illustrates the many ways that globalization permeates students' everyday lives. Topics here range from food, to clothing, to cell phones, to music, and beyond.

  - **GEOGRAPHERS AT WORK,** which highlights the research and career paths of individual geographers active in the region under consideration. Interview questions posed to the subjects help bring these portrayals to life.

- **QR links to author blogs.** These links lead readers to two blogs where authors discuss everything from current events to their travels and field research. Both blogs are graphically rich with innovative maps and photos.

Acknowledgments

We have many people to thank for the conceptualization, writing, rewriting, and production of *Diversity Amid Globalization*. First, we'd like to thank the thousands of students in our world regional geography classes who have inspired us with their energy, engagement, and curiosity; challenged us with their critical insights; and demanded a textbook that better meets their need to understand the contemporary geography of their dynamic and complex world.

Next, we are deeply indebted to many professional geographers and educators for their assistance, advice, inspiration, encouragement, and constructive criticism as we labored through the different stages of this book. Among the many who provided invaluable comments on various drafts and editions of *Diversity Amid Globalization* or who worked on supporting print or digital material are:

Gillian Acheson, *Southern Illinois University, Edwardsville*

Joy Adams, *Humboldt State University*

Dan Arreola, *Arizona State University*

Bakama Bernard BakamaNume, *Texas A&M University*

Brad Baltensperger, *Michigan Technological University*

Jessica Barnes, *Northern Arizona University*

Karen Barton, *University of Northern Colorado*

Max Beavers, *Samford University*

Laurence Becker, *Oregon State University*

Dan Bedford, *Weber State University*

James Bell, *University of Colorado*

Katie Berchak, *University of Louisiana, Lafayette*

William H. Berentsen, *University of Connecticut*

Kevin Blake, *Kansas State University*

Mikhail Blinnikov, *St. Cloud State University*

Karl Byrand, *University of Wisconsin, Sheboygan*

Michelle Calvarese, *California State University, Fresno*

Craig Campbell, *Youngstown State University*

Scott Campbell, *College of DuPage*

Elizabeth Chacko, *George Washington University*

Philip Chaney, *Auburn University*

Xuwei Chen, *Northern Illinois University*

David B. Cole, *University of Northern Colorado*

Malcolm Comeaux, *Arizona State University*

Jonathan C. Comer, *Oklahoma State University*

Catherine Cooper, *George Washington University*

Jeremy Crampton, *George Mason University*

Kevin Curtin, *University of Texas at Dallas*

James Curtis, *California State University, Long Beach*

Dydia DeLyser, *Louisiana State University*

Francis H. Dillon, *George Mason University*

Jason Dittmer, *Georgia Southern University*

Jerome Dobson, *University of Kansas*

Caroline Doherty, *Northern Arizona University*

Vernon Domingo, *Bridgewater State College*

Roy Doyon, *Ball State University*

Dawn Drake, *Missouri Western State University*

Jane Ehemann, *Shippensburg University*

Chuck Fahrer, *Georgia College and State University*

Dean Fairbanks, *California State University, Chico*

Emily Fekete, *University of Kansas*

Caitie Finlayson, *University of Mary Washington*

Doug Fuller, *George Washington University*

Gary Gaile, *University of Colorado*

Douglas Gamble, *University of North Carolina, Wilmington*

Sherry Goddicksen, *California State University, Fullerton*

Sarah Goggin, *Cypress College*

Joe Guttman, *University of Tennessee*

Reuel Hanks, *Oklahoma State University*

James Harris, *Metropolitan College of Denver*

Andrew Hillburn, *Kansas State University*

Steven Hoelscher, *University of Texas, Austin*

Erick Howenstine, *Northeastern Illinois University*

Tyler Huffman, *Eastern Kentucky University*

Peter J. Hugill, *Texas A&M University*

Eva Humbeck, *Arizona State University*

Shireen Hyrapiet, *Oregon State University*

Drew Kapp, *University of Hawaii, Hilo*

Ryan Kelly, *Bluegrass Community College*

Richard H. Kesel, *Louisiana State University*

Richard Kotula, *College of Staten Island*

Rob Kremer, *Front Range Community College*

Robert C. Larson, *Indiana State University*

Alan A. Lew, *Northern Arizona University*

Elizabeth Lobb, *Mt. San Antonio College*

Catherine Lockwood, *Chadron State College*

Max Lu, *Kansas State University*

Luke Marzen, *Auburn University*

Kent Matthewson, *Louisiana State University*

James Miller, *Clemson University*

Bob Mings, *Arizona State University*

Wendy Mitteager, *SUNY, Oneonta*

Sherry D. Morea-Oakes, *University of Colorado, Denver*

Anne E. Mosher, *Syracuse University*

Julie Mura, *Florida State University*

Tim Oakes, *University of Colorado*

Nancy Obermeyer, *Indiana State University*

Karl Offen, *University of Oklahoma*

Thomas Orf, *Las Positas College*

Kefa Otiso, *Bowling Green State University*

Joseph Palis, *University of North Carolina*

Jean Palmer-Moloney, *Hartwick College*

Bimal K. Paul, *Kansas State University*

Michael P. Peterson, *University of Nebraska, Omaha*

Richard Pillsbury, *Georgia State University*

Brandon Plewe, *Brigham Young University*

Jess Porter, *University of Arkansas at Little Rock*

Patricia Price, *Florida International University*

Erik Prout, *Texas A&M University*

Claudia Radel, *Utah State University*

David Rain, *United States Census Bureau*

Rhonda Reagan, *Blinn College*

Kelly Ann Renwick, *Appalachian State University*

Craig S. Revels, *Portland State University*

Pamela Riddick, *University of Memphis*

Scott M. Robeson, *Indiana State University*

Paul A. Rollinson, *Missouri State University*

Yda Schreuder, *University of Delaware*

Kathy Schroeder, *Appalachian State University*

Kay L. Scott, *University of Central Florida*

Patrick Shabram, *South Plains College*

J. Duncan Shaeffer, *Arizona State University*

Dmitrii Sidorov, *California State University, Long Beach*

Susan C. Slowey, *Blinn College*

Andrew Sluyter, *Louisiana State University*

Christa Smith, *Clemson University*

Joseph Spinelli, *Bowling Green State University*

Michael Strong, *Towson University*

William Strong, *University of Northern Alabama*

Philip W. Suckling, *University of Northern Iowa*

Austen Thelen, *Imperial Valley College*

Curtis Thomson, *University of Idaho*

Suzanne Traub-Metlay, *Front Range Community College*

James Tyner, *Kent State University*

Nina Veregge, *University of Colorado*

Fahui Wang, *Louisiana State University*

Gerald R. Webster, *University of Alabama*

Keith Yearman, *College of DuPage*

Emily Young, *University of Arizona*

Bin Zhon, *Southern Illinois University, Edwardsville*

Henry J. Zintambia, *Illinois State University*

Sandra Zupan, *Temple University*

In addition, we wish to thank the many publishing professionals who have made this book possible. We start with Christian Botting, Pearson's Executive Editor for Geosciences, a consummate professional and good friend whose leadership, high standards, and enduring patience has been laudable, inspiring, and necessary. Many thanks, as well, to Brett Coker, our Project Manager, Emily Bornhop, Editorial Assistant, and Veronica Jurgena, our outstanding Development Editor, whose insights, guidance, and encouragement (much of it coming from her education as a geographer) have been absolutely crucial to this revision. Also to be thanked are a number of behind-the-curtain professionals: Kristin Piljay and Dena Betz, photo researchers; Ben Ferrini, Manager, Rights Management; Michelle Gardner, SPi Global Project Manager, for somehow turning thousands of pages of manuscript into a finished product; and Kevin Lear, International Mapping Senior Project Manager, for his outstanding work updating and revising our maps. Thanks are due to Zhaohui Li for his timely production of all data tables.

Not to be overlooked are the 14 professional geographers who allowed us to pry into their personal and professional lives so we could profile them in the new *Geographers at Work* sidebars. They are:

Fenda Akiwumi, University of South Florida; Holly Barcus, Macalester College; Sarah Blue, Texas State University; Laura Brewington, East-West Center; Karen Culcasi, West Virginia University; Corrie Drummond Garcia, USAID; M. Jackson, University of Oregon; Cary Karacas, College of Staten Island; Weronika Kusek, Northern Michigan University; Chandana Mitra, Auburn University; Kelsey Nyland, Michigan State University; Diane Papineau, Montana State Library; Rachel Silvey, University of Toronto; and Susan Wolfinbarger, American Association for the Advancement of Science.

Last, the authors want to thank that special group of friends and family who were there when we needed you most—early in the morning and late at night; in foreign countries and at home; when we were on the verge of tears and rants, but needed lightness and laughter; for your love, patience, companionship, inspiration, solace, enthusiasm, and understanding. Words cannot thank you enough: Meg Conkey, Rob, Joseph, and James Crandall, Marie Dowd, Evan and Eleanor Lewis, Karen Wigen, Linda and Tom Wyckoff, and Katie Sander.

Les Rowntree

Martin Lewis

Marie Price

William Wyckoff

Les Rowntree is a Research Associate at the University of California, Berkeley, where he researches and writes about global and local environmental issues. This career change came after more than three decades teaching both Geography and Environmental Studies at San Jose State University. As an environmental geographer, Dr. Rowntree's interests focus on international environmental issues, biodiversity conservation, and human-caused global change. He sees world regional geography as a way to engage and inform students by giving them the conceptual tools needed to critically assess the contemporary world. His current research and writing projects include a natural history book on California's Coast Range and essays on Europe's environmental issues; additionally, he maintains an assortment of web-based natural history, geography, and environmental blogs and websites.

Martin Lewis is a Senior Lecturer in History at Stanford University, where he teaches courses on global geography. He has conducted extensive research on environmental geography in the Philippines and on the intellectual history of world geography. His publications include *Wagering the Land: Ritual, Capital, and Environmental Degradation in the Cordillera of Northern Luzon, 1900–1986* (1992), and, with Karen Wigen, *The Myth of Continents: A Critique of Metageography* (1997). Dr. Lewis has traveled extensively in East, South, and Southeastern Asia. His current research focuses on the geography of languages. In April 2009, Dr. Lewis was recognized by *Time* magazine as one of America's most favorite lecturers.

Marie Price is a Professor of Geography and International Affairs at George Washington University. A Latin American specialist, Dr. Price has conducted research in Belize, Mexico, Venezuela, Panama, Cuba, and Bolivia. She has also traveled widely throughout Latin America and Sub-Saharan Africa. Her studies have explored human migration, natural resource use, environmental conservation, and sustainability. In 2016 she became the President of the American Geographical Society, the oldest national geography organization in the country dedicated to the advancement of geographic research for a more sustainable future. Dr. Price brings to *Diversity Amid Globalization* a special interest in regions as dynamic spatial constructs that are shaped over time through both global and local forces. Her publications include the co-edited book *Migrants to the Metropolis: The Rise of Immigrant Gateway Cities* (2008) and numerous academic articles and book chapters.

William Wyckoff is a Professor of Geography in the Department of Earth Sciences at Montana State University, specializing in the cultural and historical geography of North America. He has written and co-edited several books on North American settlement geography, including *The Developer's Frontier: The Making of the Western New York Landscape* (1988), *The Mountainous West: Explorations in Historical Geography* (1995) (with Lary M. Dilsaver), *Creating Colorado: The Making of a Western American Landscape 1860–1940* (1999), and *On the Road Again: Montana's Changing Landscape* (2006). His most recent book, *How to Read the American West: A Field Guide*, appeared in the Weyerhaeuser Environmental Books series and was published in 2014 by the University of Washington Press. A World Regional Geography instructor for 27 years, Dr. Wyckoff emphasizes in the classroom the connections between the everyday lives of his students and the larger global geographies that surround them and increasingly shape their future.

DIGITAL & PRINT RESOURCES

This edition provides a complete world regional geography program for students and teachers.

FOR STUDENTS & TEACHERS

MasteringGeography™ with Pearson eText

The Mastering platform is the most widely used and effective online homework, tutorial, and assessment system for the sciences. It delivers self-paced coaching activities that provide individualized coaching, focus on course objectives, and are responsive to each student's progress. The Mastering system helps teachers maximize class time with customizable, easy-to-assign, and automatically graded assessments that motivate students to learn outside of class and arrive prepared for lecture. MasteringGeography offers:

- **Assignable activities** that include GIS-inspired MapMaster™ interactive map activities, *Encounter* Google Earth Explorations, Video activities, Geoscience Animation activities, Map Projections activities, GeoTutor coaching activities on the toughest topics in geography, Dynamic Study Modules that customize the student's learning experience, book questions and exercises, reading quizzes, Test Bank questions, and more.

- **A student Study Area** with GIS-inspired MapMaster™ interactive maps, videos, Geoscience Animations, web links, glossary flashcards, "In the News" articles, Google Earth Virtual Tour videos, chapter quizzes, PDF downloads of regional outline maps, an optional Pearson eText and more.

Pearson eText gives students access to the text whenever and wherever they can access the Internet. Features include:

- Now available on smartphones and tablets.
- Seamlessly integrated videos and other rich media.
- Fully accessible (screen-reader ready).
- Configurable reading settings, including resizable type and night reading mode.
- Instructor and student note-taking, highlighting, bookmarking, and search. www.masteringgeography.com

Television for the Environment *Earth Report* Geography Videos on DVD (0321662989)

This three-DVD set helps students visualize how human decisions and behavior have affected the environment and how individuals are taking steps toward recovery. With topics ranging from the poor land management promoting the devastation of river systems in Central America to the struggles for electricity in China and Africa, these 13 videos from Television for the Environment's global *Earth Report* series recognize the efforts of individuals around the world to unite and protect the planet.

FOR STUDENTS

Goode's World Atlas, 23rd Edition (0133864642)

Goode's World Atlas has been the world's premiere educational atlas since 1923. It features over 260 pages of maps, from definitive physical and political maps to important thematic maps that illustrate the spatial aspects of many important topics. The 23rd edition includes dozens of new maps, incorporating the latest geographic scholarship and technologies, with expanded coverage of the Canadian Arctic, Europe's microstates, African islands, and U.S. cities. Several new thematic maps include: oceanic environments, earthquakes and tsunamis, desertification vulnerability, maritime political claims, megacities, human trafficking, labor migration, and more. Now available in Pearson Custom Library (www.pearsoncustomlibrary.com), as well as in various eText formats, including an eText upgrade option from MasteringGeography courses.

Pearson's Encounter Series provides rich, interactive explorations of geoscience concepts through GoogleEarth™ activities, covering a range of topics in regional, human, and physical geography. For those who do not use MasteringGeography, all chapter explorations are available in print workbooks, as well as in online quizzes at www.mygeoscienceplace.com, accommodating different classroom needs. Each exploration consists of a worksheet, online quizzes, and a corresponding Google Earth™ KMZ file.

- *Encounter World Regional Geography* Workbook and Website by Jess C. Porter (0321681754)
- *Encounter Human Geography* Workbook and Website by Jess C. Porter (0321682203)
- *Encounter Physical Geography* Workbook and Website by Jess C. Porter and Stephen O'Connell (0321672526)

Dire Predictions: Understanding Climate Change, 2nd Edition (0133909778 by Michael E. Mann and Lee R. Kump)

Periodic reports from the Intergovernmental Panel on Climate Change (IPCC) evaluate the risk of climate change brought on by humans. In just over 200 pages, this practical text presents and expands upon the IPCC's essential findings in a visually stunning and undeniably powerful way to the lay reader. Scientific findings that provide validity to the implications of climate change are presented in clear-cut graphic elements, striking images, and understandable analogies. The Second Edition covers the latest climate change data and scientific consensus from the IPCC Fifth Assessment Report and integrates mobile media links to online media. The text is also available in various eText formats, including an eText upgrade option from MasteringGeography courses.

FOR INSTRUCTORS

Instructor Resource Manual (Download) (013461089X)

The *Instructor Resource Manual* follows the new organization of the main text. It includes a sample syllabus, chapter learning objectives, lecture outlines, a list of key terms, and answers to the textbook's review and end-of-chapter questions. Discussion questions, classroom activities, and advice about how to integrate MasteringGeography and Learning Catalytics resources are integrated throughout the chapter lecture outlines.

TestGen/Test Bank (Download) (0134711742)

TestGen is a computerized test generator that lets instructors view and edit Test Bank questions, transfer questions to tests, and print tests in a variety of customized formats. This Test Bank includes approximately 1,500 multiple-choice, true/false, and short-answer/essay questions. Questions are correlated with the book's learning objectives, the revised U.S. National Geography Standards, chapter-specific learning outcomes, and Bloom's Taxonomy to help teachers better map the assessments against both broad and specific teaching and learning objectives. The Test Bank is also available in Microsoft Word® and Blackboard formats.

Instructor Resource Materials (0134610873)

The Instructor Resource Materials provides a collection of resources to help instructors make efficient and effective use of their time. All digital resources can be found in one well-organized, easy-to-access place. The IRC includes:

- All textbook images as JPEGs, PDFs, and PowerPoint™ Presentations

- Pre-authored Lecture Outline PowerPoint™ Presentations, which outline the concepts of each chapter with embedded art and can be customized to fit instructors' lecture requirements

- CRS "Clicker" Questions in PowerPoint™ format, which correlate to the book's learning objectives, the U.S. National Geography Standards, chapter-specific learning outcomes, and Bloom's Taxonomy

- The TestGen software and *Test Bank* questions

- Electronic files of the *Instructor Resource Manual* and *Test Bank*

This Instructor Resource content is also available completely online via the Instructor Resources section of MasteringGeography and **www.pearsonhighered.com/irc**.

LEARNING CATALYTICS

Learning Catalytics™ is a "bring your own device" student engagement, assessment, and classroom intelligence system. With Learning Catalytics, you can

- Assess students in real time, using open-ended tasks to probe student understanding.

- Understand immediately where students are and adjust your lecture accordingly.

- Improve your students' critical thinking skills.

- Access rich analytics to understand student performance.

- Add your own questions to make Learning Catalytics fit your course exactly.

- Manage student interactions with intelligent grouping and timing.

Learning Catalytics is a technology that has grown out of 20 years of cutting-edge research, innovation, and implementation of interactive teaching and peer instruction. Available integrated with MasteringGeography.

www.learningcatalytics.com
www.masteringgeography.com

ABOUT OUR SUSTAINABILITY INITIATIVES

Pearson recognizes the environmental challenges facing this planet, as well as acknowledges our responsibility in making a difference. This book is carefully crafted to minimize environmental impact. The binding, cover, and paper come from facilities that minimize waste, energy consumption, and the use of harmful chemicals. Pearson closes the loop by recycling every out-of-date text returned to our warehouse. Along with developing and exploring digital solutions to our market's needs, Pearson has a strong commitment to achieving carbonneutrality. As of 2009, Pearson became the first carbon- and climate-neutral publishing company, having reduced our absolute carbon footprint by 22% since then. Pearson has protected over 1,000 hectares of land in Columbia, Costa Rica, the United States, the UK and Canada. In 2015, Pearson formally adopted The Global Goals for Sustainable Development, sponsoring an event at the United Nations General Assembly and other ongoing initiatives. Pearson sources 100% of the electricity we use from green power and invests in renewable energy resources in multiple cities where we have operations, helping make them more sustainable and limiting our environmental impact for local communities. The future holds great promise for reducing our impact on Earth's environment, and Pearson is proud to be leading the way. We strive to publish the best books with the most up-to-date and accurate content, and to do so in ways that minimize our impact on Earth. To learn more about our initiatives, please visit https://www.pearson.com/sustainability.html.

Diversity Amid Globalization

World Regions, Environment, Development

Lesbos, Greece

EUROPE

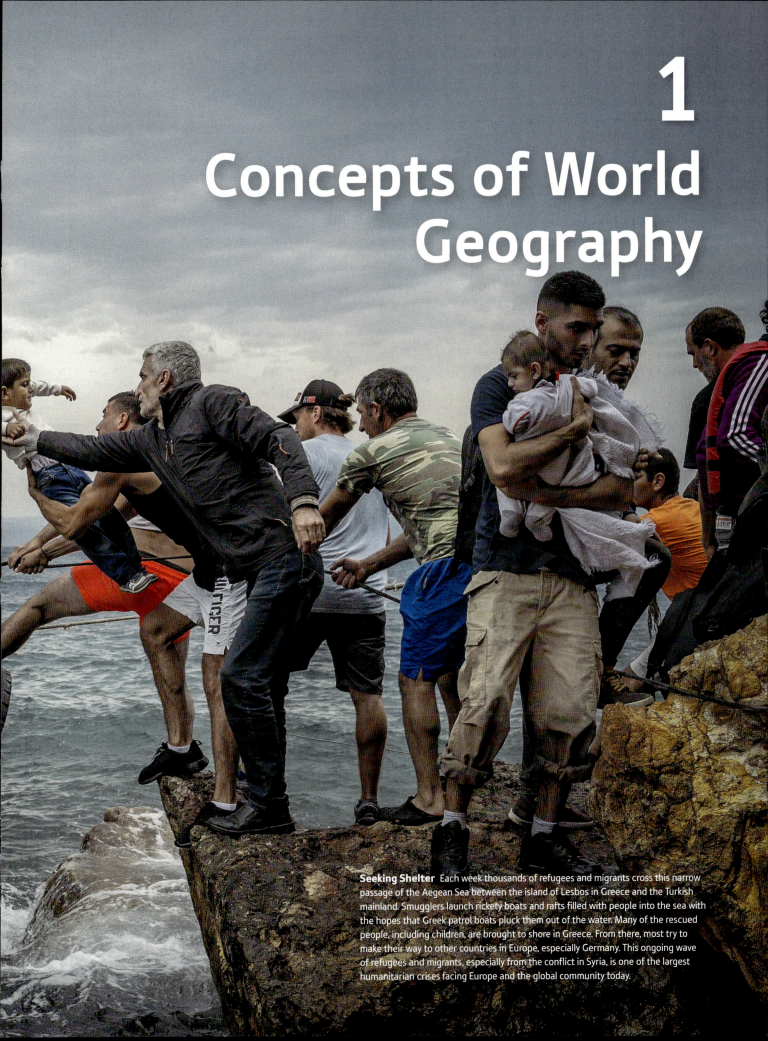

1
Concepts of World Geography

Seeking Shelter Each week thousands of refugees and migrants cross this narrow passage of the Aegean Sea between the island of Lesbos in Greece and the Turkish mainland. Smugglers launch rickety boats and rafts filled with people into the sea with the hopes that Greek patrol boats pluck them out of the water. Many of the rescued people, including children, are brought to shore in Greece. From there, most try to make their way to other countries in Europe, especially Germany. This ongoing wave of refugees and migrants, especially from the conflict in Syria, is one of the largest humanitarian crises facing Europe and the global community today.

Migration is a fundamental expression of human adaptation, a response to crisis, or an aspiration for a better or different life. Globalization, it can be argued, has contributed to increased migration as more parts of the world are connected through technology, trade, and the movement of people. Although the movement of goods and capital across international boundaries has become easier, however, the international movement of people across these same borders has been more challenging. There are stories of migrants fleeing conflict, poverty, or prejudice, be it from Burma, Haiti, or other countries around the world. Yet there is growing reluctance to allow these newcomers to settle in their destination countries.

The impasse in Europe regarding the movement and settlement of refugees fleeing the conflict in Syria is one of many such examples. Hoping for a resolution to the conflict, but tired of living in refugee camps, many Syrian refugees have taken to boats or trails to reach Western Europe, encountering both hostility and hospitality along the way. National governments have become deeply divided as to the correct response to this surge of immigrants, and in some cases, the very structure of the European Union (EU) has been tested, as demonstrated by the United Kingdom's vote to leave the EU in 2016. Despite the EU's cosmopolitan reputation and professed tolerance of cultural expression, it seems that living with diversity, a global mix of different peoples, languages, and religions, is not always easy or even desired.

Diversity Amid Globalization investigates these global patterns and interactions through the lens of geography. The analysis is by world regions, which invites consideration of long-term cultural and environmental practices that characterize and shape these distinct areas. Yet we contend that *globalization*—the increasing interconnectedness of people and places through converging economic, political, and cultural activities—is one of the most important forces reshaping the world today. Pundits say globalization is like the weather: It's everywhere, all the time. It is a ubiquitous part of our lives and landscapes that is both beneficial and negative, depending on our needs and point of view. While some people in some places embrace the changes brought by globalization, others resist and push back, seeking refuge in traditional habits and places. Thus, globalization's impact is highly uneven across space, which invites the need for a geographic (or spatial) understanding. As you will see in the pages that follow, geographers, who study places and phenomena around the globe and seek to explain the similarities and differences among places, are uniquely suited to analyze the impacts of globalization in different countries and world regions.

As a counterpoint to globalization, *diversity* refers to the state of having different forms, types, practices, or ideas, as well as the inclusion of distinct peoples, in a particular society. We live on a diverse planet with a mix of languages, cultures, environments, political ideologies, and religions that influence how people in particular localities view the world. At the same time, the intensification of communication, trade, travel, and migration that result from global forces have created many more settings in which people from vastly different backgrounds live, work, and interact. For example, in metropolitan Toronto, Canada's largest city, over half of the area's 5.5 million residents were born in another country. Increasingly, modern diversifying societies must find ways to build social cohesion among distinct peoples. Confronting diversity can challenge a society's tolerance, trust, and sense of shared belonging. Yet, diverse societies also stimulate creative exchanges and new understandings that are beneficial, building greater inclusion. The regional chapters that follow provide examples of the challenges and opportunities that diverse societies in an interconnected world experience today. We begin by introducing the discipline of geography, and then examine this ongoing diversity in the context of globalization from a geographer's perspective.

> ## Globalization's impact is highly uneven across space, which invites the need for a geographic (or spatial) understanding.

LEARNING Objectives

After reading this chapter you should be able to:

1.1 Describe the conceptual framework of world regional geography.

1.2 Identify the different components of globalization, including controversial aspects, and list several ways in which globalization is changing world geographies.

1.3 Summarize the major tools used by geographers to study Earth's surface.

1.4 Explain the concepts and metrics used to document changes in global population and settlement patterns.

1.5 Describe the themes and concepts used to study the interaction between globalization and the world's cultural geographies.

1.6 Explain how different aspects of globalization have interacted with global geopolitics from the colonial period to the present day.

1.7 Identify the concepts and data important to documenting changes in the economic and social development of more and less developed countries.

Geography Matters: Environments, Regions, Landscapes

Geography is a foundational discipline, inspired and informed by the long-standing human curiosity about our surroundings and how we are connected to the world. The term *geography* has its roots in the Greek word for "describing the Earth," and this discipline is central to all cultures and civilizations as humans explore their world, seeking natural resources, commercial trade, military advantage, and scientific knowledge about diverse environments. In some ways, geography can be compared to history: Historians describe and explain what has happened over time, whereas geographers describe and explain the world's spatial dimensions—how it differs from place to place.

Given the broad scope of geography, it is no surprise that geographers have different conceptual approaches to investigating the world. At the most basic level, geography can be broken into two complementary pursuits: *physical* and *human geography*. **Physical geography** examines climate, landforms, soils, vegetation, and hydrology. **Human geography** concentrates on the spatial analysis of economic, social, and cultural systems.

A physical geographer, for example, studying the Amazon Basin of Brazil, might be interested primarily in the ecological diversity of the tropical rainforest or the ways in which the destruction of that environment changes the local climate and hydrology. A human geographer, in contrast, would focus on the social and economic factors explaining the migration of settlers into the rainforest, or on the tensions and conflicts over resources between new settlers and indigenous peoples. Both human and physical geographers share an interest in human–environment dynamics, asking how humans transform the physical environment and how the physical environment influences human behaviors and practices. Thus, they learn that Amazon residents may depend on fish from the river and plants from the forest for food (**Figure 1.1**), but raise crops for export and grow products such as black pepper or soy rather than wheat, because wheat does poorly in humid tropical lowlands.

Another basic division in geography is the focus on a specific topic or theme as opposed to analyzing a specific place or a region. The theme approach is termed **thematic** or **systematic geography**, while the regional approach is called **regional geography**. These two perspectives are complementary and by no means mutually exclusive. This textbook, for example, utilizes a regional scheme for its overall organization, dividing Earth into 12 separate world regions. It then presents each chapter thematically, examining the topics of environment, population and settlement, cultural differentiation, geopolitics, and socioeconomic development in a systematic way. In doing so, each chapter combines four kinds of geography: physical, human, thematic, and regional geography.

Areal Differentiation and Integration

As a spatial science, geography is charged with the study of Earth's surface. A central theme of that responsibility is describing and explaining what distinguishes one piece of the world from another. The geographical term for this is **areal differentiation** (*areal* means "pertaining to area"). Why is one part of Earth humid and lush, while another, just a few hundred kilometers away, is arid (**Figure 1.2**)?

Geographers are also interested in the connections between different places and how they are linked. This concern is one of **areal integration**, or the study of how places interact with one another. An example is the analysis of how and why the economies of Singapore and the United States are closely intertwined, even though the two countries are situated in entirely different physical, cultural, and political environments. Questions of areal integration are becoming increasingly important because of the new global linkages inherent in globalization.

Scale: Global to Local All systematic inquiry has a sense of *scale*, whatever the discipline. In biology, some scientists study the very small units such as cells, genes, or molecules, while others take a larger view, analyzing plants, animals, or whole ecosystems. Geographers also work at different scales. While one may concentrate on analyzing a local landscape—perhaps a single village in southern China—another might focus on the broader regional picture, examining all of southern China. Other geographers do research on a still larger global scale, perhaps studying emerging trade networks between southern India's center of information technology in Bangalore and North America's Silicon Valley, or investigating how the Indian monsoon might be connected to, and affected by, the Pacific Ocean's El Niño phenomenon. But even though geographers may work at different scales, they never lose sight of the interactivity and

▼ **Figure 1.1 Rio Itaya Settlement in the Amazon Basin** A woman and child peer out the doorway of their newly built waterfront home near Iquitos, Peru. Settlers in the Amazon Basin have relied upon the vast forests and rivers of this region for their food, livelihood, and transport.

and practices are connected and impact each other. For example, a geographer interested in economic development may measure income inequality and examine how it differs from one location to another to better understand how poverty might be addressed. Similarly, a geographer interested in the impacts of climate change might model the effects of sea-level change on coastal settlements based on different warming scenarios. An appreciation for space and place is critical in understanding geographic change.

▲ Figure 1.2 **Areal Differentiation** This satellite photo of oasis villages on the southern slope of Morocco's Atlas Mountains is a classic illustration of areal differentiation, or how landscapes can differ significantly within short distances. The dark green bands are irrigated date palm and vegetable fields, watered by rivers that rise in the high mountains and then flow southward into the Sahara Desert. Because irrigated fields along the rivers are precious land, the village settlements are nearby in the dry areas.

connectivity among local, regional, and global scales. They will note the ways that the village in southern China might be linked to world trade patterns or how the late arrival of the monsoon could affect agriculture and food supplies in Bangladesh.

The Cultural Landscape: Space into Place

Humans transform space into distinct places that are unique and heavily loaded with significance and symbolism. **Place**, as a geographic concept, is not just the characteristics of a location but also encompasses the meaning that people give to such areas, as in the sense of place. This diverse fabric of *placefulness* is of great interest to geographers because it tells us much about the human condition throughout the world. Places can tell us how humans interact with nature and how they interact among themselves; where there are tensions and where there is peace; where people are rich and where they are poor.

A common tool for analyzing place is the concept of the **cultural landscape**, which is the tangible, material expression of human settlement, past and present. Thus, the cultural landscape visually reflects the most basic human needs—shelter, food, and work. Additionally, the cultural landscape acts to bring people together (or keep them apart) because it is a marker of cultural values, attitudes, and symbols. As cultures vary greatly around the world, so do cultural landscapes (**Figure 1.3**).

Geographers are also interested in spatial analysis and the concept of space. **Space** represents a more abstract, quantitative, and model-driven approach to understanding how objects

Regions: Formal and Functional

The human intellect seems driven to make sense of the universe by lumping phenomena together into categories that emphasize similarities. Biology has its taxa of living organisms, while history marks off eras and periods of time. Geography, too, organizes information about the world into units of spatial similarity called **regions**—each a contiguous bounded territory that shares one or many common characteristics.

Sometimes, the unifying threads of a region are physical, such as climate and vegetation, resulting in a regional designation like the *Sahara Desert* or *Siberia*. Other times, the threads are more complex, combining economic and social traits, as in the use of the term *Rust Belt* for parts of the northeastern United States that have lost industry and population. Think of a region as spatial shorthand that provides

▼ Figure 1.3 **The Cultural Landscape** Despite globalization, the world's landscapes still have great diversity, as shown by this village and its surrounding rice terraces on the island of Luzon, Philippines. Geographers use the cultural landscape concept to better understand how people interact with their environment.

▶ **Figure 1.4 U.S. Rust Belt** The rust belt is an example of a functional region. It is delimited to show an area that has lost manufacturing jobs and population over the last four decades. By constructing this region, a set of functional relationships is highlighted. **Q: In what formal and functional regions do you live?**

an area with some signature characteristic that sets it apart from surrounding areas. In addition to delimiting an area, generalizations about society or culture are often embedded in these regional labels.

Geographers designate two types of regions: formal and functional. **Formal regions** can be defined by some aspect of physical form, for example a climate type or mountain range such as Appalachia. Cultural features, such as the dominance of a particular language or religion, can also be used to define formal regions. Belgium, for example, can be divided into Flemish-speaking Flanders and French-speaking Wallonia. Many of the maps in this book denote formal regions. In contrast, a **functional region** is one where a certain activity (or cluster of activities) takes place. The earlier example of North America's Rust Belt is such a region because it encompasses a triangle from Milwaukee to Cincinnati to Syracuse, where manufacturing dominated through the 1960s and then experienced steady decline as factories shut down and people left (**Figure 1.4**). Geographers designate functional regions to show associations or activities that can change with distance, such as the spatial extent of a sports team's fan base or the commuter shed of a major metropolitan area such as Los Angeles. Delimiting such regions can be valuable for marketing, planning transportation, or simply thinking about the ways that people identify with an area.

Regions can be defined at various scales. In this book, we divide the world into 12 *world regions* based on formal characteristics such as physical features, language groups, and religious affiliations, but also defined by functional characteristics such as trade groups and regional associations (**Figure 1.5**). Some of these regional groupings are in

▼ **Figure 1.5 World Regions** The boundaries shown here are the basis for the 12 regional chapters in this book. Countries or areas within countries that are treated in more than one chapter are designated on the map with a striped pattern. For example, western China is discussed in both Chapter 10, Central Asia, and Chapter 11, East Asia. Also, three countries on the South American continent are discussed as part of the Caribbean region because of their close cultural similarities to that island region.

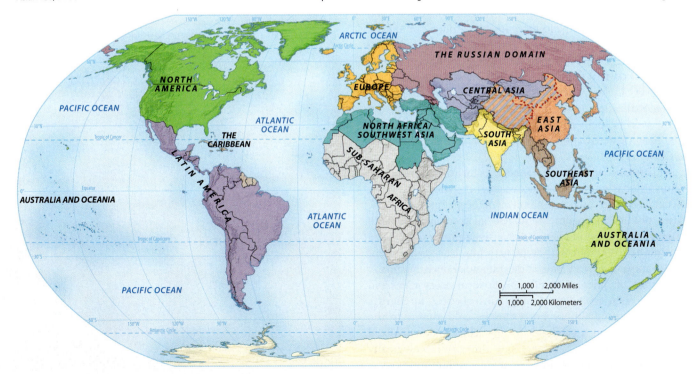

common use, such as Europe or East Asia. Ways of understanding and characterizing these regions have often evolved over centuries. But the boundaries of these regions can and do shift. For example, during the Cold War it made sense to divide Europe into east and west, with eastern Europe closely linked to the Soviet Union. With the 1991 collapse of the Soviet Union and the expansion of the European Union in the 2000s, that divide became less meaningful. Working at the world regional scale invariably creates regions that are not homogeneous, with some states fitting better into regional stereotypes than others. Yet understanding world regional formations is an important way to explore the impacts of globalization on environments, cultures, politics, and development, the focus of the next section.

REVIEW

1.1 Explain the difference between areal differentiation and areal integration.

1.2 How is the concept of place different from space in terms of geographic understanding and analysis?

1.3 How do functional regions differ from formal regions?

■ **KEY TERMS** geography, physical geography, human geography, thematic geography (systematic geography), regional geography, areal differentiation, areal integration, place, cultural landscape, space, regions, formal region, functional region

---

Converging Currents of Globalization

One of the most important features of the 21st century is **globalization**—the increased interconnectedness of people and places around the world. Once-distant regions and cultures are now more and more linked through commerce, communications, and travel. Although earlier forms of globalization existed, especially during Europe's colonial period, the current degree of planetary integration is stronger than ever. In fact, many observers argue that contemporary globalization is the most fundamental reorganization of the world's socioeconomic structure since the Industrial Revolution (see *Exploring Global Connections: A Closer Look at Globalization*).

Economic activities may be the major driving force behind globalization, but the consequences affect all aspects of land and life: human settlement patterns, cultural attributes, political arrangements, and social development are all undergoing profound change. Because natural resources are now global commodities, the planet's physical environment is also impacted by globalization. Financial decisions made thousands of miles away now affect local ecosystems and habitats, often with far-reaching consequences for Earth's health and sustainability. For example, gold mining in the Peruvian Amazon is profitable for the corporations involved and even for individual miners, but it may ruin biologically rich ecosystems and threaten indigenous communities.

The Environment and Globalization

The expansion of a globalized economy is creating and intensifying environmental problems throughout the world. Transnational firms conducting business through international subsidiaries disrupt ecosystems around the globe with their incessant search for natural resources and manufacturing sites. Landscapes and resources previously used by only small groups of local peoples are now considered global commodities to be exploited and traded in the world marketplace.

On a larger scale, globalization is aggravating worldwide environmental problems such as climate change, air and water pollution, and deforestation. Yet it is only through global cooperation, such as the United Nations treaties on biodiversity protection or greenhouse gas reductions, that these problems can be addressed. Environmental degradation and efforts to address it are discussed in detail in Chapter 2.

Globalization and Changing Human Geographies

Globalization changes cultural practices. The spread of a global consumer culture, for example, often accompanies globalization and frequently hurts local economies. It sometimes creates deep and serious social tensions between traditional cultures and new, external global culture. Television shows and movies available via satellite, along with online videos and social media such as Facebook and Twitter, implicitly promote Western values and culture that are then imitated by millions throughout the world (**Figure 1.6**).

▼ **Figure 1.6 Global Communications** The effects of globalization are everywhere, even in remote villages in developing countries. Here, in a small village in southwestern India, a rural family earns a few dollars a week by renting out viewing time on its globally linked television set.

Exploring Global Connections

A Closer Look at Globalization

Globalization comes in many shapes and forms as it connects far-flung people and places. Many of these interactions are common knowledge, such as the global reach of multinational corporations. Others are more complex and sometimes rather surprising. Who would expect to find scores of Bosnian refugees reshaping the economy and society of St. Louis, Missouri? Would you have predicted that the world's busiest international air traffic hub is in the United Arab Emirates?

Indeed, global connections are ubiquitous and often complex—so much so that an understanding of the many different shapes, forms, and scales of these interactions is a key component of the study of world geography. To complement that study, each chapter of this book contains an *Exploring Global Connections* sidebar that presents a globalization case study.

The Chapter 9 sidebar, for example, describes the potential for new trade routes through the Arctic. Other examples include international cooperation and competition at the North and South poles (Chapter 2); the growth of ecotourism in Costa Rica (Chapter 4), and Southeast Asia's resurging opium trade (Chapter 13) (**Figure 1.1.1**). A Google Earth virtual tour supplements each sidebar.

Google Earth
Virtual MG
Tour Video

http://goo.gl/Uorj2U

1. Consider complex global connections based on your own experiences. For example, what item from another part of the world did you buy today, and how did it get to your store?

2. Now choose a foreign place in a completely different part of the world, either a city or a rural village, then suggest how globalization affects the lives of people in that place.

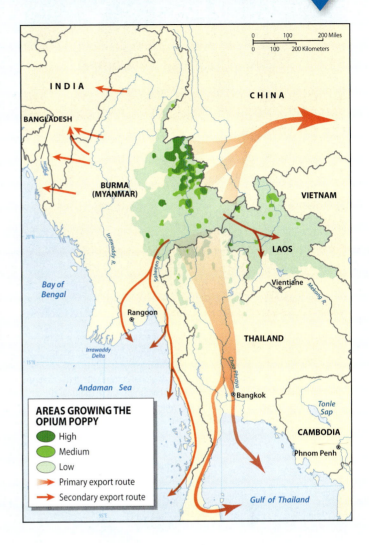

▶ **Figure 1.1.1 Golden Triangle** This area of Southeast Asia is a major production zone for opium poppies, the source of heroin. Export routes link rural villages to markets for the drug throughout the world.

Fast-food franchises are changing—some would say corrupting—traditional diets, with explosive growth in most of the world's cities. Although these foods may seem harmless to North Americans because of their familiarity, they are an expression of deep cultural changes for many societies and are also generally unhealthy and environmentally destructive. Yet some observers contend that even multinational corporations have learned to pay attention to local contexts. **Glocalization** (which combines globalization with locale) is the process of modifying an introduced globalized product or service to accommodate local tastes or cultural practices. For example, a McDonald's in Japan may serve shrimp burgers along with Big Macs.

Although the media give much attention to the rapid spread of Western consumer culture, nonmaterial culture is also dispersed and homogenized through globalization. Language is an obvious example—American tourists in far-flung places are often startled to hear locals speaking an English made up primarily of movie or TV clichés. However, far more than speech is involved, as social values also are dispersed globally. Changing expectations about human rights, the role of women in society, and the intervention of nongovernmental organizations are also expressions of globalization that may have far-reaching effects on cultural change.

In return, cultural products and ideas from around the world greatly impact U.S. culture. The large and diverse immigrant population in the United States has contributed to heightened cultural diversity and exchange. The internationalization of American food and music and the multiple languages spoken in U.S. cities are all expressions of globalization (**Figure 1.7**).

Globalization also clearly influences population movements. International migration is not new, but increasing numbers of people from all parts of the world are now crossing national boundaries, legally and illegally, temporarily and permanently. The United Nations (UN) estimates that there are nearly 250 million immigrants in the world (people who are living in a country other than their country of birth). **Figure 1.8** shows the major migration flows from

▲ **Figure 1.7 Ethiopian Culture in Washington, DC** While many think that globalization is the one-way spread of North American and European socioeconomic traits into the developing world, one needs only to look around their own neighborhood to find expressions of global culture within the United States. For example, the largest concentration of Ethiopians in the United States is in Metropolitan Washington, DC, where Ethiopian cuisine and music are a visible presence in the nation's capital.

regions of origin designated as Africa, Asia, Europe, Latin America and the Caribbean, North America, and Oceania. One of the most striking aspects of the figure is that many of the largest international flows are *intra-regional* (60 million within Asia, 40 million within Europe, and 18 million within Africa). Yet there are also substantial *inter-regional* flows, such as 26 million from Latin America and the Caribbean to North America and 20 million from Asia to Europe or 17 million from Asia to North America. Attempts to control the movement of people are evident throughout the world—much more so, in fact, than control over the movement of goods or capital. Yet this growing flow of immigrants is propelled, in part, by the uneven economic development (discussed in more detail later in the chapter) and demographic changes in regions where populations are aging and labor is needed.

Geopolitics and Globalization

Globalization also has important geopolitical components. An essential dimension of globalization is that this process is not restricted by territorial or national boundaries. For example, the creation of the United Nations following World War II was a step toward creating an international governmental structure in which all nations could find representation (**Figure 1.9**).

The simultaneous emergence of the Soviet Union as a military and political superpower led to a rigid division into Cold War blocs that slowed further geopolitical integration. However, with the peaceful end of the Cold War in the early 1990s, the former communist countries of eastern Europe and the Soviet Union were opened almost immediately to global trade and cultural exchange, changing those countries immensely. This also greatly increased the number of immigrants within Europe.

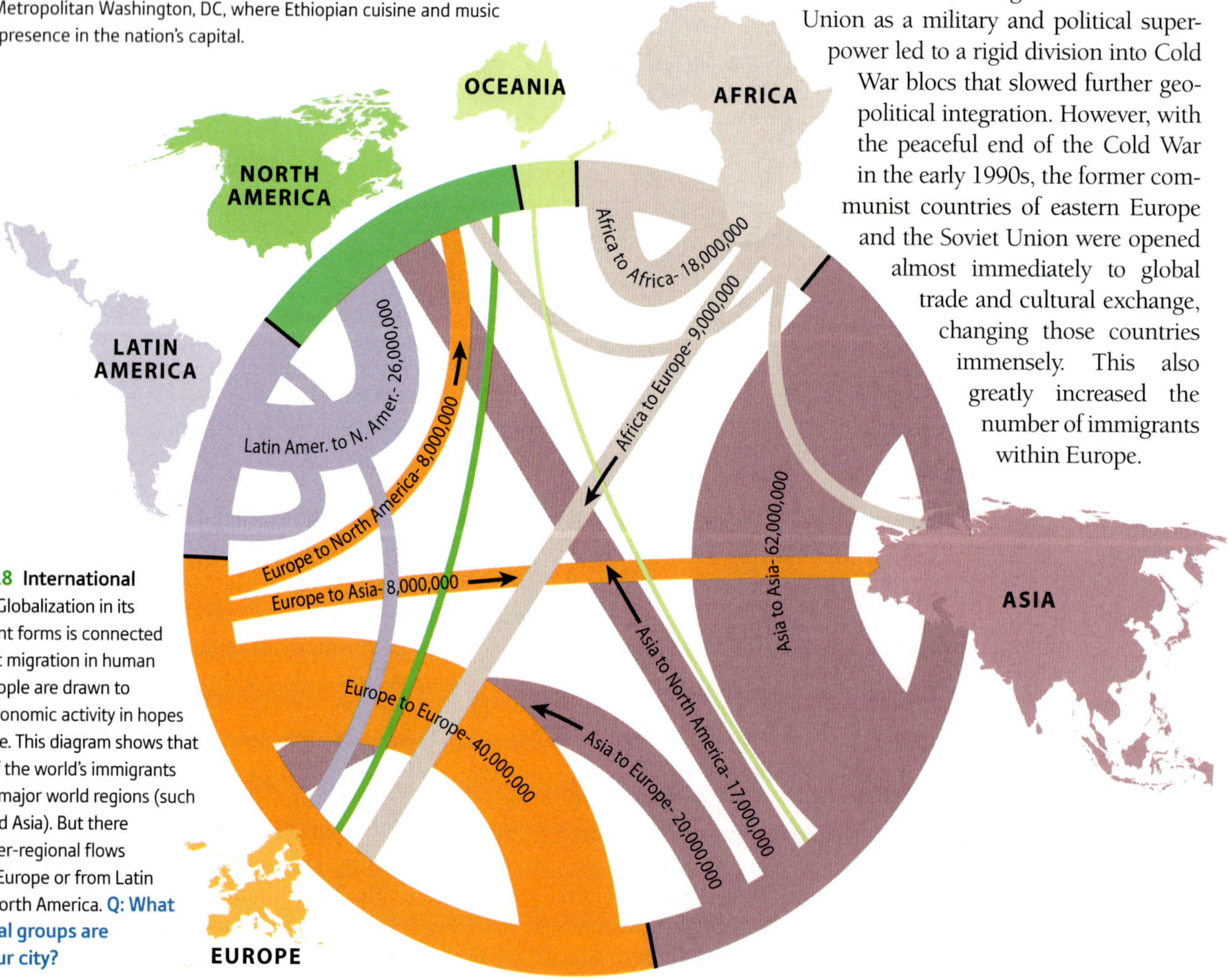

▶ **Figure 1.8 International Migration** Globalization in its many different forms is connected to the largest migration in human history as people are drawn to centers of economic activity in hopes of a better life. This diagram shows that nearly half of the world's immigrants move within major world regions (such as Europe and Asia). But there are major inter-regional flows from Asia to Europe or from Latin America to North America. **Q: What international groups are found in your city?**

OCEANIA

NORTH AMERICA

AFRICA

LATIN AMERICA

Africa to Africa- 18,000,000

Africa to Europe- 9,000,000

Latin Amer. to N. Amer.- 26,000,000

Europe to North America- 8,000,000

Europe to Asia- 8,000,000

Asia to Asia- 62,000,000

ASIA

Europe to Europe- 40,000,000

Asia to North America- 17,000,000

Asia to Europe- 20,000,000

EUROPE

trade. Even areas that do not directly produce drugs are involved in their global sale and transshipment; many Caribbean economies are becoming reoriented to drug transshipments and the laundering of drug money. Prostitution, pornography, and gambling have also emerged as highly profitable global businesses. Over the past decades, for example, parts of eastern Europe have become major sources of both pornography and prostitution, finding a lucrative, but morally questionable, niche in the new global economy.

▲ **Figure 1.9 UN Peacekeepers in Africa** A convoy rolls past displaced people walking towards a UN camp outside of Malakal, South Sudan. Conflict in South Sudan has displaced tens of thousands. The town of Malakal was destroyed by fighting; its former residents sought shelter in a UN camp.

A significant international criminal element is another globalization outcome and includes terrorism (discussed later in this chapter), drugs, pornography, slavery, and prostitution, which require international coordination and agreements to address (**Figure 1.10**). Some of the world's most remote places, such as the mountains of northern Burma and the valleys of southern Afghanistan, are thoroughly integrated into the circuits of global exchange through the production of opium that is central to the world heroin

Further, many observers argue that globalization—almost by definition—has weakened the political power of individual states by strengthening regional economic and political organizations, such as the European Union and the World Trade Organization (WTO). In some world regions, a weakening of traditional state power has led to stronger local and separatist movements, as illustrated by the turmoil on Russia's southern borders. Yet even established regional blocs such as the EU may be contested, as witnessed by the surprising result of the 2016 referendum in the United Kingdom to leave the European Union. Similarly, many view the election of U.S. President Donald J. Trump in 2016 as a vote against open trade and open borders, as he campaigned aggressively against such policies.

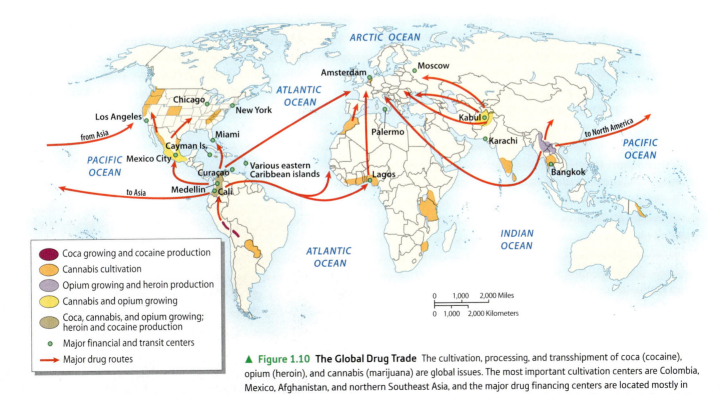

▲ **Figure 1.10 The Global Drug Trade** The cultivation, processing, and transshipment of coca (cocaine), opium (heroin), and cannabis (marijuana) are global issues. The most important cultivation centers are Colombia, Mexico, Afghanistan, and northern Southeast Asia, and the major drug financing centers are located mostly in the Caribbean, the United States, and Europe. In addition, Nigeria and Russia play significant roles in the global transshipment of illegal drugs.

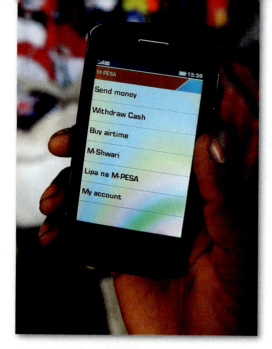

▲ **Figure 1.11 Global Use of Cell Phones** Mobile technologies have revolutionized the way people communicate, acquire information, and interact in a globalized world. Today cell phones are used for far more than simply talking. In Nairobi, Kenya, the majority of the city's adult population now uses M-Pesa, a cell-phone-based money transfer service, to pay for everything from street food to rides on the city's privately owned minibuses.

Economic Globalization and Uneven Development Outcomes

Most scholars agree that the major component of globalization is the economic reorganization of the world. Although different forms of a world economy have existed for centuries, a well-integrated, truly global economy is primarily the product of the past several decades. The attributes of this system, while familiar, are worth stating:

- Global communication systems and the digital flow of information that links all regions and most people instantaneously (**Figure 1.11**)

- Transportation systems that can quickly and inexpensively move goods by air, sea, and land

- Transnational business strategies that have created global corporations more powerful than many sovereign nations

- New and more flexible forms of capital accumulation and international financial institutions that make 24-hour trading possible

- Global and regional trade agreements that promote more free trade

- Market economies and private enterprises that have replaced state-controlled economies and services

- An emphasis on producing more goods, services, and data at lower costs to fulfill consumer demand for products and information (**Figure 1.12**)

- Growing income inequality between rich and poor, both within and between countries

This global reorganization has resulted in unprecedented economic growth in some areas of the world in recent years; China is a good example, with an average annual growth rate of 8.6 percent from 2010 to 2014. But not everyone profits from economic globalization, as the growing wage gap within China indicates, nor have all world regions shared equally in the benefits.

Thinking Critically About Globalization

Globalization, particularly its economic aspects, is one of today's most contentious issues. Supporters believe that it results in a greater economic efficiency that will eventually result in rising prosperity for the entire world. In contrast, critics claim that globalization largely benefits those who are already prosperous, leaving most of the world poorer than before as the rich and powerful exploit the less fortunate. Increasingly, scholars discuss the pros and cons of digital globalization, which is less about the movement of capital, goods, and people but instead describes the accelerated movement of data to facilitate daily demands for information, searches, financial transactions, communication, and video.

Pro-Globalization Economic globalization is generally applauded by corporate leaders and economists, and underlies the pro-market reforms and fiscal discipline as exemplified by the **Washington Consensus**, a term referring to a set of policies for economic development in which the key players include the U.S. Treasury and Federal Reserve, the World Bank, the International Monetary Fund (IMF), and the World Trade Organization. The primary function of the World Bank is to make loans to poor countries so that they can invest in infrastructure and build more modern economic foundations. The IMF makes short-term loans to countries in financial difficulty—those having trouble, for example, making interest payments on the loans that they had previously taken. The WTO, a much smaller organization than the other two, works to reduce trade barriers between countries to enhance economic globalization. The WTO also tries to mediate between countries and trading blocs that are engaged in trade disputes.

▼ **Figure 1.12 Chinese Factories** Workers sew denim jeans in the city of Shenzhen, China. Typcially from rural parts of China where wages are lower, these workers live in factory-owned dorms and work six days a week. The products they sew are shipped around the world.

Beyond North America, moderate and conservative politicians in most countries generally support free trade and other aspects of economic globalization. Advocates argue that globalization is a logical and inevitable expression of contemporary international capitalism that benefits all nations and all peoples. Economic globalization can work wonders, they contend, by enhancing competition, increasing the flow of capital and employment opportunities to poor areas, and encouraging the spread of beneficial new technologies and ideas. To support their claims, pro-globalizers argue that countries that have embraced the global economy have generally enjoyed more economic success than those that have sought economic self-sufficiency. The world's most isolated countries, Burma (Myanmar) and North Korea, are economic disasters with little growth and rampant poverty, whereas those that have opened themselves to global forces during the same period, such as Singapore and Thailand, have seen rapid growth and substantial reductions in poverty.

Anti-Globalization Opponents of globalization, such as labor and environmental groups, as well as many social justice movements, often argue that globalization is not a "natural" process. Instead, it is the product of an explicit economic policy promoted by free-trade advocates, capitalist countries (mainly the United States, but also Japan and the countries of Europe), financial interests, international investors, and multinational firms that maximize profits by moving capital and seeking low-wage labor. Further, because the globalization of the world economy is creating greater inequity between rich and poor, the trickle-down model of developmental benefits for all people in all regions has yet to be validated. On a global scale, the richest 20 percent of the world's people consume 86 percent of the world's resources, whereas the poorest 80 percent use only 14 percent. Critics also worry that a globalized economic system—with its instantaneous transfers of vast sums of money over nearly the entire world on a daily basis—is inherently unstable. The world-wide recession of 2008–2010 demonstrated that global interconnectivity can also increase economic vulnerability, as illustrated by the collapse of financial institutions in Iceland or the decline of remittances from Mexicans working in North America to their families in Mexico.

There are growing concerns that an emphasis on export-oriented economies at the expense of localized ones has led to overexploitation of resources. World forests, for example, are increasingly cut for export timber, rather than serving local needs. As part of their economic structural adjustment package, the World Bank and the IMF often encourage developing countries to expand their resource exports to earn more hard currency to make payments on their foreign debts. When commodity prices are high, this strategy can stimulate growth, but as commodity prices decline, as they did in 2014–2016, growth rates in developing countries slow. Moreover, the IMF often requires developing countries to adopt programs of fiscal austerity that entail substantial reductions in public spending for education, health, and food subsidies. Adopting such policies, critics warn, will further impoverish the people of these poor countries (**Figure 1.13**).

A Middle Position A middle-ground position argues that economic globalization is indeed unavoidable and that, despite its promises and pitfalls, globalization can be managed at both the national and the international levels to reduce economic inequalities and protect the natural environment. These experts stress the need for strong, yet efficient national governments, supported by international institutions (such as the UN, World Bank, and IMF) and globalized networks of nongovernmental environmental, labor, and human rights groups. Moreover, the global movement of goods has flattened in the last few years, whereas the digital flow of information has soared, creating new opportunities and pitfalls that require further study.

Unquestionably, globalization is one of the most important issues of the day—and certainly one of the most complicated. While this book does not pretend to resolve the controversy, nor does it take a position, it does encourage readers to reflect on these critical points as they apply to each world region.

Diversity in a Globalizing World

As globalization progresses, many observers foresee a world far more uniform and homogeneous than today's. The optimists among them imagine a universal global culture uniting all humankind into a single community untroubled by war, ethnic strife, or resource shortage—a global utopia of sorts.

▼ **Figure 1.13 Protests Against Globalization** Meetings of international groups such as the World Trade Organization (WTO) and the International Monetary Fund (IMF) commonly draw protesters against economic globalization. This group of protesters at a WTO meeting in Bali, Indonesia is demanding more local autonomy in food systems.

A more common view is that the world is becoming blandly homogeneous as different places, peoples, and environments lose their distinctive character and become indistinguishable from their neighbors. Yet even as globalization generates a certain degree of homogenization, the world is still a highly diverse place (**Figure 1.14**). We can still find marked differences in culture (language, religion, architecture, foods, and other attributes of daily life), economy, and politics—as well as in the physical environment—from place to place. Such **diversity** is so vast that it cannot readily be extinguished, even by the most powerful forces of globalization. Diversity may be difficult for a society to live with, but it also may be dangerous to live without. Nationality, ethnicity, cultural distinctiveness—all are defining expressions of humanity that are nurtured in distinct places.

In fact, globalization often provokes a strong reaction on the part of local people, making them all the more determined to maintain what is distinctive about their way of life. Thus, globalization is understandable only if we also examine the diversity that continues to characterize the world and, perhaps most important, the tension between these two forces: the homogenizing power of globalization and the reaction against it, often through demands for protecting cultural distinctiveness.

The politics of diversity demand increasing attention as we try to understand global terrorism, ethnic identity, religious practices, and political independence. Groups of people throughout the world seek self-rule of territory they can call their own. Today most wars are fought *within* countries, not *between* them. As a result, our interest in geographic diversity takes many forms and goes far beyond simply celebrating traditional cultures and unique places. People have

many ways of making a living throughout the world, and it is important to recognize this fact as the globalized economy becomes increasingly focused on mass-produced retail goods. Furthermore, a stark reality of today's economic landscape is uneven outcomes: While some people and places prosper, others suffer from unrelenting poverty (**Figure 1.15**). To analyze these patterns of unevenness and change, the next section considers the tools used by geographers to better know the world.

▼ **Figure 1.15 Landscape of Economic Inequality** The geography of diversity takes many expressions, one of which is economic unevenness, as depicted in this photo from the city of Makati in the Philippines where squatter settlements of the poor contrast with high-rise office buildings and apartments of the more affluent.

▼ **Figure 1.14 Shopping in Isfahan** Young women shop in the grand bazaar in Isfahan, Iran, in preparation for Eid al-Fitr, the celebration at the end of Ramadan. While few places are beyond the reach of globalization, it is also true that distinct cultures, traditions, and landscapes exist in the world's various regions.

REVIEW

1.4 Provide examples of how globalization impacts the culture of a place or region.

1.5 Describe and explain five components of economic globalization.

1.6 Summarize three elements of the controversy about globalization.

■ **KEY TERMS** globalization, glocalization, Washington Consensus, diversity

The Geographer's Toolbox: Location, Maps, Remote Sensing, and GIS

Geographers use many different tools to represent the world in a convenient form for examination and analysis. Different kinds of images and data are needed to study vegetation change in Brazil or mining activity in Mongolia; population density in Tokyo or language regions in Europe; religions practiced in Southwest Asia or rainfall distribution in southern India. Knowing how to display and interpret information in map form is part of a geographer's skill set. In addition to traditional maps, today's modern satellite and communications systems provide an array of tools not imagined 50 years ago.

Latitude and Longitude

To navigate their way through daily tasks, people generally use a mental map of *relative locations* that locate specific places in terms of their relationship to other landscape features. The shopping mall is near the highway, perhaps, or the college campus is along the river. In contrast, map makers use *absolute location*—often called a mathematical location—which draws upon a universally accepted coordinate system providing every place on Earth with a specific numerical address based upon latitude and longitude. The absolute location for the Geography Department at the University of Oregon, for example, has the mathematical address of 44 degrees, 02 minutes, and 42.95 seconds north and 123 degrees, 04 minutes, and 41.29 seconds west. This is written 44° 02' 42.95" N and 123° 04' 41.29" W.

Lines of **latitude**, often called parallels, run east–west around the globe and are used to locate places north and south of the equator (0 degrees latitude). In contrast, lines of **longitude**, called meridians, run from the North Pole (90 degrees north latitude) to the South Pole (90 degrees south latitude). Longitude values locate places east or west of the **prime meridian**, located at 0 degrees longitude at the Royal Naval Observatory in Greenwich, England (just east of London) (**Figure 1.16**). The equator itself divides the globe into northern and southern hemispheres, whereas the prime meridian divides the world into eastern and

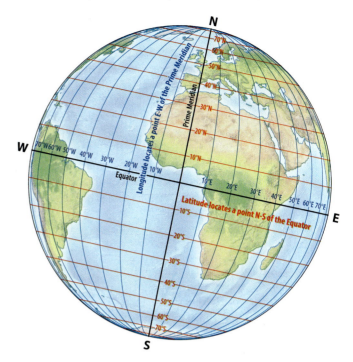

▲ **Figure 1.16 Latitude and Longitude** Latitude locates a point between the equator and the poles and is designated as so many degrees north or south. Longitude locates a point east or west of the prime meridian, which is located just east of London at the Royal Observatory in Greenwich, England. **Q: What is the latitude and longitude of your school?**

western hemispheres; these east-west hemispheres meet at 180 degrees longitude in the western Pacific Ocean. The International Date Line, where each new solar day begins, lies along much of 180 degrees longitude, deviating where necessary to ensure that small Pacific island nations remain on the same calendar day.

A degree of latitude measures 60 nautical miles or 69 land miles (111 km) and is made up of 60 minutes. Each minute is 1 nautical mile (1.15 land miles) and measures 60 seconds of distance, each of which is approximately 100 feet (30.5 meters).

From the equator, parallels of latitude are used to mathematically define the tropics: the Tropic of Cancer at 23.5 degrees north and the Tropic of Capricorn at 23.5 degrees south. These lines of latitude denote where the Sun is directly overhead at noon on the solar solstices in June and December. Similarly, the Arctic and Antarctic circles, at 66.5 degrees north and south latitude, respectively, mathematically define where these areas experience 24 hours of sunlight on the summer solstice and 24 hours of complete darkness on the winter solstice.

Global Positioning Systems (GPS)

Historically, precise measurements of latitude and longitude were determined by a complicated method of celestial navigation, based upon the observer's location relative to the Sun, Moon, planets, and stars. Today, though, absolute location on Earth (or in airplanes above Earth's surface) is achieved through satellite-based **global positioning systems (GPS)**. These systems use time signals sent from your location to a satellite and back to your GPS receiver (which can be a smart phone) to calculate precise coordinates of latitude and longitude. GPS was first used by the U.S. military in the 1960s and then made available to the public in the later decades of the 20th century. Today GPS guides airplanes across the skies, ships across the oceans, private autos on the roads, and hikers through wilderness areas. In the future such systems will guide driverless cars. While most smartphones use locational systems based on triangulation from cell phone towers, some smartphones are capable of true satellite-based GPS accurate to 3 feet (or 1 meter).

Map Projections

Because the world is spherical, mapping the globe on a flat piece of paper creates inherent distortions in the latitudinal, or north–south, depiction of Earth's land and water areas. Cartographers (those who make maps) have tried to limit these distortions by using various **map projections**, defined as the different ways to project a spherical image onto a flat surface. Historically, the Mercator projection was the projection of choice for maps used for oceanic exploration. However, a glance at the inflated Greenland and Russian landmasses shows its weakness in accurately depicting high-latitude land areas (**Figure 1.17**). Over time, cartographers have created literally hundreds of different map projections in their attempts to find the best and most accurate way of mapping different parts of the world.

For the last several decades cartographers have generally used the Robinson projection for their maps and atlases. In fact, several professional cartographic societies tried

unsuccessfully in 1989 to actually ban projections such as the Mercator because of their spatial distortions. Like many other professional publications, maps in this book utilize the Robinson projection.

Map Scale

All maps must reduce the area being mapped to a smaller piece of paper. This reduction involves the use of **map scale**, or the mathematical ratio between the map and the surface area being mapped. Many maps note their scale as a ratio or fraction between a unit on the map and the same unit in the area being mapped. An illustration is 1:63,360 or 1/63,360, which means that 1 inch on the map represents 63,360 inches on the land surface; thus, the scale is 1 inch equals 1 mile, or 1:63,360. Although this is a convenient mapping scale to understand, the amount of surface area that can be mapped on a standard-sized sheet of paper at this scale is limited to about 20 square miles. But at this scale mapping a larger area, say 100 square miles, would produce a much larger, unwieldy map. Therefore, the ratio must be changed to a larger number, such as 1:316,800, or 1 inch on the map represents 5 miles (8 km) on land.

Based upon the *representative fraction*, which is the cartographic term for the numerical value of map scale, maps are categorized as having either large or small scales (**Figure 1.18**). It may be easy to remember that large-scale maps make landscape features like rivers, roads, and cities *larger,* but because the features are larger, the maps must cover *smaller* areas. Conversely, small-scale maps cover *larger* areas, but to do so, these maps must make landscape features *smaller.* A bit harder to remember is that the larger the second

Animation MG
Map Projections

https://goo.gl/Y3z9rY

(a) Mercator projection

(b) Robinson projection

▲ **Figure 1.17 Map Projections** Cartographers have long struggled with how best to accurately map the world given the inherent distortions when transferring features on a round globe to a flat piece of paper. Early map makers commonly used the Mercator projection (a) which distorts features in the high latitudes, but worked fairly well for seagoing explorers. The map on the right (b) is the Robinson projection, which was developed in the 1960s and is now the industry standard because it minimizes cartographic distortion.

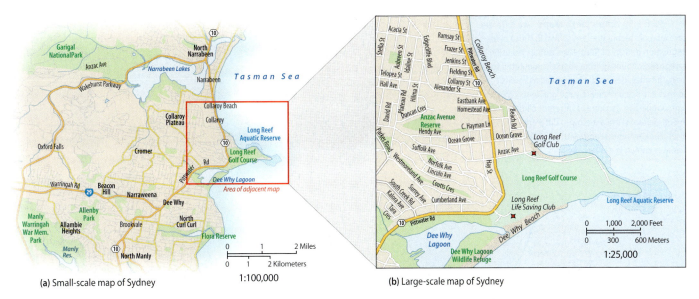

(a) Small-scale map of Sydney

1:100,000

(b) Large-scale map of Sydney

1:25,000

▲ **Figure 1.18 Small and Large Scale Maps** A portion of Australia's east coast north of Sydney is mapped at two scales, one (on the left) at a small scale and the other (on the right) at a large scale. Note the differences in distance depicted on the linear scales of the two different maps. There is more close-up detail in the large-scale map, but it covers only a small portion of the area mapped at a small scale.

number of the representative fraction—the 63,360 in the fraction 1/63,360 or the 100,000 in the fraction 1/100,000, for example—the smaller the scale of the map.

Map scale is probably easiest to interpret when shown as a **graphic** or **linear scale**, which depicts units such as feet, meters, miles, or kilometers in a horizontal bar. Most of the maps in this book are small-scale maps of large areas; thus, the graphic scale is in miles and kilometers. You can measure distances between two points on the map by making two tick marks on a piece of paper held next to the points and then measuring the distance between the two marks on the linear scale.

Map Patterns and Legends

Maps depict everything from the most basic representation of topographic and landscape features to complicated patterns of population, migration, economic conditions, and more. A map can be a simple *reference map* showing the location of certain features, or a *thematic map* displaying data such as rainfall patterns or income distribution. Most of the maps in this text are thematic maps illustrating complicated spatial phenomena. Every map has a *legend* that provides information on the categories used in the map, their values (when relevant), and other symbols that may need explanation.

One type of thematic map used often in this book is the **choropleth map**, in which color shades represent different data values, with darker shades generally showing larger average values. Per capita income and population density are often represented by these maps, with data divided into categories and then mapped by spatial units such as countries, provinces, counties, or neighborhoods. The category breaks and spatial units selected can have a dramatic impact on the patterns shown in a choropleth map (**Figure 1.19**).

Aerial Photos and Remote Sensing

Although maps are a primary tool of geography, much can be learned about Earth's surface by deciphering patterns on aerial photographs taken from airplanes, balloons, or satellites. Originally available only in black and white, today these images are digital and can exploit visible light (like a photograph) or other light wavelengths such as infrared that are not visible to the human eye.

Even more information about Earth comes from electromagnetic images taken from aircraft or satellites, termed **remote sensing** (**Figure 1.20**). Unlike aerial photography, remote sensing gathers electromagnetic data that must be processed and interpreted by computer software to produce images of Earth's surface. This technology has many applications, including monitoring the loss of rainforests, tracking the biological health of crops and woodlands, and even measuring the growing of cities. Remote sensing is also central to national defense, such as monitoring the movements of troops or the building of missile sites in hostile countries.

The Landsat satellite program launched by the United States in 1972 is a good example of both the technology and the uses of remote sensing. These satellites collect data simultaneously in four broad bands of electromagnetic energy, from visible through near-infrared wavelengths, that is reflected or emitted from Earth. The resolution on Earth's surface ranges from areas 260 square feet (80 square meters) down to 98 square feet (30 square meters).

Commercial satellites such as GeoEye and DigitalGlobe now provide high-resolution satellite imagery down to 1.5 square feet (or 0.5 square meter). This means that a car, small structure, or group of people would be easily seen, but not an individual person. Of course, cloud cover often compromises the continuous coverage of many parts of the world.

▶ **Figure 1.19 Choropleth Maps** Two different cartographic techniques are shown in these maps. (a) The population density of South Asia is mapped using different categories of density, from sparsely populated to very high densities, depicted with increasing intensity of colors so that you see immediately the gradients from low to high population density. (b) Different climate categories in Sub-Saharan Africa are given different colors, with drier climates represented with sand-like tan and wetter climates shown in darker colors.

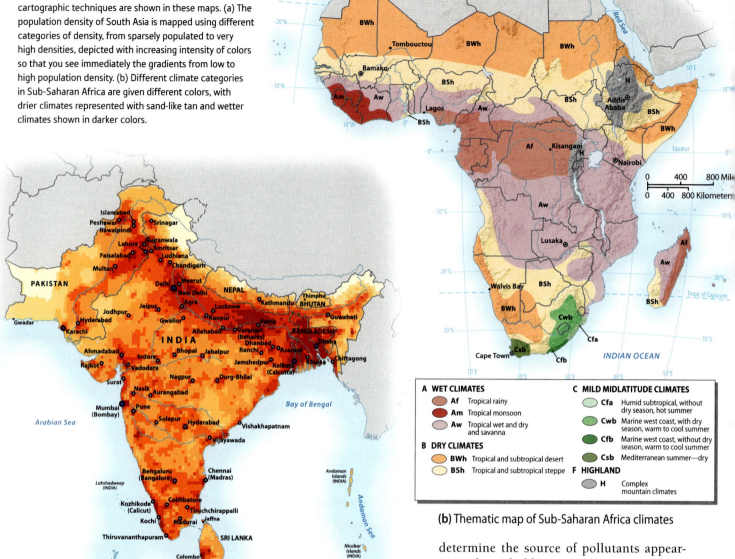

(b) Thematic map of Sub-Saharan Africa climates

A	WET CLIMATES		C	MILD MIDLATITUDE CLIMATES	
	Af	Tropical rainy		Cfa	Humid subtropical, without dry season, hot summer
	Am	Tropical monsoon		Cwb	Marine west coast, with dry season, warm to cool summer
	Aw	Tropical wet and dry and savanna		Cfb	Marine west coast, without dry season, warm to cool summer
B	DRY CLIMATES			Csb	Mediterranean summer—dry
	BWh	Tropical and subtropical desert	F	HIGHLAND	
	BSh	Tropical and subtropical steppe		H	Complex mountain climates

PEOPLE PER SQUARE KILOMETER
- Fewer than 6
- 6–25
- 26–100
- 101–250
- 251–500
- 501–1,000
- 1,001–12,801
- More than 12,800

POPULATION
- Metropolitan areas more than 20 million
- Metropolitan areas 10–20 million
- Metropolitan areas 5–9.9 million
- Metropolitan areas 1–4.9 million
- Selected smaller metropolitan areas

(a) Choropleth map of South Asia population density

Geographic Information Systems (GIS)

Vast amounts of computerized data from different sources, such as maps, aerial photos, remote sensing, and census tracts, are brought together in **geographic information systems (GIS)**. The resulting spatial databases are used to analyze a wide range of issues. Conceptually, GIS can be thought of as a computer system for producing a series of overlay maps showing spatial patterns and relationships (**Figure 1.21**). A GIS map, for example, might combine a conventional map with data on toxic waste sites, local geology, groundwater flow, and surface hydrology to

determine the source of pollutants appearing in household water systems.

Although the earliest GIS dates back to the 1960s, it is only in the last several decades—with the advent of desktop computer systems and remote sensing data—that GIS has become absolutely central to geographic problem solving. It has a central role in city planning, environmental science, public health, and real estate development, to name a few of the many activities using these systems.

REVIEW

1.7 Explain the difference between latitude and longitude.

1.8 What does a map's scale tell us? List two ways to portray map scale.

1.9 What is a choropleth map, and what might it depict?

1.10 How does remote sensing differ from aerial photography?

■ **KEY TERMS** latitude, longitude, prime meridian, global positioning systems (GPS), map projections, map scale, graphic (linear) scale, choropleth map, remote sensing, geographic information systems (GIS)

◄ **Figure 1.20 Remote Sensing of Urbanization in Dubai** This NASA satellite image of Dubai shows the extraordinary changes that have taken place along the arid gulf coast of the United Arab Emirates. Sprawling urbanization, construction of port facilities, new water features, and the creation of expensive island real estate (one shaped like a palm and the other shaped like the continents). Areas in red are irrigated green spaces for parks and golf courses.

Explore the **Sights** of Dubai's Palm Jumeirah

http://goo.gl/4akxhs

Themes and Issues in World Regional Geography

Following two introductory chapters, this book adopts a regional perspective, grouping all of Earth's countries into a framework of 12 world regions (see Figure 1.5). We begin with a region familiar to most of our readers—North America—and then move on to Latin America, the Caribbean, Sub-Saharan Africa, North Africa and Southwest Asia, Europe, the Russian Domain, and the different regions of Asia, before concluding with Australia and Oceania. Each regional chapter employs the same five-part thematic structure—physical geography and environmental issues, population and settlement, cultural coherence and diversity, geopolitical framework, and economic and social development. The concepts and data central to each theme are discussed in the following sections.

▼ **Figure 1.21 GIS Layers** Geographic information systems (GIS) maps usually consist of many different layers of information that can be viewed and analyzed either separately or as a composite overlay. This is a typical environmental planning map where different physical features (such as wetlands and soils) are combined with zoning regulations.

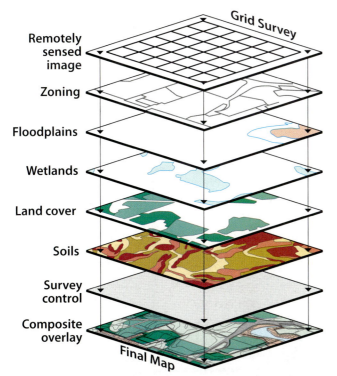

Grid Survey

Remotely sensed image

Zoning

Floodplains

Wetlands

Land cover

Soils

Survey control

Composite overlay

Final Map

Physical Geography and Environmental Issues: The Changing Global Environment

Chapter 2 provides background on world physical and environmental geography, outlining the global elements fundamental to human settlement—landforms, climate, vegetation, and water and energy resources. In the regional chapters, the physical geography sections describe the physical environment and explain the environmental issues relevant to each world region, such as climate change, sea-level rise, acid rain, energy and resource issues, deforestation and wildlife conservation. Each regional chapter also discusses policies and plans to resolve those issues (see *Working Toward Sustainability: Meeting the Needs of Future Generations*).

Meeting the Needs of Future Generations

The word *sustainability* seems to be everywhere, as we hear about sustainable cities, agriculture, forestry, businesses—even sustainable lifestyles. With so many different uses of the word, it is appropriate to revisit its original definition.

Sustainable has two meanings: The first is to endure or to maintain something at a certain level so that it lasts. The second means something that can be upheld or defended, such as a *sustainable idea* or *action.* Resource management has long used terms such as *sustained-yield forestry* to refer to timber practices in which tree harvesting is attuned to the natural rate of forest growth so that the resource is not exhausted, but can renew itself over time.

Moral and ethical dimensions were added to this traditional usage in 1987 when the UN World Commission on Environment and Development addressed the complicated relationship between economic development and environmental deterioration. The commission stated that "sustainable development is development that meets the needs of the present without compromising the ability of future generations to meet their own needs." This cautionary message expands the notion of sustainability from a narrow focus on resource management to include the whole range of human "needs," both now and in the future. Fossil fuels, for example, are finite, so we should not greedily

Google Earth Virtual (MG) Tour Video

http://goo.gl/oGTPq9

consume them now without considering their availability for future generations. Similar cautions apply to sustainable use of all other resources—air, water, genetic biodiversity, wildlife habitats, and so on.

Sustainably utilizing a specific resource, however, can be extremely difficult because it requires knowing the total amount of the resource in question and the current rate of consumption, and then estimating the needs of future generations. These challenges have led to the field of *sustainability science*, which emphasizes quantifying these factors. Because of these measurement difficulties, many researchers suggest that sustainability is better thought of as a process rather than an achievable state.

Sidebars in the following chapters explore the different ways that people around the world are working toward environmental and resource sustainability. Examples include the city of Copenhagen's bike culture (Chapter 8; see **Figure 1.2.1**); solar power in Morocco (Chapter 7); and gibbon conservation in Cambodian forests (Chapter 13). Each sidebar links to a Google Earth virtual tour video.

1. Does your college or community have a sustainability plan? If so, what are the key elements?

2. How might the concept of sustainability differ for a city or town in India or China compared to a U.S. city? Browse the Internet to see what you can learn about sustainability programs in other cities.

Population and Settlement: People on the Land

Currently, Earth has more than 7.4 billion people, with demographers—those who study population dynamics—forecasting an increase to 9.8 billion by 2050. Most of that increase will take place in Africa, particularly in Sub-Saharan Africa, North Africa and Southwest Asia, and Australia and Oceania (**Figure 1.22**). In contrast, the regions of Europe, the Russian Domain, and East Asia will likely experience no demographic growth between now and 2050. Population concerns vary, with some countries, such as Bangladesh, trying to slow population growth, while others, such as Ukraine, worry about population decline.

Population is a complex topic, but several points may help to focus the issues:

- The current rate of population growth is now half the peak rate experienced in the early 1960s, when the world population was around 3 billion. At that time, talk of a "population bomb" and "population explosion" was common, as scholars and activists voiced concern about what might happen if such high growth rates continued. Still, even with today's slower growth, demographers predict that some 2.5 billion more people will

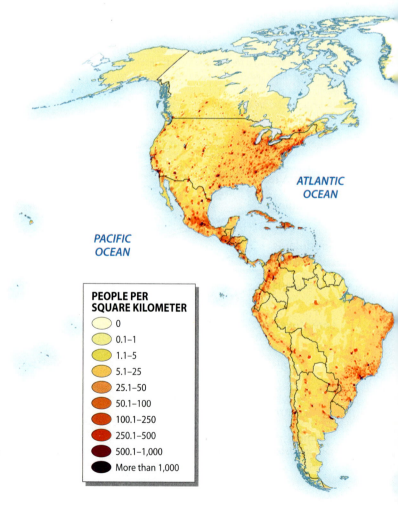

ATLANTIC OCEAN

PACIFIC OCEAN

PEOPLE PER SQUARE KILOMETER
- 0
- 0.1–1
- 1.1–5
- 5.1–25
- 25.1–50
- 50.1–100
- 100.1–250
- 250.1–500
- 500.1–1,000
- More than 1,000

▶ **Figure 1.22 World Population** This map emphasizes the world's different population densities. East and South Asia stand out as the most populated regions, with high densities in Japan, eastern China, northern India, and Bangladesh. In arid North Africa and Southwest Asia, population clusters are often linked to the availability of water, as is apparent with the population cluster along the Nile River. Higher population densities in Europe, North America, and other countries are usually associated with major metropolitan areas.

▲ **Figure 1.2.1 Copenhagen's Bike Culture** Urban planners are designing for bicycles as a key transportation mode that is clean and efficient.

be added by 2050, with much of the growth taking place in the world's poorest countries.

- Population planning takes many forms, from China's fairly rigid two-child policy intended to slow population growth, to the family-friendly policies of no-growth countries like South Korea that would like to increase their natural birth rates. Over half of the world's married women use modern contraceptive methods, which has contributed to slower growth (**Figure 1.23**).

- Not all attention should be focused on natural growth, because migration is increasingly a significant cause of growth in some countries. International migration is often driven by a desire for a better life in a new country. Although much international migration is to developed countries in Europe, North America, and Oceania, there are comparable flows of migrants moving between developing countries, such as flows from South Asia to Southwest Asia or immigration within Latin America and Sub-Saharan Africa. In addition, the UN estimates that 60 million people were displaced as a result of civil strife, political persecution, and environmental disasters in 2015, the largest number every recorded. This includes both internally displaced people and refugees who have left their country of origin.

- The greatest migration in human history is going on now as millions of people move from rural to urban places. In 2009, a landmark was reached when demographers estimated that, for the first time, more than half the world's population lived in towns and cities.

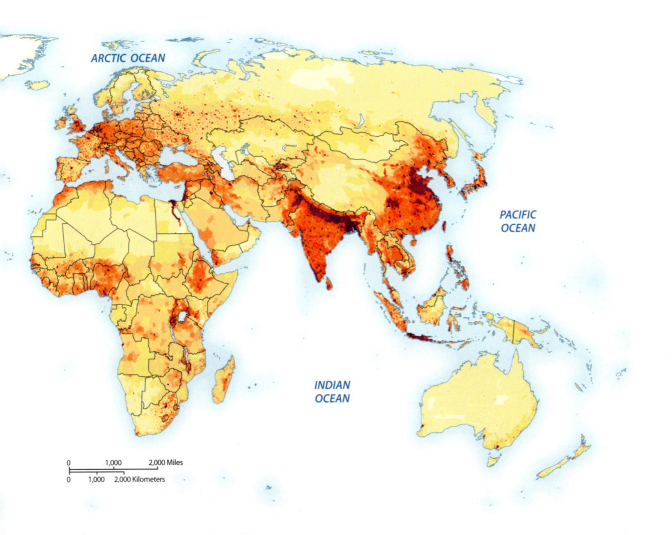

▲ **Figure 1.23 Family Planning in Cambodia** Women in a reproductive health clinic in Kampong Cham, Cambodia learn about modern birth control options. The availability of modern contraception has brought total fertility rates down throughout the world.

Population Growth and Change

Because of the central importance of population dynamics, each regional chapter includes a table of population data for the countries in that region. Table 1.1 provides the key demographic indicators for the world's ten largest countries by population size. China and India are the largest countries by far, each having one billion more people than the third largest country, the United States.

Population size alone tells only part of the story. **Population density**, for the purposes of this text, is the average number of people per square kilometer. Thus, China is the world's largest country demographically, but the population density of India, the second largest country, is more than twice that of China. Bangladesh's population

density is far greater still at over 1,200 people per square kilometer.

Population densities differ considerably across a large country and between rural and urban areas, making the gross national figure a bit misleading. Many of the world's largest cities, for example, have densities of more than 30,000 people per square mile (10,300 per square kilometer), with the central areas of São Paulo, Brazil, and Shanghai, China, easily twice as dense because of the prevalence of high-rise apartment buildings. In contrast, most North American cities have densities of fewer than 10,000 people per square mile (3,800 per square kilometer), due largely to a cultural preference for single-family dwellings on individual urban lots.

The statistics in Table 1.1 might seem daunting, but this information is crucial to understanding general population trends, overall growth rates, and patterns of settlement among the countries that make up various world regions.

Natural Population Increase A common starting point for measuring demographic change is the **rate of natural increase (RNI)**, which provides the annual growth rate for a country or region as a percentage. This statistic is produced simply by subtracting the number of deaths from the number of births in a given year. Important to remember, however, is that population gains or losses through migration are not considered in the RNI.

The RNI is a small number with major consequences. It can be positive, as in the case of Nigeria or stable as in the case of Russia, some countries even have negative rates. China's RNI is 0.5, whereas India's is 1.5. Yet if those rates are maintained, China's population will double in 140 years, whereas India's will double in 50 years. This is why demographers are confident that India will surpass China as the largest country in the next decade or so. The country with the

TABLE 1.1 Population Indicators								
Country	Population (millions) 2016	Population Density (per square kilometer)[1]	Rate of Natural Increase (RNI)	Total Fertility Rate	Percent Urban	Percent < 15	Percent > 65	Net Migration (Rate per 1000)
China	1,378.0	145	0.5	1.6	56	17	10	0
India	1,328.9	436	1.5	2.3	33	29	6	0
United States	323.9	35	0.4	1.8	81	19	15	4
Indonesia	259.4	140	1.3	2.5	54	28	5	−1
Brazil	206.1	25	0.8	1.8	86	23	8	0
Pakistan	203.4	240	2.3	3.7	39	36	4	−1
Nigeria	186.5	195	2.6	5.5	48	43	3	0
Bangladesh	162.9	1,222	1.5	2.3	34	33	6	−2
Russia	144.3	9	0.0	1.8	74	17	14	2
Mexico	128.6	65	1.4	2.2	79	28	6	−1

MasteringGeography™

Source: Population Reference Bureau, *World Population Data Sheet, 2016.*
[1] World Bank, *World Development Indicators, 2016.*

Login to MasteringGeography™ & access MapMaster to explore these data!

1) Compare the maps for the overall Population and the RNI for each county in this table. How might the top 10 rankings change in the next 20 years?
2) What is the value of the replacement rate? Which countries are currently below the replacement rate of 2.1? What does that mean for their long term growth?

highest RNI on Table 1.1 is Nigeria at 2.6. If Nigeria maintains that rate, it will double its size in 28 years. Countries with a rate close to zero are demographically stable (see Russia), but countries with persistent negative rates will experience slow declines in population unless immigration occurs. For many years Japan was one of the demographic top 10, but due to negative RNI the country has lost population and in 2015 was replaced in the top 10 by Mexico (**Figure 1.24**).

Total Fertility Rate Population change is impacted by the **total fertility rate (TFR)**, which is the average number of live births a woman has in her lifetime. The TFR is a good indicator of a country's potential for growth. A TFR of 2.1 is considered the **replacement rate** and suggests that it takes two children per woman, with a fraction more to compensate for infant and child mortality, to maintain a stable population. Where infant mortality is high, a country's actual replacement rate could be higher—say, 3.0. Clearly, women do not have 1.6 or 5.6 children; rather, women in some countries on average have 1 to 2 children versus 5 to 6 children, with the TFR being the average. In 1970, the global TFR was 4.7, but by 2015 that rate was nearly cut in half to 2.5. Around the world, total fertility rates have been coming down for the last four decades as women move to cities, become better educated, work outside the home, control their fertility with modern contraception, and receive better medical care for themselves and their infants.

Four of the countries listed in Table 1.1 have a below-replacement TFR, meaning that over time their natural growth will slow as fewer children are born, and in some cases, population will decline if immigration does not occur. Even India's current TFR is 2.3, a dramatic change from 5.5 in 1970. India will still grow for many decades to come, but the potential for growth has been reduced as Indian women have smaller families. The countries with the highest total fertility rates are in Sub-Saharan Africa, where the average is about 5. Nigeria's TFR is slightly higher at 5.5.

Population Age Structure Another important indicator of a population's relative youthfulness, and its potential for growth, is the percentage of the population under age 15. Currently, 26 percent of the world's population is younger than age 15. However, in fast-growing Sub-Saharan Africa, that figure is 43 percent. This is another indication that population growth will continue in this region for at least another generation. In contrast, only 17 percent of the population of East Asia and 16 percent of the population in Europe are under 15, suggesting slower growth and shrinking family sizes.

The other end of the age spectrum—the percentage of a population over age 65—is also important. Just 8 percent of the world's population is over age 65, yet the percentage is twice that in many developed countries. Japan distinguishes itself in this regard with 26 percent of its population over 65; in contrast, only 13 percent of its population is under 15, indicating that the average age of the population is increasing as well. An aging population is significant when calculating a country's need to provide social services for its senior citizens and pensioners. It also has implications for the size of the overall workforce that supports retired and elderly individuals.

Population Pyramids A useful graphical indicator of a population's age and gender structure is the **population pyramid**, which depicts the percentage of a population (or in some cases, the raw numbers) that is male or female in the different age classes, from young to old (**Figure 1.25**). If a country has higher numbers of young people than old, the graph has a broad base and a narrow tip, thus taking on a pyramidal shape that commonly forecasts rapid population growth. In contrast, slow-growth or no-growth populations are top-heavy, with a larger number of seniors than people of younger age.

Not only are population pyramids useful for comparing different population structures around the world at a given point in time, but also they can capture the structural changes of a country's population in time as it transitions from fast to slow growth. Population pyramids are also useful for displaying gender differences within a population, showing whether or not there is a disparity in the numbers of males and females. In the mid-20th century, for example, population pyramids for those countries that fought in World War II (such as the United States, Germany, France, and Japan) showed a distinct deficit of males, indicating those lost to warfare. Similar patterns are found today in those countries experiencing widespread conflict and civil unrest.

Cultural preferences for one sex or another, such as the preference for male infants in China and India, also show up in population pyramids. Because of their usefulness in showing different population structures, comparative population pyramids are found in many regional chapters of this book.

▼ **Figure 1.24 Smaller Families, Declining Population** Japan has seen its family size shrink to one or two children. Consequently, the total population of the country has declined as well.

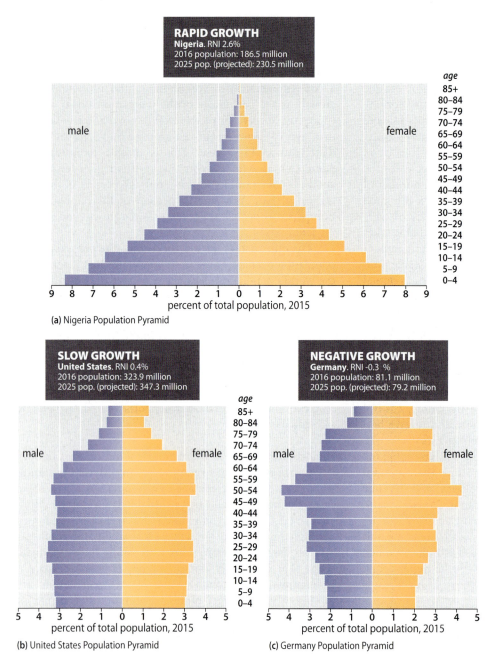

RAPID GROWTH
Nigeria. RNI 2.6%
2016 population: 186.5 million
2025 pop. (projected): 230.5 million

male   female

percent of total population, 2015

(a) Nigeria Population Pyramid

SLOW GROWTH
United States. RNI 0.4%
2016 population: 323.9 million
2025 pop. (projected): 347.3 million

male   female

percent of total population, 2015

(b) United States Population Pyramid

NEGATIVE GROWTH
Germany. RNI -0.3 %
2016 population: 81.1 million
2025 pop. (projected): 79.2 million

male   female

percent of total population, 2015

(c) Germany Population Pyramid

▲ **Figure 1.25 Population Pyramids of Nigeria, United States, and Germany** The term *population pyramid* comes from the shape of the graph assumed by a rapidly growing country such as (a) Nigeria, when data for age and sex are plotted as percentages of the total population. The broad base illustrates the high percentage of young people in the country's population, which indicates that rapid growth will probably continue for at least another generation. This pyramidal shape contrasts with the narrow bases of slow- and negative-growth countries, such as the (b) United States and (c) Germany, which have fewer people in the child-bearing years and a larger proportion of the population over age 65. **Q: Find two examples of countries that fit into each of the three categories: rapid growth, slow growth, and negative growth.**

Life Expectancy A demographic indicator containing information about a society's health and well-being is **life expectancy**, which is the average number of years a typical male or female in a specific country can be expected to live. Life expectancy generally has been increasing around the world, indicating that conditions supporting life and longevity are improving. To illustrate, in 1970 the average life expectancy figure for the world was 58 years, whereas today it is 71. Some countries, such as Bangladesh, Iran,

and Nepal, have seen average life expectancies increase by 20 years or more since 1970.

Because a large number of social factors—such as health services, nutrition, and sanitation—influence how long a person can be expected to live, many researchers use life expectancy as a surrogate measure for development. When this figure is improving, it indicates that other aspects of development are occurring. Yet life expectancy can decline, as many countries in southern Africa experienced in the early 2000s when the AIDS epidemic was at its height. Certain segments of the population may experience declines; for example, U.S. demographers in 2015 reported a decline in life expectancy for white Americans who did not complete high school. In this book, we use life expectancy as a social indicator and therefore report these data in the economic and social development tables instead of in the population data tables.

The Demographic Transition The historical record suggests that population growth rates have slowed over time. More specifically, in Europe and North America, population growth slowed as countries became increasingly industrialized and urbanized. From these historical data, demographers generated the **demographic transition model**, a conceptualization that tracks the changes in birth rates and death rates over time. Birth rates are the number of annual births in a country per 1000 people, and death rates are the annual number of deaths per 1000 people. When birth rates exceed death rates, natural increase occurs (**Figure 1.26**).

Stage 1 of the demographic transition model is characterized by both high birth rates and high death rates, resulting in a very low rate of natural increase. Historically, this stage is associated with Europe's preindustrial period, a time that predated common public health measures such as water and sewage treatment, an understanding of disease transmission, and the most fundamental aspects of modern medicine. Not surprisingly, death rates were high, life expectancy was short,

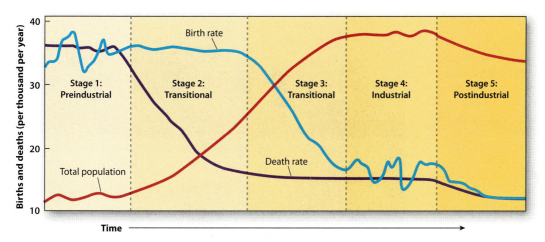

▲ **Figure 1.26 Demographic Transition** As a country goes through industrialization, its population moves through the five stages in this diagram, referred to as the *demographic transition*. In Stage 1, population growth is low because high birth rates are offset by high death rates. Currently there are no Stage 1 countries. Rapid growth takes place in Stage 2, as death rates decline. Stage 3 is characterized by a decline in birth rates. The transition was initially thought to end with low growth once again in Stage 4, resulting from a relative balance between low birth rates and low death rates. But with a large number of developed countries now showing no natural growth, demographers have added a fifth stage to the traditional demographic transition model, one that shows no or even negative natural growth.

and population growth was limited. Currently there are no Stage 1 countries in the world.

In Stage 2, death rates fall dramatically while birth rates remain high, thus producing a rapid rise in the RNI. In both historical and contemporary times, this decrease in death rates is commonly associated with the development of public health measures and modern medicine. Additionally, one of the assumptions of the demographic transition model is that these health services become increasingly available only after some degree of economic development and urbanization takes place.

However, even as death rates fall and populations increase, it takes time for people to respond with lower birth rates, which happens only in Stage 3. This, then, is the second transitional stage, in which people become aware of the advantages of smaller families in an urban and industrial setting, contrasted with the earlier need for large families in rural, agricultural settings or where children worked at industrial jobs (both legally and illegally). Then, in Stage 4, a low RNI results from a combination of low birth rates and very low death rates. Until recently, this stage was assumed to be the static end point of change of a developing, urbanizing population. However, in many highly urbanized developed countries, particularly those in Europe and Japan, the death rate now exceeds the birth rate. As a result, the RNI falls below a replacement level, expressed as a negative number. This negative growth state led to demographers adding a fifth stage to the traditional four-stage model to show that many countries have slowed to a no-growth point. Remember, though, that the RNI is just that—the rate of natural increase. Thus, it does not include a country's population growth or loss from immigration.

Global Migration and Settlement

Never before have so many people been on the move, either from rural areas to cities or across international borders. Today nearly 250 million people live outside the country of their birth and thus are officially designated as immigrants by the UN and other international agencies. Much of this international migration is directly linked to the new globalized economy because the majority of these migrants live either in the developed world or in developing countries with vibrant industrial, mining, or petroleum extraction economies. In the oil-rich countries of the United Arab Emirates, Kuwait, Qatar, and Saudi Arabia, the labor force is composed primarily of foreign migrants, especially from South Asia (**Figure 1.27**). The top six destination countries, which account for 40 percent of the world's immigrants, are major industrial or mining economies: United States, Russia, Germany, Saudi Arabia, United Kingdom, and United Arab Emirates. Within these countries, and for most other destinations, migrants are drawn to the opportunities found in major metropolitan areas (see *People on the Move: Migrants and Refugees*).

Not all migrants move for economic reasons. War, persecution, famine, and environmental destruction cause people to flee to safe havens elsewhere. Accurate data on refugees are often difficult to obtain for several reasons (such as individuals illegally crossing international boundaries or countries deliberately obscuring the number for political reasons), but UN officials estimate that there are currently 60 million refugees or internally displaced persons. More than

▼ **Figure 1.27 Global Workforce** South Asia laborers working on a construction site in Doha, Qatar. Many of the Persian Gulf countries rely upon vast numbers of contract laborers, mostly from South Asia, to provide the labor force necessary to build these modern cities and serve the populations living there.

Migrants and Refugees

Globalization has led to one of the largest migrations in human history. People are leaving their homes in search of better living conditions elsewhere, either because of desperate conditions at home—what geographers call "push forces"—or because of the lure of a better life elsewhere—the "pull forces." Often, though, it's a combination of the two that makes people move.

A third factor in the migration chain is the informational networks people draw upon to make their move. Sometimes it's family connections, as people follow relatives who have successfully migrated. Other times a paid agent or labor contractor is involved. Stories abound, unfortunately, of migrant smugglers who, after taking the migrants' money, abandon them on the open seas or in the Saharan or Mexican deserts. Increasingly, political groups in destination countries form to protest immigrant flows, fearful of the changes that newcomers might bring. Other groups organize to gain political status, such as the undocumented population in the U.S. pushing for immigration reform (**Figure 1.3.1**).

Accurate data on migration are notoriously difficult to gather because people often migrate without documentation. However, the UN estimates that roughly 3 percent of the world's people live in a country different from where they were born, be they flows of economic migrants, refugees, students, or irregular migrants. Many of these migrants have moved to cities in the developed world, but a considerable flow is drawn to magnet cities in the developing world, places such as Dubai, United Arab Emirates; Johannesburg, South Africa; Panama City, Panama; or Bangkok, Thailand. Furthermore, UN statistics do not include those people who leave home but stay within their own country, like the highland peasants of Bolivia moving to La Paz or farmers from China's Hunan Province who move to the factories of Shenzen. Internal migration in China alone is estimated at over 200 million.

Because the human geography of global migration wears many different faces and tells many different stories, each of our chapters sheds light on this complicated process of migration through our *People on the Move* sidebars. Examples include African migrants to North America

▲ **Figure 1.3.1 Immigrant Reform Rally** Latinos gather in front of the U.S. Capitol to protest for immigration reform. Record numbers of immigration deportations have driven Latinos to mobilize for change.

(Chapter 3); Portuguese workers in Angola and Mozambique (Chapter 8); and Rohingya refugees in Thailand (Chapter 13).

1. Does your community include international migrants? If so, where did they come from? What push and pull forces influenced their locational decisions?

2. Now choose a city in North America, Europe, or Asia and, using the Internet, collect information on its international migrant population.

Google Earth Virtual MG Tour Video

http://goo.gl/PCqouy

half of these people are in Africa and Southwest Asia. The conflict in Syria has displaced over half of that country's population. Most of the 11.6 million displaced people are scattered within Syria, but over 4 million live outside the territory, mostly in Turkey, Lebanon, Jordan, and Iraq (see chapter-opener photo).

Net Migration Rates The amount of immigration (people entering a country) and emigration (those leaving a country) is measured by the **net migration rate**. A positive figure means that a country's population is growing because of migration, whereas a negative number means more people are leaving. As with other demographic indicators, the net migration rate is expressed as the number of migrants per 1000 of a base population. Returning to Table 1.1, only

the United States and Russia show positive net migration rates. Five states (India, Indonesia, Pakistan, Bangladesh and Mexico) have negative rates, and three (China, Brazil, Nigeria) have rates at zero, meaning the numbers of people entering and leaving in a particular year cancel each other out. This does not mean that these countries do not produce immigrants—both China and Nigeria have large populations overseas—but for that particular year, incoming and outgoing flows equaled each other.

Countries with some of the highest net migration rates depend heavily on migrants for their labor force, such as the United Arab Emirates, Kuwait, and Oman. Countries with the highest negative migration rates include those in conflict (Syria) and Pacific island nations such as Samoa and Micronesia with relatively small populations and weak economies.

Settlement in an Urbanizing World The focal points of today's globalizing world are cities—the fast-paced centers of deep and widespread economic, political, and cultural change. This vitality, and the options that these centers offer to impoverished and uprooted rural peoples, make cities magnets for migration. The scale and rate of growth of some world cities are absolutely staggering. Between natural growth and in-migration, Mumbai, India, is expected to add over 7 million people by 2020, which, assuming growth is constant throughout the period (perhaps a questionable assumption), would mean that this urban area would add over 10,000 new people each week. The same projections show Lagos, Nigeria, which currently has the highest annual growth of any megacity, adding almost 15,000 residents per week. A **megacity** is a metropolitan area of more than 10 million people. There are approximately three dozen megacities in the world.

Based on data on the **urbanized population**, which is the percentage of a country's population living in cities, 53 percent of the world's population is urbanized. Further, demographers predict that the world will be 60 percent urbanized by 2025. Urbanization rates vary by region. Sub-Saharan Africa may be rapidly urbanizing, but in countries such as Ethiopia, Kenya, and Malawi, more than three-quarters of the population still lives in rural areas. The population of India, with its three megacities of Delhi, Mumbai, and Kolkata, is still two-thirds rural.

Generally speaking, most countries with high rates of urbanization are also more developed and industrialized because manufacturing tends to cluster around urban centers. We know that urbanization is a major demographic reality, but it is important to remember that 3 billion people live in rural settings that are also being transformed by globalization.

REVIEW

1.11 What is the rate of natural increase (RNI), and how can it be a negative number?

1.12 Explain a high versus a low total fertility rate, and give examples.

1.13 Describe the demographic transition model and explain reasons for the transitions from one stage to the next.

■ **KEY TERMS** population density, rate of natural increase (RNI), total fertility rate (TFR), replacement rate, population pyramid, life expectancy, demographic transition model, net migration rate, megacity, urbanized population

---

Cultural Coherence and Diversity: The Geography of Change and Tradition

Social scientists often say that culture binds together the world's diverse social fabric. If this is true, one glance at the daily news suggests this complex global tapestry could be unraveling because of widespread cultural tensions and conflict. As noted earlier, with the rise of digital communication, stereotypical Western culture is spreading at a rapid pace. Although some societies accept these new cultural influences willingly, others resist and push back against what they perceive as cultural imperialism with protests, censorship, and even terrorism. Still others use this technology to advance their own cultural or political agendas.

The geography of cultural coherence and diversity, then, entails an examination of both tradition and change, new cultural forms produced by interactions between cultures, gender issues, global languages, religions, and cuisines.

Culture in a Globalizing World

The dynamic changes connected with globalization have blurred traditional definitions of culture. A very basic definition provides a starting point. **Culture** is learned, not innate, behavior shared by a group of people, empowering them with what is commonly called a "way of life."

Culture has both abstract and material dimensions: language, religion, ideology, livelihood, and value systems, but also technology, housing, foods, and music. Even something like sports can have deep cultural meaning. Think of how billions of people watch the World Cup and support their "national" teams with near-religious devotion. These varied representations of culture are relevant to the study of world regional geography because they tell us much about the way people interact with their environment, with one another, and with the larger world (**Figure 1.28**). Not to be overlooked

Explore the **Sights** of Brooklyn's Williamsburg

http://goo.gl/dSOCQd

▼ **Figure 1.28 Culture as a Way of Life** The clothing, appearance, and urban setting of this young couple give clues to their membership in Brooklyn's Hipster culture.

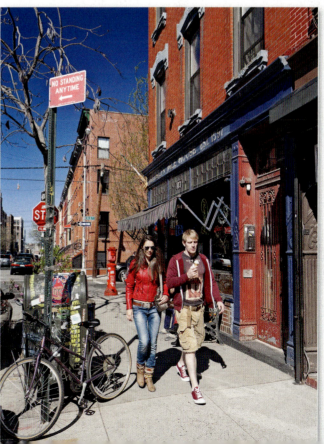

Everyday Globalization

Common Cultural Exchanges

Globalization is so ubiquitous that it's often taken for granted. What you're wearing was probably made overseas, because 98 percent of all U.S. apparel is imported. Your shirt could be made in China, Bangladesh, Thailand, Haiti, Mexico, or India, all of which are major manufacturing centers for the world's clothing. Even some "Made in the U.S.A." clothing might be pushing the truth a bit by being produced in the U.S. commonwealth countries of Puerto Rico and the Northern Mariana Islands in the far western Pacific. However, if you paid $300 or so for your jeans, they could be made in the United States—most probably in Los Angeles, where 30 different apparel firms turn out designer jeans.

The point is that globalization is not just about multinational corporations doing business all over the world; it is everywhere in your daily life, from what you eat to what you wear to the smartphone in your hand to the coffee you drink. Chances are that whatever it is, it involves a complex world geography.

A sidebar in each chapter illustrates this idea: how ethnic foods diffuse worldwide (Chapters 7 and 8); where soccer balls used on your campus are made (Chapter 12); and why coffee and wine production is shifting eastward (Chapters 13 and 14). One way that U.S. college students experience the world is through study-abroad programs, which can be important opportunities to learn about cultures other than your own (**Figure 1.4.1**).

1. How has globalization changed higher education in the United States?

2. Identify a commonplace item or activity in your life that has an interesting backstory involving globalization.

▲ **Figure 1.4.1 Cultural Exchange Through Study Abroad** College students from the United States and Panama work together on a research project investigating urban sustainability in the historic colonial core of Panama City.

is the fact that culture is dynamic and ever-changing, not static. Thus, culture is a process, not a condition—an abstract, yet useful concept that is constantly adapting to new circumstances. As a result, there are always tensions between the conservative, traditional elements of a culture and the newer forces promoting change (see *Everyday Globalization: Common Cultural Exchanges*).

When Cultures Collide Cultural change often takes place within the context of international tensions. Sometimes, one cultural system will replace another; at other times, resistance by one group to another's culture will stave off change. More commonly, however, a newer, hybrid form of culture results from an amalgamation of two or more cultural traditions (**Figure 1.29**).

The active promotion of one cultural system at the expense of another is called **cultural imperialism**. Although many expressions of cultural imperialism still exist today, the most severe examples occurred in the colonial period. In those years, European cultures spread worldwide, often

overwhelming, eroding, and even replacing indigenous cultures. During this period, Spanish culture spread widely throughout South America, French culture diffused into parts of Africa and Southeast Asia, and British culture overwhelmed South and Southwest Asia. New languages were mandated, new educational systems were implanted, and new administrative institutions replaced the old. Foreign dress styles, diets, gestures, and organizations were added to

▶ **Figure 1.29 Culture Clash in Goa, India** Beach-loving western tourists gets better acquainted with a sacred symbol of Hindu culture.

▲ **Figure 1.30 Grinding in Panama** A young man practices grinding his skateboard on a rail in Panama City. Skateboarding began in the U.S. but has made its way into popular youth culture throughout the world, especially in urban areas with paved surfaces. In the background looms a Catholic church, an indicator of the Spanish colonial influence in this region.

existing cultural systems. Many vestiges of colonial culture are still evident today.

Today's cultural imperialism is seldom linked to an explicit colonizing force, but more often comes as a fellow traveler with economic globalization. Though many expressions of cultural imperialism carry a Western (even U.S.) tone—such as McDonald's, KFC, Marlboro cigarettes, the widespread use of English as the dominant language of the Internet, or trends such as skateboarding—these facets result more from a search for new consumer markets than from deliberate efforts to spread modern U.S. culture throughout the world (**Figure 1.30**). The reaction against cultural imperialism is **cultural nationalism**. This is the process of protecting and defending a cultural system against diluting or offensive cultural expressions, while at the same time actively promoting national and local cultural values. Often cultural nationalism takes the form of explicit legislation or official censorship that simply outlaws unwanted cultural traits. Examples of cultural nationalism are common. France has long fought the Anglicization of its language by banning "Franglais" in official governmental language, thereby exorcising commonly used words such as *weekend, downtown, chat,* and *happy hour.* France has also sought to protect its national music and film industries by legislating that radio DJs play a certain percentage of French songs and artists each broadcast day. Similarly, many Muslim countries limit Western cultural influences by restricting or censoring international TV, an element they

consider the source of many undesirable cultural influences. Most Asian countries are also increasingly protective of their cultural values, and many demand changes to tone down the sexual content of MTV. In China, government censors block access to Facebook.

Cultural Hybrids As mentioned, a common product of cultural collision is the blending of forces to form a new, synergistic form of culture in a process called **cultural syncretism** or hybridization. To characterize India's culture as British, for example, is to grossly oversimplify and exaggerate England's colonial influence. Instead, Indians have adapted many British traits to their own circumstances, infusing them with their own meanings. India's use of English, for example, has produced a unique form of "Indlish" that often befuddles visitors to South Asia. Nor should we forget that India has added many words to our English vocabulary—*khaki, pajamas, veranda,* and *bungalow,* among others. Clearly, both the Anglo and the Indian cultures have been changed by the British colonial presence in South Asia. Other examples of cultural hybrids abound: Australian-rules football, hip-hop music, Tex-Mex fast food, and so on.

Popular culture linked to sports, music, and food often blends global cultural influences with local and national cultural identities. In each regional chapter we will explore the meaning of these various cultural expressions as a way to understand how globalization contributes to culture change in context.

Language and Culture in a Global Context

Language and culture are so intertwined that often language is the major characteristic that differentiates and defines one cultural group from another (**Figure 1.31**). Furthermore, because language is the primary means for communication, it folds together many other aspects of cultural identity, such as politics, religion, commerce, and customs. Language is fundamental to cultural cohesiveness and distinctiveness, for language not only brings people together, but also sets them apart from nonspeakers of that language. Therefore, language is an important component of national or ethnic identity, as well as a means for creating and maintaining boundaries for group and regional identity.

Because most languages have common historical (and even prehistorical) roots, linguists have grouped the thousands of languages spoken throughout the world into a handful of *language families.* This is simply a first-order grouping of languages into large units, based on common ancestral speech. For example, about half of the world's people speak languages of the Indo-European family, a large group that includes not only European languages such as English and Spanish, but also Hindi and Bengali, the dominant languages of South Asia.

Within language families, smaller units also give clues to the common history and geography of peoples and cultures. *Language branches and groups* (also called *subfamilies*) are closely related subsets within a language family, usually sharing similar sounds, words, and grammar. Well known

▶ **Figure 1.31 World Language Families** Most languages of the world belong to a handful of major language families. About 50 percent of the world's population speaks a language belonging to the Indo-European language family, which includes not only languages common to Europe and Russia, but also major South Asia languages, such as Hindi. They are in the same family because of their linguistic similarities. The next largest family is the Sino-Tibetan family, which includes languages spoken in China, the world's most populous country. **Q: What languages, other than English, are spoken in your community?**

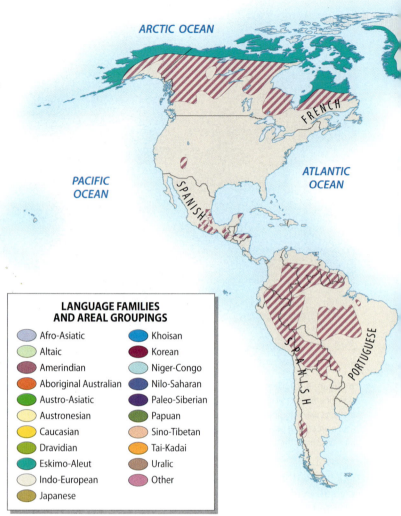

are the similarities between German and English, and between French and Spanish.

Additionally, individual languages often have very distinctive *dialects* associated with specific regions and places. Think of the distinctive differences, for example, among British, Canadian, and Jamaican English, or the city-specific dialects that set apart New Yorkers from residents of Dallas, Berliners from inhabitants of Munich, Parisians from villagers of rural France, and so on.

When people from different cultural groups cannot communicate directly in their native languages,

LANGUAGE FAMILIES AND AREAL GROUPINGS

- Afro-Asiatic
- Altaic
- Amerindian
- Aboriginal Australian
- Austro-Asiatic
- Austronesian
- Caucasian
- Dravidian
- Eskimo-Aleut
- Indo-European
- Japanese
- Khoisan
- Korean
- Niger-Congo
- Nilo-Saharan
- Paleo-Siberian
- Papuan
- Sino-Tibetan
- Tai-Kadai
- Uralic
- Other

▲ **Figure 1.32 Mandarin and English** This road sign in Shanghai, China, displays two of the world's most popular languages: Mandarin, spoken by about 12 percent of the world's population, and English, the global language of commerce, transportation, and science. An estimated 2 billion people know some English.

they often agree on a third language to serve as a common tongue, a **lingua franca**. Swahili has long served that purpose for speakers of the many tribal languages of eastern Africa, and French was historically the lingua franca of international politics and diplomacy. Today English is increasingly the common language of international communications, science, and air transportation (**Figure 1.32**).

The Geography of World Religions

Another important defining trait of cultural groups is religion (**Figure 1.33**). Indeed, in this era of a comprehensive global culture, religion is becoming increasingly important in defining cultural identity. Recent ethnic violence based upon religious differences in far-flung places such as the Balkans, Iraq, Syria, and Burma illustrates the point.

Universalizing religions, such as Christianity, Islam, and Buddhism, attempt to appeal to all peoples, regardless of location or culture. These religions usually have a proselytizing or missionary program that actively seeks new converts throughout the world. In contrast are **ethnic religions**, which are identified closely with a specific ethnic, tribal,

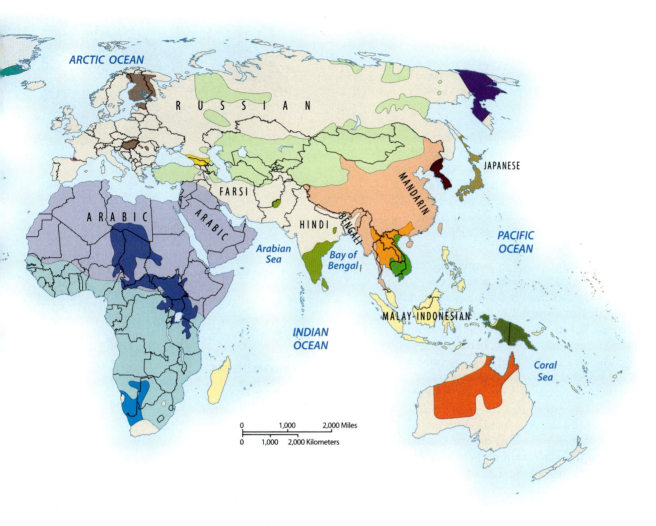

or national group. Judaism and Hinduism, for example, are usually regarded as ethnic religions because they normally do not actively seek new converts; instead, people are born into ethnic religions.

Christianity, because of its universalizing ethos, is the world's largest religion in both areal extent and number of adherents. Although fragmented into separate branches and churches, Christianity as a whole has 2.1 billion adherents, encompassing about one-third of the world's population. The largest numbers of Christians are found in Europe, Africa, Latin America, and North America.

Islam, which has spread from its origins on the Arabian Peninsula east to Indonesia and the Philippines, has about 1.8 billion members. Although not as severely fragmented as Christianity, Islam should not be thought of as a homogeneous religion because it is also split into separate groups. One of the two major branches is *Shi'a Islam*, which constitutes about 11 percent of the total Islamic population and represents a majority in Iran and southern Iraq. The other branch is the more dominant *Sunni Islam*, which is found from the Arab-speaking lands of North Africa to Indonesia. Probably in response to Western influences connected to globalization, both of these forms of Islam are currently experiencing fundamentalist revivals in which proponents are interested

in maintaining purity of faith, separate from these Western influences.

Judaism, the parent religion of Christianity, is also closely related to Islam. Although tensions are often high between Jews and Muslims, these two religions, along with Christianity, actually share historical and theological roots in the Hebrew prophets and leaders. Judaism today numbers about 14 million adherents, having lost perhaps one-third of its total population to the systematic extermination of Jews during World War II.

Hinduism, which is closely linked to India, has nearly 1 billion adherents and is the world's third largest religion. Outsiders often regard Hinduism as polytheistic because Hindus worship many deities. Most Hindus argue, however, that all of their faith's gods are merely representations of different aspects of a single divine, cosmic unity. Historically, Hinduism is linked to the caste system, with its segregation of peoples based on ancestry and occupation. However, because India's democratic government is committed to reducing the social distinctions among castes, the connections between religion and caste are now much less explicit than in the past.

Buddhism, which originated as a reform movement within Hinduism about 2500 years ago, is widespread in Asia, extending from Sri Lanka to Japan and from Mongolia to

▶ **Figure 1.33 Major Religious Traditions** The relative dominance of major religions in a particular area is shown on this map. For example, most Brazilians are Catholic, but the eastern half of the country (darker red) has a higher percentage of Catholics than the Amazonian west (lighter red). Similarly, Canada is a mix of Protestants and Catholics, with Quebec having a strong Catholic majority. Yet there are large stretches in western Canada (gray) with no dominant religion.

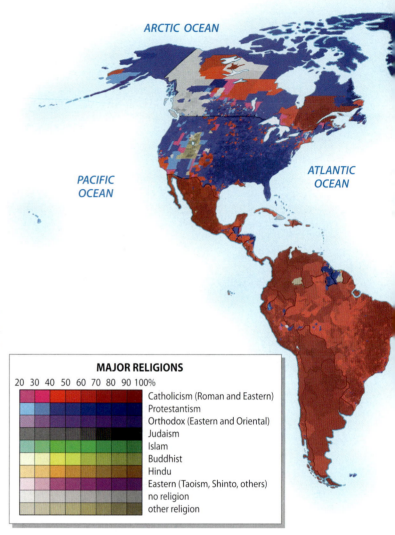

MAJOR RELIGIONS

20 30 40 50 60 70 80 90 100%

Catholicism (Roman and Eastern)
Protestantism
Orthodox (Eastern and Oriental)
Judaism
Islam
Buddhist
Hindu
Eastern (Taoism, Shinto, others)
no religion
other religion

Vietnam (**Figure 1.34**). Buddhism has two major branches: *Theravada,* found throughout Southeast Asia and Sri Lanka, and *Mahayana,* found in Tibet and East Asia. In its spread, Buddhism came to coexist with other faiths in certain areas, making it difficult to accurately estimate the number of its adherents. Estimates of the total Buddhist population range from 350 million to 900 million people.

Finally, in some parts of the world, religious practice has declined significantly, giving way to **secularism**, in which people consider themselves either nonreligious or outright atheistic. Though secularism is difficult to measure, social scientists estimate that about 1.1 billion people fit into this category worldwide. Perhaps the best example of secularism comes from the former communist lands of Russia and eastern Europe, where overt hostility occurred between government and church from the time of the Russian Revolution of 1917. Since the demise of Soviet communism in the 1990s, however, many of these countries have experienced religious revivals.

Recently, secularism has also grown more pronounced in western Europe. Although France is, historically and to some extent still culturally, a Roman Catholic country, today, between secularism and an increase of the immigrant population, there are possibly more people attending Muslim mosques on Fridays than attending Christian churches on Sundays. Japan and the other countries of East Asia are also noted for their high degree of secularization.

Culture, Gender, and Globalization

Culture includes not just the ways people speak or worship, but also embedded practices that influence behavior and values. **Gender** is a sociocultural construct, linked to the

values and traditions of specific cultural groups that differentiate the characteristics of the two biological sexes, male and female. Central to this concept are **gender roles**, the cultural guidelines that define appropriate behavior within a specific context. In traditional tribal or ethnic groups, for example, gender roles might rigidly distinguish between women's work (often domestic tasks) and men's work (done mostly outside the home). Gender roles similarly guide many other social behaviors within a group, such as child rearing, education, marriage, and even recreational activities.

The explicit and often rigid gender roles of a traditional social unit contrast greatly with the less rigid, more implicit, and often flexible gender roles of a large, modern, urban, post-industrial society. More to the point, globalization in its varied expressions is causing significant changes to traditional gender roles throughout the world. Nowhere is this more apparent than in the growing legal recognition of same-sex marriage worldwide (**Figure 1.35**). Since 2000, over 25 countries have recognized such unions, including the United States. Yet there is also a distinct geography of anti-gay legislation, especially in Africa, Southwest Asia, Russia, and South Asia. In extreme cases, gay expression can result in imprisonment and even death. Changes to the institution

▼ **Figure 1.34 Buddhist Landscapes** An array of buildings—temples, monasteries, and shrines—produces a distinctive landscape throughout Asia, as is illustrated by this photo from Chiang Mai in northern Thailand.

Explore the **Sights** of Chiang Mai Buddhist Temples

http://goo.gl/BlhtSw

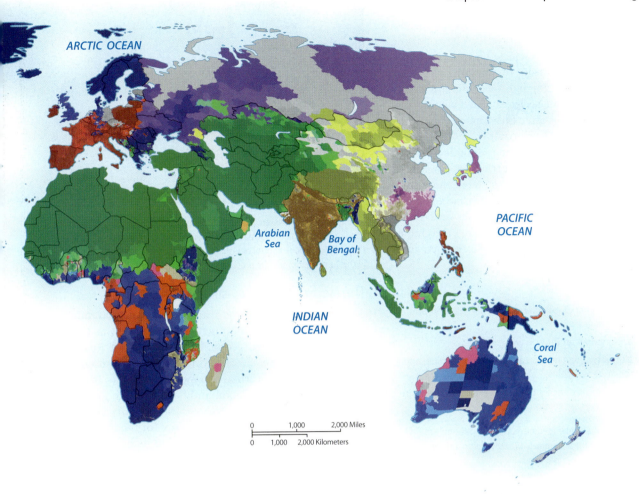

▼ **Figure 1.35 Mapping Gay Rights** Since 2000, more than 25 countries have recognized same-sex marriage. From Australia to Mexico and from South Africa to Ireland, a major cultural shift has occurred. At the same time, there are countries where gay expression is illegal and, in the most extreme cases, punishable by death.

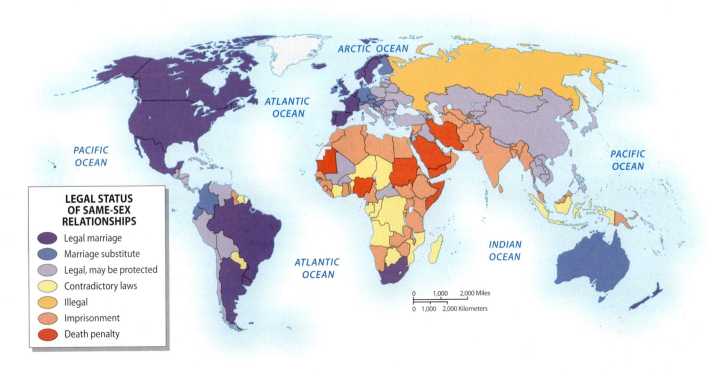

LEGAL STATUS OF SAME-SEX RELATIONSHIPS

- Legal marriage
- Marriage substitute
- Legal, may be protected
- Contradictory laws
- Illegal
- Imprisonment
- Death penalty

of marriage are part of a more globalized cultural discussion of what constitutes basic human rights. These shifting norms are embraced by some and rejected by others.

Globalization has also spread the notion of gender equality around the globe, calling into question and exposing those cultural groups and societies that blatantly discriminate against women. This topic is discussed later in the chapter as a measure of social development.

There are gender dimensions to the economic effects of globalization in many developed countries. In the United States, for example, male workers have suffered more from unemployment than have females, as industrial and technology jobs have been outsourced to China and India. Consequently, in many households women are emerging as primary income earners, while men have taken on new roles in domestic activities.

▲ **Figure 1.36 Russian Troops in Crimea** An unmarked soldier stands next to a combat vehicle on a street in Simferopol, Ukraine in the Crimea Peninsula. In 2014, Russia took control of Crimea with the support of ethnic Russians living in this territory, despite the protests of the Ukrainian government.

Geopolitical Framework: Unity and Fragmentation

The term **geopolitics** is used to describe the close link between geography and politics. More specifically, geopolitics focuses on the interactivity between political power and territory at all scales, from the local to the global. Unquestionably, one of the global characteristics of the last several decades has been the speed, scope, and character of political change in various regions of the world; thus, discussions of geopolitics are central to world regional geography.

With the demise of the Soviet Union in 1991 came opportunities for self-determination and independence in eastern Europe and Central Asia, resulting in fundamental changes to economic, political, and even cultural alignments. Religious freedom helped drive national identities in some new Central Asian republics, whereas eastern Europe was primarily concerned with new economic and political links to western Europe. Russia itself still wavers perilously between different geopolitical pathways. Russia's justification for taking over parts of Ukraine in 2014 is based on the fact that ethnic Russians live there. Meanwhile, these acts have been condemned internationally as an affront to state sovereignty (**Figure 1.36**). All of these topics are discussed further in Chapters 8, 9, and 10.

The Nation-State Revisited

A map of the world consists of an array of about 200 countries, ranging in size from the microstates like Vatican City and Andorra to the huge geopolitical expanses of Russia, the United States, Canada, and China. All of these countries are regulated by governmental systems, ranging from democratic to autocratic. Commonly, these different forms of government share a concern with **sovereignty**, which can be defined geopolitically as the ability (or the inability) of a government to control activities within its borders. Integral to the practice of sovereignty, is the concept of **territory**—the delimited area over which a state exercises control and which is recognized by other states. A sovereign state must have a territory that is recognized by other states.

One of the ways the governments maintain their sovereign territory and the unity of the people within it, is through the concept of the **nation-state**. In this hyphenated term, *nation* describes a large group of people with shared sociocultural traits, such as language, religion, and shared identity. The word *state* refers to a political entity that has a government and a clearly delimited territory that is maintained and controlled. Historically, France and England are often cited as the archetypal examples of nation-states. Contemporary countries such as Albania, Egypt, Bangladesh, and Japan are modern examples of countries that show close overlap between nation and state. The related term *nationalism* is the sociopolitical expression of identity and allegiance to the shared values and goals of the nation-state.

Globalization, however, is shifting our understanding of the nation-state concept because today most of the world's countries are culturally diverse and their sense of nationhood comes from shared political values and common experiences rather than the traditional definition of nation. In particular, international migration has led to many countries having large populations of ethnic minorities who may not share the national culture of the majority. In the United Kingdom, for example, many South Asians have formed their own communities, speak their own languages, practice their own religions, and dress to their own standards, practices that some native Britons have criticized. At the same time, a majority of Londoners elected Sadiq Khan as their mayor in 2016; Khan is of Pakistani ancestry and is Muslim. In countries with large and diverse immigrant populations, their presence over time changes the very nature of "national culture." The fact that countries such as the United States, Canada, the United Kingdom, and Germany officially embrace cultural diversity, with policies declaring themselves as multicultural states, shows how nationhood concepts can embrace diversity.

Decentralization and Devolution

Also residing within many nation-states are groups of people who seek autonomy from the central government and argue for the right to govern themselves. This autonomy can range from the simple decentralization of power from a central government to smaller governmental units, as is the case with states in the United States and French Departments. At the far end of the spectrum is outright political separation and full governmental autonomy, referred to as devolution. As an illustration, the citizens of Scotland voted in 2014 on the issue of full separation from the United Kingdom, which narrowly ended with Scotland remaining part of the union. Other separatist movements are found among French-speaking people of Quebec Province in Canada, the Catalonians and Basques of Spain, and the more radical groups of native Hawaiians who seek autonomy from the United States (**Figure 1.37**).

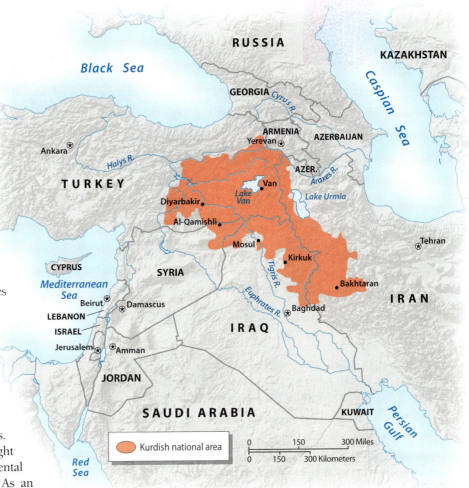

▲ **Figure 1.38 A Nation Without a State** Not all nations or large cultural groups control their own political territories. As this map shows, the Kurdish people of Southwest Asia occupy a large cultural territory that lies in four different political states—Turkey, Iraq, Syria, and Iran. As a result of this political fragmentation, the Kurds are considered a minority in each of these four countries. **Q: What kinds of issues result from the Kurds lacking a political state?**

Not to be overlooked is the fact that political organizations have eclipsed the power of traditional political states. This is certainly the case for the 28 member states of the European Union, a topic discussed in Chapter 8. Finally, some cultural groups lack political voice and representation due to the way political borders have been drawn. In Southwest Asia, the Kurdish people have long been considered a nation of people without a state because they are divided by political borders among Turkey, Syria, Iraq, and Iran (**Figure 1.38**).

Colonialism, Decolonialization, and Neocolonialism

One of the overarching themes in world geopolitics is the waxing and waning of European colonial power in South America, Asia, and Africa. **Colonialism** consists of the formal establishment of rule over a foreign population and territory. A colony has no independent standing in the world community, but instead is seen only as an appendage of the colonial power. The historical Spanish presence and rule over much of the Americas until the early 19th century is an

▼ **Figure 1.37 Basque Separatism** The Basque people of northeastern Spain and southwestern France are a distinct cultural group that has long sought autonomy and even independence from Spain and France. At times, militant Basque separatists have used violence and terrorism to further their cause. The photo is of a recent Basque demonstration in France showing support for two Basque activists accused of terrorism.

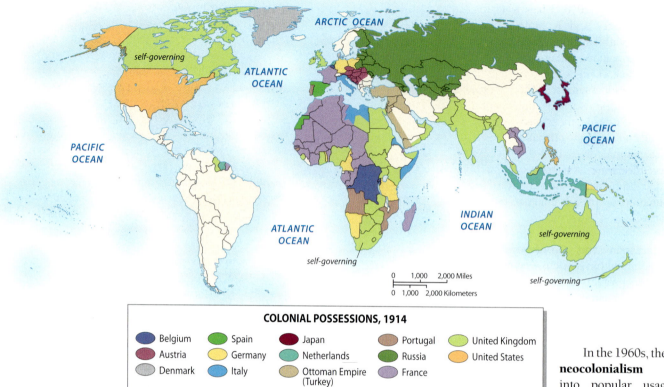

COLONIAL POSSESSIONS, 1914

- Belgium
- Austria
- Denmark
- Spain
- Germany
- Italy
- Japan
- Netherlands
- Ottoman Empire (Turkey)
- Portugal
- Russia
- France
- United Kingdom
- United States

▲ **Figure 1.39 The Colonial World, 1914** This map shows the extent of colonial power and territory just prior to World War I. At that time, most of Africa was under colonial control, as were Southwest Asia, South Asia, and Southeast Asia. Australia and Canada were very closely aligned with England. Also note that in Asia, Japan controlled the Korean Peninsula and Taiwan.

example. Generally speaking, the main period of colonialization by European countries was from 1500 through the mid-1900s, with the major players being England, Belgium, the Netherlands, Spain, Portugal, and France (**Figure 1.39**).

Decolonialization refers to the process of a colony's gaining (or, more correctly, regaining) control over its own territory and establishing a separate, independent government. As was the case with the Revolutionary War in the United States, this process often involves violent struggle. Similar wars of independence were common during the 19th century in Latin America and, in the mid-20th century, in South Asia, Southeast Asia, and Africa. Consequently, most European colonial powers recognized the inevitable and began working toward peaceful disengagement from their colonies. British rule ended in most of South Asia in 1947, and in the late 1950s and early 1960s, Britain and France accepted the independence of their former African colonies. This period of European decolonialization symbolically closed in 1997 when England turned over Hong Kong to China.

However, decades and even centuries of colonial rule are not easily erased. The influences of colonialism are still commonly found in the culture, government, educational systems, and economic life of the former colonies. Examples are the many contemporary manifestations of British culture in India and the continuing Spanish and Portuguese influences in Latin America.

In the 1960s, the term **neocolonialism** came into popular usage to characterize the many ways that newly independent states, particularly those in Africa, felt the continuing control of Western powers, especially in economic and political matters. To receive financial aid from the World Bank, for example, former colonies were often required to revise their internal economic structures to become better integrated with the emerging global system. This economic restructuring may seem warranted from a global perspective, but the dislocations caused at national and local scales led critics of globalization to characterize these external influences as no better than the formal control by historical colonial powers.

Global Conflict and Insurgency

As mentioned earlier, challenges to a centralized political state or authority have long been part of global geopolitics as rebel and separatist groups seek independence, autonomy, and territorial control. These actions are termed **insurgency**. Armed conflict has also been part of this process; the American and Mexican revolutions were both successful wars for independence fought against European colonial powers. **Terrorism**, which can be defined as violence directed at nonmilitary targets, has also been common, albeit to a far lesser degree than today.

Until the September 2001 terrorist attacks, terrorism was usually directed at specific local targets and committed by insurgents with focused goals. The Irish Republican Army (IRA) bombings in Great Britain and Basque terrorism in Spain are examples. The attacks on the World Trade Center and the Pentagon (as well as the thwarted attack on the U.S. Capitol) by Al Qaeda, however, went well beyond conventional geopolitics as a small group of religious extremists based in another part of the world attacked the symbols of Western culture, finance, and power. Experts believe that Al Qaeda's goal was less about disrupting world commerce and politics and more about displaying the strength of their own convictions and power. Regardless of motives, those acts of terrorism underscore the need to expand our conceptualization of the linkages between globalization and geopolitics.

Many experts argue that global terrorism is both a product of, and a reaction to, globalization. Unlike earlier geopolitical conflicts, the geography of global terrorism is not defined by a war between well-established political states. Instead, the Al Qaeda terrorists appear to belong to a web of small, well-organized cells located in many different countries. Boko Haram, a Muslim extremist group in Nigeria, has terrorized villages and kidnapped schoolchildren and is linked to Al Qaeda. Similarly Al-Shabaab, based in Somalia and responsible for attacks in Kenya, is affiliated with Al Qaeda.

The terrorist group Islamic State of Iraq and the Levant (ISIL; also known as ISIS or Islamic State in Iraq and Syria, or as Daesh) controls significant territory in Iraq and Syria and, unlike other terrorists groups, has the stated goal of forming a modern-day caliphate or a fundamentalist Islamic state in Southwest Asia (**Figure 1.40**). ISIL's use of terror, kidnapping, public executions, extortion, and social media has gained it international recognition and condemnation—as well as converts to its extremist cause. ISIL has taken advantage of the political vacuum left by years of conflict in Syria and Iraq that weakened those states and fostered the anti-Western, anti-secular, and anti-globalization sentiment that resonates in some parts of the world. Increasingly, geospatial tools, such as satellite imagery, are employed to track such groups and the chaos they cause to people, infrastructure, and cultural heritage sites (see *Geographers at Work: Tracking Conflict from Space*).

Although most of the terrorist organizations identified by the U.S. State Department are clustered in North Africa and Southwest and Central Asia, this list also includes insurgent groups in all other regions of the world. The military responses to global terrorism and insurgency involve several components, ranging from the neutralization of terrorist activities, known as counterterrorism, to **counterinsurgency**. The latter is a more complicated, multifaceted strategy that combines military warfare with social and political service activities, designed to win over the local population and deprive insurgents of a political base. Counterinsurgency activities include clearing and then holding territory held by insurgents, followed by building schools, medical clinics, and a viable economy. These nonmilitary activities are often referred to as nation building, because the goal is to replace the insurgency with a viable social, economic, and political fabric more complementary to the larger geopolitical state. This is the strategy recently employed by the United States in Iraq and Afghanistan.

REVIEW

1.18 Why is it common to use two different concepts—nation and state—to describe political entities?

1.19 Define, then contrast, colonialism and neocolonialism.

1.20 Describe the differences between counterterrorism and counterinsurgency.

■ **KEY TERMS** geopolitics, sovereignty, territory, nation-state, colonialism, decolonialization, neocolonialism, insurgency, terrorism, counterinsurgency

---

Economic and Social Development: The Geography of Wealth and Poverty

The pace of global economic change and development has accelerated dramatically in the past several decades. It increased rapidly at the start of the 21st century and then slowed precipitously in 2008 as the world fell into an economic recession. Most countries gradually recovered from the depths of the global recession, although overall growth rates are lower than a decade ago and major economies such as Brazil and Russia have faltered in the last couple of years. If nothing else, the recent global recession and its unsteady recovery have highlighted the overarching question of whether the benefits of economic globalization outweigh the negative aspects. Responses vary considerably, depending on one's point of view, occupation, career aspirations, and socioeconomic status. Any attempt to understand the contemporary world requires a basic understanding of global economic and social development. To that end, each regional chapter contains a substantive section on development, drawing on concepts discussed in the following paragraphs.

Economic development is considered desirable because it generally brings increased prosperity to people, regions, and nations. Following conventional thinking, this economic development usually translates into social improvements such as better health care, improved educational systems, higher wages, and longer life expectancies. One of the most troubling expressions of global economic growth,

▼ **Figure 1.40 ISIL** Heavily armed Islamic State fighters travel in the back of pickup trucks across the Iraqi desert in this shot from a propaganda video released by ISIL.

Geographers at Work

Tracking Conflict from Space

▲ **Figure 1.5.1 Susan Wolfinbarger**

As an undergraduate at Eastern Kentucky University, Susan Wolfinbarger took a world regional geography class, and was mesmerized: "There are so many things you learn in geography, and the methods of analysis can be applied to different careers and research." Years later, with a PhD in Geography from the Ohio State University, Wolfinbarger directs the Geospatial Technologies Project at the American Association for the Advancement of Science (AAAS) (**Figure 1.5.1**). Her group uses high-resolution satellite imagery to track conflicts and document issues of global concern, such as human rights abuses and damage to cultural heritage sites.

Most people have used Google Earth satellite images to look at places. Wolfinbarger's team employs a time series of such images in order to assess events such as destruction of villages. Interpreting images and quantifying findings is a challenge, but, she says, "Geography taught me not just mapping but statistics and surveying . . . it gave me a great tool kit to apply to any topic." Much of her analysis is used by human rights organizations such as the European Court of Human Rights and the Inter-American Court of Human Rights.

Wolfinbarger's team analyzed the increase in roadblocks in the Syrian city of Aleppo (**Figure 1.5.2**). Roadblocks demonstrate a decline in the circulation of people and goods in this densely settled city, which is a major problem. The Geospatial Technologies Project also documented heritage sites at risk from damage and looting, especially in Southwest Asia, and is developing training materials so that others can use this technology.

Geographers are at the cutting edge of applying satellite imagery to a broad spectrum of human rights issues. Wolfinbarger notes, "There are a lot of ways that geographers can contribute to things happening in

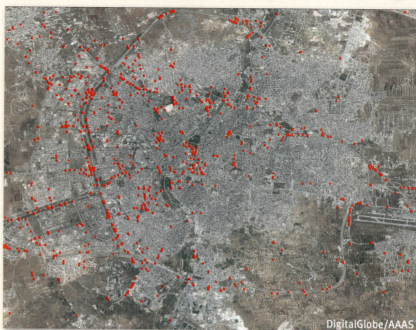

DigitalGlobe/AAAS

▲ **Figure 1.5.2 Monitoring Aleppo** This image shows the city of Aleppo in May 2013, where over 1000 roadblocks were detected. Roadblocks are an indicator of ongoing conflict and potential humanitarian concerns because they restrict the movement of people and goods throughout the city. In a nine-month period from September 2012 to May 2013, the number of roadblocks doubled.

the world, and a lot of opportunities out there other than academic jobs. Everyone wants a geographer!"

1. Suggest ways that satellite imagery could be used to document not just conflict, but environmental change.

2. Government agencies are constantly developing and using satellite technology. How might a citizen or non-governmental group in your city or state use this kind of analysis?

---

however, has been the geographic unevenness of wealth and social improvement. That is, although some regions and places in the world prosper, others languish and, in fact, apparently fall further behind the more developed countries. As a result, the gap between rich and poor regions has actually increased over the past several decades. This economic and social unevenness has, unfortunately, become one of the signatures of globalization. The numbers of people living in *extreme poverty*, defined as those living on less than $1.90 a day, have declined since the 1990s, which is largely due to the economic growth in China. Yet if one considers the population living on less than $3.10 a day, the poverty measure used by the World Bank and the UN, about 2.1 billion people (or three out of ten people in the world) still struggle for existence at this level. Many of these people live in Sub-Saharan Africa, South Asia, and Southeast Asia (**Figure 1.41**). These inequities are problematic because of their inseparable interaction with political, environmental, and social issues. For example, political instability and civil strife

▼ **Figure 1.41 Living in Extreme Poverty** The World Bank uses two measures of global poverty. Subsisting on less than $1.90 a day, the definition of extreme poverty, is a reality for roughly 1 billion people such as this Ugandan family sharing a daily meal in their mud home in Masaka. The numbers of people at this level of extreme poverty have decreased. Unfortunately, there are still 2 billion people classified as living in poverty, which is living on less than $3.10 a day.

Explore the **Sights** of Masaka, Uganda

http://goo.gl/FqjMhs

within a nation are often driven by the economic disparity between a poor peripheral area and an affluent industrial core—between the haves and have-nots. Such instability throughout a country can strongly influence international economic interactions.

More and Less Developed Countries

Until the middle of the 20th century, economic development was centered in North America, Japan, and Europe, with most of the rest of the world gripped in poverty. This uneven distribution of economic power led scholars to devise a **core–periphery model** of the world. According to this model, these countries and regions constituted the global economic *core*, centered for the most part in the Northern Hemisphere, whereas most of the areas in the Southern Hemisphere made up a less developed *periphery*. Although oversimplified, this core–periphery dichotomy does contain some truth. All the G8 countries—the exclusive club of the world's major industrial nations, made up of the United States, Canada, France, England, Germany, Italy, Japan, and Russia—are located in the Northern Hemisphere. (China—unquestionably an industrial power located in the Northern Hemisphere—is currently excluded from the G8.) Many critics contend that the developed countries achieved their wealth primarily by exploiting the poorer countries of the southern periphery, historically through colonial relationships and today through various forms of neocolonialism and foreign direct investment (**Figure 1.42**).

Following this core–periphery model, much has been made of "north–south tensions," a phrase that distinguishes the rich and powerful countries of the Northern Hemisphere from the poor and less powerful countries of the Southern Hemisphere. However, over recent decades the global economy has grown much more complicated; some former colonies of the periphery or "south"—most notably, Singapore—have become very wealthy, while a few northern countries—notably, Russia—have experienced very uneven economic growth since 1989, with some parts of Russia actually seeing economic declines. Additionally, the developed Southern Hemisphere countries of Australia and New Zealand never fit into the north–south division. For these reasons, many global experts conclude that the designation *north–south* is outdated and should be avoided.

Third world is another term often erroneously used as a synonym for the developing world. Historically, the term was part of the Cold War vocabulary used to describe countries that were not part of either the capitalist Westernized "first world" or the communist "second world" dominated by the Soviet Union and China. Thus, in its original sense *third world* signified a political and economic orientation (capitalist vs. communist), not a level of economic development. With the Soviet Union's demise and China's considerably changed economic orientation, *third world* has lost its original political meaning. In this book, we prefer relational terms that capture a complex spectrum of economic and social development—*more developed country* (*MDC*) and *less developed country* (*LDC*). Table 1.2 provides various

▼ **Figure 1.42 Patterns of Foreign Direct Investment** The World Bank tracks Foreign Direct Investment (FDI) as a development indicator. FDI is private foreign capital that enters a country for purposes of resource extraction, infrastructure development, and industrialization, among other activities. FDI dropped significantly in 2009 but was estimated at $739 billion in 2013. This figure shows net levels of FDI as a proportion of Gross Domestic Product (GDP) by country. **Q: Review the countries in which FDI as a share of GDP is 6 percent or higher. What might these countries have in common?**

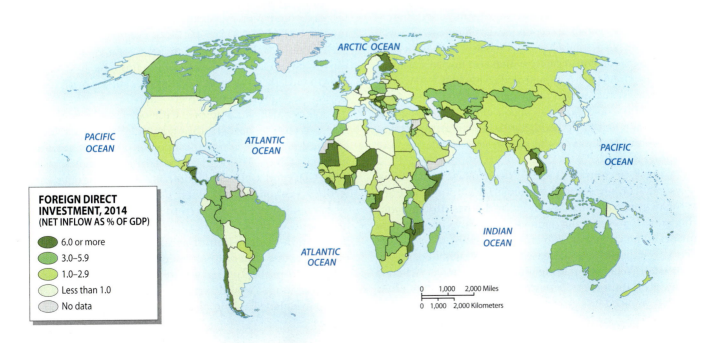

				Percent	Life Expectancy (2016)[2]			Under Age 5 Mortality Rate (2015)	Youth Literacy (% pop ages 15–24) (2005–2014)	Gender Inequality Index (2015)[3,1]
Country	GNI per capita, PPP 2014	GDP Average Annual %Growth 2009–14	Human Development Index (2015)[1]	Population Living Below $3.10 a Day	Male	Female	Under Age 5 Mortality Rate (1990)			
China	13,170	8.5	0.728	33.0	75	78	49	11	100	0.191
India	5,630	6.9	0.609	67.9	67	70	114	48	86	0.563
United States	55,900	2.2	0.915	–	76	81	11	7	–	0.280
Indonesia	10,190	5.8	0.684	54.4	69	73	82	27	99	0.494
Brazil	15,570	3.1	0.755	9.1	72	79	58	16	99	0.457
Pakistan	5,090	3.4	0.538	53.7	66	67	122	81	73	0.536
Nigeria	5,710	5.5	0.514	76.5	53	53	214	109	66	–
Bangladesh	3,330	6.2	0.570	81.5	71	73	139	38	81	0.503
Russia	22,160	2.9	0.798	‹2.0	66	77	27	10	100	0.276
Mexico	16,840	3.3	0.756	11.9	74	79	6	13	99	0.373

TABLE 1.2   Development Indicators

MasteringGeography™

Source: World Bank, *World Development Indicators, 2016.*

[1]United Nations, *Human Development Report, 2015.*

[2]Population Reference Bureau, *World Population Data Sheet, 2016.*

[3]Gender Equality Index—A composite measure reflecting inequality in achievements between women and men in three dimensions: reproductive health, empowerment, and the labor market that ranges between 0 and 1. The higher the number, the greater the inequality.

Login to MasteringGeography™ and access MapMaster to explore these data!

1) Look at the table and review the maps for the table data. Which countries you would classify as MDC or LDC, and why?
2) Which countries experienced the greatest improvement in Under Age 5 Mortality Rate?

development indicators for the world's ten most populous countries. In this book, each regional chapter includes a similar table of economic and social development indicators defined below.

Indicators of Economic Development

The terms *development* and *growth* are often used interchangeably when referring to international economic activities. There is, however, value in keeping them separate. *Development* has both qualitative and quantitative dimensions. When we talk about economic development, then, we usually imply structural changes, such as a shift from agricultural to manufacturing activity that also involves changes in the allocation of labor, capital, and technology. Along with these changes are assumed improvements in standard of living, education, and political organization. The structural changes experienced by Southeast Asian countries such as Thailand and Malaysia in the past several decades capture this process.

Growth, in contrast, is simply the increase in the size of a system. The agricultural or industrial output of a country may grow, as it has for India in the past decade, and this growth may—or may not—have positive implications for development. Many growing economies, in fact, have actually experienced increased poverty with economic expansion. When something grows, it gets bigger; when it develops, it improves. Critics of the world economy often say that we need less growth and more development.

Gross Domestic Product and Growth A common measure of the size of a country's economy is the **gross domestic product (GDP)**, the value of all final goods and services produced within its borders. Table 1.2 shows GDP average annual growth for 2009–2014. Most countries saw growth in GDP during this period, especially in the less developed world. But this period also captures growth that occurred after the recession. Compare the average annual growth of the United States with that of Indonesia. The U.S. growth rate is far lower than Indonesia's, but the U.S. economy is far larger and more diversified, and people in the high-income United States have far more resources than they do in Indonesia, a lower-middle-income country. In general, the less developed countries shown in this table have higher growth rates than the more developed countries, with China having the highest annual growth rate.

When GDP is combined with net income from outside a country's borders through trade and other forms of investment, this constitutes a country's **gross national income (GNI)**. Although these terms are widely used, both GDP and GNI are incomplete and sometimes misleading economic indicators because they completely ignore nonmarket economic activity, such as bartering and household work, and do not take into account ecological degradation or depletion of natural resources. For example, if a country were to clear-cut its forests—an activity that would probably limit future economic growth—this resource usage would actually increase the GNI for that particular year, but then GNI would likely decline.

▲ **Figure 1.43 Purchasing Power** Purchasing power parity (PPP) takes into consideration the strength or weakness of local currencies and relative value of what can be purchased with the equivalent of one international dollar in a particular country. The GNI per capita in India is less than $1,600 but the GNI PPP is over $5,000.

Diverting educational funds to purchase military weapons might also increase a country's GNI in the short run, but its economy would likely suffer in the future because of its less-well-educated population. In other words, GDP and GNI are a snapshot of a country's economy at a specific moment in time, not a reliable indicator of continued vitality or social well-being.

Comparing Purchasing Power and Poverty Because GNI data vary widely among countries, **gross national income (GNI) per capita** figures are used, allowing large and small economies to be compared, regardless of population size. An important qualification to these GNI per capita data is the concept of adjustment through **purchasing power parity (PPP)**, which takes into account the value of goods that can be purchased with the equivalent of one international dollar in a particular country. An international dollar has the same purchasing power parity over all GNI and is set at a designated U.S. dollar value. Thus, if a country's food costs are lower than the U.S. cost, the per capita purchasing power for that country increases. PPP was created to adjust comparisons between countries because (for example) an income of $5,000 in India can purchase more basic goods than the same amount of money in the United State (**Figure 1.43**).

In Table 1.2, the United States has the strongest GNI (PPP), followed by Russia. The country with the lowest GNI (PPP) is Bangladesh. In this case, purchasing power per capita GNI is greater than crude GNI per capita, which

is closer to $1000, because the cost of basic goods is far less in this developing country.

Measuring Poverty The international definition of *poverty* is living on less than $3.10 per day, and extreme poverty is living on less than $1.90 per day. While the cost of living varies greatly around the world, the UN usually uses unadjusted per capita income figures when measuring poverty. Table 1.2 shows that roughly four out of five Nigerians and Bangladeshis live in poverty. Two-thirds of all India's population live in poverty, whereas one-third of China's population live in poverty. In contrast, this measure of poverty is not used for the United States. While poverty data are usually presented at the country level, the World Bank and other agencies compile data at the sub-state level in order to better understand the poverty landscape within a country. The patterns of poverty in Morocco provide an instructive example, with areas around Rabat, Casablanca, and Tangier having lower rates of poverty compared to higher rates of poverty in the inland communities around Fez and Marrakesh (**Figure 1.44**). Poverty

▼ **Figure 1.44 Poverty Mapping** The World Bank regularly engages in poverty mapping to better understand where poverty is concentrated. This map of Morocco shows rates of poverty are higher around the inland cities of Marrakesh and Fez, and lower around the major coastal cities of Casablanca, Rabat, and Tanger. Mapping poverty at this scale is used to decide where scarce resources could be spent to improve overall development.

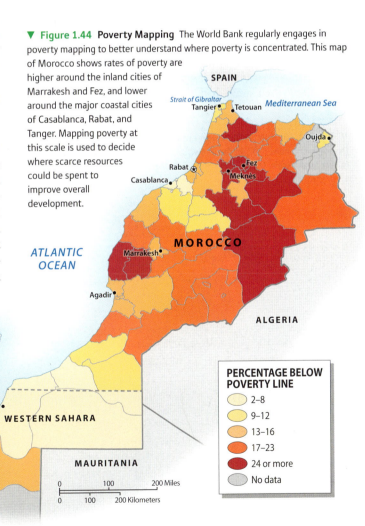

PERCENTAGE BELOW POVERTY LINE

- 2–8
- 9–12
- 13–16
- 17–23
- 24 or more
- No data

41

mapping at this scale helps governments and development agencies decide which areas of a country require more aid and investment in order to reduce poverty.

Indicators of Social Development

Although economic growth is a major component of development, equally important are indicators of the quality of human life. As noted earlier, the standard assumption is that economic development will spill over into the social infrastructure, leading to improvements in life expectancy, child mortality, gender inequality, and education. Even some of the world's poorest countries have experienced significant improvements in all these measures. Much of the foreign development aid since 2000 goes to tracking and improving these development indicators.

The Human Development Index For the past three decades, the UN has tracked social development in the world's countries through the **Human Development Index (HDI)**, which combines data on life expectancy, literacy, educational attainment, gender equity, and income (**Figure 1.45**). In a 2015 analysis, the 188 countries that provided data to the UN are ranked from high to low, with Norway achieving the highest score, Australia in second place, and the United States in sixth place. At the lowest end of the HDI are a handful of African countries, including Niger, Central African Republic, Eritrea, and Chad.

Although the HDI has been criticized for using national data that overlook the diversity of development within a country, overall the HDI conveys a reasonably accurate sense of a country's human and social development. Thus, we include HDI data in our social development table for each regional chapter.

Child Mortality Another widely used indicator of social development is data on *under age 5 mortality*, which is the number of children in that age bracket who die per 1000 of the general population. Aside from the tragedy of infant death, child mortality also reflects the wider conditions of a society, such as the availability of food, health services, and public sanitation. If those factors are lacking, children under age five suffer most; therefore, their death rate is taken as an indication of whether a country has the necessary social infrastructure to sustain life (**Figure 1.46**). In the social development tables throughout this book, child mortality data are given for two points in time, 1990 and 2015, to indicate whether the social structure has improved over the intervening years.

Youth Literacy Reading and writing are crucial skills in today's world, yet current data show that many in the developing world lack these skills. Of those who cannot read or write, two-thirds are women. The greatest disparity between male and female literacy is in South Asia, an artifact of long-standing cultural favoritism toward males. The World Bank focuses its resources on

▼ Figure 1.45 **Human Development Index** This map depicts the most recent rankings assigned to four categories that make up parts of the Human Development index (HDI). In the numerical tabulation, Norway, Australia, Switzerland and Denmark have the highest rankings, while several African countries are lowest on the scale. **Q: Why does India rank higher than Pakistan on the HDI?**

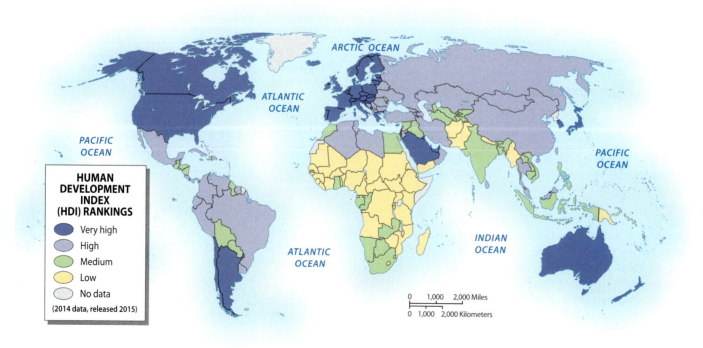

HUMAN
DEVELOPMENT
INDEX
(HDI) RANKINGS

- Very high
- High
- Medium
- Low
- No data

(2014 data, released 2015)

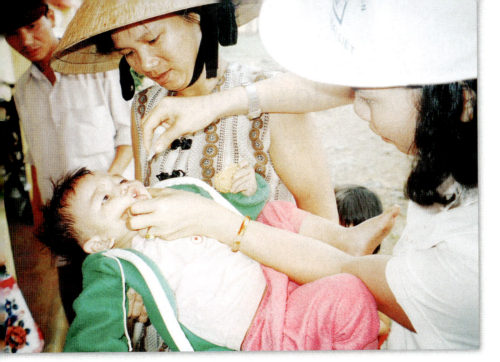

▲ **Figure 1.46 Children's Health and Mortality** The mortality rate of children under the age of 5 is an important indicator of social conditions such as food supply, public sanitation, and public health services. This child is receiving a polio vaccine in Vietnam.

UN gender inequality scores are found in the development indicators tables in the regional chapters.

Some countries may register reasonably high on the HDI, which is positive, yet also receive a relatively high gender inequality score (which is not so good). Qatar, for example, is a rich country that uses assets from its oil resources to provide many social benefits to its citizens, which explains its high HDI ranking, while at the same time its conservative Muslim culture produces a gender inequality score of .524 and a rank of 116 among countries in that index. Table 1.2 shows India's score of .563, which gives it a ranking of 130, which means relatively high gender inequality. Given the data problems in Nigeria, it is unranked in this index. Readers are advised to look carefully at the development indicators data and note these kinds of contradictions and inconsistencies.

REVIEW

1.21 Explain the difference between GDP and GNI.

1.22 What is PPP, and why is it useful?

1.23 How does the UN measure gender inequality? Explain why this is a useful metric for social development.

■ **KEY TERMS** core–periphery model, gross domestic product (GDP), gross national income (GNI), gross national income (GNI) per capita, purchasing power parity (PPP), Human Development Index (HDI), gender inequality

measuring and improving youth literacy (people ages 15–24). The hope is that literacy will increase dramatically for young adults and that disparities between males and females will disappear. Table 1.2 shows that Nigeria and Pakistan have the lowest rates of youth literacy, while in China, Indonesia, and Brazil virtually all youth are literate. Of course, literacy rates do not comment on the quality of education in a country, which can vary tremendously.

Gender Inequality Discrimination against women takes many forms, from not allowing them to vote to discouraging school attendance (**Figure 1.47**). Given the importance of this topic, the United Nations calculates **gender inequality** among countries in order to measure the relative position of women to men in terms of employment, empowerment, and reproductive health (in terns of maternal mortality and adolescent fertility). The UN index ranges from 0 to 1, which expresses the highest level of gender inequality. Slovenia, for example, has the lowest score for inequality with 0.016, while Yemen has one of the highest at 0.744. The

▶ **Figure 1.47 Women and Literacy** Gender inequities in education lead to higher rates of illiteracy for women. However, when there is gender equity in education, female literacy has several positive outcomes in a society. For example, educated women have a higher participation rate in family planning, which usually results in lower birth rates. These school children in Iran are being led by their teacher on a field trip to Isfahan. Iran has excellent youth literacy rates.

Chapter 1 Concepts of World Geography

Summary

- Geography is the study of Earth's varied and changing landscapes and environments. This study can be done conceptually in many different ways, by physical or human geography and either topically or regionally—or by using a combination of all these approaches.

- Globalization affects all aspects of world geography with its economic, cultural, and political interconnectivity. However, despite fears that globalization will produce a homogeneous world, a great deal of diversity is still apparent. Geographers use various tools that draw on information gathered on the ground and by satellites to examine the world at different scales, from an inner-city block to the entire planet.

- Human populations around the world are growing either quickly or slowly depending on natural increase and widely different migration patterns. Urbanization is also a major factor in settlement patterns as people continue to move from rural to urban locales.

- Culture is learned behavior and includes a range of tangible and intangible behaviors and objects, such as language and architecture. Globalization is changing the world's cultural geography, producing new cultural hybrids in many places. In other places, people resist change by protecting (or even resurrecting) traditional ways of life.

- Varying political systems provide the world with a dynamic geopolitical framework that is stable in some places and filled with tension and violence in others. As a result, the traditional concept of the nation-state is challenged by separatism, insurgency, and even terrorism.

- Proponents of globalization argue that all people in all places gain from expanded world commerce. But instead, there appear to be winners and losers, resulting in a geography of growing income inequality. Social development of health care and education is also highly uneven, but many key indicators are improving.

Review Questions

1. Define geography. Then define globalization and explain its relevance to understanding the world's changing geography.

2. What are the benefits of GIS, GPS, and satellite imagery in being able to monitor change and improve sustainability in a given place?

3. Summarize general migration trends around the world, and explain how these are influenced by and are impacting demographic, cultural, economic, and political change.

4. Explain the nation-state concept and provide examples. Is it still relevant in the age of globalization?

5. What is the difference between economic and social development? How might a rapidly developing country's population indicators from Table 1.2 change due to increasing well-being of its people?

Image Analysis

1. The flow of investment capital to remote parts of the planet is a feature of economic globalization. Which regions of the world receive relatively high foreign direct investment when compared with their gross domestic product? What do you think investors find attractive in these settings?

2. Imagine if you mapped which countries received the most FDI in absolute terms. What would that map look like, and why would it be different?

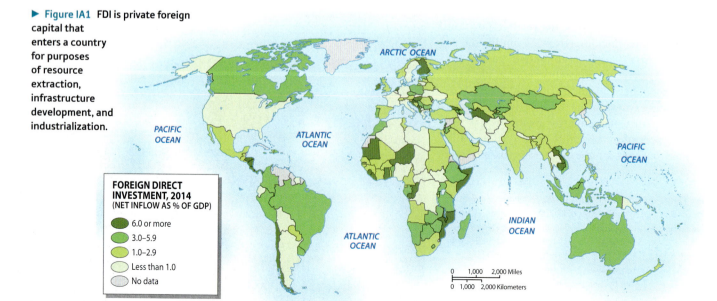

▶ **Figure IA1** **FDI is private foreign capital that enters a country for purposes of resource extraction, infrastructure development, and industrialization.**

FOREIGN DIRECT INVESTMENT, 2014 (NET INFLOW AS % OF GDP)

- 6.0 or more
- 3.0–5.9
- 1.0–2.9
- Less than 1.0
- No data

ARCTIC OCEAN

PACIFIC OCEAN

ATLANTIC OCEAN

PACIFIC OCEAN

INDIAN OCEAN

ATLANTIC OCEAN

0   1,000   2,000 Miles
0   1,000   2,000 Kilometers

JOIN THE DEBATE

Globalization is most often associated with economic activity, but impacts all aspects of the world's physical and human landscapes (Figure D1). Global linkages are complex and can result in a variety of outcomes, some unexpected. Is globalization generally good or bad for the social and economic development?

▲ **Figure D1 Global Consumers** A busy shopping mall in Guangzhou, China

Globalization Advances Social and Economic Development!

- Technological advances level the global playing field and allow more people to engage in economic activity and trade.

- With open markets there are fewer barriers, increasing the efficiency of goods production and reducing the price of goods.

- Open economies tend to be more democratic, more tolerant of diversity, and have less gender inequality.

Globalization has Negative Consequences for Development!

- As trade increases, wages decline and income inequality is exacerbated. Furthermore, digital globalization increases efficiency, but creates fewer high-skilled jobs because less labor is required.

- A growth-at-all-costs argument often accelerates depletion of natural resources and unsustainable development. Moreover, fluctuations in commodity prices can lead to economic instability.

- The speed at which capital is transferred can lead to a herd mentality with regard to financial markets, promoting instability.

KEY TERMS

areal differentiation (p. 5)
areal integration (p. 5)
choropleth map (p. 17)
colonialism (p. 35)
core–periphery model (p. 39)
counterinsurgency (p. 37)
cultural imperialism (p. 28)
cultural landscape (p. 6)
cultural nationalism (p. 29)
cultural syncretism (p. 29)
culture (p. 27)
decolonialization (p. 36)
demographic transition model (p. 24)
diversity (p. 14)
ethnic religions (p. 30)
formal regions (p. 7)
functional region (p. 7)
gender (p. 32)
gender inequality (p. 43)
gender roles (p. 32)
geographic information systems (GIS) (p. 18)
geography (p. 4)
geopolitics (p.34)
globalization (p. 8)
global positioning systems (GPS) (p. 16)
glocalization (p. 9)
graphic (linear) scale (p. 17)
gross domestic product (GDP) (p. 40)
gross national income (GNI) (p. 40)
gross national income (GNI) per capita (p. 41)
Human Development Index (HDI) (p. 42)

human geography (p. 5)
insurgency (p. 36)
latitude (p. 15)
life expectancy (p. 24)
lingua franca (p. 30)
longitude (p. 15)
map projections (p. 16)
map scale (p. 16)
megacity (p. 27)
nation-state (p. 34)
neocolonialism (p. 36)
net migration rate (p. 26)
physical geography (p. 5)
place (p. 6)
population density (p. 22)
population pyramid (p. 23)
prime meridian (p. 15)
purchasing power parity (PPP) (p. 41)
rate of natural increase (RNI) (p. 22)
regions (p. 6)
regional geography (p. 5)
remote sensing (p. 17)
replacement rate (p. 23)
secularism (p. 32)
sovereignty (p. 34)
space (p. 6)
territory (p. 34)
terrorism (p. 36)
thematic (systematic) geography (p. 5)
total fertility rate (p. 23)
universalizing religion (p. 30)
urbanized population (p. 27)
Washington Consensus (p. 12)

DATA ANALYSIS

http://goo.gl/3y1Ebm

The tables in this chapter show data for the world's 10 largest countries. But what are the world's next 10 largest countries, and where are they located? What are their per capita income levels? Are these economies growing or contracting? You can answer these questions and others by going to the website of the World Bank (http://wdi.worldbank.org) and accessing Table 1.1 for the 2016 development indicators.

1. Review the first column, and make a table of the next 10 largest countries. In which world regions are these countries located?

2. After selecting the countries, compare their gross national incomes with their purchasing power parity at a per capita basis. Based on your findings, would you consider these countries more developed or less developed, and why?

3. Compare the population densities of these countries. Some social scientists have argued that population density is a problem that can contribute to higher levels of poverty. Is there any correlation between population density and overall levels of development?

MasteringGeography™

MasteringGeography™ Looking for additional review and test prep materials? Visit the Study Area in **MasteringGeography™** to enhance your geographic literacy, spatial reasoning skills, and understanding of this chapter's content by accessing a variety of resources, including **MapMaster** interactive maps, videos, *In the News* RSS feeds, flashcards, web links, self-study quizzes, and an eText version of *Diversity Amid Globalization*.

Authors' Blogs

Scan to visit the **GeoCurrents blog**
www.geocurrents.info

Scan to visit the **Author's Blog** for field notes, media resources, and chapter updates
https://gad4blog.wordpress.com/category/globalization-and-world-geography/

Geology: A Restless Earth

Earth's surface is comprised of numerous tectonic plates that slowly move about, driven by convection cells deep within the mantle. This movement not only provides shape to the world's landscapes but also causes hazardous earthquakes and volcanoes.

Global Climates: Adapting to Change

Our world has a wide variety of climate regions, ranging from polar to tropical. These climates, however, may be changing—with problematic consequences—due to human-caused changes to Earth's atmosphere and biosphere.

Bioregions And Biodiversity: The Globalization Of Nature

Cloaks of natural vegetation vary greatly from place to place on Earth, creating novel ecosystems that have been highly altered by human activities.

Water: A Scarce World Resource

Clean freshwater is a necessity of life, but global supplies are scarce, with increasing water stress and shortages in many areas of the world.

Global Energy: The Essential Resource

Nonrenewable fossil fuels—coal, oil, and natural gas—currently dominate global energy usage, emitting climate-changing atmospheric gases. But the future may lie with renewable energy sources—water, wind, solar, and biomass.

▶ **Big Sur Coast** One of the world's more scenic coastlines, California's Big Sur coast south of Monterey embodies the interaction of ocean, atmosphere, and land. Directly along the coastline, vegetation is adapted to salty air while farther inland plants and trees draw upon ocean nutrients brought ashore in the frequent coastal fog seen in the background. Offshore marine life is protected by the Monterey Bay National Marine Sanctuary created in 1992.

Big Sur, California

NORTH AMERICA

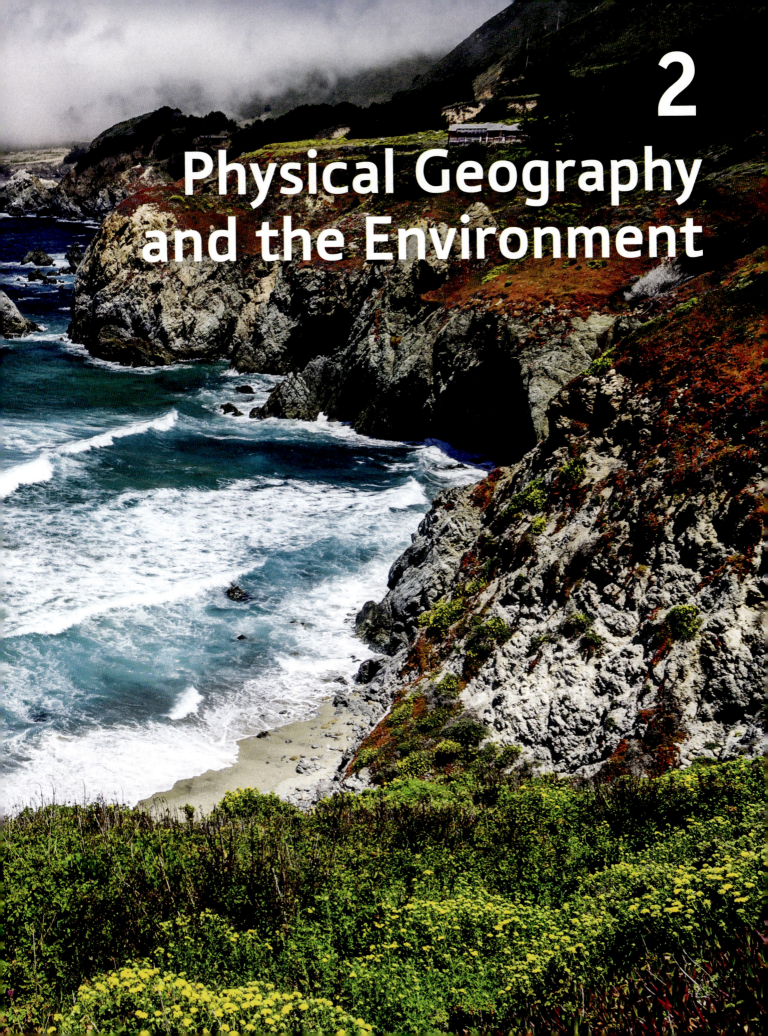

2

Physical Geography and the Environment

Earth may be misnamed because, in many ways, water dominates our planet's environmental geography. Oceans, after all, cover almost three-quarters of Earth's surface area (71 percent to be more specific), and unsurprisingly these vast areas of oceans and seas are major influences on our planet's weather and climate. These climatic effects, in turn, influence, in many ways, the geography of plants and animals on Earth's continents and islands and, to a lesser degree, the distribution of crops we humans draw upon for sustenance.

Life itself may have begun in the oceans billions of years ago, perhaps in the unique chemistry of seawater near volcanic vents on the ocean floor. Regardless of life's place of origin, today, most of Earth's known species live in the ocean. This rich array of oceanic plants and animals has long attracted the attention of humans, so much so that today over half of the world's population lives within 50 miles (80 km) of an ocean, with the other half drawing heavily on oceanic resources in one way or another. At the most basic level, marine microorganisms provide about half of the oxygen we need to survive; at the dinner table, almost a quarter of the world's protein intake comes from the seas. A far larger portion of our food needs travels across oceans when one thinks about the globalized source-to-consumer geography of chicken meat from China or beef from South America.

Oceans, and our relationship with seas and coasts, are central to understanding Earth's human geography because they sustain us in so many different ways. Also important are ocean-based environmental problems and issues, ranging from ocean acidification and the flooding of coastal cities as sea levels rise due to climate change, to water pollution and overexploitation of fisheries. Although a fuller discussion of the many problems plaguing our oceans is warranted, the best we can do in this book is to mention only the most pressing, and hope the readers' curiosity will take them further.

Studying seas and coasts is just one part of understanding our world. The immense physical diversity of Earth, from its deep oceans to its towering mountain ranges, dry deserts, and rainy tropics, have resulted in life forms of all different sorts—plant, animal, and human—that in turn interact with the physical environment to produce the diverse landscapes and habitats that make Earth our home. Thus, a necessary starting point for the study of world regional geography is knowing more about Earth's physical environment—its geology, climates, life forms, water and energy resources, and how these physical features influence and are influenced by human activities.

> ## At the most basic level, marine microorganisms provide about half of the oxygen we need to survive; at the dinner table, almost a quarter of the world's protein intake comes from the seas.

Explore the **Sounds** of Humpbacked Whales

http://goo.gl/HDcMBD

LEARNING Objectives
After reading this chapter you should be able to:

2.1 Describe those aspects of tectonic plate theory responsible for shaping Earth's surface.

2.2 Identify on a map those parts of the world where earthquakes and volcanoes are hazardous to human settlement.

2.3 List and explain the factors that control the world's weather and climate, and use these to describe the world's major climate regions.

2.4 Define the greenhouse effect and explain how it is related to anthropogenic climate change.

2.5 Summarize the major issues underlying international efforts to address climate change.

2.6 Locate on a map and describe the characteristics of the world's major bioregions.

2.7 Name some threats to Earth's biodiversity.

2.8 Identify the causes of global water stress.

2.9 Describe the world geography of fossil fuel production and consumption.

2.10 List the advantages and disadvantages of the different kinds of renewable energy.

Geology: A Restless Earth

The world's continents, separated by our vast oceans, are made up of an array of high mountains, deep valleys, rolling hills, and flat plains created over time by geologic processes originating deep within our planet and then sculpted on the surface by everyday processes such as wind, rain, and running water. Not only does this physical landscape give Earth its unique character, but this geologic fundament also affects a wide range of human activities—creating resources in many places, but posing daunting challenges in others with destructive earthquakes and volcanic eruptions

Mobile Field Trip: Introduction to Geography (MG)

http://goo.gl/5d3WtR

Plate Tectonics

The starting point for understanding geologic processes is the theory of **plate tectonics**, which states that Earth's outer layer, the **lithosphere**, consists of large geologic platforms, or plates, that move very slowly across its surface. Driving the movement of these plates is a heat exchange deep within Earth. **Figure 2.1** illustrates this complicated process.

On top of these plates sit continents and ocean basins; however, note in **Figure 2.2** that the world's continents and oceans are not identical to the underlying plates, but rather have different margins and boundaries. This is important because most earthquakes and volcanoes and their associated hazards are generated along these plate boundaries. This map also shows that there are different types of plate boundaries linked to the underlying convection cells. **Convergent plate boundaries** are those where plates move toward one another, whereas **divergent plate boundaries** are those where plates move apart. **Transform plate boundaries** are characterized by two plates grinding laterally past one another.

Along convergent boundaries, where an oceanic plate meets a continental plate, the denser seafloor plate often sinks below the lighter continental plate, creating a **subduction zone**.

▶ **Figure 2.1** **Plate Tectonics** The driving force behind plate movement is the convection cells generated by heat differences within Earth's mantle. These cells circulate slowly, and in different directions, producing surface movement in the crustal plates. New plate material reaches the surface at the mid-oceanic ridges, then moves away slowly from these divergent boundaries. As the plate material cools it tends to sink, creating subduction zones at convergent plate boundaries.

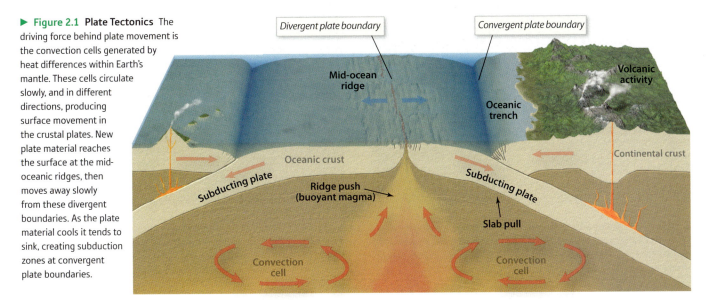

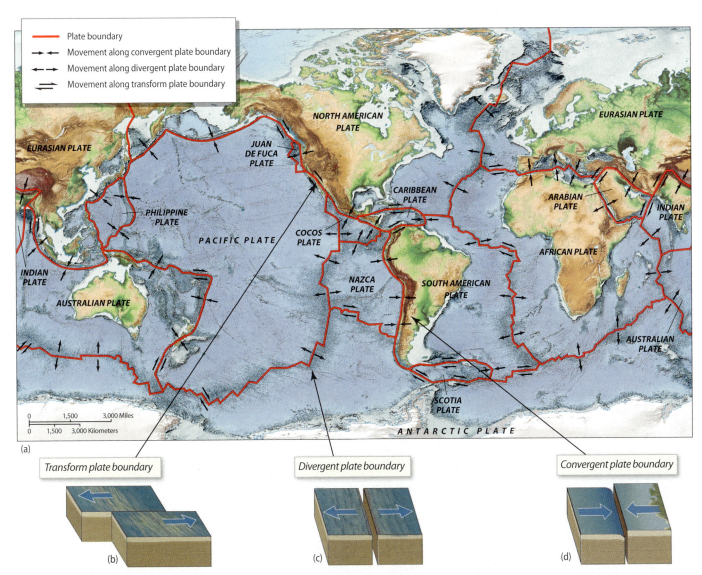

▲ **Figure 2.2** **Plate Boundaries** This world map shows the global distribution of the major plates, along with the general direction of plate movement. As well, the different categories of plate boundaries are shown. Note that continental boundaries do not always coincide with plate boundaries. Put differently, continents are not the same as tectonic plates; instead, and to simplify a bit, continents ride on top of the plates.

▲ **Figure 2.3** **Chile's Subduction Zone Earthquakes** A damaged car lies on debris after a magnitude 8.3 subduction zone earthquake hit areas of central Chile north of Santiago on September 17, 2015. Over 1 million residents lost their homes.

▲ **Figure 2.4** **Iceland's Divergent Plate Boundary** Volcanic activity, like this eruption on the Holuhraun Fissure near the Bardarbunga Volcano, is common in Iceland because of its location on the divergent boundary that bisects the Atlantic Ocean.

Explore the **Sights** of Plate Divergence in Iceland

http://goo.gl/U7JIKZ

Deep trenches characterize these zones where the ocean floor has been pulled downward by sinking plates. Subduction zones exist off the west coast of South America, off the northwest coast of North America, offshore of eastern Japan, and near the Philippines, where the Mariana Trench is the deepest point of the world's oceans at 35,000 feet (10,700 meters) below the surface. These subduction zones are also the locations of Earth's most powerful earthquakes, often with accompanying tsunamis, as evidenced by the magnitude 8.3 earthquake that displaced a million people in Chile in 2015 (**Figure 2.3**) and the magnitude 9.0 earthquake that devastated coastal Japan in 2011. These zones are also sites for many volcanoes, such as

the 130 active volcanoes in Indonesia, located along the Sunda Trench. On convergent boundaries where two continental plates collide rather than subduct, towering mountains are formed. The best known of these collision-generated ranges are the Himalayas that stretch across Asia.

Where plates diverge, magma from Earth's interior often flows to the surface, creating mountain ranges and active volcanoes. In the North Atlantic, Iceland, the product of lava flows, lies on the divergent plate boundary that bisects the Atlantic Ocean (**Figure 2.4**). But at divergent boundaries, deep depressions—called **rift valleys**—are formed. An example is the Red Sea between northern Africa and Saudi Arabia; movement of the African and Arabian plates away from one another created this body of water.

In western North America, coastal California lies atop two different plates—the Pacific and the North American plates. The infamous San Andreas Fault traversing the coast forms this plate boundary, or **transform fault** (**Figure 2.5**),

▼ **Figure 2.5** **Transform Boundary** This 3-D map shows the many different earthquake faults in the San Francisco Bay Area associated with the San Andreas fault system, which is a large transform boundary separating the oceanic Pacific Plate from the continental North American Plate.

Mobile Field Trip: San Andreas Fault

http://goo.gl/4ZNDAk

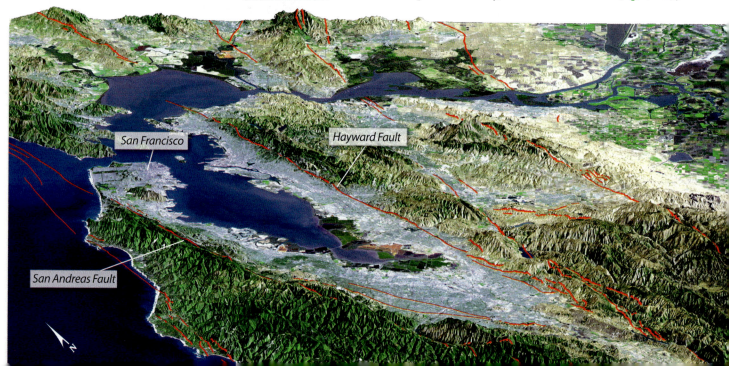

San Francisco

Hayward Fault

San Andreas Fault

with the eastern edge of the Pacific Plate moving northward at a rate of several inches each year, pushing sideways past the North American Plate. The nearness of San Francisco and Los Angeles to the San Andreas Fault makes these two urban areas—like many others around the world—vulnerable to destructive earthquakes.

Geologic evidence suggests that some 250 million years ago all the world's land masses were tightly consolidated into a supercontinent centered on present-day Africa. Over time, this supercontinent, called **Pangaea**, broke up as convection cells in the mantle moved the plates apart. A hint of this former continent can be seen in the jigsaw-puzzle fit of South America with Africa and of North America with Europe.

Geologic Hazards

Although extreme weather events like floods and tropical storms typically take a higher toll on human life each year, earthquakes and volcanoes can significantly affect human settlement and activities (**Figure 2.6**). Nearly 20,000 people perished in March 2011 from the combination of an earthquake and a tsunami in coastal Japan, and a year earlier (January 2010) as many as 300,000 people may have died

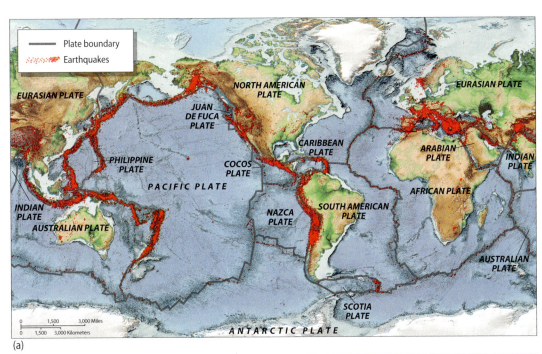

▶ **Figure 2.6 The Geography of Earthquakes and Volcanoes** (a) Most, but not all, earthquakes take place near plate boundaries. Further, most of the strongest and most devastating earthquakes are located near converging subduction-zone boundaries. (b) Although there is a strong correlation between the distribution of volcanoes, plate boundaries, and earthquakes, there are many places in the world where volcanoes are found far removed from plate boundaries. The island volcanoes of Hawaii are an example. **Q: Search the Internet for news of a recent large earthquake and locate it on this map. Near what type of plate boundary did the quake occur?**

◄ **Figure 2.7 Seattle and Mt. Rainier** Picturesque as it may be, Mt. Rainier is a classic subduction zone volcano, conveying silent warning that Seattle should expect a strong earthquake sometime in the future. The last major quake along the Cascadia subduction zone was in 1700, and was estimated to be 8.7–9.2 in magnitude. A similar earthquake today would cause considerable damage in both Seattle, Washington and Portland, Oregon.

in a magnitude 7.0 earthquake in Haiti. The vastly different effects of these two quakes underscore the fact that vulnerability to geologic hazards differs considerably around the world, depending on local building standards, population density, housing traditions, and the effectiveness of search, rescue, and relief organizations.

In addition to earthquakes, volcanic eruptions occur along both divergent plate boundaries and subduction zones (see Figure 2.6) and can also cause major destruction. But because volcanoes usually provide an array of warnings before they erupt, the loss of life from volcanoes is generally a fraction of that from earthquakes. In the 20th century, an estimated 75,000 people were killed by volcanic eruptions, whereas approximately 1.5 million died in earthquakes.

Unlike earthquakes, volcanoes provide some benefits to people. In Iceland, New Zealand, and Italy, geothermal activity produces energy to heat houses and power factories. In other parts of the world, such as the islands of Indonesia, volcanic ash has enriched soil fertility for agriculture. Additionally, local economies benefit from tourists attracted by scenic volcanic landscapes in such places as Hawaii, Japan, and the Pacific Northwest (**Figure 2.7**).

REVIEW

2.1 Sketch and describe the different kinds of tectonic plate boundaries. What causes their differences?

2.2 Where do most of the world's earthquakes and volcanoes occur? Why are they located where they are?

■ **KEY TERMS** plate tectonics, lithosphere, convergent plate boundaries, divergent plate boundaries, transform plate boundaries, subduction zone, rift valleys, transform fault, Pangaea

Global Climates: Adapting to Change

Many human activities are closely tied to weather and climate. Farming depends on certain conditions of sunlight, temperature, and precipitation to produce the world's food,

while urban transportation systems are often disrupted by extreme weather events like snowstorms, typhoons, and even heat waves. Furthermore, a severe weather event in one location can affect far-flung places. Reduced harvests due to drought in Russia's grain belt, for example, ripple through global trade and food supply systems, with serious consequences worldwide.

Aggravating these interconnections is global climate change. Just what the future holds is not entirely clear, but even if the long-term forecast has some uncertainty, there is little question that all forms of life—including humans—must adapt to vastly different climatic conditions by the middle of the 21st century (see *Geographers at Work: Studying Glaciers and Climate Change*).

Climate Controls

The world's climates differ significantly from place to place and seasonally, with highly variable patterns of temperature and precipitation (rain and snow) that can be explained by physical processes known as climate controls.

Solar Energy The Sun's heating of Earth and its atmosphere is the most important factor affecting world climates. Not only does solar energy cause temperature differences between warmer and colder regions, but it also drives other important climate controls such as global pressure systems, winds, and ocean currents.

Incoming short-wave solar energy, called **insolation**, passes through the atmosphere and is absorbed by Earth's land and water surfaces. As these surfaces warm, they **reradiate** heat back into the lower atmosphere as infrared, long-wave energy. This reradiating energy, in turn, is absorbed by water vapor and other atmospheric gases such as carbon dioxide (CO_2), creating the envelope of warmth that makes life possible on our planet. Because there is some similarity between this heating process and the way a garden greenhouse traps warmth from the Sun, this natural process of atmospheric heating is called the **greenhouse effect** (**Figure 2.8**). Without this process, Earth's climate would average about 60°F (33°C) colder, resulting in conditions much like Mars.

Studying Glaciers and Climate Change

Climate change scientist and arctic adventurer **M Jackson** is a role model for mixing serious science with global adventuring. M leads National Geographic Society expeditions (**Figure 2.1.1**), lecturing on arctic science during tours, and recently wrote a book, *While Glaciers Slept: Being Human in a Time of Climate Change.*

Glaciers and Geography After completing her master's degree in environmental science at the University of Montana, Jackson traveled to Turkey as a Fulbright Scholar to study glaciers on the Turkish/Iranian border. What she observed led to more questions—and to the study of geography. "Glaciologists told me that if I really wanted to understand glaciers within a greater system, I should do so from a geographic perspective. Geography provides an avenue to view something from many different angles: the social sciences, the humanities, the natural sciences. And it's the geographer's job to combine these different elements into a whole picture of what is happening . . . I love that."

In addition to her NGS expeditions, M is working on her PhD in geography, doing fieldwork on the impacts of glacier loss in Iceland. Both activities allow her to "sell" geography to undergraduates: "If a student is leaning more toward natural science or lab work, or if a person is interested in social theory, or wants to learn GIS or computer skills, geography is a major where the student can be selective, but add these bits into a greater toolkit. The size and strength of that toolkit will help immensely on the job market."

"If you look at something in just one way, you can't get a full view," continues M. "With geography, you can go in different directions, or just synthesize everything together."

1. Explain the impacts of human-induced climate change on glaciers, and list the possible consequences for people and their communities.

2. What are the advantages of understanding human geography when researching natural phenomena, such as glaciers, forests, or oceans?

▲ **Figure 2.1.1** M Jackson takes a well-deserved break from her glaciology work in Iceland.

▶ **Figure 2.8 Solar Energy and the Greenhouse Effect** The greenhouse effect is the trapping of solar radiation in the lower atmosphere, resulting in a warm envelope surrounding Earth. Most incoming shortwave solar radiation is absorbed by land and water surfaces, then reradiated into the atmosphere as longwave infrared radiation. It is this longwave radiation absorbed by greenhouse gases—both natural and human-generated—that warms the lower atmosphere and affects Earth's weather and climate.

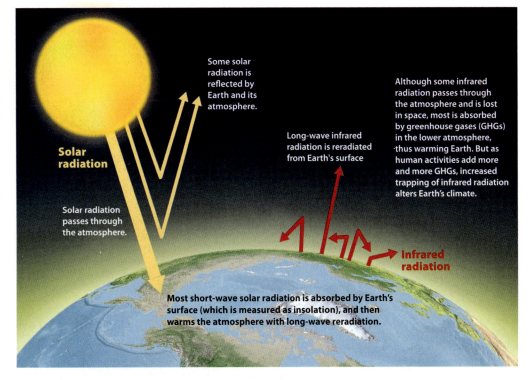

Some solar radiation is reflected by Earth and its atmosphere.

Long-wave infrared radiation is reradiated from Earth's surface

Although some infrared radiation passes through the atmosphere and is lost in space, most is absorbed by greenhouse gases (GHGs) in the lower atmosphere, thus warming Earth. But as human activities add more and more GHGs, increased trapping of infrared radiation alters Earth's climate.

Solar radiation

Solar radiation passes through the atmosphere.

Infrared radiation

Most short-wave solar radiation is absorbed by Earth's surface (which is measured as insolation), and then warms the atmosphere with long-wave reradiation.

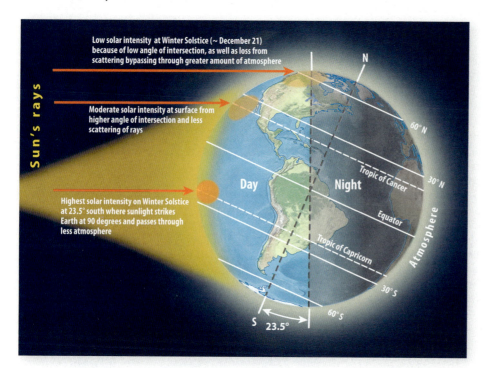

Low solar intensity at Winter Solstice (~ December 21) because of low angle of intersection, as well as loss from scattering bypassing through greater amount of atmosphere

Moderate solar intensity at surface from higher angle of intersection and less scattering of rays

Highest solar intensity on Winter Solstice at 23.5° south where sunlight strikes Earth at 90 degrees and passes through less atmosphere

Sun's rays

N

Day | Night

60° N

Tropic of Cancer

30° N

Equator

Atmosphere

Tropic of Capricorn

30° S

S 23.5°

60° S

◄ **Figure 2.9 Solar Intensity and Latitude** Because of Earth's curvature, solar radiation is more intense and more effective at warming the surface in the tropics than at higher latitudes. The resulting heat buildup in the equatorial zone energizes global wind and pressure systems, ocean currents, and creates tropical storms.

or hurricanes, and even midlatitude storms (**Figure 2.10**).

Interactions Between Land and Water Land and water areas differ in their ability to absorb and reradiate insolation; thus, the global arrangement of oceans and land areas is a major influence on world climates. More solar energy is required to heat water than to heat land, so land areas heat and cool faster than do bodies of water (**Figure 2.11**). This explains why the temperature extremes of hot summers and cold winters are found in the continental interiors, such as the Great Plains of North America, while coastal areas

Latitude Because Earth is a tilted sphere with the North Pole facing away from the Sun for half of the year and toward the Sun for the other half, maximum solar radiation at a given location occurs seasonally (with summer marking the season of peak insolation for each hemisphere), and insolation strikes the surface at a true right angle only in the tropics. Therefore, solar energy is more intense and effective at heating land and water at low latitudes than at higher latitudes. Not only does this difference in solar intensity result in warmer tropical climates, contrasted with the cooler middle and high latitudes, but also heat accumulates on a larger scale in these equatorial regions (**Figure 2.9**). This heat is then distributed away from the tropics through global pressure and wind systems, ocean currents, subtropical typhoons

▶ **Figure 2.10 Tropical Cyclone Winston** Heat buildup in equatorial zones and the lower latitudes produce massive tropical cyclones called typhoons in the Pacific and hurricanes in the Atlantic. Both are capable of widespread damage through high winds, coastal flooding from waves, and heavy rainfall. This is a satellite image of tropical cyclone Winston, a storm that devastated Fiji and adjacent islands in February 2016, and was one of the strongest tropical storms ever recorded.

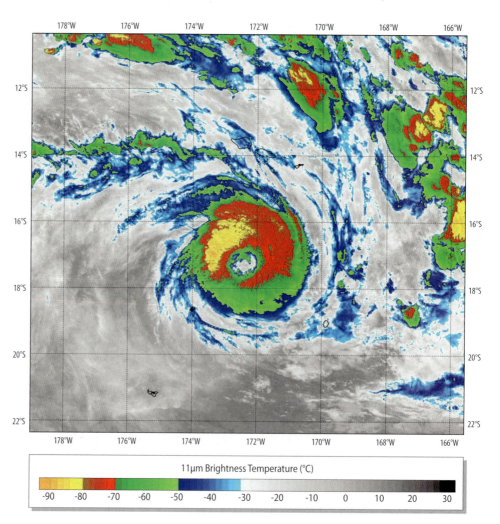

11μm Brightness Temperature (°C)

-90 -80 -70 -60 -50 -40 -30 -20 -10 0 10 20 30

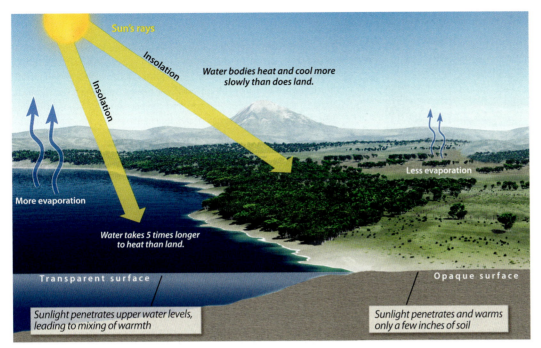

◀ **Figure 2.11**
Differential Heating of Land and Water Land heats and cools faster than does water because it takes more energy to raise the temperature of water compared to land. This is why inland temperatures are usually both warmer in the summer and colder in the winter than coastal locations.

experience more moderate winters and cooler summers. These coastal-inland temperature differences also occur at smaller scales, often within just hundreds of miles of each other. In coastal San Francisco, for example, the average July maximum temperature is 60°F (15.6°C), whereas 80 miles (129 km) away in California's inland capital, Sacramento, the average maximum July temperature is 92.4°F (33.5°C).

The term **continental climate** describes inland climates with hot summers and cold winters, while locations where oceanic influences dominate and milder temperatures are the norm are termed **maritime climates**. The island countries of Southeast Asia and the British Isles in Europe are good examples of areas with maritime climates, while interior North America, Europe, and Asia have continental climates.

Global Pressure Systems The uneven heating of Earth due to latitudinal differences and the arrangement of oceans and continents produces a regular pattern of high and low pressure cells (**Figure 2.12**). Low pressure cells form at the surface when warm, buoyant air molecules rise, while surface high pressures occur

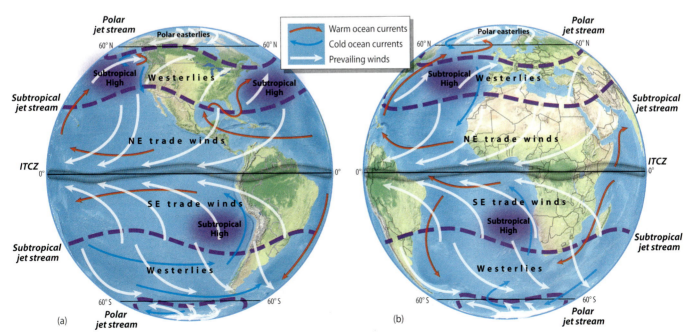

▲ **Figure 2.12 Global Pressure Systems and Winds** Two jet streams, the polar and subtropical jets, are found in each hemisphere, northern and southern. In circling Earth, these jets often change position as they steer storms and air masses. The subtropical high pressure cells are large areas of subsiding air that energize the midlatitude westerly winds and the tropical trade winds. The Inter-Tropical Convergence Zone (ITCZ) is a belt of low pressure that circles Earth and results from strong solar radiation in the equatorial zone.

International Activity at the Poles

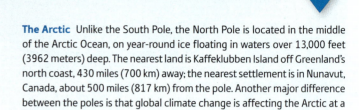

The North and South Poles are the centers of two similar—but also very different—world regions of global interest that could influence our future in dramatic ways. Let's take a closer look.

Antarctica The South Pole lies on the continent of Antarctica, the world's fifth largest land mass. Antarctica is huge, but many other superlatives apply: It's the coldest, driest, windiest, and highest continent. The average yearly temperature at the pole is −70°F (−57°C), and annual precipitation averages only 6.5 inches (16.5 cm). Winds of 50–60 miles per hour (80–95 km/h) blow constantly, producing terrifically low wind-chill factors.

Two ice sheets dominate Antarctica, separated by a large mountain range. These ice sheets have an average depth of 7900 feet (2400 meters) but reach 16,500 feet (5000 meters) in the thickest part. In addition, several giant coastal ice shelves extend into nearby ocean waters (**Figure 2.2.1**).

Laboratory on Ice With no indigenous people, Antarctica is cleaner and purer than any other place on Earth—perfect for investigating Earth's atmosphere without the effects of air pollution. Currently, 53 countries have signed an international agreement to protect the continent from economic exploitation and preserve it for scientific study.

A major focus for these studies is Earth's climate: past, present, and future. Much about past climates is learned from the chemical composition of ice cores taken from the ice sheets. In terms of present-day climate, it was in Antarctica that the notorious ozone hole in the atmosphere was discovered and then addressed in the Montreal Protocol, an international treaty banning the use of ozone-depleting chemicals. As for the future, researchers uncovered the potential for the melting and collapse of the West Antarctica Ice Sheet due to climate change (**Figure 2.2.2**). Should that happen, Earth's sea level could rise by at least 10 feet (33 meters), devastating coastal settlements worldwide.

The Arctic Unlike the South Pole, the North Pole is located in the middle of the Arctic Ocean, on year-round ice floating in waters over 13,000 feet (3962 meters) deep. The nearest land is Kaffeklubben Island off Greenland's north coast, 430 miles (700 km) away; the nearest settlement is in Nunavut, Canada, about 500 miles (817 km) from the pole. Another major difference between the poles is that global climate change is affecting the Arctic at a

▼ **Figure 2.2.1 Antarctica** The East and West Antarctica Ice Sheets, separated by the Transantarctic Mountains, make up much of the continent. McMurdo Station, the U.S. Antarctic research center, is southeast of the Ross Ice Shelf, on the small point north of Victoria Land.

where air subsides. These pressure cells drive the movement of the world's wind and storm systems because air (in the form of wind) moves from high to low pressure. The interaction between high- and low-pressure systems over the North Pacific, for example, produces storms that are carried by winds onto the North American continent. Similar processes in the North Atlantic produce winter and summer weather for Europe. Farther south, over the subtropical zones, large cells of high pressure cause very different conditions. The subsidence (sinking) of warm air moving in from the equatorial regions causes the great desert areas at these latitudes. These high-pressure areas expand during the warm summer months, producing the warm, rainless summers of Mediterranean climate areas in Europe and California. In the low latitudes, summer heating of the oceans also spawns the strong tropical storms known as typhoons or cyclones in Asia, and hurricanes in North America and the Caribbean.

Global Wind Patterns Several different wind patterns strongly influence Earth's weather and climate. At the global level are the **polar jet stream** and the **subtropical jet stream**,

powerful atmospheric rivers of eastward-moving air that affect storms and pressure systems in both the Northern and the Southern hemispheres (Figure 2.12). These jets are products of Earth's rotation and global temperature differences. The two polar jets (north and south) are the strongest and the most variable, flowing 23,000–39,000 feet (7–12 km) above the surface at speeds reaching 200 miles per hour (322 km/h). The subtropical jets are usually higher and somewhat weaker. While the northern jet stream has a major effect on the weather of North America and Europe, steering storms across the continents, the southern polar jet circles the globe in the sparsely populated areas near Antarctica (see *Exploring Global Connections: International Activity at the Poles*).

Nearer Earth's surface are continent-scale winds that, as mentioned earlier, flow from high to low pressure areas. Good examples are the **monsoon winds** of Asia and North America; summer monsoons bring welcome rainfall to the dry areas of interior India and the Southwest United States (**Figure 2.13**).

Also important is that in many parts of the world surface-level winds drive global-scale ocean currents. In the northern hemisphere equatorial belt, for example, the northeasterly trade winds energize westward-flowing ocean currents

▲ **Figure 2.2.2 West Antarctica Ice Sheet** The edge of the ice shelf is in the Ross Sea, with the Transantarctic Mountains in the background.

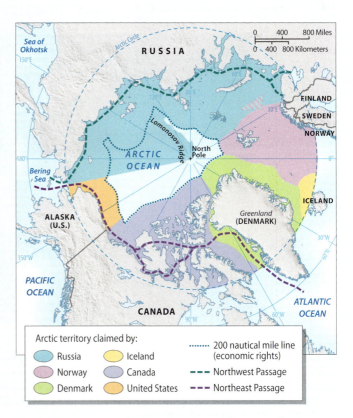

Arctic territory claimed by:

● Russia	● Iceland
● Norway	● Canada
● Denmark	● United States

········ 200 nautical mile line (economic rights)
– – – Northwest Passage
– – – Northeast Passage

▲ **Figure 2.2.3 Arctic Territorial Land Claims** The Arctic consists of land, internal waters, territorial seas, exclusive economic zones (EEZs), and high seas. All land, internal waters, territorial seas, and EEZs in the Arctic are under the jurisdiction of one of the six Arctic coastal states: Canada, Iceland, Norway, Russia, Denmark (via Greenland), or the United States.

much faster rate than the Antarctic region, and sea ice in the Arctic Ocean is diminishing extensively each year. Although the North Pole is not likely to be seasonally ice-free anytime soon, the fringes of the Arctic Ocean are free of ice for longer and longer periods. This open water has opened the door to international tensions, squabbles, even land grabs.

Territorial Claims in the Arctic Unlike Antarctica's protection under international treaty, the icy Arctic Ocean—and, importantly, the ocean floor beneath these waters—are claimed by Russia, Norway, Denmark (through its relationship with Greenland), Canada, and the United States. Under international law, countries can stake out claims for an Exclusive Economic Zone (EEZ) reaching out 200 miles from their shoreline. **Figure 2.2.3** illustrates those claims. With more open water and less ice, these claims allow deep water oil and gas mining, fishing rights, and future maritime shipping routes. More detail on these topics can be found in the North America and Russian Domain chapters.

1. In terms of their physical geography, how are the polar regions similar? How are they different?

2. Explain why international cooperation differs for each polar region.

◄ **Figure 2.13 Summer Monsoon in Southwest North America** As the U.S. Southwest warms during the northern hemisphere summer, this heating creates thermal lows over the interior that draw in moist air from the Gulfs of Mexico and California, resulting in cloudiness, thunderstorms, and much-needed rainfall from July to September.

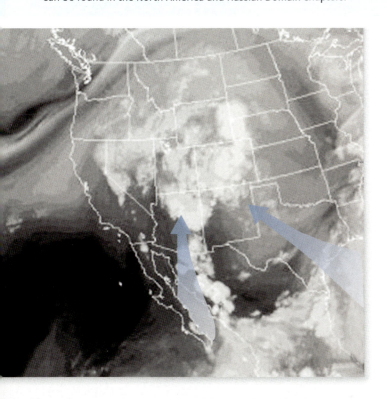

in both the Pacific and Atlantic Oceans. In the higher latitudes, prevailing northwesterly winds drive ocean currents poleward, enhancing the maritime effect to the coasts of western Europe and northwestern North America.

Topography Weather and climate are affected by topography—an area's surface characteristics—in two ways: Cooler temperatures are found at higher elevations, and precipitation patterns are strongly influenced by topography.

Because the lower atmosphere is heated by solar energy reradiated from Earth's surface, air temperatures are warmer closer to the surface and become cooler with altitude. As a general rule, the atmosphere cools by 3.5°F for every 1000 feet gained in elevation (.05°C per 100 meters). This is called the **environmental lapse rate**. To illustrate, on a typical summer day in Phoenix, Arizona, at an elevation of 1100 feet

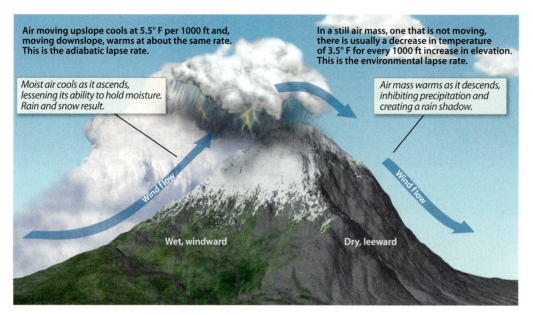

Air moving upslope cools at 5.5° F per 1000 ft and, moving downslope, warms at about the same rate. This is the adiabatic lapse rate.

In a still air mass, one that is not moving, there is usually a decrease in temperature of 3.5° F for every 1000 ft increase in elevation. This is the environmental lapse rate.

Moist air cools as it ascends, lessening its ability to hold moisture. Rain and snow result.

Air mass warms as it descends, inhibiting precipitation and creating a rain shadow.

Wind flow

Wind flow

Wet, windward

Dry, leeward

▲ **Figure 2.14** **The Orographic Effect** Upland and mountainous areas are usually wetter than the adjacent lowland areas because of the orographic effect. This results from the cooling of rising air over higher topography, and as the air mass cools it loses its ability to hold water vapor, resulting in rain and snowfall. In contrast, the leeward or downwind side of the mountains is drier because downslope air masses warm, thus increasing their ability to retain moisture. These dry areas on the downwind side of mountains are called rain shadows. **Q: After reviewing the concept of the orographic effect, look at a map of the world and locate at least five different areas where the orographic effect would be found.**

Mobile Field Trip: Clouds: Earth's (MG) Dynamic Atmosphere

https://goo.gl/2GoynZ

(335 meters), the temperature often reaches 100°F (37.7°C). Just 140 miles away, in the mountains of northern Arizona at 7100 feet (2160 meters) in the small town of Flagstaff, the temperature is a pleasant 79°F (26°C). This difference of 21°F (11.7°C) results from 6000 feet (1825 meters) of elevation; this can be easily calculated by multiplying elevation in thousands (6) by the environmental lapse rate (3.5°F).

If an area has rugged topography, this can wring moisture out of clouds when moist air masses cool as they are forced up and over mountain ranges in what is called the **orographic effect** (Figure 2.14). Cooler air cannot hold as much water vapor; this moisture condenses, resulting in precipitation. Note that the rising air mass cools (and warms upon descending) faster than a

▶ **Figure 2.15** **Global Climate Regions** A standard scheme, called the Köppen system, named after the Austrian geographer who devised the plan in the early 20th century, is used to describe the world's diverse climates. Combinations of upper- and lower-case letters describe the general climate type, along with precipitation and temperature characteristics. Specifically, the *A* climates are tropical, the *B* climates are dry, the *C* climates are generally moderate and are found in the middle latitudes, and the *D* climates are associated with continental and high-latitude locations. Closest to the poles are the *E* climates of arctic tundra and ice caps. Also on this map are two climographs that plot average month temperature and precipitation through the calendar year. Note that the seasons are reversed in the southern hemisphere from those in the northern hemisphere.

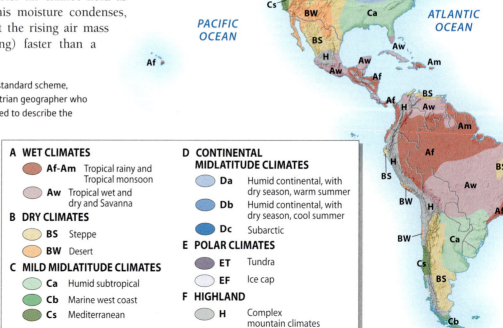

A WET CLIMATES

- **Af-Am** Tropical rainy and Tropical monsoon
- **Aw** Tropical wet and dry and Savanna

B DRY CLIMATES

- **BS** Steppe
- **BW** Desert

C MILD MIDLATITUDE CLIMATES

- **Ca** Humid subtropical
- **Cb** Marine west coast
- **Cs** Mediterranean

D CONTINENTAL MIDLATITUDE CLIMATES

- **Da** Humid continental, with dry season, warm summer
- **Db** Humid continental, with dry season, cool summer
- **Dc** Subarctic

E POLAR CLIMATES

- **ET** Tundra
- **EF** Ice cap

F HIGHLAND

- **H** Complex mountain climates

hypothetical stable or non-moving air mass, for which the change in temperature with elevation is measured by the environmental lapse rate. More specifically, an air mass moving up a mountain slope will cool at 5.5°F per 1000 feet (1°C per 100 meters) of elevation; this is referred to as the **adiabatic lapse rate**.

This process explains the common pattern of wet mountains and nearby dry lowlands. These dry areas are said to be in the **rain shadow** of the adjacent mountains. Rainfall lessens as downslope winds warm (the opposite of upslope winds, which cool), thus increasing an air mass's ability to retain moisture and deprive nearby lowlands of precipitation. Rain shadow areas are common in the mountainous areas of western North America, Andean South America, and many parts of South and Central Asia.

Climate Regions

Even though the world's weather and climate vary greatly from place to place, areas with similarities in temperature, precipitation, and seasonality can be mapped into global climate regions (**Figure 2.15**). Before going further, it is important to note the difference between these two terms. *Weather* is the short-term, day-to-day expression of atmospheric processes; that is, weather can be rainy, cloudy, sunny, hot, windy, calm, or stormy, all within a short time period. As a result, weather is measured at regular intervals each day, usually hourly. These data are then compiled over a 30-year period to generate statistical averages that describe the typical meteorological conditions of a specific place, which is the climate. Simply stated, weather is the short-term expression of atmospheric processes, and **climate** is the long-term average from daily weather measurements. As pundits like to say, climate is what you expect and weather is what you get.

We use a standard scheme of climate types throughout this text, and each regional chapter contains a map showing the different climates of that region. In addition, these maps contain **climographs**, which are graphic representations of monthly average temperatures and precipitation. Two lines for temperature data are presented on each climograph: The upper line plots average high temperatures for each month, while the lower line shows average low temperatures. Besides these temperature lines, climographs contain bar graphs depicting average monthly precipitation. The total amount of rainfall and snowfall is important, as is the seasonality of precipitation. Figure 2.15 contains climographs for Tokyo, Japan, and Cape Town, South Africa. In looking at these two climographs, remember that the seasons are reversed in the Southern Hemisphere.

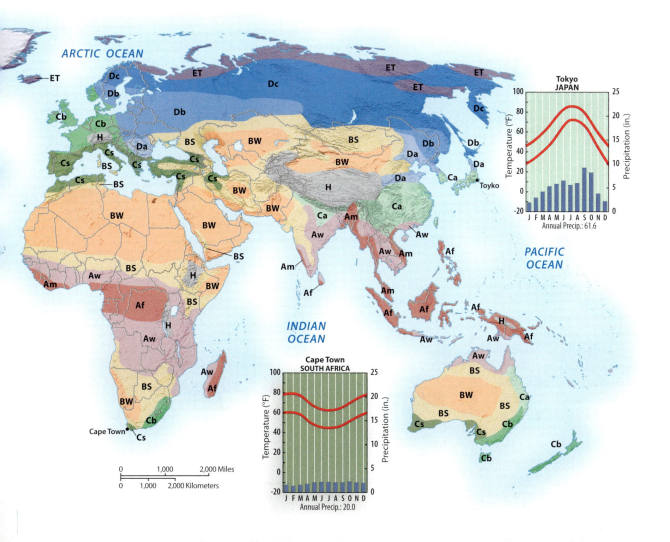

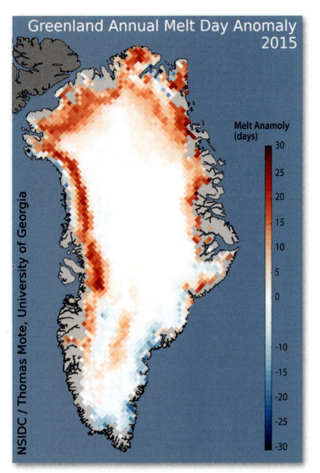

Greenland Annual Melt Day Anomaly 2015

Melt Anamoly (days)

NSIDC / Thomas Mote, University of Georgia

▲ **Figure 2.16 Melting Greenland** The areas in red on the map are those that melted at a faster than average rate during 2015, the 11th highest melt year in the 37 year record of satellite melt data. Not all years have the same pattern of more intense melting in the north and northwest. This illustrates that Arctic climate change takes place differentially with the polar region.

Explore the **Sights** of Greeland's Melting Ice Cap

http://goo.gl/ROOHtE

Global Climate Change

Human activities connected to economic development have caused significant **climate change** worldwide over the last century, resulting in warmer temperatures, melting ice caps, rising sea levels, and more extreme weather events (**Figure 2.16**). Unless international action is taken soon to limit atmospheric pollution, climate change will produce a challenging world environment by mid-century. Rainfall patterns may change, so that agricultural production in traditional breadbasket areas such as the U.S. Midwest and Canadian prairies may be threatened; low-lying coastal settlements in places like Florida and Bangladesh will be flooded as sea levels rise; increased heat waves will cause higher human death tolls in the world's cities; and clean water will become increasingly scarce in many areas of the world.

Causes of Climate Change As mentioned earlier, the natural greenhouse effect provides Earth with a warm atmospheric envelope; this warmth comes from incoming and outgoing solar radiation that is trapped by an array of such natural constituents as water vapor, carbon dioxide (CO_2), methane (CH_4), and ozone (O_3). Although the composition of these natural **greenhouse gases (GHGs)** has varied somewhat over long periods of geologic time, it has been relatively stable since the last ice age ended 20,000 years ago (more detail on GHGs is found online in MasteringGeography).

However, widespread consumption of coal and petroleum associated with global industrialization has caused a huge increase in atmospheric carbon dioxide and methane. As a result, the natural greenhouse effect has been greatly magnified by **anthropogenic**, or human-generated, GHGs, trapping increased amounts of Earth's long-wave reradiation and thus warming the atmosphere and changing our planet's climates. **Figure 2.17** shows that in 1880 atmospheric CO_2 was measured at 280 parts per million (ppm); today it is 400 ppm. More troubling is that these CO_2 emissions are forecast to reach 450 ppm by 2020, a level at which climate scientists predict irrevocable climate change.

Although the complexity of the global climate system leaves some uncertainty about exactly how the world's climates may change, climate scientists using high-powered computer models are reaching consensus on what can be expected. These computer models predicted that average global temperatures will increase 3.6°F (2°C) above the preindustrial average by 2020, a temperature change of the same magnitude as the amount of cooling that caused ice-age glaciers to cover much of Europe and North America 30,000 years ago. Further, even with international policies to limit emissions, this temperature is

▼ **Figure 2.17 Global Temperature and Atmospheric Conditions** The straight line across the graph is the average global temperature (57°F) for the 20th century. Blue bars indicate year temperatures below that average, red bars those above. The upward trending black line is the increasing amounts of CO_2 put into the atmosphere since 1880, starting at 280 parts per million (ppm) with current levels at 400 ppm.

Mobile Field Trip: Climate Change in the Arctic

https://goo.gl/4m3Py0

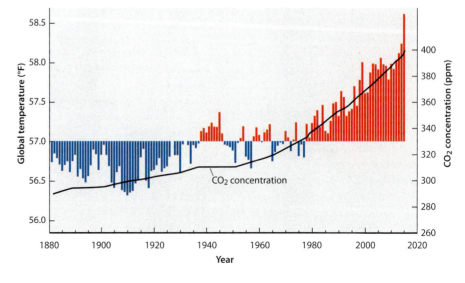

projected to increase an additional 2.7°F (1.5°C) by 2100. The resulting melting of polar ice caps, ice sheets, and mountain glaciers will cause a sea-level rise that currently is estimated to be on the order of 4 feet (1.4 meters) by century's end.

International Efforts to Limit Emissions Climate scientists have long expressed concern about fossil fuel emissions aggravating the natural greenhouse effect; as a result, in 1988 the United Nations (UN) began coordinating study of global warming by creating the International Panel on Climate Change (IPCC). This group of international scientists was (and is still) charged with providing the world with periodic Assessment Reports of climate change science. The panel's first report came out in 1990; the most recent, finalized in late 2014, included a strongly worded statement that failure to reduce emissions could threaten society with food shortages, refugee crises, the flooding of major cities and entire island nations, mass extinctions of plants and animals on land and in the oceans, and a climate so drastically altered it might become dangerous for people to work or play outside during the hottest times of the year. Further, the "continued emission of greenhouse gases will cause further warming and long-lasting changes in all components of the climate system, increasing the likelihood of severe, pervasive and irreversible impacts for people and ecosystems."

International efforts to reduce atmospheric emissions began in 1992, shortly after the IPCC's first report, when 167 countries meeting in Rio de Janeiro, Brazil, signed the Rio Convention, in which they agreed to voluntarily limit their GHG emissions. However, none of the Rio signatories reached its emission reduction targets, so a more formal international agreement came from a 1997 meeting in Kyoto, Japan. Here, 30 Western industrialized countries agreed to cut their emissions to 1990 levels by 2012. Unlike the Rio Convention, which was voluntary, this **Kyoto Protocol** had the force of international law, with penalties for those countries not reaching their emission reduction targets. At that point in time, the 30 signatories produced over 60 percent of the world's emissions, and there were no emissions limitations on the large developing economies of China and India.

But the Kyoto Protocol was not a solution. First, the world's largest polluter at that time, the United States, refused to ratify the protocol over concerns about injuring the struggling U.S. economy. Additionally, by 2005, China's yearly GHG emissions exceeded those of the United States, underscoring the need to include developing economies in international emission reduction programs (**Figure 2.18**).

By design, the Kyoto Protocol was to expire in 2012, to be replaced by a new, more inclusive treaty. But progress was slow and Kyoto was extended to the end of 2015. In 2014, the UN began a very different "bottom up" approach toward an emissions agreement that contrasted with the "top down" mandates of the protocol. All countries were asked to submit

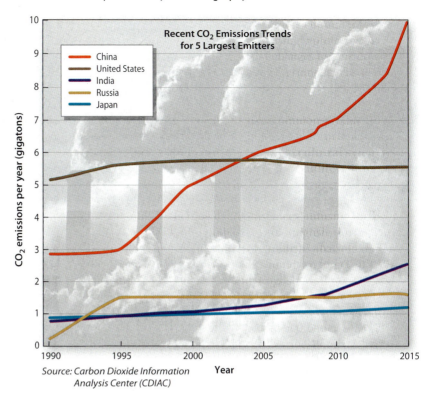

Source: Carbon Dioxide Information Analysis Center (CDIAC)

▲ **Figure 2.18 Emission Trends for the World's Largest CO₂ Emitters** China's yearly CO_2 emissions continue to grow as hundreds of new coal-fired power plants come online. In contrast, emissions in the United States have stabilized recently because more power plants have switched from coal to natural gas as the price of that cleaner source of energy has become increasingly competitive. The flat line of Russia's emissions after the collapse of the Soviet Union is somewhat of a mystery and may be a reporting problem. **Q: What are the similarities and differences between the CO₂ emission reduction plans of China, the United States, and India?**

their own strategy for addressing climate change in plans tailored to their unique national economies and social structures. This process gave countries flexibility in achieving the shared global goal of reducing GHG emissions.

These national plans, called "Intended Nationally Determined Contributions" (INDCs), were submitted to the UN Climate Change Committee during the first half of 2015, and were the basis for negotiations at the Climate Change Conference held in Paris, France that December. At that meeting a new international greenhouse gas reduction agreement was approved; this plan now serves as the world's blueprint for addressing the challenges of climate change. While the 2015 **Paris Agreement** does not solve the climate change problem, it does create a new and promising pathway for the world to follow.

- It is an inclusive, international agreement, approved by 195 countries, covering the economic spectrum from developed to developing economies. It becomes a formal international treaty in April 2017 if it is ratified either by 55 percent of the Paris signatories or by countries responsible for 55 percent of the world's atmospheric emissions. If ratified, the treaty takes effect in 2020.

- At that time, signatories are then committed to reducing their emissions as presented in their 2015 INDCs. Additionally, each country will assess and revise their INDC every five years with the goal of reducing further their GHG emissions.

◀ **Figure 2.19 India's Emissions Future** As the world's third largest CO_2 emitter, India's emissions future is a grave concern, particularly because, unlike the United States and China, the country has been deliberately vague about its plans to limit its emissions as its population grows to 1.5 billion by 2030. One low-emission option is to build more nuclear power plants such as the one pictured above. Another cheaper yet high-emission pathway is to expand the country's use of coal power plants.

- Countries commit themselves to achieving zero net emissions as soon as possible. This strategy combines the reduction of GHG emissions with carbon offsets, such as planting more trees to store carbon. Important, however, is that this component of the Paris Agreement provides flexibility in achieving zero net emissions so that developing countries, such as India and Brazil, can move at their own pace toward this goal (**Figure 2.19**).

- A fund of $100 billion dollars will be created by 2020 to assist poorer countries in meeting the challenges of climate change. Low-lying island nations suffering from sea level rise are possible candidates for this aid.

Critics of the Paris Agreement have two main concerns: first, that the total emission reduction from the INDCs currently in hand is not enough to prevent irrevocable climate change by 2100; second, beyond "naming and shaming," there is no formal regulatory structure to penalize those countries not adhering to their INDCs. Proponents respond that the agreement should be thought of as an innovative, flexible work in progress that can be easily made stricter by its signatories, which include most of the world's countries.

REVIEW

2.3 List the similarities and differences between maritime and continental climates. What causes the differences?

2.4 Provide examples of how topography affects weather and climate.

2.5 What is the natural greenhouse effect, and how has it been changed by human activities?

2.6 What are the major differences between the Paris 2015 climate agreement and the earlier Kyoto Protocol?

■ **KEY TERMS** insolation, reradiate, greenhouse effect, continental climate, maritime climates, polar jet stream, subtropical jet stream, monsoon winds, environmental lapse rate, orographic effect, adiabatic lapse rate, rain shadow, climate, climographs, climate change, greenhouse gases (GHGs), anthropogenic, Kyoto Protocol, Paris Agreement

▶ **Figure 2.20 Bioregions of the World** Although global vegetation has been greatly modified by clearing land for agriculture and settlements, as well as by cutting forests for lumber and paper pulp, there are still recognizable patterns to the world's bioregions, from tropical forests to arctic tundra. An important point is that each bioregion has its own unique array of ecosystems; also important is that these natural resources are used, abused, and conserved differently by humans throughout the world.

Mobile Field Trip: Cloud Forest MG

https://goo.gl/wxHN1z

Bioregions and Biodiversity: The Globalization of Nature

The rich diversity of plants and animals covering its continents and oceans makes Earth unique. This **biodiversity** can be thought of as the green glue that binds together geology, climate, hydrology, and life. Like climate regions, the world's biological resources can be broadly categorized into **bioregions**—areas defined by natural characteristics such as similar plant and animal life. Earth's principal bioregions, mapped in **Figure 2.20**, are illustrated in **Figures 2.21** through **2.26**.

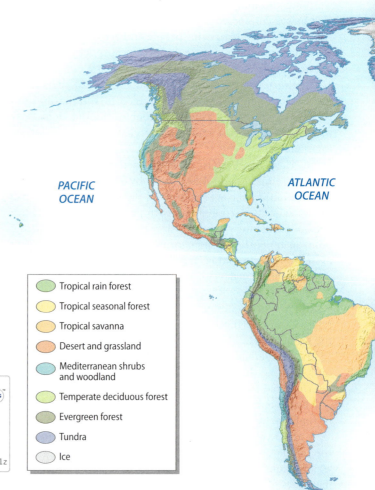

PACIFIC OCEAN

ATLANTIC OCEAN

- Tropical rain forest
- Tropical seasonal forest
- Tropical savanna
- Desert and grassland
- Mediterranean shrubs and woodland
- Temperate deciduous forest
- Evergreen forest
- Tundra
- Ice

Humans obviously are very much a part of this interaction. Not only are we evolutionary products of the African tropical savanna—a specific bioregion—but also our long human prehistory includes the domestication of plants and animals that led to modern agriculture and, later, our global food systems.

This human activity, however, combined with later urbanization and industrialization, has taken an immense toll on nature to the point where it is questionable whether natural vegetation bioregions truly exist anymore. Instead, cultivated fields have replaced grasslands and woodlands; forests have been logged for wood products (Figures 2.23 and 2.26); and wildlife has been hunted for food and fur or subjected to habitat destruction. As a result, Earth's natural world has become a distinctly human, globalized one, dominated by **novel ecosystems** that are completely new to Earth. Three pressing issues outlined below illustrate why biodiversity and bioregions are critical components of world regional geography.

Nature and the World Economy

Natural plant and animal products are an inseparable part of the world economy, be they foodstuffs, wood products, or animal meat and fur. While most of this world trade is legal and appropriately regulated, much is illegal and has detrimental environmental consequences. Examples are the illegal logging of certain rainforest trees and the poaching of protected animal species (Figure 2.25).

Climate Change and Nature

The world map of natural bioregions (see Figure 2.20) shows a close relationship with the map of global climate regions (see Figure 2.15) because temperature and precipitation—the two major components of climate—are also the two most important influences on flora and fauna. Today, however, global climate change is causing possibly irreversible changes in the world's bioregions because plants and animals cannot adapt to the rapid changes in climate. There is another important linkage as well: Vegetation takes up and stores carbon during growth and then releases CO_2 as it ages and dies. Long-lived forest trees, for example, are good storehouses of carbon, making responsible forest management essential to atmospheric emission reduction plans. Therefore, the widespread practice of cutting and burning tropical rainforests to clear land for farming or cattle pastures is a major contributor of atmospheric CO_2 and a controversial component of a developing country's economic strategy (Figure 2.26).

The Current Extinction Crisis

Not to be overlooked is the reality that climate change and habitat loss will hasten the extinction of plants and animals. There have been five major extinction events over

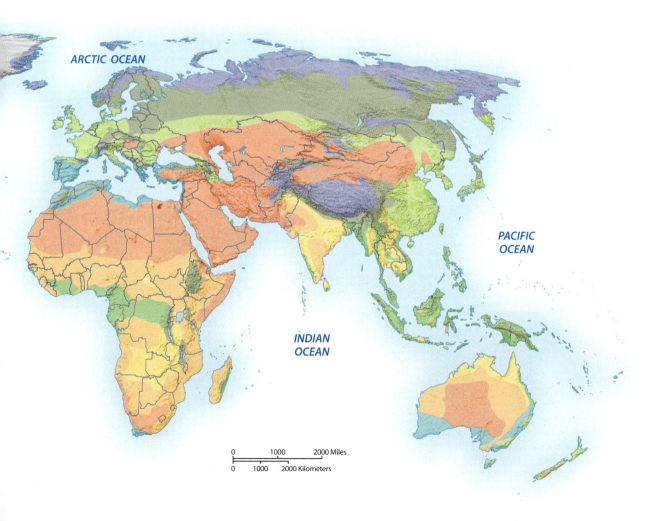

ARCTIC OCEAN

PACIFIC OCEAN

INDIAN OCEAN

0   1000   2000 Miles

0   1000   2000 Kilometers

(a)

(b)

(a)

(b)

▲ **Figure 2.21 Ecotourists in the Rainforests**
(a) A group of ecotourists and their guide examine the buttressed trunk of a rainforest tree on the Oso Peninsula of Costa Rica. Buttressed trunks (the woody outward extensions) are a feature of many rainforest trees as a way for the tree to expand its root system in shallow tropical soils where mineral nutrients lie in the upper layers. (b) The tropical savanna ecosystem is less dense than the true tropical rainforest because it is adapted to less rainfall falling in only part of the year. As a result, the vegetation is often sparse, such as these widely spaced trees in a grassland in the famed Serengeti National Park in Tanzania.

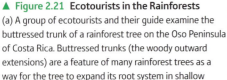
Explore the **Sounds** of Rainforest Ambience

http://goo.gl/l5pm0F

▲ **Figure 2.23 Evergreen and Deciduous Forests** (a) Evergreen trees usually have needles rather than leaves and keep their foliage throughout the year, which is why they're called "evergreen." Most evergreens, like these in British Columbia, Canada, are cone-bearing, consist of softer interior wood, and thus are called "softwoods." (b) In contrast, deciduous trees drop their leaves during the winter season, often after providing a colorful display (as in this photo from Maine) that results from the tree's slowing physiology as it prepares to hibernate.

(a)

(b)

▲ **Figure 2.22 Desert and Steppe** (a) Poleward of the tropics, in both the Northern and Southern Hemispheres, are true desert areas of sparse rainfall. On the global climate region map these are the areas of the BW climate found in Asia (such as the Gobi Desert shown here), Africa, North America, South America, and Australia. (b) Lush grasslands characterize the natural steppe bioregion, as depicted here in Mongolia, whereas in many parts of the world this environment has been converted into mechanized grain agriculture. These areas are found poleward of the true deserts where rainfall is commonly between 10 and 15 inches (254–381mm).

Explore the **Sights** of the Gobi Desert

http://goo.gl/WgVE1V

(a)

(b)

▲ **Figure 2.24 Arctic and Alpine Tundra** Tundra landscapes, comprised mainly of grasses, shrubs, and small stunted trees, are found in both high Arctic latitudes and, at lower latitudes, at high elevations in the continent's major mountain ranges. In both Arctic and Alpine environments a very short summer growing season influences what kinds of plants grow in this bioregion. (a) The Arctic tundra photo is from Canada's Yukon Territory; (b) the Alpine tundra is in the Colorado's Rocky Mountains.

▼ **Figure 2.25 Tropical Forest Destruction** Humans have cleared away much of the tropical rainforest to create grasslands for cattle, plantation crops, and, on a smaller scale, for family subsistence farms, such as this one in southern Laos. Burning the rainforest not only clears an area for crops and pasture but also transfers nutrients from the vegetation to the soil. These rainforest fires, however, add vast amounts of CO_2 to the atmosphere.

Mobile Field Trip: Forest Fires **MG** in the West

https://goo.gl/TueehL

▲ **Figure 2.26 Clear-cut Forests** The evergreen, softwood forests of the Northern Hemisphere are a major source of the world's lumber and paper pulp needs, and are often clear cut, as shown in this aerial photo from British Columbia, Canada, for the efficiency of harvesting as well as to facilitate replanting of nursery species. These forests, however, also store vast amounts of CO_2, much of which is released back into the atmosphere during cutting and milling. Further, because the visual impact of clear-cutting is often unacceptable, the protection (or harvesting) of these northern forests are the foci of numerous environmental controversies.

Earth's 4.5-billion-year history, all natural events that dramatically affected Earth's biological evolution. When climates changed naturally (such as at the end of the ice age 20,000 years ago), plants and animals migrated or adapted. Today, however, humans are causing a sixth extinction event with more rapid, human-caused climate change; highways, cities, and farmland now form barriers to plant and animal migration, further aggravating habitat loss and extinction rates. Water, air, and ground pollution and the wastes generated by human activities exacerbate habitat loss (see *Everyday Globalization: Our Plastic Bag World*). Biologists estimate that as a result of habitat destruction and other changes, we are losing several dozen species every day, resulting in an extinction rate that could see as much as 50 percent of Earth's species gone by 2050—reducing Earth's genetic resources, with potentially disastrous results.

REVIEW

2.7 Use a map to locate and describe the bioregions found in the world's tropical climates.

2.8 Describe the different kinds of forests found in the midlatitudes.

2.9 What is the difference between an evergreen and a deciduous forest?

■ **KEY TERMS** Biodiversity, bioregions, novel ecosystems

65

Our Plastic Bag World

Looking for something to celebrate? How about International Plastic Bag–Free Day— every July 3, people all over the world shop without using plastic grocery bags, pick up trash from beaches and roadsides, even commit to banning plastic bags in their communities.

One ban-the-bag organization estimates that one million plastic bags are used each minute. Besides being a pollution problem, manufacturing new bags takes millions of barrels of oil. In the United States alone, 12 million barrels of oil are used each year to make plastic bags.

True, some plastic bags can be recycled, but most are not, creating a global problem. Millions of plastic bags litter the countryside in China, where it's called "white pollution." Kenyans and South Africans refer to discarded white bags wafting about on the winds as their national flower (**Figure 2.4.1**). Even if these wayward bags were collected and buried in a landfill, they'd take several hundred years to decompose, releasing climate-changing methane gas.

Banning bags, or making them more expensive, seems to reduce usage. Some 20 U.S. states and over 200 cities have plastic bag ordinances. China claims to have an outright ban on plastic bags, but enforcement seems lax. Paris may be the biggest city with a full ban, and flood-prone Bangladesh may have the most compelling reason for a ban: engineers say that loose plastic bags clog drains, worsening flooding.

▲ **Figure 2.4.1** Waste Plastic Bags Litter the World's Landscapes

1. Does your community restrict or discourage the use of plastic shopping bags? If not, do you know of one? If so, what do people use instead?

2. Does your community recycle plastic bags? Where do the bags eventually end up?

Water: A Scarce World Resource

Water is central to all life, yet it is unevenly distributed around the world—plentiful in some areas; distressingly scarce in others. As a result, around 4 billion people, or two-thirds of the world's population, face severe water shortages during at least one month of the year. Water issues are not due simply to the geography of diverse climates that produce wet or dry conditions: they are also caused by a range of complex socioeconomic factors at all scales, local to global.

As mentioned earlier, Earth is indeed the water planet, but 97 percent of the total global water budget is saltwater, with only 3 percent freshwater. Almost 70 percent of that small amount of freshwater is locked up in polar ice caps and mountain glaciers. Additionally, groundwater accounts for almost 30 percent of the world's freshwater. This leaves less than 1 percent of the world's water in more accessible surface rivers and lakes.

Another way to conceptualize this limited amount of freshwater is to think of the total global water supply as 100 liters, or 26 gallons. Of that amount, only 3 liters (0.8 gallon) is freshwater; and of that small supply, a mere 0.003 liter, or only about half a teaspoon, is readily available to humans.

Water planners use the concept of **water stress** and scarcity to map where water problems exist, and to predict where future problems will occur (**Figure 2.27**).

These water stress data are generated by calculating the amount of freshwater available in relation to current and future population needs. Northern Africa stands out as the region of highest water stress; hydrologists predict that three-quarters of Africa's population will experience water shortages by 2025. Other problem areas are China, India, much of Southwest Asia, and even several countries in humid Europe. Although climate change may actually increase rainfall in some parts of the world, scientists forecast that global warming will, in general, aggravate global water problems because of higher evaporation rates and increased water usage due to warmer temperatures.

Water Sanitation

Where clean water is not available, people use polluted water for their daily needs, resulting in a high rate of sickness, and even death. More specifically, the UN reports that over half of the world's hospital beds are occupied by people suffering from illnesses linked to contaminated water. Further, more people die each year from polluted water than are killed in all forms of violence, including wars. This toll from polluted water is particularly high for infants and children who have not yet developed any sort of resistance to or tolerance for contaminated water. The United Nations Children's Fund (UNICEF) reports that nearly 4000 children die each day from unsafe water and lack of basic sanitation facilities.

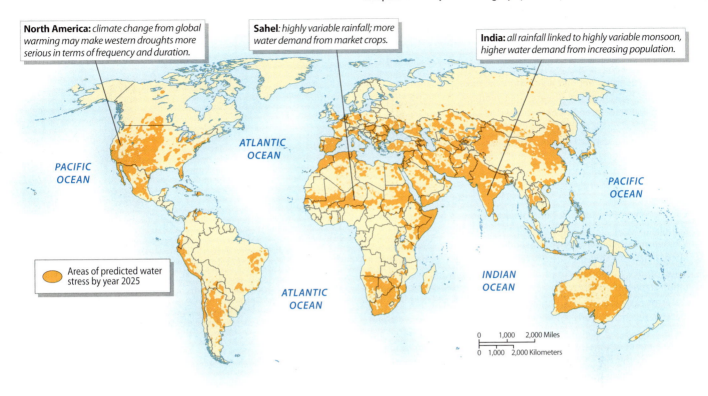

North America: *climate change from global warming may make western droughts more serious in terms of frequency and duration.*

Sahel: *highly variable rainfall; more water demand from market crops.*

India: *all rainfall linked to highly variable monsoon, higher water demand from increasing population.*

Areas of predicted water stress by year 2025

▲ **Figure 2.27 Global Water Stress Forecast for 2025** Although drought and highly variable rainfall regimes are major causes of water stress, other socioeconomic factors can limit supplies and access to water. A recent study shows that 4 billion people, or two-thirds of the world's population, are vulnerable to water scarcity. Half of those people are in either China or India. However, because of the global linkages in food systems, water stress in one part of the world can affect food supplies in many other areas. **Q: Does your community practice water conservation?**

Water Access

When a resource is scarce, access is problematic, and the hardships take many forms in the case of water. Women and children, for example, often bear the daily burden of providing water for family use, and this can mean walking long distances to pumps and wells and then waiting in long lines to draw water (see *Working Toward Sustainability: Women and Water in the Developing World*). Given the amount of human labor involved in providing water for crops, it is not surprising that some studies have shown that in certain areas people expend as many calories of energy irrigating their crops as they gain from the food itself.

Ironically, some recent international efforts to increase people's access to clean water have gone astray and have actually aggravated access problems instead. Historically, domestic water supplies have been public resources, organized and regulated—either informally by common consent, or more formally as public utilities—resulting in free or low-cost water. In recent decades, however, the World Bank and the International Monetary Fund have promoted the privatization of water systems as a condition for providing loans and economic aid to developing countries. The agencies' goals have been laudable, trying to ensure that water is clean and healthful. However, the means have been controversial because typically the international engineering firms

that have upgraded rudimentary water systems by installing modern water treatment and delivery technology have also increased the costs of water delivery to recoup their investment. Although the people may now have access to cleaner and more reliable water, in many cases the price is higher than they can afford, forcing them to either do without or go to other, unreliable and polluted sources.

REVIEW

2.10 How much water is there on Earth, and how much is available for human usage? Use in your answer the concept that Earth's water budget is just 100 liters.

2.11 Describe several major issues causing water stress.

2.12 Where in the world are the areas of the most severe water stress?

■ **KEY TERM** water stress

Global Energy: The Essential Resource

The world runs on energy. While sunlight drives the natural world, providing energy for the modern human world is much more complicated because of the uneven distribution

Women and Water in the Developing World

Women and children bear the burden of water problems in most developing countries. Not only are children the most vulnerable to waterborne diseases, but mothers, aunts, grandmothers, and older sisters are the care-givers for these sick children, adding yet another time-consuming task to their already busy days.

Google Earth Virtual MG Tour Video

http://goo.gl/Dvcsy4

Further, women and older girls are the primary conveyers of water from wells or streams to their homes. Every person requires at least 5 gallons (19 liters) of water per day for their hydration, cooking, and sanitation needs; consequently, this amount (multiplied by the number of people in a household) must be carried each day from source to residence. Women and children are also responsible for supplying water for kitchen gardens that provide the family's food. At a global level, the water source for about a third of the developing world's rural population is more than half a mile (1 km) away from residences. To meet water needs, women spend about 25 percent of their day carrying water. A recent UN study estimated that in Sub-Saharan Africa about 40 billion hours a year are spent collecting and carrying water, the same amount of time spent in a year by France's entire workforce.

Besides the time expenditure, water is heavy, and most of it is carried by hand. In Africa, 44-pound (20-liter) jerry cans are common; in northwest India, women and girls balance several 5-gallon (19-liter) containers on their heads to lessen the number of trips made (**Figure 2.3.1**). (Note that 40–45 pounds is the weight of the

▲ **Figure 2.3.2** Woman using Wello WaterWheel.

suitcase you check with the airlines on a typical trip. Try carrying it on your head through the airport parking lot someday.) After years of carrying water, women commonly suffer from chronic neck and back problems, many of which complicate childbirth. Additionally, girls' water-carrying responsibilities often interfere with their schooling, resulting in a high dropout rate and furthering female illiteracy in rural villages.

Toward a Solution: The Wello WaterWheel After studying the water-carrying issue in semiarid northwestern India, Cynthia Koenig, an engineering graduate from the University of Michigan, invented the Wello WaterWheel, a barrel-like 13-gallon (50-liter) rolling water container that greatly reduces women's water-carrying duties (**Figure 2.3.2**). Previously in that part of India, women and girls were spending 42 hours per week carrying water back and forth; with the Wello WaterWheel, that has been reduced to only 7 hours a week. Using this timesaving device has also reduced the school dropout rate for young girls in the region. Currently, Wello, which is a nonprofit organization, can deliver a WaterWheel from its factory in Mumbai to a rural Indian family for a mere $20. Thousands of Wello Water-Wheels have been purchased by international aid organizations and donated to villages in Rajasthan, moving them closer to a sustainable existence.

1. List the social costs incurred when the women and children of a village must provide water. What are the probable social benefits of readily available clean water in a rural village?

2. The average American uses 80–100 gallons of water each day. A gallon of water weighs over 8 pounds. How would your water usage habits change if you had to carry water to your home?

▲ **Figure 2.3.1** Women in India carrying water on their heads.

of energy resources; the complex technologies required to mine fuel resources; and, not least, the economic, environmental, and geopolitical dynamics involved in finding, exploiting, and transporting these essential resources (**Figure 2.28**).

Nonrenewable and Renewable Energy

Energy resources are commonly categorized as either nonrenewable or renewable. **Nonrenewable energy** is consumed at a higher rate than it is replenished, which is the case with the world's oil, coal, uranium, and natural gas resources—fossil fuels that took millennia to create. In contrast, **renewable energy** comes from natural processes that are constantly renewed—namely, water, wind, and solar energy. Currently, 90 percent of the world's energy needs is satisfied by nonrenewable sources. Although renewable energy powers the remaining 10 percent, this energy sector is increasing rapidly because of global concerns about reducing atmospheric emissions. Oil, coal, and natural gas are considered "dirty" fuels because of their carbon content, whereas renewable energy is generally "clean," with no GHG emissions resulting from its usage. Among the dirty fuels, coal adds the most harmful emissions, with natural gas emitting 60 percent less CO_2 than coal.

At a global level, oil and coal are the major fuels, currently making up 36 and 33 percent of all fossil fuel usage, respectively; natural gas trails at 26 percent, with nuclear power far behind at 5 percent. Recent trends, however, show an increase in natural gas at coal's expense because of both environmental and cost concerns.

Fossil Fuel Reserves, Production, and Consumption

As discussed at the start of this chapter, plate tectonics and other complex geologic forces have shaped ancient landscapes over millions of years. As a result, fossil fuels are not evenly distributed around the world, but instead are clustered into specific geologic formations at select locations, resulting in a complicated international pattern of supply and demand. Table 2.1 shows the varied world geography of energy reserves, production, and consumption—where the resources lie, which countries mine and produce them, and which are primary consumers of coal, oil, and gas. More details on this complex global energy geography appear in each of the regional chapters of this text.

Because of the high degree of technological difficulty involved in mining fossil fuels, the energy industry uses the concept of **proven reserves** of oil, coal, and gas to refer to deposits that can be mined and distributed under current economic and technological conditions. An important component of this definition is "current," because economic, regulatory, and technological conditions change rapidly; as they do, this expands or contracts the amount of a resource deemed feasible to mine. If, for example, the price of oil on the global market is high, then oil reserves with relatively high drilling costs (such as in the Arctic) can be produced at a profit. Conversely, if the market price of oil is low, then only easily accessible reserves qualify as "proven." This was the case during 2014–2016 when the global supply for oil far exceeded its demand, resulting in extremely low prices and the closure of many high-cost operations in the Arctic, other offshore locations, and in North America.

Not only do market prices for fossil fuels change over time, but also drilling and mining technologies become more efficient, changing the amount of proven reserves. A good example is the recent development of **hydraulic fracturing (fracking)**, a relatively new oil and gas mining technique that forces oil and gas out of shale rock (**Figure 2.29**). Currently, fracking in North America has contributed to increasing the world's oil and gas supplies to the point where traditional global oil suppliers (Russia, Saudi Arabia, and Venezuela) are suffering economically from low oil prices.

Renewable Energy

Renewable energy sources, mentioned earlier, can be naturally replenished at a faster rate than their energy is consumed. Wind and solar power are prime examples. The list of renewable energy sources includes not just wind and solar power, but also hydroelectric, geothermal (from Earth's interior heating),

▼ **Figure 2.28 North American Oil Transport** Hydraulic fracturing ("fracking") has produced a North American oil glut that benefits consumers with low fuel prices but creates transportation difficulties in moving oil safely from inland production areas to refineries located primarily in coastal areas, a locational artifact from when the United States imported most of its oil. Although pipelines carry much North American oil, railroads are moving increasing amounts, like this oil train creeping through the suburbs of Kansas City, Missouri, resulting in railroad traffic jams, oil spills, and noise and traffic problems for railroad towns.

Mobile Field Trip: Oil Sands

https://goo.gl/cIIDir

TABLE 2.1　2015 Geography of Fossil Fuels

Proven Reserves	World Share	Production	World Share	Consumption	World Share
Oil		**Oil**		**Oil**	
Venezuela	17.7%	Saudi Arabia	13%	United States	19.7%
Saudi Arabia	15.7%	United States	13%	China	12.9%
Canada	10.1%	Russia	12.4%	India	4.5%
Iran	9.3%	China	4.9%	Japan	4.4%
Iraq	8.4%	Iraq	4.5%	Brazil	3.2%
Coal		**Coal**		**Coal**	
United States	26.6%	China	47.7%	China	50%
Russia	17.6%	United States	11.9%	India	10.6%
China	12.8%	Australia	7.2%	United States	10.3%
Australia	8.6%	Indonesia	6.3%	Japan	3.1%
India	6.8%	India	7.4%	South Africa	2.2%
Natural Gas		**Natural Gas**		**Natural Gas**	
Iran	18.2%	United States	22%	United States	22.8%
Russia	17.3%	Russia	16.1%	Russia	11.2%
Qatar	13.1%	Iran	5.4%	China	5.7%
Turkmenistan	9.4%	Qatar	5.1%	Iran	5.5%
United States	5.6%	Canada	4.6%	Japan	3.3%
				Nuclear Energy	
				United States	32.6%
				France	17%
				South Korea	6.6%
				China	4.4%
				Canada	4%

Source: BP Statistics, 2015 (June 2016)

Login to MasteringGeography™ **& access** MapMaster **to explore these data!**

1) If a country produces a large amount of the world's supply of oil, coal, or natural gas, yet it is not among the world's largest consumers of that energy source then one can assume they are a major energy exporter. Which countries on Table 2.1 fall into that category?

2) Conversely, if a country is one of the world leaders with a proven reserve of an energy source but is not a world leader in production of that energy source one can assume they're either not capable of high production and/or are banking those reserves for the future. Which countries fall into that category?

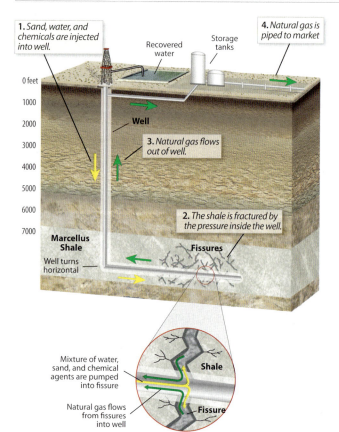

1. Sand, water, and chemicals are injected into well.

Recovered water

Storage tanks

4. Natural gas is piped to market

Well

3. Natural gas flows out of well.

0 feet
1000
2000
3000
4000
5000
6000
7000

2. The shale is fractured by the pressure inside the well.

Marcellus Shale

Fissures

Well turns horizontal

Mixture of water, sand, and chemical agents are pumped into fissure

Natural gas flows from fissures into well

Shale

Fissure

◀ **Figure 2.29 Hydraulic Fracturing** This graphic shows the different components of hydraulic fracturing (fracking) for oil and gas. What it does not show are the environmental issues associated with fracking, which include use of massive amounts of freshwater, the potential for polluting local ground water, wastewater disposal problems, local noise and nuisance issues, and the possibility that fracking could trigger earthquakes in certain geologic structures.

tidal currents, and biofuels, which use the carbon in plants as their power source.

Despite the widespread availability of sunlight, wind, and biofuels, renewable energy currently provides just 11 percent of the world's power. However, because of the emphasis on reducing atmospheric emissions in the 2015 Paris Agreement, the future for renewables is bright as countries reduce their traditional dependence on fossil fuels. Some countries already make considerable use of renewables in their power grid. Iceland, blessed with bountiful supplies of both water and thermal resources, generates fully 95 percent of its power from renewable sources.

Of the large industrial economies, Germany leads the way with over 20 percent of its power coming from renewables, primarily through extensive wind and solar power stations (**Figure 2.30**). China, the world's largest consumer of energy, claims that one-quarter of its massive economy is powered by renewable sources, and its coal usage reportedly fell in both 2014 and 2015, a significant achievement

▲ **Figure 2.30 Germany's Renewable Energy Program** Germany is on course to become the world's first large industrial economy to be powered primarily by renewable energy. The first step, taken almost a decade ago, was to subsidize its solar industry by offering attractive incentives to households that converted to solar power. Here, homeowners have used the steep roofs of traditional German homes to maximize solar panel placement. The second stage has been to expand the country's wind power sector, a technology that currently supplies most of the country's renewable energy. **Q: Check to see if your campus is partially powered by renewable energy sources.**

as China consumes half of the world's coal each year. Unlike Germany, most of China's renewable energy comes from hydropower, although the country is rapidly expanding its wind and solar facilities.

Despite the attractiveness of wind and solar power, significant issues remain before they can completely replace fossil fuels. Both wind and solar, for example, are intermittent sources of power, because the Sun does not always shine or the wind blow. To compensate for these lulls, and because energy storage remains problematic, most commercial wind and solar facilities are often mandated to have gas-fired generators as backups to produce power at a moment's notice on cloudy or windless days.

Important to note is that historically, subsidies and tax incentives have favored the development of fossil fuels over renewables—at a global level, it is estimated that fossil fuels receive six times the financial support through governmental incentives than do renewables. When new technologies serve the public interest, they usually receive considerable

government support, and while this has happened in several notable cases (China and Germany) with renewables, these incentives need to become more widespread before renewable energy can compete on a level playing field with fossil fuels. Not to be overlooked is the challenge of providing power to the 25 percent of the world's population that currently lacks access to a power grid (**Figure 2.31**). If the international community adheres to their Paris commitments on emission reduction, the role of renewables in the world's energy mix should increase dramatically in the next decades.

REVIEW

2.13 What is meant by fossil fuel proven reserves?

2.14 List the three countries using the most oil, coal, and natural gas.

2.15 What is fracking? How might fracking influence the world's energy geography?

2.16 What are some of the problems associated with renewable energy?

■ **KEY TERMS** Nonrenewable energy, renewable energy, proven reserves, hydraulic fracturing (fracking)

▼ **Figure 2.31 Small Solar in Africa** Many villages in Africa are moving to small-scale solar systems to generate energy that was formerly produced by gas or diesel generators. Not only is solar less expensive than fossil fuels, but it's more adaptable to specific needs, as illustrated here by a Ugandan mother's small business charging cell phones from a single solar panel.

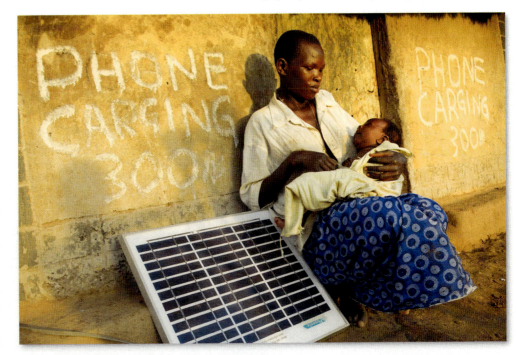

Chapter 2: Physical Geography and the Environment

Summary

- Earth is the water planet because of the large expanses of ocean covering the globe. These oceans, a critical part of Earth's physical geography, affect other physical phenomena as well as many aspects of human life.

- The arrangement of tectonic plates on Earth is responsible for diverse global landscapes, and the motion of these plates also causes earthquake and volcanic hazards that threaten the safety of millions of people.

- Earth's climate is changing because of the addition of greenhouse gases (GHGs) from human activities. As a result of both historic and current emissions, global temperatures have warmed several degrees and, despite international efforts, will likely warm several more degrees by century's end.

- Plants and animals throughout the world face an extinction crisis because of habitat destruction from a variety of human activities. Tropical forests are a focus of these problems because of their diversity and capacity to both store and emit large amounts of carbon dioxide.

- Water is becoming an increasingly scarce resource in the world and will cause serious water stress problems in Sub-Saharan Africa, Southwest Asia, and western North America.

- Fossil fuels dominate the world's energy picture, with renewable energy currently providing only a fraction of the world's energy needs. While energy demand has leveled off (and even decreased) in more developed countries, it continues to rise in China, India, and other developing economies.

Review Questions

1. Think about the different kinds of housing people build in different climates. Then, after looking at the map of geologic hazards, discuss whether or not certain climate regions of the world are more or less susceptible to earthquake damage.

2. Generally speaking, bioregions correspond closely to climate regions. However, in some parts of the world that's not the case. Where are those places, and what explains these exceptions?

3. Are water stress issues found only in dry climates? Give examples of where they are and where they are not, and defend your answers.

4. Describe the environmental, economic, and social benefits for a country increasing its share of renewable energy consumption. In what climate regions is the potential for renewable energy usage greatest?

5. Several rich countries such as Japan and Saudi Arabia have been buying up irrigated land along the Colorado River in North America's dry Southwest. Why?

Image Analysis

1. This image was taken from a plane flying over a clearcut forest, consisting primarily of Douglas firs, in British Columbia. How can you differentiate between older and newer clearcuts?

2. What other geographic patterns do you see? Are they close to or far away from settlements and major roads? What about the relationship between clearcuts and watersheds?

▶ **Figure IA2** **Aerial View of Clear Cut Forest**

JOIN THE DEBATE

Representatives of more than 190 countries participating in the 2015 United Nations Climate Change Conference voted to adopt the Paris Agreement to mitigate climate change. However, the agreement has been criticized for not going far enough to reduce greenhouse gas emissions, and for not being realistic about the financial support needed to reach this goal. Does the Paris Agreement go far enough to limit global warming?

▲ **Figure D2** **Sea Level Rise** Sea level rise from global climate change will be a serious problem in many of the world's coastal areas by 2030.

The Paris Agreement goes far enough to limit global warming

- Unlike the Kyoto Protocol, it involves many more countries, including the world's largest atmospheric polluter, China, and emphasizes limiting emissions from industrializing countries like India and Brazil, along with developed countries in Europe and North America.

- The Paris Agreement is flexible as to how countries achieve carbon reduction by emphasizing carbon management and sequestration.

- Global carbon management can be achieved by recognizing that rich countries can help poor countries through carbon offsets.

- The Paris Agreement will be revised and updated every 5 years, with reduction goals being increased regularly.

The Paris Agreement does not go far enough to limit global warming

- Are you kidding me? This is simply wishful thinking about good intentions and lacks any sort of formal legal status.

- The Agreement is too flexible, and has no standardized method for evaluating whether or not countries are achieving their carbon management goals.

- There is no mechanism other than "naming and shaming" countries that don't achieve their emission reduction goals.

- The Agreement doesn't even take effect until 2020, and by that time the world will have emitted so much more atmospheric pollution that irrevocable climate change and catastrophic sea level rise are inevitable.

■ KEY TERMS

adiabatic lapse rate *(p. 59)*
anthropogenic *(p. 60)*
biodiversity *(p. 62)*
bioregions *(p. 62)*
climate *(p. 59)*
climate change *(p. 60)*
climographs *(p. 59)*
continental climate *(p. 55)*
convergent plate boundaries *(p. 48)*
divergent plate boundaries *(p. 48)*
environmental lapse rate *(p. 57)*
greenhouse effect *(p. 52)*
greenhouse gases (GHGs) *(p. 60)*
hydraulic fracturing (fracking) *(p. 69)*
insolation *(p. 52)*
Kyoto Protocol *(p. 61)*
lithosphere *(p. 48)*
maritime climates *(p. 55)*
monsoon winds *(p. 56)*
nonrenewable energy *(p. 69)*
novel ecosystems *(p. 63)*
orographic effect *(p. 58)*
Pangaea *(p. 51)*
Paris Agreement *(p. 61)*
plate tectonics *(p. 48)*
polar jet stream *(p. 56)*
proven reserves *(p. 69)*
rain shadow *(p. 59)*
renewable energy *(p. 69)*
reradiate *(p. 52)*
rift valleys *(p. 50)*
subduction zone *(p. 48)*
subtropical jet stream *(p. 56)*
transform plate boundaries *(p. 48)*
transform fault *(p. 50)*
water stress *(p. 66)*

MasteringGeography™

Looking for additional review and test prep materials? Visit the Study Area in **MasteringGeography™** to enhance your geographic literacy, spatial reasoning skills, and understanding of this chapter's content by accessing a variety of resources, including **MapMaster** interactive maps, videos, RSS feeds, flashcards, web links, self-study quizzes, and an eText version of *Diversity Amid Globalization*.

DATA ANALYSIS

http://goo.gl/jvkljY

Emissions per capita are a useful measure of each country's contribution to climate change. Compare a rich country where people drive gas-guzzling SUVs and live in huge houses requiring lots of energy for heating and cooling with a not-so-rich country where people use public transportation and live in modest energy-efficient apartments. If both countries have the same population, then clearly the first country has higher per capita emissions, right? Reality, however, is not so simple. Go to the World Bank website (http://data.worldbank.org) and access data on CO_2 emissions (metric tons per capita). List countries emitting less than .5 ton, and those emitting over 15 tons per person. Rank the countries in both categories from lowest to highest.

1. Do rich countries emit more CO_2 than less wealthy countries? (Use GNI per capita figures from the World Bank website as a measure of wealth for each country.)

2. Does a country's energy mix—renewables vs. fossil fuels— influence emissions per capita? Do countries with abundant fossil fuel resources produce more emissions than those importing fuel? (You can find a country's energy mix and resources in The World Factbook, https://www.cia.gov)

3. Does climate make a difference? Do countries in colder climates use more energy and emit more CO_2 than those in warmer climates?

4. Write a paragraph on why per capita emissions differ so much. Then browse the Internet for articles on CO_2 emissions per capita, and refine your explanations.

Authors' Blogs

Scan to visit the
GeoCurrents blog
http://www.geocurrents.
info/category/
physical-geography

Scan to visit the
Author's Blog
for field notes, media resources, and chapter updates
https://gad4blog.wordpress.com/
category/global-environment/

Physical Geography and Environmental Issues

Stretching from Texas to the Yukon, the North American region is home to an enormously varied natural setting and to an environment that has been extensively modified by human settlement and economic development.

Population and Settlement

Settlement patterns in North American cities reflect the diverse needs of an affluent, highly mobile population. The region's sprawling suburbs are designed around automobile travel and mass consumption, whereas many traditional city centers struggle to redefine their role within the decentralized metropolis.

Cultural Coherence and Diversity

Cultural pluralism remains strong in North America. Currently, more than 49 million immigrants live in the region, more than double the total in 1990. The tremendous growth in the numbers of Hispanic and Asian immigrants since 1970 has fundamentally reshaped the region's cultural geography.

Geopolitical Framework

Cultural pluralism continues to shape political geographies in the region. Immigration policy remains hotly contested in the United States, and Canadians confront persistent regional and native peoples' rights issues.

Economic and Social Development

North America's economy recovered in many settings after the harsh economic downturn between 2007 and 2010. Still, persisting poverty and many social issues related to gender equity, aging, and health care challenge the region today.

▶ **Downtown Vancouver, British Columbia, Canada** Vancouver is Canada's largest port city, and the dynamic metropolis of 2.5 million people has forged many links to the global economy.

Vancouver, British Columbia

NORTH AMERICA

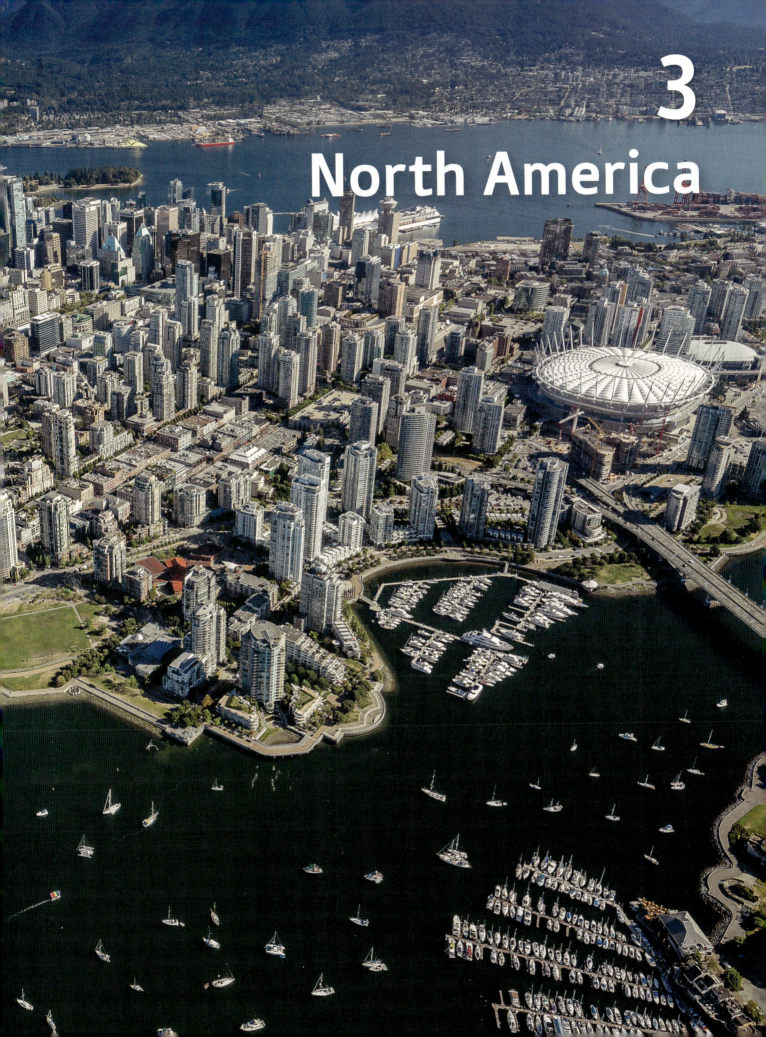

3
North America

Port Metro Vancouver is now North America's third largest port (and Canada's largest). The sprawling dock facilities, cargo and cruise terminals, and shipyards symbolize the close connections between this dynamic North American city and the global economy. Vancouver (2.5 million people) is also one of North America's most culturally diverse cities and its healthy economy has attracted a sizable foreign-born population—45 percent of its residents. The city's Chinese population is among North America's largest, along with sizable immigrant communities from South Asia and the Philippines. Future trade with Asia promises further growth and even stronger connections across the Pacific. Signed in 2016, the **Trans-Pacific Partnership (TPP)**, a comprehensive agreement among 12 countries including Canada and the United States, is designed to stimulate Pacific trade and lower tariff barriers between member states. If ratified by all participants, the TPP guarantees that Vancouver's bustling port and its regional economy will continue to grow in the 21st century.

Similar processes of globalization have fundamentally refashioned many portions of North America. Large foreign-born populations are found in many North American settings. Tourism brings in millions of additional foreign visitors and billions of dollars, which are spent everywhere from Las Vegas to Disney World. North Americans engage globally in more subtle ways: eating ethnic foods, enjoying the sounds of salsa and Senegalese music, and surfing the Internet from one continent to the next. Globalization is also a two-way street, and North American capital, popular culture, and power are ubiquitous. By any measure of multinational corporate investment and global trade, the region plays a role that far outweighs its population of 360 million residents.

> North America, one of the world's wealthiest regions with two highly urbanized, mobile populations, helps drive the processes of globalization.

Defining North America

North America is a culturally diverse and resource-rich region that has seen tremendous, sometimes destructive, modification of its landscape and extraordinary economic development over the past two centuries (**Figure 3.1**). As a result, North America is one of the world's wealthiest regions, with two highly urbanized, mobile populations that help drive the processes of globalization and have the highest rates of resource consumption on Earth. Indeed, the region exemplifies a **postindustrial economy** shaped by modern technology, innovative information services, and a popular culture that dominates both North America and the world beyond (**Figure 3.2**).

Politically, North America is home to the United States, the last remaining global superpower. In addition, North America's largest metropolitan area, New York City (20 million people), is home to the United Nations and other global political and financial institutions. North of the United States, Canada is the region's other political unit. Although slightly larger in area than the United States (3.83 million square miles [9.97 million square kilometers] versus 3.68 million square miles [9.36 million square kilometers]), Canada's population is only about 11 percent that of the United States.

The United States and Canada are commonly referred to as "North America," but that regional terminology can be confusing. As a physical feature, the North American continent commonly includes Mexico, Central America, and often the Caribbean. Culturally, however, the U.S.–Mexico border seems a better dividing line, although the growing Hispanic presence in the southwestern United States, as well as ever-closer economic links across the border, makes even that regional division problematic. In addition, while Hawaii is a part of the United States (and included in this chapter), it is also considered a part of Oceania (and discussed in Chapter 14). Finally, Greenland (population 56,000), which often appears on the North American map, is actually an autonomous territory within the Kingdom of Denmark and is mainly known for its valuable, but diminishing, ice cap.

LEARNING Objectives

After reading this chapter you should be able to:

3.1 Describe North America's major landform and climate regions.

3.2 Identify key environmental issues facing North Americans and connect these to the region's resource base and economic development.

3.3 Analyze map data to identify and describe major migration flows in North American history.

3.4 Explain the processes that shape contemporary urban and rural settlement patterns.

3.5 List the five phases of immigration shaping North America and describe the recent importance of Hispanic and Asian immigration.

3.6 Provide examples of major cultural homelands (rural) and ethnic neighborhoods (urban) within North America.

3.7 Describe how the United States and Canada developed distinctive federal political systems and identify each nation's current political challenges.

3.8 Discuss the role of key location factors that explain why economic activities are located where they are in North America.

3.9 List and explain contemporary social issues that challenge North Americans in the 21st century.

ELEVATION IN METERS

4,000+
2,000–4,000
500–1,999
200–499
0–199
Sea Level
Below sea level

NORTH AMERICA
Political & Physical Map

⊛ ● Metropolitan areas more than 20 million
⊛ ● Metropolitan areas 10–20 million
⊛ • Metropolitan areas 5–9.9 million
⊛ • Metropolitan areas 1–4.9 million
⊛ ○ Selected smaller metropolitan areas
⌐ Plate boundaries

▲ **Figure 3.1 North America** With 360 million people and extensive economic development, North America plays a pivotal role in globalization. The region also contains one of the world's most highly urbanized and culturally diverse populations, and is also one of the largest consumers of natural resources on the planet.

▲ Figure 3.2 **Googleplex Campus, Silicon Valley** Google, a multinational technology company, is headquartered in California, and its sprawling Googleplex Campus is a major Silicon Valley employer.

Physical Geography and Environmental Issues: A Vulnerable Land of Plenty

North America's physical and human geographies are enormously diverse. In the past decade, the region has witnessed a dizzying array of natural disasters and environmental hazards that suggest the close connections between the region's complex physical setting and its human population. In 2012, much of the East Coast was slammed by Hurricane Sandy. The so-called superstorm (formed by the merging of tropical and midlatitude storm systems) rearranged the coastline of New Jersey, flooded lower Manhattan, and produced $50–$60 billion in damage. For millions of coastal residents from Virginia to New England, Sandy was yet another reminder that such settings are extraordinarily vulnerable to natural disasters, especially in a time of increasingly dense coastal development and the looming possibilities of global climate change. On the West coast, California's record-breaking drought (2013–2016) showed some signs of improving thanks to more normal winter precipitation, but long-term shifts in the region's climate suggest extremely dry conditions will regularly revisit the Golden State (**Figure 3.3**).

Diverse Physical Settings

North America's complex landscape is dominated by interior lowlands bordered by mountainous topography in the western portion of the region (see Figure 3.1). In the eastern United States, extensive coastal plains stretch from southern New York to Texas and include a sizable portion of the lower Mississippi Valley. The Atlantic coastline is complex, made up of drowned river valleys, bays, swamps, and low barrier islands. The nearby Piedmont region, a transition zone between nearly flat lowlands and steep mountain slopes, consists of rolling hills and low mountains that are much older and less easily eroded than the lowlands. West and north of the Piedmont are the Appalachian Highlands, an internally complex zone of higher and rougher country, reaching altitudes of 3000–6000 feet (915–1830 meters). To the southwest, Missouri's Ozark Mountains and the Ouachita Plateau of northern Arkansas resemble portions of the southern Appalachians. Much of the North American interior is a vast lowland extending east–west from the Ohio River Valley across the Great Plains and north–south from west central Canada to the lower Mississippi near the Gulf of Mexico (**Figure 3.4**). Glacial forces, particularly north of the Ohio and Missouri rivers, once actively carved and reshaped the landscapes of this lowland zone.

In the West, mountain-building (including large earthquakes and volcanic eruptions), alpine glaciation, and erosion produce a visually spectacular regional topography quite unlike that of eastern North America. The Rocky Mountains

Explore the **Sights** of California's Shasta Lake

http://goo.gl/rGuISI

▼ Figure 3.3 **Shasta Lake, California** California's recent droughts sent Lake Shasta, a reservoir that feeds into the Sacramento River, to record low levels.

and British Columbia; the Coast Ranges of Washington, Oregon, and California; the lowlands of the Puget Sound (Washington), Willamette Valley (Oregon), and Central Valley (California); and the complex uplifts of the Cascade Range and Sierra Nevada.

Mobile Field Trip: Yosemite

https://goo.gl/jE56g2

Patterns of Climate and Vegetation

North America's climates and vegetation are highly diverse, mainly due to the region's size, latitudinal range, and varied terrain (**Figure 3.6**). Much of eastern North America south of the Great Lakes is characterized by a long growing season, 30–60 inches (75–150 cm) of precipitation annually, and a once-vast deciduous broadleaf forest (now replaced by crops). From the Great Lakes north, the coniferous evergreen or **boreal forest** dominates the continental interior. Near Hudson Bay and across the harsher northern latitudes, trees give way to **tundra**, a mixture of low shrubs, grasses, and flowering herbs that briefly flourish in the short growing season.

Drier continental climates from west Texas to Alberta feature large seasonal ranges in temperature and unpredictable precipitation that averages 10–30 inches (25–75 cm) annually (see climographs for Dallas and Cheyenne in Figure 3.6). The soils of much of this region are fertile and originally supported **prairie** vegetation, dominated by tall grasslands in the East and by short grasses and scrub vegetation in the West. Western North American climates and vegetation are greatly complicated by the region's many mountain ranges. The Rocky Mountains and the intermontane interior experience the typical midlatitude seasonal variations, but the topography greatly modifies climate and vegetation patterns.

▲ **Figure 3.4 Satellite View of the Lower Mississippi Valley** This view of the Mississippi Delta shows how sediments from the interior lowlands have accumulated in the area. Note the sediment plume (right) extending into the Gulf of Mexico.

Mobile Field Trip: Mississippi River Delta

https://goo.gl/B0kvSe

reach more than 10,000 feet (3000 meters) in height and stretch from Alaska's Brooks Range to northern New Mexico's Sangre de Cristo Mountains (**Figure 3.5**). West of the Rockies, the Colorado Plateau is characterized by highly colorful sedimentary rock eroded into spectacular buttes and mesas. Nevada's sparsely settled basin and range country features north–south mountain ranges alternating with structural basins with no outlet to the sea. North America's western border is marked by the mountainous and rain-drenched coasts of southeast Alaska

▼ **Figure 3.5 Rocky Mountains** This spectacular lake in Alberta's Banff National Park reveals the characteristic signatures of alpine glaciation that are found in many portions of the Rocky Mountain region, both in the United States and Canada.

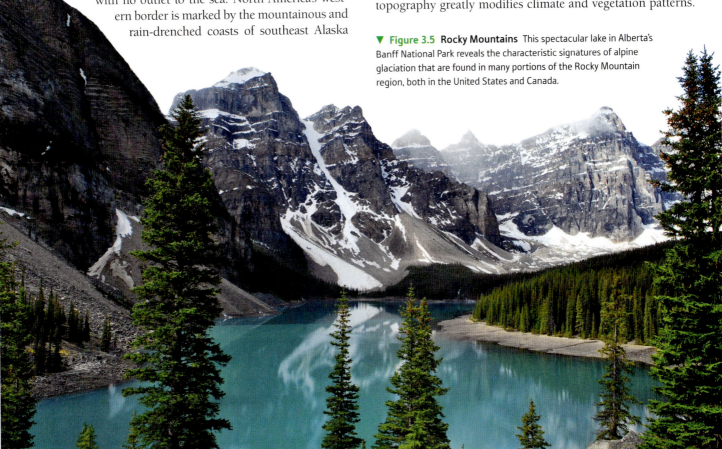

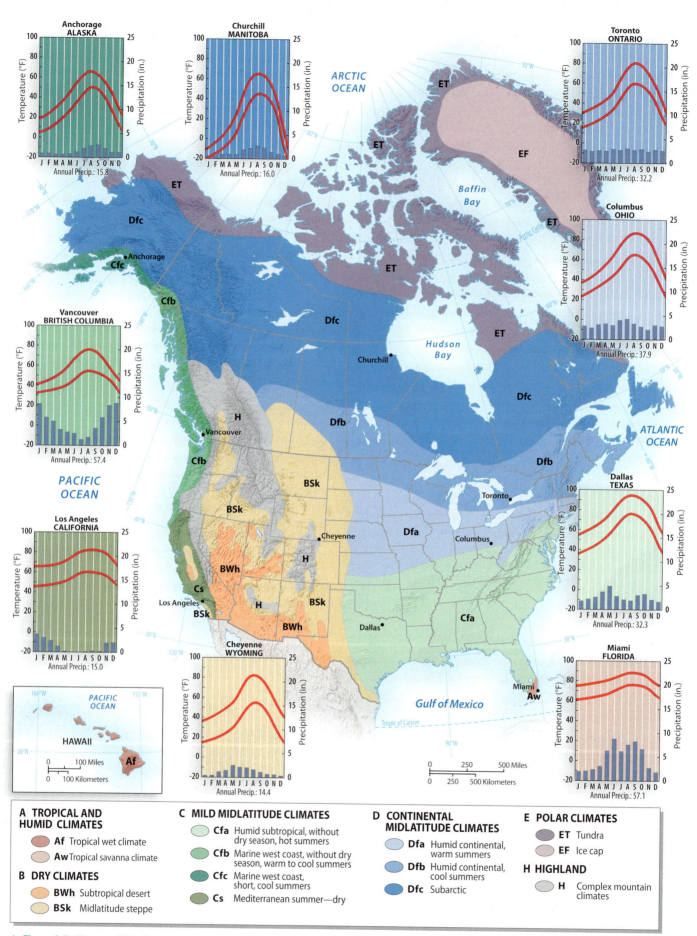

Anchorage ALASKA
Annual Precip.: 15.8

Churchill MANITOBA
Annual Precip.: 16.0

Toronto ONTARIO
Annual Precip.: 32.2

Columbus OHIO
Annual Precip.: 37.9

Vancouver BRITISH COLUMBIA
Annual Precip.: 57.4

Los Angeles CALIFORNIA
Annual Precip.: 15.0

Dallas TEXAS
Annual Precip.: 32.3

Cheyenne WYOMING
Annual Precip.: 14.4

Miami FLORIDA
Annual Precip.: 57.1

ARCTIC OCEAN

Baffin Bay

Hudson Bay

ATLANTIC OCEAN

PACIFIC OCEAN

Gulf of Mexico

Tropic of Cancer

PACIFIC OCEAN

HAWAII

0   100 Miles
0   100 Kilometers

0   250   500 Miles
0   250   500 Kilometers

A TROPICAL AND HUMID CLIMATES
- **Af** Tropical wet climate
- **Aw** Tropical savanna climate

B DRY CLIMATES
- **BWh** Subtropical desert
- **BSk** Midlatitude steppe

C MILD MIDLATITUDE CLIMATES
- **Cfa** Humid subtropical, without dry season, hot summers
- **Cfb** Marine west coast, without dry season, warm to cool summers
- **Cfc** Marine west coast, short, cool summers
- **Cs** Mediterranean summer—dry

D CONTINENTAL MIDLATITUDE CLIMATES
- **Dfa** Humid continental, warm summers
- **Dfb** Humid continental, cool summers
- **Dfc** Subarctic

E POLAR CLIMATES
- **ET** Tundra
- **EF** Ice cap

H HIGHLAND
- **H** Complex mountain climates

▲ **Figure 3.6 Climate of North America** The region's climates include everything from tropical savanna (Aw) to tundra (ET) environments. Most of the region's best farmland and densest settlements lie in the mild (C) or continental (D) midlatitude climate zones.

Legend:
- Areas affected by acid precipitation
- Desert
- Areas of groundwater depletion
- Vulnerable to sea-level rise
- Coastal pollution
- Endangered and polluted rivers
- Proposed pipeline routes
- Major hazardous waste sites
- Selected mining areas

Melting Sea Ice. *Recently, because of global climate change, the Arctic Ocean has seen dramatically reduced levels of sea ice in summer.*

Athabasca Oil Sands. *Gigantic deposits of oil sands in Alberta have generated controversial plans for long-distance pipelines, passing near some environmentally sensitive areas.*

Acid Precipitation. *Acid precipitation has devastated hundreds of sensitive lake environments across eastern Canada.*

California Drought. *In 2014 and 2015, much of California witnessed its worst drought in decades, and mountain snow packs in the Sierra Nevada remained far below average.*

Threatened Coastlines. *Increased shoreline development and recent coastal storms such as Hurricanes Katrina and Sandy have combined to threaten many low-lying portions of the East Coast as well as the Gulf of Mexico.*

Southwest Wildfires. *Recent wildfires from Arizona to Texas suggest the dangers of future drought in this part of North America.*

▲ **Figure 3.7 Selected Environmental Issues in North America** Acid rain damage is widespread in regions downwind from industrial source areas. Elsewhere, widespread water pollution, cities with high levels of air pollution, and zones of accelerating groundwater depletion pose health dangers and economic costs to residents of the region. Since 1970, however, both Americans and Canadians have become increasingly responsive to the dangers posed by these environmental challenges.

Farther west, marine west coast climates dominate north of San Francisco, whereas a dry-summer Mediterranean climate occurs across central and southern California (compare the climographs for Los Angeles and Vancouver in Figure 3.6).

The Costs of Human Modification

North Americans have modified their physical setting in many ways. Processes of globalization have brought many benefits to North America, but the accompanying urbanization,

industrialization, and heightened consumption have also transformed the region's landforms, soils, vegetation, and climate. Indeed, problems such as acid rain, nuclear waste storage, groundwater depletion, and toxic chemical spills are all manifestations of a way of life unimaginable only a century ago, and the environmental price for the region's affluence (**Figure 3.7**).

Transforming Soils and Vegetation Since prehistoric times, Native Americans substantially altered the landscape by using fire, planting crops, and hunting wild animals.

This pattern of human alteration accelerated greatly with the arrival of Europeans on the North American continent. They impacted the region's flora and fauna by introducing countless new species, including wheat, cattle, and horses. As the number of settlers increased, forest cover was removed from millions of acres. Grasslands were plowed under and replaced with grain and forage crops not native to the region. Widespread soil erosion was increased by unsustainable farming and ranching practices, and many areas of the Great Plains and South suffered lasting damage.

Managing Water North Americans also consume huge amounts of water. In the United States, conservation efforts and technology have slightly reduced per capita rates of water use over the past 25 years, but on average city dwellers still use more than 175 gallons daily. Indirectly, through other forms of food and industrial water use, Americans may consume more than 1400 gallons of water per day. While the Great Lakes region is home to more than 20 percent of the world's freshwater supply, many other places in North America are threatened by water shortages. Metropolitan areas such as New York City struggle with outdated municipal water-supply systems.

Beneath the Great Plains, the waters of the Ogallala Aquifer are being depleted. Center-pivot irrigation systems have steadily lowered water tables across much of the region by as much as 150 feet (45 meters) in the past 50 years (**Figure 3.8**). The costs of pumping are rising steadily, and 50 percent of the area's irrigated land may see wells run dry by 2020. A 2014 federal report on the aquifer revealed that southwest Kansas and north Texas had witnessed the most dramatic groundwater depletion.

Farther west, California's complex system of water management is a reminder of that state's large demands within a setting prone to periodic drought. Many of the state's irrigated crops require massive amounts of water. For example,

▼ **Figure 3.8 Center-Pivot Irrigation, Kansas** This aerial view of the Kansas landscape reveals the telltale circular signatures of center-pivot irrigation systems. Deep-water drilling in these settings has nourished crops but depleted the Ogallala Aquifer.

California's one million acres of almond trees require more than a trillion gallons of water annually, and much of the crop is shipped to hungry consumers in China. Recently, many farmers in that state's Central Valley dramatically increased their consumption of groundwater as winter snow packs in the nearby Sierra Nevada have waned (farmers use 80 percent of the state's groundwater). As a result, underground aquifers in the state are being rapidly depleted (see Figure 3.7). In the southwestern United States, the fluctuating flows of the Colorado River are chronically frustrating to that rapidly growing portion of the country, although creative approaches to water management are being utilized to improve river flow, especially in its lower reaches (see *Working Toward Sustainability: Greening the Colorado River Delta*).

Water quality is also a major issue. North Americans are exposed to water pollution every day, and even environmental laws and guidelines, such as the U.S. Clean Water Act or Canada's Green Plan, have not eliminated the problem. In 2010, the *Deepwater Horizon* rig explosion and leaking oil well in the Gulf of Mexico quickly became one of North America's greatest environmental disasters. Five years later, Flint, Michigan's tragic tale of toxic tap water (corrosive chemicals from the polluted Flint River leached lead and other contaminants into the city's water supply) was another reminder that North America's aging infrastructure and increasingly toxic environment can be a deadly combination.

Elsewhere, North America's varied fisheries illustrate the complexities of continental water-resource management. The case of the Asian carp suggests how unpredictable the process can be. Imported from East Asia by southern catfish farmers in the 1970s, the carp removed algae and suspended matter from ponds in the lower Mississippi Valley. But many carp escaped their pens during large floods in the early 1990s. Twenty-five years later, the carp have migrated northward and are now knocking on the door of the Great Lakes as they discover artificial ship canals that link the lakes with the Mississippi system. The Army Corps of Engineers has spent millions to prevent an invasion that could decimate the Great Lakes fishing industry.

Eastern Canada's cod fisheries and many of Alaska's freshwater and saltwater fishing grounds have seen their annual catches decline dramatically in recent years. Overfishing, deadly infections from commercial fish-farming operations, and global climate shifts have all been blamed for the declines. New England fisheries have had steep quotas imposed on their catch so that fish stocks can grow in the Georges Bank and Gulf of Maine areas. On Canada's west coast, University of British Columbia scientists released a report in 2016 suggesting that climate change could decimate First Nation salmon fisheries by nearly 50 percent by 2050.

Altering the Atmosphere North Americans modify the very air they breathe; in doing so, they change local and regional climates, as well as the chemical composition of the atmosphere.

Mobile Field Trip: Moving Water Across California

https://goo.gl/rxtERk

For example, built-up metropolitan areas create **urban heat islands**, in which development associated with cities often produces nighttime temperatures some 9 to 22°F (5 to 12°C) warmer than those of nearby rural areas. At the local level, industries, utilities, and automobiles contribute carbon monoxide, sulfur, nitrogen oxides, hydrocarbons, and particulates to the urban atmosphere. While U.S. cities such as Los Angeles and Houston are among the worst offenders, Canadian cities such as Toronto, Hamilton, and Edmonton experience significant air-quality problems. Although overall air quality in urban North America has improved in the past few decades, a 2014 American Lung Association report estimated that about 47 percent of the nation's residents still live in places with unsafe levels of air pollution, including high levels of particulates as well as elevated ozone levels. Adding to these chronic problems are infrastructure failures such as the natural gas leak from a huge underground storage facility in Porter Ranch, California (near Los Angeles) which for months released 65,000 pounds of methane (a potent greenhouse gas) per hour into the atmosphere. In addition to inducing environmental illnesses in nearby areas, the daily impact on the planet's atmosphere was staggering: each day of the leak warmed the atmosphere at a rate equal to driving 4.5 million automobiles.

On a broader scale, North America is plagued by **acid rain**, caused by industrially produced sulfur dioxide and nitrogen oxides in the atmosphere that combine with precipitation to damage forests, poison lakes, and kill fish. Many industrial plants and power-generating facilities are located in the Midwest and southern Ontario, and prevailing winds transport their pollutants and deposit damaging acid rain and snow across the Ohio Valley, Appalachia, the northeastern United States, and eastern Canada (see Figure 3.7).

Air pollution is also going global, freely drifting into Canada and the United States on prevailing winds. One recent estimate suggests at least 30 percent of the region's ozone comes from beyond its borders. Both China and Mexico are major contributors of airborne pollutants.

Growing Environmental Initiatives

Many U.S. and Canadian environmental initiatives have addressed local and regional problems. For example, wildlife habitats across the United States and Canada are being successfully restored in a variety of settings. Recently, a collection of public agencies and private land owners have expanded bison habitat across eastern Montana and restored these animals to their native habitat. Today, the American Prairie Preserve (APR)

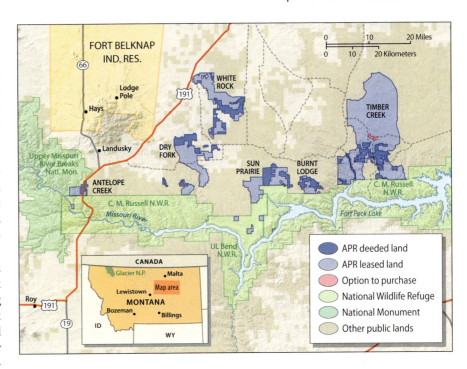

▲ **Figure 3.9 American Prairie Reserve (APR), Northeast Montana** North American bison are free to roam on lands owned by the APR. These acres are often combined with nearby public acreage, such as land allotments within the Charles M. Russell Wildlife Refuge.

owns or indirectly controls more than 270,000 acres (110,000 hectares) and plans call for 5,000–10,000 bison to be "home on the range" in the area by 2025 (**Figure 3.9**).

Conservation practices also have greatly reduced water- and wind-related soil erosion in North America, although experts estimate that one-third of U.S. and one-fifth of Canadian croplands are still at high to severe risk of future erosion. Tougher air-quality standards have also selectively reduced emissions in many North American cities. Similarly, the U.S. Superfund program (begun in 1980) and Canada's Environmental Protection Act (CEPA, begun in 1988) have significantly cleaned up hundreds of North America's toxic waste sites.

In the United States, the selective removal of dams also illustrates the trend. The 1973 passage of the Endangered Species Act mandated that federal agencies preserve threatened plants and animals along critical river habitat. The legislation became the political genesis of the dam removal movement. A varied political coalition of environmentalists, fishermen, and Native American groups has increasingly pressured governmental agencies to remove dams that many argue have outlived their usefulness and do more harm than good.

By 2015, more than 1200 dams had been removed nationwide. The world's largest dam removal project recently occurred along the Elwha River in western Washington's Olympic Mountains. Advocates of the project claim that the removal of a pair of dams (built in 1913 and 1929) has restored valuable habitat for 400,000 salmon and will provide benefits that far exceed the modest value of the

Greening the Colorado River Delta

A quiet environmental success story is unfolding along the lower Colorado River near the U.S.–Mexico border (**Figure 3.1.1**). For decades, the Colorado Delta had been a continental sacrifice zone, the victim of dam building and poor water management in the Southwest and northern Mexico. The completion of Hoover Dam (1937), Morelos Dam (1950), and Glen Canyon Dam (1963) helped dry up seasonal water flows into the delta, transforming more than 1.5 million acres (600,000 hectares) of precious wetlands into sunbaked mud flats. For decades, the delta was starved for water.

Creating a Green Coalition But thanks to a broad coalition of environmental groups and government agencies on both sides of the border, thousands of new cottonwood trees are taking root, along with willows and mesquite (**Figure 3.1.2**). Seasonal birds are revisiting the region's resurrected habitat, and there is real hope that the delta may one day return to its former splendor. Environmental groups based in the United States, such as the Sonoran Institute, the Defenders of Wildlife, and the Environmental Defense Fund, have partnered with Mexican groups, such as Pronatura, to make the case for the delta. Their actions have included legal battles (suing the U.S. government over lost habitat for endangered bird species), small-scale restoration initiatives demonstrating the potential for regreening the region (providing more than 800 acres of new trees and native plants), and intense lobbying efforts with water agencies throughout the Southwest and northern Mexico.

Google Earth Virtual (MG) Tour Video

http://goo.gl/DJ82sQ

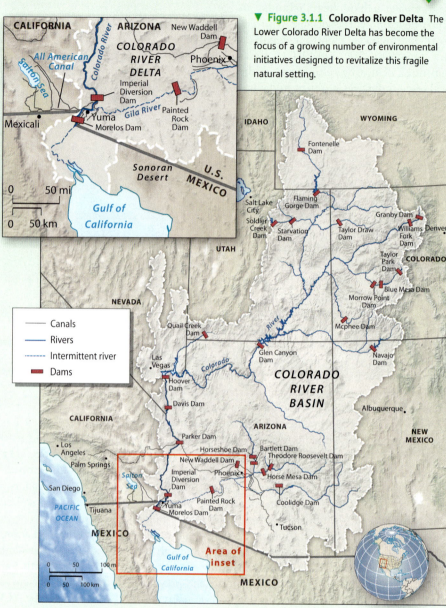

▼ **Figure 3.1.1 Colorado River Delta** The Lower Colorado River Delta has become the focus of a growing number of environmental initiatives designed to revitalize this fragile natural setting.

---

hydroelectric power generated by the dams. The Colorado River's Glen Canyon Dam is often cited as the biggest target for removal, but given the political and the logistical challenges involved in such a mammoth effort, nothing is likely to happen in the near future. Still, the deconstruction of America's dams may be just beginning, marking a policy shift intended to restore streams to some of their former environmental vitality.

The Shifting Energy Equation

Energy consumption in the region remains extremely high (the United States is still the source of almost 20 percent of Earth's greenhouse gas emissions), but growing incentives for energy efficiency may reduce rates of per capita consumption in the future. The growing technological and economic appeal of **renewable energy sources**, such as hydroelectric, solar, wind, and geothermal, are likely to fundamentally rework North America's economic geography in coming years as policymakers, industrial innovators, and consumers are attracted to their enduring availability and potentially lower environmental costs (**Figure 3.10**).

Recent evidence also suggests that North America's domestic supplies of oil, natural gas, coal-based resources, and tar sands may offer relatively abundant fossil fuel–based energy resources. New discoveries and drilling technologies have fundamentally shifted the continent's energy equation.

2013

2014

(a)   (b)

▲ **Figure 3.1.2 Restored Riparian Habitat, Colorado River Delta** Before (a) and after (b) views show the effects of a recent effort to engineer controlled flows of freshwater into the Colorado Delta to help restore riparian vegetation.

The New Agreement Those combined initiatives have paid off, even amid the ongoing reality of drought in this part of North America. Both the U.S. and Mexican governments, as well as a triumvirate of regional water agencies (the Metropolitan Water District of Southern California, the Southern Nevada Water Authority, and the Central Arizona Project), worked out an agreement to (1) increase the efficiency of existing irrigation and water storage infrastructure in the region, (2) release a huge pulse of water (more than 100,000 acre-feet) through the delta to replicate a natural flood, and (3) provide future base flows (about 50,000 acre-feet of water annually) until 2017 to maintain the ecological lifeline to delta plants and wildlife.

A Fragile Rebirth During the spring of 2014, the pulse of water released through the lower Colorado system triggered a massive ecological renaissance: Satellite imagery verified a 40 percent increase in riparian vegetation, water tables have risen, and bird populations

are increasing. But experts warn that the recovery is fragile and may be ephemeral. If severe drought conditions return to the Colorado Basin, annual base flows are not guaranteed past 2017. Furthermore, the tentative spirit of international cooperation that made recent progress possible may evaporate if more water shortages loom. But for now, the delta's pale green beauty has partly returned, eloquent testimony to what is possible when water's long-term ecological value is more fully appreciated in the complex calculus of resource management.

1. Describe another setting in North America where an international border potentially complicates an important environmental issue.

2. How might future plans for restoring the Colorado Delta go astray? Offer one scenario for an unexpected challenge, and describe how planning agencies might respond to it.

The Bakken formation in North Dakota and Montana, thanks to new oil extraction methods, may one day produce more than 15–20 billion barrels of oil, making it one of the planet's great energy reserves (on par with Alaska's North Slope field). The International Energy Agency predicts the United States will become a major exporter of natural gas by 2020. Canada alone has the world's third largest proven oil reserves: about 170 billion barrels of oil can be recovered just from its rich oil sands. Overall, even with its high rates of consumption, North America may be an oil export region by 2035. However, fluctuating global energy prices, such as the dramatic drop in crude oil prices in 2015 and 2016, makes such long-term projections more complex, especially if a slowdown in the energy economy discourages

investment in both alternative sources (such as wind and solar) and in shale-based fossil fuels (**Figure 3.11**).

Many environmental issues also complicate the clean development of North America's untapped fossil fuels. Moving fossil fuels involves huge investments and risks. Plans for transporting coal from the Rocky Mountains (especially from Wyoming and Montana) to the Pacific Coast (for export to Asia) have stirred protests. Controversial energy pipelines such as the Keystone XL project (Alberta to the Gulf of Mexico) and the Northern Gateway Pipeline (Alberta to the Pacific Coast), designed to tap into Canada's rich Athabasca oil sands, also illustrate the tensions between increasing energy production (and job creation) and the potentially dangerous environmental consequences of moving fossil fuels

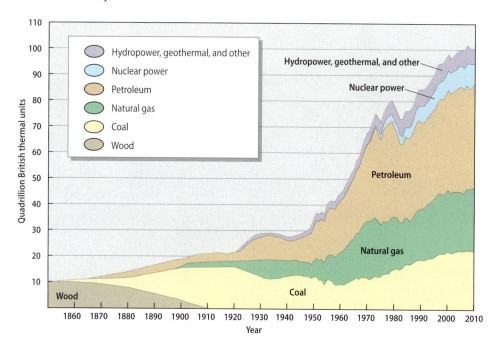

▲ **Figure 3.10 U.S. Energy Consumption** The growing popularity of fossil fuels is evident in U.S. energy consumption during the late 19th century as coal, oil, and then natural gas supplanted wood consumption. **Q: Approximately what percentage of U.S. energy consumption is currently accounted for by nuclear power and renewable energy sources?**

long distances (see Figure 3.7). In the United States, despite lower energy prices, a Republican administration may accelerate pipeline projects such as the Keystone XL initiative, but growing protests by western Canada's indigenous peoples have hampered plans for the rapid construction of the Northern Gateway Pipeline.

Mobile Field Trip: Oil Sands MG

https://goo.gl/48czBB

In addition, the growing use of hydraulic fracturing, or **fracking** (a drilling technology in which a mix of water, sand, and chemicals is injected underground to release natural gas) in settings from North Dakota to Pennsylvania has been challenged by critics claiming that the practice may lead to polluted groundwater and potentially hazardous environmental conditions for nearby residents.

Climate Change in North America

In 2014, more than 400,000 people from all around the United States participated in the People's Climate March through New York City. The marchers, including New York City Mayor Bill de Blasio, pressed for more rapid national and global responses to climate change, which has already profoundly reshaped many North American settings. After the unparalleled warmth between 2012 and 2015 (some of the hottest years on record in the United States), many experts predict that more extreme temperatures and precipitation events are likely in the near future. The accompanying droughts and wildfires cost billions.

Projections for the impacts of climate change in North America suggest varying regional patterns. In the United States, southern California, the Southwest, and parts of Texas may be drier, while Great Lakes states and the Northeast may see average precipitation, but more damaging, very heavy precipitation events. Many coastal localities along the East Coast and in the Gulf of Mexico will be especially vulnerable to rising sea levels and more intense coastal storms.

Mobile Field Trip: Forest Fires MG in the West

https://goo.gl/v5RVD1

North America's western mountains are especially sensitive to climate change. Expanding bark beetle populations can survive milder winters and are infesting pine forests. Many of the region's spectacular alpine glaciers are rapidly disappearing. Earlier spring melting of mountain snowpacks also impacts downstream fisheries, farms, and metropolitan areas that depend on these seasonal water resources.

Many government agencies are beginning to respond to the reality of climate change. Most dramatically, municipal planners are increasingly considering the likely impacts of changes on urban water supplies, flooding hazards, health risks, and disaster management plans. Gradually, states and provinces are also developing longer-term planning documents that will help guide future residents as they adjust to the reality of climate change.

▼ **Figure 3.11 Ivanpah Solar Power Facility, California** The Ivanpah Solar Electric Generating System, opened in 2014, is a solar thermal power project in the California desert, 40 miles (64 km) southwest of Las Vegas. It deploys 173,500 heliostats each with two mirrors focusing solar energy on boilers located on centralized solar towers.

Polar Transformations In North America's high latitudes, changes in arctic temperatures, sea ice, and sea levels have increased coastal erosion and affected whale and polar bear populations. Inuit residents, who have long traditions of hunting in the region, are increasingly at odds with animal-rights groups that want to more aggressively protect animals such as the polar bear.

Mobile Field Trip:
Climate Change
in the Arctic MG

https://goo.gl/sh7QgZ

At the same time, a more ice-free Arctic Ocean is also opening up potential for commercial shipping and resource development. Thanks to global climate change, beginning in 2016, large luxury cruise ships will be plying the Arctic, exploring a newly thawed and quite lucrative global connection between Anchorage and New York City. Following the so-called Northwest Passage around the northern perimeter of the continent, the high-latitude tourists—for a mere $20,000 per person—can enjoy the pleasures of the polar region, including arctic wildlife, native villages, and iceberg-studded ocean vistas (**Figure 3.12**).

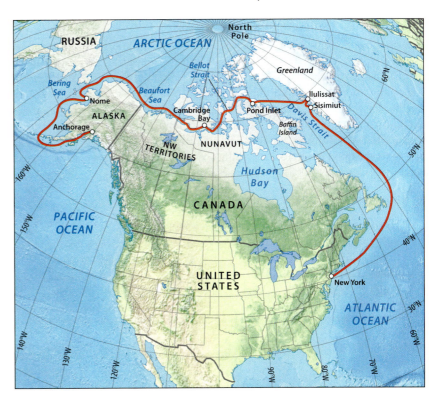

▲ **Figure 3.12   The Fabled Northwest Passage** Modern cruise ships may increasingly ply arctic waters as they follow the fabled Northwest Passage through ice-free polar seas. The Crystal Cruises line now features a high-end, 32-day journey from Anchorage to New York City.

REVIEW

3.1 Describe North America's major landform regions and climates, and suggest ways in which the region's physical setting has shaped patterns of human settlement.

3.2 Identify the key ways in which humans have transformed the North American environment since 1600.

3.3 Describe four key environmental problems that North Americans face in the early 21st century.

■ **KEY TERMS** Trans-Pacific Partnership (TPP), postindustrial economy, boreal forest, tundra, prairie, urban heat island, acid rain, renewable energy sources, fracking

Population and Settlement: Reshaping a Continental Landscape

The North American landscape is the product of human settlement extending back in time for at least 12,000–25,000 years. The pace of change for much of that period was modest and localized, but the last 400 years have witnessed an extraordinary transformation as Europeans, Africans, Asians, and Middle and South Americans arrived in the region, disrupted Native American peoples, and created dramatically new patterns of human settlement. Today 360 million people —some of the world's most affluent and highly mobile populations—live in the region (Table 3.1).

TABLE 3.1	**Population Indicators**							*Mastering*Geography™
Country	Population (millions), 2016	Population Density (per square kilometer)	Rate of Natural Increase (RNI)	Total Fertility Rate	Percent Urban	Percent <15	Percent >65	Net Migration (rate per 1000)
Canada	36.2	4	0.3	1.6	82	16	16	6
United States	323.9	35	0.4	1.8	81	19	15	4

Source: Population Reference Bureau, *World Population Data Sheet*, 2016.

[1]World Bank, *World Development Indicators*, 2016.

Login to *Mastering*Geography™ **& access** **MapMaster** **to explore these data!**

1) Compare the relative populations and the population densities for Canada and the United States. What do these data suggest about the relative sizes of the these two countries?

2) Using the demographic and urban data in the table, can you make an argument that there are differences in levels of development between these two nations?

Modern Spatial and Demographic Patterns

Metropolitan clusters dominate North America's population geography, producing strikingly uneven patterns of settlement across the region (**Figure 3.13**). About 90 percent of Canada's population is found within 100 miles (160 km) of the U.S. border. Within this broad region, Canada's "Main Street" corridor contains most of that nation's urban population, led by the cities of Toronto (6 million) and Montreal (4 million). The federal capital of Ottawa (1.3 million) and the industrial center of Hamilton (800,000) are also within the Canadian urban corridor. **Megalopolis**, the largest settlement agglomeration in the United States, includes the Washington, DC/Baltimore area (8.8 million), Philadelphia (6.1 million), New York City (20 million), and the greater Boston metropolitan area (4.7 million) (Figure 3.13). Beyond these two national core areas, other sprawling urban centers cluster around the southern Great Lakes (Chicago, 9.6 million), in various parts of the South (Dallas, 7 million), and along the Pacific Coast (Los Angeles, 17.6 million; Vancouver, 2.5 million) (**Figure 3.14**).

PEOPLE PER SQUARE KILOMETER
- Fewer than 6
- 6–25
- 26–100
- 101–250
- 251–500
- 501–1000
- 1001–12,800
- More than 12,800

POPULATION
- Metropolitan areas more than 20 million
- Metropolitan areas 10–20 million
- Metropolitan areas 5–9.9 million
- Metropolitan areas 1–4.9 million
- Selected smaller metropolitan areas

Growth in the interior West. *From Edmonton to Phoenix, many interior cities in the North American West have witnessed some of the continent's most rapid growth in the past 50 years, reflecting the region's amenities and rich base of natural resources.*

Dense populations in Megalopolis. *North America's densest regional urban populations remain in Megalopolis, stretching from Boston to Washington, DC.*

Black exodus from the rural South. *Many rural blacks left the South after 1900, seeking jobs in the urban North and West. Today, most of the growth in the southern black population is focused in urban areas.*

▶ **Figure 3.13 Population of North America** North America's geography of population reveals a strikingly clustered pattern of large cities interspersed with more sparsely settled zones. Notable concentrations are found on the eastern seaboard between Boston and Washington, DC; along the shores of the Great Lakes; and across the Sun Belt from Florida to California.

◀ **Figure 3.14 Chicago Skyline** Chicago's spectacular downtown skyline along the Lake Michigan shoreline is a classic example of a large North American central business district (CBD).

Explore the **Sights** of Chicago's "Loop"

http://goo.gl/m47nZL

Today, however, as in much of the developed world, rates of natural increase in North America are below 1 percent annually, and the overall population is growing older. Still, the region attracts many immigrants: More than 49 million foreign-born migrants now live in North America. These growing numbers, along with higher birth rates among immigrant populations, recently have led demographic experts to increase long-term population projections for the 21st century. Indeed, UN predictions that by 2050 the region's population will reach 464 million (423 million in the United States and 41 million in Canada) may prove conservative.

Occupying the Land

When Europeans began occupying North America more than 400 years ago, they were not settling an empty land; the region had been populated for at least 12,000–25,000 years by peoples as culturally diverse as the Europeans who conquered them. Native Americans migrated to North America from northeast Asia in multiple waves and dispersed across the region, adapting in diverse ways to its many natural environments. Cultural geographers estimate Native American populations in 1500 CE at 3.2 million for the continental United States and another 1.2 million for Canada, Alaska, Hawaii, and Greenland. In many areas, however, as contacts increased, European diseases, wars, and economic disruptions reduced these Native American populations by more than 90 percent.

North America's native peoples met many different fates. Some groups were exterminated by disease and war. Many others were expelled from their homelands and relocated on reservations, both in Canada and in the United States. For example, the Indian Removal Act of 1830, drove thousands of Native Americans out of the Southeast in the notorious "Trail of Tears" trek to Indian Territory (later Oklahoma). Some Native Americans also intermingled with European populations, losing parts of their cultural identity in the process. Today, the majority of the region's native peoples actually live in cities, often far removed from the land of their ancestors.

Who replaced Native North Americans? The first stage of a dramatic new settlement geography began with a series of European colonies between 1600 and 1750, mostly within the coastal regions of eastern North America (**Figure 3.15**). These regionally distinct societies were anchored in the north by the French settlement of the St. Lawrence Valley and extended south along the Atlantic Coast, including separate English colonies. Scattered developments along the Gulf Coast and in the Southwest also appeared before 1750.

Regular census data are gathered by Canada (every five years) and by the United States (every ten years). Most recently (in 2010 for the United States and 2016 for Canada), mailed questionnaires and census takers fanned out across the North American continent, gathering basic information on household size, age, and a variety of other social and economic characteristics. The U.S. Census also utilized millions of bilingual forms for the first time, in an effort to increase the response rate in Latino communities across the country.

In both countries, census data are used to make key geographical decisions. For example, census tallies are used to adjust election districts and to apportion tax dollars more efficiently. The United States allocates more than $400 billion each year to schools, hospitals, job-training centers, senior centers, and infrastructure improvements, based on census data. Businesses also use the census as North America's largest market-research survey. It helps restaurants and retailers plan locations for new stores and can suggest what products and services might do well in particular neighborhoods. In addition, the censuses are one of the most widely used tools of spatial analysis by North American human geographers interested in solving a variety of demographic, social, and economic problems. Information gathered in 2010 and 2016 is used as benchmark data for thousands of experts examining a myriad of urban, economic, and political issues in the region.

Population Change over Time Historically, North America's population has increased greatly since the beginning of European colonization. Before 1900, high birthrates produced large families. In addition, waves of foreign immigration swelled settlement, a pattern that continues today. In Canada, a population of fewer than 300,000 Indians and Europeans in the 1760s grew to an impressive 3.2 million a century later. For the United States, a late colonial (1770) total of around 2.5 million increased more than tenfold to exceed 30 million by 1860. Both countries saw even higher rates of immigration in the late 19th and 20th centuries, although birthrates gradually fell after 1900. After World War II, birthrates rose once again in both countries, resulting in the "baby boom" generation born between 1946 and 1965.

EUROPEAN SETTLEMENT EXPANSION
- Settled by 1750
- Settled 1750–1850
- Settled by 1910

▲ **Figure 3.15 European Settlement Expansion** Sizable portions of North America's East Coast and the St. Lawrence Valley were occupied by Europeans before 1750. The most remarkable surge of settlement occurred during the next century, as Europeans opened vast areas of land and dramatically disrupted Native American populations.

The second stage in the Europeanization of the North American landscape (1750–1850) was highlighted by settlement of much of the better agricultural land within the eastern half of the continent. After the American Revolution (1776) and a series of Indian conflicts, pioneers surged across the Appalachians. They found much of the Interior Lowlands region almost ideal for agricultural settlement. Much of southern Ontario, or Upper Canada, was also opened to widespread development after 1791.

The third stage in North America's settlement expansion picked up speed after 1850 and continued until just after 1910. During this period, most of the region's remaining agricultural lands were settled by a mix of native-born and immigrant farmers. Farmers were challenged and sometimes defeated by drought, mountainous terrain, and short northern growing seasons. In the American West, settlers were attracted by opportunities in California, the Oregon country, Mormon Utah, and the Great Plains. In Canada, thousands occupied southern portions of Manitoba, Saskatchewan, and Alberta. Gold and silver discoveries led to initial development in areas such as Colorado, Montana, and British Columbia's Fraser Valley.

Incredibly, in a mere 160 years, much of the North American landscape was occupied as expanding populations sought new land to settle and as the global economy demanded resources to fuel its growth. It was one of the largest and most rapid transformations of the landscape in human history. This European-led advance forever reshaped North America in its own image, and in the process it also changed the larger globe in lasting ways by creating a "New World" destined to reshape the Old.

North Americans on the Move

From the mythic days of Davy Crockett and Calamity Jane to the 20th-century sojourns of John Steinbeck and Jack Kerouac, North Americans have been on the move.

Indeed, almost one in every five Americans moves annually, suggesting that the region's people are quite willing to change residence in order to improve their income or their quality of life. Although interregional population flows are complex in both the United States and Canada, several trends dominate the picture.

Westward-Moving Populations The most persistent regional migration trend has been the tendency for people to move west. By 1990, more than half of the U.S. population lived west of the Mississippi River, a dramatic shift from colonial times. Since 1990, some of the fastest-growing states have been in the American West (including Arizona and Nevada), as well as in the western Canadian provinces of Alberta and British Columbia. This move was fueled by new job creation in high-technology, energy, and service industries, as well as by the region's scenic, recreational, and retirement attractions.

Black Exodus from the South African Americans have also generated distinctive patterns of interregional migration. Most blacks remained economically tied to the rural South after the Civil War, but by the early 20th century many African Americans migrated because of declining demands for labor in the agricultural South, and growing industrial opportunities in the North and West, particularly in cities. Boston, New York, Philadelphia, Detroit, Chicago, Los Angeles, and Oakland became key destinations for southern black migrants. Since 1970, however, more urban blacks have moved from North to South, attracted by Sun Belt jobs and federal civil rights guarantees to growing southern cities. The net result is still a major change from 1900: more than 90 percent of African Americans lived in the South at the beginning of the last century, whereas today only about 55 percent of the nation's 46 million blacks reside within those states.

Rural-to-Urban Migration Another continuing trend in North American migration has taken people from the country to the city. Two centuries ago, only 5 percent of North Americans lived in urban areas (cities of more than 2500 people), whereas today about 80 percent of the North American population is urban. Shifting economic opportunities account for much of the transformation: As mechanization on the farm reduced the demand for labor, many young people left for new employment opportunities in the city.

Sun Belt Growth Twentieth-century moves to the American South are clearly related to other dominant trends in North American migration, yet the pattern deserves closer inspection. Particularly after 1970, southern states from the Carolinas to Texas grew much

more rapidly than states in the Northeast and Midwest. Since 2010, the South has been home to most of the nation's fastest-growing counties and has remained a key regional destination for domestic migrants within the United States, adding people even more rapidly than the West. Florida, Texas, Georgia, North Carolina, and Virginia have experienced sizable population gains, with migrants heading for job-rich suburbs, as well as high-amenity retirement locations (**Figure 3.16**). Dallas–Fort Worth (23 percent), Houston (26 percent), and Atlanta (24 percent) enjoyed some of the nation's fastest metropolitan growth rates between 2000 and 2010, replacing previous high-fliers such as Las Vegas and Phoenix. Factors contributing to the South's growth are its buoyant economy, growing global exports, modest living costs, adoption of air conditioning, attractive recreational opportunities, and appeal to snow-weary retirees. Movements have been selective, however; North Carolina and Florida have seen many migrants from the Northeast, while the growth in Texas has come from high in-state birth rates, a higher foreign-born population, and in-migration from elsewhere in the South.

Nonmetropolitan Growth During the 1970s, certain areas in North America beyond its large cities began to see significant population gains, including many rural settings that had previously lost population. Selectively, this pattern of **nonmetropolitan growth**, in which people leave large cities for smaller towns and rural areas, continues today. A substantial number of migrants attracted to nonmetropolitan areas are younger, so-called *lifestyle migrants*. They find or create employment in affordable smaller cities and rural

▼ Figure 3.16 **The Villages, near Orlando, Florida** This planned retirement community near Orlando has witnessed some of Florida's most rapid population growth in the twenty-first century.

settings that are rich in amenities and often removed from the perceived problems of urban America. Nonmetropolitan population growth has exceeded metropolitan growth in many western states, and smaller communities outside the West, such as Mason City, Iowa; Mankato, Minnesota; and Traverse City, Michigan, also are seen as desirable destinations for migrants interested in downsizing from their metropolitan roots.

Settlement Geographies: The Decentralized Metropolis

North America's settlement landscape reflects the population movements, shifting regional economic fortunes, and technological innovations of the last century. The ways in which settlements are organized—the actual appearance of cities, suburbs, and farms—as well as the very ways in which North Americans socially construct their communities, have changed greatly in the past century. Today's cloverleaf interchanges, sprawling suburbs, outlet malls, and theme parks would have struck most 1900-era residents as utterly extraordinary.

North American cities boldly display the consequences of **urban decentralization**, in which metropolitan areas sprawl in all directions and suburbs take on many of the characteristics of traditional downtowns. Although both Canadian and U.S. cities have experienced decentralization, the impact is particularly profound in the United States, where inner-city problems, poor public transportation, widespread automobile ownership, and fewer regional-scale planning initiatives have encouraged many middle-class urban residents to move beyond the central city. Even beyond North America, observers note the globalization of urban sprawl: Many Asian, European, and Latin American cities are taking on attributes of their North American counterparts as they experience similar technological and economic shifts. Indeed, suburban Walmarts, semiconductor industrial parks, and shopping malls, so familiar in Seattle and Albuquerque, may become increasingly common sights on the peripheries of Kuala Lumpur or Mexico City.

Historical Evolution of the U.S. City Changing transportation technologies decisively shaped the evolution of the city in the United States (**Figure 3.17**). The pedestrian/horsecar city (pre-1888) was compact, essentially limiting urban growth to a 3- or 4-mile-diameter ring around downtown. The invention of the electric trolley in 1888 expanded the urbanized landscape farther into new "streetcar suburbs," often 5 or 10 miles from the city center. A star-shaped urban pattern resulted, with growth extending outward along and near streetcar lines. The biggest technological revolution came after 1920 with the widespread adoption of the automobile. The automobile city (1920–1945) promoted the growth of middle-class suburbs beyond the reach of the streetcar and added even more distant settlement in the surrounding countryside. Postwar growth in the outer city (1945 to the

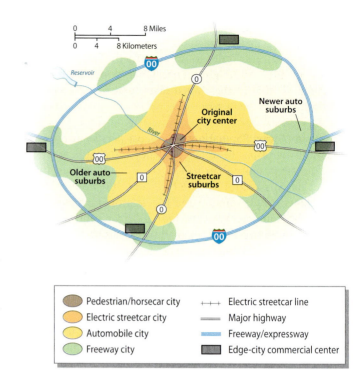

▲ **Figure 3.17 Growth of the American City** Many U.S. cities became increasingly decentralized as they moved through eras dominated by the pedestrian/horsecar, electric streetcar, automobile, and freeway. Each era left a distinctive mark on metropolitan America, including the recent growth of edge cities on the urban periphery.

present) promoted more decentralized settlement along freeways and commuter routes as built-up areas appeared 40 to 60 miles from downtown.

Urban decentralization also reconfigured land-use patterns, producing metropolitan areas today that are strikingly different from their early 20th century counterparts. In the city of 1920, urban land uses were generally organized in rings around a highly focused central business district (CBD) that contained much of the city's retail and office functions. Residential districts beyond the CBD were added as the city expanded, with higher-income groups seeking more desirable locations on the outside edge of the urbanized area.

Today's suburbs, however, feature a mix of peripheral retailing (such as commercial strips, shopping malls, and big-box stores), industrial parks, office complexes, and entertainment facilities. This larger, peripheral node of activity, called an **edge city**, has fewer functional connections with the central city than it has with other suburban centers. Southern California's Costa Mesa office and retailing district, located south of Los Angeles, is an excellent example of an edge-city landscape on the expanding periphery of a North American metropolis (**Figure 3.18**).

The Consequences of Sprawl The rapid evolution of the North American city continues to transform the urban landscape and those who live in it. As suburbanization accelerated in the 1960s and 1970s, many inner cities, especially in the Northeast and Midwest, suffered losses in population, increased

▲ **Figure 3.18 Costa Mesa, California** North America's edge-city landscape is nicely illustrated by Costa Mesa. Far from downtown Los Angeles, this sprawling complex of suburban offices and commercial activities reveals how and where many North Americans will live their lives in the 21st century.

San Francisco, and New York City) continue to post healthy population growth as new residents relocate to these vibrant, convenient downtown settings. Some experts suggest that the corporate move to suburban office complexes has also slowed, and that central cities may once again prove highly attractive to large corporations interested in convenient, prestigious downtown locations.

Many city planners and developers involved in such efforts are advocates of **new urbanism**, an urban design movement stressing higher-density, mixed-use, pedestrian-scaled neighborhoods where residents might be able to walk to work, school, and local entertainment. Pittsburgh's recent urban renaissance offers an affordable housing market, an older, highly skilled workforce, and mixed-use neighborhoods such as the SouthSide Works development, a 34-acre assortment of residences, offices, and stores created on the site of an old steel plant on the Monongahela River.

The suburbs are also changing. The edge-city construction of new corporate office centers, retail malls, and industrial facilities has created true "suburban downtowns" in suburbs that are no longer simply bedroom communities for central-city workers. Indeed, such localities have been growing players in the continent's globalization process. Many of North America's key internationally connected corporate offices (IBM, Microsoft), industrial facilities (Boeing, Cisco, Oracle), and entertainment complexes (Walt Disney World, Universal Studios, and the Las Vegas Strip) are now in such settings, and they are intimately tied to global information, technology, capital, and migration flows. Links to speedy and convenient metropolitan transit systems in many peripheral settings (such as Dallas and Washington, DC) have also blurred the lines between central cities and suburbs.

levels of crime and social disruption, and a shrinking tax base, which often brought them to the brink of bankruptcy. Today inner-city poverty rates average almost three times those of nearby suburbs. Unemployment rates remain above the national average. Central cities in the United States are also places of racial tension, the product of decades of discrimination, segregation, and poverty. Although the number of middle-class African Americans and Hispanics is growing, many exit the central city for the suburbs, further isolating the urban underclass that remains behind. In Detroit, for example, about 85 percent of inner-city residents are black, and the city's poverty rate is about triple the national norm. Racial tensions in largely black portions of Baltimore erupted in rioting and protests in 2015, partly in response to tensions between the city's police force and its poorer residents (**Figure 3.19**).

Amid these challenges, inner-city landscapes are also experiencing a selective renaissance. Termed **gentrification**, the process involves displacing lower-income residents of central-city neighborhoods by higher-income residents, rehabilitating deteriorated inner-city landscapes, and constructing new shopping complexes, entertainment attractions, or convention centers in selected downtown locations. The older and more architecturally diverse housing of the central city is also a draw, combining specialty shops and restaurants for a cosmopolitan urban clientele with residential opportunities for upscale singles who wish to live near downtown. Seattle's Pioneer Square, Toronto's Yorkville district, and Baltimore's Harborplace exemplify how such public and private investments shape the central city (**Figure 3.20**).

Recent census data also show that selected central-city settings (such as in Chicago,

▼ **Figure 3.19 Inner City Baltimore** This scene along Baltimore's blighted Pennsylvania Avenue unfolded in 2015 as protesters responded to the death of Freddie Gray, who died while in police custody.

▲ **Figure 3.20 Yorkville Neighborhood, Toronto** Toronto's fashionable Yorkville neighborhood is home to gentrified housing, upscale shops, and well-manicured public spaces.

Longer term, North Americans are likely to continue their outward drift, creating a vast suburban periphery where the boundary between country and city blurs in a mosaic of clustered housing developments, shopping complexes, and remaining open space (**Figure 3.21**).

Settlement Geographies: Rural North America

Rural North American landscapes trace their origins to early European settlement. Over time, these immigrants from Europe showed a clear preference for a dispersed rural settlement pattern as they created new farms. In portions of the United States settled after 1785, the federal government surveyed and sold much of the rural land. Surveys were organized around the simple, rectangular pattern of the federal government's township-and-range system, which offered a convenient method of dividing and selling the public domain in 6-mile-square townships (**Figure 3.22**). Canada developed a similar system of regular surveys that stamped much of southern Ontario and the western provinces with a strikingly rectilinear character.

Commercial farming and technological changes further transformed the settlement landscape. Railroads opened corridors of development, provided access to markets for commercial crops, and helped to establish towns. By 1900, several transcontinental lines spanned North America, radically transforming the farm economy and the pace of rural life. After 1920, however, even greater change accompanied the arrival of the automobile, farm mechanization, and better rural road networks. The need for farm labor declined with mechanization, and many smaller market centers declined as farmers equipped with automobiles and trucks could travel farther and faster to larger, more diverse towns.

Today many areas of rural North America face population declines as they adjust to the changing conditions of modern agriculture. Both U.S. and Canadian farm populations fell by more than two-thirds during the last half of the 20th century. Typically, fewer but larger farms dot the modern rural scene, and many young people leave the land to obtain employment elsewhere. Traditional resource-producing regions such as the Canadian Maritime provinces are also losing people, particularly from rural areas and small towns. The visual record of abandonment offers a painful reminder of the economic and social adjustments that come from population losses. Weed-choked driveways, empty farmhouses, roofless barns, and the empty marquees of small-town movie houses tell the story more powerfully than any census or government report.

Elsewhere, rural settings show signs of growth. Some places have experienced the effects of expanding edge cities. Other rural settings lie beyond direct metropolitan influence, but attract new residents who seek amenity-rich environments removed from city pressures. These trends are shaping the settlement landscape from British Columbia's Vancouver Island to Michigan's Upper Peninsula. Newly subdivided land, numerous real estate offices, and the presence of new espresso bars are all signs of growth in such surroundings.

REVIEW

3.4 Describe the dominant North American migration flows during the 20th century.

3.5 Sketch and describe the principal patterns of land use within the modern U.S. metropolis, including (a) the central city and (b) the suburbs/edge cities. How have forces of globalization shaped North American cities?

■ **KEY TERMS** Megalopolis, nonmetropolitan growth, urban decentralization, edge city, gentrification, new urbanism

▼ **Figure 3.21 Suburbs South of Denver, Colorado** These suburban homes, interspersed with planned open space, are located in Highlands Ranch, Colorado, almost 30 miles south of downtown Denver.

▲ **Figure 3.22 Iowa Settlement Patterns** The regular rectangles of this Iowa landscape are common features across the region. In the United States, the township-and-range survey system stamped such predictable patterns across vast portions of the North American interior. **Q: Can you identify any visual evidence of underlying land survey patterns in your local community?**

Cultural Coherence and Diversity: Shifting Patterns of Pluralism

North America's cultural geography is both globally dominant and internally pluralistic. History and technology have produced a contemporary North American cultural force that is second to none in the world. Many people outside the United States speak of cultural imperialism when they describe the global dominance of American popular culture, which they often see as threatening the vitality of other cultural values. Yet North America is also a home for many different peoples who retain part of their traditional cultural identities and celebrate their pluralistic roots.

The Roots of a Cultural Identity

Powerful forces formed a common dominant culture within the region. Both the United States (1776) and Canada (1867) became independent from Great Britain, but remained closely tied to their English, or Anglo, roots. Key Anglo legal and social institutions solidified the common set of core values that many North Americans shared with Britain and, eventually, with one another. Traditional Anglo beliefs emphasized representative government, separation of church and state, liberal individualism, privacy, pragmatism, and social mobility. From those shared foundations, particularly within the United States, consumer culture blossomed after 1920, producing a common set of experiences oriented around convenience, consumption, and the mass media.

However, North America's cultural unity coexists with pluralism: the persistence and assertion of distinctive cultural identities. Closely related is the concept of **ethnicity**, the shared cultural identity held by a group of people with a common background and history. For Canada, the early

and enduring French colonization of Quebec and the lasting power of the country's native peoples complicate its modern cultural geography. Canadians face the challenge of creating a truly multicultural society where issues of language and political representation are central concerns. Within the United States, given its unique immigration history, a greater diversity of ethnic groups exists, and differences in cultural geography are often found at both local and regional scales.

Peopling North America

North America is a region of immigrants. Quite literally, global-scale migrations made possible the North America we know today. Decisively displacing Native Americans in most portions of the region, immigrant populations created a new cultural geography of ethnic groups, languages, and religions. Though small in number, early migrants often had considerable cultural influence. Over time, varied immigrant groups and their changing destinations produced a culturally diverse landscape. Also varying among groups was the pace and degree of **cultural assimilation**, the process in which immigrants are absorbed into the larger host society.

Migration to the United States Variations in the number and source regions of migrants produced five distinctive chapters in U.S. history (**Figure 3.23**). In Phase 1 (prior to 1820), English influences dominated. Slaves, mostly from West Africa, contributed additional cultural influences in the South. In Phase 2 (1820–1870), Northwest Europe served as the main source region of immigrants, but in this phase Irish and Germans dominated the flow and provided more cultural variety.

As Figure 3.23 shows, immigration reached a much higher peak around 1900, when almost 1 million foreigners entered the United States *annually*. During Phase 3 (1870–1920), the majority of immigrants were southern and eastern Europeans. News of available land and expanding industrialization in the United States offered an escape from political strife and poor economies in Europe during this period. By 1910, almost 14 percent of the nation was foreign-born. Very few of these immigrants, however, targeted the job-poor U.S. South, creating a cultural divergence that still exists.

Between 1920 and 1970 (Phase 4), more immigrants came from neighboring Canada and Latin America, but overall totals fell sharply, a function of more restrictive federal immigration policies (the Quota Act of 1921 and the National Origins Act of 1924), the Great Depression, and the disruption caused by World War II.

Immigration has sharply increased since 1970 (Phase 5), and now annual arrivals surpass those of the early 20th century (see Figure 3.23). Most legal migrants since 1970 originated in Latin America or Asia. In 2000, about 60 percent of immigrants were Hispanics and only 20 percent were Asians, but by 2010 the balance shifted: 36 percent came from diverse Asian settings, while only about 30 percent were from Latin America. In percentage terms, Africans make up the fastest growing immigrant group in the United States (see *People on the Move: More Black Immigrants Arrive on American Shores*).

People on the Move

More Black Immigrants Arrive on American Shores

I t is a tasty, but perhaps surprising, dining spot: MaMa Ti's African Kitchen is located in an unassuming Minneapolis suburban shopping mall, just across the street from Walgreens. Customers come into the Liberian-themed restaurant to enjoy jollof rice and curried goat, or perhaps sample the plantains with fried fish. The restaurant is a small part of a much larger story that is quietly redefining what it means to be "African American."

Since 1980, the most rapidly growing immigrant population in the United States—at least in percentage terms—has been neither Hispanic nor Asian.

Rather, the nation's black immigrant population has steadily grown, with the largest share coming from the Caribbean and Sub-Saharan Africa (Figure 3.2.1). Today, about 9 percent of the nation's black population is foreign born (about 3.8 million immigrants), four times the total in 1980. Africans make up the fastest growing component of that immigrant population, with many new residents arriving from Nigeria, Ethiopia, Sudan, Somalia, Ghana, and Liberia.

Changing immigration laws have enabled black migrants to make the move to American shores. The Immigration and Nationality Act (1965) eased

▼ **Figure 3.2.1** **Black Immigrant Populations in the United States, 2013** The map shows the population and percent of U.S. foreign-born blacks by birth region and birth countries that contributed at least 100,000 black immigrants.

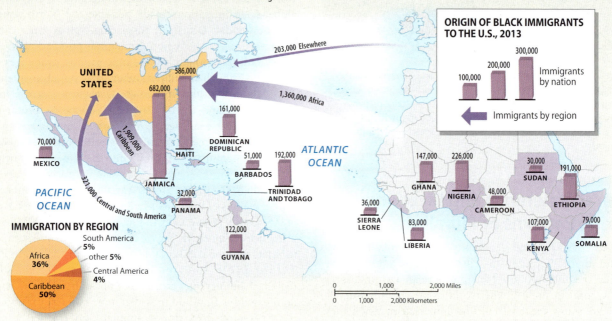

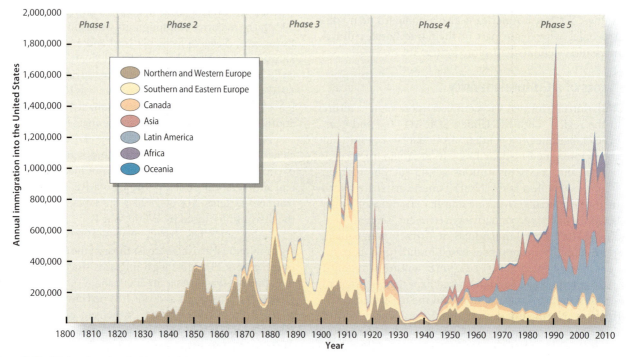

▲ **Figure 3.23** **U.S. Immigration, by Year and Group** Annual immigration rates peaked around 1900, declined in the early 20th century, and then surged again, particularly since 1970. The source areas of these migrants have also shifted. Note the decreased role of Europeans currently versus the growing importance of Asians and Latin Americans. **Q: What were dominant source regions of the U.S. population in 1850, 1910, and 1980? Why did they change?**

96

restrictions on families and skilled immigrants. More importantly, the Refugee Act of 1980 made it easier for migrants arriving from distressed political settings (many in Africa, such as Somalia and Ethiopia). The Immigration Act of 1990 mandated a more diverse migrant stream, opening the way for a variety of black immigrants worldwide. Between 2000 and 2013, almost 30 percent of Africa's black immigrants came in as refugees, and about 20 percent arrived with diversity-based visas. Once an immigrant has established legal residence, chain migration often occurs, with family members and eventually friends adding to the mix.

Many black immigrants settle in larger metropolitan areas, where jobs are available and existing immigrant communities provide support. For example, Washington, DC and its suburbs are now home to a thriving black immigrant population, including a large Eritrean and Ethiopian community living in Silver Spring, Maryland (**Figure 3.2.2**). In Miami, where more than one-third of the black population is foreign born, the mix is very different, favoring migrants from the nearby Caribbean (especially from Jamaica, Cuba, Haiti, and the Dominican Republic). In MaMa Ti's Minneapolis suburb of Brooklyn Park, black immigrants from Liberia, Somalia, and Kenya mingle with Hmong and Vietnamese from Southeast Asia and a growing number of migrants from Latin America. In 2014, almost 80 percent of Brooklyn Park's school children were non-white.

On average, these black immigrants tend to be older, wealthier, more suburban, and more likely to be college educated (35 percent hold at least a bachelor's degree) than native-born African Americans. Many migrants from the Caribbean speak English, but language can be a challenging barrier for African-born immigrants. According to the Pew Research Center, by 2060, over 16 percent of the U.S. black population is expected to be foreign born. Still unclear are the larger social and cultural implications of these people on the move, and how their communities and life experiences will be shaped by the country's culturally rich

▲ **Figure 3.2.2 Ethiopian Neighborhood, Silver Spring, Maryland** The recent influx of Ethiopian immigrants to this suburban community just north of Washington, DC has transformed the neighborhood's cultural landscape and ethnic identity.

and diverse African American communities. What is certain is that the country's black population is changing and that it will increasingly reflect influences from the Caribbean, Sub-Saharan Africa, and beyond.

1. As a young migrant from Ethiopia or Nigeria, suggest the three most important variables that shape where you settle in the United States.

2. Identify two possible consequences and/or conflicts that may result from growing black immigrant populations moving into established African American communities.

The post-1970 surge in immigration generally was made possible by economic and political instability abroad, a growing postwar American economy, and a loosening of immigration laws (the Immigration Acts of 1965 and 1990 and the Immigration Reform and Control Act of 1986). Undocumented immigration, particularly from Mexico, rose after 1970, but since 2008 the pace has slowed appreciably, mostly because of fewer job opportunities in the United States, as well as more rigorous U.S. border patrol. Today the United States is home to about 11–12 million undocumented immigrants.

The nation's Hispanic population continues to grow (**Figure 3.24**). Almost 12 million Mexican-born residents (about 10 percent of Mexico's population) now live in the United States. In the next 25 years, however, most of the projected increase in the U.S. Hispanic population will be fueled by births within the country, rather than by new immigrants. In fact, a recent report by the Pew Research Center found that between 2009 and 2014 more Mexican citizens actually returned home from the

▶ **Figure 3.24 Latino Community, Mission District, San Francisco** This bustling commercial corridor south of downtown San Francisco serves that city's largest Latino population. Many residents are recent immigrants from Mexico and Central America.

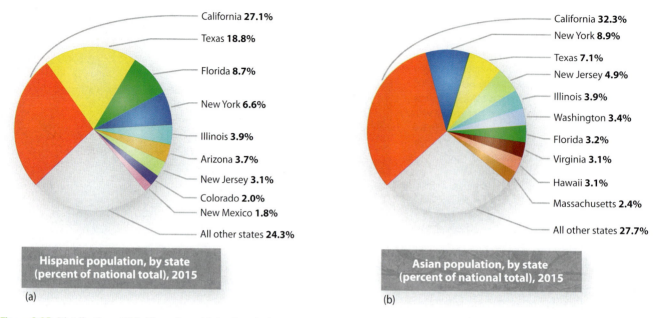

California 27.1%
Texas 18.8%
Florida 8.7%
New York 6.6%
Illinois 3.9%
Arizona 3.7%
New Jersey 3.1%
Colorado 2.0%
New Mexico 1.8%
All other states 24.3%

Hispanic population, by state
(percent of national total), 2015

(a)

California 32.3%
New York 8.9%
Texas 7.1%
New Jersey 4.9%
Illinois 3.9%
Washington 3.4%
Florida 3.2%
Virginia 3.1%
Hawaii 3.1%
Massachusetts 2.4%
All other states 27.7%

Asian population, by state
(percent of national total), 2015

(b)

▲ **Figure 3.25 Distribution of U.S. Hispanic and Asian Populations, by State, 2015** California, Texas, and Florida claim more than half of the nation's Hispanic population, and California alone is still home to almost one-third of the country's Asian population. **Q: Outside of the West and South, why do you think New York, New Jersey, and Illinois are also important destinations for these immigrants?**

United States than moved north of the border, reversing a decades-long trend.

Over 40 percent of the nation's 57 million Hispanics live in California (27 percent of California's population is foreign-born) or Texas, but they are increasingly moving to other areas (**Figure 3.25**). States such as South Carolina, Alabama, Wisconsin, Georgia, Kansas, and Arkansas have witnessed dramatic increases in Hispanic populations. Many settlements across the Great Plains are also home to Hispanic immigrants, who bring new churches, taquerías, and school-aged children to once-dying communities. The cultural implications of these Hispanic migrations are profound: Today the United States is the fifth largest Spanish-speaking nation on Earth.

In percentage terms, migrants from Asia constitute another fast-growing immigrant group, and various Asian ethnicities, both native and foreign-born, account for about 6 percent of the U.S. population. Chinese is the third most common spoken language in the United States (behind English and Spanish). California remains a key entry point for migrants and is home to almost one-third of the nation's Asian population, whereas Hawaii has the highest state-wide percentage of Asian immigrants (Figure 3.25). Asian migrants often move to large cities such as Los Angeles, San Francisco, Seattle, and New York City. Beyond these key gateway cities, a diverse array of Asian immigrants is also moving to growing communities in Washington, DC, Chicago, and Houston. The largest Asian groups in the United States include Chinese (4.0 million), Filipino (3.4 million), Asian Indian (3.2 million), Vietnamese (1.7 million), and Korean (1.7 million). Although some Asian immigrants live in poverty, on average, Asian Americans tend to have higher household incomes and are better educated than the overall U.S. population.

The future cultural geography of the United States will be dramatically redefined by these recent immigration patterns (**Figure 3.26**). By 2050, Asians may total almost 10 percent of the U.S. population, and almost one American in three will be Hispanic. Indeed, it is likely that the U.S. non-Hispanic white population will achieve minority status by that date.

The Canadian Pattern The peopling of Canada included early French arrivals who concentrated in the St. Lawrence Valley. After 1765, many migrants came from Britain, Ireland, and the United States. Canada then experienced the same surge and reorientation in migration flows seen in the United States around 1900. Between 1900 and 1920, more than 3 million foreigners moved to Canada, an immigration rate far higher than for the United States given Canada's much smaller population. Eastern Europeans, Italians, Ukrainians, and Russians dominated these later movements. Today, about 60 percent of Canada's recent immigrants are Asians, and its 21 percent foreign-born population is among the highest in the developed world. In Toronto, the city's 45 percent foreign-born population reveals a slight bias toward European backgrounds, although the city's Chinese-Canadian population remains a vibrant part of the community. On Canada's west coast, Vancouver (45 percent foreign-born) has been a key destination for Asian immigrants, particularly people from China, India, and the Philippines (**Figure 3.27**). Recently, however, Asian immigration has leveled off and British Columbia is actually seeing fewer new immigrants than it did at the turn of the 21st century.

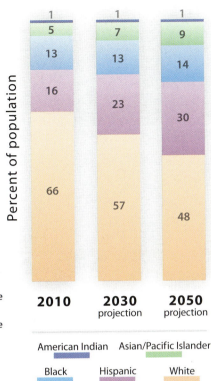

▶ **Figure 3.26 Projected U.S. Ethnic Composition, 2010 to 2050** By the middle of the 21st century, almost one in three Americans will be Hispanic, and non-Hispanic whites will achieve minority status amid an increasingly diverse U.S. population.

Culture and Place in North America

Cultural and ethnic identity is often strongly tied to place. North America's cultural diversity is expressed geographically in two ways. First, similar people congregate near one another and derive meaning from the territories they occupy in common. Second, culture marks the visible landscape with the artifacts, habits, language, and values of different groups. Boston's Italian North End simply looks different from nearby Chinatown, and rural French Quebec is a world away from a Hopi village in Arizona.

Persistent Cultural Homelands French-Canadian Quebec superbly exemplifies a **cultural homeland**: a culturally distinctive nucleus of settlement in a well-defined geographical area, where ethnicity has survived over time, stamping the cultural landscape with an enduring personality (**Figure 3.28**). Overall, about 23 percent of Canadians are French, but about 80 percent of the population of Quebec speaks French,

▶ **Figure 3.27 Vancouver's Chinatown** This traditional Chinatown is located just east of downtown Vancouver, but the city's ethnic Chinese population resides throughout the metropolitan area.

and language remains the cultural glue that unites the homeland. Indeed, policies adopted after 1976 strengthened the French language within the province by mandating French instruction in the public schools and national bilingual programming by the Canadian Broadcasting Corporation (CBC). Ironically, many Quebecois feel that the greatest cultural threat comes not from Anglo-Canadians, but rather from recent immigrants to the province. Southern Europeans or Asians in Montreal, for example, show little desire to learn French, preferring instead to put their children in English-speaking private schools.

Another well-defined cultural homeland is the Hispanic Borderlands (see Figure 3.28). It is similar in size to French-Canadian Quebec and significantly larger in total population, but not specifically linked to a single political entity such as a state or province. Historical roots of the homeland are deep, extending back to the 16th century, when Spaniards opened the region to the European world. The homeland's historical core is in northern New Mexico, including Santa Fe and much of the surrounding rural hinterlands. Spanish place-names, earth-toned Catholic churches, and traditional Hispanic settlements dot the rolling highlands of northern New Mexico and southern Colorado. From California to Texas, other historical sites, place-names, missions, and presidios also reflect this rich Hispanic heritage.

Unlike Quebec, however, massive 20th-century migrations from Latin America brought an entirely new wave of Hispanic settlement to the Southwest. More than half of the 57 million Hispanics in the United States live in California, Texas, and Florida, and by 2016, Hispanics outnumbered non-Hispanic whites in California. Within the homeland, Hispanics have created a distinctive Borderlands culture that mixes many elements of Latin and North America. These newer migrants augment the rural Hispanic presence in agricultural settings such as the lower Rio Grande Valley in Texas and the Imperial and Central valleys in California. Cities such as San Antonio, Phoenix, and Los Angeles also play leading roles in expressing Hispanic culture within the Southwest. Regionally distinctive

► **Figure 3.28 Selected Cultural Regions of North America** From northern Canada's Nunavut to the Southwest's Hispanic Borderlands, different North American cultural groups strongly identify with traditional local and regional homelands. Shaded portions of the map display a sampling of these regions across North America. Dotted areas suggest general locations for surviving ethnic islands of rural European settlement. A mother tongue other than English characterizes some cultural regions, both urban and rural.

Latin foods and music add internal cultural variety to the Borderlands. New York City, Chicago, and Miami serve as key points of Hispanic cultural influence beyond the homeland.

African Americans also retain a cultural homeland, but it has diminished in intensity because of out-migration (see Figure 3.28). Reaching from coastal Virginia and North Carolina to East Texas, the Black Belt is a zone of African American population remaining from the cotton South, when a vast majority of American blacks resided within the region. Although many African Americans have left for cities, dozens of rural counties in the region still have large black majorities. Blacks account for more than one-quarter of the populations of Mississippi (37 percent), Louisiana (31 percent), South Carolina (27 percent), Georgia (31 percent), and Alabama (27 percent). More broadly, the South is home to many black folk traditions, including music such as black spirituals and the

Explore the **Sounds** of Delta Blues

http://goo.gl/ucQKOW

blues, which have now become popular far beyond their rural origins. Beyond the South, African Americans have created large, vibrant communities in mostly urban settings in the Northeast, Midwest, and West (**Figure 3.29**).

A second rural homeland in the South is Acadiana, a zone of enduring Cajun culture in southwestern Louisiana (see Figure 3.28). This homeland was created in the 18th century when French settlers were expelled from eastern Canada (an area around the Bay of Fundy known as Acadia) and relocated to Louisiana. Nationally known today through their food and music, the Cajuns are tightly linked to the bayous and swamps of southern Louisiana.

Native American Signatures Native Americans are also strongly tied to their homelands. Indeed, many native peoples maintain intimate relationships with their surroundings, weaving elements of the natural environment into their material and spiritual lives. Over 5 million Native Americans, Inuits,

▲ **Figure 3.29 African-American Rap Artists, Denver, Colorado** This colorful mural on Denver's west side celebrates the musical contributions of (left to right) Tupac Shakur, Biggie Smalls, Easy-E, and Krayzie Bone.

and Aleuts live in North America, claiming allegiance to more than 1100 tribal bands. Although many Native Americans now live in cities, they still retain close contact with their homelands. Place-names, landscape features, and family ties cement this connection between people and place.

Particularly in the American West and the Canadian and Alaskan North, native peoples also control sizable reservations, although less than 25 percent of the overall native population resides on reservations. The largest block of native-controlled land in the lower 48 states is the Navajo Reservation in the Southwest. About 300,000 people claim allegiance to the Navajo Nation. To the north, Canada's self-governing Nunavut

Territory (population about 35,000) is another reminder of the enduring presence of native cultural influence and emergent political power within the region (see Figure 3.28).

Although these homelands preserve traditional ties to the land, they are also settings for pervasive poverty, health problems, and increasing cultural tensions (**Figure 3.30**). Within the United States, many Native American groups have taken advantage of the special legal status of their reservations and have built gambling casinos and tourist facilities that bring in much-needed capital, but also challenge traditional lifestyles. Many Native languages are also threatened: in 2015, more than four dozen recognized indigenous languages (such as Nez Perce, Klamath, Northern Paiute, and Mono) in the American West claimed fewer than 1000 speakers each, and their numbers were steadily declining.

Explore the **Sounds** of Navajo Language

http://goo.gl/pwXM9D

Explore the **Sights** of Pine Ridge Reservation

http://goo.gl/De7ZSW

▼ **Figure 3.30 Native American Poverty, South Dakota** South Dakota's Pine Ridge Indian Reservation is home to the Oglala Sioux nation. The reservation faces a severe housing shortage, made worse by recent floods and persisting poverty.

A Mosaic of Ethnic Neighborhoods North America's cultural mosaic is also enlivened by smaller-scale ethnic signatures that shape both rural and urban landscapes (see Figure 3.28). Distinctive rural communities that range from Amish settlements in Ohio and Pennsylvania to Ukrainian villages in southern Saskatchewan add cultural variety. Immigrants often established close-knit communities as they settled the agricultural interior. Among others, German, Scandinavian, Slavic, Dutch, and Finnish neighborhoods took shape, held together by common origins, languages, and religions. Although many of these ties weakened over time, the rural landscapes of Wisconsin, Minnesota, the Dakotas, and the Canadian prairies still display some of these cultural signatures, including folk architecture, distinctive settlement patterns, ethnic place-names, and the simple elegance of rural churches.

Ethnic neighborhoods also enrich the urban landscape and reflect both global-scale and internal North American migration patterns. Complex social and economic processes are clearly at work. Historically, employment opportunities fueled population growth in North American cities, but the cultural makeup of the incoming labor force varied, depending on the timing of the economic expansion and the relative accessibility of an urban area to different cultural groups. The cultural geography of Los Angeles exemplifies the interplay of the economic and cultural forces at work (**Figure 3.31a**). Because most of its economic expansion took place during the 20th century, the city's cultural makeup reflects the movements of more recent migrants. African American communities on the city's south side (Compton and Inglewood) represent the legacy of black population movements out of the South. Hispanic (East Los Angeles) and Asian (Alhambra and Monterey Park) neighborhoods are a reminder that about 40 percent of the city's population is foreign-born (**Figure 3.31b**).

Particularly in the United States, ethnic concentrations of nonwhite populations increased in many cities during the 20th century as whites exited for the perceived safety of the suburbs. In terms of central-city population, African Americans make up more than 60 percent of Atlanta, whereas Los Angeles is now more than 40 percent Hispanic (almost 5 million Hispanics reside in Los Angeles County, greater than the population of Costa Rica). Ethnic concentrations are also growing in the suburbs of many U.S. cities: Southern California's Monterey Park has been called the "first suburban Chinatown," and growing numbers of middle-class African Americans and Hispanics shape suburban neighborhoods in metropolitan settings such as Atlanta and San Antonio.

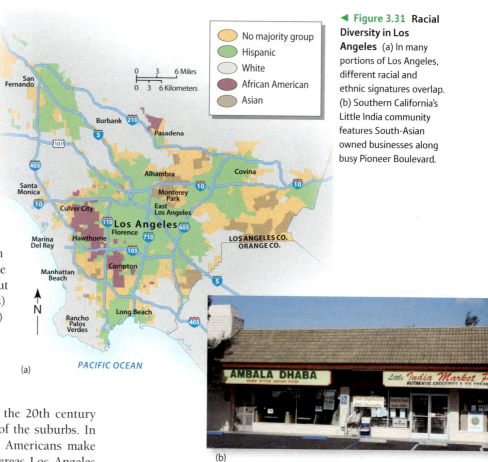

▲ **Figure 3.31** **Racial Diversity in Los Angeles** (a) In many portions of Los Angeles, different racial and ethnic signatures overlap. (b) Southern California's Little India community features South-Asian owned businesses along busy Pioneer Boulevard.

Patterns of North American Religion

Many religious traditions also shape North America's human geography. Reflecting its colonial roots, Protestantism is dominant within the United States, accounting for about 50 percent of the population (**Figure 3.32**). In some settings, hybrid American religions sprang from broadly Protestant roots. By far the most successful are the Latter-Day Saints (LDS or Mormons), regionally concentrated in Utah and Idaho and claiming more than 6 million North American members. Although many traditional Catholic neighborhoods have lost population in the urban Northeast, Catholic numbers are growing in the West and South, reflecting both domestic migration patterns and higher rates of Hispanic immigration and births. Strong Protestant religious traditions enliven many African-American communities, both in the rural South and in dozens of urban settings where churches serve as critical centers of community identity and solidarity.

Within Canada, almost 40 percent of the population is Protestant, with the United Church of Canada claiming large numbers of followers (see Figure 3.32). Roman Catholicism is important in regions that received large numbers of Catholic immigrants. French-Canadian Quebec is a bastion of Catholic tradition and makes Canada's population (39 percent) distinctly more Catholic than that of the United States (22 percent).

Millions of other North Americans practice religions outside of the Protestant and Catholic traditions. Orthodox Christians congregate in the urban Northeast, where many Greek, Russian, and Serbian Orthodox communities were established between 1890 and 1920. The telltale domes of Ukrainian Orthodox churches still dot the Canadian prairies of Alberta, Saskatchewan, and Manitoba. Among other religious traditions, more than 5 million Jews live in North America, concentrated in East and West Coast cities. In the United States, the rapidly growing organization known as the Nation of Islam also has a strong urban orientation, reflecting its appeal to many economically dispossessed African Americans. Many other Muslims (3.6 million), Buddhists (3.9 million), and Hindus (2.2 million) also live in the United States. Only about 9 percent of people in the United States classify themselves as nonbelievers, but a recent survey showed that 30 percent of the population claimed to have a largely secular lifestyle in which religion was rarely practiced.

The Globalization of American Culture

Simply put, North America's cultural geography is becoming more global at the same time that global cultures are becoming more North American (influenced particularly by the United States). But processes of cultural globalization in the

21st century are complex; rather than simple flows of foreign influences into North America or of U.S. culture invading every traditional corner of the globe, the story of cultural globalization increasingly mixes influences flowing in many directions at once, resulting in new hybrid cultural creations.

North Americans: Living Globally More than ever, North Americans in their everyday lives are exposed to people from beyond the region. With more than 49 million foreign-born migrants living across the region, diverse global influences mingle in new ways. Millions of international visitors come to North America annually, both for business and pleasure. By 2019, the United States is predicted to have about 88 million foreign visitors annually. In U.S. colleges and universities, about 975,000 international students (more than 30 percent from China) add global flavor to the classroom. Canada plays host to an additional 200,000 international students.

Globalization also presents cultural challenges for North Americans. In the United States, one key issue revolves around the English language, which some have described as the "social glue" holding the nation together. The growing popularity of **Spanglish**, a hybrid combination of English and Spanish spoken by Hispanic Americans, illustrates the complexities of North American globalization. Spanglish, an example of "code switching," where a speaker alternates between two or more languages, includes interesting hybrids such as *chatear,* which means "to have an online conversation."

North Americans are going global in other ways. By 2017, the vast majority of Americans and Canadians had Internet access, opening the door for far-reaching journeys in cyberspace. Social media such as Facebook and Twitter have for many North Americans redefined the kinds of communities and networks that shape their daily lives. Today, millions of North Americans also have access to vast online databases and GIS-based information resources that link them more effectively to their own localities or to the other side of the world (see *Geographers at Work: Building Montana's Online GIS Infrastructure*).

Many global influences have reshaped North American landscapes and cultural preferences. Ethnic restaurants pepper the region with a bewildering variety of Cuban, Ethiopian, Basque, and South Asian eateries. Americans consume imported beer in record quantities, with varieties from Mexico proving especially popular (**Figure 3.33**). In a related example of globalization, Heineken, famed for its fine Dutch brew, now owns

▼ **Figure 3.32 Christian Denominations of North America** Although many portions of North America feature great religious diversity, Roman Catholicism or various Protestant denominations dominate selected regions. Portions of rural Utah and Idaho dominated by the Mormon faith display some of the West's highest concentrations of any single religion.

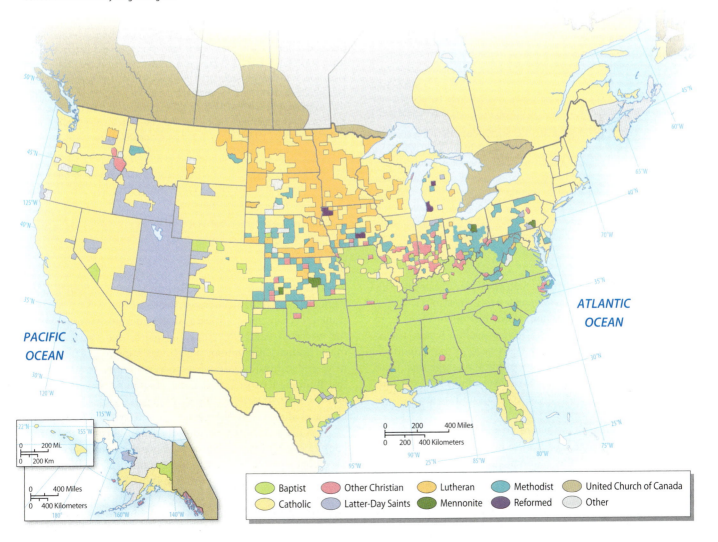

Legend: Baptist · Catholic · Other Christian · Latter-Day Saints · Lutheran · Mennonite · Methodist · Reformed · United Church of Canada · Other

Geographers at Work

Supporting Montana's Online GIS Infrastructure

▲ **Figure 3.3.1** **Diane Papineau**

Diane Papineau, a GIS Analyst for the Montana State Library in Helena, is part of a team of geographers that maintains and updates Montana's GIS Data Clearinghouse, which has provided online mapping applications and downloadable, geographic databases (land ownership, transportation, surface waters, addresses, land use, etc.) since 1994 (Figure 3.3.1). Papineau works with highly specialized GIS users as well as the general public. Her duties include creating maps from digital data representing landscape phenomena, contributing to the design of easy-to-use online map applications and data access tools, and offering outreach to those interested in mapping Montana. "The Clearinghouse makes geographic data discoverable and accessible. The geographers on staff help develop the geographic literacy of all Montanans," she explains.

GIS and historical geography Papineau also serves as Montana's State Geographic Names Advisor and Montana Geographic Names Database theme lead, responsibilities supported by her graduate work in historical geography at Montana State University. Her research examined Yellowstone's development as a key tourist destination. With her knowledge of GIS, Papineau georeferenced maps and other archival location information, permitting her to identify and illustrate the location of an historic early tourist settlement of roads, trails, overlooks, hotels, campgrounds, and other structures near the Grand Canyon of the Yellowstone. (Figure 3.3.2). History and geography have much in common, she notes: "Time is a variable that sits between lots of disciplines; so does space." Just as knowing history helps to interpret events, "understanding geography and maps can inform your interpretation of the natural and cultural landscape."

GIS in State and Local Government Diane's GIS education and graduation from Montana State University were well timed: between 1990 and 2010, most state and local governments adopted GIS technologies and integrated them into many workflows and services—the incorporation of GIS throughout local governments is still underway. Transportation and utility departments, commerce bureaus, tax assessors, and various environmental and natural resource agencies moved toward spatially-enabling their databases and planning infrastructure for a GIS-enabled world. But Papineau cautions that using

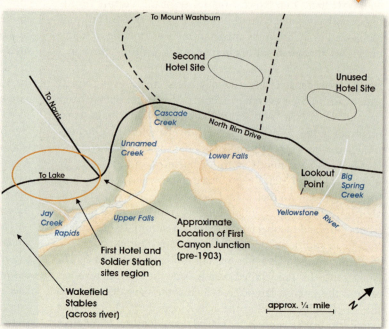

▲ **Figure 3.3.2** **Historical Evolution of the Canyon Area, Yellowstone National Park** This map, produced in a historical GIS format, shows some of the early tourist developments near the Grand Canyon of the Yellowstone River in Yellowstone National Park.

GIS without understanding geography can be risky: "Mapping technology in the hands of people who don't understand geography and cartography can result in inaccurate maps and data products that are presumed to be correct in part because they were produced using GIS. A trained geographer is better prepared to find and appraise geographic data for a project, to ensure that the spatial analysis and final products use geographic and GIS best practices, and a geographer can more clearly communicate and defend the information conveyed in maps, tables, charts, and datasets produced from a GIS."

1. Explore your state's online GIS-based resources. What kinds of information can you access that might be used to solve problems and address land-use issues in your area? What challenges might be involved in using these databases?

2. What types of online geographical information would be useful to you as an urban planner? List three examples and briefly explain why.

---

the rights to resell in the United States both the Tecate and Dos Equis brands, two popular Mexican beers. In fashion, Gucci, Brioni, and Prada are household words for millions who keep their eyes on European styles. German techno bands, Gaelic instrumentals, and Latin rhythms have become the soundtrack of daily life. Indeed, from acupuncture and massage therapy to soccer and New Age religions, North Americans tirelessly borrow, adapt, and absorb the larger world around them.

The Global Diffusion of U.S. Culture In parallel fashion, U.S. culture has forever changed the lives of billions of people beyond the region. Although the economic and military power of the United States was notable by 1900, it was not until after World War II that the country's popular culture fundamentally reshaped global human geographies. The Marshall Plan and

Peace Corps initiatives exemplified the growing presence of the United States on the world stage, even as European colonialism waned. Perhaps most critical was the marriage between growing global demands for consumer goods and the rise of the multinational corporation, which was superbly structured to meet and cultivate those needs. Global corporate advertising, distribution networks, and mass consumption bring Cokes and Big Macs to Moscow and Beijing, golf courses to Thai jungles, and Mickey and Minnie Mouse to Tokyo and Paris.

But challenges to U.S. cultural control illustrate the varied consequences of globalization. Hollywood's dominance within the global film industry has declined dramatically as filmmakers build their own movie businesses in India, Latin America, West Africa, China, and elsewhere. As worldwide use of the Internet has grown, the online dominance of English-speaking users

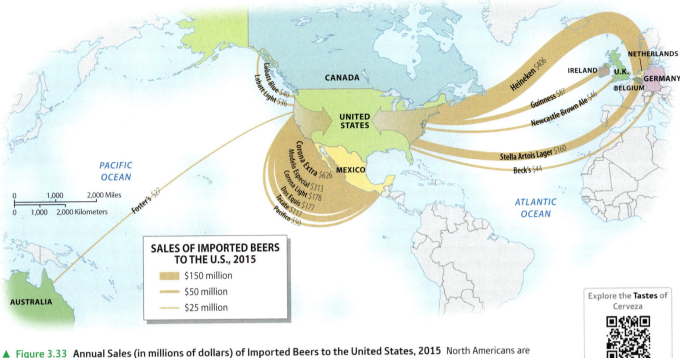

SALES OF IMPORTED BEERS TO THE U.S., 2015

- $150 million
- $50 million
- $25 million

Labatt Blue $40
Labatt Light $36
Foster's $27

Heineken $406
Guinness $87
Newcastle Brown Ale $46
Stella Artois Lager $160
Beck's $44

Corona Extra $626
Modelo Especial $313
Corona Light $178
Dos Equis $177
Tecate $117
Pacifico $50

NETHERLANDS
IRELAND
U.K.
GERMANY
BELGIUM

CANADA
UNITED STATES
MEXICO

PACIFIC OCEAN
ATLANTIC OCEAN
AUSTRALIA

0 1,000 2,000 Miles
0 1,000 2,000 Kilometers

Explore the **Tastes** of Cerveza
http://goo.gl/RSdXvv

▲ **Figure 3.33** **Annual Sales (in millions of dollars) of Imported Beers to the United States, 2015** North Americans are increasingly eating and drinking globally. Rising beer imports, including many more expensive foreign brands, exemplify the pattern. The nation's beer drinkers know no bounds to their thirsts. Mexico dominates, along with varied European, Asian, and Australian producers.

has dramatically declined, while Hindi and Mandarin users grow in number. Active resistance to U.S. cultural influence is also notable. For example, Canada routinely chastises their radio, television, and film industries for allowing too much U.S. cultural influence, and France also criticizes U.S. dominance in such media as the Internet. Elsewhere, Iran has banned satellite dishes and many U.S. films, although illegal copies of top box-office hits often find their way through national borders.

The Globalization of North American Sports Increasingly, major North American sports (hockey, baseball, basketball, football) transcend national boundaries. At the same time, sports popular in other parts of the world (particularly soccer) have become integral elements of North American culture. The so-called *global media sports complex* has facilitated this cultural and economic transformation: Today media companies broadcast sporting events by satellite and cable TV around the globe. Huge investments in league franchises, individual players, stadiums, and related sport-equipment companies, along with growing mobility among professional athletes, have all made it easier for North American sports to go global.

Undoubtedly, North American sports are reshaping the larger world (see *Everyday Globalization: Disc Golf Goes Global*). American-style golf courses are being built in exotic settings such as the United Arab Emirates (now home to the Dubai Desert Classic). Baseball is a passion from the Caribbean to Japan, where American teams have played exhibition games since 2000. Interest in American-style football has also expanded: The National Football League regularly plays exhibition or regular-season games in settings such as London, Toronto, and Mexico City. Although the NFL Europe League (1995–2007) failed financially, it signaled the growing visibility of the sport in countries such as Germany, Great Britain, and the Netherlands.

In the 1992 Olympics, the electrifying performance of the "Dream Team" (led by Michael Jordan) brought basketball onto the world stage. Today, the sport has a huge following in settings as diverse as Serbia, China, and Argentina, where viewers watch National Basketball Association (NBA) games with a passion and home-grown leagues have blossomed. In fact, since 2008, Chinese planners have mandated the building of basketball courts in every Chinese village, an initiative no doubt spurred by the popularity of the legendary Yao Ming in the NBA.

Similarly, North American sports have been transformed by the rest of the world. Within the region, Canadian ice hockey has expanded to become a truly continental sport. Soccer has also flourished on North American shores. Since 1994, the American Youth Soccer Organization has multiplied its presence from Iowa to Alabama. Millions of North American children now participate in league competition. The popularity of professional soccer teams in many U.S. cities, as well as the growing North American interest in World Cup Soccer events, also illustrates the trend. More broadly, the number of foreign players on professional American teams continues to grow rapidly. In the interconnected world of the 21st century, great talent from Nigeria to Nicaragua can more easily be discovered and imported to North America. Baseball teams feature truly globalized rosters, and many of the region's leading basketball players grew up practicing layups in another part of the world.

REVIEW

3.6 What are the distinctive eras of immigration in U.S. history, and how do they compare with those of Canada?

3.7 Identify four enduring North American cultural regions, and describe their key characteristics.

■ **KEY TERMS** ethnicity, cultural assimilation, cultural homeland, Spanglish

Disc Golf Goes Global

While pioneering examples of playing golf with a flying disc can be traced to Saskatchewan, Canada in the 1920s, the modern sport of disc golf took off along with the popular Frisbee in the 1960s. Frisbee golf tournaments blossomed in city parks, especially in sunny California and, by the 1980s, today's golf disc (for more range and accuracy) and the familiar basket-and-chain targets were established.

The sport has diffused to selected global settings, especially since 2000 (Figure 3.4.1). Today, the Professional Disc Golf Association (PDGA) boasts more than 20,000 active members in 36 countries, sanctioning almost 2,000 annual events. The PDGA's Marco Polo Program is designed to spread the sport to more countries in the years ahead. Challenges include securing the correct equipment, organizing events, and simply finding the extensive acreage required, even for a simple 9-hole course.

With more than 400 courses and 160 clubs, Finland has emerged as the European hotbed of flying discs. Most European courses are smaller than their North American counterparts, often because of a lack of available open land. Disc golf has also found its way across the Pacific: Japan, Taiwan, and South Korea have already sponsored tours, and players in Australia and New Zealand have also teed off with their favorite fairway or distance drivers. What's next? The sky's the limit. The Marco Polo program recently funded seed grants for courses in Israel, Honduras, and Barbados and disc golf, once played in a handful of North American city parks, may shape 21st-century landscapes worldwide.

▲ **Figure 3.4.1 Disc Golf in Finland** This Finnish disc-golf enthusiast aims at his target in one of that country's more than 400 outdoor courses.

1. Name another American sport that has "gone global," and identify two players from foreign countries who exemplify the pattern.

2. What sport do you enjoy playing or watching? Where did it originate and how did it spread to your community?

Geopolitical Framework: Patterns of Dominance and Division

In disarmingly simple fashion, North America's political geography brings together two of the world's largest countries. The creation of these political entities was the outcome of complex historical processes that might have created quite a different North American map. Once established, the two countries have coexisted in a close relationship of mutual economic and political interdependence. President John F. Kennedy summarized the links in a speech to the Canadian parliament in 1962: "Geography has made us neighbors, history has made us friends, economics has made us partners, and necessity has made us allies." That cozy continental relationship has not been without its tensions, and some persist today. In addition, both countries have had to deal with fundamental internal political complexities that have not only tested the limits of their federal structures, but also challenged their very existence as states.

Creating Political Space

The United States and Canada have very different political roots. The United States broke cleanly and violently from Great Britain. Canada, in contrast, was a country of convenience, born from a peaceful separation from Britain and then assembled as a collection of distinctive regional societies that only gradually acknowledged their common political destiny.

Europe imposed its own political boundaries on a future United States. The 13 English colonies, sensing their common destiny after 1750, united two decades later in the Revolutionary War. The Louisiana Purchase (1803) nearly doubled the national domain, and by the 1850s the remainder of the West had been added. The acquisition of Alaska (1867) and Hawaii (1898) rounded out what became the 50 states.

Canada was created under quite different circumstances. After the American Revolution, England's remaining territories in the region were controlled by British administrators, and in 1867, the British North America Act united the provinces of Ontario, Quebec, Nova Scotia, and New Brunswick in an independent Canadian Confederation. Soon, the Northwest Territories (1870), Manitoba (1870), British Columbia (1871),

▲ **Figure 3.34 St. Lawrence Seaway** This aerial view of the St. Lawrence Seaway shows a portion of the Welland Canal in Ontario that links Lakes Erie and Ontario.

Explore the **Sights** of the St. Lawrence Seaway

http://goo.gl/u0HP22

and Prince Edward Island (1873) joined this confederation, and the continental dimensions of the country took shape. Later infilling added Alberta, Saskatchewan, and Newfoundland. The creation of Nunavut Territory (1999) represents the latest change in Canada's political geography.

Continental Neighbors

Geopolitical relationships between Canada and the United States have always been close: Their common 5525-mile (8900-km) boundary requires both nations to pay close attention to one another. During the 20th century, the two countries lived largely in political harmony.

Sharing the Great Lakes The Great Lakes in particular have been a setting for remarkable continental cooperation. In 1909, the Boundary Waters Treaty created the International Joint Commission, an early step in the common regulation of cross-boundary issues involving Great Lakes water resources, transportation, and environmental quality. The St. Lawrence Seaway (1959) opened the Great Lakes region to better global trade connections (**Figure 3.34**). With the signing of the Great Lakes Water Quality Agreement (1972) and the U.S.–Canada Air Quality Agreement (1991), the two nations have joined more formally in cleaning

up Great Lakes pollution and in reducing acid rain in eastern North America. In 2012, these environmental agreements were updated, opening the way to new cooperative cleanup efforts.

Close Trading Connections Close political ties also have strengthened trade. The United States receives about three-quarters of Canada's exports and supplies almost two-thirds of its imports. Conversely, Canada accounts for roughly 20 percent of U.S. exports and 15 percent of its imports. A bilateral Free Trade Agreement, signed in 1989, paved the way five years later for the larger **North American Free Trade Agreement (NAFTA)**, which extended the alliance to Mexico. Paralleling the success of the European Union (EU), NAFTA has forged the world's largest trading bloc, including more than 480 million consumers and a huge free-trade zone that stretches from beyond the Arctic Circle to Latin America. In addition, the United States, Canada, and Mexico are all participants in the evolving Trans-Pacific Partnership (TPP) trade agreement, which should build trade between North America and East Asia.

Continuing Conflicts Political conflicts occasionally still divide North Americans (**Figure 3.35**). In addition to the intricate trans-boundary mingling of the Great Lakes, other regional water issues are common, because so many drainage systems cross the border. For example, Canada has protested North Dakota's plans to control the Red River (which flows northward into Manitoba), while Montana residents worry that Canadian logging and mining interests in British Columbia will increase pollution on the south-flowing North Flathead River. Long-standing agreements on dams within the shared Columbia Basin are also being renegotiated amid new demands by indigenous groups on both sides of the border for expanded salmon habitat.

More generally, tighter U.S. and Canadian regulations since 2009 have made it more difficult to cross the border in either direction. Reflecting security concerns in the United States, the world's longest "open border" now sees more surveillance drones and border agents than ever before. Persons crossing the border must now present a passport or other approved form of identification, just as they would on the border with Mexico.

Agricultural and natural resource issues occasionally cause controversy. The appearance of mad cow disease in Canadian livestock has curtailed exports to the United States and elsewhere. Problems have periodically developed when Canadian wheat and potato growers were accused of dumping their products into U.S. markets, thus depressing prices and profits for U.S. farmers. Similar issues have arisen in the logging industry, although a 2006 bilateral agreement has lessened tensions. The controversy that halted completion of the Keystone XL Pipeline has also frustrated various resource and environmental constituencies in both nations. In 2016, TransCanada, a Canadian company involved in the Keystone project, sued the U.S. government under the NAFTA treaty, claiming that the United States discriminated against the pipeline in an arbitrary manner.

Conflicting Claims. *Canada and the United States have made conflicting claims to resources and shipping lanes in the increasingly ice-free Arctic Ocean.*

An unpredictable Russian state

Keystone Pipeline. *Controversy over the Keystone XL Pipeline has resulted in a lawsuit between a Canadian energy company and the U.S. government.*

Commercial Lumbering. *U.S. loggers have protested British Columbia timber operations, arguing that the provincial government has given Canadian lumber companies special privileges and an edge in competing with their southern neighbors.*

Cooperation in Great Lakes commerce and environmental cleanup

An independent Quebec?

Dumping of Canadian wheat on U.S. markets

Undocumented immigration along U.S./Mexico border

Undocumented Immigration. *Undocumented immigration has fallen in recent years. Still, the issue emerged to dominate the U.S. presidential election in 2016.*

Improved U.S. /Cuba Relations. *Since 2014, the U.S. and Cuba have normalized political and economic ties.*

RUSSIA

ARCTIC OCEAN

GREENLAND (DENMARK)

Baffin Bay

ALASKA (U.S.)

Canadian Territory of Nunavut

C A N A D A

Hudson Bay

PACIFIC OCEAN

ATLANTIC OCEAN

U N I T E D

S T A T E S

Gulf of Mexico

MEXICO

CUBA

Puerto Rico

HAITI DOMINICAN REPUBLIC

0   250   500 Miles
0   250   500 Kilometers

▲ **Figure 3.35 Geopolitical Issues in North America** Although Canada and the United States share a long and peaceful border, many political issues still divide the two countries. In addition, internal political conflicts cause tensions, particularly in multicultural Canada.

In the far north, the two countries also disagree on the maritime boundary between the Yukon and the state of Alaska. In addition, the United States does not agree with the assertion that a potential Northwest Passage opening across a more ice-free Arctic Ocean would essentially be within Canada's territorial waters. Canadian officials are equally uneasy with the prospect of a warmer world in which "their" arctic waters become filled with unregulated oil tankers and cruise ships (**Figure 3.36**).

The Legacy of Federalism

The United States and Canada are **federal states**, meaning that both nations allocate considerable political power to units of government beneath the national level.

Other nations, such as France, have traditionally been **unitary states**, in which power is centralized at the national level. Federalism leaves many political decisions to local and regional governments and often allows distinctive cultural and political groups to be recognized as distinct entities within a country.

Both nations have federal constitutions, but their origins and evolution are very different. The U.S. Constitution (1787), created out of a violent struggle with the powerful British nation, specifically limited centralized authority, giving all unspecified powers to the states or the people. In contrast, the Canadian Constitution (1867), which created a federal parliamentary state, was an act of the British Parliament. Originally, it reserved most powers to central

▲ **Figure 3.36 Retreating Arctic Sea Ice** While annual sea ice levels can fluctuate substantially, the general trend has been for North America's polar region to become more ice free. **Q: In 2050, what do you think will be the key political and economic issues in the North American arctic?**

authorities and maintained many political links between Canada and the British Crown. Ironically, the evolution of the United States as a federal republic produced an increasingly powerful central government, whereas Canada's geopolitical balance of power shifted toward more provincial autonomy and a relatively weak national government. For example, the federal government largely controls U.S. public lands, but in Canada provincial authorities retain power over public Crown lands.

Quebec's Challenge The political status of Quebec remains a major issue within Canada (see Figure 3.35). Economic disparities between Anglo and French Canadians have reinforced cultural differences between the two groups, with the French Canadians often suffering when compared with their wealthier neighbors in Ontario. Beginning in the 1960s, a separatist political party in Quebec (the Parti Quebecois) increasingly voiced French Canadian concerns. When the party won provincial elections in 1976, it declared French the official language of Quebec. Formal provincial votes over the question of Quebec's independence were held in 1980 and 1995: Both measures failed. Since then, support for separation has ebbed in favor of a more modest strategy of increased "autonomy" within Canada.

Native Peoples and National Politics

Another challenge to federal political power in both countries comes from North American Indian and Inuit populations. Within the United States, Native Americans asserted their political power in the 1960s, marking a decisive turn away from policies of assimilation. Passage of the Indian Self-Determination and Education Assistance Act of 1975 has increased Native American control of their economic and political destiny. The Indian Gaming Regulatory Act (1988) offered potential economic independence for many tribes. Indian gaming operations (primarily gambling casinos) nationally netted tribes about $28 billion in 2015. In the western American interior, where Native Americans control roughly 20 percent of the land, tribes are also solidifying their hold on resources, reacquiring former reservation acreage, and participating in political interest groups, such as the Native American Fish and Wildlife Society and the Council of Energy Resource Tribes. In Alaska, native peoples acquired title to 44 million acres (18 million hectares) of land in 1971 under the Alaska Native Claims Settlement Act.

In Canada, ambitious challenges by native peoples have yielded dramatic results. Canada established the Native Claims Office in 1975. Agreements with native peoples in Quebec, Yukon, and British Columbia turned over millions of acres of land to aboriginal control and increased native participation in managing remaining public lands. By far the most ambitious agreement created Nunavut out of the eastern portion of the Northwest Territories in 1999 (see Figure 3.28). Home to 30,000 people (85 percent Inuit), Nunavut is the largest territorial/provincial unit in Canada. Its creation represents a new level of native self-government in North America. Agreements between the Parliament and British Columbia tribes (the Nisga'a) have made similar moves toward more native self-government (see Figure 3.28). A growing number of indigenous groups are also using this power to control the pace of natural resource development in the country. In 2016, 26 First Nations communities (Canadian indigenous groups other than Inuit or Metis), working with environmental groups, protected a Pacific Coast area (now called the Great Bear Rainforest) from commercial logging. Nearby, other coastal and inland First Nations groups collectively voiced strong opposition to the $11 billion Pacific Northwest Liquefied Natural Gas (LNG) project planned for Canada's west coast. The tribes cite studies that suggest the plant and related infrastructure could seriously harm the region's salmon fisheries.

The Politics of U.S. Immigration

Immigration policies are hotly contested in the United States, and four key issues are at the center of the debate. First, there are disagreements concerning how many legal immigrants should be allowed into the country. Some suggest that sharply reduced numbers protect American jobs and allow for gradual assimilation of existing foreigners, but others argue that looser restrictions could actually boost economic growth and business expansion.

▲ **Figure 3.37 International Border** North America's southwestern landscape is boldly divided by an increasingly hardened international border that separates the United States and Mexico. This view is south of San Diego, California.

A second major issue, particularly along the U.S.–Mexico border, is tightening daily flows of undocumented immigrants. While actual flows of undocumented workers have actually declined in recent years, the issue of "building a wall" on the southern border emerged once again in 2016 during the contentious U.S. presidential election won by Donald Trump. Many people also argue that the country's southern border is a national security issue. Federal legislation mandates increasing the number of border patrol agents, and more than 21,000 officers presently monitor the boundary. More than 700 miles (1,125 kilometers) of fencing also have been built or improved (Figure 3.37).

Third, U.S. relations with Mexico have soured with the growth of drug-related violence near the border between the two countries. Mexico remains the leading source of methamphetamine, heroin, and marijuana for the United States and is a key transit nation for northward-bound cocaine originating in South America. In addition, according to Human Rights Watch, more than 65,000 deaths, mostly in northern Mexico, were tied to the illegal drug business between 2006 and 2015. Violence has spilled north into places such as El Paso and Phoenix, causing U.S. officials to worry that Mexico's government has effectively lost control of its northern border.

Finally, there is no political consensus on a policy to deal with existing undocumented workers. While most are from Latin America, a growing number come from Asia, Africa, and Europe, often overstaying their tourist or student visas. In 2014, the Obama Administration issued a series of executive orders (bypassing Congress) that effectively delayed deportation for many undocumented immigrants and also offered a path to citizenship for immigrants who had been in the United States for at least five years and were parents of citizens or legal permanent residents. That policy remains mired in uncertainty, however, as it has been challenged by later court rulings and by the immigration policies of the Trump administration. While many Republicans have criticized the plan because it grants a path to amnesty for some undocumented residents, others felt it did not go far enough in addressing the needs of millions more who remain in the country and who desire U.S. citizenship.

A Global Reach

The geopolitical reach of the United States, in particular, has taken its influence far beyond the borders of the region. The Monroe Doctrine (1824) asserted that U.S. interests extended throughout the western hemisphere, and between 1890 and 1920, the United States became increasingly involved in both the Pacific and the Latin American regions.

World War II and its aftermath forever redefined the U.S. role in world affairs. The United States emerged from the conflict as the world's dominant political power. It also developed multinational political and military agreements, such as the North Atlantic Treaty Organization (NATO) and the Organization of American States (OAS). Conflicts in Korea (1950–1953) and Vietnam (1961–1975) pitted U.S. political interests against attempts to extend communist control beyond the Soviet Union and China. Even as the Cold War faded during the late 1980s, the global political reach of the United States expanded. Direct involvement in conflicts within Central America, the Middle East, Serbia, and Kosovo exemplified the country's global agenda, which has more recently extended into controversial wars in Iraq (2003–2011) and Afghanistan (2001–2017). The growing U.S. involvement against ISIL (Islamic State of Iraq and the Levant, also known as ISIS or IS) since 2014 demonstrates the continuing American presence in that troubled region. Tensions with Iran, Russia, and North Korea are also a reminder of how the global scene remains unpredictable, while normalized relations with Cuba since 2014 mark a profound shift from the Cold War era.

The geographic distribution of the U.S. military has also changed dramatically since 2000. Military planners argue that in the future, less emphasis should be given to housing large numbers of troops in relatively friendly foreign "hubs" such as Germany, South Korea, and Japan. A growing number of the U.S. military's 1.4 million active-duty personnel appear headed for more spartan assignments, located near source regions for terrorists and in arenas of potential conflict. From the drug- and rebel-filled jungles of Latin America to the terrorist training camps of the Philippines, areas of growing global deployment also include much of central and northern Africa, the Middle East, Central Asia, and Southeast Asia. China's growing military and political presence in East Asia and nearby portions of the western Pacific are also drawing American attention. Specifically, the United States has recently cited China's growing number of "excessive maritime claims" that have caused increased political tensions with several countries in the region. Some observers believe that China will emerge as an increasingly threatening military sea power,

Bosnian Refugees Reshape a St. Louis Neighborhood

How did the largest collection of Bosnian Americans in the United States end up in a rundown working-class neighborhood on the south side of St. Louis? The journey was long and tortuous, beginning with a dreadful war in southeastern Europe that consumed the former Yugoslavia for much of the 1990s. Bosnia-Herzegovina in particular witnessed massive genocide, and the "ethnic cleansing" practiced by multiple groups produced disruptive waves of desperate refugees. The first Bosnians arrived in St. Louis in 1993, relocated there under a federal refugee resettlement program. As the neighborhood grew and became known as "Little Bosnia," many more immigrants came in the late 1990s from elsewhere in Europe and from other U.S. cities. By 2015, the Bevo Mill community (named after an ornamental mill on a major commercial street) contained more than 70,000 Bosnian Americans (**Figure 3.5.1**).

The immigrants transformed the low-rent, crime-ridden district of abandoned buildings, older factories, and rundown housing into a vibrant and healthy ethnic enclave. Most immigrants learned English, many people created small businesses, and many more gradually invested their sweat equity to fix up their modest early 20th-century brick homes. Most of the existing residents welcomed the new arrivals, although there were a few issues with Bosnian families roasting whole lambs in backyard smokehouses. Overall, the Bosnians have thrived: their annual incomes of more than $80,000 are 25 percent higher than those of average St. Louis residents. Today, the neighborhood also has lower crime and unemployment rates than the city overall. While many immigrants are homesick, they welcome the opportunity to begin new lives in a setting where they can still enjoy elements of the old country.

What are the landscape signatures of this community? Perhaps the most visible sign of this largely Muslim population is an old branch bank building that has been converted into an Islamic Community Center with a 107-foot tall Turkish-style minaret added in 2007. Aside from the backyard smokehouses, other changes are fairly subtle. The main commercial avenue supports many businesses catering to the immigrant community. Grocery stores, restaurants, bars, and community centers often advertise with bilingual signs, and the colorful blue and gold colors of the flag of Bosnia and Herzegovina adorn many storefront windows. Many of these businesses, especially the local cafes serving Bosnian specialties such as stuffed cabbage, janjetina (lamb), and sausage, also cater to visitors from beyond the neighborhood. Women tend to dress in Western-style clothing on the street, rarely wearing a headscarf, but look carefully at the front porches of Bosnian homeowners: many residents leave their footwear at the door, an old custom adopted when Bosnians were part of the Ottoman Empire.

No surprise that the St. Louis mayor has welcomed these arrivals from southeastern Europe. In fact, he has been an outspoken advocate of resettling a newer generation of refugees from war-torn

Google Earth
Virtual **MG**
Tour Video

http://goo.gl/zSs92n

▲ Figure 3.5.1 **St. Louis Bosnian Community** This modern minaret stands outside an Islamic Community Center in the Bevo Mill community. The Ottoman-style tower was added in 2007, six years after the Community Center was established on the site of a bank branch building.

Syria in the city, no doubt figuring they might repeat the Bosnian miracle that has so impressively reinvigorated the Bevo Mill community.

1. What might be the three greatest challenges you would be confronted with as a new Bosnian immigrant to St. Louis?

2. In your community, can you identify landscape signatures of immigrant populations, either from the recent past or earlier?

---

challenging American dominance in the Pacific. U.S. defense expenditures of more than $600 billion in 2015 (nearly as much as the rest of the world combined) suggest the country will continue to play a highly visible role in global affairs.

Finally, both the United States and Canada continue to be destinations for political refugees, and questions over accepting these migrants always seems to provoke a great deal of discussion (and sometimes resistance), especially in the United States: Southeast Asia's "Boat People," periodic inflows of Cubans escaping Castro's rule, Bosnian refugees seeking new homes after the crisis in the former Yugoslavia, and most recently, Syrian refugees fleeing from that war-torn

region each demonstrate North America's enduring attraction in a politically unstable world (see *Exploring Global Connections: Bosnian Refugees Reshape a St. Louis Neighborhood*).

REVIEW

3.8 How do the political origins of the United States and Canada differ, and what issues divide these nations today?

3.9 What are the four key elements surrounding U.S. immigration policy?

■ **KEY TERMS** North American Free Trade Agreement (NAFTA), federal states, unitary states

Economic and Social Development: Patterns of Abundance and Affluence

Along with its global political clout, North America possesses the world's most powerful economy and its most affluent population. Its 360 million people consume huge quantities of global resources, but also produce some of the world's most sought-after manufactured goods and services. North America's size, geographic diversity, and resource abundance have all contributed to the region's global dominance in economic affairs. More than that, however, the region's human capital—the skills and diversity of its population—has enabled North Americans to achieve high levels of economic development (Table 3.2).

An Abundant Resource Base

North America is blessed with a varied storehouse of natural resources. The region's climatic and biological diversity, its soils and terrain, and its abundant energy, metals, and forest resources have provided a variety of raw materials for development. Indeed, the direct extraction of natural resources still makes up 3 percent of the U.S. economy and more than 6 percent of the Canadian economy. Some of these resources are then exported to global markets, while other raw materials are imported to the region.

Opportunities for Agriculture North Americans have created one of the most efficient food-producing systems in the world, and agriculture remains a dominant land use across much of the region (**Figure 3.38**). Farmers practice highly commercialized, mechanized, and specialized agriculture (**Figure 3.39**). The system also relies on efficient transportation and global agricultural markets. Today, agriculture employs only a small percentage of the labor force in both the United States (1 percent) and Canada (2 percent). At the same time, changes in farm ownership have sharply reduced the number of operating units, while average farm sizes have steadily risen.

The geography of North American farming represents the combined impacts of (1) diverse environments; (2) varied continental and global markets for food; (3) historical patterns of settlement and agricultural evolution; and (4) the role of **agribusiness**, or corporate farming. Agribusiness involves large-scale business enterprises that control closely integrated segments of food production, from farm to grocery store. In the Northeast, dairy operations and truck farms take advantage of proximity to major cities in Megalopolis and southern Canada. Corn and soybeans dominate the Midwest and western Ontario, where a tradition of mixed farming combines cultivation of feed grains with the production and fattening of livestock. To the south, only remnants of the old Cotton Belt remain, largely replaced by varied subtropical specialty crops; poultry, catfish, and livestock production; and commercial logging. Farther west, extensive, highly mechanized commercial grain-growing operations stretch from Kansas to Saskatchewan and Alberta, while irrigation offers opportunities for farming in Western North America, depending on surface and groundwater resources. Indeed, California, nourished particularly by large agribusiness operations in the irrigated Central Valley, accounts for more than 10 percent of the nation's agricultural output.

Large-scale commercial farming is also seeing some important consumer-driven shifts. The growing popularity of **sustainable agriculture** exemplifies the trend, where organic farming principles, limited use of chemicals, and an integrated plan of crop and livestock management combine to offer both producers and consumers environmentally friendly alternatives to large-scale commercial farming. In addition, the **locavore movement** has encouraged more community gardens and farmers markets, making it easier for consumers to eat locally grown food.

TABLE 3.2	Development Indicators									MasteringGeography™
Country	GNI per Capita PPP, 2014	GDP Average Annual %Growth 2009–15	Human Development Index (2015)[1]	Percent Population Living Below $3.10 a Day	Life Expectancy (2016)[2] Male	Female	Under Age 5 Mortality Rate (1990)	Under Age 5 Mortality Rate (2015)	Youth Literacy (% pop ages 15–24) (2005–2014)	Gender Inequality Index (2015)[3,1]
Canada	44,350	2.5	0.913	–	79	84	8	5	–	0.129
United States	55,900	2.2	0.915	–	76	81	11	7	–	0.280

Sources: World Bank, *World Development Indicators*, 2016.

[1]United Nations, *Human Development Report*, 2015.

[2]Population Reference Bureau, *World Population Data Sheet*, 2015.

[3]Gender Inequality Index—A composite measure reflecting inequality in achievements between women and men in three dimensions: reproductive health, empowerment, and the labor market that ranges between 0 and 1. The higher the number, the greater the inequality.

Login to MasteringGeography™ & access MapMaster to explore these data!

1) Compare GNI per Capita and Life Expectancy for these two countries. What might explain why Canadians earn less income but live longer?

2) While Gender Inequality Index values for both the United States and Canada are low by global standards, why might values for Canada suggest substantially less gender inequality in that country versus the United States?

▲ Figure 3.38 Major Agricultural and Manufacturing Activities of North America Varied environmental settings, settlement histories, and economic conditions have produced the modern map of North American agriculture and manufacturing. A growing number of major metropolitan areas play increasingly visible roles in the global economy.

Legend:
- Specialty crop or livestock farming
- Mixed farming
- Commercial wheat and other small grain farming
- Dairy farming
- General farming
- Livestock ranching
- Irrigated agriculture
- Mediterranean agriculture (with irrigation)
- Nonfarming
- Major manufacturing region
- City with key global links
- Other center of manufacturing

▼ Figure 3.39 Specialized Viticulture in California Specialized viticulture has revolutionized the agricultural economy in many portions of the American West. This commercial vineyard is in California's Napa Valley, north of San Francisco.

Energy and Industrial Raw Materials North Americans produce and consume huge quantities of other natural resources. The region consumes 40 percent more oil than all of the European Union. Key areas of oil and gas production are the Gulf Coast, the Central Interior (North Dakota became a major producing state after 2000), Alaska's North Slope, and Central Canada (especially Alberta's oil sands) (Figure 3.40). The most abundant fossil fuel in the United States is coal (27 percent of the world's total), but its relative importance in the overall energy economy has declined in the 21st century as industrial technologies changed (natural gas is cleaner and can also produce cheap electricity) and environmental concerns grew.

North America also remains a major producer of metals, although global competition, rising extraction costs, and environmental concerns pose challenges for this sector of the economy. Still, the region is endowed with more than 20 percent of the world's copper, lead, and zinc reserves, and it accounts for more than 20 percent of global gold, silver, and nickel production.

▲ **Figure 3.40 An Energy Landscape in the Permian Basin, West Texas** In addition to the visible signs of center-pivot irrigation evident in this West Texas landscape, note the numerous roads and development pads associated with oil drilling in this energy-rich portion of the Permian Basin.

Creating a Continental Economy

The timing of European settlement in North America was critical in its rapid economic transformation. The region's abundant resources came under the control of Europeans possessing new technologies that reshaped the landscape and reorganized its economy. By the 19th century, North Americans actively contributed to those technological changes. New natural resources were developed in the interior, and new immigrant populations arrived in large numbers. In the 20th century, although natural resources remained important, industrial innovations and more jobs in the service sector added to the economic base and extended the country's global reach.

Connectivity and Economic Growth North America's economic success was a function of its **connectivity**, or how well its different locations became linked with one another through improved transportation and communications networks. Those links greatly facilitated the interaction between locations and dramatically reduced the cost of moving people, products, and information, thereby laying the foundation for urbanization, industrialization, and the commercialization of agriculture.

Technological breakthroughs revolutionized North America's economic geography between 1830 and 1920. By 1860, more than 30,000 miles (48,000 kilometers) of railroad track had been laid in the United States, and the network grew to more than 250,000 miles (400,000 kilometers) by 1910. Farmers in the Midwest and Plains found ready markets for their products in cities hundreds of miles away. Industrialists collected raw materials from faraway places, processed them, and shipped manufactured goods to their final destinations. The telegraph brought similar changes to information: Long-distance messages flowed across eastern North America by the late 1840s, and 20 years later, undersea cables linked the

region to Europe, another milestone in the process of globalization.

Transportation and communications systems were modernized further after 1920. Automobiles, mechanized farm equipment, paved highways, commercial air links, national radio broadcasts, and dependable transcontinental telephone service reduced the cost of distance across North America. Perhaps most important, the region has taken the lead in the global information age, integrating computer, satellite, telecommunications, and Internet technologies in a web of connections that facilitates the flow of knowledge both within the region and beyond.

The Sectoral Transformation Changes in employment structure signaled North America's economic modernization just as surely as its increasingly interconnected society. The **sectoral transformation** refers to the evolution of a nation's labor force from one dependent on the *primary* sector (natural resource extraction) to one with more employment in the *secondary* (manufacturing or industrial), *tertiary* (services), and *quaternary* (information processing) sectors. For example, agricultural mechanization reduced demand for primary-sector workers but created new opportunities in the growing industrial sector. In the 20th century, new services (trade, retailing) and information-based activities (education, data processing, research) created other employment opportunities. Today, the tertiary and quaternary sectors employ more than 70 percent of the labor force in both Canada and the United States.

Regional Economic Patterns The locations of North America's industries show important regional patterns. **Location factors** are the varied influences that explain *why* an economic activity is located where it is and *how* patterns of economic activity are shaped. Patterns of industrial location illustrate the concept (see Figure 3.38). The historical manufacturing core includes Megalopolis (Boston, New York, Philadelphia, Baltimore, and Washington, DC), southern Ontario (Toronto and Hamilton), and the industrial Midwest. This core's proximity to *natural resources* (farmland, coal, and iron ore); its increasing *connectivity* (canals and railroad networks, highways, air traffic hubs, and telecommunications centers); its ready supply of *productive labor;* and a growing national, then global, *market demand* for its industrial goods encouraged continued *capital investment* within the area. Traditionally, the core dominated in steel, automobiles, machine tools, and agricultural equipment and played a key role in financial and insurance services.

In the last half of the 20th century, industrial- and service-sector growth shifted to the South and West. Cities of the South's Piedmont manufacturing belt (Greensboro to

▲ **Figure 3.41 Research Triangle Park, North Carolina** Located near Durham and created in 1959, North Carolina's Research Triangle Park is one of the leading centers of high-technology research and science in the Southeast.

North America and the Global Economy

Together with Europe and East Asia, North America plays a pivotal role in the global economy. In prosperous times, the region benefits from global economic growth, but in periods of international instability, globalization means that the region is more vulnerable to economic downturns. These links mean more to the region than abstract trade flow and foreign investment statistics. Increasingly, North American workers and localities find their futures directly tied to export markets in Latin America, the rise and fall of Asian imports, or the pattern of global investments in U.S. stock and bond markets.

The region is home to a growing number of truly "global cities" that serve as key connecting points and decision-making centers in the world economy (see Figure 3.38). New York City is the largest, although smaller metropolitan areas such as Toronto, Chicago, Los Angeles, and Seattle are pivotal urban players on the global stage. Centers such as Miami (links to Latin America) and Las Vegas (a global tourist destination) have carved out their own niches. The global status of these urban centers has profound local consequences. Life changes as thousands of new jobs are created; suburbs grow with a rush of new and ethnically diverse migrants; and a new, more cosmopolitan culture replaces older regional traditions.

The United States, with Canada's firm support, played a formative role in creating much of this new global economy and in shaping many of its key institutions.

Birmingham) grew after 1960, partly because lower labor costs and Sun Belt amenities attracted new investment. The North Carolina "research triangle" area, encompassing Raleigh, Durham, and Chapel Hill, has emerged as the nation's third largest biotech cluster, behind California and Massachusetts (**Figure 3.41**). The Gulf Coast industrial region is strongly tied to nearby fossil fuels that provide raw materials for the energy-refining and petrochemical industries.

The varied West Coast industrial region stretches from Vancouver, British Columbia, to San Diego, California (and beyond into northern Mexico), demonstrating the increasing importance of Pacific Basin trade. Large western aerospace firms also suggest the role of *government spending* as a location factor. Silicon Valley is now one of North America's leading regions of manufacturing exports. Its proximity to Stanford, Berkeley, and other universities demonstrates the importance of *access to innovation and research* for many fast-changing high-technology industries. Silicon Valley's location also shows the advantages of *agglomeration economies,* in which companies with similar, often integrated manufacturing operations cluster together. Smaller cities such as Provo, Utah, and Austin, Texas, specialize in high-technology industries and demonstrate the growing role of *lifestyle amenities* in shaping industrial location decisions, both for entrepreneurs and for skilled workers attracted to such amenities (**Figure 3.42**).

Explore the **Sights** of Provo, Utah

http://goo.gl/DtRfcQ

▼ **Figure 3.42 Provo, Utah** Located about 40 miles south of Salt Lake City, Provo, Utah has become a major center of technology-oriented businesses in the West. Its setting, nestled beneath the Wasatch Mountains, has attracted a large number of amenity migrants.

In 1944, allied nations met at Bretton Woods, New Hampshire, to discuss economic affairs. Under U.S. leadership, the group set up the International Monetary Fund (IMF) and the World Bank, giving these global organizations the responsibility for defending the world's monetary system and making key postwar investments in infrastructure. The United States also spurred the creation (1948) of the General Agreement on Tariffs and Trade (GATT). Renamed the **World Trade Organization (WTO)** in 1995, its 164 member states are dedicated to reducing global barriers to trade. The United States and Canada also participate in the **Group of Eight (G8)**—a collection of economically powerful countries (including Japan, Germany, Great Britain, France, Italy, and sometimes Russia)—which confers regularly on key global economic and political issues.

Patterns of Trade North America is prominent in both the sale and the purchase of goods and services within the international economy. Both countries import diverse products from many global sources, and the post-1990 growth of the Asian trade (particularly with China, South Korea, and southeast Asia) has fundamentally changed the North American economy.

Dominated overwhelmingly by the United States, Canada imports large quantities of manufactured parts, vehicles, computers, and foodstuffs. For the United States, imports continue to grow, creating a persistent global trade deficit for the country. Canada, Mexico, China, Japan, and world oil exporters supply the United States with a variety of raw materials, low-cost consumer goods, and high-quality vehicles and electronics products.

Outgoing trade flows suggest what North Americans produce most cheaply and efficiently. Canada's exports include large quantities of raw materials (energy, metals, grain, and wood products), but manufactured goods are becoming increasingly important, particularly in its pivotal trade with the United States. Since 1994, trade initiatives with Pacific Rim countries have offered Canadians new opportunities for export growth. The United States also enjoys many lucrative global economic ties, and its geography of exports reveals particularly strong links to other portions of the more developed world. Sales of automobiles, aircraft, computer and telecommunications equipment, entertainment, financial and tourism services, and food products contribute to the nation's flow of exports. Supporters of the Trans-Pacific Partnership in the United States and Canada argue that ratification of the new agreement will lead to even more dramatic growth in global trade, especially with nations in the Pacific and East Asia, but passage of the agreement in the United States remained highly doubtful following the election of Donald Trump in late 2016.

Patterns of Global Investment Patterns of capital investment and corporate power place North America at the center of global money flows and economic influence. The region's relative stability attracts huge inflows of foreign capital,

both as investments in North American stocks and bonds and as foreign direct investment (FDI) by international companies. For Canada, U.S. wealth and proximity mean that 80 percent of foreign-owned corporations in the country are based in the United States. In the United States, sustained economic growth and supportive government policies have encouraged large foreign investments, particularly since the late 1970s. Today, the United States is the world's largest destination for foreign investment, and the average annual pay of manufacturing-related jobs originating from these investments is more than $70,000, more than 30 percent higher than the average pay across the entire workforce.

Many immigrants also become urban entrepreneurs, generating another valuable form of foreign investment in both Canada and the United States (**Figure 3.43**). Whether it is the Chinese in Vancouver or the Cubans in Miami, immigrants in many of North America's largest, most global cities have made huge capital and human investments in their adopted communities. The economic consequences of these skilled people are enormous. William Frey, a demographer with the Brookings Institution, notes that "a lot of cities rely on immigration to prop up their housing market and prop up their economies." Almost 30 percent of the Korean-born and 20 percent of the Iranian-born populations in the United States are self-employed, a strong indicator of business ownership. One study of the country's 50 largest metropolitan areas found that immigrants owned 58 percent of dry cleaning and laundry businesses, 40 percent of motels, and 43 percent of liquor stores. A recent report issued by the Center for an Urban Future argues that "immigrants have been the entrepreneurial spark plugs of cities from New York to Los Angeles."

The impact of U.S. investments in foreign stock markets also suggests how outbound capital flows transform the way business is done throughout the world. Aging U.S. baby boomers have poured billions of pension fund and investment dollars into Japanese, European, and "emerging" stock markets such as Brazil, Russia, and China. In addition, U.S. investments

▼ **Figure 3.43 Immigrant Entrepreneurs** Minority and immigrant entrepreneurs have transformed the economic geographies of North American cities. This Miami woman owns Margarita Flowers, a thriving small business in this culturally diverse south Florida city.

in foreign countries flow through direct investments made by multinational corporations based in the United States.

However, the geography of 21st-century multinational corporations is changing, illustrating three recent shifts in broader patterns of globalization. These shifts have important consequences for North Americans. First, traditional U.S.-based multinational corporations are adopting a new, more globally integrated model. For example, IBM now has more than 150,000 employees in India.

Second, a growing array of multinational corporations based elsewhere in the world—especially in places such as China, India, Russia, and Latin America—is buying up companies and assets once controlled by North American or European capital. Brazilian investors, for example, spend billions annually on overseas assets (especially in North America). In Asia, China's Lenovo purchased IBM's huge personal computer business within that country. Similarly, Indian multinational companies are buying hundreds of foreign assets based in the more developed world. One sign of the times can be measured in the home country of the world's largest multinational companies. In 2015, the *Forbes* list of the "Global 2000" companies (largest as measured by their revenues, assets, profits, and market value) included 691 from Asia, 645 from North America, and 486 from Europe.

Third, many of these same multinational companies are making huge investments of their own in other portions of the less developed world, from Africa to Southeast Asia, bypassing North American control altogether. Today, more than one-third of FDI in emerging market nations comes from other emerging market nations. Simply put, the late-20th-century top-down model of multinational corporate control and investment, traditionally based in North America, Europe, and Japan, is being replaced by a more globally distributed model of corporate control. This new model has many origins, many destinations, and new patterns of labor, capital, production, and consumption.

North Americans have directly experienced the consequences of these shifts in global capitalism. For example, U.S. citizens are increasingly reacting to corporate **outsourcing**, a business practice that transfers portions of a company's production and service activities to lower-cost settings, often located overseas. In 2016, the five top countries ranked for their outsourcing potential included India, China, Malaysia, Brazil, and Indonesia.

In addition, millions of jobs in manufacturing, textiles, semiconductors, and electronics have effectively migrated to settings such as China, India, and Mexico, because those countries offer low-cost, less regulated settings for production, both for local and for foreign firms. The results are complex: North American consumers benefit from cheap imports, but may find their own jobs threatened by the corporate restructurings that make such bargains possible.

Enduring Social Issues

Profound economic and social problems shape the human geography of North America. Even with its continental wealth, great differences persist and have increased between rich and poor. High per capita incomes in the United States and Canada fail to reveal the differences in wealth within the two countries (see Table 3.2). Broader measures of social well-being suggest significant disparities in health care and education. Race, particularly within the United States, continues to be an issue of overwhelming importance. In addition, both nations face problems associated with gender inequity, education, poor diets, health, and aging populations. One consequence of globalization is that many of these economic and social challenges are increasingly defined beyond the region. Poverty in the rural American South may be related to low Asian wage rates, for example, and a viral outbreak in Hong Kong or Rio de Janeiro might be only a plane flight away from suburban Vancouver.

Wealth and Poverty   The regional landscape displays contrasting scenes of wealth and poverty. Elite northeastern suburbs, gated California neighborhoods, upscale shopping malls, and posh alpine ski resorts are all expressions of private and exclusive landscapes that characterize wealthier North American communities. In contrast, substandard housing, abandoned property, aging infrastructure, and unemployed workers are reminders of the gap between rich and poor. The problems of the rural poor remain major social issues in the Canadian Maritimes, Appalachia, the Deep South, the Southwest, and agricultural California. While many of the poorest Americans live in central cities, poverty is also moving to the suburbs. A 2013 Brookings Institution report noted that more poor people live in suburbs than in central cities, and that suburban poverty rates are growing much more rapidly than in either inner cities or rural areas. Overall, about 13–16 percent of the U.S. and Canadian populations live in poverty. In the United States, about 26 percent of the country's African American and 24 percent of the Hispanic populations live below the poverty line.

Food Deserts in a Land of Plenty   One revealing measure of poverty involves **food deserts**, where people do not have ready access to supermarkets and fresh, healthy, and affordable food. In many North American cities, for example, there are more than three times as many supermarkets in wealthier neighborhoods versus poorer areas. At the same time, unhealthy fast food restaurants are often most concentrated in low-income districts. Reduced availability of automobiles in poorer neighborhoods may also restrict access to affordable, healthy food options. The result is that urban poverty is often associated with poorer diets, more food-related diseases (such as diabetes and obesity), and relatively higher expenditures on expensive, unhealthy fast foods. The U.S. Agriculture Department considers an urban area a food desert if at least 20 percent of residents live in poverty and at least one-third of residents live more than a mile from a supermarket. A rural food desert is defined as a rural setting where residents must drive more than 10 miles to the nearest supermarket. More than 2.3 million people (often elderly and less mobile) are impacted in these settings.

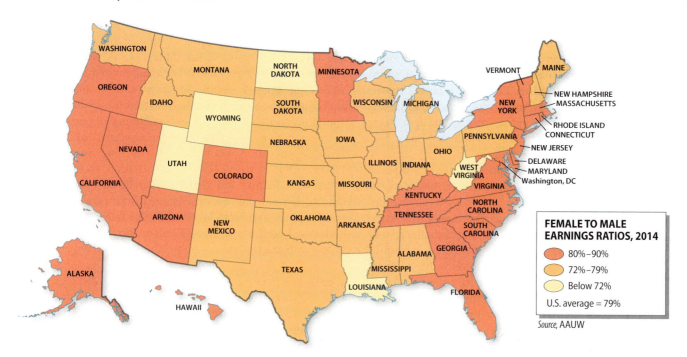

▲ **Figure 3.44 Earnings Ratios, by Gender, United States, 2014** This map shows the relative median annual earnings for women versus men, by state. Note the relatively higher earnings for women in the Northeast and in California versus selected southern and Midwestern settings. **Q: What variables might help explain how your own state fits into the larger national pattern?**

Food costs are often higher as rural residents depend on smaller, less affordable grocery stores or must travel far to obtain cheaper, healthier food. In these small-town and rural settings, even closing a single supermarket (or Walmart) can have a major impact on the surrounding area.

Access to Education Education is also a major public policy issue in Canada and the United States. Although political parties differ in their approach, most public officials agree that more investment in education can only improve North America's chances for competing successfully in the global marketplace. Canada has steadily improved its school drop-out rate since the early 1990s, and by 2015, U.S. high schools also reported their highest graduation rates (over 81 percent) in decades, although dramatically lower numbers in many poor urban and rural districts suggest ongoing challenges. In addition, race plays a key role: white Americans are two or three times more likely than African Americans or Hispanics to hold a college degree. Another challenge, particularly in the United States, is debating an effective national education policy in the face of a very strong tradition of local control of education.

Gender Equity Since World War II, both the United States and Canada have seen great improvements in the role that women play in society. However, the **gender gap** is yet to be closed when it comes to differences in salary, working conditions, and political power. Women comprise more than half of the North American workforce and are often more educated than men, but still earn only about 78 cents for

every dollar that men earn (**Figure 3.44**). Women also head the vast majority of poorer single-parent families in the United States, and more than 40 percent of all U.S. births are to unwed mothers. Canadian women, particularly single mothers working full-time, are also greatly disadvantaged, averaging only about 70 percent of the salaries of Canadian men.

Women voters have played critical roles in deciding recent national elections (they made up an essential component of President Obama's 2008 and 2012 presidential victories). In addition, Hillary Clinton's run for the presidency in 2016 signaled important changes for American politics, traditionally dominated by males. Still, political power in both nations remains largely in male hands. Canadian women have voted since 1918 and U.S. women since 1920, but females in the early 21st century remain significant minorities in both the Canadian Parliament (26 percent in 2016) and the U.S. Congress (19 percent in 2016).

Aging and Health-Care Issues Aging and other health-care issues are also key concerns within a region of graying baby boomers. A recent report on aging predicted that 20 percent of the U.S. population will be older than 65 by 2050. Today, the most elderly senior citizenry (age 85+) constitutes more than 12 percent of all seniors, and they are the fastest-growing part of the population. Poverty rates are also higher for seniors. With fewer young people to support their parents and grandparents, officials debate the merits of reforming social security programs. Whatever the outcome of such debates, the geographical consequences of aging are already abundantly clear. Whole sections of the United States—from Florida to southern Arizona—have become increasingly oriented around retirement. Communities cater to seniors with special assisted-living arrangements, health-care facilities, and recreational opportunities (Figure 3.45).

Health care remains a key issue in both countries. Both systems are costly by global standards: Canadians spend about 12 percent of their gross domestic product (GDP) on health care, and costs are even higher in the United States (over 15 percent of GDP). For several decades, Canada has offered an enviable system of government-subsidized universal health care to its residents (who pay higher taxes to fund it). In the United States, the Patient Protection and Affordable Care Act was signed into law in 2010 and has gradually moved that nation toward more universal coverage, still largely within a system of private insurers. Even with ongoing issues related to costs and access, the long life spans and low childhood mortality rates (see Table 3.2) of the region suggest that both countries reap many rewards from these modern health-care systems that offer the latest in high technology.

Still, many challenges remain. Hectic lives, often oriented around fast food, have contributed to rapidly growing rates of obesity in North America between 1975 and 2005. Caloric intake has since leveled off, but remains far above the world average. On a typical day, more than 30 percent of American children and adolescents report eating fast food. The long-term results are sobering: Almost two-thirds of adult Americans are overweight (over one-third are considered obese), contributing to higher rates of heart disease and diabetes. Combined with more sedentary lifestyles, the convenience-oriented diets that shape the everyday routine of millions of North Americans will no doubt add to the long-term cost of health care. Still, there are some bright spots, including declining rates of full-sugar soda consumption and a selective trend among more affluent and well-educated North Americans to eat more fresh and locally produced foods.

Chronic alcoholism and substance abuse are also widespread and costly. For example, about half of U.S. college students who drink engage in harmful binge drinking, more than 150,000 students develop alcohol-related health problems annually, and large numbers of students report being assaulted (about 700,000 per year) or sexually abused (about 100,000 per year) by other students who

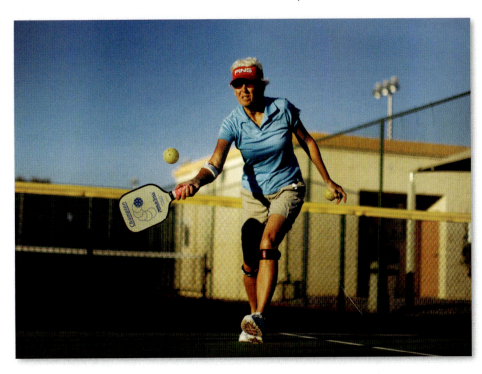

▲ **Figure 3.45 Retirement Communities** Almost 40,000 seniors live in Sun City, Arizona, near Phoenix. The retirement community, opened in 1959 by developer Del Webb, offers affordable housing, secure neighborhoods, and many recreational opportunities.

have been drinking. Similarly, an epidemic of drug-related overdoses and deaths (many from legal prescriptions of opioid painkillers) has hit many communities, including small towns and rural areas. Heroin addiction, once linked to poor inner city populations, has spread to the wealthier suburbs and to the Farm Belt in record numbers. In New Hampshire, for example, state treatment centers have seen an avalanche of new cases in the past decade (a 90 percent increase for heroin and a 500 percent increase for prescription opiate abuse), and there is simply not enough money or expertise to handle the deluge.

Another critical health-care issue has been the care and treatment of the region's 1.2 million HIV/AIDS victims. The price of the disease will be broadly borne in the 21st century, but particularly among poorer black (44 percent of AIDS cases in the United States) and Hispanic (19 percent) populations.

REVIEW

3.10 What is the sectoral transformation, and how does it help explain economic change in North America?

3.11 Cite five types of location factors, and illustrate each with examples from your local economy.

■ **KEY TERMS** agribusiness, sustainable agriculture, locavore movement, connectivity, sectoral transformation, location factors, World Trade Organization (WTO), Group of Eight (G8), outsourcing, food deserts, gender gap

Chapter 3 North America

Summary

- North America's affluence comes with a considerable price tag. Today the region's environmental challenges include air and water pollution, improving the efficiency of its energy economy, and adjusting to the realities of climate change.

- In a remarkably short time, a unique and changing mix of peoples from around the world radically disrupted indigenous populations and settled a huge, resource-rich continent that is now one of the world's most urbanized regions.

- North America is home to one of the world's most culturally diverse societies, and the region's contemporary popular culture has had an extraordinary impact on almost every corner of the globe.

- The region's two societies are closely intertwined, yet they face distinctive political and cultural issues.

- Canada's multicultural identity remains problematic, and it must deal with both the costs and benefits of living next door to its continental neighbor.

- For the United States, social and political challenges linked to its ethnic pluralism, immigration issues, and enduring poverty and racial discrimination remain central concerns, particularly in its largest cities.

Review Questions

1. Explain how "natural hazards" can be shaped by human history and settlement. In other words, what role do humans play in shaping the distribution of hazards?

2. How have the major North American migration flows since 1900 influenced contemporary settlement patterns, cultural geographies, and political issues within the region?

3. Summarize and map the ethnic background and migration history of your own family. How do these patterns parallel or depart from larger North American trends?

4. Describe the strengths and weaknesses of federalism and cite examples from both the United States and Canada.

5. The environmental price for North America's economic development has been steep. Suggest why it may or may not have been worth the price and defend your answer.

6. Who will be North America's leading trade partner in 2050? Explain the reasons for your answer.

Image Analysis

1. This chart shows annual immigration to the United States by region of origin. Note the sharp peaks clustered in Phase 3 and Phase 5. Which immigrant groups dominated immigration during these peak years? What common economic or cultural factors might explain both surges?

2. Which of the immigrant groups shown in the graphic make up important portions of your local community, and why did they settle there? Did they settle in your community as foreign-born immigrants or did they arrive in later generations?

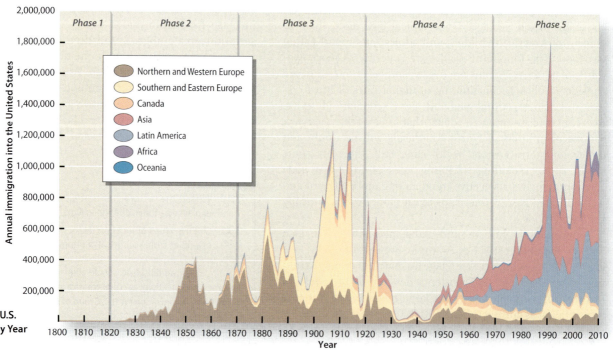

▶ Figure IA3 U.S. Immigration, by Year and Group

JOIN THE DEBATE

Fracking has revolutionized the American energy economy, but it is a highly controversial technology in which millions of gallons of water, sand, and other chemicals are injected deep into the earth. Are the economic benefits worth the potential costs?

Fracking has Brought an Amazing Set of Benefits to North America!

- Many shale-rich regions of the United States, including Texas, Oklahoma, Ohio, Pennsylvania, and North Dakota, enjoy economic growth as energy-related jobs are created and as landowners benefit from the leasing of their acreage to energy companies.

- Fracking has made the country less vulnerable to foreign energy producers, and someday the country may even be a major energy exporter.

- Fracking has dramatically lowered the cost of clean-burning natural gas, a boon to consumers and to the entire economy as it outcompetes older, dirtier coal-fired power plants.

The Costs of Fracking Far Outweigh the Benefits! It Should be Stopped Until We Know All the Long-Term Consequences.

- Fracking wells have a notoriously short life, and the drilling process at the site seriously impacts the environment.

- Much of the water injected into the shale formations remains there, forever removed from other uses. Furthermore, contaminated water from fracking waste pits has leached into groundwater and increased contaminants such as methane gas and benzene, causing serious health problems for nearby residents.

- In places such as Ohio and Oklahoma, fracking has been strongly connected to increased earthquake activity. We don't know the long-term geological consequences of this technology or who will pay for damages. We need strong, uniform federal standards.

▲ **Figure D3** **Fracking Rig in Butler Country, PA** Pennsylvania's energy-rich shales have been a major target of fracking operations such as this one in the western part of the state. While bringing new jobs and producing energy resources, the technology has also been fraught with environmental controversy.

KEY TERMS

acid rain *(p. 83)*
agribusiness *(p. 112)*
boreal forest *(p. 79)*
connectivity *(p. 114)*
cultural assimilation *(p. 95)*
cultural homeland *(p. 99)*
edge city *(p. 92)*
ethnicity *(p. 95)*
federal states *(p. 108)*
food deserts *(p. 117)*
fracking *(p. 86)*
gender gap *(p. 118)*
gentrification *(p. 93)*
Group of Eight (G8) *(p. 116)*
location factors *(p. 114)*
locavore movement *(p. 112)*
Megalopolis *(p. 88)*

new urbanism *(p. 93)*
nonmetropolitan growth *(p. 91)*
North American Free Trade
 Agreement (NAFTA) *(p. 107)*
outsourcing *(p. 117)*
postindustrial economy *(p. 76)*
prairie *(p. 79)*
renewable energy sources *(p. 84)*
sectoral transformation *(p. 114)*
Spanglish *(p. 103)*
sustainable agriculture *(p. 112)*
Trans-Pacific Partnership (TPP)
 (p. 76)
tundra *(p. 79)*
unitary states *(p. 108)*
urban decentralization *(p. 92)*
urban heat island *(p. 83)*
World Trade Organization (WTO)
 (p. 116)

MasteringGeography™

Looking for additional review and test prep materials? Visit the Study Area in **MasteringGeography**™ to enhance your geographic literacy, spatial reasoning skills, and understanding of this chapter's content by accessing a variety of resources, including **MapMaster** interactive maps, videos, *In the News* RSS feeds, flashcards, web links, self-study quizzes, and an eText version of *Diversity Amid Globalization*.

DATA ANALYSIS

http://goo.gl/GJMDyY

Every decade, the Census Bureau gathers and summarizes an enormous amount of data for the United States. These data are used by planners and government agencies to forecast future needs for public infrastructure and social services. Age and sex distributions for cities and states can provide real insights into the social and economic characteristics of these settings. Population pyramids are convenient ways to visualize these characteristics (see Figure 3.13). Go to the Census Bureau's website (www.census.gov) and access the summaries and predictions of state populations.

1. Examine the 2010 and 2030 (projected) pyramids for Florida and Utah. Describe major similarities and differences for both years. Write a paragraph that summarizes reasons for these differences.

2. Select two additional states that display quite different population structures. Write a paragraph that summarizes and explains these differences.

3. From the point of view of a planner or budget expert, explain how the different population structures in the states you selected might impact future expenditures and trends in economic development in 2030 and beyond.

Authors' Blogs

Scan to visit the
GeoCurrents blog
http://www.geocurrents.
info/category/place/
north-america

Scan to visit the
Author's Blog
for field notes, media resources, and chapter updates
https://gad4blog.wordpress.
com/category/north-america

Physical Geography and Environmental Issues

Tropical ecosystems in Latin America, are one of the planet's greatest reserves of biological diversity. A critical question is how to manage this diversity in the face of increasing pressure to extract mineral wealth, build roads and dams, and convert forests into farms or pasture.

Population and Settlement

Latin America is the most urbanized region of the developing world, with 80 percent of the population living in cities. Four megacities (>10 million people) are found here. Yet it is also a region with high rates of emigration, especially to North America.

Cultural Coherence and Diversity

Amerindian activism is on the rise in Latin America. Indigenous people from Central America to the Andes and the Amazon demand territorial and cultural recognition while Latin America is on a global stage due to Brazil hosting the 2016 Olympic Games.

Geopolitical Framework

As Latin American governments mark 200 years of independence, most are fully democratic. A new trade group, the Pacific Alliance, has joined Mercosur and CAFTA in reshaping political and economic ties in the region. Meanwhile, heightened violence and insecurity, especially in Central America, have more people living in fear and trying to leave.

Economic and Social Development

This region falls in the middle-income category, but economic downturns have slowed growth and serious income inequality persists. Government programs such as Brazil's Bolsa Familia have sought to address both social and economic development for poor families.

▶ A monumental infrastructure project completed in 2016, the expansion of the Panama Canal with a new set of locks will increase the number and size of ships that use this passage. On the left side of the image, the new far larger locks. To the right are the 100-year-old Gatun Locks that smaller ships can still use.

Colón, Panama

LATIN AMERICA

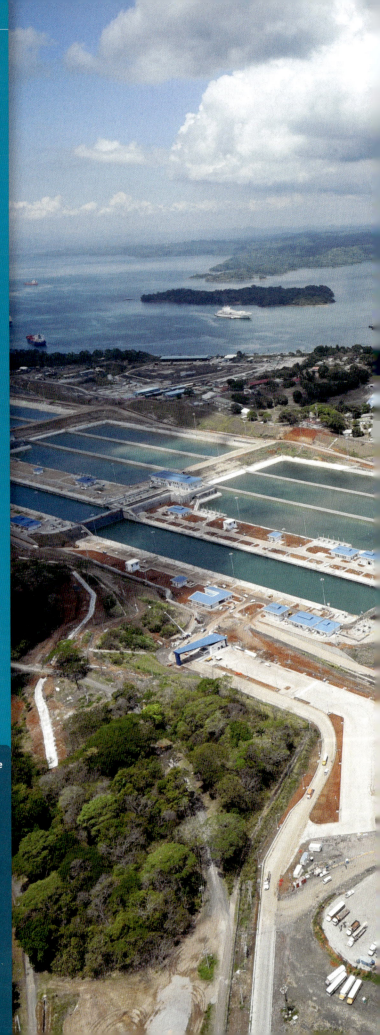

4

Latin America

A powerful symbol of Latin America's role in global trade—and of U.S. involvement in the region—is the Panama Canal. Built by the United States, it opened in 1914. Since 2000, the canal has been controlled and managed by Panamanians when the U.S.-controlled Canal Zone was returned to Panama. In 2016, a new set of larger locks opened, allowing for the passage of more and much larger ships. The size of the old locks limited access to vessels no more than 100 feet wide and 1000 feet long, the so-called Panamax vessels. The Panama Canal Authority that manages the canal estimates that the expansion will increase total cargo levels by 3 percent a year. Ports around North America and Europe have also retrofitted to adjust for the arrival of larger post-Panamax ships. The timing of this expansion is especially important, considering potential competition from Arctic sea routes opening up as a consequence of climate change.

Just 400 miles north, in Nicaragua, a new project is underway, financed by Chinese billionaire Wang Jing in close cooperation with the Nicaraguan government. In 2014, the Nicaraguan Canal was approved, and the Hong Kong Nicaraguan Canal Development Investment (HKND) Group was formed. It was estimated that canal construction would cost US$50 billion and span five years, but many experts argued that the cost and time for construction would be far more. In late 2015, Wang ran into financial troubles, having lost much of his fortune in the volatile Chinese stock market. Moreover, the environmental effects of canal construction for Lake Nicaragua and the country's coastal zones, tropical forests, and rural communities are thought to be devastating. The economic justification for a second canal is questionable, and many doubt that the project will be completed. But the political and economic power associated with an alternative canal route through the Americas has proven to be both irresistible and daunting.

Defining Latin America

The concept of Latin America as a distinct region has been popularly accepted for nearly a century. The boundaries of this region are straightforward, beginning at the Rio Grande (called the *Rio Bravo* in Mexico) and ending at Tierra del Fuego (**Figure 4.1**). French geographers are credited with coining the term *Latin America* in the 19th century to distinguish the Spanish- and Portuguese-speaking republics of the Americas plus Haiti from the English-speaking territories. There is nothing particularly "Latin" about the area, other than the predominance of Romance languages. The term stuck because it was vague enough to be inclusive of different colonial histories, while also offering a clear cultural boundary from Anglo-America, the region called North America in this book.

This chapter describes the Spanish- and Portuguese-speaking countries of Central and South America, including Mexico. This division emphasizes the important Indian and Iberian influences on mainland Latin America while separating it from the unique colonial and demographic history of the Caribbean and the Guianas, discussed in Chapter 5.

Roughly equal in area to North America, Latin America has a much larger and faster-growing population of nearly 600 million. Its most populous state, Brazil, has 206 million people, making it the world's fifth largest country by population; the next largest state, Mexico, has a population of nearly 130 million, making it the tenth largest.

Through colonialism, immigration, and trade, the forces of globalization are embedded in the Latin American landscape (**Figure 4.2**). The early Spanish Empire focused on extracting precious metals, sending galleons laden with silver and gold across the Atlantic. The Portuguese became prominent producers of dyewoods, sugar products, gold and, later, coffee. By the late 19th and early 20th centuries, exports to North America and Europe fueled the region's economy. Most countries specialized in one or two products: bananas and coffee, meats and wool, wheat and corn, petroleum and copper. Since then, Latin American states have industrialized and diversified their production, but continue to be major producers of primary goods for North America, Europe, and East Asia.

Today, neoliberal policies that encourage foreign investment, export production, and privatization have been adopted by many states. Extractive industries continue to prevail, in part because of the area's impressive natural resources. Latin America is home to Earth's largest rainforest, the greatest river by volume, and massive reserves of natural gas, oil, gold, and copper. It is also a major exporter of grains, especially soy. With its vast territory, tropical location, and relatively low population density—half the population of India in nearly seven times the area—Latin America is also recognized as one of the world's great reserves of biological diversity. How this diversity will be managed in the face of global demand for natural resources is an increasingly important question for the countries of this region.

> Latin American countries have industrialized and diversified their production, but continue to be major producers of primary goods for North America, Europe, and East Asia

LEARNING Objectives
After reading this chapter you should be able to:

4.1 Explain the relationships among elevation, climate, and agricultural production, especially in tropical highland areas.

4.2 Identify the major environmental issues of Latin America and how countries are addressing them.

4.3 Summarize the demographic issues impacting this region, such as rural-to-urban migration, urbanization, smaller families, and emigration.

4.4 Describe the cultural mixing of European and Amerindian groups in this region and indicate where Amerindian cultures thrive today.

4.5 Explain the global reach of Latino culture through immigration, sport, music, and television.

4.6 Describe the Iberian colonization of the region and how it affected the formation of today's modern states.

4.7 Identify the major trade blocs in Latin America and how they are influencing development.

4.8 Summarize the significance of primary exports from Latin America, especially agricultural commodities, minerals, wood products, and fossil fuels.

4.9 Describe the neoliberal economic reforms that have been applied to Latin America and how they have influenced the region's development.

LATIN AMERICA
Political & Physical Map

⊛ ● Metropolitan areas more than 20 million
⊛ ● Metropolitan areas 10–20 million
⊛ • Metropolitan areas 5–9.9 million
⊛ • Metropolitan areas 1–4.9 million
⊛ ∘ Selected smaller metropolitan areas
– – – Plate boundaries

ELEVATION IN METERS

4000+
2000–4000
500–1999
200–499
0–199
Below sea level
Sea Level

▶ **Figure 4.1 Latin America** Roughly equal in size to North America, Latin America supports a larger population and far greater ecological diversity. The 17 countries of this region share a history of Iberian colonization. Four-fifths of the region's 600 million people live in cities, making it the most urbanized region of the developing world. In addition to Mexico and Brazil, subregions include Central America, the Andean states, and the Southern Cone. The region is noted for its production of primary exports and manufactured goods, although rates of economic development vary greatly among states.

▲ **Figure 4.2 Spanish Colonial Influence** Throughout Latin America, the Iberian influence is still seen in towns and cities established by Spain and Portugal. Typically on a grid pattern, the center is a market plaza with nearby church and municipal buildings. This is the village of Mucuchies, Venezuela, established by the Spanish in the 17th century.

Physical Geography and Environmental Issues: Neotropical Diversity and Urban Degradation

Much of this world region is characterized by its tropicality. Travel posters of Latin America showcase verdant forests and brightly colored parrots. The diversity and uniqueness of the **neotropics** (tropical ecosystems in the Western Hemisphere) have long been attractive to naturalists eager to understand their distinct flora and fauna (**Figure 4.3**). It is no accident that Charles Darwin's theory of evolution was inspired by his two-year journey in tropical America. Even today, scientists throughout the region work to understand complex ecosystems, discover and protect new species, conserve genetic resources, and interpret the impact of human settlement, especially in neotropical forests.

Not all of the region is tropical. Important population centers lie below the Tropic of Capricorn—most notably Buenos Aires, Argentina, and Santiago, Chile. Much of northern Mexico, including the city of Monterrey, is north of the Tropic of Cancer. Highlands and deserts exist throughout the region. Yet Latin America's tropical climate and vegetation define the region's image. Given its large size and relatively low population density, Latin America has not experienced the same levels of environmental degradation witnessed in East Asia and Europe. Huge areas remain relatively untouched, supporting an incredible diversity of plant and animal life. Throughout the region, national parks offer some protection to unique plant and animal communities. A growing environmental movement in countries such as Costa Rica and Brazil has yielded both popular and political support for "green" initiatives. In short, Latin Americans have entered the 21st century with a real opportunity to avoid many of the environmental mistakes seen in

other world regions. At the same time, global market forces are driving governments to exploit minerals, fossil fuels, forests, shorelines, transportation routes, and soils. The region's biggest resource management challenge is to balance the economic benefits of extraction with the principles of sustainable development. Another major challenge is to improve the environmental quality of Latin American cities.

Western Mountains and Eastern Shields

Latin America is a region of diverse landforms, including high mountains and extensive upland plateaus. The movement of tectonic plates explains much of the region's basic topography, including the formation of its geologically young and tectonically active western mountain ranges (see Figure 4.1). In contrast, the Atlantic side of South America is characterized by humid lowlands interspersed with large upland plateaus called **shields**. Across these lowlands meander some of the great rivers of the world, including the Amazon.

Historically, the most important areas of settlement in tropical Latin America were not along the region's major rivers, but across its shields, plateaus, and fertile intermontane basins. In these localities, the combination of arable land, mild climate, and sufficient rainfall produced the region's most productive agricultural areas and its densest settlements. The Mexican Plateau, for example, is a massive upland area ringed by the Sierra Madre Mountains, with the Valley of Mexico located at the plateau's southern end. Similarly, the elevated and well-watered basins of Brazil's southern mountains provide an ideal setting for agriculture. These especially fertile areas are able to support high population densities, so it is not surprising that the region's two largest cities, Mexico City and São Paulo, emerged in these settings.

The Andes From northwestern Venezuela and ending at Tierra del Fuego, the Andes are relatively young mountains

▼ **Figure 4.3 Tropical Flora** The Osa Peninsula in Costa Rica is one of the most biodiverse places on the planet. Protected rainforest, notably in Corcovado National Park, is home to scarlet macaws, jaguars, tapir, and squirrel monkeys in addition to a vast variety of flora. With lowland tropical forest reaching to the Pacific Ocean, it is also a popular destination for tourists.

Mobile Field Trip: Cloud Forest MG

https://goo.gl/wxHN1z

▲ **Figure 4.4 Altiplano** Straddling the Bolivian and Peruvian Andes is an elevated plateau, the Altiplano. This high and windswept land is home to many Amerindian peoples and native species such as the llama and alpaca.

The Uplands of Mexico and Central America The Mexican Plateau and the Volcanic Axis of Central America are the most important Latin American uplands in terms of long-term settlement. Most major cities of Mexico and Central America are located here. The Mexican Plateau is a large, tilted block that has its highest elevations, about 8000 feet (2500 meters), in the south around Mexico City, and its lowest, just 4000 feet (1200 meters), at Ciudad Juárez. The southern end of the plateau, the Mesa Central, contains several flat-bottomed basins interspersed with volcanic peaks that have long been significant areas for agricultural production (**Figure 4.5**). It also contains Mexico's megalopolis—a concentration of the largest population centers, such as Mexico City and Puebla.

Along Central America's Pacific coast lies the Volcanic Axis, a chain of volcanoes that stretches from Guatemala to Costa Rica. It is a handsome landscape of rolling green hills, elevated basins with sparkling lakes, and conical volcanic peaks. More than 40 volcanoes, many still active, have produced a rich volcanic soil that yields a wide variety of domestic and export crops. Most of Central America's population is also concentrated in this zone, in the capital cities or surrounding rural villages. The bulk of the agricultural land is tied up in large holdings that produce beef, cotton, and coffee for export. However, in terms of numbers, most of the farms are small subsistence properties that produced corn, beans, squash, and assorted fruits.

The Shields South America has three major shields—large upland areas of exposed crystalline rock that are similar to upland plateaus found in Africa and Australia.

that extend nearly 5000 miles (8000 km). They are an ecologically and geologically complex mountain chain, with some 30 peaks higher than 20,000 feet (6000 meters). Created by the collision of oceanic and continental plates, the Andes are a series of folded and faulted sedimentary rocks with intrusions of crystalline and volcanic rock. Many rich veins of precious metals and minerals are found in these mountains. In fact, the initial economic wealth of many Andean countries came from mining silver, gold, tin, copper, and iron.

The Andes are still forming, so active volcanism and regular earthquakes are common in this zone. For example, in 2015 there were volcanic eruptions of Villarica and Calbuco in Chile, along with an 8.3M earthquake. In Ecuador, a 7.8M quake along the Nazca plate struck near the coastal city of Muisne in 2016, killing over 600 people and injuring thousands.

The lengthy Andean chain is typically divided into northern, central, and southern components. In Colombia, the northern Andes actually split into three distinct mountain ranges before merging near the border with Ecuador. High-altitude plateaus and snow-covered peaks distinguish the central Andes of Ecuador, Peru, and Bolivia. The Andes reach their greatest width here. Of special interest is the treeless high plain of Peru and Bolivia, called the **Altiplano**. The floor of this elevated plateau ranges from 11,800 feet (3600 meters) to 13,000 feet (4000 meters) in altitude, and it has limited usefulness for grazing. Two high-altitude lakes, Titicaca on the Peruvian and Bolivian border and the smaller Poopó in Bolivia, are located in the Altiplano, as are many mining sites (**Figure 4.4**). The southern Andes are shared by Chile and Argentina. The highest peaks of the Andes are found in the southern Andes, including the highest peak in the Western Hemisphere, Aconcagua, at almost 23,000 feet (7000 meters).

▼ **Figure 4.5 Mexico's Mesa Central** Mexico's elevated central plateau has long been the demographic and agricultural core of the country. This image shows a variety of agave grown in Jalisco that is used for the country's tequila production. Tequila, a traditional drink in Mexico, has a growing export market.

▶ **Figure 4.6 South American River Basins and Dams** The three great river basins of the region are the Amazon, La Plata, and Orinoco. The Amazon Basin covers 6 million square kilometers, including portions of eight countries, but the majority of the basin is within Brazil. The Amazon is the world's largest river system in terms of volume of water and area. It is currently experiencing a boom in dam construction to provide for Brazil's energy needs. The Plata Basin drains nearly 3 million square kilometers across five countries and is intensely farmed. It also contains Latin America's largest hydroelectric dam, Itaipú. The Orinoco Basin is shared by Venezuela and Colombia, covering nearly 1 million square kilometers. **Q: What are the benefits and costs of reliance upon hydroelectricity?**

Countries	Area of basin km²	%
Venezuela	639,000	77.00
Colombia	191,000	23.00
Total	830,000	

Countries	Area of basin km²	%
Brazil	3,672,600	62.61
Peru	974,600	16.61
Bolivia	684,400	11.67
Colombia	353,000	6.02
Ecuador	137,800	2.35
Venezuela	35,500	0.66
Guyana	5,200	0.09
Suriname	20	0.00
Total	5,866,100	

Countries	Area of basin km²	%
Brazil	1,379,300	46.69
Argentina	817,900	27.68
Paraguay	400,100	13.54
Bolivia	245,100	8.30
Uruguay	111,600	3.37
Total	2,954,500	

↘ Major dam
↗ Dam under construction

(Two are discussed here, while the third, the Guiana shield, is described in Chapter 5.) The Brazilian shield is larger and more important in terms of natural resources and settlement. Far from a uniform land surface, the Brazilian shield covers much of Brazil from the Amazon Basin in the north to the Plata Basin in the south. In the southeast corner of the plateau is São Paulo, the largest urban conglomeration in South America. The other major population centers are on the coastal edge of the plateau, where large protected bays made the sites of Rio de Janeiro and Salvador attractive to Portuguese colonists. Finally, the Paraná basalt plateau on the southern end of the Brazilian Shield is famous for its fertile red soils (*terra roxa*), which yield coffee, oranges, and soybeans. So fertile is this area that the economic rise of São Paulo is attributed to the expansion of commercial agriculture, especially coffee, into this area.

The vast low-lying Patagonian shield lies in the southern tip of South America. Beginning south of Bahia Blanca and extending to Tierra del Fuego, the region to this day remains sparsely settled and hauntingly beautiful. It is treeless, covered by scrubby steppe vegetation, and home to wildlife such as the condor and guanaco. Sheep were introduced to Patagonia in the late 19th century, spurring a wool boom. More recently, offshore oil production has renewed the economic importance of this area.

River Basins and Lowlands

Three great river basins drain the Atlantic lowlands of South America: the Amazon, La Plata, and Orinoco (**Figure 4.6**). Within these basins are vast interior lowlands, less than 600 feet (200 meters) in elevation, which lie over young sedimentary rock. From north to south, they are the Llanos, the Amazon lowlands, the Pantanal, the Chaco, and the Pampas. With the exception of the Pampas, most of these lowlands are sparsely settled and offer limited agricultural potential except for grazing livestock. Yet the pressure to open new areas for settlement and to exploit natural resources has created pockets of intense economic activity in the lowlands. Areas such as the Amazon and the Chaco have witnessed marked increases in resource extraction, soy cultivation, dam construction, and settlement since the 1970s.

Amazon Basin The Amazon drains an area of roughly 2.4 million square miles (6.6 million square km), making it the largest river system in the world by volume and area and the second longest by length. Everywhere in the basin, annual rainfall is more than 60 inches (150 cm), and in many places exceeds 80 inches (200 cm). The basin's largest city is Belém.

The mighty Amazon drains eight countries, but two-thirds of the watershed is within Brazil. Active settlement of the Brazilian portion of the Amazon since the 1960s has boosted the population. Today some 34 million people live in the Amazon Basin, which is equal to 8 percent of the total population in South America. The basin's development—most notably through towns, roads, dams, farms, and mines—is forever changing what was viewed as a vast tropical wilderness just a half century ago.

The Brazilian government has plans to build up to 30 new dams in its portion of the Amazon to meet the country's growing energy demands and to facilitate resource extraction in this region. Perhaps the most contested dam is Belo Monte on the Xingu River (a tributary of the Amazon) (**Figure 4.7**). When completed, Belo Monte will be the third largest hydroelectric dam in the world, generating more than 11,000 megawatts of electricity. It includes two dams, two

canals, two reservoirs, and a system of dikes. Yet the dam has its critics, as a relatively pristine river will be radically transformed, up to 20,000 people will be displaced, the river will be diverted, and forest will be flooded. The Belo Monte project has galvanized both domestic and international movements to stop construction. The project has experienced setbacks and court-ordered stoppages, but as of 2016 the construction continues.

La Plata Basin The region's second largest watershed begins in the tropics and discharges into the Atlantic in the midlatitudes near Buenos Aires. Several major rivers make up this system: the Paraná, the Paraguay, and the Uruguay.

Unlike the Amazon, much of the La Plata Basin is now economically productive through large-scale mechanized agriculture, especially soybean production. Arid areas such as the Chaco and inundated lowlands such as the Pantanal support livestock. The Plata Basin contains several major dams, including Latin America's largest hydroelectric plant, the Itaipú Dam on the Paraná, which generates electricity for all of Paraguay and much of southern Brazil. Only China's Three Gorges Dam and hydroelectric plant is larger. As agricultural output in this watershed grows, sections of the Paraná River are being canalized and dredged to enhance the river's capacity for barge and boat traffic.

Orinoco Basin The third largest river basin by area is the Orinoco in northern South America. Although it is only one-seventh the size of the Amazon watershed, its discharge roughly equals that of the Mississippi River. The Orinoco River meanders through much of southern Venezuela and part of eastern Colombia, giving character to the sparsely settled tropical grasslands called the *Llanos*. Since the colonial era, these grasslands have supported large cattle ranches. Although cattle are still important, the Llanos are also a dynamic area of petroleum production for both Colombia and Venezuela.

Climate and Climate Change in Latin America

In tropical Latin America, average monthly temperatures in settings such as Managua (Nicaragua), Quito (Ecuador), or Manaus (Brazil) show little variation (**Figure 4.8**). Precipitation patterns do vary, however, and create distinct wet and dry seasons. In Managua, for example, January is typically a dry month, and June is a wet one. The tropical lowlands of Latin America, especially east of the Andes, are usually classified as tropical humid climates that support forest or savanna, depending on the amount of rainfall. The region's desert climates are found along the Pacific coasts of Peru and Chile, Patagonia, northern Mexico, and the Bahia of Brazil. Because of the extreme aridity of the Peruvian coast, a city such as Lima, Peru, which is clearly in the tropics, averages only 1.5 inches (4 cm) of annual rainfall. Some sections of the Atacama Desert of Chile get no measurable

▼ **Figure 4.7 Belo Monte Dam** Currently the largest infrastructure project in Brazil, when completed it will be the third largest hydroelectric dam in the world. Critics claim it will do irreparable damage to the Amazonian forest systems and indigenous communities near Altamira, in Para state.

Explore the **Sights** of Amazon Basin along Xingu River

http://goo.gl/VJA9gi

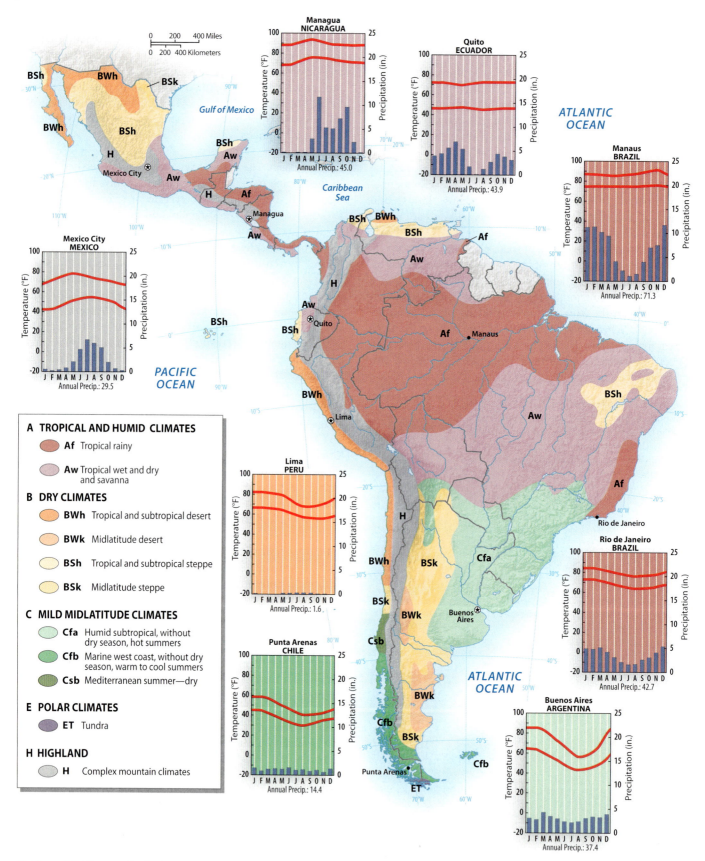

▲ **Figure 4.8 Climate of Latin America** Latin America includes the world's largest rainforest (Af) and driest desert (BWh), as well as nearly every other climate classification. Latitude, elevation, and rainfall play important roles in determining the region's climates. Note the contrast in rainfall patterns between humid Quito and arid Lima.

rainfall (**Figure 4.9**). Yet the discovery of resources such as nitrates in the 19th century and copper in the 20th century made this hyper-arid region a source of conflict among Chile, Bolivia, and Peru.

Midlatitude climates, with hot summers and cold winters, prevail in Argentina, Uruguay, and parts of Paraguay and Chile (see the climographs for Buenos Aires and Punta Arenas in Figure 4.8). Recall that the midlatitude temperature shifts in the Southern Hemisphere are the opposite of those in the Northern Hemisphere (cold Julys and warm Januarys). In the mountain ranges, complex climate patterns result from changes in elevation. To appreciate how humans adapt to tropical mountain ecosystems, it is important to understand the concept of **altitudinal zonation**—the relationship between cooler temperatures at higher elevations and changes in vegetation.

Altitudinal Zonation First described in the scientific literature by Prussian naturalist Alexander von Humboldt in the early 1800s, altitudinal zonation has practical applications that are intimately understood by all the region's native inhabitants. Humboldt systematically recorded declines in air temperature as he ascended to higher elevations, a phenomenon known as the **environmental lapse rate**. According to Humboldt's description of the environmental lapse rate, temperature declines approximately 3.5°F for every 1000 feet in elevation, or 6.5°C for every 1000 meters. Humboldt also noted changes in vegetation by elevation, demonstrating that plant communities common to the midlatitudes could thrive in the tropics at higher elevations. These different altitudinal zones are commonly referred to as the *tierra caliente* (hot land) from sea level to 3000 feet (900 meters); the *tierra templada* (temperate land) at 3001 to 6000 feet (900 to 1800 meters); the *tierra fría* (cold land) at 6001 to 12,000 feet (1800 to 3600 meters); and the *tierra helada* (frozen land) above 12,000 feet (3600 meters). Exploitation of these zones allows agriculturists, especially in the uplands, access to a great diversity of domesticated and wild plants (**Figure 4.10**).

The concept of altitudinal zonation is most relevant for the Andes, the highlands of Central America, and the Mexican Plateau. For example, traditional Andean farmers might use the high pastures of the Altiplano for grazing llamas and alpacas, the tierra fría for potato and quinoa production, and the lower temperate zone to produce corn. All the great indigenous civilizations, especially the Incas and the Aztecs, systematically extracted resources from these zones, thus ensuring a diverse and abundant resource base. Yet these complex ecosystems are extremely fragile and have become important areas of research for the effects of climate change in the tropics.

El Niño One of the most studied weather phenomena in Latin America, called **El Niño** (referring to the Christ child), occurs when a warm Pacific current arrives along the normally cold coastal waters of Ecuador and Peru in December, around Christmastime. This change in ocean temperature happens every few years and produces torrential rains, signaling the arrival of an El Niño year. The 1997–1998 El Niño was especially strong, but the 2015–2016 El Niño also proved to be a major event for Latin America. In the Plata Basin heavy rains in December and January caused the worst flooding in 50 years and displaced more than 200,000 people. Heavy rains in April flooded thousands of acres of cropland, which drastically reduced harvests of corn and soy. Moreover, greater precipitation led to heightened concern for the spread of mosquito-borne viruses such as Dengue and Zika.

The less-talked-about result of El Niño is drought. While parts of South America experienced record rainfall in 2015–2016, northern Brazil and Central America were in the grip of drought. Guatemala, Honduras, Nicaragua, and El Salvador are affected by one of the worst droughts in decades. Brazil, one of the world's leading soy producers,

▼ **Figure 4.9 Atacama Desert** One of the driest places on earth with almost no vegetation, many visitors liken it to a moonscape. Yet, the soils of the Atacama contain a wealth of copper and nitrates. Shown is an image of the Valley of the Moon in northern Chile.

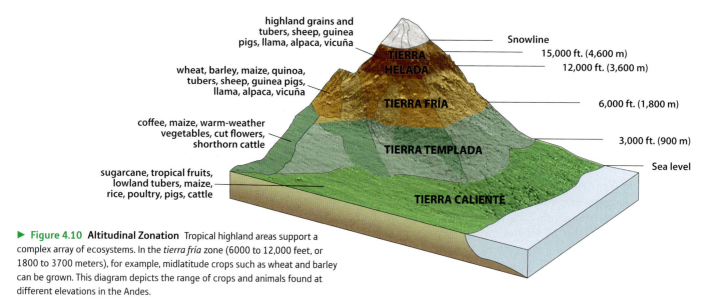

▶ **Figure 4.10 Altitudinal Zonation** Tropical highland areas support a complex array of ecosystems. In the *tierra fría* zone (6000 to 12,000 feet, or 1800 to 3700 meters), for example, midlatitude crops such as wheat and barley can be grown. This diagram depicts the range of crops and animals found at different elevations in the Andes.

saw their exports plummet due to drought. In addition to crop and livestock losses, estimated to be in the billions of dollars, some 3.5 million rural people in Central America face food shortages. Lack of rain also led to hundreds of bush and forest fires leaving their mark on the landscape.

Impacts of Climate Change for Latin America

Latin America represents about 8 percent of the world's population and produces about 6 percent of global greenhouse gas (GHG) emissions. The rate of increase in GHG emissions in Latin America is dramatically lower than in all other regions of the developing world, except for Sub-Saharan Africa. The region's relatively low emissions can be explained by lower average energy consumption, higher reliance on renewable energy (especially hydropower and biofuels), and greater dependence on public transportation. The burning of forest and brush, a common practice in the region, does produce spikes in carbon dioxide (CO_2) emissions, but the regrowth of vegetation also absorbs vast amounts of CO_2.

Global climate change has both immediate and long-term implications for Latin America. Of greatest immediate concern is how climate change is influencing agricultural productivity, water availability, changes in the composition and productivity of ecosystems, and incidence of vector-born diseases such as malaria, Dengue Fever, and now Zika. Changes attributable to climate change are already apparent in higher elevations, making these concerns more pressing. For example, coffee growers in the Colombian Andes have seen a decline in productivity over the past five years, which they attribute to higher temperatures and longer dry spells. The long-term effects of global climate change on lowland tropical forest systems are less clear; for example, some areas may experience more rainfall, others less. Other long-term impacts, such as rising sea level, will not cause the same levels of displacement in Latin America as predicted for the Caribbean or Oceania.

Climate change research indicates that highland areas are particularly vulnerable. Tropical mountain systems are projected to experience a 1–3°C increase in temperatures, as well as lower rainfall. This will raise the altitudinal limits of various ecosystems, affecting the range of crops and arable land available to farmers and ranchers. Research has documented the dramatic retreat of Andean glaciers—some no longer exist, such as Chacaltaya in Bolivia, and others have been drastically reduced, such as Argentina's Upsala glacier (**Figure 4.11**). Although this is a visible indicator of climate change, it also has pressing human repercussions. Many Andean villages get much of their water from glacial runoff. Thus, as average temperatures increase in the highlands and glaciers recede, there is widespread concern about future supplies of drinking water.

▼ **Figure 4.11 Andean Glacial Retreat** The Upsala Glacier in Glaciers National Park in southern Argentina has been in retreat for decades. The tongue of Upsala Glacier is seen in the bottom photo, lower right. It drains into Lake Argentina. (a) Upsala Glacier in 1928. (b) The same site in January 2004 (the southern hemisphere summer).

(a)

(b)

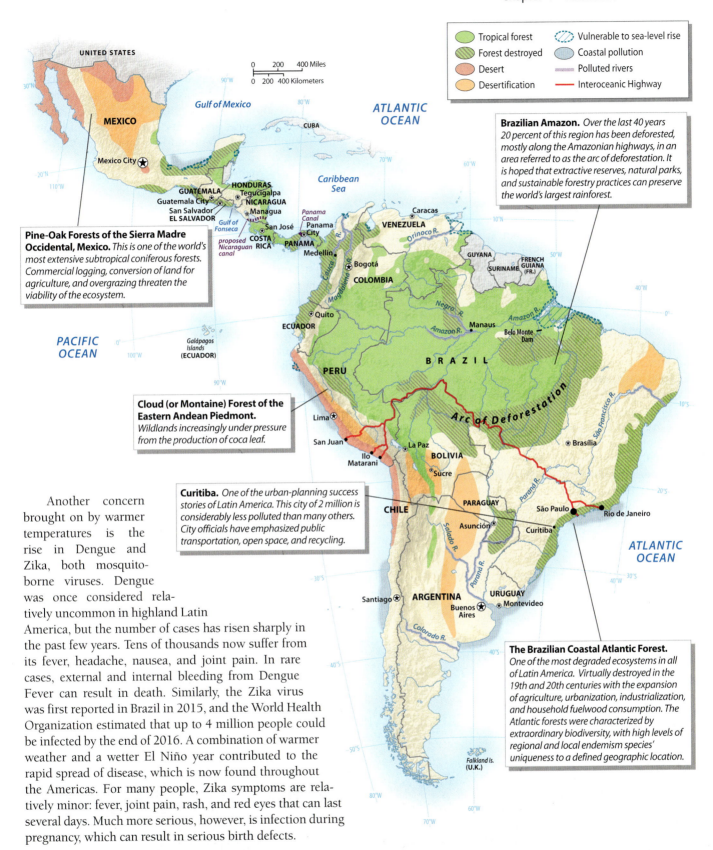

Pine-Oak Forests of the Sierra Madre Occidental, Mexico. *This is one of the world's most extensive subtropical coniferous forests. Commercial logging, conversion of land for agriculture, and overgrazing threaten the viability of the ecosystem.*

Brazilian Amazon. *Over the last 40 years 20 percent of this region has been deforested, mostly along the Amazonian highways, in an area referred to as the arc of deforestation. It is hoped that extractive reserves, natural parks, and sustainable forestry practices can preserve the world's largest rainforest.*

Cloud (or Montaine) Forest of the Eastern Andean Piedmont. *Wildlands increasingly under pressure from the production of coca leaf.*

Curitiba. *One of the urban-planning success stories of Latin America. This city of 2 million is considerably less polluted than many others. City officials have emphasized public transportation, open space, and recycling.*

The Brazilian Coastal Atlantic Forest. *One of the most degraded ecosystems in all of Latin America. Virtually destroyed in the 19th and 20th centuries with the expansion of agriculture, urbanization, industrialization, and household fuelwood consumption. The Atlantic forests were characterized by extraordinary biodiversity, with high levels of regional and local endemism species' uniqueness to a defined geographic location.*

Legend:
- Tropical forest
- Forest destroyed
- Desert
- Desertification
- Vulnerable to sea-level rise
- Coastal pollution
- Polluted rivers
- Interoceanic Highway

Another concern brought on by warmer temperatures is the rise in Dengue and Zika, both mosquito-borne viruses. Dengue was once considered relatively uncommon in highland Latin America, but the number of cases has risen sharply in the past few years. Tens of thousands now suffer from its fever, headache, nausea, and joint pain. In rare cases, external and internal bleeding from Dengue Fever can result in death. Similarly, the Zika virus was first reported in Brazil in 2015, and the World Health Organization estimated that up to 4 million people could be infected by the end of 2016. A combination of warmer weather and a wetter El Niño year contributed to the rapid spread of disease, which is now found throughout the Americas. For many people, Zika symptoms are relatively minor: fever, joint pain, rash, and red eyes that can last several days. Much more serious, however, is infection during pregnancy, which can result in serious birth defects.

Environmental Issues: The Destruction of Forests

Perhaps the environmental issue most commonly associated with Latin America is deforestation (**Figure 4.12**). The Amazon Basin and portions of the eastern lowlands

▲ **Figure 4.12 Environmental Issues in Latin America** Tropical forest destruction, desertification, water pollution, and poor urban air quality are some of the pressing environmental problems facing Latin America. Still present, however, are vast areas of tropical forest, supporting a wealth of genetic and biological diversity.

of Central America and Mexico still maintain unique and impressive stands of tropical forest. But other woodland areas, such as the Atlantic coastal forests of Brazil and the Pacific forests of Central America, have nearly disappeared as a result of agriculture, settlement, and ranching. The coniferous forests of northern Mexico are also falling, in part because of a bonanza for commercial logging stimulated by the North American Free Trade Agreement (NAFTA) (see Chapter 3). In Chile, the ecologically unique evergreen rainforest (the Valdivian forest) in the midlatitudes is being cleared for wood chip exports to Asia.

The loss of tropical rainforests is most critical in terms of biological diversity. Tropical rainforests account for only 6 percent of Earth's landmass, but at least 50 percent of the world's species are found in this biome. Moreover, the Amazon Basin contains the largest undisturbed stretches of rainforest in the world. Unlike Southeast Asian forests, where hardwood extraction drives forest clearance, Latin American forests are usually seen as an agricultural frontier that state governments divide in an attempt to give land to the landless or reward political cronies. Thus, forests are cut and burned, with settlers and politicians carving them up to create permanent settlements, slash-and-burn plots, or large cattle ranches. In addition, some tropical forest cutting has been motivated by the search for gold (Brazil, Peru, and Costa Rica) and the production of coca leaf for cocaine (Peru, Bolivia, and Colombia).

Brazil has incurred more criticism than other countries for its Amazon forest policies. During the past 40 years, one-fifth of the Brazilian Amazon forest has been cleared. Close to 60 percent of some states, such as Rondônia, have been deforested (**Figure 4.13**). What most alarms environmentalists

and forest dwellers (Indians and rubber tappers) is the dramatic increase in the rate of rainforest clearing since 2000, estimated at nearly 8000 square miles (20,000 square km) per year. The increased rates of deforestation in the Brazilian Amazon are due to the expansion of industrial mining and logging, the growth in corporate farms, the development of new road networks and dams, the incidence of human-ignited wildfires, and continued population growth. Under the Advance Brazil program started in 2000, some $40 billion will go to new highways, railroads, gas lines, hydroelectric projects, power lines, and river canalization projects that will reach into remote areas of the basin such as the Interoceanic Highway (see Figure 4.12). The Brazilian government, under President Lula da Silva, created 150,000 square kilometers of new conservation areas, many of them alongside the "arc of deforestation"—a swath of agricultural development along the southern edge of the Amazon Basin where the worst deforestation has occurred (again see Figure 4.12). However, Brazil's Forest Code, as revised in 2012, has reduced the amount of "forest reserve" that private landholders must maintain. Many conservationists fear this will lead to more forest clearing and fragmentation.

Grassification The conversion of tropical forest into pasture, called **grassification**, is another practice that has contributed to deforestation. Particularly in southern Mexico, Central America, and the Brazilian Amazon, a hodgepodge of development policies from the 1960s through the 1980s encouraged deforestation to make room for cattle. The preference for ranching as a status-conferring occupation seems to be a carryover from Iberian control. The image of the *vaquero* (cowboy) looms large in the region's history.

▼ **Figure 4.13 Tropical Forest Settlement in the Amazon** These satellite images of Rondônia, Brazil, illustrate the dramatic change in forest cover in just 10 years near the settlement of Buritis and road BR-364. Intact forest is dark green, whereas cleared areas are light green (crops) or tan (bare ground). (a) Typically, the first clearings appear off of roads, forming a fishbone pattern. (b) Over time, as more forest is cleared and settlements grow, the fishbone pattern collapses into a mosaic of pasture, farmland, and forest fragments.

(a) July 30, 2000

(b) August 2, 2010

Even poor farmers appreciate the value of having livestock. They know that cattle can be quickly sold for cash, something like having a savings account on hand.

Many natural grasslands are suitable for grazing in Latin America, such as the Llanos in Colombia and Venezuela and the Chaco and Pampas in Argentina. However, the rush to convert forest into pasture made ranching a scourge on the land. Even where domestic demand for beef has increased, ranching in remote tropical frontiers is seldom economically self-sustaining. Grassification is especially dramatic in the Central American countries of Guatemala, Costa Rica, and Panama, where huge tracts of forest have been cleared and converted to pasture (**Figure 4.14**). In the case of Panama, the Guna (an indigenous group formerly known as the Kuna) have banned cattle ranching in the Guna Yala territory in an effort to protect their forest lands.

Protecting Lands for Future Generations Latin America has more nationally protected lands than any other developing region. The areas designated as national parks, nature reserves, wildlife sanctuaries, and scientific reserves with limited public access went from 10 percent of the territory in 1990 to 21 percent in 2013, according to World Bank estimates. Brazil's protected land went from just 9 percent of the national territory to 26 percent in over 20 years. Although conservationists complain that many of these areas are "paper parks" with limited real protection, many countries in the region have used the conservation of forests and other lands as a means to attract tourists.

A variety of factors drive these conservation efforts. In some cases, such as the *páramos* (high altitude grasslands) surrounding the metropolitan area of Bogotá, these lands are critical to maintain the city's watershed and ultimately its water supply. Similarly, in Caracas, Venezuela, the forested slopes of the Avila were protected decades ago to reduce the threat of landslides and protect the watershed. Some of the

▼ **Figure 4.14 Converting Forest into Pasture** Cattle graze in northern Guatemala's Petén region. Clearing of this tropical forest lowland began in the 1960s and continues today. Ranching is a status-conferring occupation in Latin America with serious ecological costs. The beef produced from this region is for domestic and export markets.

▲ **Figure 4.15 Andean Potatoes** With some 4,000 edible varieties of potatoes in the Andes, this region is the hearth of potato domestication. Since 1971 the International Potato Center has been dedicated to protecting this vast diversity of tubers to insure food security and the well being for the rural Andean farmers who maintain these valuable varieties.

larger and more recent reserves and parks in Latin America were created explicitly to protect forests and biodiversity, including marine parks. In the process of protecting the environment, some countries saw their tourism, and even ecotourism, flourish. A regional leader in this regard is Costa Rica (see *Working Toward Sustainability: Ecotourism in Costa Rica*).

Problems on Agricultural Lands The pressure to modernize agriculture has produced a series of environmental problems. As peasants were encouraged to adopt new hybrid varieties of corn, beans, and potatoes, an erosion of genetic diversity occurred. Efforts to preserve dozens of native domesticates are under way at agricultural research centers in the central Andes and Mexico (**Figure 4.15**). Nonetheless, many useful native plants may have been lost. Modern agriculture also depends on chemical fertilizers and pesticides that eventually run off into surface streams and groundwater. Consequently, many rural areas suffer from contamination of local water supplies. Even more troublesome is the direct exposure of farm workers to toxic agricultural chemicals. Mishandling of pesticides and fertilizers can result in rashes and burns. In some areas, such as Sinaloa, Mexico, a rise in serious birth defects parallels the widespread application of chemicals.

Soil erosion and fertility decline occur in all agricultural areas. Certain soil types in Latin America are particularly vulnerable to erosion, most notably the volcanic soils and the reddish *oxisols* found in the humid tropical lowlands. The productivity of the Paraná basalt plateau in Brazil, for example, has declined over decades due to the ease with which these volcanic soils erode and the failure to apply soil conservation methods. By contrast, the oxisols of the tropical lowlands can quickly degrade into a hard-baked claypan surface when the natural cover is removed,

Ecotourism in Costa Rica

Costa Rica is a small tropical country that straddles the Pacific and the Caribbean and has impressive biodiversity in its highland and its lowland environments. It is the one Latin American country that has become nearly synonymous with the term *ecotourism*, because most of the country's tourist sector is oriented toward enjoying the natural environment. Whether you want to ride a zip-line through the cloud forest, bird watch, hike to the rim of a volcano, or explore two distinctive coastal environments, the country has a wide variety of ecologically oriented activities. Costa Rica received over 2.5 million tourists in 2015, nearly half of those from the United States. Tourism is the country's leading source of foreign exchange and the fastest growing sector of the economy.

National Parks Protect Land This approach to tourism did not happen overnight, but developed over forty years. Long known for its agricultural commodities, especially coffee and bananas, Costa Rica was like many of its Central American neighbors in the 1960s. In the 1970s, Costa Rican conservationist Mario Boza successfully lobbied for the creation of national parks in response to the rampant forest destruction occurring to expand coffee production, banana plantations, and cattle pasture. Importantly, his first park was created near the capital city of San Jose, with the intent to attract Costa Ricans to enjoy the natural environment. Today Costa Ricans still visit the country's many parks, such as Manuel Antonio (**Figure 4.1.1**), and about 20 percent of the national territory is protected from development. International tourists also visit the parks but pay higher entrance fees that go to supporting conservation and park maintenance.

The business of ecotourism requires a protected environment as well as infrastructure: hotels for accommodations, knowledgeable guides and park rangers, roads and trails to access places. Ecotourism has grown through a mix of private and public investment—the private sector creating tourist accommodations and adventure opportunities, and the public

▲ **Figure 4.1.2 Eco-lodge in Costa Rica** International tourists view wildlife from the deck of an eco-lodge on the Osa Peninsula, one of the most biodiverse locations in Central America.

▼ **Figure 4.1.1 Costa Rican National Park** Hugging the Pacific Coast, the tropical forest and beaches of Manuel Antonio National Park make it a popular destination for Costa Ricans as well as international tourists. The pressures to develop tropical coasts are real, which makes creating protected areas an urgent need.

sector investing in protected lands (**Figure 4.1.2**). Like any business there is a delicate balance, as too many tourists at a particular location can overwhelm the flora and fauna that is being protected.

Of course, there was resistance to protecting lands, especially from traditional agricultural and ranching interests. Yet today, Costa Ricans have come to embrace their status as a leading ecotourism destination. *Pura vida* is a widely used Costa Rican expression for enjoying life, happiness, and satisfaction. It literally means "pure life," but for the eco-consciously minded Costa Ricans, *pura vida* is surely reinforced by the robust national park system they have created and maintained.

Google Earth Virtual **MG** Tour Video

http://goo.gl/u8BvAw

1. What are the key components needed to foster ecotourism?

2. Why do you think Costa Rica ecotourism is more developed than that of other Central American countries?

making permanent agriculture nearly impossible. Ironically, the consolidation of the large-scale modern farms in the basins and valleys of the highlands tends to push peasant subsistence farmers into marginal areas. On these hillside farms, gullies and landslides reduce productivity. Lastly, the sprawl of Latin American cities consumes both arable land and water, eliminating some of the region's best farmland.

Urban Environmental Challenges

For most Latin Americans, air pollution, water availability and quality, and garbage removal are the pressing environmental problems of everyday life. Consequently, many environmental activists from the region focus their efforts on making urban environments cleaner by introducing "green" legislation and calling people to action. In this most urbanized region of the developing world, city dwellers do have better access to water, sewers, and electricity than their counterparts in Asia and Africa. Moreover, the density of urban settlement seems to encourage the widespread use of mass transportation; both public and private bus and van routes make getting around cities fairly easy. However, the usual environmental problems that arise from dense urban settings ultimately require expensive remedies, such as new power plants and modernized sewer and water lines. The money for such projects is never enough, due to currency devaluation, inflation, and foreign debt. Because many urban dwellers tend to reside in unplanned squatter settlements, servicing these communities with utilities after they are built is difficult and costly.

Air Pollution Air pollution is a concern for most major cities, but this is especially true for Santiago and Mexico City. A combination of geographical factors (basin settings) and meteorological factors (winter inversion layers), along with dense human settlement and automobile dependence, has led to these two capital cities having some of the highest recorded concentrations of particulate matter and ozone, major contributors to air pollution. Air pollution is not just an aesthetic issue—the health costs of breathing such contaminated air are real, as shown by elevated death rates due to heart disease, asthma, influenza, and pneumonia. The burden of air pollution is not evenly distributed among city residents, as the elderly, the very young, and the poor are more likely to suffer the negative health effects of contaminated air. Fortunately, both cities have taken steps to address this vexing problem.

Santiago, a prosperous city of nearly 7 million in Chile's Central Valley, has an elevation of 1700 feet (520 meters). Although not as high as Mexico City, this basin setting regularly produces thermal inversions, when warm air aloft traps a layer of cold air near the surface. This trapped surface layer becomes filled with engine exhaust, industrial pollution, garbage, and even fecal matter (**Figure 4.16**). The inversion layers happen year-round, but can be especially bad in the winter months from May through August, often forcing schools to suspend all sports when smog emergencies are called. Santiago officials began addressing this problem in the late 1980s by restricting vehicular traffic. On a given weekday, 20 percent of all buses, taxis, and cars are restricted from driving, based on license plate numbers. During smog emergencies, up to 40 percent of vehicles can be banned from the roads. In addition, older buses were replaced by a new fleet of cleaner-running buses with greater carrying capacity. These buses, combined with the city's subway system, can move over 2 million riders a day. By 2010, air quality had noticeably improved, and public support for restrictive measures and public transport has solidified.

The smog of Mexico City has been so bad that most modern-day visitors have no idea that mountains surround them. Air quality has been a major issue for Mexico City since the 1960s, driven in part by the city's unusually high rate of growth. (Between 1950 and 1980, the city's annual rate of growth was 4.8 percent.) It is difficult to imagine a better setting for creating air pollution. The city sits in a bowl 7400 feet (2200 meters) above sea level, and thermal inversions form

▼ **Figure 4.16 Air Pollution in Santiago** Smog blankets Santiago, with the Andes in the background. During the winter months (May through August), thermal inversion layers form that trap pollutants near ground level, causing a spike in pollution-related health problems. By reducing vehicular traffic and greatly expanding public transportation, the city has experienced improved air quality.

regularly. Steps were finally taken in the late 1980s to reduce emissions from factories and cars. Unleaded gas is now widely available, and cars manufactured for the Mexican market must have catalytic converters. Also, some of the worst polluting factories in the Valley of Mexico have closed. In the last few years, the mayor of Mexico City has expanded a low-emissions bus system, eliminating thousands of tons of carbon monoxide. In 2007, the decision was made to close the elegant Paseo de Reforma to traffic on Sunday mornings and open it to bike riders. This change was so popular that now bike lanes have been introduced to some downtown areas in an effort to encourage bike ridership. For longer commutes, a suburban train system is being built to complement the existing subway system. The payoff is real: Mexico City no longer ranks among the top polluted cities in the world, and it appears to have cut most of its pollutants by at least half.

Water Providing access to clean and reliable freshwater is also a challenge for Latin America's large cities. When Vicente Fox was president of Mexico, he declared water scarcity and water quality a national security issue—not just for the capital, but also for the entire country. Ironically, it was the abundance of water that initially made Mexico City attractive for settlement. Large shallow lakes once filled the valley, but over the centuries most were drained to expand agricultural land. As surface water became scarce, wells were dug to tap the basin's massive freshwater aquifer. Today approximately 70 percent of the metropolitan area's water is drawn from the valley's aquifer. There is troubling evidence that the aquifer is being overdrawn and at risk of contamination, especially in areas where unlined drainage canals can leak pollutants into the surrounding soil, which then leach into the aquifer. To reduce reliance on the aquifer, the city now pumps water nearly a mile uphill from more than 100 miles (160 km) away.

Andean cities such as Bogotá, Quito, and La Paz are increasingly experiencing water scarcity and rationing. Some of this is due to increased demands on aging water systems brought about by population growth. However, changes in precipitation patterns, either due to El Niño years or global climate change, make these large urban centers especially vulnerable. La Paz, for example, gets much of its water from glacial runoff. A major Bolivian glacier, Chacaltaya, is now gone. Thus, as average temperatures increase in the highlands and glaciers recede, there is widespread concern about future drinking-water supplies in this metropolitan area of nearly 2 million people.

> **REVIEW**
>
> **4.1** Describe the major ecosystems in Latin America, and explain how humans have adapted to and modified these different ecosystems.
>
> **4.2** Summarize some of the major environmental issues impacting this region and how different countries have tried to address them.

■ **KEY TERMS** neotropics, shields, altiplano, altitudinal zonation, environmental lapse rate, El Niño, grassification

Population and Settlement: The Dominance of Cities

Latin America did not have great river-basin civilizations like those in Asia. In fact, the great rivers of the region are surprisingly underutilized as areas of settlement or corridors for transportation. The major population clusters of Central America and Mexico are in the interior plateaus and valleys, whereas the interior lowlands of South America are relatively empty. Historically, the highlands supported most of the region's population during the pre-Hispanic and colonial eras, although archeological and geographical research has shown greater pre-Hispanic lowland settlement in the Amazon and Central American lowlands than previously thought. In the 20th century, population growth and migration to the Atlantic lowlands of Argentina and Brazil, along with continued growth of Andean coastal cities such as Guayaquil, Barranquilla, and Maracaibo, have reduced the demographic significance of the highlands. Major highland cities such as Mexico City, Guatemala City, Bogotá, and La Paz still dominate their national economies, but the majority of large cities are on or near the coasts (**Figure 4.17**).

Like the rest of the developing world, Latin America experienced dramatic population growth in the 1960s and 1970s. In 1950, its population totaled 150 million, which equaled the population of the United States at that time. By 1995, the population had tripled to 450 million; in comparison, the United States reached 300 million in 2006. Now nearly 600 million people live in the region. Latin America outpaced the United States because its birth rate remained consistently higher as infant mortality rates dropped and life expectancy soared. In 1950, Brazilian life expectancy was only 43 years; by the 1980s, it was 63, and now it is 75. In fact, between 1950 and 1980, most countries in the region experienced a 15- to 20-year improvement in life expectancy, which pushed up growth rates. Four countries account for over 70 percent of the region's population: Brazil with 206.1 million, Mexico with 128.6 million, Colombia with 48.8 million, and Argentina with 43.6 million (see Table 4.1).

Patterns of Rural Settlement

Although the majority of Latin Americans live in cities, some 120 million people do not. Throughout the region, a distinct rural lifestyle exists, especially among peasant subsistence farmers. In Brazil alone, more than 30 million people live in rural areas. Interestingly, the absolute number of people living in rural areas today is roughly equal to the number in the 1960s. Yet rural life has changed dramatically. In addition to subsistence agriculture, highly mechanized, capital-intensive farming occurs in most rural areas. The links between rural and urban areas are much improved, with the result that rural folks are less isolated. Also, as international migration increases, many rural communities are directly connected to cities in North America and Europe, with immigrants sending back remittances and supporting

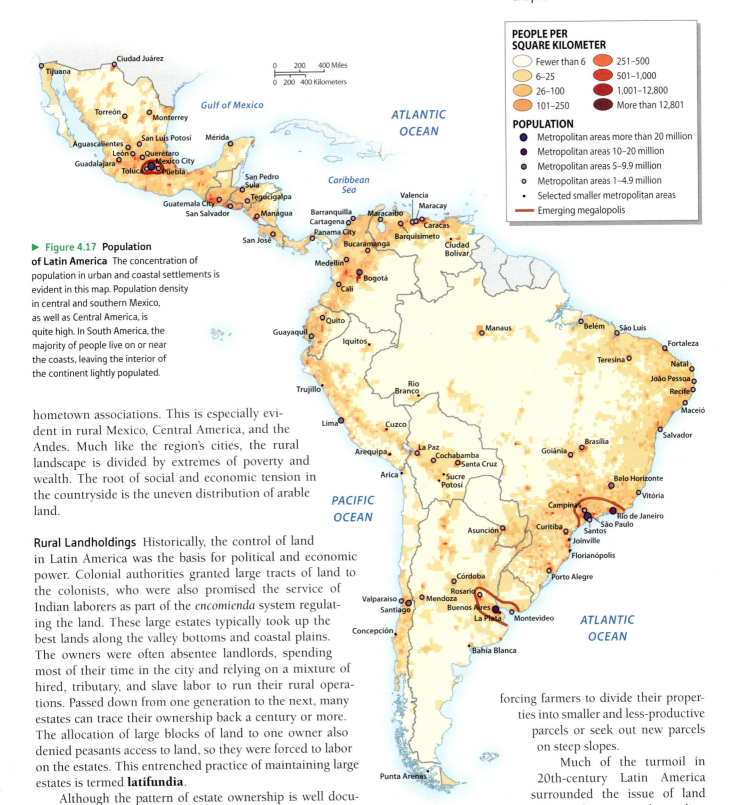

PEOPLE PER SQUARE KILOMETER

- Fewer than 6
- 6–25
- 26–100
- 101–250
- 251–500
- 501–1,000
- 1,001–12,800
- More than 12,801

POPULATION

- Metropolitan areas more than 20 million
- Metropolitan areas 10–20 million
- Metropolitan areas 5–9.9 million
- Metropolitan areas 1–4.9 million
- Selected smaller metropolitan areas
- Emerging megalopolis

▶ **Figure 4.17 Population of Latin America** The concentration of population in urban and coastal settlements is evident in this map. Population density in central and southern Mexico, as well as Central America, is quite high. In South America, the majority of people live on or near the coasts, leaving the interior of the continent lightly populated.

hometown associations. This is especially evident in rural Mexico, Central America, and the Andes. Much like the region's cities, the rural landscape is divided by extremes of poverty and wealth. The root of social and economic tension in the countryside is the uneven distribution of arable land.

Rural Landholdings Historically, the control of land in Latin America was the basis for political and economic power. Colonial authorities granted large tracts of land to the colonists, who were also promised the service of Indian laborers as part of the *encomienda* system regulating the land. These large estates typically took up the best lands along the valley bottoms and coastal plains. The owners were often absentee landlords, spending most of their time in the city and relying on a mixture of hired, tributary, and slave labor to run their rural operations. Passed down from one generation to the next, many estates can trace their ownership back a century or more. The allocation of large blocks of land to one owner also denied peasants access to land, so they were forced to labor on the estates. This entrenched practice of maintaining large estates is termed **latifundia**.

Although the pattern of estate ownership is well documented, peasants have always farmed small plots for their subsistence. This practice, called **minifundia**, can lead to permanent or shifting cultivation. Small farmers typically plant a mixture of crops for subsistence, as well as for trade. Peasant farmers in Colombia or Mexico, for example, grow corn, fruits, and various vegetables alongside coffee bushes that produce beans for export. Strains on the minifundia system occur when rural populations grow and land becomes scarce,

forcing farmers to divide their properties into smaller and less-productive parcels or seek out new parcels on steep slopes.

Much of the turmoil in 20th-century Latin America surrounded the issue of land ownership, with peasants demanding its redistribution through the process of **agrarian reform**. Governments have addressed these concerns in different ways. The Mexican Revolution in 1910 yielded a system of communally held lands called *ejidos*. In the 1950s, Bolivia crafted agrarian reform policies that led to the expropriation of estate lands and their reallocation to small farmers. As part of the Sandinista

TABLE 4.1   Population Indicators

MasteringGeography™

Country	Population (millions), 2016	Population Density (per square kilometer)[1]	Rate of Natural Increase (RNI)	Total Fertility Rate	Percent Urban	Percent < 15	Percent > 65	Net Migration (Rate per 1000)
Argentina	43.6	16	1.0	2.3	92	25	11	1
Bolivia	11.0	10	1.8	3.0	68	33	5	−2
Brazil	206.1	25	0.8	1.8	86	23	8	0
Chile	18.2	24	0.8	1.8	83	20	11	2
Colombia	48.8	43	1.0	2.3	76	27	7	−1
Costa Rica	4.9	93	1.1	1.8	77	23	7	1
Ecuador	16.5	64	1.6	2.5	64	29	7	0
El Salvador	6.4	295	1.1	2.0	67	27	8	−7
Guatemala	16.6	149	1.9	3.1	52	40	5	−1
Honduras	8.2	71	1.7	2.5	55	32	5	−2
Mexico	128.6	65	1.4	2.2	79	28	6	−1
Nicaragua	6.3	50	1.5	2.4	59	30	5	−4
Panama	4.0	52	1.4	2.4	67	27	8	2
Paraguay	7.0	16	1.6	2.6	60	32	5	−3
Peru	31.5	24	1.4	2.5	79	28	7	−1
Uruguay	3.5	20	0.4	1.9	95	21	14	−1
Venezuela	31.0	35	1.5	2.4	89	28	6	−2

Source: Population Reference Bureau, *World Population Data Sheet*, 2016.

[1] World Bank, *World Development Indicators*, 2016.

Login to Mastering Geography™ **& access** MapMaster **to explore these data!**

1) Which countries have the highest population densities, and where are they located within the region?

2) Which countries have total fertility rates below 2.1? How do you think this will impact their population growth over the next 30 years?

revolution in Nicaragua in 1979, lands were expropriated from the political elite and converted into collective farms. In 2000, President Hugo Chavez ushered in a new era of agrarian reform in Venezuela. In 2006, Bolivian President Evo Morales introduced a program aimed at giving land title to indigenous communities in the eastern lowlands, often far from their ancestral territories. These programs have met with resistance and, at times, have proven to be costly politically. Eventually, the path chosen by most governments was to make frontier lands available to land-hungry peasants. The opening of tropical frontiers, especially in South America, was a widely practiced strategy that changed national settlement patterns and began waves of rural-to-rural and even urban-to-rural migration.

Agricultural Frontiers The expansion into agricultural frontiers serves several purposes: providing peasants with land, tapping unused resources, and shoring up political boundaries. Several frontier colonization efforts in South America are noteworthy. In addition to settlement along Brazil's Trans-Amazon Highway, Peru developed its Carretera Marginal (Marginal Highway) in an effort to lure colonists into the cloud and rain forests of

eastern Peru. Most recently, the completion of the Interoceanic Highway linking Peru and Brazil has opened up new areas of settlement in the Amazonian territories of these countries (**Figure 4.18**). In Bolivia, Colombia, and Venezuela, agricultural

▼ **Figure 4.18 Puerto Maldonado, Peru** The newly constructed Puerto Maldonado Bridge in the Peruvian Amazon spans the Madre de Dios River. The bridge is a major infrastructural feature of the Interoceanic Highway, which connects Atlantic ports in Brazil with Pacific ports in Peru by traversing the Andes and the Amazon.

▶ **Figure 4.19** Major **Latin American Migration Flows** Internal, intraregional, and international migrations have opened frontier zones and created transnational communities. Over the past three decades, the flow of Latin Americans to the United States has grown. In 2015, the U.S. Census Bureau estimated there were over 57 million people of Hispanic ancestry in the United States. Most of these people either were born in or have ancestral ties to Latin America.

frontier schemes in the lowland tropical plains attracted peasant farmers and large-scale investors. Bolivians moved into the eastern plains, which spurred the growth and importance of the city of Santa Cruz. Mexico sent colonists, some displaced by dam construction, into the forests of Tehuantepec. Guatemala developed its northern Petén region. El Salvador had no frontier left, but many desperately poor Salvadorans poured into the neighboring states of Honduras and Belize in search of land. In short, although the dominant demographic trend has been a rural-to-urban movement, an important rural-to-rural flow has turned previously virgin areas into agricultural zones (see the purple arrows in **Figure 4.19**).

The opening of the Brazilian Amazon for settlement was the most ambitious frontier colonization scheme in the region. In the 1960s, Brazil began its frontier expansion by constructing several new Amazonian highways, a new capital (Brasília), and state-sponsored mining operations. The Brazilian military directed the opening of the Amazon to provide an outlet for landless peasants and to extract the region's many resources. However, the plans did not deliver as intended. Throughout the basin, thin forest soils were incapable of supporting permanent agricultural colonies and, in the worst cases, degraded into baked claylike surfaces devoid of vegetation. Government-promised land titles, agricultural subsidies, and credit were slow to reach small farmers, even if they were fortunate enough to be given land on *terra roxa*, a nutrient-rich purple clay soil. Instead, a disproportionate amount of money went to subsidizing large cattle ranches through tax breaks and improvement deals, where "improved" land meant cleared land.

Because of the many competing factions (miners, loggers, ranchers, peasants, and corporate farmers) and the uncertainty of land title and political authority, the Amazon can be a violent place. Chico Mendes, a rubber tapper and internationally recognized environmental activist, was slain in 1988 by a rancher who was eventually prosecuted and jailed. Mendes's death brought an international outpouring

of support, and eventually a large extractive reserve was created in Acre state in his memory. Yet violence toward environmental activists, religious workers, and organizers of rural workers is extremely common. In the last 20 years, more than 1000 Amazonian activists have been murdered in Brazil.

Today five times more people live in the Amazon than in the 1960s; thus, continued human modification of this region is inevitable. Yet most of the people in the Brazilian Amazon live in large cities such as Manaus and Belém.

The Latin American City

A quick glance at the population map of Latin America (see Figure 4.17) shows a concentration of people in cities. One of the most significant demographic shifts in the region has been the movement out of rural areas to cities, which began in earnest in the 1950s. Just one-quarter of the region's population was urban in 1950; the rest lived in small villages and the countryside. Today the pattern is reversed, with four-fifths of Latin Americans living in cities. In the most urbanized countries, such as Argentina, Chile, Uruguay, and Venezuela, more than 83 percent of the population lives in cities (see Table 4.1). This preference for urban life is attributed to cultural as well as economic factors. Under Iberian rule, people residing in cities had higher social status and greater economic opportunity. Initially, only Europeans were allowed to live in the colonial cities, but this exclusivity was not strictly enforced. Over the centuries, colonial cities became the hubs for transportation and communication, making them the primary centers for economic and political activities.

Latin American cities are noted for high levels of **urban primacy**, a condition in which a country has a *primate city* three to four times larger than any other city in the country. Examples of primate cities are Lima, Caracas, Guatemala City, Santiago, Buenos Aires, and Mexico City (**Figure 4.20**). Primacy is often viewed as a liability, as too many national resources are concentrated into one urban center. In an effort to decentralize, some governments have intentionally built new cities far from existing primate cities (for example, Ciudad Guayana in Venezuela and Brasília in Brazil). Despite these efforts, the tendency toward primacy remains. Moreover, the growth of urbanized regions that include several major cities has inspired the label *megalopolis* for three areas in Latin America: Mexico City–Puebla–Toluca–Cuernavaca on the Mesa Central; the Niterói–Rio de Janeiro–Santos–São Paulo–Campinas axis in southern Brazil; and the Rosario–Buenos Aires–Montevideo–San Nicolás corridor in Argentina and Uruguay's lower Rio Plata Basin (see Figure 4.17).

Urban Form Latin American cities have a distinct urban morphology that reflects both their colonial origins and their contemporary growth (**Figure 4.21**). Usually, a clear central business district (CBD) exists in the old colonial core. Radiating out from the CBD is older middle- and lower-class housing found in the zones of maturity and *in situ* accretion. In this model, residential quality declines as you move from the center to the periphery. The exception is the elite spine, a newer commercial and business strip that extends from the colonial core to newer parts of the city. Along the spine are superior services, roads, and transportation. The city's best residential zones, as well as shopping malls, are usually on either side of the spine. Close to the elite residential sector, a limited area of middle-class housing is typically found. Most major urban centers also have a *periférico* (a ring road or beltway highway) that circumscribes the city. Industry is located in isolated areas of the inner city and in larger industrial parks outside the ring road.

▼ **Figure 4.20 Primacy in Buenos Aires** This capital city with a metropolitan population of over 13 million is the economic and cultural hub of Argentina. The bustling ceremonial boulevard, 9 de Julio Avenue, cuts through the downtown and commemorates Argentine Independence Day (July 9). The obelisk built in the early 20th century is an iconic structure for the city. Buenos Aires is both a primate city and a megacity (over 10 million people).

Central Business District,
Caracas, Venezuela

Squatter Settlements,
Caracas, Venezuela

Mall Shop,
Caracas, Venezuela

Elite Housing,
Caracas, Venezuela

Industrial
Park

Disamenity   Disamenity

Market

CBD

Spine

Periférico   Periférico

Mall

Legend:
- Commercial
- Market
- Industrial
- Zone of maturity
- Zone of *in situ* accretion
- Zone of peripheral squatter settlements
- Disamenity
- Elite residential sector
- Gentrification
- Middle-class residential tract

▲ **Figure 4.21 Latin American City Model** This urban model highlights the growth of Latin American cities and the class divisions within them. The central business district (CBD), elite spine, and residential sectors may have excellent services and utilities, but life in the zone of peripheral squatter settlements is much more difficult. In many Latin American cities, one-third of the population resides in squatter settlements. **Q: How does the Latin American city model compare with that of North America in terms of where the rich and the poor live? What are the factors driving urban growth?**

In outer rings of the city (sometimes straddling the periférico) is a zone of peripheral squatter settlements where many of the urban poor live in the worst housing. Services and infrastructure are extremely limited: roads are unpaved, water is often trucked in, and sewer systems are nonexistent. The dense ring of squatter settlements (variously called *ranchos, favelas, barrios jovenes,* or *pueblos nuevos*) that encircle Latin American cities reflect the speed and intensity with which these zones were created. The squatter settlements are also found in disamenity zones near the core of the city, settings such as steep hillsides or narrow gorges prone to flooding that are considered too risky for formal housing. These can also be polluted industrial areas or even garbage dumps where the very poor reside because they might find employment there and/or a place to live. In some cities, more than one-third of the population lives in these self-built homes of marginal or poor quality. These kinds of dwellings are recognizable throughout the developing world, yet the practice of building homes on the "urban frontier" has a longer history in Latin America than in most Asian and African cities. The combination of a rapid inflow of migrants, the inability of governments to meet pressing housing needs, and the eventual official recognition of many of these neighborhoods with land titles and utilities meant that this housing strategy was rarely discouraged. In cities as diverse as Medellin, Colombia, and La Paz, Bolivia, planners have become creative in addressing the needs of urban settlers on steep hillsides by introducing gondolas to link shanty settlements with the rest of the city. Rio de Janeiro installed a six-station gondola line running over a group of favelas known as the *Complexo do Alemão* in 2011. Residents now have a 15-minute ride in the gondola versus an hour-plus-long hike to the rail station (**Figure 4.22**).

Among the inhabitants of these neighborhoods, the **informal sector** is a fundamental force that houses, services, and employs them. Definitions of the informal sector are much debated. The term usually refers to the economic sector that relies on low-wage self-employment, such as street vending, shoe shining, and artisan manufacturing, work that is virtually unregulated and untaxed. Some scholars include as part of the informal sector illegal activities such as drug smuggling, prostitution, and sale of contraband items such as illegally copied DVDs or music CDs and tapes. The informal economy is discussed in more detail later in the chapter, but one interesting expression of informality is the housing in squatter settlements. In arid Lima, Peru, an estimated 40 percent of the population lives in self-built housing, often of very poor quality. Typically, these settlements begin as illegal invasions of open spaces that are carefully planned and timed to avoid the risk of eviction by city authorities. If the hastily built communities go unchallenged, squatters steadily improve their houses. The creation of these landscapes reflects a conscious and organized effort on the part of the urban poor, many of whom have rural origins, to make a place for themselves in Latin America's cities.

Rural-to-Urban Migration As conditions in rural areas deteriorated due to the consolidation of lands, mechanization of agriculture, and increased population pressure, peasants began to pour into the cities of Latin America in the process of **rural-to-urban migration**. The strategy of

Explore the **Sights** of Rio's Sugarloaf Mountain

http://goo.gl/jYub8Q

▼ **Figure 4.22 Gondolas over Rio de Janeiro** Urban planners have creatively addressed issues of access in some of the city's densely settled favelas (informal settlements) by installing gondolas to improve transport in the Complexo do Alemão.

rural households sending family members to the cities for employment as domestics, construction workers, artisans, and vendors has been well established since the 1960s. Once in the cities, rural migrants generally found conditions better, especially in terms of access to education, health care, electricity, and clean water.

It was not poverty alone that drove people out of rural areas, but individual choice and an urban preference. Migrants believed in, and often realized, greater opportunities in cities, especially the capital cities. Those who came were usually young (in their twenties) and better educated than those who stayed behind. Women slightly outnumbered men in this migrant stream. The move itself was made easier by extended kin networks formed by earlier migrants who settled in discrete areas of the city and aided new arrivals. The migrants maintained their links to their rural communities by periodically sending remittances and making return visits.

Population Growth and Mobility

Latin America's high growth rates throughout the 20th century are attributed to natural increase, as well as immigration in the early part of the century. The 1960s and 1970s were decades of tremendous growth, resulting from high fertility rates and increasing life expectancy. In the 1960s, for example, a Latin American woman typically had six or seven children. By the 1980s, family sizes were half as large. Today the total fertility rate (TFR) for the region is 2.1, which is replacement value (see Table 4.1). Several factors explain this trend: more urban families, which tend to be smaller than rural ones; increased participation of women in the workforce; higher education levels of women; state support of family planning; and better access to birth control. Even the more rural countries with a high percentage of Amerindians are experiencing smaller families—Bolivia's TFR is 3.0, and Guatemala's is 3.1.

Even with family sizes shrinking—and in the cases of Uruguay, Chile, and Brazil, falling below replacement rates—there is built-in potential for continued growth because of the relative demographic youth of these countries. The average percentage of the population below age 15 is 27 percent. In North America, that same group is 19 percent of the population, and in Europe it is just 16 percent. This means that a proportionally larger segment of the population has yet to enter its childbearing years.

The population pyramids of Uruguay and Guatemala contrast the profile of a country with a stable population size with that of a demographically growing state (Figure 4.23). Uruguay is a small but prosperous country, with a high Human Development Index ranking and relatively little poverty. Uruguayan women average slightly less than two children (TFR of 1.9), which is below replacement level. Life expectancy is also high, but most population projections show the country's population as relatively stable between now and 2050. In contrast, Guatemala has a wide-based population pyramid and is considerably poorer. Total fertility rates have declined in Guatemala, but are still considered high at 3.1. Due to its youthful population and the increase in life expectancy, Guatemala's population is expected to double between 2015 and 2050.

In addition to natural increase, waves of immigrants into Latin America and migrant streams within Latin America have influenced population size and patterns of settlement. Beginning in the late 19th century, new immigrants from Europe and Asia added to the region's size and ethnic complexity. Important population shifts within countries have also occurred in recent decades, as witnessed by the growth of Mexican border towns and the demographic expansion of the Bolivian plains. In an increasingly globalized economy, even more Latin Americans live and work outside the region, especially in the United States and Europe.

European Migration After Latin American countries gained independence from Iberia in the 19th century, their new leaders sought to develop economically through immigration. Firmly believing that "to govern is to populate," many countries set up immigration offices in Europe to attract hardworking peasants to till the soils and "whiten" the **mestizo** (people of mixed European and Amerindian ancestry)

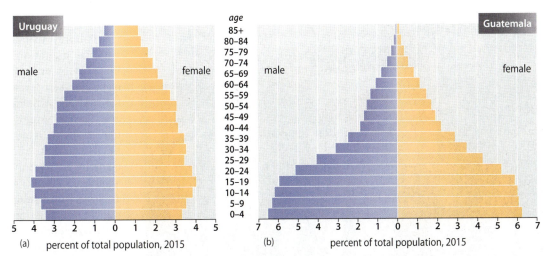

▶ **Figure 4.23 Population Pyramids of Uruguay and Guatemala** These two pyramids contrast the population structure of (a) the more developed and demographically stable Uruguay with that of (b) the youthful and rapidly growing Guatemala. The average Uruguayan woman has 2 children, whereas Guatemalan women have a total fertility rate (TFR) of 3.1; a few years ago it was closer to 4.0. Due to differences in natural increase and TFR, Uruguay is projected to have about the same population size in 2050, but Guatemala is projected to be twice as large.

population. The Southern Cone countries of Argentina, Chile, Uruguay, and southern Brazil were the most successful in attracting European immigrants from the 1870s until the depression of the 1930s. During this period, some 8 million Europeans arrived (more than came during the entire colonial period), with Italians, Portuguese, Spaniards, and Germans the most numerous. Some of this immigration was state-sponsored, such as the nearly 1 million laborers (including entire families) brought to the coffee estates surrounding São Paulo at the start of the 20th century. Other migrants came seasonally, especially the Italian peasants who left Europe in the winter for agricultural work during the Argentine summer and were thus nicknamed "the swallows." Still others paid their own passage, intending to settle permanently and prosper in the growing commercial centers of Buenos Aires, São Paulo, Montevideo, and Santiago.

Asian Migration Less well known than the European immigrants to Latin America are the Asian immigrants, who arrived during the late 19th and 20th centuries. Although considerably fewer, over time they established an important presence in the large cities of Brazil, Peru, Argentina, and Paraguay. Beginning in the mid-19th century, the Chinese and Japanese who settled in Latin America were contracted to work on the coffee estates in southern Brazil and the sugar estates and guano mines of Peru. Over time, Asian immigrants became prominent members of society. A son of Japanese immigrants, Alberto Fujimori was president of Peru from 1990 to 2000.

Between 1908 and 1978, a quarter-million Japanese immigrated to Brazil; today the country is home to more than 1.3 million people of Japanese descent (**Figure 4.24**).

▼ **Figure 4.24 Japanese Brazilians** In 1908, the first Japanese immigrants arrived as agricultural workers, choosing Brazil as a destination after countries such as the United States and Canada had banned Japanese immigration. Today there are over 1.3 million ethnic Japanese in Brazil, especially in the states of São Paulo and Parana. Here people dance samba at an event where the Japanese Brazilian community gathered to watch a FIFA 2013 soccer match between Brazil and Japan in São Paulo.

Initially, most Japanese were landless laborers, yet by the 1940s they had accumulated enough capital so that three-quarters of the migrants had their own land in the rural areas of São Paulo and Paraná states. As a group, the Japanese have been closely associated with the expansion of soybean and orange production. Today Brazil leads the world in exports of orange juice concentrate, with most of the oranges grown on Japanese-Brazilian farms. Increasingly, second- and third-generation Japanese have taken professional and commercial jobs in Brazilian cities; many of them have married outside their ethnic group and have lost their fluency in Japanese. South America's economic turmoil in the 1990s encouraged many ethnic Japanese to emigrate to Japan in search of better opportunities. Nearly one-quarter of a million ethnic Japanese left South America in the 1990s (mostly from Brazil and Peru) and now work in Japan.

The latest Asian immigrants are from South Korea. Unlike their predecessors, most of the Korean immigrants came with enough capital to invest in small businesses, and they settled in cities rather than in the countryside. According to official South Korean statistics, 120,000 Koreans emigrated to Paraguay between 1975 and 1990. Although many have stayed in Paraguay, a pattern has appeared of secondary immigration to Brazil and Argentina. Recent Korean immigrants in São Paulo have created more than 2500 small businesses. Unofficial estimates of the number of Koreans living in Brazil range from 40,000 to 120,000. As a group, they are decidedly commercial in orientation and urban in residence; their cities of choice are Asunción and Ciudad del Este in Paraguay, São Paulo in Brazil, and Buenos Aires in Argentina.

Latino Migration and Hemispheric Change Migration within Latin America and between Latin America and North America has had a significant impact on both sending and receiving communities. Within Latin America, international migration is shaped by shifting economic and political realities. For example, Venezuela's oil wealth attracted Colombian immigrants, who tended to work as domestics or agricultural laborers. Argentina has long been a destination for Bolivian and Paraguayan laborers. Nicaraguans seek employment in Costa Rica. And farmers in the United States have depended on Mexican laborers for more than a century.

Political turmoil has also sparked international migration. The bloody civil wars in El Salvador and Guatemala in the 1980s, for example, sent waves of refugees into Mexico and the United States. Violence in northern Central America is again driving people toward Mexico and the United States. In 2014, nearly 70,000 unaccompanied minors from Guatemala, El Salvador, and Honduras overwhelmed border facilities in the United States as these youth sought to reunite with family in the United States and to find refuge from gang-driven violence in Central America. By 2015 the numbers of unaccompanied minors dropped

The Impact of U.S. Deportations on Latin America

Much of the immigration debate in the United States focuses on the 11 million undocumented immigrants. Less attention is paid to the nearly 4 million Latin Americans who have been forcibly removed from the United States since 2000. More removals occurred from 2000 to 2013 than in the entire 107-year period preceding that decade, according to U.S. Department of Homeland Security data. The surge in removals does not count people turned away or detained when trying to cross the border; the deportees have lived and worked in the United States, often for many years, and were caught up in the deportation machine created through integrated databases, biometric data (such as digital fingerprints), and tougher local enforcement. The vast majority of the deported were sent to Latin America, mostly Mexico (72 percent), followed by Guatemala, Honduras, and El Salvador. And 90 percent of those removed are men. Hence, deportation is a politically charged and deeply personal issue for people in Latin America. A recent Pew survey estimated that one in three Latinos in the United States knows someone who has been deported or detained in the last 12 months.

Personal Calamity This unintended return happens suddenly and without warning, like a lightning strike. A migrant can be stopped for a traffic violation or caught up in a workplace raid and suddenly be channeled back to his or her country of origin (often after months of detention). Rather than making a triumphant return with new possessions, financial resources for investment, and even one's entire family, the unintended returnee has little to show for years of toil abroad. Moreover, the process of forced removal often results in financial calamity: Remittances are no longer sent, mortgages cannot be paid, and savings are rapidly depleted. Research on involuntary returns to Mexico shows income loss, family strain, social stigma, and difficult reintegration, especially for young children who have grown up in the United States and must now be integrated into Mexican schools.

Forced removals, especially criminal deportations, have always taken place. But the recent growth in this involuntary movement is due to noncriminal deportees (people removed because they did not have the right to stay in the United States). Mexico eclipses all other countries by its sheer number of deportations, 3 million. Living within the United States are 12 million individuals born in Mexico and some 35 million people of Mexican ancestry. This means that nearly one in four foreign-born Mexicans in the United States was forcibly removed in the 2000s.

▲ **Figure 4.2.1 Deported Guatemalans** Guatemalan men deplaning in Guatemala City after being deported from the United States. Every day, planeloads of Central Americans are returned to Guatemala, Honduras, and El Salvador by the Department of Homeland Security.

National Calamity Over 800,000 Central Americans have been removed from the United States since 2000 (Figure 4.2.1), and two to three planeloads of removals and people apprehended at the U.S.-Mexico border arrive in El Salvador, Honduras, and Guatemala every day. The economic and social impact of deportations may be greater on these countries, because Hondurans and Guatemalans removed from the United States made up a greater proportion of the remittance senders from those states. Remittances are a critical source of income for households in these poor countries, so at a time when these countries are also suffering from a surge in crime, the rise in deportations is an added strain. Unless comprehensive immigration reform is passed, forced removal of the undocumented is likely to remain a contentious issue between Latin America and the United States.

1. Why is such a large portion of the undocumented population in the United States from Latin America?

2. Describe some of the consequences of the unintended return of nearly 4 million Latinos on Latin America.

---

significantly, in part because more were apprehended on the Mexico-Guatemala border. Although some of these youth were able to remain, many were returned to Central America. Moreover, Central American countries actively dissuaded people from sending their youth to the north because of the many dangers involved in the illegal crossing. (See *People on the Move: The Impact of U.S. Deportations on Latin America*).

Presently, Mexico is the country of origin for the most legal immigrants to the United States. There are 55 million Hispanics in the United States; two-thirds of them claim Mexican ancestry, including approximately 12 million who were born in Mexico. Mexican labor migration to the United States dates back to the late 1800s, when relatively unskilled labor was recruited to work in agriculture, mining, and

railroads. Mexican immigrants are concentrated in California and Texas, but increasingly they are found throughout the country. Although Mexicans continue to have the greatest presence among Latinos in the United States, the number of immigrants from El Salvador, Guatemala, Nicaragua, Colombia, Ecuador, and Brazil has steadily grown. Most U.S. Hispanics have ancestral ties with peoples from Latin America and the Caribbean (see Chapter 5 on Caribbean migration).

Today Latin America is seen as a region of emigration, rather than one of immigration. Both skilled and unskilled workers from the region are an important source of labor in North America, Europe, and Japan, although the outflow is slowing—for example, in the early 2000s Mexico's net migration was -10, but by 2016 it was -1. Many of these immigrants send monthly **remittances** (monies sent back home) to sustain family members. Peaking at nearly US$70 billion in 2008, remittances fell to $61 billion by 2013 but climbed back to $69 billion in 2016. In particular, remittance income to Mexico (Latin America's largest recipient) declined as many Mexicans returned to their country or were removed from the United States, which has stepped up deportation efforts over the last decade. The economic significance of remittances will be discussed again at the end of the chapter.

REVIEW

4.3 What are the historical and economic explanations for urban dominance and urban primacy in Latin America?

4.4 How have policies such as agrarian reform and frontier colonization impacted the patterns of settlement and primary resource extraction in the region?

4.5 Demographically, Latin America has grown much faster than North America. What factors contribute to faster growth, and is this growth likely to continue?

■ **KEY TERMS** latifundia, minifundia, agrarian reform, urban primacy, informal sector, rural-to-urban migration, mestizo, remittances

---

Cultural Coherence and Diversity: Repopulating a Continent

The Iberian colonial experience (1492 to the 1800s) imposed a political and cultural coherence on Latin America that makes it a recognizable world region today. Yet this was not a simple transplanting of Iberia across the Atlantic. Instead, a complex process unfolded in which European and Indian traditions blended as indigenous groups were incorporated into either the Spanish or the Portuguese Empire. In some areas, such as southern Mexico, Guatemala, Bolivia, Ecuador, and Peru, Amerindian cultures have shown remarkable resilience, as evidenced by the survival of indigenous languages. Yet the prevailing pattern is one of forced assimilation in which European religion, languages, and political organization were imposed on the surviving fragments of native society. Later, other cultures—especially 10 million African slaves—added to the cultural mix of Latin America, the Caribbean, and North America. The legacy of the African slave trade will be examined in greater detail in Chapters 5 and 6. For Latin America, perhaps the single most important factor in the dominance of European culture was the demographic collapse of native populations.

Decline of Native Populations

It is difficult to grasp the enormity of cultural change and human loss due to this encounter between the Americas and Europe. Throughout the region, archaeological sites are reminders of the complexity of Amerindian civilizations prior to European contact. Dozens of stone temples found throughout Mexico and Central America, where the Mayan and Aztec civilizations flourished, attest to the ability of these societies to thrive in the area's tropical forests and upland plateaus. The Mayan city of Tikal flourished in the lowland forests of Guatemala, supporting tens of thousands, before its mysterious collapse centuries before the arrival of Europeans (**Figure 4.25**). In the Andes, stone terraces built by the Incas are still being used by Andean farmers. Cuzco—the core of the great Incan empire—and the Incan site of Machu Picchu are testaments to Incan ingenuity. The Spanish, too, were impressed by the sophistication and wealth they saw around them, especially in Tenochtitlán, where Mexico City is today. Tenochtitlán was the political and ceremonial center of the Aztecs, supporting a complex metropolitan area with some 300,000 residents. The largest city in Spain at the time was considerably smaller.

The Demographic Toll The most telling figures of the impact of European expansion are demographic. Experts believe that the precontact Americas had 54 million inhabitants; by comparison, western Europe in 1500 had approximately 42 million. Of those 54 million, about 47 million lived in what is now Latin America, and the rest were in North America and the Caribbean. The region had two major

Explore the **Sights** of Mayan Civilization

http://goo.gl/nIUhH4

▼ **Figure 4.25 Tikal, Guatemala** This ancient Mayan city, located in the lowland forests of the Petén, was part of a complex network of cities located in the Yucatan and northern Guatemala. At its height, Tikal supported over 100,000 residents before its collapse in the late 10th century. Today Tikal is a major tourist destination.

population centers: one in central Mexico with 14 million people and the other in the central Andes (highland Peru and Bolivia) with nearly 12 million. By 1650, after a century and a half of colonization, the indigenous population was one-tenth its precontact size. The human tragedy of this population loss is difficult to comprehend. The relentless elimination of 90 percent of the indigenous population was largely caused by epidemics of influenza and smallpox, but warfare, forced labor, and starvation due to a collapse of food production systems also contributed to the rapid population decline.

Interestingly, one legacy of the Amerindian collapse was an extensive recovery of forest, wildlife, and soils due to reduced human pressure. Geographer William Denevan has argued that a **pristine myth** has been perpetuated about Latin America being a sparsely settled wilderness when the Spanish arrived in 1492. In fact, there is tremendous evidence that the native population cleared forest, created grasslands, built urban centers, and experienced serious problems with soil erosion. But by 1750, as the colonial period was ending, the extensive human presence throughout Latin America was much less evident than it was when the Spanish first arrived. The population low point for Amerindians was in 1650, but the tragedy continued throughout the colonial period and to a much lesser extent continues today. After the indigenous population began its slow recovery in the central Andes and central Mexico, there were still tribal bands in southern Chile (the Mapuche) and Patagonia (the Araucania) that experienced the ravages of disease three centuries after Columbus landed. Even now, the isolation of some Amazonian tribes has made them vulnerable to disease.

The Columbian Exchange and Global Food Systems Historian Alfred Crosby likens the contact period between the Old World (Europe, Africa, and Asia) and the New World (the Americas) to an immense biological swap, which he terms the **Columbian exchange**. According to Crosby, Europeans benefited greatly from this exchange, and Amerindian peoples suffered the most from it. On both sides of the Atlantic, however, the introduction of new diseases, peoples, plants, and animals forever changed the human ecology, especially the foods we eat.

Consider, for example, the introduction of Old World crops. The Spanish brought their staples of wheat, olives, and grapes to plant in the Americas. Wheat did surprisingly well in the highland tropics and became a widely consumed grain over time. Grapes and olive trees did not fare as well, but eventually grapes were produced commercially in the temperate zones of the Americas. The Spanish grew to appreciate the domestication skills of Indian agriculturalists, who had developed valuable starch crops such as corn, potatoes, and bitter manioc, as well as condiments such as hot peppers, tomatoes, pineapple, cacao, and avocados. Corn never became a popular food for direct consumption by Europeans, but many African peoples adopted it as a vital staple. And corn as an animal grain and a sweetener (corn syrup) is an integral part of most foods today. Yet for many Mexicans and Central Americans, corn tortillas are considered a basic and daily food item (**Figure 4.26**).

After initial reluctance, Europeans and Russians widely consumed the potato as a basic food. Domesticated in the highlands of Peru and Bolivia, the humble potato plant has an impressive ability to produce a tremendous volume of food in a very small area, especially in cool climates. This root crop is credited with driving Europe's rapid population increase in the 18th century, when peasant farmers from Ireland to Russia became increasingly dependent on it as a basic food. This potato dependence also made them vulnerable to potato blight, a fungal disease that emerged in the 19th century and came close to unraveling Irish society.

Tropical crops transferred from Asia and Africa reconfigured the economic potential of the region. Sugarcane, an Asian transfer, became the dominant cash crop of the Caribbean and the Atlantic tropical lowlands of South America. With sugar production came the importation of millions of African slaves. Coffee, a later transfer from East Africa, emerged as one of the leading export crops throughout Central America, Colombia, Venezuela, and Brazil in the 19th century. Pasture grasses introduced from Africa enhanced the forage available to livestock. Rice, from Africa and Asia, also became a critical addition to Latin American diets.

The movement of Old World animals across the Atlantic had a profound impact on the Americas. Initially, these animals hastened Indian decline by introducing animal-borne diseases and by producing feral offspring that consumed everything in their paths. However, the utility of domesticated swine, sheep, cattle, and horses was eventually appreciated by native survivors. Draft animals were adopted, as was the plow, which facilitated the preparation of soil for planting. Wool became a very important fiber for indigenous communities in the uplands.

Explore the **Tastes** of Tortillas

http://goo.gl/p7XnNY

▼ **Figure 4.26 Daily Tortillas** Corn tortillas are the daily starch consumed by Mexicans and northern Central Americans. Here two women prepare tortillas for sale on the streets of Antigua, Guatemala. Corn was first domesticated in Mexico but is grown throughout the world as a food and animal grain.

Slowly, pork, chicken, and eggs added protein and diversity to the staple diets of corn, potatoes, and cassava. With the major exception of disease, many transfers of plants and animals ultimately benefited both worlds. Still, it is clear that the ecological and material basis for life in Latin America was completely reworked through the exchange process initiated by Columbus.

Amerindian Survival and Political Recognition Presently, Mexico, Guatemala, Ecuador, Peru, and Bolivia have the largest indigenous populations. Not surprisingly, these areas had the densest native populations at contact. Amerindian survival also occurs in isolated settings where the workings of national and global economies are slow to penetrate. The isolated Miskito Coast of Honduras is home to Miskito, Pech, and Garífuna. In eastern Panama, the Guna and Emberá are present. In the Brazilian state of Roraima in the northern Amazon, some 15,000 indigenous Pemon speakers organized in 2004 to create the Raposa/Serra do Sol Indian reservation. In these relatively isolated areas, small groups of people have managed to maintain a distinct way of life despite the pressures to assimilate.

In many cases, Indian survival comes down to one key resource—land. Indigenous peoples who are able to maintain a territorial home, formally through land title or informally through long-term occupancy, are more likely to preserve a distinct ethnic identity. Because of this close association between identity and territory, native peoples are increasingly insisting on a recognized space within their countries. Some of Panama's Amerindians have organized indigenous territories called *comarcas* where they assert local authority and have limited autonomy. The comarca of Guna Yala, on the Caribbean coast of eastern Panama, is the recognized territory of some 40,000 Guna. These efforts to define indigenous territory are seldom welcomed by the state, but they are occurring throughout the region.

From Amazonia to the highlands of Chiapas, many native groups are demanding formal political and territorial recognition as a means to redress centuries of injustice. The Zapatista rebellion in southern Mexico began on January 1, 1994, in Chiapas, the day NAFTA took effect. Although Zapatista supporters—largely Amerindian peasants—are mostly interested in access to land and basic services, their movement reflects a general concern about how increased foreign trade and investment hurts rural peasants. Bolivia witnessed the organized protests of Amerindians who inhabit the city of El Alto, overlooking La Paz. Angry with their pro-mining and pro-trade president, the Indians set up roadblocks in 2004 that cut off La Paz from supplies. In the end, indigenous-led protests forced two presidents to resign in two years and led to the election of Bolivia's first Amerindian president, Evo Morales, in 2005. Such experiences show the greater involvement of organized Indian groups in national politics and, some might argue, a deepening of democracy. Still others express concern that ethnically driven politics could lead to national fragmentation.

Patterns of Ethnicity and Culture

The Indian demographic collapse enabled Spain and Portugal to refashion Latin America into a European likeness. Yet instead of a neo-Europe rising in the tropics, a complex ethnic blend evolved. Beginning with the first years of contact, unions between European sailors and Indian women began the process of racial mixing that over time became a defining feature of the region. The courts of Spain and Portugal officially discouraged racial mixing, but were unable to prevent it. Spain, which had a far larger native population to oversee than did the Portuguese in Brazil, became obsessed with the matter of race and maintaining racial purity among its colonists. An elaborate classification system was constructed to distinguish emerging racial castes. Thus, in Mexico in the 18th century a Spaniard and an Indian union resulted in a *mestizo* child. A child of a mestizo and a Spanish woman was a *castizo*. However, the children from a castizo woman and a Spanish man were considered Spanish in Mexico, but a quarter mestizo in Peru. Likewise, *mulattoes* were the progeny of European and African unions, and *zambos* were the offspring of Africans and Indians.

After generations of intermarriage, such a classification system collapsed under the weight of its complexity, and four broad categories resulted: *blanco* (European ancestry), *mestizo* (mixed ancestry), *indio* (Indian ancestry), and *negro* (African ancestry). The blancos (or Europeans) continue to be well represented among the elites, yet the vast majority of Latin Americans are of mixed racial ancestry. Dia de la Raza, the region's observance of Columbus Day, recognizes the emergence of a new mestizo race as the legacy of European conquest. Throughout Latin America, more than other regions of the world, miscegenation—or racial mixing—is the norm, which makes the process of mapping racial or ethnic groups especially difficult.

Languages Roughly two-thirds of Latin Americans are Spanish speakers, and one-third speak Portuguese. These colonial languages were so prevalent by the 19th century that they were the unquestioned languages of government and instruction for the newly independent Latin American republics. In fact, until recently many countries actively discouraged, and even repressed, Indian tongues. It took a constitutional amendment in Bolivia in the 1990s to legalize native-language instruction in primary schools and to recognize the country's multiethnic heritage (more than half the population is Amerindian, and Quechua, Aymara, and Guaraní are widely spoken) (**Figure 4.27**).

Because Spanish and Portuguese dominate, there is a tendency to neglect the influence of indigenous languages in the region. Mapping the use of indigenous languages, however, reveals important pockets of Indian resistance and survival. In the central Andes of Peru, Bolivia, and southern Ecuador, more than 10 million people still speak Quechua and Aymara, along with Spanish. In Paraguay and lowland Bolivia, there are 4 million Guaraní speakers, and in southern Mexico and Guatemala at least 6 to

Explore the **Sounds** of Quechua Language

http://goo.gl/9sNLBG

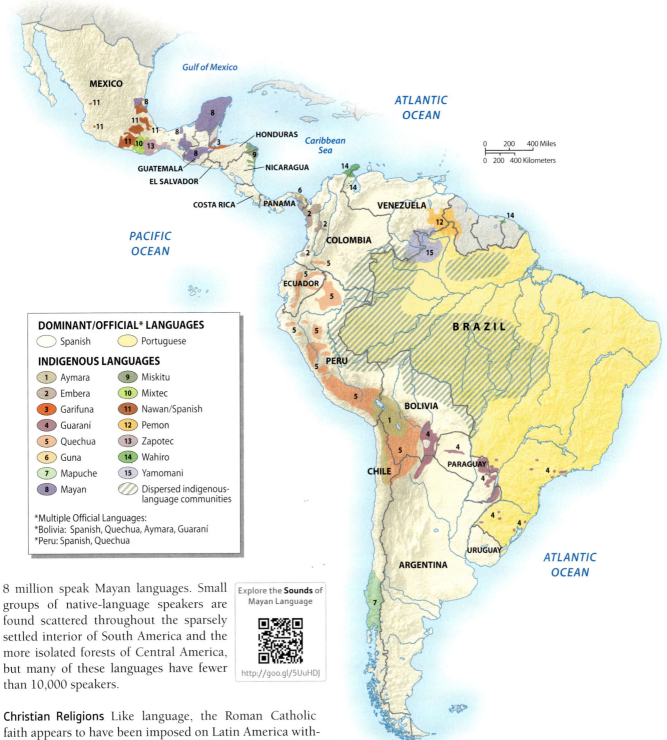

▲ Figure 4.27 Languages of Latin America The dominant languages of the region are Spanish and Portuguese. Nevertheless, there are significant areas in which indigenous languages persist and, in some cases, are recognized as official languages. Smaller language groups exist in Central America, the Amazon Basin, and southern Chile. Q: What does this language map tell us about the patterns of Amerindian survival and endurance in Latin America?

DOMINANT/OFFICIAL* LANGUAGES
- Spanish
- Portuguese

INDIGENOUS LANGUAGES
1. Aymara
2. Embera
3. Garifuna
4. Guaraní
5. Quechua
6. Guna
7. Mapuche
8. Mayan
9. Miskitu
10. Mixtec
11. Nawan/Spanish
12. Pemon
13. Zapotec
14. Wahiro
15. Yamomani
- Dispersed indigenous-language communities

*Multiple Official Languages:
*Bolivia: Spanish, Quechua, Aymara, Guaraní
*Peru: Spanish, Quechua

8 million speak Mayan languages. Small groups of native-language speakers are found scattered throughout the sparsely settled interior of South America and the more isolated forests of Central America, but many of these languages have fewer than 10,000 speakers.

Explore the **Sounds** of Mayan Language

http://goo.gl/5UuHDJ

Christian Religions Like language, the Roman Catholic faith appears to have been imposed on Latin America without challenge. Most countries report 90 percent or more of their population as Catholic. Every major city has dozens of churches, and even the smallest hamlet maintains a graceful church on its central square. Throughout Central America, Brazil, and Uruguay, however, a sizable portion of the population practice Protestant evangelical faiths.

The demographic core of the Catholic Church is in Latin America, not Europe. Worldwide, Brazil has the largest Catholic population (150 million) followed by Mexico (106 million). In 2013 a new pope was selected, and for the first time the spiritual leader of nearly 1.2 million Roman Catholics was from Latin America. Pope Francis, formerly Bishop Jorge Mario Bergoglio, is the son of Italian immigrants and was born in Buenos Aires, Argentina. As a religious leader and member of the Jesuit Order, he earned a reputation for his humility and devotion to the poor. Now, as leader of the Catholic Church, he oversees a vast global network of churches, schools, missions, and clergy that look to his guidance from Rome.

Throughout Latin America, **syncretic religions**, or blends of different belief systems, enabled pre-Hispanic religious practices to be folded into Catholic worship. These blends took hold and endured, in part because Christian saints were easy surrogates for pre-Christian gods and because the Catholic Church tolerated local variations in worship as long as the process of conversion was under way. The Mayan practice of paying tribute to spirits of the underworld seems to be replicated today in Mexico and Guatemala via the practice of building small cave shrines to favorite Catholic saints and leaving offerings of fresh flowers and fruits. One of the most celebrated religious icons in Mexico is the Virgin of Guadalupe—a dark-skinned virgin seen by an Indian shepherd boy in the 16th century—who became the patron saint of Mexico.

Syncretic religious practices also evolved and endured among African slaves. By far the greatest concentration of slaves was in the Caribbean, where they were transported to replace the indigenous population, which was wiped out by disease (see Chapter 5). Within Latin America, the Portuguese colony of Brazil received the most Africans—at least 4 million. In Brazil, where the volume and the duration of the slave trade were the greatest, the transfer of African-based religious systems is most evident. West African–based religious systems such as Batuque, Umbanda, Candomblé, and Shango are often mixed with, or ancillary to, Catholicism and are widely practiced in Brazil. In many parts of southern Brazil, Umbanda is as popular with people of European ancestry as with Afro-Brazilians. Typically, a person becomes familiar with Umbanda after falling victim to a magician's spell by having some object of black magic buried outside his or her home. To regain control of his or her life, the victim needs the help of a priest or priestess.

The syncretic blend of Catholicism with African traditions is most obvious in the celebration of carnival, Brazil's most popular festival and one of the major components of Brazilian national identity. The three days of carnival, known as the Reign of Momo, combine Christian Lenten beliefs with pagan influences and feature African musical traditions epitomized by the rhythmic samba bands. Although the street festival was banned for part of the 19th century, Afro-Brazilians in Rio de Janeiro resurrected it in the 1880s with nightly parades, music, and dancing. Within 50 years, the street festival had given rise to formalized samba schools and helped break down racial barriers. By the 1960s, carnival became an important symbol for Brazil's multiracial national identity. Today the festival—which is most closely associated with Rio de Janeiro—draws thousands of participants from all over the world (**Figure 4.28**).

▲ **Figure 4.28 Carnival in Rio de Janeiro** Samba schools, such as this one, compete each year during Carnival for the best costumes and music. Rio de Janeiro's Carnival is a spectacle that draws thousands of revelers.

The Global Reach of Latino Culture

Latin American culture, vivid and diverse as it is, is widely recognized throughout the world. Whether it is the sultry pulse of the tango or the fanaticism with which Latinos embrace soccer as an art form, aspects of Latin American culture have been absorbed into the global world culture. In the arts, Latin American writers such as Gabriel García Marquez, Jorge Luis Borges, Octavio Paz, and Isabel Allende have obtained worldwide recognition. In terms of popular culture, musical artists such as Colombia's Shakira and Brazil's hip-hop samba singer Max de Castro reach international audiences. Through music, literature, and even *telenovelas* (soap operas), Latino culture is being transmitted to an eager worldwide audience (see *Everyday Globalization: The Zumba Sensation*).

Telenovelas Popular nightly soap operas are a mainstay of Latin American television. These tightly plotted series are filled with intrigue and double dealing. Unlike their counterparts in the United States, they end, usually after 100 episodes. Once standard fare for the working class, many telenovelas take hold and absorb an entire nation. During particularly popular episodes, the streets are noticeably calm as millions of people tune in to catch up on the lives of their favorite heroines. Brazil, Venezuela, and Mexico each produce scores of telenovelas, but the Mexican ones are international mega-hits.

Televisa, a Mexican production agency, has aggressively marketed its inventory of soap operas to an eager global public. Mexican telenovelas are avidly watched in countries as diverse as Croatia, Russia, China, South Korea, Iran, the United States, and France, as well as throughout Latin America. Predictably scripted as Mexican Cinderella stories, these sagas of poor underclass women (often domestics) falling in love with members of the elite, battling jealous rivals, and ultimately emerging triumphant seem to resonate with fans around the world. In addition to their broad appeal, telenovelas are big business, perhaps Mexico's largest cultural export. While Hollywood and Mumbai grind out movies for theaters, much of Mexico's entertainment industry is geared toward producing this popular home art form.

Soccer Perhaps the quintessential global sport, soccer has a fanatical following throughout much of the world. Yet it is in

The Zumba Sensation

Go to a local gym anytime, or an urban park in the summer, and you may hear the pulsing sounds of merengue or reggaeton and see an enthusiastic, well-toned instructor leading a class in a fast-paced hip-swinging exercise dance called *zumba*. Zumba began in Colombia. The story goes that a popular fitness teacher, Alberto Perez, was leading an aerobics class in the 1990s but forgot his music. He rushed to his car and grabbed cassettes featuring cumbia and salsa. As he improvised the steps to this Latin music, the students loved it. Perez eventually moved to Miami, trademarked the class style, and called it zumba, a word with no meaning but easy to remember. Zumba took off in 2001, and hasn't slowed down.

Zumba's appeal is its vast sample of Latin music including salsa, samba, mambo, and soca. The dance moves are a mix of Latin American–inspired steps mixed with aerobics. There are even Zumba performing artists, such as pop star Cláudia Leitte from Brazil, who performs some of the most popular Zumba songs. The combination of fast tempo, upbeat music, and fun moves has made Zumba a global sensation.

While most Zumba enthusiasts are women, men are welcome to shake it as well. There are Zumba classes for all ages and ability levels. In cities all over the world people will exercise if it's fun—and Zumba fits the bill. In 2015, 13,000 Filipinos gathered for an outdoor Zumba class, breaking the Guinness World Record for the largest class size (**Figure 4.3.1**).

1. Identify factors that make this type of group exercise globally appealing.

2. Have you ever taken a Zumba class, if so, where?

▲ **Figure 4.3.1. Zumba in the Philippines** What began in Colombia as an aerobics alternative with Latin music has become a global exercise craze. Nearly 13,000 Filipinos in the town of Mandaluyong participated in a zumba session in July 2015, earning a place in the Guinness Book of World Records for the largest zumba class.

Latin America, and especially in South America, where *fútbol* is considered a cultural necessity. Still largely a male game, young boys and men are constantly seen on fields, beaches, and blacktops playing soccer, especially in late afternoons and on weekends. This is beginning to change in some countries as women take up the sport, especially in Brazil. The great soccer stadiums of Buenos Aires (Bombonera) and Rio de Janeiro (Maracaña) are regarded as shrines to the game. Many individuals use the victories and losses of their national soccer teams as the important chronological markers of their lives.

Pelé, the Brazilian soccer phenomenon of the 1960s and 1970s, introduced the free-flowing acrobatic style that became known as "the beautiful game." Today Latino soccer stars such as Argentine Lionel Messi, Uruguayan Luis Suarez, and Brazilian Neymar da Silva Santos play for corporate clubs in Europe and earn millions. Latin Americans also fill up the few slots allotted to foreign players on the U.S. Major League Soccer teams. Yet the dream of many Latin American soccer players is to be on the national team and bring home the World Cup. Visit any Latin American country when its team is playing a World Cup qualifying match, and the streets are eerily quiet. Of the 20 World Cups awarded between 1930 and 2014, South American teams have won 9. In 2014, Brazil hosted the World Cup but lost to Germany in the semifinals, in a game that few Brazilians want to remember. Germany ended up winning the cup by beating Argentina. As Latin Americans emigrate (both as players and as laborers), they bring their enthusiasm for the sport with them.

National Identities Viewed from the outside, the region displays considerable homogeneity; yet distinct national identities and cultures flourish in Latin America. Since the early days of the republics, countries celebrated particular elements from their pasts when creating their national histories. In the case of Brazil, the country's interracial characteristics were highlighted to proclaim a new society in which the color lines between Europeans and Africans ceased to matter—although racism against Afro-Brazilians does persist. Mexico celebrated the architectural and cultural achievements of its Aztec predecessors, while at the same time forging an assimilationist strategy that discouraged surviving indigenous culture and language.

Musical and dance traditions evolved and became emblematic of these new societies: The tango in Argentina, caporales in Bolivia, the vallenato and cumbia in Colombia, the mariachi in Mexico, and the samba in Brazil are easily distinguished styles that are representative of distinct national cultures. Literature also reflects the distinct identities found in Latin America. Writers such as Isabel Allende, Gabriel García Marquez, Mario Vargas Llosa, Carlos Fuentes, and Jorge Amado situate their stories in their native countries and, in so doing,

Explore the **Sounds** of Samba Music

http://goo.gl/LWiOpK

celebrate the unique characteristics of Chileans, Colombians, Peruvians, Mexicans, and Brazilians. Distinct political cultures also evolved, occasionally leading to conflict with neighbors.

REVIEW

4.6 Identify factors that contributed to racial mixing in Latin America, and locate areas of strongest Amerindian survival on a map.

4.7 What are the cultural legacies of Iberia in Latin America, and how are they expressed?

■ **KEY TERMS**   pristine myth, Columbian exchange, syncretic religions

Geopolitical Framework: From Two Iberian Colonies to Many Nations

Latin America's colonial history, more than its present condition, unifies this region geopolitically. For the first 300 years after the arrival of Columbus, Latin America was a territorial prize sought by various European countries, but effectively settled by Spain and Portugal. By the early 19th century, the independent states of Latin America had formed but continued to experience foreign influence and, at times, overt political pressure, especially from the United States. At other times, a more neutral hemispheric vision of American relations and cooperation has held sway, represented by the formation of the **Organization of American States (OAS)**. The present organization was chartered in 1948, but its origins date back to 1889. Yet there is no doubt that U.S. policies toward trade, economic assistance, political development, and at times military intervention are often seen as undermining the sovereignty of these states.

Today the geopolitical influence of the United States in the region is declining, especially in South America. In many South American countries, trade with the European Union, China, and Japan is as important as, if not more important than, trade with the United States. For example, Brazil's largest trading partner is now China. Brazil's influence in the region and the world is rising as Latin America's largest economy and host of the 2016 Summer Olympics. Another sign of Brazil's economic clout is the appointment of Brazilian Roberto Azevêdo as the Director of the World Trade Organization in 2013.

Within Latin America, cycles of intraregional cooperation and antagonism have occurred. Neighboring countries have fought over territory, closed borders, imposed high tariffs, and cut off diplomatic relations. Even today a dozen long-standing border disputes in Latin America have the potential to erupt into conflict. The last two decades or so have witnessed a revival in the trade bloc concept, with the formation of Mercosur (the Southern Common Market) in South America, NAFTA in North America and Mexico, and CAFTA, the Central American Free Trade Association. In 2008, Brazil proposed the formation of the **Union of South American Nations (UNASUR)**, which includes all the states of South America except French Guiana. Unlike NAFTA and CAFTA, UNASUR is more a political project

than a trade group, and its influence remains to be seen. In 2011, however, the trade-oriented **Pacific Alliance** was formed, and includes Mexico, Colombia, Peru, and Chile with Costa Rica and Panama in the process of joining. Both groups mark a significant change in the region's geopolitical orientation.

Iberian Conquest and Territorial Division

When Christopher Columbus claimed the Americas for Spain, the Spanish became the first active colonial agents in the Western Hemisphere. In contrast, the Portuguese presence in the Americas was the result of the **Treaty of Tordesillas** in 1493–1494. At that time, Portuguese navigators had charted much of the coast of Africa in an attempt to find an ocean route to the Spice Islands (Moluccas) in Southeast Asia. With the help of Columbus, Spain sought a western route to the Far East. When Columbus landed in the Americas, Spain and Portugal asked the Pope to settle how these new territories should be divided. Without consulting other European powers, the Pope divided the Atlantic world in half—the eastern half, containing the African continent, was awarded to Portugal, and the western half, with most of the Americas, was given to Spain. The line of division established by the treaty actually cut through the eastern edge of South America, placing it under Portuguese control. The treaty was never recognized by the French, English, or Dutch, who also asserted territorial claims in the Americas, but it did provide the legal apparatus for the creation of a Portuguese territory in America—Brazil—which would later become the largest and most populous state in Latin America (**Figure 4.29**).

Six years after the treaty was signed, Portuguese navigator Alvares Cabral inadvertently reached the coast of Brazil on a voyage to southern Africa. The Portuguese soon realized that this territory was on their side of the Tordesillas line. Initially, they were unimpressed by what Brazil had to offer; there were no spices or major indigenous settlements. Over time, they came to appreciate the utility of the coast as a provisioning site, as well as a source for brazilwood, used to produce a valuable dye. Portuguese interest in the territory intensified in the late 16th century with the development of sugar estates and the expansion of the slave trade, and in the 17th century with the discovery of gold in the Brazilian interior.

Spain, in contrast, aggressively pursued the conquest and settlement of its new American territories from the very start. After discovering little gold in the Caribbean, by the mid-16th century Spain's energy was directed toward developing the silver resources of central Mexico and the central Andes (most notably Potosí in Bolivia). Gradually, the economy diversified to include some agricultural exports, such as cacao (for chocolate) and sugar, as well as a variety of livestock. In terms of foodstuffs, the colonies were virtually self-sufficient. Some basic manufactured items, such as crude woolen cloth and agricultural tools, were also produced but, in general, manufacturing was forbidden in the Spanish American colonies to keep them dependent on Spain.

Revolution and Independence Not until the rise of revolutionary movements between 1810 and 1826 was Spanish authority on the mainland challenged. Ultimately, elites born

Latin America in 1650

ATLANTIC OCEAN

SANTO DOMINGO 1511

New Spain

MEXICO 1529

GUATEMALA 1544

NUEVA GALICIA 1549

SANTA FE 1549

PANAMA 1538 & 1567

New Granada

Unexplored Spanish Territory

Treaty of Tordesillas, 1494

QUITO 1563

Unexplored Spanish Territory

PACIFIC OCEAN

LIMA 1542

CHARCAS 1559

La Plata

CHILE 1565 & 1609

Viceroyalty of New Spain
Viceroyalty of Peru
Brazil

0   500   1000 Miles
0   500   1000 Kilometers

Unexplored Spanish Territory

(a)

Latin America in 1830
States with date of independence

ATLANTIC OCEAN

MEXICO 1821

Mexico City   Veracruz

UNITED PROVINCES OF CENTRAL AMERICA 1823–1839

Caracas

Bogotá

GRAN COLOMBIA 1819–1830

Quito

Manaus

Natal

BRAZIL 1822

PERU 1821

Lima

La Paz

BOLIVIA 1825

Salvador

PACIFIC OCEAN

PARAGUAY 1811

Asunción

Río de Janeiro

São Paulo

CHILE 1817

Santiago

URUGUAY 1828

Buenos Aires   Montevideo

UNITED PROVINCES OF LA PLATA 1816

0   500   1000 Miles
0   500   1000 Kilometers

(b)

▲ **Figure 4.29 Shifting Political Boundaries of Latin America** The evolution of political boundaries in Latin America began with the Treaty of Tordesillas, which gave much of the Americas to Spain and a slice of South America to Portugal. The larger Spanish territory was gradually divided into viceroyalties and audiencias, which formed the basis for many modern national boundaries. The 1830 borders of these newly independent states were far from fixed. Bolivia would lose its access to the coast; Peru would gain much of Ecuador's Amazon; and Mexico would be stripped of its northern territory by the United States.

in the Americas gained control, displacing the representatives of the crown. In Brazil, the evolution from Portuguese colony to independent republic was a slower and less violent process that spanned eight decades (1808–1889). In the 19th century, Brazil was declared a separate kingdom from Portugal with its own monarch, and later it became a republic.

The territorial division of Spanish and Portuguese America into administrative units provided the legal basis for the modern states of Latin America (see Figure 4.29). The Spanish colonies were first divided into two viceroyalties (the administrative units of New Spain and Peru), and within these were various subdivisions that later became the basis for the modern states. (In the 18th century, the viceroyalty of Peru, which included all of Spanish South America, was divided into the viceroyalties of La Plata, Peru, and New Granada.) Unlike Brazil, which evolved from a colony into a single republic, the former Spanish colonies fragmented in the 19th century. Prominent among revolutionary leaders was Venezuelan-born Simon Bolívar (**Figure 4.30**), who advocated

▼ **Figure 4.30 Simon Bolívar** Heroic likenesses of Simon Bolívar (the Liberator) are found throughout South America, especially in his native country of Venezuela. This statue stands in the central plaza of the Andean city of Mérida, Venezuela.

his vision for a new and independent state of Gran Colombia. For a short time (1822–1830), Bolívar's vision was realized, as Colombia, Venezuela, Ecuador, and Panama were combined into one political unit. Similarly, in 1823 the United Provinces of Central America was formed to avoid annexation by Mexico. By the 1830s, this union also broke apart into the states of Guatemala, Honduras, El Salvador, Nicaragua, and Costa Rica.

Today the former Spanish mainland colonies include 16 Latin American states plus 3 Caribbean ones, with a total population of over 400 million. If the Spanish colonial territory had remained a unified political unit, it would now have the third largest population in the world, following China and India.

Persistent Border Conflicts As the colonial administrative units turned into states, it became clear that the territories were not clearly delimited, especially the borders that stretched into the sparsely populated interior of South America. This would later become a source of conflict as the new states struggled to demarcate their boundaries. Numerous border wars erupted in the 19th and 20th centuries, and the map of Latin America has been redrawn many times. Some notable conflicts were the War of the Pacific (1879–1882), in which Chile expanded to the north and Bolivia lost its access to the Pacific; warfare between Mexico and the United States in the 1840s, which resulted in the present border under the Treaty of Hidalgo (1848); and the War of the Triple Alliance (1864–1870), the bloodiest war of the postcolonial period, which occurred when Argentina, Brazil, and Uruguay allied themselves to defeat Paraguay in its claim to control the upper Paraná River Basin. It is estimated that 90 percent of Paraguay's adult males died in this conflict. Sixty years later, the Chaco War (1932–1935) resulted in a territorial loss for Bolivia in its eastern lowlands and a gain for Paraguay. In the 1980s, Argentina lost a war with Great Britain over control of the Falkland, or Malvinas, Islands in the South Atlantic. As recently as 1998, Peru and Ecuador skirmished over a disputed boundary in the Amazon Basin.

Outright war in the region is less common than inactive, but unresolved disputes over international boundaries. Any of a dozen dormant claims can erupt from time to time given the political climate between neighbors. For example, every March in Bolivia, the Dia del Mar (Day of the Sea) recognizes the day that Bolivia lost its coast to Chile to inspire support for regaining it. Other disputed boundaries include the Venezuela–Guyana border, the Peruvian eastern lowlands claimed by Ecuador, and the maritime boundary between Venezuela and Colombia (**Figure 4.31**). Many of these disputes are based on territorial claims dating back to poorly delimited boundaries during the colonial period.

The Mexico–U.S. border is the most fortified border of the Americas. Since 1996, the United States has poured resources into more fencing, surveillance, and biometric forms of identification for all recipients of visas and green cards. Part of the rationale is to prevent illegal border crossings—which have declined dramatically since the late 2000s—in an effort to cut down on the number of undocumented people in the United States. Ironically, scholars who track the flow of undocumented entrants say that intense border security may have had the unintended consequence of reducing circular labor flows because recrossing the border has become so difficult and expensive. As a result, undocumented laborers who used to leave the United States are now staying. In the last couple years, the United States has worked with Mexico to better patrol the Mexico-Guatemala border in the hopes of reducing the number of Central Americans entering Mexico to either work in Mexico or attempt to enter the United States. The other strategy has been to remove or deport undocumented migrants—over 4 million Latin Americans since 2000.

The Trend Toward Democracy Most of the 17 countries in Latin America have celebrated, or will soon celebrate, their bicentennials. Compared with most of the rest of the developing world, Latin Americans have been independent for a long time. Yet political stability is not a hallmark of the region. Among the countries in the region, some 250 constitutions have been written since independence, and military takeovers have been alarmingly frequent. Since the 1980s, however, the trend has been toward democratically elected governments, the opening of markets, and broader popular participation in the political process. Where dictators once outnumbered elected leaders, by the 1990s each country in the region had a democratically elected president. (Cuba, the one exception, will be discussed in Chapter 5.)

Democracy may not be enough for the millions frustrated by the slow pace of political and economic reform. In survey after survey, Latin Americans register their dissatisfaction with politicians and governments. Many of the democratic leaders have also been free-market reformers who are quick to eliminate state-backed social safety nets, such as food subsidies, government jobs, and pensions. Many of the poor and middle class have grown doubtful about whether this brand of neoliberal policy can make their lives better. Even in prosperous Chile, widespread student protests in 2011–2015 demanding lower-cost university tuition and fairer access to and funding for secondary education have left politicians scrambling. Under such conditions, it is not surprising the left-leaning politicians have won presidential elections in Mexico, Brazil, Bolivia, Chile, Nicaragua, Ecuador, and Venezuela. These leaders have not rolled back neoliberal trade reforms, but many have tried to improve social services and attempted to reduce income inequalities.

Regional Organizations

At the same time that democratically elected leaders struggle to address the pressing needs of their countries, political developments at the supranational and subnational levels pose new challenges to their authority. The most discussed **supranational organizations** (governing bodies that include several states) are the trade blocs, the newest ones being UNASUR and the Pacific Alliance. **Subnational organizations** (groups that represent areas or people within a state) form along ethnic or ideological lines. These organizations can have positive or destabilizing impacts. Examples include native groups that seek territorial recognition (such as the Guna in Panama), insurgent groups (such as the FARC [Revolutionary Armed Forces of Colombia] or the Zapatistas in Mexico) that

United States (top left, map area)

Chiapas. *Territorial dispute between Amerindian groups and the Mexican state.*

Guatemala/Mexico Border. *Increased border enforcement to deter Central Americans from migrating north.*

Gulf of Venezuela. *Maritime boundary and resource conflict between Colombia and Venezuela.*

U.S.–Mexican Border. *Intensified border security since 1996 over contraband trade and undocumented migration.*

Essequibo. *Boundary dispute between Venezuela and Guyana.*

Cordillera del Condor. *Boundary dispute between Peru and Ecuador, fighting in the 1990s.*

ECONOMIC TRADE BLOCS
- CAFTA-DR
- Mercosur
- NAFTA
- Pacific Alliance
- UNASUR

▶ **Figure 4.31 Geopolitical Issues in Latin America** Of the five economic trade blocs depicted, Mercosur, NAFTA and the Pacific Alliance are the most dynamic. As UNASUR develops, it could unite all of South America into a single common market but that seems unlikely at this point. Members of the Central American Common Market signed an agreement in 2004 to form CAFTA (Central American Free Trade Association), which also includes the Dominican Republic. **Q: How could the growth and strength of trade blocs impact how Latin America functions as a region?**

Falkland/Malvinas Islands. *Territorial dispute between the United Kingdom and Argentina.*

have challenged the authority of the state and, more recently, drug cartels such as the *Zetas* and *Sinaloa,* which have terrorized Mexican society with extreme violence since 2006.

Trade Blocs In the 1990s, Mercosur and NAFTA emerged as supranational structures that could influence development in the region (see Figure 4.31). For Latin America, the lessons from Mercosur in particular led Brazil to propose UNASUR in 2008, uniting virtually all of South America, and the creation of the Pacific Alliance in 2011 to expand trade with Asia.

NAFTA took effect in 1994 as a free-trade area that would gradually eliminate tariffs and ease the movement of goods among the member countries (Mexico, the United States, and Canada). NAFTA has increased intraregional trade, but has provoked considerable controversy about costs to the environment and to employment (see Chapter 3). NAFTA did prove, however, that a free-trade area combining industrialized and developing states was possible. In 2004, the United States, five Central American countries—Guatemala, El Salvador, Nicaragua, Honduras, and Costa Rica—and the Dominican Republic signed CAFTA. CAFTA, like NAFTA, aims to increase trade and reduce tariffs among member

countries. The treaty was fully ratified in 2009, but whether such a treaty will lead to more economic development in Central America is still debated.

Mercosur was formed in 1991 with Brazil and Argentina, the two largest economies in South America, and the smaller states of Uruguay and Paraguay as members. Since its formation, trade among these countries has grown, so much so that Venezuela joined as a full member in 2012. This success is significant in two ways: It reflects the growth of these economies and the willingness to put aside old rivalries (especially long-standing antagonisms between Argentina and Brazil) for the economic benefits of cooperation. Yet after initial trade expansion, trade within the group accounted for only 14 percent of its members' total trade in 2014 and the bloc has yet to be able to sign a trade agreement with the EU.

In 2008, Brazil initiated the formation of the 12-member UNASUR. Some see this as an assertion of Brazil's greater political and economic clout in the region, since it is Latin America's largest country and the world's ninth largest economy. UNASUR includes all countries in South America, minus French Guiana, which is a territory of France. It has formally organized with a permanent secretariat, and has responded to political crises in Bolivia (2008), Ecuador (2010), and Paraguay (2012). Significantly, UNASUR was a Brazilian-led effort, not one led by the United States. Brazil's lead in this initiative underscores its larger geopolitical ambitions to influence South American development and to secure a permanent seat on the United Nations Security Council.

In contrast, the Pacific Alliance formed in 2011 but has made many concrete moves to promote cooperation and trade among the member states and to increase bargaining power when creating trade deals with Asia, especially China (see Figure 4.31). The alliance is also seen as a geopolitical counterweight to Brazil's influence in the region. Led by Mexico, Colombia, Peru, and Chile, the members of the Pacific Alliance represent nearly half of the region's world trade. The much smaller economies of Costa Rica and Panama are in the process of joining.

Insurgencies and Drug Cartels Guerilla groups such as the FARC in Colombia have controlled large territories of their countries through the support of those loyal to the cause, along with theft, kidnapping, and violence. The FARC, along with the ELN (National Liberation Army), gained wealth and weapons through the drug trade. The level of violence in Colombia escalated further with the rise of paramilitary groups—armed private groups that terrorize those sympathetic to insurgency. The paramilitary groups have been blamed for hundreds of politically motivated murders each year. As many as 2.5 million Colombians have been internally displaced by violence since the late 1980s, most fleeing rural areas for towns and cities. Fortunately, after more than a decade of negotiations, in 2014 unilateral cessation of FARC hostilities was agreed to, and the situation has improved considerably. While the level of violence has fallen, Colombia remains the world's largest cocaine producer, followed by Peru and Bolivia (see *Geographers at Work: Development Work in Post-Conflict Colombia*).

Geographers at Work

Development Work in Post-Conflict Colombia

Corrie Drummond Garcia, who works for a U.S. Agency for International Development (USAID) program working on post-conflict issues in Colombia, says that her geography studies "provided a background in several disciplines that can be applicable in many types of jobs" (**Figure 4.4.1**). What fascinates Drummond Garcia about Colombia is its regional diversity: "It is a country with a strong sense of national pride, but also strong regional identities."

Building a Career Drummond Garcia discovered geography as an undergraduate at Bucknell University where she had a professor who could "take what is happening in the world on any particular day and bring it to your attention in ways you never thought about . . . it made me totally rethink what geography was and how it affected my analysis of the world." She later brought these analytical skills to a stint with the Peace Corps in El Salvador, where she found that she enjoyed working in Latin America with marginalized groups. After earning her MA in Geography at George Washington University and working for the Pan American Development Foundation, she joined USAID, working in Washington DC, Haiti, and now Colombia.

Post-Conflict Development After decades of violence, Colombia is in a post-conflict phase where many rural areas need development or redevelopment. An overarching concern is Colombia's displaced population, estimated to be at least 2.5 million. Legislation enacted in 2011 assists victims of violence and helps those uprooted by violence return to their land, though many have been displaced for more than a decade.

Political dysfunction was high during the conflict years, and some groups thrived in the intentional chaos. The challenge now is to build trust and find local solutions. One creative development strategy is to design educational or cultural parks on reclaimed sites once associated with violent episodes. Actively involving local groups in planning and reclaiming sites is the key to community healing and development.

▲ **Figure 4.4.1 Development worker, Corrie Drummond Garcia, in the field** Corrie is wearing protective netting as she visits a beekeeping project in rural Colombia.

Drummond Garcia says, "With regard to conflict, each community has experienced it differently . . . we go into a community without any preconceived notions, and build solutions together."

1. What is unique about development efforts in a post-conflict situation?

2. What skills do geographers bring to the field of development?

▲ **Figure 4.32 The Influence of Drug Cartels in Mexico** This map shows the major areas of influence of Mexican Cartels in 2015 according to the U.S. Drug Enforcement Administration. Cartel influence is fluid, and groups often compete for territories. The Sinaloa, Gulf, and Zeta cartels control the drug trade in large swaths of Mexico, and their influence also crosses into Central America as well.

Drug cartels and gangs in states as diverse as Mexico, Guatemala, El Salvador, Honduras, and Brazil have been blamed for increases in violence and lawlessness. The spike in violence and corruption in Mexico has been especially destabilizing. Profiting from the illegal production and/or shipment of cocaine, marijuana, methamphetamine, and heroin, the cartels generate billions of dollars. Beginning in 2006, the Mexican government brought in the army to quell the violence, kidnapping, and intimidation brought on by cartel groups, especially in the border region, but now extending throughout Mexico and into Central America (**Figure 4.32**). The Mexican government reported some 151,000 murders from 2006 to 2015 and over 20,000 more "disappeared" persons; many of these murders are at the hands of the cartels. Finding ways to stem the violence (rather than the flow of drugs) was one of the biggest issues in the 2012 Mexican election. Enrique Peña Nieto promised a shift toward police work and judicial reform to reduce violence but accusations of corruption and a lack of progress on crime have marred his presidency.

Tragically, the most violent Latin American states on a per capita basis are in Central America. Honduras, El Salvador, and Guatemala have some of the highest homicide rates in the world outside of war zones. These impoverished countries have long been in the transshipment zone of cocaine produced in Colombia, Peru, and Bolivia. More recently, however, Mexico's Sinaloa, Gulf, and Zetas cartels have been active in the isthmus, paying locals in drugs,

creating drug-processing centers and, in the process, driving up the murder rate. The high levels of violence in Central America contributed to the surge in unaccompanied minors crossing through Mexico and into the United States in 2014.

REVIEW

4.8 How did the colonization of Latin America by Iberia lead to the formation of the modern states of Latin America?

4.9 How are trade blocs reshaping the region's geopolitics and development?

■ **KEY TERMS** Organization of American States (OAS), Union of South American Nations (UNASUR), Pacific Alliance, Treaty of Tordesillas, supranational organizations, subnational organizations

Economic and Social Development: From Dependency to Neoliberalism

Most Latin American economies fit into the broad middle-income category set by the World Bank. Clearly part of the developing world, Latin American people are much better off than those in Sub-Saharan Africa, South Asia, and most of China. Still, the economic contrasts are sharp, both between states and within them. Generally, the Southern Cone countries (including southern Brazil and excluding Paraguay) and

TABLE 4.2   Development Indicators

MasteringGeography™

Country	GNI per capita, PPP 2014	GDP Average Annual %Growth 2009-14	Human Development Index (2015)[1]	Percent Population Living Below $3.10 a Day	Life Expectancy (2016)[2] Male	Life Expectancy (2016)[2] Female	Under Age 5 Mortality Rate (1990)	Under Age 5 Mortality Rate (2015)	Youth Literacy (% pop ages 15-24) (2005-2014)	Gender Inequality Index (2015)[3,1]
Argentina	22,500*	4.1	0.836	3.6	73	80	28	13	99	0.376
Bolivia	6,290	5.4	0.662	13.4	68	74	120	38	99	0.444
Brazil	15,570	3.1	0.755	9.1	72	79	11	16	99	0.457
Chile	21,320	4.8	0.832	2.1	77	82	19	8	99	0.338
Colombia	12,910	4.9	0.720	13.8	72	79	34	16	98	0.429
Costa Rica	14,420	4.4	0.766	1.4	77	82	17	10	99	0.349
Ecuador	11,190	5.3	0.732	11.6	73	79	52	22	99	0.407
El Salvador	8,000	1.9	0.666	11.5	68	78	60	17	97	0.427
Guatemala	7,250	3.6	0.627	26.5	69	76	78	29	92	0.533
Honduras	4,570	3.5	0.606	34.6	71	76	55	20	96	0.480
Mexico	16,840	3.3	0.756	10.3	74	79	49	13	99	0.373
Nicaragua	4,790	4.9	0.631	25.2	72	78	66	22	-	0.449
Panama	19,930	8.7	0.780	8.0	75	81	33	17	98	0.454
Paraguay	8,470	6.3	0.679	6.3	71	75	53	21	99	0.472
Peru	11,440	5.9	0.734	9.7	72	77	75	17	99	0.406
Uruguay	20,220	4.8	0.793	<2	74	81	23	10	99	0.313
Venezuela	17,700	1.9	0.762	14.9	72	78	31	15	98	0.476

Source: World Bank, *World Development Indicators, 2016.*

[1]United Nations, *Human Development Report, 2015.*

[2]Population Reference Bureau, *World Population Data Sheet, 2016.*

[3]Gender Inequality Index—A composite measure reflecting inequality in achievements between women and men in three dimensions: reproductive health, empowerment and the labor market that ranges between 0 and 1. The higher the number, the greater the inequality.

*additional data from the CIA Factbook.

Login to MasteringGeography™ **& access** MapMaster **to explore these data!**

1) Rank the countries in this region by GNI per capita PPP and their Human Development Index figure. Compare the maps and describe the relationship between these two measures.

2) Compare the figures for Youth Literacy with Gender Inequality. What patterns do you see?

Mexico are the richest states. The poorest countries in terms of per capita purchasing power parity (PPP) are Nicaragua, Bolivia, Honduras, and Guatemala. Although per capita incomes in Latin America are well below levels of developed countries, the economies are growing, and the region has witnessed steady improvements in various social indicators such as life expectancy, child mortality, and literacy. This is reflected in the relatively strong rankings of Latin American countries in the Human Development Index, most notably Argentina and Chile (Table 4.2). Yet Brazil and Venezuela have experienced serious economic downturns in the last two years, especially with lower commodity prices coupled with political instability. And, in general, growth rates are slowing in most states in the region as the demand for primary commodities, especially from China, is declining.

The economic engines of Latin America are the two largest countries, Brazil and Mexico. According to the International Monetary Fund, Brazil's economy in 2015 was the 9th largest in the world (down from 6th in 2012), and Mexico's was the 15th largest, based on gross domestic product. The region has also seen a reduction in extreme poverty, as the percentage of people living on less than $2 per day dropped from 22 percent in 1999 to 12 percent in 2008. The World Bank

recently adjusted its poverty marker to less than $3.10 a day. At this level some of the poorest Latin American countries such as Nicaragua, Honduras, Guatemala, and even Venezuela have one-quarter of their populations living in poverty.

The path toward economic development in Latin America has been a volatile one. In the 1960s, Brazil, Mexico, and Argentina all seemed poised to enter the ranks of the developed world. Multilateral agencies such as the World Bank and the Inter-American Development Bank loaned money for big development projects: continental highways, dams, mechanized agriculture, and power plants. All sectors of the economy were radically transformed. Agricultural production increased with the application of "green revolution" technology and mechanization (see Chapter 12). State-run industries reduced the need for imported goods, and the service sector ballooned as a result of new government and private-sector jobs. In the end, most countries in Latin America made the transition from predominantly rural and agrarian economies, dependent on one or two commodities, to more economically diversified and urbanized countries with mixed levels of industrialization.

The modernization dreams of Latin American countries were trampled in the 1980s, when debt, currency devaluation, hyperinflation, and falling commodity prices undermined the aspirations of the region. By the 1990s, most Latin American governments had radically changed their economic development strategies. State-run national industries and tariffs were jettisoned for policy reforms that emphasized privatization, direct foreign investment, and free trade, collectively labeled **neoliberalism**. Through tough fiscal policy, increased trade, privatization, and reduced government spending, most countries saw their economies grow and poverty decline. In aggregate, the economies of the region averaged an annual growth rate of 3.5 percent from 2009 to 2014. However, sporadic economic downturns have made these neoliberal policies highly unpopular with the masses, at times causing major political and economic turmoil. Much of the economic growth in the last few years is attributed to a boom in primary exports, driven by an increased demand from China. With demand softening, causing a drop in commodity prices, growth rates have also declined.

Primary Export Dependency

Historically, Latin America's abundant natural resources were its wealth. In the colonial period, silver, gold, and sugar generated tremendous riches for the colonists. With independence in the 19th century, the region began a series of export booms to an expanding world market, including commodities such as bananas, coffee, cacao, grains, tin, rubber, copper, wool, and petroleum. One of the legacies of this export-led development was a tendency to specialize in one or two major commodities, a pattern that continued into the 1950s. During that decade, Costa Rica earned 90 percent of its export earnings from bananas and coffee; Nicaragua earned 70 percent from coffee and cotton; 85 percent of Chilean export income came from copper; half of Uruguay's export income came from wood. Even Brazil generated 60 percent of its export earnings from coffee in 1955; by 2000, coffee accounted for less than 5 percent of the country's exports, yet Brazil remained the world leader in coffee production.

Explore the **Sights** of Brazilian Agriculture

http://goo.gl/Am2ekd

▼ **Figure 4.33 Soy Production in Brazil** Fartura Farm in the state of Mato Grosso, Brazil, embodies the large-scale industrial agriculture that has transformed much of South America into one of the world's largest producers and exporters of soy products.

Agricultural Production

Since the 1960s, the trend in Latin America has been to mechanize agriculture and expand the range of crops produced. Nowhere is this more evident than in the Plata Basin, which includes southern Brazil, Uruguay, northern Argentina, Paraguay, and eastern Bolivia. Soybeans, used for oil and animal feed, transformed these lowlands in the 1980s and 1990s. Brazil is now the second largest soy producer in the world (following the United States) and the world's largest exporter of soy. Argentina is the third largest producer, and production is still increasing: between the late 1990s and 2010, its soy production tripled. The speed with which the Plata and Amazon basins are being converted into soy fields alarms many, as forests and savannas are eliminated, negatively impacting biodiversity and increasing greenhouse gas emissions (**Figure 4.33**). In addition to soy, acres of rice, cotton, and orange trees, as well as the more traditional wheat and sugar, continue to be planted in the Plata Basin.

Coffee looms large in Latin America's connection to global trade. For over a century, Brazil has been the world leader in coffee production. But with over 200 million people, it is also a major coffee consumer as well. Up until 1990, Colombia was the second largest global producer, but in the past two decades Colombia's production has declined and the output of two Southeast Asian countries—Vietnam and Indonesia—has surpassed it. Today, Latin America accounts for 58 percent of the world's coffee production with Brazil, Colombia, and Mexico the leading regional producers (**Figure 4.34**). Over 145 million 60-kilogram bags were produced in the 2015–2016 harvest. Most of the exported coffee is bound for Europe, North America, and Japan.

Other large-scale agricultural production zones include the piedmont of the Venezuelan Llanos (mostly grains), the Pacific slope of Central America (cotton and some tropical fruits), and the Central Valley of Chile and the foothills of Argentina (wine and fruit production). In northern Mexico, water supplied from dams along the Sierra Madre Occidental has turned the valleys in Sinaloa into intensive producers of fruits and vegetables for consumers in the United States. The relatively mild winters in northern Mexico allow growers to produce strawberries and tomatoes during the winter months.

In these cases, the agricultural sector is capital-intensive and dynamic. By using machinery, high-yielding hybrids, chemical fertilizers, and pesticides, many corporate farms are extremely productive and profitable. What these operations

▲ **Figure 4.34 Harvesting Colombian Coffee** Two coffee pickers in Risaralda Department harvest coffee by hand. Noted for its quality coffee, harvests in Colombia have declined in the last decade due to climate change and rural violence.

fail to do is employ many rural people, which is especially problematic in countries where a quarter or more of the population depends on agriculture for its livelihood. Thus, the modernization of agriculture has left behind many subsistence producers who make up the ranks of Latin America's most impoverished people. Interestingly, a few traditional Amerindian foods, such as quinoa, are gaining consumers thanks to a growing appetite for organic and healthy foods. Recently, Peru and Bolivia have experienced a boom in quinoa production and exports, with much of the crop being grown in small and medium-sized highland farms. Bolivian President Evo Morales even declared 2012 the "year of quinoa" in an effort to promote traditional "Indian" foods.

Mining and Forestry

The mining of silver, zinc, copper, iron, bauxite, and gold is an economic mainstay for many countries in the region. Many commodity prices reached record levels in the last decade, boosting foreign exchange earnings, but prices began to fall in 2013. Chile is the world leader in copper production, far surpassing the next two largest producers, Peru and the United States. Mexico and Peru were top silver producers in 2015, but Chile and Bolivia were also top 10 producers. Peru was also Latin America's top gold producer in 2015, followed by Mexico.

Lithium is gaining worldwide interest. This soft, silver-white metal is used to make lightweight batteries, like those in cell phones and laptops. It is also a key metal for electric car batteries. Today the largest producer of lithium is Chile, but the world's largest reserves are in Bolivia, under the Salar de Uyuni in the Altiplano. These reserves are so immense that Bolivia has been dubbed the Saudi Arabia of lithium. But it remains to be seen how and under what terms this critical resource will be extracted from this remote region (**Figure 4.35**).

Like agriculture, mining has become more mechanized and less labor-intensive. Even Bolivia, a country long dependent on tin production, cut 70 percent of its miners from the

payrolls in the 1990s. The measure was part of a broad-based austerity program, yet it suggests that the majority of the miners were not needed. Similarly, the vast copper mines of northern Chile are producing record amounts of copper with few miners. In contrast, gold mining continues to be labor-intensive, offering employment for thousands of prospectors.

Logging is another important, and controversial, extractive activity in the region. Ironically, many of the forest areas cleared for cattle were not systematically harvested. More often than not, all but the most valuable trees were burned. Logging concessions are commonly awarded to domestic and foreign timber companies, which export boards and wood pulp. These one-time arrangements are seen as a quick means for foreign exchange, particularly if prized hardwoods such as mahogany are found. Logging can mean a short-term infusion of cash into a local economy. Yet rarely do long-term conservation strategies exist, making this system of extraction unsustainable. Interest is growing in certification programs to designate wood products that have been produced sustainably. This is due to consumer demand for certified wood, mostly in Europe. Unfortunately, such programs are small, and the lure of profit usually overwhelms the impulse to conserve for future generations.

Explore the **Sights** of Mining in Bolivia

http://goo.gl/r0bzsj

▼ **Figure 4.35 Lithium Mining in Bolivia** The Solar de Uyuni is a salt flat in the Bolivian Andes. Here a drilling operation taps into the brine beneath the surface in search of lithium, a key mineral for battery production.

Several countries rely on plantation forests of introduced species of pines, teak, and eucalyptus to supply domestic fuelwood, pulp, and board lumber. These plantation forests grow single species and fall far short of the complex ecosystems occurring in natural forests. Nonetheless, growing trees for paper or fuel reduces the pressure on other forested areas. Leaders in plantation forestry are Brazil, Venezuela, Chile, and Argentina. Considered Latin America's economic star in the 1990s, Chile relied on timber and wood chips to boost its export earnings. Thousands of hectares of nonnative trees (eucalyptus and pine) have been planted, systematically harvested, and cut into boards or chipped for wood pulp. The expansion of the wood chip business, however, has led to a significant increase in the logging of native forests as well.

The Energy Sector

The oil-rich nations of Venezuela, Mexico, and now Brazil are able to meet their own fuel needs and also earn vital state revenues from oil exports. In 2014, Brazil was the world's 9th largest producer of oil, Mexico was 10th, and Venezuela was 12th. These three countries account for 75 percent of the region's oil production. The largest new oil discovery in recent years has been off the coast of Brazil, and in the past decade oil production has soared, making Brazil Latin America's top oil producer and drawing increased foreign investment. Venezuela's overall production has declined, yet it still earns up to 90 percent of its foreign exchange from petroleum and natural gas products. The Venezuelan oil company, PDVSA, became highly politicized in the 2000s with revenues going to national and international development projects of then President Hugo Chavez. Even though revenues were up, basic investment in extractive infrastructure deteriorated, negatively affecting overall production levels. Although Latin American oil producers do not receive as much attention as those in the Middle East, they are significant. Venezuela was one of the five founding members of the Organization of the Petroleum Exporting Countries (OPEC).

Natural gas production is also increasing in the region. Venezuela and Bolivia have the largest proven reserves of natural gas in Latin America, but Mexico and Argentina are by far the largest producers. In recent years, Argentina's natural gas production has been boosted by new finds in Patagonia. In 2012, Argentina made headlines when then President Christina Fernandez de Kirchner seized majority control of YPF, the major producer of oil and natural gas in Argentina, from a Spanish-owned company, claiming frustration over unnecessary declines in output. Argentina is especially dependent on natural gas for its urban markets and exports; the production declines produced shortages and price increases in the domestic liquefied natural gas market that required action.

In the area of biofuels, Brazil offers a story of sweet success. In the 1970s, when oil prices skyrocketed, then oil-poor Brazil decided to convert its abundant sugarcane harvest into ethanol. Over the years, even when oil prices plummeted, Brazil continued to invest in ethanol, building mills and a distribution system that delivered ethanol to gas stations. One of Brazil's major technological successes was inventing flex-fuel cars that

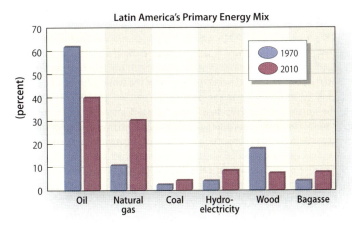

▲ **Figure 4.36** **Latin America's Energy Mix, 1970–2010** As the region's energy sources have increased and diversified, there is much less reliance on wood, and more reliance on natural gas. **Q: What would explain the decline in wood fuel for this region?**

run on any combination of ethanol and gasoline. At the time, Brazil was motivated by its limited oil reserves, but today, with interest in biofuels growing as a way to reduce CO_2 emissions, Brazil's support of its ethanol program looks visionary.

Due to these innovations, Latin America's energy mix has changed considerably since the 1970s (**Figure 4.36**). Nearly 50 years ago, 60 percent of the region's energy was supplied by oil and 20 percent by wood fuel. By 2010, the region's energy supply was more diverse and cleaner, most notably in the growth in natural gas, hydroelectricity, and biofuels (or bagasse). That said, Latin America's energy consumption has also increased fivefold from 1970 to 2010 due to population increase, urban growth, improved transportation, and greater economic activity.

Latin America in the Global Economy

During the 1990s, Latin American governments and the World Bank became champions of neoliberalism as a sure path to economic development. Neoliberal policies accentuate the forces of globalization by turning away from policies that emphasize state intervention and self-sufficiency. Most Latin American political leaders have embraced neoliberalism and the benefits that come with it, such as increased trade, greater foreign direct investment, and more favorable terms for debt repayment. A few countries have rejected these policies, most notably oil-rich Venezuela during the Chavez (1999–2013) and Maduro (2013–present) administrations. Yet a combination of low oil prices, virtually no foreign direct investment, and limited hard currency has created an economic and political crisis for this country where poverty and hunger are on the rise. Despite the turmoil in Venezuela, signs of discontent with neoliberalism are visible throughout the region, in part due to frustration with political corruption and declining commodity prices (see *Exploring Global Connections: Brazil in Crisis*).

Maquiladoras and Foreign Investment Increased foreign investment and the presence of foreign-owned factories are examples of neoliberalism. **Maquiladoras**, the Mexican

Exploring Global Connections

Brazil in Crisis

By all accounts, 2016 was supposed to be a culminating year for Brazil: the country anticipated garnering the global spotlight as host of the summer Olympics, basking in the accomplishments of poverty reduction, and being recognized as the world's 7th largest economy. Yet after hitting an eye-popping 7.6 percent growth rate in 2010, the government of President Dilma Rousseff introduced government control over banks and energy companies, which may have undermined growth rather than supporting it. A series of economic and political crises followed, and in 2016 Rousseff was impeached, the economy was at negative or zero growth, and the Olympic glory was undercut by austerity and concern over the spread of the Zika virus (**Figure 4.5.1**). How did things go so wrong?

An Economic Crisis Global and domestic factors contributed to Brazil's economic recession. The drop in global commodity prices began in 2013 for exports such as iron ore, soy, oil, and metals. These price declines, coupled with slumping demand from China, hurt the country's overall growth. Moreover, an El Niño–induced drought in the northeast negatively impacted agricultural harvests and hydroelectricity production.

Domestically, in order to propel further growth, Rousseff pressured the central bank to reduce interest rates, which fueled more consumer spending and debt. She cut taxes for certain domestic industries and implemented price controls on gasoline and electricity, which led to losses for the public utilities. As her administration attempted to correct for energy price controls, Brazilians found their domestic energy costs increasing while world prices fell. This led to massive street protests in 2013 and 2014 over increased bus fares, the most important form of transportation for the urban poor and middle class. Electricity prices surged more than 40 percent in 2015, and unemployment and inflation rates rose to about 10 percent by 2016. Under these conditions, the country's fragile middle class fear that many of the gains of the past decade could be lost.

And Political Uncertainty Brazil is a large country that has always had serious income inequality. In 2003 with the election of President Lula da Silva of the Workers Party, it was believed that the country would finally work toward fairness and the poor could benefit from Brazil's incredible natural resource base. For 13 years the Worker's Party was in power. One of its signature programs, the Bolsa Familia, is a monthly conditional cash transfer that kept children in school with regular health check-ups, and is credited for substantially reducing poverty levels in Brazil.

Yet corruption of government officials at all levels seems to have undercut Brazil's ability to navigate its way through the current recession. As President Rousseff narrowly won re-election in 2014, a massive bribery scandal involving her popular predecessor, President

▲ **Figure 4.5.1 Brazil's Olympic Games** People take selfies before a set of Olympic rings, created from recycled materials, on Copacabana Beach in Rio de Janeiro. Brazil committed to hosting the Olympics when its economy was booming. By the 2016 games, economic growth was near zero and the president was impeached.

Lula da Silva, and Petrobas (the Brazilian national oil company) rocked the party establishment and infuriated citizens. President Rousseff was impeached and on August 31, 2016 Vice President Michel Temer became the President of Brazil.

Global Impact Part of Brazil's soaring economic performance from 2010 to 2013 was its growing trade with China, its largest trading partner, accounting for half of Brazil's foreign trade. As the Chinese economy slowed, so inevitably, did the Brazilian economy. One of the BRICS, Brazil went from being the 7th largest economy in 2014 to the 9th in two short years. Fixing Brazil's issues could take years, including addressing corruption and increasing exports of manufactured goods. But Brazilians also have the desire to shape their future and to lead, both within Latin America and globally.

1. Describe Brazil's linkages to the global economy.

2. Which domestic practices or policies contributed to the current economic and political crisis?

---

assembly plants that line the border with the United States, are characteristic of manufacturing systems in an increasingly globalized economy. These plants were first constructed in the 1960s as part of a border industrialization program, and by 2000 there were over 3,000 maquiladoras along the border. With NAFTA, foreign-owned manufacturing plants are no longer restricted to the border zone and are increasingly being built near the population centers of Monterrey, Puebla, and Veracruz. Aguascalientes in central Mexico has emerged as the country's auto city, although most of the cars are produced by foreign companies and are destined for export. The city of Chihuahua, a four-hour drive from the U.S.–Mexico border, is Mexico's aerospace center (**Figure 4.37**). In the last

decade, over three dozen aerospace plants have opened there, providing parts for the booming U.S. airplane manufacturing business. Labor costs average US$6 per hour in Chihuahua, far cheaper than U.S. labor. And although Mexican wages are higher than those in China, northern Mexico is still an attractive location because of its proximity to the U.S. border and its membership in NAFTA.

Considerable controversy on both sides of the border surrounds this form of industrialization. Organized labor in the United States complains that well-paying manufacturing jobs are being lost to low-cost competitors, whereas environmentalists point out serious industrial pollution resulting from lax government regulation. Mexicans worry that these

▲ **Figure 4.37 Chihuahua's Aerospace Industry** Mexican workers at the Hawker Beechcraft plant in Chihuahua assemble jet airplane parts for export to the United States. Chihuahua has the largest concentration of aerospace engineers and technicians in Mexico.

plants are poorly integrated with the rest of the economy and that many of the factories choose to hire young, unmarried women because they are viewed as docile laborers.

Other Latin American states are attracting foreign companies through tax incentives and low labor costs. Assembly plants in Honduras, Guatemala, and El Salvador are drawing foreign investors, especially in the apparel industry. A recent report from El Salvador claims that not one of its apparel factories has a union. Making goods for American labels such as the Gap, Liz Claiborne, and Nike, many Salvadoran garment workers complain they do not make a living wage, work 80-hour weeks, and face mandatory pregnancy tests. With the signing of CAFTA, Central American states are hopeful that more foreign investment will flow into their countries and that wages and conditions will improve.

The situation in Costa Rica is different. The country had been a major semiconductor chip manufacturer for Intel since 1998, but Intel closed its plant in 2014, deciding to relocate to Vietnam. With a well-educated population, low crime rate, and stable political scene, Costa Rica wants to attract other high-tech firms and repurpose the old Intel plant as a high-tech incubator. Hopeful officials claim that Costa Rica is transitioning from a so-called banana republic (bananas and coffee were the country's long-standing exports) to a high-tech manufacturing center. As a result, the Costa Rican economy averaged 4.4 percent annual growth from 2009 to 2014.

Uruguay is another small country with a well-educated population that has recently emerged as the leader in Latin American **outsourcing** operations. Outsourcing, most commonly associated with India, is the practice of moving service jobs such as tech support, data entry, and programming to cheaper locations. Partnered with the Indian multinational company Tata, in the last few years TCS Iberoamerica in Uruguay has created the largest outsourcing operation in the region. Uruguay takes advantage of being in a similar time zone as the eastern United States. While India's top

engineers sleep, Uruguayan engineers and programmers can serve their customers from Montevideo.

Latin America is linked to the world economy in ways other than trade. **Figure 4.38** shows the changes in foreign direct investment (FDI) as a percentage of gross domestic product (GDP) from 1990 and 2014. For nearly every country in the region, the value of FDI in terms of the percentage of GDP went up. In 1995, Brazil's FDI was less than $5 billion, and Mexico's was $9.5 billion. By 2014, FDI in Brazil had soared to $97 billion, and Mexico's was $24 billion. Much of this foreign investment was from Europe and Asia. In 2012, China became Brazil's largest trading partner. China and Brazil—the so-called rapidly developing BRICS nations, along with Russia, India and South Africa—have recognized their global strategic partnership. Not only is Brazil exporting grains, minerals, and energy resources to China, but it has also signed an agreement that will allow Brazilian airplane maker Embraer to manufacture and sell its regional jets in China.

Remittances Another important indicator that reflects the integration of Latin American workers into labor markets around the world is remittances. Scholars debate whether this flow of capital can actually lead to sustained development or whether it is simply a survival strategy of last resort. World Bank research shows that remittances sharply dropped during the global economic recession that began in 2008, but by 2014 had reached $66 billion (below the 2008 peak of $69 billion). Many economists project remittances will continue to be a major source of capital for the region.

The economic impact of remittances shown on a per capita basis is real (see Figure 4.38). Mexico is the regional leader, receiving over $24 billion in remittance income in 2014 (which is equal to over $200 per capita). But for smaller countries such as El Salvador and Honduras, remittances contribute far more to the domestic economy. El Salvador, a country of about 6 million people, received $4 billion in remittances in 2014, which is nearly $630 per capita. For many Latinos, remittances are the surest way to alleviate poverty, although they depend upon an international migration system that is constantly changing and includes both legal and illegal channels of movement.

Dollarization During the 1990s, as Latin American governments faced various financial crises, many began to consider the economic benefits of **dollarization**, a process by which a country adopts—in whole or in part—the U.S. dollar as its official currency. In a totally dollarized economy, the U.S. dollar becomes the only medium of exchange, and the country's national currency ceases to exist. This was the radical step taken by Ecuador in 2000 to address the dual problems of currency devaluation and hyperinflation rates of more than 1000 percent annually. El Salvador adopted dollarization in 2001 as a means to reduce the cost of borrowing money. Dollarization is not a new idea; back in 1904, Panama dollarized its economy the year after it gained independence from Colombia. Until 2000, however, Panama was the only fully dollarized state in Latin America.

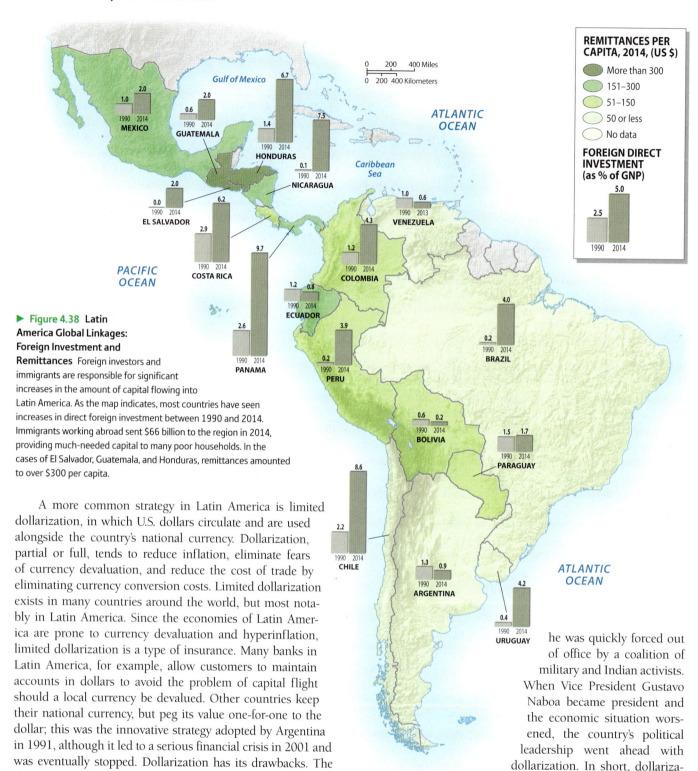

▶ **Figure 4.38 Latin America Global Linkages: Foreign Investment and Remittances** Foreign investors and immigrants are responsible for significant increases in the amount of capital flowing into Latin America. As the map indicates, most countries have seen increases in direct foreign investment between 1990 and 2014. Immigrants working abroad sent $66 billion to the region in 2014, providing much-needed capital to many poor households. In the cases of El Salvador, Guatemala, and Honduras, remittances amounted to over $300 per capita.

A more common strategy in Latin America is limited dollarization, in which U.S. dollars circulate and are used alongside the country's national currency. Dollarization, partial or full, tends to reduce inflation, eliminate fears of currency devaluation, and reduce the cost of trade by eliminating currency conversion costs. Limited dollarization exists in many countries around the world, but most notably in Latin America. Since the economies of Latin America are prone to currency devaluation and hyperinflation, limited dollarization is a type of insurance. Many banks in Latin America, for example, allow customers to maintain accounts in dollars to avoid the problem of capital flight should a local currency be devalued. Other countries keep their national currency, but peg its value one-for-one to the dollar; this was the innovative strategy adopted by Argentina in 1991, although it led to a serious financial crisis in 2001 and was eventually stopped. Dollarization has its drawbacks. The obvious one is that a country no longer has control of its monetary policy, making it reliant on the decisions of the U.S. Federal Reserve. Foreign governments do not have to ask permission to dollarize their economies. At the same time, the United States insists that all its monetary policies be based exclusively on domestic considerations, regardless of the impact such decisions may have on foreign countries. The political impact of eliminating a national currency is serious. The case of Ecuador is instructive. In 1999, when President Jamil Mahuad announced his plan to dollarize the economy to head off hyperinflation, he was quickly forced out of office by a coalition of military and Indian activists. When Vice President Gustavo Naboa became president and the economic situation worsened, the country's political leadership went ahead with dollarization. In short, dollarization may help in a time of economic duress, but it is not a popular policy.

The Informal Sector Even in prosperous capital cities, a short drive to the urban periphery shows large neighborhoods of self-built housing filled with street traders and family-run workshops. Such activities make up the informal sector, the provision of goods and services without the benefit of government regulation, registration, or taxation. Most people in the informal economy are self-employed and receive no wages or benefits except the profits they clear. The most common informal activities are housing construction (in many cities, as many as half of all residents live in self-built housing), manufacturing in small workshops, street vending, transportation services (messenger services, bicycle delivery, and collective taxis), garbage picking, street performing, and even line-waiting (**Figure 4.39**). These activities are legal. Illegal informal activities also exist: drug trafficking, prostitution, and money laundering, for example. The vast majority of people who rely on informal livelihoods produce legal goods and services.

No one is sure how big this economy is, in part because separating formal activities from informal ones is difficult. Visitors to Lima, Belém, Guatemala City, or Guayaquil could easily get the impression that the informal economy *is* the economy. From self-help housing that dominates the landscape to hundreds of street vendors that crowd the sidewalks, it is impossible to avoid. There are advantages in the informal sector—hours are flexible, children can work with their parents, and there are no bosses. Peruvian economist Hernando de Soto even argues that this most dynamic sector of the economy should be encouraged and offered formal lines of credit. As important as this sector may be, however, widespread dependence on it signals Latin America's poverty, not its wealth. It reflects the inability of the formal economies of the region, especially in industry, to absorb labor. For millions of urban dwellers, formal employment that offers benefits, safety, and a living wage is still a dream. With nowhere else to go, the numbers of the informally employed are substantial.

Social Development

Over the past three decades, Latin America has experienced marked improvements in life expectancy, child survival, and educational equity. One telling indicator is the steady decline between 1990 and 2014 in mortality rates for children below age five (see Table 4.2). This indicator is important because an increase in the number of children younger than five years surviving suggests that basic nutritional and health-care needs are being met. We can also conclude that resources are being used to sustain women and their children. Despite economic downturns, the region's social networks have been able to lessen the negative effects on children.

A combination of government policies and grassroots and nongovernmental organizations (NGOs) plays a fundamental role in contributing to social well-being. In the past few years, conditional cash transfer programs, such as **Bolsa Familia**, have reduced extreme poverty. Poor Brazilian families who qualify for Bolsa Familia receive a monthly check from the state, but are required to keep their children in school and take them to clinics for health checkups. Such programs have both the immediate impact of giving poor families cash and the long-term impact of improving the educational attainment and health of their children. Mexico has adopted a similar program. For states with far fewer resources than Brazil or Mexico, international humanitarian organizations, church organizations, and community activists provide many services that state and local governments cannot. Catholic Relief Services and Caritas, for example, work with rural poor throughout the region to improve their water supplies, health care, and education. Other groups lobby local governments to build schools and recognize squatters' claims. Grassroots organizations also develop cooperatives that market everything from sweaters to cheeses.

Other important gauges for social development are life expectancy, gender educational equity, and access to improved water sources. In aggregate, 95 percent of Latin Americans have access to an adequate amount of water from an improved source, and 83 percent have access to improved sanitation facilities. Life expectancy (men and women) is 75 years and nearly all youth in the region are literate (see Table 4.2). Masked by this aggregate data are extreme variations between rural and urban areas, between regions, and along racial and gender lines (**Figure 4.40**).

Within Mexico and Brazil, tremendous internal differences exist in socioeconomic indicators. The northeastern part of Brazil lags behind the rest of the country in every social indicator. The country has an overall literacy rate of over 85 percent, but in the northeast it is only 60 percent. Moreover, within the northeast, literacy for city residents is 70 percent, but for rural residents it is only 40 percent. In Mexico, the levels of poverty are highest in the more

▼ **Figure 4.39 Peruvian Street Vendors** Street vendors sell produce in Huancayo, Peru. Street vending plays a critical role in distributing goods and generating income. It is representative of the informal sector in Latin America.

▲ **Figure 4.40 School Children in Panama** Uniformed public school children walk to school in Panama City. Latin American states have seen steady improvements in youth literacy, with 97 percent of youth (between the ages of 15 and 24) being literate. Per capita expenditures on education and access to postsecondary education still lag behind levels in Europe and North America.

Indian south. In contrast, Mexico City and the states of Nuevo Leon (Monterrey is the capital), Quintana Roo (home to Mexico's largest resort, Cancun), and Campeche have the highest GDP per capita. Noting this north–south economic divide, former President Felipe Calderón explained, "there is one Mexico more like North America and another Mexico more like Central America. It is a very clear challenge for me to make them more alike." All countries have spatial inequities regarding income and availability of services, but the contrasts tend to be sharper in the developing world. In the cases of Mexico and Brazil, it is hard to ignore ethnicity and race when explaining these patterns.

Race and Inequality There is much to admire about race relations in Latin America. The complex racial and ethnic mix that was created in Latin America fostered tolerance for diversity. That said, Indians and blacks are more likely to be counted among the poor of the region. More than ever, racial discrimination is a major political issue in Brazil. Reports of organized killings of street children, most of them Afro-Brazilian, make headlines. For decades, Brazil put forward its vision of a color-blind racial democracy. True, residential segregation by race is rare in Brazil, and interracial marriage is common, but certain patterns of social and economic inequity seem best explained by race.

Assessing racial inequities in Brazil is problematic. The Brazilian census asks few racial questions, and all are based on self-classification. In the 2000 census, less than 11 percent of the population called itself black. Some Brazilian sociologists, however, claim that more than half the population is of African ancestry, making Brazil the second largest "African state" after Nigeria. Racial classification is always highly subjective and

relative, but certain patterns support the existence of racism. Evidence from northeastern Brazil, where Afro-Brazilians are the majority, shows death rates approaching those of some of the world's poorest countries. Throughout Brazil, blacks suffer higher rates of homelessness, landlessness, illiteracy, and unemployment. To address this problem, various affirmative action measures have been implemented (along with the Bolsa Familia program). From federal ministries to public universities, various quota systems are being tried to improve the condition of Afro-Brazilians.

In areas of Latin America where Indian cultures are strong, indicators of low socioeconomic position are also present. In most countries, areas where native languages are widely spoken invariably correspond with areas of persistent poverty. In Mexico, the Indian south lags behind the booming northern states and Mexico City. Prejudice is embedded in the language. To call someone an *indio* (Indian) is an insult in Mexico. In Bolivia, women who dress in the Indian style of full, pleated skirts and bowler hats are called *cholas*, a descriptive term referring to the rural mestizo population that suggests backwardness and even cowardice. No one of high social standing, regardless of skin color, would ever be called a *chola* or *cholo*.

It is difficult to separate status divisions based on class from those based on race. From the days of conquest, being European meant an immediate elevation in status over the Indian, African, and mestizo populations. Class awareness is very strong. Race does not necessarily determine a person's economic standing, but it certainly influences it. Most people recognize the power of the elite and envy their lifestyle.

A growing middle class also exists, and its members are formally employed, aspire to own a home and car, and strive to give their children a university education. The vast majority of people, however, are the working poor who struggle to meet basic food, shelter, clothing, and transport needs. These class differences are evident in the landscape. Go to any large Latin American city, and you will find handsome suburbs, country clubs, and trendy shopping centers. High-rise luxury apartment buildings with beautiful terraces offer all the modern amenities, including maids' quarters. The elite and the middle class even show a preference for decentralized suburban living and dependence on automobiles, as do North Americans. Yet near these same residences are shantytowns where urban squatters build their own homes, create their own economy, and eke out a living.

The Status of Women Many contradictions exist with regard to the status of women in Latin America. Many Latina women work outside the home. In most countries, the formal figures are between 30 and 40 percent of the workforce, not far off from many European countries, but lower than in the United States. Legally speaking, women can vote, own property, and sign for loans, although they are less likely to do so than men, reflecting the patriarchal (male-dominated) tendencies in the society. Even though Latin America is predominantly Catholic, divorce is legal, and family planning is promoted. In most countries, however, abortion remains illegal.

PERCENT OF SEATS HELD BY WOMEN IN NATIONAL PARLIAMENTS, 2014

- 5–14.9
- 15–24.9
- 25–34.9
- 35–44.9
- 45–54.9

● Countries that have or have had women presidents

In Latin America 28.5% of seats are held by women in national parliaments. By comparison, in the United States 19.3% of congressional seats are held by women. In Canada the figure is 25.1%.

▶ **Figure 4.41 Development Issues in Latin America: Female Participation in Politics** Women are increasingly playing vital roles in Latin American politics. On average, 25 percent of seats in national parliaments are held by women. (In contrast, the figure for the United States is 19 percent.) In Brazil, only 9 percent of the seats were held by women, but in Mexico and Argentina 37 percent were held by women in 2014. Moreover, six countries have or have had women presidents. (World Bank Development Indicators, 2015).

Overall, access to education in Latin America is good, compared to other developing regions, and thus illiteracy rates tend to be low. The rates of adult illiteracy are slightly higher for women than for men, but usually by only a few percentage points. Throughout higher education in Latin America, male and female students are equally represented today. Consequently, women are regularly employed in the fields of education, medicine, and law.

The biggest changes for women are the trends toward smaller families, urban living, and educational parity with men. These factors have greatly improved the participation of women in the labor force. In the countryside, however, serious inequalities remain. Rural women are less likely to be educated and tend to have larger families. In addition, they are often left to care for their families alone, as husbands leave in search of seasonal employment. In most cases, the conditions facing rural women have been slow to improve.

Women are increasingly playing an active role in politics. In 1990, Nicaragua elected the first woman president in Latin America, Violeta Chamorro, the owner of an opposition newspaper. Nine years later, Panamanians voted Mireya Moscoso into power. In 2005, South America had its first woman president when Michelle Bachelet, a pediatrician and single mother, took the oath of office in Chile. She was re-elected president for a second term in 2013. Brazilian President Dimla Rousseff took office in 2011, so Latin America's largest economy was run by a woman. As shown in **Figure 4.41**, many Latin American states have larger percentages of women in their national parliaments than does the United States.

Across the region, women and indigenous groups are active organizers and participants in cooperatives, small businesses, and unions and are elected to national office. In a relatively short period, they have won a formal place in the economy and a political voice. Moreover, evidence suggests that this trend will continue.

REVIEW

4.10 How has the export of primary products (food, fiber, and energy) shaped the economies of Latin America?

4.11 What explains some of the positive indicators of social development in Latin America?

■ **KEY TERMS** neoliberalism, maquiladoras, outsourcing, dollarization, Bolsa Familia

Chapter 4 Latin America

Summary

- Latin America is still rich in natural resources and relatively lightly populated. Yet as populations continue to grow and trade in natural resources increases, environmental problems multiply. Tropical forest clearing and dam construction are particular concerns.

- Unlike in other developing areas, 80 percent of Latin Americans live in cities. This shift started early and reflects a cultural bias toward urban living with roots in the colonial past. Cities are large and combine aspects of the formal industrial economy with the informal one.

Grinding poverty in rural areas drives people to cities or to overseas employment.

- Perhaps 90 percent of Latin America's native population died from disease, cruelty, and forced resettlement when colonized by Europe. The slow demographic recovery of native peoples and the continual arrival of Europeans and Africans resulted in an unprecedented level of racial and cultural mixing. Today Amerindian activism is on the rise as indigenous groups seek territorial and political recognition.

- Most Latin American states have been independent for 200 years, yet remained politically and economically dependent on Europe and North America. Today Latin

America, especially Brazil, is exerting more geopolitical influence. Most Latin American states have been independent for 200 years, yet remained politically and economically dependent on Europe and North America. Today Latin America, especially Brazil, is exerting more geopolitical influence. New political actors—from indigenous groups to women—are challenging old ways of doing things.

- Latin American governments were early adopters of neoliberal economic policies. Some states prospered whereas others faltered, sparking popular protests against the effects of neoliberalism and globalization. Extreme poverty has declined, and social indicators of development are improving.

Review Questions

1. How do different countries in the region address environmental issues, and why might approaches differ? How might resolving environmental issues affect the economies of countries in the region?

2. How does Latin America's history and geography explain why the region

is much more urbanized than other developing world regions? With its vast lightly inhabited areas, why does Latin America have so many megacities?

3. In what ways do Latin American countries influence popular culture around the world? Which countries are the most influential, and why?

4. NAFTA and Mercosur are established trade blocs in the region, whereas

the Pacific Alliance is relatively new. What are the benefits of forming trade blocs? How have the trade blocs in Latin America impacted the region's economic growth and geopolitics?

5. How has the export of primary products (food, fiber, and energy) shaped the economies of Latin America, and why does the region continue to produce so many raw materials for the world?

Image Analysis

1. Two important sources of foreign capital in Latin America are foreign direct investment (FDI) and remittances. Remittances are often a sign of a country's dependence on emigration as a livelihood strategy. Which countries in the region are most dependent on remittances? What generalizations can you make about these countries?

2. The largest recipient of FDI in absolute dollars is Brazil, and yet FDI accounts for a relatively small portion of Brazil's economy. Which countries have seen the largest increase in FDI? What might explain that increase?

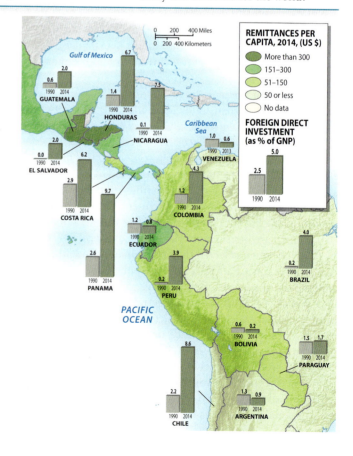

▶ **Figure IA4** **Latin America Foreign Direct Investment and Remittances**

JOIN THE DEBATE

Dams are necessary for flood control, irrigation, and energy production, but are often controversial, especially large daxms built in wilderness areas. Examine Figure 4.6 to see the major existing and planned dams in the Amazon Basin. Also consider that the Amazon Basin contains the world's largest tropical forest with immense biodiversity and thousands of acres of undisturbed land. Are dams beneficial or disastrous for Amazonian peoples?

▲ **Figure D4** Construction Site of the Belo Monte Hydroelectric Dam on the Xingu River in Brazil

Dam construction benefits the people of the Amazon Basin!

- With the growing demand for energy in the region, dam construction creates a renewable and clean source of electricity.

- The Amazon Basin is vast and for many South American countries it represents a resource and settlement frontier. By building roads and constructing dams, the countries of this region can better benefit from their natural resources.

- Dam construction creates jobs, advances in engineering abilities, and enhances overall infrastructural development. It can be done safely with limited environmental impact.

Dams construction is a poor choice for the Amazon Basin!

- Dams are expensive projects that have limited life spans, especially in a rainforest environment. The costs of construction can drive countries into debt with limited long-term benefits.

- The ecological costs of building dams in remote tropical forests are serious; many species can be lost and the negative impact of the dam goes far beyond the footprint of the actual project.

- For indigenous people in the region, dam construction inevitably results in displacement.

■ KEY TERMS

agrarian reform *(p. 139)*
Altiplano *(p. 127)*
altitudinal zonation *(p. 131)*
Bolsa Familia *(p. 167)*
Columbian exchange *(p. 149)*
dollarization *(p. 165)*
El Niño *(p. 131)*
environmental lapse rate *(p. 131)*
grassification *(p. 134)*
informal sector *(p. 144)*
latifundia *(p. 139)*
maquiladoras *(p. 163)*
mestizo *(p. 145)*
minifundia *(p. 139)*
neoliberalism *(p. 161)*
neotropics *(p. 126)*

Organization of American States (OAS) *(p. 154)*
outsourcing *(p. 165)*
Pacific Alliance *(p. 154)*
pristine myth *(p. 149)*
remittances *(p. 148)*
rural-to-urban migration *(p. 144)*
shields *(p. 126)*
subnational organizations *(p. 156)*
supranational organizations *(p. 156)*
syncretic religions *(p. 152)*
Treaty of Tordesillas *(p. 154)*
Union of South American Nations (UNASUR) *(p. 154)*
urban primacy *(p. 142)*

Mastering Geography™

Looking for additional review and test prep materials? Visit the Study Area in **MasteringGeography**™ to enhance your geographic literacy, spatial reasoning skills, and understanding of this chapter's content by accessing a variety of resources, including interactive maps, videos, RSS feeds, flashcards, web links, self-study quizzes, and an eText version of *Diversity Amid Globalization*.

DATA ANALYSIS

http://goo.gl/LErlnD

Coffee production is extremely important for many Latin American countries, and the region produces more than half of the world's coffee on large estates and small farms in the tropics. For many rural families, the coffee bean provides access to international markets and income. The International Coffee Organization (ICO) maintains statistics on coffee production worldwide. Consult the ICO's website (http://www.ico.org) and access the coffee production figures from 2012 to 2015.

1. Which countries were the top producers in 2012? Which countries were the top producers in 2015? What political or economic factors might explain the changes?

2. Which world regions have seen the biggest increase in coffee production? Which world regions have seen the biggest decline?

3. Write a paragraph explaining how significant increases or decreases in coffee production could impact the physical environment and economic development potential of these regions.

Authors' Blogs

Scan to visit the
GeoCurrents blog
http://geocurrents.info/
category/place/latin-america

Scan to visit the
Author's Blog
for field notes, media resources, and chapter updates
http://gad4blog.wordpress.com/
category/latin-america

Physical Geography and Environmental Issues

Climate change threatens the Caribbean, with the potential for stronger and more frequent hurricanes, loss of territory due to rising sea level, and destruction of coral reefs.

Population and Settlement

Having experienced its demographic transition, the region now has slow population growth. In addition, large numbers of Caribbean people have emigrated from the region, leaving in search of economic opportunity and sending back billions of dollars.

Cultural Coherence and Diversity

Creolization—the blending of African, European, and Amerindian elements—has resulted in unique Caribbean expressions of culture, such as rara, reggae, and steel drum bands. Caribbean-styled celebrations of carnival have diffused with the movement of people from the region to Europe and North America.

Geopolitical Framework

The first area of the Americas to be extensively explored and colonized by Europeans, the region has seen many rival European claims and, since the early 20th century, has experienced strong U.S. influence. In 2015 the United States restored diplomatic relations with Cuba, a major geopolitical shift for the region.

Economic and Social Development

Environmental, locational, and economic factors make tourism a vital component of this region's economy. Offshore manufacturing and banking are also significant in the region's modern economic development.

▶ **Surf and Sun** Cabarete, on the north shore of the Dominican Republic, is an international destination for wind and kite surfers due to its protected bay and nearly constant breeze. Like many Caribbean countries, the Dominican Republic relies upon a constant stream of tourists.

Cabarete beach, Dominican Republic

THE CARIBBEAN

Tourism is big business for the Caribbean, and with over 5 million visitors in 2014, the Dominican Republic is the single largest destination for international tourists in the region. The island offers world heritage monuments, stunning palm-lined beaches, tropical mountain forests, and a pulsing merengue-driven beat heard almost everywhere. It is a destination that offers a full range of options, from high-end packaged resorts to budget accommodations. The Dominican diaspora, migrants and their descendants, also regularly visit the island. And because prostitution is legal, there is a significant sex industry. International tourism accounts for $5 billion a year, making it the largest earner of foreign exchange for the country.

Yet the Dominican Republic is not just a place for a beach holiday. Its nearly 11 million people share the island of Hispaniola with Haiti, a far poorer but equally large country. Seeking economic growth and resilience, the Dominican Republic struggles to expand its manufacturing by offering cheap labor to firms that outsource production, modernize its agricultural sector, and accommodate an increasingly urbanized population. The capital of Santo Domingo, settled in 1496, is now a bustling metropolis of 3 million people, the largest in the Caribbean. This cosmopolitan city was once linked to the world economy through slavery, sugar, and rum. Today planes filled with sun-seeking tourists drive the island's economic growth.

The Caribbean was the first region of the Americas to be extensively explored and colonized by Europeans. Yet its modern regional identity is unclear—often merged with Latin America, but also viewed as apart from it. Today the region is home to 45 million inhabitants scattered across 26 countries and dependent territories, ranging from the small British dependency of the Turks and Caicos, with 36,000 people, to the island of Hispaniola, with 22 million. In addition to the Caribbean islands, Belize of Central America and the three Guianas—Guyana, Suriname, and French Guiana—of South America are often included as part of the Caribbean. For historical and cultural reasons, the peoples of these mainland states identify with the island nations and are thus included in this chapter (**Figure 5.1**).

The basis for treating the Caribbean as a distinct area lies within its particular economic and cultural history. As in many developing areas, external control of the Caribbean produced highly dependent and inequitable economies, accentuated by the area's reliance on slave labor, plantation agriculture, and the prosperity that sugar promised in the 18th century. Culturally, this region can be distinguished from the largely Iberian-influenced mainland of Latin America because of its more diverse European colonial history, and its strong African imprint due to the region's historical reliance upon slavery.

Still a developing, most Caribbean states have achieved life expectancies in the 70s, low child mortality rates, and high literacy rates. Millions of tourists view the Caribbean as an international playground for sun, sand, and fun. However, there is another Caribbean that is far poorer and economically more dependent than the one portrayed on travel posters. Haiti, by most measures, is the Western Hemisphere's poorest country. In fact, the majority of Caribbean people, living in the shadow of North America's vast wealth, suffer from serious economic problems and widespread poverty.

Nevertheless, the Caribbean has evolved as a distinct, but economically marginal world region (**Figure 5.2**). This status expresses itself today as workers leave the region in search of better wages, while foreign companies are attracted to the Caribbean for its cheaper labor costs. The economic well-being of most Caribbean countries is precarious. Despite such uncertainty, an enduring cultural richness and an attachment to place are present here that may explain a growing countercurrent of immigrants back to the region.

> ## This region can be distinguished from the largely Iberian-influenced Latin America because of its more diverse European colonial history and strong African imprint.

Physical Geography and Environmental Issues: Paradise Undone

Tucked between the Tropic of Cancer and the equator, with year-round temperatures averaging in the high 70s, the hundreds of islands and picturesque waters of the Caribbean have often inspired comparisons to paradise. Columbus began the tradition by describing the islands of the New World as the most marvelous, beautiful, and fertile lands he had ever known, filled with flocks of parrots, exotic plants, and friendly natives. Writers today are still lured by the sea, sands, and swaying palms of the Caribbean.

Ecologically speaking, it is difficult to picture a landscape that has been more completely altered than that of the Caribbean. For over five centuries, the destruction of forests

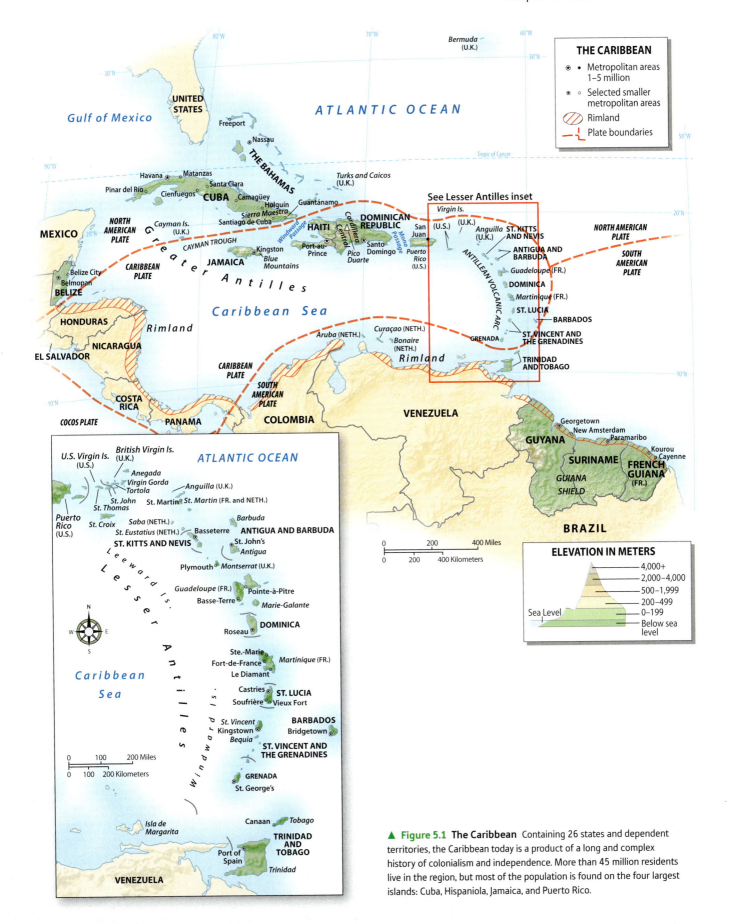

▲ **Figure 5.1 The Caribbean** Containing 26 states and dependent territories, the Caribbean today is a product of a long and complex history of colonialism and independence. More than 45 million residents live in the region, but most of the population is found on the four largest islands: Cuba, Hispaniola, Jamaica, and Puerto Rico.

▲ **Figure 5.2 Carnival Drummer** A steel pan drummer performs while his drum cart is pushed through the streets during carnival in Port of Spain, Trinidad. Steel drums were created in Trinidad in the 1940s from discarded oil drums from a U.S. military base. They have become an iconic sound for the region.

and unrelenting cultivation of soils resulted in the extinction of many endemic (native) Caribbean plants and animals, including various kinds of shrubs and trees, songbirds, large mammals, and monkeys. This severe depletion of biological resources helps explain some of the present economic and social instability of the region. Most Caribbean environmental problems are associated with agricultural practices, soil erosion, excessive reliance on wood and charcoal for fuel, and the threat of global climate change. The devastating Port-au-Prince earthquake of 2010 underscored how quickly a place such as Haiti can become undone due to a major natural disaster. However, because many countries rely upon tourism as a vital source of income, the region has also experienced a growth in protected areas, both on the land and in maritime locales.

Island and Rimland Landscapes

The Caribbean Sea itself—that body of water enclosed between the Antillean islands (the arc of islands that begins with Cuba and ends with Trinidad) and the mainland of Central and South America—links the states of the Caribbean region. Historically, the sea connected people through its trade routes and sustained them with its marine resources of fish, green turtle, manatee, lobster, and crab. The Caribbean Sea is noted for its clarity and biological diversity, yet the quantity of any one species is not great, so the region has never supported large commercial fishing.

Sea surface temperatures range from 73° to 84°F (23° to 29°C), over which forms a warm tropical marine air mass that influences daily weather patterns. This warm water and tropical setting continue to be key resources for the region, attracting millions of tourists to the Caribbean each year (**Figure 5.3**).

The arc of islands that stretches across the sea is the region's most distinguishing feature. The Antillean islands are divided into two groups: the Greater and Lesser Antilles. The **rimland** (the Caribbean coastal zone of the mainland) includes Belize and the Guianas, as well as the Caribbean shoreline of Central and South America stretching between them (see Figure 5.1). In contrast to the islands, the rimland has low population densities.

Most of the islands, with the exception of Cuba, are on the Caribbean tectonic plate, wedged between the South American and North American plates. Generally, this is not one of the most tectonically active zones, although earthquakes and volcanic eruptions do occur. In January 2010, a magnitude 7.0 earthquake leveled Port-au-Prince in one of the most tragic natural disasters to strike the Caribbean.

The Enriquillo Fault, near Haiti's densely settled and extremely poor capital city, had been inactive for more than a century. When it violently shifted in 2010, the epicenter of the resulting earthquake was just a few miles away from the city. The disaster affected nearly 3 million people, as homes were rendered unsafe and water and electricity supplies were disrupted. Shockingly, the earthquake left over 200,000 people dead and 1 million more homeless. Many of the

▼ **Figure 5.3 Caribbean Sea** Noted for its calm, turquoise waters, steady breezes, and treacherous shallows, the Caribbean Sea has both sheltered and challenged sailors for centuries. Recreational sailors enjoy the waters around Tabago Cays in the Grenadines.

government agencies that would have assisted in the relief response were also destroyed. The tragedy of the Haitian earthquake was compounded by the state's poverty and corruption, as most buildings were not built to standards that could withstand an earthquake of this magnitude. The international community, as well as the large diaspora of Haitians living abroad, immediately offered financial aid and assistance. More than US$12 billion in humanitarian and development aid and debt relief were pledged in the wake of the crisis. Slowly rebuilding has occurred, including a new two-lane highway between Port-au-Prince and Gonaives, a new airport, and hundreds of new schools (**Figure 5.4**). The last of the tent cities that housed over 1 million people immediately after the earthquake were finally taken down. But housing remains a problem, and Haiti's large impoverished population still struggles.

Greater Antilles The four large islands of Cuba, Jamaica, Hispaniola (shared by Haiti and the Dominican Republic), and Puerto Rico make up the **Greater Antilles**. These islands have most of the region's population, arable lands, and large mountain ranges. Given the popular interest in the Caribbean coasts, it still surprises many people that Pico Duarte on the Dominican Republic is more than 10,000 feet (3000 meters), Jamaica's Blue Mountains top 7000 feet (2100 meters), and Cuba's Sierra Maestra is more than 6000 feet (1800 meters). The mountains of the Greater Antilles were of little economic interest to plantation owners, who preferred the coastal plains and valleys. However, the mountains were an important refuge for runaway slaves and subsistence farmers and thus figure prominently in the cultural history of the region.

The best farmlands are found in the central and western valleys of Cuba, where a limestone base contributes to the formation of a fertile red clay soil (locally called *matanzas*) and a gray or black soil called *rendzinas* (also found in Antigua, Barbados, and lowland Jamaica). The rendzinas soils, which consist of a gravelly loam with a high organic content—ideal for sugar production—are actively exploited for agriculture wherever found. Surprisingly, given the area's agricultural orientation, many of the soils are nutrient-poor, heavily leached, and acidic. These poor, *ferralitic* soils are found in the wetter areas where crystalline base rock exists (as in parts of Hispaniola, the Guianas, and Belize). They are characterized by heavy accumulations of red and yellow clays and offer little potential for permanent intensive agriculture.

Lesser Antilles The **Lesser Antilles** form a double arc of small islands stretching from the Virgin Islands to Trinidad. Smaller in size and population than the Greater Antilles, they were important early footholds for rival European colonial powers. The islands from St. Kitts to Grenada form the inner arc of the Lesser Antilles. These mountainous islands, with peaks ranging from 4000 to 5000 feet (1200 to 1500 meters), have volcanic origins. In this subduction zone, the heavier North and South American plates sink beneath the Caribbean Plate, producing volcanic activity and earthquakes. Erosion of the island peaks and the accumulated ash from eruptions have created small pockets of arable soils, although the steepness of the terrain limits agricultural development.

Just east of this volcanic arc are the low-lying islands of Barbados, Antigua, Barbuda, and the eastern half of Guadeloupe. Covered in limestone that overlays volcanic rock, these lands were inviting for agriculture, especially sugarcane. Barbados was the experimental hub for colonial British sugar production in the Caribbean; innovations from that intensely cultivated island diffused throughout the region. Trinidad and Tobago are on the South American Plate and consist of sedimentary rather than volcanic rock. These islands include alluvial soils and, more important, sedimentary basins that contain oil and natural gas reserves.

The Rimland The rimland consists of the coastal zone of the mainland, beginning with Belize and extending along the coast of Central America to northern South America. Unlike the rest of the Caribbean, the rimland states of Belize and the Guianas still contain significant amounts of forest cover. As on the islands, agriculture in these states is closely tied to local geology and soils. Much of low-lying Belize is limestone. Sugarcane is the dominant crop in the drier north, whereas citrus is produced in the wetter central portion of the state. The Guianas are characterized by the rolling hills of the Guiana Shield, whose crystalline rock is responsible for the area's overall poor soil quality. Thus, most agriculture in the Guianas occurs on the narrow coastal plain, where sugar and rice are produced. Timber continues to be an important export for these rimland states.

▼ **Figure 5.4** **Post-earthquake Reconstruction in Haiti** The magnitude 7.0 earthquake that struck Port-au-Prince on January 12, 2010, was one of the worst natural disasters in the region's history, killing over 200,000 and leveling over 150,000 structures. The Iron Market in downtown Port-au-Prince, where thousands of vendors work, was one of the first major public structures to be rebuilt after the earthquake.

Explore the **Sights** of Iron Market in Port-au-Prince

https://goo.gl/MlLody

Metal extraction (bauxite and gold) also is vital to the economies of Guyana and Suriname. French Guiana, which is an overseas territory of France, relies mostly on French subsidies, but exports shrimp and timber. It is also home to the European Space Center at Kourou (**Figure 5.5**).

Caribbean Climates and Climate Change

Much of the Antillean islands and rimland receives more than 80 inches (200 cm) of rainfall annually, which is enough to support tropical forests. Average temperatures are typically highs of 80°F and lows of 70°F (27°C and 21°C) (Figure 5.6). Seasonality in the Caribbean is defined by changes in rainfall more than temperature. Although some rain falls throughout the year, the rainy season is from July to October. This is when the Atlantic high-pressure cell is farthest north and easterly winds generate moisture-laden and unstable atmospheric conditions that sometimes yield hurricanes and heavy rainfall in the fall (see **Figure 5.6**). During the slightly cooler months of December through March, rainfall declines. This time of year corresponds with the peak tourist season.

The Guianas have a different rainfall cycle. These territories, on average, receive more rain than the Antillean islands. In Cayenne, French Guiana, an average of 126 inches (320 cm) falls each year (see Figure 5.6). Unlike the Antilles, the Guianas experience a brief dry period in late summer (September to October). Also, January tends to be a wet period for the mainland, while it is a dry time for the islands. Climatically, the Guianas also are distinguishable from the rest of the region because they are not affected by hurricanes.

Hurricanes Each year several **hurricanes** pound the Caribbean, as well as Central and North America, with heavy rains and fierce winds. Beginning in July, westward-moving low-pressure disturbances form off the coast of West Africa and pick up moisture and speed as they move across the Atlantic Ocean. Usually no more than 100 miles across,

these disturbances achieve hurricane status when wind speeds reach 74 miles per hour. Hurricanes may take several paths through the region, but they typically enter through the Lesser Antilles. They then arc north or northwest and collide with the Greater Antilles, Central America, Mexico, or southern North America before moving to the northeast and dissipating in the Atlantic. The hurricane zone lies just north of the equator on both the Pacific and Atlantic sides of the Americas. Typically, a half-dozen to a dozen hurricanes form each season and move through the region, causing limited damage.

There are, of course, exceptions, and most longtime residents of the Caribbean have felt the full force of at least one major hurricane. The destruction caused by these storms is not just from the high winds, but also from the heavy downpours, which can cause severe flooding and deadly coastal storm surges. Modern tracking equipment has improved hurricane forecasting and reduced the number of fatalities, primarily through early evacuation of areas in a hurricane's path. This has saved lives but cannot reduce the damage to crops, forests, or infrastructure. In 2012, Hurricane Sandy struck Jamaica and eastern Cuba, removing roofs, downing power lines, and flooding rivers before moving up the U.S. East Coast (**Figure 5.7**). Similarly, in 2010, Antigua in the Lesser Antilles was hit hard by Hurricane Earl, which destroyed homes and caused serious flooding. Many atmospheric scientists and geographers believe that we have entered a more active hurricane period. Since 1995, a change in the multiyear cycle of sea surface temperatures in the Atlantic Ocean may be contributing to more-intense tropical depressions.

Climate Change Of all the issues facing the Caribbean, climate change is one of the most serious. The Caribbean has not been a major contributor of greenhouse gases (GHGs), but this maritime region is extremely vulnerable to the negative impacts of climate change, including sea-level rise, increased intensity of storms, variable rainfall leading to both floods and droughts, and loss of biodiversity (both in forests and in coral reefs). The scientific consensus is that surface temperatures would increase between 1.2° and 2.3°C across the Caribbean in this century and that rainfall will decrease by 5 to 6 percent. Perhaps most destructive of all, the latest climate change models show sea levels rising by 0.6 meters by the end of the century. In terms of land loss due to inundation, the low-lying Bahamas would be the most affected country in the region. In terms of people affected by inundation, Suriname, French Guiana, Guyana, Belize, and The Bahamas would be the most severely impacted: A 3-foot (1-meter) sea-level rise would be

▼ **Figure 5.5 European Space Center** An Ariane 5 rocket carrying telecommunications and meteorology satellites launches from Kourou, French Guiana. The European Space Agency uses this French territory as it is near the equator, on the coast, and lightly populated.

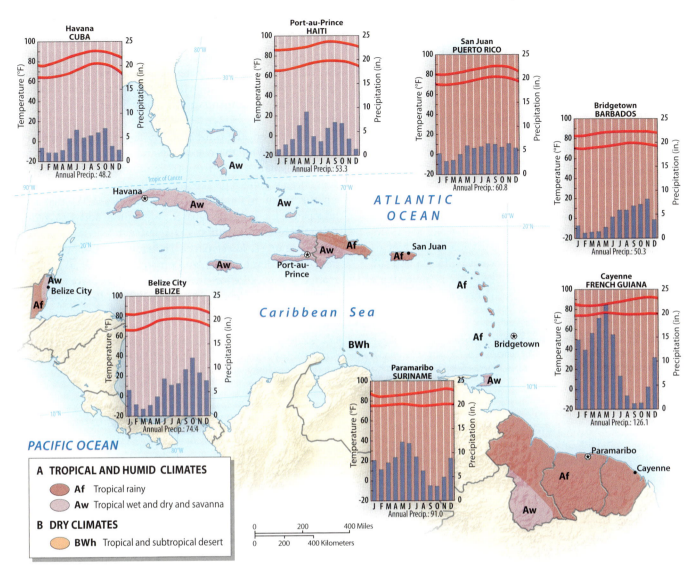

▲ **Figure 5.6 Climate of the Caribbean** Most of the region is classified as having either a tropical wet (Af) or a tropical savanna (Aw) climate. Temperature varies little over the year, as shown by the relatively straight temperature lines. Important differences in total rainfall and the timing of the dry season distinguish different localities. **Q: What are the wettest and driest months for each of these Caribbean cities? Is there a general pattern between the northern vs. the southern Caribbean?**

▶ **Figure 5.7 Hurricane Sandy** Residents walk through the wreckage of Santiago de Cuba after Hurricane Sandy moved through this east Cuban city in October 2012. In addition to fallen trees and missing roofs, 11 Cubans were killed by this fierce storm, which later pounded the New York metropolitan area.

devastating because most of the population lives near the coast (**Figure 5.8**).

In addition to land loss and population displacement due to sea-level rise, other concerns include changes in rainfall patterns, leading to declines in agricultural yields and freshwater supplies, and increases in storm intensity—especially hurricanes—that cause destruction of infrastructure and other problems. All of these changes would negatively affect tourism and thus the gross domestic income of countries in the region. Some of the worst-case scenarios are catastrophic.

In terms of biodiversity, continued ocean temperature increases will further negatively affect the Caribbean's coral reefs, which are the most biologically diverse ecosystems of the marine world. These reefs, particularly those of the rimland, are already threatened by water pollution and subsistence fishing practices. Recently, evidence is mounting of coral bleaching and die-off due to higher sea temperatures.

Coral reefs are diverse and productive ecosystems that function as nurseries for many marine species. Healthy reefs, along with mangrove swamps and wetlands, also serve as barriers to protect populated coastal zones. As the reefs become more ecologically vulnerable, so, too, do the human populations that depend on the many benefits that the reefs provide.

Throughout the Caribbean, protecting the environment and preparing for the effects of climate change are considered fundamental to the region's economic livelihood. In fact, Caribbean nations represent half of the 41 countries that make up the Small Island Development States (SIDS) that have advocated for ecosystem-based adaptation and risk reduction strategies. Adaptation strategies such as conservation of water resources, lower Green House Gas emissions, renewable energy development, and protection of coastal reefs and mangroves are all efforts to reduce the risks associated with climate change. Keeping in line with the Paris Agreement

▼ **Figure 5.8 Environmental Issues in the Caribbean** It is hard to imagine a region in which the environment has been so completely transformed. Most of the island forests were removed long ago for agriculture or fuel, and soil erosion is a chronic problem. Coastal pollution is serious around the largest cities and industrial zones. The forest cover of the rimland states, however, is largely intact and is attracting the interest of environmentalists.

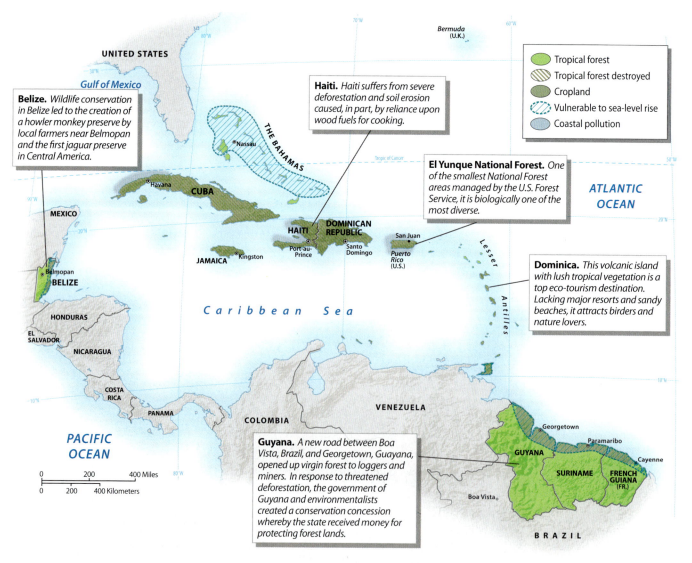

Belize. Wildlife conservation in Belize led to the creation of a howler monkey preserve by local farmers near Belmopan and the first jaguar preserve in Central America.

Haiti. Haiti suffers from severe deforestation and soil erosion caused, in part, by reliance upon wood fuels for cooking.

El Yunque National Forest. One of the smallest National Forest areas managed by the U.S. Forest Service, it is biologically one of the most diverse.

Dominica. This volcanic island with lush tropical vegetation is a top eco-tourism destination. Lacking major resorts and sandy beaches, it attracts birders and nature lovers.

Guyana. A new road between Boa Vista, Brazil, and Georgetown, Guayana, opened up virgin forest to loggers and miners. In response to threatened deforestation, the government of Guyana and environmentalists created a conservation concession whereby the state received money for protecting forest lands.

Legend:
- Tropical forest
- Tropical forest destroyed
- Cropland
- Vulnerable to sea-level rise
- Coastal pollution

on Climate Change, the government of Japan launched a $15 million partnership with eight Caribbean nations to support 50 coastal communities in adaptation efforts.

Environmental Issues

Climate change is both a medium- and a long-term concern for the Caribbean region. However, other environmental issues, such as soil erosion and deforestation, have preoccupied the region due to its long-standing dependence on agriculture. Also, as the Caribbean has become more urbanized and more reliant on tourism, governments have realized that protection of local ecosystems is not just good for the environment, but also good for the overall economy of the region (see Figure 5.8).

Agriculture's Legacy of Deforestation Prior to the arrival of Europeans, much of the Caribbean was covered in tropical forests. The great clearing of these forests began on European-owned plantations on the smaller islands of the eastern Caribbean in the 17th century and spread westward. The island forests were removed not only to make room for sugarcane, but also to provide the fuel necessary to turn the cane juice into sugar and the lumber for housing, fences, and ships. Primarily, however, tropical forests were removed because they were seen as unproductive; the European colonists valued cleared land. The newly exposed tropical soils easily eroded and ceased to be productive after several harvests, a situation that led to two distinct land-use strategies. On the larger islands of Cuba and Hispaniola, as well as on the mainland, new lands were constantly cleared and older ones abandoned or fallowed in an effort to keep up sugar production. On the smaller islands where land was limited, such as Barbados and Antigua, labor-intensive efforts to conserve soil and maintain fertility were employed. In either case, the island forests were replaced by a landscape devoted to crops for world markets.

Haiti's problems with deforestation were accentuated throughout the 20th century as a destructive cycle of economic and environmental impoverishment was established. Half of Haiti's people are peasants who work small hillside plots and seasonally labor on large estates. As the population grew, people sought more land. They cleared the remaining hillsides, subdivided their plots into smaller units, and abandoned the practice of fallowing land in an effort to eke out an annual subsistence. When the heavy tropical rains came, the exposed and easily eroded mountain soils washed away. As sediment collected in downstream irrigation ditches and behind dams,

▲ **Figure 5.9 Differences in Deforestation** Hispaniola has serious deforestation problems. Hillsides are denuded to make room for agricultural production or to secure wood fuel. **Q: Which side of the yellow border is Haiti? How would you compare land use in Haiti with that in the Dominican Republic?**

agriculture suffered, electricity production declined, and water supplies were degraded throughout the country. Deforestation was further aggravated by reliance of many poor Haitians on charcoal (made from trees) for household needs. Less than 3 percent of Haiti remains forested (**Figure 5.9**).

While Haiti has lost most of its forest cover, on Jamaica and Cuba nearly one-third of the land is still forested. About 40 percent of the Dominican Republic remains forested, as is more than half of Puerto Rico. On these more-forested islands, the decline in agricultural production overall has allowed forests to recover. In the case of Puerto Rico, a territory of the United States, its national forests—such as El Yunque on the eastern side of the island, which is protected by the U.S. Forest Service—contribute greatly to the island's biological diversity (**Figure 5.10**).

Energy Needs and Innovations With the exception of the country of Trinidad and Tobago, which exports oil and liquefied natural gas, the Caribbean states are net importers of oil and highly dependent on foreign sources for their energy needs, which results in disproportionally high energy costs. The region has some oil refineries that process crude oil shipments into petroleum for domestic consumption and even some for export, but dependence on foreign energy and volatile oil pricing make the small economies of this region vulnerable.

Not surprisingly, Caribbean nations have a growing interest in renewable energy. In many ways, wind energy has long been important for the Caribbean economy. After all, the entire colonial enterprise depended on the trade

▲ **Figure 5.10 El Yunque National Forest** Puerto Rico's largest remaining rainforest is in El Yunque National Forest. Visitors enjoy the park's hiking trails and its rich biological diversity.

winds to move commodities and people across the Atlantic. Commercial wind energy is gaining popularity. Puerto Rico just opened a major new wind farm on its southern coast near the city of Ponce, and the government intends to generate 12 percent of its energy from renewable sources. Similarly, the Los Cocos wind farm in the Dominican Republic is a major investment in wind power to serve the country's growing electricity needs. The region also has excellent potential for solar power. Nearly half the households in Barbados use solar water heaters, and solar panels are increasingly seen around government buildings, hospitals, businesses, and private homes. Some of the volcanically active Caribbean islands, especially in the Lesser Antilles, are tapping into clean **geothermal** energy (see *Working Toward Sustainability: Geothermal Energy for the Lesser Antilles*). For example, western Guadeloupe's La Bouillante power station generates 15 megawatts of geothermal power.

Conservation Efforts In general, biological diversity and stability are less threatened in the rimland states than in the rest of the Caribbean. Thus, current conservation efforts could produce important results. Even though much of

Belize was selectively logged for mahogany in the 19th and 20th centuries, healthy forest cover still supports a diversity of mammals, birds, reptiles, and plants. Public awareness of the negative consequences of deforestation is also greater now. Many protected areas have been established in Belize; for example, in the mid-1980s villagers in Bermudian Landing established a community-run sanctuary for black howler monkeys (locally referred to as baboons). The villagers banded together to maintain habitat for the monkeys and commit to land management practices that accommodate this gregarious species. The success of the project has resulted in tourists visiting the villages to see these indigenous primates up close (**Figure 5.11**). In 1986, a jaguar preserve was established in the Cockscomb Basin in southern Belize, the first of its kind in Central America.

Explore the **Sounds** of Howler Monkeys

https://goo.gl/KWDQLu

Slowly, the territorial waters surrounding the Caribbean nations have gained protection, although more could be done. Here again, Belize has been a leader in creating over a dozen marine reserves and national parks along its barrier reef and outer atolls. The country has also created a substantial coastal wildlife sanctuary to protect mangrove swamps. The island of Bonaire, a municipality of The Netherlands, attracts large numbers of scuba divers and maintains the Bonaire Marine Park, recognized as one of the most effectively managed marine reserves in the Caribbean.

REVIEW

5.1 Describe the locational, environmental, and climatic factors that together help make the Caribbean a major international tourist destination.

5.2 What environmental issues currently impact the Caribbean? Describe the risks and possible solutions.

■ **KEY TERMS** rimland, Greater Antilles, Lesser Antilles, hurricanes, geothermal

▼ **Figure 5.11 Protecting Habitat and Wildlife** Tourists visit the community "Baboon" Sanctuary in Bermudian Landing, Belize. The sanctuary is a community-run project to preserve the habitat and increase the number of black howler monkeys (locally referred to as baboons). The sanctuary, established in 1985, attracts domestic and foreign visitors.

Geothermal Energy for the Lesser Antilles

As Caribbean states strive for greener and less expensive energy sources, geothermal energy has become an attractive alternative for the islands of the Lesser Antilles. Many of these islands lie in a subduction zone where the Caribbean plate collides with the North American or South American plates. Such locations are ideally positioned for drilling geothermal wells that tap into subsurface heat for electricity production. Geothermal energy is the intense heat deep within the Earth, felt at the surface in geysers, hot springs, and volcanoes. Now several Caribbean islands, with international grants and technological help from experts in Iceland, New Zealand, and Japan, are drilling geothermal wells and constructing the first power stations. Not many countries in the world can exploit geothermal energy, but those that do benefit from a reliable and renewable energy alternative.

For most of the residents in the Lesser Antilles, energy costs are three times that of wealthy mainland countries. Many of these islands burn oil or gas to generate electricity, which is expensive and produces greenhouse gases (GHGs). In addition to energy conservation, such as using LED light bulbs or highly efficient air conditioners, geothermal energy represents a clean, local, and renewable alternative for power generation. Seven island nations are exploring or developing geothermal energy sources: Guadeloupe, Dominica, Grenada, Montserrat, St. Lucia, St. Kitts and Nevis, and St. Vincent and the Grenadines. Currently Guadeloupe is the only Caribbean island with an active geothermal plant (**Figure 5.1.1**).

▲ **Figure 5.1.2 Green Energy for Guadeloupe** Workers tend to the turbines at the new geothermal plant in Bouillante. Geothermal energy, widely used in Iceland, Japan, and New Zealand, is being developed in the Lesser Antilles, including the French territory of Guadeloupe.

▲ **Figure 5.1.1 Powering the Lesser Antilles** The islands that arc from St. Kitts and Nevis to Grenada are part of the Volcanic Axis. On fault lines and with geothermal activity, these islands are seeking to develop renewable energy from the earth.

With the international focus on reducing GHG emissions, there are new sources of funding to support geothermal infrastructure in the region. After a grant from the European Union resulted in identifying drill sites, Dominica now has a loan from the World Bank to build two geothermal plants that will supply much of the island's energy needs and even energy for export. The smaller plant, in the Roseau Valley, will soon be operational (**Figure 5.1.2**).

Researchers from New Zealand are mapping potential well sites in Montserrat, an island that was largely depopulated in 1997 after a volcanic eruption destroyed the capital city of Plymouth. The northern end of the island has gradually been resettled, and geothermal power is an important part of the energy strategy for this island's few thousand residents.

Geothermal energy does have its risks. Tapping the wells can be costly and time consuming. Also, hot water from geothermal sources can contain trace amounts of toxic metals such as mercury and arsenic, which evaporate into the air as the water cools. Consequently, there have been local protests against geothermal plants, especially from nearby farmers and landowners. But countries such as Iceland, Japan, and New Zealand that have exploited geothermal energy for years demonstrate the enormous potential of this resource to produce electricity without using fossil fuels.

Google Earth Virtual Tour Video

https://goo.gl/4YgtUP

1. What are the potential benefits of geothermal energy for the Lesser Antilles?

2. Consider Figure 2.2, which shows Earth's plate boundaries. Where else might geothermal energy be developed?

Population and Settlement: Densely Settled Islands and Rimland Frontiers

Caribbean population density is generally quite high and, as in neighboring Latin America, increasingly urban. Eighty-five percent of the region's population is concentrated on the four islands of the Greater Antilles (**Figure 5.12**). Add to this Trinidad and Tobago's 1.4 million and Guyana's 800,000, and most of the population of the Caribbean is accounted for by six countries and one U.S. territory (Puerto Rico). Of these, Puerto Rico has the greatest population density with 399 people per square kilometer (1,011 per square mile), followed by Haiti with 384 people per square kilometer (1,007 per square mile).

In absolute numbers, few people inhabit the Lesser Antilles; nevertheless, some of these microstates are densely settled. The small island of Barbados has only 166 square miles (430 square kilometers) of territory, but averages 1,712 people per square mile (659 people per square kilometer). Bermuda is even more densely settled. Just one-third the size of the District of Columbia, Bermuda has 3380 people per square mile (1303 people per square kilometer). Population densities on St. Vincent and Grenada, while not as high, are still more than 700 people per square mile (280 people per square kilometer). If you take into consideration the scarcity of arable land on some of these islands, it is clear that access to land is a basic resource problem for many inhabitants of the Caribbean. The growth in the region's population, coupled with its scarcity of land, has forced many people into the cities or abroad. It also has forced many Caribbean states to be net importers of food.

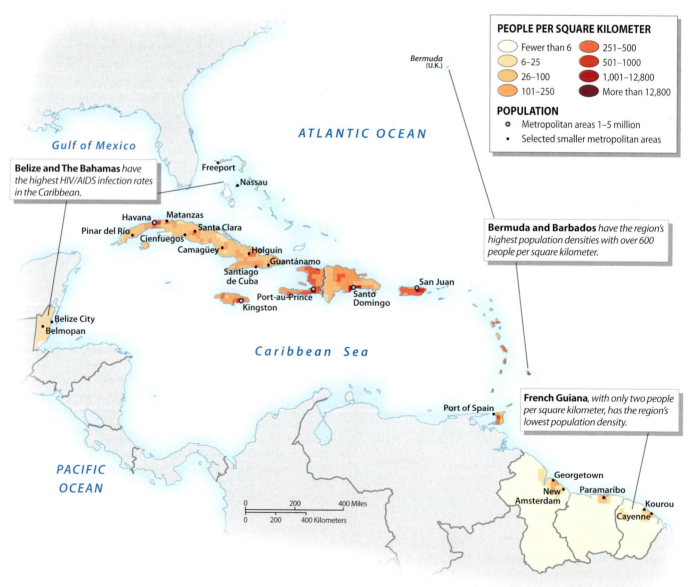

PEOPLE PER SQUARE KILOMETER

- Fewer than 6
- 6–25
- 26–100
- 101–250
- 251–500
- 501–1000
- 1,001–12,800
- More than 12,800

POPULATION
- ⊙ Metropolitan areas 1–5 million
- • Selected smaller metropolitan areas

Belize and The Bahamas *have the highest HIV/AIDS infection rates in the Caribbean.*

Bermuda and Barbados *have the region's highest population densities with over 600 people per square kilometer.*

French Guiana, *with only two people per square kilometer, has the region's lowest population density.*

▲ Figure 5.12 **Population of the Caribbean** The major population centers are on the islands of the Greater Antilles. The tendency here, as in the rest of Latin America, is toward greater urbanism. The largest metropolitan areas in the region are San Juan, Santo Domingo, and Havana; each has over 2 million residents. In comparison, the rimland states are very lightly settled.

TABLE 5.1	Population Indicators							MasteringGeography™
Country	Population (millions) 2016	Population Density (per square kilometer)[1]	Rate of Natural Increase (RNI)	Total Fertility Rate	Percent Urban	Percent < 15	Percent > 65	Net Migration (Rate per 1000)
Anguilla*	0.02	173	–	1.8	100	24	8	13
Antigua and Barbuda	0.1	207	0.6	1.5	24	24	7	0
Bahamas	0.4	38	0.8	1.8	83	26	6	4
Barbados	0.3	659	0.3	1.7	32	19	14	2
Belize	0.4	15	1.6	2.5	45	36	4	4
Bermuda*	0.07	1,303		2.0	100	18	16	2
Cayman*	0.05	247	–	1.9	100	19	11	15
Cuba	11.2	107	0.2	1.7	77	16	14	−2
Dominica	0.1	96	0.5	2.1	70	22	10	−1
Dominican Republic	10.6	215	1.4	2.4	79	30	7	−3
French Guiana	0.3	–	2.3	3.5	84	34	5	−2
Grenada	0.1	313	0.9	2.1	36	26	7	−2
Guadeloupe	0.4	–	0.4	2.2	98	22	15	−1
Guyana	0.8	4	1.2	2.6	29	29	5	−7
Haiti	11.1	384	1.8	3.2	59	35	4	−3
Jamaica	2.7	251	0.8	2.0	55	24	9	−5
Martinique	0.4	–	0.3	1.9	89	19	17	−10
Montserrat*	0.005	51		1.3	14	46	6	0
Puerto Rico	3.4	399	0.1	1.4	99	17	18	−19
St. Kitts and Nevis	0.1	211	0.6	1.8	32	21	8	7
St. Lucia	0.2	301	0.6	1.5	18	22	9	2
St Vincent and the Grenadines	0.1	280	0.9	2.0	51	25	7	−8
Suriname	0.5	3	1.1	2.3	66	27	7	−2
Trinidad and Tobago	1.4	264	0.6	1.7	8	21	9	−1
Turks and Caicos*	0.05	50	–	1.7	93	22	4	15

Source: Population Reference Bureau, World Population Data Sheet, 2016.

* Additional data from the CIA World Factbook.

[1] World Bank, World Development Indicators, 2016.

Login to MasteringGeography™ **& access** MapMaster **to explore these data!**

1) Which countries in this region have a positive net migration rate, and which ones are negative? What might explain these differences?

2) Which countries have "older populations" (a higher percentage of the population over 65 years of age) and which have younger ones (a higher percentage of population under 15)?

In contrast to the islands, the mainland territories of Belize and the Guianas are lightly populated; Guyana averages 10 people per square mile (4 people per square kilometer), Suriname only 8 (3 per square kilometer), and Belize 36 (15 per square kilometer). These areas are sparsely settled in part because the relatively poor quality and accessibility of arable land made them less attractive to colonial enterprises (Table 5.1).

Demographic Trends

Prior to European contact with the New World, diseases such as smallpox, influenza, and malaria did not exist in the Americas. As discussed in Chapter 4, these diseases contributed to the demographic collapse of Amerindian populations. In the Caribbean, epidemics spread quickly, and within 50 years of Columbus' arrival, the indigenous population was virtually gone. Only the name Caribbean suggests that a Carib people once inhabited the region. Initially, European planters experimented with white indentured labor to work on sugar plantations. However, newcomers from Europe were especially vulnerable to malaria in the lowland Caribbean; typically, half died during the first year of settlement. Those that survived were considered "seasoned." In contrast, Africans had prior exposure to malaria and thus some immunity. They, too, died from malaria, but at much lower rates. This is not to argue that malaria caused slavery in the region, but it did strengthen the economic rationale for it.

During the years of slave-based sugar production, mortality rates were extremely high because of disease, inhumane treatment, and malnutrition. Consequently, the only way population levels could be maintained was through the continual importation of African slaves. With the end of slavery in the mid- to late 19th century and the gradual improvement of health and sanitary conditions on the islands, natural population increase began to occur. In the 1950s and 1960s, many states achieved peak growth rates of 3.0 or higher, causing population totals and densities to soar. Over the past 30 years, however, growth rates have steadily come down and stabilized. As noted earlier, the current population of the Caribbean is 45 million. The population is now growing at an annual rate of 1.1 percent, and projected population in 2025 is 49 million.

Fertility Decline and Longer Lives The most significant demographic trends in the Caribbean are the decline in fertility and the increase in life expectancy. Most countries of the region have low rates of natural increase and are below replacement values for total fertility. Puerto Rico's rate of natural increase is only 0.1, and it has the region's highest negative net migration rate (–19), which has risen sharply as the island's economy has struggled. In socialist Cuba, the RNI is only 0.2 and the average woman has 1.7 children. In general, educational improvements, urbanization, availability of contraception, and a preference for smaller families have contributed to slower growth rates in the region, regardless of political ideology. Even states with relatively high total fertility rates (TFR), such as Haiti, have seen a decline in family size. Haiti's TFR fell from 6.0 in 1980 to 3.2 in 2016. Currently French Guiana has the region's highest TFR at 3.5.

Figure 5.13 provides a stark contrast in the population profiles of Cuba and Haiti. Although both are poor Caribbean countries, Haiti has the more classic, broad-based pyramid of a developing country, where more than one-third of the population is under the age of 15. Also, there are very few old people, due to the relatively low life expectancy (64 years) in Haiti. In contrast, Cuba's population pyramid is more diamond-shaped, bulging in the 35- to 49-year-old age cohort and tapering down after that. Here the impact of the Cuban revolution and socialism is evident. Family size came down sharply after education improved and modern contraception became readily available. With better health care, Cuba's population also lives longer, having nearly the same life expectancy as those in the United States (78 years in Cuba vs. 79 years in the United States). Cuba has 14 percent of its population over 65 and just 16 percent under 15; thus, it has an extremely low rate of natural increase, similar to many developed countries.

The Impact of HIV/AIDS The rate of HIV/AIDS infection in the Caribbean has come down in the past decade, but it is still twice that of North America, making the disease an important regional issue. Although the infection rates are nowhere near those in Sub-Saharan Africa (see Chapter 6), 1.0 percent of the Caribbean population between the ages of 15 and 49 was living with HIV/AIDS in 2012—or roughly 250,000 people. In Haiti, one of the locations where AIDS was detected earliest, 1.8 percent of the population between the ages of 15 and 49 is infected with the virus. The infection rate is 1.8 percent in Jamaica, 2.3 percent in Belize, and 2.8 percent in The Bahamas.

Poverty, gender inequality, misinformation, and stigma attached to people infected with HIV contributed to the spread of the disease in the 1990s and 2000s. Added to this are the considerable movement among the island nations and the commercial sex trade, encouraged by a strong tourism economy. Now nearly every country in the region has launched educational campaigns to bring awareness up and infection rates down. Reflecting global patterns, heterosexual sex is the main route of HIV transmission in the Caribbean, often associated with paid sex. More than half of the infected population is female. In 2001, the Pan-Caribbean Partnership Against HIV/AIDS (PANCAP) formed to help

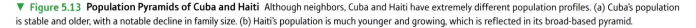

▼ **Figure 5.13 Population Pyramids of Cuba and Haiti** Although neighbors, Cuba and Haiti have extremely different population profiles. (a) Cuba's population is stable and older, with a notable decline in family size. (b) Haiti's population is much younger and growing, which is reflected in its broad-based pyramid.

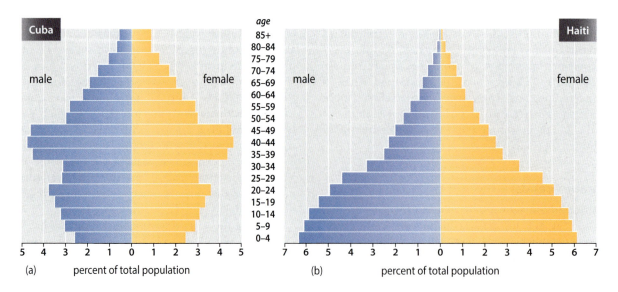

address the spread of the disease. PANCAP negotiated for lower-cost antiretroviral drugs, so that now two-thirds of the infected population receives these life-prolonging therapies. Through state and regional efforts, mother-to-child transmission prevention is the norm, condoms are widely available, and testing is easily done.

Emigration Driven by the region's limited economic opportunities, a pattern of emigration to other Caribbean islands, North America, and Europe began in the 1950s. For more than 50 years, a **Caribbean diaspora**—the economic flight of Caribbean peoples across the globe—has defined existence and identity for much of the region (**Figure 5.14**). Barbadians generally choose England, most settling in the London suburb of Brixton with other Caribbean immigrants. In contrast, one out of every three Surinamese has moved to the Netherlands, with most residing in Amsterdam. As for

Puerto Ricans, only slightly more live on the island than reside on the U.S. mainland. About 700,000 Jamaicans currently live in the United States, which is equal to one-quarter of the island's population. Cubans have made the city of Miami their destination of choice since the 1960s. Today they are a large percentage of that city's population.

Intraregional movements also are important. Perhaps one-fifth of all Haitians do not live in their country of birth. Their most common destination is the neighboring Dominican Republic, followed by the United States, Canada, and French Guiana (see *People on the Move: Haitian Struggles in the Dominican Republic*). Dominicans are also on the move; the vast majority come to the United States, settling in New York City, where they are the single largest immigrant group. Others, however, simply cross the Mona Passage and settle in Puerto Rico. As a region, the Caribbean has one of the highest annual rates of net migration in the world at −4.0 per

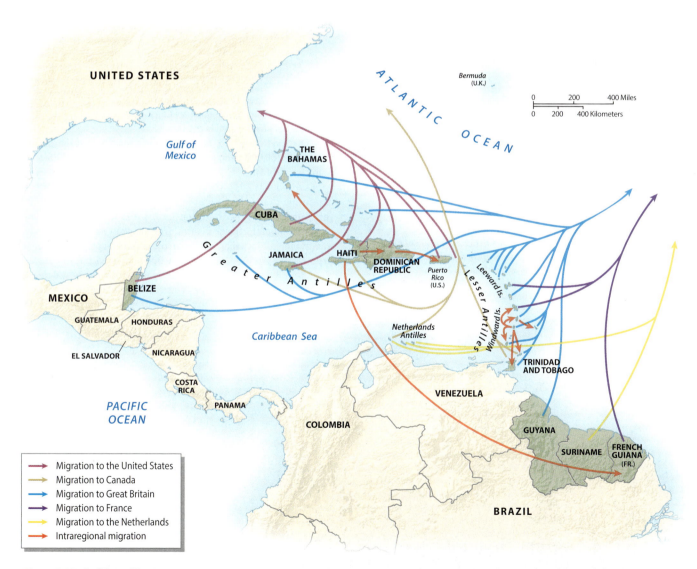

▲ **Figure 5.14 Caribbean Diaspora** Emigration has long been a way of life for Caribbean peoples. With relatively high education levels but limited professional opportunities, migrants from the region head to North America, Great Britain, France, and the Netherlands. Intraregional migration between Haiti and the Dominican Republic and between the Dominican Republic and Puerto Rico also occurs. **Q: Compare this map with Figure 5.23. What is the relationship between migration flows and the languages spoken in origin and destination countries?**

Haitian Struggles in the Dominican Republic

For nearly a century, Haitians have worked the difficult and dirty jobs in the Dominican Republic. Initially they came to cut sugar cane and many still work in agriculture, but are more likely to work in all-inclusive resorts, construction jobs, or housekeeping. Haitians generally work the low-wage and low-skilled jobs that are not taken by Dominicans (**Figure 5.2.1**).

Exactly how many *Haitianos* (the term used in the Dominican Republic) live across the border is unclear. Current estimates range from a low of 400,000 to nearly 750,000, counting the children of Haitian immigrants. Certainly after the devastating earthquake in 2010, there was an uptick in newcomers from Haiti into the Dominican Republic and a corresponding spike in resentment by Dominicans. Still, *Haitianos* represent, at most, 7 percent of the Dominican Republic's total population.

Race and Identity Dominican attitudes toward Haitianos can be xenophobic and even racist. While most Dominicans are of Afro-Caribbean ancestry (much like Haitians), many refer to themselves as *indio-oscuro* (dark Indian), which is a reference to the ancient Taino people who were wiped out within 50 years of Columbus' discovery of the island. Dominicans reserve the term *negro* (black) exclusively for Haitians. Much of Dominican racial identity is about not being Haitian, which is distinguished not just by an acculturated understanding of skin color but by language (Dominicans speak Spanish and Haitians speak French Creole), hairstyle, and dress.

Even now, there are documented abuses of Haitians living in near slave-like conditions, not being regularly paid, harassed by police, and even subject to lynchings—abuses that have been denounced by human rights organizations. It is common to see anti-Haitian graffiti in the countryside.

▲ **Figure 5.2.2 Protests in Santo Domingo** Haitians protest in the capital of the Dominican Republic in June 2015 to extend the registration deadline for legal status. Thousands of Haitians had to reapply for citizenship in the Dominican Republic after a law stripped it away in 2013.

Mass Deportation Threats The situation for Haitians and their descendants became worse under President Danilo Medina (elected in 2012 and reelected in 2016), who promised mass deportations of *Haitianos*. In 2013 the Dominican Constitutional Court ruled that the citizenship of anyone born in the Dominican Republic to parents deemed "foreigners in transit" (e.g., Haitian workers) could be revoked. Suddenly, more than 200,000 people were stateless, and nearly all of them were Dominican-Haitians.

A subsequent 2014 law created a program for those who lost their citizenship to reapply, by presenting birth certificates to authorities. Yet many Haitians reported that local authorities were unwilling or unable to supply birth certificates. The following year all Haitians were told to register for legal residence in the Dominican Republic or face deportation (**Figure 5.2.2**). Over 290,000 registered, but very few were granted legal status. In June 2015 when the registration deadline approached and the specter of mass deportations loomed, some 30,000 Haitians self-deported. There was also widespread international condemnation of this policy as well as threats to pull back on foreign investment and aid. In the end, the mass removal did not happen.

President Medina won his reelection in May 2016. The official Dominican policy is that mass deportations are planned. As such, many Haitians and Dominican-Haitians live in fear of a sudden knock on the door and their forced removal to a country that many have never known.

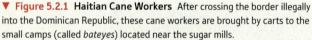

▼ **Figure 5.2.1 Haitian Cane Workers** After crossing the border illegally into the Dominican Republic, these cane workers are brought by carts to the small camps (called *bateyes*) located near the sugar mills.

1. What forces are driving Haitians to work in the Dominican Republic?

2. How do Dominican Republic policies to register and/or deport Haitians compare with anti-immigrant policies in your home country?

thousand. That means for every 1000 people in the region, 4 leave each year. Individual countries have much higher rates (see Table 5.1).

Crucial in this exchange of labor from south to north is the counterflow of cash remittances. Immigrants are expected to send something back, especially when immediate family members are left behind. Collectively, remittances add up; it is estimated that US$3.5 billion are sent annually to the Dominican Republic by immigrants in the United States, making remittances the country's second leading source of income. Jamaicans and Haitians remit nearly US$2 billion annually to their countries. Governments and individuals alike depend on these transnational family networks. Families carefully select the household member most likely to succeed abroad in the hope that money will flow back and a base for future immigrants will be established. A Caribbean nation in which no one left would soon face crisis.

Most migrants are part of a **circular migration** flow. In this type of migration, a man or woman typically leaves children behind with relatives in order to work hard, save money, and return home. Other times a **chain migration** begins, in which one family member at a time is brought over to the new country. In some cases, large numbers of residents from a Caribbean town or district send migrants to a particular locality in North America or Europe. Thus, chain migration can account for the formation of immigrant enclaves. Caribbean immigrants have increasingly practiced **transnational migration**—the straddling of livelihoods and households between two countries. Dominicans are probably the most transnational of all the Caribbean groups. They regularly move back and forth between two islands: Hispaniola and Manhattan. Former Dominican President Leonel Fernandez was first elected in 1996 for a four-year term and was reelected in 2004 and in 2008. He grew up in New York City and for many years held a green card.

The Rural–Urban Continuum

Initially, plantation agriculture and subsistence farming shaped Caribbean settlement patterns. Low-lying arable lands were dedicated to export agriculture and controlled by wealthy colonial landowners. Only small amounts of land were set aside for subsistence production. Over time, villages of freed or runaway slaves were established, especially in remote areas of the interior. But the vast majority of people continued to live on estates as owners, managers, or slaves. Cities were formed to serve the administrative and social needs of the colonizers, but most were small, containing a small fraction of a colony's population, and often defensive. The colonists who linked the Caribbean to the world economy saw no need to develop major urban centers.

Plantation America Anthropologist Charles Wagley coined the term **plantation America** to designate a cultural region that extends from midway up the coast of Brazil through the Guianas and the Caribbean into the southeastern United States. Ruled by a European elite dependent on an African labor force, this society was primarily coastal and produced agricultural exports. It relied upon **monocrop production** (a single commodity, such as sugar) under a plantation system that concentrated land in the hands of elite families. Such a system created rigid class lines, as well as formed a multiracial society in which people with lighter skin were privileged. The term *plantation America* is not meant to describe a race-based division of the Americas, but rather a production system that relied on export commodities, coerced labor, and limited access to land (**Figure 5.15**).

Even today, the structure of Caribbean communities reflects the plantation legacy. Many of the region's subsistence farmers are descendants of former slaves who continue to work their small plots and seek seasonal wage-labor on estates. The social and economic patterns generated by slavery still mark the landscape. Rural communities tend to be loosely organized; labor is transient; and small farms are scattered on available pockets of land. Because men have tended to leave home for seasonal labor, matriarchal family structures and female-headed households are common.

Caribbean Cities Since the 1960s, the mechanization of agriculture, offshore industrialization, and rapid population growth have caused a surge in rural-to-urban migration. Cities have grown accordingly, and today 66 percent of the region is classified as urban. Of the large islands, Puerto Rico is the most urban, and Haiti and Jamaica are the least urban (see Table 5.1). Caribbean metropolitan areas are not large by world standards, as only five have more than 1 million residents: Santo Domingo, Port-au-Prince, San Juan, Havana, and Kingston.

Like their counterparts in Latin America, the Spanish Caribbean cities were laid out on a grid with a central plaza. Vulnerable to raids by rival European powers and

▼ **Figure 5.15 Sugar Plantation** Commodities such as tobacco and sugar were profitable; but the work was arduous. Several million Africans were enslaved and forcibly relocated to the region to produce these commodities. This illustration from 1823 depicts slaves planting sugarcane in Antigua.

pirates, these cities were usually walled and extensively fortified. The oldest continually occupied European city in the Americas is Santo Domingo in the Dominican Republic, settled in 1496. Merengue—a fast-paced, highly danceable music that originated in the Dominican Republic—is the soundtrack that pulses through the metropolis day and night. As rural migrants poured into the city over the last four decades in search of employment and opportunity, the city steadily grew, and now is home to 3 million. In 2009 a high-speed Metro opened in Santo Domingo, and it now has two lines and 30 stations to reduce the city's crushing traffic (**Figure 5.16**). The country has experienced solid growth in the 2000s, but there is still inadequate housing, electricity, employment, and schooling for a large portion of Santo Domingo's residents. Some critics argue that an expensive underground Metro was ill-advised given the city's other pressing needs. Yet it is also a sign of both big-city status and modernity that Dominicans have embraced.

The region's second largest city is metropolitan San Juan, estimated at 2.6 million. It, too, has a renovated colonial core that is dwarfed by the modern sprawling city, which supports the island's largest port (**Figure 5.17**). San Juan is the financial, political, manufacturing, and tourism hub of Puerto Rico. With its highways, high rises, shopping malls, and ever-present shoreline, it is an interesting blend of Latin American, North American, and Caribbean urbanism.

Port-au-Prince, Haiti's capital, is now the third largest Caribbean city with 2.5 million residents in 2016. Much of the city was leveled in the 2010 earthquake, but even before that event the city's basic infrastructure, especially access to plumbing and electricity, was poor. While new roads and

▲ **Figure 5.17 Old San Juan** Tourists roam the cobbled streets of Old San Juan, a UNESCO World Heritage Site. Many of the 18th- and 19th-century structures of this port city have been handsomely restored.

Explore the **Sights** of Old San Juan

https://goo.gl/TM9gS7

▼ **Figure 5.16 Santo Domingo Metro** Passengers load onto metro cars in downtown Santo Domingo, Dominican Republic. The Metro, which opened in 2009, received technical support from the Metro in Madrid, Spain.

homes have been built, rural migrants have poured into the city in the last few years and services are once again strained.

Havana emerged as the most important colonial city in the region, serving as a port for all incoming and outgoing Spanish galleons. Strategically situated on Cuba's north coast at a narrow opening to a natural deep-water harbor, Havana became an essential city for the Spanish empire. Consequently, Old Havana possesses a handsome collection of colonial architecture, especially from the 18th and 19th centuries, and is also a UNESCO World Heritage Site. The modern city of 2 million is more sprawling, with a mix of Spanish colonial and Soviet-inspired concrete apartment blocks. It is also a city that had to reinvent itself when subsidies from the former Soviet Union stopped flowing. Now that Americans can freely travel to Cuba, many believe that the number of tourists bound for Havana, which is already large, will surge.

Other colonial powers left their mark on Caribbean cities. For example, Paramaribo, the capital of Suriname, has been described as a tropical, tulipless extension of Holland. In the British colonies, a preference for wooden whitewashed cottages with shutters is evident. Yet the British and French colonial cities tended to be unplanned afterthoughts; these port cities were built to serve the needs of the rural estates, rather than the needs of all residents. Most have grown dramatically over the past 40 years. These cities are no longer small ports for agricultural exports; increasingly, their focus is on welcoming cruise ships and sun-seeking tourists.

Caribbean cities and towns do have their charms and reflect a variety of cultural influences. Throughout the region, houses are often simple structures (made of wood, brick, or stucco), raised off the ground a few feet to avoid flooding, and painted in pastels. Most residents still get around by foot, bicycle, motorbike, or public transportation; neighborhoods are filled with small shops and services within easy walking distance (**Figure 5.18**). Streets are narrow, and the pace of life is markedly slower than in North America and Europe. Even when space is tight in town, most settlements are close to the sea and its cooling breezes. An afternoon or evening stroll along the waterfront is a common activity.

REVIEW

5.3 What are the major demographic trends for this region, and what factors explain these patterns?

5.4 How did the long-term reliance on a plantation economy influence settlement patterns and livelihoods in the Caribbean?

■ **KEY TERMS:** Caribbean diaspora, circular migration, chain migration, transnational migration, plantation America, monocrop production

Cultural Coherence and Diversity: A Neo-Africa in the Americas

Linguistic, religious, and ethnic differences abound in the Caribbean. The presence of several former European colonies, millions of descendants of ethnically distinct Africans and indentured workers from India and China, and isolated Amerindian communities on the mainland challenge any notion of cultural coherence.

Common historical and cultural processes hold the region together. In particular, this section focuses on three cultural influences shared throughout the Caribbean: the European colonial presence, African influences, and the mix of European and African cultures, termed *creolization*.

The Cultural Imprint of Colonialism

The arrival of Columbus in 1492 triggered a devastating chain of events that depopulated the region within 50 years. A combination of Spanish brutality, enslavement, warfare, and disease reduced the densely settled islands, which supported up to 3 million Caribs and Arawaks, into an uninhabited territory ready for the colonizer's hand. The demographic collapse of Amerindian populations occurred throughout the Americas (see Chapter 4), but the death rates were highest in the Caribbean. Only fragments of Amerindian communities survive, mostly on the rimland.

By the mid-16th century, as rival European states vied for Caribbean territory, the lands they fought for were virtually empty. In many ways, this simplified their task, as they did not have to acknowledge indigenous land claims or work amid Amerindian societies. Instead, the colonizers reorganized the Caribbean territories to serve a plantation-based production system. The critical missing element was labor. Once slave labor from Africa, and later contract labor from Asia, were secured, the small Caribbean colonies became surprisingly profitable.

Creating a Neo-Africa The introduction of African slaves to the Americas began in the 16th century and continued into the 19th century. This forced migration of mostly West Africans to the Americas was only part of a much more complex **African diaspora**—the forced removal of Africans from their native

▼ **Figure 5.18 Caribbean Motorbikes** A woman rides her motorbike past the whitewashed Dutch colonial-styled buildings in Paramaribo, Suriname. Across the Caribbean, motorbikes and bicycles are popular urban transportation options.

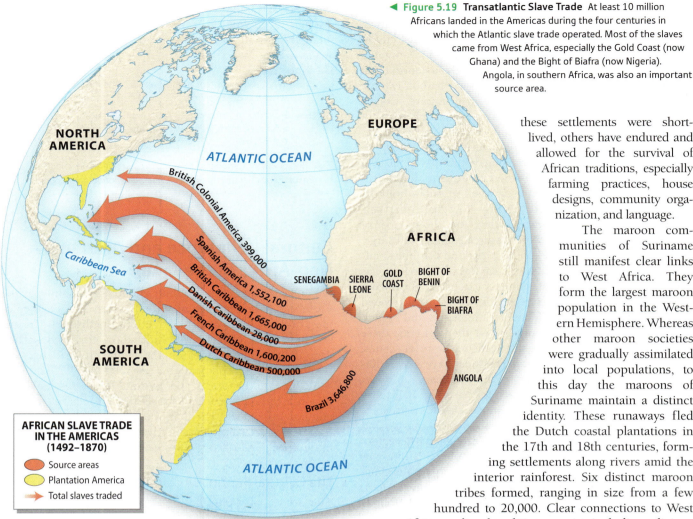

◄ **Figure 5.19** **Transatlantic Slave Trade** At least 10 million Africans landed in the Americas during the four centuries in which the Atlantic slave trade operated. Most of the slaves came from West Africa, especially the Gold Coast (now Ghana) and the Bight of Biafra (now Nigeria). Angola, in southern Africa, was also an important source area.

AFRICAN SLAVE TRADE IN THE AMERICAS (1492–1870)

- Source areas
- Plantation America
- Total slaves traded

these settlements were short-lived, others have endured and allowed for the survival of African traditions, especially farming practices, house designs, community organization, and language.

The maroon communities of Suriname still manifest clear links to West Africa. They form the largest maroon population in the Western Hemisphere. Whereas other maroon societies were gradually assimilated into local populations, to this day the maroons of Suriname maintain a distinct identity. These runaways fled the Dutch coastal plantations in the 17th and 18th centuries, forming settlements along rivers amid the interior rainforest. Six distinct maroon tribes formed, ranging in size from a few hundred to 20,000. Clear connections to West African cultural traditions persist, including religious practices, crafts, patterns of social organization, agricultural systems, and even dress (**Figure 5.20**). Living relatively

area. The slave trade also crossed the Sahara to include North Africa and linked East Africa with a slave trade in Southwest Asia (see Chapter 6). The best-documented slave route is the transatlantic one; at least 10 million Africans landed in the Americas, and it is estimated that another 2 million died en route. More than half of these slaves were sent to the Caribbean (**Figure 5.19**).

This influx of slaves, combined with the extermination of nearly all the native inhabitants, recast the Caribbean as the area with the greatest concentration of relocated African people in the Americas. The African source areas extended from Senegal to Angola, and slave purchasers intentionally mixed tribal groups in order to dilute ethnic identities. Consequently, intact transfer of African religions and languages into the Caribbean did not occur; instead, languages, customs, and beliefs were blended.

Maroon Societies Communities of runaway slaves—termed **maroons** in English, *palenques* in Spanish, and *quilombos* in Portuguese—offer the most compelling examples of African cultural diffusion across the Atlantic. Hidden settlements of escaped slaves existed wherever slavery was practiced. While many of

▼ **Figure 5.20** **Maroons in Suriname** Maroons (descendants of runaway slaves) dance in Paramaribo at the annual Black People's Day celebration, the first Sunday in January. Living in Suriname for over two centuries, Maroons maintain West African cultural traditions.

Explore the **Sights** of Maroon Communities in Suriname

https://goo.gl/EBNSbc

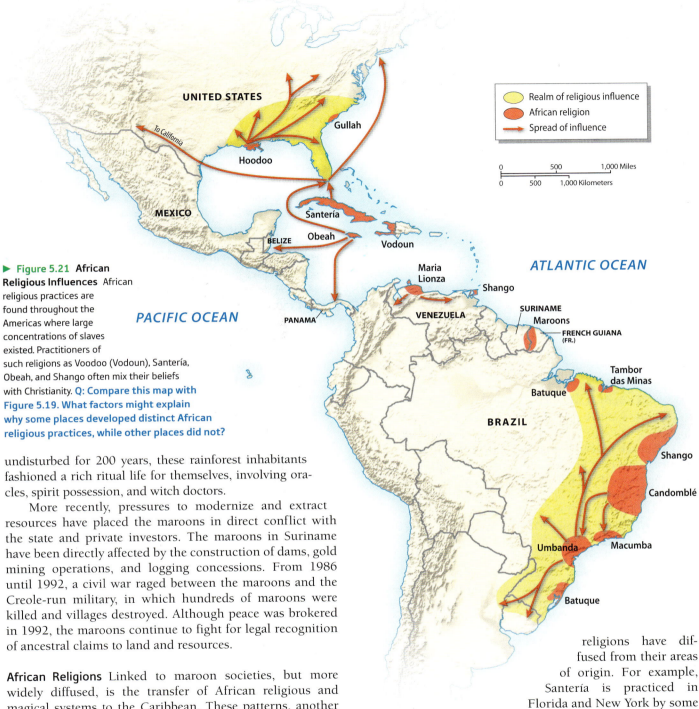

▶ **Figure 5.21 African Religious Influences** African religious practices are found throughout the Americas where large concentrations of slaves existed. Practitioners of such religions as Voodoo (Vodoun), Santería, Obeah, and Shango often mix their beliefs with Christianity. **Q: Compare this map with Figure 5.19. What factors might explain why some places developed distinct African religious practices, while other places did not?**

undisturbed for 200 years, these rainforest inhabitants fashioned a rich ritual life for themselves, involving oracles, spirit possession, and witch doctors.

More recently, pressures to modernize and extract resources have placed the maroons in direct conflict with the state and private investors. The maroons in Suriname have been directly affected by the construction of dams, gold mining operations, and logging concessions. From 1986 until 1992, a civil war raged between the maroons and the Creole-run military, in which hundreds of maroons were killed and villages destroyed. Although peace was brokered in 1992, the maroons continue to fight for legal recognition of ancestral claims to land and resources.

African Religions Linked to maroon societies, but more widely diffused, is the transfer of African religious and magical systems to the Caribbean. These patterns, another reflection of neo-Africa in the Americas, are most closely associated with northeastern Brazil and the Caribbean. Millions of Brazilians practice the African-based religions of Umbanda, Macuba, and Candomblé, along with Catholicism. Likewise, Afro-religious traditions in the Caribbean have evolved into unique forms that have clear ties to West Africa.

The most widely practiced religions are Voodoo (also Vodoun) in Haiti, Santería in Cuba, and Obeah in Jamaica. These religions have their own priesthoods and unique patterns of worship. As **Figure 5.21** shows, many of these religions have diffused from their areas of origin. For example, Santería is practiced in Florida and New York by some Cuban immigrants. Likewise, belief in Obeah diffused when Jamaicans migrated to Panama and Los Angeles. The impact of these belief systems is considerable; the father and son dictators of Haiti, the Duvaliers, were known to hire Voodoo priests to scare off government opposition during their rule from 1957 to 1986.

Indentured Labor from Asia By the mid-19th century, most colonial governments in the Caribbean had begun to free their slaves. Fearful of labor shortages, they sought

indentured labor (workers contracted to labor on estates for a set period of time, often several years) from South, Southeast, and East Asia.

The legacy of these indentured arrangements is clearest in Suriname, Guyana, and Trinidad and Tobago. In Suriname, a former Dutch colony, more than one-third of the population is of South Asian descent, and 16 percent is Javanese (from Indonesia, another former Dutch colony). Guyana and Trinidad were British colonies, and most of their contract labor came from India. Today half of Guyana's population and 40 percent of Trinidad and Tobago's claim South Asian ancestry. Hindu temples are found in the cities and villages, and many families speak Hindi at home. In 2015 Moses Nagamootoo, of Indian ancestry, was elected Prime Minister of Guyana (**Figure 5.22**).

Most of the former English colonies have Chinese populations of not more than 2 percent. Once these East Asian immigrants fulfilled their agricultural contracts, they often became merchants and small-business owners, positions they still hold in Caribbean society. Cuba and Suriname have the largest ethnic Chinese populations in the region. Moreover, Suriname has experienced a substantial surge in Chinese immigration, with some reports suggesting that recent Chinese arrivals may account for 10 percent of the country's population.

Creolization and Caribbean Identity

Creolization refers to the blending of African, European, and even some Amerindian cultural elements into the unique sociocultural systems found in the Caribbean. The Creole identities that have formed over time are complex; they illustrate the cultural and national identities of the region. Today Caribbean writers (V. S. Naipaul, Derek Walcott, and Jamaica Kincaid), musicians (Bob Marley, Ricky Martin, and Juan Luis Guerra), and athletes (Dominican baseball player David Ortiz or Jamaican sprinter Usain Bolt) are internationally recognized. Collectively, these artists and athletes represent their individual islands and Caribbean culture as a whole.

The story of the Garifuna people illustrates creolization at work. Now settled along the Caribbean rimland from the southern coast of Belize to the northern coast of Honduras, the Garifuna (formerly called the *Black Carib*) are descendants of African slaves who speak an Amerindian language. Unions between Africans and Carib Indians on the island of St. Vincent produced an ethnic group that was predominantly African, but spoke an Indian language. In the late 18th century, Britain forcibly resettled some 5000 Garifuna from St. Vincent to the Bay Islands in the Gulf of Honduras. Over time, the Garifuna settled along the Caribbean coast of Central America, living in isolated fishing communities in Belize, Guatemala, and Honduras. In addition to maintaining an Indian language, the Garifuna are the only group in Central America who regularly eat bitter manioc—a root crop common in lowland tropical South America. It is assumed that they acquired their taste for manioc from their exposure to Carib culture. The Afro-Indian blend that the Garifuna manifest is unique, but the process of creolization is recognizable throughout the Caribbean, especially in language and music.

Explore the **Sounds** of Garifuna Music

https://goo.gl/Zrrshk

Language The dominant languages in the region are European: Spanish (25 million speakers), French (11 million), English (7 million), and Dutch (0.5 million) (**Figure 5.23**). However, these figures tell only part of the story. In Cuba, the Dominican Republic, and Puerto Rico, Spanish is the official language, and it is universally spoken. As for the other countries, colloquial variants of the official language exist, especially in spoken form, which can be difficult for a nonnative speaker to understand. In some cases, completely new languages have emerged. In the islands of Aruba, Bonaire, and Curaçao, Papiamento (a trading language that blends Dutch, Spanish, Portuguese, English, and African languages) is the *lingua franca,* with usage of Dutch declining. In Suriname, the vernacular language is Sranan Tongo (an amalgam of Dutch and English with many African words). Similarly, French Creole or *patois* in Haiti has constitutional status as a distinct language. In practice, French is used in higher education, government, and the courts, but patois (with clear African influences) is the language of the street, the home, and oral tradition. Most Haitians speak patois, but only the formally educated know French.

Explore the **Sounds** of Papiamento

https://goo.gl/8cDbS9

With the independence of Caribbean states from European colonial powers in the 1960s, Creole languages became politically and culturally charged with national meaning. Most formal education is taught using standard language forms, but the richness of vernacular expression and its ability to instill a sense of identity are appreciated. As linguists began to study these languages, they found that, although the vocabulary came from Europe, the syntax or semantic structure had other origins, notably from African language families. Locals rely on their ability to switch from standard to vernacular forms of speech. Thus, a Jamaican can converse with a tourist in standard English and then switch to a Creole variant when a friend walks by, effectively excluding the

▼ **Figure 5.22 South Asians in the Caribbean** Guyana's Prime Minister, Moses Nagamootoo (in white), congratulates Veerasammy Permaul, a cricket player on the Guyana Amazon Warriors team, for being player of the match. Both men share South Asian ancestry.

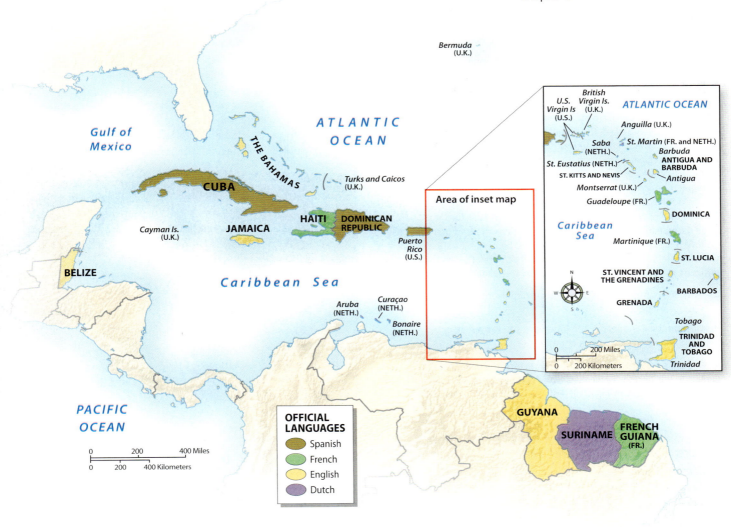

▲ **Figure 5.23 Languages of the Caribbean** Because this region has no significant Amerindian population (except on the mainland), the dominant languages are European: Spanish (25 million), French (11 million), English (7 million), and Dutch (0.5 million). However, many of these languages have been creolized, making it difficult for outsiders to understand them.

outsider from the conversation. This ability to switch is evident in many cultures, but is widely used in the Caribbean.

Music The rhythmic beats of the Caribbean might be the region's best-known product. This small area is the birthplace of reggae, calypso, merengue, rumba, zouk, and scores of other musical forms. The roots of modern Caribbean music reflect a combination of African rhythms with European forms of melody and verse. These diverse influences, coupled with a long period of relative isolation, sparked distinct local sounds. As movement among the Caribbean inhabitants increased, especially during the 20th century, musical traditions were blended, but characteristic sounds remained.

The famed steel pan drums of Trinidad were created from oil drums discarded from a U.S. military base there in the 1940s. The bottoms of the cans are pounded with a sledge hammer to create a concave surface that produces different tones. During carnival (a pre-Lenten celebration),

racks of steel pans are pushed through the streets by dancers while the drummers play (see Figure 5.2). So skilled are these musicians that they even perform classical music, and government agencies encourage troubled teens to learn steel pan (see *Everyday Globalization: Caribbean Carnival*)

The distinct sound and ingenious rhythms have made Caribbean music very popular. When Jamaican Bob Marley and the Wailers crashed the international music scene with their soulful reggae sound, it was the lyrics about poverty, injustice, and freedom that resonated with the world. More than good dance music, Caribbean music can be closely tied to the Afro-Caribbean religions and is a popular form of political protest. In Haiti, rara music mixes percussion instruments, saxophones, and bamboo trumpets, weaving in funk and reggae base lines. The songs are always performed in French Creole and typically celebrate Haiti's African ancestry and the use of Voodoo. The lyrics often address difficult issues, such as political oppression or poverty (**Figure 5.24**).

Caribbean Carnival

Modern carnival originated in Christian Lenten beliefs and African musical traditions. The term is Latin in origin, a reference to giving up eating meat (*carnivale* means to put meat away) in observance of Lent (the 40-day period prior to Easter). In the Caribbean, former slaves imbued carnival with special meaning because it became their opportunity to break the monotony of their daily lives.

Today carnival is celebrated on nearly every Caribbean island as a national street party that can go on for weeks and attract thousands of tourists. Many official carnivals are no longer tied to Lent, but are scheduled at other times of the year. Carnival celebrations have also followed the Caribbean diaspora to new settings in North America and Europe. One of North America's biggest carnivals is in Toronto, where a large Caribbean immigrant population maintains the tradition every July (**Figure 5.3.1**). In London, Birmingham, and Leicester, Caribbean carnivals are celebrated annually. As a cultural measure of globalization, the diffusion of carnival shows that a great party transfers easily!

1. Why do you think carnival became an important expression of Caribbean identity?

2. Outside the Caribbean where else have you witnessed or experienced this kind of celebration?

▶ **Figure 5.3.1** Carnival in Toronto revelers march in the Toronto Caribbean Carnival Parade held each summer in Canada's largest and most diverse city. Toronto has long been an important destination for Caribbean migrants.

Sports: Caribbean Olympic Glory

Two Caribbean countries dominate in producing Olympic-caliber athletes: Jamaica and Cuba. Each country brought home 11 medals during the 2016 Olympic Games in Rio de Janeiro. While other Caribbean countries direct athletes toward soccer, baseball, or cricket, Jamaicans prefer track and field, especially sprinting. Jamaica's 11 medals were all in track and field, which is particularly impressive considering it is a nation of less than 3 million. An important individual in their success is Usain Bolt, a sprinter who has dominated his events since exploding onto the scene in 2008 at the Beijing games. In the last three Olympics, Bolt won gold in the 100 and 200 meter sprints and the 4 x100 meter relay. His charismatic personality on the track has made him an international sensation and a much-admired figure at home (**Figure 5.25**). Jamaican women sprinters are also world class; at the 2016 games Elaine Thompson won gold in the 100 meter and 200 meter sprints, and was part of the 4 x100 women's team that took the silver medal.

For Cubans, however, their Olympic glory has come through boxing. As in many socialist countries, promising athletes are nurtured from an early age. In the Rio de Janeiro games, six of Cuba's 11 medals were earned in the boxing ring. There was also three medalists in wrestling, one in judo, and one in track. Cubans enjoy boxing, but as a nation they are passionate about baseball, something they share with Puerto Rico, the Dominican Republic, and the United States. The importance of baseball in the Caribbean is a byproduct of early U.S. influence in the region, and will be discussed in the next section.

▼ **Figure 5.24 Haiti's Rara Music** Performed in procession, rara music is sung in patois. Considered the music of the poor, it is used to express risky social commentary. This rara band performs at a folk festival in Washington, D.C.

REVIEW

5.5 What kinds of neo-African influences exist in the Caribbean, and how do they express themselves?

5.6 What is meant by creolization, and how does it explain different cultural patterns found in the Caribbean?

■ **KEY TERMS** African diaspora, maroons, indentured labor, creolization

▲ **Figure 5.25 Jamaican Sprinter** Usain Bolt is considered one of greatest sprinters in history. Wearing Jamaican yellow, he celebrates gold after winning the 200m race at the Olympics in Rio de Janeiro. Track and field is a national obsession in Jamaica, and the country produces many world class athletes.

Geopolitical Framework: Colonialism, Neocolonialism, and Independence

Caribbean colonial history is a patchwork of competing European powers fighting over profitable tropical territories. By the 17th century the Caribbean had become an important proving ground for European ambitions. Spain's grip on the region was slipping, and rival European nations felt confident that they could gain territory by gradually pushing Spain out. Many territories, especially the islands in the Lesser Antilles, changed hands several times.

Europeans viewed the Caribbean as a strategically located and profitable region in which to produce sugar, rum, and spices. Geopolitically, rival European powers also felt that their presence in the Caribbean limited Spanish authority there. However, Europe's geopolitical dominance in the Caribbean began to diminish by the mid-19th century, just as the U.S. presence increased.

Inspired by the **Monroe Doctrine**, which claimed that the United States would not tolerate European military involvement in the Western Hemisphere, the U.S. government made it clear that it considered the Caribbean to be within its sphere of influence. The view was highlighted during the Spanish-American War in 1898. Even though several English, Dutch, and French colonies persisted after this date, the United States indirectly (and sometimes directly) asserted its control over the region, ushering in a period of **neocolonialism**. Neocolonialism is the indirect control of one country or region by another through economic and cultural domination, rather than through direct military or political control as occurs under colonialism.

In an increasingly global age, however, even neocolonial interest can be short-lived or sporadic. The Caribbean has not attracted the level of private foreign investment seen in other regions, and as its strategic importance in a post–Cold War era fades, new geopolitical ties are shaping the region. Taiwan began wooing small Caribbean islands in the 1990s with strategic investments, in the hopes of winning United Nations votes for its cause. Not surprisingly, China has invested even greater amounts of money in the region, in part to convince nations that supported Taiwan, such as Dominica and Grenada, to switch their support to China.

Life in the "American Backyard"

To this day, the United States exerts considerable influence in the Caribbean, which was commonly referred to as the "American backyard" in the early 20th century. The stated foreign policy objectives were to free the region from European authority and encourage democratic governance. Yet time and again, American political and economic ambitions undermined those goals. President Theodore Roosevelt made his priorities clear with imperialistic policies that extended the influence of the United States beyond its borders. Policies and projects such as the construction of the Panama Canal and the maintenance of open sea-lanes benefited the United States, but did not necessarily support social, economic, or political gains for the Caribbean people. The United States later offered benign-sounding development packages such as the Good Neighbor Policy (1930s), the Alliance for Progress (1960s), and the Caribbean Basin Initiative (1980s). The Caribbean view of such initiatives has been wary at best. Rather than feeling liberated, many residents believe that one kind of political dependence was being traded for another—colonialism for neocolonialism (see *Exploring Global Connections: From Baseball to Béisbol*).

In the early 1900s, the role of the United States in the Caribbean was overtly military and political. The Spanish-American War (1898) secured Cuba's freedom from Spain and also resulted in Spain ceding the Philippines, Puerto Rico, and Guam to the United States; the latter two are still U.S. territories. The U.S. government also purchased the Danish Virgin Islands in 1917, renaming them the U.S. Virgin Islands and developing the harbor of St. Thomas. French, English, and Dutch colonies were tolerated as long as these allies recognized U.S. supremacy in the region. Avowedly against colonialism, the United States had become much like an imperial force.

One privilege of empire is the ability to impose one's will, by force if necessary. When a Caribbean state refused to abide by American trade rules, U.S. Navy vessels would block its ports. U.S.-backed governments were installed throughout the Caribbean Basin. These were not short-term engagements: U.S. troops occupied the Dominican Republic from 1916 to 1924, Haiti from 1915 to 1934, and

From Baseball to Béisbol

The popularity of baseball in the Caribbean is attributed to U.S. influence in the region in the early 20th century as American businesses and military personnel introduced the game to the region. In particular, Cubans, Puerto Ricans, Venezuelans, and Dominicans adore the game (Figure 5.4.1). Recently, several Cuban baseball stars defected to the United States; players such as Jose Abreu, Alex Guerrero, and Aroldis Chapman landed multimillion-dollar contracts and appreciative fans. Over one-quarter of Major League Baseball (MLB) players are born outside the United States, mostly from the Caribbean or Latin America. The Dominican Republic sends the most players to the major leagues, accounting for 10 percent of all players in 2016. After

▲ **Figure 5.4.2 Dominican Training Camps** Major League hopefuls train at the Washington Nationals Dominican baseball training complex in Boca Chica. Many MLB teams have training facilities in the Dominican Republic, anchored near the city of San Pedro de Macoris.

▼ **Figure 5.4.1 Caribbean Baseball** A young Cuban batter takes aim during a pick-up game in rural Cuba. Several Caribbean islands have adopted baseball as their national sport—most notably the Dominican Republic, Cuba, and Puerto Rico.

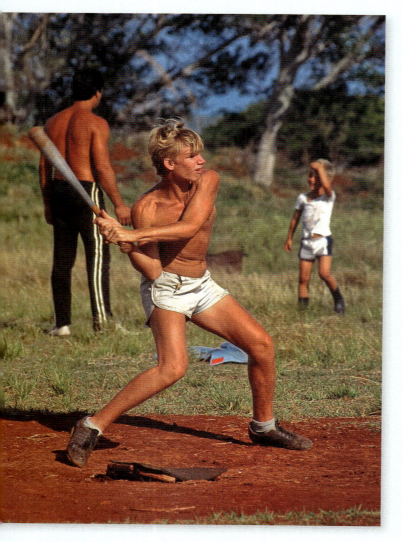

Dominicans, the top sources for international players are Venezuela, Cuba, and Puerto Rico.

The Dominican Republic became an MLB talent pipeline due to a complex mix of boyhood dreams, economic inequality, and greed. This small country has produced many baseball legends, and over the decades franchises have invested millions in training camps there. In the past two decades, however, Dominican pride in its baseball prowess has been tinged by the realities of a merciless feeder system that depends on impoverished kids, performance-enhancing drugs, fake documents, and scouts who skim a percentage of the signing bonuses. Yet the reality is that more and more young boys, who can sign contracts at age 16, see their future in baseball rather than in schooling. Even a modest signing bonus of $10,000 to $20,000 can build a nice home for a boy's family.

San Pedro de Macoris, not far from Santo Domingo, epitomizes this field of dreams (Figure 5.4.2). It is a place of cane fields, kids on bicycles with bats and gloves, sugarcane factories, dusty baseball diamonds, and large homes of former players, such as George Bell, Pedro Guerrero, and Sammy Sosa. These houses are silent testaments to what is possible through baseball. In an effort to clean up baseball's image, MLB has officials in the Dominican Republic investigating drug use and fraudulent papers. However, as long as there are families pushing their teenage boys and a talent pool that delivers, this transnational system is self-perpetuating.

Google Earth Virtual **MG** Tour Video

https://goo.gl/llvstr

1. What has attributed to the popularity of baseball in the Caribbean?

2. In what other countries is baseball popular?

Cuba from from 1898 to 1902 (**Figure 5.26**). Even today the United States maintains several important military bases in the region, including Guantánamo in eastern Cuba. There is greater reluctance to commit troops in the area now, but as recently as 1994 and 2004, U.S. troops were sent to Haiti to suppress political violence and prevent a mass exodus of Florida-bound refugees. Also, after the Haitian earthquake in 2010, U.S. naval vessels and troops were deployed to assist in the international relief effort.

Critics of U.S. policy in the Caribbean complain that business interests overwhelm democratic principles when foreign policy is determined. The American banana companies that settled the coastal plain of the Caribbean rimland operated as if they were independent states. Sugar and rum manufacturers from the United States bought the best lands in Cuba, Haiti, and Puerto Rico. Meanwhile, truly democratic institutions remained weak, and there was little improvement in social development. True, exports increased, railroads were built, and port facilities were improved, but levels of income, education, and health remained abysmally low throughout the first half of the 20th century.

The Commonwealth of Puerto Rico Because of its status as a commonwealth of the United States, Puerto Rico is both within the Caribbean and apart from it. Throughout the 20th century, various Puerto Rican independence movements sought to uncouple the island from the United States. Even today residents of the island are divided about their political future. At the same time, Puerto Rico depends on U.S. investment and welfare programs; U.S. food stamps are a major source of income for many Puerto Rican families. Commonwealth status also means that Puerto Ricans can freely move between the island and the U.S. mainland, a right they actively assert. In other ways, Puerto Ricans symbolically manifest their independence; for example, they support their own "national" sports teams and send a Miss Puerto Rico to international beauty pageants. The dispute over the use of Vieques Island (a small island off the east coast) as a naval testing ground for bombing exercises became a flashpoint in U.S.–Puerto Rican relations. Ultimately, the political pressure from Puerto Ricans led President George W. Bush to close the facility in 2003. The island now supports an upscale resort.

▼ **Figure 5.26 Geopolitical Issues in the Caribbean** The Caribbean was labeled the geopolitical backyard of the United States, and U.S. military occupation was a common occurrence in the first half of the 20th century. Border and ethnic conflicts also exist, most notably in the Guianas.

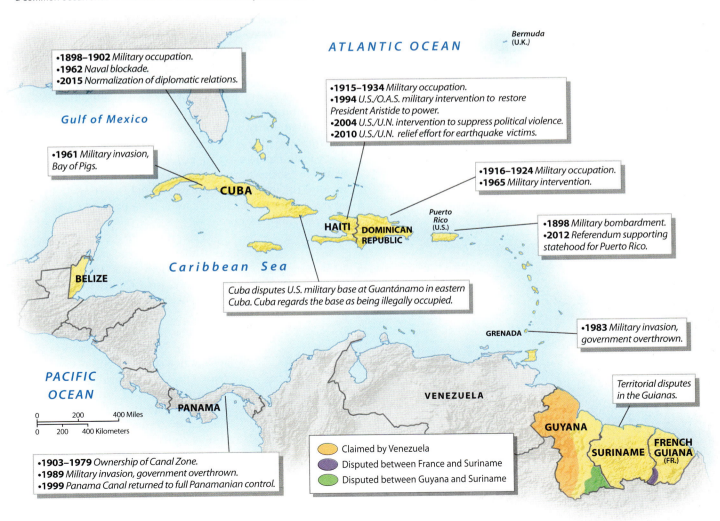

Beginning in the 1950s, Puerto Rico led the Caribbean in the transition from an agrarian economy to an industrial one. For some U.S. officials, Puerto Rico became the model for the rest of the region. Puerto Rican President Muñoz Marín championed an industrialization program called "Operation Bootstrap." Drawn by tax incentives and cheap labor, hundreds of U.S. textile and apparel firms relocated to Puerto Rico. Over the next two decades, 140,000 industrial jobs were added, resulting in a marked increase in per capita gross national income (GNI). In the 1970s, when Puerto Rico faced stiff competition from Asian apparel manufacturers, the government encouraged petrochemical and pharmaceutical plants to relocate to the island. By the 1990s, Puerto Rico was one of the most industrialized places in the region, with a significantly higher per capita income than its neighbors. But by the 2000s, many of the tax benefits associated with establishing factories in Puerto Rico had ceased to exist, and manufacturing declined. Consequently, there are growing signs of unemployment, extensive out-migration, and serious debt problems. In 2012, a controversial island referendum resulted in the majority of residents voting for a change in political status, with their preference being for statehood. There are no plans in the U.S. Congress to change the political status of the island (**Figure 5.27**).

Cuba and Geopolitics The most profound challenge to U.S. authority in the region came from Cuba and its superpower ally, the former Soviet Union. In the 1950s, a revolutionary effort led by Fidel Castro began in Cuba against the pro-American Batista government. Cuba's economic productivity had soared under Batista, but its people were still poor, uneducated, and increasingly angry. The contrast between the lives of average cane workers and the foreign elite was stark. Castro tapped a deep vein of Cuban resentment against six decades of U.S. neocolonialism. In 1959, Castro took power.

After Castro's government nationalized American industries and took ownership of all foreign-owned properties, the United States responded by refusing to buy Cuban sugar and ultimately ending diplomatic relations with the state. Various U.S. trade embargoes against Cuba have existed for nearly five decades. What sealed Cuba's fate as a geopolitical enemy was its establishment of diplomatic relations with the Soviet Union in 1960, during the height of the Cold War. With the Soviet Union financially and militarily backing Castro, a direct U.S. invasion of Cuba was too risky. Instead, CIA-trained paramilitaries attempted the Bay of Pigs invasion in the spring of 1961, but failed. The fall of 1962 produced one of the most dangerous episodes of the Cold War when Soviet missiles were discovered on Cuban soil. Ultimately, the Soviet Union removed its weapons; in return, the United States promised not to invade Cuba.

Even with the end of the Cold War, when Cuba lost its financial support from the Soviet Union, the country managed to reinvent itself by growing its tourism sector and courting foreign investment, especially from Spain. Castro and the late Venezuelan President Hugo Chavez also became close political allies. In 2004, they signed an important exchange agreement in which Cuba provided Venezuela with doctors, while Venezuela shipped much-needed oil to Cuba.

In many ways, Cuba is entering a new political era. In 2008, Fidel Castro, 82 and in poor health, left office and his younger brother, Raúl Castro, assumed the duties of president. Raúl seems more willing to encourage small private enterprise in Cuba as he expanded licenses to individuals and small businesses. In 2013, Raúl announced that he would not seek a third term in office and praised the appointment of the first vice president, Miguel Díaz-Canel. A much younger man in his 50s, many believe that Díaz-Canel will lead Cuba in the future.

▼ **Figure 5.27 Puerto Rican Referendum for Statehood** Supporters of the pro-statehood New Progressive party hold placards supporting the vote for Puerto Rico to become the 51st state. The referendum passed in November 2012, but the island is still deeply divided over its relationship with the United States. Meanwhile, the U.S. Congress has yet to take any action regarding Puerto Rican statehood.

The most dramatic change in U.S.-Cuban relations came in December 2014 when President Obama and President Castro agreed to restore full diplomatic relations between the two countries. Through executive order, restrictions on trade, remittances, and travel have been eased. The first regular commercial flights from the United States to Cuba resumed in 2016, and American tourists in Cuba are expected to increase dramatically. That same year Fidel Castro died at the age of 90. Interestingly, Cuba has been importing some food from the United States for over a decade, based on chronic food scarcity. Currently, Cuba's main sources of agricultural imports are the European Union,

Brazil, and the United States. The major destinations for Cuban exports are China, the EU, and Russia. This was expected to change as the U.S. restrictions ease and Cuban sugar, oranges, and cigars find their way directly into U.S. markets. Although the Trump Administration has promised to reverse many of these political openings with Cuba.

Independence and Integration

Given the repressive colonial history of the Caribbean, it is no wonder that the struggle for political independence began more than 200 years ago. Haiti was the second colony in the Americas to gain independence in 1804, after the United States in 1776. However, political independence in the region has not guaranteed economic independence. Many Caribbean states struggle to meet the basic needs of their people. Today some Caribbean territories maintain their colonial status as an economic asset. For example, Martinique, Guadeloupe, and French Guiana are overseas departments of France; residents have full French citizenship and social welfare benefits.

Independence Movements Haiti's revolutionary war began in 1791 and ended in 1804. Spanish, French, and British forces were involved, as well as factions within Haiti that had formed along racial lines. During this conflict, the island's population was cut in half by casualties and emigration; ultimately, the former slaves became the rulers. Independence, however, did not allow this jewel of the French Caribbean to prosper. Plantation America watched in horror as Haitian slaves used guerrilla tactics to gain their freedom. Fearing that other colonies might follow Haiti's lead, plantation owners in other countries were on guard for the slightest hint of revolt. For its part, Haiti did not become a leader in liberation. Slowed by economic and political problems, it was shunned by the European powers and never fully accepted by the Latin American countries on the mainland once they gained their own political independence in the 1820s.

Several revolutionary periods followed in the 19th century. In the Greater Antilles, the Dominican Republic finally gained independence in 1844 after taking control of the territory from Spain and Haiti. Cuba and Puerto Rico were freed from Spanish colonialism in 1898, but their independence was compromised by greater U.S. involvement. The British colonies also faced revolts, especially in the 1930s, yet it was not until the 1960s that independent states emerged from the English Caribbean. First, the larger colonies of Jamaica, Trinidad and Tobago, Guyana, and Barbados gained their independence. Other British colonies followed throughout the 1970s and early 1980s. Suriname, the only Dutch colony on the rimland, became an autonomous territory in 1954, but remained part of the Kingdom of the Netherlands until 1975, when it declared itself an independent republic.

Present-Day Colonies Britain still maintains several crown colonies in the region: the Cayman Islands, the Turks and Caicos, Anguilla, Montserrat, and Bermuda. The combined population of these islands is about 200,000 people, yet their standard of living is high, due in part to their specialization in the recently developed industry of offshore financial services. French Guiana, Martinique, and Guadeloupe are French departments and thus, technically speaking, not colonies. Together, they total over 1 million people. The Dutch islands in the Caribbean were once part of the federation with the Kingdom of the Netherlands. Today Aruba, Curaçao, and St. Maarten are completely independent while Bonaire, Saba, and St. Eustatius continue ties with the Netherlands. Together, the population of the Dutch islands is a quarter of a million people.

Limited Regional Integration Perhaps the most difficult task facing the Caribbean is to increase economic integration. Scattered islands, a divided rimland, different languages, and limited economic resources hinder the formation of a meaningful regional trade bloc. Economic cooperation is more common between groups of islands with a shared colonial background than between, for example, former French and English colonies.

During the 1960s, the Caribbean began to experiment with regional trade associations as a means to improve its economic competitiveness. The goals of regional cooperation were to improve employment rates, increase intraregional trade, and ultimately reduce economic dependence. The countries of the English Caribbean took the lead in this development strategy. In 1963, Guyana proposed an economic integration plan with Barbados and Antigua. In 1972, the integration process intensified with the formation of the **Caribbean Community and Common Market (CARICOM)**. Representing the former English colonies, CARICOM proposed an ambitious regional industrialization plan and the creation of the Caribbean Development Bank to assist the poorer states. CARICOM also oversees the University of the West Indies, with campuses in Trinidad, Jamaica, and Barbados. As important as this trade group is as an institutional symbol of collective identity, it has produced limited improvements in intraregional trade.

Today CARICOM has 15 full-member states—all of the English Caribbean, French-speaking Haiti, and Dutch-speaking Suriname. Other dependencies, such as Anguilla, Bermuda, the Turks and Caicos, the Cayman Islands, and the British Virgin Islands, are associate members. As CARICOM's membership grows, so does its services. In 2005, CARICOM passports began to be issued to facilitate travel between member nations, but efforts to create a unified economic development policy have been challenging. CARICOM has been an important voice in global climate negotiation treaties to reduce GHG emissions and to respond to the worst impacts of climate change.

The dream of regional integration as a way to produce a more stable, self-sufficient Caribbean has never been realized. One scholar of the region argues that a factor limiting this regional integration is a "small-islandist ideology."

Islanders tend to keep their backs to the sea, oblivious to the needs of neighbors. At times, such isolationism results in suspicion, distrust, and even hostility toward nearby states. Yet economic necessity dictates engagement with partners outside the region.

REVIEW

5.7 Which countries have had colonial or neocolonial influences in the Caribbean, and why have they engaged with the region?

5.8 What are the obstacles to Caribbean political or economic integration?

■ **KEY TERMS** Monroe Doctrine, neocolonialism, Caribbean Community and Common Market (CARICOM)

Economic and Social Development: From Cane Fields to Cruise Ships

Collectively, Caribbean peoples, although poor by U.S. standards, are economically better off than most of Sub-Saharan Africa, South Asia, and even China. Despite periods of economic stagnation in the Caribbean, social gains in education, health, and life expectancy are significant (Table 5.2). Historically, the Caribbean's links to the world economy were tropical agricultural exports, yet several specialized industries—such as tourism, offshore financial services, and assembly plants—have challenged agriculture's dominance. These industries grew because of the region's proximity to North America and Europe, abundant cheap and educated labor, and policies that created a nearly tax-free environment for foreign-owned companies. Unfortunately, growth in these sectors does not employ all the region's displaced rural workers, so the lure of jobs outside the region remains strong.

From Fields to Factories and Resorts

Agriculture used to dominate the economic life of the Caribbean. However, decades of turbulent commodity prices and the decline of preferential trade agreements with former colonial states have produced more hardship than prosperity. Ecologically, the soils are overworked, and there are no frontiers into which to expand production, except for areas of the rimland. Moreover, agricultural prices have not kept pace with rising production costs, so wages and profits remain low. With the exception of a few mineral-rich territories, such as Trinidad, Guyana, Suriname, and Jamaica, most countries have systematically tried to diversify their economies, relying less on their soils and more on manufacturing and services.

Comparing export figures over time demonstrates the shift away from monocrop dependence. In 1955, Haiti earned more than 70 percent of its foreign exchange through the export of coffee; by 1990, coffee accounted for only 11 percent of its export earnings and today it is less than 1 percent. Similarly, in 1955 the Dominican Republic earned close to 60 percent of its foreign exchange through sugar, but 35 years later sugar earned less than 20 percent of the country's foreign exchange, and today it earns about 2 percent.

The Land of Sugarcane and Rum The economic history of the Caribbean cannot be separated from the production of sugarcane. Even relatively small territories such as Antigua and Barbados yielded fabulous profits because there was no limit to the demand for sugar in the 18th century. Once considered a luxury crop, it became a popular necessity for European and North American laborers by the 1750s. It sweetened tea and coffee and made jams a popular spread for stale bread. In short, it made the meager and bland diets of ordinary people tolerable, and it also boosted caloric intake. Distilled into rum, sugar produced a popular intoxicant. Though it is hard to imagine today, individual consumption of a pint of rum a day was not uncommon in the 1800s. The Caribbean and Latin America still produce the majority of the world's rum (Figure 5.28).

Sugarcane is still grown throughout the region for domestic consumption and export. Its economic importance has declined, however, mostly because of increased competition from corn and sugar beets grown in the mid-latitudes. The Caribbean and Brazil are the world's major sugar exporters. Until 1990, Cuba alone accounted for more than 60 percent of the value of world sugar exports, and the country earned 80 percent of its foreign exchange through sugar production. However, Cuba's dominance in sugar exports had more to do with its subsidized and guaranteed markets in eastern Europe and the Soviet Union than with exceptional productivity. Since 1990, the value of the Cuban sugar harvest has plummeted, although there is discussion of importing Cuban organic sugar into U.S. markets now that trade restrictions have been lifted.

The Caribbean also has other agricultural commodities. Several small states of the Lesser Antilles reply on banana exports, although reduced preferential trade with the European Union has hurt this industry. Tobacco (for cigars) has long been a staple for Cuba and the Dominican Republic. There is also specialty production in nontraditional export commodities such as flowers or spices. A not inconsiderable cash crop is marijuana; illegal, but tolerated, it is grown for local consumption, along with some for export.

Assembly-Plant Industrialization Another important Caribbean development strategy has been to invite foreign investors to set up assembly plants and thus create jobs. This was first tried successfully in Puerto Rico in the 1950s and was copied throughout the region. During Puerto Rico's "Operation Bootstrap," island leaders encouraged U.S. investment by offering cheap labor, local tax breaks, and, most

TABLE 5.2   Development Indicators

MasteringGeography™

Country	GNI per capita, PPP 2014	GDP Average Annual % Growth 2009–15	Human Development Index (2015)[1]	Percent Population Living Below $3.10 a Day	Life Expectancy (2016)[2] Male	Female	Under Age 5 Mortality Rate (1990)	Under Age 5 Mortality Rate (2015)	Youth Literacy (% pop ages 15–24) (2005–2014)	Gender Inequality Index (2015)[3,1]
Anguilla*	12,200	–	–	–	79	84	–	–	–	–
Antigua and Barbuda	21,370	0.1	0.783	–	74	80	27	8	–	–
Bahamas	22,290	1.1	0.790	–	72	78	22	12	–	0.298
Barbados	15,190	0.3	0.785	–	73	78	18	13	–	0.357
Belize	7,590	2.8	0.715	26.0	71	77	44	17	–	0.426
Bermuda	66,560	–3.4	–	–	78	84	–	–	–	–
Cayman*	43,800	–	–	–	78	84	–	–	–	–
Cuba	10,200*	2.8	0.769	–	78	82	13	6	100	0.356
Dominica	10,480	0.5	0.724	–	72	77	17	21	–	–
Dominican Republic	12,600	4.6	0.715	9.1	71	77	58	31	97	0.477
French Guiana	–	–	–	–	77	83	–	–	–	–
Grenada	11,720	1.1	0.750	–	74	79	21	12	–	–
Guadeloupe	–	–	–	–	76	83	–	–	–	–
Guyana	6,940	5.0	0.636	28.3	64	69	63	39	93	0.515
Haiti	1,730	2.5	0.483	71.0	61	65	143	69	–	0.603
Jamaica	8,640	0.2	0.719	8.2	73	78	35	16	96	0.430
Martinique	–	–	–	–	78	84	–	–	–	–
Montserrat*	8,500	–	–	–	76	73	–	–	–	–
Puerto Rico	23,960	–2.0	–	–	76	83	–	6	99	–
St. Kitts and Nevis	22,600	1.9	0.752	–	73	78	28	11	–	–
St. Lucia	10,540	–0.3	0.729	61.8	75	82	23	14	–	–
St Vincent and the Grenadines	10,730	0.2	0.720	–	71	75	27	18	–	–
Suriname	17,040	3.6	–	40.2	68	74	52	21	98	0.463
Trinidad and Tobago	31,970	0.9	0.772	12.2	71	78	37	20	100	0.371
Turks and Caicos*	29,100*	–	–	–	77	83	–	–	–	–

Source: World Bank, *World Development Indicators,* 2016.

[1] United Nations, *Human Development Report,* 2015.

[2] Population Reference Bureau, *World Population Data Sheet,* 2016.

[3] Gender Inequality Index—A composite measure reflecting inequality in achievements between women and men in three dimensions: reproductive health, empowerment, and the labor market that ranges between 0 and 1. The higher the number, the greater the inequality.

* Additional data from the CIA World Factbook.

Login to MasteringGeography™ **& access** MapMaster **to explore these data!**

1) Identify some of the fastest growing economies in the Caribbean. What might attribute to their growth?

2) What generalizations can you make about youth literacy in this region? Which countries are lagging in this indicator?

importantly, federal tax exemptions (something only Puerto Rico can do because of its special status as a U.S. commonwealth). The manufacturing sector accounts for nearly half of the island's GDP. However, competition from other states with even lower wages, and the U.S. Congress' 1996 decision to phase out many of the tax exemptions, has undermined Puerto Rico's ability to maintain its specialized industrial base. Moreover, with mechanization, manufacturing now

employs a smaller percentage of the workforce than in the past.

Through the creation of **free trade zones (FTZs)**— duty-free and tax-exempt industrial parks for foreign corporations—the Caribbean is an increasingly attractive location for assembling goods for North American consumers. The Dominican Republic took advantage of tax incentives and guaranteed access to the U.S. market offered through the

▲ **Figure 5.28 Bacardi Rum and San Juan** Rum is the quintessential Caribbean beverage. Made from sugar cane, it has been an important regional export for five centuries. The Bacardi factory, shown here in San Juan, Puerto Rico, is a popular tourist destination.

Explore the **Tastes** of Rum

https://goo.gl/9fS7Fs

and corporations by offering specialized services that are confidential and tax exempt. Places that provide offshore banking make money through registration fees, not taxes. The Bahamas was so successful in developing this sector that by 1976, the country was the world's third largest banking center. Its dominance began to decline due to corruption concerns linked to money laundering and competition from the Caribbean, Hong Kong, and Singapore. In the 1990s, the Cayman Islands emerged as the region's leader in financial services—and one of the leading financial centers globally (**Figure 5.30**). With a population of 56,000, this crown colony of Britain has many more registered corporations than residents, including banks, insurers, and mutual funds. The per capita income purchasing power parity in 2016 was US$44,000. Yet with growing competition from Asia, the Pacific Islands, and even Europe, the Caymans has slipped from one of the top 5 banking centers to one of the top 50.

It is estimated that US$20–$30 trillion are hidden in offshore tax havens all over the world; the Caribbean is just one location where corporations and rich individuals park their money. Each offshore banking center in the Caribbean tries to develop specialized financial services to attract clients, such as banking, functional operations, mutual funds, hedge funds, insurance, or trusts. Bermuda, for example, is the global leader in the reinsurance business, which makes money from underwriting part of the risk of other insurance

Caribbean Basin Initiative. There are now 50 Dominican FTZs, with the majority clustered around the outskirts of Santo Domingo and Santiago, the country's two largest cities (**Figure 5.29**). Firms from the United States and Canada are the most frequent investors in these zones, followed by Dominican, South Korean, and Taiwanese companies. Traditional manufacturing on the island was tied to sugar refining and rum production, whereas production in the FTZs focuses on garments, medical instruments, and textiles. These manufacturing centers now account for half of the country's exports.

Growth in manufacturing depends on national and international policies that support export-led development through foreign investment. Certainly, new jobs are being created, and national economies are diversifying in the process, but critics believe that foreign investors gain more than the host countries. Because most goods are assembled from imported materials, there is little development of national suppliers. Although wages are often higher than local averages, they are still low compared to those in the developed world—sometimes just a few dollars a day.

Offshore Banking and Online Gambling The rise of **offshore banking** in the Caribbean is most closely associated with The Bahamas in the 1920s. Offshore banking centers appeal to foreign banks

▲ **Figure 5.29 Free Trade Zones in the Dominican Republic** A sign of globalization is the increase in duty-free and tax-exempt industrial parks in the Caribbean. The Dominican Republic, which is also a member of the Central American Free Trade Association, has 50 FTZs with foreign investors from the United States, Canada, South Korea, and Taiwan. **Q: Where are FTZs clustered, and what might explain this pattern?**

▲ **Figure 5.30 Financial Services in the Cayman Islands** Upland House has the offices of Maples and Calder, the largest law firm in the Cayman Islands specializing in financial services. It is also the official address of nearly 13,000 companies. Offshore financial services are vital to the Cayman Islands economy.

companies. The Caribbean is an attractive location for such services because of its closeness to the United States (home of many of their registered firms), client demand for these services in different countries, and the steady improvement in telecommunications that make this industry possible. The resource-poor islands of the region see providing financial services as a way to bring foreign capital to state treasuries. Envious of the economic success of The Bahamas, Bermuda, and the Cayman Islands, countries such as Antigua, Aruba, Barbados, and Belize have also developed offshore sectors. Barbados, for example, has been the preferred tax haven for Canadians who desire to bank offshore.

Online gambling is the newest industry for the microstates of the Caribbean. Antigua and St. Kitts were the leaders of the region, beginning legal online gambling services in 1999. Other states soon followed; as of 2003, Dominica, Grenada, Belize, and the Cayman Islands had gambling domain sites. In 2007, the World Trade Organization deemed restrictions imposed on overseas Internet gambling sites by the United States to be illegal. The tiny nation of Antigua is currently seeking US$3 billion from the United States as compensation for lost revenue due to illegal restrictions placed on Antigua's business.

Meanwhile, sensing a lucrative business opportunity, efforts to legalize Internet gambling in the United States moved into full gear. By 2013, the governors of Delaware, New Jersey, and Nevada all signed laws to allow online gambling in their states. Other cash-strapped states have followed suit, seeing the taxation of online gambling as an attractive new revenue source. For the Caribbean, however, this suggests that its place as the premier site for online gambling will soon be eclipsed.

Tourism Environmental, locational, and economic factors converge to support tourism in the Caribbean. The earliest visitors to this tropical sea admired its clear and sparkling turquoise waters. By the 19th century, wealthy North Americans were fleeing winter to enjoy the healing warmth of the Caribbean during its dry season. Developers later realized that the simultaneous occurrence of the Caribbean dry season and the Northern Hemisphere winter was ideal for beach resorts. Tourism was well established by the 20th century, with both destination resorts and cruise lines. By the 1950s, the leader in tourism was Cuba, and The Bahamas was a distant second. Castro's rise to power, however, eliminated this sector of the island's economy for nearly three decades and opened the door for other islands to develop tourism economies.

Six countries or territories hosted two-thirds of the 21.5 million international tourists who came to the Caribbean in 2014: the Dominican Republic, Puerto Rico, Cuba, Jamaica, The Bahamas, and Aruba (**Figure 5.31**). Puerto Rico saw its tourist sector begin to grow with common-wealth status in 1952. San Juan is now the largest home port for cruise lines and the second largest cruise-ship port in the world in terms of total visitors. The Bahamas attributes most of its economic development and high per capita income to tourism. With 1.4 million stay-over visitors in 2014, it is another major hub for tourism in the region. Some 30 percent of the Bahamian population is employed in tourism, and tourism represents nearly half the country's GDP.

The Dominican Republic is the region's largest tourist destination, receiving over 5 million visitors in 2014, many of them Dominican nationals who live in the United States. Since 1980, tourist receipts have increased 20-fold, making tourism the leading foreign-exchange earner at more than US $5.6 billion. Jamaica has become similarly dependent on tourism for hard currency. The vast majority of tourists to Jamaica are from the United States and the United Kingdom. While drawing over 2 million stay-overs in 2014, Jamaica was a port of call for another half-million cruise-ship passengers. Tourism receipts for Jamaica totaled US$2.2 billion in 2014.

After years of neglect, Cuba has revived tourism in an attempt to earn badly needed hard currency. Tourism represented less than 1 percent of the national economy in the early 1980s. By 2014, 2.9 million tourists (mostly Canadians and Europeans) poured onto the island, bringing in US$2.5 billion in tourism receipts. Conspicuous in their absence were travelers from the United States, forbidden to travel to Cuba as "tourists" because of the U.S.-imposed sanctions until now. Normalization of U.S.–Cuban relations in 2014 increased the number of educational tours to Cuba, and more conventional tourism is picking up. How Cuba will deal with the expected surge in American tourists remains to be seen. Neighboring islands worry that they may see fewer tourists as a result of interest in Cuba (see *Geographers at Work: Educational Tourism in Cuba*).

As important as tourism is for the larger islands, it is often the principal source of income for smaller ones. The Virgin Islands, Barbados, the Turks and Caicos, and, recently, Belize all greatly depend on international tourists. To show how quickly this sector can grow, consider this example: When Belize began promoting tourism in the early 1980s, it had just 30,000 arrivals per year. An English-speaking country close to North America, Belize specialized in ecotourism that showcased its interior tropical forests and coastal barrier reef. By the mid-1990s, the number of land-based tourists topped 300,000, and tourism was credited with employing one-fifth of the workforce. Belize City became a port of call for day visitors from cruise ships in 2000, making it the fastest-growing tourist port in the Caribbean. Yet the influx of day visitors to

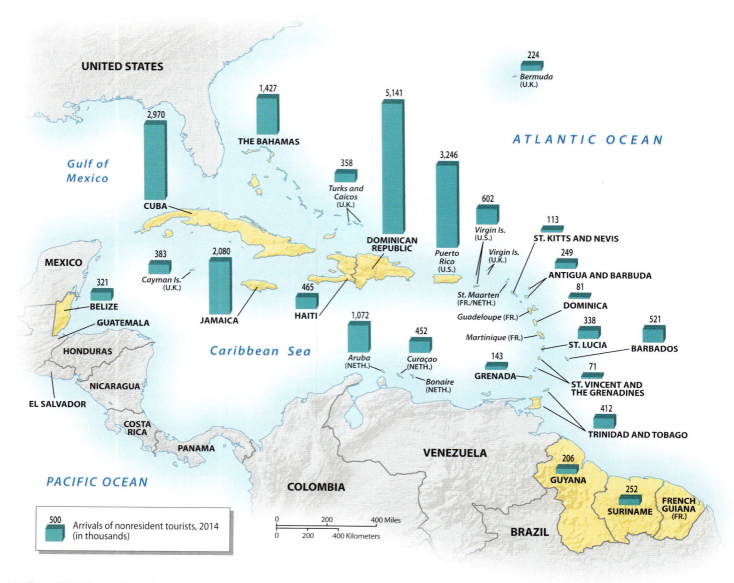

▲ **Figure 5.31** **The Caribbean's Global Linkages: International Tourism** The Caribbean is directly linked to the global economy through tourism. Each year more than 21 million tourists come to the islands, mostly from North America, Latin America, and Europe. The most popular destinations are the Dominican Republic, Puerto Rico, Cuba, Jamaica, The Bahamas, and Aruba. **Q: What might explain the lower levels of tourism in the Guianas compared to the rest of the region?**

this impoverished coastal town of 60,000 has done little to improve the city's infrastructure or high unemployment.

For more than four decades, tourism has been the foundation of the Caribbean economy. However, this regional industry has grown more slowly in recent years, compared to other tourism destinations in the Middle East, southern Europe, and even Central America. It seems that Americans are favoring domestic destinations, such as Hawaii, Florida, and Las Vegas, or are going to more "exotic" locales, such as Costa Rica. European tourists also seem to be staying closer to home or venturing to new locations, such as Dubai on the Persian Gulf or Goa in India. Increasingly, tourists are opting to experience the Caribbean from the decks of cruise ships, rather than land-based resorts. This trend undermines the local benefits of tourism, directing capital to large cruise lines rather than island economies (**Figure 5.32**).

Tourism-led growth has detractors for other reasons. It is subject to the overall health of the world economy and current political affairs. Thus, if North America experiences a recession or international tourism declines due to heightened fears of terrorism or health concerns such as the Zika virus, the flow of tourist dollars to the Caribbean dries up. Where tourism is on the rise, local resentment may build as residents confront the disparity between their own lives and those of the tourists. There is also a serious problem of **capital leakage**, which is the huge gap between gross receipts and the total tourist dollars that actually remain in the Caribbean. Because many guests stay in hotel chains or on cruise ships with corporate headquarters outside the region, leakage of profits is inevitable. On the plus side, tourism tends to promote stronger environmental laws and regulation. Countries quickly learn that their physical environment is the foundation for success. Also, although tourism does have its costs

Educational Tourism in Cuba

Unlike many people who discovered geography later on, Sarah Blue, an associate professor at Texas State University, says that she started college as a geography major: "I wanted to do something about pollution. Then I realized that geography was so much more than that." She first visited Cuba in 1996 to improve her Spanish-language skills and learn to dance salsa, then returned to study topics such as the role of remittances in the Cuban economy and Cuba's influence on medical practices throughout the developing world.

When the Obama administration allowed people-to-people tourism in 2013, Blue started Candela Cuba Tours, a company that leads educational trips to Cuba, as a way to give back to Cuban colleagues. Her tours incorporate food production, culture, music, education, health care, and the economy (**Figure 5.5.1**). Blue's clients stay in family-run establishments, eating breakfast with their Cuban hosts. (These *casas particulares*, licensed by the Cuban government, allow Cubans to rent rooms in their homes to benefit directly from international tourism and educational tours.)

Changing Perceptions The recent change in U.S.–Cuba diplomatic relations means that U.S. visitors to Cuba will likely increase, especially through educational tourism. Blue notes that these developments "will boost Cuba's economy, but it's not going to change things radically. Really, what's opening up are opportunities for Americans." Her geography background helps her analyze possible changes to Cuban society. Says Blue, "A colleague once said to me, 'History is about the past, but geography is about the future.' I love that!"

Blue's American clients are surprised by how open, friendly, and happy Cubans are; they also appreciate the different ways that Cubans approach problems and view the role of government. These insights offer a poignant juxtaposition, leading visitors to meaningful reflection and deeper understanding about society and place. "Traveling is fundamental to geography and understanding different places," notes Blue.

▲ **Figure 5.5.1** **Cuba Tourism** Sarah Blue discusses food production on the island with a group of American tourists

"The more you see other places in the world, the more you can reflect on your own home."

1. How is people-to-people tourism different from regular tourism? Have you or someone you know engaged in this kind of tourist experience?

2. Consider how globalization is changing Cuba today. Do you think Cuba can keep its distinctive qualities in the future?

Explore the **Sights** of St. John's Harbor, Antigua

https://goo.gl/X1HoOC

▼ **Figure 5.32** **Caribbean Cruising** A massive cruise ship docks in St. John's harbor in Antigua, an island of the Lesser Antilles. The cruise business brings many visitors to these small islands, but they spend relatively little time there.

(higher energy and water consumption, as well as demand for more imports), it is environmentally less destructive than traditional export agriculture and, at present, more profitable.

Social Development

The record of economic growth in the region is inconsistent, but measures of social development are generally strong. For example, Caribbean peoples have an average life expectancy of 73 years (see Table 5.2). Literacy levels are high, and there is near parity in terms of school enrollment by gender. Indeed, high levels of educational attainment and out-migration have contributed to a marked decline in the natural increase rate over the past 30 years, which hovers around 1 percent.

These demographic and social indicators explain why Caribbean nations fare well in the Human Development Index (**Figure 5.33**). Almost all ranked states (territories are not ranked) are in the high human development category. Only Guyana was in the medium category. Haiti has the lowest ranking at 163rd in the world, placing it in the category of low human development. Figure 5.33 also shows that many of the well-ranked states, especially Jamaica, Curacao, the Dominican Republic, and St. Kitts and Nevis, have significant annual per capita flows of **remittances** (monies

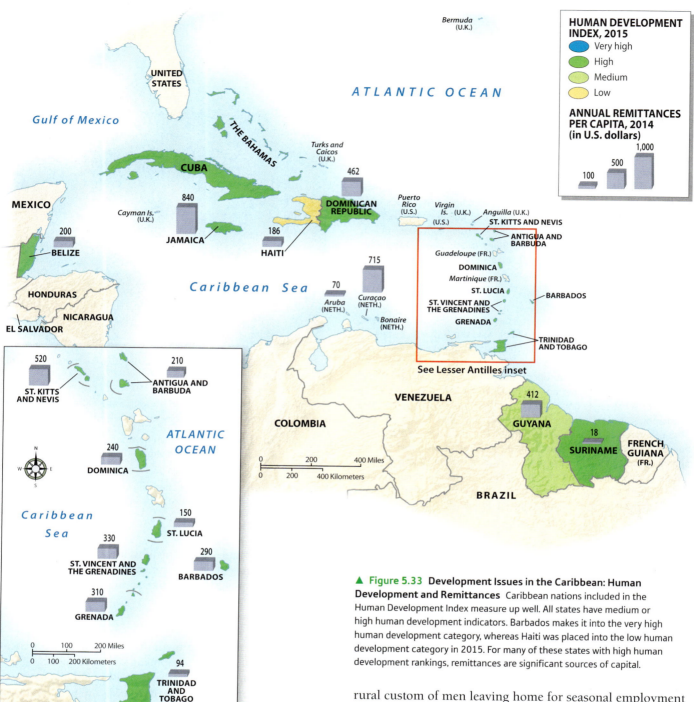

HUMAN DEVELOPMENT INDEX, 2015
- Very high
- High
- Medium
- Low

ANNUAL REMITTANCES PER CAPITA, 2014 (in U.S. dollars)
100 500 1,000

▲ **Figure 5.33 Development Issues in the Caribbean: Human Development and Remittances** Caribbean nations included in the Human Development Index measure up well. All states have medium or high human development indicators. Barbados makes it into the very high human development category, whereas Haiti was placed into the low human development category in 2015. For many of these states with high human development rankings, remittances are significant sources of capital.

sent back home by migrants working overseas) entering the economy. It has been argued that remittances have become extremely important in boosting the overall level of social and economic development in the region. Despite real social gains, many inhabitants are chronically underemployed, poorly housed, and perhaps overly dependent on foreign remittances. For rich and poor alike, the temptation to leave the region in search of better opportunities remains.

Gender, Politics, and Culture The matriarchal (female-dominated) basis of Caribbean households is often singled out as a distinguishing characteristic of the region. The

rural custom of men leaving home for seasonal employment tends to nurture strong and self-sufficient female networks. Women typically run the local street markets. With men absent for long periods of time, women tend to make household and community decisions. Although giving women local power, this position does not always confer status. In rural areas, female status is often undermined by the relative exclusion of women from the cash economy—men earn wages, while women provide subsistence.

As Caribbean society urbanizes, more women are being employed in assembly plants (the garment industry, in particular, prefers to hire women), in data-entry firms, and in tourism. With new employment opportunities, female labor-force participation has surged; in countries such as Barbados, Haiti, and Jamaica, more than 45 percent of the workforce is female. Increasingly, women are the principal earners of cash, and are more likely than men to complete secondary education. There are also signs of greater political involvement by women. In recent years, Jamaica, Dominica, Trinidad and Tobago, and Guyana have all elected women prime ministers.

Today slightly more women than men migrate to the United States, and many seek employment in the globalized care economy as health-care workers, nannies, and eldercare aides. Many feminist scholars argue that the care industry has increasingly become the domain of immigrant women of color throughout North America and Europe. This segmentation of the labor market is driven by a complex mix of income inequalities, racial and gender preferences, and the demands and relatively low status of care work in general.

An estimated 300,000 to 500,000 Caribbean-born health-care professionals, mostly women, work in the United States. More than half are home health-care aides, and another third are registered nurses or technicians (**Figure 5.34**). Many of these women are reliable senders of remittances. Yet it is also true that these female migrants may leave their own families and children behind due to work demands and visa stipulations. Many Caribbean governments worry about this trend, but it is not one that can be easily addressed.

Education Many Caribbean states have excelled in educating their citizens. Literacy is the norm, and the expectation is for most people to receive at least a high school degree. In many respects, Cuba's educational accomplishments are the most impressive given the size of the country and its high illiteracy rates in the 1960s. Today nearly all adults are literate. Haiti is the obvious contrast to Cuba's success. Among youth (ages 15 to 24), 74 percent of males and 70 percent of females are literate. When considering all Haitians over the age of 15, less than half are literate.

Education is expensive for these nations, but it is considered essential for development. Ironically, many states express frustration about training professionals for the benefit of developed countries, a phenomenon called **brain drain**. Brain drain occurs throughout the developing world,

especially between former colonies and the mother countries. In the early 1980s, Jamaica's prime minister complained that 60 percent of his country's newly university-trained workers left for the United States, Canada, and Britain, representing a subsidy to these economies far greater than the foreign aid Jamaica received from them. A study by the World Bank of skilled migrants revealed that 40 percent of Caribbean immigrants living abroad were college educated. In countries such as Guyana, Grenada, Jamaica, St. Vincent and the Grenadines, and Haiti, over 80 percent of the college-educated population will emigrate. No other region in the world has this many educated people leaving. To be fair, some of these migrants moved when they were young and received their education in North America and Europe. Still, other immigrants, especially health professionals, received their educations in the Caribbean and were recruited abroad because of higher wages and better opportunities. Given the small population of many Caribbean territories, each professional person lost to emigration can negatively impact local health care, education, and enterprise.

The Brain Gain Although the outflow of professionals continues to be high, many countries are experiencing a return migration of Caribbean peoples who left for work in North America and Europe in the 1960s and 1970s and are now returning after decades overseas. This **brain gain**, some argue, offers the potential for returnees to contribute to the social and economic development of a home country with the experiences they have gained abroad. Throughout the English-speaking Caribbean, crates filled with household possessions are arriving, new houses are being built, and associations of returnees are being formed. Many returnees are finally living a long-deferred dream of return, bringing with them cash, new skills, distinct life experiences, and new expectations. Sometimes, they bring adult children with them who have never lived in the Caribbean but are "returning" nevertheless to a place that they were told was home. Economic inequalities may drive people to emigrate in search of work, but the emotional attachment to places persists and explains, in part, the Caribbean emigrants' return.

Many Caribbean countries are also reaching out to their diasporas, in an effort to forge transnational linkages that can foster greater investment in development for the region. In countries as diverse as Haiti and Barbados, the combination of brain gain and the potential of the diaspora to stimulate social and economic development are strategies being exploited to improve the region's future.

▼ **Figure 5.34 Caribbean Health-care Worker** A Jamaican nurse with her colleagues in a Massachusetts hospital. Up to half a million Caribbean health-care professionals work in the United States.

REVIEW

5.9 As the Caribbean has shifted out of dependence on agriculture, what other economic sectors have emerged in the region?

5.10 What explains the relatively high levels of social development in the Caribbean given the region's relative poverty?

■ **KEY TERMS** free trade zones (FTZs), offshore banking, capital leakage, remittances, brain drain, brain gain

Chapter 5 The Caribbean

Summary

- This tropical region was exploited to produce export commodities such as sugar, tobacco, and bananas. The region's warm waters and mild climate attract millions of tourists, yet serious problems with deforestation and soil erosion have degraded urban and rural environments. Global warming poses a serious threat to the region, with the likelihood of more-intense hurricanes and sea-level rise.

- Population growth in the Caribbean has slowed over the past two decades; the average woman now has two to three children. Life expectancy is quite high, as are literacy rates. Most Caribbean people live in cities, but Caribbean cities are not large by world standards. The Caribbean region is noted for high rates of emigration, especially among the highly skilled who settle in North America and Europe.

- The Caribbean was forged through European colonialism and the labor of millions of Africans. The blending of African and European elements, termed creolization, has resulted in many unique cultural expressions in music, language, and religion. Others view the Caribbean as a neo-Africa, in which African peoples, cultures, and even some agricultural practices dominate, especially in isolated maroon communities.

- Today the region contains 26 independent countries and territories. Recent developments in U.S.–Cuban relations are bringing political and economic change to one of the Caribbean's largest countries. While U.S. influence is strong in this region, other countries such as Brazil, China, and the United Kingdom also exert influence in the Caribbean.

- The Caribbean has gradually shifted from being an exporter of primary agricultural resources (especially sugar) to a service and manufacturing economy. Employment opportunities in assembly plants, tourism, and offshore banking have replaced jobs in agriculture. The region's strides in social development, especially in education, health, and the status of women, distinguish it from other developing areas.

Review Questions

1. What are the historical, cultural, and resource differences between the Caribbean islands and the rimland, and what environmental issues affect these distinct areas?

2. What is the relationship between the shift from agrarian to urban settlement and the reliance on emigration in assessing the long-term population growth of this region?

3. What are the demographic and cultural implications of the forced transfer of African people to the Caribbean and the creation of a neo-African society in the Americas?

4. Why did European colonists so aggressively seek control of the Caribbean and why did independence in the region come about more gradually than in neighboring Latin America?

5. How is the contemporary Caribbean linked to the world economy through emigration, off-shore banking, free trade zones, and tourism? What changes might occur over the next twenty years?

Image Analysis

1. Look at the population pyramid for Barbados. Based on the pyramid, what percentage of the population is under 15? Over 65?

2. When did the TFR begin to decline? Is the population growing, declining, or stable? What factors might account for demographic trends in Barbados?

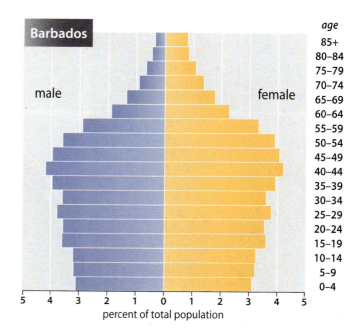

▶ **Figure IA5** Barbados Population Pyramid

JOIN THE DEBATE

For over five decades the United States had a trade embargo with Cuba in an effort to isolate this socialist country and to force political change. In 2015 relations between the two countries began to normalize due to the executive action of the Obama Administration, yet Raúl Castro is still in power and the island is still socialist, although some private sector activities and foreign investments are allowed. Did the economic isolation of Cuba for 50 years support the status quo, or help to bring about change?

The embargo did have a positive impact overall

- The trade embargo kept pressure on the Cuban government and discouraged other Caribbean and Latin American countries from following Cuba's path.
- By forcing Cuba to develop other trade partners and alternative sources of revenue, the country became an innovator in health care and biomedical research.
- Limited interaction with the United States over 50 years allowed for a distinct Cuban way of life and identity to form.

The embargo did not serve U.S. citizens nor the citizens of Cuba

- The embargo resulted in needless suffering of the Cuban people and did not change the political system.
- The United States would have had more influence on Cuba earlier if it had stayed economically and politically engaged in the country, as the U.S. did with communist China.
- By maintaining the embargo for so long, foreign investors from Europe and Asia have more economic influence in the country.

▲ **Figure D5** **The United States and Cuba** A Cuban gives the thumbs up from his Havana balcony decorated with US and Cuban flags. Normalization of relations between the two countries began in 2015.

■ KEY TERMS

African diaspora *(p. 191)*
brain drain *(p. 209)*
brain gain *(p. 209)*
capital leakage *(p. 206)*
Caribbean Community and Common Market (CARICOM) *(p. 201)*
Caribbean diaspora *(p. 187)*
chain migration *(p. 189)*
circular migration *(p. 189)*
creolization *(p. 194)*
free trade zones (FTZs) *(p. 203)*
geothermal *(p. 182)*
Greater Antilles *(p. 177)*
hurricanes *(p. 178)*

indentured labor *(p. 194)*
Lesser Antilles *(p. 177)*
maroons *(p. 192)*
monocrop production *(p. 189)*
Monroe Doctrine *(p. 197)*
neocolonialism *(p. 197)*
offshore banking *(p. 204)*
plantation America *(p. 189)*
remittances *(p. 207)*
rimland *(p. 176)*
transnational migration *(p. 189)*

MasteringGeography™

Looking for additional review and test prep materials? Visit the Study Area in **MasteringGeography**™ to enhance your geographic literacy, spatial reasoning skills, and understanding of this chapter's content by accessing a variety of resources, including **MapMaster** interactive maps, videos, RSS feeds, flashcards, web links, self-study quizzes, and an eText version of *Diversity Amid Globalization*.

DATA ANALYSIS

https://goo.gl/qEpia

Remittances are an important source of revenue for many Caribbean countries. It is important to know not only the total amount a country receives from remittances, but also where the money comes from. Visit the World Bank website (http://econ.worldbank.org) and search for data on migration and remittances. Select the most recent bilateral remittance matrix to download a spreadsheet listing countries receiving remittances across the top row and source countries for remittances down the first column.

1. Select five or six Caribbean countries. Be sure to mix both large and small countries.

2. Create a table with the major source countries and the amounts of remittances sent to each selected country. Now map these patterns.

3. Which countries are the most important sources of remittances, and why? Which countries receive relatively few remittances, and why?

4. What does your map tell you about the major destinations for Caribbean migrants? Write a summary that compares two or more of your selected countries and outlines cultural, geopolitical, and economic factors explaining the patterns you identified.

Authors' Blogs

Scan to visit the
GeoCurrents Blog
http://www.geocurrents.info/category/place/caribbean

Scan to visit the
Author's Blog
for field notes, media resources, and chapter updates
https://gad4blog.wordpress.com/category/the-caribbean/

Physical Geography and Environmental Issues

Much of Southern and East Africa were gripped by El Niño–induced drought in 2015–2016, causing food and water shortages. In addition, biofuels (especially wood) is a main source of energy for this region. Tree planting through the Green Belt Movement is critical.

Population and Settlement

Sub-Saharan Africa is demographically young and growing. With nearly 1 billion people, its rate of natural increase is 2.7, making it the fastest-growing world region in terms of population. It is also the region hit hardest by HIV/AIDS, which has lowered overall life expectancies in many countries.

Cultural Coherence and Diversity

The region is culturally complex, with dozens of languages spoken in some countries. Religious life is also important, with large and growing numbers of Muslims and Christians. Religious diversity and tolerance generally have been distinctive features of this region, but religious conflict, especially in the Sahel region, has been on the rise.

Geopolitical Framework

Most countries gained their independence in the 1960s. Since then, many ethnic conflicts have taken place, as governments have struggled for national unity within the boundaries drawn by European colonialists. There are significant numbers of internally displaced people and refugees.

Economic and Social Development

Extreme poverty is all too common in this region. The UN's Sustainable Development Goals to reduce extreme poverty by 2030 are the new benchmarks to measure progress. Fortunately, significant improvements in education, life expectancy, and economic growth are occurring.

▶ **Roses from Kenya** Highland East Africa has become a major flower exporter to Europe and SW Asia. Here women prepare cut roses for shipping. They work for Simbi Roses, a fair trade flower farm in Thika, Kenya.

Thika, Kenya

SUB-SAHARAN AFRICA

GROUP 2 CELL E

Sub-Saharan Africa

6

Highland African countries, especially Kenya and Ethiopia, are emerging leaders in the global cut flower industry. This is a multibillion-dollar business with The Netherlands as its distribution hub, but new growers are cutting into Dutch dominance as a grower and re-exporter of flowers to the world's major markets. Critical to Sub-Saharan Africa's floriculture is the highland tropical setting at elevations of 6,000 to 9,000 feet (1,800 to 2,700 meters), year-round growing conditions with 12 hours of daily sunlight, and temperatures in the 70s Fahrenheit, perfect for growing roses. Access to international airports is also critical, as cut flowers are flown to Europe, Asia, and North America. Kenya is the regional leader, although Ethiopia's floral business is catching up quickly with significant investment from the Netherlands, India, and Ecuador.

Flowers are an attractive export due to relatively high value and growing global demand. The comparative advantages for African producers are the right climate and low wages; the disadvantage is distance, although refrigerated containers on ships could cut transport costs in half. A capital-intensive crop requiring greenhouses, irrigation systems, pack houses, and cold storage to be successful, floriculture is not suitable for small farmers. Nevertheless, investors in Tanzania, South Africa, and even Rwanda are also betting on blooms to diversify their economies.

The downside of floriculture is a heavy reliance on pesticides to grow the perfect rose, and the risk of chemical exposure to workers not properly trained and protected. Moreover, for a country suffering food shortages such as Ethiopia, redirecting limited arable land and water for a non-food export is controversial. Foreign investors receiving government incentives regularly displace small farmers. Yet in an increasingly complex production chain, investors from India, using floral breeds from Colombia and Ecuador, are expanding Ethiopia's floriculture by opening new markets in the Middle East and Russia that bypass the Netherlands. With billions at stake, flowers are a serious business.

Defining Sub-Saharan Africa

Sub-Saharan Africa—that portion of the African continent lying south of the Sahara Desert—is a commonly accepted world region (**Figure 6.1**). No shared religion, language, philosophy, or political system has ever united the area. Instead, the region's unity stems from loose cultural bonds developed from a variety of lifestyles and idea systems that evolved here. The impact of outsiders also shaped the region's identity. Slave traders from Europe, North Africa, and Southwest Asia treated Africans as chattel; until the mid-1800s, millions of Africans were taken from the region and sold into slavery. In the late 1800s, the entire continent was divided by European colonial powers, imposing political boundaries that, for the most part, remain to this day. In the postcolonial period, Sub-Saharan African countries shared many of the same economic and political challenges.

When defining the region, the major question is whether to include North Africa and treat the entire continent as one world region. However, North Africa is considered to be more closely linked, both culturally and physically, to Southwest Asia, as Arabic is the dominant language and Islam the dominant religion. Consequently, North Africa is discussed along with Southwest Asia in Chapter 7. In this chapter the new country of South Sudan, along with the Sahelian states of Mauritania, Mali, Niger, and Chad, form the region's northern boundary.

The people of Africa south of the Sahara are generally poorer, more rural, and much younger than those in Latin America and the Caribbean. Nearly 1 billion people reside in this region, which includes 48 states and 1 territory (Reunion off the coast of Madagascar). Demographically, this is the fastest-growing world region; in many countries, nearly half the population is younger than 15 years. Income levels are extremely low: 43 percent of the people live in extreme poverty (less than $1.90 per day). Such statistics and the all-too-frequent negative headlines about violence, disease, and poverty might lead to despair. Yet many Africans are optimistic about the future. Local and international nongovernmental organizations, diaspora-led groups, and government agencies have worked to reduce infant mortality, expand basic education, and increase food production in the past two decades. Most national economies are growing faster than their populations, another positive indicator for the region. One transformational change is the rapid diffusion of cell phones, along with innovative applications that improve communication, commerce, and information sharing. Sub-Saharan Africa is a vast region with abundant natural resources and a growing young population that needs better education. Sustained investments and infrastructural improvements are also needed. To better appreciate these opportunities and challenges, one must consider the region's physical geography.

> ## Most African economies are growing much faster than their populations, a positive indicator for the region.

LEARNING Objectives
After reading this chapter you should be able to:

6.1 Describe the major ecosystems in the region and how humans have adapted to living in them.

6.2 Outline the environmental issues that challenge Africa south of the Sahara.

6.3 Explain the region's rapid demographic growth and describe the differential impact of HIV/AIDS and Ebola on the region.

6.4 Connect ethnicity to conflicts in this region and identify strategies for maintaining peace.

6.5 Summarize the various cultural influences of African peoples within the region and globally.

6.6 Trace the colonial history of Sub-Saharan Africa and link colonial policies to the region's postindependence conflicts.

6.7 Assess the roots of African poverty and explain why many of the fastest-growing economies in the world today are in Sub-Saharan Africa.

6.8 List the region's major resources, especially metals and fossil fuels, and describe how they are being developed.

6.9 Explain how reductions in conflicts and increased investment can improve educational and social development outcomes in the region.

▲ **Figure 6.1 Sub-Saharan Africa** This vast world region includes 48 states and 1 territory, and is often divided into western, eastern, central, and southern subregions. The rainforests, tropical savannas, and deserts of Africa South of the Sahara are home to nearly 1 billion people.

Much of the region consists of broad plateaus ranging in elevation from 1600 to 6500 feet (500 to 2000 meters). Although the population is growing rapidly, overall population density of Sub-Saharan Africa is low. Considered one of the least developed world regions, it remains an area rich in natural resources.

Physical Geography and Environmental Issues: The Plateau Continent

Sub-Saharan Africa is the largest landmass straddling the equator (**Figure 6.2**). Its physical environment is remarkably beautiful, dominated by extensive elevated plateaus. Some 250 million years ago the forces of continental drift began the breakup of **Gondwana**, the ancient megacontinent that included Africa, South America, Antarctica, Australia, and the Arabian Peninsula. As this process unfolded, the African landmass experienced a series of continental uplifts that left much of the area with vast plateaus. The highest areas are found on the eastern edge of the continent, where the **Great Rift Valley** forms a complex upland area of lakes, volcanoes, and deep valleys. In contrast, lowlands prevail in West Africa (see Figure 6.1).

The landscape of Sub-Saharan Africa offers a palette of intense colors: deep red soils studded with bright green food crops; the blue tropical sky; golden savannas that ripple with the movement of animal herds; dark rivers meandering through towering rainforests; and sun-drenched deserts. Amid this beauty, however, are relatively poor soils,

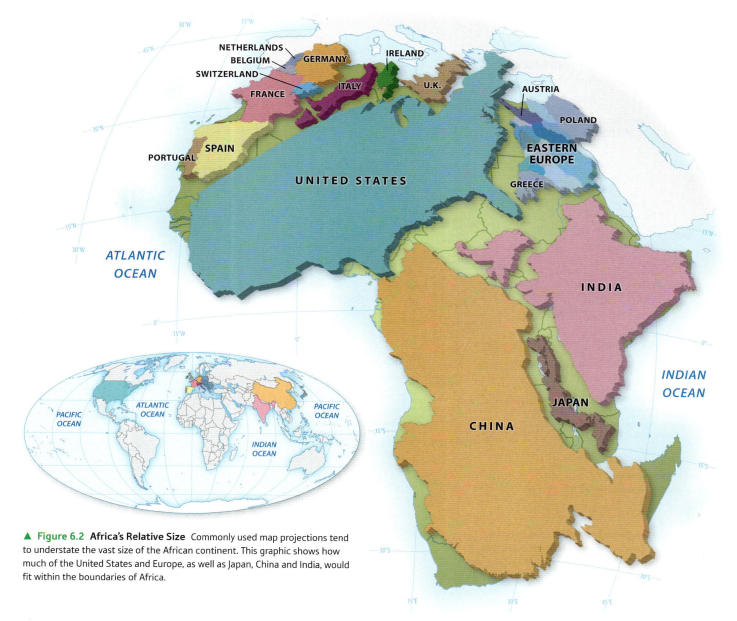

▲ **Figure 6.2 Africa's Relative Size** Commonly used map projections tend to understate the vast size of the African continent. This graphic shows how much of the United States and Europe, as well as Japan, China and India, would fit within the boundaries of Africa.

persistent tropical diseases, and frequent droughts. Large areas exist with great potential for agricultural development, especially in southern Africa, and throughout the continent significant water resources, biodiversity, and mineral wealth abound.

Plateaus and Basins

A series of plateaus and elevated basins dominates the African interior and explains much of the region's unique physical geography. Generally, elevations increase toward the south and east of the continent. Most of southern and eastern Africa lies well above 2000 feet (600 meters), and sizable areas sit above 5000 feet (1500 meters). These areas are typically referred to as High Africa; Low Africa includes western and much of central Africa. Steep escarpments form where the plateaus abruptly end, as illustrated by the majestic Victoria Falls on the Zambezi River (**Figure 6.3**). Much of

southern Africa is rimmed by a landform called the **Great Escarpment** (a steep cliff separating the coastal lowlands from the plateau uplands), which begins in southwestern Angola and ends in northeastern South Africa. South Africa's Drakensberg Range (with elevations reaching 10,000 feet, or 3100 meters) rise up from the Great Escarpment. Because of this escarpment, coastal plains tend to be narrow with few natural harbors, and river navigation is impeded by a series of falls. Such landforms proved to be a barrier for European colonial settlement in the interior of the continent and, in part, explain the prolonged colonizing period.

Though Sub-Saharan Africa is an elevated landmass, it has few significant mountain ranges. The one extensive area of mountainous topography is in Ethiopia, in the northern portion of the Rift Valley zone. Yet even there the dominant features are high plateaus intercut with deep valleys,

▲ **Figure 6.3 Victoria Falls** The Zambezi River descends over Victoria Falls. A fault zone in the African plateau explains the existence of a 360-foot (110-meter) drop. The Zambezi has never been important for navigation, but is a vital supply of hydroelectricity for Zimbabwe, Zambia, and Mozambique.

rather than actual mountain ranges. Receiving heavy rains in the wet season, the Ethiopian Plateau is densely settled and forms the headwaters of several important rivers—most notably, the Blue Nile, which joins the White Nile at Khartoum, Sudan.

A discontinuous series of volcanic mountains, some of them quite tall, is associated with the southern half of the Rift Valley. Kilimanjaro at 19,000 feet (5900 meters) is the continent's tallest mountain (**Figure 6.4**), and nearby Mount Kenya (17,000 feet, or 5200 meters) is the second tallest. The Rift Valley itself reveals the slow, but inexorable progress of geological forces. Eastern Africa is slowly being torn away from the rest of the continent, and within some tens of millions of years it will form a separate landmass. Such motion has already produced a great gash across the uplands of eastern Africa, much of which is occupied by elongated and extremely deep lakes (most notably, Nyasa, Malawi, and Tanganyika). In central East Africa, this rift zone splits into two separate valleys, each flanked by volcanic uplands. Between the eastern and western rifts lies a bowl-shaped depression containing Lake Victoria—Africa's largest body of water. Not surprisingly, some of the densest areas of settlement are found amid the fertile and well-watered soils that border the Rift Valley.

▶ **Figure 6.4 Mount Kilimanjaro** Rising above the tropical plateau and capped in snow, Africa's highest peak is a popular destination for tourists who aspire to reach its lofty heights.

Watersheds Africa south of the Sahara conspicuously lacks the broad alluvial lowlands that influence patterns of settlement throughout other regions. The four major river systems are the Congo, Nile, Niger, and Zambezi. Smaller rivers—such as the Orange in South Africa; the Senegal dividing Mauritania and Senegal; and the Limpopo in Mozambique—are locally important, but drain much smaller areas. Ironically, most people think of this region as suffering from water scarcity and tend to discount the size and importance of the watersheds (or catchment areas) that these river systems drain.

The Congo (or Zaire) River is the region's largest watershed in terms of drainage and water volume. It is second only to South America's Amazon River in terms of annual flow. The Congo flows across a relatively flat basin that lies more than 1000 feet (300 meters) above sea level, meandering through Africa's largest tropical forest, the Ituri (**Figure 6.5**). Entry from the Atlantic into the Congo Basin is prevented by a series of rapids and falls, making the river only partially navigable. Despite these limitations, the Congo River has been the major travel corridor within the Republic of the Congo and the Democratic Republic of the Congo (DRC; formerly Zaire); the capitals of both countries, Brazzaville and Kinshasa, rest on opposite shores of the river to form a metropolitan area of nearly 10 million people.

The Nile River, the world's longest, is the lifeblood of Egypt and Sudan. Yet this river originates in the highlands of the Rift Valley zone and is an important link between North and Sub-Saharan Africa. The Nile begins in the lakes of the Rift Valley zone (Victoria and Edward) before descending into a vast wetland in South Sudan known as the Sudd. One of the great wetlands of the world, the Sudd averages over 30,000 square kilometers, but can expand to four times that size during the rainy season. Navigation in this region is tricky, but settlers have long been drawn to this area's abundant water supply and aquatic life (**Figure 6.6**). Development projects in the 1970s increased the agricultural potential of the Sudd, especially its peanut crop. Unfortunately, three decades of civil war in what is now South Sudan ravaged this area, turning farmers and herders into refugees and undermining the productive capacity of this important ecosystem. One of the development objectives in South Sudan is to harness the potential of the Sudd. A more extensive discussion of the Nile River appears in Chapter 7.

For the energy-poor nation of Ethiopia, harnessing the hydroelectric potential of the Blue Nile, a major tributary of the Nile, is key to the country's power needs. The country opened the Tekeze Dam in 2009 on a tributary of the Nile in northern Tigray Province to provide water for irrigation and electricity. In 2011, the ambitious Grand Renaissance Dam project began on the Blue Nile. When completed, it will be the region's largest hydroelectric project. But because the dam is located near the border with Sudan, Ethiopia's downstream neighbors, Sudan and Egypt, are concerned about how this project will affect their overall supply of water. Ethiopia's response is that hydropower along the Blue Nile is an essential and sustainable approach to supplying the energy needs of the country's 90 million people.

Like the Nile, the Niger River is the critical source of water for two otherwise arid countries: Mali and Niger. Originating in the humid Guinea highlands, the Niger flows first to the northeast and then spreads out to form a huge inland delta in Mali before making a great bend southward at the margins of the Sahara near Gao. On the banks of the Niger River are the capitals of Mali (Bamako) and Niger (Niamey), as well as the historic city of Tombouctou (Timbuktu). After flowing through the desert, the Niger River returns to the humid lowlands of Nigeria, where the Kainji Reservoir temporarily blocks its flow to produce electricity for Africa's most populous state. At the river's end lies the Niger Delta, which is also the center of Nigeria's oil industry. The fertile delta region, home to ethnic groups such as the Igbo and Ogoni, is extremely poor. For decades, conflict in the delta has centered on who benefits from the region's oil and on the serious environmental degradation resulting from oil and gas extraction.

The considerably smaller Zambezi River originates in Angola and flows east, spilling over an escarpment at Victoria Falls and finally reaching Mozambique and the Indian Ocean. More than other rivers in the region, the Zambezi has been a major supplier of

▼ **Figure 6.5 Congo River** Africa's largest river by volume, the mighty Congo River flows through the Ituri rainforest in the Democratic Republic of the Congo.

▲ **Figure 6.6 Settlement in the Sudd** The Sudd in South Sudan is one of the region's largest wetlands, and during the rainy season the White Nile River can expand the wetlands to four times its normal size. This is a traditional village near the Bor, South Sudan.

Explore the **Sights** of the Sudd

http://goo.gl/hqVsra

Climate and Vegetation

Most of Sub-Saharan Africa lies in the tropical latitudes. Only the far south of the continent extends into the subtropical and temperate belts. Much of the region averages high temperatures from 70 to 80°F (22 to 28°C) year-round (**Figure 6.7**). The seasonality and amount of rainfall, more than temperature, determine the different vegetation belts characterizing the region. Addis Ababa, Ethiopia, and Walvis Bay, Namibia, have similar average temperatures (see the climographs in Figure 6.7), but Addis Ababa is in the moist highlands and receives nearly 50 inches (127 cm) of rainfall annually, whereas Walvis Bay lies on the Namibian Desert and receives less than 1 inch (2.5 cm).

The three main biomes of the region are tropical forests, savannas, and deserts.

Tropical Forests The core of Sub-Saharan Africa is remarkably moist. The world's second largest expanse of humid equatorial rainforest, the Ituri, lies in the Congo Basin, extending from the Atlantic coast of Gabon two-thirds of the way across the continent, including the Republic of the Congo and northern portions of the Democratic Republic of the Congo. The conditions here are constantly warm to hot, and precipitation falls year-round (see the climograph for Kisangani in Figure 6.7).

Commercial logging and agricultural clearing have degraded the western and southern fringes of this vast forest, but much of the northeastern section is still intact. The Ituri has, so far, fared much better than the forests of Southeast Asia and Latin America in terms of tropical deforestation. Major national parks such as Okapi and Virunga have been created in the Democratic Republic of the Congo. Virunga National Park—one of the oldest in Africa—is on the eastern fringe of the Ituri and is home to endangered mountain gorillas. Poor infrastructure and political chaos in the DRC over the past 20 years have made large-scale logging impossible, but have also made conservation difficult. Due to regional conflict, parks such as Virunga have been repeatedly taken over by rebel groups, and park rangers have been killed; poaching has become a means of survival for people in the region. In the future, it seems likely that Central Africa's rainforest, and the wildlife within it, could suffer the same kind of degradation experienced in other tropical forests, especially with illegal logging on the rise.

Savannas Wrapped around the Central African rainforest belt in a great arc lie Africa's vast tropical wet and dry savannas. Savannas are dominated by a mixture of trees and tall grasses in the wetter zones immediately adjacent to the forest belt and shorter grasses with fewer trees in the drier zones.

commercial energy for decades. Two of Sub-Saharan Africa's early hydroelectric installations are here: The Kariba Dam is on the Zambia–Zimbabwe border, and the Cabora Bassa Dam is in Mozambique.

Soils With a few major exceptions, Sub-Saharan Africa's soils are relatively infertile. Generally speaking, fertile soils are young soils, deposited in recent geological time by rivers, volcanoes, glaciers, or windstorms. In older soils—especially those located in moist tropical environments—natural processes tend to wash out most plant nutrients over time. Over most of Sub-Saharan Africa, the agents of soil renewal have largely been absent.

Portions of Sub-Saharan Africa are, however, noted for their natural soil fertility, and, not surprisingly these areas support denser settlement. Some of the most fertile soils are in the Rift Valley, enhanced by the volcanic activity associated with the area. The population densities of rural Rwanda and Burundi, for example, are partially explained by the highly productive volcanic soils. The same can be said for highland Ethiopia, which supports the region's second largest population. The Lake Victoria lowlands and central highlands of Kenya also are noted for their sizable populations and productive agricultural bases.

The drier grasslands and semidesert areas have a soil type called *alfisol*. High in aluminum and iron, these red soils are much more fertile than comparable soils found in wetter zones. This helps to explain why farmers tend to plant in drier areas such as the Sahel, even though they risk exposure to drought. With irrigation, many agronomists suggest that the southern African countries of Zambia and Zimbabwe could greatly increase commercial grain production on these soils.

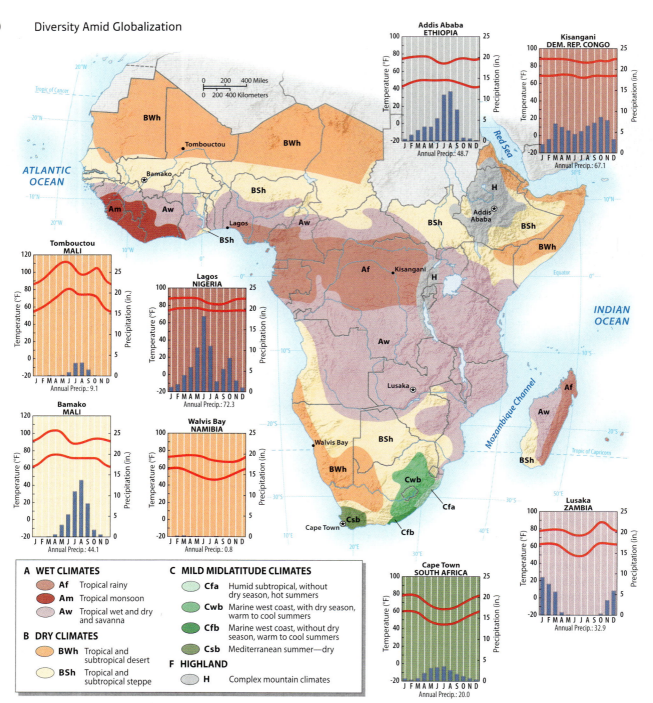

▲ **Figure 6.7 Climate of Sub-Saharan Africa** Much of the region lies within the tropical humid and tropical dry climatic zones; thus, the seasonal temperature changes are not great. Precipitation, however, varies significantly. Compare the distinct rainy seasons in Lusaka and Lagos: Lagos is wettest in June, and Lusaka receives most of its rain in January. Although West and Central Africa have important tropical forests, much of the territory is tropical savanna.

North of the equatorial belt, rain generally falls only from May to October. The farther north one travels, the less the total rainfall and the longer the dry season. Climatic conditions south of the equator are similar, only reversed, with the wet season occurring between October and May and precipitation generally decreasing toward the south (see the climograph for Lusaka in Figure 6.7). A larger area of wet savanna exists south of the equator, with substantial woodlands in southern portions of the Democratic Republic of the Congo, Zambia, northern Zimbabwe, and eastern Angola. These savannas are a critical habitat for the region's large fauna (**Figure 6.8**).

Deserts Major deserts exist in the southern and northern boundaries of the region. The Sahara, the world's largest desert and one of its driest, spans the landmass from the Atlantic coast of Mauritania all the way to the Red Sea coast of Sudan. A narrow belt of desert extends to the south and east of the Sahara, wrapping around the **Horn of Africa** (the northeastern corner that includes Somalia, Ethiopia, Djibouti,

▲ **Figure 6.8** **Savanna Wildlife** A herd of elephants gather at a riverbank in Krueger National Park in South Africa. The savannas of southern Africa are a noted habitat for the region's larger mammals, including buffalo, impala, wildebeest, zebra, lions, and elephants.

and Eritrea) and pushing as far as eastern and northern Kenya. An even drier zone is found in southwestern Africa. In the striking red dunes of the Namib Desert of coastal Namibia, rainfall is a rare event, although temperatures are usually mild (**Figure 6.9**). Inland from the Namib lies the Kalahari Desert. Most of the Kalahari is not dry enough to be classified as a true desert, as it receives slightly more than 10 inches (25 cm) of rain a year. Its rainy season, however, is brief, and most of the precipitation is immediately absorbed by the underlying sands. Surface water is thus scarce, giving the Kalahari a desertlike aspect for most of the year.

Africa's Environmental Issues

The prevailing perception of Africa south of the Sahara is one of environmental scarcity and degradation, no doubt fostered by televised images of drought-ravaged regions and starving children. Single explanations such as rapid population growth or colonial exploitation cannot fully capture the complexity of Africa's environmental issues or the ways that people have adapted to living in marginal ecosystems with debilitating tropical diseases. Because Sub-Saharan Africa's people are largely rural, earning their livelihood directly from the land, sudden environmental changes can have devastating effects on household income and food consumption.

As **Figure 6.10** illustrates, **desertification**—the expansion of desertlike conditions as a

result of human-induced degradation— is commonplace. Sub-Saharan Africa is also vulnerable to drought, most notably in the Horn of Africa, parts of southern Africa, and the Sahel. Many scientists fear that climate change will lead to more frequent and more prolonged droughts. During the El Niño event of 2015–2016, rainfall levels were abnormally low in southern and eastern Africa, contributing to water and food scarcity. Yet the region is also home to some of the most impressive wildlife populations in the world, which is a source of pride and revenue for many African nations.

The Sahel and Desertification The **Sahel** is a zone of ecological transition between the Sahara in the north and the wetter savannas and forest of the south (Figure 6.10). In the 1970s, the Sahel became a symbol for the dangers of unchecked population growth and human-induced environmental degradation when a relatively wet period came to an abrupt end. Six years of drought (1968–1974) were followed by a second prolonged drought during the mid-1980s, ravaging the land. During these droughts, rivers in the area diminished, and desertlike conditions began to move south. Unfortunately, tens of millions of people lived in this area, and farmers and pastoralists whose livelihoods had come to depend on the more abundant precipitation of the relatively wet period were temporarily forced out.

Life in the Sahel depends on a delicate balance of limited rain, drought-resistant plants, and a pattern of **transhumance** (the movement of animals between

▼ **Figure 6.9** **Namib Desert** Arid Namibia has one of the region's lowest population densities, yet its coastal dunes attract foreign tourists. Hikers are seen scaling the orange-colored dunes in Namib-Naukluft National Park.

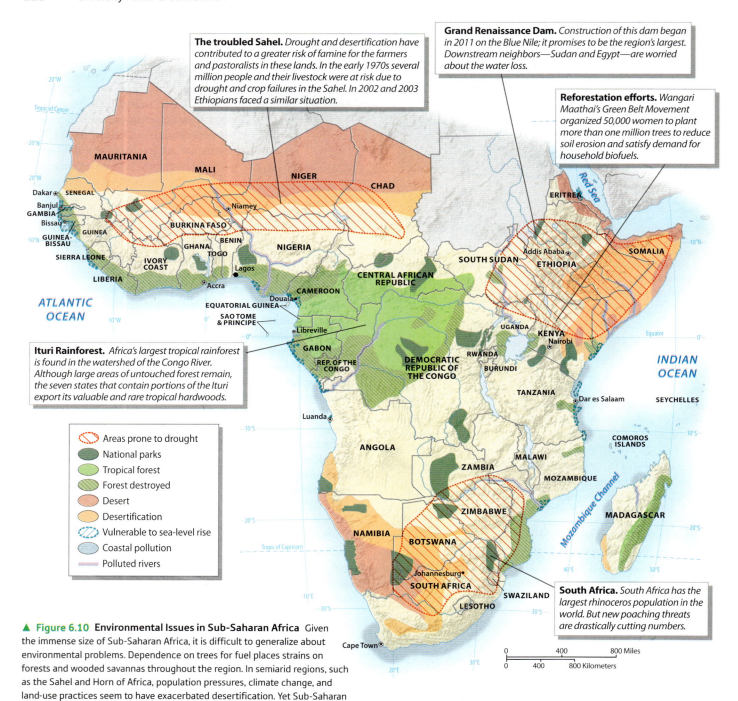

The troubled Sahel. *Drought and desertification have contributed to a greater risk of famine for the farmers and pastoralists in these lands. In the early 1970s several million people and their livestock were at risk due to drought and crop failures in the Sahel. In 2002 and 2003 Ethiopians faced a similar situation.*

Grand Renaissance Dam. *Construction of this dam began in 2011 on the Blue Nile; it promises to be the region's largest. Downstream neighbors—Sudan and Egypt—are worried about the water loss.*

Reforestation efforts. *Wangari Maathai's Green Belt Movement organized 50,000 women to plant more than one million trees to reduce soil erosion and satisfy demand for household biofuels.*

Ituri Rainforest. *Africa's largest tropical rainforest is found in the watershed of the Congo River. Although large areas of untouched forest remain, the seven states that contain portions of the Ituri export its valuable and rare tropical hardwoods.*

South Africa. *South Africa has the largest rhinoceros population in the world. But new poaching threats are drastically cutting numbers.*

Legend:
- Areas prone to drought
- National parks
- Tropical forest
- Forest destroyed
- Desert
- Desertification
- Vulnerable to sea-level rise
- Coastal pollution
- Polluted rivers

▲ **Figure 6.10 Environmental Issues in Sub-Saharan Africa** Given the immense size of Sub-Saharan Africa, it is difficult to generalize about environmental problems. Dependence on trees for fuel places strains on forests and wooded savannas throughout the region. In semiarid regions, such as the Sahel and Horn of Africa, population pressures, climate change, and land-use practices seem to have exacerbated desertification. Yet Sub-Saharan Africa also supports the most impressive array of wildlife, especially large mammals, on Earth.

wet-season and dry-season pasture). What appears to be desert wasteland in April or May transforms into a lush garden of millet, sorghum, and peanuts after the drenching rains of June. Relatively free of the tropical diseases found in the wetter zones to the south, the Sahel also has soils that are quite fertile, which helps to explain why people continue to live there despite the unreliable rainfall (**Figure 6.11**).

Considerable disagreement continues over the basic causes of desertification and drought in the Sahel: Is it too

many humans degrading their environment, or unsound settlement schemes encouraged by European colonizers, or a failure to understand global atmospheric cycles? Certainly, human activity in the region greatly increased in the mid-20th century, making the case for human-induced desertification more likely. But parts of the Sahel were important areas of settlement long before European colonization.

The main practices cited in the desertification of the Sahel are the expansion of agriculture and overgrazing, leading to the loss of natural vegetation and declines in soil fertility. For example, French colonial authorities forced villagers to grow peanuts as an export crop, a policy continued

▲ **Figure 6.11 Sahel in Bloom** A woman prepares millet grains grown near the city of Maradi, Niger. The soils of the Sahel are fertile, and peasant farmers can produce a surplus when adequate rain falls. Yet in times of drought, crop failures can lead to famine in this region.

by the newly independent states of the region. However, peanuts tend to deplete several key soil nutrients, which means that peanut fields are often abandoned after a few years as cultivators move on to fresh sites. Harvesting this crop also turns up the soil at the onset of the dry season, leading to accelerated wind erosion of valuable topsoil.

Overgrazing by livestock, another traditional product of the region, is also implicated in Sahelian desertification. Development agencies, hoping to increase livestock production, introduced deep wells into areas previously unused by herders through most of the year. The new supplies of water, in turn, allowed year-round grazing in places that, over time, could not withstand it. Large barren circles around each new well began to appear even on satellite images.

Some Sahelian areas are experiencing some vegetative recovery thanks to simple actions taken by farmers, changes in government policy, and better rainfall. In the Sahelian portion of Niger, local agronomists recently documented an unanticipated increase in tree cover over the past 35 years. More interesting still, increases in tree cover have occurred in some of the most densely populated rural areas. After a drought in 1984, farmers began to actively protect saplings instead of clearing them from their fields, including the nitrogen-fixing *goa* tree, which had disappeared from many villages. During the rainy season, the goa tree loses its leaves, so it does not compete with crops for water or sun. The leaves themselves fertilize the soil. Sahelian farmers also use branches, pods, and leaves from trees for fuel and for animal fodder.

Until the 1990s, all trees were considered the property of the state of Niger, thus giving farmers little incentive to protect them. Since then, the government has recognized the value of allowing individuals to own trees. Not only can farmers sell branches, pods, or fruit—they can also

conserve trees to ensure sustainable rural livelihoods. The villages that protect their trees are much greener than those that do not. The Sahel is still poor and prone to drought, but as the case of Niger shows, relatively simple conservation practices can have a positive impact.

Deforestation Although Sub-Saharan Africa still contains extensive forests, much of the region is either grasslands or agricultural lands that were once forest. Lush forests that existed in places such as highland Ethiopia were long ago reduced to a few remnant patches. Throughout history, local populations have relied on such woodlands for their daily needs. Tropical savannas, which cover large portions of the region to the north and south of the tropical rainforest zone, are dotted with woodlands. For many people of the region, deforestation of the savanna woodlands is of greater local concern than the commercial logging of the rainforest. This is because of the importance of **biofuels**—wood or charcoal used for household energy needs, especially cooking—as the leading source of energy for many rural settlements. Loss of woody vegetation has resulted in extensive hardship, especially for women and children who must spend many hours a day looking for wood.

In some countries, village women have organized into community-based nongovernmental organizations (NGOs) to plant trees and create green belts to meet ongoing fuel needs. One of the most successful efforts is the Green Belt Movement in Kenya, started by Wangari Maathai. The Green Belt Movement has 15,000 members, mostly women, who planted half a million trees in 2015 alone. Since the movement's beginning in 1977, millions of trees have been planted. In Green Belt areas, village women now spend less time collecting fuel, and local environments have improved. Kenya's success has drawn interest from other African countries, spurring a Pan-African Green Belt Movement largely organized through NGOs interested in biofuel generation, protection of the environment, and the empowerment of women. In 2004, Maathai was awarded a Nobel Peace Prize for her contribution to sustainable development, democracy, and peace. She died in 2011, but the Green Belt Movement, currently led by her daughter, Wanjira Maathai, remains a powerful force in the region (*Working Toward Sustainability: Reforesting a Continent*).

Destruction of tropical rainforests by logging is most evident in the fringes of Central Africa's Ituri (see Figure 6.10). Given the vastness of this forest and the relatively small number of people living there, however, it is less threatened than other forest areas. Two smaller rainforests—one along the Atlantic coast from Sierra Leone to western Ghana, and the other along the eastern coast of the island of Madagascar—have nearly disappeared. These rainforests have been severely degraded by commercial logging and

Reforesting a Continent

A significant outcome from the 2015 UN Climate Change Conference held in Paris was a pledge by a dozen Sub-Saharan African nations to restore their natural forests, even though the developing region is the least responsible for greenhouse gas (GHG) emissions. The African Forest Landscape Restoration Initiative, or AFR100, is the first regional effort at forest restoration, with the goal of replanting 100 million hectares (about 386,000 square miles) of forest by 2030. The World Bank, the German government, and private funders committed $1.5 billion toward restoration, and the World Resources Institute, a U.S.-based NGO, will monitor the effort.

A New Greenbelt Movement Reforestation has many benefits for rural Africa, including reduced soil erosion, improved soil fertility, and greater food security. Trees can slow desertification and supply animal feed and household cooking needs. If successful, such a project will increase biodiversity, create jobs, reduce food insecurity, and increase capacity for climate change resilience and adaptation. Growing trees also absorbs large amounts of carbon dioxide. Wanjira Maathai, President of the Green Belt Movement and daughter of the late Kenyan Nobel Peace Prize Laureate Wangari Maathai, supports this initiative, saying that "I have seen restoration in communities both large and small across Africa. . . . Restoring landscapes will empower and enrich rural communities while providing downstream benefits to those in cities" (**Figure 6.1.1**).

Ethiopia, Kenya, Uganda, Madagascar, Burundi, and Rwanda have pledged millions of acres to the project. The Democratic Republic of the Congo alone has pledged 8 million hectares (20 million acres) to forest restoration. West African nations in the Sahel have also pledged to plant more trees to slow desertification.

Turning Talk into Action Countries pledge to implement these kinds of programs because they hope to get funding. In the worst cases, however, the promised funds go to government agencies and never reach the intended communities. Tree planting programs have been tried before in Sub-Saharan Africa, with mixed results. The difference this time is the scale of the AFR100 program and the financial support. Yet the challenges of having people on the ground with the training and the tools to do the work are real, and there must be local community buy-in to the benefits of the program—something that the Green Belt Movement has

▲ **Figure 6.1.1 Reforestation Efforts** Women tend seedlings in a tree nursery in Meru, Kenya. Beginning with the Green Belt Movement, Kenyan women have played a leadership role in planting trees for fuel needs and ecological restoration.

worked to build. Finally, any reforestation program, just like the seedlings they plant, take years to mature.

The need for wood fuel, the demand for farmland, and the abundance of free-grazing animals put enormous pressure on the forests of Sub-Saharan Africa. At the same time, forests provide valuable ecosystem services such as reducing erosion and absorbing carbon. It is hoped that AFR100 can meet its goals, and that millions of hectares of forest can be restored.

Google Earth Virtual Tour Video

http://goo.gl/jLlISM

1. What services do tropical forests provide to rural residents?

2. What strategies should be employed to make this program work?

agricultural clearing. Madagascar's eastern rainforests, as well as its western dry forests, have suffered serious degradation in the past three decades. Deforestation in Madagascar is especially worrisome because the island forms a unique environment with a large number of native species—most notably, the charismatic lemurs. In order to address biofuel needs in areas of deforestation, NGOs are experimenting with the introduction of bamboo as a fast-growing alternative for construction material and fuel.

Energy Issues The people of this region suffer from serious energy shortages; one of the justifications for the Great Renaissance Dam project is that, when completed, it will meet Ethiopia's electricity needs. At the same time, foreign investors are actively developing the region's oil and natural gas supplies, mostly for export. Many Sub-Saharan states have oil and natural gas; major producers such as Nigeria and Angola are even members of OPEC (**Figure 6.12**). More recently, countries such as Ivory Coast, Tanzania, and Mozambique have developed their natural gas reserves for domestic consumption and export. Yet for most Africans, wood and charcoal account for the majority of total energy production. Figure 6.12 shows the 21 states for which the World Bank estimates the percentage that these biofuels (labeled as combustible renewables) contribute to national energy production. Even though Angola is a major

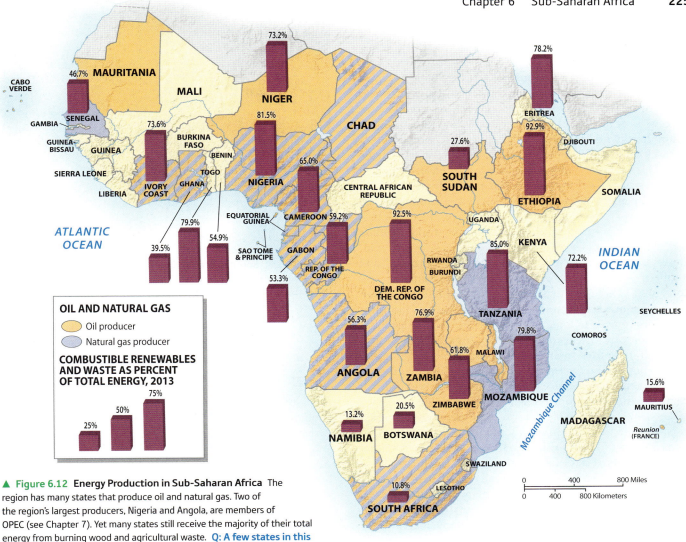

OIL AND NATURAL GAS

◯ Oil producer

◯ Natural gas producer

COMBUSTIBLE RENEWABLES AND WASTE AS PERCENT OF TOTAL ENERGY, 2013

25%   50%   75%

▲ **Figure 6.12** **Energy Production in Sub-Saharan Africa** The region has many states that produce oil and natural gas. Two of the region's largest producers, Nigeria and Angola, are members of OPEC (see Chapter 7). Yet many states still receive the majority of their total energy from burning wood and agricultural waste. **Q: A few states in this figure are much less dependent upon biofuels. Why might that be?**

oil producer, biofuels supply more than half of the country's energy, and biofuels account for 84 percent of oil-rich Nigeria's national energy supply. In large countries such as Ethiopia or the Democratic Republic of the Congo, the figure is over 90 percent. This is why energy production places tremendous strain on forests and vegetation in the region, and why so many countries are developing alternatives such as hydroelectricity and solar power. Another environmental issue for the majority of Africans who rely on biofuels is the smoke that fills homes, causing respiratory problems.

Developing oil and natural gas reserves is not a sure path to economic development, and some have even called it a curse for Sub-Saharan Africa. In Nigeria, politicians and oil executives have prospered from oil profits. But in the Niger Delta, where oil was first extracted over 50 years ago, many places lack roads, electricity, and schools. Moreover, careless and unregulated oil extraction has grossly degraded the delta ecosystem (**Figure 6.13**). As geographer Michael Watts has observed about the delta, oil has been "a dark tale of neglect and unremitting misery." Not all oil production leads

▼ **Figure 6.13** **Oil Pollution in the Niger Delta** Thousands of Nigerians engage in the practice of hacking into oil pipelines to steal crude oil, refine it, and then sell it locally or abroad. This informal practice leaves the delta horribly polluted and cuts deeply into Nigeria's national oil production.

to such misery, but Nigeria is a cautionary tale about the limits of oil's ability to foster development.

Wildlife Conservation Sub-Saharan Africa is famous for its wildlife. No other world region has such abundance and diversity of large mammals. The survival of wildlife here reflects, to some extent, the historically low human population density and the fact that sleeping sickness and other diseases have kept people and their livestock out of many areas. In addition, many African peoples have developed various ways of successfully coexisting with wildlife, and about 12 percent of the region is included in nationally protected lands.

However, as is true elsewhere in the world, wildlife is declining in much of the region. The most noted wildlife reserves are in East Africa (Kenya and Tanzania) and southern Africa (South Africa, Zimbabwe, Namibia, and Botswana). These reserves are vital for wildlife protection and are major tourist attractions. Wildlife reserves in southern Africa now seem to be the most secure. In fact, elephant populations are considered to be too large for the land to sustain in countries such as Zimbabwe. Yet throughout the region, population pressure, political instability, and poverty make the maintenance of large wildlife reserves difficult even though many countries benefit from wildlife tourism.

In 1989, a worldwide ban on ivory trade was imposed as part of the Convention on International Trade in Endangered Species (CITES). Although several African states, such as Kenya, lobbied hard for the ban, others, such as Zimbabwe, Namibia, and Botswana, complained that their herds were growing and the sale of ivory helped to fund conservation efforts. Conservationists feared that lifting the ban would bring on a new wave of poaching and illegal trade. However, in the late 1990s, the ban was lifted so that some southern African states could sell off their inventories of elephant ivory confiscated from poachers, and limited sales have continued. The last legal auction of elephant ivory was in 2008; officials have been reluctant to hold more auctions, as there has been a corresponding spike in poaching.

Today the rhinoceros is especially threatened. The illegal market for rhino horn is lucrative, with most of the demand coming from Vietnam and China. Wildly valued for its questionable medicinal properties, ground rhino horn can fetch $65,000 per kilo in the black market. According to geographer Elizabeth Lunstrum, 80 percent of Sub-Saharan Africa's rhinos are in South Africa, and half of those are in one place—Kruger National Park (**Figure 6.14**), which borders Mozambique. In 2008, rangers noted a spike in shot and dehorned rhinos, and by 2014, more than 1,200 rhinos were reported killed in South Africa. Poaching decreased only slightly in 2015, with 826 rhinos killed in Kruger National Park alone. Many poachers stage their attacks from Mozambique, which makes this an international as well as a domestic issue. With poaching rates still rising, South African officials are deploying drones and military personnel in an attempt to stop the slaughter. Even with these resources, protecting these endangered animals is proving difficult.

▲ **Figure 6.14 South African Wildlife** A white rhinoceros moves across the savannas of Kruger National Park in South Africa, home to more than half of the world's rhinos. A sudden spike in poaching since 2008 has led to dramatic losses. In 2015 it was estimated that four rhinos a day were being killed. The fate of the rhino, a protected and endangered species, in the wild is being challenged by this latest wave of poaching.

Climate Change and Vulnerability in Sub-Saharan Africa

Global climate change poses extreme risks for Sub-Saharan Africa due to the region's poverty, recurrent droughts, and overdependence on rain-fed agriculture. Sub-Saharan Africa is the world's lowest emitter of greenhouse gases, but it is likely to experience greater-than-average human vulnerability to climate change because of the region's limited resources to both respond and adapt to environmental changes. The areas most vulnerable are arid and semiarid regions such as the Sahel and the Horn of Africa, some grasslands, and the coastal lowlands of West Africa and Angola.

Climate change models suggest that parts of highland East Africa and equatorial Central Africa may receive more rainfall in the future. Thus, some lands that are currently marginal for farming might become more productive. These effects are likely to be offset, however, by the decline in agricultural productivity in the Sahel, as well as in the grasslands of southern Africa, especially in Zambia and Zimbabwe. Drier grassland areas could deplete wildlife populations, which are a major factor behind the growing tourist economy. As in Latin America, higher temperatures in the tropics could result in the expansion of vector-borne diseases such as malaria and dengue fever into the highlands, where these have until now been relatively rare. Given the relatively high elevations in the region, the negative consequences of a rising sea level would be mostly felt on the West African coast (Senegal, Gambia, Sierra Leone, Nigeria, Cameroon, and Gabon).

Even without the threat of climate change, famine stalks many areas of Africa. A severe drought, related to El Niño, occurred in southern Africa in 2015–2016 where rainfall already had been well below normal. The Famine

Early Warning Systems (FEWS) network is in place to monitor food insecurity throughout the developing world, but especially in Sub-Saharan Africa. **Food insecurity** measures both daily food intake and irreversible coping strategies—for example, selling assets such as livestock or machinery—that will lead to food consumption gaps. By tracking rainfall, vegetation cover, food production, food prices, and conflict, the FEWS network maps food insecurity along a continuum from food secure to famine. Figure 6.15 shows the areas of concern as of March 2016, especially with regard to corn production—the staple for this part of Africa. While conditions in Tanzania and Angola were rated as favorable, much of Zimbabwe, Swaziland, Lesotho, and even portions of South Africa and Mozambique were rated poor to failure with regard to basic food production. What is especially troubling is that these areas are usually the most important production zones for all of southern Africa. Corn prices have risen, and government agencies are worried about possible famine conditions, especially in rural areas.

REVIEW

6.1 What economic and environmental factors contribute to the region's reliance on biofuels?

6.2 Summarize the factors that make Sub-Saharan Africans especially vulnerable to climate change.

■ **KEY TERMS** Gondwana, Great Rift Valley, Great Escarpment, Horn of Africa, desertification, Sahel, transhumance, biofuels, food insecurity

Population and Settlement: Young and Restless

Sub-Saharan Africa's population is growing quickly. By 2050, the projected population for Africa south of the Sahara is 2 billion, nearly double the current population. It is also a very young population, with 43 percent of the people younger than age 15, compared to just 16 percent for developed countries. Only 3 percent of the region's population is over 65 years of age, whereas in Europe 16 percent is over 65. Families tend to be large, with a woman averaging five births (Table 6.1). However, child and maternal mortality rates are also high, although child mortality rates have declined substantially in the past two decades. The most troubling indicator for the region is its low life expectancy, which dropped to 50 years in 2008 (in part due to the AIDS epidemic), but is currently estimated at 57 years. Life expectancy in other developing nations is much better: India's is 68 years and China's, 75 years. The growth of cities is also a major trend. In 1980, an estimated 23 percent of the population lived in cities; now the figure is 38 percent (Figure 6.16).

Behind these demographic facts lie complex differences in settlement patterns, livelihoods, belief systems, and access to health care. Despite its rapid population growth, Sub-Saharan Africa is not densely populated. The entire region holds 975 million persons—roughly half the population that is crowded into the much smaller land area of South Asia. Just six states account for over half of the region's population: Nigeria, Ethiopia, the Democratic Republic of the Congo, South Africa,

▼ **Figure 6.15 Food Insecurity in Southern Africa** Anticipating areas of food insecurity, based upon the timing and amount of rainfall and changes in vegetation/crop cover, is the mission of the FEWS network. FEWS has existed since the late 1980s to map areas of potential famine. The 2015–2016 drought in southern Africa was a major area of concern for FEWS.

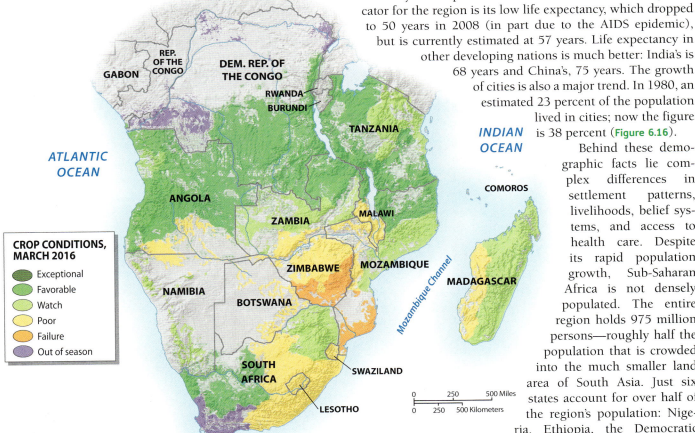

CROP CONDITIONS, MARCH 2016
- Exceptional
- Favorable
- Watch
- Poor
- Failure
- Out of season

TABLE 6.1   Sub-Saharan Africa Population Indicators   MasteringGeography™

Country	Population (millions) 2016	Population Density (per square kilometer)[1]	Rate of Natural Increase (RNI)	Total Fertility Rate	Percent Urban	Percent <15	Percent >65	Net Migration (Rate per 1000)
Western Africa								
Benin	10.8	94	2.7	4.7	44	45	3	0
Burkina Faso	19.0	64	3.1	5.7	30	49	3	−1
Cape Verde	0.5	128	1.6	2.3	66	28	6	−4
Gambia	2.1	191	3.2	5.6	60	46	2	−1
Ghana	28.2	118	2.5	4.2	54	39	5	−1
Guinea	11.2	50	2.7	5.1	37	43	3	0
Guinea–Bissau	1.9	64	2.5	4.9	49	43	3	−1
Ivory Coast	23.9	70	2.4	4.9	54	42	3	0
Liberia	4.6	46	2.6	4.7	50	42	3	−1
Mali	17.3	14	3.1	6.0	40	47	3	−3
Mauritania	4.2	4	2.3	4.2	60	40	3	−1
Niger	19.7	15	4.0	7.6	22	50	3	0
Nigeria	186.5	195	2.6	5.5	48	43	3	0
Senegal	14.8	76	3.2	5.0	45	44	4	−1
Sierra Leone	6.6	87	2.3	4.9	40	42	3	−1
Togo	7.5	131	2.7	4.7	38	42	3	0
Eastern Africa								
Burundi	11.1	421	3.2	6.1	12	46	2	0
Comoros	0.8	414	2.6	4.3	28	40	3	−3
Djibouti	0.9	38	1.6	3.2	77	33	4	−3
Eritrea	5.4	–	2.7	4.2	23	43	3	−5
Ethiopia	101.7	97	2.3	4.2	20	41	4	0
Kenya	45.4	79	2.4	3.9	26	42	3	0
Madagascar	23.7	41	2.6	4.3	35	41	3	0
Mauritius	1.3	621	0.2	1.4	41	20	9	−1
Reunion	0.8	–	1.2	2.5	95	24	10	−7
Rwanda	11.9	460	2.7	4.2	29	41	3	−1
Seychelles	0.1	201	1.0	2.4	54	22	8	1
Somalia	11.1	17	3.2	6.4	40	47	3	−6
South Sudan	12.7	–	2.5	6.7	19	42	3	10
Tanzania	54.2	59	3.0	5.2	30	45	3	−1
Uganda	36.6	188	3.3	5.8	20	48	3	−4
Central Africa								
Cameroon	24.4	48	2.8	4.9	54	43	3	0
Central African Republic	5.0	8	2.0	4.4	40	39	4	0
Chad	14.5	11	3.3	6.4	22	48	2	1
Congo	4.9	13	2.7	4.7	65	41	3	−7
Dem. Rep. of Congo	79.8	33	3.0	6.5	42	46	3	0
Equatorial Guinea	0.9	29	2.4	4.9	40	39	3	5
Gabon	1.8	7	2.3	4.1	87	37	5	1
São Tomé and Principe	0.2	194	2.6	4.4	67	42	4	−6
Southern Africa								
Angola	25.8	19	3.1	6.0	62	48	2	1
Botswana	2.2	4	1.7	2.8	57	33	5	2
Lesotho	2.2	69	1.4	3.3	27	36	5	−2

TABLE 6.1   Sub-Saharan Africa Population Indicators (*continued*)

Country	Population (millions) 2016	Population Density (per square kilometer)[1]	Rate of Natural Increase (RNI)	Total Fertility Rate	Percent Urban	Percent <15	Percent >65	Net Migration (Rate per 1000)
Malawi	17.2	177	2.4	4.4	16	40	4	0
Mozambique	27.2	35	3.1	5.9	32	45	3	0
Namibia	2.5	3	2.2	3.6	47	34	4	0
South Africa	55.7	45	1.2	2.4	65	30	5	3
Swaziland	1.3	74	1.5	3.3	21	37	4	−1
Zambia	15.9	21	3.0	5.3	40	46	3	0
Zimbabwe	16.0	39	2.6	4.0	33	42	3	−3

Source: Population Reference Bureau, *World Population Data Sheet,* 2016.

[1]World Bank, *World Development Indicators,* 2016.

*Combined data from the *World Population Data Sheet,* 2016 and the *World Development Indicators,* 2015.

Login to Mastering Geography™ **& access** MapMaster **to explore these data!**

1) Which countries have the highest RNI? What is the relationship between RNI and TFR? The doubling of a population in years can be calculated using the formula 70/RNI. Use this fraction to determine the countries with the shortest doubling time.

2) Which five countries have the highest population density in this region? Which countries have the lowest? What characteristics do each group have in common?

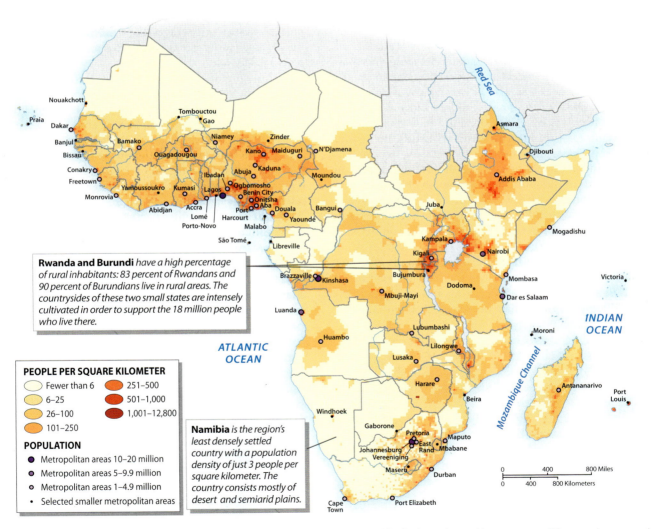

Rwanda and Burundi *have a high percentage of rural inhabitants: 83 percent of Rwandans and 90 percent of Burundians live in rural areas. The countrysides of these two small states are intensely cultivated in order to support the 18 million people who live there.*

PEOPLE PER SQUARE KILOMETER

- Fewer than 6
- 6–25
- 26–100
- 101–250
- 251–500
- 501–1,000
- 1,001–12,800

POPULATION

- Metropolitan areas 10–20 million
- Metropolitan areas 5–9.9 million
- Metropolitan areas 1–4.9 million
- Selected smaller metropolitan areas

Namibia *is the region's least densely settled country with a population density of just 3 people per square kilometer. The country consists mostly of desert and semiarid plains.*

▲ **Figure 6.16 Population of Sub-Saharan Africa** The majority of people in Sub-Saharan Africa live in rural areas. However, some of these rural zones—such as West Africa and the East African highlands—are densely settled. Major urban centers, especially in South Africa and Nigeria, support millions. Lagos, Nigeria, is the one megacity in the region (with over 10 million people), but more than three dozen cities have more than 1 million residents. **Q: What factors contribute to the extremely low population density in the southwest corner of the continent?**

Tanzania, and Kenya (see Table 6.1). Some states have very high population densities (such as Rwanda or Mauritius), whereas others are sparsely settled (such as Namibia and Botswana).

Crude population density is an imperfect indicator of whether a country is overpopulated. Geographers are often more interested in **physiological density**, or the number of people per unit of arable land. The physiological density in Chad, where only 3 percent of the land is arable, is 300 people per square kilometer (188 per square mile), which is much higher than its crude population density of 11 people per square kilometer. Perhaps a more telling indicator of population pressure and potential food shortages is **agricultural density**, the number of farmers per unit of arable land. Because the majority of people in Sub-Saharan Africa earn their livings from agriculture, agricultural density indicates the number of people who directly depend on each arable square kilometer. The agricultural density of many Sub-Saharan countries is 10 times greater than their crude population density.

Demographic Trends and Disease Challenges

The demographic profile of the region is changing. One positive change is the decline in child mortality due to greater access to primary health care and disease prevention efforts. Gone are the days when 1 in 5 children did not live past his or her fifth birthday. Today the child mortality figure is closer to 1 in 10—still high by world standards, but a considerable improvement. Also, life expectancy figures bottomed out in the 2000s due to the devastating impact of HIV/AIDS and are now on the rise. Finally, like people in other world regions, Africans are moving to cities, which in most cases results in fertility declines. The family size in South Africa, one of the more urbanized large countries, is half the regional average.

Figure 6.17 compares the population pyramids of Ethiopia and South Africa. Ethiopia has the classic broad-based pyramid of a demographically growing and youthful country. In Ethiopia, most women have four children, and the rate of natural increase is 2.3 percent. There are nearly even numbers of men and women, and only 3 percent of the people are over the age of 65 (life expectancy is 64). In contrast, the South African population pyramid tapers down, reflecting the country's smaller family size (the average woman has two or three children). One unusual aspect in this graphic is the smaller number of women in their 30s and early 40s when compared to men. This is due to the disproportionate impact of AIDS on women in Africa (which will be discussed later). South Africa has more people over the age of 65 (5 percent), but its life expectancy is only 62 years, less than that of Ethiopia. This, too, is the legacy of the AIDS epidemic, which hit southern Africa with deadly force in the 1990s; thankfully, infection rates are now on the decline.

Family Size A continued preference for large families is the basis for the region's demographic growth. In the 1960s, many areas in the developing world had a total fertility rate (TFR) of 5.0 or higher. Today Sub-Saharan Africa, at 5.0, is the only region with such a high TFR. A combination of cultural practices, rural lifestyles, child mortality, and economic realities encourages large families (**Figure 6.18**). Yet the average family size is decreasing; as recently as 1996, the regional TFR was 6.0.

Throughout the region, large families guarantee a family's lineage and status. Even now most women marry young, typically when they are teenagers, maximizing their childbearing years. Demographers often point to the limited formal education available to women as another factor contributing to high fertility. Religious affiliation seems to have little bearing on the region's fertility rates; Muslim, Christian, and animist communities all have similarly high birth rates.

The everyday realities of rural life make large families an asset. Children are an important source of labor; from tending crops and livestock to gathering wood, they add more to the household economy than they take. Also, for the poorest places in the developing world such as Sub-Saharan Africa, children are seen as social security: When parents' health falters, grown children are expected to care for them.

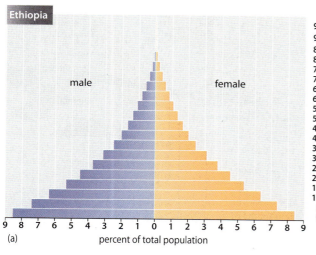

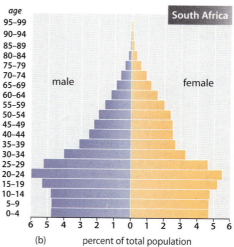

(a) percent of total population

(b) percent of total population

◀ **Figure 6.17 Population Pyramids of Ethiopia and South Africa** These two pyramids show the contrasting demographic profiles of (a) the more rural and rapidly growing Ethiopia and (b) the more urbanized and slower-growing South Africa. A lost generation of South African women in their 30s and 40s is also evident, due to the disproportionate impact of HIV/AIDS on women in South Africa.

▲ **Figure 6.18 Large Families** Large families are still common in Sub-Saharan Africa. The average total fertility rate for the region is five children per woman. Above is a large family in front of their home in Kolahun, Liberia.

foreign aid sources have increased spending to reduce the threat of infection. Presently, a malaria vaccine does not exist, but research in this area is promising. Medication helps in many cases but is not reliable over the long term, and insecticide use has led to mosquito resistance to the chemicals. The most effective tool to reduce infection has been the distribution of insecticide-treated mosquito nets to millions of African homes. That, along with rapid diagnostic tests and access to medication once infected, has cut infections and related deaths by one-third since 2000.

Malaria and poverty are closely related in Sub-Saharan Africa, with many of the poorest tropical countries experiencing higher infection rates. West and Central Africa are the areas hardest hit. The Democratic Republic of the Congo and Nigeria, along with India, account for 40 percent of worldwide malaria cases, and 40 percent of malaria deaths occur in these two large African nations.

HIV/AIDS Now in its fourth decade, HIV/AIDS has been one of the deadliest epidemics in modern human history, yet it is beginning to be tamed. This is especially welcome news for Sub-Saharan Africa, home to 70 percent of the 35 million people living with HIV/AIDS. Human immunodeficiency virus (HIV) is the virus that can lead to acquired immunodeficiency syndrome (AIDS). The human body cannot get rid of HIV, but antiretroviral drugs can suppress it and keep it from becoming AIDS.

The HIV/AIDS virus is thought to have originated in the forests of the Congo, possibly crossing over from chimpanzees to humans sometime in the late 1950s. Yet it was not until the 1980s that the impact of the disease was first felt. In Sub-Saharan Africa, as in much of the developing world, the virus is transmitted primarily by unprotected heterosexual activity or from mother to child during the birth process or through breastfeeding. A long-standing pattern of seasonal male labor migration helped to spread the disease. So, too, did lack of education, inadequate testing early on in the epidemic, and disempowerment of women. Consequently, women bear a disproportionate burden of the HIV/AIDS epidemic, accounting for approximately 60 percent of HIV infections, and serving as caregivers for those who are infected. Until the late 1990s, many African governments were unwilling to acknowledge publicly the severity of the situation or to discuss frankly the measures necessary for prevention.

Southern Africa is ground zero for the AIDS epidemic; the countries with the highest HIV prevalence (South Africa, Swaziland, Lesotho, Botswana, Namibia, Mozambique, Zambia, Zimbabwe, and Malawi) are all located there. In South Africa, the most populous state in southern Africa, 7 million people (nearly one person in five between the ages of 15 and 49) are infected with HIV/AIDS. Rates of infection in neighboring Botswana, Lesotho, and Swaziland are even higher. Infection rates in other African states are lower, but

When most African states gained their independence in the 1960s, population growth was not perceived as a problem and many equated demographic growth with development. National policies began shifting in the 1980s. For the first time, government officials argued that smaller families and slower population growth were needed for social and economic development. Following the 1994 United Nations International Conference on Population and Development held in Cairo, Egypt, African governments announced their intent to cut natural increase rates to 2.0 and to increase the rate of contraceptive use to 40 percent by 2020. They may reach their goal: As of 2016, the region's rate of natural increase was 2.7 percent, and 30 percent of married women use contraception. Other factors are bringing down the rate of natural increase, such as the decline in family size among urban populations—a pattern seen throughout the world. Tragically, declines in natural increase were also occurring as a result of AIDS.

The Disease Factor: Malaria, HIV/AIDS, and Ebola

Historically, the hazards of malaria and other tropical diseases such as sleeping sickness limited European settlement in the tropical portions of Sub-Saharan Africa. It was only in the 1850s, when European doctors discovered that a daily dose of quinine could protect against malaria, that the balance of power in Africa radically shifted. Explorers immediately began to penetrate the interior of the continent, while merchants and expeditionary forces moved inland from the coast. The first imperial claims soon followed, culminating in colonial division of Africa in the 1870s (discussed later in this chapter).

Malaria A scourge in this region for centuries, malaria is transmitted from infected individuals to others via the anopheles mosquito, causing high fever, severe headache, and in the worst cases, death. The World Health Organization estimates that 200 million people contract malaria each year, resulting in 600,000 deaths. The majority of infections and deaths occur in Sub-Saharan Africa. Since 2000, African governments, NGOs, and still high by world standards. In Kenya, an estimated 6 percent of the 15–49 age group has HIV or AIDS. By comparison, only 0.6 percent of the same age group in North America is infected.

The social and economic implications of this epidemic have been profound. Life expectancy rates tumbled—in a few places even dropping to the early 40s. AIDS typically hits the portion of the population that is most economically productive. The time lost to care for sick family members and the outlay of workers' compensation benefits have reduced economic productivity and overwhelmed public services in hard-hit areas. The disease makes no class distinctions: Countries are losing both peasant farmers and educated professionals (doctors, engineers, and teachers). Many countries struggle to care for millions of children orphaned by AIDS.

After three devastating decades, there are finally hopeful signs. Prevention measures are widely taught, and treatment with a mix of available drugs means that HIV infection is now manageable and not a death sentence. Due to international financial support and national outreach efforts, Sub-Saharan Africa now has many more health facilities that offer HIV testing and counseling. Some 15 million people receive life-prolonging drugs. Prevention services in prenatal clinics provide the majority of pregnant HIV-positive women with antiretroviral drugs to prevent transmission of the virus to their babies. Political activism and changes in sexual practices, driven by educational campaigns, more condom use, and higher rates of male circumcision, have prevented hundreds of thousands of new HIV cases (**Figure 6.19**).

Ebola The 2014–2015 Ebola outbreak in West Africa attracted international attention due to the highly contagious and deadly nature of this disease—once infected, an individual may live only a week or two without intensive medical care. The 2014 outbreak was the largest Ebola epidemic in history, affecting multiple countries in West

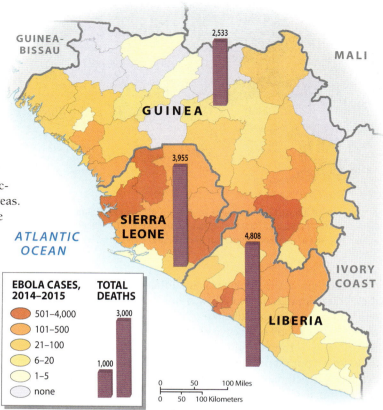

▲ **Figure 6.20 Ebola Outbreak in West Africa** Liberia, Sierra Leone, and Guinea were the states most impacted by the 2015 Ebola outbreak. Cases were found throughout Liberia but especially in the capital city of Monrovia. Sierra Leone's cases were also more concentrated near the cities of Freetown and Kenama. Guinea, which had fewer deaths, had a concentration of rural cases near the Liberia border.

Africa and resulting in some 30,000 confirmed cases. A rare disease, Ebola was first identified in 1976 along the Ebola River in the Democratic Republic of the Congo. There had been isolated outbreaks before, but the 2014 outbreak was by far the largest and deadliest (**Figure 6.20**), although as of late 2015 there were no new cases in Liberia and Sierra Leone.

International organizations such as Médecins Sans Frontières (Doctors Without Borders) and various government agencies contributed money, medical personnel, and equipment to fight this epidemic, fearing that it could spread and potentially infect millions. There is no medication to treat Ebola, and it is critical that people not come into contact with the blood or bodily fluids of those infected. This often meant quarantining infected people, developing stringent procedures for health-care personnel, conducting education efforts, and abandoning traditional burial practices to reduce the spread of infection. International cooperation and the efforts on the ground in Sierra Leone, Liberia, and Guinea worked in the 2014–2015 outbreak, and worst-case scenarios were not realized. How such diseases spread and are treated has much to do with settlement patterns, quality of infrastructure, and availability of health care in this poor but developing region.

▼ **Figure 6.19 Combating HIV/AIDS in South Africa** To address high infection rates, southern African countries have created explicit campaigns to educate the public on the need to use condoms to prevent sexually transmitted diseases such as AIDS. This billboard is in Johannesburg, South Africa's largest city.

Patterns of Settlement and Land Use

Because of the dominance of rural settlements in Sub-Saharan Africa, people are widely scattered throughout the region (see Figure 6.16). Population concentrations are highest in West Africa, highland East Africa, and the eastern half of South Africa. The first two areas have some of the region's best soils, and indigenous systems of permanent agriculture developed there. In South Africa, the more densely settled east results from an urbanized economy based on mining, as well as the forced concentration of black South Africans into eastern homelands.

West Africa is more heavily populated than most of Sub-Saharan Africa, although the actual distribution pattern is patchy. Density in the far west, from Senegal to Liberia, is moderate. This area is characterized by broad lowlands with decent soils, and in many areas the cultivation of rice has enhanced agricultural productivity. Greater concentrations of people are found along the Gulf of Guinea, from southern Ghana through southern Nigeria, and again in northern Nigeria along the southern fringe of the Sahel. Nigeria is moderately to densely settled through most of its extensive territory; with over 187 million inhabitants, it stands as the demographic core of Sub-Saharan Africa. The next largest country, Ethiopia, has a population of nearly 100 million.

As more Africans move to cities, settlement patterns are becoming more concentrated. Towns that were once small administrative centers for colonial elites grew into major cities. The region has one megacity, Lagos, estimated at 13 to 17 million residents. Throughout the continent, African cities are growing faster than rural areas. But before examining the Sub-Saharan urban scene, a more detailed discussion of rural subsistence is needed.

Agricultural Subsistence and Food Ways The staple crops over most of Sub-Saharan Africa are millet, sorghum, rice, and corn (maize), as well as a variety of tubers and root crops such as yams. Irrigated rice is widely grown in West Africa and Madagascar. Geographer Judith Carney in her book *Black Rice* documents how African slaves introduced rice cultivation to the Americas. Corn, in contrast, was brought to Africa from the Americas by the slave trade and quickly grew to become a basic food. In higher elevations of Ethiopia and South Africa, wheat and barley are grown. Intermixed with subsistence foods are a variety of export crops—coffee, tea, rubber, bananas, cocoa, cotton, and peanuts—that are grown in distinct ecological zones and often in some of the best soils.

In areas that support annual crop yields, population densities are greater. In parts of humid West Africa, for example, the yam became the king of subsistence crops. The mastery of yam production in earthen mounds throughout West Africa enhanced food production so that people could live in denser permanent settlements (Figure 6.21). Nigerian novelist Chinua Achebe described the intricacies of yam production in his novel *Things Fall*

Explore the **Tastes** of Yams

http://goo.gl/SweKlx

▲ **Figure 6.21** **Seeding Yams** A couple plants yams in earthen mounds in Benin, West Africa. A native crop to West Africa, yams are a daily starch for people of this sub-region. In Benin, yams form the basis of the agricultural economy.

Apart: "The young tendrils were protected from earth-heat with rings of sisal leaves. As the rains became heavier, the women planted maize, melons, and beans between the yam mounts. The yams were then staked, first with little sticks and later with tall and big tree branches."

Over much of the continent, African agriculture remains relatively unproductive, and rural population densities tend to be low. Growing crops in poorer tropical soils usually entails shifting cultivation, or **swidden**. This process involves burning the natural vegetation to release fertilizing ash and planting crops such as maize, beans, sweet potatoes, bananas, papaya, manioc, yams, melon, and squash. Each plot is temporarily abandoned once its source of nutrients has been exhausted. Swidden cultivation often is a very finely tuned adaptation to local environmental conditions, but it cannot support high population densities. Women are the subsistence farmers of the region, producing for their household needs as well as for local and foreign markets.

Export Agriculture Agricultural exports, whether from large estates or small producers, are critical to the economies of many Sub-Saharan states. If African countries are to import the modern goods and energy resources they require, they must sell their own products on the world market. Because the region has few competitive industries, the bulk of its exports are primary products derived from farming, mining, and forestry.

In the densely settled country of Rwanda, most farms are small, but the highland volcanic soil is ideal for growing coffee. However, for decades the country's coffee production languished, and farmers earned very little for their low-quality beans. Yet across Rwanda's hillsides were older varieties of coffee plants with high value in today's premium coffee market. Farmers just needed a better way to prepare the beans and market them. In an effort to rebuild the country after the ethnic genocide in the 1990s, the government targeted improving the quality of Rwandan coffee by forming cooperatives that would bring ethnic groups together and raise farmers' incomes. Through community-run cooperatives, the quality of washed and sorted beans improved. So, too, did the farmers' ability to bargain with major buyers such as Starbucks and Green Mountain, leaving out the middleman. For many small farmers in Rwanda, their premium coffee is now a source of pride. The cooperatives have taken off, and farmers, many of them war widows, have seen their incomes improve.

Several African countries rely heavily on one or two export crops. Coffee is vital for Ethiopia, Kenya, Rwanda, Burundi, and Tanzania. Peanuts have historically been the primary source of income in the Sahel, whereas cotton is tremendously important for the Central African Republic and South Sudan. Ghana and Ivory Coast have long been the world's main suppliers of cacao (the source of chocolate), Liberia produces plantation rubber, and many farmers in Nigeria specialize in palm oil. The export of such products can bring good money when commodity prices are high, but when prices collapse, as they periodically do, economic hardship follows.

Nontraditional agricultural exports that depend on significant capital inputs and refrigerated air transport have emerged in the last two to three decades. One such industry is floriculture in the highlands of Kenya, Ethiopia, and South Africa as discussed at the beginning of this chapter. Similarly, the European market for fresh vegetables and fruits in the winter is being met by some producers in West and East Africa.

Pastoralism Animal husbandry (the care of livestock) is extremely important in Sub-Saharan Africa, particularly in the semiarid zones. Camels and goats are the principal animals in the Sahel and the Horn of Africa, but farther south cattle are king (**Figure 6.22**). Many African peoples have traditionally specialized in cattle raising and are often tied into mutually beneficial relationships with neighboring farmers. Such **pastoralists** typically graze their stock on the stubble of harvested fields during the dry season and then move them to drier uncultivated areas during the wet season, when the pastures turn green. Farmers thus have their fields

▼ **Figure 6.22 Pastoralists in the Sahel** Longhorn cattle and goats are led by a pastoralist on horseback in Ngalwa, Niger. Throughout the Sahel, the movement of animals to seasonal grazing areas is critical for economic livelihoods.

fertilized by the manure of the pastoralists' stock, while the pastoralists find good dry-season grazing. At the same time, the nomads can trade their animal products for grain and other goods of the sedentary world. Several pastoral peoples of East Africa, such as the Masai of the Tanzanian-Kenyan borderlands, are noted for their extreme reliance on cattle and general (but never complete) independence from agriculture. The Masai derive a large percentage of their nutrition from drinking a mixture of milk and blood, the latter obtained by periodically tapping the animal's jugular vein, a procedure that evidently causes little harm.

Large expanses of Sub-Saharan Africa have been off-limits to cattle because of infestations of *tsetse flies*, which spread sleeping sickness to cattle, humans, and some wildlife. In environments containing brush or woodland (necessary for tsetse fly survival), wild animals that harbor the disease but are immune to it might be present in large numbers, but cattle simply could not be raised. Some evidence suggests that tsetse fly infestations dramatically increased in the late 1800s, greatly harming African societies dependent on livestock but benefiting wildlife populations. At present, tsetse fly eradication programs have reduced the threat, and cattle ranching is spreading into areas where it was previously unsuitable. This process is beneficial for African livestock, but when people move large numbers of cattle into new areas, wildlife almost inevitably declines.

Sub-Saharan Africa thus presents a difficult environment for raising livestock because of the virulence of its animal diseases. In the tropical rainforests of Central Africa, cattle have never survived well, and the only domestic animal that thrives is the goat. Raising horses, moreover, has historically been feasible only in the Sahel and in South Africa.

Urban Life

Sub-Saharan Africa and South Asia are the least urbanized world regions, although in both regions cities are growing at twice the national growth rates. More than one-third of Sub-Saharan people live in urbanized areas. One consequence of this surge in city living is urban sprawl. Rural-to-urban migration, industrialization, and refugee flows are forcing the cities of the region to absorb more people and use more resources. As in Latin America, the tendency is toward urban primacy—the condition in which one major city dominates and is at least three times larger than the next largest city.

Nairobi, Kenya's capital, was a city of 250,000 in 1960 but now has 3.5 million residents and is considered the hub of transportation, finance, and communication for all of East Africa. In the last decade, Nairobi has become a high-tech superstar, with half the city's population using the Internet and many start-up companies being created. Despite such a robust embrace of technology, Nairobi still has many unemployed and impoverished residents. Close to the city center is the vast slum of Kibera. Here garbage lines the streets, crime is rampant, and housing is crude and crowded. And yet for many years this was a blank spot on city maps (see *Exploring Global Connections: The Map Kibera Initiative*). Municipal

Exploring Global Connections

The Map Kibera Initiative

Near downtown Nairobi, bordering a golf course, the slum of Kibera is home to some 250,000 people. One of the oldest and most studied slums, today Kibera is experiencing the dual processes of upgrading and resident relocation (**Figure 6.2.1**). Residents are also using social media and open source mapping technologies to make the invisible visible by literally inserting their place on Nairobi's map. Through the creative use of digital tools and citizen reporting and training, the issues of slum dwellers can be recorded and widely shared.

Filling in the Map Map Kibera was formed in 2009 as a not-for-profit organization established to train slum residents in mapping technology and digital advocacy. Throughout the developing world, informal or slum settlements are often densely settled, but appear as blank spots on city maps. Key to the success of Map Kibera was open-source software that is free and available to anyone with Internet access. Working in Open Street Map (OSM) and using GPS devices, Kibera residents began mapping clinics, water sources, pathways, markets, pharmacies, and services in their community. As more information was gathered, Quantum GIS (now

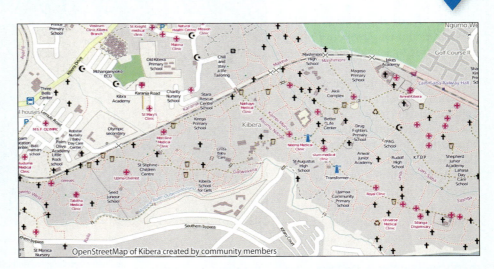

OpenStreetMap of Kibera created by community members

▲ **Figure 6.2.2** **Community Mapping** A map of Kibera produced by its residents through Open Street Map. This version of the map shows clinics and places of worship along with pathways through the neighborhood. Before the Map Kibera project, this huge slum was a blank spot on the city map.

QGIS, a free GIS application) and satellite imagery were used to analyze the data and make specialty maps that could be printed and distributed to residents and city officials (**Figure 6.2.2**). The success of Map Kibera expanded to other Nairobi slums, Mathare and Mukuru, which are also mapped using the same methodology.

Giving Voice Mapping gave young residents the tools and the skills necessary to assert their presence and their stories in other ways. The "Voice Kibera" program uses blogging through WordPress and Ushahidi software to report on issues that impact the neighborhood such as crime, water quality, or even garbage. Using Ushahidi, designed in Kenya, residents can text in news to an editorial team who can then geo-locate an event and blog about it, such as a breaking news story about a fire or mudslide. Logistically, Kibera and other mapped slums also receive better service provision; police and firefighters now know where to go when needed. Such volunteered geographic information and citizen reporting is made possible with software and technologies that did not exist in a meaningful way a decade ago. Together, mapping and blogging have helped to build a sense of community.

Connected and Technically Sophisticated Map Kibera is the exception for slums in the region, but with the growth of digital technology and open source platforms, the hope is that more urban communities can emulate this model. In July 2016, when the White House invited volunteer mappers to a Mapathon focusing on malaria prevention in Sub-Saharan Africa, mappers from Kibera joined through a video conference and traced areas of concern in OpenStreet Map using satellite imagery that was part of the event. With support from organizations such as UN Habitat and Ushahidi, citizen movements such as Map Kibera are creating a set of tools so that "all people have equal ability to create and share information to influence their future" (see http://www.mapkibera.org/).

▲ **Figure 6.2.1** **Kibera Slum** One of Nairobi's oldest and largest slums, Kibera is home to at least 250,000 people. The Map Kibera project is based here.

Google Earth Virtual **MG** Tour Video

http://goo.gl/4TkJC9

1. Why do you think informal settlements such as Kibera are often off the map?

2. Using Open Street Map, search for your community and see how it appears on the map.

officials throughout the region struggle to build enough roads and provide electricity, water, trash collection, and employment for so many people in such rapidly growing places.

European colonialism greatly influenced urban form and development in the region. Although a very small percentage of the population lived in cities, Africans did have an urban tradition prior to the colonial era. Ancient cities, such as Axum in Ethiopia, thrived 2000 years ago. Similarly, prominent trans-Saharan trade centers in the Sahel, such as Tombouctou (Timbuktu) and Gao, have existed for more than a millennium. In East Africa, an urban mercantile culture rooted in Islam and the Swahili language emerged. The prominent cities of Zanzibar, Tanzania, and Mombasa, Kenya, flourished by supporting a trade network that linked the East African highlands with the Persian Gulf. The stone ruins of Great Zimbabwe in southern Africa are testimony to the achievements of stone working, metallurgy, and religion achieved by Bantu groups in the 14th century. West Africa, however, had the most developed precolonial urban network, reflecting both indigenous and Islamic traditions. It also supports some of the region's largest cities today.

West African Urban Traditions The West African coastline is dotted with cities, from Dakar, Senegal, in the far northwest to Lagos, Nigeria, in the east. Half of Nigerians live in cities, and in 2015 the country had eight metropolitan areas with populations of more than 1 million. Historically, the Yoruba cities in southwestern Nigeria have been the best documented. Developed in the 12th century, cities such as Ibadan were walled and gated, with a palace encircled by large rectangular courtyards at the city center. An important center of trade for an extensive hinterland, Ibadan also was a religious and political center. Lagos was another Yoruba settlement. Founded on a coastal island in the Bight of Benin, most of the modern city has spread onto the nearby mainland. Its coastal setting and natural harbor made this relatively small, indigenous city attractive to colonial powers. When the British took control in the mid-19th century, the city grew in size and importance. Lagos was a city of 1 million when Nigeria gained its independence in 1960; today it is Sub-Saharan Africa's largest city (**Figure 6.23**).

Most West African cities are hybrids, combining Islamic, European, and national elements such as mosques, Victorian architecture, and streets named after independence leaders. Metropolitan Accra is Ghana's capital and home to 4 million people. Originally settled by the Ga people in the 16th century, it became a British colonial administrative

▲ **Figure 6.23 Downtown Lagos, Nigeria** A sea of shoppers, vendors, and buses fills the streets of the Oshodi Market, the commercial hub of Lagos. Lagos is the largest city in Sub-Saharan Africa, with at least 13 million inhabitants.

Explore the **Sights** of Oshodi Market in Lagos

http://goo.gl/gmcUzC

center by the late 1800s. The modern city is being transformed through neoliberal policies introduced in the 1980s that attracted international corporations. Increased foreign investment in financial and producer services led to the creation of a "Global Central Business District (CBD)" on the east side of the city, away from the "National CBD." Here foreign companies clustered in areas with secure land title, new modern roads, parking, and airport access. Upper-income gated communities have also formed near the Global CBD, with names such as Trasacco Valley, Airport Hills, Buena Vista, and Legon (**Figure 6.24**). This influx of foreign capital is rapidly changing Accra and other cities in the region. The result is highly segregated urban spaces that reflect a global phase in urban development, in which world market forces, rather than colonial or national ones, are driving the change.

Urban Industrial South Africa The major cities of southern Africa, unlike those of West Africa, are colonial in origin. Cities such as Lusaka, Zambia, and Harare, Zimbabwe, grew as administrative or mining centers. South Africa is one of the most urbanized states in the entire region, and it is certainly the most industrialized. The foundations of South Africa's urban economy rest largely on its incredibly rich mineral resources (diamonds, gold, chromium, platinum, tin, uranium, coal, iron ore, and manganese). Seven of its metropolitan areas have more than 1 million people; the largest of these are Johannesburg, Durban, and Cape Town.

The form of South African cities continues to be imprinted with the legacy of **apartheid**, an official policy

▲ **Figure 6.24 Elite Accra Neighborhood** In the Legon area east of downtown Accra one finds the city's nicest homes, private schools, and the University of Ghana.

of racial segregation that shaped social relations in South Africa for nearly 50 years. Even though apartheid was abolished in 1994, it is still evident in the landscape. Apartheid rules divided all cities into residential areas based on racial categories: white, **coloured** (a South African term describing people of mixed African and European ancestry), Indian (South Asian), and African (black). Whites resided in the most appealing and spacious portions of the cities, whereas blacks were crowded into the least desired areas, forming squatter settlements called **townships**. Today blacks, coloureds, and Indians are legally allowed to live anywhere they want. Yet the economic differences among racial groups, as well as deep-rooted animosity, hinder residential integration.

Even in relatively affluent South Africa, the challenge of accommodating new urban residents is constant. The largely black community of Diepsloot, north of Johannesburg, illustrates how even planned communities can become densely settled in short order. Diepsloot is a post-apartheid project for which the government divided land into lots and eventually built several thousand houses. **Figure 6.25** illustrates the increased settlement density of a Diepsloot neighborhood over a decade. The relatively large lots were gradually filled in with small shacks built with scrap metal, wood, and plastic. The initial homes had access to running water and sewage, but the informal dwellings surrounding them do not. Later, small shops were added, and trees were planted, but much of the population lacks formal employment. This community, initially planned for a few thousand people, now numbers about 200,000.

Still, Johannesburg is the African metropolitan area that most consciously aspires to global city status, a dream underscored by its hosting of soccer's World Cup in 2010. However, as apartheid ended, Johannesburg became infamous for its high crime rate. Many white-owned businesses fled the CBD for the northern suburb of Sandton, and by the late 1990s, Sandton was the new financial and business hub for Johannesburg. It epitomizes the modern urban face of South Africa: While it is racially mixed, affluent whites are overrepresented. When the province of Gauteng invested $3 billion in a new high-speed commuter rail called Gautrain, it was no accident that the first functioning line linked Sandton with Tambo International Airport (**Figure 6.26**).

REVIEW

6.3 How have infectious diseases impacted population trends, and what are governments and international organizations doing to fight diseases such as HIV/AIDS and malaria?

6.4 What are the major rural livelihoods in this world region?

6.5 Explain the factors contributing to the region's high population growth rates and rapid urbanization.

■ **KEY TERMS** physiological density, agricultural density, swidden, pastoralists, apartheid, coloured, townships

▼ **Figure 6.25 Satellite Images of Diepsloot, South Africa** Over a decade, black settlers poured into the planned community of Diepsloot, filling in single-family-home lots (a) with multiple structures. They also added businesses, planted trees, and vastly increased the population density of the area (b).

(a)

(b)

Cultural Coherence and Diversity: Unity Through Adversity

No world region is culturally homogeneous, but most have been partially unified in the past by widespread systems of belief and communication. Traditional African religions, however, were largely limited to local areas, and the religions that did become widespread—namely, Islam and Christianity—are primarily associated with other world regions. A handful of African trade languages have long been understood over vast territories (Swahili in East Africa, Mandingo and Hausa in West Africa), but none spans the entire Sub-Saharan region. Sub-Saharan Africa also lacks a history of widespread political union or even an indigenous system of political relations. The powerful African kingdoms and empires of past centuries were limited to distinct subregions of the landmass.

The lack of traditional cultural and political coherence across Sub-Saharan Africa is not surprising if you consider the region's huge size—more than four times larger than Europe or South Asia (see Figure 6.2). Had foreign imperialism not impinged on the region, it is quite possible that West Africa and southern Africa would have developed into their own distinct world regions.

An African identity south of the Sahara was created through a common history of slavery and colonialism, as well as struggles for independence and development. More telling, the people of the region often define themselves as African, especially to the outside world. No one will argue the fact that Sub-Saharan Africa is poor. And yet the cultural expressions of its people—its music, dance, and art—are joyous. Africans share a resilience and optimism that visitors to the region often comment on. The cultural diversity of the region is obvious, yet there is unity among the people, drawn from surviving many adversities.

Language Patterns

In most Sub-Saharan countries, as in other former colonies, people speak multiple languages that reflect tribal, ethnic, colonial, and national affiliations. Indigenous languages, many from the Bantu subfamily, often are localized to relatively small rural areas. More widely spoken African trade languages, such as Swahili or Hausa, serve as lingua francas over broader areas. Overlaying native languages are Indo-European (French and English) and Afro-Asiatic

▲ **Figure 6.26 Sandton, Johannesburg** Just north of Johannesburg's central business district, Sandton has emerged as the financial and business center of the new South Africa. Many businesses relocated here, fleeing high crime rates in the old central business district. With world-class shopping, hotels, a convention center, and even Nelson Mandela Square, Sandton has become the part of Johannesburg that most tourists and businesspeople experience.

(Arabic) languages. **Figure 6.27** illustrates the complex pattern of language families and major languages found in the region today. A comparison of the larger map with the inset of current "official" languages shows that most African countries are multilingual, which can be a source of tension within states. In Nigeria, for example, the official language is English, yet there are millions of Hausa, Yoruba, Igbo (or Ibo), Ful (or Fulani), and Efik speakers, as well as speakers of dozens of other languages.

African Language Groups Three of the six language groups mapped in Figure 6.27 are unique to the region (Niger-Congo, Nilo-Saharan, and Khoisan), while the other three (Afro-Asiatic, Austronesian, and Indo-European) are more closely associated with other parts of the world. Afro-Asiatic languages, especially Arabic, dominate North Africa and are understood in Islamic areas of Sub-Saharan Africa as well. Amharic in Ethiopia and Somali of Somalia are also Afro-Asiatic languages. The Austronesian language family is limited to the island of Madagascar, which many believe was first settled by seafarers from present-day Indonesia some 1500 years ago. Indo-European languages, especially French, English, Portuguese, and Afrikaans, are a legacy of colonialism and are widely used today.

Of the three language groups found exclusively in the region, the Niger-Congo language group is by far the most influential. This linguistic group originated in West Africa and includes Mandingo, Yoruba, Ful(ani), and Igbo, among others. Around 3000 years ago a people of Niger-Congo stock began to expand out of western Africa into the equatorial zone (see arrows on Figure 6.27). This group, called the Bantu, commenced one of the most far-ranging migrations in human history, which introduced agriculture into large areas of central and southern Africa. One Bantu group migrated east across the fringes of the rainforest to settle in the Lake Victoria basin in East Africa, where they formed an eastern Bantu core that later pushed south all the way to South Africa. Another group moved south, into the rainforest proper. The equatorial rainforest belt immediately adjacent to the original Bantu homeland had been very sparsely settled by the ancestors of the modern pygmies (a distinct people noted for their short stature and hunting skills). Pygmy groups, having entered into close trading relations with the Bantu newcomers, eventually came to speak Bantu languages as well.

Once the Bantu migrants had advanced beyond the rainforest into the savannas and woodlands, their agricultural techniques proved highly successful and their influence expanded. Over the centuries, the various languages and dialects of the many Bantu-speaking groups, which often were separated from each other by considerable distances, gradually diverged from each other. Today there are several hundred distinct languages in the Bantu subfamily of the great Niger-Congo group. All Bantu languages, however, remain closely related to each other, and a speaker of one can generally learn any other without undue difficulty.

One language in the Bantu subfamily, Swahili, eventually became the most widely spoken Sub-Saharan language. Swahili originated on the East African coast, where several merchant colonies from Arabia were established around 1100 CE. A hybrid society grew up in a narrow coastal band of modern Kenya and Tanzania, speaking a language of Bantu structure enriched with many Arabic words. Swahili became the primary language only in the narrow coastal belt, but it spread far into the interior as the language of trade. After independence, both Kenya and Tanzania adopted Swahili as an official language. With some 100 million speakers, Swahili is the lingua franca of East Africa and parts of Central Africa. It has generated a fairly extensive literature and is often studied in other regions of the world.

Explore the **Sounds** of Swahili Language

http://goo.gl/oJo9k0

Language and Identity Historically ethnic identity, as well as linguistic affiliation, have been highly unstable over much of Sub-Saharan Africa. The tendency was for new groups to form when people threatened by war fled to less settled areas, where they often mixed with peoples from other places. In such circumstances, new languages arise quickly, and divisions between groups become blurred. Nevertheless, distinct **tribes** formed that initially consisted of a group of families or clans with a common kinship, language, and definable territory. European colonial administrators were eager to establish a fixed indigenous social order to control native peoples. During this process, a flawed cultural map of Sub-Saharan Africa evolved. Some tribes were artificially divided, meaningless names were applied, and territorial boundaries were often misinterpreted.

Social boundaries between different ethnic and linguistic groups have become more stable in recent years, and some individual languages have become particularly important for communication on a national scale. Wolof in Senegal; Mandingo in Mali; Mossi in Burkina Faso; Yoruba, Hausa, and Igbo in Nigeria; Kikuyu in central Kenya; and Zulu, Xhosa, and Sesotho in South Africa are all nationally significant languages spoken by millions of people. None, however, has the status of being the official language of any country. With the end of apartheid in South Africa, the country officially recognized 11 languages, although English is still the lingua franca of business and government. Indeed, a single language has a clear majority status in only a handful of Sub-Saharan countries. The more linguistically homogeneous states include Somalia (where virtually everyone speaks Somali) and the very small states of Rwanda, Burundi, Swaziland, and Lesotho.

European Languages In the colonial period, European countries used their own languages for administrative purposes in their African empires. Education in the colonial period also stressed literacy in the language of the imperial power. In the postindependence period, most Sub-Saharan African countries have continued to use the languages of their former colonizers for government and

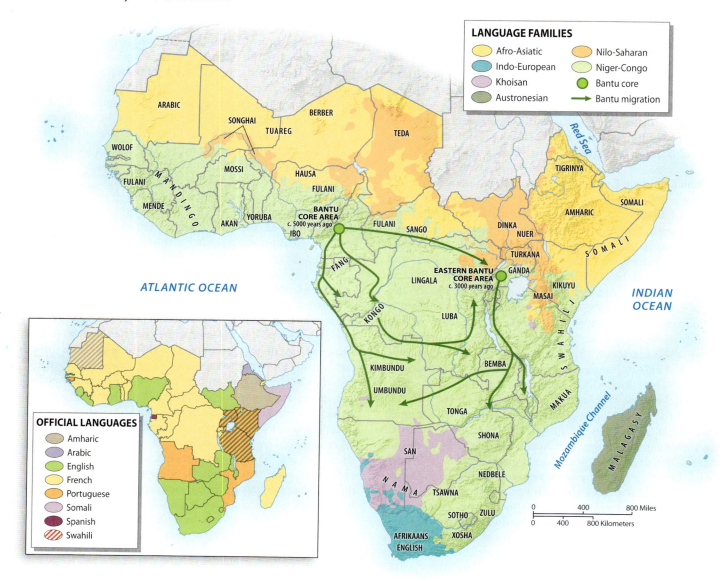

▲ **Figure 6.27 Languages of Sub-Saharan Africa** Mapping language is a complex task for Sub-Saharan Africa. There are languages with millions of speakers, such as Swahili, and languages spoken by a few hundred people living in isolated areas. Six language families are represented in the region. Among these families are scores of individual languages (see the labels on the map). Because most modern states have many indigenous languages, the colonial language often became the "official" language because it was less controversial than picking from among several indigenous languages. English and French are the most common official languages in the region (see inset).

higher education. Few of these new states had a clear majority language that they could employ, and picking any minority tongue would have aroused the opposition of other peoples. The one exception is Ethiopia, which maintained its independence during the colonial era. The official language is Amharic, although other indigenous languages are also spoken.

Two vast blocks of European languages exist in Africa today: Francophone Africa, encompassing the former colonies of France and Belgium, where French serves as the main language of administration; and Anglophone Africa, where the use of English prevails (see inset, Figure 6.27). Early Dutch settlement in South Africa resulted in the use of Afrikaans (a Dutch-based language) by several million South Africans. Portuguese is spoken in Angola and Mozambique, former colonies of Portugal. Interestingly, when South Sudan

gained its independence from Sudan in 2011, it changed its official language from Arabic to English.

Religion

Indigenous African religions generally are classified as *animist,* a somewhat misleading catchall term applied to all local faiths that do not fit into one of the handful of "world religions." Most animist religions are centered on the worship of nature and ancestral spirits, but the internal diversity within the animist tradition is vast. Classifying a religion as animist says more about what it is not than what it actually is.

Both Christianity and Islam actually entered the region early in their histories, but they advanced slowly for many centuries. Since the beginning of the 20th century, both religions have spread rapidly—more rapidly, in fact, than in any other part of

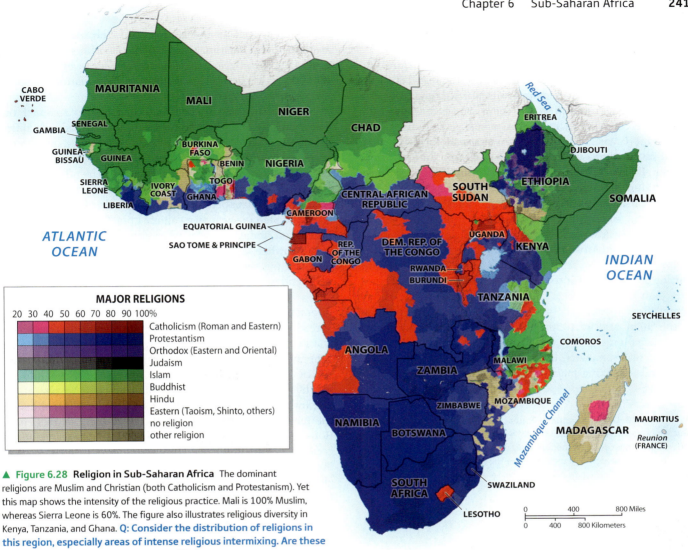

▲ **Figure 6.28 Religion in Sub-Saharan Africa** The dominant religions are Muslim and Christian (both Catholicism and Protestanism). Yet this map shows the intensity of the religious practice. Mali is 100% Muslim, whereas Sierra Leone is 60%. The figure also illustrates religious diversity in Kenya, Tanzania, and Ghana. **Q: Consider the distribution of religions in this region, especially areas of intense religious intermixing. Are these areas more prone or less prone to conflict?**

the world (**Figure 6.28**). But tens of millions of Africans still hold animist beliefs, and many others combine animist practices and ideas with their observances of Christianity and Islam.

The Introduction and Spread of Christianity Christianity came first to northeast Africa. Kingdoms in both Ethiopia and central Sudan were converted by 300 CE—the earliest conversions outside of the Roman Empire. The peoples of Ethiopia and Eritrea adopted the Coptic form of Christianity and have thus historically looked to Egypt's Christian minority for their religious leadership (**Figure 6.29**). Today, roughly half of the population of both Ethiopia and Eritrea follows Coptic Christianity; most of the rest are Muslim, but there are still some animist communities.

European settlers and missionaries introduced Christianity to other parts of Sub-Saharan Africa beginning in the 1600s. The Dutch, who began to colonize South Africa at this time, brought their Calvinist Protestant faith. Later, European immigrants to South Africa brought Anglicanism and other Protestant creeds, as well as Catholicism. A substantial Jewish community also emerged, concentrated in the Johannesburg area. Most black South Africans eventually

converted to one or another form of Christianity as well. In fact, churches in South Africa were instrumental in the long fight against white racial supremacy. Religious leaders such as Bishop Desmond Tutu were outspoken critics of

▼ **Figure 6.29 Eritrean Christians at Prayer** Coptic Christians gather for an Easter celebration in Asmara, Eritrea. Half of the populations in Eritrea and Ethiopia belong to the Coptic church, which has ties to Egypt's Christian minority.

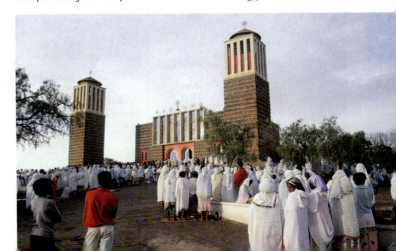

the injustices of apartheid and worked to bring down the system.

Elsewhere in Africa, Christianity came with European missionaries, most of whom arrived after the mid-1800s. As was true in the rest of the world, missionaries had little success where Islam had preceded them, but eventually made numerous conversions in animist areas. As a general rule, Protestant Christianity prevails in areas of former British colonization, while Catholicism is more important in former French, Belgian, and Portuguese territories. There are nearly 150 million Catholics in the region. In the postcolonial era, African Christianity has diversified, at times taking on a life of its own, independent from foreign missionary efforts. Increasingly active in the region are various Pentecostal, Evangelical, and Mormon missionary groups.

The Introduction and Spread of Islam Islam began to advance into Sub-Saharan Africa 1000 years ago. Berber traders from North Africa and the Sahara introduced the religion to the Sahel, and by 1050 the Kingdom of Tokolor in modern Senegal emerged as the first Sub-Saharan Muslim state (**Figure 6.30**). Somewhat later, the ruling class of the powerful Mande-speaking mercantile empires of Ghana and Mali

▼ **Figure 6.30 Grand Mosque of Touba** One of the largest mosques in Africa, Touba is a holy city for Muslims and residents there dedicate themselves to devotion and scholarship. An annual pilgrimage, called the Grant Magal, brings 1 to 2 million faithful each year. Senegal, like much of the Sahel, converted to Islam more than six centuries ago.

Explore the **Sights** of Islam in Senegal

http://goo.gl/KbU9S3

converted as well. In the 14th century, the emperor of Mali astounded the Muslim world when he and his huge entourage made the pilgrimage to Makkah (Mecca), bringing with them so much gold that they set off an inflationary spiral throughout Southwest Asia.

Mande-speaking traders, whose networks spanned the area from the Sahel to the Gulf of Guinea, gradually introduced the religion to other areas of West Africa. Many peoples remained committed to animism, and Islam made slow and fitful progress. Even in the Sahel, syncretic (blended) forms of Islam prevailed through the 1700s. In the early 1800s, however, the pastoral Fulani people launched a series of successful holy wars designed to shear away animist practices and to establish pure Islam. Today orthodox Islam prevails through most of the Sahel. Farther south, Muslims are mixed with Christians and animists, but their numbers continue to grow, and their practices tend to be orthodox as well (see Figure 6.28).

Interaction Between Religious Traditions The southward spread of Islam from the Sahel, coupled with the northward dissemination of Christianity from the port cities, has generated a complex religious frontier across much of West Africa. In Nigeria, the Hausa are firmly Muslim, while the southeastern Igbo are largely Christian. The Yoruba of the southwest are divided between Christians and Muslims. In the more remote parts of Nigeria, moreover, animist traditions remain vital. Despite this religious diversity, religious conflict in Nigeria has been relatively rare until recently. In 2000, seven Nigerian states imposed Islamic *sharia* (religious) law, which has triggered intermittent violence ever since, especially in the northern cities of Kano and Kaduna. In the last few years, an armed jihadist group called Boko Haram formed in northeastern Nigeria, escalating the levels of religious violence in the north and in the south through kidnapping and killing. Initially claiming affiliation with Al Qaeda, in 2013 Boko Haram claimed a formal allegiance with the Sunni Muslim extremist group, Islamic State of Iraq and the Levant (ISIL, also abbreviated ISIS or IS; see Chapter 7). The Nigerian military launched an offensive to dislodge the group from Borno Province in northeastern Nigeria, but Boko Haram has been a difficult organization to control.

Religious conflict historically has been far more acute in northeastern Africa, where Muslims and Christians have struggled against each other for centuries. Such a clash eventually led to the creation of the region's newest state when South Sudan separated from the country of Sudan in 2011. Islam was introduced to Sudan in the 1300s by an invasion of Arabic-speaking pastoralists who destroyed the indigenous Coptic Christian kingdoms of the area. Within a few hundred years, northern and central Sudan had become completely Islamic. The southern equatorial province of Sudan, where tropical diseases and extensive wetlands prevented Arab advances, remained animist or converted to Christianity under British colonial rule.

In the 1970s, the Arabic-speaking Muslims of northern and central Sudan began to build an Islamic state.

Experiencing both religious discrimination and economic exploitation, the Christian and animist peoples of the south launched a massive rebellion. In the 1980s, fighting became intense, with the government generally controlling the main towns and roads and the rebels maintaining power in the countryside. A peace was brokered in 2003, and as part of the peace agreement, southern Sudan was promised an opportunity to vote on secession from the north in 2011. The vote took place, and the new nation was formed with Juba as its capital. Yet this landlocked territory is still not at peace, as discussed later in the chapter.

Sub-Saharan Africa is a land of religious vitality. Both Christianity and Islam are spreading rapidly, and devotional activities are part of the daily flow of life in cities and rural areas. Animism continues to hold widespread appeal as well, so that new and syncretic forms of religious expression are also emerging. With such a diversity of faiths, it is fortunate that religion is not typically the cause of overt conflict.

Globalization and African Culture

The slave triangle that linked Africa to the Americas and Europe set in process patterns of cultural diffusion that transferred African peoples and practices across the Atlantic. Tragically, slavery damaged the demographic and political strength of African societies, especially in West Africa, from where most slaves were taken. An estimated 12 million Africans were shipped to the Americas as slaves from the 1500s until 1870 (see Figure 5.19 in Chapter 5). Slavery impacted the entire region, sending Africans not just to the Americas, but also to Europe, North Africa, and Southwest Asia. The vast majority, however, worked on plantations across the Americas.

Out of this tragic displacement of people came a blending of African cultures with Amerindian and European ones. African rhythms are at the core of various musical styles, from rumba to jazz, the blues, and rock and roll. Brazil, the largest country in Latin America, is claimed to be the second largest "African state" (after Nigeria) because of the huge Afro-Brazilian population. Thus, the forced migration of Africans as slaves had an enormous cultural influence on many areas of the world.

So, too, have contemporary movements of Africans influenced the cultures of many world regions. Perhaps one of the most celebrated persons of African ancestry today is former U.S. President Barack Obama, the son of a Kenyan man and a woman from Kansas. Obama's heritage and upbringing embody the forces of globalization. In Kenya, he is hailed as part of the modern African diaspora—young professionals (and their offspring) who leave the continent and make their mark somewhere else. In television, South African comedian Trevor Noah was selected to take over Jon Stewart's *Daily Show* in 2015 when Stewart retired. This comedy "news" show's brand of political and social commentary will inevitably become more international with a

South African at the helm. And Academy Award–winning actress Lupita Nyong'o, raised in Kenya, is also a global style icon (see *Everyday Globalization: Africa's Fashion Industry*).

Popular culture in Africa, like everywhere else in the world, is a dynamic mixture of global and local influences. Kwaito, a popular musical form in South Africa, sounds a lot like rap from the United States. A closer listen, however, reveals an incorporation of local rhythms, lyrics in Zulu and Xhosa, and themes about life in the post-apartheid townships. Zola, a kwaito superstar, also costarred in the Oscar-winning film *Tsotsi,* about gang life and hardship in Johannesburg. In films, music, and sports, this region continues to influence global culture.

Nollywood   Africa's undisputed film capital is Lagos, Nigeria. Called Nollywood, the Nigerian movie industry makes more films than Hollywood, currently grinding out as many as 2500 films a year (50 films a week) and employing more than 1 million people. Relying on relatively inexpensive digital video technology, most of these movies are shot in a few days and with budgets of $10,000 to $20,000. The typical themes of religion, ethnicity, corruption, witchcraft, the spirit world, violence, and injustice resonate with African audiences. The films are almost always shot on location—in city streets, office buildings, and homes or in the countryside. Nollywood films can be bloody and exploitive; they can also be overtly evangelical, promoting Christianity over indigenous faiths. Many are acted in English, but there are also movies for Yoruba, Igbo, and Hausa speakers (**Figure 6.31**).

Rather than theatrical release, most Nollywood movies go directly to DVD and are rarely viewed beyond Africa. The shelf life of these movies is rather short; production companies need to make their money back quickly, before pirated copies undermine profits. However, a $20,000 film

▼ **Figure 6.31 Location Filming in Nigeria**   A Nigerian film crew prepares to shoot a scene in Lagos. Nicknamed Nollywood, Nigeria produces more films each year than Hollywood.

Africa's Fashion Industry

West Africa has long been a major center of textile production, but by the early 2000s, runway models were strutting in West African designs for a growing market. While European designers have taken inspiration from African fabrics, the region has not had major fashion houses of their own. But this is changing. Beginning in Dakar, Senegal, major urban centers such as Lagos, Nairobi, and Johannesburg now host Fashion Weeks that are equal to the glitz and serious business associated with Paris and Milan.

Known for daring use of color, geometric designs, and comfortable cottons, African fabrics are easily recognized. With investment in new designers and Fashion Week formats, the exuberant fashion industry is catching on, especially in West Africa. Celebrities such as Beyonce, Lupita Nyong'o, and former First Lady Michelle Obama have worn designs by Nigerian Maki Oh. Adama Paris, a Senegalese designer, sells her line to the region's largest musical artists and to shops in France. Paris notes that African fashion is also about changing notions of beauty, and taking pride in one's ethnic community through design (**Figure 6.3.1**).

The region's apparel and footwear business is valued at over US$30 billion, and a handful of big-name designers are beginning to enter European and North American markets. Move over, Raf Simons—Adebayo Oke-Lawal, the Lagos-based creative director of the menswear line Orange Culture, wants to dress you.

▲ **Figure 6.3.1 African Fashion** A model presents a creation by Ivorian designer Zak Kone during Dakar Fashion Week in the Senegalese capital. West Africa has a long textile tradition, but the area's designers have only recently begun to influence global fashion trends.

1. What factors contribute to the burgeoning fashion industry in the region?

2. Do you see African fashion influences when you shop for clothing?

---

can earn $500,000 in DVD sales in just a couple of weeks. Consequently, film distribution remains tightly controlled by Igbo businessmen, sometimes called the Alaba cartel for their distribution center on the outskirts of Lagos. But this is changing as middle-class Africans want to go to theaters and directors want to create higher-quality films for Nigeria and beyond.

Music in West Africa Nigeria is the music center of West Africa, with a well-developed and cosmopolitan recording industry. Modern Nigerian styles such as juju, highlife, and Afro-beat are influenced by jazz, rock, reggae, and gospel, but they are driven by an easily recognizable African sound. Yet one of the continent's biggest stars is Youssou N'Dour, a singer and percussionist from Senegal, who performs to packed venues around the world. He is also a politician, having been appointed Senegal's Minister of Tourism and Culture in 2012.

Farther up the Niger River lies Mali; Bamako, the capital, is also a music center that has produced scores of recording artists. Many Malian musicians descend from a traditional caste of musical storytellers performing on either the traditional kora (a cross between a harp and a lute) or the guitar. The musical style is strikingly similar to that of the blues from the Mississippi Delta—so much so that the late Ali Farka Touré, from northern Mali, was known as the Bluesman of Africa because of his distinctive, yet familiar, guitar work (**Figure 6.32**).

Explore the **Sounds** of Kora Music

http://goo.gl/i0Er50

Contemporary African music can be both commercially and politically important. Nigerian singer Fela Kuti became a voice of political conscience for Nigerians struggling for true democracy. Born in an elite family and educated in England, Kuti borrowed from jazz, traditional, and popular music to produce the Afro-beat sound in the 1970s. The music was irresistible, but his searing lyrics also attracted attention. Acutely critical of the military government, he sang of police harassment, the inequities of the international economic order, and even Lagos' infamous traffic. Singing in English and Yoruba, his message was transmitted to a larger audience, and he became a target

▲ **Figure 6.32** **Ali Farka Touré** Known as the Bluesman of Africa, Touré was an extremely popular and influential musician. Here he performs at the famous Desert Festival near Timbouctou, Mali. His musical style and guitar abilities earned worldwide praise.

of state harassment. Kuti died in Lagos in 1997 from complications related to AIDS; yet his music and politics later became the subject of the awarding-winning Broadway musical *Fela!*

Pride in East African Runners Ethiopia and Kenya have produced many of the world's greatest distance runners. Marathoner Abebe Bikila won Ethiopia's and Africa's first Olympic gold medal, running barefoot at the Rome games in 1960. Since then, nearly every Olympic Games has yielded medals for Ethiopia and Kenya. At the 2016 Rio Olympics, Kenyan runners won twelve medals and Ethiopians, eight.

Running is a national pastime in both countries, where elevation—Addis Ababa sits at 7300 feet (2200 meters) and Nairobi at 5300 feet (1600 meters)—increases oxygen-carrying capacity. Past medalists Haile Gebrselassie and Derartu Tulu are national celebrities in Ethiopia, where they are idealized by the country's youth. Tulu, the first black African woman to win a gold medal in distance running, is a forceful voice for women's rights in a country where women are discouraged from putting on running shorts. In the 2016 Olympics, Ethiopian Almaz Ayana won gold in the women's 10,000 meter race, setting a new world record (**Figure 6.33**).

▼ **Figure 6.33** **Ethiopian Running Star Almaz Ayana** Kneeling beside her world record time and draped in the Ethiopian flag, Almaz Ayana was one of several East African medalists at the 2016 Olympic Games in Rio de Janeiro.

REVIEW

6.6 What are the dominant religions of Sub-Saharan Africa, and how have they diffused throughout the region?

6.7 Describe the ways in which African peoples have influenced world regions beyond Africa.

■ **KEY TERMS** tribes

Geopolitical Framework: Legacies of Colonialism and Conflict

The duration of human settlement in Sub-Saharan Africa is unmatched by that of any other region. Evidence shows that humankind originated here, evidently evolving from a rather apelike *Australopithicus* all the way to modern *Homo sapiens*. Over millennia, many diverse ethnic groups formed in the region. Although conflicts among these groups have occurred, cooperation and coexistence among different peoples have also continued over centuries.

Some 2000 years ago, the Kingdom of Axum arose in northern Ethiopia and Eritrea, strongly influenced by political models derived from Egypt and Arabia. The first wholly indigenous African states were founded in the Sahel around 700 CE. Kingdoms such as Ghana, Mali, Songhai, and Kanem-Bornu grew rich by exporting gold to the Mediterranean and importing salt from the Sahara, and they maintained power over lands to the south by monopolizing horse breeding and mastering cavalry warfare (**Figure 6.34**).

Over the next several centuries, a variety of other states emerged in West Africa. Some were large, but diffuse empires, organized through elaborate hierarchies of local kings and chiefs; others were centralized states focused on small centers of power. The Yoruba of southwestern Nigeria, for example, developed a city-state form of government, and their homeland is still one of the most urbanized and densely populated parts of Africa. The most powerful Sub-Saharan states continued to be located in the Sahel until the 1600s, when European coastal trade undercut the lucrative trans-Saharan networks. The focus of power subsequently moved to the Gulf of Guinea, where well-organized African states (such as Dahomey and Ashanti) as well as the Europeans took advantage of lucrative opportunities presented by the slave trade and in the process increased their military and economic power (see Figure 6.34).

Thus, prior to European colonization, Sub-Saharan Africa presented a complex mosaic of kingdoms, states, and tribal societies. However, the arrival of Europeans forever changed patterns of social organizations and ethnic relations. As the European powers rushed to carve up the continent to serve their imperial ambitions, they set up various administrations that heightened ethnic tensions and promoted hostility. Many of the region's modern conflicts can

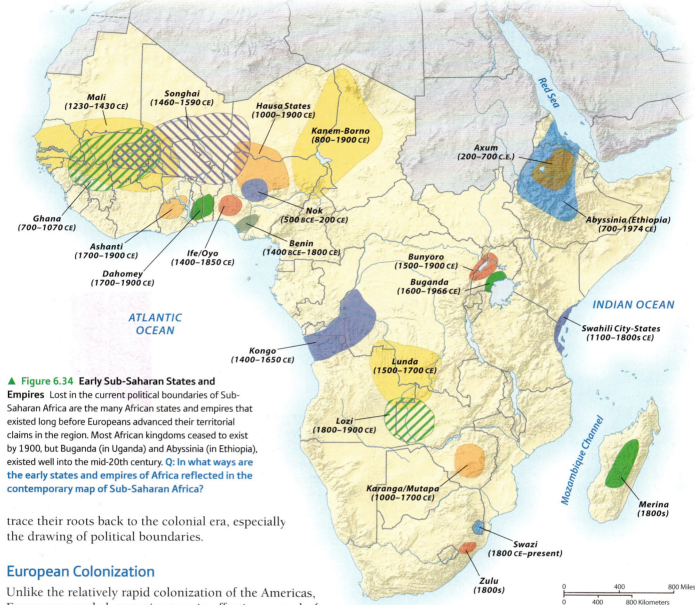

▲ **Figure 6.34 Early Sub-Saharan States and Empires** Lost in the current political boundaries of Sub-Saharan Africa are the many African states and empires that existed long before Europeans advanced their territorial claims in the region. Most African kingdoms ceased to exist by 1900, but Buganda (in Uganda) and Abyssinia (in Ethiopia), existed well into the mid-20th century. **Q: In what ways are the early states and empires of Africa reflected in the contemporary map of Sub-Saharan Africa?**

trace their roots back to the colonial era, especially the drawing of political boundaries.

European Colonization

Unlike the relatively rapid colonization of the Americas, Europeans needed centuries to gain effective control of Sub-Saharan Africa. Portuguese traders landed along the coast of West Africa in the 1400s, and by the 1500s were well established in East Africa as well. Initially, the Portuguese made large profits, converted a few local rulers to Christianity, established several fortified trading posts, and acquired dominion over the Swahili trading cities of the east. They stretched themselves too thin, however, and failed in many of their colonizing activities. Only along the coasts of modern Angola and Mozambique, where a sizable population of mixed African and Portuguese descent emerged, did Portugal maintain power. Along the Swahili, or eastern coast, the Portuguese were eventually expelled by Arabs from Oman, who then established their own mercantile empire in the area.

Another reason for the Portuguese failure was Africa's disease environment. With no resistance to malaria and other tropical diseases, roughly half of all Europeans who

remained on the African mainland died within a year. Protected both by their formidable armies and by the diseases of their native lands, African states were able to maintain an upper hand over European traders and adventurers well into the 1800s. Unlike the Americas, where European conquest was facilitated by the introduction of Old World diseases that devastated native populations (see Chapters 4 and 5), Sub-Saharan Africa's native diseases limited European settlement until the mid-19th century.

Also in the early 1800s, two small territories were established in West Africa so that freed and runaway slaves would have a place to return to in Africa. The territory that was to become Liberia was set up by the American Colonization Society in 1822 to settle former African American slaves. By 1847, it was the independent and free state of Liberia. Sierra Leone served a similar function for ex-slaves from the

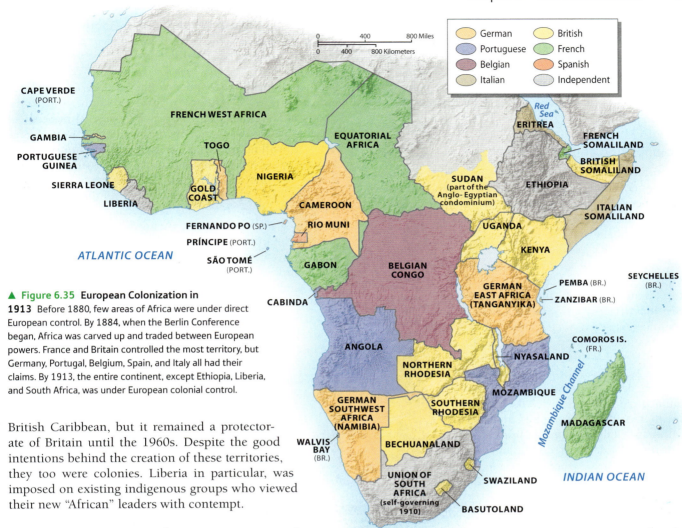

Legend:
- German
- Portuguese
- Belgian
- Italian
- British
- French
- Spanish
- Independent

CAPE VERDE (PORT.)

FRENCH WEST AFRICA

GAMBIA

PORTUGUESE GUINEA

SIERRA LEONE

LIBERIA

TOGO

GOLD COAST

NIGERIA

EQUATORIAL AFRICA

FERNANDO PO (SP.)

PRÍNCIPE (PORT.)

ATLANTIC OCEAN

SÃO TOMÉ (PORT.)

CAMEROON

RIO MUNI

GABON

CABINDA

BELGIAN CONGO

ANGOLA

NORTHERN RHODESIA

GERMAN SOUTHWEST AFRICA (NAMIBIA)

WALVIS BAY (BR.)

BECHUANALAND

SOUTHERN RHODESIA

UNION OF SOUTH AFRICA (self-governing 1910)

BASUTOLAND

SWAZILAND

MOZAMBIQUE

NYASALAND

Red Sea

ERITREA

FRENCH SOMALILAND

BRITISH SOMALILAND

SUDAN (part of the Anglo-Egyptian condominium)

ETHIOPIA

ITALIAN SOMALILAND

UGANDA

KENYA

GERMAN EAST AFRICA (TANGANYIKA)

PEMBA (BR.)

ZANZIBAR (BR.)

SEYCHELLES (BR.)

COMOROS IS. (FR.)

Mozambique Channel

MADAGASCAR

INDIAN OCEAN

▲ **Figure 6.35 European Colonization in 1913** Before 1880, few areas of Africa were under direct European control. By 1884, when the Berlin Conference began, Africa was carved up and traded between European powers. France and Britain controlled the most territory, but Germany, Portugal, Belgium, Spain, and Italy all had their claims. By 1913, the entire continent, except Ethiopia, Liberia, and South Africa, was under European colonial control.

British Caribbean, but it remained a protectorate of Britain until the 1960s. Despite the good intentions behind the creation of these territories, they too were colonies. Liberia in particular, was imposed on existing indigenous groups who viewed their new "African" leaders with contempt.

The Scramble for Africa   In the 1880s, European colonization of the region quickly accelerated, leading to the so-called scramble for Africa. By this time, due to the invention of the machine gun, no African state could long resist European forces.

As the colonization of Africa intensified, tensions among the colonizing forces of Britain, France, Belgium, Germany, Italy, Portugal, and Spain mounted. Rather than risk war, 13 countries convened in Berlin at the invitation of German Chancellor Bismarck in 1884 in a gathering known as the **Berlin Conference**. During the conference, which no African leaders attended, rules were established as to what constituted "effective control" of a territory, and Sub-Saharan Africa was carved up and traded like properties in a game of Monopoly (**Figure 6.35**). Exact boundaries in the interior, which was still poorly known, were not determined, and a decade of "orderly" competition followed as imperial armies marched inland.

Although European weapons in the 1880s were far superior to anything found in Africa, several indigenous states did mount effective resistance campaigns. For example, in South Africa, Zulu warriors resisted British invasion into their lands in what has been termed the Anglo-Zulu Wars (1879–1896). Eventually, European forces prevailed

everywhere except Ethiopia. The Italians had conquered the Red Sea coast and the far northern highlands (modern Eritrea) by 1890, and they quickly set their sights on the large Ethiopian kingdom, called Abyssinia, which had been vigorously expanding for several decades. In 1896, however, Abyssinia defeated the invading Italian army, earning the respect of the European powers. In the 1930s, fascist Italy launched a major invasion of the country, by this time renamed Ethiopia, to redeem its earlier defeat, and with the help of poison gas and aerial bombardment, it quickly prevailed. However, by 1942 Ethiopia had regained its freedom.

Although Germany was a principal instigator of the scramble for Africa, it lost its own colonies after suffering defeat in World War I. Britain and France then partitioned most of Germany's African empire between themselves. Figure 6.35 shows the colonial status of the region in 1913, prior to Germany's territorial loss.

While the Europeans were cementing their rule over Africa, South Africa was inching toward political independence, at least for its white population. One of the oldest colonies in Sub-Saharan Africa, in 1910 South Africa became the first to obtain its political independence from

Europe. However, because of its formalized system of discrimination and racism, it was hardly a symbol of liberty. Ironically, as the Afrikaners tightened their political and social control over the nonwhite population through their policy of apartheid, introduced in 1948, the rest of the continent was preparing for political independence from Europe.

Decolonization and Independence

Decolonization of Sub-Saharan Africa began in 1957. Independence movements, however, had sprung up throughout the continent, some dating back to the early 1900s. Workers' unions and independent newspapers became voices for African discontent and the hope for freedom. Black intellectuals, who had typically studied abroad, were influenced by the ideas of the **Pan-African Movement** led by W. E. B. Du Bois and Marcus Garvey in the United States. Founded in 1900, the movement's slogan of "Africa for Africans" encouraged a trans-Atlantic liberation effort. Nevertheless, Europe's hold on Africa remained secure through the 1940s and early 1950s, even as other colonies in South and Southeast Asia gained their independence.

By the late 1950s, Britain, France, and Belgium decided that they could no longer maintain their African empires and began to withdraw. (Italy had already lost its colonies during World War II, and Britain gained Somalia and Eritrea.) Once begun, the decolonization process moved rapidly and by the mid-1960s virtually the entire region had achieved independence. In most cases, the transition was relatively peaceful and smooth, with the exception of southern Africa.

Dynamic African leaders put their mark on the region during the early independence period. Men such as Kenya's Jomo Kenyatta, Ivory Coast's Felix Houphuët-Boigney, Tanzania's Julius Nyerere, and Ghana's Kwame Nkrumah became powerful father figures who molded their new nations (**Figure 6.36**). President Nkrumah's vision for Africa was the most expansive. After helping to secure independence for Ghana in 1957, his ultimate aspiration was the political unity of Africa. Although his dream was never realized, it set the stage for the founding of the Organization of African Unity (OAU) in 1963, which was renamed the **African Union (AU)** in 2002. The AU is a continent-wide organization headquartered in Addis Ababa, Ethiopia, whose main role has been to mediate disputes between neighbors. Certainly, in the 1970s and 1980s it was a constant

▶ **Figure 6.36 Julius Nyerere Monument** An independence leader from 1961 until he retired from the presidency in 1985, Julius Nyerere is the founding father of Tanzania.

voice of opposition to South Africa's minority rule, and the AU intervened in some of the more violent independence movements in southern Africa.

Southern Africa's Independence Battles Independence did not come easily to southern Africa. In Southern Rhodesia (modern-day Zimbabwe), the problem was the presence of some 250,000 white residents, most of whom owned large farms. Unwilling to see power pass to the country's black majority, then some 6 million strong, these settlers declared themselves the rulers of an independent, white-supremacist state in 1965. The black population continued to resist, however, and in 1978 the Rhodesian government was forced to give up power. The renamed country of Zimbabwe was, henceforth, ruled by the black majority, although the remaining whites still formed an economically privileged community. Since the mid-1990s, disputes over government land reform (splitting up the large commercial farms, mostly owned by whites, and giving the land to black farmers) and President Robert Mugabe's strongman politics have resulted in serious racial and political tensions, as well as the collapse of the country's economy.

In the former Portuguese colonies, independence came violently. Unlike the other imperial powers, Portugal refused to relinquish its colonies in the 1960s. As a result, the people of Angola and Mozambique turned to armed resistance. The most powerful rebel movements adopted a socialist orientation and received support from the Soviet Union and Cuba. A new Portuguese government came to power in 1974, however, and it withdrew abruptly from its African colonies. At this point, Marxist regimes quickly came to power in both Angola and Mozambique. The United States, and especially South Africa, responded to this perceived threat by supplying arms to rebel groups that opposed the new governments. Fighting dragged on for nearly three decades in Angola and Mozambique. The countryside in both states became heavily laced with land mines, a threat that rural residents still face if they unknowingly walk or farm in a mined area. Efforts to clear the mines and make the land usable are ongoing, but uneven. With the end of the Cold War, however, outsiders lost interest in continuing these conflicts, and sustained efforts to negotiate a peace settlement began. Mozambique has been at peace since the mid-1990s. After several failed attempts at peace in Angola, the Angolan army signed a peace treaty with rebels in 2002 that ended a 27-year conflict in which more than 300,000 people died and 3 million Angolans were displaced.

Apartheid's Demise in South Africa While fighting continued in the former Portuguese zone, South Africa underwent a remarkable transformation. From 1948 through the 1980s, the ruling Afrikaners' National Party was firmly committed to apartheid. Only whites enjoyed real political freedom, while blacks were denied even citizenship in their own country—technically, they were citizens of segregated **homelands.** Likened to reservations created for Native Americans in the United States, homelands were established on marginal lands that were rural and overcrowded. Moreover, to uphold the notion that every black South African had a homeland, some 3 million blacks were forcibly relocated into homelands, and residence outside of a homeland was strictly regulated.

Opposition to apartheid began in the 1960s, intensifying and becoming more violent by the 1980s. Blacks led the opposition, but coloureds and Asians (who suffered severe, but less extreme, discrimination) also opposed the Afrikaner government. As international pressure mounted, white South Africans found themselves ostracized. Many corporations refused to do business there, and South African athletes (regardless of color) were banned from most international competitions, such as the Olympics and World Cup Soccer. Increasing numbers of whites also opposed the apartheid system, and many businesspeople began to believe that apartheid threatened to undermine their economic endeavors.

The first major change came in 1990 when South Africa withdrew from Namibia, which it had controlled as a protectorate since the end of World War I. South Africa now stood alone as the single white-dominated state in Africa. A few years later, the leaders of the Afrikaner-dominated National Party decided they could no longer resist the pressure for change. In 1994, free elections were held in which Nelson Mandela (1918–2013), a black leader who had been imprisoned for 27 years by the old regime, emerged as the new president. Black and white leaders pledged to put the past behind them and work together to build a new, multiracial country. The homelands themselves were the first to be eliminated from the political map of the new South Africa. Since Mandela's presidency, orderly elections have been held, and South Africans elected Thabo Mbeki for two terms (1999–2009) and Jacob Zuma in 2009 and again in 2014.

Unfortunately, the legacy of apartheid is not so easily erased. Residential segregation is officially illegal, but neighborhoods are still sharply divided along racial lines. A black middle class emerged under the multiracial political system, but most blacks remain extremely poor (and most whites remain prosperous). Violent crime has increased, and rural migrants and immigrants have poured into South African cities, producing a xenophobic anti-immigrant backlash. Because the political change was not matched by a significant economic transformation, the hopes of many people are frustrated (see *People on the Move: The Chinese in Africa*).

Persistent Conflict

Although most Sub-Saharan countries made a relatively peaceful transition to independence, virtually all of them immediately faced a difficult set of institutional and political problems. In several cases, the old authorities had done virtually nothing to prepare their colonies for independence. Lacking an institutional framework for independent government, countries such as the Democratic Republic of the Congo confronted a chaotic situation from the beginning. Only a handful of Congolese had received higher education, let alone been trained for administrative posts. The indigenous African political framework had been essentially destroyed by colonization, and in most cases very little had been built in its place.

Even more problematic in the long run was the political geography of the newly independent states. Civil servants could always be trained and administrative systems built, but little could be done to rework the region's basic political map. The problem was the fact that the European colonial powers had essentially ignored indigenous cultural and political patterns, both in dividing Africa among themselves and in creating administrative subdivisions within their own imperial territories.

The Tyranny of the Map All over Africa, different ethnic groups found themselves forced into the same state with peoples of different linguistic and religious backgrounds, many of whom had recently been their enemies. At the same time, some of the larger ethnic groups of the region found their territories split between two or more countries. The Hausa people of West Africa, for example, were divided between Niger (formerly French) and Nigeria (formerly British), each of which they had to share with several former rivals.

Given the imposed political boundaries, it is no wonder that many African countries struggled to generate a common sense of national identity or establish stable political institutions. **Tribalism**, or loyalty to the ethnic group rather than to the state, has emerged as the bane of African political life. Especially in rural areas, tribal identities usually supersede national ones. Because virtually all of Africa's countries inherited an inappropriate set of colonial borders, you might assume that they would have been better off redrawing a new political map based on indigenous identities. However, such a strategy was impossible, as all the leaders of the newly independent states realized. Any new territorial divisions would have created winners and losers, and thus would have resulted in more conflict. Moreover, because ethnicity in Sub-Saharan Africa was traditionally fluid, and because many groups were territorially intermixed, it would have been difficult to generate a clear-cut system of division. Finally, most African ethnic groups were considered too small to form viable countries. With these complications in mind, the new African leaders, meeting in 1963 to form the OAU, agreed that colonial boundaries should remain. The violation of this principle, they argued, would lead to

The Chinese in Africa

The rise of China as the largest trading partner of, and investor in, Sub-Saharan Africa has generated much geopolitical discussion, but less talked about are the growing number of Chinese workers and economic migrants in the region. In Kenya, Angola, Tanzania, Mozambique, and South Africa, they are a growing presence. In the last two decades, over 1 million Chinese are estimated to have settled, at least temporarily, in the region. Most agree that the top destination is South Africa, where Chinese migrants number between 200,000 and 400,000.

Life as a Chinese Migrant in Africa On the ground in African cities and villages, the increase in Chinese merchants and small businesses evokes both admiration and resentment, especially in southern Africa (**Figure 6.4.1**). Chinese merchants are an important component of this new migration; a survey by a South African foundation revealed important insights into who these people are and local perceptions about them.

The majority of the Chinese emigrating to Africa in the past decade are from the southern province of Fujian, long an important source for Chinese traders and emigrants. Informal networks from Fujian facilitate this migration, as well as wholesale connections for the goods that are sold. This is not a state-directed flow, but the result of individuals and households seeing economic potential in emigration and small trade.

Generally, these are not the most educated Chinese, and southern Africa was usually not their first choice, but was selected because of its minimal entry requirements. The move also makes economic sense, as shopkeepers can earn perhaps three times more in southern Africa than in their home country. In interviews, Chinese migrants see themselves as working hard, forging opportunities in places neglected by others, creating new supply chains, and outcompeting local merchants. Most do not see themselves staying in their new homes—which is typical for most economic migrants. Yet they also express fears of local resentment, language difficulties, and cultural isolation as they struggle to build new businesses and lives in this region. Only Chinese traders in South Africa expressed any sense of attachment to their country of settlement.

Local Resentments On the other hand, these close-knit and ethnically derived Chinese business networks are seen as exclusionary by Africans. Isolated xenophobic incidents against some merchants have

▲ **Figure 6.4.1 Chinese Merchants in Sub-Saharan Africa** A man browses through pairs of Chinese-made shoes in a shop owned by a Chinese merchant in Kampala, Uganda.

occurred, most recently in the Democratic Republic of the Congo. Although many African consumers enjoy the wider range of goods, they complain about low quality and price fixing. A number of Sub-Saharan governments welcome large-scale Chinese investment, but many farmers, retailers, and petty traders complain about unfair competition from Chinese newcomers. In Malawi, Tanzania, Uganda, and Zambia, new rules have restricted the industries or sectors in which Chinese migrants can operate.

1. What factors contributed to the growth of Chinese migration to this region in the past two decades?

2. Thinking globally, what are the other major destinations of Chinese immigrants today? How do you think Sub-Saharan Africa compares?

---

pointless wars between states and endless civil struggles within them.

Despite the determination of Africa's leaders to build their new nations within existing boundaries, challenges to the states began soon after independence. **Figure 6.37** shows the ethnic and political conflicts that have disabled parts of Africa since 2005. The human cost of this turmoil is several million refugees and internally displaced persons. **Refugees** are people who flee their state because of a well-founded fear of persecution based on race, ethnicity, religion, or political orientation. Nearly 5 million Africans were considered refugees at the end of 2015. Added to this figure are nearly 11 million **internally displaced persons (IDPs)**. IDPs have fled from conflict, but still reside in their country of origin. Nigeria has the largest number of IDPs (2.2 million) followed by South Sudan (1.8 million), Democratic Republic of the Congo (1.5 million), and Somalia (1.1 million). These

populations are not technically considered refugees, making it difficult for humanitarian NGOs and the UN to assist them. After a dip in the number of refugees and IDPs in 2011, the numbers are up again due to conflict exacerbated by drought.

Of course, many extremely poor African states are also the top recipients of refugees. This includes Ethiopia, Kenya, Uganda, the Democratic Republic of the Congo, and Chad. The challenges of accommodating such large numbers of refugees are real, even with the support of the United Nations and other NGOs. Consequently, many countries have been reluctant to accept refugees, resulting in larger numbers of IDPs. In the context of long-standing conflicts, refugees are often warehoused for years in camps. In 2016 Kenya announced the closing of Dabaab Refugee Camp, home to 300,000 Somalis, based on security concerns. The camp, first opened in 1991, has served many people fleeing conflict in the Horn of Africa. At its peak it held over

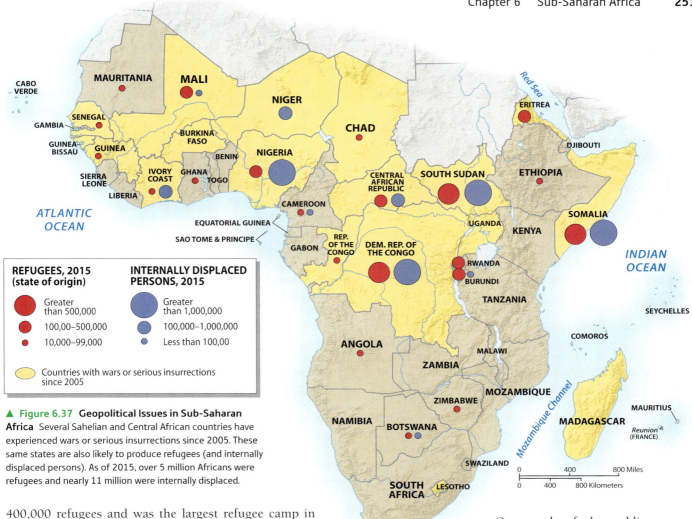

REFERENCES, 2015
(state of origin)

- Greater than 500,000
- 100,00–500,000
- 10,000–99,000

INTERNALLY DISPLACED PERSONS, 2015

- Greater than 1,000,000
- 100,000–1,000,000
- Less than 100,00

Countries with wars or serious insurrections since 2005

▲ **Figure 6.37   Geopolitical Issues in Sub-Saharan Africa**   Several Sahelian and Central African countries have experienced wars or serious insurrections since 2005. These same states are also likely to produce refugees (and internally displaced persons). As of 2015, over 5 million Africans were refugees and nearly 11 million were internally displaced.

400,000 refugees and was the largest refugee camp in the world (**Figure 6.38**).

Ethnic Conflicts   In the 1990s, nearly two-thirds of the states in the region were experiencing serious ethnic conflict and, in the case of the Rwanda, even genocide. As Figure 6.37 shows, most of the conflict since 2005 occurred in the Sahelian states and Central Africa. Fortunately, in the past few years peace has returned to Sierra Leone, Liberia, Ivory Coast, and Angola, states that produced large numbers of refugees in the 1990s.

Many attributed the cycles of violence in Sierra Leone and Liberia to the availability of diamonds as a means of financing the conflict. Although the relationship between resources and conflict is complex, the term **conflict diamonds** was employed when discussing the diamond trade in West Africa in the 1990s.

One result of the public concern about conflict diamonds was a certification scheme adopted in 2002, called the Kimberly Process. Its aim is to keep conflict diamonds out of the global market and thus avoid tainting the image of the diamond business. In Ivory Coast, where conflict began as violence spilled over from Liberia in 2002, a peace deal was brokered in 2007 between the New Forces rebel group in the north and the government-controlled south,

▶ **Figure 6.38   Kenyan Refugee Camp**   This is just one camp (called Ifo) in the larger complex of Dabaab in northern Kenya. Dabaab has housed refugees from the Horn of Africa for 25 years. In 2016, the Kenyan government announced closure of the camp for security reasons.

although there are still many Ivorian IDPs and refugees, as shown in Figure 6.37.

The deadliest ethnic and political conflict in the region has been in the Democratic Republic of the Congo. It is estimated that 5.4 million people died there between 1998 and 2010, although many of the deaths were from war-induced starvation and disease rather than bullets or machetes. In addition, half a million refugees are living outside the country and there are 1.5 million IDPs within it (Figure 6.37). In 1996 and 1997, a loose alliance of armed groups from Rwanda (led by Tutsis) and Uganda joined forces with other militias in the DRC and marched their way across the country, installing Laurent Kabila as president. Under Kabila's rocky and ruthless leadership, which ended in assassination in 2001, rebel groups from Uganda and Rwanda again invaded and soon controlled the northern and eastern portions of the country, while the Kinshasa-based government loosely controlled the western and southern portions.

With Kabila's death, his son Joseph took power and signed a peace accord with the rebels in 2002. In 2003, rebel leaders were made part of a transitional government, and an unsteady peace was in place, with help from the UN, the AU, and Western donors. Remarkably, when elections were held in 2006, Joseph Kabila was elected president, and he was reelected in 2011. Yet Sub-Saharan Africa's largest state in terms of territory has only limited experience with democracy. Its civil service barely functions, corruption is rampant, and there are few roads and little working infrastructure for the 79 million people who live there. Moreover, armed groups and the DRC army scattered throughout the country continue to commit serious crimes, including mass rapes, torture, and murders (**Figure 6.39**). Due to years of conflict, the formal economy is small, and the informal economy dominates. And yet elections are scheduled for the end of 2016 to select a new president.

▼ **Figure 6.39 Protest in the DRC** People hold flags and banners for the Reformist Democratic Movement party at a Kinshasa rally in May 2016. With growing political dissatisfaction, opposition parties pushed for elections in 2016. President Kabila proposed elections in 2018.

Another area experiencing ethnic-religious conflict is northeast Nigeria near the Lake Chad basin, where over 2 million Nigerians were displaced at the end of 2015. Here the clash is with Boko Haram, a violent Muslim sect that has terrorized the region's different ethnic groups. The conflict has spilled over into Niger, Chad, and Cameroon—all bordering Lake Chad.

Secessionist Movements Problematic African political boundaries have occasionally led to attempts by territories to secede and form new states. The Shaba (or Katanga) Province in what was then the state of Zaire (now the DRC) tried to leave its national territory soon after independence. The rebellion was crushed a couple of years after it started with the help of France and Belgium. Similarly, the Igbo in oil-rich southeastern Nigeria declared an independent state of Biafra in 1967. After a short but brutal war, during which Biafra was essentially starved into submission, Nigeria was reunited.

In 1991 the government of Somalia disintegrated, and the territory has been in civil war for nearly three decades. The lack of political control facilitated the rise of piracy; Somali pirates raid vessels and extort ransom from ships from the Gulf of Aden to the Indian Ocean. The territory has been ruled by warlords and their militias, who have informally divided the country into clan-based units. **Clans** are social units that are branches of a tribe or an ethnic group larger than a family. Early in the conflict, the northern portion of the country declared its independence as a new country—Somaliland. Somaliland has a constitution, a functioning parliament, government ministries, a police force, a judiciary, and a president. The territory produces its own currency and passports. Yet no country has recognized this territory, in part because no government exists in Somalia to negotiate the secession. In 1998, neighboring Puntland also declared itself a semi-autonomous state. Although it does not seek outright independence like Somaliland, Puntland has created its own administration. In 2012, the Somalian government was finally able to reform, with the help of the African Union along with the Ethiopian military in Mogadishu. Meanwhile, well-armed Islamic insurgents (led by Al-Shabab) are present in the south. The political instability has been exacerbated by several years of drought, creating a humanitarian emergency (**Figure 6.40**).

Only two territories in the region have successfully seceded. In 1993, Eritrea gained independence from Ethiopia after two decades of civil conflict. This territorial secession is striking because Ethiopia gave up its access to the Red Sea, making it landlocked. Yet the creation of Eritrea still did not bring about peace. After years of fighting, the transition to Eritrean independence began remarkably well. Unfortunately, border disputes between

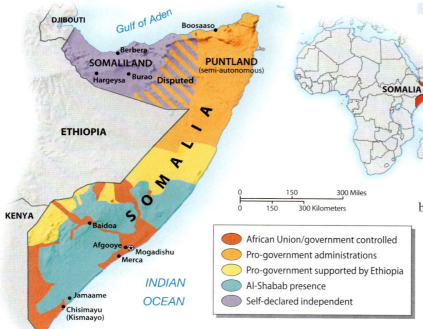

▲ **Figure 6.40 Divided Somalia** For three decades conflict has raged in Somalia, challenging every aspect of governance. The territory is fractured with one area seeking independence, another seeking semi-autonomy, and pro-government and pro-Al Shabab groups, along with the Ethiopian military, vying for control.

the two countries erupted in 1998, resulting in the deaths of some 100,000 troops. In 2000, a peace accord was reached and the fighting stopped, but in 2016 border clashes erupted again.

The second example is South Sudan, which gained its independence from Sudan in 2011 after some three decades of violent conflict between the largely Arab and Muslim north and the Christian and animist south. But peace has not come to this new territorial state either. A power struggle between rival factions of Dinka and Nuer began in 2013 and has led to some 2 million displaced people, 800,000 refugees, and thousands dead. In 2015, after two years of civil war, President Salva Kiir signed a peace accord, yet this ethnically driven conflict continues as of 2016. The difficulties experienced by the newly created states of South Sudan and Eritrea suggest that major changes to Africa's political map should not be expected.

REVIEW

6.8 What are the processes behind Sub-Saharan Africa's political map, and why have there been relatively few boundary changes since the 1960s?

6.9 What are the present major conflicts in this region, and where are they occurring?

■ **KEY TERMS** Berlin Conference, Pan-African Movement, African Union (AU), homelands, tribalism, refugees, internally displaced persons (IDPs), conflict diamonds, clans

Economic and Social Development: The Struggle To Develop

By almost any measure, Sub-Saharan Africa is the poorest world region. According to World Bank estimates, 43 percent of the population lives in extreme poverty, surviving on under $1.90 per day. Poverty and low life expectancy puts most of the states in the region at the bottom of the Human Development Index. Some demographically small or resource-rich states, such as Botswana, Equatorial Guinea, Mauritius, the Seychelles, and South Africa, have much higher per capita gross national income, adjusted for purchasing power parity (GNI-PPP); the average for the region was about $3400 in 2014. By way of comparison, the figure for South Asia, the next poorest region, was $5300 (See Table 6.2).

Since 2000 there have been signs of economic growth. The average annual growth rate for the region from 2000 to 2009 was an impressive 5.4 percent, although from 2009 to 2014 the growth rate slipped to 4.3 percent. Even with declines in commodity prices, several countries have average annual growth rates of 5 percent or more; Ethiopia's annual growth rate averaged a stunning 10.5 percent from 2009 to 2014. Over the past 10 years, real income per person grew (20 to 30 percent), whereas in the previous 20 years it actually decreased. The most optimistic views of the region see strengthened democracies, greater civic engagement, less violence, and growing investment both from within and outside the region. American economist Jeffrey Sachs argues that in order for the region to get out of the poverty trap, it will need substantial sums of new foreign aid and investment.

Roots of African Poverty

In the past, observers often attributed Africa's poverty to its colonial history, poorly conceived development policies, and/or corrupt governance. Those who favored environmental explanations pointed to the region's infertile soils, erratic patterns of rainfall, lack of navigable rivers, and virulent tropical diseases as reasons for underdevelopment. The best explanations for the region's poverty tend to focus on historical and institutional factors rather than environmental circumstances.

Numerous scholars have singled out the slave trade for its debilitating effect on Sub-Saharan African economic life. Large areas of the region were depopulated, and many people were forced to flee into poor, inaccessible refuges. Colonization was another blow to Africa's economy. European powers invested little in infrastructure, education, and public health and were instead interested

TABLE 6.2	Sub-Saharan Africa Development Indicators

MasteringGeography™

Country	GNI per capita, PPP 2014	GDP Average Annual % Growth 2009–2014	Human Development Index (2015)[1]	Percent Population Living Below $3.10 a Day	Life Expectancy (2016)[2] Male	Life Expectancy (2016)[2] Female	Under Age 5 Mortality Rate (1990)	Under Age 5 Mortality Rate (2015)	Youth Literacy (% pop ages 15–24) (2005–2014)	Gender Inequality Index (2015)[3,1]
WESTERN AFRICA										
Benin	2,020	4.7	0.480	75.6	58	61	177	100	–	0.614
Burkina Faso	1,600	5.7	0.402	80.5	57	60	208	89	–	0.631
Cape Verde	6,200	2	0.646	–	71	80	58	25	98	–
Gambia	1,580	2.6	0.441	–	59	62	165	69	71	0.622
Ghana	3,900	8.9	0.579	49.0	60	63	121	62	86	0.554
Guinea	1,130	2.8	0.411	68.7	58	59	228	94	31	–
Guinea–Bissau	1,380	2.5	0.420	83.6	54	57	210	93	75	–
Ivory Coast	3,130	5.2	0.462	–	51	53	151	93	48	0.679
Liberia	700	6.9	0.430	89.6	60	62	241	70	–	0.651
Mali	1,510	2.8	0.419	77.7	54	54	257	115	47	0.677
Mauritania	3,710	5.4	0.506	32.5	62	65	125	85	–	0.610
Niger	910	6.7	0.348	81.8	61	62	314	96	24	–
Nigeria	5,710	5.5	0.514	76.5	53	53	214	109	66	–
Senegal	2,300	3.6	0.466	66.3	65	68	136	47	56	0.528
Sierra Leone	1,770	11	0.413	80.0	50	52	267	120	64	0.650
Togo	1,290	5.2	0.484	74.5	59	61	147	78	80	0.588
EASTERN AFRICA										
Burundi	770	4.2	0.400	92.2	58	61	183	82	89	0.492
Comoros	1,430	2.8	0.503	11.1	62	65	122	74	87	–
Djibouti	–	4.8	0.470	37.0	60	64	122	65	–	–
Eritrea	–	–	0.391	–	62	66	138	47	92	–
Ethiopia	1,500	10.5	0.442	71.3	62	66	198	59	–	0.558
Kenya	2,940	5.8	0.548	58.9	60	65	98	49	82	0.552
Madagascar	1,400	2.1	0.510	92.9	64	67	161	50	65	–
Mauritius	18,150	3.5	0.777	3.0	71	78	24	14	98	0.419
Reunion	–	–	–	–	77	84	–	–	–	–
Rwanda	1,630	7.2	0.483	80.7	62	66	156	42	82	0.400
Seychelles	24,810	6.2	0.772	< 2.0	70	79	17	14	99	–
Somalia	–	–	–	–	54	57	180	137	–	–
South Sudan	1,800	–	0.467	–	55	57	217	93	37	–
Tanzania	2,510	6.7	0.521	76.1	64	66	158	49	86	0.547
Uganda	1,720	5.5	0.483	63.0	62	64	178	55	84	0.538
CENTRAL AFRICA										
Cameroon	2,950	4.7	0.512	54.3	56	59	145	88	81	0.587
Central African Republic	600	–7.6	0.350	82.3	49	53	169	130	36	0.655
Chad	2,070	6.5	0.392	64.8	51	53	208	139	50	0.706
Congo	5,200	4.7	0.591	–	57	60	119	45	81	0.593
Dem. Rep. of Congo	650	7.7	0.433	–	49	52	181	98	84	0.673
Equatorial Guinea	17,660	–0.3	0.587	–	56	59	190	94	98	–
Gabon	17,200	6.2	0.684	7.2	64	65	94	51	89	0.514
São Tomé and Principe	3,140	4.5	0.555	–	64	68	96	47	80	–

TABLE 6.2   Sub-Saharan Africa Development Indicators (*continued*)   MasteringGeography™

Country	GNI per capita, PPP 2014	GDP Average Annual % Growth 2009–2014	Human Development Index (2015)[1]	Percent Population Living Below $3.10 a Day	Life Expectancy (2016)[2] Male	Female	Under Age 5 Mortality Rate (1990)	Under Age 5 Mortality Rate (2015)	Youth Literacy (% pop ages 15–24) (2005–2014)	Gender Inequality Index (2015)[3,1]
SOUTHERN AFRICA										
Angola	–	–	0.532	22.5	51	54	243	157	73	–
Botswana	16,030	6.6	0.698	35.7	62	67	53	44	98	0.480
Lesotho	3,150	4.9	0.497	77.3	50	50	88	90	83	0.541
Malawi	790	4.4	0.445	87.6	62	64	227	64	72	0.611
Mozambique	1,120	7.1	0.416	87.5	52	56	226	79	67	0.591
Namibia	9,810	5.5	0.628	45.7	62	67	73	45	–	0.401
South Africa	12,700	2.5	0.666	34.7	60	64	62	41	99	0.407
Swaziland	7,880	2.3	0.531	63.1	50	48	83	61	94	0.557
Zambia	3,690	7	0.586	78.9	51	56	193	64	–	0.587

Source: World Bank, *World Development Indicators,* 2016.

[1]United Nations, *Human Development Report,* 2015.

[2]Population Reference Bureau, *World Population Data Sheet,* 2016.

[3]Gender Inequality Index—A composite measure reflecting inequality in achievements between women and men in three dimensions: reproductive health, empowerment, and the labor market that ranges between 0 and 1. The higher the number, the greater the inequality.

Login to MasteringGeography™ **& access** MapMaster **to explore these data!**

1) Which countries rank the highest with regard to the Human Development Index? What factors account for these relatively high rankings?

2) Look at the literacy rankings for the various subregions of Sub-Saharan Africa. Which subregion has the lowest youth literacy and what might explain it? What could be done to improve literacy in these areas?

mainly in developing mineral and agricultural resources for their own benefit. Several plantation and mining zones did achieve some prosperity under colonial regimes, but strong national economies failed to develop. In almost all cases, the basic transport and communications systems were designed to link administration centers and zones of extraction directly to the colonial powers, rather than to their own surrounding areas. As a result, after achieving independence, Sub-Saharan African countries faced economic and infrastructural challenges that were as daunting as their political problems.

Failed Development Policies The first decade or so of independence was a time of relative prosperity and optimism for many countries of the region. Most of them relied heavily on the export of mineral and agricultural products, and through the 1970s commodity prices generally remained high. Some foreign capital was attracted to the region, and in many cases the European economic presence actually increased after decolonization.

In the 1980s, as most commodity prices began to decline, foreign debt began to weigh down many Sub-Saharan countries. By the 1990s, most states were registering low or negative growth rates. Not only was the AIDS crisis raging, but also an economic and debt crisis in the 1980s and 1990s prompted the introduction of **structural adjustment programs** by the International Monetary Fund

(IMF) and the World Bank. These programs typically reduce government spending, cut food subsidies, and encourage private-sector initiatives. Yet these same policies caused immediate hardships for the poor, especially women and children, and led to social protest, most notably in cities. Although the region's debt was low compared to that of other developing regions (such as Latin America), as a percentage of economic output, Sub-Saharan Africa's debt was the highest in the world.

Many economists argue that the region's governments enacted counterproductive economic policies and thus brought some of their misery on themselves. Eager to build their own economies and reduce dependency on the former colonial powers, most African countries followed a course of economic nationalism. More specifically, they set about building steel mills and other forms of heavy industry that were simply not competitive. Local currencies were often maintained at artificially elevated levels, which benefited the elite who consumed imported products, but undercut exports.

The largest blunders made by Sub-Saharan leaders in the postindependence period were in agricultural and food policies. The main objective was to maintain a cheap supply of staple foods in urban areas. Yet the majority of Africans are farmers, who could not make money from their crops because prices were artificially low. Thus, they opted to grow food mainly for subsistence, rather than selling at a loss to national marketing boards. The end result was the failure to meet staple food needs at a time when the population was growing explosively. Many of these price subsidies have since been removed as part of structural adjustment policies.

Reflecting a neoliberal turn toward agricultural production, several Sub-Saharan states—Uganda, Tanzania, Somalia, and Mozambique—have experienced a 21st-century land grab driven by food-insecure governments. Dutch geographer Annelies Zoomers describes food-insecure governments, such as China and the Gulf States, as those that seek to outsource their domestic food production by buying or leasing vast areas of farmland abroad for their own offshore food production. Ironically, the places that these countries invest in tend to be poor countries with their own food insecurity issues.

Corruption Although prevalent throughout the world, corruption also seems to have been particularly rampant in several African countries. Part of this is driven by a lack of transparent and representative governance and by a civil service that lacks both resources and professionalism. According to a recent poll of international businesspeople, Nigeria ranks as the world's most corrupt country. (Skeptical observers, however, point out that several Asian nations with highly successful economies, such as China, also are noted for high levels of corruption, so that corruption alone may not be an explanation for Africa's economic problems.)

With millions of dollars in loans and aid pouring into the region, officials at various levels have been tempted to take something for themselves. Some African states, such as the Democratic Republic of the Congo, were dubbed kleptocracies. A **kleptocracy** is a state in which corruption is so institutionalized that politicians and government bureaucrats siphon off a huge percentage of the country's wealth. President Mobutu of the DRC was a legendary kleptocrat. While his country was saddled with an enormous foreign debt, he reportedly skimmed several billion dollars and deposited them in Belgian banks during his presidency from 1965 to 1997.

Signs of Economic Growth

Most of the economies of the region are growing at 4 to 5 percent, even with the recent dip in commodity prices. And there is a small but growing middle class of people who aspire to own homes and cars, and consume more than basic needs. This growth is not just tied to natural resources; for example, mobile telephones have transformed commerce throughout the region. The percentage of people living in extreme poverty has steadily declined over the last 20 years. In addition, more children are in school, and tremendous strides have been made in combating HIV/AIDS and malaria.

One bright spot in the region's economy is the growth in cellular and digital technology. Admittedly, fixed telephone lines are scarce; the regional average is 1 line per 100 people. Cell-phone usage, however, has soared. In 2007, the World Bank estimated 23 cell-phone subscriptions per 100 people in Sub-Saharan Africa; by 2014, the figure tripled to 71 per 100 people. Multinational providers now compete for mobile-phone customers. Development specialists and entrepreneurs are exploring many new uses for cell phones and smartphones, with applications to secure not only micro-finance, but also educational tools and updates about health issues or weather patterns (**Figure 6.41**). In Kenya, one of the most wired countries in the region, 43 percent of the population uses the Internet; the average for the region as a whole is 19 percent. And Kenya's cell phone–based money transfer service, M-Pesa, allows residents to pay for goods and services from street food to bus fare with their phone. A recent report indicated that mobile technologies and services accounted for 6.7 percent of Africa's total Gross Domestic Product (GDP) and supported 3.8 million jobs in 2015.

Sustainable Development Goals It can be argued that domestic and international aid aimed at reducing extreme poverty, as outlined in the **Sustainable Development Goals (SDGs),** has had a positive impact. The SDGs resulted from a UN–led effort to end extreme poverty by focusing on 17 key indicators. The top five are no poverty, zero hunger, good health, quality education, and gender equality, with key benchmarks for 2030.

In 2000 the Millennium Development Goals (predecessors of the SDGs) were announced with a set of targets to be reached by 2015. Most Sub-Saharan African countries did not achieve their specific goals by that date, but the approach of articulating specific goals and targeting foreign aid towards reaching them has gained momentum. This is still the world's poorest region, with serious problems, but its connections with the world are deepening and, in many cases, proving beneficial.

▼ **Figure 6.41 Mobile Phones for Africa** A woman on her cell phone in downtown Monrovia, Liberia. Cell phone subscriptions in Sub-Saharan Africa tripled between 2007 and 2014, greatly improving communication.

A Surge in New Infrastructure Limited paved roads and railroads inhibit national economies, but large-scale infrastructural programs are in the works. South Africa is the only African state with a fully developed modern road network. Recently, Kenya inaugurated its first superhighway, an eight-lane, 42-kilometer road from Nairobi to Thika. The highway not only has transformed many of the towns along its route, but also is an expression of growing Chinese investment in this region, because the Chinese firm Wu Yi Company Ltd. performed much of the engineering and construction work.

Two major projects to improve regional train networks also are under way. In East Africa, billions are being invested to renovate existing railroads, standardize gauges, and construct new lines to link East African cities to the port of Mombasa, Kenya. A new line from Kigali, Rwanda, is now being constructed and will connect to Uganda and then to Kenya. There are also plans to extend the East African rail network into South Sudan, Ethiopia, and the eastern Democratic Republic of the Congo. Much of the engineering and some of the financing for this 3000-mile (4800-km) network comes from China (**Figure 6.42**). In 2015, seven West African states (Ivory Coast, Ghana, Togo, Benin, Nigeria, Burkina Faso, and Niger) announced plans to renovate, build, and integrate 1860 miles (3000 km) of railroads to facilitate the export of minerals and other primary products. For landlocked Burkina Faso and Niger, this project could greatly reduce transportation costs.

Finally, major water and energy projects are under way in the region. Ethiopia is constructing the Renaissance Dam on the Blue Nile. When completed, it will be Africa's largest dam. A project of this scale is not without controversy; the downstream nation of Egypt is deeply concerned that the dam will reduce total water flow into the Nile River. Others complain that the scale of the dam is more than Ethiopia needs—annual energy production is estimated to be 15,000 gigawatt-hours—and that

▼ **Figure 6.42** **East African Railroad Construction** Kenyan workers lay new standard gauge railroad track as part of the Chinese-funded regional railway project that will modernize the rail link between Kenya's port of Mombasa and its capital, Nairobi. Plans to improve railroad links throughout East Africa are under way.

Explore the **Sights** of Mombasa's Port Facilities

http://goo.gl/BzSe41

▲ **Figure 6.43** **Solar Power** Workers assemble panels on a 8.5 megawatt utility-scale solar farm in Rwamagana, Rwanda. Opened in 2015, it is one of the first solar farms in East Africa. Gigawatt Global, a multinational renewable energy corporation, led the project.

the dam's environmental impacts are not fully understood. The region's potential for solar energy is also great. Sub-Saharan Africa's largest photovoltaic solar power project opened near Kimberly, South Africa, in 2014. The Jasper plant can produce 180,000 megawatt-hours of energy a year, enough to power 80,000 homes (**Figure 6.43**). This renewable energy project was developed by Solar Reserve, a California-based company that builds solar power stations for utilities around the world. Many other states in the region are interested in following South Africa's lead in solar power and are exploring other ways to encourage development that takes into account indigenous practices (see *Geographers at Work: Vision for Sustainable Development in West Africa*).

Links to the World Economy

Sub-Saharan Africa's trade connection with the world is limited, accounting for just over 2 percent of global trade. The level of trade is low both within the region and outside it. Traditionally, most exports went to the European Union (EU), especially the former colonial powers of England and France. The United States is the second most common destination. That pattern is changing fast; China is now the region's single largest trading partner, although collective trade between the EU nations and Sub-Saharan Africa is greater. Throughout the decade of the 2000s, China's trade with Sub-Saharan Africa grew on average 30 percent per year. During the same decade, India's and Brazil's levels of trade with Sub-Saharan Africa annually grew more than 20 percent.

The rise of China as the largest trading partner of, and investor in, Sub-Saharan Africa has generated much geopolitical discussion about China's influence in this resource-rich, but still developing, world region. In 2013, China's President Xi Jinping visited Tanzania, South Africa, and the Democratic Republic of the Congo, promoting the mutual benefits of Sino-African relations. And in 2016 China built its first overseas base anywhere in the world—in the tiny nation of Djibouti in East Africa. Located between the critical choke point between the Gulf of Aden and the Red Sea, the base is supposedly a logistic hub for anti-piracy efforts.

Geographers **at Work**

Vision for Sustainable Development in West Africa

Fenda Akiwumi was always interested in the environment and the development of her country, Sierra Leone. Trained as a hydrogeologist in the United Kingdom, Akiwumi worked for over a decade in Sierra Leone's Ministry of Agriculture and Forestry and as a consultant developing and managing water resources. Yet her "eureka moment" came much later when she found a way to bridge the gap between the physical sciences and the social sciences—by studying geography. Says Akiwumi, "You have to understand why geography is so important: no matter what we do, no matter what field we're in, ultimately it all boils down to people . . . you have a great opportunity to help people in whatever it is you're doing."

The one thing Akiwumi loved most was fieldwork: constructing wells, regulating wetlands for rice growing, and training mostly male technicians in basic hydrogeology. Because mining is a vital part of Sierra Leone's economy, she also worked as a consultant at mine sites. During those years, she often wondered what villagers thought about these projects: "Our countries are embracing these investments without regard to people and places."

The blood-diamond war forced Akiwumi to migrate to the United States in 1991. She taught for a while, but wanted a more holistic approach toward development that included the physical environment and cultural geography. Eventually she earned a PhD in geography at Texas State University.

Now an associate professor at the University of South Florida, Akiwumi continues to work and conduct research in West Africa (**Figure 6.5.1**). She says that her geographical training encourages her to examine human–environment dynamics, particularly the interconnections between customary livelihoods and sacred places (rivers, lakes, and springs with great cultural meaning) that local communities use in ceremonies and rituals. Sierra Leone's government recognizes these spaces, but when conflict over mining rights ensues, these indigenous claims are often ignored. "The mining industry must respect cultural heritage and promote community participation for sustainable, conflict-free mining to exist," argues Akiwumi.

▲ **Figure 6.5.1 Discussing Sustainability** Professor Fenda Akiwumi (in yellow and green dress, fourth from the right) speaks with village women in rural Sierra Leone about natural resource practices. Akiwumi now teaches geography at the University of South Florida and conducts research in West Africa.

One challenge facing West Africa is developing practices that take into account indigenous systems, because "colonial policies are still embedded in post-colonial policies and laws," notes Akiwumi. Yet she is optimistic about the resilience of the rural people she works with: "Despite external influences and pressures, they find ways to sustain important aspects of their culture."

1. How does development conflict with cultural practices in West Africa?

2. Provide an example of a development project that affects local practices in your area.

China is not alone; the largest U.S. military base in Africa is in Djibouti, and France has its largest military presence abroad there as well.

Aid Versus Investment In many ways, Sub-Saharan Africa is more tightly linked to the global economy through the flow of financial aid and loans than through the flow of goods. As **Figure 6.44** shows, for several states (Central African Republic, Liberia, Malawi, Sierra Leone, and Somalia) foreign aid accounted for more than 20 percent of GNI in 2014. Most of the aid comes from a handful of developed regions (Europe, North America, and Japan). Other countries, including Botswana, South Africa, Angola, and Nigeria, receive relatively little aid because their economies are larger and they have mineral or oil wealth. In 2014 the total value of foreign assistance was $33 billion, whereas remittances to Sub-Saharan Africa countries from migrants were valued at $34 billion.

Although aid is extremely important for many African states, foreign direct investment in the region substantially increased from only $4.5 billion in 1995 to $46 billion in 2014. Yet the overall level of foreign investment remains low when compared to other developing regions. In 2014, the largest recipients of foreign investment were Nigeria, Mozambique, South Africa, and the Republic of the Congo. China has been the leading investor in the region at a time when the United States and the EU are more focused on offering aid or fighting terrorism. China wants to secure the oil and ore it needs for its massive industrial economy. In exchange, it offers Sub-Saharan nations money for roads, railways, housing, and schools, with relatively few strings attached.

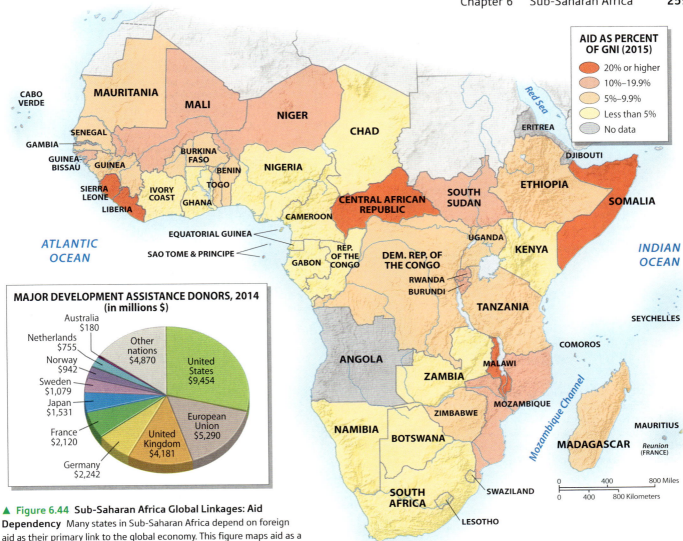

▲ **Figure 6.44** **Sub-Saharan Africa Global Linkages: Aid Dependency** Many states in Sub-Saharan Africa depend on foreign aid as their primary link to the global economy. This figure maps aid as a percentage of GNI, which ranges from less than 1 percent to 44 percent in Liberia. **Q: Compare the most aid-dependent states in the region with the least aid-dependent ones. What are the differences between these two groups of states?**

Some African leaders see China as a new kind of global partner, one that wants straight commercial relations without an ideological or political agenda. Angola, a country in which China has invested heavily, is now one of China's top suppliers of oil.

Economic Differentiation Within Africa

As in most other regions, considerable differences in levels of economic and social development persist in Sub-Saharan Africa. In many respects, the small island nations of Mauritius and the Seychelles have little in common with the mainland. With high per capita GNI, life expectancies averaging in the low 70s, and economies built on tourism, they could more easily fit into the Caribbean were it not for their Indian Ocean location.

Two small African states noted for oil wealth are Gabon (population of 2 million) and Equatorial Guinea (less than

1 million), which began producing oil in the 1990s. In 2014, the GNI-PPP levels of these two states were $17,200 and $17,660, respectively. Yet after two decades of oil production, these revenues have not been invested in the country's citizens; rather, as is often the case, they seem to have fallen into the pockets of a few members of the elite. Only a few states, mostly in southern Africa, have per capita GNI-PPP levels over $5000 (see Table 6.2).

Given the scale of the African continent, it is not surprising that groups of states have formed trade blocs to facilitate intraregional exchange and development (**Figure 6.45**). The two most active regional organizations are the **Southern African Development Community (SADC)** and the **Economic Community of West African States (ECOWAS)**. Both were founded in the 1970s but became more important in the 1990s, and each is anchored by one of the region's two largest economies: South Africa and Nigeria. Other regional trade blocs include the Economic Community of Central African States (ECCAS) and the smaller, but more effective East African Community (EAC).

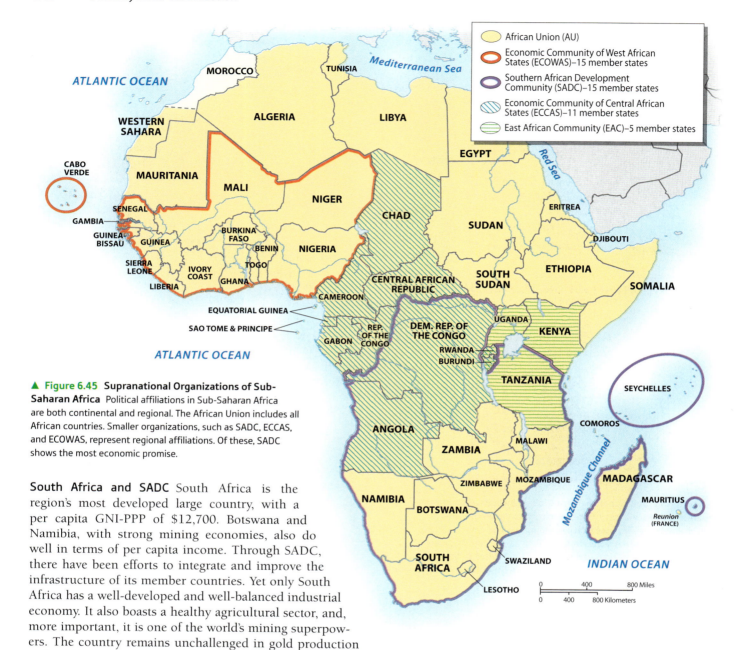

▲ Figure 6.45 Supranational Organizations of Sub-Saharan Africa Political affiliations in Sub-Saharan Africa are both continental and regional. The African Union includes all African countries. Smaller organizations, such as SADC, ECCAS, and ECOWAS, represent regional affiliations. Of these, SADC shows the most economic promise.

South Africa and SADC South Africa is the region's most developed large country, with a per capita GNI-PPP of $12,700. Botswana and Namibia, with strong mining economies, also do well in terms of per capita income. Through SADC, there have been efforts to integrate and improve the infrastructure of its member countries. Yet only South Africa has a well-developed and well-balanced industrial economy. It also boasts a healthy agricultural sector, and, more important, it is one of the world's mining superpowers. The country remains unchallenged in gold production and is a leader in many other minerals and precious gems, including diamonds. In 2010, South Africa hosted the World Cup, the first African country to do so, symbolizing its arrival as a developed and modern nation.

South Africa is undeniably a wealthy country by African standards, but although its white minority is prosperous by any standard, it is also a country beset by severe and widespread poverty. In the townships lying on the outskirts of the major cities and in the rural districts of the former homelands, unemployment is high and living standards marginal. Despite the end of apartheid, South Africa continues to suffer from one of the most unequal distributions of income in the world.

The Leaders of ECOWAS The largest economy and the most populous country in Africa, Nigeria is the core member of the Economic Community of West African States. Nigeria

has the region's largest oil reserves and is a member of the Organization of Petroleum Exporting Countries (OPEC). Yet despite its natural resources, its per capita GNI-PPP is only $5710. It has been argued that oil money has helped to make Nigeria notoriously inefficient and corrupt. A small minority of its population has grown fantastically wealthy, more by manipulating the system than by engaging in productive activities. Three-quarters of Nigerians, however, remain trapped in poverty, earning less than $3.10 per day in 2014. Oil money led to the explosive growth of the former capital city of Lagos, which by the 1980s had become one of the most expensive—and least livable—cities in the region. As a result, the Nigerian government built a new capital city in Abuja, near the country's center, a move that has proved tremendously expensive. In 1991, Abuja became the national capital, and it now has over 2 million residents.

The southwestern corner of the country, in the Niger Delta where most of the oil is produced, has seen few of the benefits of oil production but bears the burden of the environmental costs and related social unrest.

The second and third most populous states in ECOWAS, Ivory Coast and Ghana, are also important West African commercial centers. These states rely on a mix of agricultural and mineral exports. In the mid-1990s, the Ivorian economy began to take off. Boosters within the country called it an emerging "African elephant" (comparing it to the successful "economic tigers" of eastern Asia). But in 2002 a destructive civil war resulted in rebel forces taking control of the northern half of the country, displacing over half a million Ivorians. Today the country is doing much better, and although its GNI-PPP is lower than Nigeria's it does not have as high a poverty rate. Ghana, a former British colony, began its economic recovery in the 1990s. In 2001, it negotiated with the IMF and World Bank for debt relief to reduce its nearly US$6 billion foreign debt. Ghana maintained a high annual growth rate of nearly 9 percent and by 2011, it became an emerging oil producer for Africa, with offshore wells being pumped near the city of Takoradi.

East Africa Long the commercial and communications center of East Africa, Kenya experienced economic decline and political tension throughout the 1990s. From 2009 to 2014, the economy has averaged 5.8 percent annual growth, and per capita GNI-PPP is at $2940. Kenya boasts good infrastructure by African standards, and over 1 million foreign tourists come each year to marvel at its wildlife and natural beauty. Traditional agricultural exports of coffee and tea, as well as nontraditional exports such as cut flowers, dominate the economy.

Kenya is also East Africa's technological leader. In 2009, the government launched plans for Konza City, a walkable technology-oriented development 60 kilometers (37 miles) southeast of Nairobi. The intent is to capture the growing business in outsourced information technology services; Kenya is an English-speaking country where 82 percent of youth (ages 15–24) are literate. If Kenya can avoid political unrest due to ethnic rivalries, it could lead East Africa toward better economic integration.

The political and economic indicators for Kenya's neighbors, Uganda and Tanzania, are also improving, with average annual growth rates at 5.5 and 6.7 percent, respectively (see Table 6.2). Both countries rely heavily on agricultural exports and mining (especially gold), and both benefited from debt reduction agreements in the 2000s that redirected debt repayment funds to education and health care.

Measuring Social Development

By global standards, measures of social development in Africa are extremely low. Yet some positive trends, especially with regard to child survival and literacy, are cause for hope (see Table 6.2). Many governments in the region have reached out to the modern African diaspora (economic migrants and refugees who now live in Europe and North America). In other parts of the world, African immigrant organizations have worked to improve schools and health care, and former emigrants have returned to invest in businesses and real estate. The economic impact of remittances in this region is small, but growing.

Child Mortality and Life Expectancy Reductions in child mortality are a surrogate measure for improved social development, because if most children make it to their fifth birthday, it usually indicates adequate primary care and nutrition. A child mortality rate of 200 per thousand means that one child in five dies before his or her fifth birthday. As Table 6.2 shows, most of the states in the region saw modest to significant improvements in child survival between 1990 and 2015. Eritrea, Liberia, Madagascar, Malawi, and Rwanda actually experienced dramatic gains. However, countries with prolonged conflict (such as Somalia and the Democratic Republic of the Congo) have seen little improvement. In 1990, the regional child mortality rate was 175 per thousand; in 2015, it was down to 83. Thus, high child mortality is still a major concern for the region, but steady reductions in this rate are significant for African families.

Life expectancy for Sub-Saharan Africa is only 57 years. Countries hit hard by HIV/AIDS or conflict have seen life expectancies tumble into the 40s. Despite these statistics, there are indications that access to basic health care is improving and, eventually, so will life expectancies. Keep in mind that high infant- and child-mortality figures depress overall life expectancy figures; average life expectancies for people who make it to adulthood are much better.

Low life expectancies are generally related to extreme poverty, environmental hazards (such as drought), and various environmental and infectious diseases (malaria, cholera, AIDS, and measles). Often these factors work in combination. Malaria, for example, kills half a million African children each year. The death rate is also affected by poverty, as undernourished children are the most vulnerable to the effects of high fevers. Tragically, preventable diseases such as measles occur when people have no access to, or cannot afford, vaccines. National and international health agencies, along with NGOs such as the Gates Foundation, are working to improve access to vaccines, bed nets (to prevent malaria), and primary health care. These efforts are making a difference.

Meeting Educational Needs Basic education is another challenge for the region. The goal of universal access to primary education is a daunting one for a region where 43 percent of the population is under 15 years old. The UN estimates that 75 percent of African children are enrolled in primary school, but only 23 percent of the relevant population is in secondary school (high school or its equivalent). Sub-Saharan Africa is home to one-sixth of the world's children under 15, but to half of the world's uneducated children. Girls are still less likely than boys to attend school. In West African countries such as Chad, Niger, and Ivory Coast, girls are decidedly underrepresented.

A renewed focus on education since 2000 has been attributed to UN efforts to reduce extreme poverty by focusing on basic education, health care, and access to clean water through the Sustainable Development Goals. More government and nonprofit organization resources have been directed toward education: More schools are being built across the region, and more children are attending them (**Figure 6.46**).

Women and Development

Development gains cannot be achieved in the region unless the economic contributions of African women are recognized. Officially, women are the invisible contributors to local and national economies. In agriculture, women account for 75 percent of the labor that produces more than half the food consumed in the region. Tending subsistence plots, taking in extra laundry, and selling surplus produce in local markets all contribute to household income. Yet because many of these activities are considered part of the informal sector, they are not counted. For many of Africa's poorest people, however, the informal sector is the economy, and within this sector women dominate.

Status of Women The social position of women is difficult to measure for Sub-Saharan Africa. Female traders in West Africa, for example, have considerable political and economic power. By such measures as female labor force participation, many Sub-Saharan African countries show relative gender equality. Also, women in most Sub-Saharan societies do not suffer the kinds of traditional social restrictions encountered in much of South Asia, Southwest Asia, and North Africa; in Sub-Saharan Africa, women work outside the home, conduct business, and own property. In 2006, Ellen Johnson-Sirleaf was sworn in as Liberia's president, making her Africa's first elected female leader. In 2012, she was joined by Joyce Banda, the recently elected president of Malawi. In fact, throughout the region, women occupied 22 percent of all seats in national parliaments in 2014. In Rwanda, over half the parliamentary seats are filled by women (**Figure 6.47**).

By other measures, however, such as the prevalence of polygamy, the practice of the "bride-price," and the tendency for males to inherit property over females, African women do suffer discrimination. Perhaps the most controversial issue regarding women's status is the practice of female circumcision, or genital mutilation. In Ethiopia, Somalia, and Eritrea, as well as parts of West Africa, the majority of girls are subjected to this practice, which is extremely painful and can have serious health consequences. Yet because the practice is considered traditional, most African states are unwilling to ban it.

Regardless of their social position, most African women still live in remote villages where educational and wage-earning opportunities remain limited and caring for large families is time-consuming and demanding labor. As educational levels increase and urban society expands—and as reduced infant mortality provides greater security—we can expect fertility in the region to gradually decrease. Governments can speed up the process by providing birth control information and cheap contraceptives—and by investing more money in health and educational efforts aimed at women. As the economic importance of women receives greater attention from national and international organizations, more programs are being directed exclusively toward them.

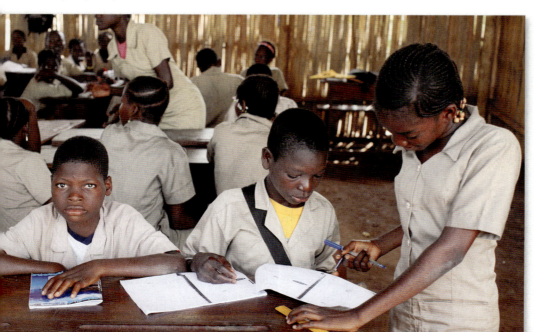

▼ **Figure 6.46 Educating African Youth** A high school classroom in Hevie, Benin. One of the greatest challenges for this region is educating its youth beyond primary school.

Building from Within

Surveys reveal that the majority of people in Sub-Saharan Africa are optimistic about their future. This surprises outside observers, who look at the region's many development hurdles. Yet considering the levels of conflict, food insecurity, and neglect that Sub-Saharan Africa experienced during the 1990s, perhaps the developments of the past two decades are a cause for hope. Most African states have been

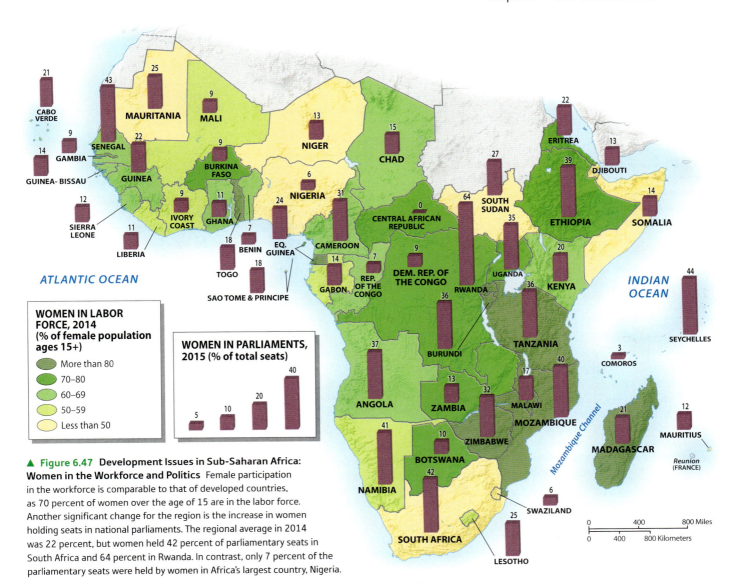

▲ **Figure 6.47** **Development Issues in Sub-Saharan Africa: Women in the Workforce and Politics** Female participation in the workforce is comparable to that of developed countries, as 70 percent of women over the age of 15 are in the labor force. Another significant change for the region is the increase in women holding seats in national parliaments. The regional average in 2014 was 22 percent, but women held 42 percent of parliamentary seats in South Africa and 64 percent in Rwanda. In contrast, only 7 percent of the parliamentary seats were held by women in Africa's largest country, Nigeria.

independent for only half a century. During that time, the governance of these countries has shifted from one-party authoritarian states to multiparty democracies. Targeted aid projects, aimed at addressing particular critical indicators such as mother and child mortality or combating malaria, have proven effective. Civil society is also vigorous, from raising the status of women to supporting small businesses with micro-credit loans. Even members of the African diaspora are beginning to return and invest in their countries. The region's rapid adoption and adaptation of mobile phone technology demonstrates its desire to be better connected with each other and the world.

Whether inspired by the free market, African socialism, or feminism, various internal and international organizations are finding ways to meet basic needs and innovate in the future. No doubt some projects will

fall short of their objectives. And for areas such as the Sahel, the long-term implications of climate change may overwhelm efforts to develop and adjust to severe water shortages. Yet for many people, the message of forming local networks to solve community problems is an empowering one.

REVIEW

6.10 What are the environmental, historical, structural, and institutional reasons offered to explain poverty in the region?

6.11 What technological and investment changes are impacting the region's social development?

■ **KEY TERMS** structural adjustment programs, kleptocracy, Sustainable Development Goals (SDGs), Southern African Development Community (SADC), Economic Community of West African States (ECOWAS)

Chapter 6 Sub-Saharan Africa

Summary

- The largest landmass straddling the equator, Africa is called the plateau continent because it is dominated by extensive uplifted plains. Key environmental issues facing this tropical region are desertification, deforestation, and drought. At the same time, the region supports a tremendous diversity of wildlife, especially large mammals.

- With nearly 1 billion people, Sub-Saharan Africa is the fastest-growing world region in terms of population; the average woman has five children. Yet it is also the poorest region, with 43 percent of the population living in extreme poverty (less than $1.90 per day). In addition, it has the lowest average life expectancy, at 57 years. HIV/AIDS has hit this region especially hard.

- Culturally, Sub-Saharan Africa is extremely diverse, where multiethnic and multireligious societies are the norm. With a few exceptions, religious diversity and tolerance have been distinctive features of the region. Most states have been independent for 50 years, and in that time pluralistic, but distinct, national identities have been forged. Many African cultural expressions, such as music, dance, and religion, are influential beyond the region.

- In the 1990s, many bloody ethnic and political conflicts occurred in the region. Peace now exists in many conflict-ridden areas, such as Angola, Sierra Leone, and Liberia, but ongoing ethnic and territorial disputes in Somalia, the Democratic Republic of the Congo, and South Sudan, and more recently in Mali and Nigeria, have produced millions of internally displaced persons and refugees.

- Widespread poverty and limited infrastructure are the region's most pressing concerns. Since 2000, Sub-Saharan economies have grown, led in part by higher commodity prices, greater investment, foreign assistance, and the end of some of the longest-running regional conflicts. Social indicators of development are also improving, due to greater attention from the international community and the formulation of Sustainable Development Goals.

Review Questions

1. Sub-Saharan Africa is noted for its wildlife, especially its large mammals. What environmental, historical, and institutional processes explain the existence of so much fauna? How important is wildlife for this region?

2. Consider how disease impacted colonization in Latin America and the Caribbean (demographic decline). How did the disease environment differ in Sub-Saharan Africa, and how did this delay colonization of this region? How do various diseases still influence the region's development?

3. Rates of urbanization are on the rise everywhere, including Sub-Saharan Africa. What are the urbanization challenges facing this region? How might a more urbanized region impact demographic and economic trends?

4. Compare and contrast the role of tribalism in Sub-Saharan Africa with that of nationalism in Europe. How might competing forms of social loyalties explain some of the development challenges and opportunities for this region?

5. Consider the development model put forward by the United States and Europe, with an emphasis on foreign assistance, and China's emphasis on foreign investment. Will Chinese influence in the region alter the course of development for Sub-Sahara Africa?

Image Analysis

Consider this image of per capita income for selected states in Africa South of the Sahara.

1. Compare the data for the three African subregions pictured and describe the income patterns for each. What demographic, geopolitical, and socioeconomic factors explain these patterns?

2. How might this map appear if data were shown at the sub-state level? For example, look at Figure 6.1 and suggest which areas of South Africa have the highest and lowest incomes, and why.

GNI PER CAPITA INCOME, 2014
- High $12,736 or more
- Upper middle $4,126–$12,735
- Lower middle $1,046–$4,125
- Low $1,045 or less

▶ **Figure IA6 Income Level by State** World Bank per capita income levels (PPP) by state shows relative poverty and wealth in Southern, Central, and East Africa.

JOIN THE DEBATE

The United Nations recently implemented 17 Sustainable Development Goals that focus on ending poverty and addressing inequalities while promoting sustainable growth. SDGs replace Millennium Development Goals that targeted only developing countries, by covering more ground and calling on developed countries to help attain benchmarks. Can SDGs effectively promote sustainable development?

SDGs will reduce poverty in Sub-Saharan states

- Since the creation of the Millennium Development Goals in 2000, more people have been lifted out of extreme poverty. The SDGs hope to continue that trend.

- Focusing on specific targets, such as decreasing maternal mortality or increasing youth literacy, will contribute to the region's overall social and economic development.

- The SDGs are broadly conceived benchmarks to support a holistic approach toward development, and do not focus exclusively on economic growth as a measure of success.

SDGs are unrealistic and do not impact development

- Such metrics are often abused, and countries manipulate the data to look good rather than address development concerns.

- Much of the development that occurred globally in the 2000s was due to factors other than foreign assistance to support these goals.

- National-level data mask the complex patterns of poverty and underdevelopment, especially for large countries.

▲ **Figure D6** To read more about the SDGs, go to http://www.un.org/sustainabledevelopment/sustainable-development-goals/. Pearson Education supports the Sustainable Development Goals.

■ KEY TERMS

African Union (AU) (p. 248)
agricultural density (p. 230)
apartheid (p. 236)
Berlin Conference (p. 247)
biofuels (p. 223)
clans (p. 252)
coloured (p. 237)
conflict diamonds (p. 251)
desertification (p. 221)
Economic Community of West African States (ECOWAS) (p. 259)
food insecurity (p. 227)
Gondwana (p. 215)
Great Escarpment (p. 216)
Great Rift Valley (p. 215)
homelands (p. 249)
Horn of Africa (p. 220)
internally displaced persons (IDPs) (p. 250)
kleptocracy (p. 256)
Pan-African Movement (p. 248)
pastoralists (p. 234)
physiological density (p. 230)
refugees (p. 250)
Sahel (p. 221)
Southern African Development Community (SADC) (p. 259)
structural adjustment programs (p. 255)
Sustainable Development Goals (SDGs) (p. 256)
swidden (p. 233)
townships (p. 237)
transhumance (p. 221)
tribalism (p. 249)
tribes (p. 239)

Mastering Geography™

Looking for additional review and test prep materials? Visit the Study Area in **MasteringGeography**™ to enhance your geographic literacy, spatial reasoning skills, and understanding of this chapter's content by accessing a variety of resources, including **MapMaster** interactive maps, videos, RSS feeds, flashcards, web links, self-study quizzes, and an eText version of *Diversity Amid Globalization*.

DATA ANALYSIS

http://goo.gl/Ga02UX

Sub-Saharan Africa has experienced more AIDS-related deaths and holds more people living with HIV/AIDS than any other world region. Country-specific data shed light on the impact of this disease and its geography. Go to the United Nations AIDS info website (http://aidsinfo.unaids.org) and access an interactive map of country-level data since 1990. Scroll over a country to see the data, and adjust the time bar along the bottom to retrieve data from a particular year. The column on the left allows you to select different variables. Pick six to eight countries in Sub-Saharan Africa, including at least one country from its western, eastern, central, and southern subregions.

1. Contrast the number of male adults and the number of female adults living with HIV/AIDS. Where are female numbers higher, and why?

2. New infections are declining, but not in all places. Look at new HIV infections, and compare figures in 1990 and 2000 with figures in 2015. Describe the trends. Now explore AIDS-related deaths (perhaps comparing males and females) and also estimates of AIDS orphans.

3. Use your data to write a couple of paragraphs about how gender, treatment, and loss of life have impacted each subregion and the region as a whole.

Authors' Blogs

Scan to visit
the GeoCurrents blog
http://geocurrents.info/category/place/subsaharan-africa

Scan to visit the
Author's Blog
for field notes, media resources, and chapter updates
http://gad4blog.wordpress.com/category/sub-saharan-africa/

Physical Geography and Environmental Issues

The region's vulnerability to water shortages is increasing as growing populations, rapid urbanization, increasing demands for agricultural land, and climate change pressure limited supplies.

Population and Settlement

Many settings within the region continue to see rapid population growth. These demographic pressures are visible in fragile, densely settled rural zones as well as in large, fast-growing cities.

Cultural Coherence and Diversity

Islam continues to be a vital cultural and political force within the region, but increasing fragmentation within that world has led to more culturally defined political instability.

Geopolitical Framework

Failed reforms jolted the status quo in Tunisia, Egypt, Libya, Yemen, and Bahrain. Internal instability and the growth of ISIL have led to violence in Syria and Iraq. Prospects for peace between Israel and the Palestinians remain murky, and Iran's growing political role is viewed as a threat both within and beyond the region.

Economic and Social Development

Unstable world oil prices and unpredictable geopolitical conditions have discouraged investment and tourism in many countries. The pace of social change, especially for women, has quickened, stimulating diverse regional responses.

▶ Beddawi Palestinian Refugee Camp Near Tripoli, Lebanon A family walks through the streets of Beddawi, near Tripoli in Lebanon. Several generations of refugees now call Beddawi home.

Tripoli, Lebanon

NORTH AFRICA SOUTHWEST ASIA

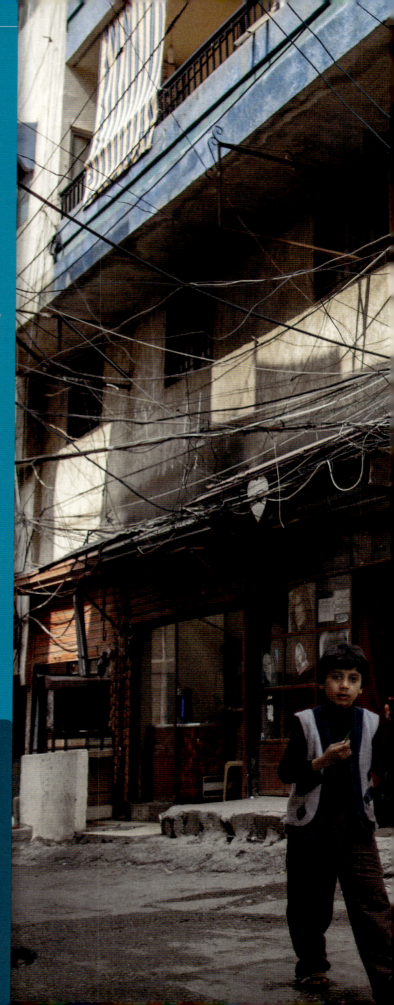

7

Southwest Asia and North Africa

Because of the ongoing violence in neighboring Syria, Lebanon is once again home to huge and growing refugee populations. The tiny country is littered with places that have been refugee camps for decades, and the newest wave of displaced people are moving into localities where generations of desperate migrants have sought refuge from earlier conflicts. Take the refugee camp of Baddawi in northern Lebanon, created more than 60 years ago as a temporary refuge for displaced Palestinians from nearby Israel; today, Baddawi is home to somewhere between 25,000 and 40,000 Palestinians. But recently, these established residents have seen their streets flooded by new waves of Syrian and Iraqi refugees. The result? Refugees helping refugees, or what some experts call "refugee-refugee humanitarianism."

Despite occasional tensions between established refugees and the recent arrivals, most residents of the camp feel their common plight. Many new refugees are given food and clothing. They are provided with temporary shelter. Seasoned refugees show them how to navigate the intricacies of everyday life in unfamiliar surroundings. And the mix of old-timers and newcomers also shapes the distribution of relief services to the refugees. The United Nations Relief and Works Agency (UNRWA) has long had a mandate to support Palestinian refugees, but not other groups. On the other hand, the UN High Commissioner for Refugees (UNHCR) now provides relief for displaced Syrians and Iraqis. While these international agencies have certainly helped in the streets of Baddawi, often it is one generation of refugees offering help to another that makes the biggest difference.

Defining Southwest Asia and North Africa

Climate, culture, and oil help define Southwest Asia and North Africa. Located where Europe, Asia, and Africa meet, the region includes thousands of square miles of parched deserts, rugged plateaus, and oasis-like river valleys (**Figure 7.1**).

"Southwest Asia and North Africa" is both an awkward term and a complex region. The area is often called the Middle East, but some experts exclude the western parts of North Africa, as well as Turkey and Iran, from such a region. Moreover, "Middle East" suggests a European point of view, reflecting the nations that colonized the region and still shape the names we give the world today.

Diverse languages, religions, and ethnic identities have molded land and life within the region, strongly wedding people and place. One traditional zone of conflict surrounds Israel, where Jewish, Christian, and Islamic peoples have yet to resolve long-standing differences. In the early 2010s, **Arab Spring** movements—a series of public protests, strikes, and rebellions calling for fundamental government and economic reforms—toppled several governments in the region and pressured others to accelerate political and economic change. Overall, however, the uprisings failed to produce many democratic reforms or more stable political regimes, and some observers suggest that Arab Spring has been replaced by "Arab Winter."

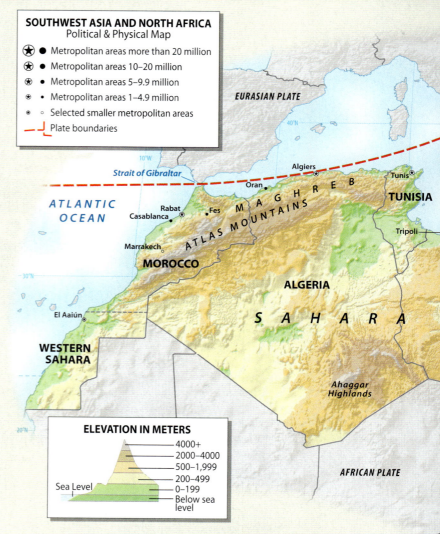

SOUTHWEST ASIA AND NORTH AFRICA
Political & Physical Map

- ★ ● Metropolitan areas more than 20 million
- ✳ ● Metropolitan areas 10–20 million
- ✱ • Metropolitan areas 5–9.9 million
- ✲ • Metropolitan areas 1–4.9 million
- ✳ ○ Selected smaller metropolitan areas
- ⌐ Plate boundaries

ELEVATION IN METERS

- 4000+
- 2000–4000
- 500–1,999
- 200–499
- 0–199
- Below sea level
- Sea Level

▲ **Figure 7.1 Southwest Asia and North Africa** This vast region extends from the shores of the Atlantic Ocean to the Caspian Sea. Within its boundaries, major cultural differences and globally important petroleum reserves have contributed to recent political tensions.

At the same time, cycles of **sectarian violence**—conflicts dividing people along ethnic, religious, and sectarian lines—plague the region. For example, enduring differences have led to clashes between Jews and Muslims and among varying factions of Islam. Syria and Iraq have been especially violent settings for recent instability brought about by the Sunni extremist organization **ISIL (Islamic State of Iraq and the Levant)**; (also known as ISIS or as IS). ISIL has further destabilized an already war-torn region as it attempts to create a new religious state (a *caliphate*) in the area.

Islamic fundamentalism in the region more broadly advocates a return to traditional practices within Islam. A related political movement within Islam, known as **Islamism**, challenges the encroachment of global popular culture and blames colonial, imperial, and Western elements for the region's political, economic, and social problems.

No world region better exemplifies globalization than Southwest Asia and North Africa. The region is a key global **culture hearth**, producing religions and civilizations of global significance.

> ## No world region better exemplifies globalization than Southwest Asia and North Africa

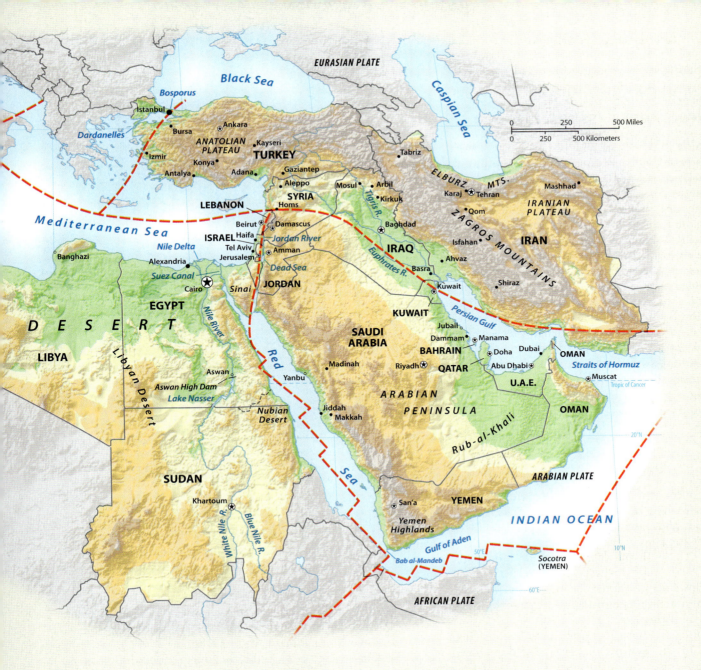

Particularly within the past century, the region's strategic importance increasingly opened Southwest Asia and North Africa to outside influences. The 20th-century development of the petroleum industry, largely initiated by U.S. and European investment, had enormous consequences for the region. Many key members of the **Organization of the Petroleum Exporting Countries (OPEC)** are found within the region, and greatly influence global prices and production levels for petroleum.

LEARNING Objectives

After reading this chapter you should be able to:

7.1 Explain how latitude and topography produce the region's distinctive patterns of climate.

7.2 Describe how the region's fragile, often arid, setting shapes contemporary environmental challenges.

7.3 Describe four distinctive ways in which people have learned to adapt their agricultural practice to the region's arid environment.

7.4 Summarize the major forces shaping recent migration patterns within the region.

7.5 List the major characteristics and patterns of diffusion of Islam.

7.6 Identify the key modern religions and language families that dominate the region.

7.7 Identify the role of cultural variables in understanding key regional conflicts in North Africa, Israel, Syria, Iraq, and the Arabian Peninsula.

7.8 Summarize the geography of oil and gas reserves in the region.

7.9 Describe traditional roles for Islamic women and provide examples of recent changes.

Physical Geography and Environmental Issues: Life in a Fragile World

In the popular imagination, much of Southwest Asia and North Africa is a land of shifting sand dunes, searing heat, and scattered oases. Although examples of those stereotypes certainly can be found across the region, the actual physical setting, in both landforms and climate, is considerably more complex (see Figure 7.1). In reality, the regional terrain varies greatly, with rocky plateaus and mountain ranges more common than sandy deserts. Even the climate, although dominated by aridity, varies remarkably from the dry heart of North Africa's Sahara Desert to the well-watered highlands of northern Morocco, coastal Turkey, and western Iran. One theme is pervasive, however: A lengthy legacy of human settlement has left its mark on a fragile environment, and the entire region is faced with increasingly daunting ecological problems in the decades ahead.

Regional Landforms

A quick tour of Southwest Asia and North Africa reveals diverse environmental settings (see Figure 7.1). In North Africa, the **Maghreb** (meaning "western island") includes the nations of Morocco, Algeria, and Tunisia and is dominated near the Mediterranean coastline by the Atlas Mountains. The rugged flanks of the Atlas rise like a series of islands above the narrow coastal plains to the north and the vast stretches of the lower Saharan deserts to the south (**Figure 7.2**). South and east of the Atlas Mountains, interior North Africa varies between rocky plateaus and extensive desert lowlands. In northeast Africa, the Nile River shapes regional drainage patterns as it flows north through Sudan and Egypt (**Figure 7.3**).

Southwest Asia is more mountainous than North Africa. In the **Levant**, or eastern Mediterranean region, mountains rise within 20 miles (30 km) of the sea, and

▲ **Figure 7.3 Nile Valley** This satellite image dramatically reveals the impact of water on the North African desert. Cairo lies at the southern end of the Nile Delta, where it begins to widen toward the Mediterranean Sea. The coastal city of Alexandria sits on the northwest edge of the low-lying delta. Lake Nasser is visible toward the bottom (center) of the image.

the highlands of Lebanon reach heights of more than 10,000 feet (3000 meters). Farther south, the Arabian Peninsula forms a massive tilted plateau, with western highlands higher than 5000 feet (1500 meters) gradually sloping eastward to extensive lowlands in the Persian Gulf area (**Figure 7.4**). North and east of the Arabian Peninsula lie the two great upland areas of Southwest Asia: the Iranian and Anatolian plateaus (*Anatolia* refers to the large peninsula of Turkey, sometimes called Asia Minor; see Figure 7.1). Both plateaus, averaging 3000–5000 feet (1000–1500 meters) in elevation, are geologically active and prone to earthquakes.

▼ **Figure 7.2 Atlas Mountains** The rugged Atlas Mountains dominate a broad area of interior Morocco. This view shows a landscape with cultivated fields and a small settlement in the Ait Bouguemez Valley.

▼ **Figure 7.4 Yemen Highlands** Steep highlands and narrow valleys dominate much of Yemen's rugged interior.

▲ **Figure 7.5** **Jordan Valley** This view of the fertile Jordan Valley shows a mix of irrigated vineyards and date palm plantations.

Explore the **Sights** of the Jordan Valley

http://goo.gl/bTPRv5

One quake near the Iranian city of Bam (2003) claimed more than 30,000 lives, and several devastating earthquakes have rocked western Turkey in the past 30 years.

Smaller lowlands characterize other portions of Southwest Asia. Narrow coastal strips are common in the Levant, along the southern (Mediterranean Sea) and the northern (Black Sea) Turkish coastlines, and north of Iran's Elburz Mountains near the Caspian Sea. Iraq contains the most extensive alluvial lowlands in Southwest Asia, dominated by the Tigris and Euphrates rivers, flowing southeast to empty into the Persian Gulf. The much smaller Jordan River Valley is a notable lowland that straddles the strategic borderlands of Israel, Jordan, and Syria and drains southward to the Dead Sea (**Figure 7.5**).

Patterns of Climate

Although the region of Southwest Asia and North Africa is often termed the "dry world," a closer look reveals more complex patterns (**Figure 7.6**). Both latitude and altitude come into play. Aridity dominates large portions of the region (see the climographs for Cairo, Riyadh, Baghdad, and Tehran in Figure 7.6). A nearly continuous belt of desert land stretches eastward across interior North Africa, through the Arabian Peninsula, and into central and eastern Iran (**Figure 7.7**). Throughout this zone, plant and animal life adapts to extreme conditions. Deep or extensive root systems allow desert plants to benefit from the limited moisture they receive, and animals efficiently store water, hunt at night, or migrate seasonally to avoid the worst of the dry cycle.

Elsewhere, altitude and latitude produce a surprising variety of climates. The Atlas Mountains and nearby lowlands of northern Morocco, Algeria, and Tunisia experience

a Mediterranean climate, in which dry summers alternate with cooler, wet winters (see the climographs for Rabat and Algiers). In these areas, the landscape resembles those in nearby southern Spain or Italy (**Figure 7.8**). A second zone of Mediterranean climate extends along the Levant coastline, into the nearby mountains, and northward across sizable portions of northern Syria, Turkey, and northwestern Iran (see the climographs for Jerusalem and Istanbul).

Legacies of a Vulnerable Landscape

The larger environmental history of Southwest Asia and North Africa contains many examples of the clever, but short-sighted legacies of its human occupants. Lengthy human settlement in a marginal land has resulted in deforestation, soil salinization and erosion, and depleted water resources (**Figure 7.9**). Indeed, the pace of population growth and technological change during the past century suggests that the region's vulnerable environment is destined to face even greater challenges in the 21st century.

Deforestation and Overgrazing Deforestation is an ancient problem in Southwest Asia and North Africa. Although much of the region is too dry for trees, the more humid and elevated lands that ring the Mediterranean once supported heavy forests. Included in these woodlands were the cedars of Lebanon, cut during ancient times and now reduced to a few scattered groves in a largely denuded landscape. In many settings, growing demands for agricultural land caused upland forests to be removed and replaced with grain fields, orchards, and pastures.

Human activities have conspired with natural conditions to reduce most of the region's forests to grass and scrub. Mediterranean forests grow slowly, are highly vulnerable to fire, and usually fare poorly if subjected to heavy grazing. Browsing by goats in particular often has been blamed for much of the region's forest loss, but other livestock have also impacted the vegetative cover, especially in steeply sloping semiarid settings that are slow to recover from grazing. Deforestation also has resulted in a millennia-long deterioration of the region's water supplies and in accelerated soil erosion.

Some forests survive in the mountains of northern and southern Turkey, and northern Iran also retains considerable tree cover. Scattered forests can be found in western Iran, the eastern Mediterranean, and the Atlas Mountains. Moreover, several governments have launched reforestation drives and forest preservation efforts in recent years. For example, both Israel and Syria have expanded their coverage of wooded lands since the 1980s. In nearby Lebanon, more than 5 percent of the country's total area was included in the Al-Shouf Cedar Reserve in 1996 to protect old-growth cedars, junipers, and oak forests. An even larger area (including the Al-Shouf Cedar Reserve) was declared the Shouf Biosphere Reserve by the UN in

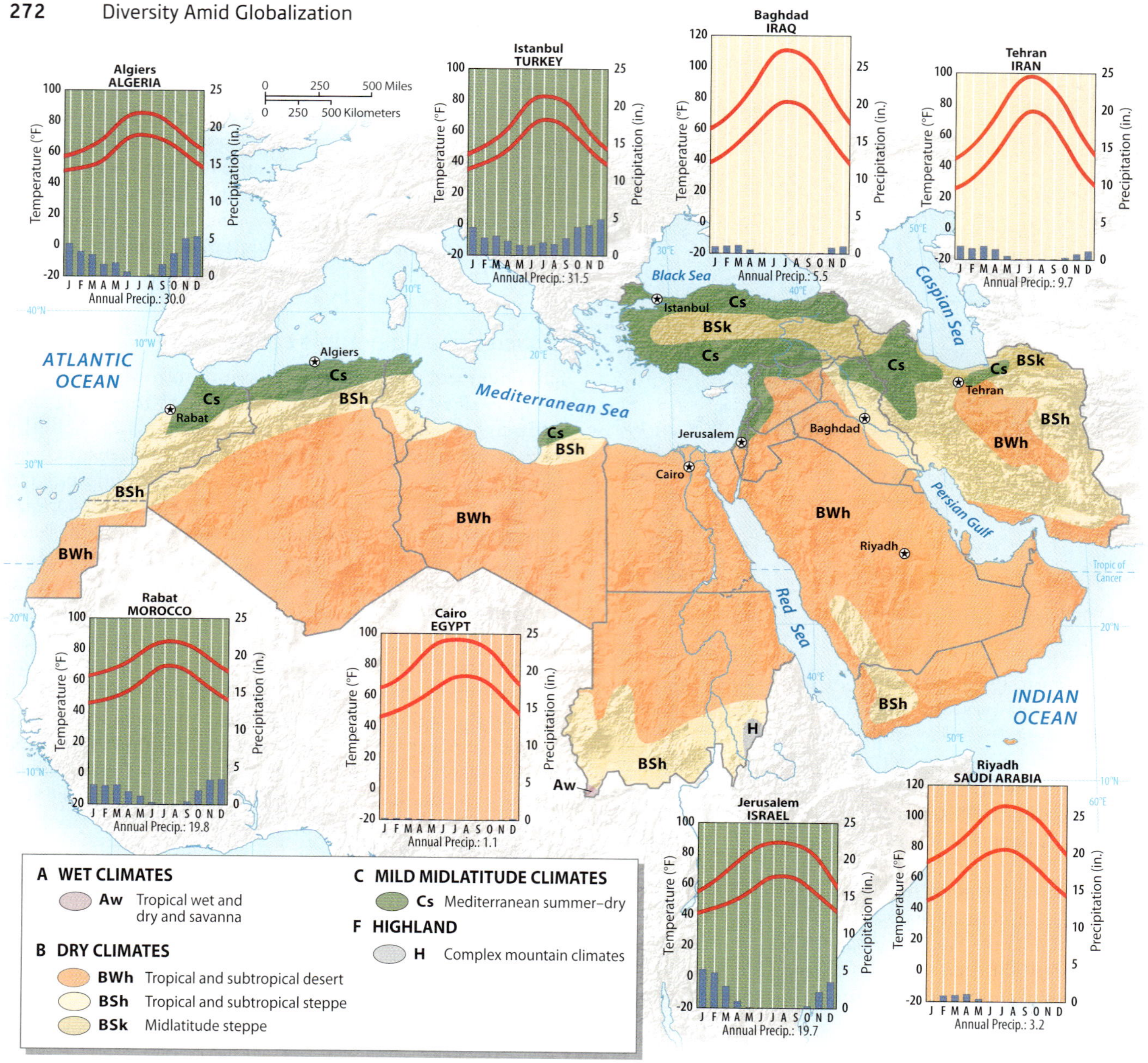

**Algiers
ALGERIA**

Annual Precip.: 30.0

**Istanbul
TURKEY**

Annual Precip.: 31.5

**Baghdad
IRAQ**

Annual Precip.: 5.5

**Tehran
IRAN**

Annual Precip.: 9.7

0   250   500 Miles
0   250   500 Kilometers

*ATLANTIC
OCEAN*

Mediterranean Sea

Black Sea

Caspian Sea

Algiers

Cs

BSh

Rabat

Cs

BSh

Istanbul

Cs

BSk

Cs

Cs

Cs   BSk

Tehran

BSh

BWh

Jerusalem

Baghdad

Cairo

Riyadh

Persian Gulf

BWh

BWh

Red Sea

BSh

BSh

*INDIAN
OCEAN*

H

BSh

Aw

BWh

BSh

Tropic of
Cancer

**Rabat
MOROCCO**

Annual Precip.: 19.8

**Cairo
EGYPT**

Annual Precip.: 1.1

**Jerusalem
ISRAEL**

Annual Precip.: 19.7

**Riyadh
SAUDI ARABIA**

Annual Precip.: 3.2

A   WET CLIMATES

Aw   Tropical wet and
dry and savanna

B   DRY CLIMATES

BWh   Tropical and subtropical desert

BSh   Tropical and subtropical steppe

BSk   Midlatitude steppe

C   MILD MIDLATITUDE CLIMATES

Cs   Mediterranean summer–dry

F   HIGHLAND

H   Complex mountain climates

▲ **Figure 7.6   Climate of Southwest Asia and North Africa**   Dry climates dominate from western Morocco to eastern Iran. Within these zones, persistent subtropical high-pressure systems offer only limited opportunities for precipitation. Elsewhere, mild midlatitude climates with wet winters are found near the Mediterranean Basin and Black Sea.

◀ **Figure 7.7   Arid Iran**   Much of Iran is arid or semiarid. These camel herders are trekking across Isfahan province in the central part of the country.

◀ **Figure 7.8 Mediterranean Landscape, Northern Algeria** The Mediterranean moisture in northern Algeria produces an agricultural landscape similar to that of southern Spain or Italy. Winter rains create a scene that contrasts sharply with deserts found elsewhere in the region.

2005 and is designed to protect the rare trees as well as endangered mammals, such as the wolf and Lebanese jungle cat.

Salinization The buildup of toxic salts in the soil, termed **salinization**, is another ancient environmental issue in a region where irrigation has been practiced for centuries (see Figure 7.9). The accumulation of salt in the topsoil is a common problem wherever desert lands are subjected to extensive irrigation. All freshwater contains a small amount of dissolved salt, and when water is diverted from streams into fields, salt remains in the soil after the moisture is absorbed by the plants and evaporated by the sun. In humid climates, accumulated salts are washed away by saturating rains, but in arid climates this rarely occurs. Where irrigation is practiced, salt concentrations build up over time, leading to lower crop yields and eventually to abandoned lands.

Hundreds of thousands of acres of once-fertile farmland within the region have been destroyed or degraded by salinization. The problem has been particularly acute in Iraq, where centuries of canal irrigation along the Tigris and Euphrates rivers have seriously degraded land quality. Similar conditions plague central Iran, Egypt, and irrigated portions of the Maghreb.

▼ **Figure 7.9 Environmental Issues in Southwest Asia and North Africa** Growing populations, pressures for economic development, and pervasive aridity combine to create environmental hazards across the region. Long human occupancy has contributed to deforestation, irrigation-induced salinization, and expanding desertification. **Q: Compare this map with Figure 7.6. What climate types are most strongly associated with desertification?**

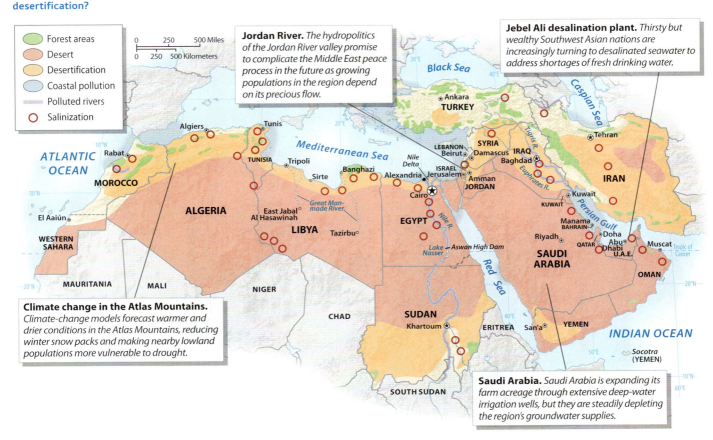

Jordan River. *The hydropolitics of the Jordan River valley promise to complicate the Middle East peace process in the future as growing populations in the region depend on its precious flow.*

Jebel Ali desalination plant. *Thirsty but wealthy Southwest Asian nations are increasingly turning to desalinated seawater to address shortages of fresh drinking water.*

Climate change in the Atlas Mountains. *Climate-change models forecast warmer and drier conditions in the Atlas Mountains, reducing winter snow packs and making nearby lowland populations more vulnerable to drought.*

Saudi Arabia. *Saudi Arabia is expanding its farm acreage through extensive deep-water irrigation wells, but they are steadily depleting the region's groundwater supplies.*

Legend:
- Forest areas
- Desert
- Desertification
- Coastal pollution
- Polluted rivers
- Salinization

Water Management Water has been manipulated and managed in the region for thousands of years. Traditional systems emphasized directing and conserving surface and groundwater resources at a local scale, but in the past half-century the scope of environmental change has been greatly magnified. One remarkable example is Egypt's Aswan High Dam, completed in 1970 on the Nile River south of Cairo (see Figure 7.9). Increased storage capacity in the upstream reservoir made more water available for agriculture and generates clean electricity. But irrigation has increased salinization because water is not rapidly flushed from fields. The dam has led to more use of costly fertilizers, the infilling of Lake Nasser behind the dam with accumulating sediments, and the collapse of the Mediterranean fishing industry near the Nile Delta, an area previously nourished by the river's silt.

Additional water-harvesting strategies have also proven useful. **Fossil water**, or water supplies stored underground during earlier and wetter climatic periods, is being mined. For example, Saudi Arabia has invested huge sums to develop deep-water wells, allowing it to greatly expand its food output. Unfortunately, underground supplies are being depleted more rapidly than they are recharged, limiting their long-term sustainability.

Desalination of seawater is another partial solution to this water crisis in the dry world. Wealthier countries such as Saudi Arabia, the United Arab Emirates, and Israel are making huge investments in new and emerging desalination technologies, which can more affordably and efficiently transform seawater into potable drinking water. The region already accounts for half of the world's desalination plants, and that total will grow in the next 20 years. In Dubai (a part of the United Arab Emirates), one of the world's largest combined desalination and power plants opened in 2013 along the shores of the Persian Gulf at Jebel Ali (**Figure 7.10**). Huge pipes extending into the Gulf can draw in up to a

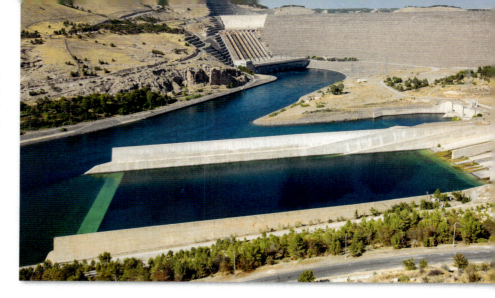

▲ **Figure 7.11 Southeast Anatolia Project** Turkey's Ataturk Dam, one of the world's largest, was completed in 1990 and is a centerpiece of the region's Southeast Anatolia Project.

billion gallons of water a day. The seawater is heated by efficient natural gas and diesel fuel boilers to produce steam and provide more than 2000 megawatts of electricity daily. Recent new additions to the plant (to be completed in 2019) by Siemens, a German company, will increase that capacity to more than 2700 megawatts. Plant operators note that the thermal efficiency of their technology is among the best in the world.

Most dramatically, **hydropolitics**, or the interplay of water resource issues and politics, has raised tensions between countries that share drainage basins. For example, with the help of Chinese capital and engineering expertise, Ethiopia recently built Tekeze Dam on a tributary of the Nile. Already referred to by promoters as "China's Three Gorges Dam in Africa," the controversial project threatens to disrupt downstream fisheries and irrigation in North Africa. Similarly, Sudan's Merowe Dam project (on another stretch of the Nile) has raised major concerns in nearby Egypt.

In Southwest Asia, Turkey's growing development of the upper Tigris and Euphrates rivers (the Southeast Anatolia Project, or GAP), complete with 22 dams and 19 power plants, has raised issues with Iraq and Syria, who argue that capturing "their" water might be considered a provocative political act (**Figure 7.11**). Turkey has

◄ **Figure 7.10 Jebel Ali Desalination Plant** The United Arab Emirates and many other countries in the region have made large investments in desalination technology. The Jebel Ali plant also supplies electricity to the area.

periodically withheld water from Syria by controlling flows along the Euphrates, provoking protests. In addition, a 2013 study based on new NASA satellite data measured the overall water flows within the Tigris–Euphrates Basin and found accelerating losses since 2003. As surface water flows from neighboring nations have declined, Iraqi farmers have dug over 1000 new wells, severely impacting groundwater supplies and lowering regional water tables.

Hydropolitics also has played into negotiations among Israel, the Palestinians, and other neighboring states, particularly in the Golan Heights area (an important headwaters zone for multiple nations) and within the Jordan River drainage (**Figure 7.12**). Israelis worry about Palestinian and Syrian pollution; nearby Jordanians argue for more water from Syria; and all regional residents must deal with the uncomfortable reality that, regardless of their political differences, they must drink from the same limited supplies of freshwater. A recent study by the International Institute of Sustainable Development is not encouraging; it estimated that the Jordan River may shrink by 80 percent by 2100, given increased demands and reduced flows.

Water is not only a resource, but also a vital transportation link in the area. The region's physical geography created enduring **choke points**, where narrow waterways are vulnerable to military blockade or disruption. For example, the Strait of Gibraltar (entrance to the Mediterranean), Turkey's Bosporus and Dardanelles, and the Suez Canal have all been key historical choke points within the region (see Figure 7.1). In addition, the narrow Bab-el-Mandeb, located at the southern entrance to the Red Sea, has gained strategic importance, given Yemen's increasingly unstable political situation (**Figure 7.13**). Finally, Iran's periodic threat to close the Straits of Hormuz (at the eastern end of the Persian Gulf) to world oil shipments suggests the strategic role water continues to play in the region.

Climate Change in Southwest Asia and North Africa

The 2014 report of the 10th Intergovernmental Panel on Climate Change (IPCC) suggested that 21st-century climate changes in Southwest Asia and North Africa will aggravate already existing environmental issues. Temperature changes are predicted to have a greater impact on the region than changes in precipitation. The already arid and semiarid region will probably remain

▲ **Figure 7.12 Hydropolitics in the Jordan River Basin** Many water-related issues complicate the geopolitical setting in the Middle East. The Jordan River system has been a particular focus of conflict.

▲ **Figure 7.13 Straits of Bab al-Mandeb** This satellite view shows the entrance to the Red Sea at the Straits of Bab al-Mandeb, one of the region's key choke points.

relatively dry, but warmer average temperatures are likely to have several major consequences:

- Higher overall evaporation rates and lower overall soil moisture across the region will stress crops, grasslands, and other vegetation. Semiarid lands in North Africa's Maghreb region are particularly vulnerable, especially dryland cropping systems that cannot depend on irrigation.

- Warmer temperatures will reduce runoff into rivers, reducing hydroelectric potential and water available for the region's increasingly urban population. Less snow in the Atlas Mountains will stress nearby farmers who depend on meltwater for irrigation.

- More extreme, record-setting summertime temperatures will lead to more heat-related deaths, particularly in cities.

A NASA study made public in 2016 found that the region's current drought is the worst in the past 900 years. Using a region-wide variety of tree-ring data (used to measure historical moisture and growth-rate patterns), the NASA scientists discovered that the current episode of intensified drought was "significantly drier" than ordinary dry periods, and that this evidence was the "smoking gun for climate change" affecting the region.

Changes in sea level also pose special threats to the Nile Delta. This portion of northern Egypt is a vast, low-lying landscape of settlements, farms, and marshland. Studies that model sea-level changes suggest much of the delta could be lost to inundation, erosion, or salinization. Farmland losses of more than 250,000 acres (100,000 hectares) are quite possible with even modest sea-level changes. The IPCC estimates that a sea-level rise of 3.3 feet (1 meter) could affect 15 percent of Egypt's habitable land and displace 8 million Egyptians in coastal and delta settings. Total losses of $30 billion have been projected for Alexandria because sea-level changes will devastate the city's huge resort industry as well as nearby residential and commercial areas.

Experts also estimate broader political and economic costs associated with potential climate changes. Given the region's political instability, even small changes in water supplies, particularly where they might involve several nations, could add to potential conflicts. In addition, wealthier nations such as Israel and Saudi Arabia may have more available resources to plan, adjust, and adapt to climate shifts and extreme events than poorer, less developed countries such as Yemen, Syria, and Sudan. A study by the World Resources Institute, for example, suggests that Syria's civil war that began several years ago was at least partly triggered by drought and water shortages.

Population and Settlement: Changing Rural and Urban Worlds

The geography of human population across Southwest Asia and North Africa demonstrates the intimate tie between water and life in this part of the world. The pattern is complex: Large areas of the population map remain almost devoid of permanent settlement, while more moisture-favored lands suffer increasingly from problems of crowding and overpopulation (**Figure 7.14**). Almost everywhere across the region, humans have uniquely adapted themselves to living within arid or semiarid settings.

The Geography of Population

Today about 500 million people live in Southwest Asia and North Africa (Table 7.1). The distribution of that population is strikingly varied (see Figure 7.14). In countries such as Egypt, large zones of almost empty desert land stand in sharp contrast to crowded, well-watered locations, such as those

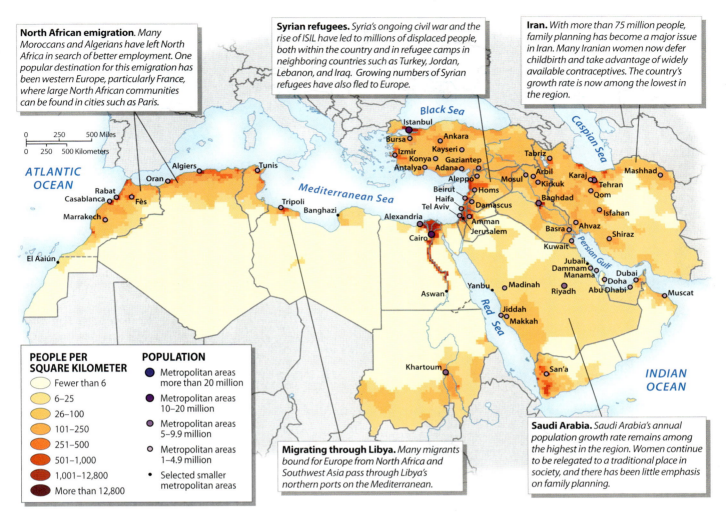

North African emigration. *Many Moroccans and Algerians have left North Africa in search of better employment. One popular destination for this emigration has been western Europe, particularly France, where large North African communities can be found in cities such as Paris.*

Syrian refugees. *Syria's ongoing civil war and the rise of ISIL have led to millions of displaced people, both within the country and in refugee camps in neighboring countries such as Turkey, Jordan, Lebanon, and Iraq. Growing numbers of Syrian refugees have also fled to Europe.*

Iran. *With more than 75 million people, family planning has become a major issue in Iran. Many Iranian women now defer childbirth and take advantage of widely available contraceptives. The country's growth rate is now among the lowest in the region.*

Saudi Arabia. *Saudi Arabia's annual population growth rate remains among the highest in the region. Women continue to be relegated to a traditional place in society, and there has been little emphasis on family planning.*

Migrating through Libya. *Many migrants bound for Europe from North Africa and Southwest Asia pass through Libya's northern ports on the Mediterranean.*

PEOPLE PER SQUARE KILOMETER

- Fewer than 6
- 6–25
- 26–100
- 101–250
- 251–500
- 501–1,000
- 1,001–12,800
- More than 12,800

POPULATION

- Metropolitan areas more than 20 million
- Metropolitan areas 10–20 million
- Metropolitan areas 5–9.9 million
- Metropolitan areas 1–4.9 million
- Selected smaller metropolitan areas

▲ **Figure 7.14 Population of Southwest Asia and North Africa** The striking contrasts are clearly evident between large, sparsely occupied desert zones and much more densely settled regions where water is available. The Nile Valley and the Maghreb region contain most of North Africa's people, whereas Southwest Asian populations cluster in the highlands and along the better-watered shores of the Mediterranean.

along the Nile River. Although the overall population density in such countries appears modest, the **physiological density**, which is the number of people per unit of arable land, is among the highest on Earth. Patterns of urban geography also are highly uneven: Less than two-thirds of the overall population is urban, but many nations are overwhelmingly dominated by huge and sprawling cities that produce the same problems of urban crowding found elsewhere in the developing world. Rates of recent urban growth have been phenomenal: Cairo, a modest-sized city of 3.5 million people in the 1960s, has more than quadrupled in population in the past 50 years.

Across North Africa, two dominant clusters of settlement, both shaped by the availability of water, account for most of the region's population (see Figure 7.14). In the Maghreb, the moister slopes of the Atlas Mountains and nearby better-watered coastal districts have accommodated denser populations for centuries. Today concentrations of both rural and urban settlement extend from south of Casablanca in Morocco to Algiers and Tunis on

the shores of the southern Mediterranean. Indeed, most of the populations of Morocco, Algeria, and Tunisia crowd into this crescent, a stark contrast to the almost empty lands south and east of the Atlas Mountains. Casablanca and Algiers are the largest cities in the Maghreb, with rapidly growing metropolitan populations of 3–4 million residents each.

Farther east, much of Libya and western Egypt is very thinly settled. Egypt's Nile Valley, however, is home to the other great North African population cluster. The vast majority of Egypt's 93 million people live within 10 miles of the river (see Figure 7.3). Optimistic politicians and planners are creating a second corridor of denser settlement in Egypt's "New Valley" west of the Nile by diverting water from Lake Nasser into the Western Desert.

Most Southwest Asian residents are clustered in favored coastal zones, moister highland settings, and desert localities where water is available from nearby rivers or subsurface aquifers. High population densities are found in better-watered portions of the eastern Mediterranean (Israel,

TABLE 7.1   Population Indicators

MasteringGeography™

Country	Population (millions) 2016	Population Density (per square kilometer)[1]	Rate of Natural Increase (RNI)	Total Fertility Rate	Percent Urban	Percent <15	Percent >65	Net Migration (Rate per 1000)
Algeria	40.8	16	2.1	3.1	71	29	6	0
Bahrain	1.4	1,769	1.4	2.1	100	17	2	3
Egypt	93.5	90	2.5	3.5	43	31	4	−1
Gaza and West Bank	4.8	713	–	4.1	75	40	3	−2
Iran	79.5	48	1.4	1.8	72	24	5	−1
Iraq	38.1	81	2.8	4.2	70	40	3	2
Israel	8.2	380	1.6	3.1	92	28	11	1
Jordan	8.2	84	2.3	3.5	84	37	3	−13
Kuwait	4.0	211	1.5	2.2	98	22	2	22
Lebanon	6.2	549	1.0	1.7	88	25	7	−1
Libya	6.3	4	1.5	2.4	79	30	5	−10
Morocco	34.7	76	1.4	2.4	60	25	6	−2
Oman	4.4	14	1.9	2.9	75	21	3	39
Qatar	2.5	187	1.1	2.0	99	15	1	75
Saudi Arabia	31.7	14	1.8	2.8	83	30	3	3
Sudan	42.1	22	2.9	5.2	34	43	3	−3
Syria	17.2	102	1.8	2.7	58	32	4	−20
Tunisia	11.3	71	1.3	2.4	68	24	8	−1
Turkey	79.5	101	1.2	2.1	73	24	8	7
United Arab Emirates	9.3	109	0.9	1.8	86	14	1	8
Western Sahara	0.6	–	1.2	2.1	81	26	3	8
Yemen	27.5	50	2.6	4.2	35	40	3	−1

Source: Population Reference Bureau, *World Population Data Sheet*, 2016.

[1] World Bank, *World Development Indicators*, 2016

Login to MasteringGeography™ & access MapMaster to explore these data!

1) Explain why the overall Population Densities of Egypt and Algeria are so low, while the population densities of Lebanon and Gaza and West Bank are so high. Why might these dramatic differences be misleading?

2) What are the three youngest populations (highest Percent < 15) and the three oldest (highest Percent > 65) in the region? What connections can you make to levels of economic development, if any?

Lebanon, and Syria) and Turkey. Nearby Iran is home to almost 80 million residents, but population densities vary considerably, from thinly occupied deserts in the east to more concentrated settlements near the Caspian Sea and across the more humid highlands of the northwest. Tehran, Iran's capital, is now home to 9 million people.

The region's largest city is Istanbul (formerly Constantinople), with a metropolitan population of about 15 million (**Figure 7.15**). At current rates of growth, Istanbul may surpass both London and Moscow in metropolitan population by 2018. Elsewhere, sizable populations are scattered through the Tigris and Euphrates river valleys, in the Yemen Highlands, and near oases where groundwater can be tapped to support agricultural or industrial activities.

Shifting Demographic Patterns

High population growth remains a critical issue throughout Southwest Asia and North Africa, but the demographic picture is shifting. Uniformly high growth rates in the 1960s

Explore the **Sights** of Istanbul

http://goo.gl/LPKR26

▼ **Figure 7.15 Downtown Istanbul, Turkey** Istanbul's bustling Taksin Square attracts many shoppers and tourists in Turkey's largest city.

have been replaced by more varied regional patterns (**Figure 7.16**). For example, women in Tunisia and Turkey now average fewer than three births, representing a large decline in total fertility rates (see Table 7.1). Various factors explain these changes. More urban, consumer-oriented populations opt for fewer children. Many Arab women now delay marriage into their middle 20s and early 30s. Family-planning initiatives are expanding in many countries; programs in Tunisia, Egypt, and Iran have greatly increased access to contraceptive pills, IUDs, and condoms.

Intriguingly, fundamentalist Iran has witnessed a rapid decline in fertility in the past two decades. Fertility has fallen since the mid-1970s from an average of 6.6 births to about 1.8 births per woman. While Iran's family-planning program was initially dismantled after the fundamentalist revolution in 1979 (it was seen as a Western idea), recent leaders have recognized the wisdom in containing the country's large population.

Still, areas such as the West Bank, Gaza, and Yemen are growing much faster than the world average. Poverty and traditional ways of rural life contribute to large rates of population increase, and even in more urban Saudi Arabia, growth rates remain between 1.5 and 2 percent. The increases result from high birth rates combined with low death rates. In Egypt, even though birth rates may decline, the labor market will need to absorb more than 500,000 new workers annually over the next 10 to 15 years just to keep up with the country's large youthful population (see Figure 7.16). Given the recent political instability in the country, many of Egypt's family planning priorities have fallen off the government's agenda. Some experts predict that if the country's population grows to 160 million people by 2050, there could be dire consequences for Egypt's economy and quality of life.

Water and Life: Rural Settlement Patterns

Water and life are closely linked across rural settlement landscapes of Southwest Asia and North Africa (**Figure 7.17**). Indeed, Southwest Asia is one of the world's earliest hearths of **domestication**, where plants and animals were purposefully selected and bred for their desirable characteristics. Beginning around 10,000 years ago, increased experimentation with wild varieties of wheat and barley led to agricultural settlements that later included domesticated animals, such as cattle, sheep, and goats. Much of the early agricultural activity focused on the **Fertile Crescent**, an ecologically diverse zone stretching from the Levant inland through the fertile hill country of northern Syria into Iraq. Between 5000 and 6000 years ago, improved irrigation techniques and increasingly powerful political states encouraged the spread of agriculture into nearby lowlands, such as the Tigris and Euphrates valleys (Mesopotamia) and North Africa's Nile Valley.

Pastoral Nomadism In the drier portions of the region, **pastoral nomadism**, where people move livestock seasonally, is a traditional form of subsistence agriculture. The settlement landscape of pastoral nomads reflects their need for mobility and flexibility as they move

▼ **Figure 7.16 Population Pyramids: Egypt, Iran, and United Arab Emirates, 2015** Three distinctive demographic snapshots highlight regional diversity: (a) Egypt's above-average growth rates differ sharply from those of (b) Iran, where a focused campaign on family planning has reduced recent family sizes. (c) Male immigrant laborers play a special role in skewing the pattern within the United Arab Emirates. **Q: For each example, cite a related demographic or cultural issue that you might potentially find in these countries.**

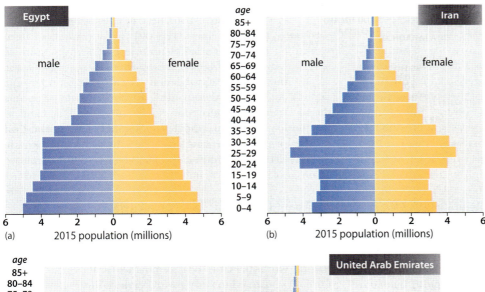

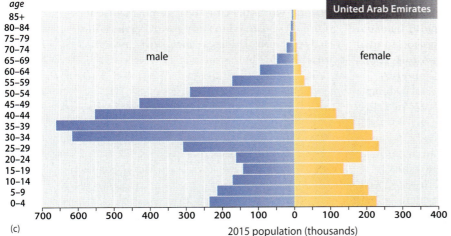

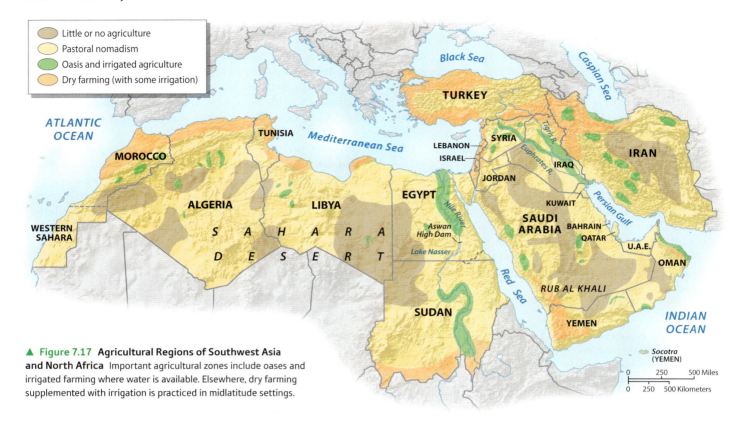

●	Little or no agriculture
●	Pastoral nomadism
●	Oasis and irrigated agriculture
●	Dry farming (with some irrigation)

▲ **Figure 7.17 Agricultural Regions of Southwest Asia and North Africa** Important agricultural zones include oases and irrigated farming where water is available. Elsewhere, dry farming supplemented with irrigation is practiced in midlatitude settings.

camels, sheep, and goats from place to place. Near highland zones such as the Atlas Mountains and the Anatolian Plateau, nomads practice **transhumance**—seasonally moving livestock to cooler, greener high-country pastures in the summer and returning them to valley and lowland settings for fall and winter grazing. Elsewhere, seasonal movements often involve huge areas of desert that support small groups of a few dozen families. Fewer than 10 million pastoral nomads remain in the region today.

Oasis Life Permanent oases exist where high groundwater levels or modern deep-water wells provide reliable water (Figures 7.17 and **7.18**). Tightly clustered, often walled villages sit next to small, intensively utilized fields where underground water is applied to tree and cereal crops. In newer oasis settlements, concrete blocks and prefabricated housing add a modern look. Traditional oasis settlements contain families that

work their own irrigated plots or, more commonly, work for absentee landowners. While oases are usually small, eastern Saudi Arabia's sprawling Hofuf Oasis covers more than 30,000 acres (12,000 hectares). Although some crops are raised for local consumption, expanding world demand for products such as figs and dates increasingly draws even these remote locations into the global economy as products end up on the tables of hungry Europeans or North Americans.

▶ **Figure 7.18 Oasis Settlement** This view of Morocco's Tinghir Oasis features small cultivated fields and date palms in the foreground.

Exotic Rivers For centuries, the densest rural settlement of Southwest Asia and North Africa has been tied to its great irrigated river valleys and their seasonal floods of water and fertile nutrients. In such settings, **exotic rivers** transport much-needed water from more humid areas to drier regions suffering from long-term moisture deficits (**Figure 7.19**). The Nile and the combined Tigris and Euphrates rivers are the largest regional examples of such activity, and both systems have large, densely settled deltas. Similar settlements are found along the Jordan River in Israel and Jordan, in the foothills of the Atlas Mountains, and on the peripheries of the Anatolian and Iranian plateaus. These settings, although capable of supporting sizable populations, are also vulnerable to overuse and salinization. Rural life is also changing in such settings. New dam- and canal-building schemes in Egypt, Israel, Syria, Turkey, and elsewhere are increasing the storage capacity of river systems, allowing for more year-round agriculture.

The Challenge of Dryland Agriculture Mediterranean climates in the region permit dryland agriculture that depends largely on seasonal moisture. These zones include better-watered valleys and coastal lowlands of the northern Maghreb, lands along the shore of the eastern Mediterranean, and favored uplands across the Anatolian and Iranian plateaus. A mix of tree crops, grains, and livestock is raised in these settings. More mechanization, crop specialization, and fertilizer use are also transforming such agricultural settings, following a pattern set earlier in nearby areas of southern Europe. One commercial adaptation of growing regional and global importance is Morocco's flourishing hashish crop. More than 200,000 acres (80,000 hectares) of cannabis are cultivated in the hill country

▼ **Figure 7.19 Nile Valley Agriculture** Irrigated rice is a major staple in Egypt's fertile Nile Valley.

near Ketama in northern Morocco, generating more than US$2 billion annually in illegal exports (mostly to Europe).

Many-Layered Landscapes: The Urban Imprint

Cities have played a pivotal role in the region's human geography. Indeed, some of the world's oldest urban places are located in the region. Today enduring political, religious, and economic ties link the city and countryside.

A Long Urban Legacy Urbanization in Mesopotamia (modern Iraq) began by 3500 BCE, and cities such as Eridu and Ur reached populations of 25,000 to 35,000 residents. Similar centers appeared in Egypt by 3000 BCE, with Memphis and Thebes assuming major importance amid the dense populations of the middle Nile Valley. These ancient cities functioned as centers of political and religious control. Temples, palaces, tombs, and public buildings dominated the urban landscapes of such settlements, and surrounding walls (particularly in Mesopotamia) offered protection from outside invasion. By 2000 BCE, however, a different kind of city was emerging, particularly along the shores of the eastern Mediterranean and at the junction points of important caravan routes. Centers such as Beirut, Tyre, and Sidon, all in modern Lebanon, as well as Damascus in nearby Syria, exemplified the growing role of local and long-distance trade in creating urban landscapes. Expanding port facilities, warehouse districts, and commercial thoroughfares suggested how trade and commerce shaped these urban settlements, and many of these early Middle Eastern trading towns have survived to the present.

Islam also left a lasting mark because cities traditionally served as centers of Islamic religious power and education. By the 8th century, Baghdad had emerged as a religious center, followed soon thereafter by the appearance of Cairo as a seat of religious authority and expansion. Urban settlements from North Africa to Turkey felt the influences of Islam. Indeed, the Islamic Moors carried the Muslim culture to Spain, where it shaped urban centers such as Córdoba and Málaga. Islam's impact on the settlement landscape merged with older urban traditions across the region and established a characteristic Islamic cityscape that exists to this day. Its traditional attributes include a walled urban core, or **medina**, dominated by the central mosque and its associated religious, educational, and administrative activities (**Figure 7.20**). A nearby bazaar, or *suq,* functions as a marketplace where products from city and countryside are traded. Housing districts feature an intricate maze of narrow, twisting streets that maximize shade and accentuate the privacy of residents, particularly women (**Figure 7.21**). Houses have small windows, frequently are situated on dead-end streets, and typically open inward to private courtyards, which are often shared by extended families with similar ethnic or occupational backgrounds.

More recently, European colonialism added another layer of urban landscape features in selected cities. Particularly in North Africa, coastal and administrative centers during the

▲ **Figure 7.20 Mosque in Qom, Iran** Qom's Hazrati Masumeh Shrine is visited annually by thousands of faithful Shiites in this sacred Iranian city south of Tehran.

late 19th century added dozens of architectural features from Britain and France. Victorian building blocks, French mansard roofs, suburban housing districts, and wide European-style commercial boulevards complicated the settlement landscapes of dozens of cities, both old and new. Centers such as Algiers (French), Fes (French), and Cairo (British) vividly displayed the effects of colonial control, and many of these urban signatures remain today (see Figure 7.21).

Signatures of Globalization Since 1950, dramatic new forces have transformed the urban landscape. Cities have become key gateways to the global economy. As the region has been opened to new investment, industrialization, and tourism, the urban landscape reflects the fundamental changes taking place. Expanded airports, commercial and financial districts, industrial parks, and luxury tourist facilities all mark the imprint of the global economy.

Further, as urban centers become focal points of economic growth, surrounding rural populations are drawn to the new employment opportunities, thus fueling rapid population increases. The results are both impressive and problematic. Many traditional urban centers, such as Algiers and Istanbul, have more than doubled in size in recent years. Booming demand for homes has produced ugly, cramped high-rise apartment houses in some government-planned neighborhoods. Elsewhere, sprawling squatter settlements provide little in the way of quality housing or municipal services.

Crowded Cairo is now home to more than 15 million people, making it North Africa's largest city. Founded more than 1000 years ago along the shores of the Nile River, Cairo emerged as a key Islamic city in the Muslim world. Its universities cultivated Islamic scholarship, and its commercial bazaars served as a regional crossroads between northern Africa and western Asia. Modern Cairo began taking shape in the later 19th century when Egyptian rulers reimagined major sections of downtown, inspired by the European Renaissance. This section of the city—sometimes called the Paris of the Nile—was redesigned with distinctive open spaces, grand boulevards, and key government buildings. On its southwestern edge, the area's most famous landmark is undoubtedly Tahrir (Liberation) Square, which served as the public rallying point in 2011 during Egypt's Arab Spring rebellions that overthrew the regime of Hosni Mubarak. Some of Cairo's most interesting neighborhoods are found east and south of downtown in Old Cairo (**Figure 7.22**). These southern districts grew slowly and haphazardly, and some streets are remnants of cities that even predate Cairo's official founding. The city's Coptic population and Christian churches are also concentrated here.

▼ **Figure 7.21 Map of Fes, Morocco** The tiny neighborhoods and twisting lanes of the old walled city reveal features of the traditional Islamic urban center. To the southwest, however, the rectangular street patterns, open spaces, and broad avenues suggest colonial European influences.

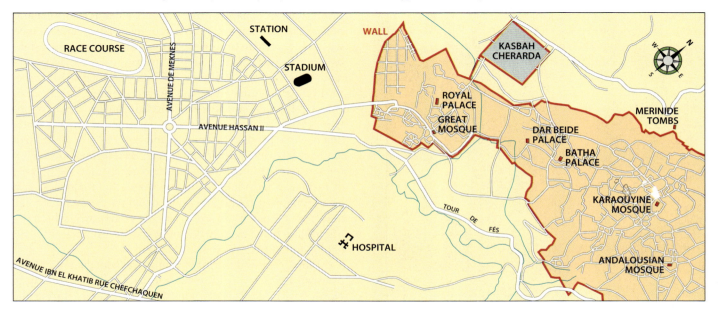

Huge suburban developments west of the river (in Giza) and east of downtown (Nasr City and beyond) have spilled into former farmland and desert waste. Nearby, one of the city's prime burial grounds—the famous City of the Dead—is home to almost 1 million residents (mostly poor), who live in cheap homes and apartment houses erected amid a sprawling cemetery. Another satellite community called New Cairo City, about 20 miles east of downtown, features planned boulevards, affluent housing, and new university facilities. However, the development's long-term prospects remain uncertain amid Egypt's current economic and political challenges.

Certainly, the oil-rich states of the Persian Gulf display the most extraordinary changes in the urban landscape. Before the 20th century, urban traditions were relatively weak in the area, and even as late as 1950 only 18 percent of Saudi Arabia's population lived in cities. All that changed, however, as the global economy's demand for petroleum mushroomed. Today the Saudi Arabian population is more urban than those of many industrialized nations, including the United States, and the capital city of Riyadh has grown to over 5 million people. Particularly after 1970, other cities, such as Dubai (UAE), Doha (Qatar), Kuwait City (Kuwait), and Manama (Bahrain), grew in size and took on modern Western characteristics, including futuristic architecture and new transportation infrastructure. Dubai's skyline is now home to the soaring Burj Khalifa (completed in 2010), a needle-like, 160-story building that ascends a half mile into the desert sky (**Figure 7.23**).

▲ **Figure 7.23 Burj Khalifa, United Arab Emirates** Soaring more than 2,700 feet into the desert sky, Dubai's Burj Khalifa is home to more than 50 elevators that whisk visitors to a variety of hotel rooms, restaurants, residences, offices, and shops. It is the world's tallest building.

▼ **Figure 7.22 Old Cairo** The Egyptian capital's narrow, twisting streets are often crowded with shoppers and pedestrians.

▼ **Figure 7.24 Masdar City, United Arab Emirates** Masdar City is being laid out near the Abu Dhabi Airport.

In addition, investments in petrochemical industries have fueled the creation of new urban centers, such as Jubail along Saudi Arabia's Persian Gulf coastline. Masdar City, one of the world's most carefully planned and energy-efficient cities, is also taking shape in this part of the world, within the United Arab Emirates (**Figure 7.24**).

A Region on the Move

Although nomads have crisscrossed the region for ages, entirely new patterns of migration reflect the global economy and recent political events. The rural-to-urban shift seen widely in the less developed world is reworking population patterns across Southwest Asia and North Africa. The Saudi Arabian example is echoed in many other countries. Cities from Casablanca to Tehran are experiencing phenomenal growth, spurred by in-migration from rural areas.

Foreign workers have also migrated to areas within the region that have large labor demands. In particular, Persian Gulf nations support immigrant workforces that often comprise large proportions of the overall population (see Figure 7.16c). Over 40 percent of the Gulf nations' total population is made up of foreign-born workers. The influx has major economic, social, and demographic implications in nations such as the United Arab Emirates, where more than 90 percent of the country's private workers are immigrants. Source regions for these immigrants vary, but most come from South Asia and other Muslim countries within and beyond the region (**Figure 7.25**). In Dubai,

Pakistani cab drivers, Filipino nannies, and Indian shop clerks typify a foreign workforce that currently makes up a large majority of the city's 1.4 million inhabitants. Many of these foreign workers in the Gulf send their wages home: India, Egypt, the Philippines, Pakistan, and Bangladesh are often key destinations for these remittances. Recent reports of guest-worker abuse—especially of women—in settings such as Saudi Arabia have cast a shadow on these migrant flows, however. Some countries such as Indonesia and India are now putting limits on the number and types of guest workers that can travel to the region.

On the other hand, other residents from the region migrate to jobs elsewhere in the world. Because of its strong economy and close location, Europe is a powerful draw. More than 2 million Turkish guest workers live in Germany. Algeria and Morocco also have seen large out-migrations to western Europe, particularly France.

Political instability has also sparked migration. Wealthier residents, for example, have fled nations such as Lebanon, Syria, Iraq, and Iran since the 1980s and today live in cities such as Toronto, Los Angeles, and Paris. More recent political instability has provoked other refugee movements. Since 2003, huge numbers of people in western Sudan's unsettled Darfur region have moved to dozens of refugee camps in nearby Chad. Elsewhere, thousands of displaced Afghans have moved to nearby Iran and elsewhere across Southwest Asia and beyond.

Syria's civil war and sectarian conflicts produced a massive refugee crisis that has displaced over half the country's population. By 2016, almost 7 million internally displaced persons (IDPs) had left their homes and were living elsewhere in the country. About 4.8 million Syrians were refugees and were living in other countries. The nearby nations of Turkey, Jordan, Lebanon, and Iraq house the vast majority of these

▼ **Figure 7.25 Migrant Workers in Saudi Arabia, 2014** Most migrant workers in Saudi Arabia are Muslims. Many come from nearby areas in North Africa and Southwest Asia, but some workers also relocate from far-off Indonesia and the Philippines.

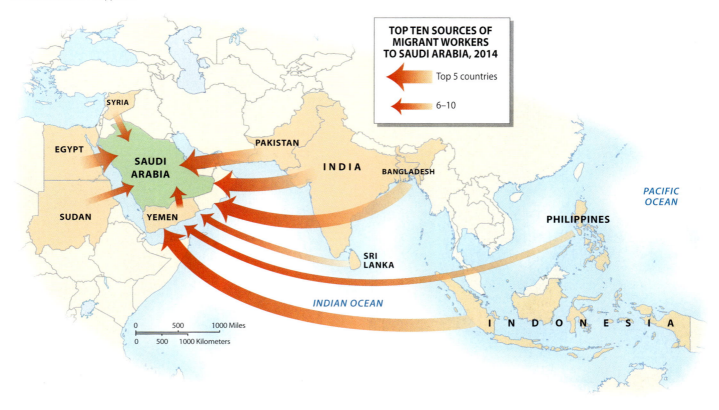

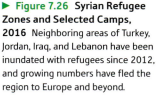

▶ **Figure 7.26   Syrian Refugee Zones and Selected Camps, 2016** Neighboring areas of Turkey, Jordan, Iraq, and Lebanon have been inundated with refugees since 2012, and growing numbers have fled the region to Europe and beyond.

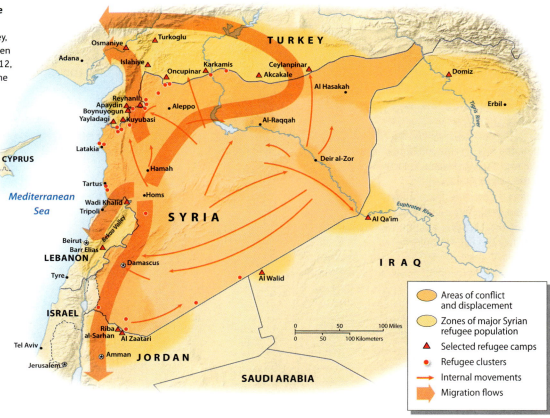

refugees. The burden can be huge: for example, Jordan's modest population of 8 million people is being swamped by about 700,000 refugees, most from Syria (**Figure 7.26**).

Growing numbers of Syrians were also fleeing the region, bound for opportunities in Europe and beyond. By early 2016, more than 500,000 Syrians, for example, had arrived in Germany (the largest destination in Europe) and many applied for asylum. Depending on conditions in Syria, more than 3 million additional refugees may arrive in Europe by 2018. Large numbers of refugees travel via the eastern Mediterranean and through Turkey and Greece to the rest of Europe.

Elsewhere, North Africa has been another major focus for refugee populations. Many African and Southwest Asian migrants have traveled through war-torn Libya on their desperate search to find better lives in Europe and beyond (see *People on the Move: The Libyan Highway to Europe*).

REVIEW

7.3 Discuss how pastoral nomadism, oasis agriculture, and dryland wheat farming represent distinctive adaptations to the regional environments of Southwest Asia and North Africa. How do these rural lifestyles create distinctive patterns of settlement?

7.4 Describe the contributions of (a) Islam, (b) European colonialism, and (c) recent globalization to the region's urban landscape.

7.5 Summarize the key patterns and drivers of migration in and out of the region.

■ **KEY TERMS** physiological density, domestication, Fertile Crescent, pastoral nomadism, transhumance, exotic rivers, medina

Cultural Coherence and Diversity: A Complex Cultural Mosaic

Southwest Asia and North Africa clearly define the heart of the "Islamic" and "Arab" worlds, but a surprising degree of cultural diversity characterizes the region. Muslims practice their religion in varied ways, often disagreeing profoundly on basic religious views, as well as on how much of the modern world and its mass consumer culture should be incorporated into their daily lives. In addition, diverse religious minorities complicate the region's contemporary cultural geography. Linguistically, Arabic languages form an important cultural core historically centered on the region. However, different Arab dialects can mean that a Syrian may struggle to comprehend an Algerian. In addition, many non-Arab peoples, including Persians, Kurds, and Turks, populate important homelands in the region and historically have dominated large portions of it. Understanding these varied patterns of cultural geography is essential to comprehending many of the region's political tensions, as well as appreciating why many of its residents resist processes of globalization and celebrate the lasting cultural identity of their home neighborhoods and communities.

Patterns of Religion

Religion permeates the lives of most people within the region. Its centrality stands in sharp contrast to largely secular cultures in many other parts of the world. Whether it is

The Libyan Highway to Europe

Revolutions bring many unintended consequences. When Libyan dictator Muammar al-Qaddafi was overthrown in 2011, few experts believed it would dramatically reorient and enhance one of the world's most diverse flows of refugees. The newly formed Libyan Highway has truly international implications that reach from Syria and Nigeria to Italy and Sweden (**Figure 7.1.1**).

A Highway for Refugees All of the critical variables in the creation of the highway fell into place in 2014. First and foremost, Libya itself ceased to truly exist as multiple political forces vied for power, essentially ending

Google Earth Virtual (MG) Tour Video

http://goo.gl/aJaZC0

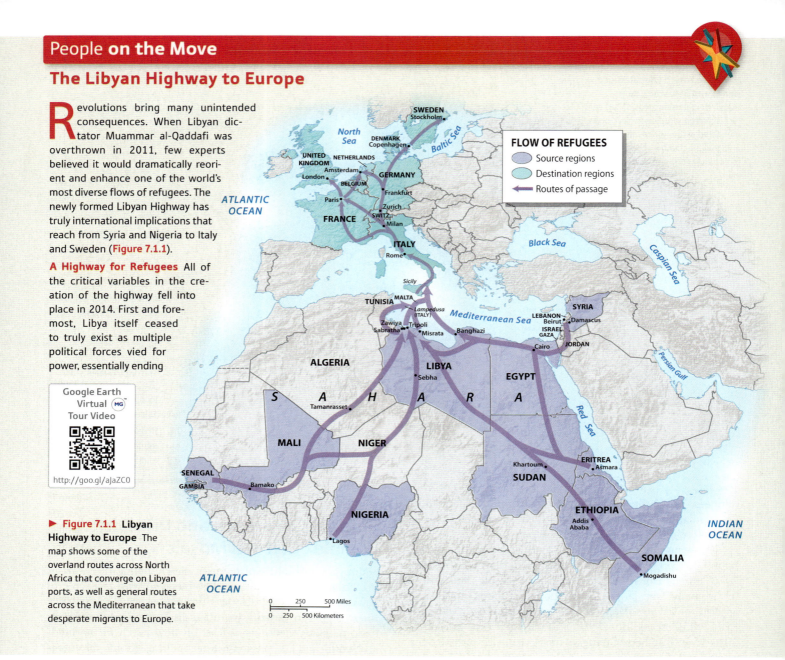

► **Figure 7.1.1 Libyan Highway to Europe** The map shows some of the overland routes across North Africa that converge on Libyan ports, as well as general routes across the Mediterranean that take desperate migrants to Europe.

FLOW OF REFUGEES
- Source regions
- Destination regions
- Routes of passage

the quiet ritual of morning prayers or profound discussions about contemporary political and social issues, religion is part of the daily routine of most regional residents from Casablanca to Tehran. The geographies of religion—their points of origin, paths of diffusion, and patterns of modern regional distribution—are essential elements in understanding cultural and political conflicts within the region.

Hearth of the Judeo-Christian Tradition Both Jews and Christians trace their religious roots to the eastern Mediterranean, and while neither group is numerically dominant across the area, each plays a key cultural role. The roots of Judaism lie deep in the past: Abraham, an early patriarch in the Jewish tradition, lived some 4000 years ago and led his people from Mesopotamia to Canaan

Explore the Sounds of Jewish Folk Music

http://goo.gl/lgO9FS

(modern-day Israel), near the shores of the Mediterranean. From Jewish history, recounted in the Old Testament of the Holy Bible, springs a rich religious heritage focused on a belief in one God (or **monotheism**), a strong code of ethical conduct, and a powerful ethnic identity that continues to the present. Around 70 CE, during the time of the Roman Empire, most Jews were forced to leave the eastern Mediterranean after they challenged Roman authority. The resulting forced migration, or diaspora, of the Jews took them to the far corners of Europe and North Africa. Only in the past century have many of the world's far-flung Jewish populations returned to the religion's hearth area, a process that accelerated greatly with the formation of the Jewish state of Israel in 1948.

Christianity also emerged in the vicinity of modern-day Israel and has left a lasting legacy across the region. An outgrowth of Judaism, Christianity was based on the teachings any effective control over the country. Migrants and smugglers were free to make trip arrangements without much fear of government interference.

Second, an unregulated extralegal industry designed around transporting desperate refugees expanded across the region. Former North African tour guides who owned a jeep or a van could make a new, often lucrative living by driving refugees. This new geography includes overland travel routes across Libya (with checkpoints, smuggler transfer agents, and staging areas) as well as a casual, often chaotic collection of boats (everything from unsafe inflatable rafts to rickety fishing vessels) available to make the short but hazardous journey to Europe (most commonly to southern Italy and Malta) (**Figure 7.1.2**). The Italian island of Lampedusa (south of Sicily) has been an initial destination for many migrants.

Third, Europe shone as an ever-brighter destination (Germany, Italy, Sweden, and Switzerland have accepted the most asylum seekers) as prospects for decent lives dimmed in West Africa (Mali, Gambia, and Nigeria), the African Horn (Eritrea and Somalia), Southwest Asia (Syria and Gaza), and Libya itself. No surprise that desperate Syrian refugees have dominated recent flows as that country disintegrates. Experts estimate that more than 219,000 people migrated from North Africa to Europe in 2014, most of them along the Libyan Highway. Even reports of disastrous journeys (in 2015, more than 900 people died as an overloaded boat sank) have failed to quell the human migration north. In one week in 2016, more than 13,000 refugees were plucked from the sea on their way from North Africa to Europe.

Addressing the Challenge Some officials within the European Union (EU) have asked how to throw an effective roadblock across the Libyan Highway—or at least how to safely control the traffic flow. In the short term, EU nations have agreed to improve patrols and rescue efforts in the Mediterranean to reduce the risk of disasters like the one that took almost 1000 lives in 2015. There is also movement toward a broader resettlement plan for refugees once they fall within EU jurisdiction, an initiative pushed by Italians overwhelmed by migrants. Still, many European interests—especially anti-immigrant political parties—simply want

▲ **Figure 7.1.2 Refugees on the Mediterranean Sea, 2015** These refugees, rescued in the Straits of Sicily in 2015, were only 30 miles off the coast of Libya. Most of the migrants on this boat were from Eritrea or Syria.

to shut the highway down. In the longer term, however, the Libyan Highway can be controlled only when stable civil authority returns to Libya and when conditions improve in the far-flung localities generating the refugees in the first place. Neither appears likely soon, so for the foreseeable future the hazardous route north will continue to swell with desperate people on the move.

1. From the diverse list of migrant source areas mentioned above, choose two countries and write a paragraph on each that explains why residents of these areas are willing to make the journey.

2. Should Europe welcome or curtail these diverse migrants? Defend your answer.

---

of Jesus and his disciples, who lived and traveled in the eastern Mediterranean about 2000 years ago. Although many Christian traditions became associated with European history, some forms of early Christianity remained strong near the religion's original hearth. To the south, one stream of Christian influences linked with the Coptic Church diffused into northern Africa, shaping the culture of places such as Egypt and Ethiopia. In the Levant, another group of early Christians, known as the Maronites, retained a separate cultural identity that survives today. The region remains the "Holy Land" in the Christian tradition and annually attracts millions of the faithful as they visit the most sacred sites in the Christian world.

The Emergence of Islam Islam originated in Southwest Asia in 622 CE, forming yet another cultural hearth of global significance. Muslims can be found today from North America

to the southern Philippines; however, the Islamic world is still centered on its Southwest Asian origins. Most Southwest Asian and North African peoples still follow its religious teachings and moral doctrines. Muhammad, the founder of Islam, was born in Makkah (Mecca) in 570 CE and taught in nearby Madinah (Medina) (**Figure 7.27**). In many respects, his creed parallels the Judeo-Christian tradition. Muslims believe that both Moses and Jesus were true prophets and that both the Hebrew Bible (or Old Testament) and the Christian New Testament, while incomplete, are basically accurate. Ultimately, however, Muslims hold that the **Quran** (or Koran), a book of revelations received by Muhammad from Allah (God), represents God's highest religious and moral revelations to humankind.

The basic teachings of Islam offer an elaborate blueprint for leading an ethical and a religious life. Islam literally means "submission to the will of God," and the creed rests on five essential

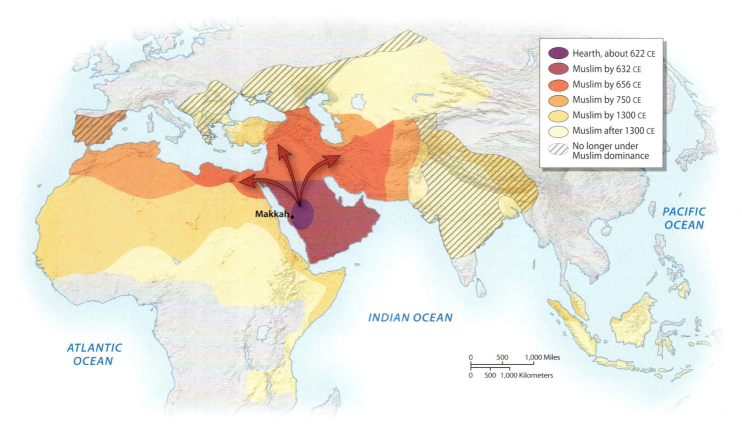

Hearth, about 622 CE	
Muslim by 632 CE	
Muslim by 656 CE	
Muslim by 750 CE	
Muslim by 1300 CE	
Muslim after 1300 CE	
No longer under Muslim dominance	

▲ **Figure 7.27 Diffusion of Islam** The rapid expansion of Islam that followed its birth is shown here. From Spain to Southeast Asia, Islam's legacy remains strongest nearest its Southwest Asian hearth. In some settings, its influence has ebbed or has come into conflict with other religions, such as Christianity, Judaism, and Hinduism.

pillars: (1) repeating the basic creed ("There is no god but God, and Muhammad is his prophet"); (2) praying facing Makkah five times daily; (3) giving charitable contributions; (4) fasting between sunup and sundown during the month of Ramadan; and (5) making at least one religious pilgrimage, or **Hajj**, to Muhammad's birthplace of Makkah (**Figure 7.28**).

Islam is a more austere religion than most forms of Christianity, and its modes of worship and forms of organization are generally less ornate. Muslims avoid the use of religious images, and the strictest interpretations even forbid the depiction of the human form. Followers of Islam are instructed to lead moderate lives, avoiding excess. Many Islamic fundamentalists still argue for a **theocratic state**, such as modern-day Iran, in which religious leaders (ayatollahs) guide public policy.

A major religious schism divided Islam early on, and the differences endure to the present. The breakup occurred almost immediately after the death of Muhammad in 632 CE. Key questions surrounded the succession of religious power. One group, now called the **Shiites**, favored passing on power within Muhammad's own family—specifically to Ali, his son-in-law. Most Muslims, later known as **Sunnis**, advocated passing power down through the

▼ **Figure 7.28 Makkah** Thousands of faithful Muslims gather at the Grand Mosque in central Makkah, part of the pilgrimage to this sacred place that draws several million visitors annually.

established clergy. This group emerged victorious. Ali was killed, and his Shiite supporters went underground. Ever since, Sunni Islam has formed the mainstream branch of the religion, to which Shiite Islam has presented a recurring, and sometimes powerful, challenge (**Figure 7.29**). The Shiites argue that a successor to Ali will someday return to reestablish the pure, original form of Islam. The Shiites also are more hierarchically organized than the Sunnis; Iran's Shiite ayatollahs, for example, have overriding religious and political power.

In a very short period of time, Islam diffused widely from its Arabian hearth, often following camel caravan routes and Arab military campaigns as it expanded its geographical range and converted thousands to its beliefs. By the time of Muhammad's death in 632 CE, the peoples of the Arabian Peninsula were united under its banner. Shortly thereafter, the Persian Empire fell to Muslim forces, and the Eastern Roman (or Byzantine) Empire lost most of its territory to Islamic influences. By 750 CE, Arab armies had swept across North Africa, conquered most of the Iberian Peninsula (modern Spain and Portugal), and established footholds in Central and South Asia. At first, only the Arab conquerors followed Islam, but the diverse inhabitants of Southwest Asia and North Africa gradually were absorbed into the religion, although with many distinct local variants. By the 13th century, most people in the region were Muslims, while older religions such as Christianity and Judaism became minority faiths or disappeared altogether.

Between 1200 and 1500, Islamic influences expanded in some areas and contracted in others. The Iberian Peninsula returned to Christianity by 1492, although many Moorish (Islamic) cultural and architectural features remained behind and still shape the region today. At the same time,

Muslims expanded their influence southward and eastward into Africa. In addition, Muslim Turks largely replaced Christian Greek influences in Southwest Asia after 1100. One group of Turks moved into the Anatolian Plateau and finally conquered the last vestiges of the Byzantine Empire in 1453. These Turks soon created the vast **Ottoman Empire** (named after one of its leaders, Osman), which included southeastern Europe (including modern-day Albania, Bosnia, and Serbia) and most of Southwest Asia and North Africa. The legacy of the Ottoman Empire was considerable. It offered a new, distinctly Turkish interpretation of Islam, and it provided a focus of Muslim political power within the region until the empire's disintegration in the late 19th and early 20th centuries.

Modern Religious Diversity Today Muslims form the majority population in all the countries of Southwest Asia and North Africa except Israel, where Judaism dominates (see Figure 7.29). Still, divisions within Islam create regional cultural differences. Many of the region's recent conflicts are defined along Sunni–Shiite lines, although specific issues focus more on power, politics, and economic policy than on theological differences.

The region is dominated by Sunni Muslims (73 percent), but Shiites (23 percent) remain important elements in the contemporary cultural mix. In Iraq, for example, southern Shiites (around Najaf, Karbala, and Basra) asserted their cultural and political power following the fall of Saddam Hussein in 2003. Shiites also claim majorities in Iran and Bahrain, form substantial minorities in Lebanon, Yemen, and Egypt, and comprise about 10 percent of Saudi Arabia's population

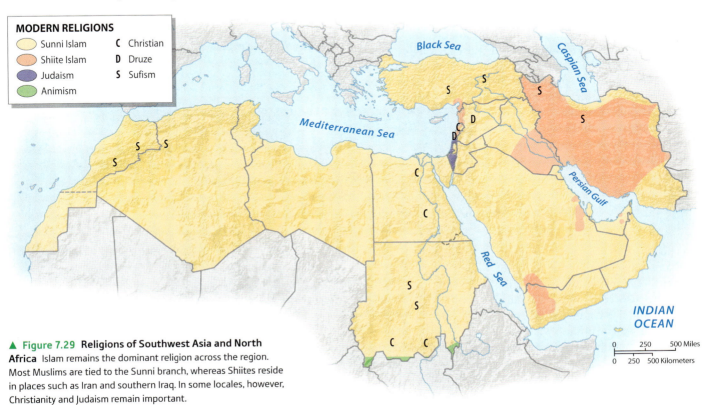

▲ **Figure 7.29 Religions of Southwest Asia and North Africa** Islam remains the dominant religion across the region. Most Muslims are tied to the Sunni branch, whereas Shiites reside in places such as Iran and southern Iraq. In some locales, however, Christianity and Judaism remain important.

(see Figure 7.29). Since 1980, many radicalized Shiite groups have pushed a cultural and political agenda of Islamic fundamentalism across the region. Cultural and political tensions are often evident within Shiite communities as fundamentalist elements clash with more secular Muslims. In Iran, for example, many younger, urban, and more affluent Muslims reject much of the extremist rhetoric of Shiite clerics who continue to entirely reject Western values and cultural practices.

Recently, Sunni fundamentalism has also been on the rise, most evident in the growth of ISIL in Iraq, Syria, and beyond. More mainstream Sunnis still make up the majority of the region's population and reject its more radical cultural and political precepts, arguing for a more modern Islam that accommodates some Western values and traditions. Even in a moderate Islamic country such as Turkey, however, tensions between more secular and more fundamentalist Sunnis have been on the rise in the past 15 years.

While the Sunni–Shiite split is the Muslim world's great divide, other variations of Islam are also practiced in the region. One division separates the mystically inclined form of Islam known as *Sufism* from mainstream traditions. Sufism is prominent in the peripheries of the region, including the Atlas Mountains and across northwestern Iran and portions of Turkey. Elsewhere, the Salafists and Wahhabis, numerous in both Egypt and Saudi Arabia, are austere, conservative Sunnis who adhere to what they see as an earlier, purer form of Islamic doctrine. The Druze of Lebanon practice yet another variant of Islam.

Southwest Asia also is home to many non-Islamic communities. Israel's dominant Jewish population (77 percent) is divided between Jewish fundamentalists and more reform-minded Jews. The country also has an important Muslim minority (16 percent), and Christians (including Armenians) make up another 2 percent of the total.

Jerusalem (Israel's capital) holds special religious significance for several groups and also stands at the core of the region's political problems (**Figure 7.30**). Indeed, the sacred space of this ancient city remains deeply scarred and divided. Considering just the 220 acres of land within the Old City, Jews pray at the old Western Wall (the site of a Roman-era Jewish temple); Christians honor the Church of the Holy Sepulchre (the burial site of Jesus); and Muslims hold sacred rites in the city's eastern quarter (including the place from which the prophet Muhammad reputedly ascended to heaven). Nearby suburban communities are also contested real estate as Arab and Israeli neighborhoods (including newly built Jewish settlements) uneasily sit next to one another.

Neighboring Lebanon also has a complex religious geography.

The nation had a slight Christian (Maronite and Orthodox) majority as recently as 1950. Christian out-migration and differential birth rates, however, have created a nation that today is about 60 percent Muslim. Christians also form approximately 5 percent of Syria's population; Iraqi Christians, concentrated mostly in the rugged northern uplands, make up about 2 percent of its population.

Important Jewish and Christian communities also have left a long legacy across North Africa. Roman Catholicism was once dominant in much of the Maghreb, but it disappeared several hundred years after the Muslim conquest. The Maghreb's Jewish population, on the other hand, remained prominent until the post–World War II period, when most of the region's Jews migrated to the new state of Israel.

In Egypt, Coptic Christianity has maintained a presence through the centuries and today includes approximately 8 percent of the country's population. In earlier years, the Coptic community had a secure place in Egyptian society, and numerous Copts held high-level posts in government and business. Today, however, Egypt's Christians are being increasingly marginalized, and some of their communities have been put under pressure, even subjected to physical attack, by extremist Islamic elements.

Overall, there has been a gradual decline in Christian populations across most countries within the region. A century ago, about 14 percent of the region's population was Christian; today it is only 4 percent. Egypt, Syria, and Iraq have witnessed the largest recent declines and, given current political uncertainties, the exodus is likely to continue.

Finally, a smaller area of diverse, locally based animist religions is found across southern Sudan along the border with South Sudan.

▼ **Figure 7.30 Old Jerusalem** The historic center of Jerusalem reflects its varied religious legacy. Sacred sites for Jews, Christians, and Muslims are all located within the Old City. The Western Wall, a remnant of the ancient Jewish temple, stands at the base of the Dome of the Rock and Islam's al-Aqsa Mosque.

Geographies of Language

Although the region is often referred to as the "Arab World," linguistic complexity creates many important cultural divisions across Southwest Asia and North Africa (**Figure 7.31**). The geography of language offers insights into regional patterns of ethnic identity and potential cultural conflicts that exist at linguistic borders. The language map is useful in identifying the major families found across the region, but it is important to remember that many local variations in language, representing distinctive dialects and well-defined islands of cultural and ethnic identity, are not seen on the map. For example, more than 70 separate languages are recognized in Iran, 36 in Turkey, and 23 in Iraq.

Semites and Berbers Afro-Asiatic languages dominate much of the region. Within that family, Arabic-speaking Semitic peoples can be found from Morocco to Saudi Arabia. Before the expansion of Islam, Arabic was limited to the Arabian Peninsula. Today, however, Arabic is spoken from the Persian Gulf to the Atlantic and reaches southward into Sudan, where it borders the Nilo-Saharan-speaking peoples of Sub-Saharan Africa. As the language has diffused, it has slowly diverged into local dialects. As a result, the everyday Arabic spoken on the streets of Fes, Morocco, is quite distinct from the Arabic spoken in the UAE. The Arabic language also has a special religious significance for all Muslims because it was the sacred language in which God delivered his message to Muhammad. Although most of the world's Muslims do not speak Arabic, the faithful often memorize certain prayers

in the language, and many Arabic words have entered other important languages of the Islamic world. Advanced Islamic learning, moreover, demands competence in Arabic.

Hebrew, another traditional Semitic language of Southwest Asia, was recently reintroduced into the region with the creation of Israel. Hebrew originated in the Levant and was used by the ancient Israelites 3000 years ago. Today its modern version survives as the sacred tongue of the Jewish people and is the official language of Israel, although the country's non-Jewish population largely speaks Arabic. English is also widely used as a second or third language throughout the country.

While Arabic eventually spread across North Africa, several older languages survive in more remote areas. Older Afro-Asiatic tongues endure in the Atlas Mountains and in certain parts of the Sahara. Collectively known as Berber, these languages are related to each other, but are not mutually intelligible. Most Berber languages have never been written, and none has generated a significant literature. Indeed, a Berber-language version of the Quran was not completed until 1999. The decline in pastoral nomadism and pressures of modernization threaten the integrity of these Berber languages. Scattered Berber-speaking communities are found as far to the east as Egypt, but Morocco is the center of this language group, where it plays an important role in shaping that nation's cultural identity.

Persians and Kurds Although Arabic spread readily through portions of Southwest Asia, much of the Iranian Plateau and nearby mountains are dominated by older Indo-European languages. Here the principal tongue remains

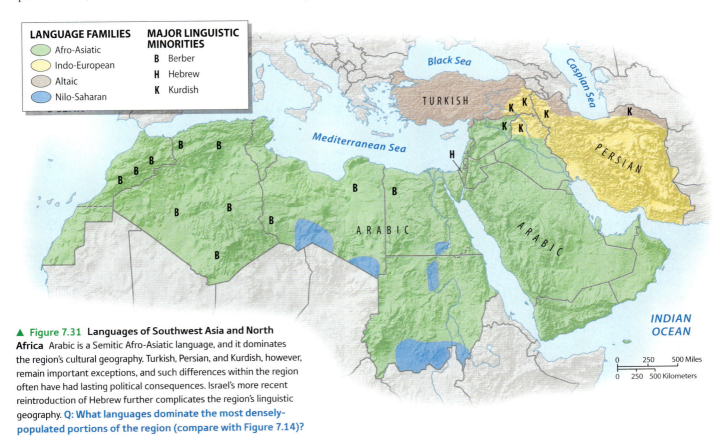

▲ **Figure 7.31 Languages of Southwest Asia and North Africa** Arabic is a Semitic Afro-Asiatic language, and it dominates the region's cultural geography. Turkish, Persian, and Kurdish, however, remain important exceptions, and such differences within the region often have had lasting political consequences. Israel's more recent reintroduction of Hebrew further complicates the region's linguistic geography. **Q: What languages dominate the most densely-populated portions of the region (compare with Figure 7.14)?**

Persian, although, since the 10th century, the language has been enriched with Arabic words and written in the Arabic script. Persian, like other languages, has developed distinct local dialects. Today Iran's official language is called *Farsi*, which denotes the form of Persian spoken in Fars, the area around the city of Shiraz. Thus, although both Iran and neighboring Iraq are Islamic nations, their ethnic identities spring from quite different linguistic and cultural traditions.

The Kurdish speakers of northern Iraq, northern Iran, and eastern Turkey add further complexity to the regional pattern of languages. Kurdish, also an Indo-European language, is spoken by 10–15 million people in the region (**Figure 7.32**). Kurdish has not historically been a written language, but the Kurds do have a strong sense of shared cultural identity. Indeed, "Kurdistan" has sometimes been called the world's largest nation without its own political state because the group remains a minority in several countries of the region. In the 2003 war in Iraq, the Kurds emerged as a cohesive group in the northern part of the country and opposed Saddam Hussein. Iraqi Kurds have gained more political autonomy in a post-Saddam Iraq state and are also intent on maintaining control of important oil resources found in their portion of the country. Their leading city of Kirkuk has been called the "Kurdish Jerusalem" because its history and settlement so richly capture the group's cultural identity. Recently, however, the Iraqi Kurds have suffered in the battle with ISIL and thousands have left the country since 2014. Nearby Kurds in eastern Turkey are also under siege. They complain that their ethnic identity is frequently challenged by the majority Turks, leaving some to wonder if they will attempt to join forces with their Iraqi neighbors.

The Turkic Imprint Turkic languages provide more variety across much of modern Turkey and in portions of far northern Iran. The Turkic languages are a part of the larger Altaic language family that originated in Central Asia. Turkey remains the largest nation in Southwest Asia dominated by that family. Tens of millions of people in other countries of Southwest and Central Asia speak related Altaic languages, such as Azeri, Uzbek, and Uighur. During the era of the Ottoman Empire, Turkic speakers ruled much of Southwest Asia and North Africa, but Iran is the only other large country in the region today where Turkic languages persist, particularly in the northwest part of the country.

Regional Cultures in the Global Context

Many cultural connections tie the region with the world beyond. These global connections are nuanced and complex and are expressed in fascinating, often unanticipated ways.

Religion and Dietary Preferences For example, religiously-based dietary preferences (known as *halal* in Islam and *kashrut* [or kosher-style food preparation] in Judaism) shape food consumption patterns, both within the region and globally. From Indonesia to the United States, traditional Muslims and Jews incorporate dietary preferences that are based on holy teachings within the Quran and Torah. Certain practices are shared by these two religious faiths: pork and the consumption of animal blood are avoided, and foods are traditionally prepared using prescribed methods of slaughter (**Figure 7.33**). In addition, alcohol consumption is prohibited in the Islamic world, and orthodox Jews carefully avoid consuming meat and dairy together.

In addition to traditional food preferences, many younger, more urban residents in the region are growing more accustomed to American-style fast food, sushi, and

▼ **Figure 7.32 Kurdish Women** Eastern Turkey is home to these young Kurdish women. Many Kurds in this region face discrimination.

▼ **Figure 7.33 Ritual Food Preparation, Israel** An ultra-Orthodox Jew twirls a live chicken over the head of his wife as he recites a prayer. Religious Jews believe the practice transfers sins into the bird. The bird is then butchered in a kosher manner and the meat usually donated to the poor.

Explore the **Tastes** of Kosher Meats

http://goo.gl/MkRk2M

Everyday **Globalization**
Falafel Round the World

Falafel has arrived on the global stage. This ancient Middle Eastern food (perhaps originating among Coptic Christians in Egypt) is becoming a trendy, easy-to-prepare, vegetarian alternative with worldwide appeal. From London to Los Angeles, these tasty deep-fried balls (or sometimes patties) are filled with some combination of ground chickpeas and fava beans, stuffed in a pita, and topped with green garnish, vegetables, and tasty tahini sauce (**Figure 7.2.1**). Long a national dish in both Egypt and Israel (falafel spans religious and political divides), it is well-designed as street food and as fast food (for a while, some McDonalds even marketed the McFalafel).

In 2016, London's inaugural Falafel Festival was marked by a fierce but nonviolent competition between Israeli, Lebanese, Palestinian, and Egyptian chefs. Judges finally awarded the Egyptian chef the top prize, eking out a victory over Israel and prompting many lighthearted tweets celebrating Egypt's culinary conquest of its neighbor.

Falafel may be coming soon to a community near you. A global franchiser, Just Falafel, opened in 2007 in Abu Dhabi and the business is rapidly expanding. Company owners want to do for deep-fried chick peas what Starbucks did for Frappuccinos. Numerous restaurants already have sprouted in the United Kingdom, plans are afoot for 70 branches in the United States, more than 60 in Australia, and additional outlets in India.

1. As a marketing director for your new local falafel fast-food restaurant, make a case for the food's appeal in your community.

2. Identify another authentic "foreign" food that is popular in your community, and explain why.

▲ **Figure 7.2.1 Falafel** These deep-fried delicacies, usually made of ground chickpeas and fava beans, have become a popular Middle Eastern food choice around the world.

Explore the **Tastes** of Falafel

http://goo.gl/QydFqE

other examples of changing global dietary preferences. It is also a two-way street: the growing popularity of Middle Eastern food in Europe, North America, and beyond represents another example of globalization (see *Everyday Globalization: Falafel Round the World*).

Islamic Internationalism Islam, in particular, has also become a globally dispersed religion. All Muslims recognize the fundamental unity of their religion, which extends far beyond Southwest Asia and North Africa. Islamic communities are well established in such distant places as central China, European Russia, central Africa, Indonesia, and the southern Philippines. Today Muslim congregations also are expanding rapidly in the major urban areas of western Europe and North America. Even with its global reach, however, Islam remains centered on Southwest Asia and North Africa, the site of its origins and its holiest places. As Islam expands in number of followers and geographic scope, the religion's tradition of pilgrimage ensures that Makkah will become a city of increasing global significance in the 21st century. Today, the Hajj is a huge business for Makkah and authorities are increasingly enlarging local infrastructure to cope with the crowds and imposing limits on how many people can visit the city's sacred sites.

The recent global growth of Islamist fundamentalism and Islamism also focuses attention on the region. In addition, the oil wealth accumulated by many Islamic nations is used to sustain and promote the religion. Countries such as Saudi Arabia invest in Islamic banks and economic ventures and make donations to Islamic cultural causes, colleges, and hospitals worldwide. In Southeast Asia, for example, money from Saudi Arabia is funding a growing number of new mosques being built in several Muslim portions of that world region.

Globalization and Technology Technology also shapes cultural and political change. Particularly among the young, millions within and beyond the region found themselves linked by the Internet, cell phones, and various forms of social media during the Arab Spring uprisings of 2011 and 2012. Cell phones, blogs, email, and tweets facilitated the flow of information that helped protesters plan events and coordinate strategies with their allies. Local videos from smartphones and pinhole cameras documented government abuse and often provided (in settings such as Syria) the only proof of widespread state-supported violence. Also, the global diffusion of this information promoted the internationalization of political discourse and made it easier to spread the word about local conflicts and to identify common threads among different protest movements.

In a related fashion, the fundamentalist ISIL movement claims considerable success in recruiting new global converts to its cause (including those from Europe and the United States) via the digital world through the use of social media such as Facebook and Twitter. Often, the group communicates to its adherents, potential recruits, and its enemies through a sophisticated digital communications strategy that involves online videos, Twitter, and Facebook. A related point is that about 60 percent of the region's population is under 30—precisely the group most inclined to use these technologies and often the one most frustrated by unresponsive governments that refuse to change their ways. The battle against ISIL, however, is also gaining technological traction.

▲ **Figure 7.34 UAE Soccer Star, Ahmed Khalil** London's Wembley Stadium plays host to UAE soccer superstar Ahmed Khalil, as he matches up against British players in a preliminary contest during the 2012 Olympic Games.

For example, the U.S. Department of State has sponsored several hundred YouTube videos to counter the group's terrorist and extremist messages. In addition, Kurdish media outlets—strongly opposed to ISIL—have published online videos of military operations that have garnered millions of views.

The Role of Sports Soccer plays a hugely important cultural role in everyday life within the region, both as a spectator sport and as an activity that many young people enjoy. Most countries have national football (soccer) associations that also participate in regional and global (FIFA) league competitions. The Union of Arab Football Associations (UAFA), headquartered in Riyadh, Saudi Arabia, offers many opportunities for regional competition, often carried on the Al Jazeera Sports network. A smaller Gulf Cup of Nations competition also sparks spirited rivalries. Large soccer stadiums are a common part of modern urban landscapes, including Kuwait City's Sabah Al-Salem Stadium, which comfortably seats more than 28,000 fans. Recently, Algeria, Tunisia, Iran, and Israel have supported strong teams with high global rankings. Excitement is already growing about the potential for Qatar to host the 2022 FIFA World Cup Soccer Championships. As in other parts of the world, regional players can earn superstar status, such as the United Arab Emirate's goal-shooting forward, Ahmed Khalil (**Figure 7.34**). In 2015, Ahmed won the Asian Footballer of the Year award after leading his team to several major victories in regional play.

REVIEW

7.6 Describe the key characteristics of Islam, and explain why distinctive Sunni and Shiite branches exist today.

7.7 Identify three examples that illustrate how modern communication and transportation technologies have promoted cultural change and globalization within the region.

■ **KEY TERMS** monotheism, Quran, Hajj, theocratic state, Shiites, Sunnis, Ottoman Empire

Geopolitical Framework: Never-Ending Tensions

Geopolitical tensions remain very high in Southwest Asia and North Africa (**Figure 7.35**). While occurring more than five years ago, the Arab Spring rebellions remain a key turning point in marking this era of heightened regional tensions. Governments fell in Tunisia (where the regional movement began in late 2010), Egypt, Libya, and Yemen; widespread protests shook once-stable states such as Bahrain; and a more protracted civil war erupted in Syria, producing a huge and ongoing refugee crisis. Other countries witnessed shorter, more intermittent demonstrations against state authority in a region that has a long tradition of authoritarian political regimes. To varying degrees, these uprisings focused broadly on (1) charges of widespread government corruption; (2) limited opportunities for democracy and free elections; (3) rapidly rising food prices; and (4) the enduring reality of widespread poverty and high unemployment, especially for people under 30.

More recently, sectarian conflicts between Sunnis and Shiites have dominated the geopolitical map. Iran (mainly Shiite) and Saudi Arabia (mainly Sunni) have each played major regional roles in bankrolling their respective supporters. For example, Iran supports the Shiite-dominated Iraq government in its struggles against ISIL Sunni extremists as well as Syria's Assad regime (largely made up of Alawites, an offshoot of Shiite Islam). Saudi Arabia, from its perspective, has made it clear that it does not want to see nearby Yemen led by Shiite extremists, and began intervening directly in that conflict in 2015.

In addition to these recent conflicts, ongoing issues include Israeli-Palestinian relations. Some of these tensions relate to age-old patterns of cultural geography or to European colonialism because modern boundaries were formed by colonial powers. In addition, geographies of wealth and poverty enter the geopolitical mix: Some residents profit from petroleum resources and industrial expansion, while others struggle to feed their families. The result is a region where the political climate is charged with tension, and the sounds of bomb blasts and gunfire remain all-too-common characteristics of everyday life.

The Colonial Legacy

European colonialism arrived relatively late in Southwest Asia and North Africa, but that era left an important imprint on the region's modern political geography. Between 1550 and 1850, the region was dominated by the Ottoman Empire, which expanded from its Turkish hearth to engulf much of North Africa as well as nearby areas of the Levant, the western Arabian Peninsula, and modern-day Iraq. After 1850, Ottoman influences waned, and European colonial dominance grew after the dissolution of the Ottoman Empire in World War I (1918).

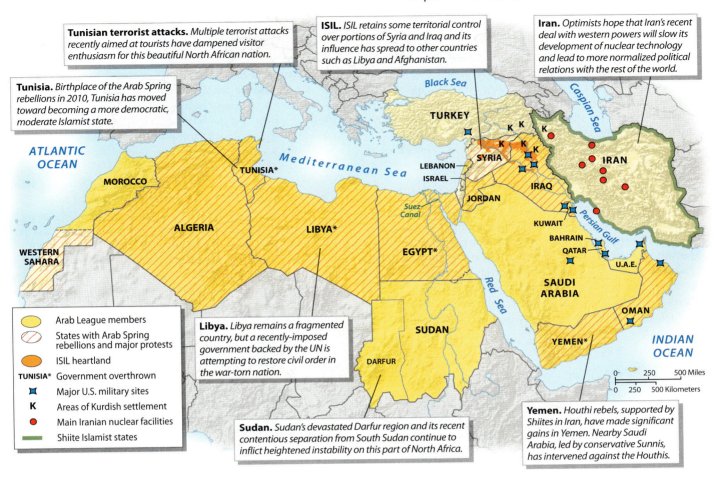

Tunisian terrorist attacks. *Multiple terrorist attacks recently aimed at tourists have dampened visitor enthusiasm for this beautiful North African nation.*

ISIL. *ISIL retains some territorial control over portions of Syria and Iraq and its influence has spread to other countries such as Libya and Afghanistan.*

Iran. *Optimists hope that Iran's recent deal with western powers will slow its development of nuclear technology and lead to more normalized political relations with the rest of the world.*

Tunisia. *Birthplace of the Arab Spring rebellions in 2010, Tunisia has moved toward becoming a more democratic, moderate Islamist state.*

Libya. *Libya remains a fragmented country, but a recently-imposed government backed by the UN is attempting to restore civil order in the war-torn nation.*

Sudan. *Sudan's devastated Darfur region and its recent contentious separation from South Sudan continue to inflict heightened instability on this part of North Africa.*

Yemen. *Houthi rebels, supported by Shiites in Iran, have made significant gains in Yemen. Nearby Saudi Arabia, led by conservative Sunnis, has intervened against the Houthis.*

Legend:
- Arab League members
- States with Arab Spring rebellions and major protests
- ISIL heartland
- TUNISIA* Government overthrown
- Major U.S. military sites
- K Areas of Kurdish settlement
- Main Iranian nuclear facilities
- Shiite Islamist states

▲ **Figure 7.35 Geopolitical Issues in Southwest Asia and North Africa** Political tensions continue across much of the region. The Arab Spring rebellions shaped subsequent political changes in several settings, and the rise of ISIL has disrupted life in Iraq, Syria, and elsewhere. The Israeli–Palestinian conflict also remains pivotal.

Both France and Great Britain were major colonial players within the region. Beginning around 1800, France became committed to a colonial presence in North Africa, including an expedition to Egypt by Napoleon. By the 1830s, France moved more directly into Algeria. French interests in North Africa also included Tunisia and Morocco. These **protectorates** in Tunisia and Morocco retained some political autonomy, but remained under a broader sphere of French influence and protection from other competing colonial powers.

During World War I, France and Great Britain signed a secret pact known as the **Sykes-Picot Agreement** that in effect proposed different spheres of European colonial influence across the region as the Ottoman Empire disintegrated. France added more colonial territories in the Levant (Syria and Lebanon). The British loosely incorporated what are now Kuwait, Bahrain, Qatar, the United Arab Emirates, and Aden (in southern Yemen) into their empire to help control sea trade between Asia and Europe. Nearby Egypt also caught Britain's attention. Once the European-engineered **Suez Canal** linked the Mediterranean and Red seas in 1869, European banks and trading companies gained more influence over the Egyptian economy. In Southwest Asia, British

and Arab forces joined to force out the Turks during World War I. The Saud family convinced the British that a country should be established on the Arabian Peninsula, and Saudi Arabia became fully independent in 1932. Britain divided its other territories into three entities: Palestine (now Israel) along the Mediterranean coast; Transjordan to the east of the Jordan River (now Jordan); and a third zone that became Iraq.

Persia and Turkey were never directly occupied by European powers. In Persia, the British and Russians agreed to establish two spheres of economic influence in the region (the British in the south, the Russians in the north), while respecting Persian independence. In 1935, Persia's modernizing ruler, Reza Shah, changed the country's name to Iran. In Turkey, European powers attempted to divide up the Ottoman Empire following World War I. The successful Turkish resistance to European control was based on new leadership provided by Kemal Ataturk. Ataturk decided to imitate the European countries and establish a modern, culturally unified, secular state.

European colonial powers began withdrawing from Southwest Asian and North African colonies before World War II. By the 1950s, most countries in the region were independent.

In North Africa, Britain withdrew troops from Sudan and Egypt in 1956. Libya (1951), Tunisia (1956), and Morocco (1956) achieved independence peacefully during the same era, but the French colony of Algeria became a major problem. Several million French citizens resided there, and France had no intention of simply withdrawing. A bloody war for independence began in 1954, and France finally agreed to an independent Algeria in 1962.

Southwest Asia also lost its colonial status between 1930 and 1960. Iraq became independent from Britain in 1932, but its later instability resulted in part from its imposed borders, which never recognized its cultural diversity. Similarly, the French division of the Levant into Syria and Lebanon (1946) greatly angered local Arab populations and set the stage for future political instability. As a favor to the Lebanese Maronite Christian majority, France carved out a separate Lebanese state from largely Arab Syria, even guaranteeing the Maronites constitutional control of the government. The action created a culturally divided Lebanon as well as a Syrian state that has repeatedly asserted its influence over its Lebanese neighbors.

Modern Geopolitical Issues

The geopolitical instability in Southwest Asia and North Africa continues today. A quick regional transect from the shores of the Atlantic to the borders of Central Asia suggests how these forces are playing out in different settings early in the 21st century.

Across North Africa Varied North African settings have recently witnessed dramatic political changes (see Figure 7.35). In Tunisia, birthplace of the Arab Spring, a moderate Islamist government was elected to replace deposed dictator Zine el-Abidine Ben Ali, but the country has not been immune to terrorist attacks by jihadist extremists, including multiple attacks aimed at tourists in 2015 and 2016.

In nearby Libya, while many cheered the end of Colonel Muammar al-Qaddafi's rule in 2011, several rival militia alliances initially split the nation into several dysfunctional fragments. Libya's political power vacuum has also made it a North African stronghold for ISIL sympathizers. In 2016, some hope for a more unified Libya emerged with a UN-backed government that controlled the country's national oil firm and its central bank. Still, separatists in the east and ISIL's continuing presence may derail any hopes for stability within the war-torn country.

Next door, following the 2011 overthrow of Hosni Mubarak, Egypt remained politically unstable. Parliamentary and presidential elections in 2012 ushered in a brief period of rule by the Muslim Brotherhood. In 2013, the Egyptian military staged a coup and ousted the Brotherhood. Since then, some political stability has returned to the country within the shadows of a regime strongly influenced by the military. Egyptian President Abdel-Fattah al-Sisi has insisted that a strong authoritarian hand (including restrictions on demonstrations and limits to a free press) is necessary to preserve the nation's relative stability amid growing regional threats, including the rise of ISIL.

Sudan also faces daunting political issues. A Sunni Islamist state since a military coup in 1989, Sudan imposed Islamic law across the country, antagonizing both moderate Sunni Muslims and the large non-Muslim (mostly Christian and animist) population in the south. Civil war between the north and south produced more than 2 million casualties (mostly in the south) between 1988 and 2004. A tentative peace agreement was signed in 2005, which opened the way for a successful vote on independence in South Sudan. Even though the two nations officially split in 2011, tensions remain, especially focused on an oil-rich and contested border zone between the two countries.

In addition, Sudan's western Darfur region remains in shambles (see Figure 7.35). Ethnicity, race, and control of territory seem to be at the center of the struggle in the largely Muslim region, as a well-armed Arab-led militia group with links to the central government in Khartoum has attacked hundreds of black-populated villages, killing more than 300,000 people (through violence, starvation, and disease) and driving 2.5 million more from their homes.

The Arab–Israeli Conflict The 1948 creation of the Jewish state of Israel produced another enduring zone of cultural and political tensions within the eastern Mediterranean (**Figure 7.36**). Jewish migration to Palestine increased after the defeat of the Ottoman Empire in World War I. In 1917, Britain issued the Balfour Declaration, a pledge to encourage the creation of a Jewish homeland in the region. After World War II, the UN divided the region into two states, one to be predominantly Jewish, the other primarily Muslim. Indigenous Arab Palestinians rejected the partition, and war erupted. Jewish forces proved victorious, and by 1949 Israel had actually grown in size. Hundreds of thousands of Palestinian refugees fled from Israel to neighboring countries, where many of them remained in makeshift camps. Under these conditions, Palestinians nurtured the idea of creating their own state on land that had become part of Israel.

Israel's relations with neighboring countries remained poor. Supporters of Arab unity and Muslim solidarity sympathized with the Palestinians, and their antipathy toward Israel grew. Israel fought additional wars in 1956, 1967, and 1973. In territorial terms, the Six-Day War of 1967 was the most important conflict (see Figure 7.36). In this struggle against Egypt, Syria, and Jordan, Israel occupied substantial new territories in the Sinai Peninsula, the Gaza Strip, the West Bank, and the Golan Heights. Israel also annexed the eastern (Muslim) part of the formerly divided city of Jerusalem, arousing particular bitterness among the Palestinians. A peace treaty with Egypt resulted in the return of the Sinai Peninsula in 1982, but tensions focused on other occupied territories under Israeli control. To strengthen its geopolitical claims, Israel built additional Jewish settlements in the West Bank and in the Golan Heights, further angering Palestinian residents.

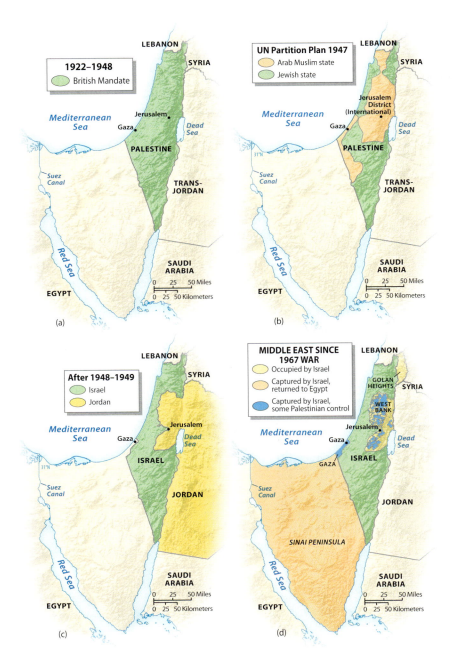

▲ **Figure 7.36 Evolution of Israel** Modern Israel's complex evolution began with (a) an earlier British colonial presence and (b) a UN partition plan in the late 1940s. (c) Thereafter, multiple wars with nearby Arab states produced Israeli territorial victories in Gaza, the West Bank, and the Golan Heights. (d) Each of these regions continues to be important in Israel's recent relations with nearby states and with resident Palestinian populations.

Palestinians and Israelis began to negotiate a settlement in the 1990s. Preliminary agreements called for a quasi-independent Palestinian state in the Gaza Strip and across much of the West Bank. A tentative agreement late in 1998 strengthened the potential control of the ruling **Palestinian Authority (PA)** in the Gaza Strip and portions of the West Bank (**Figure 7.37**). But a new cycle of heightened violence

erupted late in 2000 as Palestinian attacks against Jews increased and the Israelis continued to build new settlements in occupied lands (especially in the West Bank). Indeed, the Israeli government has pledged to continue its support for new construction, including developments in and near Jerusalem and Bethlehem (**Figure 7.38**). Palestinian authorities continue to strongly protest these new West Bank Jewish settlements near Israel's capital, especially in an area (known as E1) that sits on hills just east of Jerusalem.

Adding to the friction is the ongoing construction of an Israeli security barrier, a partially completed series of concrete walls, electronic fences, trenches, and watchtowers designed to effectively separate the Israelis from Palestinians across much of the West Bank region (see Figure 7.37). Israeli supporters of the barrier (to be more than 400 miles long when completed) see it as the only way to protect their citizens from terrorist attacks. Palestinians see it as a land grab, an "apartheid wall" designed to isolate many of their settlements along the Israeli border.

Political fragmentation of the Palestinians adds further uncertainty. In 2006, control of the Palestinian government was split between the Fatah and Hamas political parties. Israelis have long regarded Hamas as an extremist political party, whereas Fatah has shown more willingness to work peacefully with Israel. Hamas gained effective control of the PA within Gaza, and Fatah maintains its greatest influence across the West Bank. Rockets launched into Israel from Hamas-controlled Gaza have repeatedly provoked Israeli counterattacks, decimating the Gaza economy.

One thing is certain: Geographical issues will remain at the center of the conflict. Israelis continue their search for secure borders to guarantee their political integrity. Most Palestinians still call for a "two-state solution," in which their autonomy is guaranteed, but a growing minority of Palestinians, frustrated with the stalemate, suggest considering a "one-state solution," in which Israel would be compelled to recognize the Palestinians as equals.

Devastation in Syria and Iraq Elsewhere in the region, political instability in Syria erupted into civil war in 2011. Rebel (mostly Sunni Muslim) protests against the autocratic regime of President Bashar Hafez al-Assad (a member of the minority Alawite sect) reached a fever pitch, and government soldiers killed thousands of civilians and used chemical weapons in a series of violent confrontations. The larger regional Arab community reacted against Assad,

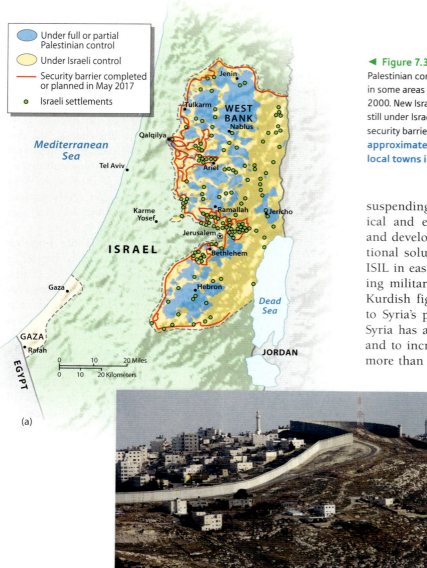

(a)

(b)

Figure 7.37 West Bank (a) Portions of the West Bank were returned to Palestinian control in the 1990s, but Israel has partially reasserted its authority in some areas and has expanded the construction of its security barrier since 2000. New Israeli settlements are scattered throughout the West Bank in areas still under Israel's nominal control. (b) The photo shows a segment of the Israeli security barrier. **Q: Look carefully at the scale of the map. Measure the approximate distance between Jerusalem and Hebron, and find two local towns in your area that are a similar distance apart.**

suspending Syria from the **Arab League** (a regional political and economic organization focused on Arab unity and development; see Figure 7.35) and urging an international solution to the crisis. Since 2014, the presence of ISIL in eastern Syria (see Figure 7.35)—as well as a growing military response to ISIL from moderate Arab states, Kurdish fighters, and U.S.-led bombing raids—has added to Syria's political disintegration. Russian involvement in Syria has also grown, mostly to bolster the Assad regime and to increase Russian influence in the region. By 2015, more than 220,000 deaths were directly related to the violence, and millions of people had fled their homes (see *Geographers at Work: How Do We Define "Middle East"?*) (**Figure 7.39**). In addition, many of Syria's great cultural landmarks—such as the ancient city of Palmyra—have been devastated as ISIL has pressed its case to erase many of these historic sites from the landscape. Many observers now see the Syrian civil war as one of the world's worst humanitarian crises since World War II.

Neighboring Iraq, another multinational state born during the colonial era, has yet to escape the consequences of its geopolitical origins. When the country was

Explore the **Sights** of West Bank Jewish Settlement

http://goo.gl/kPKByi

Figure 7.38 Jewish Settlement, West Bank This new housing is part of the Jewish settlement of Givat Ze'ev located just a few miles northwest of Jerusalem.

Figure 7.39 Syrian Refugee Camp in Turkey A Kurdish refugee woman from the Syrian town of Kobani walks with her baby at a refugee camp in the border town of Suruc, Sanliurfa province.

Geographers at Work

How Do We Define "Middle East"?

▲ Figure 7.3.1 Karen Culcasi

Karen Culcasi is no stranger to the Middle East (Figure 7.3.1). A geographer at the University of West Virginia, Culcasi regularly teaches courses on the Arab World and takes her students to the region, including explorations of Jordan and the United Arab Emirates. This gives students the opportunity to experience both the region's diverse natural setting as well as its complex cultural mosaic. The geographic perspective is important, stresses Culcasi: "The spatial element that you don't get [in other disciplines] is powerful." This is particularly true when looking at the politics of this region. "Geopolitics doesn't just happen in a place; the place affects geopolitics," she explains.

Culcasi explores how the "Middle East" has evolved as a regional idea and how it is represented on maps. She has examined old maps and atlases to discover where the term "Middle East" originated (largely from imperial Great Britain), and traveled to the region to ask local residents and experts if they use the term ("Arab Homeland" is more commonly used). She concludes that the "Middle East" is largely an imposed term popularized by European and American politicians to compartmentalize and simplify a complicated mix of peoples and places.

Refugee Experiences More recently, Culcasi has examined the challenges faced by refugees, especially women. She interviews Palestinians, both in the United States and in the region, about their homeland. How do they map its location and describe its character? She is also spending time with Syrian refugee women in Jordan, trying to understand how their lives have been suddenly transformed by this incredible disruption (Figure 7.3.2). Whether exploring archives or working with people in a

▲ Figure 7.3.2 **Refugee Camp** Professor Culcasi's recent research has taken her to Syrian refugee camps in Jordan where she has interviewed women about their recent experiences.

refugee camp, Culcasi credits her geography studies for her research approach: "My professors helped me cultivate a critical perspective and to question assumptions. And on the undergrad level, the diversity and breadth of geography is a benefit."

1. On a blank map of Southwest Asia and North Africa, draw a line around your definition of the "Middle East" and then write a paragraph defending your answer. Compare your map with those of classmates.

2. Select a regional term used locally ("New England," "Southern California," "the Panhandle," etc.). Have five friends/classmates identify the area on a blank map of the region and defend their answers. Then summarize and explain their responses.

carved out of the British Empire in 1932, it contained the cultural seeds of its later troubles. Iraq remains culturally complex today (Figure 7.40). Most of the country's Shiites live in the lower Tigris and Euphrates river valleys near and south of Baghdad. Indeed, the region near Basra contains some of the world's holiest Shiite shrines. In northern Iraq, the Kurds have their own ethnic identity and political aspirations. Many Kurds want complete independence from Baghdad and have managed to establish a federal region that already enjoys some autonomy from the central Iraqi government. A third major subregion, traditionally dominated by the Sunnis, encompasses part of the Baghdad area as well as territory to the north and west that includes strongholds such as Fallujah and Tikrit. The country's oil fields are mainly located in Shiite- and Kurdish-controlled portions of Iraq, an uneasy fact of resource geography that has long troubled the nation's sizable Sunni population.

When Iraqi leaders assumed control of their new state in 2004, growing sectarian violence between different Iraqi factions threw portions of the nation into civil war. Rival Sunni and Shiite groups forced many Iraqis from their communities. Before largely leaving Iraq in 2011, American troops successfully worked with Iraqi officials to reduce the level of violence. But the growth of ISIL in 2014 (as in nearby Syria) disrupted Iraq's regional balance of power. As ISIL enlarged its sphere of influence in the north and west, increased military responses from the United States and others (especially Shiite-led militias supported by Iran) suggest that Iraq will remain a political battleground haunted by sectarian violence and terrorism for the foreseeable future.

Politics in the Arabian Peninsula Change has also rocked the Arabian Peninsula. In Saudi Arabia, the Al Saud royal family retains its conservative control of the country, although the regime is gradually passing into the hands of younger family members who might be more inclined to democratize the nation's political structure. Saudi Arabia officially supports U.S. efforts to provide stable flows of petroleum and fight

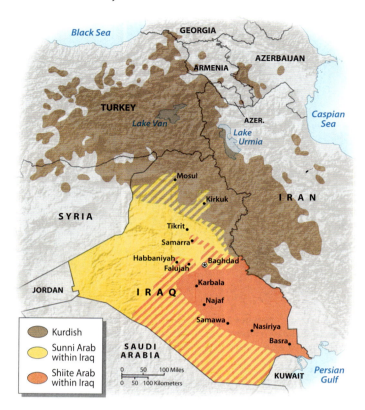

▲ **Figure 7.40 Multicultural Iraq** Iraq's complex colonial origins produced a state with varying ethnic characteristics. Shiites dominate south of Baghdad, Sunnis hold sway in the western triangle zone, and Kurds are most numerous in the north, near oil-rich Kirkuk and Mosul.

terrorism, but beneath the surface, certain elements of the regime may have financed radically anti-American groups such as Al Qaeda. The Saudi people themselves, largely Sunni Arabs, are torn among an allegiance to their royal family (and the economic stability it brings); the lure of a more democratic, open Saudi society; and an enduring distrust of foreigners, particularly Westerners. The Sunni majority also includes Wahhabi sect members, whose radical Islamist philosophy has fostered anti-American sentiment. In addition, the large number of foreign laborers and the persistent U.S. military and economic presence within the country (a chief complaint of former Al Qaeda leader Osama bin Laden) create a setting ripe for political instability.

Nearby Yemen has been torn apart by political conflict. President Ali Abdullah Saleh was forced from office, and elections were held in 2012. Calls for democratic reforms have been complicated by ongoing factionalism within the country, including the presence of Shiite militants (the Houthis) who maintain close connections with Iran. Houthi political gains in Yemen (including their taking control of the capital, San'a) provoked a military response in 2015 from Saudi Arabia, which fears growing Iranian interests in the region. Over 9000 civilian deaths or injuries have been attributed to the Saudi air campaign, which has garnered increasing international disapproval. Added to the mix is Al Qaeda's influence (including terrorist training camps) in other portions of the country.

Iran Ascendant? Iran increasingly attracts global notice. Islamic fundamentalism dramatically appeared on the political scene in 1978 as Shiite Muslim clerics overthrew Mohammad Reza Pahlavi, an authoritarian, pro-Western ruler friendly to U.S. interests. The new leaders proclaimed an Islamic republic in which religious officials ruled both clerical and political affairs.

Today Iran's influence has grown across the region. The country supports Shiite-allied interests throughout the region (including the Houthis in Yemen, sympathetic regimes in Iraq and Syria, and the Hezbollah movement in Lebanon) and has repeatedly threatened Israel. Adding uncertainty has been Iran's ongoing nuclear development program, an initiative its government claims is related to the peaceful construction of power plants (see Figure 7.35). Both Israel and Arab states such as Saudi Arabia, the United Arab Emirates, and Egypt fear Iran's ascendance. Others in the West (including the United States) negotiated a settlement with Iran in 2015 that allows for more limited development of its nuclear capabilities and an end to many of the economic sanctions against the country. Optimists hope that this leads to more normalized political relations between Iran and the rest of the world.

Within Iran, varied political and cultural impulses are also evident. Many younger, wealthier, more cosmopolitan Iranians are hopeful that the country will become less isolated on the world stage. After all, more than 70 percent of the country's population was not even born at the time of the Iranian Revolution in the late 1970s. Popular interest in fundamentalism has waned, and many Iranians have actually moved toward a more secular lifestyle. At the same time, most Iranians support the nuclear program, arguing that they have the same right as Pakistan or India to develop this resource. Hard-line religious extremists also maintain control of key government positions, and these Shiite clerics continue to harshly criticize remaining Western sanctions and suspected interference in their country.

Tensions in Turkey Turkey has also emerged as a geopolitical question mark, as it is strategically positioned between diverse, often contradictory geopolitical forces. Many pro-Westerners within Turkey, for example, are committed to joining the EU. To do so, the country has embarked on an active agenda of reforms designed to demonstrate its commitment to democracy. On the other hand, Islamist elements (mainly Sunni) within the country are wary of moving too close to Europe. In addition, Turkey's President, Recip Erdogan, responding to a failed military coup within the country in 2016, cracked down on dissidents who were challenging his increasingly broad assertion of presidential powers.

Regional issues within the country also remain important. In the east, the Kurds (a key cultural minority in Turkey) have continued to press for more recognition and regional autonomy from the Turkish government (see Figure 7.40). The militant Kurdistan Workers' Party (PKK) and their allies are frequently the target of government air raids, while more moderate Kurdish groups have pressed for closer ties with the Turkish government. Turkey's government also fears that Turkish Kurds will unite with their counterparts in nearby Syria and Iraq, and press for a larger independent state.

In addition, Syria's political fragmentation has produced a huge refugee problem in the southern part of the country, and Turkey has closed its border with its troubled neighbor to control the flow of desperate Syrian migrants (see Figure 7.26).

REVIEW

7.8 Describe the role played by the French and British in shaping the modern political map of Southwest Asia and North Africa. Provide specific examples of their lasting legacy.

7.9 Discuss how the Sunni–Shiite split has recently played out in sectarian violence across the region.

7.10 Explain how ethnic differences have shaped Iraq's political conflicts in the past 50 years.

■ **KEY TERMS** protectorates, Sykes-Picot Agreement, Suez Canal, Palestinian Authority (PA), Arab League

Economic and Social Development: Lands of Wealth and Poverty

Southwest Asia and North Africa constitute a region of incredible wealth and discouraging poverty (Table 7.2). While some countries enjoy prosperity, due mainly to rich reserves of petroleum and natural gas, other nations are among the world's least developed. Continuing political instability contributes to the region's economic challenges. Civil wars and internal conflicts within Syria, Iraq, and Libya have devastated their economies. Palestinians living in the Gaza and West Bank regions also suffer as political minorities within Israel. Elsewhere, recent economic sanctions have hurt the Iranian economy. Petroleum will

TABLE 7.2   Development Indicators

MasteringGeography™

Country	GNI per capita, PPP 2014	GDP Average Annual % Growth 2009–15	Human Development Index (2015)[1]	Percent Population Living Below $3.10 a Day	Life Expectancy (2016)[2] Male	Life Expectancy (2016)[2] Female	Under Age 5 Mortality Rate (1990)	Under Age 5 Mortality Rate (2015)	Youth Literacy (% pop ages 15–24) (2005–2014)	Gender Inequality Index (2015)[3,1]
Algeria	13,880	3.2	0.736	–	73	78	66	26	–	0.413
Bahrain	37,680	3.9	0.824	–	76	78	21	6	98	0.265
Egypt	10,280	2.5	0.690	–	70	73	86	24	92	0.573
Gaza and West Bank	5,000	4.4	–	2.6	71	75	43	21	99	–
Iran	16,590	0.2	0.766	‹2.0	74	77	61	16	98	0.515
Iraq	15,100	7.3	0.654	–	67	72	46	32	82	0.539
Israel	33,300	3.8	0.894	–	80	84	12	4	–	0.101
Jordan	11,910	2.7	0.748	‹ 2.0	73	76	37	18	99	0.473
Kuwait	79,850	3.5	0.816	–	74	76	17	9	99	0.387
Lebanon	17,590	2.6	0.769	–	76	78	33	8	–	0.385
Libya	16,000	-9.8	0.724	–	69	75	44	13	100	0.134
Morocco	7,290	3.9	0.628	15.5	73	75	81	28	82	0.525
Oman	33,690	3.5	0.793	–	75	78	48	12	99	0.275
Qatar	134,420	8.6	0.850	–	77	80	20	8	99	0.524
Saudi Arabia	51,320	5.4	0.837	–	73	75	43	15	99	0.284
Sudan	3,920	0.6	0.479	38.9	61	64	123	70	88	0.591
Syria	–	–	0.594	–	64	77	36	13	96	0.533
Tunisia	11,020	2.1	0.721	8.4	73	77	51	14	97	0.240
Turkey	19,560	5.2	0.761	3.1	75	79	72	14	99	0.359
United Arab Emirates	67,720	4.8	0.835	–	76	79	22	7	–	0.232
Western Sahara*	2,500	–	–	–	60	65	–	–	–	–
Yemen	3,650	-2.7	0.498	–	63	67	126	42	88	0.744

Sources: World Bank, *World Development Indicators*, 2016.

[1] United Nations, *Human Development Report*, 2015.

[2] Population Reference Bureau, *World Population Data Sheet*, 2016.

[3] Gender Inequality Index—A composite measure reflecting inequality in achievements between women and men in three dimensions: reproductive health, empowerment, and the labor market that ranges between 0 and 1. The higher the number, the greater the inequality.

*Additional data from the *CIA World Factbook*

Login to Mastering**Geography**™ **& access** **MapMaster** **to explore these data!**

1) Look at the HDI map and find the three countries with the *lowest* Human Development Index (HDI). What other development indicators correlate well with the HDI data? Why?

2) Where might you expect to live longest in this region? What are the top five countries in terms of life expectancy, and how do they measure up in terms of GNI per capita? Why might there be differences in these two measures?

no doubt figure significantly in the region's future economy, but some countries in the area have also focused on increasing agricultural output, investing in new industries, and promoting tourism to broaden their economic base.

The Geography of Fossil Fuels

The striking global geographies of oil and natural gas reveal the region's continuing importance in the world economy as well as the extremely uneven distribution of these resources within the region (**Figure 7.41**). North African settings (especially Algeria and Libya), as well as the Persian Gulf region, have large sedimentary basins containing huge reserves of oil and gas, whereas other localities (for example, Israel,

▼ **Figure 7.41 Crude Petroleum and Natural Gas Production and Reserves** The region plays a central role in the global geography of both (a) crude petroleum production and (b) natural gas production. Abundant regional reserves of both (c) crude petroleum and (d) natural gas suggest that the pattern will continue.

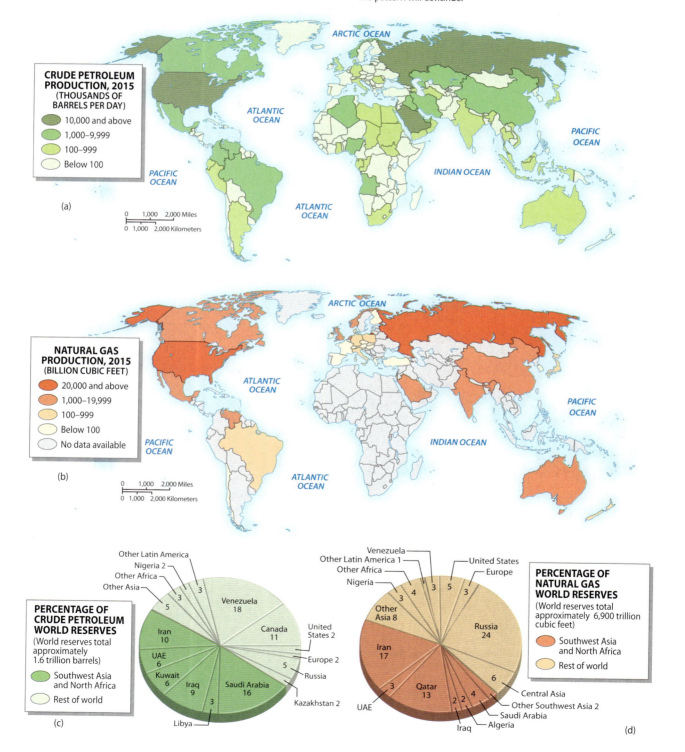

Jordan, and Lebanon) lie outside zones of major fossil fuel resources. Saudi Arabia, Iran, Iraq, Kuwait, and the United Arab Emirates hold large petroleum reserves, while Iran and Qatar possess the largest regional reserves of natural gas. The distribution of fossil fuel reserves suggests that regional supplies will not be exhausted anytime soon. Overall, with only 7 percent of the world's population, the region holds over half of the world's proven oil reserves. Saudi Arabia's pivotal position, both regionally and globally, is clear: Its 30 million residents live atop almost 20 percent of the planet's known oil supplies.

Intriguingly, the region's abundant sunshine and expansive desert landscapes also make it an ideal setting for solar energy. Recently, one of the world's largest and most modern solar power facilities opened in Morocco (see *Working Toward Sustainability: Noor 1 Shines Brightly in the North African Desert*).

Global Economic Relationships

Southwest Asia and North Africa share close economic ties with the rest of the world. While oil and gas remain critical commodities that dominate international economic linkages, the growth of manufacturing and tourism is also redefining the region's role in the world.

OPEC's Changing Fortunes Although OPEC does not control global oil and gas prices, it still influences the cost and availability of these pivotal products. While the United States has made some progress toward greater energy independence from the Mideast (via its larger domestic oil and natural gas output), western Europe, Japan, China, and many less industrialized countries still depend on the region's fossil fuels.

Lower energy prices in 2015 and 2016, however, may be signaling changes for OPEC's fortunes. Falling prices indicate increased global energy production by non-OPEC producers, such as Canada and the United States, suggesting more competition that might decrease OPEC's importance on the global stage. OPEC's move to keep production high indicates the organization (especially Saudi Arabia) is interested in driving higher-cost producers (such as North American fracking operations) out of business.

In the meantime, however, lower prices have already exerted pressure on the budgets of many OPEC nations within the region. Government programs are being cut, and many residents resent paying more for many services. The International Monetary Fund estimated that major Gulf producers recently ran a combined budget deficit of more than 6 percent of gross domestic product (compared with surpluses of more than 10 percent in recent years), and Saudi Arabia was forced to reduce government spending by more than 10 percent. Even the United Arab Emirates, one of the region's wealthiest and most diverse economies, operated with a deficit in 2016, the first in decades. Foreign workers have been hit hard in the slowdown, with many layoffs in the construction and service sectors leaving some

workers with large debts and no way to return to their home countries.

Other Global and Regional Linkages Beyond key OPEC producers, other trade flows also contribute to global economic integration. Turkey, for example, ships textiles, food products, and manufactured goods to its principal trading partners: Germany, the United States, Italy, France, and Russia. Tunisia sends more than 60 percent of its exports (mostly clothing, food products, and petroleum) to nearby France and Italy. Israeli exports emphasize the country's highly skilled workforce: Products such as cut diamonds, electronics, pharmaceuticals, and machinery parts go to the United States, western Europe, and Japan.

Future interconnections between the global economy and Southwest Asia and North Africa may depend increasingly on cooperative economic initiatives far beyond OPEC. Relations with the European Union (EU) are critical. Since 1996, Turkey has enjoyed close ties with the EU, but recent attempts at full membership in the organization have failed. Other so-called Euro-Med agreements have been signed between the EU and countries across North Africa and Southwest Asia that border the Mediterranean Sea.

Most Arab countries, however, are wary of too much European dominance. In 2005, 17 Arab League members established the **Greater Arab Free Trade Area (GAFTA)**, an organization designed to eliminate all intraregional trade barriers and spur economic cooperation. In addition, Saudi Arabia plays a pivotal role in regional economic development through organizations such as the Islamic Development Bank and the Arab Fund for Economic and Social Development. Many of these financial organizations offer services compliant with Islamic law (*sharia law*). In fact, these Islamic banking assets are projected to grow about 20 percent annually between 2015 and 2018.

Tourists are another link to the global economy. Traditional magnets such as ancient historical sites and globally significant religious localities draw millions of visitors annually. In addition, as the developed world becomes wealthier, a growing demand arises for recreational spots that offer beaches, sunshine, and novel entertainment. Indeed, many miles of the Mediterranean Sea, Black Sea, and Red Sea coastlines are now lined with upscale, but often ticky-tacky, landscapes of resort hotels and condominiums dedicated to serving travelers' needs. More adventurous travelers seek ecotourist activities such as snorkeling in Naama Bay on Egypt's Sinai coast or four-wheeling among the Berbers in the Moroccan backcountry. Endangered wildlife beckons photographers and poachers hoping to catch a glimpse of a South Arabian grey wolf, a Nubian ibex, or a darting Persian squirrel. Localities such as Dubai within the United Arab Emirates have even become globally significant transportation hubs, attracting air travelers from around the world (see *Exploring Global Connections: Dubai's Role as a Global Travel Hub*).

Noor 1 Shines Brightly in the North African Desert

Morocco's king performed the honors early in 2016 when he switched on the Noor 1 facility in the Moroccan desert near the town of Ouarzazate (Figure 7.4.1). The first of three phases, Noor 1 is a huge concentrated solar power plant that sprawls across more than 1,100 acres (450 hectares) of the North African desert (Figure 7.4.2). It is already plainly visible from space. The final facility will encompass more than 7400 acres (3000 hectares). When completed in 2018, the US$9 billion Noor complex will be the world's largest concentrated solar power facility, designed to provide electricity for 1.1 million people. Initially, consumers will be in the local region, but ultimately Morocco's leaders and other investors hope that this and similar North African solar projects can light the living rooms of Europeans living far to the north in less sunny climes. The completed Noor project will have the capacity to generate 580 megawatts of electricity, saving hundreds of thousands of tons of carbon emissions annually and reducing Morocco's energy dependence by an equivalent of about 2.5 million tons of oil per year.

Shifting Solar Technologies Noor 1, similar to several new facilities in the American Southwest, relies on so-called concentrated solar technology, which is quite different from traditional (and still very competitive) photovoltaic (or pv) technology. Photovoltaic solar cells generate power with sensitized panels that absorb sunlight and convert it directly to electricity (that typically must be transformed from DC [direct current] to AC [alternating current]), while

▲ **Figure 7.4.2 Solar Facilities at Noor 1** Solar panels shimmer in the desert sun near Ouarzazate at the newly-opened Noor 1 facility in central Morocco.

concentrated solar technology systems use parabolic mirrors to capture and concentrate heat that can then be used to generate steam to power electricity-generating turbines. The Noor 1 facility contains an array of 500,000 solar mirrors that efficiently follow the sun along its diurnal journey. One of the current advantages of the latter technology is its ability to store the heat (and the ability to generate electricity) for several hours, typically in tanks containing molten salt.

New Regional Initiatives With Noor 1's success and the prospects for the full project coming online soon, Morocco has taken a remarkable leap into the age of renewable energy. In fact, the country, poorly endowed with fossil fuels (it imports 97 percent of its traditional energy needs), hopes to generate more than 40 percent of its energy from home-grown renewables by 2020 and potentially more than half by 2030. Morocco's feat mirrors a growing regional realization that solar energy's future shines bright in a part of the world endowed with some of the highest rates of annual sunshine on the planet. Other massive solar projects are gaining traction in Egypt, Saudi Arabia, the United Arab Emirates, and Kuwait. Even Dubai—the global epicenter of petroleum wealth—has developed its Clean Energy Strategy 2050 initiative that envisions the Gulf city generating 75 percent of its electricity with renewable sources by the middle of the century. Dubai's urban leaders hope to become "the city with the smallest carbon footprint in the world by 2050," echoing Morocco's commitment to a brighter, cleaner energy future.

Google Earth Virtual (MG) Tour Video

http://goo.gl/ODMV32

▼ **Figure 7.4.1 Noor 1, Morocco** The Noor 1 solar plant is located in central Morocco.

1. Look at Figure 7.41a (Crude Petroleum Production) and identify other nations within the region, in addition to Morocco, that would logically be strong advocates for increasing their investment in renewable solar energy.

2. Find an online map of potential sunshine/solar potential for North America and describe the patterns you find. How much potential does your surrounding area have for solar energy?

Dubai's Role as a Global Travel Hub

The spidery web of the Air Emirates route map tells it all (**Figure 7.5.1**). Anchored strategically midway between Europe and Asia, Dubai's International Airport, the main route hub for Air Emirates, has blossomed into the planet's busiest center of global air travel. It is about six flying hours from Central Europe to the Middle East hub and another seven flying hours to East Asia.

A Growing Air Hub It wasn't so long ago that the airport claimed a single lonely runway in the desert, but in the past 20 years—thanks to the petroleum economy and the efficient management of the state-run facility and its flagship airline—Dubai has strikingly emerged to rule long-distance air traffic that connects many of the globe's most important places to one another. While Atlanta's Hartsfield airport still claims the most daily traffic, Dubai is the world's busiest international travel hub, surpassing London Heathrow in 2014.

In February 2016, the Dubai Airport opened a massive new $1.2 billion concourse that boosted the facility's annual capacity from 75 to 90 million passengers. Its high-tech infrastructure and upscale passenger services has made it by far the busiest, most modern airport in the Middle East (**Figure 7.5.2**). Here, passengers from Hyderabad, Amsterdam, Cape Town, and Osaka mingle with one another as they grab a meal and pass the time between flights.

Success in the Skies Air Emirates, based in Dubai, now boasts the world's largest fleet of Boeing 777s and Airbus A380s. In 2013, in a single staggering purchase, Air Emirates ordered $99 billion worth of new airplanes from the dominant commercial manufacturers (Boeing and Airbus), more wide-body jets than are flown by American Airlines and United Airlines combined. The airline's route map reveals its dominance of long-distance global travel with some nonstop flights to Latin America lasting more than 17 hours. The carrier also has a great global reputation for excellent service, on-time arrivals, and competitive prices.

Part of Dubai's success as an aviation hub is simply its location. Long-distance travelers often change planes there on trips from Europe and eastern North America as they make their way to India, China, and other Asian destinations. Many competing U.S. carriers (such as Delta and United) also claim Dubai and its state-owned airline have unfair advantages because the government controls the facility, the airplanes, and all the regulations surrounding their integrated operation. In other words, the airline and related facilities can do the job cheaper and more efficiently, out-competing

▲ **Figure 7.5.2 Dubai Airport** The Dubai Airport, located in the United Arab Emirates, is one of the world's busiest air traffic hubs.

privately-owned operators. Some see Dubai already emerging as a global *aerostate*, in which a country derives a significant portion of its economic muscle from controlling a key airport and associated air carriers. Indeed, it looks as if the airport and its associated airline will increasingly dominate long-distance travel in that part of the world.

Ironically, the recent downturn in oil prices—mostly a drag on the petro-rich regional economy—has produced even more business profitability for the commercial airline business and for Air Emirates in particular.

Google Earth
Virtual
Tour Video

http://goo.gl/JK4v8m

1. Looking at the global route map (Figure 7.5.1) for Air Emirates, which parts of the wealthier, more developed world appear to be most attractive targets for the airline's future expansion plans? Why?

2. Identify a nearby example of an airline "hub" in your area. What makes it a hub? What airlines dominate traffic?

▶ **Figure 7.5.1 Air Emirates Route Map for Dubai** Dubai's striking global centrality is apparent as Air Emirates has successfully expanded to be a major international carrier.

Emirates routes, April 2015

Political instability, however, can dampen demand for travel to areas perceived to be volatile and dangerous. In the aftermath of the disruptive Arab Spring rebellions, for example, countries such as Tunisia have redoubled their efforts to boost their tourist economies, trying to assure potential visitors that all is well. But recently several major attacks in the country were aimed at tourists, including one at a museum in Tunis, and another at a beach resort on the Mediterranean. Egypt, in particular, has seen tourist travel decline, and many hotels near historic sites and resort areas sit half empty. In 2016, the whole Sinai Peninsula was basically closed to outside visitors due to security concerns. Widespread violence in Libya, Syria, and Iraq has made travel in those areas extremely hazardous. Even Turkey's multibillion-dollar tourist economy has been hit, given growing political instability in that country and several high-profile terrorist incidents that have even hit the country's flagship city of Istanbul.

Regional Economic Patterns

Remarkable economic differences characterize Southwest Asia and North Africa (see Table 7.2). Some oil-rich countries have prospered greatly since the early 1970s, but in many cases fluctuating oil prices, political disruptions, and rapidly growing populations threaten future economic growth. Regional levels of military spending have accelerated in many nations

since 2010, suggesting that future conflicts may further disrupt the region's path to economic stability (**Figure 7.42**). Many countries in the region (Saudi Arabia and Israel, for example), on a per capita basis, are some of the world's largest spenders on national defense, surpassing even the United States.

Higher-Income Oil Exporters The richest countries of Southwest Asia and North Africa owe their wealth to massive oil reserves. Nations such as Saudi Arabia, Kuwait, Qatar, Bahrain, and the United Arab Emirates benefit from fossil fuel production as well as from their relatively small populations. Large investments in transportation networks, urban centers, and other petroleum-related industries have reshaped the cultural landscape. The petroleum-processing and shipping centers of Jubail (on the Persian Gulf) and Yanbu (on the Red Sea) are examples of this commitment to expand the Saudi economic base beyond simple extraction of crude oil (**Figure 7.43**). Billions of dollars have poured into new schools, medical facilities, low-cost housing, and modernized agriculture, significantly raising the standard of living in the past 40 years.

Still, problems remain. Dependence on oil and gas revenues produces economic pain when prices fall as they did in 2015 and 2016. Such fluctuations in world oil markets will inevitably continue in the future, disrupting construction projects, producing large layoffs of immigrant populations, and slowing investment in the region's

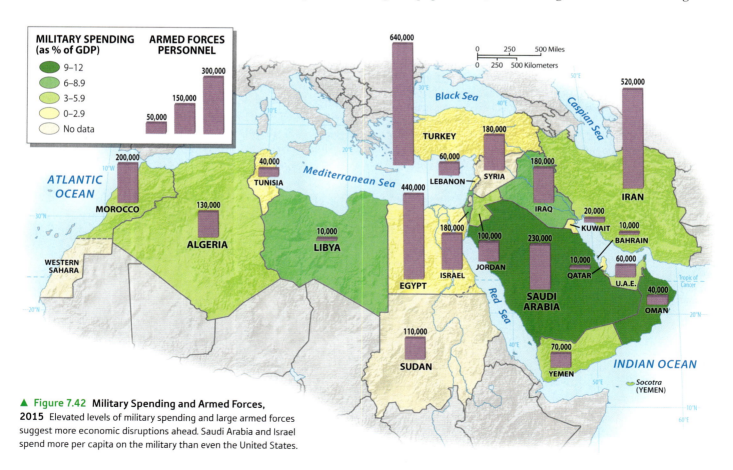

▲ **Figure 7.42 Military Spending and Armed Forces, 2015** Elevated levels of military spending and large armed forces suggest more economic disruptions ahead. Saudi Arabia and Israel spend more per capita on the military than even the United States.

▲ **Figure 7.43 Saudi Arabian Oil Refinery** These oil storage tanks are a part of Saudi Arabia's sprawling Ras Tanura refinery, located in the eastern part of the country.

Explore the **Sights** of Oil Refineries in Yanbu

http://goo.gl/tN8GWJ

region around Basra, have been leaving the country, signaling more economic instability ahead.

The situation in Iran is also challenging. The country's oil and gas reserves are huge, but Iran is relatively poor, burdened with a stagnating standard of living. Since 1980, fundamentalist leaders have limited international trade in consumer goods and services, fearing the import of unwanted cultural influences. International sanctions on purchases of Iranian oil (related to its nuclear development program) have depressed the economy until recently. With sanctions reduced, many hope for more economic activity and the government has targeted a growth rate of 8 percent between 2015 and 2020. More international travel to the country may stimulate a tourist industry. Some optimistic government officials would like to increase annual visitation from 4 million to 25 million by 2025. Increased revenues from global oil sales should also stimulate the economy, but inflation and unemployment remained stubbornly high in 2016.

economic and social infrastructure. In 2016, Saudi Arabia released its Vision 2030 economic agenda to address some of these issues. It proposes 1) diversifying its national wealth by selling off a small part of its state-owned oil assets and then investing the proceeds in varied global investments; 2) holding down domestic spending and making its economy more competitive; and 3) generating more public revenues from fees, an expanded tax base, and more tourism.

Lower-Income Oil Exporters Some countries possess fossil fuel reserves, but different political and economic variables have hampered sustained economic growth. For example, Algerian oil and natural gas overwhelmingly dominate the country's exports, but the past 20 years have brought political instability and shortages of consumer goods. Nearby Libya's political disintegration has had profound economic consequences, sharply reducing oil and gas output and causing severe economic disruptions.

Iraq faces huge challenges. War has crippled much of its already deteriorated infrastructure, and political instability has made rebuilding its economy more difficult. Iraq suffers from high unemployment, more than 20 percent of the population remains malnourished, and only 25 percent of the country is served by dependable electricity. Oil output has increased since 2009, however, suggesting the potential for economic recovery if sectarian conflicts can be contained. Many middle-class Iraqis, however, even in the oil-rich

Prospering Without Oil Some countries, while lacking petroleum resources, have still found paths to economic prosperity. Israel, for example, supports one of the highest standards of living in the region, even with its political challenges (see Table 7.2). Israelis and many foreigners have invested large amounts of capital to create a highly productive industrial base, which produces many products for the global marketplace. The country is also a global center for high-tech computer and telecommunications products, known for its fast-paced and highly entrepreneurial business culture that resembles California's Silicon Valley. Israel also has daunting economic problems. Its struggles with the Palestinians and with neighboring states have sapped potential vitality. Defense spending absorbs a large share of total gross national income (GNI), necessitating high tax rates. Poverty and unemployment among Palestinians also remain unacceptably high, both in Gaza (recently devastated by more violence) and in the West Bank.

Turkey has a diversified economy, even though incomes are modest by regional standards. Lacking petroleum, Turkey produces varied agricultural and industrial goods for export. About 20 percent of the workforce remains in agriculture, and principal commercial products include cotton, tobacco, wheat, and fruit. The industrial economy has grown since 1980, including exports of textiles, processed food, and chemicals. Turkey has also gone high tech: About half the Turkish population uses the Internet, and the country has been a fertile ground for dozens of global Internet startup companies that connect well (many are online or virtual gaming enterprises) with younger Turks as well as with the global economy. Turkey also remains a major tourist destination in the region, attracting more than 6 million visitors annually in recent years. Still, many economists argue the country is performing well below its potential as unemployment and inflation both hover around 10 percent and as the manufacturing economy contracted between 2013 and 2016.

Regional Patterns of Poverty Poorer countries of the region share the problems of the less developed world. For example, Sudan, Egypt, Syria, and Yemen each face unique economic challenges. Sudan's continuing political problems stand in the way of progress. Political disruptions have resulted in major food shortages. The country's transportation and communications systems have seen little new investment, and secondary school enrollments remain very low. On the other hand, Sudan's fertile soils could support more farming, and its share of regional oil prospects suggests petroleum's expanding role in the economy.

Egypt's economic prospects are unclear. Since Mubarak's overthrow in 2011, unemployment has risen, tourism has declined, and foreign investors have remained wary of the nation's uncertain political environment. While leaders have recently been promoting the country's relatively stable political situation, many Egyptians still live in poverty, and the gap between rich and poor continues to widen. Illiteracy is widespread, and the country suffers from the **brain drain** phenomenon as some of its brightest young people leave for better jobs in western Europe or the United States.

Syria once enjoyed both a growing economy and a stable political regime. Now it has neither, as its ongoing civil war and sectarian conflicts have decimated the economy. Millions of people who considered themselves part of the Syrian middle class have been thrown into utter poverty, often as desperate refugees. It will be many years before more normal economic conditions can be restored (**Figure 7.44**).

Yemen remains the poorest country on the Arabian Peninsula. Positioned far from the region's principal oil fields, Yemen's low per capita GNI puts it on par with nations in impoverished Sub-Saharan Africa. The largely rural country relies mostly on marginally productive subsistence agriculture, and much of its mountain and desert interior lacks effective links to the outside world. Coffee, cotton, and fruits are commercially grown, and modest oil exports bring in needed foreign currency. Overall, however, widespread unemployment and high

childhood mortality suggest the failure of the region's inadequate health-care system combined with challenging social and economic conditions and complex environmental determinants. Recent political upheavals have only further dimmed the nation's economic prospects.

Gender, Culture, and Politics: A Woman's Changing World

The role of women in largely Islamic Southwest Asia and North Africa remains a major social issue. Female participation rates in the workforce are among the world's lowest,

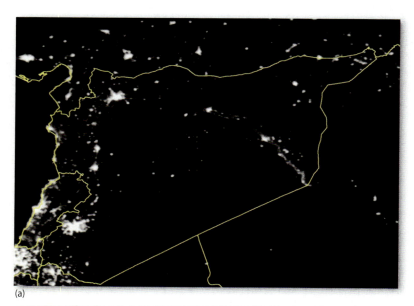
(a)

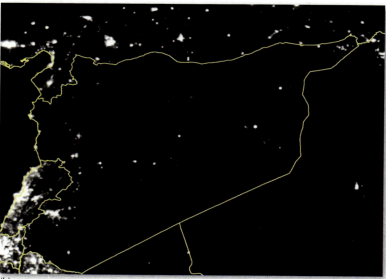
(b)

▲ **Figure 7.44 Syria Sinks into Darkness** These comparative nighttime satellite views reveal the devastating political and economic impact of the Syrian civil war and the ongoing presence of ISIL. Photo (a) was taken before the conflicts began, photo (b) after hostilities began to rock the country.

▲ **Figure 7.45 Iranian Women** These fashionably dressed young women in Tehran suggest how Iran's more urban and affluent residents have embraced many elements of Western culture. **Q: What unique challenges confront educated women in Iran? Unique opportunities?**

and large gaps typically exist between levels of education for males and females. In conservative parts of the region, few women work outside the home. Even in parts of Turkey, where Western influences are widespread, it is rare to see rural women selling goods in the marketplace or driving cars in the street. A recent poll on women's rights in the region ranked Egypt, Iraq, and Saudi Arabia among the lowest and Oman, Kuwait, Jordan, and Qatar among the highest.

More orthodox Islamic states impose legal restrictions on the activities of women. In Saudi Arabia, for example, women are not allowed to drive, although growing protests among many younger, educated Saudi females may overturn that ban. In neighboring Qatar, women can vote (and hold public office), but they still need a husband's consent to obtain a driver's license. In Iran, full veiling remains mandatory in more conservative areas, but many wealthier Iranian women have adopted Western dress, reflecting a more secular outlook, especially among the young (**Figure 7.45**). Today, 60 percent of Iranian university students are female. Generally, Muslim women still lead more private lives than men: Much of their domestic space is shielded from the world by walls and shuttered windows, and their public appearances are filtered through the use of the *niqab* (face veil) or *chador* (full-body veil).

In some settings, women's lives are changing, even within norms of more conservative Islamist societies. From Tunisia to Yemen, women widely participated in the Arab Spring rebellions, asserting their new political visibility in very public ways. Kurdish women have also volunteered to fight ISIL, both in Iraq and in Syria. The local responses to these political and social changes are complex, however: In liberated Libya, young Islamic women rejoice in their freedom to wear the niqab in public, a practice banned under Qaddafi's rule.

Even in conservative Saudi Arabia, women have enjoyed increased freedom, thanks to reforms passed under former King Abdullah. Some women have been encouraged to play more active roles in political affairs, including running for office. Many women are also attending universities, making up 52 percent of the nation's student body. Women are pouring into the workplace, with female employment growth skyrocketing by 50 percent between 2010 and 2016. Two-income families are becoming the norm as women's roles change and as the country's economy struggles amid lower oil prices.

Algerian women also demonstrate the pattern. Most studies suggest those in the younger generation are more religious than their parents and more likely to cover their heads and bodies with traditional clothing. At the same time, they are more likely to be educated and employed. Today 70 percent of Algeria's lawyers and 60 percent of its judges are women. A majority of university students are women, and women dominate the health-care field. These new social and economic roles help explain why birth rates are declining. Women also have a more visible social position in Israel, except in fundamentalist Jewish communities, where conservative social customs limit women to traditional domestic roles.

Even with these changes, many argue that the region still has a long way to go and that women, especially in the Arab World, are only in the early stages of a social transformation that will still take decades to unfold.

REVIEW

7.11 Describe the basic geography of oil reserves across the region, and compare the pattern with the geography of natural gas reserves.

7.12 Identify different strategies for economic development recently employed by nations such as Saudi Arabia, Turkey, Israel, and Egypt. How successful have they been, and how are they related to globalization?

■ **KEY TERMS** Greater Arab Free Trade Area (GAFTA), brain drain

Summary

- Many nations within Southwest Asia and North Africa suffer from significant environmental challenges. Twentieth-century population growth across the region was dramatic, but expanding the region's limited supplies of agricultural land and water resources is costly and difficult. The eroded soils of the Atlas Mountains to overworked garden plots along the Nile illustrate the environmental price paid when population growth outstrips the ability of the land to support it.

- The population geography of the region is strikingly uneven. Areas with higher rainfall or access to exotic water often have high physiological population densities, whereas nearby arid zones remain almost empty of settlement.

- Culturally, the region remains the hearth of Christianity, the spatial and spiritual core of Islam, and the political and territorial focus of modern Judaism. In addition, important sectarian divisions within religious traditions, as well as long-standing linguistic differences, continue to shape the area's local cultural geographies and regional identities.

- Political conflicts have disrupted economic development. Civil wars, sectarian violence, conflicts between states, and regional tensions hamper initiatives for greater cooperation and trade. Perhaps most important, the region must deal with the conflict between modernity and more fundamentalist interpretations of Islam.

- Abundant reserves of oil and natural gas, coupled with the world's continuing reliance on fossil fuels, ensure that the region will remain prominent in world petroleum markets. Yet economic prospects remain muted by fluctuating oil prices and never-ending political instability, which challenge progress in modernizing the region's social development, including expanded opportunities for women.

Review Questions

1. Discuss five important human modifications of the Southwest Asian and North African environment, and assess whether these changes have benefited the region.

2. Why are birthrates declining in many countries of the region? Despite cultural differences with North America, what common processes seem to be at work in both regions that have contributed to this demographic transition?

3. Compare the modern maps of religion and language for the region, and identify three major examples where Islam dominates non-Arab-speaking areas. Explain why this is the case.

4. What economic changes could occur if Israel and the Palestinians achieved peace and if the Syrians could unite behind a single leader? What kinds of general connections might be found between political conflict and economic conditions throughout the region?

5. Describe some of the similarities and differences in how lower oil prices have recently affected life in a) France, b) Saudi Arabia, and c) Texas.

Image Analysis

1. This simple bar graph shows the Percentage of Female Labor Force Participation in selected Middle Eastern North African (MENA) countries for 2016. Why might rates be relatively high in Qatar and the United Arab Emirates (UAE), but so low in Jordan and the West Bank and Gaza?

2. What underlying economic, cultural, and political variables are really being measured by Female Labor Force Participation rates?

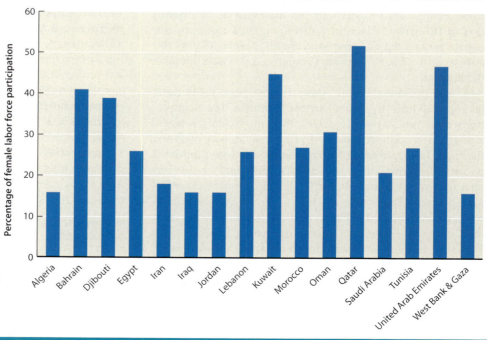

▶ Figure IA7 Percentage of Female Labor Force Participation Rates in MENA Countries

JOIN THE DEBATE

Historically, Southwest Asia's Kurdish population has been a "nation without a state." An enduring "Kurdistan" almost came to fruition after World War I, but never became a reality, due to resistance from newly emerged Turkey as well as a lack of support from leading Western European nations. Is the time now right for an independent Kurdish state?

The time is right to create a Kurdish Republic out of portions of Iraq, Turkey, Syria, and Iran.

- Kurdish-speaking peoples make up a large and distinctive cultural group in Southwest Asia, with a well-defined homeland that would define the limits of a new political entity within the region.

- If they were at peace and united, the Kurds would cease to be a constant irritant as a cultural and political minority in nations such as Turkey and Iraq, increasing stability in those countries.

- A Kurdish Republic would quickly become a real economic bright spot within the region, thanks to lessened hostilities, a solid economic base (including rich petroleum reserves), and a skilled, entrepreneurial population.

Although attractive in the abstract, a Kurdish nation would only disrupt an already unstable region further.

- Nations such as Turkey and Iraq would strongly resist giving up large, resource-rich portions of their national territory, and there would be endless arguments over boundaries that would define the new country.

- Massive, potentially violent migrations of both Kurds and non-Kurds could result from poorly-drawn borders, similar to what happened with the partition of South Asia in the late 1940s.

- While not ideal, Kurds can work toward greater representation and increased autonomy within established nations such as Turkey and Iraq, and the past decade suggests they can make further progress in these efforts.

▲ **Figure D7** Kurdish Women in Turkey

KEY TERMS

Arab League (p. 298)
Arab Spring (p. 268)
brain drain (p. 308)
choke points (p. 275)
culture hearth (p. 268)
domestication (p. 279)
exotic rivers (p. 281)
Fertile Crescent (p. 279)
fossil water (p. 274)
Greater Arab Free Trade Area (GAFTA) (p. 303)
Hajj (p. 288)
hydropolitics (p. 274)
ISIL (Islamic State of Iraq and the Levant) (p. 268)
Islamic fundamentalism (p. 268)
Islamism (p. 268)
Levant (p. 270)

Maghreb (p. 270)
medina (p. 281)
monotheism (p. 286)
Organization of the Petroleum Exporting Countries (OPEC) (p. 269)
Ottoman Empire (p. 289)
Palestinian Authority (PA) (p. 297)
pastoral nomadism (p. 279)
physiological density (p. 277)
protectorates (p. 295)
Quran (p. 287)
salinization (p. 273)
sectarian violence (p. 268)
Shiites (p. 288)
Suez Canal (p. 295)
Sunnis (p. 288)
Sykes-Picot Agreement (p. 295)
theocratic state (p. 288)
transhumance (p. 280)

Mastering Geography™

Looking for additional review and test prep materials? Visit the Study Area in **MasteringGeography**™ to enhance your geographic literacy, spatial reasoning skills, and understanding of this chapter's content by accessing a variety of resources, including **MapMaster** interactive maps, videos, RSS feeds, flashcards, web links, self-study quizzes, and an eText version of *Diversity Amid Globalization*.

DATA ANALYSIS

http://goo.gl/0cwTYO

Health care is often considered a basic human right in more developed portions of the world, but large parts of Southwest Asia and North Africa are poorly served by health-care providers. The World Health Organization (WHO) gathers data on the number of physicians per 1000 population, which can be used as a measure of access to health care as well as social development. According to recent data, the United States had about 2.5 physicians per 1000 and Germany about 3.9. Go to the WHO website (www.who.int) and access the data/interactive atlas page on physicians per 1000 population.

1. Make your own data table and map showing the regional pattern of health-care access across Southwest Asia and North Africa.

2. In a few sentences, summarize the general patterns and trends you see. How would you explain some of the major variations you observe across the region?

3. Compare the pattern you see for physicians with the data in the text on childhood mortality (Table 7.2). What similarities and differences do you see? How might these two indicators be a good measure of future social development? How might they predict political stability?

Physical Geography and Environmental Issues

Diverse European environments range from subtropical Mediterranean lands to the arctic tundra, and from the moderate Atlantic coast to inland continental climates. One of the "greenest" world regions, Europe enacts strong measures on pollution, recycling, and renewable energy.

Population and Settlement

Europe has very low rates of natural growth and very high rates of internal mobility and international in-migration. Most international migration consists of refugees from strife-torn Africa and Southwest Asia, an issue causing numerous problems for European countries.

Cultural Coherence and Diversity

Europe has a long history of cultural tensions linked to internal differences in language and religion; however, today's tensions are primarily connected with immigration from other world regions.

Geopolitical Framework

Two world wars and a lengthy Cold War divided 20th-century Europe into warring camps, producing an ever-changing map of new states. Although Europe is now largely an integrated and peaceful region, devolution tensions linked to micro-nationalism and anti-European Union (EU) feeling threaten this unity.

Economic and Social Development

For half a century, the European Union (EU) has worked successfully to integrate the region's diverse economies and political systems, making Europe a global superpower. Today, however, internal economic and social issues challenge this unity.

▶ In June 2016, Britain startled Europe by approving a referendum to withdraw from the European Union (EU), a process now known as "Brexit" (from Britain + exit). The 52% of the voters who approved this measure thought withdrawing from the EU would be a cure for all sorts of issues ranging from unemployment to foreign immigration. Those voting against, such as these demonstrators, feel the benefits of EU membership achieved over the last 50 years far outweigh any costs.

London, England

EUROPE

Europe today: a continent in crisis, a casualty of its own economic and political success, some say, a region whose hard-won unity and vitality is now threatened by internal struggles over an agreed-upon future, complicated by external pressures from millions of strangers looking to Europe for safety and solace.

What is at risk is the postwar dream of a unified Europe consisting of a mosaic of independent countries with a common political, economic, and cultural agenda. While many of these goals have been achieved through the **European Union (EU)**, the alliance of 28 of the region's countries that form the world's largest economic and political bloc, success has come with an array of serious problems that threaten the very nature of this newfound unity. Here is a simplified list, with more details provided throughout the chapter:

- Although nationalism was put on the back shelf temporarily, it is alive and well in various forms, politically and culturally. Several subregions of Spain seek independence from the mother country, as do the two distinct cultural groups that comprise Belgium. Scotland is not sure it wants to remain part of the historic United Kingdom, and smaller regions of France periodically agitate for more autonomy.

- At a larger level, there is a backlash against perceived overreach and dysfunction by a hyperbureaucratic EU by resisting its stated goal of "an ever-closer union." This anti-EU feeling was fueled across Europe by the United Kingdom's June 2016 referendum to leave the European Union, the so-called British exit, or "**Brexit**."

- The EU's common monetary policy forces richer countries to bail out and subsidize poorer countries. While this seemed like a good idea in boom times, it's less well received as Europe's economy stagnates, as is evident from the disagreement over how to handle Greece's debt crisis of 2015.

- A cherished EU goal was the "softening" of national boundaries to facilitate freer movement of people and goods within a unified Europe. These soft boundaries, however, became problematic in 2015 as Europe was flooded by over a million refugees from war-torn Syria, Iraq, Somalia, and Afghanistan. Once-open borders were closed, causing chaos to the region's transportation and economic systems.

- Adding to the cultural and political unrest resulting from the migrant crisis were terrorist attacks from radical Islamists, exploiting Europe's experiment in political and social openness and furthering the continent's crisis (**Figure 8.1**).

▲ **Figure 8.1 Terrorism in Belgium** Mourners gather in central Brussels to pay tribute to the 32 people killed by the terrorist attacks of March 27, 2016 committed by Islamic State members.

Defining Europe

Europe, of course, encompasses much more than the EU member states. One of the world's most diverse regions, Europe is home to a wide assortment of people and places in an area considerably smaller than North America. More than half a billion people reside in 42 countries ranging in size from France, Spain, and Germany to microstates such as Andorra and Monaco (**Figure 8.2**). The region is commonly divided into the four general subregions of western, eastern, southern (or Mediterranean), and northern (or Scandinavian) Europe, terms we use throughout this chapter. These subregions are characterized by shared physical and human geographies, but individual countries have distinctive cultural characteristics, including language and ethnic or religious affiliation. Their political and economic trajectories vary widely as well, as do the extent of their linkages with other European countries and world regions.

> EU success has come with an array of serious problems that threaten the very nature of European unity.

LEARNING Objectives

After reading this chapter you should be able to:

8.1 Describe the topography, hydrology, and climate of Europe.

8.2 Identify the major environmental issues in Europe as well as measures taken to resolve those problems.

8.3 Provide examples of countries with different rates of natural growth.

8.4 Describe both the patterns of internal migration within Europe and the geography of immigration to the region.

8.5 Discuss the origins and configurations of Europe's ongoing migration crisis.

8.6 Describe the major languages and religions of Europe and locate these on a map.

8.7 Summarize how the map of European states has changed in the last 100 years.

8.8 Explain why and how Europe was divided during the Cold War and what the geographic implications are today.

8.9 Describe Europe's economic and political integration as driven by the EU, along with the devolution forces that currently threaten this Union.

8.10 Identify the major characteristics of Europe's ongoing fiscal crisis.

ELEVATION IN METERS

- 4000+
- 2000–4000
- 500–1999
- 200–499
- 0–199
- Below sea level

Sea Level

EUROPE
Political & Physical Map

- ★ ● Metropolitan areas more than 20 million
- ✦ ● Metropolitan areas 10–20 million
- ✦ ● Metropolitan areas 5–9.9 million
- ✦ ● Metropolitan areas 1–4.9 million
- ✦ ○ Selected smaller metropolitan areas
- — — Plate boundaries

NORTH AMERICAN PLATE

EURASIAN PLATE

Reykjavik

ICELAND

Jan Mayen (NORWAY)

Arctic Circle

Faroe Islands (DENMARK)

Shetland Islands (U.K.)

Orkney Islands (U.K.)

Norwegian Sea

FINLAND

Fenno-Scandian Shield

Helsinki

NORWAY

SWEDEN

Bergen

Oslo

Stockholm

Goteborg

ESTONIA

Tallinn

Riga

LATVIA

LITHUANIA

Vilnius

North Sea

Skagerrak

Kattegat

Baltic Sea

Glasgow

Belfast

UNITED KINGDOM

Dublin

Leeds

Manchester

IRELAND

Birmingham

Thames R.

London

Copenhagen

Malmo

DENMARK

Hamburg

North European Lowland

Elbe R.

Berlin

POLAND

Vistula R.

Warsaw

Lodz

Oder R.

Wroclaw

Katowice

Krakow

Southern limit of Pleistocene glaciation

NETHERLANDS

Amsterdam

The Hague

Rotterdam

Antwerp

Brussels

Lille

BELGIUM

Cologne

GERMANY

Frankfurt

Prague

CZECHIA

SLOVAKIA

CARPATHIANS

Iasi

English Channel

ATLANTIC OCEAN

LUXEMBOURG

Luxembourg

Seine R.

Paris Basin

Paris

Nantes

Loire R.

Rhine R.

Stuttgart

Munich

Danube

Bratislava

Vienna

LIECHTENSTEIN

AUSTRIA

Budapest

HUNGARY

Hungarian Basin

ROMANIA

Zurich

Bern

SWITZERLAND

ALPS

SLOVENIA

Ljubljana

Zagreb

Bucharest

Valachian Plain

FRANCE

Lyon

Milan

Po Valley

CROATIA

BOSNIA & HERZEGOVINA

Belgrade

Danube R.

Basin of Aquitaine

Turin

Po R.

Sarajevo

SERBIA

Sofia

Burgas

Bay of Biscay

Garonne R.

Rhone R.

Nice

Florence

ITALY

SAN MARINO

MONTENEGRO

KOSOVO

Pristina

Podgorica

Skopje

BULGARIA

RHODOPE MTS.

Bilbao

Toulouse

ANDORRA

Marseille

MONACO

DINARIC ALPS

Tirana

MACEDONIA

Thessaloniki

Porto

PYRENEES

Ebro R.

VATICAN CITY

Rome

APENNINE MOUNTAINS

Adriatic Sea

ALBANIA

PORTUGAL

SPAIN

Madrid

Tagus R.

Barcelona

Corsica (FRANCE)

Naples

GREECE

Aegean Sea

Lisbon

Balearic Islands (SPAIN)

Sardinia (ITALY)

Valencia

Seville

Mediterranean Sea

EURASIAN PLATE

AFRICAN PLATE

Strait of Gibraltar

Palermo

Sicily (ITALY)

MALTA

Valletta

Crete (GREECE)

Athens

TURKEY

Mediterranean Sea

Nicosia

CYPRUS

0   150   300 Miles

0   150   300 Kilometers

▲ **Figure 8.2 Europe** Stretching east from Iceland in the Atlantic to Russia, Europe includes 42 countries, ranging in size from large states such as France and Germany to the microstates of Liechtenstein, Andorra, San Marino, and Monaco. Currently, the population of the region is about 540 million. Europe is commonly divided into the four subregions of western, eastern, southern (or Mediterranean), and northern (or Scandinavian) Europe. Tables 8.1 and 8.2 use these subregions for their organizational format.

Physical Geography and Environmental Issues: Human Transformation of a Diverse Landscape

Despite Europe's small size, its environmental diversity is extraordinary. A startling array of landscapes is found within its borders: the semiarid hills of the Mediterranean countryside, explosive volcanoes in southern Italy, wide lowland plains, glaciated seacoasts in Norway and Iceland, and the arctic tundra of northern Scandinavia. Three factors explain this impressive environmental diversity: the complex geology of this western edge of the Eurasian landmass; the broad latitudinal range extending from the Mediterranean to the Arctic seas, which influences climate, vegetation, and hydrology patterns (**Figure 8.3**); and the region's long history of human settlement, which has transformed the landscape in fundamental ways.

Landform Regions

Europe can be organized into four general landform regions (see Figure 8.2): the European Lowland, forming an arc from southern France to the northeast plains of Poland, but also including southeastern England; the Alpine mountain system, extending from the Pyrenees in the west to the Balkan Mountains of southeastern Europe; the Central Uplands, positioned between the Alps and the European Lowland; and the Western Highlands, which include mountains in Spain and portions of the British Isles and the highlands of Scandinavia. Iceland, unquestionably a part of Europe yet lying 900 miles (1500 km) west of Norway, has its own unique landforms, straddling as it does two tectonic plates.

The European Lowland Also known as the North European Plain, this lowland is the unquestioned economic focus of western Europe, with its high population density, intensive agriculture, large cities, coastal ports, and major industrial regions (**Figure 8.4**). Though not completely flat, most of this lowland lies below 500 feet (150 meters) in elevation. Many of Europe's major rivers (the Rhine, Loire, Thames, and Elbe) meander across this lowland and form broad estuaries before emptying into the Atlantic. Several of Europe's busiest ports, including London, Le Havre, Rotterdam, and Hamburg, are located in these lowland settings.

The Rhine River delta conveniently divides the unglaciated southern European Lowland from the glaciated plains to the north, which were covered by a Pleistocene ice sheet until about 15,000 years ago. Rocky clay materials in Scandinavia were eroded and transported south by glaciers. As the climate warmed and the glaciers retreated, piles of glacial debris were left on the plains of northern Germany and Poland, which are far less fertile than the unglaciated portions of Belgium and France.

The Alpine Mountain System The Alpine ranges form the topographic spine of Europe and consist of a series of mountains running west to east from the Atlantic to the Black Sea and the southeastern Mediterranean. These mountain ranges carry distinct regional names, such as the Pyrenees, Alps, Apennines, Carpathians, Dinaric Alps, and Balkans, even though they share a similar geology.

The Pyrenees mark the political border between Spain and France and include the microstate of Andorra. This rugged range extends almost 300 miles (480 km) from the Atlantic to the Mediterranean, where glaciated peaks reaching to 11,000 feet (3350 meters) alternate with broad glacier-carved valleys.

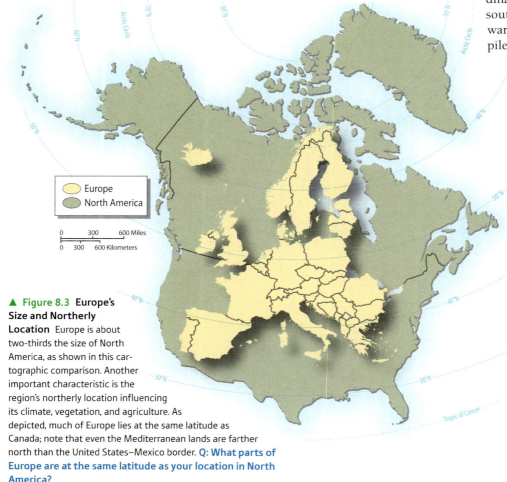

▲ **Figure 8.3 Europe's Size and Northerly Location** Europe is about two-thirds the size of North America, as shown in this cartographic comparison. Another important characteristic is the region's northerly location influencing its climate, vegetation, and agriculture. As depicted, much of Europe lies at the same latitude as Canada; note that even the Mediterranean lands are farther north than the United States–Mexico border. **Q: What parts of Europe are at the same latitude as your location in North America?**

▲ **Figure 8.4 The European Lowland** Also known as the North European Plain, this large lowland extends from southwestern France to northern Germany and eastward into Poland. Numerous rivers drain interior Europe by crossing this lowland; many of Europe's port cities are located on the lower reaches of these rivers. This photo shows the Normandy region of western France; the small island is Mont Saint-Michel, a historic, fortified monastery.

The centerpiece of this geologic system is the prototypical mountain range, the Alps, running more than 500 miles (800 km) from France to eastern Austria. These impressive mountains are highest in the west, rising to more than 15,000 feet (4600 meters) at Mt. Blanc on the French-Italian border. In Austria, the Alps are much more subdued, with few peaks exceeding 10,000 feet (3000 meters). Though easily crossed today by car or train through long tunnels and valley-spanning bridges, these mountains have historically formed an important cultural divide between the Mediterranean lands to the south and central and western Europe to the north.

The Apennine Mountains are located south of the Alps, mainly in Italy; the two ranges, however, are physically connected by the hilly coastline of the French and Italian Riviera. The Apennines are lower and lack the spectacular glaciated peaks and valleys of the true Alps, but take on their own distinctive character farther south with the explosive volcanoes of Mt. Vesuvius (just over 4000 feet, or 1200 meters) outside Naples and the much higher Mt. Etna (almost 11,000 feet, or 3350 meters) on the island of Sicily.

The Carpathian Mountains define the eastern limits of the Alpine system in Europe. They are a plow-shaped upland area extending from eastern Austria to the Iron Gate gorge, which is a narrow passage for Danube River traffic where the borders of Romania and Serbia meet.

Central Uplands A much older highland region occupies an arc between the Alps and the European Lowland in France and Germany. These mountains are much lower in elevation than the Alpine system, with their highest peaks around 6000 feet (1800 meters). Much of these uplands are characterized by rolling landscapes 3000 feet (1000 meters) above sea level, and are important to western Europe because they contain the raw materials for Europe's industrial areas. In Germany and France, for example, they provided the iron and coal central to each country's steel industry. To the east, mineral resources from the Bohemian highlands fuel major industrial areas in Germany, Poland, and the Czech Republic.

Western Highlands Defining the western edge of the European subcontinent are the Western Highlands, extending from Portugal in the south, through portions of the British Isles, to Norway, Sweden, and Finland. These are Europe's oldest mountains, formed about 300 million years ago.

As with other upland areas that traverse Europe, specific names for these mountains differ from country to country. A portion of the Western Highlands forms the highland spine of England, Wales, and Scotland, where picturesque glaciated landscapes are found at modest elevations of 4000 feet (1200 meters) or less. These U-shaped glaciated valleys also appear in Norway's uplands, where they produce a spectacular coastline of **fjords**, or flooded valley inlets, similar to the coastlines of Alaska and New Zealand (**Figure 8.5**).

Geologically, the far western edge of Europe is found in Iceland, which is divided by the Eurasian and North American tectonic plates. Like other divergent plate boundaries, Iceland has many active volcanoes that occasionally spew ash into the atmosphere, sometimes causing serious problems for the heavy airline traffic between Europe and North America.

Seas, Rivers, and Ports

In many ways, Europe is a maritime region with strong ties to its surrounding seas. Even landlocked countries such as Austria, Hungary, Serbia, and the Czech Republic have access to the ocean and seas through extensive networks of navigable rivers and canals.

Europe's Ring of Seas Four major seas and the Atlantic Ocean encircle Europe. In the north, the Baltic Sea separates Scandinavia from north-central Europe. Denmark and Sweden have long controlled the narrow Skagerrak and Kattegat straits that connect the Baltic to the North Sea, which is both a major fishing ground and a principal source of Europe's oil and gas, mined from deep-sea drilling platforms.

The English Channel (in French, *La Manche*) separates the British Isles from continental Europe. At its narrowest point, the Dover Straits, the channel is only 20 miles (32 km) wide.

▲ **Figure 8.5 Fjords in Norway** During the Pleistocene epoch, continental ice sheets carved deep U-shaped valleys along what is now Norway's coastline. As the ice sheets melted around 20,000 years ago, the world's oceans rose, flooding these valleys. This is the Geiranger Fjord in western Norway.

Explore the **Sights** of Norway's Fjorded Coast

http://goo.gl/zBIDX9

Although England has regarded the channel as a protective moat, it has primarily been a symbolic barrier as it deterred neither the French Normans from the continent nor the Viking raiders from the north. Only Nazi Germany found it a formidable barrier during World War II. Since 1993, and after decades of resistance by the English, the British Isles have been connected to France through the 31-mile (50-km) Eurotunnel, with a high-speed rail system carrying passengers, autos, and freight.

Gibraltar guards the narrow straits between Africa and Europe at the western entrance to the Mediterranean Sea, and Britain's stewardship of this passage remains an enduring symbol of a once great sea-based empire. Finally, on Europe's southeastern flanks are the straits of Bosporus and the Dardanelles, the narrows connecting the eastern Mediterranean with the Black Sea. Disputed for centuries, these pivotal waters are now controlled by Turkey. Though these straits are often thought of as the physical boundary between Europe and Asia, they are easily bridged in several places that facilitate truck and train transportation within Turkey and between Europe and Southwest Asia.

Rivers and Ports Europe is a region of navigable rivers, connected by a system of canals and locks that allow inland barge travel from the Baltic and North Seas to the Mediterranean and between western Europe and the Black Sea (**Figure 8.6**). Many rivers on the European Lowland—the Loire, Seine, Rhine, Elbe, and Vistula—flow into Atlantic and Baltic waters. However, the Danube, Europe's longest river, flows east and south, rising in the Black Forest of Germany only a few miles from the Rhine River and running to the Black Sea, offering a connecting artery between central and eastern Europe. Similarly, the Rhône headwaters rise close to those of the Rhine in Switzerland, yet the Rhône flows southward into the Mediterranean.

Major ports are found at the mouths of most western European rivers, where they serve as transshipment points for inland waterways and rail and truck networks. From south to north, these ports include Bordeaux at the mouth of the Garonne, Le Havre on the Seine, London on the Thames, Rotterdam (the world's largest port in terms of tonnage) at the mouth of the Rhine, Hamburg on the Elbe, and, to the east in Poland, Szczecin on the Oder and Gdansk on the Vistula.

Europe's Climate

Three major climate types characterize Europe (**Figure 8.7**). Along the Atlantic coast, a moderate and moist **marine west coast climate** modified by oceanic influences dominates. Farther inland, **continental climates** prevail, with hotter summers and colder winters. Finally, a dry-summer

▶ **Figure 8.6 Inland Barge Traffic** Europe's river systems are connected by an extensive inland canal system so that barge traffic can readily move throughout the region, connecting ports on the Atlantic Ocean with those on the Baltic, Mediterranean, and Black Seas. This photo is of a barge on north Germany's Elbe Lateral Canal.

Explore the **Sights** of Elbe Canal near Brandenburg

http://goo.gl/ryzGLv

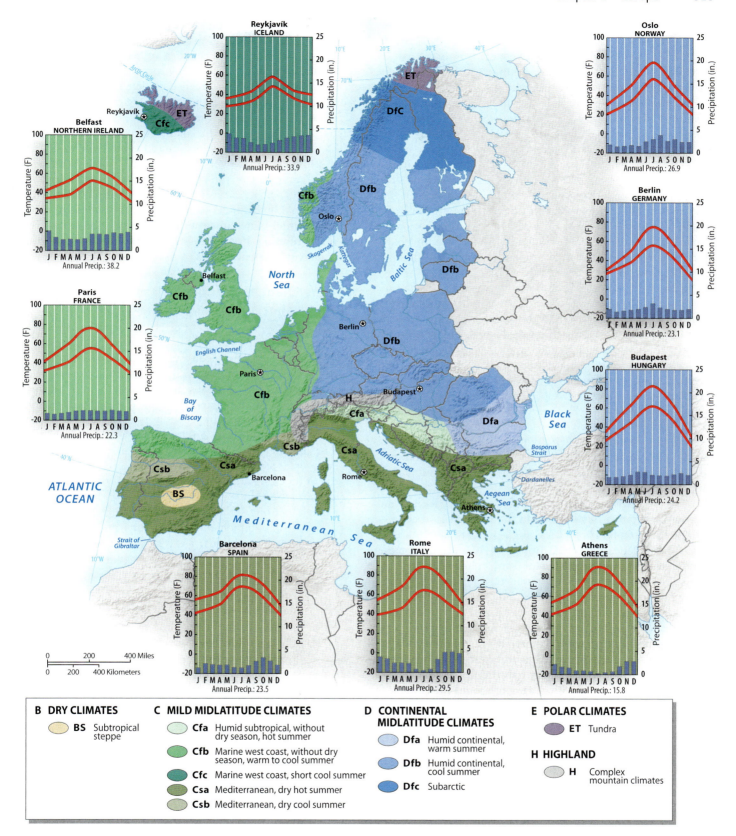

B DRY CLIMATES

    BS Subtropical steppe

C MILD MIDLATITUDE CLIMATES

    Cfa Humid subtropical, without dry season, hot summer

    Cfb Marine west coast, without dry season, warm to cool summer

    Cfc Marine west coast, short cool summer

    Csa Mediterranean, dry hot summer

    Csb Mediterranean, dry cool summer

D CONTINENTAL MIDLATITUDE CLIMATES

    Dfa Humid continental, warm summer

    Dfb Humid continental, cool summer

    Dfc Subarctic

E POLAR CLIMATES

    ET Tundra

H HIGHLAND

    H Complex mountain climates

▲ **Figure 8.7 Climates of Europe** Three major climate zones dominate Europe. Close to the Atlantic Ocean, the marine west coast climate has cool seasons and steady rainfall throughout the year. Farther inland, continental climates have at least one month averaging below freezing, as well as hot summers, with a precipitation maximum occurring during the warm season. Southern Europe has a dry-summer Mediterranean climate. **Q: Where in Europe are there climates similar to the climate where you live in North America?**

Mediterranean climate is found in southern Europe, from Spain to Greece.

One of the most important climate controls is the Atlantic Ocean. Even though much of Europe is at relatively high latitudes (London, England, for example, is slightly farther north than Vancouver, British Columbia), the mild North Atlantic Current, which is a continuation of the warmer Atlantic Gulf Stream, moderates coastal temperatures from Iceland and Norway south to Portugal. This maritime influence gives western Europe a climate 5–10°F (3–6°C) warmer than regions at comparable latitudes that lack the moderating influence of a warm ocean current. As a result, in the marine west coast climate region no winter months average below freezing, even though cold rain, sleet, and an occasional blizzard can be common winter events. Summers are often cloudy and overcast, with frequent drizzle and rain as moisture flows in from the ocean.

Inland, far removed from the ocean (or where a mountain chain limits the maritime influence, as in Scandinavia), landmass heating and cooling becomes a strong climatic control, producing hotter summers and colder winters. Indeed, all continental climates average at least one month below freezing during the winter.

In Europe, the transition between maritime and continental climates takes place close to the Rhine River border of France and Germany. Farther north, although Sweden and other nearby countries are close to the moderating influence of the Baltic Sea, their higher latitude coupled with the blocking effect of the Norwegian mountains produces cold winter temperatures characteristic of true continental climates.

The Mediterranean climate has a distinct dry season during the summer, resulting from the warm-season expansion of the Atlantic (or Azores) high-pressure area. As this warm air descends between latitudes of 30 and 40 degrees, it inhibits summer rainfall. This same phenomenon also produces the Mediterranean climates of California, western Australia, parts of South Africa, and Chile. While these rainless summers attract sun-starved tourists from northern Europe, this seasonal drought is problematic for agriculture. It is no coincidence that traditional Mediterranean cultures, such as the Arab, Moorish, Greek, and Roman civilizations, were major innovators of irrigation technology.

Environmental Issues: Local and Global

Because of its long history of agriculture, resource extraction, industrial manufacturing, and urbanization, Europe has its share of environmental issues (**Figure 8.8**). Compounding the situation is the fact that pollution rarely stays within political boundaries. Air pollution from England, for example, creates serious acid rain problems in Sweden. Similarly, water pollution from Swiss factories on the upper Rhine River creates major problems downstream for the Netherlands, where Rhine River water is commonly used for urban drinking supplies. As a result of these numerous trans-boundary environmental problems, the EU has taken the lead in addressing the region's environmental issues, and Europe is today probably

one the "greenest" of the major world regions, surpassing even North America in its environmental sensibilities.

Until recently, the countries of eastern Europe were plagued by far more serious environmental problems than their western neighbors because of that region's history of Soviet control, where economic planning emphasized short-term industrial output at the expense of environmental protection (see Chapter 9). In Poland, for example, industrial effluents reportedly had wiped out all aquatic life in 90 percent of the country's rivers, with damage from air pollution affecting over half of the country's forests. Similar legacies from the Soviet period were reported in the Czech Republic, Romania, and Bulgaria. Today, however, most of these environmental issues in eastern Europe have been resolved through funding from the EU and the strengthening of national environmental laws.

Climate Change in Europe

The fingerprints of global warming are everywhere in Europe, from dwindling sea ice, melting glaciers, and sparse snow cover in arctic Scandinavia to more frequent droughts in the water-starved Mediterranean subregion. Furthermore, the projections for future climate change are ominous: World-class ski resorts in the Alps are forecast to have warmer winters with less snow pack, while in the lowlands, higher summer temperatures will probably produce more frequent heat waves like those in 2003 and 2015, affecting farmers and urban dwellers alike. In addition, rising sea levels from melting polar ice sheets will threaten the Netherlands, where much of the population lives in diked lands that are actually below sea level (**Figure 8.9**). Because of these threats, Europe has taken a strong stand in addressing climate change and, as a result, has implemented numerous policies and programs to reduce greenhouse gas (GHG) emissions.

The EU entered the 1997 Kyoto Protocol climate negotiations with an innovative scheme that reinforced its philosophy that regional action was superior to that of individual countries. Specifically, the EU set a target of an 8 percent reduction below 1990 GHG emission levels for the EU as a whole. Under this umbrella scheme, several EU member states were required to make significant emissions cuts, whereas others like Greece and Spain were actually allowed increased emissions. The point was to promote growth and industrial development in Europe's poorer countries while at the same time requiring emission reductions in the traditional industrial core of Germany, France, and the UK. Noteworthy is that this umbrella approach has remained in place as the EU has grown from 15 members in 1997, when it made its original Kyoto commitment, to its current 28 member states. Under the 2015 Paris Agreement the EU is committed to a 40 percent reduction over 1990 levels by the year 2030, and an impressive 80–95 percent reduction by 2050.

The EU's Emission-Trading Scheme As part of its emissions reduction strategy, the EU inaugurated the world's first carbon-trading scheme in 2005. Under this plan, specific

Areas affected by acid precipitation

Vulnerable to sea-level rise

Coastal pollution

Polluted rivers

Area of worst air pollution

Dutch coastline. *Low-lying coastal settlements and farmlands are threatened by sea-level rise from global warming.*

Acid precipitation. *Half of Poland's forests and three-quarters of those in the Czech Republic are damaged from acid precipitation.*

Global warming in the Alps. *Warmer temperatures have caused Alpine glaciers to retreat, and sparse snowfall threatens the economic vitality of Alpine ski resorts.*

▲ **Figure 8.8 Environmental Issues in Europe** Although western Europe has worked energetically over the past 50 years to solve environmental problems such as air and water pollution, eastern Europe lags a bit behind because environmental protection was not a high priority during the postwar communist period, 1945–1989. Current efforts, however, show great promise.

yearly emission caps were set for the EU's largest GHG emitting firms. Emitters exceeding those caps must either purchase carbon emission equivalences from a source below their own cap or, alternatively, buy credits from the EU carbon market. The goal of this cap-and-trade system was to make business more expensive for companies that pollute, while rewarding those staying under their carbon quota. Although there were considerable problems

with this trading scheme in its first decade, these issues had been largely resolved by 2016, making the EU's carbon-trading scheme the world's largest and most successful cap-and-trade plan.

Energy and Emissions Greenhouse gas emissions are closely linked to a country's energy mix and population size. Not surprisingly, within the EU, the highest emissions come

◄ **Figure 8.9** **Protecting Low-Country Europe from Sea Level Rise** Conceived more than half a century ago, the Dutch Deltaworks were originally built to keep ocean storm surges and Rhine River flooding from the southwestern Netherlands. But now, given the forecasts for sea-level rise from global warming, the Deltaworks must be reengineered and made higher to protect the 50 percent of the Netherlands that lies below the current sea level.

fastest-growing renewable energy segment with both offshore and onshore wind turbine sites (**Figure 8.10**). Denmark, a wind energy pioneer, currently produces 43 percent of its energy from wind, with a goal of increasing this to 50 percent by 2020.

REVIEW

8.1 Name and locate on a map the major lowland and mountainous areas of Europe.

8.2 What and where are the three major climate regions of Europe?

8.3 Describe how inland barge traffic would get from the mouth of the Rhine to the delta of the Danube.

8.4 Explain why and how Europe has been so successful in reducing its CO_2 emissions over the last 20 years.

■ **KEY TERMS** European Union (EU), Brexit, fjords, marine west coast climate, continental climates, Mediterranean climate

from countries having the largest populations and burning the most fossil fuels. Germany, Europe's most populated country with almost 83 million people, emits over 900 million metric tons of pollutants each year. It is followed by the UK, Italy, and France, all with populations of about 60 million and emissions about half that of Germany. Noteworthy though, is that Germany's current CO_2 emissions are almost 30 percent less than in 1990, the Kyoto Protocol and Paris Agreement's baseline year. Most of this reduction has come from shutting down coal-fired power plants and greatly expanding its renewable energy sector.

Despite these advances in reducing emissions, generally speaking, Europe still runs on fossil fuels: More specifically, on imported coal from North America and Russia, oil from Southwest Asia, and natural gas from Russia. France is the only country to depend on nuclear energy for more than half its energy supply. In contrast to France, both Germany and England are committed to shutting down their nuclear plants because of safety and nuclear waste disposal issues. Ironically, as Germany shuts down its nuclear plants the country has had to use more coal—resulting in higher emissions—during their transition to more renewable energy.

Complementing EU emission reduction goals is a policy to increase the region's renewable energy resources; by 2020 the EU as a whole is expected to generate 20 percent of its power from hydropower, wind, solar, and biofuels. This goal seems well within reach given existing hydropower facilities in the Alpine and Scandinavian countries, coupled with the recent expansion of wind and solar power throughout the EU. Germany, Italy, and Spain already generate well over 20 percent of their energy from renewable resources. Throughout Europe, wind power is the

▶ **Figure 8.10** **Wind Power in Europe** Large wind energy turbines are now a common feature of the European landscape, both on land and offshore. This offshore wind farm is in Denmark, a country whose goal is to meet half of its energy needs through wind power by the year 2020.

Population and Settlement: Slow Growth and Problematic Migration

Several themes dominate Europe's population and settlement geography: very low rates of natural growth; aging populations, and, perhaps most pressing, large streams of legal and extralegal migration, mainly coming from Africa and Southwest Asia. Western Europe's highly urbanized and relatively wealthy core area, which includes southern England, northern France, Belgium, the Netherlands, and western Germany, is the focus of most migration, both internal and international (**Figure 8.11**).

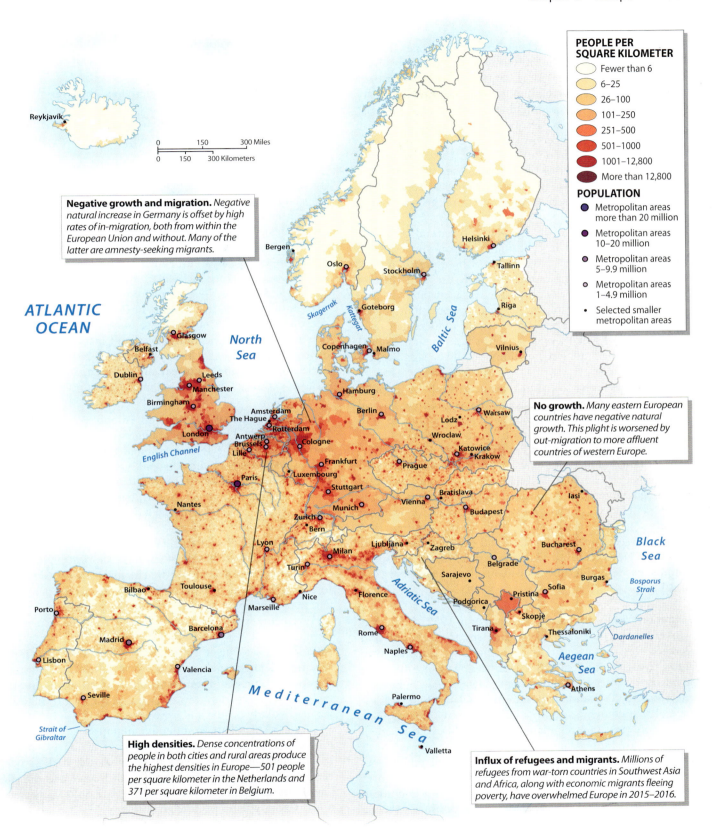

PEOPLE PER SQUARE KILOMETER

- Fewer than 6
- 6–25
- 26–100
- 101–250
- 251–500
- 501–1000
- 1001–12,800
- More than 12,800

POPULATION

- Metropolitan areas more than 20 million
- Metropolitan areas 10–20 million
- Metropolitan areas 5–9.9 million
- Metropolitan areas 1–4.9 million
- Selected smaller metropolitan areas

Negative growth and migration. *Negative natural increase in Germany is offset by high rates of in-migration, both from within the European Union and without. Many of the latter are amnesty-seeking migrants.*

No growth. *Many eastern European countries have negative natural growth. This plight is worsened by out-migration to more affluent countries of western Europe.*

High densities. *Dense concentrations of people in both cities and rural areas produce the highest densities in Europe—501 people per square kilometer in the Netherlands and 371 per square kilometer in Belgium.*

Influx of refugees and migrants. *Millions of refugees from war-torn countries in Southwest Asia and Africa, along with economic migrants fleeing poverty, have overwhelmed Europe in 2015–2016.*

▲ **Figure 8.11 Population of Europe** The European region includes about 540 million people, many of them clustered in large cities in both western and eastern Europe. As can be seen on this map, the most densely populated areas are in England, the Netherlands, Belgium, western Germany, northern France, and south across the Alps to northern Italy. **Q: What best explains the different population densities between eastern and western Europe?**

Low (and No) Natural Growth

Probably the most striking characteristic of Europe's demography is the lack of natural growth as death rates exceed birth rates (Table 8.1). Several large countries, notably Germany and Italy, have negative natural growth, meaning that without in-migration their populations would actually decrease in size over the next decades. Germany is forecast to lose 6 million people by mid-century, and Italy, 2 million. Numerous smaller European countries such as Latvia, Lithuania, Bulgaria, Romania, Serbia, and Portugal are also projected to have smaller populations in the coming decades. Again, these forecasts are based only upon natural growth figures and do not consider migration.

The reason behind this low and no natural growth is that Europe's population, like that of Japan and even the United States, is characterized by the fifth, or postindustrial, stage of the demographic transition (discussed in Chapter 1), in which fertility falls below the replacement level. The consequences of shrinking national populations—labor shortages, smaller internal markets, and reduced tax revenues to support social services (such as retirement pensions) essential for their aging populations—could be significant.

Pro-growth Policies To address concerns about population loss, many European countries try to promote natural growth through various programs and policies ranging from bans on abortion and the sale of contraceptives (Hungary) to what are commonly called **family-friendly policies** (Germany, France, and Scandinavia). In these countries, pro-growth policies include full-pay maternity and paternity leaves, guarantees of continued employment once these leaves conclude, extensive child-care facilities for working parents, outright cash subsidies for having children, and free or low-cost public education and job training for their offspring. However, even with these policies, no European country has a total fertility rate above the replacement level of 2.1.

Migration Within Europe Since its origin in 1957, the EU has worked toward the goal of free movement of both people and goods within the larger European community. Consequently, today residents of EU member countries can generally move about as they please. And they are doing just that. For example, some 16,000 Lithuanians moved to Ireland to take advantage of that country's then booming economy, leaving their home country (total population of 2.9 million) to languish economically. More recently, when the Irish boom went bust, almost half of these migrants either moved back to Lithuania or went elsewhere in the EU. Recent net migration figures show Ireland is continuing to lose people to out-migration, as are Lithuania and the other two Baltic countries, Latvia and Estonia. Other areas of significant out-migration are the Balkans and the Mediterranean countries; Spain and Romania are two of the largest countries losing population to emigration. As for receiving European migrants, Germany is a favored destination, as are the UK and several smaller countries, including Norway, Luxembourg, and Austria

(see *Geographers at Work: Migrants in the Digital Age*). The UK's status as a major destination for EU citizens, however, may change dramatically because of the Brexit vote.

The Schengen Agreement Underlying this intra-Europe mobility is a formal legislative act that eroded Europe's historical national borders: the **Schengen Agreement**, named after the small town in Luxembourg where it was signed in 1985.

Before Schengen, crossing a European border always involved showing passports and auto insurance papers, car inspection records, and so on. Until recently, however, there were either no border stations at all or only the most cursory formalities when traveling between Schengen countries.

However, in 2015 free movement across the borders of Schengen Europe was suspended due to a combination of terrorist acts (mainly in France) and more than a million extralegal migrants flooding in from Africa and Southwest Asia. Instead of the free movement that defined Europe for decades, temporary stations were established at most borders to check identification papers and inspect the innards of trucks and buses. At this writing (late 2016) the future of Schengen Europe is unclear, and will remain so until the issue of extralegal international migration is resolved.

Legal and Extralegal Migration to Europe Western Europe has long accepted international migrants into its population, particularly into the former colonial powers of Spain, the Netherlands, France, and England, countries that willingly provided their overseas citizens with visas and residential permits. Thus, historically, South Asians came to England, Indonesians to the Netherlands, North and West Africans to France, and South Americans to Spain.

Europe's doors to foreigners were opened further to solve postwar labor shortages as cities and factories rebuilt from the destruction of World War II. The former West Germany, for example, drew heavily on workers from Europe's rural and poorer periphery—Italy, the former Yugoslavia, Greece, and even Turkey—to fill industrial, construction, and service jobs. Later, with the collapse of the Soviet Union, emigrants from former satellite countries poured into western Europe, seeking relief from the economic chaos in Russia and eastern European countries. This post–Cold War immigration also included refugees from war-torn areas of the former Yugoslavia, particularly Bosnia and Kosovo.

As a result of these different migration streams, legal migrants now make up about 5 percent of the EU population; Germany, the largest country, has the highest percentage of foreigners (10 percent), with France and the UK having about 5 percent each. (For comparison, 11.7 percent of the U.S. population is foreign born.)

Extralegal Migration While at one level the distinction between legal and extralegal migration may seem clearcut—a migrant either does or does not have the proper entry papers—the situation today is much more complex because of Europe's long-standing **asylum laws** designed to protect

TABLE 8.1   Population Indicators   MasteringGeography™

Country	Population (millions) 2016	Population Density (per square kilometer)[1]	Rate of Natural Increase (RNI)	Total Fertility Rate	Percent Urban	Percent ‹15	Percent ›65	Net Migration (Rate per 1000)
Western Europe								
Austria	8.8	103	0.0	1.5	66	14	18	13
Belgium	11.3	371	0.1	1.7	98	17	18	4
France	64.6	121	0.3	1.9	80	18	18	1
Germany	82.6	232	−0.2	1.5	75	13	21	14
Ireland	4.7	67	0.8	1.9	63	22	13	−3
Luxembourg	0.6	215	0.4	1.5	90	16	14	20
Netherlands	17.0	501	0.1	1.7	90	17	18	3
Switzerland	8.4	207	0.2	1.5	84	15	18	9
United Kingdom	65.6	267	0.3	1.8	83	18	17	3
Southern Europe								
Albania	2.9	106	0.3	1.7	58	18	13	−6
Bosnia & Herzegovina	3.7	75	−0.1	1.3	40	15	14	0
Croatia	4.2	76	−0.4	1.5	59	15	19	−4
Cyprus	1.2	125	0.6	1.4	67	17	12	−12
Greece	10.8	85	−0.2	1.3	78	15	19	1
Italy	60.6	207	−0.3	1.4	69	14	22	2
Kosovo	1.8	167	1.2	2.3	38	24	8	−31
Macedonia	2.1	82	0.1	1.5	57	17	13	0
Montenegro	0.6	46	0.2	1.6	64	18	14	−3
Portugal	10.3	114	−0.3	1.3	64	14	20	−1
Serbia	7.1	82	−0.6	1.5	60	14	18	2
Slovenia	2.1	102	0.0	1.6	50	15	18	0
Spain	43.3	93	0.0	1.3	80	15	18	−1
Northern Europe								
Denmark	5.7	133	0.1	1.7	88	17	19	9
Estonia	1.3	31	−0.1	1.6	69	16	19	2
Finland	5.5	18	0.0	1.6	84	16	20	2
Iceland	0.3	3	0.5	1.8	94	20	14	3
Latvia	2.0	32	−0.3	1.6	68	15	19	−4
Lithuania	2.9	47	−0.3	1.6	67	15	19	−8
Norway	5.2	14	0.3	1.7	81	18	16	6
Sweden	9.9	24	0.3	1.8	86	17	20	8
Eastern Europe								
Bulgaria	7.1	67	−0.6	1.5	73	14	20	−1
Czech Republic	10.6	136	0.0	1.6	73	15	18	2
Hungary	9.8	109	−0.4	1.4	71	14	18	0
Poland	38.4	124	0.0	1.3	60	15	16	0
Romania	19.8	87	−0.3	1.2	55	16	15	−27
Slovakia	5.4	113	0.0	1.4	54	15	14	1
Micro States								
Andorra	0.1	155	0.5	1.2	85	15	14	8
Liechtenstein	0.0	233	0.3	1.6	14	15	16	4
Malta	0.4	1,336	0.2	1.4	95	14	18	7
Monaco	0.0	18,812	0.1	1.5	100	13	24	21
San Marino	0.0	527	0.1	1.5	94	15	18	6
Vatican City	−	−	−	−	−	−	−	−

Source: Population Reference Bureau, *World Population Data Sheet*, 2016.

[1] World Bank, *World Development Indicators*, 2016.

Login to MasteringGeography™ **& access** MapMaster **to explore these data!**

1) Use MapMaster to map all of those countries with a negative migration rate. Then write a short essay discussing what this distribution says about the traits shared by these countries. You could also draw upon socioeconomic data in Table 8.2 for your discussion.

2) Using the percentage of the population over age 65 data, make a map that shows which European countries have the highest number of older people. Draw upon relevant data from Table 8.1 and Table 8.2 for your analysis.

Geographers at Work

Migrants in the Digital Age

▲ **Figure 8.1.1** Dr. Weronika Kusek

Weronika Kusek, an assistant professor of geography at Northern Michigan University, grew up in Poland but did her college and graduate work in the United States (**Figure 8.1.1**). During those years, she thought of herself not as a migrant, but as an international student who would surely return to Poland someday. But after 10 years living and working in the United States, she's recalibrating her identity, in part because she studies migrants and how migrant networks have changed dramatically in the digital age.

Everyday Geography Kusek's interest in geography started early, based on a Polish school curriculum that introduces geography in the 4th grade, and reinforced by parents who valued travel and global experiences. She says, "Geography is very applicable to everyday life . . . you interact with people from other cultures, other parts of the world, so a basic understanding of other regions and cultures is extremely important for any career."

Kusek's family hosted an American student while she was in high school; this, in turn, led to her extending a summer visit to Ohio into a trial semester at a college in Toledo. Although she initially planned to return to Poland for college, Kusek received her BA and MA from the University of Toledo, then went on to earn her PhD in Geography from Kent State University.

Ties to Home Based on her fieldwork in England, Kusek says that some migrant groups maintain such strong ties to their homeland through daily phone calls, Skype, and other media, that they challenge the traditional process of assimilation into the host culture (**Figure 8.1.2**). Many Polish migrants in London, for example, spend every holiday in Poland reinforcing ties with family and friends. These new cultural behaviors and geographies contrast significantly

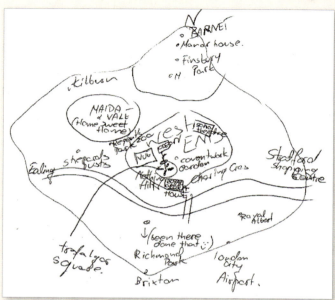

▲ **Figure 8.1.2** **Mental Map of London** One way of learning about a migrant's adaptation to a new place is to ask them to draw a mental map of the area, noting the important parts where they work and live, shop, and recreate. Here's an example from one of Dr. Kusek's London subjects.

with how pre-digital migrant communities interacted with their host cultures.

And what about her own ties to Poland? Just like the people she studies, Kusek Skypes often and makes at least one trip back to Poland each year.

1. Do you know people who are pre-Internet migrants? If so, ask them how they stayed in touch with their home and family.

2. Are you a "digital migrant" of some sort? That is, do you use the Internet to keep in touch with your home?

---

international refugees from political and ethnic persecution. These asylum laws stem from Europe's post–World War II humanitarian efforts to care for the region's refugees displaced by the war as well as those displaced by the subsequent Cold War that divided the continent politically from 1945 to 1991.

In 2015, however, Europe was inundated with asylum seekers from afar, primarily from war-torn Syria, Iraq, Somalia, and Afghanistan (**Figure 8.12**). Although it is theoretically possible to apply for asylum in Europe from one's home country, the very nature of persecution often prevents this, leading most asylum seekers to try to enter Europe extralegally. Once (or if) they reach European soil, refugees can then ask authorities for political asylum.

Getting to European soil, however, is not simple, cheap, or safe. Refugees reportedly pay thousands of dollars to smugglers for a risky land or sea journey to Europe (**Figure 8.13**). Many die in the process. Even if they reach Europe, as one million did in 2015, they will spend months (often years) in overcrowded relocation camps waiting to make their legal

case to European authorities (Figure 8.13). Years may pass before a decision is reached, because separating asylum seekers with legitimate claims of persecution from the so-called economic migrants simply looking for work is a difficult and time-consuming task. If the refugees are granted asylum, they will be sent to an accepting European country, where they will face the additional challenges of adapting to a new culture and environment; if they are denied asylum, they will be summarily sent back to their home country.

Sweden and Norway, on a per capita basis (of the native population), have accepted the largest number of migrants granted asylum. Germany has accepted the largest total number (more than 140,000 in 2016), with France accepting around 15,000 and the UK about 10,000. As of this writing, the EU is developing a quota system that will require all 28 member states to accept a specific number of refugees each year.

To help the EU perimeter countries of Greece, Italy, Malta, and Spain police their borders, the EU provides funds to strengthen their borders with guards and, in some places,

▲ **Figure 8.12 Migration Into and Within Europe** In 2015, over a million extralegal migrants entered Europe, with about half that number in 2016. Despite the reduction in numbers, the migrant crisis remains one of Europe's most pressing issues.

with physical border barriers to inhibit illegal entry. Even Turkey, a non-EU country but a popular point of departure for migrants into nearby Greece and Bulgaria, has been provided EU funding for more accommodating refugee camps in hopes of keeping potential migrants there instead of attempting entry to Europe.

Today, some observers describe Europe as being divided into a geographical system in which its perimeter consists of hard borders—a "Fortress Europe," as critics (and anti-immigrant groups) call the plan—while its internal borders are deliberately soft and porous due to the Schengen Agreement (**Figure 8.14**). However, until the illegal international migration issue is resolved, those soft internal Schengen borders will become increasingly controversial, challenging earlier political and economic goals of a "Europe without borders."

◄ **Figure 8.13 Migrants Walking to Europe** In 2015, one of the major migration pathways into Europe was from Greece into Macedonia and other Balkan countries, with the goal of reaching Austria and Germany. These migrants are walking through Macedonia.

▲ **Figure 8.14 Greece-Macedonia Border** Two young Syrian refugees play alongside the razor-wire fence erected along the Greece-Macedonian border. This fortified border was created in 2016 to limit the flow of migrants into the Balkan countries. Currently, only a handful of refugees are allowed into Macedonia each day to continue their journey to the heart of Europe.

Urban Europe

Europe is highly urbanized, with almost three-quarters of the region's people living in cities. Further, Europe has several microstates that are essentially city-states: Monaco (aka Monte Carlo), Malta, and Vatican City. High rates of urbanization, however, produce very different landscapes. Belgium, for example, is 98 percent urban, but this comes more from a landscape of connected mid-sized towns rather than from huge megacities. And no traveler in Iceland would think of that country as primarily urban since the country has only one true city, Reykjavik. At the other end of the scale are Bosnia–Herzegovina and Kosovo, two Balkan countries that are Europe's only two countries less than 50 percent urban. Indeed, the dominant rural landscape reinforces that fact.

Europe's largest countries are more typical of much of the region: nearly 80 percent urban, with numerous large cities scattered among expansive rural landscapes. With the exception of Germany, Europe's countries are dominated by one primate city, usually its historic capital. London, England; Paris, France; Madrid, Spain; Sofia, Bulgaria; and Stockholm, Sweden are a few examples.

London easily heads the list of Europe's largest cities at 8.6 million, followed by Berlin (3.5 million), Madrid (3.1 million), Rome (2.8 million), and Paris (2.2 million) within the true city limits. However, when each city's suburbs are counted the metro population size is usually doubled. Not to be overlooked are the urban agglomerations of several cities grown together.

▶ **Figure 8.15 London, Europe's World City** London's booming growth over the last years has been closely tied to the city's central role in global finance. That role, however, may be jeopardized by an uncertain future because of Brexit.

Explore the **Sights** of London

http://goo.gl/pzWJt3

Two stand out: the Rhine-Ruhr area centered on Cologne, where some 11 million people are clustered together in one megacity, and not far away the so-called *Randstadt* ("Ring City") of the Netherlands, consisting of Rotterdam, Amsterdam, Utrecht, and the Hague, where 7.1 million live.

Not only is London Europe's largest city, but it is also Europe's leading world city in terms of its interaction with the global economy, competing with New York and Hong Kong as one of the world's most influential banking centers (**Figure 8.15**). A distant second place in terms of its interaction with the global economy is Zurich, Switzerland. Interesting to note is that despite London's vitality as a world city, its population only recently attained pre–World War II levels. That is, since 1945 London actually lost population as the British Empire dissolved and the UK sought resolution to both internal and international problems. But by the 1970s it was clear that London, determined to reclaim its status as a world city, implemented an ambitious (and, at times, highly contested) building program that reshaped the city's landscape. With many of London's new developments mixing residential spaces with nearby office space, the population slowly rebounded to the point where in 2015 London achieved its pre-war population.

A significant part of London's regained vitality results from its international population, with thousands upon thousands of inhabitants coming from countries both within Europe and from the world at large. There are so many French people living and working in London that it's said that London is now France's second-largest city. Unknown, however, is how the UK's departure from the European Union (Brexit) will affect London's cosmopolitan population.

Urban Planning Issues European cities face a number of difficult problems, some of them similar to those in North America, others unique to the region. Perhaps the most vexing is the problem of auto congestion in central cities. Even though most European cities have public transportation systems far superior to those in North America, auto traffic remains a problem because European city planners have been reluctant to sacrifice the historical central city

Working Toward Sustainability

Copenhagen, Where Bikes Rule

Google Earth Virtual Tour Video

http://goo.gl/WnFW5Z

All European cities are trying to reduce traffic congestion in their central areas, and one common strategy is to encourage bicycling rather than automobile transport. The acknowledged leader of this movement is Copenhagen, Denmark, an attractive and vibrant capital of 591,000 people, and a place commonly considered the world's most bike-friendly city. And bike-friendly it is indeed, with over half of the city's inhabitants commuting by bike every day, rain or shine. For the larger Copenhagen metro area, almost a third of commuters travel daily by bicycle.

What's Copenhagen's secret? At the higher municipal planning level, it's treating bikes as a legitimate form of transportation, not a recreational add-on, and integrating biking needs into every kind of planning decision. Not that this happened overnight: Copenhagen has been working hard at becoming bike-friendly for over 40 years, nor has it come easily; many decisions are made at the expense of car drivers, many of whom are in open battle with what they call the country's plague of "two-wheeled locusts."

Cycle Tracks The major feature of Copenhagen's bike friendliness are the 220 miles (330 km) of what are called cycle tracks, where the bike paths are physically separated from auto traffic by a curb and sidewalk. Each side of main streets has a cycle track so bike traffic flows in the same direction as auto traffic. On these routes streetside parking is on the outside of the dividing sidewalk, creating an additional barrier between bikes and cars. True, adding these dedicated cycle tracks to Copenhagen's streets does hinder auto traffic—which is part of the plan.

Complementing the cycle tracks are 14 miles (23 km) of on-street bike lanes, where cars and bikes are separated only by the conventional painted lines, along with 42 miles (67 km) of off-street bike lanes running through parks and other green spaces. Most recently, city planners have designated 310 miles (500 km) of dedicated super bikeways connecting the suburbs to the inner city, hoping to lure more suburban commuters from their cars to bike.

But it's not just about bike lanes. With more bikes than people, Copenhagen has created large bike parking areas; as well, all forms of public transportation are bike-friendly, with racks on buses and dedicated bike cars on the subway (**Figure 8.2.1**).

Bike Culture Physical cycle tracks and bike racks are one thing, but central to Copenhagen's cycling story is the development over the years of a strong bike culture that goes far beyond the mere enjoyment of recreational bike riding to incorporate new technology (cargo bikes started in Copenhagen), fashion ("bike chic" is closely linked to Copenhagen), behavior (a "bike date" is common), and on and on (**Figure 8.2.2**). Perhaps nothing captures Danish bike culture quite like the fact that the former prime minister rode her bike to the Queen's palace for her ceremonial inauguration.

1. What is your town or college campus doing to become more bike-friendly? What can they do that they haven't done already?

2. Have you experienced some bike-friendly measures that are auto-unfriendly? Describe them.

▲ **Figure 8.2.1 Which Bike is Mine?** With so many residents of Copenhagen riding their bikes to work and school, providing adequate bike parking spaces in the inner city has become a major problem.

▲ **Figure 8.2.2 Bike Chic** Except in rainy weather when Danish bikers break out their rain pants and parkas, everyday work apparel is standard dress for Copenhagen's urban bikers. Bike helmets seem to be used far less often than in the United States.

landscape to accommodate cars and trucks. Even London, which suffered major bomb damage in World War II, chose to recreate the historical urban fabric of pre-automobile narrow streets and alleys. One common solution has been to create car-free pedestrian zones in historic city centers (**Figure 8.16**). Another is to make it outrageously expensive to drive (and park) autos in central districts; this strategy is often complemented by reducing public transportation costs. Still another is to make central cities more bike friendly, a model adopted by Copenhagen, Denmark several decades ago (see *Working Toward Sustainability: Copenhagen, Where Bikes Rule*).

▲ **Figure 8.16 Central City Pedestrian Districts** Many European cities have created historical preservation areas in their central cities. Besides preserving the historical landscape of the area, a common feature is to create pedestrian-friendly areas where auto traffic is severely limited. This photo is from the Marais historical district in central Paris.

Suburban sprawl and preserving open space—familiar North American planning problems—are also issues in Europe. London first addressed these in 1935 with a protective green-belt surrounding the urbanized area. This notion has been furthered by several postwar planning policies to strengthen open space protection; this English greenbelt model is now a common planning practice throughout Europe.

REVIEW

8.5 Which European countries have the highest and lowest rates of natural population increase? Which have the highest rates of out-migration and in-migration?

8.6 What is the Schengen Agreement, and how is it related to population movement?

8.7 Discuss the reasons behind the current flood of illegal migration to Europe.

8.8 Explain why European cities are so congested, and give an example of a policy to counter such congestion.

■ **KEY TERMS** family-friendly policies, Schengen Agreement, asylum laws

---

Cultural Coherence and Diversity: A Mosaic of Differences

The rich cultural geography of Europe demands our attention for several reasons. First, the highly varied mosaic of languages, customs, religions, and ways of life that characterize Europe not only strongly shaped regional identities, but have also stoked the fires of ethnic conflict. Embers from those historical conflicts still smolder today in different ways and in different parts of Europe.

Second, European cultures played leading roles in globalization as European colonialism brought about changes in languages, religions, political systems, economies, and social customs in every corner of the globe. Examples include cricket

games in Pakistan, high tea in India, Dutch architecture in South Africa, the millions of Spanish-speaking inhabitants of Latin America, and French and Italian cuisine around the globe (see *Everyday Globalization: The Cultural Heritage of Pizza*).

Today, however, waves of global culture are spreading back into Europe, and while some Europeans embrace (or passively condone) these changes, others actively resist. France, for example, struggles against both global popular culture as well as the multicultural influences of its large Muslim migrant population (**Figure 8.17**).

Geographies of Language

Language has always been an important component of nationalism and group identity in Europe. Today, while some small ethnic groups such as the Irish and the Bretons work hard to preserve their local language, millions of other Europeans are busy learning multiple languages—primarily English—so they can better communicate across cultural and national boundaries.

Explore the **Sounds** of Irish Folk Music

http://goo.gl/6pUYpx

At the broadest scale, most Europeans speak a language of the **Indo-European language family**; only in Finland, Estonia, and Hungary are the native languages not Indo-European (**Figure 8.18**). As their first language, 90 percent of Europe's population speaks a Germanic, Romance, or Slavic language, all linguistic groups within the Indo-European family. Germanic and Romance speakers each number almost 200 million in the region. Although Slavic languages are spoken by 400 million when Russia and its immediate neighbors are included, there are only 80 million Slavic speakers within Europe proper.

Germanic Languages Germanic languages dominate Europe north of the Alps. Today German, claimed by about 90 million people as their mother tongue, is spoken in

▼ **Figure 8.17 Muslims in Europe** Europe has long had a small Muslim population, historically in the Balkans and Spain but more recently in western Europe, because of Europe's colonial ties to Asia and Africa. Moreover, the postwar guest worker program led to the creation of Turkish communities in many German cities. These two Muslim schoolgirls in Germany are of Turkish descent, and perhaps are even German citizens. Currently, nationalistic, anti-migrant groups have created concerns about the "Islamization" of Europe from the large numbers of extralegal migrants from Southwest Asia and Africa.

The Cultural Heritage of Pizza

Although eating local food is always an exciting part of traveling, some people prefer to avoid some unfamiliar items commonly lurking on European menus. After all, unless you know the local cuisine, you could end up with a plate of cow's tongue in France (*langue de boeuf*), sheep's stomach in Scotland (*haggis*) or fried carp sperm (*smazeny kapr spermie*) in the Czech Republic.

But if you tire of that sort of culinary adventure there's always pizza, ubiquitous in Europe just as it is in North America, found in restaurants and at street stands everywhere throughout the region. It may not be the California, Chicago, Detroit, or New Jersey version you seek, but it'll most likely be a version of the authentic thin-crust pie served for centuries in Naples, Italy. Although pizza historians say that the earliest pizza-like food was brought by the Romans from Egypt to southern Italy, it was Naples' medieval trade ties that brought together the unique mixture of ingredients that make up the authentic and traditional Neapolitan pizza.

Explore the **Tastes** of Neapolitan Pizza

http://goo.gl/HZEpbQ

Perhaps because this mother-of-all-pizzas is now found worldwide in all sorts of flavors and forms, the Neapolitan Pizza Association has upped the ante in protecting the traditional form by requesting it be put on UNESCO's Intangible Cultural Heritage list. If approved, Neapolitan pizza will join two other important culinary practices on the UNESCO list, the Mediterranean Diet, and the French Gastronomic Meal. So try a slice of history when you tire of animal parts. *Bon appetit*!

By the way: When in Naples, you can dine at what pizza historians say is the world's first pizzeria, the *Antica Pizzeria Port' Alba*, which has been serving authentic Neapolitan pizza since 1830 (**Figure 8.3.1**).

1. What other national dishes have become so mainstream that they can be considered "world food"?

2. How does your favorite pizza differ from classic Neapolitan pizza? How do you suppose it evolved?

▲ **Figure 8.3.1 The World's First Pizzeria** Shown here, the Antica Pizzeria Port'Alba, founded in 1830 in Naples, Italy, is considered to be the world's oldest pizza place.

Germany, Austria, Liechtenstein, Luxembourg, eastern Switzerland, and several small areas in Alpine Italy.

English is the second-largest Germanic language, with about 60 million speakers learning it as their first language. In addition, a large number of Europeans learn English as a second language, with many as fluent in English as are native speakers. Linguistically, English is closest to the Low German spoken along the North Sea coastline, reinforcing the theory that an early form of English evolved in the British Isles through contact with the coastal peoples of northern Europe. One distinctive trait of English that sets it apart from German, however, is that almost one-third of the English vocabulary is made up of Romance words brought to England during the Norman French conquest of the 11th century.

Elsewhere in the Germanic linguistic region, Dutch (in the Netherlands) and Flemish (in northern Belgium) together account for another 20 million speakers, with roughly the same number of Scandinavians speaking the closely related languages of Danish, Norwegian, and Swedish. Icelandic is a more distinctive language because of that country's geographic isolation from its Scandinavian roots.

Romance Languages Romance languages, including French, Spanish, and Italian, evolved from the vulgar (or everyday) Latin spoken within the Roman Empire. Today, Italian is the most widely used of these Romance languages, with about 60 million Europeans speaking it as their first language. In addition to being spoken in Italy, Italian is an official language of Switzerland and is also spoken on the French island of Corsica.

French is spoken in France, western Switzerland, and southern Belgium (where it is known as *Walloon*). Today, there are about 55 million native French speakers in Europe with almost three times that many speakers in former French colonies in Africa, the Caribbean, and North America. As with other languages, French, both in Europe and in former colonial countries throughout the world, has very strong regional dialects.

Spanish also has very strong regional variations. About 25 million people speak Castilian Spanish, Spain's official language, which dominates the interior and northern areas of that large country. However, Catalan, which many argue is a completely separate language, is spoken along the eastern coastal fringe,

▲ **Figure 8.18 Languages of Europe** Ninety percent of Europeans speak an Indo-European language from one of the three major categories of Germanic, Romance, and Slavic languages. Ninety million Europeans speak German as a first language, which places it ahead of the 60 million who list English as their native language. However, given the large number of Europeans who speak fluent English as a second language, one could make the case that English is the dominant language of modern Europe.

centered on Barcelona, Spain's second-largest city. This distinctive language reinforces a strong sense of cultural separateness—also evident in unique Catalan architecture, traditions, and cuisine—that has led to Spain designating the state of Catalonia an autonomous region (**Figure 8.19**). At the global scale, Spanish is the world's fourth largest language, primarily because of the millions

of native speakers in South and Central America, Mexico, and the United States.

Portuguese is spoken by 12 million in Portugal and in the northwestern corner of Spain, although considerably more people speak the language in Brazil, a former Portuguese colony. Finally, Romanian represents an eastern outlier of the Romance language family, spoken by

(a)

(b)

▲ **Figure 8.19 Catalonian Autonomy** (a) People wave their *Esteladas*, the Catalonian separatist flag, during a recent Catalan pro-independence demonstration in Barcelona. (b) Complementing its distinct culture is Catalonian cuisine, illustrated here by this seafood-rich paella.

24 million people in Romania. Though unquestionably a Romance language, Romanian also contains many Slavic words.

The Slavic Language Family Slavic speakers are traditionally separated into northern and southern groups, divided by the non-Slavic speakers of Hungary and Romania. To the north, Polish is spoken by 35 million, while Czech and Slovakian speakers total about 15 million. As noted earlier, these numbers pale in comparison to the number of northern Slav speakers in nearby Ukraine, Belarus, and Russia, which easily total more than 150 million. Southern Slav languages include three groups: 14 million speakers of Serbian and Croatian (now considered separate languages because of the strong political and cultural differences between Serbs and Croats), 11 million Bulgarian or Macedonian speakers, and 2 million Slovenian speakers.

The use of two distinct alphabets further complicates the geography of Slavic languages. In countries with a strong

Roman Catholic heritage, such as Poland and the Czech Republic, the **Latin alphabet** is used. In contrast, countries with close ties to the Orthodox Church—Bulgaria, Montenegro, Macedonia, parts of Bosnia–Herzegovina, and Serbia—use the Greek-derived **Cyrillic alphabet** (**Figure 8.20**).

Geographies of Religion, Past and Present

Religion is an important component of the geography of cultural coherence and diversity in Europe because many of today's ethnic tensions result from historical religious events. To illustrate, significant cultural borders in the Balkans and eastern Europe are based on the 11th-century split of Christianity into eastern and western churches, as well as on the division between Christianity and Islam. Much of the terrorism termed "ethnic cleansing" in the former Yugoslavia during the 1990s was based on these religious differences.

In western Europe, blood is still occasionally shed in Northern Ireland over the tensions resulting from the 17th-century split of Christianity into Catholicism and Protestantism, and clashes between Islamic fundamentalists and everyone else. Understanding these contemporary tensions involves

▼ **Figure 8.20 Cyrillic Alphabet** A directional sign in downtown Sofia, Bulgaria, uses both the Cyrillic and the Roman alphabets to guide locals and visitors alike.

taking a brief look at the historical geography of Europe's religions (**Figure 8.21**).

The Schism Between Western and Eastern Christianity

In southeastern Europe, early Greek missionaries spread Christianity throughout the Balkans and into the lower reaches of the Danube. Because these Greek missionaries refused to accept the control of Roman Catholic bishops in western Europe, there was a formal split with western Christianity in 1054 CE.

This eastern church subsequently splintered into Orthodox sects closely linked to specific nations and states. Today we find Greek Orthodox, Bulgarian Orthodox, and Russian Orthodox churches, all with slightly different rites and rituals.

The Protestant Revolt

Besides the division into western and eastern churches, the other great split within Christianity occurred between Catholicism and Protestantism. This division arose in Europe during the 16th century and has divided the region ever since. However, with the exception of "the Troubles" in Northern Ireland, tensions today between these two major groups are far less problematic than in the past, when these religious differences led to several long wars.

Historical Conflicts with Islam

Both the eastern and western Christian churches struggled with challenges from the Islamic empires to Europe's south and east. Even though historical Islam was reasonably tolerant of Christianity in its conquered lands, Christian Europe was far less accepting of Muslim imperialism. The first crusade to reclaim Jerusalem from the Turks took place in 1095. After the Ottoman Turks conquered Constantinople in 1453 and gained control over the Bosporus Strait and the Black Sea, they moved rapidly to spread a Muslim empire throughout the Balkans, arriving at the gates of Vienna in the middle of the 16th century. There, Christian Europe stood firm militarily and stopped Islam from expanding into western Europe.

Ottoman control of southeastern Europe, however, lasted until the empire's demise in the early 20th century. This historical presence of Islam explains the current coexistence of religions in the Balkans, with intermixed areas of Muslims, Orthodox Christians, and Roman Catholics.

Islam was also the dominant religion and culture of Portugal and most of Spain from the 8th to the 17th century, when the Catholic kingdoms in Spain's northeast expanded their control of the Iberian Peninsula.

A Geography of Judaism

Europe has long been a difficult home for the Jews forced to leave Palestine during the Roman Empire. At that time, small Jewish settlements were located in cities throughout the Mediterranean.

Later, by 900 CE, about 20 percent of the Jewish population was clustered in the Muslim lands of the Iberian Peninsula, where Islam showed greater tolerance for Judaism than did Christianity. After the Christian reconquest of Iberia, however, Jews once more faced severe persecution and fled from Spain to more tolerant countries in western and central Europe.

One focus for this exodus was the area in eastern Europe that became known as the Jewish Pale (see Figure 8.21). In the late Middle Ages, at the invitation of the Kingdom of Poland, Jews settled in cities and small villages in what are now eastern Poland, Belarus, western Ukraine, and northern Romania. Jews gathered in this region for several centuries in the hope of establishing a true European homeland.

Until emigration to North America began in the 1890s, 90 percent of the world's Jewish population lived in Europe, and most were clustered in the Pale region. Tragically, Nazi Germany devastated this ethnic cluster by focusing its extermination activities on the Pale.

In 1939, on the eve of World War II, 9.5 million Jews, or about 60 percent of the world's Jewish population, lived in Europe. During the war, German Nazis murdered some 6 million Jews in the horror of the Holocaust. Today, fewer than 1.5 million Jews, about 10 percent of the world population, live in Europe.

Patterns of Contemporary Religion

Estimates of religious adherence in contemporary Europe suggest there are 277 million Roman Catholics, 100 million Protestants, and 13 million Muslims. Generally, Catholics live in the southern half of the region, except for the significant numbers in Ireland and Poland. Protestantism is most widespread in northern Germany (with Catholicism stronger in southern Germany), the Scandinavian countries, and England, and it is intermixed with Catholicism in the Netherlands, Belgium, and Switzerland. Muslims historically are found in Albania, Kosovo, Bosnia–Herzegovina, and Bulgaria. Adding to Europe's Muslim population are recent migrants from Southwest Asia and northern Africa, resulting in around 5 million Muslims today in both Germany and France. Because of migration and a generally higher natural birth rate, combined with a stagnating number of practicing Christians, Islam is Europe's fastest-growing religion.

Because of Europe's long history of religious wars and tensions, the EU's agenda of European unity is explicitly secular, a position that causes its own set of contemporary cultural tensions. For example, the euro, the EU's common currency, has purged all national symbols of Christian crosses and saints, much to the chagrin of countries like Hungary and Poland whose sense of nationalism is inseparable from Catholicism. Perhaps understandably, EU secularism is particularly difficult to accept in those former Soviet satellites where their churches were closed and boarded up during

▲ **Figure 8.21 Religions of Europe** This map shows the divide in western Europe between the Protestant north and the Roman Catholic south. Historically, this distinction was much more important than it is today. Note also the location of the former Jewish Pale, the area that was devastated by the Nazis during World War II. Today, ethnic tensions with religious overtones are found primarily in the Balkans, where adherents to Roman Catholicism, Eastern Orthodoxy, and Islam are found in close proximity. **Q: After comparing this map to the one of Europe's languages (Figure 8.18), list those areas where language families and religion appear to be related.**

the communist period and have only recently reopened as places of gathering and worship.

Despite these tensions, Europe's religious landscape, with its diverse array of cathedrals, churches, monasteries, nunneries, and other religious sites and structures, is a major attraction for travelers and tourists, both international and intra-European.

Because of historical Protestantism's reaction against the ornate cathedrals and statues of the Catholic Church, the landscape of Protestantism in northern Europe is rather sedate and subdued, in contrast to the numerous religious sites of Catholic Europe. The large cathedrals and religious monuments in Britain are associated primarily with the Church of England, which originated with strong historical ties to Catholicism; St. Paul's Cathedral and Westminster Abbey in London are examples. Newer additions to Europe's religious landscape are the Muslim mosques now found in many western European cities (**Figure 8.22**), with many more being planned to accommodate growing numbers of adherents.

Migrants and Culture

New migration streams from Africa, Asia, and South America are profoundly influencing the dynamic cultural geography of Europe. Unfortunately, in some areas of Europe the products of this recent cultural exchange are highly troubling.

Immigrant clustering, leading to the formation of ethnic neighborhoods and even ghettos, is now common in the cities and towns of western Europe. The high-density apartment buildings of suburban Paris, for example, are home to large numbers of French-speaking Africans and Arab Muslims caught in a web of high unemployment, poverty, and racial discrimination. As a result, cultural struggles, both on the streets and in the courtrooms, are now common in many European countries. For example, in 2004 French leaders drew on the country's constitutional separation of state and religion to ban a key symbol of conservative Muslim life—the head scarf (*hijab*)—for female students in public schools because, officials argued, it interfered with the educational process. In 2010, full-face veils were banned in public places, the rationale being that traditional Muslim dress inhibits assimilation into contemporary French society. Other European countries have made their own attempts at restricting Muslim dress with a portfolio of rationalizations ranging from concerns about terrorism to the unsafe driving conditions resulting from headscarves and body coverings.

What is clearly at issue here is the social unease about Muslim immigrants in what until recently was a relatively homogeneous European culture. A recent Pew Foundation survey asked Europeans about their attitudes toward Muslims. Surprisingly, three-quarters of the French population actually holds favorable views of Muslims. In Germany, home to Europe's largest Muslim population, favorable views are held by 60 percent of the population. But in Spain, Greece, and Poland, less than half the population has positive attitudes toward Muslims. Interestingly, attitudes toward Muslims reflected general political orientation—people associated with the political right were significantly more unfriendly to Muslims than those aligned with the political center or left. This political distinction is not surprising, because far-right, neonationalistic political parties throughout Europe share anti-migrant, anti-immigration, anti-asylum positions. Further, European attitudes toward Muslims are also linked to the presence or absence of terrorist activity.

Popular Culture

Sports Soccer (which Europeans call football) is unquestionably Europe's national sport, played everywhere from sandlots to stadiums, by both women and men, at all levels from family picnics to multiple-level professional leagues. At the highest pro level, soccer teams draw crowds into stadiums holding 100,000 people. Smaller soccer stadiums seating 30,000–40,000 are common in every European town.

Like many sports throughout the world, soccer is irrevocably linked to globalized culture, with fanatical fans rooting for place-based teams constituted largely of international players lacking any local allegiance. But this contradiction doesn't keep soccer fans from taking their local fandom across Europe's borders to rival towns and cities, where team loyalties sometimes turn violent. Soccer hooliganism, unfortunately, has become a common outlet for Europe's anti-migrant racism and xenophobia.

Aside from homegrown sports like soccer and rugby, Europe has shown some interest in the North American sports of basketball, baseball, and American football. Basketball is unquestionably Europe's favorite American sport, with hoops and courts increasingly common in the region's gyms and playgrounds. Pro leagues at all levels abound for both men and women, with most European cities supporting at least one pro team (**Figure 8.23**). It is now common for U.S. Women's National Basketball Association (WNBA) players to spend their off-season playing for a European pro team to augment their modest WNBA salaries. Also common is

▼ **Figure 8.22 Muslim Mosque in Germany** The Merkez Mosque in Duisburg, Germany, is said to be the largest mosque in a non-Muslim country. Duisburg, a city of half a million people in the Ruhr industrial area of northwestern Germany, has a population of 85,000 people of Turkish origin.

Explore the **Sights** of Duisburg's Merkez Mosque

http://goo.gl/dMCEbE

▲ **Figure 8.23 European Women's Basketball** Kristi Toliver (center), in action here in a European Final Four game, played her college basketball at the University of Maryland and currently is a member of the WNBA's Los Angeles Sparks. In the WNBA offseason she plays for a Slovakian team.

the increasing number of European basketball players (both male and female) who play for North American college and pro teams.

Baseball is fairly popular in Europe, having grown from seeds planted by postwar U.S. servicemen into several professional leagues. American football remains a novelty, however; with its frequent timeouts, the sport fails to capture the attention of cultures that thrive on soccer's nonstop action.

Many winter sports, primarily ice hockey and skiing, originated in Europe and are now enjoyed world-wide, with ski resorts found in the Alborz mountains of Iran and professional ice hockey leagues (complete with Russian and Canadian players) found in Dubai and other Gulf States.

The Sounds of Europe: Eurovision While the geography of European music is often dominated by discussing differences within the classical forms of, say, French versus Italian operas or the emotion (or lack of) expressed in northern versus eastern European symphonic composers, in contemporary globalized Europe it's all about the two weeks in May when Europe—along with much of the world—gets caught up in the Eurovision Song Contest. Begun in 1956 in the spirit of bridging Europe's cultural differences, it's now the world's most watched non-sport TV show, as songsters compete from most every European country (including an ever-changing list of invited guest countries such as Australia, Ukraine, and even Russia). Originally performers had to sing in their native language, a rule that was quickly amended in the early1960s with guidelines suggesting English as the appropriate language. Although today most songs are sung in English, current guidelines allow performances in any language, native or not. Frankly, critics and inveterate watchers both agree, it's not about the music but more about the campy

▲ **Figure 8.24 Eurovision: Music and More** Singer Jamie-Lee, representing Germany, performs on stage during the Grand Final of the 61st annual Eurovision Song Contest.

personalities, costumes, and stage performances (**Figure 8.24**).

During the two-week Eurovision period, performers work their way through regional and national competitions until the final round, where one singer from each country performs. Voting takes place by live audiences, a panel of judges, and—here's where it gets really interesting—by telephone. Although the Eurovision Song Contest was invented to bring Europe together, various kinds of nationalism are still present. The booing of a Russian performer resulted in Eurovision using a machine that erases the sound of booing before the live performance is transmitted from the performance hall to the outside world.

Explore the **Sounds** of Jamala's Eurovision Hit "1944"

http://goo.gl/83xN3s

REVIEW

8.9 Describe the general location within Europe of the three major language groups: Germanic, Romance, and Slavic.

8.10 Summarize the historical distribution within Europe of Catholicism, Protestantism, Judaism, and Islam.

8.11 Which countries have the highest numbers of Muslims? What cultural conflicts have resulted?

■ **KEY TERMS** Indo-European language family, Latin alphabet, Cyrillic alphabet

---

Geopolitical Framework: A Dynamic Map

One of Europe's unique characteristics is its dense fabric of 42 independent states within a relatively small region. The ideal of democratic **nation-states** (see Chapter 1 for a discussion of the nation-state concept) arose in Europe and, over time, replaced the fiefdoms and empires ruled by autocratic royalty. France, Italy, Germany, and the United Kingdom are major examples.

In many ways, however, Europe's unique geopolitical landscape has been as much problem as promise. Twice in the past century, Europe shed blood to redraw its political borders, and within the past several decades nine new states have appeared—more than half through violent wars. Today, Europe's geopolitical tensions are less about creating new nation-states and more about achieving smaller-scale regional autonomy, concurrent with safety from terrorism. (**Figure 8.25**).

Redrawing the Map of Europe Through War

Two world wars radically reshaped the geopolitical map of 20th-century Europe (**Figure 8.26**). Although World War I was known as the "war to end all wars," it fell far short of solving Europe's geopolitical problems. Instead, the resulting peace treaty actually made another world war unavoidable.

When Germany and Austria–Hungary surrendered in 1918, the Treaty of Versailles set about redrawing the map of Europe with two goals in mind: first, to punish the losers through loss of territory and severe financial reparations and, second, to recognize the nationalistic aspirations of unrepresented peoples by creating several new nation-states. As a result, the new states of Czechoslovakia and Yugoslavia were created. Poland was reestablished, as were the Baltic states of Finland, Estonia, Latvia, and Lithuania.

Though the goals of the treaty were admirable, few European states were satisfied with the resulting map. New states were resentful when some of their citizens were left

outside the redrawn borders. This created an epidemic of **irredentism**, state policies directed toward reclaiming lost territory and peoples.

These imperfect geopolitical solutions were greatly aggravated by the global economic depression of the 1930s, which brought high unemployment, food shortages, and even more political unrest to Europe. With industrial unemployment at record highs in western Europe, public opinion fluctuated wildly between the extremist solutions of far-right fascism and far-left communism and socialism. In 1936, Italy and Germany joined forces through the Rome–Berlin "axis" agreement. When an imperialist Japan signed a pact with Germany, the scene was set for a second global war.

Nazi Germany tested western European resolve in 1938 by annexing Austria, the country of Hitler's birth, and then

▶ **Figure 8.25 Geopolitical Issues in Europe** Although the major geopolitical issue of the early 21st century remains the integration of eastern and western Europe into the EU, numerous issues of micro- and ethnic nationalism also engender geopolitical fragmentation. In other parts of Europe, such as Spain, France, and Great Britain, questions of local ethnic autonomy within the nation-state structure challenge central governments.

North Atlantic Treaty Organization (NATO) member

Former Warsaw Pact member

NATO headquarters

Note: The United States and Canada are also members of NATO.

Scotland. *In 2014 Scots narrowly rejected a referendum on independence from the United Kingdom; however, separatist sentiment remains strong.*

United Kingdom. *June 2016 vote to leave the European Union destabilizes Europe's agenda for political and economic cooperation*

Catalonia. *Separatists have a plan for secession from Spain by 2017 based upon regional elections in late 2015. Spain's national government, however, strongly resists this action.*

Peace at last. *After a decade of ethnic cleansing during the 1990s following the breakup of the former Yugoslavia, relative peace has settled over the Balkans as the new independent states turn their attention to joining the European Union.*

A new Cold War? *Tensions between Europe and NATO have increased recently because of Russia's aggressive actions in Ukraine and Crimea, along with provocative military activities in international waters and airspace.*

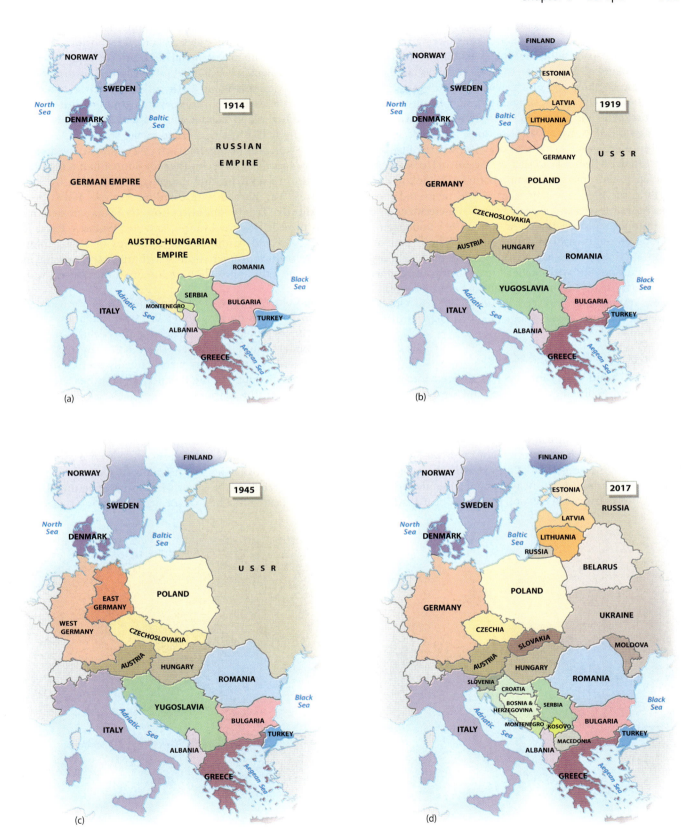

▲ **Figure 8.26 A Century of Geopolitical Change** (a) At the outset of the 20th century, central Europe was dominated by the German, Austro-Hungarian (or Hapsburg), and Russian empires. (b) Following World War I, these empires were largely replaced by a mosaic of nation-states. (c) More border changes followed World War II, largely as a result of the Soviet Union's turning the area into a buffer zone between itself and western Europe. (d) With the demise of Soviet hegemony in 1990, further political change took place. **Q: Where are the strongest relationships between political change and cultural factors such as language and religion?**

Czechoslovakia under the pretense of providing protection for ethnic Germans located there. After Germany signed a nonaggression pact with the Soviet Union, Hitler's armies invaded Poland on September 1, 1939. Two days later, France and Britain declared war on Germany.

In 1941, the war took several startling new turns. In June, Hitler broke the nonaggression pact with the Soviet Union and, catching its Red Army by surprise, took the Baltic states and then drove deep into Soviet territory. When Japan attacked the American naval fleet at Pearl Harbor, Hawaii, in December 1941, the United States entered the war in both the Pacific and Europe.

By early 1944, the Soviet army had recovered most of its territorial losses and moved against the Germans in eastern Europe, reaching Berlin in April 1945 and beginning the long communist domination of the eastern part of the region. Germany surrendered on May 8, 1945, ending the war in Europe. But with Soviet forces firmly entrenched in the eastern part of Europe, the military battles of World War II were immediately replaced by an ideological **Cold War** between communism and democracy that lasted until 1991.

A Divided Europe, East and West

From 1945 until 1991, Europe was divided into two geopolitical and economic blocs, east and west, separated by the infamous **Iron Curtain** that descended shortly after the peace agreement ending World War II.

The larger geopolitical issue was the Soviet desire for a **buffer zone** between its own territory and western Europe. This buffer zone consisted of an extensive bloc of satellite countries, dominated politically and economically by the Soviet Union, that could cushion the Soviet heartland against possible attack from western Europe. The Soviet Union took control of the Baltic states, Poland, Czechoslovakia, Hungary, Bulgaria, Romania, Albania, and, briefly, Yugoslavia. Austria and Germany were divided into occupied sectors by the four Allied victors and, in both cases, the Soviet Union dominated the eastern portions, which contained the capital cities of Berlin and Vienna. Both capitals, in turn, were divided into French, British, U.S., and Soviet sectors.

Along the border between east and west, two hostile military forces faced each other for almost half a century. Both sides prepared for and expected an invasion by the other across the barbed wire dividing Europe. Both the **North Atlantic Treaty Organization (NATO)**, created in 1949 to protect western Europe, and the **Warsaw Pact**, formed in 1955 and comprised of countries of Soviet-dominated eastern Europe, were armed with nuclear weapons, making Europe a tinderbox for a nightmarish third world war.

The Cold War Thaw The symbolic end of the Cold War in Europe came on November 9, 1989, when East and West Berliners joined forces to rip apart the Berlin Wall with jackhammers and hand tools (**Figure 8.27**). By October 1990, East and West Germany were officially reunified into a single nation-state. During this period, all other Soviet satellite states from the Baltic Sea to the Black Sea, also underwent major geopolitical changes that have resulted in a mixed bag of benefits and problems. The Cold War ended completely with the breakup of the Soviet Union at the end of 1991.

The Cold War's end came as much from a combination of problems within the Soviet Union (discussed in Chapter 9) as from rebellion in eastern Europe. By the mid-1980s, the Soviet leadership had advocated for an internal economic

(a)

(b)

Explore the **Sights** of Former Location of Berlin Wall

http://goo.gl/DtjjMv

▲ **Figure 8.27 The Berlin Wall** In August 1961, East Germany built a concrete and barbed wire structure along the border of East Berlin to stem the flow of refugees fleeing communist rule. The wall was the most visible symbol of the Cold War division between east and west until November 1989, when the failing Soviet Union renounced its control over eastern Europe. (a) The extent of the wall zone at the Brandenburg Gate (East Berlin is to the left); (b) Berliners celebrate the end of the wall with East German police, who previously guarded the border zone with shoot-to-kill orders.

Exploring **Global Connections**

The New Cold War

Connect the dots: Russia annexes Crimea, and U.S. fighter jets appear in Lithuania; Norway sells a former naval base to the highest bidder to be used for arctic exploration, and Russian submarines move in; Greece can't pay its fiscal bailout debt to Germany, and Russian banks offer to help; Britain protests the Russian army's meddling in Ukraine, and Russian bombers hold "exercises" off England's coast; fictitious Twitter reports warning Louisiana residents to shelter in place because of a huge chemical spill are tied to Russian hackers in St. Petersburg; Swiss authorities arrest corrupt world soccer executives, and the Kremlin accuses the United States of plotting to discredit Russia's hosting of the 2018 World Cup; hundreds of Syrian war refugees suddenly appear one day at the Russia-Finland border, then the flow stops as suddenly as it started.

What's going on? It's the new Cold War between Russia and the West, a possible return to the scary days of the 1960s when the world's superpowers pushed each other to the brink of global destruction.

Expansion of NATO After the thawing of the original Cold War that saw several decades of demilitarization in Europe and the former Soviet Union, tensions between Russia and the West increased considerably from 2009 to 2013 as nine former Soviet satellite countries (Poland, the Czech Republic, Slovakia, Hungary, Bulgaria, Romania, Lithuania, Latvia, and Estonia) joined the EU. Even more distressing to Russia was the fact that all nine of these former Warsaw Pact countries joined the North Atlantic Treaty Organization (NATO), the international military organization created in 1949 to contain Soviet ambitions in Europe (**Figure 8.4.1**). Instead of Russia's longed-for buffer zone between the West and its heartland, it now shared a border with three Baltic NATO countries (Estonia, Latvia, and Lithuania) and one formerly trusted but now West-leaning ally, Ukraine. Despite considerable cooperation and even joint military exercises between NATO and Russia in previous decades, this interaction was "suspended" in April 2014 in response to the Russian army's intervention in Ukraine (see Chapter 9).

Recent Incidents Since then, Cold War 2.0 has escalated, with "dangerous or sensitive" incidents reaching levels not seen since the 1960s; Russian aircraft are testing western Europe's air defenses, and the Russian navy is asserting its rights in international waters off Norway's coast and in the Baltic Sea. In November 2015, NATO member country Turkey shot

▲ **Figure 8.4.1** **New Cold War** U.S. soldiers unload after arriving at a Lithuanian air force base as part of NATO's response to Russia's increasing hostile moves in the Baltic region.

down a Russian bomber for ostensibly violating its national airspace; then, following strong Russian protests, Macedonia and several other Balkan countries were invited to become new NATO members. In the Baltic Sea Russian ships assembled in the construction area of the new underwater cable between Lithuania and Sweden, threatening the cable by "inadvertently" dragging ship anchors across the area. That Lithuania is building this new link to Sweden's hydropower while cutting back on its consumption of Russian natural gas is indicative of the many different aspects of Cold War 2.0.

1. In what European countries or areas would the people in the street be affected by the new Cold War? In what ways?

2. In what other world regions are there expressions of the new Cold War?

---

restructuring and also recognized the need for a more open dialogue with the West. More recently, unfortunately, this era of cooperation with the West seems to have passed (see *Exploring Global Connections: The New Cold War*).

As a result of the earlier Cold War thaw, the map of Europe changed once again, with the unification of Germany, the peaceful "Velvet Divorce" of Czechs and Slovaks, and the reemergence of the Baltic states (see Figure 8.25). More troublesome was the Balkan area, where the former Yugoslavia violently fractured into a handful of independent states.

The Balkans: Waking from a Geopolitical Nightmare

The Balkans have long been a troubled area, with a complex mixture of languages, religions, and ethnic allegiances. Throughout history, these allegiances have led to an often-changing geography of small countries (**Figure 8.28**). Indeed, the term **balkanization** is used to describe the geopolitical processes of small-scale independence movements based on ethnic fault lines.

Following the fall of the Austro-Hungarian and Ottoman empires in the early 20th century, much of the region was unified under the political umbrella of the former Yugoslavian state. In the 1990s, however, Yugoslavia broke apart as ethnic factionalism and nationalism produced a decade of violence, turmoil, and wars of independence, creating a geopolitical nightmare for Europe, NATO, and the world. Today, despite lingering tensions in several areas, there are signs that the Balkan countries are moving toward a new era of peace and stability, although the logjam of migrants from war-torn Southwest Asia passing through on their way to western Europe in late 2015 strained resources and led to border tensions. Two Balkan countries, Slovenia and Croatia, are EU members, and several others are candidates for EU membership.

Balkan Wars of Independence In 1990, elections were held in Yugoslavia's different republics over the issue of secession from the mother state. Secessionist parties gained control in Slovenia and Croatia, but Serbian voters opted for continued

▲ Figure 8.28 Ethnicity in the Balkans The varied and complicated pattern of ethnic diversity in the Balkans has led to geopolitical fragmentation in recent decades. Not only is the area a meeting ground for Roman Catholicism, Eastern Orthodoxy, and Islam, but also complex linguistic boundaries complicate ethnic and national identity. Unfortunately, a long history of discrimination and retaliation between ethnic groups is embedded in contemporary ethnic identities.

SLOVENIA: 2.0 million
EU: Member
Ethnicity: 83% Slovene, 2% Serb, 2% Croat, 1% Bosniak
Religion: 58% Roman Catholic, 2% Muslim, 2% Orthodox
Language: 91% Slovenian, 5% Serbian or Croatian

CROATIA: 4.3 million
EU: Member
Ethnicity: 90% Croat, 4% Serb
Religion: 86% Roman Catholic, 4% Orthodox, 2% Muslim
Language: 96% Croatian 1% Serbian

BOSNIA AND HERZEGOVINA: 3.9 million
EU: Potential applicant
Ethnicity: 50% Bosniak, 31% Serb, 15% Croat
Religion: 51% Muslim, 31% Orthodox, 15% Roman Catholic
Language: 53% Bosnian, 31% Serbian, 15% Croatian

MONTENEGRO: 0.6 million
EU: Potential applicant
Ethnicity: 45% Montenegrin, 29% Serb, 9% Bosniak, 5% Albanian
Religion: 72% Orthodox, 19% Muslim, 3% Roman Catholic
Language: 43% Serbian, 37% Montenegrin, 5% Bosnian, 5% Albanian

ALBANIA: 3.0 million
EU: Potential applicant
Ethnicity: 83% Albanian, 1% Greek
Religion: 57% Muslim, 7% Albanian Orthodox, 10% Roman Catholic
Language: 99% Albanian, 1% Greek

SERBIA: 7.1 million
EU: Candidate
Ethnicity: 83% Serb, 3% Hungarian, 2% Bosniak, 2% Roma
Religion: 85% Serbian Orthodox, 5% Roman Catholic, 3% Muslim, 1% Protestant
Language: 88% Serbian, 3% Hungarian, 2% Bosniak, 1% Romany

KOSOVO: 1.8 million
EU: Potential applicant
Ethnicity: 92% Albanian, 2% Bosniak, 2% Serb
Religion: 96% Muslim, 1% Serbian Orthodox, 2% Roman Catholic
Language: 95% Albanian, 2% Serbian, 2% Turkish

MACEDONIA: 2.1 million
EU: Candidate
Ethnicity: 64% Macedonian, 25% Albanian, 4% Turkish, 3% Roma, 2% Serb
Religion: 65% Macedonian Orthodox, 33% Muslim
Language: 66% Macedonian, 25% Albanian, 4% Turkish, 2% Romany, 1% Serbian

Boundary of the Former Yugoslavia

Yugoslav unification, in what observers considered to be a government-controlled election. Nonetheless, Slovenia and Croatia declared independence in 1991 and Macedonia and Bosnia–Herzegovina in 1992. When the Yugoslavian army attacked Slovenia, the Balkan situation got Europe's full attention, and a negotiated settlement resulted in Slovenia's independence. In Bosnia–Herzegovina, however, Serb paramilitary units waged a ferocious war of ethnic cleansing against both Muslims and Croats in a war that lasted until 1995. At that time, a complex political arrangement created a Serb republic and a Muslim–Croat federation, both ruled by the same legislature and president. Croatia also fought a devastating but successful war of independence against Serbian nationalists.

Kosovo, in the south of Serbia (as the former Yugoslavia was now called), was another trouble spot, with long-standing tensions between Serbs and Muslims. Although Kosovo had enjoyed differing degrees of autonomy within the former Yugoslavia, this autonomy was withdrawn by Belgrade in 1990 to protect the minority Serb population. Not surprisingly, Muslim Kosovar rebels responded by proclaiming Kosovo's independence in 1991. This act was resisted vigorously by Serbia, which responded with a violent ethnic-cleansing program designed to oust the Muslims and

make Kosovo a pure Serbian province. As warfare escalated in Kosovo, NATO forces (including the United States) began bombing Belgrade in 1999 to force Serbia to accept a negotiated settlement. From 1999 until 2008, Kosovo was administered by the UN as a protectorate, an arrangement enforced by some 50,000 peacekeepers from 30 different countries.

Although Serbia remains steadfast about reclaiming Kosovo and only grudgingly recognizes its independence, in most other matters its government is more moderate and less nationalistic. This has led to Serbia's being reinstated in the UN and the Council of Europe and becoming an official EU candidate. However, in Kosovo itself, political strife and tensions continue, aggravated by high unemployment and a stagnant economy.

Devolution in Contemporary Europe

As noted in Chapter 1, geopolitical **devolution** refers to a decentralization of power away from a central authority. In contemporary Europe devolution takes many forms, ranging from a central government sharing power with small states within a larger union to full-on calls for separatism and independence from a larger political entity. Germany and France, which share power between their national and regional governments (in France) and *Länder* (Germany's equivalent to U.S. states), illustrate one end of the spectrum. Spain is another example of power sharing; its 1978 constitution recognizes 17 autonomous communities within the nation-state. Somewhere in between (or perhaps outside of) these two points lies Belgium, where so much power and authority lies with the two regional cultures, the French-speaking Walloons in the south and the Flemish culture in the north, that for several years now there's been no effective central government. As a result, pundits often refer to Belgium as the "world's wealthiest failed state." More seriously, after the March 2016 terrorist attacks at the Brussels Airport and on its subway system there was much concern expressed that Belgium's devolved vacuum created a weak intelligence and security environment in which a terrorist network could exist.

Also defining part of the devolution spectrum was Scotland's historic 2014 referendum on independence from the United Kingdom. In this election, Scottish voters were asked to vote on the question, "Should Scotland be an independent country?" (**Figure 8.29**). With the highest voter turnout in the UK's long election history, the no vote won by 55 percent. Notable is that the opposition to Scottish independence received a last-minute boost from London when all three major UK parties pledged to devolve "extensive new powers" to the Scottish Parliament by the end of 2015. In Spain, Catalonia—which accounts for 19 percent of the country's economy—is also calling for a referendum on independence, a move strongly opposed by the Spanish government.

Different, but similarly complex, is the movement in many European countries to withdraw—or at least regain powers—from the European Union. The United Kingdom's June 2016 vote to leave the EU could have profound implications for the geopolitical fabric of Europe, starting with the makeup of the UK itself. Scotland and Northern Ireland, two countries that voted overwhelmingly to remain in the EU, may seek independence from England, with Northern Ireland possibly joining the Republic of Ireland and Scotland becoming an independent country. In both cases a motivating factor for breaking away from the UK is to remain part of the EU.

Beyond the UK, Brexit has emboldened right-wing Euroskeptics, anti-immigrant parties in many other European countries, most notably in France, Italy, Spain, and Hungary. Whether these anti-EU sentiments are temporary or permanent may depend on how painful the Brexit process is to the UK's economy and social structure.

REVIEW

8.12 Describe briefly how the map of Europe changed with the Treaty of Versailles in 1918.

8.13 What European countries were considered Soviet satellites during the Cold War?

8.14 What countries made up the former Yugoslavia?

8.15 Discuss, with examples, the different forms of political devolution in contemporary Europe.

■ **KEY TERMS** nation-states, irredentism, Cold War, Iron Curtain, buffer zone, North Atlantic Treaty Organization (NATO), Warsaw Pact, balkanization, devolution

---

Economic and Social Development: Integration and Transition

As the acknowledged birthplace of the Industrial Revolution, Europe in many ways invented the modern economic system of industrial capitalism that dominates the global economy. Although Europe was the world's industrial leader in the early 20th century, by mid-century it was eclipsed by both Japan and the United States as the region struggled to cope with the effects of two world wars, a decade of global depression, and the Cold War. Despite these hardships, the rebuilding of Europe's postwar economy was considered something of an economic and political miracle as former enemies put aside historical animosities to form innovative industrial and trade alliances. By 1960, the term "economic miracle" was commonly used to describe Europe's recovery.

While the last half-century of economic recovery and integration has been largely successful, Europe is currently suffering from a period of prolonged stagnation aggravated by an internal fiscal crisis due to its transition to a common currency. This internal crisis is coupled with reduced exports to developing economies from Germany, the economic engine that drives much of Europe. As a result, the immediate economic future appears mixed, with traditional industrial areas losing jobs, while newer ones—including many in eastern Europe—show increased vitality (Table 8.2).

▼ **Figure 8.29 Brexit** Supporters of the United Kingdom's June 2016 referendum to leave the European Union celebrate its passage. Once Britain begins the formal act of withdrawing from the EU, which is scheduled for March 2017, it will have two years to complete the process.

TABLE 8.2 Development Indicators

MasteringGeography™

Country	GNI per capita, PPP 2014	GDP Average Annual % Growth 2009–15	Human Development Index (2015)[1]	Percent Population Living Below $3.10 a Day	Life Expectancy (2016)[2] Male	Life Expectancy (2016)[2] Female	Under Age 5 Mortality Rate (1990)	Under Age 5 Mortality Rate (2015)	Youth Literacy (% pop ages 15–24) (2005–2014)	Gender Inequality Index (2015)[3,1]
Western Europe										
Austria	47,380	1.2	0.885	–	79	84	9	4	98	0.053
Belgium	44,090	1	0.890	–	79	84	10	4	–	0.063
France	40,100	1	0.888	–	79	85	9	4	–	0.088
Germany	47,460	1.8	0.916	–	78	83	9	4	–	0.041
Ireland	42,830	1.7	0.916	–	79	83	9	4	–	0.113
Luxembourg	65,570	2.7	0.892	–	80	85	8	2	–	0.100
Netherlands	48,860	0.3	0.922	–	80	83	8	4	–	0.062
Switzerland	59,160	1.8	0.930	–	81	85	8	4	–	0.028
United Kingdom	39,500	1.9	0.907	–	79	83	9	4	–	0.177
Southern Europe										
Albania	10,980	2	0.733	6.8	76	80	41	14	99	0.217
Bosnia & Herzegovina	10,550	0.8	0.733	–	74	79	19	5	100	0.201
Croatia	20,910	−1.2	0.818	<2.0	74	81	13	4	100	0.149
Cyprus	29,190	−1.9	0.850	–	78	82	11	3	100	0.124
Greece	27,050	−5.4	0.865	–	78	84	13	5	99	0.146
Italy	35,450	−0.8	0.873	–	80	85	10	4	100	0.068
Kosovo	9,300	3.2	–	<2.0	74	79	–	–	–	–
Macedonia	13,170	2	0.747	–	73	77	38	6	99	0.164
Montenegro	15,250	1.4	0.802	–	74	79	18	5	99	0.171
Portugal	28,370	−1.3	0.830	–	77	83	15	4	99	0.111
Serbia	13,040	0.5	0.771	–	73	78	29	7	99	0.176
Slovenia	30,360	−0.2	0.880	<2.0	78	84	10	3	100	0.016
Spain	33,490	−1.1	0.876	–	80	85	11	4	100	0.095
Northern Europe										
Denmark	46,850	0.4	0.923	–	79	82	9	4	–	0.048
Estonia	27,490	4.2	0.861	<2.0	72	82	20	3	100	0.164
Finland	40,630	0.3	0.883	–	78	85	7	2	–	0.075
Iceland	41,800	1.4	0.899	–	81	84	6	2	–	0.087
Latvia	23,360	2.9	0.819	<2.0	69	80	21	8	100	0.167
Lithuania	26,390	3.8	0.839	<2.0	69	80	17	5	100	0.125
Norway	67,100	1.5	0.944	–	80	84	8	3	–	0.067
Sweden	46,870	2	0.907	–	80	84	7	3	–	0.055
Eastern Europe										
Bulgaria	16,840	0.9	0.782	4.7	71	78	22	10	98	0.212
Czech Republic	28,740	0.7	0.870	<2.0	76	82	14	3	–	0.091
Hungary	23,960	1	0.828	<2.0	72	79	19	6	99	0.209
Poland	24,430	2.8	0.843	<2.0	74	82	17	5	100	0.138
Romania	19,950	1.5	0.793	4.1	72	79	37	11	99	0.333
Slovakia	27,410	2.4	0.844	–	73	80	18	7	–	0.164
Micro States										
Andorra	–	−3.1	0.845	–	–	–	8	3	–	
Liechtenstein	–	–	0.908	–	81	84	10	–	–	
Malta	27,390	2.2	0.839	–	80	84	11	6	99	0.227
Monaco	–	–	–	–	–	–	8	4	–	
San Marino	–	–	–	–	84	89	12	3	–	
Vatican City	–	–	–	–	–	–	–	–	–	

Source: World Bank, *World Development Indicators*, 2016.

[1] United Nations, *Human Development Report*, 2015.

[2] Population Reference Bureau, *World Population Data Sheet*, 2016.

[3] Gender Inequality Index—A composite measure reflecting inequality in achievements between women and men in three dimensions: reproductive health, empowerment, and the labor market that ranges between 0 and 1. The higher the number, the greater the inequality.

Login to MasteringGeography™ & access MapMaster to explore these data!

1) Using MapMaster, map the changes in under age 5 mortality between 1990 and 2015 for the eastern European and Balkan countries. Then list the countries with the most and least changes in infant mortality. Finally, discuss the reasons behind these changes (or lack of).

2) Make a map of the 10 countries with the lowest Human Development Index scores. What does the resulting distribution tell you? Drawing upon other data from Table 8.2, discuss whether these countries share other traits.

Background: Europe's Industrial Revolution

Two fundamental innovations made the **Industrial Revolution** possible: First, machines replaced human labor in many manufacturing processes, and, second, inanimate energy sources (water, steam, electricity, and petroleum) powered these new machines. More specifically, England's textile industry on the flanks of the Pennine Mountains was the birthplace of this new system in the years between 1730 and 1850 where waterwheels were used to power mechanized looms (**Figure 8.30**). Later, when coal-fired steam engines replaced water power, these early textile factories expanded into the lowlands to the west, closer to trading ports and global markets.

Industrial districts in continental Europe began appearing in the 1820s, located close to coalfields. The first was the Sambre-Meuse region, named for the two river valleys straddling the French-Belgian border. Like the English Midlands, it had a long history of cottage-based woolen textile manufacturing that quickly converted to the new technology of steam-powered mechanized looms.

By 1850, the world's dominant industrial area was the Ruhr district in northwestern Germany, near the Rhine River. Rich coal deposits close to the surface powered the Ruhr's transformation from a small textile region to one of heavy industry, particularly of iron and steel manufacturing. Decades later, the Ruhr industrial region became synonymous with the industrial strength behind Nazi Germany's war machine, resulting in heavy bombing in World War II.

Rebuilding Postwar Europe

The leader of the industrial world in 1900, Europe produced, at that time, 90 percent of the world's manufactured output. However, by 1945, after four decades of war and economic chaos, industrial Europe was in shambles, with many of its cities and industrial areas in ruins and much of the region's population dispirited, homeless, and hungry. Clearly, a new pathway for postwar Europe had to be forged to provide economic, political, and social security.

Evolution of the European Union In 1950, the leaders of western Europe began discussing a new form of economic integration that would avoid the historical pattern of nationalistic independence that duplicated industrial effort. In 1952, France, Germany, Italy, the Netherlands, Belgium, and Luxembourg ratified a treaty that joined them together in the European Coal and Steel Community (ECSC). Five years later, because of the resounding success of the ECSC, these six states agreed to further integration by creating a larger European common market that would encourage the free movement of goods, labor, and capital. In 1957, the Treaty of Rome was signed, establishing the European Economic Community (EEC).

The EEC expanded its goals again and in 1991 became the European Union (EU), at which time the supranational organization moved even further into the region's affairs by establishing regulations on agricultural products and pricing with the Common Agricultural Policy (CAP). In 2004, the EU expanded beyond its core membership in western Europe by adding 10 new states, including a cluster of former Soviet-controlled communist satellites from eastern Europe, bringing the total to 25. Bulgaria and Romania were admitted in 2007 and Croatia in 2013, resulting in the current 28 EU countries. Serbia, Montenegro, Macedonia, Albania, and Turkey have all been accepted as official applicants to the EU, while Kosovo and Bosnia-Herzegovina are potential candidates (**Figure 8.31**).

Economic Disintegration and Recovery in Eastern Europe

As discussed in the previous section, the former Soviet Union controlled the political and economic life of eastern Europe, and although Soviet economic planning was ostensibly an attempt to develop eastern Europe's economy by coordinating resource usage, these efforts were in fact designed primarily to serve Soviet homeland interests. This Communist system produced mixed results for more than 40 years; however, the collapse of the Soviet Union in 1991 plunged eastern European countries into economic and social chaos. Recovery and development since that time have been difficult,

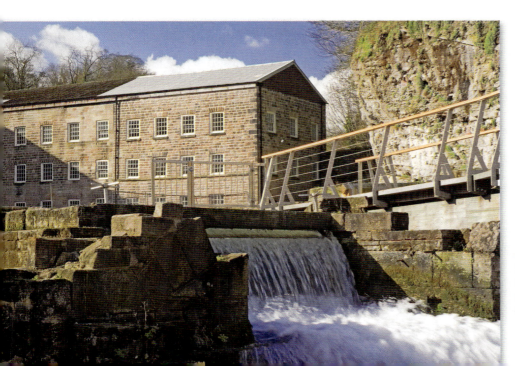

◀ **Figure 8.30   Water Power and England's Early Industrialization** Cromford Mill in Derbyshire was built in 1771 and was the world's first water powered cotton spinning mill. It is now preserved as a World Heritage Site.

▲ **Figure 8.31** **The European Union** The driving force behind Europe's economic and political integration has been the EU, which was formed in the 1950s as an organization with six members focused solely on rebuilding the region's coal and steel industries. As of late 2016, the EU has 28 members, with membership formal applications from Turkey, Serbia, and Macedonia. Iceland has withdrawn its application, and the United Kingdom will end its membership in 2019. **Q: Why is Switzerland not a member of the EU?**

with some countries making the transition more rapidly and more fully than others, resulting in a geographic patchwork of wealth and hardship throughout eastern Europe.

Change Since 1991 In place of Soviet coordination and subsidies came a painful period of economic transition that was

outright chaotic in many eastern European countries. As the Soviet Union turned its attention to its own economic and political turmoil, it stopped exporting cheap natural gas and petroleum to eastern Europe. Instead, Russia sold these fuels on the open global market to gain hard currency. Without cheap energy, many eastern European industries could not operate and were forced to close, idling millions of workers. For example, in the first two years of the transition (1990–1992), industrial production fell 35 percent in Poland and 45 percent in Bulgaria. In addition, the guaranteed Soviet markets for eastern European products simply evaporated, further aggravating the collapse of eastern European economies.

To recover, these countries began redirecting their economies toward western Europe. This meant moving from a socialist-based economy of state ownership and control to a capitalist economy of private ownership and free markets. Without Soviet price supports and subsidies, eastern European countries had to construct completely new economic systems, ones that could compete favorably in the new global marketplace. While some countries—Poland, the Czech Republic, Slovenia, and Slovakia—made the transition quickly, others—primarily the Balkan states—are taking longer (**Figure 8.32**).

Europe's Changing Industrial Heartland Until the 1990s, the industrial belt of Europe was commonly described as running eastward from the Midlands of England through the Netherlands and northern France to the Ruhr district of what was then West Germany. Its northward extension included the shipyards of northeast Britain and the Volvo factories of Sweden, while to the south were the collection of industries in the Paris Basin, pharmaceutical factories around Basel, Switzerland, and various Italian car manufacturers in the Po Valley. Other industrial centers, such as those clustered around Madrid and along Spain's north coast, as well as those in southeastern Italy, were outliers of the main heartland.

A very different geography has emerged with the opening of eastern Europe, as many western European firms opened factories in Poland, the Czech Republic, Slovakia, and Romania in order to take advantage of lower labor costs, weaker unions, and proximity to the emerging markets of eastern Europe and Russia (**Figure 8.33**). The western European auto industry is a good example, with German manufacturers such as Volkswagen and Opel GM building eastern European assembly sites, and higher-end auto firms following. In 2015, to illustrate, the traditional British high-end car firm Jaguar Land Rover wished to build a completely new manufacturing plant in eastern Europe to serve growing global demand for their product. Their search for a new factory location resulted in furious competition between two countries, Poland and Slovakia, with each offering attractive incentives such as subsidized infrastructure costs (land, highways, worker housing) and tax incentives. Slovakia won out, reportedly

(a)

(b)

◀ **Figure 8.32 The Old and the New in Poland** (a) Warsaw's Old Town, on the banks of the Vistula River, was originally established in the 13th century. Today, it is not only a symbol of historical Poland but also testimony to the country's pride and resiliency. In 1938, the Old Town was destroyed by Nazi German bombers during Hitler's invasion of Poland. Polish citizens rebuilt the Old Town as a symbol of resistance, but then it was systematically destroyed once again by the Nazis after the Warsaw Uprising of 1944. Most recently, the Old Town was rebuilt during the post-war period and became a World Heritage site in 1994. (b) Ten minutes walking distance from Warsaw's Old Town is the country's new commercial hub in Warsaw's new downtown. On the right is the Warsaw Financial Center, built in 1999; in the center is the 48-floor hotel Warsaw Intercontinental; and on the left is Zlota 44, a 54-floor residential building completed in 2014. Three other major skyscrapers are currently under construction in the downtown area, indicative of economic vitality in this former Soviet satellite country.

Explore the **Sights** of Warsaw's Old Town

http://goo.gl/T9jPQZ

North Sea

Skagerrak

Kattegat

Baltic Sea

Helsinki

Stockholm

Glasgow

Belfast

Dublin

Leeds

Manchester

Birmingham   *The English Midlands*

London   Rotterdam   *Europoort*

Lille   *Sambre-Meuse*   *The Ruhr*   *Saxon Triangle*   Lodz

Wrocław

Upper Silesia

Katowice

Paris   *Saar-Lorraine*   Stuttgart   Prague

ATLANTIC OCEAN

Munich   Vienna   Budapest

Bay of Biscay

Swiss Plateau   Zurich   Bern

Lyon   Milan   Turin   *Po Valley*

Bilbao   Toulouse   Belgrade   Bucharest   Black Sea

Burgas

Sofia   Bosporus Strait

Barcelona

Valencia

Bari

Adriatic Sea

Aegean Sea

Dardanelles

M e d i t e r r a n e a n   S e a

Strait of Gibraltar

| Older industrial areas |
| Newer industrial areas |

0 — 200 — 400 Miles
0 — 200 — 400 Kilometers

English Channel

▲ **Figure 8.33 Europe's Changing Industrial Geography** This map shows both the historic 19th and 20th century industrial areas and the newer 21st century industrial areas, many of which are in eastern Europe because of lower labor costs.

because of the factory site's proximity to the Danube River, which is important for shipping vehicles both west into the heart of Europe and southeast to Russia and Asia.

Several other components of Europe's new economic geography demand attention. First, despite the United Kingdom's loss of factory workers as industry moves eastward, an increase in finance, technology, and service workers has made it Europe's second-largest economy after Germany, moving France into third place. Whether this vitality will continue after the UK leaves the EU remains to be seen. Also notable are the many new tech centers that are part of this new industrial heartland—in southeastern Ireland, in and around Copenhagen, Berlin, Stockholm, and Helsinki.

Promise and Problems of the Eurozone

A traditional aspect of a country's sovereignty is the ability to control its own monetary system and fiscal policy, and indeed this was the case for most European countries until two decades ago. Historically, the currency of Germany was the deutschmark, with francs in France, lira in Italy, pesetas in Spain, and so on. Today, however, most EU countries have replaced national currencies with a common currency, the *euro*. This shift began in 1999 when 11 of the then 15 EU member states joined together to form the **Economic and Monetary Union (EMU)**. In 2002, new euro coins and bills replaced national currencies in the EMU countries, creating an economic subregion of Europe commonly called the **Eurozone**. Today, 19 of the EU's 28 member states use the euro, with some refusing to join the Eurozone, while others await resolution of ongoing tensions between rich and poor euro countries.

By adopting a common currency, EU members sought to increase the efficiency of both domestic and international business by eliminating the costs associated with payments made in different currencies. Although many traditional economists had (and still have) misgivings about this common-currency system, the political goal of enhancing European unity through a common currency won out. In 1999, several EU member countries resisted relinquishing control over their national monetary system and opted out of the Eurozone; consequently, the UK still uses its traditional currency, the pound, as does Denmark with the krone, and Sweden with the krona.

The non-euro EU member states of Bulgaria, Croatia, the Czech Republic, Hungary, Poland, and Romania are legally obligated to join the Eurozone at some future date. However, this expansion process has slowed recently because of a lengthy and controversial fiscal crisis in 2015 that has illuminated EMU shortcomings.

The 2015 Greek Debt Crisis Europe has long consisted of a diverse mosaic of richer and poorer countries. Historically, this disparity was between the richer industrial heartland—England, France, Germany, the Netherlands, Belgium, and northern Italy—and the poorer Mediterranean agricultural periphery—Spain, Portugal, southern Italy, the Balkans, and Greece (**Figure 8.34**). To lessen this economic disparity, a founding assumption of the EU was that unity would emerge only if the richer countries helped the poorer ones by providing financial loans and subsidies. The EMU's creation in 1999 furthered this rich-helping-the-poor process by offering relatively cheap loans available in the new common currency.

Initially, the process seemed to be working as Greece, Portugal, and Spain borrowed money to pay existing debts and start new projects that reduced unemployment and expanded the consumer base. But as shockwaves from the 2008 global financial crisis rippled through the Eurozone, it became painfully apparent that these borrowing countries were hopelessly in debt. At that point, the lending countries—led by Germany and the Netherlands, two rich countries noted for their thrifty cultures—demanded that the borrowing countries implement programs of austerity by reducing government spending and raising revenue through increased taxes. As a result, what had previously been a banker's financial abstraction now became a humanitarian crisis as jobs disappeared, pensions shrank, social services evaporated, taxes increased, and prices for food and other necessary goods soared.

Before the EMU, countries that found themselves in financial trouble would adjust their currency through devaluation, a common pathway out of indebtedness. But this was not allowed under the EMU, because the euro was a common currency controlled by the European Central Bank (ECB). Consequently, the suffering continued. Thousands of Portuguese and Spaniards took advantage of the new mobility under Schengen and left their home countries for jobs in Germany, the Netherlands, the United Kingdom, and Belgium (see *People on the Move: Portugal's Quirky Emigration Issues*). In Greece, people took to the streets in riots and demonstrations and changed governments frequently (**Figure 8.35**). By early summer 2015, Greece was on the brink of economic collapse and became the focus of the euro crisis. With 3.5 billion euros in payments due in July and no way for the government to pay, banks closed, ATMs ran dry, crucial imports like food and medicine stalled, and tourists, a mainstay of Greece's economy, stayed away. The 2015 influx of migrants attempting to make their way to western Europe has only compounded Greece's woes.

Greece's options were limited: The country could either borrow more money from the ECB and hope for the best or, alternatively, declare bankruptcy, leave the Eurozone, reinvent its traditional currency (the drachma), and muddle through somehow. Negotiations to borrow more money from the ECB were ugly: Germany, which had loaned Greece 57 billion euros, was furious and accused the Greeks of profligacy (a polite word for squandering money) and not only refused to lend more money, but also suggested Greece be thrown out of the Eurozone; France, to whom Greece owed 43 billion euros, was more

GNI/PPP IN $

- More than 50,000
- 40,000–50,000
- 30,000–39,999
- 20,000–29,999
- Less than 20,000
- Received Eurozone bailout funds between 2009–2016
- **5.2** Percent unemployed, 2015 or latest available
- *$2.82* Hourly minimum wage (US$)

ICELAND
4.0
none

Norwegian Sea

FINLAND
9.4
none

RUSSIA

Skagerrak *Kattegat*

North Sea

DENMARK
6.2
$18.00

NORWAY
4.4
none

SWEDEN
7.4
none

ESTONIA
6.2 *$3.60*

Baltic Sea

LATVIA
9.9 *$2.83*

LITHUANIA
9.1
$2.83

RUSSIA

BELARUS

IRELAND
9.4
$8.70

UNITED KINGDOM
5.3
$8.20

NETHERLANDS
6.9 *$9.60*

POLAND
7.5
$5.30

UKRAINE

ATLANTIC OCEAN

English Channel

BELGIUM
8.5 *$10.16*

GERMANY
4.6
$10.20

CZECHIA
5.1 *$3.80*

SLOVAKIA
11.5 *$3.40*

MOLDOVA

LUXEMBOURG
6.4 *$11.20*

LIECHTENSTEIN

AUSTRIA
5.7
varies

HUNGARY
6.8 *$4.20*

ROMANIA
6.8
$2.15

Bay of Biscay

FRANCE
10.4
$10.90

SWITZERLAND
none

SLOVENIA
9.0 *$6.80*

CROATIA
16.3 *$3.13*

ANDORRA

SAN MARINO

Adriatic Sea

BOSNIA & HERZEGOVINA
27.9
$1.25

SERBIA
22.2
$1.84

BULGARIA
9.2
$1.70

MONACO

KOSOVO
32.1 *$1.30*

ITALY
11.9
none

MONTENEGRO
19.1 *$1.48*

MACEDONIA
27.9 *$1.99*

PORTUGAL
12.6
$4.30

SPAIN
22.1
$5.00

ALBANIA
16.1 *$1.20*

GREECE
24.9
$4.80

Aegean Sea

Strait of Gibraltar

Mediterranean   Sea

CYPRUS
15.1
none

MALTA
5.4
$5.90

| 0 | 200 | 400 Miles |
| 0 | 200 | 400 Kilometers |

▲ **Figure 8.34** **Development Issues in Europe: Economic Disparity**   This map shows that economic disparity between rich and poor countries still exists in contemporary Europe, with stark differences between northern Europe and the Mediterranean (compare Norway and Portugal), and traditional western Europe and the former Soviet satellite countries of eastern Europe (compare France and Germany with Romania and Bulgaria). However, note also that some former Soviet satellite countries (Poland, Czech Republic, Slovenia) are doing better than others (Latvia, Bulgaria, Romania). The minimum hourly wage figure (in red) gives clues to how labor costs might differ between countries. **Q. What factors could explain the contemporary economic disparity between the former Soviet satellite countries?**

Portugal's Quirky Emigration Issues

While the free movement of labor around Europe is a cherished EU goal, allowing workers to move from areas of high unemployment to places where jobs are available, Portugal's current emigration pattern has some quirky details illustrating how that policy works in reality. Let's take a closer look.

While many Portuguese emigrants stay in Europe, with the largest number of migrants favoring Britain and Switzerland for work, equally large numbers emigrate to former Portuguese colonies in Africa, creating a reverse migration of sorts after decades of African migration to Portugal. Oil-rich Angola draws the most Portuguese workers, with Mozambique a close second. Most emigrants to Angola find construction jobs in the oil industry but many take low-skill service jobs, creating cultural tensions with the native Angolans who previously did this kind of work (Figure 8.5.1). One observer noted going into an upscale restaurant where the clientele were all black Angolans, while the waiters and kitchen workers were Portuguese migrants—a situation still rare in Africa. Adding to the irony of this reverse migration are reports of Angolan oil millionaires buying up choice real estate in coastal Portugal.

The second quirky pattern is a bit more complex. In an effort to stanch population loss by facilitating in-migration from former colonies, Portugal recently passed a law granting a Portuguese passport to anyone born before Portugal's colonial ties were formally ended with that particular country. In the case of the small country of Goa on India's west coast, that date is 1961. So in recent years, at least 20,000 Goans granted Portuguese passports have headed for Europe. But have they settled in Portugal? No—they've used their new EU citizenship to settle in the United Kingdom, mostly in the small town of Swindon, west of London. This perfectly legal entry into Britain has infuriated England's xenophobic right wing and furthered controversy over Britain's EU membership due to concerns about hundreds of thousands of new Portuguese passport holders from Brazil, East Timor, Mozambique, and Angola arriving in the United Kingdom.

Of course, Portugal is under pressure to require these new Portuguese citizens to at least spend a certain amount of time in Portugal itself before moving on to other EU countries. And maybe that's not

▲ **Figure 8.5.1 Portugal Welcomes Syrian Refugees** While many other European countries are uncomfortable with the EU's quota for accepting Syrian refugees, Portugal would welcome even more refugees to help stanch the country's population loss from out-migration.

a bad thing, because Portugal is having difficulties even luring Syrian migrants to the country. While other European countries balk at accepting a quota of amnesty migrants, Portugal has upped its own quota to 10,000 per year. The migrants, however, are not interested. While over a million migrants have flooded into Germany, at last count only 430 have accepted Portugal's generous offer.

1. What obstacles might Portuguese migrants experience in EU countries such as the United Kingdom or Belgium? What different drawbacks might they experience when emigrating to former Portuguese colonies?

2. North America has millions of people of Portuguese ancestry. Are there Portuguese communities near you? If so, how do these communities celebrate their ancestry? If not, search the Internet to learn how Portuguese communities reinforce their ethnic identity.

understanding, suggesting things could be worked out; the Netherlands, Finland, and Ireland—countries that had successfully balanced their books through austerity—supported Germany's hardline stance; Poland, Bulgaria, and Romania—potential euro countries—put their Eurozone membership on hold; and even Russia got in the act by offering to bail out Greece (even though experts said Russia lacked the financial means), an offer that sent chills through NATO headquarters because of the geopolitical implications.

Finally, after months of contentious negotiations, Greece and the Eurozone came to an agreement in August 2015 for a third round of bailout funds; an agreement that many believe is only a temporary solution. Regardless, these drawn-out negotiations have changed the conversation about the Eurozone's future in several ways:

- The assumption that Europe's richer countries will subsidize the region's poorer countries needs to be revised.
- If the Eurozone is to survive, it, too, must change to allow individual countries more flexibility in managing their internal finances.
- The International Monetary Fund (IMF), a persistent critic of Eurozone indebtedness in the past, became the strongest advocate for actually reducing Greece's debt because it sees the current solution as unsustainable. This new stance by the major international lender may significantly influence future Eurozone actions on indebtedness.

▲ **Figure 8.35 Greece Protests Against Austerity** A farmer uses his tractor to block the border road between Greece and Turkey as part of a nationwide protest against pension reform mandated by Europe's Central Bank as part of releasing another round of bailout funds.

Social Development in Europe

Europe is renowned for its strong system of public welfare services such as free medical care and higher education, job security, maternity and paternity benefits, subsidized housing, unemployment coverage, and even longer guaranteed vacation time. Some of these, such as child labor laws, are historical reactions to abuses of the early industrial period while others are postwar products of liberal and social democratic political parties. Conventional wisdom attributes these social welfare programs to extraordinarily high personal income taxes, but that's true in only a few countries, Denmark's 51 percent income tax being one of them. More common is the combination of a strong progressive tax system for the highest income levels; a flat tax for most workers; and high corporate taxes. Regardless of the source of public monies, a significant factor is that European governments traditionally spend more money on public welfare than does the United States.

Despite these social programs being taken for granted by Europe's population, they are constantly under attack as governments and corporations seek to cut costs. While governmental cost-cutters seek balanced budgets, corporations complain that these programs add to the costs that make them uncompetitive in the global marketplace.

Europe's social welfare programs are also one of the attractions that make the region a destination for the world's migrants. And, not surprisingly, the fact that migrants to Europe can often qualify for free housing, medical care, and education adds more fuel to the fire of controversy surrounding migrants in Europe.

Gender Issues Despite the visibility of female political leaders and the fact that Europe is considered one of the world's most developed regions, gender equity issues persist in government, business, and domestic life. For the EU countries as a whole, for example, male employment is 21 percent higher than female employment, and women who work generally make 25 percent less than men.

However, given the complexity of Europe, with its mixture of national, urban, rural, and migrant cultures, the nature and extent of gender issues differ widely among

◀ **Figure 8.36 Women in Europe's Business World** The employment of women in Europe's workforce differs significantly among different countries and regions. Scandinavia, for example, has the highest percentage of women in upper management positions; this contrasts with the lowest percentage in the Mediterranean countries. This photo was taken at a business meeting in Berlin, Germany.

countries and regions. To illustrate, within the 28 EU countries about a quarter of the parliamentary offices are held by women. Sweden has the highest representation, with more than half of its ministers being female, whereas Cyprus has absolutely none. Similarly, in the business world, only 11 of Europe's largest companies have women in top management, yet women make up almost a third of top management in Norway, compared to just 1 percent in Luxembourg (**Figure 8.36**).

One interesting pattern is that female participation in the workforce is generally higher in the countries of eastern Europe and the Balkans. Two interrelated factors explain this. First, women were expected to work in the communist economies of these countries from 1945 to 1990. Second, families often needed two incomes to survive during the difficult economic transition that followed the Soviet Union's collapse in 1991. Regardless of cause, the results are startling. Today, Bulgaria has the highest percentage of female CEOs (21 percent) of any EU country, and Slovenia, formerly part of socialist Yugoslavia, is the country with the least income disparity between men and women.

Women are well represented in both government and business in the Scandinavian countries, but for very different reasons than in eastern Europe and the Balkans. It is generally agreed that the foundation of Scandinavia's gender equity comes from a combination of comprehensive child care, liberal family benefits that guarantee job security and career advancement after maternity and paternity leaves, and a tax code that does not punish dual-income families. As a result, Norway and Sweden have the highest percentage of females in the workforce. Portugal has the

third-highest number at 71 percent; because of its struggling economy, women usually work out of necessity rather than choice, with grandparents and other family members providing child care—unlike the government-sponsored child care common in Scandinavia.

Not to be overlooked are the extraordinarily complex gender issues within Europe's large migrant cultures and their host cultures. We mentioned earlier how France's national policies have become entangled with Muslim gender and cultural preferences. Other examples of these complexities can be found in Germany as the state finds itself embroiled in cultural tensions. These range from increasing women's freedom beyond the family household to prosecuting Turkish honor killings, where young women have paid with their lives for behaviors, such as dating and marrying without parental consent, that are common in German culture but unacceptable in traditional Turkish culture.

REVIEW

8.16 What geographic factors explain the locations of early industry in Europe?

8.17 Describe the origin and evolution of the European Union in terms of its goals.

8.18 Why and how has eastern Europe reinvented its economy in the last two decades?

8.19 Discuss the factors explaining the highly variable geography of women in Europe's workforce.

■ **KEY TERMS** Industrial Revolution, Economic and Monetary Union (EMU), Eurozone

Chapter 8 Europe

Summary

- Europe's physical geography ranges from arctic to subtropical, marine to deep continental, lowlands to high mountains, producing a range of habitats that have been transformed by human settlement over thousands of years. Its long history of environmental interaction makes Europe today one of the greenest and most environmentally conscious world regions.

- In Europe, any sort of natural population increase is lacking because of low birth rates. As a result, any population growth will come from in-migration, legal and extralegal. In 2015–2016 over a million extralegal migrants arrived in Europe from Africa and Southwest Asia, many seeking political asylum and others seeking jobs and welfare.

- While historically Europe is a region of varied languages and religions, individual countries were largely culturally homogeneous, with a shared language, religion, and values. Today, however, migrants bring new cultural geographies to Europe, leading to considerable tension and resistance.

- Europe's national boundaries result from a war-plagued 20th century that included two world wars, an ideological Cold War, and a series of Balkan Wars that spilled over into the current century. Despite this long geopolitical process of rearranging borders, significant minority groups still seek autonomy and independence from larger units.

- Reconstructing postwar Europe was done largely through the supraregional organization known today as the European Union (EU) that began in the 1950s with economic integration goals, then gradually added political, financial, and social concerns to its agenda. While largely successful in achieving integration, this process has not been—nor is it today—problem free.

Review Questions

1. Compare the maps of Europe at the beginning of the chapter—landforms, rivers, and different climate regions—with the population, language, and geopolitical maps to identify any significant associations. Discuss your findings in a short essay.

2. Use the information about which countries have the highest percentage of foreign-born people coupled with the language and religion maps to discuss where the influence of migrants might be greatest and where it would not be great.

3. Look through other chapters of the textbook for examples of European culture that were diffused during the historical colonial period. Then make a list of cultural traits diffused to Europe from other parts of the world.

4. How does Cold War 2.0 compare with the original Cold War in terms of countries involved, military strategies, and effects on economies and peoples' lives?

5. Are there any shared economic and/or social traits of those countries not belonging to the European Monetary Union? Explain your answer.

Image Analysis

1. What regional patterns do you see in this map of income disparity? Are there any geographic generalizations you can make from the larger patterns to explain this?

2. The three richest countries in Europe are Norway, Luxembourg, and Switzerland. What traits (other than wealth) do they share? What other factors are important in explaining their wealth?

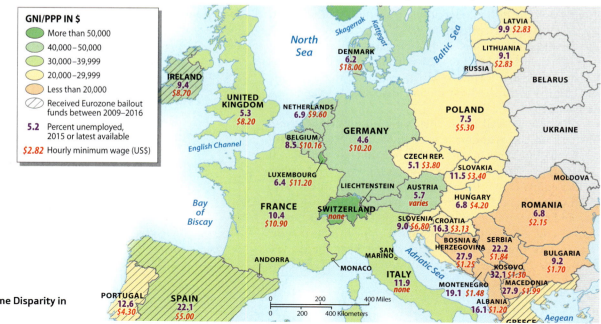

▶ **Figure IA8** Income Disparity in Europe

JOIN THE DEBATE

Free movement of people and goods across Europe's national boundaries has long been a goal of the European Union, and was furthered by the landmark Schengen Agreement of 1985 and its more recent conventions. However, recent events have called into question the dream of a borderless Europe.

▲ **Figure D8** Greek-Macedonian Border Fence at Refugee Camp Near Idomeni, Greece

A borderless Europe is crucial for a unified Europe—keep Schengenland!

- Europe's national borders are an artifact of an archaic political system, thus this must be eroded or even erased for a new unified system to thrive.

- The very nature of a unified economy demands the rapid and free flow of labor and goods across Europe's borders.

- Concerns about terrorism and illegal migration must be addressed by a common policy by the EU, not at each and every national border station.

The dream of a borderless Europe has become a nightmare in this contemporary world—discard it!

- There is no way to keep terrorists from infiltrating Europe's unprotected periphery; therefore, a strong system of internal borders is a necessary component of winning the war against terrorism.

- Massive global migration is a fact of life, and unless Europe has a strong system of internal checkpoints at each and every national border, those countries with unprotected borders will be overwhelmed with illegal migrants.

- What's wrong with a little nationalism? Europe recognizes dozens of languages, why not recognize redesigned border stations that meet the complex needs of the 21st century?

■ KEY TERMS

asylum laws (*p. 324*)
balkanization (*p. 341*)
Brexit (*p. 314*)
buffer zone (*p. 340*)
Cold War (*p. 340*)
continental climates (*p. 318*)
Cyrillic alphabet (*p. 333*)
devolution (*p. 342*)
Economic and Monetary Union (EMU) (*p. 349*)
European Union (EU) (*p. 314*)
Eurozone (*p. 349*)
family-friendly policies (*p. 324*)
fjords (*p. 317*)
Indo-European language family (*p. 330*)
Industrial Revolution (*p. 345*)
Iron Curtain (*p. 340*)
irredentism (*p. 338*)
Latin alphabet (*p. 333*)
marine west coast climate (*p. 318*)
Mediterranean climate (*p. 320*)
nation-states (*p. 337*)
North Atlantic Treaty Organization (NATO) (*p. 340*)
Schengen Agreement (*p. 324*)
Warsaw Pact (*p. 340*)

MasteringGeography™

Looking for additional review and test prep materials? Visit the Study Area in **MasteringGeography**™ to enhance your geographic literacy, spatial reasoning skills, and understanding of this chapter's content by accessing a variety of resources, including **MapMaster** interactive maps, videos, RSS feeds, flashcards, web links, self-study quizzes, and an eText version of *Diversity Amid Globalization*.

DATA ANALYSIS

http://goo.gl/8MmbpB

The flood of extralegal migrants seeking political asylum in Europe is overwhelming the region. This exercise is designed to give you a better understanding of this migration crisis.

1. Go to EU's Eurostat page (http://ec.europa.eu), access the asylum statistics, and acquaint yourself with the data tables in the right-hand column and the text content on the left. Note that asylum seekers come from many different countries and file their asylum applications in an array of different EU countries.

2. Now make either a map or a bar chart linking the different migrant source countries to the different European countries. This can be done most simply by mapping one or two leading source countries for each EU country. Summarize your findings in a paragraph or two.

3. Next, go to the data table that shows whether the migrant's asylum application was successful or not—that is, whether the migrant was granted asylum and stayed in Europe or was sent back to his or her country of origin. Once again, you can present your findings in either map or table form.

4. Which countries are most likely and which least likely to approve asylum applications from which source countries? Consider social factors such as the country's wealth (measured as gross domestic product per capita), unemployment rates, natural birth rates, and cultural factors such as each country's religion and language. Write a summary essay explaining your findings and reasoning.

Authors' Blogs

Scan to visit the
GeoCurrents blog
http://www.geocurrents.info/category/place/europe

Scan to visit the
Author's Blog
for field notes, media resources, and chapter updates
https://gad4blog.wordpress.com/category/europe/

Physical Geography and Environmental Issues

Many areas within the Russian domain suffered severe environmental damage during the Soviet era (1917–1991). Today, air, water, toxic chemical, and nuclear pollution plague large portions of the region.

Population and Settlement

Urban landscapes within the Russian domain reflect a fascinating mix of imperial, socialist, and post-communist influences. Recently, many larger urban areas within the region are showing similar trends toward sprawl and decentralization as those seen in North America and western Europe.

Cultural Coherence and Diversity

Although Slavic cultural influences dominate the region, many non-Slavic minorities, including a variety of indigenous peoples in Siberia and a complex collection of ethnic groups in the Caucasus Mountains, shape the cultural and political geography of the domain.

Geopolitical Framework

The centralization of Russian political power under the direction of President Vladimir Putin has had widespread consequences within the region, sparking ongoing tensions with neighboring Ukraine as well as limiting democratic freedoms within Russia itself.

Economic and Social Development

The region's economy has recently been hard hit by falling energy prices, global economic sanctions against Russia, and war within Ukraine.

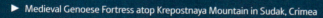

► Medieval Genoese Fortress atop Krepostnaya Mountain in Sudak, Crimea

Sudak, Crimea, Ukraine

THE RUSSIAN DOMAIN

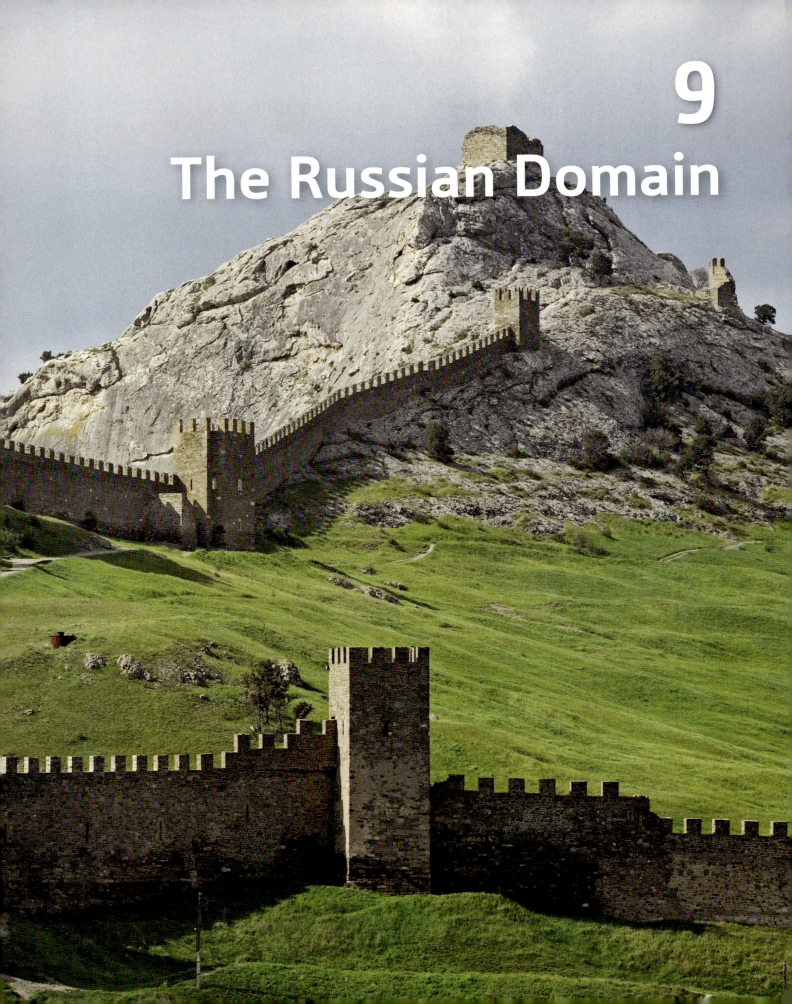

The Russian Domain

t is often called "Putin's Bridge." The huge US$3.2 billion project, designed to link southern Russia's Krasnodar region with the Russian-occupied Crimean Peninsula, is a high priority for the Russian president. Long seen as a strategic link between east and west, the Kerch Strait Bridge will make it much easier for Russia to directly supply and cement its political hold on Crimea, a region it occupied (and now controls) in its recent conflict with neighboring Ukraine. Many of Crimea's two million residents (mostly Russian) hope that the bridge, to be completed by 2020, will spur the region's struggling economy and end its isolation. Ukrainian tourists no longer flock to the region's warm beaches, but if Russians from nearby Krasnodar and beyond can use the bridge, the regional economy might revive. President Putin also sees the obvious geopolitical advantages to forging a more direct connection to his recently claimed territory.

The bridge itself is an ambitious engineering project, spanning 12 miles (19 kilometers) of often storm-tossed waters in the Kerch Strait. Designed to include both a 4-lane highway and a 2-track railroad, it will be the longest bridge in Russia and one of the longest in Europe when completed. The idea for a bridge here is nothing new: Adolf Hitler ordered the German military to build the span in World War II, but a hasty German retreat from southern Russia ended the effort before it was finished. A more recent initiative to build the bridge began as a joint effort between Russia and Ukraine, but Ukraine withdrew once Russia occupied the Peninsula in 2014. President Putin took personal responsibility for fast-tracking the project, declaring, "Let's hope we can fulfill this historic mission." He selected his boyhood friend, Arkady Rotenberg, to serve as the lead contractor.

Will it be a bridge to nowhere? That may depend on Crimea's political stability in the coming years. If the embattled region remains firmly within Russia's orbit, it could be a critical economic connection and as a powerful political symbol linking the once-isolated region to the country that now claims it.

Defining the Russian Domain

The Russian domain is rich with superlatives: Endless Siberian spaces, unlimited natural resources, legends of ruthless Cossack warriors, and tales of epic wars and revolutions are all part of the region's geographical and historical mythology. Indeed, the rise of Russian civilization remarkably parallels the story of the United States. Both cultures grew from small beginnings to become imperial powers that benefited from the fur trade,

ELEVATION IN METERS

4000+
2000–4000
500–1999
200–499
0–199
Below sea level
Sea Level

The annexation of Crimea by Russia in March of 2014 was declared invalid in a resolution passed by the United Nations General Assembly.

RUSSIAN DOMAIN
Political & Physical Map
⊛ ● Metropolitan areas more than 20 million
⊛ ● Metropolitan areas 10–20 million
⊛ ● Metropolitan areas 5–9.9 million
⊛ • Metropolitan areas 1–4.9 million
⊛ ○ Selected smaller metropolitan areas
– – Plate boundaries

gold rushes, and transcontinental railroads during the 19th century. And both were dramatically transformed by industrialization in the 20th century. Recently, however, the region has witnessed breathtaking change. In 1991, the region witnessed the complete collapse of the **Soviet Union** (Union of Soviet Socialist Republics [USSR]), a sprawling communist state that had dominated the region since 1917. In its place stood 15 former "republics" once united under the USSR. Some of the former Soviet republics are now part of greater Europe (Chapter 8), while other former Central Asian republics (Chapter 10) try to find their own footing as independent states.

Today, the remaining Russian domain extends across the vast northern half of Eurasia and includes not only Russia itself, but several surrounding nations (**Figure 9.1**). Slavic Russia (population 144 million) dominates the region. The country's European front borders Finland and Poland, while far to the east, sprawling Siberia shares a sparsely populated boundary with Mongolia and China. Russia's 6.6 million square miles (17 million square kilometers) dwarfs even Canada, and its nine time zones are a reminder that dawn in Vladivostok on the Pacific Ocean is still only evening in Moscow.

The border states of Ukraine, Belarus, Moldova, Georgia, and Armenia are inevitably linked to the evolution of their giant neighbor.

Will it be a bridge to nowhere?

Bering
Sea

Anadyr'

Northeast
Highlands

Kamchatka
Peninsula

Verkhoyansk

Norilsk
Central
Siberian
Uplands

Petropavlovsk-
Kamchatskiy

Verkhoyansk Range

Lena R.

NORTH
AMERICAN
PLATE

S I B E R I A

Yakutsk
Basin

Yakutsk

EURASIAN
PLATE

Sea of
Okhotsk

R U S S I A

Sakhalin
Island

Kuril Islands

PACIFIC
OCEAN

Yenisey R.

NORTH
AMERICAN
PLATE

Baikal-Amur Mainline

(BAM) Railroad

Krasnoyarsk

Trans-Siberian Railroad

Lake
Baikal

Amur R.

Khabarovsk

PACIFIC
PLATE

Novosibirsk

Novokuznetsk

Irkutsk

Ussuri R.

EURASIAN
PLATE

Vladivostok

◄ **Figure 9.1** **The Russian Domain** Russia and its neighboring states of Belarus, Ukraine, Moldova, Georgia, and Armenia make up a dynamic and unpredictable world region. Sprawling from the Baltic Sea to the Pacific Ocean, the region includes huge industrial centers, vast farmlands, and almost-empty stretches of tundra.

Emerging from the shadows of Soviet dominance has been difficult. Ukraine, in particular, has the size, population, and resource base to become a major European nation (**Figure 9.2**), with 43 million people in an area about the size of France. Years of political instability recently escalated in the country, however, prompting Russia's takeover of Crimea and fomenting a rebel-led civil war (with Russia's support) in eastern Ukraine.

Nearby Belarus is smaller (80,000 square miles or 208,000 square kilometers), and its population of 10 million remains closely tied

economically and politically to Russia. Moldova, with 4 million people, shares many cultural links with Romania, but its economic and political connections tie it more closely to the Russian domain than to central Europe. South of Russia and beyond the Caucasus Mountains, Armenia and Georgia are similar in size to Moldova, but their people differ culturally from that of their Slavic neighbor. Still, these small nations seem enduringly tied to Russia, although dissenting elements in both countries argue for closer ties with the West.

LEARNING Objectives
After reading this chapter you should be able to:

9.1 Explain the close connection among latitude, regional climates, and agricultural production in Russia.

9.2 Describe the major environmental issues affecting the region, and suggest how climate change might impact the Russian domain.

9.3 Summarize major migration patterns, both in the Soviet and post-Soviet eras.

9.4 Explain major urban land-use patterns in a large city such as Moscow.

9.5 Outline the major phases of Russian expansion across Eurasia.

9.6 Identify the key regional patterns of linguistic and religious diversity.

9.7 Summarize the historical roots of the region's modern geopolitical system.

9.8 Provide examples of recent geopolitical conflicts in the region and explain how these reflect persistent cultural differences.

9.9 Identify key ways in which natural resources, including energy, have shaped economic development in the region.

9.10 Describe the key sectors of the Soviet and post-Soviet era economy and discuss how recent geopolitical events have impacted prospects for future economic growth.

▲ **Figure 9.2 Kiev and the Dnieper River** The ornate domes of the Kiev Petchersk Lavra (one of the historical centers of Eastern Orthodox Christianity) frame this view of the city, which includes many new buildings (in the distance) constructed on the east bank of the Dnieper River.

Explore the **Sights** of Kiev'a Pechersk Lavra

http://goo.gl/IzNe1c

Physical Geography and Environmental Issues: A Vast and Challenging Land

The region's physical geography continues to shape its economic prospects in fundamental ways. For example, Russia's vast size poses special challenges, but its rich store of natural resources has offered unique economic benefits. At the same time, the Soviet period (1917–1991) also witnessed unparalleled, unrestrained economic development that damaged the region's environment in enduring ways. Today, more than two decades after the collapse of the Soviet Union, the environment still bears the scars of that legacy.

A Diverse Physical Setting

The Russian domain's northern latitudinal position shapes its basic geographies of climate, vegetation, and agriculture (see Figure 9.1). Indeed, the Russian domain provides the world's largest example of a high-latitude continental climate, where seasonal temperature extremes and short growing seasons limit human settlement (**Figure 9.3**). In terms of latitude, Moscow is positioned as far north as Ketchikan, Alaska, and even the Ukrainian capital of Kiev (Kyiv) sits farther north than the Great Lakes in Canada. Thus, apart from a subtropical zone near the Black Sea, the region experiences a classic continental climate with hard, cold winters and marginal agricultural potential.

The European West An airplane flight over the western portions of the Russian domain would reveal a vast, barely changing landscape. European Russia, Belarus, and Ukraine cover the eastern portions of the vast European Plain, which runs from southwest France to the Ural Mountains. One major geographic advantage of European Russia is that different river systems, all now linked by canals, flow into four separate drainage basins. The result is that trade goods can easily flow in many directions. The Dnieper and Don rivers flow into the Black Sea; the West and North Dvina rivers drain into the Baltic and White seas, respectively; and the Volga River (the longest river in Europe) runs to the Caspian Sea.

Most of European Russia experiences cold winters and cool summers by North American standards. Moscow, for example, is about as cold as Minneapolis in January, but it is not nearly as warm in July. In Ukraine, Kiev is milder, however, and Simferopol, near the Black Sea, offers wintertime temperatures that average more than 20°F (11°C) warmer than those of Moscow (see the climographs in Figure 9.3).

Three distinctive environments shape agricultural potential in the European west (**Figure 9.4**). North of Moscow and St. Petersburg, poor soils and cold temperatures severely limit farming. The region's boreal forests have been extensively logged. Belarus and central portions of European Russia possess longer growing seasons, but acidic **podzol soils**, typical of northern forest environments, limit output and thus the ability of the region to support a productive agricultural economy. The diversified agriculture that does occur includes grain (rye, oats, and wheat) and potato cultivation, swine and meat production, and dairying.

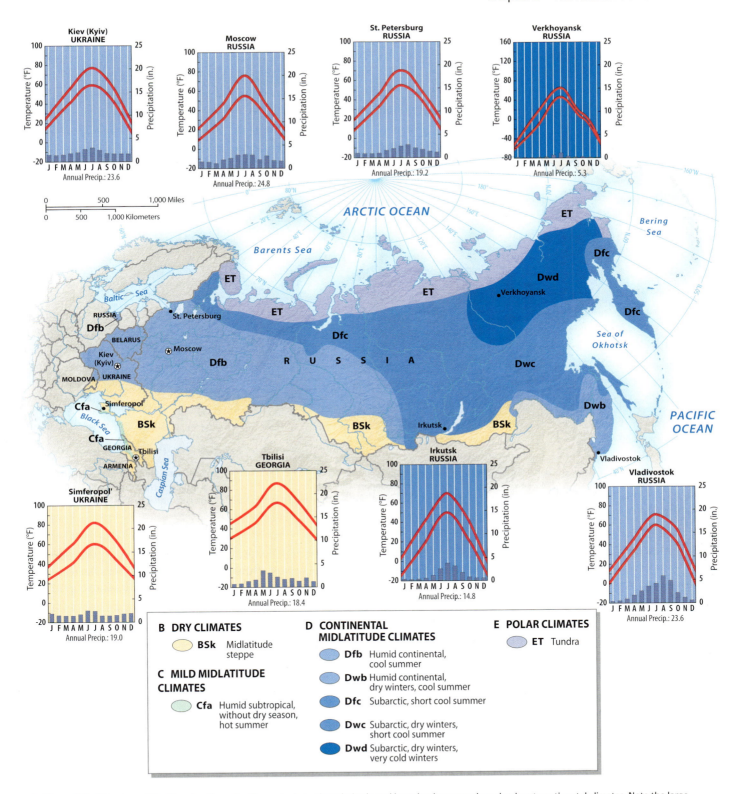

▲ **Figure 9.3 Climates of the Russian Domain** The region's northern latitude and large landmass produce dominant continental climates. Note the large seasonal ranges in temperature. Farming is limited by short growing seasons across the region. Aridity imposes limits elsewhere, especially across the southern Russian interior. Small zones of mild climate are found on the warming shores of the Black Sea in the far southwest corner of the region, producing subtropical conditions in Georgia.

▲ **Figure 9.4 Agricultural Regions** Harsh climates and poor soils combine to limit agriculture across much of the Russian domain. Better farmlands are found in Ukraine and in European Russia south of Moscow. Portions of southern Siberia support wheat production, but yield marginal results. In the Russian Far East, warmer climates and better soils translate into higher agricultural productivity. **Q: Describe the relationships between major agricultural zones and patterns on the climate map (Figure 9.3).**

South of 50° latitude, agricultural conditions improve across much of southern Russia and Ukraine. Forests gradually give way to steppe environments dominated by grasslands and by fertile "black earth" **chernozem soils**. These have proven valuable for commercial wheat, corn, and sugar beet cultivation and for commercial meat production (**Figure 9.5**).

The Ural Mountains and Siberia The Ural Mountains (see Figure 9.1) mark European Russia's eastern edge, separating it from Siberia. Topographically, the Urals are not a particularly impressive range. Still, the ancient rocks of these mountains contain valuable mineral resources, and the Urals traditionally marked European Russia's eastern cultural boundary.

East of the Urals, Siberia unfolds across the landscape for thousands of miles. The great Arctic-bound Ob, Yenisey, and Lena rivers (see Figure 9.1) drain millions of square miles of northern country, including the flat West Siberian Plain, the hills and plateaus of the Central Siberian Uplands, and the rugged and isolated

Northeast Highlands. Siberian vegetation and agriculture reflect the climatic setting. The north is too cold for tree growth and instead supports tundra vegetation,

▼ **Figure 9.5 Ukrainian Steppe** Akin to North America's Great Plains, much of southern Ukraine is a region of low relief and extensive commercial grain production.

▲ Figure 9.6 **Kamchatka Peninsula** Spawning sockeye salmon offer a ready meal for this brown bear on the shoreline of Kurile Lake, located beneath the towering Ilyinsky volcano in the southern portion of the Kamchatka Peninsula.

container ships plying nearby seas as an ice-free Arctic spurs more polar trade between Europe and China.

A concerted global effort is now under way to prevent development in seven sensitive spawning areas of Kamchatka wilderness. One Russian-supported plan calls for protecting an area nearly triple the size of Yellowstone National Park, and a recent consolidation of several national parks on the peninsula should make management of the region's natural resources more efficient and less costly.

The Russian Far East The Russian Far East is a distinctive subregion characterized by proximity to the Pacific Ocean, more southerly latitude, and fertile river valleys, such as the Amur and the Ussuri. Located at about the same latitude as New England, the region features longer growing seasons and milder climates than those found to the west or north. Here the continental climates of the Siberian interior meet the seasonal monsoon rains of East Asia. It is a fascinating zone of ecological mixing: Conifers of the taiga mingle with Asian hardwoods, and reindeer, Siberian tigers, and leopards also find common ground.

The Caucasus and Transcaucasia In European Russia's extreme south, the Caucasus Mountains stretch between the Black and Caspian seas (**Figure 9.7**). They mark Russia's southern boundary and are characterized by major earthquakes. Farther south lies Transcaucasia and the distinctive natural settings of Georgia and Armenia. Patterns of both climate and terrain in the Caucasus and Transcaucasia are very complex. Rainfall is higher in the west, while eastern

characterized by mosses, lichens, and a few ground-hugging flowering plants. Much of the tundra region is associated with **permafrost**, a cold-climate condition of unstable, seasonally frozen ground that limits vegetation and causes problems for railroad construction. South of the tundra, the Russian **taiga**, or coniferous forest zone, dominates a large portion of the Russian interior (see Figure 9.4). With a huge demand for lumber from nearby Japan and China, the eastern taiga zone has been threatened by both authorized logging and illegal timber poaching.

Along the Pacific, the Kamchatka Peninsula dangles dramatically into the waters of the North Pacific Ocean, and the region offers spectacular volcanic landscapes (**Figure 9.6**). Unlike much of the Russian domain, which has been poisoned by almost a century of reckless development and environmental exploitation, the Kamchatka region remains relatively untouched by the outside world. Six native species of Pacific salmon thrive on the peninsula, spawning in the millions in the area's free-flowing rivers. The salmon are at the center of a complex environmental web: They provide food for the brown bears, seals, and Stellar's sea eagles that are abundant here; they also remain part of the subsistence diet of the Koryak and other native peoples of these coastal zones; and they offer ecotourism and sports fishing opportunities found nowhere else in the world.

But Kamchatka's relative isolation is ending: The local government is trying to boost tourism, new oil exploration is underway in the nearby Sea of Okhotsk, a dozen new gold mines are being developed, and the region will likely see more

▼ Figure 9.7 **Caucasus Mountains** This scene, near the resort town of Dombai, reveals some of the spectacular glaciated terrain encountered in the higher reaches of the Caucasus Mountains.

valleys are semiarid. In areas of adequate rainfall or where irrigation is possible, agriculture can be quite productive. Georgia in particular produces fruits, vegetables, flowers, and wines (**Figure 9.8**).

A Devastated Environment

The Russian domain has no shortage of fragile and endangered environments. The breakup of the Soviet Union and subsequent opening of the region to international public scrutiny revealed some of the world's most severe environmental degradation (**Figure 9.9**). Even official studies commissioned by the Russian government estimate that almost two-thirds of Russians live in an environment harmful to their health and that the country's environmental problems continue to worsen. The studies also suggest that nearly 65 million Russians live in areas of chronically poor air quality and that the drinking water is unsafe in half of the country.

▲ **Figure 9.8 Subtropical Georgia** The moderating influences of the Black Sea and a more southerly latitude produce a small zone of humid subtropical agriculture in Georgia. This elderly peasant woman is picking grapes.

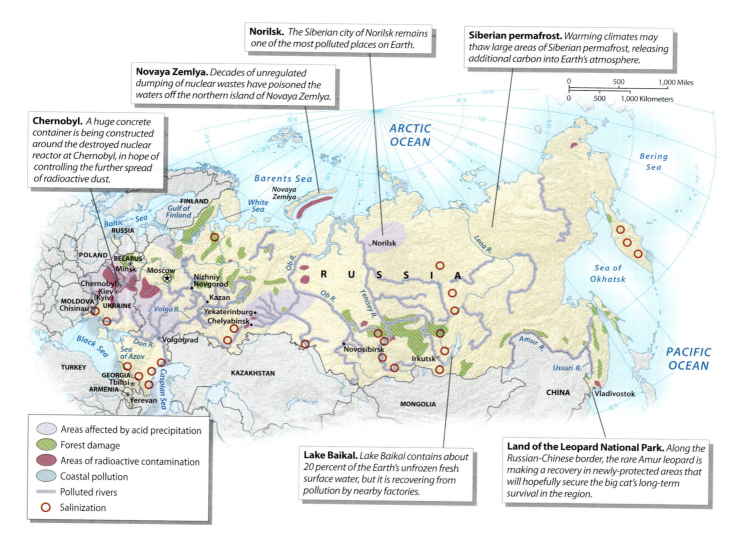

Norilsk. *The Siberian city of Norilsk remains one of the most polluted places on Earth.*

Novaya Zemlya. *Decades of unregulated dumping of nuclear wastes have poisoned the waters off the northern island of Novaya Zemlya.*

Siberian permafrost. *Warming climates may thaw large areas of Siberian permafrost, releasing additional carbon into Earth's atmosphere.*

Chernobyl. *A huge concrete container is being constructed around the destroyed nuclear reactor at Chernobyl, in hope of controlling the further spread of radioactive dust.*

Lake Baikal. *Lake Baikal contains about 20 percent of the Earth's unfrozen fresh surface water, but it is recovering from pollution by nearby factories.*

Land of the Leopard National Park. *Along the Russian-Chinese border, the rare Amur leopard is making a recovery in newly-protected areas that will hopefully secure the big cat's long-term survival in the region.*

Legend:
- Areas affected by acid precipitation
- Forest damage
- Areas of radioactive contamination
- Coastal pollution
- Polluted rivers
- ○ Salinization

▲ **Figure 9.9 Environmental Issues in the Russian Domain** Varied environmental hazards have left a devastating legacy across the region. The landscape has been littered with nuclear waste, heavy metals, and air pollution. Fouled lakes and rivers pose additional problems in many localities. Present economic difficulties and political uncertainties only add to the costly challenge of improving the region's environmental quality in the 21st century.

The frenetic pace of seven decades of Soviet industrialization took its toll across the region. Even in some of the most remote reaches of Russia, careless mining and oil drilling, the spread of nuclear contamination, and rampant forest cutting have resulted in frightening environmental damage. New Russian environmental and antinuclear movements have protested these ecological disasters, but to date these movements remain a minor political voice in a region dominated by the desire for economic growth. Indeed, the magnitudes of many of these environmental challenges are so great that they have global implications and may affect world climate patterns, water quality, and nuclear safety. For example, since the 1980s, the global environmental costs of Siberian forests lost to lumbering and pollution may have exceeded those of the more widely publicized destruction of the Brazilian rainforest.

Air and Water Pollution Poor air quality plagues hundreds of cities and industrial complexes throughout the region. The traditional Soviet practice of building large clusters of industrial processing and manufacturing plants in concentrated areas, often with minimal environmental controls, has produced an ongoing legacy of fouled air that stretches from Belarus to Siberia. A traditional reliance on abundant, but low-quality coal also contributes to pollution problems. The air quality in dozens of cities within the region typically fails to meet health standards, particularly in the winter when cold-air inversions trap the polluted atmosphere for days on end. Large numbers of urban residents across the region suffer from chronic respiratory problems.

Siberia's northern mining and smelting city of Norilsk is one of Russia's dirtiest urban areas, earning it the dubious distinction of appearing on the Blacksmith Institute's list of "ten most polluted places in the world" (**Figure 9.10**). In addition, a large swath of larch-dominated forest has died due to air pollution and acid precipitation in a huge zone of contamination that stretches more than 75 miles (120 kilometers) east of the city (see Figure 9.9). Norilsk Nickel (the major industrial polluter in the area) hopes to cut harmful sulfur dioxide emissions dramatically between 2015 and 2020. Elsewhere, however, growing rates of private car ownership have greatly increased automobile-related pollution. Today, 90 percent of Moscow's air pollution has been linked to the city's growing automobile traffic.

Degraded water is another hazard that residents of the region must cope with daily (see Figure 9.9). Oil spills have harmed thousands of square miles in the tundra and taiga of the West Siberian Plain and along the Ob River. A recent study of the Russian petroleum industry estimated that 5 million tons of oil (1 percent of Russia's annual oil production) are spilled every year, and the actual total could be higher.

Municipal water supplies are constantly vulnerable to industrial pollution, flows of raw sewage, and demands that increasingly exceed capacity. For example, the Baltic Sea near the city of St. Petersburg has reached a critical level of pollution that has killed fish and threatens to permanently damage the region's ecosystem. The biggest problem is that 30 percent of all the residential and industrial waste that enters the sea via the nearby Neva River is unfiltered raw sewage, a toxic mix of heavy metals and human waste that is rapidly killing nearby portions of the Baltic.

Elsewhere, the extensive industrialization and dam-building along Russia's Volga Valley have produced a corridor of degraded water that stretches for hundreds of miles. Water pollution has also affected much of the northern Black Sea, large portions of the Caspian Sea shoreline, and even Arctic Ocean waters off Russia's northern coast. Although Siberia's Lake Baikal has also suffered from industrial pollution, its future prospects appear brighter.

The Nuclear Threat The nuclear era brought its own particularly deadly dangers to the region. The Soviet Union's aggressive nuclear weapons and nuclear energy programs expanded greatly after 1950, and issues of environmental safety were often ignored. Northeast Siberia's Sakha region, for example, suffered regular nuclear fallout in the era of above-ground nuclear testing. In other areas, nuclear explosions were widely utilized for seismic experiments, oil exploration, and dam-building projects. The once-pristine Russian Arctic has also been poisoned. During the Soviet era, the area around the northern island of Novaya Zemlya served as a huge and unregulated dumping ground for nuclear wastes. Nearby, dozens of atomic submarines have been abandoned to rust away among the fjords of the Kola Peninsula. Aging nuclear reactors also dot the region's landscape, often contaminating nearby rivers with plutonium leaks.

▼ **Figure 9.10 Norilsk** The sprawling facilities of the Norilsk Nickel Plant dominate this portion of the city of Norilsk. Extensive air and water pollution are undesirable consequences of this industrial operation.

Nuclear pollution is particularly pronounced in northern Ukraine, where the Chernobyl nuclear power plant suffered a catastrophic meltdown in 1986. Large areas of nearby Belarus were also devastated in the Chernobyl disaster. Fallout contaminated fertile agricultural lands across much of the southern part of the country and rendered about 20 percent of the nation unhealthy to even live in. The reactor burned for 16 days, pouring smoke 2 miles into the sky and spreading nuclear contaminants from southern Russia to northern Norway. Thousands were directly killed in the accident, and millions of other residents across the region and in nearby affected portions of Europe suffered long-term health problems. It proved to be the world's worst nuclear accident and one of the greatest environmental disasters of the modern age, and it will continue to impact the ecological health of the region for decades to come.

The region's involvement with the nuclear age continues to take new forms. In 2001, Russia's Rostov Nuclear Energy Station went online, making it the region's first nuclear power plant since the Soviet era. Growing blackouts and energy shortages have produced a new government drive to revive Russia's nuclear industry. Recently, an ambitious plan to build 26 new reactors by 2020 produced a rash of protests across the country. At one gathering, protestors held signs simply declaring, "One Chernobyl was plenty."

Addressing the Environmental Crisis

Regional leaders are beginning to respond to the environmental crisis. In the case of Chernobyl, plans call for completing a huge protective roof over the entire reactor complex by 2018 (see *Working Toward Sustainability: Putting a Lid on Chernobyl*). In Siberia, the successful cleanup of Lake Baikal is a sign of greater environmental awareness in the region (**Figure 9.11**). The lake, home to about 20 percent of Earth's unfrozen fresh surface water, suffered during the later Soviet period. Large pulp and paper mills were located along the lakeshore in the 1950s and 1960s. Unfortunately, these industries discharged pollutants into the lake and surrounding atmosphere. Since the early 1990s, stricter regulations have reduced industrial pollution and the lake's water quality has improved. The lake became a UNESCO World Heritage Site in 1996, and three years later the Russian government formally created legislation designed to protect the lake. Indeed, the lake has become *the* national "poster child" of the Russian environmental movement.

Elsewhere, in the Land of the Leopard National Park, along Russia's Amur River, another success story is

▲ **Figure 9.11 Lake Baikal** Southern Siberia's Lake Baikal is one of the world's largest deep-water lakes. Industrialization devastated water quality after 1950 as pulp and paper factories poured wastes into the lake. Recent cleanup efforts have helped, but environmental threats remain.

unfolding. A 2015 census of Amur leopards, the world's rarest big cat, revealed a remarkable comeback since they faced extinction less than a decade ago. Animal counts of the rare cat more than doubled in eight years, suggesting the Russian national park is providing valuable protected habitat for breeding (**Figure 9.12**). The next step may be the creation of a jointly-managed trans-boundary nature reserve that involves similar habitat in China.

Climate Change and the Russian Domain

Given its latitude and continental climates, the Russian domain is often cited as a world region that would benefit from a warmer global climate. But such an interpretation

▼ **Figure 9.12 Amur Leopard** Eastern Siberia's endangered Amur leopard has staged a recent comeback as Russia's Land of the Leopard National Park offers increased protection for the big cats.

Working Toward Sustainability

Putting a Lid on Chernobyl

It is a sobering remembrance: April 26, 2016, marked the 30th anniversary of the Chernobyl nuclear plant disaster in northern Ukraine (**Figure 9.1.1**). Chernobyl was one of the world's worst nuclear nightmares and greatest environmental disasters ever. The explosion at the power plant and the meltdown of the reactor exposed tens of millions of Europeans to elevated doses of radiation, killed thousands of nearby residents, and contaminated a vast landscape surrounding the facility. Long-term effects on soils, livestock, wildlife, and human health are still being tabulated. A study published in 2010 by the New York Academy of Sciences, for example, estimates that the cumulative health effects of Chernobyl may have contributed to more than 985,000 deaths worldwide. An "exclusion zone" of about 1000 square miles (2600 square kilometers) still exists around the plant, where weed-filled city streets and abandoned farmland remain a 21st-century no-man's-land. Right after the disaster, a concrete structure (called the *sarcophagus*) was hastily constructed to contain the radioactive materials on site, but it was never meant to be a long-term solution to the problem.

Finally, however, a more sustainable approach is being taken to confine the long-term risks at the site (**Figure 9.1.2**). A remarkable 32,000-ton

Google Earth Virtual (MG) Tour Video

http://goo.gl/aq5Abr

▲ **Figure 9.1.2. Arch Construction near the Reactor Site** Chernobyl's new Safe Confinement structure is being assembled in two halves. This view (in 2015) shows the arch being constructed near the reactor site.

steel and concrete arch, costing more than $1.5 billion, is being assembled nearby. If construction plans remain on schedule, the otherworldly structure, more than 300 feet (90 meters) high, will be moved over the damaged plant in 2017 or 2018. Once in place, the ends of the arch will be closed off, effectively containing future radioactive dust, especially if the unsafe and aging structure within collapses. The safeguarded plant grounds will also allow for a more comprehensive cleanup of areas surrounding the site. Engineers hope that the arch can stand for at least 100 years.

Many challenges remain. Providing safe conditions for construction workers is a high priority. Maintaining the arch is also problematic. Most steel-supported structures such as this are painted every 15 years to protect them from rusting, but that task would introduce additional health risks. Instead, expensive rustproof stainless steel is being used, and special dehumidifiers will keep bolts and key parts from becoming too moist. Longer term, remaining fuel at the site must be removed to ensure radiation does not leak into groundwater, potentially endangering nearby Kiev. Perhaps most daunting is the country's unstable political environment, making it more difficult and potentially dangerous for global companies to participate in the containment effort. Many ask who will pay for maintaining the arch if Ukraine suffers an economic meltdown, a real possibility given its current challenges. Still, the arch is a testament to human ingenuity and the ability to address a catastrophe that reshaped this corner of the world three decades ago.

1. Is nuclear power a safe energy source today? Defend your answer.

2. Find a map of your local area focused on the college campus, and draw a 1000-square-mile (10-mile x 10-mile) zone around the facility to understand the size of Chernobyl's "exclusion zone." What local areas would be included?

▼ **Figure 9.1.1. Chernobyl Region** Chernobyl's close proximity to the Dnieper River and to Kiev resulted in a disaster affecting millions.

Exploring **Global Connections**

New Opportunities in the Russian Arctic

Russia's northern sea route (NSR) may be a big winner in the era of global warming (**Figure 9.2.1**). The route (also referred to as the *Northeast Passage*) connects northern Europe with northern Asia and lies within Russia's Exclusive Economic Zone. The northern

shortcut between Europe and Asia is increasingly seen as a viable commercial route that can shave more than 3000 miles (5000 kilometers) from traditional corridors around southern Africa or even through the Suez Canal (**Figure 9.2.2**). Shipping volumes within the NSR grew

▲ **Figure 9.2.1. Russia's Northern Sea Route (NSR)** The map shows the NSR that links Europe with the Far East. With continued global warming, more ice-free arctic days suggest the growing commercial potential of this high-latitude link.

oversimplifies the complex natural and human responses to global climate change, some of which are already occurring across the region.

High-Latitude Benefits At first glance, Russia's high latitude and polar regions seem likely to benefit from global warming. Optimists point to economic benefits that may result from warmer Eurasian climates. Some models predict the northern limit of spring cereal cultivation in northwestern Russia will shift 60–90 miles (100–150 kilometers) poleward for every 1°C (1.8°F) of warming. Less severe winters may make energy and mineral development in arctic settings less costly. About 15 percent of the world's undiscovered oil reserves (and 30 percent of its undiscovered natural gas reserves) are probably located in these settings, and Russia has staked large claims to the region. In

the Arctic Ocean and Barents Sea, warmer temperatures and less sea ice are translating into better commercial fishing, easier navigation, more high-latitude commerce, and more ice-free days in northern Russian ports. Since 2010, a growing number of commercial vessels have negotiated the **northern sea route** along Siberia's northern coast (see *Exploring Global Connections: New Opportunities in the Russian Arctic*). Northern ports such as Murmansk may reap the benefits of this arctic warming.

Potential Hazards Even with such rosy scenarios, might long-term regional and global costs outweigh the benefits? First, hotter summers may increase the risk of wildfire. In what could be a sign of things to come, hundreds of blazes broke out in Russia in the summer of 2010, mostly south and east of Moscow. The fires scorched more than 484,000 acres

▲ **Figure 9.2.2.** **Russian Icebreaker** The nuclear-powered *Yamal* is a state-of-the-art icebreaker in the Russian Arctic.

Google Earth Virtual MG Tour Video

http://goo.gl/EgIndu

from 3.9 million tons in 2013 to 5.4 million tons in 2015, and high-latitude enthusiasts imagine annual shipping volumes of more than 100 million tons by 2030.

How would the NSR contribute to the regional and global economy? Optimists point to three major contributions of an expanded NSR. First, it would benefit *global transit traffic* between Europe and Asia, especially to and from China. Clearly, the NSR will need to prove its reliability and its competitiveness against traditional routes to make this possible. With Russian support, a series of demonstration voyages were completed between 2010 and 2013, mainly to convince global shipping firms and insurance companies that the NSR

was indeed reliable and safe. For now, however, the NSR remains a highly seasonal option, predictably accessible only between July and November. But global warming optimists hope that more open seas lie ahead.

Second, Russia's *arctic natural-resource projects* would clearly be spurred by an expanded NSR. Currently, a good deal of existing traffic along the route is related to nickel and other metals shipments from Norilsk and to the expansion of three arctic energy projects already under development (one for liquefied natural gas, two for crude oil production). Other oil, coal, and mining projects are planned, and would greatly benefit from a more ice-free arctic and a better-developed NSR.

Finally, long-term Russian planners envision huge economic potential for *expanded Siberian trade* with a more viable NSR. Major Siberian rivers such as the Ob, the Yenisey, and the Lena (see Figure 9.2.1) flow north, and a more highly commercialized NSR could open up these once inaccessible regions to more global trade.

Still, more cautious enthusiasts suggest a slow and careful development plan is needed. Climate change is not a linear, always predictable process. What if huge investments are made in the NSR and the ice doesn't cooperate? Fluctuating global economic demand for shipping is another concern: a sustained Chinese downturn, for example, would slacken demand for the NSR. Another challenge is the lack of Russian icebreakers and ice pilots to provide competent guidance and support. More generally, a safe and reliable NSR requires better surveys of the region, expanded arctic ports, and a series of globally approved rules. Russia's Development Plan, established in 2015, seems a positive step in that direction and hopefully can provide a centralized, standardized set of protocols as well as a competitive pricing system that will attract global interest. Perhaps one day soon—in a warmer, more ice-free world—the Russian arctic will be a bustling focus of global trade and economic activity.

1. For a Chinese manufacturer, list the key advantages and disadvantages that might be associated with shipping their finished goods to western Europe via the NSR.

2. In the next century, do you think that the NSR or the Northwest Passage (around northern North America) will have the greatest economic potential? Defend your answer.

(196,000 hectares), burning wheat fields and hundreds of structures and filling the skies of Moscow with smoke as visitors and residents experienced record heat.

Second, changes in ecologically sensitive arctic and subarctic ecosystems are already leading to major disruptions in wildlife and indigenous human populations in those settings of northern Russia. Take the example of the polar bear. The shrinking volume of arctic sea-ice habitat for the bears has meant that they are forced to widen their search for food. This has brought them into closer contact with arctic villages and also disrupted traditional hunting practices. Poachers have also profited, increasing their illegal harvests.

Third, rising global sea levels will hit low-lying areas of the Black and Baltic seas particularly hard. Officials in St. Petersburg, Russia's second largest city, are already

contemplating significant costs associated with controlling the Baltic's rising waters.

Finally, the largest potential change, with global implications, relates to the thawing of the Siberian permafrost. Substantial areas of northern Russia are covered with permafrost that is already close to thawing. This same region of the world has witnessed some of the largest, most persistent global warming since 1950. Thus, even minor increases in temperature could have large and irreversible consequences for the region. Major changes in topography (mud flows, slumping, erosion, and craters caused by the release of methane), drainage (lake coverage and rivers), and vegetation will be the result. Existing fish and wildlife populations will need to adjust in order to survive. Human infrastructure such as buildings, roads, and pipelines will also require substantial modification.

The greatest global impact may come with the huge release of carbon that is currently stored in existing permafrost environments. The soils frozen in permafrost contain large amounts of organic material that decomposes quickly when thawed. Most of the planet's permafrost could release its carbon reservoir within the next century, the equivalent of 80 years of burning fossil fuels. Such a contribution to the world's carbon budget, which would likely further warm Earth, is only beginning to be incorporated into models of global climate change. Methane, a much more potent greenhouse gas than carbon dioxide, is also frozen in the permafrost. Since 2014, a growing number of mysterious Siberian craters have appeared, and one scientific guess is that they represent sudden explosive releases of thawing methane gas. Thus, the survival of the Siberian permafrost may hold one of the keys to slowing or quickening further global warming.

Russia's northern boreal forests also play a pivotal role in global climate change. The country possesses about 20 percent of world forest reserves, which absorb huge amounts of carbon dioxide. But currently, unsustainable forest management policies and illegal timber poaching are contributing to expanded logging. By 2045, the beneficial impacts of these forests may be negligible, leading to the more rapid build-up of carbon dioxide in the atmosphere.

REVIEW

9.1 Compare the climate, vegetation, and agricultural conditions of Russia's European West with those of Siberia and the Russian Far East.

9.2 Describe the high environmental costs of industrialization within the Russian domain and cite a recent effort to address some of these problems.

■ **KEY TERMS** Soviet Union, podzol soils, chernozem soils, permafrost, taiga, northern sea route

Population and Settlement: An Urban Domain

The six states of the Russian domain are home to about 200 million residents (Table 9.1). Although they are widely dispersed across a vast Eurasian land mass, most live in cities. The region's distinctive distributions of natural resources and changing migration patterns continue to shape its population geography. Government policies have encouraged migration into the eastern portions of the domain, but the population remains strongly concentrated in the traditional centers of the European West. Relatively low birth rates and higher death rates also remain a critical concern, as these trends threaten to shrink future regional populations.

Population Distribution

The favorable agricultural setting of the European West has offered a home to more people than live in the inhospitable conditions found across the larger areas of central and northern Siberia. Although Russian efforts over the past century have encouraged a wider dispersal of the population, it remains heavily concentrated in the west (**Figure 9.13**). European Russia is home to more than 100 million people, whereas Siberia, although far larger, holds only some 35 million. When you add in the 60 million inhabitants of Belarus, Moldova, and Ukraine, the imbalance between east and west becomes even more striking (see Table 9.1).

The European Core The region's largest cities, biggest industrial complexes, and most productive farms are located in the European Core, a subregion that includes Belarus, much of Ukraine, and Russia west of the Urals (see Figure 9.1). Sprawling Moscow and its nearby urbanized region clearly dominate the settlement landscape with a metropolitan area

TABLE 9.1	Population Indicators								Mastering Geography™
Country	Population (millions) 2016	Population Density (per square kilometer)[1]	Rate of Natural Increase (RNI)	Total Fertility Rate	Percent Urban	Percent < 15	Percent > 65	Net Migration (Rate per 1000)
Armenia	3.0	106	0.5	1.6	64	19	11	−1
Belarus	9.5	47	0.0	1.8	78	16	14	2
Georgia	4.0	65	0.2	1.7	54	17	14	−1
Moldova	3.6	124	0.0	1.3	42	16	10	1
Russia	144.3	9	0.0	1.8	74	17	14	2
Ukraine	42.7	78	−0.4	1.5	70	15	16	0

Source: Population Reference Bureau, *World Population Data Sheet,* 2016.
[1] World Bank, *World Development Indicators,* 2016.

1) Use MapMaster to visualize Population and Population Density patterns. Why does Russia have the largest population but the lowest population density? Why is tiny Moldova so densely settled compared to Russia?

2) Which country has the highest Total Fertility Rate? The lowest? Do you expect these rates to change much in the next ten years? Why or why not?

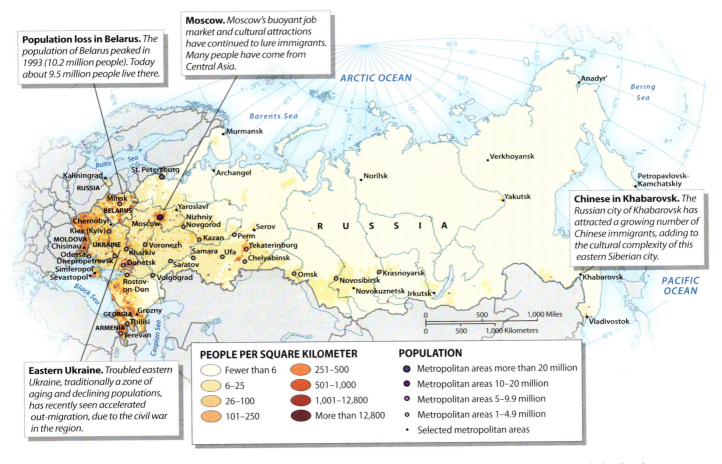

Population loss in Belarus. *The population of Belarus peaked in 1993 (10.2 million people). Today about 9.5 million people live there.*

Moscow. *Moscow's buoyant job market and cultural attractions have continued to lure immigrants. Many people have come from Central Asia.*

Chinese in Khabarovsk. *The Russian city of Khabarovsk has attracted a growing number of Chinese immigrants, adding to the cultural complexity of this eastern Siberian city.*

Eastern Ukraine. *Troubled eastern Ukraine, traditionally a zone of aging and declining populations, has recently seen accelerated out-migration, due to the civil war in the region.*

PEOPLE PER SQUARE KILOMETER
- Fewer than 6
- 6–25
- 26–100
- 101–250
- 251–500
- 501–1,000
- 1,001–12,800
- More than 12,800

POPULATION
- Metropolitan areas more than 20 million
- Metropolitan areas 10–20 million
- Metropolitan areas 5–9.9 million
- Metropolitan areas 1–4.9 million
- Selected metropolitan areas

▲ **Figure 9.13 Population of the Russian Domain** The region's people are strongly clustered west of the Ural Mountains. Dense agricultural settlements, extensive industrialization, and large urban centers are found in Ukraine, much of Belarus, and across western Russia south of St. Petersburg and Moscow. A narrower chain of settlements follows the better lands and transportation corridors of southern Siberia, but most of Russia east of the Urals remains sparsely settled.

containing more than 16 million people (**Figure 9.14**). In 2014, government officials embraced an ambitious urban plan that emphasized continued decentralization and automobile-oriented sprawl, especially to the south (**Figure 9.15**). Officials hope this takes pressure off the city's crowded core, making it more liveable. Clearly Russia's primate city, Moscow has a population that is more than double the country's next largest metropolitan area (St. Petersburg), and it produces almost 20 percent of the entire nation's wealth. Looking ahead to 2020 and beyond, Moscow is projected to grow at a healthy pace, attracting domestic migrants from the country's less vibrant smaller cities and rural areas, as well as foreign immigrants from regions such as Central Asia.

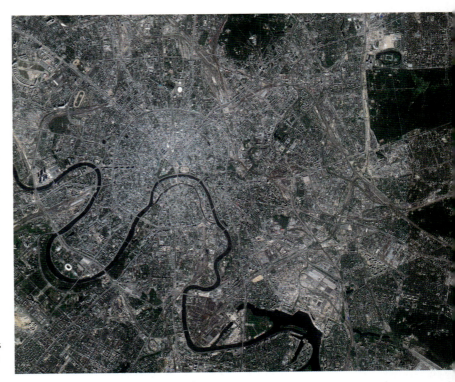

▶ **Figure 9.14 Metropolitan Moscow** Sprawling Moscow extends more than 50 miles (80 kilometers) beyond the city center, on both sides of the Moscow River. The larger metropolitan area is home to more than 16 million people, and the relative strength of its urban economy continues to attract migrants from elsewhere in the country, thus putting more pressure on its infrastructure.

▶ **Figure 9.15 Moscow Traffic** The capital's Ring Road (the MKAD) features North American–style interchanges and traffic jams.

On the shores of the Baltic, St. Petersburg (Leningrad in the Soviet period; 4.9 million people) has traditionally had a great deal of contact with western Europe. Literally rising from swamplands in 1703, St. Petersburg grew from the creative imagination of its founder, Tsar Peter the Great. It became the westward-looking tsarist capital until the Soviet takeover in 1917 (when the capital returned to Moscow). Peter's imagination helped create a Renaissance-style city built around grand avenues, palaces, and a network of canals that still draws comparisons with Venice and Amsterdam (**Figure 9.16**). Near the city center, elaborate baroque and neoclassical mansions (many now converted to other uses) and commercial buildings recall earlier days when Russians such as Tchaikovsky and Dostoyevsky walked the streets. Cutting through the heart of the city and bordered by the bright lights of shops, cafes, and theaters, Nevsky Prospekt ends near the Neva River, just two blocks from the world-famed Hermitage Museum, home to 2.8 million pieces of art.

Other urban clusters are oriented along the lower and middle stretches of the Volga River, including the cities of Kazan, Samara, and Volgograd. Industrialization within the region accelerated during World War II, as this area lay somewhat removed from German advances in the west. Today, the highly commercialized river corridor, also containing important petroleum reserves, supports a diverse industrial base strategically located to serve the large populations of the European Core. Nearby, the resource-rich Ural Mountains include the gritty industrial landscapes of Yekaterinburg (1.4 million) and Chelyabinsk (1.1 million).

▼ **Figure 9.16 St. Petersburg** Often called Russia's most beautiful city, St. Petersburg's urban design features a varied mix of gardens, open space, waterways, and bridges. This view shows the ornate architecture (center) of the Church of the Savior of Spilled Blood, completed in 1907.

Beyond Russia, major population clusters within the European Core are also found in Belarus and Ukraine (see Table 9.1). The Belorussian capital of Minsk (1.8 million) is the dominant urban center in that country, and its landscape recalls the drab Soviet-style architecture of an earlier era. In nearby Ukraine, the capital of Kiev (Kyiv, 2.8 million) straddles the Dnieper River. The city's old and beautiful buildings on the west side of the river contrast markedly with much of the Soviet-era growth evident on the east side of the city (see Figure 9.2).

Siberian Hinterland Leaving the southern Urals city of Yekaterinburg on a Siberia-bound train, you become aware that the land ahead is ever more sparsely settled (see Figure 9.13). The distance between cities grows, and the intervening countryside reveals a landscape shifting gradually from farms to forest. The Siberian hinterland is divided into two characteristic zones of settlement, each linked to railroad lines. To the south, a collection of isolated, but sizable urban centers follows the **Trans-Siberian Railroad**, a key rail passage to the Pacific completed in 1904 (see Figure 9.1). The eastbound traveler encounters Omsk (1.1 million) as the rail line crosses the Irtysh River, Novosibirsk (1.5 million) at its junction with the Ob River, and Irkutsk (600,000) near Lake Baikal. At the end of the line, the port city of Vladivostok (600,000) provides access to the Pacific. To the north, a thinner sprinkling of settlement appears along the more recently completed (1984) **Baikal–Amur Mainline (BAM) Railroad**, which parallels the older line, but runs north of Lake Baikal to the Amur River. From the BAM line to the Arctic, the almost empty spaces of central and northern Siberia dominate the scene,

Exploring High-Latitude Siberia

Kelsey Nyland's fascination for polar regions—and geography—blossomed when she was an undergraduate at George Washington University. Initially a geology major, Nyland "didn't know what geography was," but an arctic environments course and subsequent fieldwork in Alaska hooked her. She found that in geography, you "get to see places, not just learn about them in classrooms abstractly," and eventually became an undergraduate research assistant for Dr. Nikolay Shiklomanov, who leads a program that monitors permafrost change throughout the Arctic and Antarctic. As permafrost warms and thaws, it can undermine roads, damage infrastructure, and release additional greenhouse gasses. "What's interesting is how [the thawing] permafrost actually affects infrastructure, how it affects people," Nyland notes. Her interest in both physical and human environments made geography an obvious fit—"everything I was doing was lining up with geography"—and she found that this is true for many students.

▲ **Figure 9.3.1.** **Geographer Kelsey Nyland** observes material transport by Romantic Glacier in the Polar Ural Mountains, Russia during an expedition after the Tenth International Conference on Permafrost (July 2012).

Siberian Research Nyland continued in the Master's program in Geography at George Washington and her interests returned her to Siberia (**Figure 9.3.1**). By 2015, Nyland had traveled to Siberia for five different field courses and learned all about using Landsat images to remotely monitor changing permafrost conditions. She also developed an additional research project (using historical photographs) examining how depopulating Siberian cities look decades after people had left.

Currently a doctoral student in Geography at Michigan State University, Nyland has contributed directly to research initiatives quantifying changes in Siberia's natural and urban permafrost landscapes. "First we look at the natural landscapes, at how they develop, and then how it impacts the surprising numbers of people who live in Siberia, some in pretty big cities no less … pretty neat that we can do that," she says. For example, her master's thesis documented how changing

Russian economic and urban policies impact the natural environments surrounding Siberian cities. Nyland's diverse interests in both physical and human geography demonstrate how good research benefits from a thorough background in both parts of the discipline.

1. In addition to the effects of human-induced climate change, how else might permafrost in arctic settlements become unstable or damaged?

2. Describe an issue or problem in your local setting that requires a thorough knowledge of *both* physical and human geography to address.

interrupted only rarely by small settlements, often oriented around points of natural resource extraction (see *Geographers at Work: Exploring High-Latitude Siberia*).

The Demographic Crisis

Russia has identified population loss as a key issue of national importance. Some government and United Nations (UN) estimates have predicted that Russia's population could fall by a startling 45 million by 2100. Similar conditions are affecting the other countries within the region (see Table 9.1). Beginning with World War II, large numbers of deaths combined with low birth rates to produce sizable population losses. Although population increased in the 1950s, growth slowed by 1970, and death rates began exceeding birth rates in the early 1990s.

President Putin has declared that demographic decline was Russia's "most acute problem." He has pushed for programs aimed at raising birth rates and challenging the one-child family norm that has become widely accepted. Under the plan, mothers of multiple children receive cash payments, extended maternity leave, and extensive day-care subsidies. Some Russian cities have even sponsored competitions, encouraging couples to have babies in the hopes of winning prizes such as automobiles.

Recently, birth rates have risen slightly within Russia, perhaps a function of changing government policies. In particular, rates among ethnic non-Russians in the country (for example, in the Caucasus region and portions of Siberia) have been significantly above those of ethnic Russians. Russia's birth rate is now significantly higher than those of Germany, Italy, and Japan.

Still, many uncertainties remain concerning future demographic growth. The current economic decline and political instability in the region may depress birth rates. In addition, residents still face significantly higher death rates than much of the developed world, and they also have daunting health-care challenges. While global life expectancies have grown by an impressive 6.2 years since 1990, Russian lifespans have only eked out a gain of 1.8 years.

The region's demographic structure also forecasts fewer women of child-bearing age in the near future, making the recent improvements difficult to maintain. A report issued in 2015 by the Russian Presidential Academy of National Economy and Public Administration predicted that the number of Russian women aged 20 to 29 will shrink by almost 50 percent by 2025, further pressuring growth rates.

Regional Migration Patterns

Over the past 150 years, millions of people within the Russian domain have been on the move. These major migrations, both forced and voluntary, reveal sweeping examples of human mobility that rival the great movements from Europe and Africa or the transcontinental spread of settlement across North America.

Eastward Movement Just as settlers of European descent moved west across North America, exploiting natural resources and displacing native peoples, European Russians moved east across the vast Siberian frontier. Although the deeper historical roots of the movement extend back several centuries, the pace and volume of the eastward drift accelerated in the late 19th century as portions of the Trans-Siberian Railroad were being completed. Peasants were attracted to the region by its agricultural opportunities (in the south) and by greater political freedoms than they traditionally enjoyed under the **tsars** (or czars; Russian for "Caesar"), the authoritarian leaders who dominated politics during the pre-1917 Russian Empire. Almost 1 million Russian settlers moved into the Siberian hinterland between 1860 and 1914.

The eastward migration continued during the Soviet period, once communist leaders consolidated power during the late 1920s and saw the economic advantages of developing the region's rich resource base. The German invasion of European Russia during World War II demonstrated that there were also strategic reasons for settling the eastern frontier, and this propelled further migrations during and following the war. Indeed, by the end of the communist era, 95 percent of Siberia's population was classified as Russian (including Ukrainians and other western immigrants). With completion of the BAM Railroad in the 1980s, another corridor of settlement opened in Siberia, prompting new migrations into a region once remote from the outside world.

Political Motives Political motives also have shaped migration patterns. Particularly in the case of Russia, leaders from both the imperial and the Soviet eras saw advantages in moving selective populations to new locations. The infilling of the southern Siberian hinterland had a political as well as an economic rationale. Both the tsars and the Soviet leaders saw their political power grow as Russians moved into the resource-rich Eurasian interior. For some, however, the move to Siberia was not voluntary as the region became a repository for political dissidents and troublemakers. The communist regime of Joseph Stalin (1928–1953) was particularly noted for its forced migrations, including the removal of thousands of Jews to the Russian Far East after 1928. Overall, during the Soviet period, uncounted millions were forcibly relocated to the region's infamous **Gulag Archipelago**, a vast collection of political prisons in which inmates often disappeared or spent years far removed from their families and home communities.

Russification, the Soviet policy of resettling Russians into non-Russian portions of the Soviet Union, also changed the region's human geography. Millions of Russians were given economic and political incentives to move elsewhere in the Soviet Union in order to increase Russian dominance in many of the outlying portions of the country. The migrations were geographically selective in that most of the Russians moved either to administrative centers, where they took government positions, or to industrial complexes focused on natural resource extraction. As a result, by the end of the Soviet period, Russians made up significant minorities within former Soviet republics (now independent nations) such as Kazakhstan (30 percent Russian), Latvia (30 percent), and Estonia (26 percent).

Since 2014, another wave of politically motivated migration has rocked the region. War-torn areas of eastern Ukraine, caught in the crossfire between Russian rebel and Ukrainian forces, have lost about one-third of their population. About half of these displaced peoples have moved elsewhere within Ukraine and the other half (largely ethnic Russians) have spilled into nearby Russia, overwhelming border zones with new refugee populations. Some observers estimate that more than one million people have flowed into Russia, many ending up in poorly provisioned refugee camps.

New International Movements Other recent migrations have also crossed international borders (**Figure 9.17**). In the post-Soviet era, Russification has often been reversed. Several of the newly independent non-Russian countries imposed rigid language and citizenship requirements, which encouraged Russian residents to leave. In other settings, ethnic Russians experienced varied forms of discrimination. The Russian government also has promoted a repatriation program for ethnic Russians worldwide, offering incentives for Russian-speaking migrants to return or move to their cultural homeland. As a result, the Central Asia and Baltic regions,

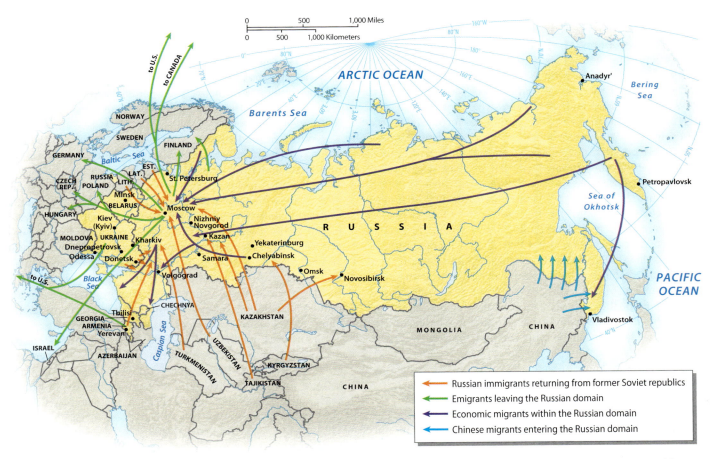

	Russian immigrants returning from former Soviet republics
	Emigrants leaving the Russian domain
	Economic migrants within the Russian domain
	Chinese migrants entering the Russian domain

▲ **Figure 9.17 Recent Migration Flows in the Russian Domain** Recent events are encouraging the return of ethnic Russians from former Soviet republics, while other Russians are emigrating from the domain for economic, cultural, and political reasons. Within Russia, both political and economic forces are also at work, encouraging people to be on the move.

once a part of the Soviet Union, have seen their Russian populations decline significantly since 1991, often by 20 to 35 percent.

Russia has also experienced a growing immigrant population, many of them undocumented migrants drawn to the region for work. The story has a familiar ring to it. More than 11 million undocumented immigrants are suspected to be in the country (the world's second largest total, behind that of the United States). Most are young, upwardly mobile people, predominantly male, who come for better-paying jobs. Many migrants send money they earn in Russia back home to their native lands. For example, one estimate suggests about 20–30 percent of Tajikistan's economy is based on funds sent there from émigrés living and working in Russia. Recently, the government has implemented tighter controls to restrict border crossings and enacted tougher penalties against businesses that hire illegal immigrants. There is also a growing national debate concerning how many legal foreign workers should be allowed in the country and whether or not undocumented immigrants should be granted amnesty.

The actual flows of undocumented immigrants are complex. Most immigrants—legal and undocumented—come from portions of the former Soviet Union, especially from ethnically non-Slavic regions in Central Asia (**Figure 9.18**). Almost one-third of the country's undocumented immigrants may live in the job-rich Moscow metropolitan region.

▼ **Figure 9.18 Central Asian Immigrants in Moscow** These migrants from Tajikistan work at an outdoor market near Moscow's Kiev Railway Station.

As is true in the United States, tougher economic times in Russia—as has been the case recently—often leads to reduced flows of immigrants into the country.

In addition, immigration into Russia's Far East, principally from northern China, is reshaping the economic and cultural geographies of that region. Walk through the Russian cities of Vladivostok and Khabarovsk, and you will see street signs featuring both Russian and Chinese lettering. Entire neighborhoods are dominated by Chinese immigrant populations. Chinese children are learning Russian in school, and Russians find themselves working for Chinese entrepreneurs. A significant nationalist backlash has occurred: Chinese are sometimes attacked, Chinese shopkeepers complain of being rousted by Russian police, and recent legislation has made it harder for Chinese to operate businesses. On the other hand, many younger Russians have welcomed their Chinese counterparts and a growing number of joint Russian-Chinese companies have appeared in the region. In addition, with a persisting shortage of healthy Russian men in the region, hundreds of younger Russian women have enrolled in Mandarin language schools, and there is more intermarrying between the two groups.

The domain's more open borders also have made it easier for other residents to leave the region (see Figure 9.17). Deteriorating economic conditions and the region's unpredictable politics have encouraged many to emigrate. The "brain drain" of young, well-educated, upwardly mobile Russians has been considerable. Sometimes, ethnic links also play a part. For example, many Russian-born ethnic Finns have moved to nearby Finland, much to the consternation of the Finnish government. Russia's Jewish population also continues to fall, a pattern begun late in the Soviet period. These emigrants have flocked mostly to Israel or the United States.

The Urban Attraction Regional residents also have been bound for the cities. The Marxist philosophy embraced by Soviet planners encouraged urbanization. In 1917, the Russian Empire was still overwhelmingly rural and agrarian; 50 years later, the Soviet Union was primarily urban. Planners saw great economic and political advantages in efficiently clustering the population, and Soviet policies dedicated to large-scale industrialization obviously favored an urban orientation. Today, Russian, Ukrainian, and Belorussian rates of urbanization are comparable to those of the industrialized capitalist countries (see Table 9.1).

Soviet cities grew according to strict governmental plans. Planners selected different cities for different purposes. Some were designed for specific industries, while others had primarily administrative roles. A system of internal passports prohibited people from moving freely from city to city. Instead, people generally went where the government assigned them jobs. Moscow, the country's leading administrative city, thrived under the Soviet regime. It formed the undisputed core of Soviet bureaucratic power, as well as the center of education, research, and the media. Specialized industrial cities also grew at a rapid pace. In the mining and metallurgical zone of the southern Urals, centers such as Yekaterinburg and Chelyabinsk mushroomed into major urban centers. Another cluster of specialized industrial cities, including Kharkiv and Donetsk, emerged in the coal districts of eastern Ukraine.

With the end of the Soviet Union, however, people gained basic freedoms of mobility. A more open economy, especially in Russia, also shifted urban employment opportunities, which increasingly reflected how effectively local economies could compete in the global market. This new economic reality led to the depopulating of many older industrial areas, as they simply cannot produce raw materials or finished industrial goods at competitive global prices. Since 1989, for example, the Russian Northeast has seen its population decline as many workers have left harsh, unemployment-prone industrial centers for better opportunities elsewhere. In many cases, people are freely gravitating toward growth opportunities in locations of new foreign investment, principally in larger urban areas in western and southern Russia. Some also are receiving relocation assistance from a Russian government that admits many of these outlying settlements simply cannot be justified in a more market-oriented economy.

Inside the Russian City

Today, most people in the region live in a city, the product of a century of urban migration and growth (see Table 9.1). Large Russian cities possess a core area, or center, that features superior transportation connections; the best-stocked, upscale department stores and shops; the most desirable housing; and the most important offices (both governmental and private) (**Figure 9.19**). The largest urban centers such as Moscow and St. Petersburg also feature extensive public spaces and examples of monumental architecture at the city center. Within the city, there is usually a distinctive pattern of circular land-use zones, each of which was built at a later date moving outward from the center. Such a ringlike urban morphology is not unique. However, as a result of the extensive power of government planners during the Soviet period, this urban form is probably more highly developed here than in most other parts of the world.

At their very center, the cores of many older cities predate the Soviet Union. Pre-1900 stone buildings often dominate older city centers. Some of these are former private mansions that were turned into government offices or subdivided into apartments during the communist period, but are now being privatized again. Many of these older buildings, however, are being leveled in rapidly growing urban settings such as downtown Moscow. Urban preservation experts estimate that since 1992, several thousand

▲ **Figure 9.19 Downtown Moscow** This outdoor art show and sale attracts busy shoppers along Moscow's bustling Arbat Street.

historic buildings have been destroyed, including many structures that had supposedly received official protection. Retail malls replace many of these older structures. Nearby, nightclubs and bars are filled with pleasure seekers as the city's professional elite mingles with foreign visitors and tourists.

Farther out from the city centers are **mikrorayons**, large, Soviet-era housing projects of the 1970s and 1980s. Mikrorayons are typically composed of massed blocks of standardized apartment buildings, ranging from 9 to 24 stories in height. The largest of these supercomplexes contain up to 100,000 residents. Soviet planners hoped that mikrorayons would foster a sense of community, but most now serve as anonymous bedroom communities for larger metropolitan areas.

Some of Russia's most rapid ub
an growth has occurred on the metropolitan periphery, paralleling the North American experience. Moscow, for example, has seen its urban reach expand far beyond the city center. The city's towering International Business Center now dominates the city's west-side skyline and features some of Europe's tallest buildings (Figure 9.20). Suburban shopping malls and housing districts featuring single-family homes, like their North American counterparts, are also popping up on the urban fringe, allowing upscale

residents to live and shop without having to visit the city center (Figure 9.21).

Elsewhere on Moscow's urban fringe, elite **dachas**, or cottage communities, appeal to many more well-to-do residents, particularly during the summer months. The tradition of rural retreats, dating back to the Russian Empire, also thrived during the Soviet era as Communist Party officials sought an escape from the dreary bureaucratic chores of the city. Today, about 300 cottage settlements appear on the Moscow periphery, including many new elite privatized developments northwest and southeast of the city that cater to the country's business class.

REVIEW

9.3 Discuss how major river and rail corridors have shaped the geography of population and economic development in the region. Provide specific examples.

9.4 Contrast Soviet and post-Soviet migration patterns within the Russian domain, and explain the changing forces at work.

9.5 Describe the major land-use zones in the modern Russian city, and suggest why it is important to understand the impact of Soviet-era planning within such settings.

■ **KEY TERMS** Trans-Siberian Railroad, Baikal–Amur Mainline (BAM) Railroad, tsars, Gulag Archipelago, Russification, mikrorayons, dachas

Explore the **Sights** of Moscow's Business Center

http://goo.gl/S5QHKX

▼ **Figure 9.20 Moscow's International Business Center** Also known as Moscow City, the International Business Center dominates the capital's western skyline and features some of Europe's tallest buildings.

▲ **Figure 9.21 Single-Family Housing, Moscow Suburbs** This newer upscale housing development in the Moscow suburbs was built by INKOM-Nedvizhimost. The large, neatly fenced lots and spacious homes build on North American traditions of suburban taste and design. **Q: Why has Moscow's urban landscape become more similar to that of North America in the past 30 years?**

Cultural Coherence and Diversity: The Legacy of Slavic Dominance

The Russian domain remains at the heart of the Slavic world. For hundreds of years, Slavic peoples speaking the Russian language expanded their influence from an early homeland in central European Russia. Eventually, this Slavic cultural imprint spread north to the Arctic Sea, south to the Black Sea and Caucasus, west to the shores of the Baltic, and east to the Pacific Ocean. In this process of diffusion, Russian cultural patterns and social institutions spread widely, influencing many non-Russian ethnic groups that continued to live under the rule of the Russian Empire. The legacy of that Slavic expansion continues today. It offers Russians a rich historical identity and sense of nationhood. It also provides a meaningful context in which to understand how present-day Russians are dealing with forces of globalization and how non-Russian cultures have evolved within the region.

The Heritage of the Russian Empire

The expansion of the Russian Empire paralleled similar events in western Europe. As Spain, Portugal, France, and Britain carved out empires in the Americas, Africa, and Asia, the Russians expanded eastward and southward across Eurasia. Unlike other European empires, however, the Russian Empire formed one single territory, uninterrupted by oceans or seas. Only with the fall of the Soviet Union in 1991 did this transformed empire finally begin to dissolve.

Origins of the Russian State The origin of the Russian state lies in the early history of the **Slavic peoples**, defined linguistically as a distinctive northern branch of the Indo-European language family. The Slavs originated in or near the Pripyat marshes of modern Belarus. Some 2000 years ago they began to migrate to the east, reaching as far as modern Moscow by 200 CE. Slavic political power grew by 900 CE as Slavs intermarried with southward-moving warriors from Sweden known as *Varangians,* or *Rus*. Within a century, the state of Rus extended from Kiev (the capital) in modern Ukraine to Lake Ladoga near the Baltic Sea. The new Kiev–Rus state interacted with the rich and powerful Byzantine Empire of the Greeks, and this influence brought Christianity to the Russian realm by 1000 CE. Along with the new religion came many other aspects of Greek culture, including the Cyrillic alphabet. Even as groups such as the Russians and Serbs converted to **Eastern Orthodox Christianity**—a form of Christianity historically linked to eastern Europe and church leaders in Constantinople (modern Istanbul)—their Slavic neighbors to the west (the Poles, Czechs, Slovaks, Slovenians, and Croatians) accepted Catholicism. The resulting religious division split the Slavic-speaking world into two groups, one oriented to the west, the other to the east and south. This early Russian state soon faltered and split into several principalities that were then ruled by invading Mongols and Tatars (a group of Turkish-speaking peoples).

Growth of the Russian Empire By the 14th century, however, northern Slavic peoples overthrew Tatar rule and established a new and expanding Slavic state (**Figure 9.22**). The core of the new Russian Empire lay near the eastern fringe of the old state of Rus. The former center around Kiev was now a war-torn borderland (or "Ukraine" in Russian) contested by the Orthodox Russians, the Catholic Poles, and the Muslim Turks. Gradually, this area's language diverged from that spoken in the new Russian core, and *Ukrainians* and Russians developed into two separate peoples. A similar development took place among the northwestern Russians, who experienced several centuries of Polish rule and over time were transformed into a distinctive group known as the *Belorussians*.

The Russian Empire expanded remarkably in the 16th and 17th centuries (see Figure 9.22). Former Tatar territories in the Volga Valley (near Kazan) were incorporated into the Russian state in the mid-1500s. The Russians also allied with the seminomadic **Cossacks**, Slavic-speaking Christians who had earlier migrated to the region to seek freedom in the ungoverned steppes. The Russian Empire granted them considerable privileges in exchange for their military service, an alliance that facilitated Russian expansion into Siberia during the 17th century. Premium furs were the chief lure of this immense northern territory. By the 1630s, Russian power was entrenched in central Siberia, and by the end of the century, it had reached the Pacific Ocean. Chinese resistance, however, delayed Russian occupation of the Far East region until 1858, and the imperial designs of the Japanese halted further expansion to the southeast when the Russians lost the Russo-Japanese War in 1905.

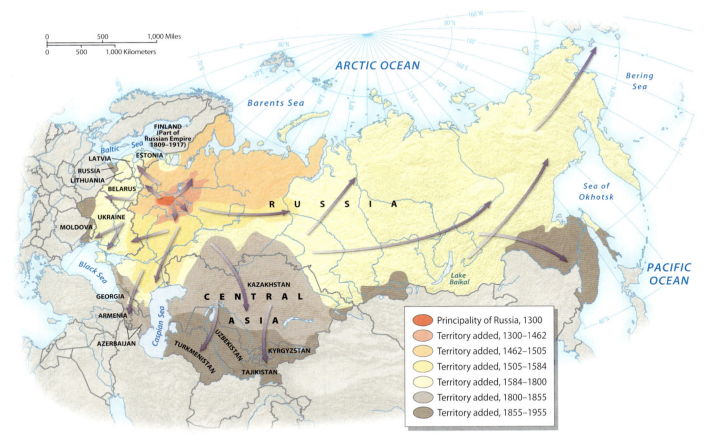

▲ **Figure 9.22** **Growth of the Russian Empire** Beginning as a small principality in the vicinity of modern Moscow, the Russian Empire took shape between the 14th and 16th centuries. After 1600, Russian influence stretched from eastern Europe to the Pacific Ocean. Later, the empire added lands in the Far East, in Central Asia, and near the Baltic and Black seas.

Although the Russian Empire expanded to the east with great rapidity, its westward expansion was slow and halting. In the 1600s, Russia still faced formidable enemies in Sweden, Poland, and the Ottoman (Turkish) Empire. By the 1700s, however, all three of these states had weakened, allowing the Russian Empire to gain substantial territories. After defeating Sweden in the early 1700s, Tsar Peter the Great (1682–1725) obtained a foothold on the Baltic, where he built the new capital city of St. Petersburg (see Figure 9.16). Later in the 18th century, Russia defeated both the Poles and the Turks and gained all of modern-day Belarus and Ukraine. Tsarina Catherine the Great (1762–1796) was particularly pivotal in colonizing Ukraine and bringing the Russian Empire to the warm-water shores of the Black Sea.

The 19th century witnessed the Russian Empire's final expansion. Large gains were made in Central Asia, where a group of once-powerful Muslim states was no longer able to resist the Russian army. The mountainous Caucasus region proved a greater challenge, as the peoples of this area had the advantage of rugged terrain in defending their lands. South of the Caucasus, however, the Christian Armenians and Georgians accepted Russian power with little struggle

because they found it preferable to rule by the Persian or Ottoman empires.

The Legacy of Empire The expansion of the Russian people was one of the greatest human movements Earth has ever witnessed. By 1900, a traveler going from St. Petersburg on the Baltic to Vladivostok on the Sea of Japan would everywhere encounter Russian peoples speaking the same language, following the same religion, and living under the rule of the same government. Nowhere else in the world did such a tightly integrated cultural region cover such a vast space.

The history of the Russian Empire also reveals points of ongoing tension with the world beyond. One of these tensions centers on Russia's ambivalent relationship with western Europe. Russia shares with the West the historical legacy of Greek culture and Christianity. Since the time of Peter the Great, Russia has undergone several waves of intentional Westernization. At the same time, however, Russia has long been suspicious of—even hostile to—European culture and social institutions. Although elements of this debate were transformed during the Soviet period, the central tension remains, and it influences

Russia to this day. As Europe further unifies through the expansion of the EU, this gap between East and West may take new forms, but it is unlikely to disappear anytime soon.

Geographies of Language

Slavic languages dominate the region (**Figure 9.23**). The distribution of Russian-speaking populations is complicated. Russian, Belorussian, and Ukrainian are closely related languages. Some linguists argue that they ought to be considered separate dialects of a single Russian language, as they are all mutually intelligible. Most Ukrainians, however, insist that Ukrainian is a distinct language in its own right, and there is a well-developed sense of national distinction between Russians and Ukrainians. Belorussians, on the other hand, are more inclined to stress their close kinship with the Russians.

Patterns in Belarus, Ukraine, and Moldova The geographic pattern of the Belorussian people is relatively simple: The vast majority of Belorussians reside in Belarus, and most people in Belarus are Belorussians (see Figure 9.23). The country does, however, contain scattered Polish and Russian

minorities. For the Russians, this presents few problems, because Russians and Belorussians can relatively easily assume each other's ethnic identity.

The situation in Ukraine, however, is much more complex. Only about 67 percent of the population is Ukrainian. Russian speakers make up almost 25 percent of the population, but are strongly concentrated in eastern Ukraine, while they make up a much smaller portion of western Ukraine's population (**Figure 9.24**). Not surprisingly, the Russian speakers in eastern Ukraine have recently formed the foundation of rebel resistance forces in that part of the country (along with Russian military support). Similarly, the Crimean Peninsula, while still considered part of Ukraine, has long ethnic and historical connections to Russia, facilitating that country's occupation of the peninsula in 2014. Conversely, many Ukrainians born in the west never learned Russian. Kiev (Kyiv), the national capital, has been described as bilingual, "a Russian-speaking city whose people know how to speak Ukrainian."

In nearby Moldova, Romanian (a Romance language) speakers are dominant, although ethnic Russians and Ukrainians each make up about 13 percent of the country's population. Use of the official Romanian language has been criticized by many Slavic speakers, particularly in the

▲ **Figure 9.23 Languages of the Russian Domain** Slavic Russians dominate the region, although many linguistic minorities are present. Siberia's diverse native peoples add cultural variety in that area. To the southwest, the Caucasus Mountains and the lands beyond contain the region's most complex linguistic geography. Ukrainians and Belorussians, while sharing a Slavic heritage with their Russian neighbors, add further variety in the west.

▼ **Figure 9.24 The Russian Language in Ukraine** Both eastern Ukraine and the Crimean Peninsula retain large numbers of Russian speakers, a function of long cultural and political ties that continue to complicate Ukraine's contemporary human geography.

PERCENT SPEAKING RUSSIAN AS A NATIVE LANGUAGE
- More than 60
- 46–60
- 31–45
- 15–30
- Less than 15

of northeast Siberia also represent Turkish speakers within the Altaic family. Other Altaic speakers in Russia belong to the Mongol group. In the west, examples include the Kalmyk-speaking peoples of the lower Volga Valley who migrated to the region in the 1600s. To the east, seminomadic Tuvans live near the Mongolian border and still practice *khoomei*, a traditional form of throat singing said to resemble the sound of wind swirling among rocks (**Figure 9.25**). Nearby, over 400,000 Buryats live in the vicinity of Lake Baikal and represent an indigenous Siberian group closely tied to the cultures and history of Central Asia.

Explore the **Sounds** of Tuvan Throat Singing
http://goo.gl/VEh04k

The plight of many native peoples in central and northern Siberia parallels the situation in the United States, Canada, and Australia. Rural indigenous peoples in each of these settings remain distinct from dominant European cultures. Such groups also are internally diverse and often are divided into several unrelated linguistic clusters. One entire linguistic grouping of Eskimo-Aleut speakers is limited to approximately 25,000 people, who are widely dispersed through northeast Siberia. Another indigenous Siberian group is the Altaic-speaking Evenki, whose historical territory covers a large portion of central and eastern Siberia. At home in the taiga, traditional Evenki hunted animals (such as elk and wild reindeer) and raised livestock (horses and reindeer). Many of these Siberian peoples have seen their customs challenged by the pressures of Russification, just as indigenous peoples elsewhere in the world have been subjected

region east of the Dniester River (Transdniester) that borders Ukraine. During the Soviet period, such geographical complexities had little consequence because the distinctions among Russians, Ukrainians, and Moldovans were not viewed as important in official circles. Now that Russia, Ukraine, and Moldova are separate countries with a heightened sense of national distinction, linguistic issues have become significant sources of tension across the region.

Patterns Within Russia Approximately 80 percent of Russia's population claims a Russian linguistic identity. Russian speakers inhabit most of European Russia, but large enclaves of other peoples are also present. The Russian linguistic zone extends across southern Siberia to the Sea of Japan. In sparsely settled lands of central and northern Siberia, Russians are numerically dominant in many areas, but they share territory with varied indigenous peoples.

Finno-Ugric peoples, though small in number, dominate sizable portions of the non-Russian north. The Finno-Ugric (part of the Uralic language family) speakers make up an entirely different language family from the Indo-European Russians. Although many have been culturally Russified, distinct ethnic groups such as the Karelians, Komi, and Mordvinians remain a part of Russia's modern cultural geography. Karelians, in particular, identify strongly with their Finnish neighbors to the west.

Altaic speakers also complicate the country's linguistic geography. This language family includes the Volga Tatars, whose territory is centered on the city of Kazan in the middle Volga Valley. While retaining their ethnic identity, the Turkish-speaking Tatars have extensively intermarried with their Russian neighbors. Bilingual education is now common in the schools, with ethnic Tatars free to use their traditional language. Yakut peoples

▼ **Figure 9.25 Tuvan Throat Singer** These throat singers are members of Huun-Huur-Tu, a popular Tuvan musical group that is helping to preserve this distinctive cultural tradition. **Q: What social and economic problems might the Tuvans share with many indigenous peoples in North America and Australia?**

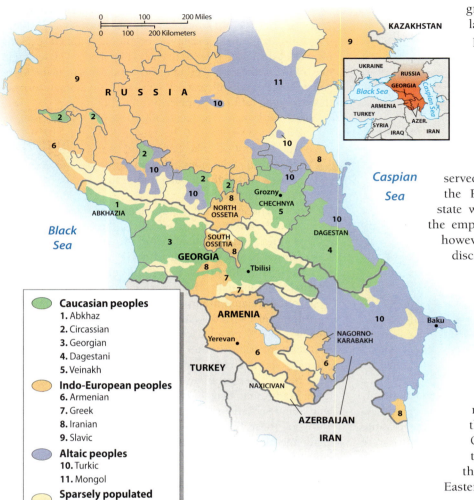

▲ **Figure 9.26** **Languages of the Caucasus Region** A complicated mosaic of Caucasian, Indo-European, and Altaic languages characterizes the Caucasus region of southern Russia and nearby Georgia and Armenia. Persistent political problems have erupted periodically in the region as local populations struggle for more autonomy. Recent examples include independence movements in Chechnya and in nearby Dagestan.

groups (**Figure 9.27**). Not surprisingly, language remains a pivotal cultural and political issue within the fragmented Transcaucasian subregion.

Geographies of Religion

Most Russians, Belorussians, and Ukrainians share a religious heritage of Eastern Orthodox Christianity. For hundreds of years, Eastern Orthodoxy served as a central cultural presence within the Russian Empire. Indeed, church and state were tightly fused until the demise of the empire in 1917. Under the Soviet Union, however, religion in all forms was severely discouraged and actively persecuted. Most monasteries and many churches were converted into museums or other kinds of public buildings, and schools disseminated the doctrine of atheism.

Contemporary Christianity With the downfall of the Soviet Union, however, a religious revival has swept much of the Russian domain. In the past 20 years, more than 12,000 Orthodox churches have been returned to religious uses. Over 70 percent of the Russian population identifies as Eastern Orthodox, up from 31 percent in

▼ **Figure 9.27** **Dagestani Women** These Dagestani women are participating in a folk festival that is being held in the region's largest city, Makhachkala.

Explore the Sights of Dagestan

http://goo.gl/CMcDZe

to similar pressures of cultural and political assimilation. Unfortunately, other common traits seen within such settings are low levels of education, high rates of alcoholism, and widespread poverty.

Transcaucasian Languages Although small in size, Transcaucasia offers a bewildering variety of languages (**Figure 9.26**). From Russia, along the north slopes of the Caucasus, to Georgia and Armenia east of the Black Sea, a complex history and a fractured physical setting have combined to produce some of the most complicated language patterns in the world. No fewer than three language families (Caucasian, Altaic, and Indo-European) are spoken within a region smaller than Ohio, and many individual languages are represented by small, isolated cultural

1991. Within Russia, the Orthodox Church appears headed toward a return to its former role as official state church. The government increasingly is using church officials to sanction various state activities, which is particularly ironic because many of Russia's current leaders played roles in the earlier Soviet period when religious observances were banned. In addition, various forms of evangelical Protestantism (especially from the United States) have been on the rise in Russia since the disintegration of the Soviet Union.

Other forms of western Christianity are also present in the region. For example, the people of western Ukraine, who experienced several hundred years of Polish rule, eventually joined the Catholic Church. Eastern Ukraine, on the other hand, remained fully within the Orthodox framework. This religious split reinforces the cultural differences between eastern and western Ukrainians. Western Ukrainians have generally been far more nationalistic, and hence more firmly opposed to Russian influence, than eastern Ukrainians.

Elsewhere, Christianity came early to the Caucasus, but modern Armenian forms—their roots dating to the 4th century CE—differ slightly from both Eastern Orthodox and Catholic traditions (see *People on the Move: Following the Armenian Diaspora*). Georgian Christianity, however, is more closely tied to the Orthodox faith.

Non-Christian Traditions Religions other than Christianity are also practiced and, along with language, shape ethnic identities and tensions within the region. Islam is the largest non-Christian religion. Russia has some 7000 mosques and approximately 20 million adherents. Most are Sunni Muslims, and they include peoples in the North Caucasus, the Volga Tatars, and Central Asian peoples near the Kazakhstan border. Growth rates among Russia's Muslim population are three times that of the non-Muslim population. Moscow alone may be home to 2–3 million Muslims, mostly Central Asian immigrants. Many of these residents filled city streets in the fall of 2015 celebrating the inauguration of Europe's grandest mosque, a sprawling white marble structure topped with golden and emerald minarets and domes (**Figure 9.28**).

Islamic fundamentalism also has been on the rise, particularly among Muslim populations in the Caucasus region, who increasingly resist what they see as strong-arm tactics and repressive actions on the part of the Russian government. Elsewhere, a growing number of young, conservative Muslim Salafists living in Russian Tatarstan also draw sharp distinctions between themselves and traditional Sunnis in the region, raising ethnic and religious tensions. Local authorities, as well as Moscow leaders, have drafted laws designed to limit the rising influence of the Salafists, but many observers feel this will only add to their appeal among the region's disaffected youth.

Russia, Belarus, and Ukraine are also home to more than 500,000 Jews, who are especially numerous in the larger cities of the European West. Jews suffered severe persecution under both the tsars and the communists. Recent out-migrations, prompted by new political freedoms, have further reduced their numbers in Russia, Belarus, and Ukraine. Buddhists also are represented in the region, associated with the Kalmyk and Buryat peoples of the Russian interior. Indeed, Buddhism has witnessed a recent renaissance and now claims approximately 1 million practitioners, mostly in Asiatic Russia.

Russian Culture in Global Context

Russian culture has enriched the world in many ways. Russian cultural norms have for centuries embodied both an inward orientation toward traditional forms of expression (and nationalism) and an outward orientation directed primarily toward western Europe. By the 19th century, even as Russian peasants interacted rarely with the outside world, Russian high culture had become thoroughly Westernized, and Russian composers, novelists, and playwrights gained considerable fame in Europe

▼ **Figure 9.28 New Moscow Mosque** Muslims bow in prayer in front of Moscow's new Cathedral Mosque, opened in 2015.

Following the Armenian Diaspora

The Armenian *diaspora* (the global scattering of a people) has been unfolding for centuries. Remarkably, Armenia's population of about 3 million is now dwarfed by perhaps 6 or 7 million people of Armenian descent who live in far-flung communities around the world (**Figure 9.4.1**).

Armenians—both native and globalized—share a powerful cultural identity rooted in four characteristics. First, the *Armenian language*

(and its special 39-letter alphabet) is a small but distinctive branch of the Indo-European family that has survived in one of the most linguistically complex parts of the world. Armenians living abroad continue to strongly identify with their linguistic roots. Second, the *Armenian Apostolic Church*, broadly a part of Eastern Orthodox Christianity, has retained its special character since the early 4th century, making it one of the world's oldest established churches. Third, Armenians worldwide

▼ **Figure 9.4.1. The Armenian Diaspora** The map shows the major countries that now host people of Armenian descent. Russia, the United States, and France are home to the largest numbers of Armenians.

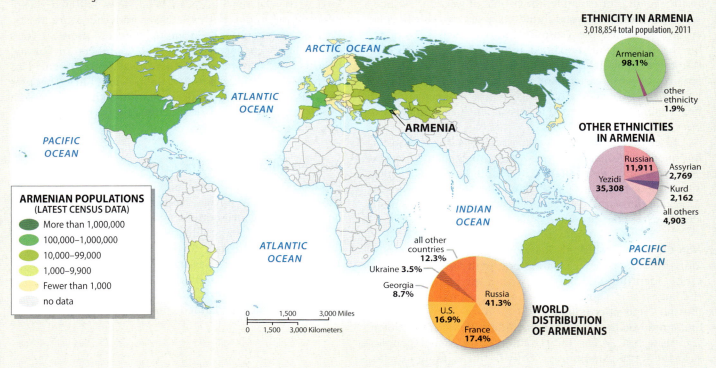

and the United States. Composers such as Tchaikovsky, Rachmaninoff, and Stravinsky stand as hugely important figures in the world of classical music. In literature, high school and college students around the world still bury themselves in the enduring stories told by legendary writers such as Leo Tolstoy (*War and Peace, Anna Karenina*) and Fyodor Dostoyevsky (*Crime and Punishment, The Brothers Karamazov*). Even in Russia today, there is great reverence and appreciation for the arts (the Bolshoi Ballet remains one of the world's leading dance companies) and for the ornate, baroque-style buildings that adorn older city centers such as St. Petersburg (see Figure 9.16).

Soviet Days During the Soviet period, a new mixture of cultural relationships unfolded within the socialist state. Initially, European-style modern art flourished in the Soviet Union, encouraged by the radical ideas of the new rulers.

By the late 1920s, however, Soviet leaders turned against modernism, which they viewed as the decadent expression of a declining capitalist world. Many Soviet artists fled to the West, and others were exiled to Siberian labor camps. Increasingly, state-sponsored Soviet artistic productions centered on **socialist realism**, a style devoted to the realistic depiction of workers harnessing the forces of nature or struggling against capitalism. Still, traditional high arts, such as classical music and ballet, continued to receive lavish state subsidies, and to this day, Russian artists regularly achieve worldwide fame.

Turn to the West By the 1980s, it was clear that the attempt to fashion a new Soviet culture based on communist ideals had failed. The younger generation was more inspired by fashion and rock music from the West, and American mass-consumer culture proved particularly attractive.

share a deep and enduring place identity with the ancient *Armenian homeland*, a sprawling area that includes much of eastern Turkey, Iran, modern-day Armenia, and portions of adjacent countries. Finally, the drama of the *Armenian diaspora* itself, historically rooted in violence, oppression, and genocide within the homeland, forms a set of stories, family sagas, and shared memories that bind together this globally dispersed nation.

Armenian life has been violently disrupted multiple times, no surprise given the homeland's pivotal setting in the midst of numerous powerful empire states. Still, the defining chapters that shaped the contemporary diaspora came in the 20th century. Most notorious was a mass extermination and expulsion of Armenians (still disputed by the Turks) from what is today eastern Turkey as the Ottoman Empire disintegrated. Between 1915 and 1922, about 1.5 million Armenians died in war or starved to death, and many others ended up in Russia, the Middle East, Europe, North America, and beyond. A secondary exodus accelerated in several waves during the Soviet period (Soviet Armenia was formed in 1922). Armenians bound for the United States came after World War II, followed by another surge after 1965 with the passage of the Immigration and Nationality Act. These immigrants include people from the Soviet Union as well as Armenians flowing in from Europe and the Middle East.

The Armenian diaspora has left a truly global imprint, both in numbers and in the distinctive communities and individuals that make up this dispersed nation. Russia (2.5 million), the United States (1.5 million), and France (500,000) are home to the largest Armenian populations outside the country, but many other states including Australia (50,000), Brazil (35,000), and The Netherlands (12,000) contain well-established Armenian communities. War-torn Syria—once home to a large Armenian population—has seen a recent exodus, with some migrants returning to their ancient homeland.

Many Armenian communities in the United States are found in northeastern states such as New York and Massachusetts, but Southern California (Glendale, a suburb of Los Angeles, is 27 percent Armenian) has attracted the largest number of migrants. Many West Coast Armenians are entrepreneurs, establishing small businesses in the dynamic California economy. Of course, some of the best-known celebrities of Armenian descent include the Kardashian family. Recently Kim

▲ **Figure 9.4.2. Kim Kardashian Visits Yerevan, Armenia** On a visit to Armenia in 2015, Kim Kardashian and her family pose for pictures with local residents.

Kardashian, making connections to her ancient homeland, has been a spokesperson for Armenian issues and identity and, not surprisingly, has caught the eye of the press and of the Armenian people in the process (**Figure 9.4.2**).

1. As an Armenian refugee, what might be the advantages of moving to nearby Georgia versus western Europe? What about disadvantages?

2. Do some detailed research on an Armenian community beyond the U.S. example. Where else do Armenian migrants live? What do they do for a living?

After the fall of the Soviet Union, global cultural influences grew, particularly in larger cities such as Moscow. Western books and magazines flooded shops, residents sought advice about home mortgages and condominium purchases, and they enjoyed the newfound pleasures of fake Chanel handbags and McDonald's hamburgers. People embraced the world that their former leaders had warned them about for generations. Cultural influences streaming into the country were not all Western in inspiration. Films from Hong Kong and Mumbai (Bombay), as well as televised romance novels (*telenovelas*) of Latin America, for example, proved even more popular in the Russian domain than in the United States.

Moscow's Changing Foodways As Russia's most globalized and cosmopolitan city, Moscow's foodways have experienced a profound transformation since the old days of the Soviet Union. Western-style fast food is popular among many urban residents: today the city is home to dozens of McDonalds restaurants (more than 450 are in Russia overall). On the high end of the food chain, some of the world's most elegant and expensive restaurants also can be found in the Russian capital, remaining busy even in tough economic times. Chefs at Ginza Project travel the world in search of tasty recipes, blending Russian, Italian, French, and Asian flavors for high-end customers. Private police will guard your Ferrari when you dine at Restaurant Mario, where fresh ingredients from Italy are flown in twice a week. Ordinary diets have changed as well, thanks to the city's large, mostly Central Asian immigrant population. Both Russians and non-Russians, for example, can dash into a trendy neighborhood Uzbeki-style eatery and enjoy a piece of *lepioshka* (the Russian name for Uzbek bread baked in clay ovens) or perhaps a bowl of *osh*, a savory mix of rice, meat, vegetables, and spices (**Figure 9.29**).

▲ **Figure 9.29 Osh** This savory Central Asian dish—a mix of rice, meat, vegetables, and spices—is an increasingly common sight in Moscow as immigrants from Uzbekistan, Tajikistan, and Kyrgyzstan arrive in large numbers.

Explore the **Tastes** of Uzbek Food

http://goo.gl/3ltQhc

The Music Scene Younger residents have embraced popular music, and their enthusiasm for American and European performers, as well as their support of a budding home-grown music industry, symbolizes the changing values of an increasingly post-Soviet generation. Today, Russian MTV reaches most of the nation's younger viewers. Major global media companies such as Sony Music Entertainment established Russian operations, and Universal also has opened the way for Russian performers to go global.

Moscow has hosted the Eurovision awards, a regional competition highlighting new musical talent from varied European countries. In 2016, the latest generation of regional musicians made major appearances at the 61st annual Eurovision Song Contest in Stockholm, Sweden. Armenia's representative, Iveta Mukuchyan, was born in Yerevan when the Armenian capital was still part of the Soviet Union (**Figure 9.30**). After the fall of the Soviet Union, her family moved to Germany in 1992, where Iveta grew up before returning to Yerevan to study jazz-vocal music in 2009. Her song for the awards, entitled "LoveWave," was cited for its broad global appeal and the associated video became widely popular and was seen by millions.

Explore the **Sounds** of Iveta Mukuchyan's Eurovision Hit "LoveWave"

http://goo.gl/fx5bRO

A Revival of Russian Nationalism

Even as the Russian domain finds itself increasingly swept up in external cultural influences, countertrends that emphasize Russian nationalism persist. Extremely conservative nationalists resist foreign influences such as Western-style music, and they have often been supported by more traditional elements of the Russian Orthodox Church. Even President Putin and the Russian government have carefully resurrected numerous symbols from the Russian past in order to cultivate a renewed sense of Russian identity within the region. Recently, Russian leaders proposed a return to the tsarist-era coat of arms (the double-headed eagle) to symbolize the ongoing connections between Russian society today and the earlier culture of Dostoyevsky and Tolstoy. The glory of Soviet communist days (1917–1991) also has been revived, with a strong Russian tone. The Soviet-era national anthem was reintroduced (with new lyrics) in 2001, and several Soviet holidays, celebrations, and national awards have been reinstated. Recent Russian textbooks have even resurrected the virtues of Joseph Stalin. Once discredited as a brutal, cruel leader, Stalin is now seen as an "effective manager" of Soviet affairs in an earlier era.

Sports and National Identity The Soviet era also revealed a close and powerful connection between popular sports and Soviet nationalism. Russia's famed love affair with chess, kindled during tsarist times, received renewed attention during the Soviet period. Lenin's army commander opened chess schools and promoted the game after 1921. Stalin played often, believing that the game embodied the Soviet knack for intellect and strategy. Indeed, the board game's popularity outlived the Soviet era, and 21st-century Russians still embrace the game.

Both the summer and the winter Olympic Games have also been signature events to display Soviet superiority

▲ **Figure 9.30 Iveta Mukuchyan** Armenia's talented Iveta Mukuchyan performs her best-selling single, *LoveWave*, at the Eurovision Song Contest held in Stockholm, Sweden, in 2016.

over their American and western European rivals. Indeed, from figure skating to gymnastics, Soviet-era athletes often dominated particular events for years. American fans were occasionally on the winning side as well: The miracle victory over the Soviet team in ice hockey at the 1980 Winter Olympics (in Lake Placid, New York) remains one of the most memorable moments in modern global sports history.

In the post-Soviet era, popular sports remain hugely important throughout the region, and serve as signatures of national pride and identity. Whether it is Moldovan Trânta (a form of wrestling), Ukrainian soccer, or Russian ice hockey, enthusiasm for winning teams and leading players is equal to anything an American basketball or football league could hope for. One of the reasons the Russians (including President Putin) prepared so carefully for the 2014 Winter Olympics in Sochi (in southern Russia) was to demonstrate the country's ability to once again host that important global sports event. More recently, however, Russia's sports reputation on the global stage has suffered. Dozens of Russian athletes were banned from many global competitions in 2016 after the World Anti-Doping Agency outlawed the use of meldonium, a substance that boosts blood flow and enhances performance.

REVIEW

9.6 What were the key phases of colonial expansion during the rise of the Russian Empire, and how did each enlarge the reach of the Russian state?

9.7 Identify some of the key ethnic minority groups (as defined by language and religion) within Russia and neighboring states.

■ **KEY TERMS** Slavic peoples, Eastern Orthodox Christianity, Cossacks, socialist realism

Geopolitical Framework: Growing Instability Across the Region

The geopolitical legacy of the former Soviet Union still weighs on the Russian domain. After all, the bold lettering of the "Union of Soviet Socialist Republics" dominated the Eurasian map for much of the 20th century, and the country's political influence affected every part of the world. Under President Putin, Russia's post-2000 political resurgence signals a return to its Soviet-era status of geopolitical dominance within the region (**Figure 9.31**). Recent events in Ukraine and in neighboring countries demonstrate Russian willingness to reassert itself on the regional stage, but sharp global reaction in the West suggests the country could isolate itself from mainstream global political initiatives and institutions. In response, Russia may pivot to the east, forging closer relations with China.

▲ **Figure 9.31 Vladimir Putin** Since 2000, Vladimir Putin's influence across the Russian domain has been immense. Within Russia, Putin (as both president and prime minister) has managed an impressive economic recovery, while limiting civil liberties. Beyond Russia, Putin has reestablished the region's geopolitical presence on the global stage.

Geopolitical Structure of the Former Soviet Union

The Soviet Union rose from the ashes of the Russian Empire, which collapsed abruptly in 1917. The Russian tsars did little to modernize the country or improve the lives of the peasant population. After the tsars fell, a government representing several political groups assumed authority. Soon, however, the **Bolsheviks**, a faction of Russian communists representing industrial workers, seized power. They espoused the doctrine of **communism**, a belief that was based on the writings of Karl Marx and that promoted the overthrow of capitalism by the workers, the large-scale elimination of private property, state ownership and central planning of major sectors of the economy (both agricultural and industrial), and one-party authoritarian rule. The leader of these Russian communists was Vladimir Ilyich Ulyanov, usually known by his self-selected name, Lenin. Lenin became the architect of the Soviet Union, the state that replaced the Russian Empire. The new socialist state radically reconfigured Eurasian political and economic geography. When the Soviet Union emerged in 1917, Lenin and other communist leaders were aware that they faced a major challenge in organizing the new state.

Creating a Political Structure Soviet leaders designed a geopolitical solution that maintained their country's territorial boundaries and recognized, in theory, the rights of non-Russian citizens. Each major nationality received its own "union republic," provided it was situated on one of the nation's external borders (**Figure 9.32**). Eventually, 15 such republics were established, creating the Soviet Union. So-called **autonomous areas** within these republics gave special recognition to smaller ethnic homelands. Even with these

▲ **Figure 9.32 Soviet Geopolitical System** During the Soviet period, the boundaries of the country's 15 internal republics often reflected major ethnic divisions. Ultimately, however, many of the ethnically non-Russian republics pressured the Soviet government for more political power. As the Soviet Union disintegrated, the former republics became politically independent states and now form an uneasy ring of satellite nations around Russia.

internal republics and autonomous areas, the Soviet Union emerged by the late 1920s as a highly centralized state, with important decisions made in the Russian capital of Moscow.

The chief architect of this political consolidation was Joseph Stalin, who did everything he could to centralize power and assert Russian authority. The Stalin period (1922–1953) also saw the enlargement of the Soviet Union. Victorious in World War II, the country acquired Pacific islands from Japan, the Baltic republics (Estonia, Latvia, and Lithuania), and portions of eastern Europe.

After World War II, the Soviet Union expanded its influence across eastern Europe. In the words of British leader Winston Churchill, the Soviets extended an **Iron Curtain** between their eastern European allies and the more democratic nations of western Europe. As eastern Europe retreated behind the Iron Curtain, the Soviet Union and the United States became antagonists in a global **Cold War** of military competition that lasted from 1948 to 1991.

End of the Soviet System Ironically, Lenin's system of republics based on cultural differences sowed the seeds of the Soviet Union's demise. Even though the republics were never allowed real freedom, they provided a political framework that encouraged the survival of distinct cultural identities. Contrary to the expectations of Soviet leaders, ethnic nationalism intensified in the post–World War II era as the

Soviet system grew less repressive. When Soviet President Mikhail Gorbachev initiated his policy of **glasnost**, or greater openness, during the 1980s, several republics—most notably the Baltic states of Lithuania, Latvia, and Estonia—demanded independence. In addition, Gorbachev's policy of **perestroika**, or restructuring of the planned centralized economy, was an admission that domestic economic conditions increasingly lagged behind those of western Europe and the United States. A failed war in Afghanistan and increasing political protests in eastern Europe added to Gorbachev's problems.

By 1991, Gorbachev saw his authority slip away amid rising pressures for political decentralization and economic reforms. During that summer, Gorbachev's regime was further endangered by the popular election of reform-minded Boris Yeltsin as head of the Russian Republic. By late December, all of the country's 15 constituent republics had become independent states, and the Soviet Union ceased to exist.

Current Geopolitical Setting

The political geography of post-Soviet Russia and the nearby independent republics has changed dramatically since the collapse of the Soviet Union in 1991 (**Figure 9.33**). All of the former republics have struggled to establish stable political relations with their neighbors.

Moldova. *Recent financial turmoil has put more political pressure on officials to improve the economy. Russian troops remain within a rebel Slavic region (Transdniester) in the eastern part of the region.*

Fractured Ukraine. *Russia's illegal occupation of Crimea and the presence of Russian rebels in eastern Ukraine have fragmented this southern Slavic nation since 2014.*

Nagorno-Karabakh. *In 2016, troubled Nagorno-Karabakh, occupied by Armenia, but claimed by Azerbaijan, erupted in violence once again.*

Chechnya. *Relative stability has allowed for the rebuilding of the Chechen capital of Grozny, but Chechen leader Ramzan Kadyrov, an ally of Vladimir Putin, continues to foster anti-Western sentiment in the region.*

Kuril Islands. *The Russians and Japanese have yet to resolve their dispute over the southernmost Kuril Islands. Japan demands return of the islands, which were seized by the Russians at the end of World War II.*

Members of the Commonwealth of Independent States (CIS)
Internal Republics of the Russian Federation

▲ **Figure 9.33 Geopolitical Issues in the Russian Domain** The Russian Federation Treaty of 1992 created a new internal political framework that acknowledged many of the country's ethnic minorities. Recently, however, Russian authorities have moved to centralize power and limit regional dissent. Russia's relations with several nearby states remain strained. **Q: Cite some similarities between Russia's internal republics shown here and the regional map of languages (Figure 9.23).**

Russia and the Former Soviet Republics For a time, it seemed that a looser political union of most of the former republics, called the *Commonwealth of Independent States (CIS)*, would emerge from the ruins of the Soviet Union. But the CIS faded in importance. More recently, Russia has backed the growth of the **Eurasian Economic Union (EEU)**, a customs union (parallelling the European Union [EU]) designed to encourage trade as well as closer political ties between member states. Formed in 2015, the EEU contains five member states (Russia, Belarus, Kazakhstan, Armenia, and Kyrgyzstan). Some observers argue Russia will use the organization to reconstruct in more formal ways many elements of a larger "Soviet-style" empire that extends from the Baltic to Central Asia.

The Crisis in Ukraine Since 2014, the region's geopolitical map has been dominated by the crisis in Ukraine (**Figure 9.34**). Why did Ukraine unravel? Three reasons help explain its plight. First, since its emergence as an independent country in the 1990s, Ukraine has been *a divided, politically unstable state,* torn between its long-time political and economic connections with Russia (and the former Soviet Union) and a burgeoning desire to drift westward into a stronger relationship with NATO and the EU. In 2010, Viktor Yanukovich, who had close ties to Russia and President Putin, came to power. When Yanukovich refused to consider a closer relationship with the European Union in 2013, protestors forced him from power the following year. Subsequent elections brought President Petro Poroshenko into office; he advocated a turn away from too much Russian influence.

Second, the country's longstanding *regional cultural divide* between its Ukrainian and Russian populations, broadly defined (see Figure 9.24), have often reinforced and reflected these political differences, with ethnic Russians in Crimea and eastern Ukraine favoring closer ties with Moscow. Many residents in these settings grew increasingly dissatisfied as events in Kiev seemed to signal a turn to the West.

Third, *Russia's expansive geopolitical ambitions* in the region, reflected by Putin's bellicose response to the

depopulated as residents have fled the violence (see Figure 9.34). Several cease-fire agreements have failed to fully end the conflict and it appears likely that eastern Ukraine will remain a zone of political instability and conflict for the foreseeable future.

Other Geopolitical Hotspots In Belarus, leaders have been slow to open political and economic opportunities in western and central Europe. The country remains firmly within Russia's political orbit. In 2010, the two countries pledged to move toward a "union state" and expanded their common military maneuvers. In 2012, Belorussian President Alexander Lukashenko (often called Europe's last dictator) adopted one of the world's most restrictive policies on free use of the Internet within the nation, strongly discouraging the use of any "foreign" websites. Many political dissidents inside the country remain imprisoned.

Tiny Moldova has also witnessed political tensions in the post-Soviet era. Conflict has repeatedly flared in the Transdniester region in the eastern part of the country, where Russian troops remain and where Slavic separatists have pushed for independence from a central government dominated by Romanian-speaking Moldovans. Complicating matters, the country witnessed a massive $1 billion theft from its leading banks in 2014, forcing a taxpayer bailout, the threat of economic collapse, and more political instability.

Transcaucasia also remains unstable. Since 2003, Georgia's government has moved toward closer ties with the West. In 2008, Russia invaded the country when Georgia attempted to reassert its control over Abkhazia and South Ossetia, two breakaway regions dominated by Russian sympathizers. Almost 1000 people died in the conflict, and more than 30,000 people were displaced from their homes. The situation remains tense today, with Georgia still claiming control over Abkhazia and South Ossetia but Russia declaring both microstates independent.

In nearby Armenia, the territories of the Christian Armenians and the Muslim Azeris interpenetrate each other in a complex fashion. The far southwestern portion of Azerbaijan (Naxicivan) is actually separated from the rest of the country by Armenia, while the important Armenian-speaking district of Nagorno-Karabakh is officially an autonomous portion of Azerbaijan. After Armenia successfully occupied much of Nagorno-Karabakh in 1994, fighting between the countries diminished, but renewed and deadly violence in 2016 (involving tanks and artillery fire) threatened to erupt into another full-scale war.

▲ **Figure 9.34** **Ukraine's Geopolitical Hot Spots** This map shows where Russian troops occupied the Crimean Peninsula in 2014. Disputed areas in eastern Ukraine, under the influence of Russian-backed rebel forces, are also shown.

Ukrainian situation, prompted Russia's illegal occupation of the Crimean peninsula in 2014 and its ongoing economic and military support (as many as 12,000 Russian troops are operating in the country) of rebel forces in eastern Ukraine who are advocating separation from Kiev, either to create independent states or to be a part of an expanded Russian union (see Figure 9.34).

What has happened? The Crimean Peninsula has been occupied by Russian forces since 2014, an illegal action not recognized by most of the world. The Russian government gains improved access to the Black Sea and its resources, and because Crimea is mostly populated by ethnic Russians (who largely voted to accept Russian annexation), the move away from Ukrainian control has met with approval from most residents. There are frustrations, however. Many Russians in the region have suffered from a Ukrainian trade blockade, rolling electricity blackouts, and a lack of economic support from Moscow. In addition, the region's Tatar minority has seen increased discrimination by the Russians, and thousands have abandoned their homes and left the region.

The situation in eastern Ukraine has been much worse. A violent, economically catastrophic civil war erupted in the region between the mixed Ukrainian and Russian-speaking populations. By 2016, more than 6,000 people had died in the conflict and about 1.5 million residents, mostly in far eastern Ukraine, had been forced from their homes. Russian rebel-held zones in the east, especially in the disputed Luhansk and Donetsk areas, have been partially

Geopolitics Within Russia Within Russia, further pressures for devolution, or more localized political control, led to the signing of the Russian Federation Treaty in 1992. The treaty granted Russia's internal autonomous republics and its lesser administrative units greater political, economic, and cultural freedoms, including more control of their own natural resources and foreign trade (see the map of internal republics, Figure 9.33). Conversely, it weakened Moscow's centralized authority to collect taxes and to shape policies within its varied hinterlands. Defined essentially along ethnic lines, 21 regions possess status as republics within the federation and now have constitutions that often run counter to national mandates.

Since 2000, Russian leaders, especially Vladimir Putin, have pushed for more centralized control in the country. Putin's prominence has been enduring: He served two terms as president (2000–2008) and one term as prime minister (2008–2012); then he was reelected for a six-year term as president in March 2012. Putin, a former internal security agent with the KGB during the Soviet period, has consolidated power in the country, pushed for strong economic growth oriented around Russia's energy economy, and reasserted Russia's political and military role, both on the world stage and as a dominant regional power. Putin's occupation of Crimea in 2014 proved widely popular in Russia, boosting his domestic popularity even as the country slipped into an economic recession.

Russian Challenge to Civil Liberties Still, public protests since 2009 have periodically challenged Putin's authority. After his 2012 election, thousands of protesters marched in the streets of Moscow and St. Petersburg, and hundreds were arrested. Disgruntled members of the urban middle class (Putin received less than 50 percent of the vote in Moscow), human rights groups, and opposition political parties resent Putin's grip on power. Protestors demand a freer press, more democracy, more open elections, and a broader commitment to economic growth. They have criticized Putin's strong-arm leadership style. They also note Putin's increasingly close ties to the **siloviki**, members of the nation's military and security forces.

Protests against the Russian central government are in part a response to the government's crackdown on civil liberties. Immediately after the fall of the Soviet Union, Russia enjoyed a genuine flowering of democratic freedoms. A multiparty political system, independent media, and a growing array of locally and regionally elected political officials signaled real change from the authoritarian legacy of the Soviet period.

Since 2002, however, many hard-won civil liberties have slipped away, victims of Putin's campaign to consolidate political power, increase the authority of the central government, limit press freedoms, and silence critics who disagreed with his policies. For example, the Russian president now has more direct control over nominating candidates for dozens of Russian governorships and mayoral positions. Putin has also forged a close political relationship with conservative Islamist strongman Ramzan Kadyrov in the Chechen Republic (a part of southern Russia), pouring aid into that once war-torn region to rebuild the capital of Grozny (**Figure 9.35**). In return, Kadyrov has endorsed Putin's more conservative, authoritarian policies.

Many of Russia's media outlets have also lost their autonomy and are now under more direct government ownership and influence. Outspoken journalists critical of the government (such as Anna Politkovskaya) have died under suspicious circumstances or been murdered. Just as disturbing, one of Putin's leading political critics, Boris Nemtsov, was gunned down in Moscow near Red Square in early 2015, sparking more protests across the country. Recently, Russian officials have also increased their surveillance and regulation of the Internet, another move to silence opposition.

In 2012, Pussy Riot, a Russian female punk rock band, made global headlines with its performance of a song critical of President Putin in Moscow's Christ Savior Cathedral, one of Russia's largest houses of worship. Band members—who were decked out in fluorescent stockings as they sang "Mother of God, blessed Virgin, drive out Putin!"—were accused of "hooliganism" and jailed. Although the women claimed their actions were a political protest, they were given multiple-year sentences, igniting a flurry of public reactions against the government. One of the singers

▶ **Figure 9.35 Grozny in Russian Chechnya** Large government-managed investments have helped rebuild the war-torn Chechen capital, even though anti-Russian sentiment still runs high in the countryside.

▲ **Figure 9.36 Pussy Riot's Nadia Tolokonnikova** This Russian punk rock star went to jail for hooliganism when she appeared in a prominent Russian Orthodox Church and sang a song protesting President Putin's rule. Recently, she has taken up the cause of prison reform in the country.

strongly opposed the addition of the Baltic republics (Estonia, Latvia, and Lithuania) to the increasingly powerful organization. Kaliningrad is a related sore point and has become a focus of east–west tensions in the region. After World War II, the former Soviet Union claimed this small, but strategic territory on the Baltic Sea. It was the northern portion of East Prussia (now the port of Kaliningrad), previously part of Germany. It still forms a small Russian **exclave**, a portion of a country's territory that lies outside its contiguous land area. Kaliningrad (population 950,000) remains a part of Russia, but since 2004 the area has been surrounded by NATO and EU nations (Poland and Lithuania), much to Russia's frustration (**Figure 9.37**). Recently, Russia has increasingly militarized the strategic exclave, building up bases and weapons in the area. Talks in 2016 between NATO diplomats and Russian officials failed to

(Nadia Tolokonnikova), recently released from prison, has led the cause for prison reform in Russia (**Figure 9.36**). In 2016, the group produced a satirical video condemning Russia's criminal justice system, suggesting that the government selectively prosecuted individuals for political reasons and routinely engaged in torture and in the mistreatment of prisoners. The country has the second highest incarceration rate in the world (after the United States) and many prisoners are mistreated and deprived of basic social services and health care.

Explore the **Sights** of Port of Kaliningrad

http://goo.gl/5UMKsf

▼ **Figure 9.37 Kaliningrad** A Russian exclave on the Baltic Sea, Kaliningrad is now surrounded by Poland and Lithuania, both EU- and NATO-member states.

The Shifting Global Setting

Since 1991, complicated regional political relationships challenge the Russians, both in the east and the west. In East Asia, Russia is working to build a closer political relationship with China to counter faltering ties to Europe. Russia has also played an important role in containing North Korea's nuclear ambitions in the region, pressuring its Far East neighbor to limit uranium enrichment and weapons development projects. Territorial disagreements—specifically, a dispute over the Kuril Islands—continue to challenge Russia's relationship with Japan.

To the west, Russia worries about the expansion of NATO. Although most Russian leaders accepted the inevitable inclusion of Poland, Hungary, and the Czech Republic in NATO, they diffuse the growing tensions between the two sides. Many Kaliningrad residents are nervous about becoming part of a Russian fortress on the nation's far western flanks, especially given recent events in Ukraine. They also are suffering economically with reduced European tourism, a weak currency, and international sanctions.

Globally, President Putin is attempting to reassert his nation's political status. Russia retains a permanent seat on the UN Security Council. The country's nuclear arsenal, while reduced in size, remains a powerful counterpoint to American and western European interests. Russia often acts as a counterweight to the United States in international maneuverings, and Putin has taken an increasingly anti-American slant in his foreign policy, a political strategy that has proven quite popular with Russians. Indeed, the long geopolitical history of the region suggests that Russia's recent reemergence as a more powerful, centralized state is a sign of things to come. Russia's recent involvement in the Syrian crisis (to support its ally, President Assad) demonstrates Putin's willingness to use force to reinforce his rhetoric. One thing is clear: Putin's global muscle flexing seems to play well at home where he enjoys broad support for reasserting Russia's presence on the world stage.

REVIEW

9.8 How do current geopolitical conflicts reflect long-standing cultural differences within the region?

9.9 Describe how Vladimir Putin has played a key role in consolidating Russia's power since 2000, both within the country and beyond.

■ **KEY TERMS** Bolsheviks, communism, autonomous areas, Iron Curtain, Cold War, glasnost, perestroika, Eurasian Economic Union (EEU), siloviki, exclave

Economic and Social Development: Coping with Growing Regional Challenges

The economic future of the Russian domain remains difficult to predict (see Table 9.2). With the dissolution of the Soviet Union and its state-controlled economy in the early 1990s, the region has experienced incredible economic changes in the past 25–30 years. Recent political instability has only added to the region's economic woes. In the late 2010s, much of the region faced enormous economic headwinds and a steep recession.

The Legacy of the Soviet Economy

Much of the Russian domain's present economic infrastructure was established during the communist era, including new urban centers, industrial developments, and the modern network of transportation and communication linkages. As communist leaders such as Stalin consolidated power in the 1920s and 1930s, they nationalized Soviet industries and agriculture, creating a system of **centralized economic planning**, in which the state controlled production targets and industrial output. The Soviets emphasized heavy basic industries (steel, machinery, chemicals, and electricity generation) rather than consumer goods. By the late 1920s, Stalin shifted agricultural land into large collectives and state-controlled farms.

Much of the region's basic infrastructure—its roads, rail lines, canals, dams, and communications networks—originated during the Soviet period (**Figure 9.38**). Dam and canal construction turned many rivers into a virtual network of interconnected reservoirs. The Volga–Don Canal (completed in 1952) connected those two river systems and greatly

TABLE 9.2   Development Indicators

Country	GNI per capita, PPP 2014	GDP Average Annual % Growth 2009–15	Human Development Index (2015)[1]	Percent Population Living Below $3.10 a Day	Life Expectancy (2016)[2] Male	Life Expectancy (2016)[2] Female	Under Age 5 Mortality Rate (1990)	Under Age 5 Mortality Rate (2015)	Youth Literacy (% pop ages 15–24) (2005–2014)	Gender Inequality Index (2015)[3,1]
Armenia	8,450	4.5	0.733	17.0	72	78	47	14	100	0.318
Belarus	17,610	3.3	0.798	< 0.5	68	78	17	5	100	0.151
Georgia	9,080	5.6	0.754	28.6	71	79	47	12	100	0.382
Moldova	5,500	5	0.693	< 2.0	68	76	35	16	100	0.248
Russia	22,160	2.9	0.798	< 2.0	66	77	27	10	100	0.276
Ukraine	8,560	0.8	0.747	< 2.0	66	76	19	9	100	0.286

Source: World Bank, *World Development Indicators*, 2016.

[1]United Nations, *Human Development Report*, 2015.

[2]Population Reference Bureau, *World Population Data Sheet*, 2016.

[3]Gender Inequality Index—A composite measure reflecting inequality in achievements between women and men in three dimensions: reproductive health, empowerment, and the labor market that ranges between 0 and 1. The higher the number, the greater the inequality.

1) Visualize Under Age 5 Mortality Rates since 1990, using MapMaster. Which countries show the most and least improvement in reducing child mortality, and why?

2) Examine the maps for GNI per Capita and Life Expectancy. Combined with Under Age 5 Mortality Rates, what do the maps suggest about levels of development in the region? Do your findings parallel or differ from the HDI (Human Development Index) data in the table?

▲ **Figure 9.38 Major Natural Resources and Industrial Zones** The region's varied natural resources and chief industrial zones are widely distributed. In southern Siberia, rail corridors offer access to many mineral resources. In the mineral-rich Urals and eastern Ukraine, proximity to natural resources sparked industrial expansion, while Moscow's industrial power is related to its proximity to markets and capital. **Q: Looking at the map, why might it be argued that Russia's size is both a blessing and a curse?**

eased the movement of raw materials and manufactured goods (**Figure 9.39**). The Trans-Siberian line was modernized and complemented by the addition of the BAM link across central Siberia. Farther north, the Siberian Gas Pipeline was built to link the Arctic's energy-rich gas fields with growing demand in Europe. Overall, the postwar period produced real economic and social improvements for the Soviet people.

Despite the successes, problems increased during the 1970s and 1980s. Soviet agriculture remained inefficient. Manufacturing quality failed to match Western standards. Equally troubling, the Soviet Union failed to participate fully in the technological revolutions transforming the United States, Europe, and Japan. Disparities also grew between the Soviet elite and ordinary people who enjoyed few personal freedoms. By the late 1980s, the Soviet Union had reached both an economic and a political impasse.

The Post-Soviet Economy

Fundamental economic changes have shaped the Russian domain since 1991. Particularly within Russia itself, much of the highly centralized state-controlled economy has been replaced by a mixed economy of state-run operations and private enterprise. The collapse of the communist state also meant that economic relationships between the former Soviet republics were no longer controlled by a single, centralized government.

Redefining Regional Economic Ties Since the breakup of the Soviet Union, Russia has worked to maintain many economic ties with other former Soviet republics. The recent expansion of the EEU (Eurasian Economic Union) in 2015 is designed to counterbalance the growth of the European Union, reduce trade barriers within the region, and encourage more economic cooperation among its five member states. Still, the potential for such an organization seems modest, especially if it fails to build ties to the European Union.

Ukraine has followed a different trajectory, turning to the West for more economic aid and trading opportunities. In 2014 it secured a 5-year, $22 billion lending guarantee from the International Monetary Fund in order to stabilize its economy. Some optimists also think a longer-term, more formal relationship with the European Union is possible. Georgia has followed a similar path, minimizing its ties to Russia and actively exploring membership in the EU.

Privatization and State Control The post-Soviet era has brought a great deal of economic uncertainty. The Russian government initiated a massive program to transform its economy in 1993, opening some economic sectors to more private initiative and investment. Unfortunately, the lack of legal and financial safeguards invited abuses and often resulted in mismanagement and corruption in the new system.

▲ **Figure 9.39 Volga–Don Canal**   Built during the Soviet era, the Volga-Don canal remains a key commercial link that facilitates the economic integration of southern Russia. This view is near Volgograd.

Explore the **Sights** of Volga-Don Canal

http://goo.gl/0GdVGg

Almost 90 percent of Russia's farmland was privatized by 2003, with many farmers forming voluntary cooperatives to work the same acreage as under the Soviet system. Thousands of private retail establishments also appeared, and they now dominate that portion of the economy. In addition, the long-established "informal economy" continues to flourish. Even during the Soviet era, millions of citizens earned extra money by informally selling Western consumer goods, manufacturing food and vodka, and providing skilled services such as computer and automobile repairs. Today, these barter transactions and informal cash deals form a huge part of an economy never reported to government authorities.

The natural resource and heavy industrial sectors of the economy were initially privatized in Russia, but in recent years, under Putin's management, state-run enterprises took back more control of the nation's energy assets and infrastructure. Gazprom, the huge Russian natural gas company, was privatized in 1994, but since 2005 its activities have increasingly been controlled by the state.

Especially in Russia—and particularly in that country's cities—the successes of the new economy are increasingly visible on the landscape. Luxury malls, office buildings, and more fashionable housing subdivisions are now part of the urban scene as the middle class grows in settings such as Moscow. On the other hand, the gap has grown between increasing urban affluence and grinding rural poverty. In 2015, the *Wall Street Journal* reported that 110 people control 35 percent of Russia's wealth, while half the population had a total average household wealth of less than $875. Clearly, the country's **oligarchs**, a small and private group of politically well-connected businessmen who control (along with organized crime) important aspects of the Russian economy, have done very well in comparison to the average Russian citizen.

The Challenge of Corruption   Throughout the Russian domain, corruption also remains widespread. Doing business often means lining the pockets of government officials, company insiders, or trade union representatives. Organized crime remains pervasive in Russia. Many ties also remain between organized crime and Russian intelligence agencies. Much of the country's real wealth has been exported to foreign bank accounts. Various local and regional crime organizations divide up much of the economy and have links to illegal global business operations. Violence and gangland-style murders still unfold on the streets of Moscow, much to the embarrassment of government officials.

Regional Economic Challenges   The economies of the countries of the Russian domain continue to struggle. Both Moldova (still trying to recover from the $1 billion theft from its banks) and Ukraine (dealing with the enormous costs of its intermittent civil war) face special challenges that have thrown their economies into steep economic declines. Russia has not fared much better. Its energy-based export economy (the source of about half the government's overall revenue) has had to adjust to lower prices, sharply lower revenues, and a weakened currency. In addition, widespread economic sanctions and growing political tensions with the West over Ukraine have hurt trade and curtailed foreign investment in the region. The Russian economy shrank by 4 percent in 2015 and its weakened currency meant that real wages fell by more than 9 percent. A recent survey reported that the share of Russian families that periodically lack funds for food or clothing has risen to almost 40 percent of the population. Still, the country has substantial foreign reserves and the Central Bank of Russia has worked carefully with the nation's private banks to ensure they survive these tough economic times.

The Russian Domain in the Global Economy

The relationship between the Russian domain and the world beyond has shifted greatly since the end of communism. During the Soviet era, the region was relatively isolated

Russia's Global High-Tech Ambitions

The Russian Venture Company (RVC) is hardly a household name to most Americans, but the ambitious goal of this government-supported development institute and venture capital initiative, created in 2006, is to integrate Russian technologies with the worldwide marketplace. The company's CEO, an applied mathematician named Igor Agamirzyan, argues that Russia's economy is too dependent on natural resources (especially oil and gas) (**Figure 9.5.1**). Between 2015 and 2035, Agamirzyan and the RVC hope to vastly increase Russian innovation and participation in technology companies, making products and services in demand globally.

The initiative builds on a Soviet-era legacy of technological accomplishment. From fashioning modern industrial operations to pioneering space travel, Russian scientists and engineers have played innovative roles in the high-tech world. The RVC wants to provide seed capital for varied tech-related exports. In 2014, the development institute evaluated over 1,800 projects, providing funding, product development ideas, and expert mentoring designed to create globally competitive products. The RVC sees great potential in supplying information technology (IT) support, in developing biotech innovations in the health-care industry, and in adding competitive products in nuclear- and radiation-related technologies. The goal is to grow today's US$4 billion in annual tech exports to more than $11 billion by 2020. Working in today's challenging political environment won't make achieving that goal easy, but Russia's innovators and entrepreneurs clearly are targeting a huge competitive marketplace that could fundamentally diversify the country's participation in the global economy.

1. As an innovator in Russia's high-tech world, what are the three greatest challenges that you might face as you prepare to take your product/service to the global marketplace?

2. Select a technology-related product or service you use every day and briefly trace its geographical origins.

▲ **Figure 9.5.1. Igor Agamirzyan** CEO of the Russian Venture Company, Agamirzyan is a major advocate for the global spread of Russian technology.

from the world economic system. But links with the global economy multiplied after the downfall of the Soviet Union. Recent economic sanctions against Russia, however, have thrown the path of its future global economic connections into doubt.

More Globalized Consumers Most visibly, since the fall of the Soviet Union, a barrage of consumer imports has transformed the lives of regional residents. All of the symbols of global capitalism are visible in the heart of Moscow and, increasingly, in many other settings throughout the region. Luxury goods from the West have also found a small, but enthusiastic and highly visible market among the Russian elite, a group noted for its devotion to BMW automobiles, Rolex watches, and other status emblems. It is also a two-way street: for example, the region's video game developers have had a large impact on these industries around the world. From *Age of Pirates* to *World of Tanks*, global video game enthusiasts are entertained with innovations from Russian, Ukrainian, and Belarusian software engineers (see *Everyday Globalization: Russia's Global High Tech Ambitions*).

Changing Flows of Foreign Investment During the post-Soviet era, foreign investment has ebbed and flowed with the region's political stability. After President Putin took office in 2000, he successfully encouraged growing foreign investment, especially from the United States, Japan, and western Europe. For many years, the success of Russia's equity markets and the relative stability of its financial sector were encouraging. Since 2014, however, a sliding currency in Russia (the ruble), war in eastern Ukraine, and a growing list of economic sanctions have dramatically slowed foreign investment. Foreign direct investment into Russia peaked at $40 billion early in 2013 and had fallen to an annual rate of $3 billion by late 2015. Sanctions related to Ukraine included freezing overseas Russian bank assets and forbidding the sale of many European and American goods to Russia (especially in energy- and technology-related industries). Capital flight from Russia also dramatically increased in 2014 as many Russian investors feared more instability ahead. As a response to sanctions, Russia banned many imported goods from the United States and Europe, including many food and consumer items. Falling currencies in the region have also made the purchase of any dollar- or euro-denominated items more expensive, further dampening consumer demand and stoking inflation.

One relatively bright spot has been Chinese investment in the Russian Far East. For example, Vladivostok's new Tigre de Cristal casino, financed with capital from Macao, features Russian card dealers and Chinese gamblers (**Figure 9.40**). Many smaller-scale Chinese entrepreneurs have also located businesses in the region, catering both to resident Russian populations as well as growing numbers of Chinese immigrants.

Globalization and Russia's Petroleum Economy Russia's enormous oil and gas industry has not escaped tougher economic times. For much of the post-Soviet era, the energy economy was a real boon to the region. The statistics are impressive: Russia's energy production makes up more than one-quarter of its economic output and two-thirds of its exports. Russia has 26 percent of the world's natural gas reserves (mostly in Siberia) and is the world's largest gas exporter. As for oil, Russia possesses more than 75 billion barrels of proven reserves and has large producing fields in Siberia, the Volga Valley, the Far East, and the Caspian Sea regions. Over the past two decades, the primary destination for Russian petroleum products has overwhelmingly shifted to western Europe.

The Siberian Gas Pipeline already connects distant Siberian fields with western Europe via Ukraine. Those connections are supplemented by lines through Belarus (the Yamal–Europe Pipeline) and Turkey (the Blue Stream Pipeline). Underwater pipelines beneath the Baltic Sea (Nord Stream—completed in 2012) deliver gas to northern Europe. To the south, a large petroleum export terminal opened at Novorossiysk (on the Black Sea) in 2001, delivering Caspian Sea oil supplies to the world market via a pipeline passing through troubled Chechnya. Nearby, oil pipelines between Baku (on the Caspian Sea) and the Black and Mediterranean seas cross Azerbaijan and Georgia.

But Russia's energy orientation to the west may be changing: Russia and China have also committed to greatly expanding their energy connections, forging major deals in 2014 that are designed to build new pipeline links (with the help of Chinese capital) between the two nations and ship large quantities of energy (mostly in the form of natural gas) from Russian Siberia to the Chinese heartland (**Figure 9.41**). The 30-year deal may be worth more than $400 billion for the Russian government and a consortium of private energy companies. The two countries will also explore offshore oil prospects in the Barents Sea and jointly develop Russian coal deposits for export to China. It is a logical marriage for one of the world's largest energy producers and the world's largest energy consumers. Still, uncertainties remain. Lower energy prices and slower growth in China could delay full-scale construction plans, or the volume of natural gas shipments might be lower than initially planned.

Enduring Social Challenges

The region's recent economic downturn has had profound negative social consequences. Less money is available for investments into social infrastructure. Moldova's economic crisis has precipitated steep cutbacks in many social programs. Ukraine is depending on a growing flow of foreign aid and international loans to maintain social programs at reduced levels. In the case of Russia, President Putin's huge new investments in the military (he plans to spend $600 billion between 2016 and 2020 to upgrade the armed forces and modernize weapons systems) will likely mean lower levels of spending for other portions of the economy.

Problems of Health Care and Alcoholism Health care remains a major social problem within the Russian domain. Health-care expenditures are only a fraction of what they were during the Soviet period. While average per capita health care expenditures are more than $9000 in the United States, they are only about $300 per year in Ukraine and around $1000 in Russia. Mortality rates for Russian men are especially grim. One in three Russian men dies before retirement. Cardiovascular disease, often related to high-fat diets and physical inactivity, is a key contributor to these elevated death rates. Smoking also remains widely popular (54 percent of physicians smoke). HIV-AIDS is also a major problem. More than 700,000 Russians live with the disease, and Russia faces the fastest growing HIV-AIDS epidemic in Europe.

Alcohol use across the region remains far above the global annual average of about 6 liters (1.6 gallons). Russians consume more than 15 liters (4 gallons) of pure alcohol per person annually and these estimates are probably low. Russian leaders initiated an anti-drinking campaign in 2010, calling their country's plight "a national disaster." An ambitious goal was set to reduce alcohol consumption by 50 percent within the next decade. Meanwhile, however, binge drinking and chronic high levels of alcohol consumption continue to be life-threatening problems for millions of people throughout

▼ **Figure 9.40 Tigre de Cristal Casino, near Vladivostok** Opening its doors in 2015, Russia's posh Tigre de Cristo Casino is located near Vladivostok. In this image, a Russian blackjack dealer tests the gambling skills of Chinese visitors.

the region. Regional patterns of alcoholism reveal higher rates of abuse in thinly populated zones dominated by natural-resource economies (in the north) and in areas dominated by poor indigenous populations (especially in northeast Siberia) (**Figure 9.42**). Relatively low rates of alcoholism in the Caucasus region may be culturally shaped by higher rates of wine consumption, usually with meals, a traditional practice less associated with chronic abuse. Regional patterns of reported drug abuse are more difficult to explain, but the highest rates seem to be found in older Soviet-era industrial districts that have experienced prolonged economic decline since the 1990s and in selected areas still experiencing rapid natural-resource development today (such as energy-rich Sakhalin Island) (see Figure 9.42).

▲ **Figure 9.41 Russian-Chinese Pipelines** The map shows some of the growing connections between Russian Siberia and China. A series of energy deals between the two countries will promote more trade between 2015 and 2030.

Gender, Culture, and Politics Women still struggle for basic rights within the conservative, patriarchal societies that characterize the Russian domain. While often better educated than men, they earn substantially less money performing the same work. Women are also underrepresented in positions of corporate or political power, often faring worse than their western European or American counterparts. Violence against women has been widely reported in the post-Soviet era. Beatings and rapes are common.

In addition, **human trafficking** (a practice in which women are lured or abducted into prostitution) is a widespread problem. Armenia, Ukraine, Moldova, and rural districts of Russia are major sources of young women who are forced into prostitution in Europe and the Middle East. It is a multi-billion-dollar business, involving the large-scale participation of organized crime and hundreds of thousands of young women. Some estimates suggest that in addition to Ukraine's large domestic sex tourism industry, thousands more women have emigrated as sex workers, both to western Europe and the United States. Since 2008, the Ukrainian feminist organization known as FEMEN has made headlines around the world with its members'

high-profile topless protests of that country's sex industry (their slogan is "Ukraine is not a brothel"). The region is also a major and notorious source for Internet brides and dating, practices that invite additional violence against women.

REVIEW

9.10 Describe how centralized planning created a new economic geography across the former Soviet Union. What is its lasting impact?

9.11 Briefly summarize the key strengths and weaknesses of the post-Soviet Russian economy and suggest how globalization has shaped its evolution.

9.12 Identify three key social problems that may have worsened during the post-Soviet period. Why might this be the case?

■ **KEY TERMS** centralized economic planning, oligarchs, human trafficking

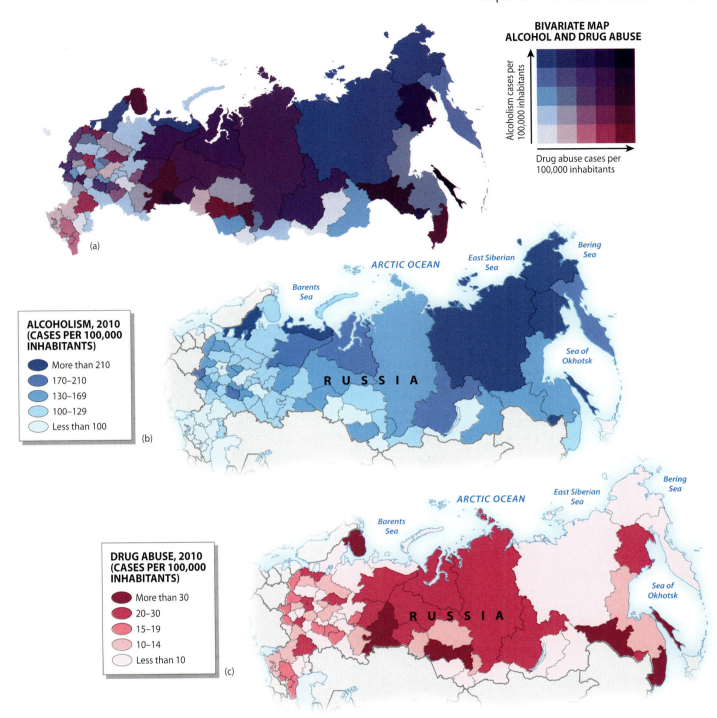

▲ **Figure 9.42 Development Issues in the Russian Domain: Alcoholism and Drug Use in Russia** Higher rates of alcoholism and drug abuse within Russia are often found in older declining industrial zones, regions with large employment in natural resource extraction, and in localities with highly impoverished indigenous populations.

Chapter 9 The Russian Domain

Summary

- Huge environmental challenges remain for the Russian domain. The legacy of the Soviet era includes polluted rivers and coastlines, poor urban air quality, and a frightening array of toxic waste and nuclear hazards.

- Declining and aging populations are a sobering reality for much of the region. While some urban areas enjoy modest growth, mostly related to immigration and domestic in-migration, many rural areas and less competitive industrial zones may see continued outflows of people and very low birthrates.

- The Russian domain's cultural geography was formed centuries ago from the complex mix of Slavic languages, Orthodox Christianity, and numerous ethnic minorities that still complicate the scene today. Further changing the region are new global influences—products, technologies, and attitudes that sometimes clash with traditional cultural values.

- Much of the region's political legacy is rooted in the colonial expansion of the Russian Empire after 1600, and then in the Soviet Union, which expanded its influence in the 20th century. Since 2000, Russian President Vladimir Putin has centralized political power and reasserted Russia's geopolitical presence in the region and around the world.

- The Russian domain's future economic success remains uncertain, related both to the unpredictable energy economy (especially in the case of Russia) and to ongoing political instability that has harmed economic growth and reduced global investment.

Review Questions

1. Join with a group of students to debate another student group on the question of whether Russia's natural environment is one of its greatest assets or one of its greatest liabilities.

2. Describe the key phases of colonial expansion during the rise of the Russian Empire. In what ways was this similar to U.S. expansion across North America? In what ways did it differ?

3. On a base map of the Russian domain, suggest possible political boundaries 20 years from now and discuss your choices in a short essay that accompanies your map. What forces might work for a larger Russian state? A smaller Russian state?

4. Why is organized crime such a critical problem in this part of the world?

5. From the perspective of a 22-year-old college student in Moscow, write a letter to a relative in the United States suggesting how political and economic changes over the past 15 years have changed your life.

Image Analysis

1. Describe the three different migration patterns shown on the map. Write a short paragraph for each pattern, briefly explaining the key drivers (demographic, economic, political, cultural, etc.) at work for each flow.

2. Why is the Moscow region such a key player in *both* incoming and outgoing waves of migration?

▶ **Figure IA9** **Recent Migration Flows in the Russian Domain**

Legend:
- → Russian immigrants returning from former Soviet republics
- → Emigrants leaving the Russian domain
- → Economic migrants within the Russian domain

JOIN THE DEBATE

Complex changes have shaped the Russian domain since the Soviet Union's disintegration in 1991. Many experts have pointed both to the advantages and disadvantages of the old Soviet system compared to what the region is witnessing today. Would people be better off or worse off if the Soviet Union had remained intact?

The Soviet era may have been authoritarian and inefficient, but economic and social indicators were higher.

- Under centralized Soviet control, huge and successful investments were made into the region's basic urban and transportation infrastructure and careful planning allowed for effective long-term projects.

- Social services—especially education and health care—were notably superior during the Soviet era versus what is available today.

- Contrary to high hopes, democracy has not flowered in the region. Corruption remains rampant. Is President Putin (a former Soviet-era KGB agent!) any better than a Soviet dictator?

Although life remains tough in the region, people have far more freedom and choice now.

- Although press and media freedoms are limited, regional societies are far more open today than under Soviet control, and important social and political changes have been shaped by elections and popular protests.

- More open societies and improved access to the Internet have allowed residents to share more fully in cultural fruits of globalization. From music to consumer goods, people have greater access to what the world has to offer.

- People can migrate much more freely today, both within and beyond the region than they ever could during the Soviet era. Furthermore, private businesses and agricultural operations have flowered, allowing people to improve their lives through innovation and hard work.

▲ **Figure D9** Vladimir Putin

KEY TERMS

autonomous areas *(p. 387)*
Baikal–Amur Mainline (BAM) Railroad *(p. 372)*
Bolsheviks *(p. 387)*
centralized economic planning *(p. 393)*
chernozem soils *(p. 362)*
Cold War *(p. 388)*
communism *(p. 387)*
Cossacks *(p. 378)*
dachas *(p. 377)*
Eastern Orthodox Christianity *(p. 378)*
Eurasian Economic Union (EEU) *(p. 389)*
exclave *(p. 392)*
glasnost *(p. 388)*

Gulag Archipelago *(p. 374)*
human trafficking *(p. 398)*
Iron Curtain *(p. 388)*
mikrorayons *(p. 377)*
northern sea route (NSR) *(p. 368)*
oligarchs *(p. 395)*
perestroika *(p. 388)*
permafrost *(p. 363)*
podzol soils *(p. 360)*
Russification *(p. 374)*
siloviki *(p. 391)*
Slavic peoples *(p. 378)*
socialist realism *(p. 384)*
Soviet Union *(p. 358)*
taiga *(p. 363)*
Trans-Siberian Railroad *(p. 372)*
tsars *(p. 374)*

DATA ANALYSIS

http://goo.gl/ywNduL

Foreign Direct Investment (FDI) can be a valuable gauge of economic activity in a country. Data are often gathered for both incoming (investment capital entering a country) and outgoing (investment capital leaving a country) FDI. The OECD (Organization for Economic Co-operation and Development) keeps annual statistics on FDI. Go to their website for the Russian Federation.

1. Make a simple chart showing incoming and outgoing flows of FDI for the years shown in the data table.

2. In a few sentences, summarize the general patterns and trends you observe. How would you explain some of the major changes you observe during the period?

3. Given Russia's current economic and political situation, what are likely patterns of incoming and outgoing FDI for the current year and the next three years? Explain your answer.

MasteringGeography™

Looking for additional review and test prep materials? Visit the Study Area in **MasteringGeography**™ to enhance your geographic literacy, spatial reasoning skills, and understanding of this chapter's content by accessing a variety of resources, including **MapMaster** interactive maps, videos, RSS feeds, flashcards, web links, self-study quizzes, and an eText version of *Diversity Amid Globalization*.

Authors' Blogs

Scan to visit the
GeoCurrents blog
http://geocurrents.info/category/place/russia-ukraine-and-caucasus

Scan to visit the
Author's Blog
for field notes, media resources, and chapter updates
http://gad4blog.wordpress.com/category/the-russian-domain/

Physical Geography and Environmental Issues

Central Asia includes some of the world's most extreme deserts, as well as some of its highest mountains. Intensive agriculture along rivers flowing from the highlands to the deserts has caused water shortages and the dessication of many lakes and wetlands.

Population and Settlement

Pastoral nomadism, the traditional way of life across much of Central Asia, is gradually disappearing as people settle in towns and cities.

Cultural Coherence and Diversity

In much of eastern Central Asia, the growing Han Chinese population threatens the indigenous Tibetan and Uyghur cultures. In the west, the role of Islam in social and political life remains a major issue.

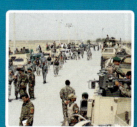

Geopolitical Framework

Afghanistan and its neighbors to the north are frontline states in the struggle between radical Islamic fundamentalism and secular governments.

Economic and Social Development

Despite its abundant resources, Central Asia remains relatively poor, although much of the region does enjoy relatively high levels of social development.

▶ **Baikonur Cosmodrome** Located in south-central Kazakhstan, Baikonur is the world's oldest and largest space launch facility. Here we see the July 7, 2016 launch of the Soyuz MS-01 spacecraft that took a three-person crew—one Russian, one American, and one Japanese—to the International Space Station.

Baikonur, Kazakhstan

CENTRAL ASIA

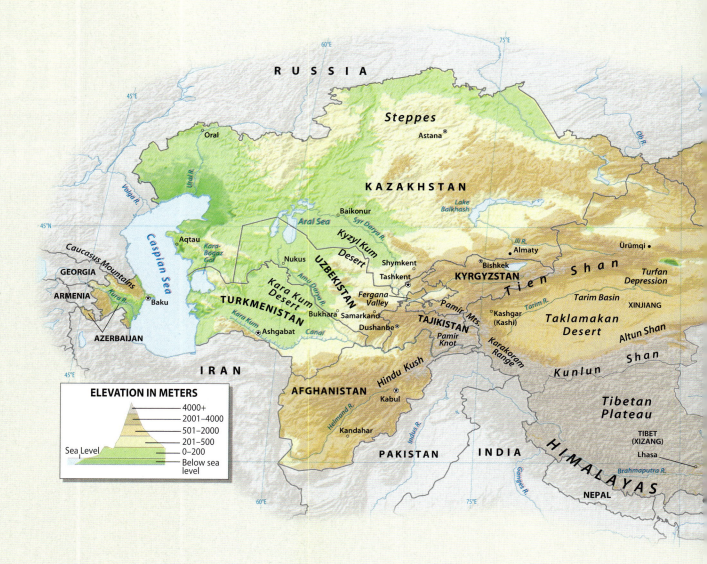

ELEVATION IN METERS

4000+
2001–4000
501–2000
201–500
0–200
Sea Level
Below sea level

A stronauts visiting the International Space Station must first travel to Baikonur, located in a remote Central Asian desert. Although situated in Kazakhstan, Baikonur functions more as part of Russia, which leases the land surrounding the city and encompassing the Baikonur Cosmodrome, the world's first—and largest—space launch facility. The Cosmodrome does much more than blast off the Soyuz rockets carrying astronauts to the space station. The busy facility launches numerous scientific, commercial, and military rockets.

Baikonur is profitable for Kazakhstan, which receives US$115 million in annual rent from Russia for the spaceport. But Kazakhstan is not pleased with this arrangement, in part because its rental income is fixed until 2050, when the lease expires. Environmental damage is another source of tension. The launching of satellite-carrying Proton rockets spreads toxic waste over a large expanse of land. Demography is another factor. Baikonur has changed drastically over the past several decades as Russians have moved out

and Kazakhs have moved in. The new Kazakh majority tends to resent living in what amounts to an island of Russia within their own country.

Concerned about Baikonur's long-term future, Moscow is building a new launch facility in far eastern Russia. This project, however, has been plagued by scandals, and it is uncertain whether it will make the 2025 deadline for its first human-occupied space flight. As a result, Russia will probably continue to use Baikonur for many years. And even if Russia does eventually pull out, other countries will still use the cosmodrome's superb facilities to launch their own satellites, generating both revenues and environmental problems for Kazakhstan.

Baikonur's ambiguous situation derives in part from Central Asia's geopolitical situation. Before 1991, most of the region was either part of the Soviet Union or was dominated by it, and almost all of the rest belonged to China. When the Soviet Union collapsed and several new countries appeared on the map, geographers began to debate their placement. Should they remain part of a Russian-centered region, should they

> **Although Russian influence and cultural ties with Southwest Asia remain pronounced, Central Asian countries are linked by a number of distinctive features**

◀ **Figure 10.1** **Central Asia** A vast, sprawling region in the center of the Eurasian continent, Central Asia is dominated by arid plains and basins in the west and by lofty mountain ranges and plateaus in the east. Eight independent countries—Kazakhstan, Turkmenistan, Uzbekistan, Kyrgyzstan, Tajikistan, Azerbaijan, Afghanistan, and Mongolia—form Central Asia's core. This book places China's lightly populated far west and north within Central Asia as well, due to patterns of cultural and physical geography.

CENTRAL ASIA
Political & Physical Map

⬢ ● Metropolitan areas more than 20 million

⬢ ● Metropolitan areas 10–20 million

⊛ ● Metropolitan areas 5–9.9 million

⊛ • Metropolitan areas 1–4.9 million

⊛ ○ Selected smaller metropolitan areas

Defining the Region

Distinctive though it may be, Central Asia is difficult to delineate precisely. Five post-Soviet republics—Kazakhstan, Kyrgyzstan, Uzbekistan, Tajikistan, and Turkmenistan—anchor the region, but it is not obvious what other places should be included. Here we adopt an unconventional strategy, adding another former Soviet state, Azerbaijan, and the long-independent countries of Mongolia and Afghanistan, as well as a large expanse of western China (**Figure 10.1**). We thus count the large autonomous Chinese regions of Tibet and Xinjiang as part of both East Asia and Central Asia, and several other parts of western China, such as Nei Mongol (Inner Mongolia), are also discussed in both chapters.

Including these additional territories in Central Asia is controversial. Azerbaijan is often classified with its neighbors in the Caucasus (Georgia and Armenia), western China fully belongs to East Asia on political grounds, and Mongolia is also often considered part of East Asia due to its location and historical connections with China. Afghanistan, moreover, could just as easily be located in either South Asia or Southwest Asia.

However, there are solid reasons for defining Central Asia as we have. The region has deep historical bonds and similar environmental and economic conditions. Azerbaijan, for example, has more in common culturally and economically with Central Asia than with its Caucasus neighbors. Central Asia is also increasingly seen as a geopolitical unit, as its various countries face similar political challenges. But whether Central Asia as we have defined it will remain a coherent world region is unclear. Continuing Han Chinese migration into Tibet and especially into Xinjiang, for example, increasingly places these areas in an East Asian cultural framework.

be linked with Southwest Asia due to cultural similarities, or should they instead become the focus of a new world region? Although both Russian influence and cultural ties with Southwest Asia remain pronounced, Central Asian countries are linked by a number of distinctive features. We therefore favor the third option.

LEARNING Objectives
After reading this chapter you should be able to:

10.1 Describe the key environmental differences among Central Asia's desert areas, its mountain and plateau zone, and its steppe belt, and how these influence human settlement and economic development.

10.2 Identify the reasons for the Aral Sea's disappearance and outline the economic and environmental consequences of the loss of this once-massive lake.

10.3 Summarize the reasons why water resources are so important to Central Asia and describe how people respond to water shortages.

10.4 Explain why Central Asia's population is so unevenly distributed, with some areas densely settled and others essentially uninhabited.

10.5 Contrast Central Asia's historical cities with those that have been established within the past 100 years.

10.6 Outline the ways in which religion divides Central Asia and describe how religious diversity has influenced the history of the region.

10.7 Identify distinct ways in which cultural globalization affects different parts of Central Asia and explain why it is controversial in much of the region.

10.8 Describe the geopolitical roles played in Central Asia by Russia, China, and the United States, and explain why the region has been the site of pronounced geopolitical tension over the past several decades.

10.9 Explain how ethnic conflict has contributed to instability in Afghanistan and assess the potential of ethnic tension to destabilize the rest of the region.

10.10 Link oil and natural gas production and transport, as well as oil price fluctuations, to uneven levels of economic and social development across Central Asia.

Physical Geography and Environmental Issues: Steppes, Deserts, and Threatened Lakes

Central Asia forms a large region in the center of the Eurasian landmass. Alone among the world regions, it lacks ocean access. Owing to its continental position in the center of the world's largest landmass, Central Asia is noted for its harsh climate. High mountains, deep basins, and extensive plateaus magnify its climatic extremes. The region's aridity has also contributed to some of the most severe environmental problems in the world. To understand the variation in Central Asia's environmental circumstances, we need to examine its physical geography.

Central Asia's Physical Regions and Climate

Central Asia is dominated by grassland plains (or **steppes**) in the northern area, desert basins in the southwestern and central areas, and high plateaus and mountains in the south-central and southeastern areas. Several mountain ranges extend into the heart of the region, dividing the desert zone into a series of basins and giving rise to the rivers that flow into the deserts.

Central Asian Highlands The highlands of Central Asia originated in one of the great tectonic events of Earth's history: the collision of the Indian subcontinent into the Asian mainland. This ongoing impact has created the highest mountain range in the world, the Himalayas, located along the boundary of South Asia and Central Asia.

Yet the Himalayas are merely part of a larger network of high mountains and plateaus. To the northwest, they merge with the Karakoram Range and then the Pamir Mountains (**Figure 10.2**). From the so-called Pamir Knot—a tangle of mountains where Pakistan, Afghanistan, China, and Tajikistan converge—other towering ranges radiate outward. The Hindu Kush sweeps to the southwest through central Afghanistan; the Kunlun Shan extends to the east along the northern border of the Tibetan Plateau; and the Tien Shan swings out to the northeast into China's Xinjiang Province. These ranges all have peaks higher than 20,000 feet (6000 meters). Much lower but still significant ranges are found along Turkmenistan's boundary with Iran and Azerbaijan's boundaries with Russia, Armenia, and Iran.

More extensive than these ranges is the Tibetan Plateau, which extends some 1250 miles (2000 km) from east to west and 750 miles (1200 km) from north to south. More remarkable than its size is its elevation; virtually the entire area is higher than 12,000 feet (3700 meters) above sea level. Most of the large rivers of South, Southeast, and East Asia originate on the Tibetan Plateau, including the Indus, Ganges, Brahmaputra, Mekong, Yangtze, and Huang He.

The average elevation of the Tibetan Plateau is 14,800 feet (4500 meters), near the maximum height at which human life can exist. Rather than forming a flat, table-like surface, the plateau is punctuated with east–west-running ranges alternating with basins. Although the southeastern sections of the plateau receive ample precipitation, most of Tibet is arid. Cut off from any source of moisture by high ranges, large areas of the plateau receive only a few inches of rain a year. Winters in Tibet are cold, and although summer afternoons can be warm, summer nights remain chilly (**Figure 10.3**; see the climograph for Lhasa).

Plains and Basins Although the mountains of Central Asia are higher and more extensive than those found anywhere else, most of the region is characterized by plains and basins of low and intermediate elevation. This lower-lying zone can be divided into two main areas: a central belt of deserts punctuated by lush river valleys, and a northern swath of semiarid steppe.

Central Asia's desert belt is itself divided into two segments by the Tien Shan and Pamir mountains (Figure 10.2). To the west lie the arid plains of the Caspian and Aral basins, located primarily in Turkmenistan, Uzbekistan, and southern Kazakhstan. Most of this area is relatively flat and low; the surface of the Caspian Sea lies 92 feet (28 meters) below sea level. The climate here is characterized by pronounced seasonal differences; summers are dry and hot, whereas winter temperatures average well below freezing (see the climographs

◀ **Figure 10.2 The Pamir Mountains** Spanning the boundaries of Tajikistan, Kyrgyzstan, Afghanistan, and China, the rugged Pamir Mountains include peaks as high as 25,000 feet (7,600 meters). Here we see the remains of Yamchun Fort, situated near the base of the range in a remote area of southeastern Tajikistan. The fort is believed to date back to the first century BCE, and once played a key role in defending a branch of the Silk Road.

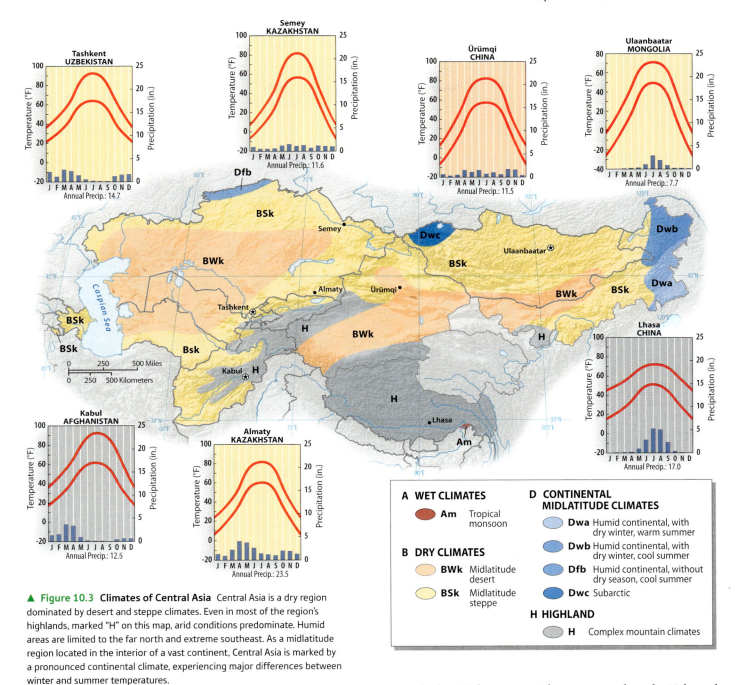

▲ **Figure 10.3 Climates of Central Asia** Central Asia is a dry region dominated by desert and steppe climates. Even in most of the region's highlands, marked "H" on this map, arid conditions predominate. Humid areas are limited to the far north and extreme southeast. As a midlatitude region located in the interior of a vast continent, Central Asia is marked by a pronounced continental climate, experiencing major differences between winter and summer temperatures.

for Tashkent and Almaty in Figure 10.3). Central Asia's eastern desert extends for almost 2000 miles (3200 km) from far western China to the southeastern edge of Mongolia. It is conventionally divided into two deserts: the Taklamakan, in the Tarim Basin of Xinjiang, and the Gobi, which runs along the border between Mongolia proper and the Chinese autonomous region of Inner Mongolia.

Western Central Asia is differentiated from the east in part by its larger rivers. More snow falls on the western than the eastern slopes of the Pamir Mountains, giving rise to abundant runoff. The largest of these rivers historically flowed into sizable lakes, although several of them have almost disappeared

since the late 20th century. Other rivers, such as the Helmand of Afghanistan, end in shallow lakes or extensive marshes. In Xinjiang, only one river, the Tarim, flows out of the highlands onto the basin floor. Before the completion of irrigation projects in the 1960s, the Tarim terminated in a salty lake called Lop Nor. Subsequently, the dried-out Lop Nor salt flat was used periodically by China for testing nuclear weapons.

North of the arid zone, rainfall increases and desert gradually gives way to the steppe of northern Central Asia. Near the region's northern boundary, trees begin to appear in favored locales, outliers of the great Siberian taiga (coniferous forest). A nearly continuous swath of grasslands extends some 4000 miles (6400 km) east to west across the region, only partially broken by Mongolia's Altai Mountains.

Summers on the northern steppe are usually pleasant, but winters can be brutally cold (see the climographs for Ürümqi and Ulaanbaatar in Figure 10.3).

Central Asia's Environmental Challenges

Much of Central Asia has a fairly clean environment, mainly due to low population density. Industrial pollution, however, is a serious problem in the larger cities, such as Tashkent in Uzbekistan, Baku in Azerbaijan, and Almaty in Kazakhstan. According to a recent study, Almaty is one of the 10 most heavily polluted cities in the world, with a particularly high concentration of airborne particulates. Mongolia's capital, Ulaanbaatar, has a similar air-pollution problem, exacerbated by its rapid growth, use of coal for heating, and stagnant wintertime air masses.

Over the past several decades, mining has emerged as a major industry in parts of Central Asia, and it too is associated with high levels of pollution and land degradation. Southern Mongolia and China's Inner Mongolia have some of the largest mines and worst environmental damage. The most controversial operation, however, is the massive Kumtor gold mine in Kyrgyzstan, which generates more that 7 percent of the country's economic output (**Figure 10.4**). Mineral extraction at high-elevation Kumtor requires "ice mining," or tunneling through glaciers to reach the veins below, a process that disrupts natural water flow and releases toxic chemicals. Large protests over such practices erupted in 2013, but a 2015 study concluded that Kumtor would not be economically viable if ice mining were discontinued.

Across much of Central Asia, the typical environmental dilemmas of arid environments plague the region: **desertification** (the spread of deserts); **salinization** (the

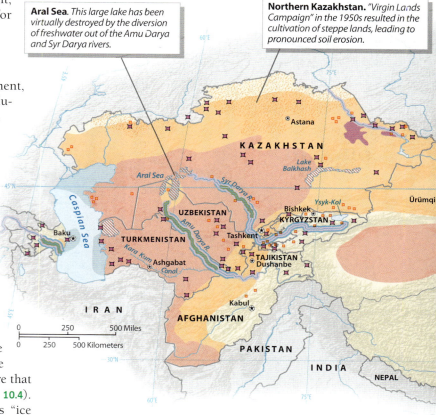

Aral Sea. *This large lake has been virtually destroyed by the diversion of freshwater out of the Amu Darya and Syr Darya rivers.*

Northern Kazakhstan. *"Virgin Lands Campaign" in the 1950s resulted in the cultivation of steppe lands, leading to pronounced soil erosion.*

▲ **Figure 10.5 Environmental Issues in Central Asia** Central Asia has experienced some of the world's most severe desertification problems. Soil erosion and overgrazing have led to the advance of desertlike conditions in much of western China and Kazakhstan. In western Central Asia, the most serious environmental problems are associated with the diversion of rivers for irrigation and the corresponding desiccation of lakes. Oil pollution is a particularly serious issue in the Caspian Sea area.

accumulation of salt in the soil); and **desiccation** (the drying up of lakes and wetlands) (**Figure 10.5**). The destruction of the Aral Sea, located on the boundary of Kazakhstan and Uzbekistan in western Central Asia, has been particularly tragic.

▼ **Figure 10.4 Kumtor Mine in Kyrgyzstan** Located at 13,100 feet (4,000 meters) above sea level in the Tien Shan Mountains, the Kumtor mine is vital to the economy of Kyrgyzstan. Owned by the Canadian firm Centerra Gold, Kumtor generates as much as 40 percent of Kyrgyzstan's exports in some years.

The Shrinking Aral Sea Until the late 20th century, the Aral Sea was the world's fourth largest lake. Its only significant sources of water are the Amu Darya and Syr Darya rivers, which flow out of the Pamir Mountains, some 600 miles (960 km) to the southeast. Water diversions for irrigation have long lowered flows in both rivers, but after 1950 the scale of diversion vastly expanded. The Amu Darya and Syr Darya valleys formed the southernmost farming districts of the Soviet Union, and the rivers supplied water for warm-season crops, especially cotton.

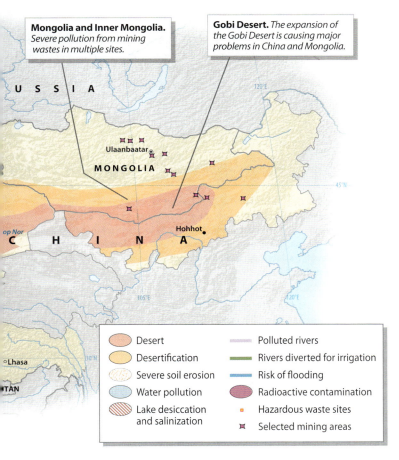

Mongolia and Inner Mongolia. *Severe pollution from mining wastes in multiple sites.*

Gobi Desert. *The expansion of the Gobi Desert is causing major problems in China and Mongolia.*

Legend:
- Desert
- Desertification
- Severe soil erosion
- Water pollution
- Lake desiccation and salinization
- Polluted rivers
- Rivers diverted for irrigation
- Risk of flooding
- Radioactive contamination
- Hazardous waste sites
- Selected mining areas

The largest Soviet-era project here was the Kara Kum Canal, which extends from the Amu Darya to the deserts of southern Turkmenistan.

Unfortunately, the more river water was diverted for crop production, the less freshwater was available for the lake. The Aral proved to be particularly sensitive because it is relatively shallow. By the 1970s, the shoreline was retreating at an unprecedented rate; eventually, several "seaside" villages found themselves stranded more than 40 miles (64 km) inland. As salinity levels increased, most fish species disappeared. New islands began to emerge, and by the 1990s the Aral Sea had been divided into two separate lakes.

The destruction of the Aral Sea resulted in extensive economic, environmental, and cultural damage. Fisheries were destroyed and even local agriculture suffered. The retreating lake left large salt flats on its exposed beds; windstorms picked up the salt, along with the agricultural chemicals that had accumulated in the lake's shallows, and deposited it in nearby fields. Crop yields have thus declined; desertification

has accelerated; and public health has been undermined. Particularly hard hit are the Karakalpak people, members of a marginalized ethnic group who inhabit the formerly rich delta of the Amu Darya River.

In 2001, the oil-rich government of Kazakhstan decided to save what was left of its portion of the Aral Sea. By reconstructing canals, sluices, and other waterworks along the Syr Darya River, it doubled the flow of water reaching the northern Aral, which correspondingly began to rise. A dike extending across the lakebed prevented the extra water from flowing onto the salt flats of the southern Aral Sea. As a result, improved water quality has allowed the revival of wildlife and even commercial fishing. In 2014, 5,595 metric tons of fish were harvested in the "Small Aral Sea" of the north. The southern Aral Sea, however, continues to shrink, and its water quality continues to deteriorate (**Figure 10.6**). It has also divided into two water bodies, and in 2014 the southeastern basin went completely dry.

The governments of Kazakhstan, Tajikistan, Uzbekistan, and Turkmenistan have been working together to try to solve this problem, meeting periodically with the International Fund for Saving the Aral Sea. Thus far, however, agriculturally dependent Uzbekistan has done little to salvage its portion of the Aral Sea. Uzbekistan is worried, moreover, that Tajikistan will divert more water from local rivers, resulting in tense relations between the two countries.

Other Fluctuating Lakes Despite its aridity, Central Asia contains large lakes because it forms a low-lying basin, without drainage to the ocean, surrounded by mountains and other more humid areas. The world's largest lake by a huge margin is the Caspian Sea, located along the region's western boundary; the 15th largest is Lake Balkhash in eastern Kazakhstan. Several other large lakes, such as Kyrgyzstan's spectacular

▼ **Figure 10.6 The Shrinking Aral Sea** Satellite images show the steady shrinkage of the Aral Sea, which was a massive lake as recently as the 1970s. The Aral is now divided into several much smaller lakes.

(a) 1987   (b) 2004   (c) 2014

Issyk Kul, the Earth's second largest alpine lake, are also located here.

Because few of Central Asia's lakes drain to the sea, most have accumulated salt. But the Caspian is less salty than the ocean, particularly in the north, while until the 1970s the Aral was only slightly brackish. Lake Balkhash is almost fresh in the west, but its long eastern extension is quite salty. The lack of drainage also means that these lakes naturally fluctuate, depending on how much precipitation falls in their drainage basin in a given year. Several Central Asian lakes have suffered from reduced water flow and hence increasing salinity, including Balkhash.

The story of the Caspian is more complicated. The Caspian Sea receives most of its water from the large rivers of the north, the Ural and the Volga. The expansion of irrigation in the lower Volga reduced freshwater reaching the Caspian by the mid-20th century. As a result, the level of the lake dropped, exposing as much as 15,000 square miles (39,000 square km) of new land, and its salinity level increased. After reaching a low point in the late 1970s, the Caspian began to rise, due mostly to higher precipitation in the north. This enlargement, too, caused problems, inundating some newly reclaimed farmlands. But then from 2007 to 2015 it began to shrink again, dropping by roughly half a meter. Currently, however, the Caspian's most serious environmental threat is probably pollution from the oil industry rather than fluctuation in size.

Water development projects are now beginning to impact the Tibetan Plateau. Concerned about shortages in the North China Plain, China is diverting water from the headwaters of the Yangtze River into those of the Huang He (or Yellow River). This and other transfer schemes threaten several large Tibetan wetlands, havens for many rare species of migrating waterfowl. Overgrazing on the Tibetan Plateau has also resulted in reduced stream flow in some areas.

Desertification and Deforestation Desertification is another concern in Central Asia. In the east, the Gobi and Taklamakan deserts have spread southward, encroaching on settled lands in northeastern China. Some farmers must relocate their houses multiple times to avoid the sands. The Chinese government is trying to stabilize dunes with massive tree- and grass-planting campaigns, but only some of these efforts have been successful (see *Working Toward Sustainability: The Greening of the Inner Mongolian Desert*). As a result, sand and dust storms have increased in frequency across much of northern China (**Figure 10.7**).

The former Soviet Central Asia has also seen extensive desertification. Northern Kazakhstan underwent the ambitious Soviet "Virgin Lands Campaign" of the 1950s, in which semiarid grasslands were plowed and planted with wheat. Many of these lands have since returned to native grasses, but not before erosion stripped away much of their

▲ **Figure 10.7 Dust Storm in the Gobi Desert** Straddling the border between Mongolia and China, the Gobi Desert is noted for its harsh climate, with frequent droughts and dust storms. The region's indigenous Mongolian inhabitants raise livestock and live in felt-covered tents. **Q: How have dust storms from Central Asia historically impacted East Asia, particularly China's Loess Plateau?**

productivity. Some reports claim that up to 50 percent of Kazakhstan's farmland has been abandoned since 1990, due in part to desertification. In the irrigated lands of Uzbekistan, salinization has resulted in extensive cropland abandonment. Throughout the region, however, efforts are being made to address such problems. In early 2013, for example, Turkmenistan initiated a massive tree-planting campaign aimed at stopping the spread of desert sands.

Deforestation has also harmed the region, resulting in wood shortages and reduced dry-season water flow. Although most of Central Asia is too dry for forests, many of its mountains were once well wooded. Today extensive forests can be found only in the wild gorge country of the eastern Tibetan Plateau; in some of the more remote west- and north-facing slopes of the Tien Shan, Altai, and Pamir mountains; and in the mountains of northern Mongolia. Some Central Asian woodlands, particularly the walnut forests of Kyrgyzstan and the apple forests of southeastern Kazakhstan, are noted for their high levels of biodiversity.

Energy Issues in Central Asia Central Asia contains several vast oil and natural gas fields. Increasingly, the region's energy is being exported to eastern China. Xinjiang holds most of China's fossil fuel reserves, which are linked to the country's major population centers by large pipelines. The first West-East Gas Pipeline, operational since 2005, stretches 2,485 miles (4000 km) from the Tarim Basin to Shanghai. It has more recently been upstaged by a second 5,650-mile (9100-km) West-East Gas Pipeline, which connects with the Central Asia-China Gas Pipeline, originating in Turkmenistan. A third West-East Gas Pipeline, linking Xinjiang to Fujian Province in southern China, was completed in 2015.

Former Soviet Central Asia is equally dependent on selling fossil fuels. In 2015, roughly 90 percent of Azerbaijan's exports were derived from oil and natural gas.

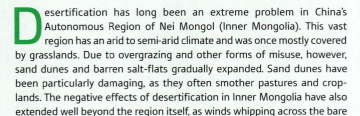

The Greening of the Inner Mongolian Desert

Desertification has long been an extreme problem in China's Autonomous Region of Nei Mongol (Inner Mongolia). This vast region has an arid to semi-arid climate and was once mostly covered by grasslands. Due to overgrazing and other forms of misuse, however, sand dunes and barren salt-flats gradually expanded. Sand dunes have been particularly damaging, as they often smother pastures and croplands. The negative effects of desertification in Inner Mongolia have also extended well beyond the region itself, as winds whipping across the bare lands generate damaging dust storms as far away as Beijing and beyond.

Halting Desertification in Inner Mongolia In recent years, China's government, along with local authorities and businesspeople, have turned the corner on desertification across much of the region. From 2005 to 2010, the area covered by extreme deserts in northern China shrank by an estimated 663 square miles (1,717 square kilometers) a year. The key process here has been the planting of drought-adapted vegetation on the dunes. If successful, the plants will send out roots to hold the sand in place, preventing dune movement. Such plants, however, usually need some help to become established. The best technique is to cover the sand with a mesh of wheat straw, creating squares roughly the size of a house. The straw temporarily stabilizes the sand until the new plants take root, and provides essential plant nutrients as it decays.

Success in the Hobq Desert One of the most successful examples of reversing desertification is found in the Hobq Desert of Inner Mongolia (**Figure 10.1.1**). A key participant is the private Chinese company Elion Resources Group, which has been working with local and regional governments since the mid-1990s to revegetate vast expanses of barren land. The company originally extracted minerals found in salt flats, but the roads to these mineral deposits kept being covered by blowing sand, halting transportation. After much trial and error, Elion researchers learned how to stabilize the sands by planting drought-adapted shrubs.

One plant that was soon thriving on the former dunes was licorice. Before long, Elion Resources learned that it could profit from the sale of licorice root, used in the pharmaceutical industry as well as in candy making. Eventually the firm began to invest in a wide array of sustainable projects in the area, using solar power extensively and building greenhouses, efficient drip-irrigation systems, and research laboratories. Even a successful tourism industry was established. An estimated 100,000 jobs have emerged, increasing the annual income of

▲ **Figure 10.1.1. Fighting Desertification in Inner Mongolia** In northern China's Inner Mongolia Autonomous Region, extensive tree-planting campaigns attempt to halt the spread of sand dunes and thus preserve agricultural fields and grazing grounds. Here villagers are planting trees to stabilize sand dunes in the Hobq Desert.

local residents tenfold. In 2012, the company's chair, Wang Wenbiao, was honored with the United Nations Environment and Development Award.

1. What other parts of the world experience severe desertification? Can the techniques pioneered in China be successfully applied in other world regions?

2. List some of the advantages and disadvantages of relying on private firms to combat desertification.

Google Earth Virtual MG Tour Video

http://goo.gl/ZeAkiH

Oil fields near Baku were among the earliest to be exploited in modern times, the first derrick going up in 1871. Kazakhstan exports even more oil than Azerbaijan, and Turkmenistan has the fourth largest reserves of natural gas in the world.

Several Central Asian countries, especially Tajikistan and Afghanistan, lack significant oil or gas reserves and must use other sources of energy. Tajikistan relies on hydropower for over 90 percent of its electricity, taking advantage of the Vakhsh and Panj rivers, major tributaries of the Amu Darya. The Nurek Dam, completed in 1980, stands as the world's second tallest dam at 997 feet (304 meters) (**Figure 10.8**). The Rogun Dam, currently under construction, will be even taller. These dams allow Tajikistan to export excess electricity to Uzbekistan, from which it imports oil and natural gas. Uzbekistan strongly objects, however, to Tajikistan's

dam-building program, fearing that it will diminish the flow of water into its farmlands.

Climate Change in Central Asia

Most climate experts predict that Central Asia will be hard hit by global warming, largely because it depends heavily on snow-fed rivers. The Tibetan Plateau has already seen marked increases in temperature, resulting in the drawback of mountain glaciers. Eighty percent of Tibet's glaciers are presently retreating, some at rates of up to 7 percent a year. A 2015 study found that Central Asia's glaciers have been vanishing four times faster than the global average, losing 27 percent of their ice since the early 1960s. Tree-ring data from Mongolia indicate that the last century has been the warmest in more than a millennium. Some researchers claim that Mongolia

▲ **Figure 10.8.** **Nurek Dam** The world's second-highest dam, Nurek Dam in Tajikistan stands at 984 feet (300 meters). Completed in 1980 under the Soviet Union, Nurek Dam is designed primarily to supply hydroelectric power, but it is also used for irrigation and for flood control.

Explore the **Sights** of Tajikistan's Nurek Dam

http://goo.gl/g28sQG

▲ **Figure 10.9.** **Flood Damage in Tajikistan** Although most of Central Asia is arid or semi-arid, many of the region's mountains can receive heavy precipitation, resulting in periodic floods, mudslides, and other natural disasters. Earthquakes can compound the risks by setting off debris flows, one of which is evident in this photo. Such hazards could be exacerbated by climate change in the near future.

REVIEW

10.1 How does the location of Central Asia, near the center of the world's largest landmass and at the junction of several large tectonic plates, influence the region's climate and landforms?

10.2 Why does Central Asia have such large lakes, and why are many of these lakes so deeply threatened?

10.3 Why does Central Asia figure so prominently in many discussions of climate change?

■ **KEY TERMS** steppes, desertification, salinization, desiccation

has experienced more warming over the past half-century, about 4°F (2°C), than any other part of the world.

The retreat of Central Asia's glaciers is especially worrisome because of their role in providing dependable flows of water. In some areas, ice melting has resulted in the temporary flooding of mountain valleys and lowland basins, sometimes accompanied by mudslides, but the long-term result will be the further reduction of freshwater resources (**Figure 10.9**). Climate change could also reduce precipitation in the lowlands of western Central Asia, compounding the problem. Prolonged and devastating droughts have recently struck Afghanistan and several of its neighbors, indicating a possible shift to a drier climate. As a result, the UN Intergovernmental Panel on Climate Change (IPCC) has predicted a 30 percent crop decline for Central Asia as a whole by mid-century.

Climate change will not affect all parts of Central Asia in a uniform manner. Some areas, including the Gobi Desert and parts of the Tibetan Plateau, could see a long-term increase in precipitation. Such forecasts are scant consolation, however, in a region that is experiencing rampant desertification. Some global climate models also indicate increased precipitation in the boreal forest zone of Siberia that extends into the mountains of northern Mongolia. But as the rivers coming out of these mountains generally flow north, few benefits would be reaped by Central Asia.

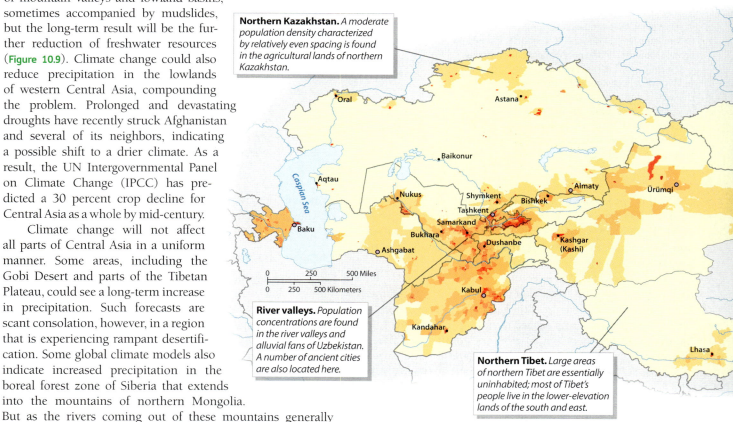

Northern Kazakhstan. *A moderate population density characterized by relatively even spacing is found in the agricultural lands of northern Kazakhstan.*

River valleys. *Population concentrations are found in the river valleys and alluvial fans of Uzbekistan. A number of ancient cities are also located here.*

Northern Tibet. *Large areas of northern Tibet are essentially uninhabited; most of Tibet's people live in the lower-elevation lands of the south and east.*

Population and Settlement: Densely Settled Oases Amid Vacant Lands

Most of Central Asia is sparsely populated (**Figure 10.10**). Large areas are essentially uninhabited, either too arid or too high to support human life. Other expanses are populated by widely scattered groups of nomadic **pastoralists** (people who raise livestock for subsistence purposes). Mongolia, which is more than twice the size of Texas, has only 3.1 million inhabitants. But as is common in arid environments, those few lowland areas with fertile soil and dependable water supplies are thickly settled. Despite its overall aridity, Central Asia is well endowed with perennial rivers and productive oases. The nomadic pastoralists of the steppe and desert zones have dominated the history of Central Asia, but the sedentary farming peoples of the river valleys have always been more numerous.

Highland Population and Subsistence Patterns

The environment of the Tibetan Plateau is particularly harsh. Not only is the climate cold and water often scarce or brackish, but also ultraviolet radiation in this high-elevation area is always pronounced. Over most of the plateau only sparse grasses and herbaceous plants can survive, and human subsistence is obviously difficult under such conditions. The only feasible way of life over much of the Tibetan Plateau is nomadic pastoralism based on the yak, an altitude-adapted relative of the cow. Several hundred thousand people manage to make a living in such a manner, roaming with their herds over vast distances.

Farming in Tibet is possible only in a few favorable locations, generally those that are *relatively* low in elevation and have good soils and either adequate rain or a dependable irrigation system. The main zone of agricultural settlement lies in the far south and east, where protected valleys offer favorable conditions. The population of Tibet proper (the Chinese autonomous region of Xizang) is only 3.2 million, a small number considering the vast size of the area.

Population densities are also low in the other highlands of Central Asia, although densely settled farming communities can be found in protected valleys. The complex topography of the Pamir Range offers a large array of small and nearly isolated valleys that are suitable for agriculture. Not surprisingly, this area is marked by great cultural and linguistic diversity. Many villages here are noted for their agricultural terraces and well-tended orchards.

Central Asia's mountains are vitally important for people living in the adjacent lowlands, whether they are settled farmers or migratory pastoralists. Many herders use the highlands for summer pasture; when the lowlands are parched, the high meadows provide rich grazing. The Kyrgyz (of Kyrgyzstan) are noted for their traditional economy based on **transhumance**, moving their flocks from lowland pastures in the winter to highland meadows in the summer. The farmers of Central Asia rely on the highlands for their wood supplies and, more important, for their water.

Pastoralism and Farming in the Lowlands

Most of the inhabitants of Central Asia's desert zone live in narrow belts where the mountains meet the basins and plains. Here water supplies are adequate, and soils are not contaminated with salt or alkali, as is often the case in the basin interiors. The population distribution pattern of China's Tarim Basin in Xinjiang thus forms a ringlike structure (**Figure 10.11**). Streams flowing from the mountains are diverted to irrigate fields and orchards in the narrow band of cropland situated between the steep slopes of the mountains and the nearly empty deserts of the basin.

To the east, the Gobi Desert has few permanent water sources. Rivers draining Mongolia's highlands flow northward or terminate in interior basins, whereas only a few of the streams from the Tibetan Plateau reach the Gobi proper. (The Huang He, however, does swing north to reach the desert edge before turning south to flow through China's Loess Plateau.) Owing to this scarcity of **exotic rivers** (those originating in more humid areas) and to its own aridity, the Gobi remains one of Asia's least-populated areas.

▼ **Figure 10.10 Population of Central Asia** Central Asia as a whole remains one of the world's least densely populated regions, although it does contain distinct clusters of higher population density. Most of Central Asia's large cities are located near the region's periphery or in its major river valleys.

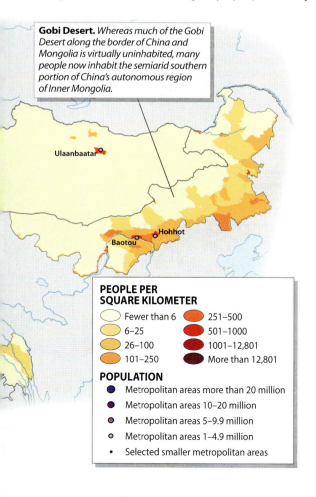

Gobi Desert. *Whereas much of the Gobi Desert along the border of China and Mongolia is virtually uninhabited, many people now inhabit the semiarid southern portion of China's autonomous region of Inner Mongolia.*

PEOPLE PER SQUARE KILOMETER

Fewer than 6	251–500
6–25	501–1000
26–100	1001–12,801
101–250	More than 12,801

POPULATION

- Metropolitan areas more than 20 million
- Metropolitan areas 10–20 million
- Metropolitan areas 5–9.9 million
- Metropolitan areas 1–4.9 million
- Selected smaller metropolitan areas

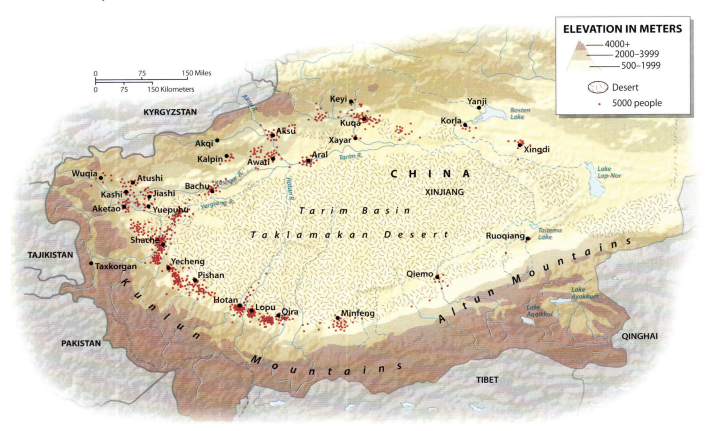

▲ **Figure 10.11 Population Patterns in Xinjiang's Tarim Basin** The central portion of the Tarim Basin is a virtually uninhabited expanse of sand dunes and salt flats. Along the edge of the basin, however, dense agricultural and urban settlements are located where streams running out of the surrounding mountains allow for intensive irrigation. The largest of these oasis communities are found along the southwestern fringe of the basin.

Agricultural Lands As is true of Xinjiang, much of the population in former Soviet Central Asia is concentrated in the transitional zone nestled between the highlands and the plains. A series of **alluvial fans**—fan-shaped deposits of sediments dropped by streams flowing out of the mountains—has long been devoted to intensive cultivation (**Figure 10.12**). Fertile **loess**, a silty soil deposited by the wind, is widespread, and in a few favored areas winter precipitation is high enough to allow rainfed agriculture. Several large valleys in this area, such as the densely populated Fergana Valley of the upper Syr Darya River, offer fertile and easily irrigated farmland. In the far west of the region, Azerbaijan's Kura River Basin also supports intensive agriculture and concentrated settlement.

Northern Kazakhstan is another important farming zone. Once covered in grasslands, the Soviets converted the most productive pastures into farmland in the mid-1900s. Some of these lands have since reverted to steppe, but large areas remain under the plow.

▼ **Figure 10.12 Densely Settled Alluvial Fan** The Tente River in Kazakhstan forms an alluvial fan where it emerges from the mountains and begins to flow across a relatively flat landscape. The fertile soils of this broad fan, measuring 12 miles (20 kilometers) across at its widest point, are intensively cultivated. Several towns and villages are visible along the fan's outer edge. Railroad tracks form a straight feature that cuts across the northeastern portion of the fan.

Kazakhstan is the world's sixth largest exporter of wheat. Consequently, the northern strip of the country has the highest population density of the steppe belt.

The Pastoral Zone The steppes of northern Central Asia are the classical land of nomadic pastoralism. To this day, pastoralism remains common across the grasslands stretching from central Kazakhstan across Mongolia. In many areas, however, most nomads have been forced to adopt sedentary lifestyles. National governments often find migratory people hard to control and difficult to provide with social services. Migratory herders are also highly vulnerable to natural disasters. Severe cold in the winter of 2010, for example, killed a quarter of the domesticated animals in Mongolia, resulting in a subsistence crisis. Although roughly one-third of Mongolians still practice nomadic pastoralism (**Figure 10.13**), the country's prime minister recently announced that he expected this way of life to diminish over the next few decades.

Changing Foodways of Pastoral Peoples The food of the Mongols and other traditional pastoral people of Central Asia has not made much of a global impact. Although "Mongolian barbeque" restaurants are found in the United States, the dishes that they serve are generally of Taiwanese origin. Traditional Mongolian cuisine is heavily focused on meat, which is abundantly produced in the country's pastoral economy. The average Mongol eats more than 200 pounds (90 kilograms) of meat each year. Milk and other dairy products are also extremely important. Dried yoghurt is an old staple food that can be easily ported from camp to camp. Kumis, or fermented mare's milk, is the traditional Mongolian

▲ **Figure 10.14 Mongolian Cuisine** *Bantan*, a traditional Mongolian dish, is a simple soup made with meat (usually mutton), dumplings, and chunks of fat.

Explore the **Tastes** of Mongolian Food

http://goo.gl/ZR1AGm

drink. Although slightly alcoholic, kumis is widely consumed by Muslim Kyrgyz and Kazakh herders as well.

Mutton is the most common meat among Central Asian pastoralists. It is usually served as a stew along with a few vegetables, and is traditionally cooked in a pot suspended over an open fire. In the traditional Mongolian *boodog* cooking style, hot stones are inserted into the carcass of a small animal, often a marmot (woodchuck), which is then roasted over open flames and sometimes singed with a blowtorch. Such meals are easily prepared outdoors with minimal equipment (**Figure 10.14**).

Mongolian food is not limited to animal products. Dumplings and noodles are popular, although they were originally borrowed from China. Intriguingly, a few Mongolians are even turning to vegetarianism, a radical departure from traditional practices. Roughly 1 percent of the people of Mongolia now avoid meat. Vegetarianism in Mongolia is associated with the revival of Buddhism, a religion that discourages the taking of life. Until recently, however, Mongolian Buddhists ignored their faith's stress on vegetarianism out of necessity, as meat was often the only food available. In the modern globalized world, however, a meat-free diet is feasible in urban areas across the planet.

Population Issues

The demographic data on Central Asia show overall fertility rates near the global middle, slightly above the replacement level (Table 10.1). Afghanistan, however, has a much higher birth rate, as might be expected based on its lack of social development, religious conservatism, and male domination.

▼ **Figure 10.13 Steppe Pastoralism** The northern and central Mongolian steppes offer lush pastures during the summer. Mongolians, some of the world's most skilled horse riders, have traditionally followed their herds of sheep and cattle, living in collapsible, felt-covered yurts. Many Mongolians still follow this way of life. **Q: Why is nomadic pastoralism so much more widespread in Mongolia than in most other countries?**

TABLE 10.1   Population Indicators

Country	Population (millions) 2015	Population Density (per square kilometer)[1]	Rate of Natural Increase (RNI)	Total Fertility Rate	Percent Urban	Percent <15	Percent >65	Net Migration (Rate per 1,000)
Afghanistan	33.4	48	2.9	5.3	27	44	2	1
Azerbaijan	9.8	115	1.1	2.1	53	22	6	0
Kazakhstan	17.8	6	1.5	2.6	53	25	7	−1
Kyrgyzstan	6.1	30	2.2	3.8	36	33	4	−1
Mongolia	3.1	2	2.1	3.1	67	27	4	0
Tajikistan	8.6	59	2.5	3.6	26	35	3	−3
Turkmenistan	5.4	11	1.3	2.2	50	28	4	−1
Uzbekistan	31.9	72	1.8	2.5	36	29	5	−1

Source: Population Reference Bureau, *World Population Data Sheet,* 2015.

[1] World Bank, *World Development Indicators,* 2016

Login to MasteringGeography™ **& access** MapMaster **to explore these data!**

1) Map the Total Fertility Rates for this region. Do the TFR values correlate with other data categories in this table or in Table 10.2?

2) Now map the Net Migration data. What correlations can you make with other data categories in this table or in Table 10.2?

Tajikistan, another relatively poor country, also has a somewhat elevated birth rate. The fertility rate of Kazakhstan, on the other hand, was below replacement level as recently as 2000, but has since rebounded and is now over 2.6.

Fertility rates are also relatively low in western China, but this region is still experiencing substantial population growth due to immigration. Most migrants in Tibet and especially Xinjiang are Han Chinese from eastern China. The indigenous inhabitants tend to resent this influx, but the Chinese government claims that it is necessary for economic development.

Migration has also complicated demographic patterns in the former Soviet zone. After the breakup of the Soviet Union, millions of Russians and Ukrainians left Central Asia to return to their homelands. Russia's economic boom in the early 2000s led many Azerbaijanis, Tajiks, Uzbeks, and other Central Asians to seek employment in Moscow and St. Petersburg, but Russia's recent economic downturn has led many labor migrants to seek opportunities elsewhere (see *People on the Move: Central Asian Labor Migrants Seek Alternatives to Russia*).

Urbanization in Central Asia

Although the steppes of northern Central Asia had no real cities before the modern age, the river valleys have been partially urbanized for thousands of years. Cities such as Samarkand and Bukhara in Uzbekistan have long been famous for their lavish architecture (**Figure 10.15**). This early urban development was built on the region's economic and political prominence. The Amu Darya and Syr Darya valleys lay near the midpoint of the trans-Eurasian silk route, and they formed the core of several ancient empires based on the cavalry forces of the steppe.

The conquest of Central Asia by the Russian and Chinese empires ushered in a new wave of urban formation.

Cities began to appear for the first time on the steppe. The Manchu and Chinese conquerors of Xinjiang built new administrative and garrison cities, often placing them a few miles from indigenous urban sites. The old cities of the area are characterized by complex networks of streets and alleyways, whereas the neighboring Chinese cities were constructed according to a strict geometrical order. The continuing growth of urban populations, along with the establishment of new policies, is now obscuring this old dualism. In 2009, Beijing announced that the old city of Kashgar (Kashi) in western Xinjiang would be rebuilt, and by 2014 more than two-thirds of it had been demolished, forcing residents to move to new apartment complexes. Although some of Kashgar's historic buildings were saved, most local people view such reconstruction as an assault on their cultural identity.

▼ **Figure 10.15 Traditional Architecture in Samarkand** Famous for its lavish Islamic architecture—some of it dating back to the 1400s—Samarkand, Uzbekistan, owes its rich architectural heritage in part to the fact that it was the capital of the empire created by the great medieval conqueror Tamerlane.

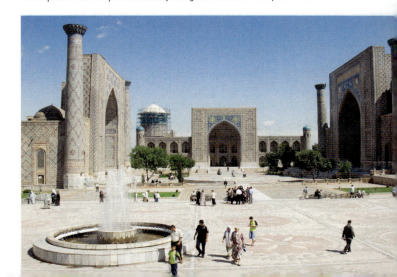

People on the Move

Central Asian Labor Migrants Seek Alternatives to Russia

After the collapse of the Soviet Union, the poorer countries of former Soviet Central Asia—Uzbekistan, Tajikistan, and Kyrgyzstan—came to rely heavily on seasonal labor migration to Moscow and other large Russian cities. Russia's booming economy during the first decade of the 2000s generated abundant jobs, especially in the construction industry. Russian-speaking Central Asians filled many of these positions, and the money that they sent to their families helped support the economies of their home countries. A 2011 World Bank report estimated that 47 percent of Tajikistan's GDP derived from funds sent back from its workers living abroad, mostly in Russia.

The situations faced by Central Asian laborers in Russia are often grim. Working conditions are often arduous and wages low, even if they are much higher than what one could expect at home. Tajiks and Uzbeks working in Russia also face increasing discrimination on racial and religious grounds. Attacks by extreme Russian nationalists on foreign workers are not uncommon.

Growing Obstacles in Russia Beginning in 2014, labor migration from Central Asia to Russia slowed considerably. Owing primarily to the global decline in energy prices, the Russian economy suffered a sharp recession, eliminating many jobs. Galloping inflation, moreover, has reduced the purchasing power of migrants still working in Russia. Russian hostility toward Central Asians is also increasing, in part because migrants tend to be blamed for the heightened crime rate in Russia's major cities. The Russian bureaucracy has responded by placing obstacles on labor migrants, forcing them to be tested for drug addiction, HIV, tuberculosis, and other diseases (**Figure 10.2.1**).

Due to these issues, the number of Uzbeks working in Russia dropped by 15 percent during 2015, while remittances to Uzbekistan fell from US$5.6 billion to US$3.0 billion in the same period. But despite the decline, almost 2 million Uzbek citizens still work in Russia, and Uzbekistan continues to rely on the money that they send home. In 2015, Uzbekistan's president negotiated with Moscow to simplify the rules restricting labor immigration, but without success. Instead, the Russian government offered to import more Uzbek products.

New Destinations As work opportunities in Russia diminish, would-be labor migrants are seeking alternative destinations. Some are looking to other parts of Central Asia, particularly Kazakhstan. Although Kazakhstan has also suffered from the global downturn in the oil market, it remains much more prosperous than its Central Asian neighbors, and massive construction projects are still under way in its burgeoning capital, Astana. Uzbek and Tajik workers generally find Kazakhstan to be much more culturally familiar than Russia, and more tolerant as well.

Other Central Asian labor migrants are finding work much further away. The United Arab Emirates, particularly Dubai, has recently emerged

▲ **Figure 10.2.1. Central Asian Workers Being Harassed in Russia** Large numbers of young men from Tajikistan and other former Soviet countries of Central Asia work for low wages in the major cities of Russia. Russian hostility toward Central Asian migrant workers, however, is pronounced and has increased in recent years as Russia's economic growth has stalled out. Here we see Russia police detaining migrant workers during a raid on a vegetable warehouse complex in Moscow in 2013.

as a favorite destination. Turkey has attracted as many as 50,000 Uzbek workers, but its growing political instability and lagging economy are now reducing the flow. A less-expected alternative has emerged in South Korea. As of early 2016, more than 16,500 Uzbeks had officially registered for work in the country. In the same year, South Korea and Uzbekistan signed a provisional memorandum of understanding that would allow in larger numbers of Uzbek workers. South Korea may be a distant and culturally unfamiliar destination, but its vibrant economy does offer many relatively high-paying jobs.

1. What are some of the potential advantages and disadvantages of working in Dubai for Central Asian labor migrants?

2. What other countries in different world regions face similar issues with labor migration and remittance-dependent economies? How do these countries deal with these issues?

Although it is possible to distinguish Russian/Soviet cities from indigenous cities in the former Soviet zone, this dichotomy is not clear-cut. In Uzbekistan, Tashkent is largely a Soviet creation, while many parts of Bukhara and Samarkand still reflect the older urban patterns and architectural forms. Several major cities, such as Kazakhstan's former capital of Almaty, did not exist before Russian colonization. A more recently established city, Astana, was designated Kazakhstan's capital in 1997 (**Figure 10.16**). Noted for its futuristic architecture, Astana is an example of a **forward capital**,

which is a new capital established in what had previously been a peripheral area in order to incorporate it more fully into the national territory.

Urbanization is gradually but unevenly spreading across Central Asia. Afghanistan has recently seen a huge expansion of Kabul, its capital and major city. Kabul now boasts upscale neighborhoods with trendy shopping streets and multistory malls with explosive-resistant windows and metal detectors. Even Mongolia, long a land virtually without permanent settlements, now has more people living in cities

▲ **Figure 10.16** **Astana, Capital of Kazakhstan** A forward capital, Astana has grown rapidly in recent years, emerging as a new metropolis near the middle of the vast country. Oil wealth has supported Astana's expansion.

Explore the **Sights** of Astana, Kazakhstan

http://goo.gl/s9BQeQ

than in the countryside. Roughly 1.4 million people, almost half of Mongolia's population, now reside in Ulaanbaatar, the capital. Ulaanbaatar has grown so rapidly that many of its residents still live in traditional felt-covered tents, enjoying few urban services.

In some parts of Central Asia, however, cities remain relatively few and small. Only 26 percent of the people of Tajikistan, for example, are urban residents. Tibet similarly remains a predominantly rural society, although the influx of Han Chinese into the region is creating a much larger urban system. To understand this movement and its broader ramifications, we must examine Central Asia's patterns of cultural geography.

REVIEW

10.4 Why are large parts of Central Asia so sparsely settled, yet others have dense populations?

10.5 Why is the urban environment of Central Asia changing so rapidly?

10.6 Why is nomadic pastoralism so historically important in Central Asia, and how is this adaptation currently changing?

■ **KEY TERMS** pastoralists, transhumance, exotic rivers, alluvial fans, loess, forward capital

Cultural Coherence and Diversity: A Meeting Ground of Different Traditions

Although Central Asia has a degree of environmental unity, its cultural coherence is more questionable. The western half of the region is largely Muslim and is sometimes grouped with Southwest Asia. In the east, Mongolia and Tibet are

historically characterized by Tibetan Buddhism. Tibet is culturally linked to both South and East Asia, and Mongolia is intimately associated with China, but neither fits easily within any world region.

Historical Overview: Steppe Nomads and Silk Road Traders

The river valleys and oases of Central Asia supported agricultural communities dating to the Neolithic period (beginning circa 8000 BCE). After the domestication of the horse around 4000 BCE, nomadic pastoralism emerged in the steppe belt. Eventually, pastoral peoples gained power over the entire region, transforming not only the history of Central Asia, but also that of virtually all of Eurasia. Before the development of gunpowder, highly mobile pastoral nomads enjoyed major military advantages over sedentary societies despite their much smaller populations.

Until the modern era of steamships, Central Asia was also a crucial location for long-distance trade. Much of the economic exchange between East Asia and both Europe and Southwest Asia passed along several parallel routes traversing the region known as the **Silk Road**. Religious practices also moved along these trade routes. Buddhism spread from India to China, for example, by way of Central Asia.

The linguistic geography of Central Asia has undergone major changes over the ages. Into the first millennium CE, most inhabitants of the region spoke Indo-European languages closely related to Persian. More than a thousand years ago, these languages began to be replaced on the steppe by Turkic languages (belonging to the Altaic family) spread by nomadic pastoralists. This process was never completed, however, and southwestern Central Asia remains a meeting ground of the Indo-European and Turkic languages.

The Turks were eventually replaced on the eastern steppes by another group of Altaic speakers, the Mongols. In the late 1100s, the Mongols, under the leadership of Genghis Khan, united the pastoral peoples of Central Asia and used the resulting force to conquer sedentary societies. By the late 1200s, the Mongol Empire had grown into the largest contiguous empire the world had ever seen, stretching from southern China to eastern Europe (**Figure 10.17**). In today's Mongolia, Genghis Khan has reemerged as the national hero.

Protected by mountain barriers and the rigorous climate, Tibet took a different course from the rest of Central Asia. It emerged as a strong kingdom around 700 CE, but Tibetan unity and power did not persist and the region reverted to semi-isolation. Tibet was briefly incorporated in the 1200s into the Mongol Empire, and in later centuries other Mongol states occasionally exercised limited power over the Tibetans. These interactions resulted in the establishment

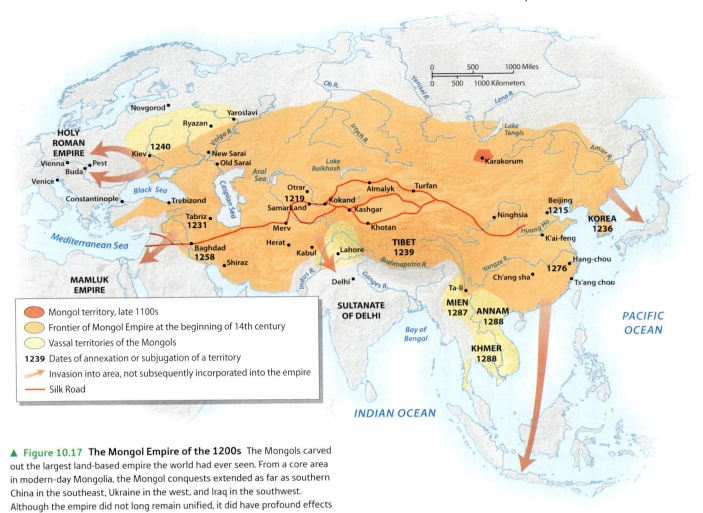

▲ **Figure 10.17 The Mongol Empire of the 1200s** The Mongols carved out the largest land-based empire the world had ever seen. From a core area in modern-day Mongolia, the Mongol conquests extended as far as southern China in the southeast, Ukraine in the west, and Iraq in the southwest. Although the empire did not long remain unified, it did have profound effects on the subsequent political and economic history of Eurasia.

of Mongolian communities in the northeastern part of the plateau and in the eventual conversion of the Mongolians to Tibetan Buddhism.

Contemporary Linguistic and Ethnic Geography

Today most of Central Asia is inhabited by peoples speaking Altaic languages, but patterns of linguistic geography remain complex (**Figure 10.18**). A few indigenous Indo-European languages are confined to the southwest, and Tibetan remains the main language of the Tibetan Plateau. Russian is also widely spoken in the west, and retains an official status in most former Soviet countries. Similarly, Mandarin Chinese is increasingly important in the east. In both Tibet and Xinjiang, Mandarin Chinese is now the language of higher education. Most urban merchants in Tibet are Chinese-speaking immigrants, causing much concern among the indigenous people.

Tibetan   Tibetan is divided into several distinct dialects that are spoken over almost the entire Tibetan Plateau. Approximately 6 million people speak Tibetan; of these, roughly 2.5 million live in Tibet proper, with most of the rest residing in China's

provinces of Qinghai and Sichuan. Tibetan is usually placed in the Sino-Tibetan family, implying a shared linguistic ancestry between the Chinese and the Tibetan peoples, but some scholars argue that there is no definite relationship between the two. Tibetan has an extensive literature written in its own script, mainly devoted to religious topics.

Mongolian   Mongolian is composed of a cluster of closely related dialects spoken by approximately 5 million people. The standard Mongolian of both the independent country of Mongolia and China's Inner Mongolia is called Khalkha; other Mongolian dialects include Buryat (found in southern Siberia) and Kalmyk (found to the northwest of the Caspian Sea). Mongolian has its own distinctive script, but Mongolia itself adopted the Cyrillic alphabet of Russia in 1941. In China's Autonomous Region of Inner Mongolia the old script is still widely used. Today, however, only about 4 million of the 25 million residents of Inner Mongolia are Mongol speakers, as the majority population is Han Chinese.

Turkic Languages   Far more Central Asians speak Turkic languages than Mongolian and Tibetan combined. Central Asia's Turkic linguistic sphere extends from Azerbaijan in

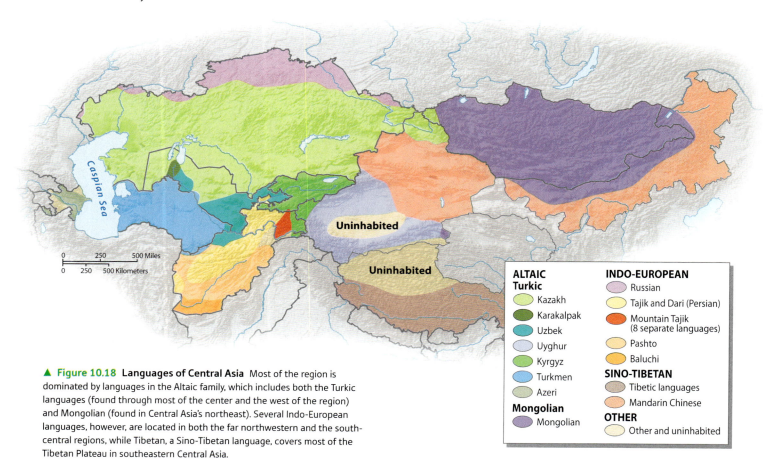

▲ **Figure 10.18 Languages of Central Asia** Most of the region is dominated by languages in the Altaic family, which includes both the Turkic languages (found through most of the center and the west of the region) and Mongolian (found in Central Asia's northeast). Several Indo-European languages, however, are located in both the far northwestern and the south-central regions, while Tibetan, a Sino-Tibetan language, covers most of the Tibetan Plateau in southeastern Central Asia.

the west through China's Xinjiang Province in the east. The various Turkic languages are not as closely related to each other as are the dialects of Mongolian, but they are still kindred tongues. Six main Turkic languages are found in Central Asia; five are associated with former Soviet republics of the west, and the sixth, Uyghur, is the main indigenous language of northwestern China.

Uyghur is an old and important language. The Uyghur number about 10 million, most of whom live in Xinjiang. As recently as 1949, Uyghurs comprised about 90 percent of the population of Xinjiang, but due to Han Chinese immigration that figure has dropped to 44 percent. The eastern and northern areas of Xinjiang are now largely Chinese-speaking. China is now discouraging the use of the indigenous language in schools in favor of Mandarin Chinese.

Five of the countries of the former Soviet zone—Kazakhstan, Uzbekistan, Turkmenistan, Kyrgyzstan, and Azerbaijan—are named after the Turkic languages of their dominant populations. In three of these countries, the indigenous people form a substantial majority. Over 90 percent of the people of Azerbaijan are classified as Azeri (there are more Azeris in northern Iran, however, than there are in Azerbaijan); some 80 percent of the people of Uzbekistan are classified as Uzbek; and between 80 and 90 percent of the people in Turkmenistan are classified as Turkmen. Most of these people speak their indigenous languages at home,

although many Kazakhs are more comfortable with Russian than with Kazakh.

With more than 27 million speakers, Uzbek is the most widely spoken Central Asian language. In the Amu Darya delta in northwestern Uzbekistan, however, people speak a Turkic language called Karakalpak. Kazakh herders are also widely scattered across Uzbekistan's sparsely populated deserts. In Uzbekistan's older cities, such as Samarkand, many residents still speak Tajik, a dialect of Persian. Tajik activists, moreover, claim that the number of Tajik speakers in Uzbekistan is much greater than official statistics indicate.

In the two other Turkic republics, the nationality for which the country is named is not so dominant. At the time of independence in 1991, the Kyrgyz formed only a bare majority in Kyrgyzstan. Due to Russian emigration and differential birth rates, however, roughly 72 percent of the country's inhabitants now identify themselves as Kyrgyz. The situation is similar in Kazakhstan, where Kazakhs and Russians each constituted about 40 percent of the population in 1991. Today the figures are 63 percent and 23 percent, respectively. (There are also about 1.8 million Kazakh speakers in northern Xinjiang in China and another 100,000 in western Mongolia; see *Geographers at Work: Kazakh Migration in Mongolia*.) In general, people of European descent live in the agricultural districts of northern Kazakhstan and in the cities of the southeast. Roughly a quarter million Central Asians are of ethnic Korean background, a legacy of the

Kazakh Migration in Mongolia

Geography's links with other disciplines comes to life in Mongolia, notes Holly Barcus, a geographer at Macalester College in Minnesota. Barcus studies the minority Kazakh population of Mongolia in collaboration with anthropologist Cynthia Werner of Texas A&M University. "Geography allows conversations, not only across the globe, but across disciplines," says Barcus. "Geographers enter these conversations with an understanding of history, politics, the natural environment, drawing on our background in each area to pursue a question." Barcus and Werner want to understand why roughly half of Mongolia's Kazakhs migrated to Kazakhstan after the end of the Cold War, why many later returned to Mongolia, and why still others simply stayed put (**Figure 10.3.1**).

Studying Migration Barcus's passion for geography was sparked by "the real-world linkages between topics discussed in class and being able to understand the process creating those outcomes—but also my own role in that process. At that moment, what I was studying became active rather than passive learning—something I enjoyed outside the classroom as well as in the classroom."

In 2004 Barcus initiated a long-term project on transnational migration in western Mongolia. She and Werner found that ethnic Kazakhs were not pushed out of Mongolia after 1991, but were instead enticed by the Kazakh government, which wanted to increase the proportion of ethnic Kazakhs in its population and had enough oil wealth to subsidize migrants. Migration later became more circular, as roughly a third of the 60,000-strong group that had left in the first wave returned to Mongolia. Some returnees complained of discrimination for being too rural and for not speaking Russian, still an important language in Kazakhstan. More recently, younger Kazakhs have left Mongolia for Kazakhstan for better education and career opportunities.

Barcus and Werner found that Mongolian Kazakhs generally frame their decisions around geographically informed narratives. The desire to stay in Mongolia is reinforced by ties to one's place of birth, while movement to Kazakhstan stems from the notion of the "original homeland." "The environment is so intrinsically related to our self-identity and how that identity plays into nationalism and migration decisions," says Barcus.

Barcus encourages her own students to travel: "People are moving around, and changes are occurring in the global economy. There are endless possibilities for going out and engaging." She also tells students that geography "is a practical skill that complements a whole range of

▲ **Figure 10.3.1. The Kazakhs of Mongolia** Geographer Holly Barcus visits an ethnic Kazakh family in western Mongolia. Dr. Barcus has been studying migration in this region since 2004.

occupations. At the undergraduate level you can't go wrong with geography. It forms the baseline for anything you want to do."

1. What cultural, historical, and environmental factors might be linked to the relative ease of movement of people from Mongolia to Kazakhstan, and back again?

2. Do you know immigrants who have returned or want to return to their home country? List push or pull factors that might influence their decision.

Soviet Union's Stalinist period, when Koreans were deported from the Russian Far East.

The Indo-European Zone The sixth republic of the former Soviet Central Asia, Tajikistan, is dominated by people who speak an Indo-European language. Tajik is so closely related to Persian (or Farsi) that it is considered a Persian dialect. Roughly 84 percent of the people of Tajikistan identify themselves as Tajiks, with most of the rest being Uzbeks. The remote mountains of eastern Tajikistan, however, are populated by peoples speaking a variety of distinctive Indo-European languages, sometimes collectively called "Mountain Tajik."

The linguistic geography of Afghanistan is more complex (**Figure 10.19**). Its two main languages are Pashto, the tongue of the Pashtun ethnic group that spans the border between Afghanistan and Pakistan, and Dari, which is another dialect of Persian. Estimates of the proportion of the populace speaking Pashto vary from 38 to 50 percent, most of whom live to the south of the Hindu Kush Mountains. A similar proportion of Afghans speak Dari, most of whom live in the north and west. Three separate ethnicities are ascribed to the Dari-speaking people: the settled farmers and townspeople in the west and north are classified as Tajiks; the traditionally seminomadic people of the west-center as Aimaks; and villagers of the central mountains as Hazaras, who are East Asian in appearance and reputed to be descendants of medieval Mongol conquerors. Although historically forming something of an underclass, Afghanistan's Hazaras are now showing more

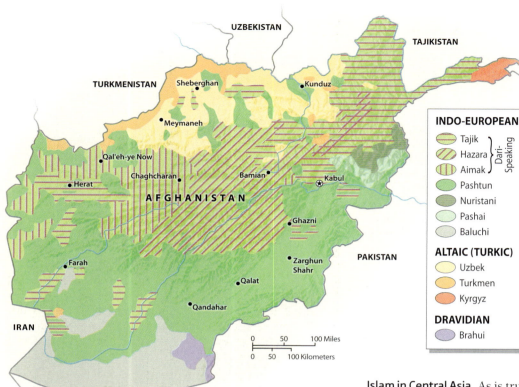

◀ Figure 10.19 Afghanistan's
Ethnolinguistic Patchwork
Afghanistan is one of the world's more ethnically complex countries. Its largest ethnic group is the Pashtuns, a people who inhabit most of the southern portion of the country, as well as the adjoining borderlands of Pakistan. Northern Afghanistan is inhabited mostly by Uzbeks, Tajiks, and Turkmens, whose main population centers are located in Uzbekistan, Tajikistan, and Turkmenistan, respectively. The Hazaras of Afghanistan's central mountains, like the Tajiks, speak a form of Persian, but are considered to be a separate ethnic group, in part because they, unlike other Afghans, follow Shiite rather than Sunni Islam. Smaller groups are found elsewhere in the country.

dedication to modern education than the country's other ethnic groups and as a result are beginning to gain some power.

Although Pashto and Dari are Afghanistan's official languages, a variety of other languages are spoken, creating a complex ethnic patchwork. Approximately 10 percent of Afghanistan's people speak Turkic languages, including both Uzbek and Turkmen. Other ethnolinguistic groups include the Baluchis, the Nuristanis (who speak five separate languages), the Pashai, all of whose languages are Indo-European, and the Brahui, who speak a Dravidian language related to the tongues of southern India. Such complex ethnic divisions give Afghanistan a weak national foundation, presenting serious challenges for the country's government.

Geography of Religion

An assortment of religions characterized ancient and medieval Central Asia. The major overland trading routes of premodern Eurasia crossed the region, giving easy access to merchants and missionaries. Varieties of Buddhism, Zoroastrianism, Islam, Christianity, Judaism, and other faiths thrived at various times and places. Eventually, however, religious lines hardened, and Central Asia was divided into two spiritual camps: Islam triumphed in the west and center, while Tibetan Buddhism prevailed in Tibet and Mongolia.

Islam in Central Asia   As is true elsewhere in the Muslim world, different Central Asian peoples hold different interpretations of Islam. The Pashtuns of Afghanistan are noted for their strict Islamic ideals—although critics contend that many Pashtun practices, such as forbidding women from showing their faces in public, are based more on their own customs. By contrast, the traditionally nomadic groups of the steppes, such as the Kazakhs and Kyrgyz, have historically been considered lax in their religious practices. Although most of the region's Muslims are Sunnis, Shiism dominates among the Hazaras of central Afghanistan, the Azeris of Azerbaijan, and the mountain dwellers of eastern Tajikistan (**Figure 10.20**).

Under communist rule in China, the Soviet Union, and Mongolia, all forms of religion were discouraged.

▼ Figure 10.20. **Hazara Mosque** Unlike most Afghans, the Hazara, whose ancestors came mostly from East Asia, follow Shia rather than Sunni Islam. As a result, they have been persecuted by the Taliban, an extremist Sunni organization. Here we see an elderly Hazara man on the roof of a Shia Mosque that is being repaired from the damage it incurred during the Afghan civil war of the 1990s.

Chinese authorities and student radicals severely suppressed Islam in Xinjiang during the Cultural Revolution of the late 1960s and early 1970s. Mosques were destroyed or converted to museums, and religious schools were closed. Periodic persecution of Islam also occurred in Soviet Central Asia, and until the 1970s many observers thought that Islam was slowly diminishing throughout the region.

Religion was not so easily discouraged, however, and interest in Islam began to grow in the former Soviet Central Asia in the 1980s. In the post-Soviet period, Islam continues to revive as people reassert their indigenous heritage and identity. Thus far, signs of Islamic political extremism are few in most areas, although the movement does have local adherents, especially in the Fergana Valley. An estimated 1400 to 4000 militants from former Soviet Central Asia have gone to Syria to fight with ISIS and other extreme Islamist groups.

Concerned about the situation in Afghanistan and the possible lure of radical Islamism among young people, governments of the former Soviet republics have clamped down on religious practices. Tajikistan, for example, prohibits teachers under age 50 from growing beards, commonly regarded as a symbol of fundamentalism, and has closed unregistered mosques. (In response to criticism from its own religious leaders, Tajikistan also banned revealing Western clothing and other practices.) Such repression led the United States to place Tajikistan on its list of violators of religious freedom in 2016. Turkmenistan and Uzbekistan were placed on the same list a decade earlier.

China has been restricting the religious practices of Muslims in Xinjiang, closely monitoring mosques and religious schools out of fear that they are entangled with political separatist movements. Although Islam has served as a focal point of an anti-Chinese movement among the Uyghurs, most Uyghur leaders insist that their program is essentially secular. Chinese fears were heightened, however, in 2009 when ethnic rioting pitting Uyghurs against Han Chinese took as many as 200 lives in Ürümqi, the capital of Xinjiang Province, and again in 2015 when more than 50 Chinese miners were stabbed in their dormitory beds by militants. In response, harsh anti-Muslim measures were instituted. In the city of Kashgar, mosques are prevented from broadcasting the call to prayer, and after-school religious classes have been banned. In nearby Hotan, more than 20 personal names have been banned for being "too Muslim," and in many areas men are prevented from growing beards and women from veiling their faces.

In Afghanistan, radical Islamic fundamentalism has emerged as a powerful political movement. From the mid-1990s until 2001, most of the country was controlled by the **Taliban**, an extremist organization that insists that all aspects of society conform to its harsh version of Islamic orthodoxy. This commitment was demonstrated in early 2001 when Afghanistan's religious authorities oversaw the destruction of Buddhist statues in the country, including some of the world's largest and most magnificent works of art. Although the Taliban regime fell in late 2001, Taliban insurgents continue to control territory, hoping eventually to rule the entire country.

Islam is not the only religion found in western Central Asia. Many Russians belong to the Russian Orthodox Church, and Uzbekistan has a small Jewish population. Kazakhstan also counts some 100,000 Catholics, mostly people of Polish or German descent.

Tibetan Buddhism   Mongolia and Tibet stand apart from the rest of Central Asia in their practice of Tibetan Buddhism. Buddhism entered Tibet from India many centuries ago, where it merged with the indigenous religion, called Bon. The resulting faith is more oriented toward mysticism than other forms of Buddhism, and it is more hierarchically organized. Standing at the apex of Tibetan Buddhism is the **Dalai Lama** (a term derived from the words for "Great Teacher"), considered to be a reincarnation of the Bodhisattva of Compassion (a Bodhisattva is a spiritual being who helps others attain enlightenment). Ranking below him is the Panchen Lama, followed by other religious officials. Until the Chinese conquest, Tibet was essentially a **theocracy**, or religious state, with the Dalai Lama enjoying political as well as religious authority (**Figure 10.21**).

Tibetan Buddhists suffered severe persecution in the 1960s. The Dalai Lama fled to India with many of his followers after China unleashed a military crackdown on Tibet in 1959 and has since been a powerful advocate for the Tibetan cause in international circles. From 1960 to 1979, an estimated 6000 Tibetan monasteries were destroyed, and thousands of monks were killed. Although China later allowed many monasteries to reopen, the number of active monks today is only about 5 percent of what it was before the Chinese occupation. Still, the Buddhist faith continues to form the basis of Tibetan identity. In 2008, Tibetan monks led a series of protests against the Chinese government, and soon afterward ethnic riots between Tibetans and Han Chinese broke out across the Tibetan Plateau. The Chinese government reacted harshly and later blamed the Dalai Lama for inciting the unrest. More recently, Tibetan monks and nuns have taken to burning themselves to death to protest Chinese policies. As of February 2016, 145 Tibetans have engaged in such acts of self-immolation.

In Mongolia, the downfall of communism led to a revival of Tibetan Buddhism. Several monasteries have been refurbished, and many people are returning to their national religion. Buddhist belief, however, is not as intense in Mongolia as it is in Tibet. About half of the people of Mongolia actively follow Buddhism, whereas about 40 percent identify themselves as nonreligious.

Central Asian Culture in Global Context

Although remote and not closely integrated into global economic and cultural circuits, Central Asia is hardly immune to the forces of globalization. Certainly the

▲ **Figure 10.21 Tibetan Buddhist Monastery** Tibet is well known for its large Buddhist monasteries, which at one time served as seats of political as well as religious authority. The Ganden Monastery is one of the three great "university monasteries" of Tibet.

Explore the **Sights** of Tibet's Ganden Monastery

http://goo.gl/Yarzyo

hyper-modern buildings appearing in Astana, Kazakhstan and Baku, Azerbaijan show strong global influences (see **Figure 10.22**). Even the tensions existing throughout the region between deeply religious and largely secular orientations and between ethnic nationalism and multiethnic inclusion are aspects of the current global condition. So, too, the increased usage of English throughout Central Asia shows that it is not cut off from global culture. Such

influences are especially marked in the oil cities of the Caspian Basin, such as Baku.

The Russian Language Issue During the Soviet period, the Russian language spread widely through western Central Asia. Russian served as both a common language and a means of instruction in higher education. People had to be fluent in Russian in order to reach any position of responsibility. The Cyrillic (Russian) script, moreover, replaced the modified Arabic script previously used for the indigenous languages.

After the fall of the Soviet Union, however, most Russians migrated back to Russia, especially from the region's poorer countries. As a result, the use of Russian in education, government, business, and the media has declined in favor of local languages, and the Cyrillic alphabet has been replaced by the Latin one in Turkmenistan and Uzbekistan. According to a 2013 survey, fewer than 20 percent of young people in Turkmenistan and Tajikistan speak Russian. But Russian still has the status of the "language of inter-ethnic communication" in

► **Figure 10.22 Baku, Azerbaijan** Baku is a vibrant and increasingly cosmopolitan city closely linked to global economic and cultural networks. Baku is now well known for its new buildings and modern architecture.

Tajikistan and Turkmenistan, irritating local nationalists. In Kyrgyzstan and Kazakhstan, Russian continues to be widely used in the media and education, generating protests by anti-Russian Kazakh student groups in 2010. Meanwhile, the number of ethnic Russians declines each year, due both to low birth rates and emigration to Russia. Kyrgyzstan, for example, saw its Russian population drop from 419,600 in 2009 to 364,600 in 2015.

An Alternative Turkic World? Many Turkic-speaking Central Asians want to move away from the Russian world and instead emphasize their own cultural and historical commonalities. This idea, once called "pan-Turkism," dates back to the 1880s when it was used to bolster the Ottoman Empire and resist that of the Russians. Today the movement operates mostly in the cultural sphere, its main vehicle being the annual "Turkvision" Song Contest, modeled on the popular Eurovision Song Contest. All countries and regions with sizable Turkic-speaking ethnic groups are eligible to enter. In 2016, however, the contest was marred by geopolitical rivalry as Russia forbade entry by people from its own Turkic-speaking regions due to its geopolitical tensions with Turkey.

Some of the more traditional forms of music are shared among the nomadic Turkic and Mongolian peoples of Central Asia. One prominent example is overtone singing (also known as "throat singing"; see Chapter 9), in which a person produces more than one pitch at the same time. Believed to have originated in southwestern Mongolia, overtone singing is still widely practiced in the area.

Globalization and Sports Central Asia has been relatively slower to embrace the globalization of sports. Mongolia's athletic culture is perhaps the most distinctive, as traditional forms of wrestling and horseback riding remain the marquee sporting events. Every summer, the three-day Naadam festival brings thousands of competitors to Ulaanbaatar to engage in wrestling, horse racing, and archery.

Horse racing is popular through most of Central Asia, which is not surprising considering the region's nomadic past. Mongolian racehorses are usually bred for endurance, as most races are cross-country events that span over 10 miles. In repressive Turkmenistan, horse races are one of the few occasions in which citizens can gather in large numbers. The country has even raised the Akhal-Teke, considered to be the world's oldest horse breed, to the status of national symbol. In Afghanistan, Tajikistan, Kazakhstan, and Kyrgyzstan, the sport of Buzkashi, in which a horse-mounted team competes to drag a goat carcass across a goal line, remains popular (**Figure 10.23**). Such equestrian pursuits—along with archery and wrestling—help link Central Asian cultures to their historical pasts.

▲ **Figure 10.23** **The Wild Sport of Buzkashi** The ancient sport of Buzkashi, played in many parts of Central Asia, entails horse-riders pulling a goat or calf carcass over a goal line. Traditionally, matches could last several days, but new rules now limit the playing time. Widely regarded as the national sport of Afghanistan, Buzkashi was outlawed by the Taliban when they ruled most of the country in the late 1990s. **Q: What historical factors might help explain the popularity of Buzkashi in Central Asia?**

REVIEW

10.7 What was the historical significance of the Silk Road for Central Asia?

10.8 How have the patterns of linguistic geography in Central Asia been transformed over the past two decades?

10.9 How do patterns of religious affiliation divide Central Asia into distinct regions, and why is religion the source of increasing tensions in the region?

■ **KEY TERMS** Silk Road, Taliban, Dalai Lama, theocracy

Geopolitical Framework: Political Reawakening in a Power Void

Central Asia has played a minor role in global political affairs for the past several hundred years. Before 1991 all of the region except Mongolia and Afghanistan was under direct Soviet or Chinese control. Mongolia, moreover, was a close Soviet ally, and even Afghanistan came under Soviet domination in the late 1970s. Today, southeastern Central Asia remains part of China, which is increasingly cementing its power over the area. And although the breakup of the Soviet Union saw the emergence of six new Central Asian countries, most of them retain close political ties to Russia (**Figure 10.24**).

Partitioning of the Steppes

Before 1500, Central Asia was a power center, a region whose mobile armies threatened the far more populous, sedentary states of Asia and Europe. The development of new weapons changed the balance of power, however, allowing the wealthier agricultural states to conquer the nomads. By the 1700s, the nomad armies had been defeated and their lands taken. The winners in this struggle were the two largest states bordering the steppes: Russia and China.

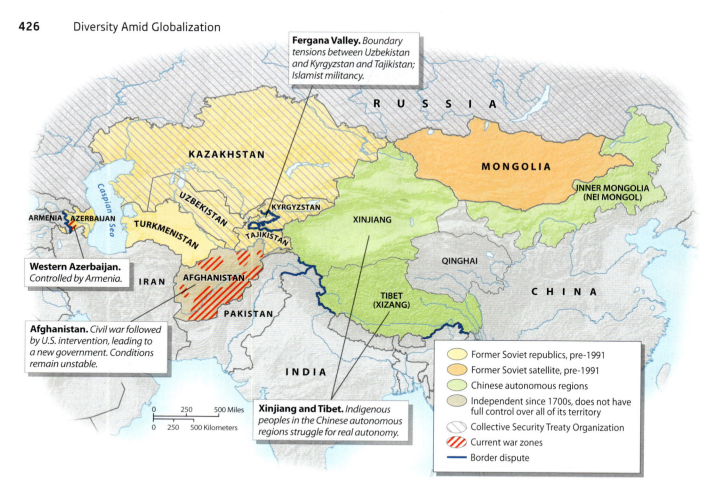

Fergana Valley. *Boundary tensions between Uzbekistan and Kyrgyzstan and Tajikistan; Islamist militancy.*

Western Azerbaijan. *Controlled by Armenia.*

Afghanistan. *Civil war followed by U.S. intervention, leading to a new government. Conditions remain unstable.*

Xinjiang and Tibet. *Indigenous peoples in the Chinese autonomous regions struggle for real autonomy.*

Former Soviet republics, pre-1991
Former Soviet satellite, pre-1991
Chinese autonomous regions
Independent since 1700s, does not have full control over all of its territory
Collective Security Treaty Organization
Current war zones
Border dispute

▲ **Figure 10.24 Geopolitical Issues in Central Asia** Six of the eight independent states of Central Asia came into existence in 1991 with the dissolution of the Soviet Union. In eastern Central Asia, the most serious difficulties stem from China's maintenance of control over areas in which the indigenous peoples are not Chinese. Afghanistan, scene of a prolonged and brutal civil war, has experienced the most extreme forms of geopolitical tension in the region.

By the mid-1700s the Chinese empire (under the Manchu, or Qing, dynasty) stood at its greatest territorial extent, including Mongolia, Xinjiang, Tibet, and a slice of modern Kazakhstan. But China weakened rapidly after 1800. When the Manchu dynasty fell in 1912, Mongolia became independent, functioning as a **buffer state** between China and Russia, although China still ruled the extensive borderlands of Inner Mongolia. Tibet also gained effective independence, although China did not recognize that status.

In the mid-1800s Russia began to push south of the steppe zone, in part to keep the British out, taking over most of western Central Asia by 1900. Britain attempted to conquer Afghanistan, but failed. Subsequently, Afghanistan's position as an independent buffer state between the Russian Empire (later the Soviet Union) and British India remained secure.

Central Asia Under Communist Rule

Western Central Asia came under communist rule not long after the emergence of the Soviet Union in 1922. Mongolia followed in 1924. After the Chinese revolution of 1949, a communist system was also imposed on Xinjiang and Tibet.

Soviet Central Asia The Soviet Union retained all the Central Asian territories of the Russian Empire. The new regime sought to build a Soviet society that would eventually knit together all the territories of the vast country. Central Asia's leaders were replaced by Communist Party officials loyal to the new state, Russian immigration was encouraged, and local languages had to be written in Cyrillic (Russian) rather than Arabic script.

Although the early Soviet leaders thought that a new Soviet nationality would eventually emerge, they realized that local ethnic diversity would not disappear overnight. They therefore divided the Soviet Union into a series of nationally defined "union republics," in which a certain degree of cultural autonomy was allowed. In the 1920s, Kazakhstan, Kyrgyzstan, Tajikistan, Uzbekistan, Turkmenistan, and Azerbaijan assumed their present shapes as Soviet republics. In some areas, particularly around the densely populated Fergana Valley, the resulting boundaries were extremely complicated, generating geopolitical problems that persist to the present (**Figure 10.25**).

Contrary to official intentions, the new republics encouraged the development of local national identities more than a broader Soviet identity. Also undercutting Soviet unity was

KAZAKHSTAN

Fergana Valley

Bishkek
Tokmok
Cholpon-Ata
Talas
Balykchy
Ysyk-Köl
Karakol
Toktogul
Tashkent
Songköl
Kara-Say
Jangy-Bazar
Tash Kömür
Olmaliq
KYRGYZSTAN
Namangan
Jalal-Abad
Naryn
Fergana Valley
Andijon
Özgön
At-Bashy
Guliston
Konsoy
Quqon
Osh
Jizzakh
Fergana
Khujand
Konibodom
Kyzyl-Kyya
Uroteppa
Sülüktü
UZBEKISTAN
Sary Tash
CHINA
Samarkand
Panjakent
Zarafobod
Jirgatol
Navabad
TAJIKISTAN
Tursunzoda
Dushanbe
Qal'ai Khum
Dinau
Dangara
Morghob
Kulob
Qurghonteppa
Khorugh
Termiz
Panj
Amu Darya R.
Syr Darya R.

AFGHANISTAN

0   50   100 Miles
0   50   100 Kilometers

◀ **Figure 10.25 Political Boundaries Around the Fergana Valley** Some of the world's most convoluted political boundaries can be found in the vicinity of the Fergana Valley. The central portion of the valley belongs to Uzbekistan, which is otherwise separated from it by high mountains. Part of the lower valley, on the other hand, is part of Tajikistan, the core area of which is likewise separated from the valley by highlands. The Fergana's upper periphery belongs to Kyrgyzstan. Note also the small exclaves of Uzbekistan within Kyrgyzstan.

Explore the **Sights** of Political Geography of Fergana Valley

http://goo.gl/WK1RcM

the fact that cultural and economic gaps separating Central Asians from Russians did not decrease as much as planned. In addition, the region's higher birthrates led many Russians to fear that the Soviet Union risked being dominated by Turkic-speaking Muslims, contributing to the breakup of the country.

The Chinese Geopolitical Order After decades of political and economic chaos, China reemerged as a united country in 1949. Its new communist government quickly reclaimed most of the Central Asian territories that had slipped away from China in the early 1900s. China's new leaders promised the non-Chinese peoples a significant degree of political self-determination and thus found much local support in Xinjiang. Gaining control over Tibet was more difficult. China occupied Tibet in 1950, but the Tibetans rebelled in 1959. When this rebellion was crushed, the Dalai Lama and some 100,000 followers found refuge in India.

Loosely following the Soviet model, China established autonomous regions in areas occupied primarily by non–Han Chinese peoples, including Xinjiang, Tibet proper (called *Xizang* in Chinese), and Inner Mongolia. Such autonomy, however, often meant little, and certainly did not prevent Han Chinese immigration to these areas. Nor were all parts of Chinese Central Asia granted autonomous status. The large and historically Tibetan and Mongolian province

of Qinghai, for example, remained an ordinary Chinese province.

Current Geopolitical Tensions

Although the former Soviet portion of Central Asia weathered the post-1991 transition to independence relatively smoothly, the region currently suffers from several ethnic conflicts, as well as from the struggle between radical Islam and secular governments. Much of China's Central Asian territory is also troubled, but Beijing retains a firm grip. Such problems are insignificant, however, compared to the situation in Afghanistan, where brutal warfare has been waged for decades.

Independence in Former Soviet Lands After 1991, the six newly independent countries of the former Soviet Union had some difficulties charting their own political courses. They still had to cooperate with Russia on security issues, and all initially joined the Commonwealth of Independent States, the hollow successor of the Soviet Union. In most cases authoritarian rulers, rooted in the old order, retained power and sought to undermine opposition groups.

All told, democracy has made less progress in Central Asia than in other parts of the former Soviet Union. Kyrgyzstan forms a partial exception, as it generally runs open elections. Such a policy has not, however, generated stability. In 2010 demonstrators brought down the government, forcing Kyrgyzstan's president to flee. Three months later, massive ethnic riots broke out in southern Kyrgyzstan as supporters of the country's ousted president attacked ethnic Uzbeks.

In Tajikistan, war broke out almost immediately after independence. Many members of the smaller ethnic groups of the mountainous east resented the authority of the Tajiks and rebelled. They were joined by Islamist groups seeking to overthrow the secular state. The civil war ended in 1997 when the Islamists agreed to work through normal political channels. However, continuing instability led Tajikistan to increase its pressure on the Islamic Renaissance Party, the only legal Islamist party in the former Soviet zone. Tajikistan's government has also enhanced its own power. In 2016, its leader was declared "president for life" after winning a referendum by a suspiciously large margin.

Azerbaijan also experienced strife following the breakup of the Soviet Union. Armenia invaded and occupied a large portion of western Azerbaijan, allowing the Armenian-speaking highlands to form the "breakaway republic" of Nagorno-Karabakh. Hostile relations with Armenia make it difficult for Azerbaijan to govern its **exclave** of Nakhchivan, a piece of territory separated from the rest of the country by Armenia and Iran. Although the Nagorno-Karabakh conflict had been viewed as inactive, it reignited in April 2016 when Azerbaijani forces moved into Armenian-controlled territory, resulting in the deaths of 27 Armenian and 28 Azerbaijani troops (**Figure 10.26**). A Russian-brokered truce quickly brought an end to the fighting with no significant territorial changes. This conflict will likely continue to fester, however, as neither side is willing to compromise.

Uzbekistan and Turkmenistan had relatively easy transitions out of the Soviet Union, but both countries are ruled by repressive governments that allow little personal freedom or opposition. According to a 2015 Human Rights Watch report, authorities in Uzbekistan "repress all forms

of freedom of expression and do not allow any organized political opposition, independent media, free trade unions, independent civil society organizations, or religious freedom." Turkmenistan is even more repressive, ranking third from last in the 2016 Press Freedom Index. As is the case in North Korea, images of its leader are ubiquitous; in 2015, Turkmen authorities erected a 21-meter-tall statue of President Gurbanguly Berdymukhammedov riding a horse on top of a massive marble pedestal.

Strife in Western China   Local opposition to Chinese rule is pronounced among the region's indigenous communities. Protests by Tibetans have not been successful, but have brought the world's attention to their struggle. China maintains several hundred thousand troops in an area with only some 3 million civilian inhabitants. Such an overwhelming military presence is considered necessary because of both Tibetan resistance and the strategic importance of the region. Massive Tibetan protests in 2008 were handled harshly by the Chinese government. Protests broke out again in 2016, focused this time on the expansion of mining operations on a mountain considered sacred by Tibetan Buddhists.

China also maintains tight control over Xinjiang. Although fragmented by deserts and mountains, its terrain is less forbidding than that of Tibet, and it is now home to millions of Han Chinese immigrants. Xinjiang is also an economically vital part of China. It contains rich mineral deposits and has been the site of nuclear weapons tests. Many Uyghurs, not surprisingly, oppose such uses of their homeland, and resent the suppression of their religion by the regime. Strife between Han Chinese and Uyghurs resulted in more than a thousand deaths between 2007 and 2015. In 2016, China launched a campaign to promote ethnic understanding and unity, encouraging companies to hire Uyghurs. Critics complain, however, that continuing restrictions on religion undermine the initiative.

Other parts of western China have experienced ethnic tensions, but always in a violent manner. In late 2015, for example, ethnic Mongolians wearing colorful traditional costumes rode horses and camels though the town of Eznee in Inner Mongolia to protest an alleged attack by Han Chinese migrants on their grazing lands.

China's position is that its Central Asian lands are integral portions of the national territory. Those advocating independence are viewed as traitors and are often considered to be in league with Western political forces that have sought to keep China weak and divided for the past 200 years.

War in Afghanistan   No other Central Asian conflict compares in intensity to the struggle being waged in Afghanistan. Afghanistan's descent began in 1978, when a Soviet-supported "revolutionary council" seized power. The new government began to suppress religion, leading to widespread resistance. To prevent its

▼ **Figure 10.26** **2016 Armenia-Azerbaijan Conflict** The so-called frozen war between Armenia and Azerbaijan, which dates from the break-up of the Soviet Union in 1991, occasionally reignites into active fighting. Here we see a soldier from the Armenian Nagorno Karabakh defense army walking past tanks at a field position near the front line on April 6, 2016.

collapse, the Soviet Union launched a massive invasion. Despite its power, the Soviet military was never able to control the more rugged parts of the country. Pakistan, Saudi Arabia, and the United States, moreover, ensured that anti-Soviet forces remained well armed. When the exhausted Soviets finally withdrew in 1989, brutal local warlords grabbed power in most areas.

In 1995, a movement called the Taliban arose in Afghanistan. Founded by religious students, the Taliban advocated the strict enforcement of Islamic law. The new model attracted militants, and by September 2001 only far northeastern Afghanistan lay outside Taliban power. But at the same time, most Afghans were turning against the group, primarily because of the severe restrictions that it imposed on daily life. These constraints were most pronounced for women, but even men were compelled to obey the Taliban's numerous decrees. Most forms of recreation were simply outlawed, including television, films, music—even kite flying.

The attacks of September 11, 2001 in the United States changed the balance of power in Afghanistan. The United States and Britain, working with Afghanistan's anti-Taliban forces, quickly overthrew the Taliban government. Although a democratic government was established within a few years, peace did not return. The new Afghan government proved corrupt and ineffective, failing to establish security in most parts of the country.

By 2004, the Taliban had regrouped, operating from safe havens in Pakistan. Afghanistan's new government has had to rely on the military power of an international coalition led by the North Atlantic Treaty Organization (NATO). But despite its strength, this force was unable to stop the Taliban resurgence. In 2009, the United States sent an additional 17,000 troops to Afghanistan. U.S. forces have continued to target high-level Taliban leaders, often by bombing their compounds, many of which are located in Pakistan. This strategy weakened the Taliban command structure, but also generated numerous civilian casualties. At the end of 2014, NATO ended formal military operations, planning to pull out remaining U.S. troops by 2016.

But as has often happened in Afghanistan, events on the ground undermined such plans. As efforts to reach a ceasefire with the Taliban collapsed in 2015, fighting intensified. In early 2016, the Taliban even managed to besiege the supposedly secure city of Kunduz in northern Afghanistan, and experts estimated that Taliban forces had regained control over one-third of the country (**Figure 10.27**). As a result, by mid-2016 the United States backtracked on its plans to continue pulling out its forces. Some evidence indicates that the Afghan military is finally turning itself into a capable fighting force, but the government remains weak and plagued by corruption. Although Kabul, the capital, is usually peaceful, explosions

Explore the **Sounds** of War in Afghanistan

http://goo.gl/3YOmfx

▲ **Figure 10.27 Battle of Kunduz, 2016** The important northern Afghan city of Kunduz was besieged by the Taliban twice during 2016. Here we see Afghan security personnel standing alert as they prepare for an operation against Taliban militants near the city on June 12, 2016. **Q: What ethnic factors may have contributed to the fact that the Kunduz area temporarily fell to the Taliban, despite being considered relatively secure? Refer to Figure 10.19 to help answer this question.**

are not uncommon and the sounds of helicopters and military planes remind people that their country is still at war.

Global Dimensions of Central Asian Geopolitics

As the previous discussion indicates, Central Asia has emerged as a key arena of geopolitical tension. Vying for power and influence are several important countries, including China, Russia, Pakistan, India, Iran, Turkey, and the United States. In addition, the revival of Islamic fundamentalism has generated international geopolitical repercussions.

A Continuing U.S. Role Afghanistan is not the only Central Asian country to have experienced a U.S. armed presence. After 9/11, the United States established military bases in Uzbekistan, Tajikistan, and Kyrgyzstan, countries eager to host the U.S. military because they feared that radical Islamists could overthrow their governments. In the late 1990s, the Islamic Movement in Uzbekistan (IMU) was engaging the armies of Uzbekistan, Tajikistan, and Kyrgyzstan. As a result, military budgets increased, straining government finances but allowing Central Asian governments to get an upper hand on the IMU.

Since 2005, U.S. influence has declined across the former Soviet zone. In that year, Uzbekistan shut down an American military base after the United States criticized the Uzbek government for human rights abuses. In 2014, the United States closed its air base in Kyrgyzstan due to pressure from the Kyrgyz government, which in turn was pressured by Russia.

U.S. relations with Mongolia, in contrast, have strengthened in recent years. In 2012 Mongolia signed an agreement of cooperation with NATO, and in 2014 the U.S. Secretary of Defense thanked the country for its military help in Afghanistan and Iraq. Mongolia seeks close ties not only with the United States but also with South Korea and Japan, due largely to its concerns about the power of China and Russia.

Central Asian Relations with China and Russia After the breakup of the Soviet Union, relations between the newly independent countries of Central Asia and China remained tense for several years. China objected to the boundaries separating it from Tajikistan and Kyrgyzstan. These disputes were peacefully settled shortly after 2000, with Kyrgyzstan ceding 222,400 acres (90,000 hectares) of land. China subsequently initiated a series of diplomatic maneuvers designed to quell tensions, fight Islamic separatism, and gain access to the natural resources of Central Asia.

After the fall of the Soviet Union, Russia continued to regard the former Soviet territories of Central Asia as lying within its sphere of influence, and resented U.S. military initiatives. Russia has, moreover, retained its own military bases in both Tajikistan and Kyrgyzstan. Economics and infrastructure also tie Russia to the former Soviet republics. One of the main transportation lines linking European Russia to central Siberia cuts across northern Kazakhstan, and the area's rail and pipeline links to the outside world are still partly oriented toward Russia.

In the early 2000s, most Central Asian leaders concluded that the potential economic and political advantages from cooperation with Russia and China outweighed the disadvantages. One consequence was the formation of the **Shanghai Cooperation Organization (SCO)**, composed of China, Russia, Kazakhstan, Kyrgyzstan, Tajikistan, and Uzbekistan. (Turkmenistan refused to join, as it remains somewhat isolationistic.) The SCO seeks cooperation on terrorism and separatism and aims to enhance trade. Recently, however, its geopolitical significance has diminished, in part because of disagreements between Russia and China, but also due to the organization's own expansion. India and Pakistan—which remain bitter enemies—are scheduled to join the SCO in 2017.

Russia has sought to enhance its power more directly through the **Collective Security Treaty Organization (CSTO)**, a military alliance that links Russia to Kazakhstan, Kyrgyzstan, Tajikistan, Armenia, and Belarus. An attack on any CSTO member state is to be regarded as an attack on all others, and the organization holds annual military exercises. But Uzbekistan's exit in 2012 weakened the CSTO, and it has not emerged as a powerful force. Due to its own financial problems, Russia is having difficulties maintaining its status in the region, while the economic power of China continues to increase.

The Roles of Iran, Pakistan, and Turkey Iran is a major trading partner of several Central Asian countries and offers a good route to the ocean. Since the completion of a rail link between Iran and Turkmenistan in the late 1990s, some of Central Asia's global trade has been conducted through Iran's ports. Iran's cultural links with the region are old and deep, particularly in Tajikistan, northern Afghanistan, and Azerbaijan. The Iranian government has gained influence in Afghanistan, especially in the western region of Herat and among the Shiite Hazaras, but problems with Afghan refugees in Iran generate tensions between the two

countries. Iran's relations with Azerbaijan remain tense, in part because many Azerbaijanis regard northwestern Iran—an Azeri-speaking region—as "Southern Azerbaijan."

Pakistan has also strived to gain influence in the region, hoping that pipelines will deliver Central Asian fossil fuels to its new deep-water port at Gwadar. During the Taliban period, Pakistan enjoyed close relations with Afghanistan. The aftermath of 9/11 thus put Pakistan in a serious bind. In response to both pressure and promises from the United States, Pakistan joined the anti-Taliban coalition. The subsequent retreat of the Taliban into northwestern Pakistan, however, harmed relations between the two countries. Another complication is Afghanistan's refusal to recognize the legitimacy of the border separating the two countries, claiming that it was a British imposition. Relations further deteriorated in 2016 after Pakistan began restricting movement across the boundary and building hardened border terminals. Two days of military clashes at one border crossing resulted in several deaths and at least 23 injuries.

Turkey's connections with Central Asia also run deep. Most Central Asians speak Turkic languages, and Turkey also offers itself as the model of a modern, secular state of Muslim heritage. It is unclear, however, whether the Turkish system can be maintained even in Turkey itself, much less exported to other Turkic-speaking states. Turkish interest in Central Asia declined, moreover, after 2011 when the Syrian war forced it to refocus its efforts on its own neighborhood. Turkey's increasingly tense relationship with Russia further complicates its relations with Kazakhstan, Kyrgyzstan, and Tajikistan, all of which are Russian allies.

REVIEW

10.10 How did the 1991 collapse of the Soviet Union change the geopolitical structure of Central Asia, and how has Russia attempted to maintain influence in the region?

10.11 How does the geopolitical situation of Afghanistan differ from those of the other countries of the region, and why is Afghanistan's political history so different from those of its neighbors?

10.12 What role do international treaty organizations play in the geopolitics of Central Asia?

■ **KEY TERMS** buffer state, exclave, Shanghai Cooperation Organization (SCO), Collective Security Treaty Organization (CSTO)

Economic and Social Development: Abundant Resources, Struggling Economies

By most measures, Central Asia is a poor region. Afghanistan is near the bottom of almost every list of economic and social indicators. Other Central Asian countries, however, have moderately high levels of health and education, a legacy of the social programs of their former communist regimes.

Unfortunately, these same governments built inefficient industrial systems, and after the fall of the Soviet Union, western Central Asia experienced a spectacular economic decline.

Although most of Central Asia experienced solid economic growth in the first 15 years of the new millennium, expansion was based primarily on the extraction of natural resources, particularly oil and natural gas. When global energy prices tumbled in 2015, the region's economy faltered as well. Further problems resulted from the simultaneous downturns in the Chinese and especially Russian economies, as China and Russia are major markets for Central Asian exports. The region's landlocked position poses additional challenges for economic development, as it increases transportation costs.

Economic Development in Central Asia

Soviet planners sought to spread the benefits of economic development across their vast country. This required building large factories, even in remote areas such as Tajikistan, regardless of the costs. Such Central Asian industries relied on subsidies from the central government. When those subsidies ended, the industrial base began to collapse, leading to plummeting living standards. But as is true elsewhere in the former Soviet Union, certain well-connected individuals grew very wealthy after the fall of communism.

Post-Communist Economies   As Table 10.2 shows, no Central Asian country or region could be considered prosperous by global standards. Kazakhstan stands as the most developed and probably has the best prospects. The country has two of the world's largest underutilized deposits of oil and natural

gas—the vast Tengiz and Kashagan fields in the northeastern Caspian Basin—and contains numerous fossil-fuel pipelines (**Figure 10.28**). Kazakhstan has signed agreements with Western oil companies to exploit its oil reserves, including a US$37 billion deal with Chevron in 2016, and has received major investments from China. Between 2001 and 2008, Kazakhstan's oil and gas wealth made it one of the world's fastest-growing economies, and after a brief recession in 2009 it quickly bounced back. But after the 2015 drop in global energy prices, the Kazakh economy again faltered. But in the same year, Kazakhstan joined the World Trade Organization, boosting its long-term prospects.

Owing to its sizable population, Uzbekistan is the region's second largest economy. Uzbekistan did not decline as sharply as its neighbors after the fall of the Soviet Union, largely because it retained many aspects of the old command economy (that is, an economy run by government planners rather than private firms responding to the market). Uzbekistan is one of the world's largest cotton exporters, and has significant gold and natural-gas deposits. However, environmental degradation threatens cotton production, and both inefficient state management and widespread corruption continue to hamper production, resulting in widespread poverty.

Kyrgyzstan, in contrast, aggressively privatized its industries after the fall of the Soviet Union. However, Kyrgyzstan's economy is heavily agricultural, few of its industries are competitive, and political strife discourages foreign investment. Kyrgyzstan does enjoy the largest supply of freshwater in Central Asia, which will likely become increasingly valuable, and its mineral reserves are substantial. A single facility, the Kumtor Gold Mine, accounts for roughly half of Kyrgyzstan's industrial output, and when production

TABLE 10.2   Development Indicators

Country	GNI per capita, PPP 2014	GDP Average Annual % Growth 2009–15	Human Development Index (2015)[1]	Percent Population Living Below $3.10 a Day	Life Expectancy (2016)[2] Male	Life Expectancy (2016)[2] Female	Under Age 5 Mortality Rate (1990)	Under Age 5 Mortality Rate (2015)	Youth Literacy (% pop ages 15–24) (2005–2014)	Gender Inequality Index (2015)[3,1]
Afghanistan	2,000	6.8	0.465	–	59	62	192	91	47	0.693
Azerbaijan	16,920	2.9	0.751	‹2.0	72	77	95	32	100	0.303
Kazakhstan	21,710	6	0.788	‹2.0	68	77	57	14	100	0.267
Kyrgyzstan	3,220	4.2	0.655	–	67	74	70	21	100	0.353
Mongolia	11,120	11.7	0.727	4.0	65	75	107	22	98	0.325
Tajikistan	2,660	7.2	0.624	23.4	66	73	114	45	100	0.357
Turkmenistan	14,520	11.3	0.688	69.1	62	70	94	51	100	–
Uzbekistan	5,830	8.2	0.675	87.8	71	76	75	39	100	–

Source: World Bank, *World Development Indicators,* 2016.

[1] United Nations, *Human Development Report,* 2015.

[2] Population Reference Bureau, *World Population Data Sheet,* 2016.

[3] Gender Inequality Index—A composite measure reflecting inequality in achievements between women and men in three dimensions: reproductive health, empowerment and the labor market that ranges between 0 and 1. The higher the number, the greater the inequality.

Login to MasteringGeography™ **and access** MapMaster **to explore these data!**

1) How closely do under age 5 mortality figures correlate with the GNI per capita figures across Central Asia?

2) Which Central Asian country stands out from the others across these different developmental measures? Provide reasons for the disparities.

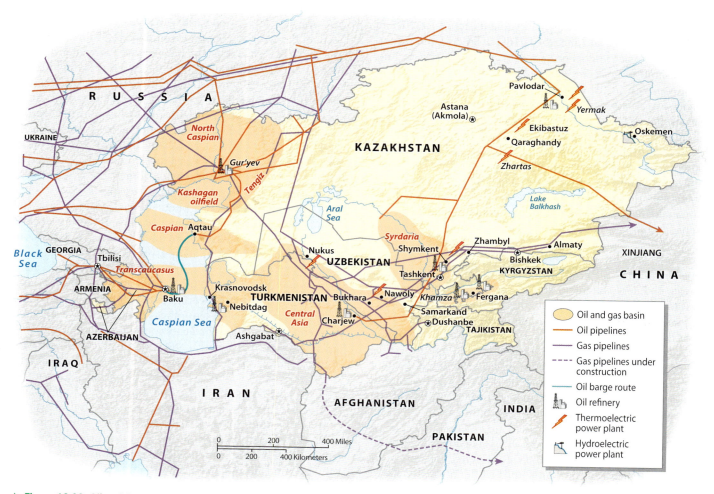

▲ **Figure 10.28 Oil and Gas Pipelines** Central Asia has some of the world's largest oil and gas deposits and recently has emerged as a major center for drilling and exploration. Because of its landlocked location, Central Asia cannot easily export its petroleum products. Pipelines have been built to solve this problem, and others are currently being planned.

tumbles due to strikes or other problems, as occurred in 2012, the entire country suffers. Kyrgyzstan depends heavily on remittances, deriving over 30 percent of its GDP from this source. The recent decline in the Russian economy has thus been a major blow to Kyrgyzstan.

Turkmenistan is another major cotton exporter, but its economy rests mostly on natural gas. A new gas pipeline to China gave the country an economic boost, but worries mounted in 2015 when Russia ceased importing its gas. Turkmenistan is now focusing on the potentially huge South Asian market, initiating in 2015 the massive Turkmenistan-Afghanistan-Pakistan-India (TAPI) pipeline. Some experts fear, however, that instability in Afghanistan along with tensions between India and Pakistan could halt the project. On the down side, Turkmenistan retains an inefficient state-run economy, has resisted pressure for economic liberalization, and remains cut off from many global economic institutions.

Central Asia's oldest fossil fuel industry is located in Azerbaijan. Azerbaijan has attracted a great deal of international investment, helping to revitalize its oil industry (**Figure 10.29**). Through the 1990s, its economy responded slowly, but in the first decade of the new century it surged

▼ **Figure 10.29 Oil Development in Azerbaijan** Oil has brought a certain amount of wealth to Azerbaijan, but it has also produced extensive pollution. Because most of the petroleum is located either near or under the Caspian Sea, this sea—actually the world's largest lake—is becoming increasingly polluted, and its fisheries are declining.

Explore the **Sights** of Azerbaijan's Oil Industry

http://goo.gl/cJuUDc

ahead. Since then, Azerbaijan's economy has been highly volatile. Critics contend that the country has not enacted adequate market reforms and remains too dependent on oil. Deteriorating economic conditions in 2016 resulted in a sharp drop in the value of Azerbaijan's currency, leading in turn to extensive protests, some of which were shut down by the country's military.

The most economically troubled of the former Soviet republics is Tajikistan. With a per capita gross national income (GNI) of only some $2660, Tajikistan is a poor country. It is burdened by its remote location far from the main roads and railways, rugged topography, lack of natural resources, and political tensions. Almost half of the country's labor force lives abroad, mostly in Russia and Kazakhstan, making Tajikistan the world's most remittance-dependent country. Russia's recent economic decline has therefore generated severe problems. According to one estimate, Tajikistan saw a 65 percent decline in the value of its remittances from Russia in 2015.

Tajikistan's government hopes to revive its fortunes by opening new transportation routes to China and by generating more hydroelectricity. A large new dam was finished in 2009, another was completed in 2011, and others are under construction. Enhanced electricity supplies are allowing Tajikistan to expand its aluminum smelting, one of its few competitive industries. The biggest of these projects is the Rogun Dam on the Vakhsh River. Although construction continues, Tajikistan has not yet gained the financing necessary to complete this huge dam.

Although never part of the Soviet Union, Mongolia was a close Soviet ally, run by the Communist Party. It, too, suffered an economic collapse in the 1990s when Soviet subsidies ended. During the early 2000s, however, mining investments by Chinese and Western firms resulted in rapid expansion, with total GDP increasing by 17.5 percent in 2011. During this same period, Mongolia began to develop the Oyu Tolgoi copper–gold project, one of the largest mining operations in the world. But in 2015, disputes between the government and foreign investors resulted in a sharp drop in funding, which, compounded by the global drop in mineral prices, stalled out the Mongolian economy. Mongolia's government is currently trying to reduce dependence on the Chinese market and seek alternative sources of foreign investment.

The Economy of Western China The Chinese portions of Central Asia did not suffer the economic crash that visited other parts of the region with the fall of the Soviet Union. Although China's growth rate slowed considerably after 2011, it is still expanding at a relatively rapid rate. The country's centers of dynamism are located in the distant coastal zone, but even the remote western parts of the country have experienced a significant degree of development. From 2002 to 2012, Inner Mongolia's economy expanded at a scorching annual rate of around 17 percent, owing in large part to its extensive deposits of rare earth minerals and other mineral resources (see *Everyday Globalization: Rare Earths from Inner Mongolia*).

Everyday Globalization

Rare Earths from Inner Mongolia

Rare earth elements form crucial components of the high-tech economy, impacting daily life across the world in many ways. Dysprosium, for example, improves high-powered magnets and lasers, yttrium is used in microwave filters for communications equipment and in energy-efficient light bulbs, and gadolinium has medical and computer memory applications.

As their name suggests, the 17 rare earth elements are not easy to find or extract. India, Brazil, South Africa, and the United States were once major producers, but by 2010 China accounted for 95 percent of total world output. Most Chinese rare earths come from Inner Mongolia's Bayan Obo Mining district near the Mongolian border (**Figure 10.4.1**). Bayan Obo is regarded as an environmental disaster zone, as huge quantities of dust, toxins, and radioactive waste are produced in the mining process. China's near monopoly on rare earth production has led other countries, including the United States, Canada, Brazil, Tanzania, Australia, Vietnam, and Malaysia, to try to develop their own reserves.

1. Is it important for the United States to mine its own rare earth deposits?

2. What objects do you own that probably contain rare earth elements?

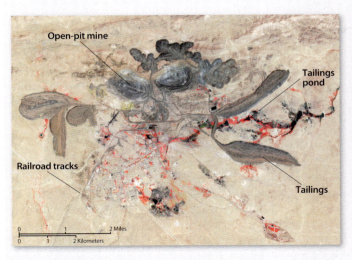

▲ **Figure 10.4.1. Rare Earth Mining in Inner Mongolia** China dominates the mining of rare earth elements, several of which are extremely important in high-tech manufacturing. Roughly half of China's rare earth production comes from Inner Mongolia's Bayan Obo mines. In this false-color satellite image of Bayan Obo, vegetation appears red, grassland is light brown, rocks are black, and water surfaces are green. Two circular open-pit mines are visible, as well as a number of tailing ponds and tailing piles.

Tibet remains burdened by poverty. Most of the plateau is relatively cut off from the Chinese economy and is even more isolated from the global economy. China, however, has been investing large amounts of money in infrastructure. In 2006, it inaugurated the monumentally expensive Qinghai–Tibet Railway, linking Tibet's capital city of Lhasa to the rest of the country. Reaching heights of over 16,400 feet (5000 meters), its trains must be equipped with supplemental oxygen (**Figure 10.30**). In 2016, plans were released for a second, more direct railway that will

▲ **Figure 10.30 Qinghai–Tibet Railway** China is engaged in a massive program to extend roads and railroads into the western half of the country. The main connection with the Tibetan Plateau is the recently completed Qinghai–Tibet Railway. This railway has greatly increased tourism in Tibet. **Q: Why have the railroad tracks been elevated so high above the valley floor?**

connect Tibet with the major city of Chengdu in Sichuan province.

As a result of these and other infrastructural projects, Tibet's tourism industry is booming. Its economy overall grew more quickly than that of China as a whole from 1995 to 2015. Many Tibetans, however, argue that most of the benefits are flowing to Han Chinese immigrants rather than to the indigenous peoples, and economic critics contend that subsidies from Beijing have unsustainably bolstered the Tibetan economy. China hopes that rapid economic development in Tibet, along with continuing Han Chinese migration, will reduce Tibetan resistance to Chinese rule.

Xinjiang has tremendous economic potential. It boasts significant mineral wealth and has benefited from recent development projects. The completion of oil and natural gas pipelines from Kazakhstan and Turkmenistan has bolstered its petrochemical industries. New infrastructural projects, including the Silk Road initiative and the China-Pakistan Corridor that extends to Pakistan's port of Gwadar, promise enhanced connections with the rest of the world. But as is the case in Tibet, many indigenous peoples of Xinjiang believe their region's wealth is being monopolized by the Chinese state and Han Chinese immigrants. Some point to the Xinjiang Production and Construction Corps, a quasi-military firm that runs hundreds of factories, farms, schools, and hospitals. This huge company was initially formed by the army that rejoined Xinjiang to China in 1949, and it is still run almost entirely by Han Chinese.

Economic Misery in Afghanistan Afghanistan is Central Asia's poorest country by a wide margin, with the lowest level of per capita production in Asia. Since the late 1970s, it has suffered nearly continuous war, thwarting most economic activities. Even before the war, it had experienced

little industrial or commercial development. Much of its exports are traditional products such as hand-woven carpets, fruits, nuts, and semiprecious gemstones.

The international community devoted roughly US$75 billion in development aid to Afghanistan between 2003 and 2015, hoping to jump-start its economy. Although official statistics indicate that the Afghan economy expanded rapidly for a few years after 2011, war, corruption, crime, and poor infrastructure have prevented most people from experiencing any gains. In 2015, moreover, the Afghan economy went into recession, due in part to cuts in foreign aid and investment. Yet Afghanistan is economically competitive in one highly problematic area: the production of illicit drugs for the global market.

By the late 1990s, Afghanistan had emerged as the world's leading producer of opium, used to produce heroin. In much of southwestern Afghanistan, opium is the main cash crop (**Figure 10.31**). Not only is opium highly profitable and easy to transport, but it also requires relatively little water. As much as US$100 million a year in narcotics profits have supposedly gone to the Taliban, which operates mainly in the opium-growing areas. As a result, both NATO and the Afghan government have tried to convince villagers to grow alternative crops, leading to a production decline from 2007 to 2010. But it soon rose again, with the 2014 harvest of 6400 metric tons setting a new record.

If Afghanistan solved its security problems, it could undergo a mining boom. China has invested heavily in the Ainak Copper Mine, projected to become one of the world's

▼ **Figure 10.31 Afghan Farmer in His Poppy Field** Most of the world's opium is produced in southern and eastern Afghanistan. Opium-poppy farmers can earn an adequate income from a small plot of land. **Q: Why is opium such an attractive crop, particularly for farmers in southern Afghanistan?**

Exploring **Global Connections**

The New Silk Road

From ancient times through the Middle Ages, East Asia was linked to Europe and Southwest Asia through the so-called Silk Road, a network of caravan routes crossing Central Asia. Such Central Asian cities as Samarkand and Bukhara, now in Uzbekistan, prospered as way stations and regional markets in the Silk Road system. But European maritime trade networks bypassed the Silk Road in the early modern period, resulting in a gradual decline in Central Asia's economic status and global significance.

Chinese Investments Over the past several years, a Chinese-led effort to reinvent the Silk Road as a modern transportation route has taken root. China is concerned about the security of the maritime trade networks that deliver its energy and raw materials, and it is keen to develop new overland connections. Central Asia itself is rich in minerals, oil, and natural gas, and Chinese leaders see the region as both a potential supplier of resources and a market for Chinese goods.

For these reasons, Chinese leader Xi Jinping announced the creation of the so-called Silk Road Economic Belt in 2013, promising massive investments. A Chinese state-owed investment firm pledged US$110 billion for various projects associated with this initiative. Current plans call for several road and rail corridors stretching from western China through Central Asia. One route will pass through Russia on its way to Europe, another will head to the Persian Gulf, and a third will terminate in South Asia.

The Development of Khorgas The current showcase of this New Silk Road is the formerly insignificant city of Khorgas (alternately, Qorğas or Horgos), located on the China–Kazakhstan border. Khorgas has been designated a free-trade zone, its businesses provided with lucrative tax breaks (**Figure 10.5.1**). Kazakhstan hopes to turn its side of the city into a "mini Dubai," complete with a dry port that will facilitate the storage and transshipment of goods, allowing containers to be quickly transferred across different railroad gauges.

Thus far, however, the development of the New Silk Road has been somewhat disappointing. Few of the facilities planned for Khorgas have actually been built, especially on the Kazakh side of the border. The environs are sparsely populated, feeder roads remain rough, and the larger Kazakh economy has faltered as oil prices have fallen. Ethnic tensions, moreover, are growing, as many Kazakhs fear that their country could become economically subordinated to China. In July 2015, a massive brawl broke out between Chinese and Kazakh workers at a copper mine in northern Kazakhstan, sending 65 men to a hospital.

Complications with Russia The New Silk Road project is also complicating the relations between Russia and China. Russia regards the former Soviet republics as part of its sphere of influence, and has been

▲ **Figure 10.5.1. China-Kazakhstan Border at Khorgas** The railway crossing from Kazakhstan to China at the small city of Khorgas is marked by a Chinese flag. China and Kazakhstan hope that Khorgas will emerge as a major station along the so-called New Silk Road, but thus far development has been limited.

pushing economic and geopolitical initiatives to enhance their connections with Moscow. But Russia, unlike China, does not have the financial power to develop the region's infrastructure. The Russian government initially tried to prevent the funding of the Silk Road project, but such efforts failed after China created the Asian Infrastructural Investment Bank in 2014. China then sought to reassure Russian leader Vladimir Putin that it was not intending to undermine Russia's political role in the region, and in 2015 Russia itself joined the Bank.

1. How much risk is entailed in the so-called New Silk Road project? What are some possible negative ramifications of the project?

2. Are there any infrastructural projects similar to the New Silk Road being carried out in other parts of the world? If so, how are they being financed?

Google Earth Virtual (MG) Tour Video

http://goo.gl/Vti2ly

---

largest mines. When fully operational, Ainak could employ as many as 20,000 Afghans. Taliban fighters, however, often target such foreign operations, reducing international investment in the country.

Central Asian Economies in Global Context Even Afghanistan —despite its poverty and relative isolation—is thoroughly embedded in the global economy, albeit partly through illicit products. But with the exceptions of the drug trade, the oil and natural gas industries, and other mining ventures, the overall extent of economic globalization in Central Asia remains low.

In the former Soviet republics, business connections with China are gradually replacing those with Russia. As early as 2012, Kazakhstan was sending 21 percent of its exports to China and only 10 percent to second-place Russia. Connections with China have been greatly enhanced by the completion of oil and gas pipelines, as well as by the transportation initiatives of the so-called New Silk Road, part of China's "One Belt, One Road" initiative (see Chapter 11) (See *Exploring Global Connections: The New Silk Road*).

The United States and other Western countries are attracted to Central Asia's oil and natural gas deposits and other mineral resources. Switzerland, surprisingly, is Kyrgyzstan's largest export partner, as it is the main destination of the country's gold. Many Western oil companies operate extensively in Azerbaijan and Kazakhstan. The delivery of Central Asian oil to the West was enhanced in 2006 with the opening of the Baku–Tbilsi–Ceyhan Pipeline—the second longest in the world—which runs from the Caspian Sea to the Mediterranean at the Turkish port of Ceyhan.

Central Asia's spectacular natural scenery and generally low population density present opportunities for international ecotourism. Kyrgyzstan hopes to build a sustainable tourism sector focused on Issyk Kul, the world's second largest alpine lake. Ecotourism companies in Mongolia offer tourists the opportunity to experience the nomadic lifestyle of traditional pastoral people, sleeping in felt tents, drinking fermented mare's milk, and riding horses over the great expanses of the steppe.

Social Development in Central Asia

Social conditions in Central Asia vary more than economic conditions (**Figure 10.32**). Afghanistan falls at the bottom of almost every measure of social development, but in the former Soviet territories levels of health and education

remain relatively high. Social conditions have improved recently in Tibet and Xinjiang, although probably not keeping pace with the progress made in the rest of China. Mongolia is noted for its relatively high levels of social development, as well as for its pronounced gender equity. Women have outnumbered men in Mongolian universities since the 1980s, and many have reached high positions in the government.

Social Conditions and the Status of Women in Afghanistan

Social conditions remain grim in most parts of Afghanistan. In 2015, the CIA ranked Afghanistan last in its global ranking of infant mortality. The country endures almost constant warfare, and its rugged topography hinders the provision of basic social and medical services. Illiteracy is commonplace, especially for women. According to 2015 UNESCO data, Afghanistan has the world's sixth lowest literacy rate (38 percent) and third lowest female literacy rate (24.2 percent).

Women in traditional Afghan society—especially Pashtun society—have very little freedom. Restrictions on female activities intensified in the 1990s under Taliban control. Taliban forces prohibited women from working, attending school, and even restricted medical care. After the Taliban lost power, its fighters continued to target female

▼ Figure 10.32 **Human Development in Central Asia** Levels of social development vary significantly across Central Asia. Afghanistan ranks as one of the least-developed countries in the world, whereas both Azerbaijan and Kazakhstan rank relatively high.

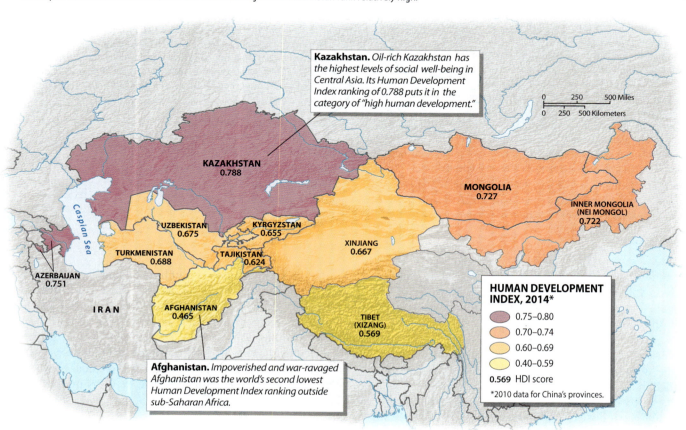

Kazakhstan. *Oil-rich Kazakhstan has the highest levels of social well-being in Central Asia. Its Human Development Index ranking of 0.788 puts it in the category of "high human development."*

KAZAKHSTAN
0.788

MONGOLIA
0.727

INNER MONGOLIA (NEI MONGOL)
0.722

UZBEKISTAN
0.675

KYRGYZSTAN
0.655

TURKMENISTAN
0.688

TAJIKISTAN
0.624

XINJIANG
0.667

AZERBAIJAN
0.751

IRAN

AFGHANISTAN
0.465

TIBET (XIZANG)
0.569

Caspian Sea

0   250   500 Miles
0   250   500 Kilometers

Afghanistan. *Impoverished and war-ravaged Afghanistan was the world's second lowest Human Development Index ranking outside sub-Saharan Africa.*

HUMAN DEVELOPMENT INDEX, 2014*

- 0.75–0.80
- 0.70–0.74
- 0.60–0.69
- 0.40–0.59

0.569 HDI score

*2010 data for China's provinces.

▲ **Figure 10.33** **Protest Against "Bride Abduction" in Kyrgyzstan** The kidnapping of young women by men who hope to marry them is an old problem in some parts of Central Asia, particularly Kyrgyzstan. In this photo we see Kyrgyz gynecologist Tukan Orunbayeva listening to women recount their abduction stories at a meeting organized to resist the practice. Orunbayeva is the founder of an NGO called Bakubat, which is devoted to protecting women's rights and delivering medical assistance.

education, destroying girls' schools throughout southern Afghanistan.

Under Afghanistan's 2004 constitution, 25 percent of parliamentary seats are reserved for women. Much evidence, however, indicates that the social position of women has improved little outside Kabul, the capital. According to a 2014 report by the U.S. Institute for Peace, 87 percent of Afghan women have been victimized by at least one episode of domestic violence. Throughout Afghanistan and especially in Kabul, many women fear the Taliban will again take control, eliminating the few gains that they have made.

Gender Issues in Mongolia and Former Soviet Central Asia In historical Central Asian societies, the social position of women varied considerably. Women generally had more autonomy in the pastoral societies of the steppe zone than in the agricultural and urban societies. In traditional Mongolian, Kazakh, and Kyrgyz societies, women usually worked closely with men and often had almost equal status. To some extent, such patterns persist to this day.

But even in the relatively gender-egalitarian societies of the steppes, women still suffer forms of discrimination. In traditional Kyrgyz society, for example, they have often been victimized by the practice of "bride abduction," in which kidnapped women are forced to marry their abductors. Although many such kidnappings took place with the consent of the bride-to-be, others were true abductions. The practice has recently resurged, leading to outrage by Kyrgyz women's groups (**Figure 10.33**). In early 2013, Kyrgyzstan passed a law to increase the maximum prison term for bride kidnapping to seven years, but one report notes that some 12,000 Kyrgyz young women and girls were abducted for this purpose in 2014 alone.

During the communist period, efforts were made to educate Central Asian women and place them in the workforce, although many encountered limited career advancement. Women in these countries still have relatively high levels of education. According to the World Economic Forum's *Global Gender Gap Report 2015*, female professional and technical workers outnumber male ones in Kazakhstan, Kyrgyzstan, Mongolia, and Azerbaijan. The same report also indicates that Mongolia has one of the world's highest percentages of women legislators, senior officials, and managers, surpassing even Sweden, Norway, and Iceland. A notable example is Sanjaasuren Oyun, Mongolia's environment minister, who was recently named president of the UN Environment Assembly.

Much of this achievement can be attributed to communist-era educational investment, leading to doubts about whether the region's relative gender equality can persist into the future. Several Central Asian countries have attempted to address such concerns by installing quotas for women in the government. By law, 30 percent of Kyrgyz and Uzbek parliament members must be female. Local women's groups complain, however, that their actual power is usually limited.

Demographic issues in Central Asia also influence the position of women. In Tajikistan, so many men have sought work in Russia that a large proportion of households are now led by women. At the same time, women are under pressure to follow more traditional gender roles due both to the rise of nationalism and the resurgence of fundamentalist Islam. In Kazakhstan, nationalist parties have tried to ban contraception for Kazakh women, hoping to increase both the country's population and the proportion of its ethnic Kazakh population. In Kyrgyzstan, feminists are now concerned about the increasing numbers of child brides, even though the country's legal age of marriage is 18. Many were disappointed in 2016 when the Kyrgyz parliament rejected a proposal that would have outlawed the religious consecration of marriages in which the bride is a minor.

REVIEW

10.13 Explain how the distribution of fossil fuel deposits influences economic and social development in Central Asia.

10.14 Why does the social position of women vary so much across Central Asia, and why is Afghanistan particularly problematic in this regard?

10.15 How is the Soviet legacy reflected in the social and economic development of western Central Asia?

Chapter 10 Central Asia

Summary

- The arid climate of Central Asia poses severe challenges for the region. Many lakes have diminished and a few have disappeared completely, and desert conditions have spread over many areas. Global climate change could further reduce the flow of the region's rivers.

- Pastoral nomadism is the traditional way of life over much of the region, but it is gradually diminishing as people move into permanent settlements. New cities have emerged in the grasslands of Central Asia, and older urban areas are being transformed by modernizing governments.

- Central Asia experienced a revival of religion—Islam in the west and Tibetan Buddhism over much of the east—after the fall or decline of communist ideology in the late twentieth century. Han Chinese migration into China's portion of Central Asia is culturally transforming the area, while the former Soviet republics of the region are gradually moving out of the Russian cultural orbit.

- Russia retains much geopolitical influence over the former Soviet portion of Central Asia, but the influence of China is growing. Conflicts connected with radical interpretations of Islam are a potential threat over much of the region, and have turned much of Afghanistan into a battleground.

- Fossil fuels and other mineral sources generated rapid economic growth in many Central Asian countries in the first decade and a half of the new millennium. More recently, the global drop in fossil fuel and commodity prices has resulted in a major economic downturn in the region.

Review Questions

1. How has the continental climate of Central Asia historically influenced the ways of life and modes or adaptations found across the region?

2. How have changing economic and political conditions in both Russia and former Soviet Central Asia influenced migration patterns in the region, considering both ethnic Russians and indigenous Central Asian peoples?

3. In what ways do cultural patterns, particularly those of language and religion, historically connect Central Asia to other world regions?

4. How have Russia and China sought to maintain or enhance their political power in the region, and in what ways have other major powers, including the United States and India, responded to such efforts?

5. Can Central Asia move beyond mining and fossil-fuel extraction to find a sustainable place in the global economy? Might it be possible for the region to use its position in the center of the world's largest landmass to become a major transportation hub?

Image Analysis

1. Examine the shrinking of the Aral Sea as revealed in these successive satellite images. After the Aral Sea split into two portions, why has the much smaller north lake retained its water level while the larger southern lake has continued to diminish? Why does the western part of the southern lake still exist in the final image, whereas the eastern part has disappeared completely?

2. Are there any lakes in the United States that have, or potentially could have, problems similar to those of the Aral Sea? If so, in what part of the country are they located, and why would that be the case?

▶ **Figure IA10** The Shrinking Aral Sea

(a) 1987   (b) 2004   (c) 2014

JOIN THE DEBATE

Mongolia has based its economic development program largely on producing minerals for the global market. Such policies resulted in rapid growth during the early 2000s, but the country experienced a major setback when the prices of most minerals tumbled in 2015. Has Mongolia been wise to rely so heavily on mining?

▲ **Figure D10** The Oyu Tolgoi Copper-Gold Mine in Khanbogd, Mongolia

Mining is Mongolia's best development strategy.

- Mongolia has large deposits of minerals vital for the world economy, and although prices have recently slumped, they will eventually recover.

- Mongolia has few people and is burdened by its remote location. As a result, it cannot depend on labor-intensive forms of development.

- Although mineral extraction often results in pronounced environmental degradation, such problems are highly concentrated in mining areas, bypassing most parts of the country.

Mining is a poor long-term strategy for Mongolia.

- Mongolia's mineral reserves are large, but they are still limited, and when the ores are exhausted the country will have to turn to other sources of income.

- Mining in Mongolia is highly dependent on China for both markets and capital, and as a result it puts the country in a precarious geopolitical situation.

- Mining not only causes intense local environmental degradation, but it also brings about pronounced health risks for mine workers.

■ KEY TERMS

alluvial fans *(p. 414)*
buffer state *(p. 426)*
Collective Security Treaty Organization (CSTO) *(p. 430)*
Dalai Lama *(p. 423)*
desertification *(p. 408)*
desiccation *(p. 408)*
exclave *(p. 428)*
exotic rivers *(p. 413)*
forward capital *(p. 417)*
loess *(p. 414)*
pastoralists *(p. 413)*
salinization *(p. 408)*
Shanghai Cooperation Organization (SCO) *(p. 430)*
Silk Road *(p. 418)*
steppes *(p. 406)*

Taliban *(p. 423)*
theocracy *(p. 423)*
transhumance *(p. 413)*

MasteringGeography™

Looking for additional review and test prep materials? Visit the Study Area in **MasteringGeography**™ to enhance your geographic literacy, spatial reasoning skills, and understanding of this chapter's content by accessing a variety of resources, including **MapMaster** interactive maps, videos, RSS feeds, flashcards, web links, self-study quizzes, and an eText version of *Diversity Amid Globalization*.

DATA ANALYSIS

http://goo.gl/yHXzJ7

Spectacular mountain scenery and rich history provides Central Asia with great tourism potential. The well-preserved cities of the Silk Road, such as Uzbekistan's Samarkand, are particularly well suited for tourism. However, international tourism has been slow to take off in Central Asia. The World Bank maintains data on international tourism receipts for most countries: http://data.worldbank.org/indicator/ST.INT.RCPT.CD

1. Write down the tourism-receipt figure for each Central Asian country for both 2010 and the most recent year available. Note if data is unavailable for any particular country. Use the data to construct a bar graph for the Central Asian countries for 2010 and the most recent year. Do the same for Turkey, a country outside the region that shares historical and cultural features with several Central Asian countries.

2. Suggest reasons for the differences you see on the graphs, based on environmental, population, cultural, geopolitical, and economic factors.

3. Could Central Asian countries significantly increase their tourism potential? How could they make this possible?

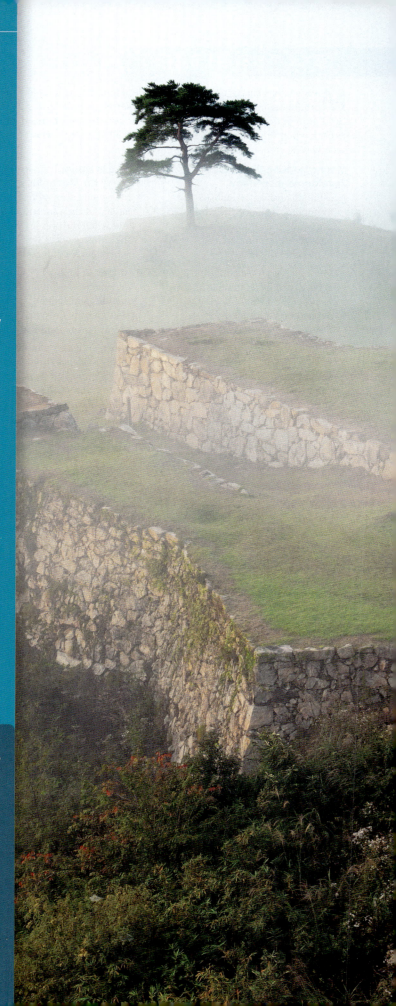

Environmental Geography

China has long experienced severe deforestation and soil erosion, and its current economic boom is generating the world's worst pollution problems. Japan, South Korea, and Taiwan all maintain extensive forests and relatively clean environments.

Population and Settlement

China is currently undergoing a major transformation as tens of millions of peasants move from poor villages in the interior to booming cities along the coastal region. Birth rates are low, and populations are aging throughout the region.

Cultural Coherence and Diversity

Historically, the region has been tied together by Mahayana Buddhism, Confucianism, and the Chinese writing system. Despite these unifying cultural features, East Asia in general and China in particular are divided along several striking cultural lines.

Geopolitical Framework

China's growing military power is generating tension with other East Asian countries, while Korea remains a divided nation. Japan, South Korea, and Taiwan have responded to China's growth by strengthening ties with the United States.

Economic and Social Development

Over the past several decades, East Asia has become a core area of the world economy, with China experiencing one of the most rapid economic expansions the world has ever seen. Japan, Taiwan, and South Korea are prosperous states, but North Korea remains desperately poor.

▶ Population decline in Japan and other parts of East Asia has led to the desertion and neglect of many buildings in recent years. The process of building abandonment, however, is nothing new. During the unification of Japan in the early 1600s, for example, many castles fell into ruins as they were no longer necessary and posed a possible threat to the new regime. Here we see the remains of Takeda Castle in western Japan, which was built in 1441 and abruptly abandoned in 1600. Now a popular tourist destination, Takeda Castle is sometimes referred to as "the Machu Picchu of Japan.

Asago, Japan

EAST ASIA

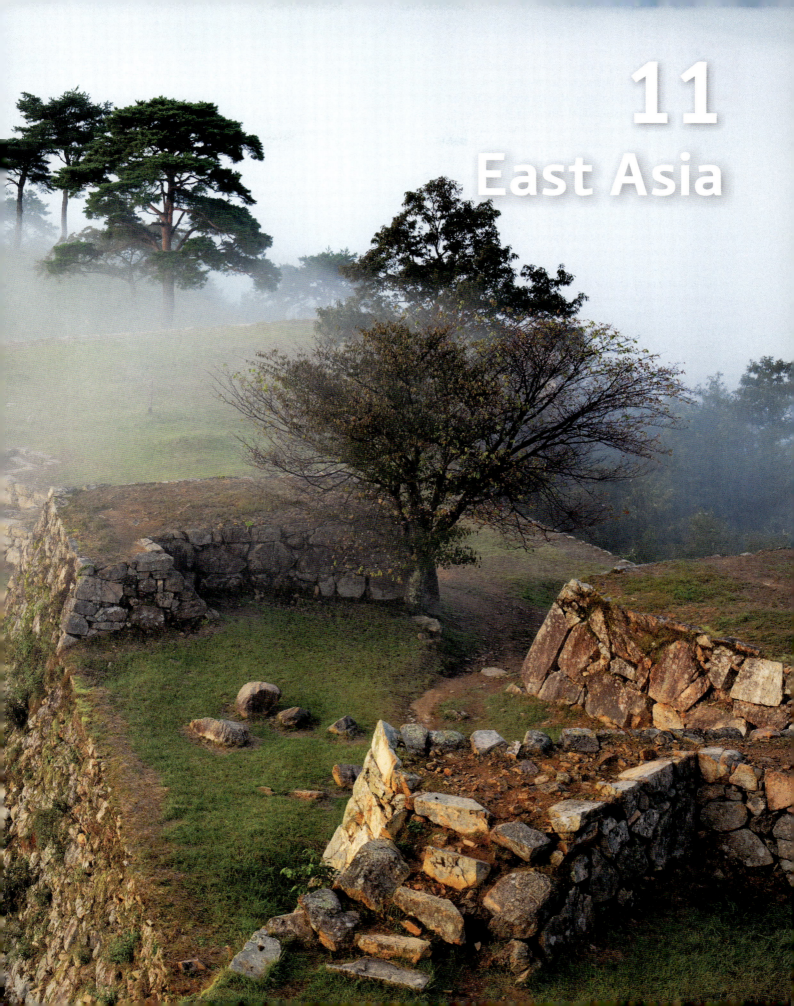

11
East Asia

E ast Asia is famous for its huge modern cities, such as Tokyo, with 37 million people in its greater metropolitan area, Shanghai with 34 million, and Seoul with 25 million. But East Asia is also a region of diminishing towns, disappearing villages, and depopulating countrysides, due both to migration to the big cities and extremely low birth rates. Consider, for example, Gunwi, a county located in a scenic region of central South Korea. As recently as the 1980s, more than 80,000 people lived in Gunwi, but by 2016 that number had dropped to 24,000, some 40 percent of whom are more than 65 years of age. Gunwi has closed more than 20 schools, and births are so infrequent that the county no longer has any obstetricians.

Officials in both Gunwi and Seoul, South Korea's capital, have tried to arrest rural depopulation and promote fertility, but with little success. South Korea spent more than US$68.3 billion from 2006 to 2016 to encourage people to marry young and have larger families, yet during that period its fertility rate barely budged. Gunwi's leaders even offered cash incentives for childbearing along with free school lunches, but could neither increase the county's birth rate nor attract people from other parts of the country.

Gunwi may be an extreme case, but the pattern that it represents is by no means unique. Across South Korea, more than 3,700 schools have been shuttered since the 1990s, and many more will soon follow. Japan has so many abandoned towns, factories, schools, and hospitals that a new hobby of "ruins exploration," or *haikyo*, has emerged, showcased in numerous websites. China has not yet reached this stage of population loss and town abandonment, but if current demographic trends persist it eventually will. Unless birthrates dramatically increase or in-migration surges, East Asia will generate many more ghost towns and ruins over the next several decades.

Defining East Asia

The contrast between booming Seoul and dying Gunwi reveals some critical processes currently impacting East Asia, a region composed of China, Japan, South Korea, North Korea, and Taiwan (**Figure 11.1**). Massive cities mushroomed as East Asia emerged as a core of the global system, noted for its economic, technological, and cultural power and linkages. But not all parts of the region have been so dynamic. Seoul itself sits only 31 miles (50 km) from the North Korean border, beyond which conditions are very different. In stark contrast to the south, North Korea is noted for its poverty, economic stagnation, and resistance to globalization.

> ## Massive cities mushroomed as East Asia emerged as a core of the global system, but not all parts of the region have been so dynamic.

Although historically unified by cultural features, in the second half of the 20th century East Asia was politically divided, with the capitalist economies of Japan, South Korea, Taiwan, and Hong Kong separated from the communist bloc of China and North Korea. As Japan became a leader in the global economy, much of China remained poor. By the early 21st century, however, divisions within East Asia had been

▲ **Figure 11.1 East Asia** This region includes China, Japan, North Korea, South Korea, and Taiwan. The physical geography of mainland East Asia varies widely, from the high plateaus and desert basins of western China to the broad river valleys and vast plains of eastern China. In the island region, landscapes are shaped by the convergence of three major tectonic plates: the Eurasian, Philippine, and Pacific.

reduced. China is still governed by the Communist Party, but has taken a path of mostly capitalist development since the 1980s. Yet relations between North Korea and South Korea remain hostile, and China's rapid rise is generating concerns across the rest of the region.

In certain respects, East Asia can be easily defined by the territorial extent of its constituent countries. Taiwan's political status, however, is ambiguous: although Taiwan is an independent country, China (officially, the People's Republic of China) claims it as part of its own territory. As a result, Taiwan (officially, the Republic of China) is recognized as a sovereign state by only a handful of countries. In cultural terms, defining East Asia is more complicated, especially with regard to the western half of China. This is a huge, but

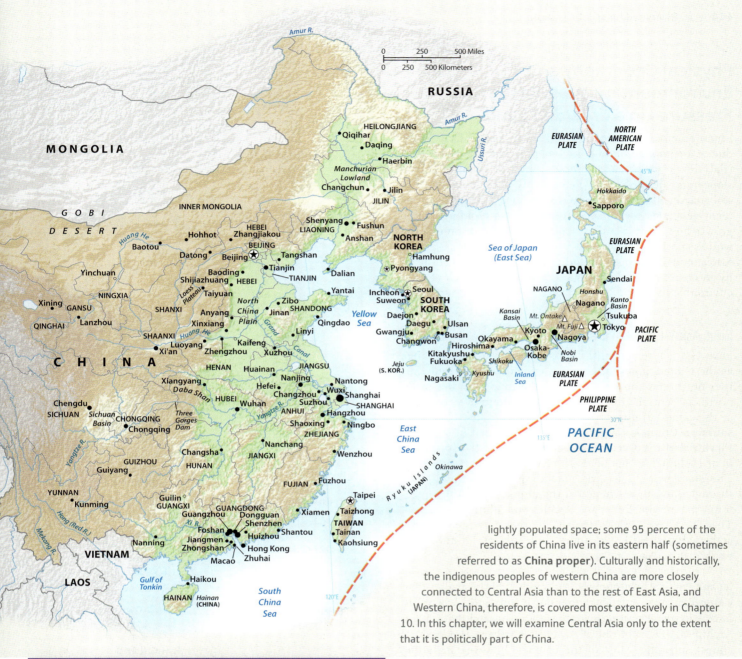

lightly populated space; some 95 percent of the residents of China live in its eastern half (sometimes referred to as **China proper**). Culturally and historically, the indigenous peoples of western China are more closely connected to Central Asia than to the rest of East Asia, and Western China, therefore, is covered most extensively in Chapter 10. In this chapter, we will examine Central Asia only to the extent that it is politically part of China.

LEARNING Objectives

After reading this chapter you should be able to:

11.1 Identify the key environmental differences between East Asia's island countries (Japan and Taiwan) and the mainland.

11.2 Describe China's main environmental problems and compare them with the environmental challenges faced by Japan, South Korea, and Taiwan.

11.3 Summarize the relationships among topography, climate, rice cultivation, and population density across East Asia.

11.4 Explain why China's population is so unevenly distributed, with some areas densely settled and others almost uninhabited.

11.5 Outline the distribution of major urban areas across East Asia and explain why continued growth of the region's largest cities is often viewed as a problem.

11.6 Describe the ways in which religion and other belief systems both unify and divide East Asia.

11.7 Distinguish between the Han Chinese and other ethnic groups of China, paying particular attention to language.

11.8 Describe the geopolitical division of East Asia during the Cold War period and explain how that division still influences East Asian geopolitical relations.

11.9 Identify the main reasons behind East Asia's rapid economic growth in recent decades and discuss possible limits to continued expansion at such a rate.

11.10 Describe the differences in economic and social development found across China and, more generally, across East Asia as a whole.

Physical Geography and Environmental Issues: Resource Pressures in a Crowded Land

Environmental problems in East Asia are particularly severe due to a combination of a huge population, massive industrial development, and unique physical features. Steep slopes and heavy rainfall make many areas vulnerable to soil erosion and mudslides, and a seismically active environment generates earthquakes. Owing to its rapid economic growth and lax regulations, China suffers from some of the world's most severe air and water pollution. Japan, South Korea, and Taiwan, on the other hand, have invested heavily in environmental protection, resulting in much cleaner environments.

East Asia's Physical Geography

East Asia is situated in the same general latitudinal range as the United States, although it extends considerably farther north and south. China's northernmost tip lies as far north as central Quebec, and its southernmost point is at the same latitude as Mexico City. Thus, the climate of southern China is roughly comparable to that of southern Florida and the Caribbean, whereas northern China's climate is similar to that of south-central Canada (**Figure 11.2**). The island belt of East Asia, extending from northern Japan through Taiwan, is situated at the intersection of three tectonic plates: the Eurasian, Pacific, and Philippine plates. This area, particularly Japan, is therefore geologically active, dotted with volcanoes and subject to numerous earthquakes (**Figure 11.3**).

Japan's Physical Environment Although slightly smaller than California, Japan extends farther north and south. As a result, Japan's extreme south, in southern Kyushu and the Ryukyu Archipelago, is subtropical, whereas northern Hokkaido is almost subarctic. Most of the country, however, including the main island of Honshu, is temperate. The climate of Tokyo is not unlike that of Washington, DC, although Tokyo is distinctly rainier.

Japan's climate varies not only from north to south, but also from southeast to northwest, across the main axis of the archipelago. In winter, the area facing the Sea of Japan receives much more snow than the Pacific Ocean coastline, because cold winds from the Asian mainland blow across the relatively warm waters of the Sea of Japan. The air picks up moisture over the sea and deposits snow when it hits land (**Figure 11.4**). Japan's Pacific coast, on the other hand, is far more vulnerable to typhoons (hurricanes), which occasionally strike the country.

Japan is one of the world's most rugged countries, with mountainous terrain covering some 85 percent of its territory. Most of these uplands are thickly wooded. Japan owes its lush forests both to its mild, rainy climate and

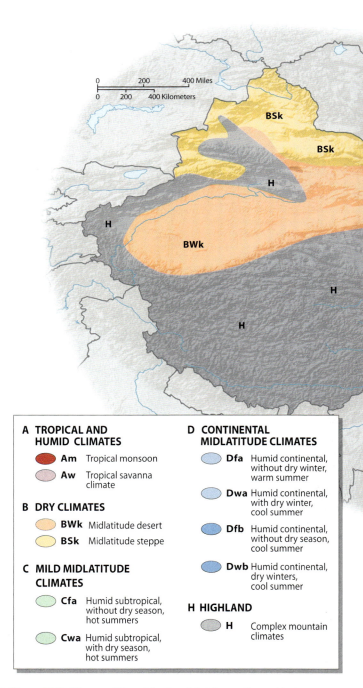

A TROPICAL AND HUMID CLIMATES

- **Am** Tropical monsoon
- **Aw** Tropical savanna climate

B DRY CLIMATES

- **BWk** Midlatitude desert
- **BSk** Midlatitude steppe

C MILD MIDLATITUDE CLIMATES

- **Cfa** Humid subtropical, without dry season, hot summers
- **Cwa** Humid subtropical, with dry season, hot summers

D CONTINENTAL MIDLATITUDE CLIMATES

- **Dfa** Humid continental, without dry winter, warm summer
- **Dwa** Humid continental, with dry winter, cool summer
- **Dfb** Humid continental, without dry season, cool summer
- **Dwb** Humid continental, dry winters, cool summer

H HIGHLAND

- **H** Complex mountain climates

▲ **Figure 11.2 Climate of East Asia** East Asia is located in roughly the same latitudinal zone as North America, and climatic parallels exist between the two world regions. The northernmost tip of China lies at about the same latitude as Quebec and shares a similar climate, whereas southern China approximates the climate of Florida. In Japan, maritime influences produce a milder climate.

to its long history of forest conservation. For hundreds of years, both the Japanese state and its village communities have enforced strict conservation rules, ensuring that timber and firewood extraction is balanced by tree growth.

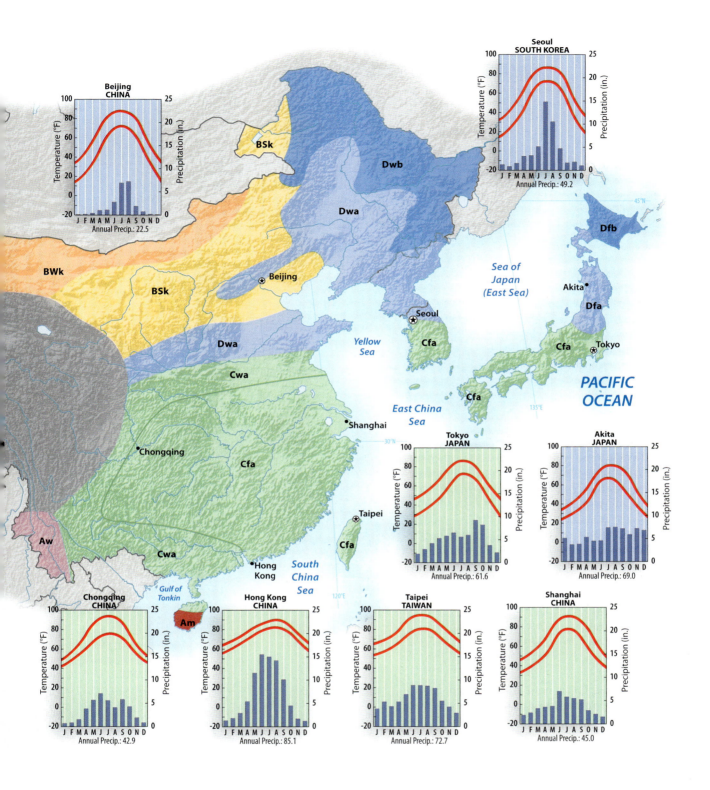

Beijing CHINA — Annual Precip.: 22.5

Seoul SOUTH KOREA — Annual Precip.: 49.2

Tokyo JAPAN — Annual Precip.: 61.6

Akita JAPAN — Annual Precip.: 69.0

Chongqing CHINA — Annual Precip.: 42.9

Hong Kong CHINA — Annual Precip.: 85.1

Taipei TAIWAN — Annual Precip.: 72.7

Shanghai CHINA — Annual Precip.: 45.0

Along Japan's coastline and interspersed among its mountains are limited areas of alluvial plains. Although these lowlands were once covered by forests and wetlands, they have long since been cleared and drained for intensive agriculture. The largest Japanese lowland is the Kanto Plain to the north of Tokyo, but it is only some 80 miles wide and 100 miles long (130 by 160 km). The country's other main lowland basins are the Kansai, located around Osaka, and the Nobi, centered on Nagoya. In the mountainous province of Nagano, smaller basins are sandwiched between the imposing peaks of the Japanese Alps.

The Tsunami Threat Japan's location makes it vulnerable to earthquakes, just as its extensive coastlines make it susceptible to **tsunamis**—large ocean waves, usually caused by earthquakes, that can be very destructive when reaching the coast. The power of such events was demonstrated in March 2011, when northeastern Japan was devastated by one of the largest and most destructive natural events of modern

▶ **Figure 11.3 Japan's Physical Geography** Japan has several sizable lowland plains, primarily along the coastline, but they are interspersed among rugged mountains and uplands. Its location at the convergence of three major tectonic plates means that Japan experiences numerous earthquakes. Volcanic eruptions can also be hazardous and are linked directly to Japan's location on tectonic plate boundaries. Much of Japan's coast is also vulnerable to devastating tsunamis caused by earthquakes in the Pacific Basin.

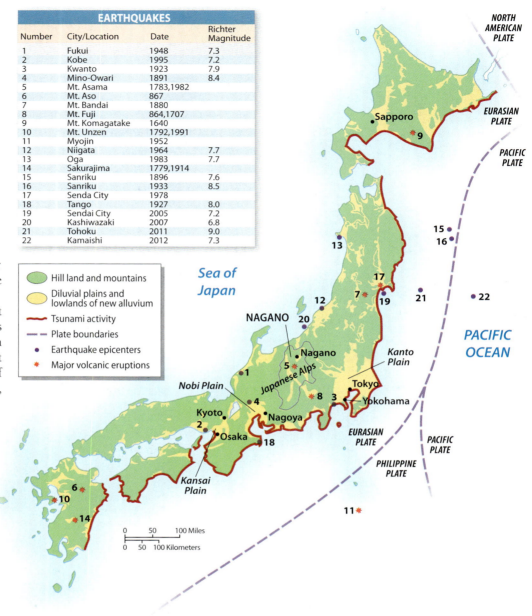

EARTHQUAKES			
Number	City/Location	Date	Richter Magnitude
1	Fukui	1948	7.3
2	Kobe	1995	7.2
3	Kwanto	1923	7.9
4	Mino-Owari	1891	8.4
5	Mt. Asama	1783,1982	
6	Mt. Aso	867	
7	Mt. Bandai	1880	
8	Mt. Fuji	864,1707	
9	Mt. Komagatake	1640	
10	Mt. Unzen	1792,1991	
11	Myojin	1952	
12	Niigata	1964	7.7
13	Oga	1983	7.7
14	Sakurajima	1779,1914	
15	Sanriku	1896	7.6
16	Sanriku	1933	8.5
17	Senda City	1978	
18	Tango	1927	8.0
19	Sendai City	2005	7.2
20	Kashiwazaki	2007	6.8
21	Tohoku	2011	9.0
22	Kamaishi	2012	7.3

Legend:
- Hill land and mountains
- Diluvial plains and lowlands of new alluvium
- Tsunami activity
- Plate boundaries
- Earthquake epicenters
- Major volcanic eruptions

times, the Tohoku earthquake and tsunami (**Figure 11.5**). Over 15,000 lives were lost and the local economy was ruined.

The Japanese government had invested billions of dollars in anti-tsunami seawalls, which stretched along almost 40 percent of its coastline. The surge of water from the Tohoku tsunami, however, washed over the tops of the barriers, collapsing many in the process. Japan, an energy-poor country, had also invested heavily in nuclear power, which was widely regarded as relatively safe and secure. The earthquake and tsunami, however, caused severe damage to the Fukushima Daiichi nuclear power plant, forcing the evacuation of over 200,000 people.

Korean Landscapes Korea forms a well-demarcated peninsula, partially separated from northeastern China by rugged mountains and sizable rivers (**Figure 11.6**). The far north, which just touches Russia's Far East, has a climate not unlike that of Maine, whereas the southern tip is more reminiscent of the Carolinas. Korea is also a mountainous land with scattered alluvial basins. The lowlands of the southern portion of the peninsula are more extensive than those of the north, giving South Korea an agricultural advantage over North Korea. However, North Korea has more abundant mineral deposits and hydroelectric resources.

▶ **Figure 11.4 Heavy Snow in Japan's Mountains** Cold, moist air moving off the Sea of Japan produces heavy snows in northwestern Japan and along the country's mountainous spine. Numerous major ski areas dot the Japanese Alps, several of which have hosted world-class sports competitions, including the 1998 Winter Olympics.

◀ **Figure 11.5 2011 Tohoku Earthquake and Tsunami** The 2011, Tohoku earthquake and tsunami constituted one of the most devastating natural disasters of recent times. At its extreme, the tsunami wave was over 133 feet (40 meters) high, giving it enough power to wash away entire villages. More than 15,000 people died as a result. **Q: Why did the Tohoku earthquake and tsunami generate such profound consequences for the Japanese economy and political system?**

In the early 20th century, the northern half of the Korean Peninsula also had much more extensive forests than the south. Over the past several decades, however, South Korea has reforested many of its mountain areas, while impoverished North Korea has seen an increase in deforestation, resulting in a starkly contrasting landscape along the Demilitarized Zone (DMZ) that separates the two states (**Figure 11.7**).

Taiwan's Environment Taiwan, an island about the size of Maryland, sits at the edge of the continental land mass. To the west, the Taiwan Strait is only about 200 feet (60 meters) deep; to the east, ocean depths of many thousands of feet are found 10 to 20 miles (16 to 32 km) offshore.

Taiwan itself forms a large tilted block. Its central and eastern regions are rugged and mountainous, whereas the west is dominated by an alluvial plain. Bisected by the Tropic of Cancer, Taiwan has a mild winter climate, but it is sometimes battered by typhoons in the early autumn. Unlike nearby areas of China proper, Taiwan still has extensive forests concentrated in its eastern uplands (**Figure 11.8**).

China's Varied Environments Even without its Central Asian regions of Tibet and Xinjiang, China is a vast country with diverse environmental regions. We can divide China proper into two main areas: Northern China lies to the north of the Yangtze River Valley, and southern China includes the Yangtze and all areas to the south.

Southern China is a land of rugged mountains and hills interspersed with large lowland basins. One of the most distinctive regions is the former lake bed of central Sichuan Province (see Figure 11.1). Protected by imposing mountains, Sichuan is noted for its mild winter climate. East of Sichuan, the Yangtze River passes through several other broad basins (in the Middle Yangtze region), partially

▲ **Figure 11.6 The Mountains and Lowlands of Korea** Like Japan, Korea has extensive uplands interspersed with lowland plains. The country's highest mountains are in the north, whereas its most extensive alluvial plains are in the south. South Korea's provinces, shown here, are culturally as well as physically distinctive.

and Guizhou, the former noted for its perennial springlike weather (**Figure 11.9**). Finally, northeast of Guangdong lies the rugged coastal province of Fujian. With its narrow coastal plain and deeply indented coastline, Fujian has long supported a maritime way of life.

Northward of the Yangtze Valley, the climate becomes both colder and drier. Summer rainfall is generally abundant except along the edge of the Gobi Desert, but the other seasons are dry. Desertification is a major threat in parts of the North China Plain, which have experienced prolonged droughts in recent years. Seasonal water shortages are growing increasingly severe as water withdrawals for irrigation and industry increase, leading to concerns about an impending crisis. With the exception of a few low mountains in Shandong Province, the entire area is a virtually flat plain.

West of the North China Plain sits the Loess Plateau, a fairly rough upland of moderate elevation and uncertain precipitation. It does, however, have fertile soil—as well as huge coal deposits. Farther west are the semiarid plains and uplands of Gansu Province, situated at the foot of the great Tibetan Plateau.

China's far northeastern region, called *Dongbei* in Chinese and Manchuria in English, is dominated by a broad, fertile lowland basin sandwiched between mountains stretching along China's borders with North Korea, Russia, and Mongolia. Although winters here can be brutal, summers are usually warm and moist. Manchuria's peripheral uplands have some of China's best-preserved forests and wildlife refuges.

East Asia's Environmental Challenges

Deforestation is a major problem not just in North Korea, but in much of eastern China as well. Pollution and reduced biological diversity are also major issues in China and other parts of East Asia (**Figure 11.10**). The region as a whole, moreover, has a shortage of energy resources.

Deforestation and Desertification Many of the uplands of China and North Korea support only grass, meager scrub, and stunted trees. China lacks the historical tradition of forest conservation that characterizes Japan. During earlier periods of Chinese history, hillsides were often cleared of wood for fuel, and in some instances entire forests were burned for ash that could be used as fertilizer. In much of southern China, sweet potatoes, maize, and other crops have been grown on steep and easily eroded hillsides for hundreds of years. After centuries of exploitation, many hillslopes are now so degraded that they cannot easily regenerate forests.

separated from each other by hills and low mountains, before flowing into a large delta near Shanghai.

South of the Yangtze Valley, the mountains are higher (up to 7000 feet, or 2150 meters, in elevation), but they are still interspersed with alluvial lowlands. Sizable valleys are found in the far south, such as the Xi Basin in Guangdong Province. Here the climate is tropical, free of frost. West of Guangdong lie the moderate-elevation plateaus of Yunnan

▼ **Figure 11.8 Mountains of Taiwan** The eastern and central areas of Taiwan are dominated by rugged and heavily forested mountains. The Taroko Gorge in Taroko National Park, seen in this photograph, is one of Taiwan's top tourist destinations.

Over the past 50 years, China has initiated several successful reforestation efforts. The government has mobilized millions of people to plant trees. A 2016 remote-sensing study by NASA scientists found that China had added 61,000 square miles (160,000 square kilometers) of forested land since 2000. It will take many years, however, before these new forests can produce marketable timber. At present, substantial timber reserves are found only in China's far northeast, where a cold climate limits tree growth, and along the eastern slopes of the Tibetan Plateau, where rugged terrain restricts commercial forestry. As a result, China suffers a severe shortage of forest resources.

Desertification is also a problem for China. In many areas, sand dunes have pushed south from the Gobi Desert, smothering farmlands. Some reports claim that more desertification occurs in China than in any other country. The government has responded with massive dune-stabilization schemes, which involve planting grass, shrubs, and trees to stop sand from blowing away (**Figure 11.11**). The most ambitious such project involved planting hundreds of millions of trees along a 3000-mile (4800-km) swath of northern China, the so-called great green wall. Although some success has been realized, one study estimated that only around one-third of the trees planted since the 1970s were still alive

▲ Figure 11.9 **Yunnan Plateau** Noted for its mild climate and scenic landscapes, the Yunnan plateau in south-central China is becoming a major tourist destination. Here we see the road heading to the city of Shangri-La in northern Yunnan, formerly called Zhongdian. To encourage tourism, Zhongdian was renamed "Shangri-La" in 2001 in reference to a mythical paradise supposedly located in the general area.

Explore the **Sights** of Dali, China

http://goo.gl/30gTH4

▼ Figure 11.10 **Environmental Issues in East Asia** This vast world region has been almost completely transformed from its natural state and continues to have serious environmental problems. In China, some of the more pressing environmental issues involve deforestation, flooding, water control, and soil erosion.

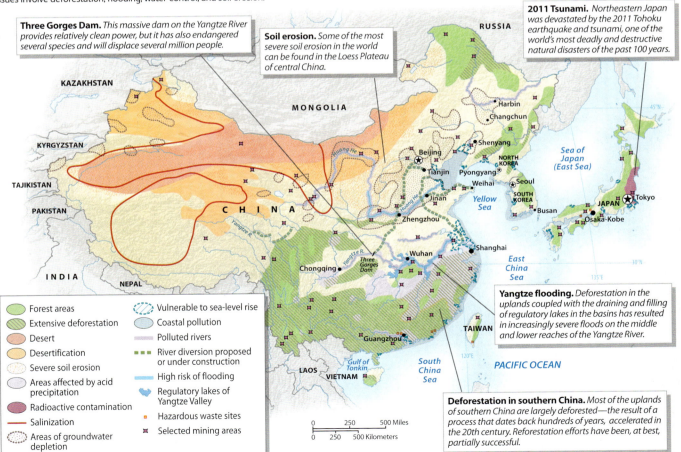

▲ **Figure 11.11 Dune Stabilization in China** The spread of sand dunes is a significant hazard in dry areas of northern China. By spreading dried vegetation and then planting grass seeds, the dunes can be stabilized.

▼ **Figure 11.12 Air Pollution in China** Major Chinese cities, such as Shanghai shown here, suffer from some of the worst air pollution in the world. The resulting haze can make it difficult to see for more than a few city blocks.

in 2015. An official 2011 government report found that the overall rate of desertification in China declined moderately after 2000, but that in some areas the problem continues to intensify.

Mounting Pollution As China's industrial base expands, environmental problems such as water pollution and toxic-waste dumping are growing more acute. The burning of high-sulfur coal generates severe air pollution, a problem aggravated by the growing number of automobiles (**Figure 11.12**). According to a 2016 study presented at the American Association for the Advancement of Science, air pollution resulted in 1.6 million deaths in China in 2013 alone. In January of the same year, a smog crisis in Beijing forced the government to restrict outdoor activities and suspend certain forms of industrial production. Pollutants from China, moreover, regularly reach not only Japan and Korea, but also the U.S. West Coast. Water pollution, another huge problem, is reportedly responsible for some 60,000 deaths a year.

China is, however, making environmental progress. Sulfur dioxide emissions peaked in 2006, and China's major cities

are now less polluted than those of India. A hard-hitting video about air pollution in the country, *Under the Dome*, went viral in 2015, with several hundred million views within a few days. The fact that this highly critical video was not initially blocked by China's censors convinced some observers that the country's leaders were finally regarding pollution as a national emergency. The documentary's popularity, however, generated its own official concerns, and within a week it was removed from all Chinese video sites.

Considering Japan's large population and intensive industrialization, its environment is relatively clean. Japan's population density has some environmental advantages, such as allowing a highly efficient public transportation system. In the 1950s and 1960s, Japan did suffer from some of the world's worst pollution, but soon afterward the government passed stringent environmental laws covering both air and water pollution.

Japan's cleanup was aided by its location, as winds usually carry smog-forming chemicals out to sea. Equally important has been the phenomenon of **pollution exporting**. Because of Japan's high cost of production and strict environmental laws, many Japanese companies have moved their dirtier factories overseas. In effect, Japan's pollution has been partially displaced to poorer countries. Taiwan and South Korea have followed suit, setting up new factories in countries—including China—that have less strict environmental standards.

Endangered Species The growing number of endangered species in East Asia has been linked to the region's economic rise. Many forms of traditional Chinese medicine are based on products derived from rare and exotic animals. Deer antlers, bear gallbladders, snake blood, tiger penises, and rhinoceros horns are believed to have medical effectiveness, and certain individuals will pay fantastic sums of money to purchase

them. As wealth has accumulated, trade in such substances has expanded. China itself has relatively little remaining wildlife, but other areas of the world supply its demand.

China has, however, moved strongly to protect some of its remaining habitats. Among the most important of these are the high-altitude forests and bamboo thickets of western Sichuan Province, home of the panda. Efforts are also being made to preserve wilderness in northern Manchuria, where the population of Siberian tigers, long thought to be locally extinct, had rebounded to around 40 by 2013. Another area of extensive nature preservation is the canyonlands of northwestern Yunnan, which were declared off-limits to loggers in 1999.

Wildlife is also scarce in Korea, but ironically, the Demilitarized Zone that separates North from South Korea functions as an unintentional wildlife sanctuary, supporting populations of several endangered species, including the Asiatic black bear. In much of Japan, wildlife is rapidly rebounding in depopulating rural areas.

Dams, Flooding, and Soil Erosion in China

Most of China's environmental problems also characterize other densely populated, rapidly industrializing countries. Other issues, however, are unique to China. Many of these problems are focused on dams, flooding, and soil erosion.

The Three Gorges Dam Controversy The Yangtze River (or Chang Jiang) is one of the most important physical features of East Asia. This river, the third largest (by volume) in the world, emerges from the Tibetan highlands onto the rolling lands of the Sichuan Basin and passes through what used to be a magnificent canyon in the Three Gorges area (**Figure 11.13**). It then meanders across the lowlands of central

▼ **Figure 11.13 The Three Gorges of the Yangtze** (a) The Three Gorges on the Yangtze River was once one of China's most scenic natural attractions. (b) It is now flooded by a reservoir created by the Three Gorges Dam, which produces more hydroelectricity than any other facility in the world.

(a)   (b)

China before entering the sea near the city of Shanghai. The Yangtze has historically been the main avenue of entry into China's interior and is celebrated in Chinese literature for its beauty and power. Since the 1990s, however, it has been the focal point of an environmental controversy of global proportions.

Finished in 2006, the Three Gorges Dam—600 feet (180 meters) high and 1.45 miles (2.3 km) long—is the world's largest hydroelectric dam. The reservoir it created is 350 miles (563 km) long, has displaced an estimated 1.2 million people, and has inundated a major scenic attraction. The dam also traps sediment and pollutants in the reservoir and disrupts habitat for several endangered species, including the Yangtze River dolphin. The Chinese government, however, argues that the clean power generated by the dam offsets its problems, claiming that it prevented the release of 100 million tons of carbon into the atmosphere in 2014 alone.

Chinese planners also praise the water-control benefits provided by the dam, claiming that it held back 7.5 billion cubic meters (260 billion cubic feet) of potential floodwater during a period of drenching rain in July 2016. Critics argue, however, that flooding was by no means eliminated. They also note that the flood risk in the Yangtze Valley has been exacerbated by the draining of lakes. East of Sichuan, the Yangtze passes through several basins (in Hunan, Hubei, and Jiangxi provinces), each containing a group of lakes. During periods of high water, river flows are diverted into these **regulatory lakes**, thus reducing the flow downstream. After the flood season is over, water drains from the lakes and therefore helps maintain the water level in the river downstream. In the 1950s and 1960s, however, the Chinese government drained many of these lakes to increase the extent of farmland.

Flooding in Northern China The North China Plain, which has been deforested for thousands of years, is plagued by both drought and flood. This area is dry most of the year, yet often experiences heavy downpours in summer. Since ancient times, large-scale hydraulic engineering projects have both controlled floods and allowed irrigation. But no matter how much effort has been put into water control, disastrous flooding has never been completely prevented. New anti-flooding measures, however, continue to be developed (see *Working Toward Sustainability: Flood-Prone China to Develop New "Sponge Cities."*)

The worst floods in northern China are caused by the Huang He, or Yellow River, which cuts across the North China Plain. Upstream erosion contributes to the Huang He's huge **sediment load**, or suspended clay, silt, and sand, making it the world's muddiest major river (**Figure 11.14**). When the river enters the low-lying plain, its velocity slows

▲ **Figure 11.14 China's Muddy Yellow River** The Yellow River, which cuts through the rapidly eroding Loess Plateau, carries a tremendous amount of sediment. Here we see the muddy river as it passes by the Incense Burner Temple in Shaanxi province.

and its sediments begin to settle and accumulate in the riverbed. As a result, the level of the riverbed gradually rises above that of the surrounding lands. Eventually, the river must break free of its course to find a new route to the sea over lower-lying ground. Twenty-six such course changes have been recorded for the Huang He throughout Chinese history.

Through the process of sediment deposition and periodic course changes, the Huang He has created the vast North China Plain. In prehistoric times, the Yellow Sea extended far inland. Even today the sea is retreating as the Huang He's delta expands. Such a process occurs in other alluvial plains, but nowhere else in the world is it so pronounced. Nowhere else, moreover, is it so destructive. The world's two deadliest recorded natural disasters, each killing between 1 and 2 million people, involved flooding of the Yellow River (in 1887 and 1931). Since ancient times, the Chinese have attempted to keep the river within its banks by building ever larger dikes. Eventually, however, the riverbed rises too high to contain the flow and catastrophic flooding results. The river has not changed its course since the 1930s, but most geographers think that another course correction is inevitable.

Although flooding is occasionally devastating along the Huang He, the more common problem is a lack of water. The lower course of the river sometimes dries out completely (**Figure 11.15**). As a result, China is building a huge diversion system, the South–North Water Transfer Project, which will channel some 44.8 billion cubic meters of fresh water each year from the Yangtze River to the Yellow River. Diversion canals designed to link the two rivers are being constructed both in the headwaters area on the Tibetan Plateau and on the North China Plain. By 2014, the project had cost some US$79 billion, making it one of the world's most expensive engineering schemes.

Working Toward **Sustainability**

Flood-Prone China to Develop New "Sponge Cities"

China has long been burdened by devastating flooding. Six of the word's eight deadliest recorded floods have occurred in China. In 1975, more than 230,000 people perished as a result of Typhoon Nina. Although recent events have not been so devastating, they remain common—and deadly. Flooding in the summer of 2016 took more than 150 lives, caused some 73,000 buildings to collapse, and forced almost 2 million people to seek safety on higher ground.

Flooding, Infrastructure, and Urbanization Until recently, China has approached flooding mainly through conventional civil engineering, building dams, dikes, and drainage canals. The country now has more than 87,000 dams and the world's largest network of dikes and other water barriers. Such efforts, however, have never solved the problem. One study found that the number of Chinese cities affected by floods doubled from 2008 to 2016.

Although climate change may be responsible for some of this recent increase in flooding, urban development is the more immediate cause. As China's cities mushroom, more and more land is being covered by impermeable surfaces, preventing excess water from soaking into the ground. As Bill Gates famously observed, China poured more concrete between 2011 and 2013 than the United States did over the entire 20th century.

A New Model In 2015, China opted for a new urban flood-control strategy based on absorbing rather than shunting away runoff. Government planners designated 16 model urban areas that will receive substantial funding to turn themselves into "sponge cities." Some of these cities are relatively small, but massive Chongqing on the Yangtze River, with more than 10 million inhabitants, is included as well.

Many of the ideas in the "sponge city" model are relatively simple. Sizable areas must be set aside so that they can fill up with water during floods. After the rains subside, the water will slowly infiltrate back into the ground, recharging aquifers and reducing potential damage from

▲ Figure 11.1.1 **Chinese Urban Wetland** China is currently designing "sponge cities" that allow run-off to be absorbed and stored in wetland environments, reducing the risk of flooding. Here we see an opera house in the large northern city of Harbin, which is situated within a large semi-urban wetland.

future droughts. Such water-management areas also serve as green spaces in otherwise crowded urban environments, providing wildlife habitat as well as seasonal recreation grounds (**Figure 11.1.1**).

Other aspects of the sponge city model entail technological developments and changes to the built environment. New forms of paving material, for example, allow water to infiltrate through the surface. Such permeable pavement, however, costs up to three times as much as conventional asphalt or concrete and requires more frequent maintenance. Planners hope that technological advances and large-scale applications will lower costs and thus help create "spongier" cities throughout China.

Google Earth Virtual Tour Video
http://goo.gl/1wgN9F

1. What parts of China are particularly vulnerable to flooding?

2. What advantages other than flood control might be expected from the "sponge city" model?

Erosion on the Loess Plateau The Huang He's sediment burden is derived from the eroding soils of the Loess Plateau, located to the west of the North China Plain. **Loess** consists of fine, windblown sediment that was deposited on this upland area during the last ice age, accumulating in some places to depths of several hundred feet.

Loess forms fertile soil, but it washes away easily when exposed to running water. At the dawn of Chinese civilization, the semiarid Loess Plateau was covered with tough grasses and scrubby forests that retained the soil. Chinese

◀ Figure 11.15 **Drought in Northern China** The Yellow River (or Huang He) is sometimes called the "cradle of Chinese civilization" owing to its historical importance. However, due to the increasing extraction of water for agriculture and industry, the river now often runs dry in its lower reaches. Two children play on a boat in the dried-up riverbed in Henan Province.

453

▲ **Figure 11.16 Reforestation on the Loess Plateau** Large-scale eco-development projects funded by the World Bank and other organizations have finally begun to reduce and even reverse the environmental degradation of China's vulnerable Loess Plateau. Terracing and tree-planting efforts have been especially successful. **Q: Why have such large ridges been constructed around every tree planted on this slope, and why are they shaped this way?**

farmers, however, began to clear the land for the abundant crops that it yields when rainfall is adequate. Cultivation required plowing, which, by exposing the soil to water and wind, exacerbated erosion. As the population of the region gradually increased, the remaining areas of woodland diminished, leading to faster rates of soil loss. As the erosion process continued, great gullies cut across the plateau, steadily reducing the extent of arable land.

Today the Loess Plateau is one of the poorest parts of China. Population is only moderately dense by Chinese standards, but good farmland is limited, and drought is common. China's government has long encouraged the construction of terraces to conserve the soil, but such efforts have not always been effective. More recently, the World Bank and the government of China have supported two Loess Plateau Watershed Rehabilitation Projects, which have achieved considerable success (**Figure 11.16**). An estimated 3861 square miles (10,000 square km) of degraded land have been restored.

Climate Change in East Asia

East Asia occupies a central position in discussions of climate change, largely because of China's rapid increases in carbon emissions. China's total production of greenhouse gases (GHGs) surpassed that of the United States in 2007. This rise has been caused both by China's explosive economic growth and its reliance on coal to meet most of its energy needs.

The potential effects of climate change in East Asia have serious implications for human populations. A 2016 World Bank report concluded that the region could see its economic growth rate decline by as much as 6 percent of GDP by 2050 as a result of water-related impacts from climate change on agriculture, health, and income. According to China's 2015 National Climate Change Assessment, the country's average temperatures have increased faster than the global average. Increased evaporation rates, coupled with the melting of glaciers on the Tibetan Plateau, could intensify local water shortages, while the wet zones of southeastern China could see more storms and flooding.

Responses to Climate-Change Risks In 2007 China released its first national plan on climate change, calling for major gains in energy efficiency and a partial transition to renewable energy sources. China's government is subsidizing the manufacture of solar panels, which it sees as a key energy technology of the 21st century. China's climate change strategy also includes a major expansion of both nuclear and wind power, along with ambitious reforestation efforts.

In late 2014, China reached a climate deal with the United States, promising that its carbon emissions would peak around 2030 and then begin to decline. The Chinese government also pledged to fill 20 percent of its energy needs from renewable sources by the same year. China's recent economic slowdown has helped its drive to reach these targets. In 2016, for example, Beijing halted the construction of dozens of new coal-burning plants, claiming that the electricity that they would provide was no longer necessary.

Japan, South Korea, and Taiwan are also major emitters of GHGs, but have energy-efficient economies, releasing far less carbon than the United States on a per capita basis. Japan generally supports international treaties designed to force carbon reductions. In 2009, for example, it pledged to reduce its carbon footprint by 25 percent by 2020. The 2011 earthquake and tsunami, however, undermined such plans. After shutting down its nuclear power plants following the disaster, Japan was forced to burn more fossil fuels, increasing its GHG emissions by 4 percent (**Figure 11.17**). By 2016 several nuclear facilities

▼ **Figure 11.17 Japan's Shuttered Daiichi Nuclear Power Plant** The huge earthquake and tsunami of March 11, 2011 crippled the Daiichi Nuclear Power Plant, seen here two weeks after the disaster. Daiichi remains shuttered, like most other nuclear power plants in Japan.

had reopened, but public resistance led the Japanese government to promise to generate no more than 15 percent of its electricity from nuclear plants. As a result, Japan's use of coal and liquefied natural gas is expected to increase over the next several decades.

East Asia's Energy Mix One reason for Japan's increasing use of coal is its lack of other energy resources, an issue that confronts the rest of East Asia. Japan lacks significant oil and natural gas deposits and depends on imports. Fossil fuels provide almost all of the country's energy, with oil accounting for almost 50 percent. Hydroelectric plants supply around 4 percent of Japan's total energy, but other renewables remain insignificant. South Korea faces similar energy challenges, but it reaffirmed its commitment to nuclear energy in the wake of Fukushima. South Korea's goal is to generate almost half its electricity from nuclear reactors by 2024.

China's rapid industrialization has forced it to expand its electricity generation at breakneck speed, as it is now the world's largest energy consumer. According to an official 2015 report, the country derives 64 percent of its energy from coal (**Figure 11.18**). China consumes more oil than any other country except the United States. Owing to both climate concerns and pollution, China is investing heavily in renewable energy. Roughly 1.7 percent of its electricity comes from the Three Gorges Dam. China has been building many other large dams, especially in the rugged lands of Tibet and Yunnan, causing concern in the downstream countries of South and Southeast Asia. China has also become an investment leader in other forms of renewable energy. Official statistics indicate that in 2015 China's grid-connected solar power grew 73.7 percent, and its grid-connected wind power grew 33.5 percent.

▼ **Figure 11.18 Energy Use in China** Most of China's surging energy demand is being met by burning coal and oil.

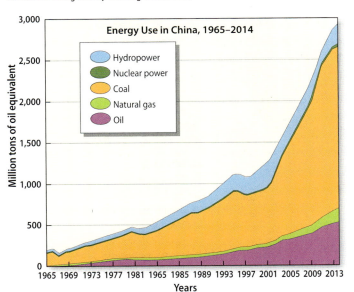

Energy Use in China, 1965–2014

Legend:
- Hydropower
- Nuclear power
- Coal
- Natural gas
- Oil

Y-axis: Million tons of oil equivalent (0 to 3,000)
X-axis: Years 1965 1969 1973 1977 1981 1965 1985 1989 1993 1997 2001 2005 2009 2013

REVIEW

11.1 Why is China the world's largest emitter of GHGs, and what is its government doing about this problem?

11.2 Why is Japan so much more heavily forested than China?

11.3 Explain how China's physical geography is linked to its historical problems with floods and droughts.

■ **KEY TERMS** China proper, tsunamis, pollution exporting, regulatory lakes, sediment load, loess

Population and Settlement: A Realm of Crowded Lowland Basins

East Asia vies with South Asia as the world's most populated region (Table 11.1). The lowlands of Japan, Korea, and China are among the most intensely used portions of Earth, containing East Asia's major cities and most of its agricultural lands (**Figure 11.19**). Despite East Asia's extremely high population density, however, the region's growth rate has plummeted since the 1970s.

East Asia's Population Dilemma

As we saw earlier, one of the biggest problems facing East Asia is its extremely low birth rates, leading to an aging population that will need to be supported by a shrinking number of workers. Japan is already losing population, and both South Korea and Taiwan will soon be in the same condition. China's population is younger than those of Japan or South Korea (**Figure 11.20**), but it is aging quickly. Only North Korea approaches replacement-level fertility.

Japan and especially South Korea have tried to increase birth rates, but with little success. Another alternative to counter shrinking and aging populations is to import workers from other countries. Japan has been reluctant to take this path, largely out of fear of reducing cultural homogeneity. But in 2015 an opinion poll found that a narrow majority (51 percent) of the Japanese public had come to favor increasing immigration. Still, governmental projections released in 2015 estimated that the number of workers in the country will probably fall by 7.9 million by 2030, and that Japan's total population will likely decline to 86 million by 2060.

South Korea has been more willing to accept immigrants than Japan, but it also maintains firm limits. By 2015 foreign residents accounted for 2.8 percent of the country's total population. South Korea's government is more interested in increasing the native birthrate, and in 2015 released a 200-page report on how to accomplish this, proposing official match-making services, state subsidies for fertility treatments, and reserving units in public housing for young couples with children.

The demographic issue in China is more complicated. A rapidly rising population from the 1950s through the 1970s led to widespread concern that the country would no

TABLE 11.1 Population Indicators

Country	Population (millions) 2016	Population Density (per square kilometer)[1]	Rate of Natural Increase (RNI)	Total Fertility Rate	Percent Urban	Percent <15	Percent >65	Net Migration (Rate per 1000)
China	1,378.0	145	0.5	1.6	56	17	10	0
Hong Kong	7.4	6,897	0.2	1.2	100	11	16	0
Japan	125.3	349	−0.2	1.5	94	13	27	0
North Korea	25.1	208	0.6	2.0	61	21	10	0
South Korea	50.8	517	0.4	1.2	82	14	14	0
Taiwan	23.5	652	0.2	1.2	77	14	13	0

Source: Population Reference Bureau, *World Population Data Sheet,* 2016.

[1] World Bank, *World Development Indicators,* 2016.

Login to MasteringGeography™ **& access** MapMaster **to explore these data!**

1) How do basic demographic patterns differ in North Korea and South Korea?

2) The rate of natural increase and the total fertility rate have different patterns in East Asia. What other demographic factors found in this data would help explain this lack of correlation?

longer be able to feed itself. By 1978 China's government had become so concerned that it instituted the infamous "one-child policy." Under this plan, couples in normal circumstances were expected to have only one offspring and could suffer financial and other penalties if they did not comply. Forced abortions and other human rights abuses followed.

A combination of economic development and the one-child policy successfully reduced China's fertility, which currently stands at 1.6, well below the replacement level. Now the government is worried about its aging population and possible future labor shortages. Such concerns led China to relax its one-child policy in 2013, allowing couples to have two children if one parent is an only child. In 2015, the government announced that it would gradually phase out the policy

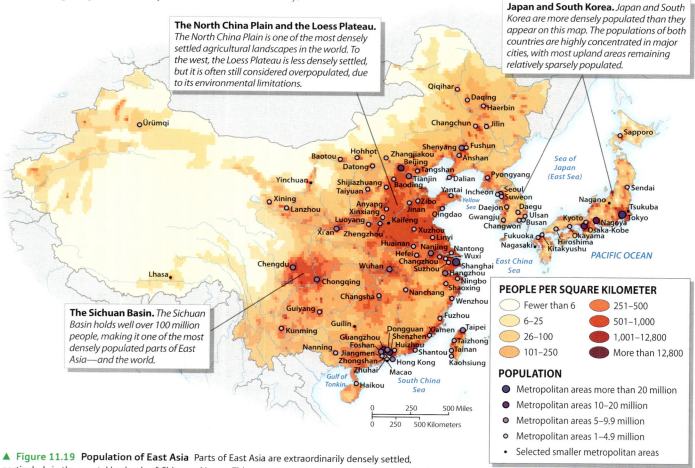

The North China Plain and the Loess Plateau. *The North China Plain is one of the most densely settled agricultural landscapes in the world. To the west, the Loess Plateau is less densely settled, but it is often still considered overpopulated, due to its environmental limitations.*

Japan and South Korea. *Japan and South Korea are more densely populated than they appear on this map. The populations of both countries are highly concentrated in major cities, with most upland areas remaining relatively sparsely populated.*

The Sichuan Basin. *The Sichuan Basin holds well over 100 million people, making it one of the most densely populated parts of East Asia—and the world.*

PEOPLE PER SQUARE KILOMETER
- Fewer than 6
- 6–25
- 26–100
- 101–250
- 251–500
- 501–1,000
- 1,001–12,800
- More than 12,800

POPULATION
- Metropolitan areas more than 20 million
- Metropolitan areas 10–20 million
- Metropolitan areas 5–9.9 million
- Metropolitan areas 1–4.9 million
- Selected smaller metropolitan areas

▲ **Figure 11.19 Population of East Asia** Parts of East Asia are extraordinarily densely settled, particularly in the coastal lowlands of China and Japan. This contrasts with the sparsely settled lands of western China, North Korea, and northern Japan. Although the total population of this world region is high, as is the overall density, the rate of natural population increase has slowed rather dramatically in the last several decades.

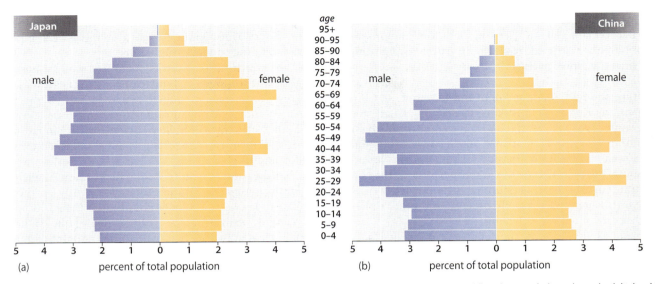

▲ Figure 11.20 Population Pyramids of China and Japan (a) Japan has one of the world's oldest and most rapidly aging populations, due to both its low birth rate and its high life expectancy. The large Japanese population in the over-60 category places a heavy burden on the country's economy. (b) China has a balanced demographic profile, but it also has a low birth rate, which could result in problems similar to Japan's in another 10 to 20 years.

and instead aim for a two-child program. Even so, experts predict that China's population will fall below 1 billion by 2100.

Agriculture and Settlement in Japan, Korea, and Taiwan

East Asia's maritime fringe—Japan, Korea, and Taiwan—is highly urbanized, supporting three of the largest urban agglomerations in the world: Tokyo, Osaka, and Seoul. Yet it is also one of the world's most mountainous areas, with lightly inhabited uplands. Agriculture must therefore share the limited lowlands with cities and suburbs, resulting in extremely intensive farming practices.

Japan's Urban-Agricultural Dilemma Japanese agriculture is largely limited to the country's coastal plains and interior basins. Rice farming, concentrated along the Sea of Japan coast in Honshu, has long been extremely productive. Vegetables and even rice crops are cultivated intensively on tiny patches within suburban and urban neighborhoods (**Figure 11.21**). The valleys of central and northern Honshu are famous for their temperate-climate fruit, while citrus comes from the country's milder southwestern reaches. Crops that thrive in cooler climates, such as potatoes, are produced mainly in Hokkaido and northern Honshu.

Japanese cities are located in the same lowlands that support the country's agriculture. Not surprisingly, the three largest metropolitan areas—Tokyo, Osaka, and Nagoya—sit near the centers of the three largest plains.

The fact that Japan's settlements are largely restricted to roughly 15 percent of its land area means that the country's effective population density is one of the highest in the world. This is especially notable in the main industrial belt extending from Tokyo through Nagoya and Osaka and then along the Inland Sea (sandwiched between Shikoku, western Honshu, and Kyushu) to the northern coast of Kyushu. In major urban areas, the amount of available living space is highly restricted for all but the most affluent families. Critics have long argued that Japan should allow its cities and suburbs to expand into nearby rural areas. However, because most uplands are too steep for residential use, such expansion would have to come at the expense of farmland.

As it now stands, the wholesale conversion of farmland to neighborhoods would be difficult. Although most farms are extremely small (the average is only several acres) and only marginally viable, farmers are politically powerful, and croplands are protected by tax codes. Moreover, most

▼ Figure 11.21 Japanese Urban Farms In Japan's limited lowland areas, farms and cities are often jumbled together in complex mosaics. Here we see lush rice fields within the general confines of the city of Mino, located in central Japan.

Japanese citizens believe that it is vitally important for their country to remain self-sufficient in rice. Rice imports are therefore highly restricted, resulting in prices several times higher than the global average.

Agriculture and Settlement in Korea and Taiwan Like China and Japan, the Korean peninsula is densely populated, containing some 76 million people (25 million in the north and 51 million in the south) in an area smaller than Minnesota. South Korea's population density is higher than Japan's. Most South Koreans live in the alluvial basins of the west and south. South Korean agriculture is dominated by rice, while North Korea relies heavily on corn and other upland crops that do not require irrigation.

Taiwan is the region's most densely populated state. Roughly the size of the Netherlands, it contains more than 23.5 million inhabitants. Because mountains cover most of central and eastern Taiwan, the population is concentrated in the narrow lowland belt in the north and west. Large cities and numerous factories are scattered amid lush farmlands. Despite its highly productive agriculture, Taiwan, like other East Asian countries, is forced to obtain much of its food from abroad.

Agriculture and Settlement in China

Unlike Japan, Korea, and Taiwan, China has almost as many people living in rural areas as in its cities. But the Chinese government predicts that 70 percent of its population will be urbanized by 2035. Chinese cities are rather evenly distributed across the plains and valleys of eastern China. As a result, the overall pattern of population distribution in China closely follows the geography of agricultural productivity.

China's Agricultural Regions A line drawn just to the north of the Yangtze Valley divides China into two main agricultural regions. To the south, rice is the dominant crop. To the north, wheat, millet, and sorghum are the most common.

In southern and central China, the population is concentrated in the broad lowlands, which are famous for their fertile soil and intensive agriculture. More than 100 million people live in the Sichuan Basin, and more than 70 million reside in and near the Yangtze River Delta in Jiangsu, a province smaller than Ohio. Crop growing occurs year-round in southern and central China; summer rice alternates with winter barley or vegetables in the north, whereas two rice crops can be harvested in the far south. Southern China also produces

a wide variety of tropical and subtropical crops, and moderate slopes throughout the area supply sweet potatoes, corn, and other upland products.

The North China Plain is one of the world's most thoroughly **anthropogenic landscapes**; that is, it has been heavily transformed by human activities. Virtually its entire extent is either cultivated or occupied by houses, factories, and other structures of human society. Too dry in most areas for rice, the North China Plain is largely devoted to wheat, millet, and sorghum.

Northeastern China (called Manchuria in English) was a lightly populated frontier zone as recently as the mid-1800s. Today, with a population of more than 100 million, its central plain is thoroughly settled. Still, Manchuria remains less crowded than many other parts of China, and it is one of the few parts of China to produce a consistent food surplus.

The Loess Plateau is more thinly settled yet, supporting only some 70 million inhabitants. However, considering the area's aridity and soil erosion, this is a high figure. Like other portions of northern China, the Loess Plateau produces wheat and millet. One of its unique settlement features is subterranean housing. Although loess erodes quickly under the impact of running water, it holds together well in other circumstances. For millennia, villagers have excavated pits on the surface of the plateau, from which they have tunneled into the earth to form underground houses (**Figure 11.22**). These subterranean dwellings are cool in the summer and warm in the winter. Unfortunately, they tend to collapse during earthquakes. One quake in 1920 took the lives of an estimated 100,000 persons; another in 1932 killed some 70,000.

▼ **Figure 11.22 Loess Plateau Settlement** In the Loess Plateau of central China, houses are often constructed underground. These dwellings are cool in the summer and warm in the winter, but tend to collapse easily during earthquakes.

Agriculture and Resources in a Global Context

Japan, South Korea, and Taiwan are major food importers, and China has recently moved in the same direction. Other resources also are being drawn in from all quarters of the world by the powerful economies of East Asia.

Global Dimensions of Japanese Agriculture and Forestry

Japan may be mostly self-sufficient in rice, but it is still one of the world's largest food importers. As the Japanese have grown more prosperous over the past 60 years, their diet has grown more diverse. That diversity is made possible largely by imports.

Japan imports food from a wide array of other countries. It procures both meat and the feed used in its domestic livestock industry from the United States, Canada, and Australia, which also supply wheat needed to produce bread and noodles. Even soybeans, long a staple of the Japanese diet, must be purchased from Brazil, the United States, and other countries. Japan has one of the world's highest rates of fish consumption, and its fishing fleet scours the world's oceans to meet demand. Japan also purchases prawns and other seafood, much of it farm-raised in former mangrove swamps, from Southeast Asia and Latin America.

Japan also depends on imports to supply forest products. Although its own forests produce high-quality cedar and cypress logs, it obtains most of its construction lumber and pulp (for papermaking) from western North America and Southeast Asia. As the rainforests of Southeast Asia diminish, Japanese interests are turning to Latin America and Africa for tropical hardwoods. Japanese and South Korean firms are also extracting resources from eastern Russia, a nearby and previously little-exploited forest zone. As a result, the environmental demands generated by Japan's large economy have become highly globalized.

Korean Agriculture in a Global Context

Like Japan and Taiwan, South Korea has also made the transition to a global food and resource procurer. From 2010 to 2014, South Korea's food imports increased by 45 percent. Although the country aims to be self-sufficient in rice and vegetables, it is a major importer of feed grains, wheat, and soybeans. South Korea also imports large quantities of beef, particularly from Australia.

North Korea, on the other hand, has pursued a goal of self-sufficiency. This policy was relatively successful for several years, but in the mid-1990s a series of floods, followed by drought, destroyed most of the country's rice and corn crops, leading to widespread famine. Since then North Korea has relied on international aid—much of it from South Korea—to feed its people, but malnutrition remains widespread. According to a 2015 United Nations report, 70 percent of North Korea's people are food insecure, and almost one-third of children under the age of five are stunted. The country's food situation further worsened in 2016, due mainly to inadequate rain, leading the government to warn of a possible famine (**Figure 11.23**). As a result, the UN concluded that North Korea would have to import 694,000 tons

▲ **Figure 11.23 Hardships in North Korea** North Korea has a difficult time feeding its population, resulting in great hardship especially for its rural population. Here we see two farmers laboriously moving topsoil from one part of a field to another as they prepare to apply precious fertilizer.

of food while cautioning that the government would only bring in 300,000 tons.

Global Dimensions of Chinese Agriculture

Until the late 1990s China was self-sufficient in food, despite its huge population and crowded lands. But rapid economic growth has led to increased consumption of meat, which requires large amounts of feed grain. Growth also has sacrificed agricultural lands to residential and industrial development. As a result, China is now the world's largest importer of soybeans and vegetable oil, and it imports significant quantities of corn and other grains in most years. In 2014 alone, China imported US$122 billion worth of agricultural products.

Although a net food importer, China exports many high-value crops, such as garlic, apples, farm-raised fish, and vegetables. Roughly half of the apple juice consumed in the United States now comes from China. Farmers in both wealthy countries such as the United States and poor countries such as the Philippines often complain that China exports its fruits and vegetables at below-market rates, seeking to build dominance. In addition, consumers in both China and foreign markets are concerned about the safety of Chinese food products. In 2014, for example, a major Chinese food company was shown to have sold tainted and expired meat to such companies as McDonald's, KFC, Pizza Hut, Starbucks, and Burger King.

East Asia's Urban Structure

China has one of the world's oldest urban foundations, dating back more than 3500 years. In medieval and early modern times, East Asia supported a well-developed system of cities. In the early 1700s, Tokyo, then called Edo, probably overshadowed all other cities, with a population of more than 1 million.

Despite this early start, East Asia was largely rural at the end of World War II, with some 90 percent of China's people living in the countryside, and Japan only about

China Reforms Its Residency Registration System

China is currently undergoing the largest internal migration in world history. As of 2014, an estimated 274 million people had moved from their official residences in the countryside to fast-growing cities. Owing to China's *hukou* system of individual registration in a particular place, such migrants are classified as temporary workers and have few rights in their new homes. They are often forced to leave their children behind, and their opportunities remain limited in the burgeoning cities that they help build (**Figure 11.2.1**). As a result, social tension is mounting.

Pressure for Change Most experts agree that the *hukou* system places large burdens on both China's economy and its social system. If internal immigration restrictions were relaxed, more people would leave areas of limited economic prospects and instead find higher-paying jobs in more prosperous urban cores. If such migrants could gain official residency, they would be able to improve their skill levels and bring their families with them, thus reducing social isolation and enhancing community connections.

With these considerations in mind, China's government has begun to liberalize the *hukou* system. In early 2016, it announced that it would phase out temporary permits for migrant workers and replace them with some form of permanent residency. Such residency would allow migrants to buy cars and homes. The details of this new plan have not been fully spelled out, and it is clear changes will be gradual. The government worries that a rapid end of the *hukou* system would generate chaos, both in depopulating rural areas and mushrooming cities. But China's state council does intend to grant urban residency permits to some 100 million migrant workers by 2020, focusing on individuals considered most "desirable" based on their employment history, education level, and housing situation.

Those Left Behind Reforming China's residency registration system could bring profound benefits to rural areas abandoned by labor migrants and especially to the children left behind. An estimated 61 million Chinese children fall into this category, 40 percent of whom are younger than five. Many live with their grandparents, but others have been farmed out to more distant relatives or other caretakers, and few see their parents more often than once a year. One study found that almost 60 percent of these children suffer from some sort of psychological problems, and official statistics

▲ **Figure 11.2.1 Undocumented Internal Migrants in Chinese Cities** As China's economy grows, tens of millions of people are moving from the countryside to urban areas. Many of these people, however, do not have official permission to live in cities, and are thus undocumented migrants in their own country.

indicate that they account for more than two-thirds of juvenile delinquency cases.

If the *hukou* system could be meaningfully reformed, many of these left-behind children would be able to join their parents and take advantage of the educational opportunities found in urban environments. In the meantime, however, Chinese authorities are seeking to ease the burdens that they currently face. One strategy currently being tested involves training local social workers to help such children cope with psychological stress.

1. Why is China so determined to manage and restrict migration to major cities?

2. What possible problems could emerge if the *hukou* system were completely eliminated?

---

50 percent urbanized. However, as the region's economies began to grow after the war, so did its cities. Japan, Taiwan, and South Korea are now between 78 and 86 percent urban, which is typical for advanced industrial countries. Almost half of China's people still live in rural areas, but the country is urbanizing rapidly. Rapid urbanization, however, has generated serious social problems (see *People on the Move: China Reforms Its Residency Registration System*).

Chinese Cities Traditional Chinese cities were separated from the countryside by defensive walls. Most were planned in accordance with strict geometric principles that were thought to reflect the cosmic order. The old-style Chinese city was dominated by low buildings and characterized by straight streets. Houses were typically built around courtyards, and narrow alleyways served both commercial and residential functions.

China's urban fabric began to change during the colonial period. Several port cities were taken over by European interests, which built Western-style buildings and modern business districts. The most important of these semicolonial cities was Shanghai, built near the mouth of the Yangtze River, the main gateway to interior China. When the communists came to power in 1949, Shanghai, with a population of more than 10 million, was the second largest city in the world. The communist authorities, however, viewed it as a decadent foreign creation, and milked it for taxes, which they invested elsewhere. As a result, the city began to decline.

Since the late 1980s, however, Shanghai has experienced a major revival and is again in many respects China's premier city. Migrants are pouring into the city even though the state tries to restrict the flow, and building cranes crowd the skyline. Official statistics put the municipality's population at more than 24 million and that of the metropolitan area at around 34 million. The new Shanghai is a city of massive high-rise apartments and concentrated industrial developments (**Figure 11.24**). In 2010, Shanghai surpassed Singapore to become the world's busiest container port. However, despite Shanghai's revived economic fortunes, the city remains politically secondary to Beijing.

Beijing was China's capital during the Manchu period (1644–1912), a status it regained in 1949. Under communist rule, Beijing was radically transformed; old buildings were razed, and broad avenues plowed through neighborhoods. Crowded residential districts gave way to large blocks of apartment buildings and massive government offices to accommodate the burgeoning population. Some historically significant structures were saved; the buildings of the Forbidden City, for example, where the Manchu rulers once lived, survived as a complex of museums. The area immediately in front of the Palace Museum, however, was cleared. The resulting plaza, Tiananmen Square, is reputed to be the largest open square in any city of the world.

The extremely rapid but uneven growth of Chinese cities over the past several decades has resulted in housing problems. In some secondary cities, apartment construction has outpaced population growth, generating unoccupied "ghost towns" along urban fringes (**Figure 11.25**). In the biggest cities, especially Beijing and Shanghai, the opposite problem has occurred, with construction failing to meet demand. Migrants are often forced to live in crowded conditions, often in poorly ventilated and unsanitary basements. People living in such conditions are sometimes termed the "rat tribe." An official 2013 report estimated the population of this group at over 250,000 in Beijing alone.

The Chinese Urban System On the whole, China's urban system is fairly well balanced, with sizable cities relatively evenly spaced across the landscape and no city overshadowing all others. This balance stems from China's heritage of urbanism, its vast size and distinctive physical-geographical regions, and its legacy of socialist planning. In 2015,

▲ Figure 11.25 **Chinese "Ghost City"** Chinese authorities have built huge new apartment blocks in several interior cities in recent years. Many of these settlements are still unoccupied, resulting in largely empty urban environments. Here we see the Kangbashi district of Ordos in Inner Mongolia, which is often described as China's largest "ghost city."

39 Chinese cities had urban-area populations of over 3 million.

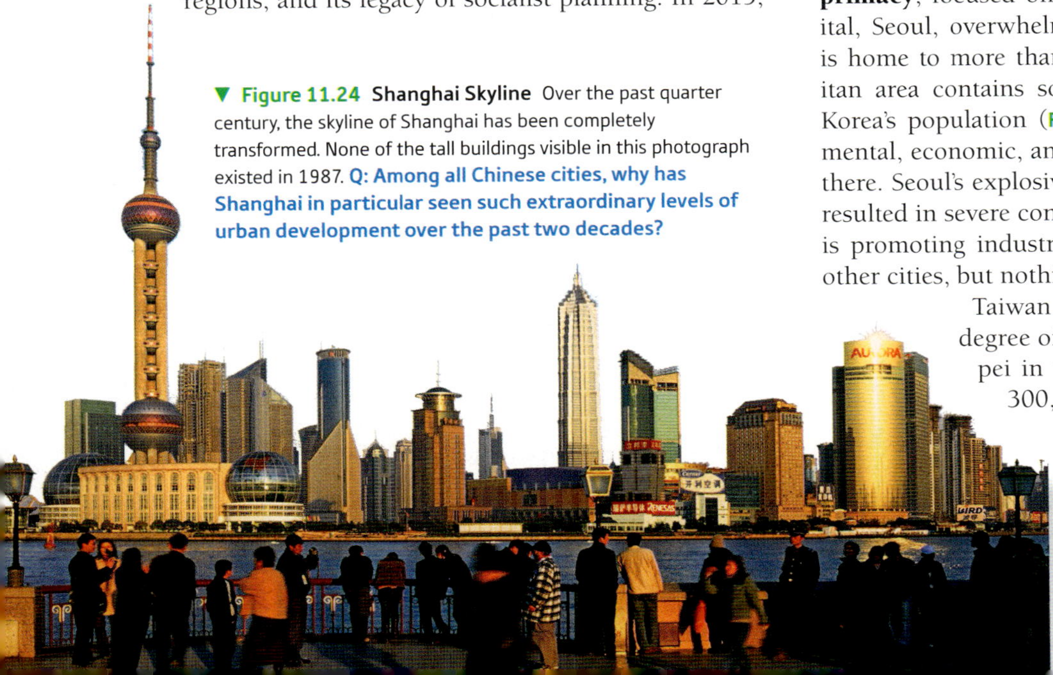

In the 1990s, Beijing and Shanghai vied for the first position among Chinese cities, with Tianjin, serving as Beijing's port, coming in third. All three cities, along with Chongqing in the Sichuan Basin, are exempt from the regular provincial structure of the country and have been granted their own metropolitan governments. In 1997 another major city, Hong Kong, passed from British to Chinese control. Rather than becoming an ordinary part of China, Hong Kong became an autonomous Special Administrative Region (SAR), allowed to manage its own affairs. Although not as populous as Beijing or Shanghai, Hong Kong is far wealthier. The greater metropolitan area of the Xi Delta, composed of Hong Kong, Shenzhen, and Guangzhou (called Canton in the West), forms one of China's premier urban areas.

Urban Patterns in South Korea, Taiwan, and Japan Unlike China, South Korea and Taiwan are noted for their **urban primacy**, focused on a single huge city. South Korea's capital, Seoul, overwhelms all other Korean cities. Seoul itself is home to more than 10 million people, and its metropolitan area contains some 25 million, almost half of South Korea's population (**Figure 11.26**). Almost all major governmental, economic, and cultural institutions are concentrated there. Seoul's explosive and generally unplanned growth has resulted in severe congestion. The South Korean government is promoting industrial and administrative development in other cities, but nothing has challenged Seoul's primacy.

Taiwan is similarly characterized by a high degree of urban primacy. The capital city of Taipei in the far north mushroomed from some 300,000 people during the Japanese colonial period (from 1895 to 1945) to more than 8 million in the metropolitan area today.

Japan has traditionally been characterized by urban "bipolarity," rather

▼ Figure 11.24 **Shanghai Skyline** Over the past quarter century, the skyline of Shanghai has been completely transformed. None of the tall buildings visible in this photograph existed in 1987. **Q: Among all Chinese cities, why has Shanghai in particular seen such extraordinary levels of urban development over the past two decades?**

▲ **Figure 11.26 Seoul's Gangnam District** Seoul is the capital city, economic hub, and cultural center of South Korea. Its Gangnam District is noted for expensive shops and luxury apartments.

Explore the Sights of Seoul's Gangnam-gu

http://goo.gl/wHu8mU

primacy of Tokyo, the Japanese government has been trying to steer new development to other parts of the country.

Japanese cities sometimes strike foreign visitors as rather gray and monotonous places, lacking historical interest. Little of the country's older architecture remains intact. Traditional Japanese buildings were made of wood, which survives earthquakes much better than stone or brick. Fires have therefore been a long-standing hazard, and in World War II, U.S. forces firebombed most Japanese cities (see *Geographers at Work: Cary Karacas Examines the Firebombing of Japan*). In addition, Hiroshima and Nagasaki were completely destroyed by atomic bombs. The one exception was Kyoto, the old imperial capital, which was spared devastation. As a result, Kyoto is famous for its beautiful (wooden) Buddhist monasteries and Shinto than urban primacy. Until relatively recently, Tokyo, the capital and main business and educational center, together with the neighboring port of Yokohama, were balanced by the mercantile center of Osaka and its port of Kobe. Kyoto, the former imperial capital and the traditional cultural center, is also situated in the Osaka region. A host of secondary and tertiary cities balance Japan's urban structure. Nagoya, with a metropolitan area of almost 9 million people, remains the center of the automobile industry.

As Japan's economy boomed in the 1960s, 1970s, and 1980s, so did Tokyo, outpacing all other urban areas in almost every urban function. Tokyo itself has about 13 million inhabitants, but the Greater Tokyo area now contains more than 37 million people, making it the world's most populous urbanized area. The Osaka–Kobe–Kyoto metropolitan area stands second, with 19 million inhabitants. Other large cities are sandwiched between these two metropolitan hubs (**Figure 11.27**). Concerned about the increasing

▶ **Figure 11.27 Urban Concentration in Japan** The inset map shows the rapid expansion of Tokyo in the postwar decades. Today the Greater Tokyo Metropolitan Area is home to roughly 37 million people. The larger map shows the cluster of urban settlements along Japan's southeastern coast. The major area of urban concentration is between Tokyo and Osaka, a distance of some 300 miles (482 km), known as the Tokkaido corridor. By some accounts, 65 percent of Japan's population lives in this area.

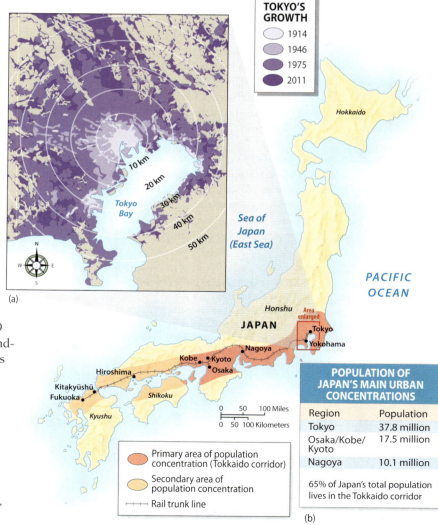

TOKYO'S GROWTH
- 1914
- 1946
- 1975
- 2011

POPULATION OF JAPAN'S MAIN URBAN CONCENTRATIONS	
Region	Population
Tokyo	37.8 million
Osaka/Kobe/Kyoto	17.5 million
Nagoya	10.1 million
65% of Japan's total population lives in the Tokkaido corridor	

Primary area of population concentration (Tokkaido corridor)

Secondary area of population concentration

Rail trunk line

Cary Karacas Examines the Firebombing of Japan

▲ **Figure 11.3.1** Geographer Cary Karacas

Hiroshima's and Nagasaki's destruction by atomic bombs at the end of World War II are well-known events. Often forgotten, however, is the U.S. fire-bombing of 66 other Japanese cities during the final months of the war. Operation Meeting House, conducted on Tokyo on March 9 and 10, 1945, is often considered the most destructive air raid in history, destroying 15.8 square miles and taking over 100,000 lives. Geographer Cary Karacas, who teaches at the College of Staten Island, CUNY, is determined to keep the memory of these events alive through mapping and other geographical techniques that explore the bombing strategies and their effects on civilian populations (**Figure 11.3.1**).

Combining Geography and History Examining old maps and documents might seem to be the work of a historian, but the process benefits from a geographer's eye, notes Karacas, who discovered geography during his first quarter at UCLA. "In classes, historians on the whole weren't really sensitive to concepts of place or space ... which is central to how geographers examine whatever is in front of them," he says. It was the "breadth and depth that could be had in the discipline" that drew him to geography. Working with historian David Fedman, Karacas found a shift in U.S. military strategy: military mapping in the early years of the war focused on high-level military targets, but by late 1944, military cartographers were instead mapping the flammability of urban neighborhoods (**Figure 11.3.2**). By the war's end, many maps simply showed urban areas obliterated by incendiary bombs and areas that remained more or less intact.

In addition to examining maps, Karacas co-created a large digital archive—JapanAirRaids.org—focused on primary documents, analyses, and oral accounts of the firebombing of Japan's cities. He is currently preserving firsthand accounts of the destruction—translating the writings of Japanese civilians who survived the firestorms, but analyzing these with a geographer's sensibility. "The possibilities are up to the individual geographer as to what he or she could focus on," he explains.

Beyond the Great Destruction Karacas' research extends beyond the bombing raids. He is also co-editor of a recently published

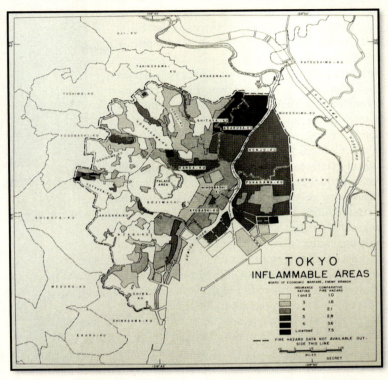

▲ **Figure 11.3.2** World War II U.S. Military Map of Tokyo's Inflammable Areas

comprehensive work on the historical geography of the Japanese archipelago, *Cartographic Japan: A History in Maps*. Lavishly illustrated, the volume is both aesthetically pleasing and deeply informative, showcasing the long history of high-level mapmaking in Japan.

1. Why do you think that the dropping of atomic bombs on Hiroshima and Nagasaki has attracted vastly more attention than the firebombing of Tokyo and many other Japanese cities?

2. Other than map analysis, what are some of the ways in which geographers have been involved with military campaigns over the past 100 years?

---

temples. Other Japanese cities were largely reconstructed in the 1950s and 1960s, when Japan was still relatively poor and could afford only inexpensive concrete buildings. In the boom years of the 1980s, however, modern skyscrapers rose in many of the larger cities, and postmodernist architecture now adds variety, especially to the wealthier neighborhoods.

REVIEW

11.4 Why does East Asia get so much of its food and so many of its natural resources from other parts of the world?

11.5 How is the urban landscape of East Asia, and particularly that of China, currently changing?

11.6 How have East Asia's patterns of population density influenced the region's agricultural and settlement systems?

■ **KEY TERMS** anthropogenic landscapes, urban primacy

Cultural Coherence and Diversity: The Historical Influence of Confucianism and Buddhism

East Asia is in some respects one of the world's most unified cultural regions. Although different parts of East Asia have their own unique cultural features, the entire region shares certain historically rooted ways of life and systems of ideas. Most of these commonalities can be traced back to ancient Chinese civilization.

Unifying Cultural Characteristics

The most important unifying cultural characteristics of East Asia are related to religious and philosophical beliefs. Throughout the region, Buddhism and Confucianism have shaped not only individual beliefs, but also social and political structures. Although the role of traditional belief systems has been seriously challenged over the past 50 years, especially in China, historically rooted cultural patterns have not disappeared.

Writing Systems The clearest distinction between East Asia and the world's other cultural regions is found in written language. Existing writing systems elsewhere in the world are based on the alphabetic principle, in which each symbol represents a distinct sound. East Asia, in contrast, evolved an entirely different system, known as **ideographic writing**. Ideographic symbols (commonly called characters) primarily represent ideas rather than sounds (although the symbols can denote sounds in certain circumstances). As a result, ideographic writing requires a large number of distinct symbols.

Because ideographic writing is largely divorced from spoken language, the Chinese ideographic writing system is difficult to learn compared to alphabetic systems; to be literate, a person must memorize thousands of characters. The advantage is that two literate people do not have to speak the same language to be able to communicate, because the written symbols that they use to express their ideas are the same. Therefore, cultures that share this writing system can still read and understand the same newspapers, books, and websites.

Korea adopted Chinese characters at an early date and used these exclusively for hundreds of years. In the 1400s, however, Korean officials decided that Korea needed its own alphabet to allow more widespread literacy, and to more clearly differentiate Korean culture from that of China. The new script spread quickly throughout the country, but Korean scholars and officials continued to use Chinese characters, regarding their own script as suitable only for popular writings. Today the Korean script is used for almost all purposes, but scholarly works in South Korea still contain scattered Chinese characters.

Japan's writing system is even more complex. Initially, the Japanese simply borrowed Chinese characters, called kanji in Japanese. Grammatical differences between the Japanese and Chinese languages meant that the exclusive use of kanji resulted in awkward sentence construction. The Japanese solved this quandary by developing a quasi-alphabet known as hiragana to express words and parts of speech not easily represented by Chinese characters. (A different but parallel system, called katakana, is used in Japan to spell words of foreign origin.) Eventually, the two styles of writing merged, and written Japanese now employs a complex mixture of symbols. Japanese kanji today differ slightly from Chinese characters.

The Confucian Legacy Another historically unifying feature in East Asia is **Confucianism**, the philosophy developed by the premier philosopher of Chinese history, Kongfuzi, known as Confucius in the West. Confucius himself lived during the 6th century BCE, a period of marked political instability, and as a result, sought to establish a more stable social order.

▲ **Figure 11.28 Confucian Temple in China** Confucianism is usually considered to be a philosophy rather than a religion, but it does have religious aspects. In Confucian temples, which are found through many parts of East Asia, the spirit and philosophy of Confucius are honored and often worshipped. **Q: To what extent can Confucianism be said to unify East Asia, both in the past and in the contemporary period?**

Although Confucianism is often considered to be a religion, Confucius himself was mostly interested in how to lead an ethical life and organize a proper society (**Figure 11.28**).

Confucius stressed deference to the proper authority figures, but he also thought that authority has a responsibility to act in a benevolent manner. The most basic level of the traditional Confucian moral order is the family unit, considered the bedrock of society. The ideal family structure is patriarchal, and children are told to obey and respect their parents—especially their fathers. Confucian philosophy also stresses the need for a well-rounded and broadly humanistic education. Confucianism holds that an individual should be judged on the basis of behavior and education rather than on family background. The high officials of Imperial China (pre-1912)—the powerful **Mandarins**—were thus selected by competitive examinations. Only wealthy families, however, could afford to give their sons the education needed for success on those grueling tests.

In Japan, Confucianism was never as important as it was on the mainland. Japanese officials were actually able to exclude certain Confucian beliefs that they considered dangerous, such as the revocable "mandate of heaven." According to this notion, the emperor of China derived his authority from the principle of cosmic harmony, but such a mandate could be withdrawn if he failed to fulfill his duties. This idea was used both to explain and to legitimize the rebellions that occasionally resulted in a change of China's ruling dynasty. In Japan, on the other hand, a single imperial dynasty has persisted throughout the entire period of written history. Although the emperor of Japan has had little real power for more than a thousand years, the sanctity of his family lineage continues to form a basic principle of Japanese society.

Over the past 100 years, Confucianism has lost much of the hold that it once had on public morality throughout East Asia, especially in China. After the communist revolution,

China's rulers sought for decades to eliminate Confucian thought. Currently, however, Chinese officials are pushing for a revival of the philosophy, hoping that it will enhance social stability. Confucian schools are multiplying, and recent books, films, and television shows about Confucius have been popular. The Chinese government maintains "Confucius Institutes" in more than 90 countries, designed to promote Chinese culture and language. Some critics, however, accuse these institutes of acting as propaganda arms of the Chinese government.

Religious Unity and Diversity

Certain religious beliefs have worked alongside Confucianism to unify East Asian cultures. The most important of these are associated with Mahayana Buddhism. Other religious practices, however, have had a more divisive role.

Mahayana Buddhism Originating in India in the 6th century BCE, Buddhism stresses escape from an endless cycle of rebirths to reach union with the divine cosmic principle (or nirvana). By the 2nd century CE Buddhism reached China, and within a few hundred years had expanded throughout East Asia. Today Buddhism remains widespread everywhere in the region, although it is far less significant here than in mainland Southeast Asia and Sri Lanka.

The variant of Buddhism practiced in East Asia—Mahayana, or Greater Vehicle—is distinct from the Theravada Buddhism of South and Southeast Asia (**Figure 11.29**). Mahayana Buddhism simplifies the quest for nirvana, in part by suggesting the existence of bodhisattvas who refuse divine union for themselves in order to help others spiritually. Mahayana Buddhism is also nonexclusive, meaning that one may follow it while simultaneously practicing other faiths. Thus, many Chinese consider themselves to be both Buddhists and Taoists, whereas most Japanese are at some level both Buddhists and followers of Shinto.

As Mahayana Buddhism spread through East Asia in the medieval period, many different sects emerged. At one time, Buddhist monasteries were rich and powerful. In all East Asian countries, however, periodic reactions against Buddhism resulted in the persecution of monks and the suppression of monasteries. Despite such hardships, East Asian Buddhism was never extinguished, although it never became the focal point of society. In Japan, that position was partially captured by a different religion, Shinto.

Shinto The practice of Shinto is so closely bound to the idea of Japanese nationality that it is questionable whether a non-Japanese person can follow it. Shinto began as the worship of nature spirits, but it was gradually refined into a subtle set of beliefs about the harmony of nature and its connections with human existence. Until the late 1800s, Buddhism and Shinto were complexly intertwined. Subsequently, the Japanese government began to disentangle the two faiths, while elevating Shinto into a nationalistic cult focused on the divinity of the Japanese imperial family. After World War II, the more excessive aspects of nationalism were removed from the religion.

Shinto is still a place- and nature-centered religion. Certain mountains, particularly the volcanic Mount Fuji, are considered sacred and are thus climbed by large numbers of people. Major Shinto shrines, often located in scenic places, attract numerous pilgrims; most notable is the Ise Shrine south of Nagoya, site of the cult of the emperor. Local Shinto shrines, as well as Buddhist temples, offer leafy oases in otherwise largely treeless Japanese urban neighborhoods.

Taoism and Other Chinese Belief Systems The Chinese religion of Taoism (or Daoism) is similarly rooted in nature worship. Like Shinto, it stresses the acquisition of spiritual harmony and the pursuit of a balanced life. Taoism is indirectly associated with feng shui, also called **geomancy**, the Chinese and Korean practice of designing buildings in accordance with the spiritual powers that supposedly course through the local topography. Even in hypermodern Hong Kong, skyscrapers worth millions of dollars have occasionally gone unoccupied because people believed that their construction failed to accord with geomantic principles. The government of China officially recognizes Taoism as a religion and seeks to regulate it through the state-run China Taoist Association.

Although both Taoism and Buddhism were historically followed throughout the country, traditional religious practice in China embraced the unique attributes of particular places. Many minor gods were traditionally associated with single cities or other specific areas. In modern-day rural China, village gods are often still honored.

Minority Religions Followers of virtually all world religions can be found in the increasingly cosmopolitan cities of East Asia. Over a million Japanese belong to Christian churches, whereas South Korea is roughly 30 percent Christian (mostly Protestant). South Korea reportedly sends more missionaries abroad

▼ **Figure 11.29 The Buddhist Landscape**
Mahayana Buddhism has been traditionally practiced throughout East Asia. This Golden Buddha statue is located in Baomo Garden in Chi Lei Village, Guangdong Province, China. **Q: What is the symbolic significance of the flowers growing in the water around the statue of the Buddha?**

Explore the **Sights** of Baomo Garden

http://goo.gl/ztcmBE

than any country except the United States. Some reports indicate that Christianity is growing rapidly in China despite state discouragement. Officially, China acknowledges a Christian population of 26 million, but this figure excludes children as well as all members of unregistered "house churches." A 2012 Chinese survey estimated the Christian population at up to 40 million, but some sources believe it could be significantly higher.

China also has a long-established Muslim community, estimated in 2009 to exceed 21 million. Most Chinese Muslims are members of minority ethnic groups living in the far west. The roughly 10 million Chinese-speaking Muslims, called *Hui*, are concentrated in Gansu and Ningxia in the northwest and in Yunnan Province in the south. The only Muslim congregations in Japan and South Korea, on the other hand, are associated with recent and often temporary immigrants from South, Southeast, and Southwest Asia.

Secularism in East Asia For all of these varied forms of religion, East Asia is one of the most secular regions of the world. Roughly half of the South Koreans are not religious. A small portion of Japan's population is highly religious, but most people only occasionally observe Shinto or Buddhist rituals. As a result, statistics on religious affiliation in Japan vary tremendously, with estimates of the followers of Shinto ranging between 4 million and 119 million! Japan also has several "new religions," a few of which are noted for their strong beliefs (**Figure 11.30**). But for Japanese society as a whole, religion is not tremendously important.

▼ **Figure 11.30 Perfect Liberty's Peace Tower** Japan is the home of a number of new religions, including PL Kyodan, or the Church of Perfect Liberty, which aims to establish world peace. This photo shows PK Kyodan's 600-foot-tall (180 m) Peace Tower, located near Osaka, Japan.

After the communists took power in China in 1949, all forms of religion and traditional philosophy were discouraged and sometimes severely repressed. Under the new regime, the atheistic philosophy of **Marxism**—the communistic belief system developed by Karl Marx—became the official ideology. In the 1960s, many observers thought that traditional forms of Chinese religion would survive only in overseas Chinese communities. With the easing of Marxist orthodoxy during the 1980s and 1990s, however, many forms of spiritual expression began to revive.

More recently, however, China has sought to reimpose Marxism and to restrict the growth of religious organizations, particularly Christian ones. In 2016, Chinese leader Xi Jinping urged China to return to its socialist ideals, while arguing that it needed to adapt Marxist ideas to the country's current realities by encouraging innovation. China's government has also demolished unregistered churches, confiscated church properties, and even arrested pastors. Although state-recognized congregations are generally secure, members of underground churches complain of persecution.

North Korea also adopted communist ideology after World War II. North Korean leaders supplemented Marxism with a unique official ideology called **juche**, or "self-reliance." Ironically, juche demands absolute loyalty to North Korea's political leaders. It also involves intense nationalism based on the idea of Korean racial purity. In North Korea's revised 2009 constitution, all references to communism were removed. North Korea has also deviated from Marxism by instituting the principle of **dynastic succession**, with power flowing from father to son. When the leader Kim Jong-il died in late 2011, he was immediately replaced by his untested 28-year-old son, Kim Jong-un.

Linguistic and Ethnic Diversity

Japanese and Mandarin Chinese may partially share a system of writing, but the two languages have no direct relationship (**Figure 11.31**). In their grammatical structures, Chinese and Japanese are more different from each other than are Chinese and English. Japanese, however, has adopted many words of Chinese origin.

Language and National Identity in Japan According to most scholars, Japanese is not related to any other language. From several perspectives, the Japanese form one of the world's most homogeneous peoples, and tend to regard themselves as such. To be sure, minor cultural and linguistic characteristics distinguish the people of western Japan (centered on Osaka) and eastern Japan (centered on Tokyo), but only the people of the Ryukyu Islands of the far south speak variants of Japanese that can be considered separate languages. Many Ryukyu people believe that they have not been treated as full members of the Japanese nation and have suffered from discrimination.

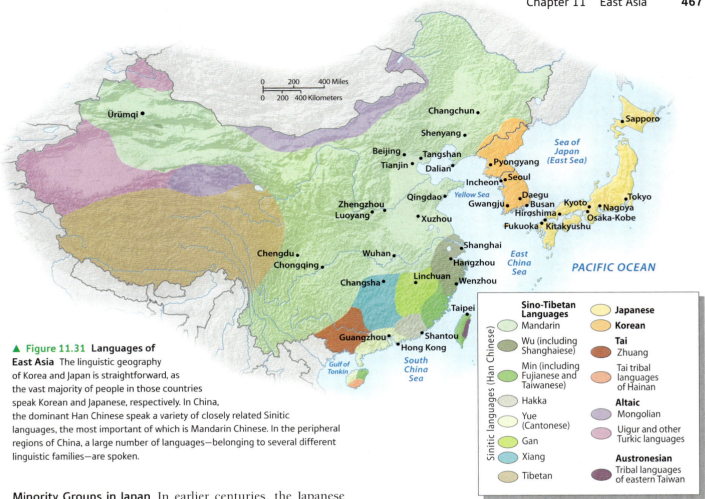

▲ **Figure 11.31 Languages of East Asia** The linguistic geography of Korea and Japan is straightforward, as the vast majority of people in those countries speak Korean and Japanese, respectively. In China, the dominant Han Chinese speak a variety of closely related Sinitic languages, the most important of which is Mandarin Chinese. In the peripheral regions of China, a large number of languages—belonging to several different linguistic families—are spoken.

Minority Groups in Japan In earlier centuries, the Japanese archipelago was divided between two very different peoples: the Japanese living to the south and the Ainu inhabiting the north. The Ainu possess their own language and have a distinctive physical appearance. Over time, the Ainu were driven north by Japanese expansion, and today only around 24,000 remain, mostly on the northern island of Hokkaido. As few as 10 people still speak the Ainu language, although efforts are now under way to revive its use.

The approximately 800,000 people of Korean descent living in Japan today also have felt discrimination. Many were born in Japan (their parents and grandparents having left Korea in the early 1900s) and speak Japanese, rather than Korean, as their primary language. Yet only around 300,000 people of Korean background have been able to gain Japanese citizenship, the rest being treated as resident aliens. Perhaps as a result of such treatment, many Japanese Koreans hold radical political views and support North Korea.

Other immigrants began to arrive in Japan in the 1980s, mostly from the poorer countries of Asia. Most do not have legal status. Men from China and southern Asia typically work in the construction industry and in other dirty and dangerous jobs; women from Thailand and the Philippines often work as entertainers and sometimes as prostitutes. However, immigration is less pronounced in Japan than in most other wealthy countries, and relatively few migrants acquire permanent residency, let alone citizenship.

The most victimized people in Japan could be the **Burakumin**, or *Eta*, an outcast group whose ancestors worked in "polluting" industries such as leathercraft. Discrimination is now illegal, but the Burakumin, who usually live in separate neighborhoods and are concentrated in the Osaka region, are still among the poorest people in Japan. Private detective agencies do a brisk business checking prospective marriage partners and employees for possible Burakumin ancestries. The Burakumin, however, have banded together to demand their rights; the Buraku Liberation League is politically powerful and is reputed to have close connections with the Yakuza (the Japanese criminal organization) (**Figure 11.32**).

Language and Identity in Korea The Koreans, like the Japanese, are a relatively homogeneous people. The vast majority of people in both North and South Korea speak Korean and unquestioningly consider themselves to be members of the Korean nation.

South Korea does, however, have a strong sense of regional identity, which can be traced back to the medieval period when the peninsula was divided into three separate kingdoms. The people of southwestern South Korea, centered on the city of Gwangju, tend to be viewed as distinctive, and many southwesterners

▲ **Figure 11.32 Men of the Yakuza** Members of Japan's organized-crime syndicate, known as the Yakuza, are noted for their elaborate tattoos. Here we see Yakuza members showing off their bodily decorations during the Sanja Matsuri festival, a Shinto religious observation, in Tokyo's Asakusa district.

believe that they have suffered periodic discrimination. In contrast, the Kyongsang region of southeastern South Korea has supplied most of the country's political leaders and has received more than its share of development funds.

Korean identity also has an international component. Several million Korean speakers reside directly across the border in northern China. Because of deportations ordered by the Soviet Union in the mid-20th century, substantial Korean communities also can be found in Kazakhstan. Over the past several decades, a Korean **diaspora**—the scattering of a particular group of people over a vast geographical area—has brought hundreds of thousands of Koreans to the United States, Canada, parts of South America, Australia, and New Zealand. South Korea's prosperity has also attracted hundreds of thousands of immigrants into the country. Particularly notable are the roughly 100,000 Vietnamese migrants, many of whom are women married to Korean farmers who have difficulty attracting local wives.

Language and Ethnicity Among the Han Chinese The geography of language and ethnicity in China is far more complex than that of Korea or Japan. This is true even if we consider only the eastern half of the country, so-called China proper. The most important distinction is that separating the Han Chinese from the non-Han peoples. The Han, who form the vast majority, are those people who have long been incorporated into the Chinese cultural and political systems and whose languages are expressed through Chinese writing. The Han do not, however, all speak the same language.

The northern, central, and southwestern regions of China—a vast area extending from Manchuria through the middle and upper Yangtze Valley to the valleys and plateaus of Yunnan in the far south—constitute a single linguistic zone. The spoken language here is generally called *Mandarin Chinese* in English. Mandarin is divided into several dialects, but the speech of the Beijing area is gradually spreading. Standard Mandarin—called *Pǔtōnghuà, or* "common language," locally—is China's official national language (see Figure 11.31).

In southeastern China, from the Yangtze Delta to China's border with Vietnam, several separate languages are spoken by peoples who are Han Chinese. Traveling from south to north, one encounters Cantonese (or Yue), in Guangdong; Fujianese (alternatively Hokkienese, or Min, locally), in Fujian; and Shanghaiese (or Wu), in and around the city of Shanghai and in Zhejiang Province. Linguistically speaking, these are true languages, because they are not mutually intelligible. They are usually called dialects, however, because they have no distinctive written form.

The Hakka, a group of people speaking a southern Chinese language, are occasionally not classified as Han Chinese. Evidently, their ancestors fled northern China roughly a thousand years ago to settle in the rough upland area where the Guangdong, Fujian, and Jiangxi provinces meet. Later migrations took them throughout southern China, where they typically settled in hilly wastelands. The Hakka traditionally made their living by growing upland crops such as sweet potatoes and by working as loggers, stonecutters, and metalworkers. Today they form one of the poorest communities of southern China.

Despite their many differences, all of the languages of the Han Chinese (including Hakka) are closely related to each other and belong to the same Sinitic language subfamily. Because their basic grammar and sound systems are similar, a person speaking one of these languages can learn another relatively easily. All Sinitic languages are **tonal** and monosyllabic; their words are all composed of a single syllable (although compound words can be formed from several syllables), and the meaning of each basic syllable changes according to the pitch in which it is uttered.

China's government, concerned about the country's linguistic diversity, encourages the use of Mandarin and discourages other languages and dialects. As part of this process, it is pressuring foreign companies to conduct business and marketing in the official national language. Such policies are resented, particularly in the far south where Cantonese is firmly established. In May 2016, demonstrators gathered outside of the Japanese consulate in Hong Kong to protest an announcement by Nintendo that the names of Pokémon characters would henceforth be written in characters that follow Mandarin rather than Cantonese pronunciation.

The Non-Han Peoples Many of the more remote upland districts of China proper are inhabited by non-Han peoples speaking non-Sinitic languages. According to the government, China contains 55 such ethnic groups. Most of these peoples are classified as tribal, implying that they have a traditional

▲ **Figure 11.33** **Tribal Villages in South China** Non-Han people are usually classified as "tribal" in China, which assumes they have a traditional social order based upon autonomous village communities. The photo shows a Miao village in the Xiangxi Tujia and Miao Autonomous Prefecture, Hunan Province.

social order based on self-governing village communities (**Figure 11.33**). Such a view is not entirely accurate, however, because some of these groups once had their own kingdoms.

Over the course of many centuries, the territory occupied by these non-Han communities has steadily declined in size, due both to continued Han expansion and non-Han emigration. Acculturation into Chinese society and intermarriage with the Han also have reduced many non-Han groups, which are primarily concentrated today in the rougher lands of the far north and the far south.

As many as 11 million Manchus live in the more remote portions of Manchuria, but only a handful of Manchus still speak their own language, the rest having abandoned it for Mandarin. This is ironic because the Manchus ruled the Chinese Empire from 1644 to 1912. For centuries the Manchus prevented the Han from settling in central and northern Manchuria, which they hoped to preserve as their homeland. Once Chinese were allowed to settle in Manchuria in the 1800s—in part to prevent Russian expansion—the Manchus soon found themselves vastly outnumbered. As they intermarried with and adopted the language and customs of the newcomers, their own culture began to disappear.

Larger and more secure communities of non-Han peoples are found in the far south, especially in Guangxi. Most of the inhabitants of Guangxi's more remote areas speak languages of the Tai family, closely related to those of Thailand. Because as many as 18 million non-Han people live in Guangxi, it has been designated an **autonomous region**. Such autonomy was designed to allow non-Han peoples to experience "socialist modernization" at a different pace from that expected of the rest of the country. Critics contend that very little real autonomy has ever existed. (In addition to Guangxi, four other autonomous regions exist in China. Three of these—Xizang [Tibet], Nei Monggol [Inner Mongolia], and Xinjiang—are located in Central Asia and are discussed in Chapter 10. The fourth, Ningxia in northwestern China, is distinguished by its large concentration of Hui, who are Mandarin-speaking Muslims.)

Other areas with sizable numbers of non-Han peoples are Yunnan and Guizhou in southwestern China, and western Sichuan. Most tribal peoples here practice swidden agriculture (also called "slash and burn") on rough slopes; flatter lands are generally occupied by rice-growing Han Chinese. A wide variety of separate languages, falling into several linguistic families, is found among the ethnic groups living in the uplands. **Figure 11.34** shows that in Yunnan, the resulting ethnic mosaic is staggeringly complex.

Language and Ethnicity in Taiwan Taiwan is also noted for its linguistic and ethnic complexity. In the island's mountainous eastern region, several small groups of "tribal" peoples speak languages related to those of Indonesia and the Philippines. These peoples resided throughout Taiwan before the 16th century, when Han migrants began to arrive in large numbers. Most of the newcomers spoke the Fujianese (or Hokkien) dialect, which evolved into the distinctive language of Taiwanese.

Taiwan was transformed almost overnight in 1949, when China's nationalist forces, defeated by the communists, sought refuge on the island. Most of the nationalist leaders spoke Mandarin, which they made the official language. Taiwan's new leadership discouraged Taiwanese, viewing it as a mere local dialect. As a result, tension developed between the Taiwanese- and the Mandarin-speaking communities. Only in the 1990s did Taiwanese speakers begin to reassert their language rights, resulting in the use of Taiwanese in schools alongside Mandarin Chinese. Today most Taiwanese citizens are fluent in both languages. In Taiwan's television industry, dramas and variety shows tend to be in Taiwanese, whereas game shows and documentaries are more often filmed in Mandarin.

East Asian Cultures in a Global Context

East Asia has long exhibited tensions between an internal orientation and more cosmopolitan tendencies. Until the mid-1800s, East Asian countries attempted to insulate themselves from Western cultural influences. Japan subsequently opened its doors, but remained ambivalent about foreign ideas. Only after its defeat in 1945 did Japan really opt for a globalist orientation. It was followed in this regard by South Korea, Taiwan, and Hong Kong (then a British colony). The Chinese and North Korean governments, to the contrary, decided during the early **Cold War** decades of the 1950s and 1960s to isolate themselves as much as possible from Western and global culture.

The Cosmopolitan Fringe Japan, South Korea, Taiwan, and Hong Kong are characterized by a vibrant internationalism that coexists with strong national and local cultural identities. Virtually all Japanese, for example, study English for 6 to 10 years. Business meetings among Japanese, Chinese, and Korean firms are often conducted in English. Relatively large numbers of advanced students, especially from Taiwan, study in the United States and other

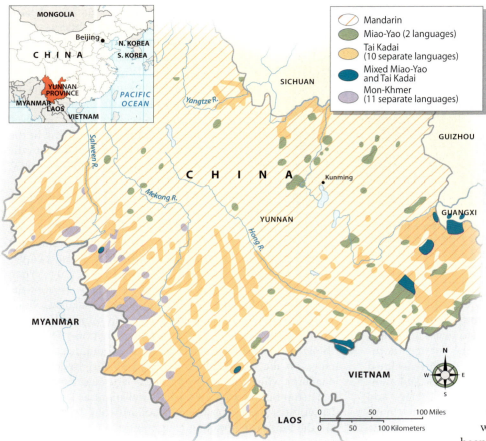

Mandarin
Miao-Yao (2 languages)
Tai Kadai (10 separate languages)
Mixed Miao-Yao and Tai Kadai
Mon-Khmer (11 separate languages)

▲ **Figure 11.34 Language Groups in Yunnan** China's Yunnan Province is the most linguistically complex area in East Asia. In Yunnan's broad valleys, its relatively level plateau areas, and its cities, most people speak Mandarin Chinese. In the hills, mountains, and steep-sided valleys, however, people speak a wide variety of tribal languages, falling into several linguistic families. In certain areas, several different languages can be found in very close proximity.

English-speaking countries and thus bring home a kind of cultural bilingualism.

The current cultural flow is not merely from a globalist West to a previously isolated East Asia, but is rather more reciprocal. Hong Kong's action films are popular throughout most of the world and have influenced filmmaking techniques in Hollywood. Over the past few years, South Korean popular culture has spread throughout East Asia and beyond. Japan almost dominates the world market in video games, and its ubiquitous comic-book culture and animation techniques are now following its karaoke bars in their overseas march. The films of Hayao Miyazaki and other directors associated with Japan's Studio Ghibli have been highly popular and influential across the world. Employment in Japan's *anime* industry, however, peaked around 2005 and has been slowly declining ever since. Critics claim that the industry is too focused on niche audiences and thus suffers from competition from South Korea.

South Korea's popular culture industry continues to thrive, as Korean music, movies, and television shows have become popular around the world, a phenomenon known as **hallyu**, or "Korean wave." K-pop, a South Korean musical genre based on rock, hip-hop, and electropop styles, has been especially fashionable abroad, its global reach enhanced by its fans' use of Facebook, Twitter, and YouTube to publicize their favorite artists. Some scholars think that the popularity of South Korean pop culture, and that of Japan as well, is reinforcing the old cultural connections that made East Asia a coherent world region.

Explore the **Sounds** of K-Pop Hit "Girls' Generation"

http://goo.gl/6KfiKT

The Chinese Heartland In one sense, Japan is more culturally predisposed to cosmopolitanism than is China. The Japanese have always borrowed heavily from other cultures (particularly from China itself), whereas the Chinese have historically been more self-sufficient. However, the southern coastal Chinese have long tended to be more international in their orientation, with close links to the Chinese diaspora communities of Southeast Asia and ultimately to maritime trading circuits extending over much of the globe.

In most periods of Chinese history, the interior orientation of the center prevailed over the external orientation of the southern coast. After the communist victory of 1949, only the small British enclave of Hong Kong was able to pursue international cultural connections. In the rest of the country, a dour and puritanical cultural order was rigidly enforced.

After China began to liberalize its economy and open its doors to foreign influences in the late 1970s and early 1980s, the southern coastal region suddenly assumed a new prominence. Through this gateway, global cultural patterns began to penetrate the rest of the country. The result has been the emergence of a vibrant, but somewhat showy urban popular culture throughout China, replete with such global features as nightclubs, karaoke bars, fast-food franchises, and theme parks. China now has some 300 theme parks, and the number continues to grow. In 2016, Disney opened its first mainland Chinese park, the US$5.5 billion Disney Resort (**Figure 11.35**). China does, however, continue to reject certain aspects of global culture. It strictly limits, for example, the number of foreign films that can be distributed in the country. Chinese censors banned the 2016 film *Ghostbusters*, supposedly due to a law that forbids the promotion of superstition.

is very popular in Japan, where it is often topped with seafood. Traditional Japanese dishes, most notably sushi, have likewise spread across much of the world.

Of China's many highly developed regional cuisines, that of Sichuan is probably best known globally. The people of Sichuan are noted in China for their dedication to the art of cooking and their love of restaurants, both cheap and expensive. For these reasons, the province's main city, Chengdu, was declared by UNESCO to be an official "city of gastronomy" in 2011 (**Figure 11.36**).

▲ **Figure 11.35 China's New Disney Resort** Theme parks have become very popular in China over the past two decades, resulting in extensive construction projects. This photograph shows the massive Shanghai Disney Resort on June 15, 2016, shortly before it was opened to the public.

The traditional dishes of Sichuan tend to differ from those of other parts of China in several respects. Beef is more commonly eaten, although pork remains the primary meat, and rabbit is widespread as well. Internal organs are especially prized. The most distinctive feature of Sichuan food, however, is its spiciness. Chili peppers and garlic are used liberally, but what really stands out is the so-called Sichuan pepper, a plant unrelated to other kinds of pepper. Sichuan pepper has a strong citrusy flavor, but more important is the tingling or numbing sensation that it brings to the mouth.

As Sichuan food has gained popularity and spread outside of the province and across the world, controversies have erupted. Prominent food critics have long scoffed at the transformation of traditional dishes in foreign lands, but noted chefs now complain about the debasement of the cuisine in Sichuan itself by less knowledgeable cooks. Many old-school dishes are disappearing, such as sliced pig kidneys fried in fermented bean paste, and younger cooks are

Globalization and Sports in East Asia Although the sporting culture of East Asia is highly varied, the last half-century has seen increasing domination by popular international sports. Japan's national sport of Sumo wrestling dates back to at least the 8th century and holds an important place in the country's psyche. The country's top spectator sports—baseball and soccer (association football)—trace their Japanese roots back to the Meiji period (1868–1912), when foreign advisors were invited into the country with the goal of modernizing Japanese society. In technology-oriented South Korea, video games have emerged as a significant professional activity. The annual World Cyber Games competition is organized by a South Korean company and is jointly sponsored by Samsung and Microsoft.

Baseball is played at a very high level in Japan, and Japan's main league—Nippon Professional Baseball—is viewed by scouts as the world's best league outside of Major League Baseball (MLB) in the United States. Despite encouragement to remain in Japan, many Japanese stars have had illustrious MLB careers. South Korea also has a strong baseball tradition, winning the Olympic gold medal in 2008.

The People's Republic of China excels in Olympic sports, hosting the 2008 Summer Games and finishing in second place in the total medal count at both the 2012 and 2016 Summer Olympics. Chinese divers consistently dominate at the Olympics and other world championships. China's biggest sport appears to be basketball, with a fan base estimated generously at 450 million. China native and former Houston Rockets all-star Yao Ming, whose U.S. debut in the National Basketball Association (NBA) attracted 200 million Chinese television viewers, did much to popularize the game in the early 2000s.

The Controversial Globalization of Sichuan Cuisine Like sports, food in East Asia is highly globalized. Foreign dishes have been embraced across the region, where they have often been modified by local preferences. Pizza, for example,

▼ **Figure 11.36 Traditional Restaurant in Chengdu** Officially recognized for its fine traditional Sichuan cuisine, the city of Chengdu in central China supports a large array of eateries. Pictured here is a plush restaurant combined with an upscale boutique located in a restored building in the old quarter of Chengdu.

Explore the **Tastes** of Sichuan Peppercorn

http://goo.gl/tMUAFw

often said to rely on overpowering spices rather than harmonious blends of distinct flavors. Upstart chefs respond by arguing that Sichuan food has always evolved, with one of its signature dishes, kung pao chicken, dating back only to the early 1800s. The provincial government decided to step into this debate in 2016, issuing guidelines for traditional dishes and devising a rating system, based on gold, silver, and bronze pandas, to recognize especially fine food.

> **REVIEW**
>
> **11.7** What features mark the Han Chinese as an ethnic group?
>
> **11.8** How has the geography of religion changed in East Asia since the end of World War II?
>
> **11.9** Describe the main historical factors in the creation of an East Asian world region.
>
> ■ **KEY TERMS** ideographic writing, Confucianism, Mandarins, geomancy, Marxism, *juche*, dynastic succession, Burakumin, diaspora, tonal language, autonomous region, Cold War, *hallyu*

Geopolitical Framework: Struggles for Regional Dominance

Much of the political history of East Asia revolves around the centrality of China and the ability of Japan to remain outside China's grasp. The traditional Chinese conception of geopolitics was based on the idea of a universal empire: All foreign territories either were part of the Chinese Empire, paying tribute to it and acknowledging its supremacy, or stood outside the system altogether. Until the 1800s, the Chinese government would not recognize any other as its diplomatic equal. When China could no longer maintain its power in the face of European aggression, the East Asian political system fell into disarray. As European power declined in the 1900s, China and Japan contended for regional leadership. After World War II, East Asia was split by larger Cold War rivalries (**Figure 11.37**).

▼ **Figure 11.37 Geopolitical Issues in East Asia** This region remains one of the world's geopolitical hot spots. Tensions are particularly severe between capitalist, democratic South Korea and the isolated communist regime of North Korea, and also between China and Taiwan. China has had several border disputes, whereas Japan and Russia have not been able to resolve their quarrel over the southern Kuril Islands.

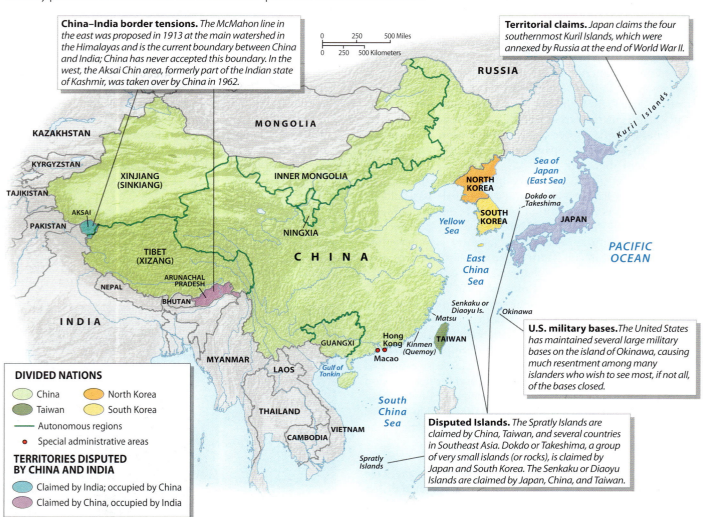

China–India border tensions. *The McMahon line in the east was proposed in 1913 at the main watershed in the Himalayas and is the current boundary between China and India; China has never accepted this boundary. In the west, the Aksai Chin area, formerly part of the Indian state of Kashmir, was taken over by China in 1962.*

Territorial claims. *Japan claims the four southernmost Kuril Islands, which were annexed by Russia at the end of World War II.*

U.S. military bases. *The United States has maintained several large military bases on the island of Okinawa, causing much resentment among many islanders who wish to see most, if not all, of the bases closed.*

Disputed Islands. *The Spratly Islands are claimed by China, Taiwan, and several countries in Southeast Asia. Dokdo or Takeshima, a group of very small islands (or rocks), is claimed by Japan and South Korea. The Senkaku or Diaoyu Islands are claimed by Japan, China, and Taiwan.*

DIVIDED NATIONS
- China
- Taiwan
- North Korea
- South Korea
- Autonomous regions
- ● Special administrative areas

TERRITORIES DISPUTED BY CHINA AND INDIA
- Claimed by India; occupied by China
- Claimed by China, occupied by India

The Evolution of China

The original core of Chinese civilization was the North China Plain and the Loess Plateau. For centuries, periods of unification alternated with times of division into competing states. The most important episode of unification occurred in the 3rd century BCE. Once political unity was achieved, the Chinese Empire began to expand vigorously to the south of the Yangtze Valley. Subsequently, the ideal of the imperial unity of China triumphed, with periods of division seen as indicating disorder. This ideology helped cement the Han Chinese into a single people.

Several Chinese dynasties rose and fell between 219 BCE and 1912, most of them controlling roughly the same territory. The core of the Chinese Empire remained China proper, excluding Manchuria. Other lands, however, were often ruled as well. Until around 1800, the Chinese Empire was the world's wealthiest and most powerful state. Its only real threat came from the pastoral peoples of Mongolia and Manchuria. Although vastly outnumbered by China, these societies were organized on a highly effective military basis. The Chinese and the Mongols usually enjoyed mutually beneficial trading relationships, but periodically waged war, and on several occasions the northern nomads conquered China (see Chapter 10). In time, the conquering armies adopted Chinese customs in order to govern the far more numerous Han people.

The Manchu Qing Dynasty The final and most significant conquest of China occurred in 1644, when the Manchus toppled the Ming Dynasty and replaced it with the Qing (or Ch'ing) Dynasty. Like earlier conquerors, the Manchus retained the Chinese bureaucracy and made few institutional changes. Their strategy was to adapt themselves to Chinese culture, yet at the same time to preserve their own identity as an elite military group. Their system functioned well until the mid-19th century, when the empire began to crumble before the onslaught of European and, later, Japanese power.

China's most significant legacy from the Manchu Qing Dynasty was the extension of its territory to include much of Central Asia. The Manchus subdued the Mongols and eventually established control over eastern Central Asia, including Tibet. Even the states of mainland Southeast Asia sent tribute and acknowledged Chinese supremacy. Never before had the Chinese Empire been so extensive or so powerful.

The Modern Era From its height of power and extent in the 1700s, the Chinese Empire descended rapidly in the 1800s as it failed to match the technological progress of Europe. Threats had always come from the north, and Chinese and Manchu officials saw little peril from European merchants operating along their coastline. The Europeans were distressed by the amount of silver needed to obtain Chinese silk, tea, and porcelain and by the Chinese disdain for their manufactured goods. In response, the British began to sell opium, which

Chinese authorities viewed as a threat. When the imperial government tried to suppress the opium trade in 1839, the British attacked and quickly prevailed (**Figure 11.38**).

This first "opium war" ushered in a century of political and economic chaos. The British demanded free trade in selected Chinese ports and in the process overturned the traditional policy of managed trade. As European enterprises penetrated China and undermined local economic interests, anti-Manchu rebellions began to break out. In the 1800s, all such uprisings were crushed, but not before causing tremendous destruction. Meanwhile, European power continued to advance. In 1858, Russia annexed the northernmost reaches of Manchuria, and by 1900 China had been divided into separate **spheres of influence**, in which the different European powers had no formal political authority, but enjoyed influence and economic clout (**Figure 11.39**).

A successful rebellion in 1911 finally toppled the Manchus, but subsequent efforts to establish a unified Chinese Republic were not successful. In many parts of the country local military leaders, or "warlords," grabbed power for themselves. By the 1920s, it appeared that China might be completely dismembered. The Tibetans had regained autonomy; Xinjiang was under Russian influence; and in China proper Europeans and local warlords vied with the weak Chinese republic for power. In addition, Japan was increasing its demands and seeking to expand its territory.

The Rise of Japan

Japan did not emerge as a strong, unified state until the 7th century. From its earliest days, Japan looked to China (and, at first, to Korea as well) for intellectual and political models. Its offshore location insulated Japan from the threat of rule by the Chinese Empire. At the same time, the Japanese conceptualized their islands as a separate empire, equivalent in certain respects to China. Between 1000 and 1580, however, Japan was usually divided into mutually hostile feudal

▼ **Figure 11.38 Opium War** Great Britain humiliated China in two "opium wars" in the early 1800s, forcing the much larger country to open its economy to foreign trade and to grant Europeans extraordinary privileges. The East India Company steamer *Nemesis* destroyed Chinese war junks in January 1841.

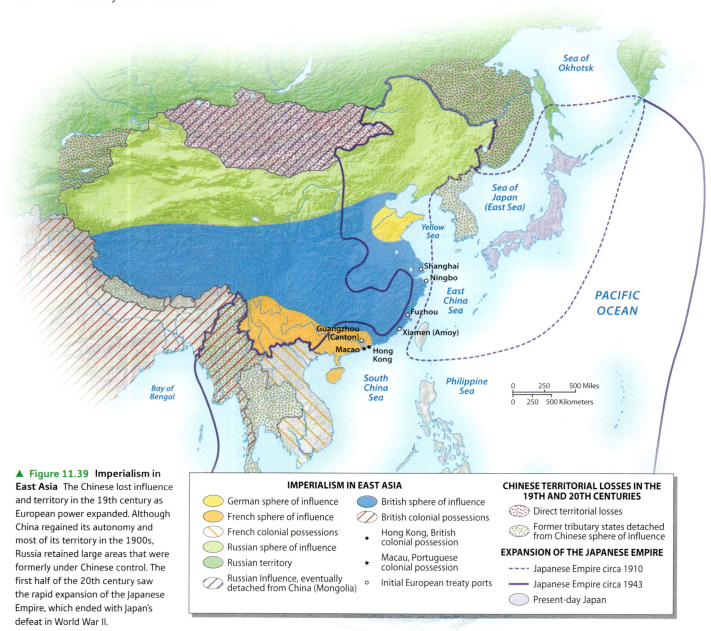

▲ Figure 11.39 Imperialism in East Asia The Chinese lost influence and territory in the 19th century as European power expanded. Although China regained its autonomy and most of its territory in the 1900s, Russia retained large areas that were formerly under Chinese control. The first half of the 20th century saw the rapid expansion of the Japanese Empire, which ended with Japan's defeat in World War II.

IMPERIALISM IN EAST ASIA

- German sphere of influence
- French sphere of influence
- French colonial possessions
- Russian sphere of influence
- Russian territory
- Russian Influence, eventually detached from China (Mongolia)
- British sphere of influence
- British colonial possessions
- Hong Kong, British colonial possession
- ★ Macau, Portuguese colonial possession
- ○ Initial European treaty ports

CHINESE TERRITORIAL LOSSES IN THE 19TH AND 20TH CENTURIES

- Direct territorial losses
- Former tributary states detached from Chinese sphere of influence

EXPANSION OF THE JAPANESE EMPIRE

- ---- Japanese Empire circa 1910
- —— Japanese Empire circa 1943
- Present-day Japan

realms. Around 1600, it was reunited by the armies of the Tokugawa **Shogunate** (a shogun was a supreme military leader who serves the emperor in name only). Shortly afterward, Japan attempted to isolate itself from the rest of the world, limiting trade with outsiders.

Japan's restrictions on foreign commerce ended in 1853 when U.S. gunboats sailed into Tokyo Bay to demand access. Aware that they could no longer keep the Westerners out, Japanese leaders set about modernizing their economic, administrative, and military systems. This effort accelerated when the Tokugawa Shogunate was toppled in 1868 by the **Meiji Restoration**, so called because it restored the emperor to the throne, although it did not give the emperor real power. Unlike China, Japan successfully accomplished most of its reform efforts.

The Japanese Empire Japan's new rulers were still worried about European imperial expansion. They decided that the only way to meet the challenge was to industrialize and become expansionistic themselves. Japan, therefore, took control over Hokkaido and began to move north into the Kuril Islands and Sakhalin. In 1895, it tested its newly modernized army against China, winning a quick and profitable victory that gave it control of Taiwan. Tensions then mounted with Russia as the two countries vied for power in Manchuria and Korea. The Japanese defeated the Russians in 1905, giving them influence in northeastern China. With no strong rival in the area, Japan annexed Korea in 1910. Alliances with Britain, France, and the United States during World War I brought further gains, as Japan was awarded Germany's island colonies in Micronesia.

The 1930s brought a global depression, putting resource-dependent Japan in a difficult situation. The country's leaders sought a military solution, and, in 1931, Japan conquered Manchuria. In 1937, Japanese armies moved south, occupying the North China Plain and the coastal cities of southern China. The Chinese government withdrew to the relatively inaccessible Sichuan Basin to continue the struggle. During this period, Japan's relations with the United States deteriorated. In 1941, Japan's leaders decided to destroy the U.S. Pacific fleet in order to clear the way for the conquest of resource-rich Southeast Asia and unite East and Southeast Asia into a "Greater East Asia Co-Prosperity Sphere." This "sphere" was to be ruled by Japan and was designed to keep the Americans and Europeans out. Japanese forces in China and Southeast Asia sometimes engaged in brutal acts, resulting in serious tensions between the Japanese and the other Asians—tensions that persist to this day.

Postwar Geopolitics

With Japan's defeat at the end of World War II, East Asia became an arena of rivalry between the United States and the Soviet Union. Initially, American interests prevailed in the maritime fringe, whereas Soviet influence advanced on the mainland. In time, however, East Asia experienced its own revival.

Japan's Revival Japan's colonial empire crumbled when it lost World War II. Its territory was reduced to the four main islands plus the Ryukyu Archipelago and a few minor outliers. In general, the Japanese accepted this loss of land. The only remaining territorial conflict concerns the four southernmost islands of the Kuril chain, which were taken by the Soviet Union in 1945. Although Japan still claims these islands, Russia refuses to discuss relinquishing control, straining Russo-Japanese relations.

After losing its overseas possessions, Japan turned to trade to obtain the resources needed for its economy. The country's military power was strictly limited by the constitution imposed on it by the United States, forcing Japan to rely on the U.S. military for much of its defense needs. The U.S. Navy continues to patrol many of Japan's vital sea lanes, and U.S. armed forces maintain bases within the country. This military presence, however, is controversial. Particularly contentious are the U.S. bases on Okinawa, which take up much of the island's territory. Large-scale public protests and pressure from the Japanese government forced the U.S. military to agree in 2016 to relinquish one-sixth of the area under its control in Okinawa.

Japan's own military has gradually emerged as a strong regional force despite the constitutional limits imposed on it. In recent years, fears about North Korea's nuclear program and China's military growth have led the Japanese government to reconsider its limitations on defense spending. In 2010, Japan announced that it would change its military focus away from the Cold War struggle with Russia and toward the rising threat posed by China. Japan has also been developing closer military ties with the United States and Australia. In 2015, new laws allowed the Japanese military to help defend allies in the case of war and to support them in international combat operations.

China is concerned about the threat posed by a potentially remilitarized Japan. Such a perception was strengthened in the early 2000s when Japan's former prime minister visited the Yasukuni Shrine, which contains a military cemetery in which several war criminals from World War II are buried. Anti-Japanese sentiments in China—which have occasionally boiled over into huge public protests—have also been encouraged by the publication of Japanese textbooks that minimize the country's atrocities during the war.

The Division of Korea The aftermath of World War II brought much greater changes to Korea than to Japan. As the end of the war approached, the Soviet Union and the United States agreed to divide Korea at the 38th parallel. Soon two separate regimes were established, the northern one allied with the Soviet Union, the southern one with the United States. In 1950, North Korea invaded South Korea, seeking to reunify the country. The United States, with support from the United Nations, supported the south, while China aided the north. The war ended in a stalemate, and Korea has remained a divided country, its two governments locked in a continuing Cold War (**Figure 11.40**).

▼ **Figure 11.40 The Demilitarized Zone in Korea** North and South Korea were divided along the 38th parallel after World War II. Today, even after the conflict of the early 1950s, these states are separated by the demilitarized zone, or DMZ, which runs near the parallel and is actively patrolled by armed forces.

Large numbers of U.S. troops remained in the south after the war. South Korea in the 1960s was a poor, agrarian country that could not defend itself. Over the past 30 years, however, the south has emerged as a wealthy trading nation, while the fortunes of the north have plummeted. The southerners' fear of the north gradually lessened, especially among the younger generation. Many South Korean students resent the presence of U.S. forces. Changing military priorities, meanwhile, led the United States to reduce its presence in Korea, dropping its troop strength in 2004 from 37,000 to 28,500.

In the late 1990s, the South Korean government began to pursue better relations with North Korea, giving it food aid and industrial investment in exchange for reduced hostility. In 2008, however, a new South Korean government concluded that this so-called Sunshine Policy had not worked. North Korean nuclear testing further strained relations, as did the March 2010 sinking of a South Korean naval vessel. Efforts to reduce tensions have since alternated with periods of belligerency. In 2015, North Korea fired an artillery blast at the South Korean city of Yeoncheon, forcing the evacuation of civilians. South Korea responded by launching several shells on the North. Although no one was hurt in these exchanges, they served to remind the world that the Korean War never actually ended.

The Division of China World War II brought tremendous destruction and loss of life to China. Even before the war began, China had been engaged in a civil conflict between nationalists and communists. The communists had originally been based in the middle Yangtze region, but in 1934 nationalist pressure forced them out. Under the leadership of Mao Zedong, they retreated in the "Long March," which took them to the Loess Plateau, an area close to the power center of northern China and the industrialized zones of Manchuria. After Japan invaded China proper in 1937, the two camps cooperated, but as soon as Japan was defeated, China returned to civil war. In 1949, the communists proved victorious, forcing the nationalists to retreat to Taiwan.

A latent state of war has persisted ever since between China and Taiwan. Although no battles have been fought, gunfire has periodically been exchanged over Matsu, several small Taiwanese islands just off the mainland. The Beijing government still claims Taiwan as an integral part of China and vows eventually to reclaim it. The nationalists who gained power in Taiwan long maintained that they represented the true government of China.

The idea of the intrinsic unity of China continues to be influential. In the 1950s and 1960s, the United States recognized Taiwan as the only legitimate government of China, but its policy changed in the 1970s when U.S. leaders decided that it would be more useful to recognize mainland China. Soon China entered the United Nations, and Taiwan found itself diplomatically isolated. As of 2016, only 21 UN members, mostly small African, Pacific, and Caribbean states, continue to recognize Taiwan and, in return, to receive

Taiwanese aid. In reality, however, Taiwan is a fully separate country. In 2001, both China and Taiwan entered the World Trade Organization (WTO), but Taiwan was forced to join under the awkward name of "Chinese Taipei" to avoid the suggestion that it is a sovereign country.

The geopolitical status of Taiwan continues to be a controversial issue in Taiwan itself. When a Taiwanese nationalist was elected president in 2000, China threatened to invade if the island declared formal independence. Military tensions remained high for several years, but during the same period economic connections strengthened. Taiwanese voters then grew dissatisfied with an independence movement that had generated tensions without delivering benefits, and in the presidential elections of 2008 and 2012 the Chinese nationalist party won clear victories after promising good relations with the mainland. The growing power of mainland China, however, revitalized the independence movement. In 2016 Taiwan elected a harsh critic of mainland China, Tsai Ing-Wen, as president. Although Tsai does not advocate formal independence, she rejects the idea that Taiwan and China form a single nation that is temporarily disunited, a stance that infuriates Beijing. As a result, China has increased its pressure on international organizations to reject Taiwan as a separate member.

Chinese Territorial Issues Other than being unable to regain Taiwan, China has successfully retained the Manchu territorial legacy. In the case of Tibet, this has required considerable force; resistance by the Tibetans compelled China to launch a full-scale invasion in 1959. The Tibetans, however, have continued to struggle for real autonomy if not actual independence, as they fear that the Han Chinese now moving to Tibet will eventually outnumber them (see Chapter 10).

China claims several other areas that it does not control. It contends that former Chinese territories in the Himalayas were illegally annexed by the British Empire in South Asia, resulting in a border dispute with India. The two countries went to war in 1962 when China occupied an uninhabited highland district in northeastern Kashmir. Tensions between India and China have eased over the past decade, but occasionally flare up as the underlying territorial disagreements remain unresolved.

One territorial issue was finally resolved in 1997 when China reclaimed Hong Kong. In the 1950s, 1960s, and 1970s, Hong Kong acted as China's window on the outside world, and it grew prosperous as a capitalist enclave. As Chinese international relations opened in the 1980s, Britain decided to honor its treaties and return Hong Kong to China. China, in turn, promised that Hong Kong would retain its capitalist economic system and its partially democratic political system for 50 years under a model known as **"one country, two systems."** Civil liberties not enjoyed in China itself were to remain protected in Hong Kong.

Hong Kong's autonomy, however, is insecure, as China has thwarted attempts to install a more representative government. Interference by Beijing in Hong Kong's local elections in 2014 generated a massive student-led protest movement. At its height, more than 100,000 people occupied the streets of the city to demand democracy. Tensions again flared in 2016 when a prominent Hong Kong bookseller was jailed in mainland China for selling books critical of the country's leaders. Many of Hong Kong's major media outlets have been acquired by firms located elsewhere in China, reducing their independence. Such developments have led a majority of Hong Kong residents to express doubt about the "one country, two systems" model.

Explore the **Sounds** of Hong Kong Mong Kok Protest

http://goo.gl/iAplkZ

In 1999, Macau, the last colonial territory in East Asia, was returned to China (**Figure 11.41**). Like Hong Kong, it is a special administrative region with its own autonomous government and subject to its own laws. This small former Portuguese enclave, located across the estuary from Hong Kong, has functioned largely as a gambling refuge. Now the largest betting center in the world, Macau derives 40 percent of its gross domestic product from gambling. When China began to crack down seriously on corruption in 2015, Macau experienced a sharp recession. New casinos continue to be constructed, however, leading to fears of over-building.

The Senkaku/Diaoyu Crisis The most serious territorial dispute in East Asia concerns a group of islets to the northeast of Taiwan, called the Senkaku Islands in Japanese and the Diaoyu Islands in Chinese (**Figure 11.42**). Although the five uninhabited islands total only 2.7 square miles (7 square km), they are probably situated over substantial oil and natural gas reserves. Japan has controlled the islands since 1895, except for the period after World War II when they were under U.S. authority, but China maintains that it has

▲ **Figure 11.42 Senkaku/Diaoyu Islands** These rugged and uninhabited islands in the East China Sea are administered by Japan but claimed by both China and Taiwan. As China has increasingly emphasized its claims to these rocky islands, tensions with Japan have mounted. **Q: Why have these small and relatively insignificant islands generated such potentially dangerous conflict over the past several years?**

better historical claims. Tensions heated up in 2012 when the Japanese government purchased three of the islands from a private Japanese family. China objected strenuously, viewing this act as an infringement on its sovereignty. In 2016, China sent warships and large numbers of fishing vessels into disputed waters around the islands, infuriating Japan.

Global Dimensions of East Asian Geopolitics

In the early 1950s, East Asia was divided into two hostile camps: China and North Korea were allied with the Soviet Union, while Japan, Taiwan, and South Korea were linked to the United States. The Chinese-Soviet alliance soon deteriorated into mutual hostility, and in the 1970s China and the United States found that they could accommodate each other. In contrast, North Korea's relations with the United States and many other countries have only grown more heated during the same period. As China's economy and military have steadily expanded, however, international geopolitical tensions across the region have again intensified.

▼ **Figure 11.41 Macau Casino** Macau, reclaimed by China from Portugal in 1999, retains a unique mixture of Chinese and Portuguese cultural influences. Its economic mainstay remains gambling, which is prohibited in the rest of China.

Explore the **Sights** of Macau

http://goo.gl/iAplkZ

The Ongoing North Korean Crisis In 1994, North Korea refused to allow international inspections of its nuclear power facilities, provoking an international crisis. It eventually relented, however, in exchange for energy assistance from the United States. But by 2002, the agreement had fallen apart, as evidence mounted that North Korea was continuing to pursue nuclear weapons. Other complaints by the international community focused on North Korea's trafficking in illegal drugs and the fact that it holds hundreds of thousands of its own citizens in brutal labor camps.

In response to North Korea's nuclear ambitions, the United States has advocated multilateral negotiations involving South Korea, China, Japan, and Russia. North Korea, however, insisted for years that it would negotiate only with the United States. Owing in part to pressure from China, its main trading partner, North Korea agreed to the broader framework, participating in five rounds of talks between 2003 and 2007. Little progress was made, however, and in 2009 North Korea pulled out of the negotiations, detonated a nuclear weapon, and tested a series of ballistic missiles. Four years later it detonated another nuclear weapon. In 2016, a U.S.–South Korea agreement on deployment of advanced missile-defense systems in the south prompted a further round of ballistic missile testing. Some of these missiles landed in the sea within Japan's exclusive maritime economic zone, infuriating the Japanese government.

Such actions have put China in a difficult diplomatic situation. China officially objects to North Korea's nuclear tests and has agreed to limited UN sanctions on the country, but it continues to support the Pyongyang government in other ways. Many experts think that China is concerned that North Korea could collapse, leading to reunification with South Korea. This event would result in a refugee crisis, as well as a shared border with a U.S. ally.

China on the Global Stage The end of the Cold War, coupled with China's rapid economic growth, reconfigured the balance of power in East Asia. The U.S. military, like China's neighbors, became increasingly worried about China's growing power. China has the world's second largest military budget, as well as nuclear capability and sophisticated missile technology. It is rapidly building up its navy, buying advanced ships from Russia and constructing its own. The country has also invested heavily in port facilities around the Indian Ocean in Burma (Myanmar), Sri Lanka, and especially Pakistan that could potentially be used as military facilities. Through its "One Belt, One Road" (or "new Silk Road") initiative, China is attempting to reorient global trade networks so that they focus on the Eurasian landmass rather than the Atlantic and Pacific ocean basins, a move that could have major geopolitical ramifications (see Chapter 10).

China's most controversial international action has been its claiming of almost the entire South China Sea along with construction of artificial islands in the area, an issue discussed in Chapter 13. In 2016, an international tribunal ruled that China has no historical basis for these claims. Both China and Taiwan rejected the ruling, and an angry Chinese government responded by flying military aircraft over the disputed islands and sending fishing boats into waters claimed by other countries. China maintains, however, that it is willing to negotiate a regional code of conduct to cover all activities taking place in the South China Sea.

China is thus coming of age as a major force in global politics. Whether it is a force to be feared by other countries is a matter of considerable debate. Chinese leaders insist that they have no expansionist designs and no intention of interfering in the internal affairs of other countries. They regard concerns expressed by the United States and other countries about their human rights record, as well as their activities in Tibet, as undue meddling in their own internal affairs.

REVIEW

11.10 Explain how geopolitical issues in East Asia changed since the end of the Cold War.

11.11 How did the decline of China during the 1800s affect the geopolitical structure of East Asia?

11.12 To what extent do conflicts over natural resources influence geopolitical tensions across East Asia?

■ **KEY TERMS** spheres of influence, shogunate, Meiji Restoration, "one country, two systems"

Economic and Social Development: A Core Region of the Global Economy

East Asia exhibits a wide range of economic and social development. Japan's urban belt contains one of the world's greatest concentrations of wealth, whereas North Korea is one of the world's most impoverished societies. Overall, East Asia has experienced rapid economic growth since the 1970s (Table 11.2), gaining global domination in certain key industries, such as shipbuilding (see *Everyday Globalization: East Asia's Domination of Shipbuilding*). Since 2000, China has experienced the most rapid economic expansion in the region, turning it into the world's second largest economy.

Japan's Economy and Society

Japan was the pacesetter of the world economy from the 1960s through the 1980s. In the early 1990s, however, the Japanese economy experienced a major setback, and growth has been slow ever since. Despite this slowdown, Japan is still one of the world's largest economic powers.

Japan's Boom and Bust Although Japan's heavy industrialization began in the late 1800s, most of its people remained poor. The 1950s, however, saw the beginnings of the Japanese

TABLE 11.2   Development Indicators

Country	GNI per capita, PPP 2014	GDP Average Annual % Growth 2009–15	Human Development Index (2015)[1]	Percent Population Living Below $3.10 a Day	Life Expectancy (2016)[2]		Under Age 5 Mortality Rate (1990)	Under Age 5 Mortality Rate (2015)	Youth Literacy (% pop ages 15–24) (2005–2014)	Gender Inequality Index (2015)[3,1]
					Male	Female				
China	13,170	8.5	0.728	27.2	75	78	49	11	100	0.191
Hong Kong	56,570	3.5	0.910	–	81	87	–	2	–	
Japan	38,120	1.4	0.891	–	80	87	6	3	–	0.133
North Korea	–	–	–	–	66	74	45	25	100	
South Korea	33,650	3.5	0.898	–	79	86	8	3	–	0.125
Taiwan	–	–	–	–	77	83	–	–	–	

Source: World Bank, *World Development Indicators,* 2016.

[1]United Nations, *Human Development Report,* 2015.

[2]Population Reference Bureau, *World Population Data Sheet,* 2016.

[3]Gender Inequality Index—A composite measure reflecting inequality in achievements between women and men in three dimensions: reproductive health, empowerment, and the labor market that ranges between 0 and 1. The higher the number, the greater the inequality.

Login to MasteringGeography™ **& access** MapMaster **to explore these data!**

1) Rank the countries by income. Why are the figures for percent of the population living on less than $2 a day missing for most East Asian countries?
2) Use the maps to see how well GNI figures correlate with other developmental indicators across East Asia.

"economic miracle." Shorn of its empire, Japan was forced to export manufactured products. Beginning with inexpensive consumer goods, Japanese industry moved to more sophisticated materials, including automobiles, cameras, electronics, machine tools, and computer equipment. By the 1980s, it was the leader in many segments of the global high-tech economy.

In the early 1990s, Japan's inflated real estate market collapsed, leading to a banking crisis. At the same time, many Japanese companies discovered that producing labor-intensive goods at home had become too expensive, and began relocating factories to Southeast Asia and China. Because of these and related difficulties, Japan's economy slumped through the 1990s and into the new millennium,

Everyday **Globalization**

East Asia's Domination of Shipbuilding

Economic globalization depends heavily on maritime shipping, as the vast majority of exported and imported goods—whether oil or iPhones—travel by sea. Today almost all large ships, with the exception of those used by the military, are made in East Asia. China has the lead with 45 percent of the global market, while South Korea is second at 29 percent and Japan follows with 18 percent.

Sixty years ago, shipbuilding was concentrated in North America and Europe, with Scotland and Northern Ireland occupying prime positions. Japan then took the lead, outcompeting its Western rivals, but by the 1980s South Korea gained the top position. Although China is now the largest producer, South Korea still makes most of the world's truly massive ships, such as cruise liners, supertankers, and large cargo carriers. The world's biggest shipyard is located in Ulsan, South Korea, which is also the headquarters of the world's largest shipbuilding company, Hyundai Heavy Industries (**Figure 11.4.1**). But shipbuilding is vulnerable to fluctuations in the global economy. South Korea's economy took a further hit in 2016 due to the global decline in ship orders.

1. List some factors that helped South Korea develop such a large and profitable shipbuilding industry.

2. What percentage of the items you are now wearing or carrying were shipped to the United States from an overseas location?

▲ **Figure 11.4.1 Hyundai Heavy Industries Ulsan Shipyard** Hyundai Heavy Industries Co., Ltd., is headquartered in Ulsan, South Korea, where it builds a range of container vessels and tankers. Other South Korean shipbuilders include Samsung Heavy Industries and Daewoo Shipbuilding & Marine Engineering.

and was further undermined by the earthquake and tsunami of 2011. In 2013, the Japanese government tried to bolster its economy by reducing interest rates below zero and weakening the value of its currency to boost exports. Such policies had little effect, however, nor did a further round of stimulus spending in 2016.

Despite its problems, Japan remains a core country of the global economic system. Its economy spans the globe as Japanese multinational firms invest heavily in production facilities in North America and Europe, as well as in poorer countries. Japan is a world leader in a number of high-tech fields, including optics, machine tools for the semiconductor industry, and especially robotics (**Figure 11.43**). It is also one of the world's largest creditor nations, despite its own debts, owning a large percentage of U.S. government bonds.

Living Standards in Japan Despite its affluence, living standards in Japan remain somewhat lower than those of the United States. Housing, food, transportation, and services are very expensive in Japan. But the Japanese also enjoy many benefits unknown in the United States. Unemployment remains low, health care is provided by the government, and the crime rate is extremely low. By such social measures as literacy, infant mortality, and average longevity, Japan surpasses the United States.

Japan, of course, has its share of social problems. Koreans and alien residents from other Asian countries suffer discrimination, as do members of the Burakumin underclass. Japan's more remote rural areas have few jobs, and many have seen population decreases. In many small villages, most of the remaining people are elderly. Farming is an increasingly marginal occupation, and many farm families survive only because one family member works in a factory or office. Professional and managerial occupations in Japan's cities are notable for their long hours and high levels of stress.

Korea's Divergent Development

In the 1960s, 1970s, and 1980s, the Japanese path to development was successfully followed by its former colonies, South Korea and Taiwan. North Korea, however, undertook a completely different and far less successful economic path.

The Rise of South Korea The postwar rise of South Korea was even more remarkable than that of Japan. Before independence, Korean industrial development was concentrated in the north, an area rich in natural resources. The south, in contrast, remained densely populated, poor, and agrarian. In the 1960s, the South Korean government initiated a program of export-led economic growth. It guided the economy with a heavy hand and denied basic political freedom to the Korean people. By the 1970s, such policies had proven highly successful. Huge Korean industrial conglomerates, called *chaebol*, moved from manufacturing inexpensive consumer goods to heavy industrial products and then to high-tech equipment.

By the 1990s, South Korea emerged as one of the world's main producers of semiconductors and other high-tech goods. South Korean wages also rose at a rapid clip. The country has invested heavily in education, which has served it well in the global economy. Large South Korean companies are themselves now multinational, building new factories and making other investments in Southeast Asia, Latin America, and Africa, as well as in the United States and Europe (see *Exploring Global Connections: South Korean Investments and Aid in Africa*). South Korea is also noted for its ready embrace of cutting-edge technology. By 2010, it was widely regarded as being the world leader in Internet connectivity and speed.

Contemporary South Korea The political development of South Korea has not been nearly as smooth as its economic progress. Throughout the 1960s and 1970s, student-led protests against the authoritarian government were brutally repressed. As the South Korean middle class

▲ **Figure 11.43 Japanese Robot** Japan is a world leader in robotics. Japanese researchers are now working hard to create useful humanoid robots like the one that serves as a receptionist at the upscale Mitsukoshi department store in Tokyo.

Exploring **Global Connections**

South Korean Investments and Aid in Africa

China's massive presence in Sub-Saharan Africa has gained much attention. Less often noted is South Korea's growing engagement in the region. South Korean firms not only have increased their exports to Africa, but also have been investing, building local factories and distribution centers. The South Korean government is also enhancing its presence in the region. The Korean Initiative for African Development has overseen a major increase in official developmental assistance, and the Korea–Africa Industry Cooperation Forum encourages industrial cooperation. Diplomatic ties are also growing; between 2007 and 2014, Kenya, Angola, Senegal, Rwanda, Ethiopia, Sierra Leone, and Zambia all opened embassies in Seoul.

Products for the African Market From 2000 to 2010, South Korea's investments in Sub-Saharan Africa increased roughly tenfold, with exports surging ahead as well. More than 75 percent of exports to the region are sophisticated manufactured products, including electronic equipment, phones, vehicles, and construction equipment. Samsung Corporation has been a leading player. From 2010 to 2012, it invested over US$150 million in the region.

Samsung is now pursuing a "build for Africa" strategy based on making products for African consumers (**Figure 11.5.1**). Its smartphones are a key component and now command over half of the regional market. Samsung also plans to build local assembly plants for refrigerators, washing machines, and other consumer goods. Other South Korean firms involved include LG Electronics, which recently announced that it will invest more than US$2 billion to develop products for East and Central Africa.

Some South Korean firms focus instead on construction, building essential infrastructure in Sub-Saharan Africa. Daewoo Corporation, for example, recently announced the construction of a major power plant in Kenya, and another Korean firm is working on an oil refinery in Gabon. Overall, South Korea's construction contracts in the region rose from US$1.5 billion in 2008 to US$2.2 billion in 2011.

Local Opposition Not all of South Korea's ventures in Sub-Saharan Africa have been welcomed. Several South Korean firms have leased huge tracts of African farmland in order to enhance Korean food security. Such projects often anger local people, who typically receive few benefits and may even be forced off their lands. In 2009, an agreement allowing Daewoo to lease over a million hectares (2,470,000 acres) of

▲ **Figure 11.5.1 Samsung's Africa Initiative** Pedestrians pass a large advertisement for Samsung Electronics Co. mobile phones in Nairobi, Kenya in 2013. Samsung, like several other South Korean companies, is making major efforts to reach the African market.

farmland in Madagascar led to the downfall of the country's government and the ouster of its president.

Another problem in South Korean–African relations is that of Korean fleets illegally fishing off the shores of West Africa. In 2013, a ship was caught off-loading some 4000 boxes of illicit fish from Sierra Leone in the South Korean port of Busan. Liberia recently detained a number of South Korean ships for illegally fishing in its waters. Unless such problems can be solved, South Korea's engagement with Africa may remain relatively limited.

Google Earth Virtual **MG** Tour Video

http://goo.gl/Rufneu

1. List some of the reasons why South Korean firms are so interested in investing in Sub-Saharan Africa. What features particular to this region might make this investment attractive?

2. Is it a sound policy for South Korea to fish extensively in foreign waters and acquire farmland abroad? What are the advantages and disadvantages of such strategies?

---

expanded and prospered, pressure for democratization grew, and by the late 1980s it could no longer be denied. But even though democratization has been successful, political tension has not disappeared. Critics contend that the South Korean economy needs substantial reforms—in particular, the breaking up of the *chaebol*. In recent years, however, South Korea's economy has been characterized by steady growth, moderate inflation, low unemployment, and large export surpluses. But concerns for its economic future are mounting due mainly to its aging population, low birth rate, and continuing tension with North Korea.

Between 1998 and 2008, South Korea tried to bolster the North Korean economy and thereby reduce tensions through its so-called Sunshine Policy. South Korean companies were encouraged to invest in jointly operated factories in North Korea, and a sizable industrial complex emerged in

the Kaesong Industrial Zone, located 10 kilometers (6 miles) north of the border (**Figure 11.44**). After 2008, however, periodic political crises between the two countries limited expansion, and in 2016 the facility was closed, at least temporarily, and all South Korean managers and workers were forced to return home.

Contemporary North Korea At the end of the Korean War in 1953, North Korea had a higher level of industrial development than South Korea. But as South Korea experienced export-led industrialization starting in the 1970s, North Korea retained its state-led economy, rejecting globalization. North Korea did maintain close ties with the Soviet Union, however, and when Soviet power collapsed in 1991, it experienced a severe blow. No longer able to afford sufficient fertilizer, its agriculture system sharply declined, leading to

▲ **Figure 11.44** **Kaesong Industrial Zone** A joint South Korean-North Korean venture, the Kaesong industrial complex is located a few miles north of the demilitarized zone (DMZ) separating the two countries. In 2016, the facility was shut down, perhaps temporarily, due to increasing tensions between North Korea and South Korea.

Explore the MG
Sights of Kaesong
Industrial Zone

http://goo.gl/aeMk7r

famine. A 2013 UN report found that roughly 25 percent of North Korean children suffer from chronic food insecurity and hunger.

North Korea's government has made limited efforts to enhance economic growth. In 2002, it legalized farmers' markets, but several years later it shut most of them down. A more ambitious move toward private farming and food selling began in 2012, with some success. Following China's lead, North Korea opened a number of Special Economic Zones (to be discussed in detail later) in which foreign firms, mostly South Korean and Chinese, produce export goods using inexpensive local labor. In 2013–2014 alone, 20 such zones were established, bringing the total up to 25. North Korea is also increasingly exporting its own people to work in foreign countries under official government contracts. Partly as a result of such policies, the country's economy finally seemed to turn a corner. But a subsequent drop in the prices of its main exports, coal and iron ore, resulted in yet another decline in 2015.

North Korea's erratic policies, harshly repressive government, and hostile relations with most of the rest of the world undermine its development policies. New human rights abuses have been uncovered as the country attempts to boost its growth. A 2015 UN report claimed that tens of thousands of North Koreans work under slavelike conditions in labor camps in China, Russia, and the Arabian Peninsula. The North Korean government has also increasingly shunted its limited development funds into Pyongyang, resulting in a capital city that has far higher living standards than the rest of the country.

Continuing Development in Taiwan and Hong Kong

Like South Korea, Taiwan and Hong Kong have experienced rapid economic growth since the 1960s. The Taiwanese government, like those of South Korea and Japan, has guided the economic development of the country. Hong Kong, unlike its neighbors, was long characterized by a **laissez-faire** economic system, one with minimal government interference (*laissez-faire* is a French term meaning "let it be"). Hong Kong traditionally functioned as a trading center, but in the 1960s and 1970s it emerged as a major producer of textiles, toys, and other consumer goods.

By the 1980s, however, such cheap products could no longer be profitably manufactured in a city as expensive as Hong Kong. Local industrialists began to move their plants to nearby areas in southern China, while Hong Kong increasingly specialized in business services, banking, telecommunications, and entertainment. Surging property values and the high cost of living, however, are now putting pressure on Hong Kong's economy. Mounting political tensions with Beijing cloud the city's economic future, and a reduction in tourists from the mainland in 2016 brought further concerns. But Hong Kong is still one of the world's most prosperous places, with a level of per capita economic output comparable to that of the United States.

Taiwan is also a prosperous place, with a level of economic production similar to that of Germany on a per capita basis. But its economic growth stalled out in 2015, leading to fears of economic stagnation similar to that of Japan. China's rapid development of its own information technology industry is a particular concern for Taiwan, as is its own rapidly aging population. Optimists, however, were heartened by signs of renewed Taiwanese expansion in mid-2016, arguing the island's economic foundations remain strong.

Both Taiwan and Hong Kong have close overseas economic connections, with particularly tight linkages with Chinese-owned firms in Southeast Asia. Taiwan's high-technology businesses are also intertwined with those of the United States, particularly those of Silicon Valley. Hong Kong's economy is more closely connected with the rest of China. Taiwan is moving in the same direction; China is now its largest export market and its second largest source of imports after Japan.

Chinese Development

China dwarfs all the rest of East Asia in both expanse and population. Its economic takeoff is thus reconfiguring the economy of the world as a whole. China now has a vast middle class that is able to afford a broad array of consumer goods. As that class is expected to grow to 600 million by

2020, both the Chinese economy and that of the world as a whole are being transformed. But despite its recent success, China's economy has several weaknesses that generate major uncertainties for both the East Asian and the global economies.

China Under Communism More than a century of war, invasion, and chaos in China ended in 1949 when the communist forces led by Mao Zedong seized power. The new government immediately set about nationalizing private firms and building heavy industries. Some successes were realized, especially in Manchuria, where a large amount of heavy industrial equipment was inherited from the Japanese colonial regime.

In the late 1950s, however, China experienced an economic disaster ironically called the "Great Leap Forward." A principle idea behind this scheme was that small-scale village workshops could produce the large quantities of iron needed for sustained industrial growth. Communist Party officials required these inefficient workshops to meet unreasonably high production quotas. In some cases, the only way they could do so was to melt peasants' agricultural tools. Peasants also were forced to contribute a large percentage of their crops to the state. The result was a horrific famine that may have killed 20 million people.

The early 1960s saw a return to more pragmatic policies, but toward the end of the decade a new wave of radicalism swept through China. This "Cultural Revolution" was aimed at mobilizing young people to stamp out the remaining vestiges of capitalism. Thousands of experienced industrial managers and college professors were expelled from their positions. Many were sent to villages to be "reeducated" through hard physical labor; others were simply killed. The economic consequences of such policies were devastating.

Toward a Post-communist Economy When Mao Zedong died in 1976, China faced a crucial turning point. A political struggle ensued between pragmatists hoping for change and dedicated communists. The pragmatists emerged victorious, and by the late 1970s it was clear that China would embark on a new economic path. The country would now seek closer connections with the world economy and allow private enterprise to take root. China did not, however, transform itself into a fully capitalist country. The state continued to run most heavy industries, and the Communist Party retained a monopoly on political power.

One of China's first capitalist openings was in agriculture, which had previously been dominated by large communal farms. Individuals were suddenly allowed to act as agricultural entrepreneurs, selling produce in the open market. This change dramatically increased the income of many farmers. By the late 1980s, however, the focus of growth had shifted to the urban-industrial sector. As the government became concerned about inflation, it placed price caps on food and increased taxes on farmers. By the early years of the new millennium, many rural areas in China's interior provinces were experiencing economic distress, encouraging many people to move to China's coastal cities.

The Era of Rapid Growth One of China's early industrial reforms involved opening **Special Economic Zones (SEZs)**, in which foreign investment was welcome and state interference minimal. The Shenzhen SEZ, adjacent to Hong Kong, proved particularly successful after Hong Kong manufacturers found it a convenient source of cheap land and labor (**Figure 11.45**). Additional SEZs were soon opened, mostly in the coastal region. The basic strategy was to attract foreign investment that could generate exports and thus supply China with the capital it needed to build its infrastructure. Other market-oriented reforms followed. From 1980 to 2010, China's economy grew at an average rate of roughly 10 percent a year, perhaps the fastest rate of expansion the world has ever seen. Economic reforms, however, were not accompanied by political reforms, and as a result China remains a relatively authoritarian state, with vast amounts of power concentrated in the Communist Party and the central government.

More recently, China's economic growth rate has dropped, averaging only around 7 percent annually from 2011 to 2016. Several factors are responsible for the slowdown, including high levels of debt from a 2008–2009 stimulus plan, industrial overcapacity, and overinvestment in housing and infrastructure in second-tier cities. China's leaders have responded by trying to increase domestic consumption and by engaging in a drive against corruption. Another strategy is the creation of Free Trade Zones, which are similar to the SEZs but characterized by more extensive deregulation and global access. The first such zone was created in Shanghai in 2013, and three more followed in 2015 in Tianjin, Guangdong, and Fujian. Some evidence suggests, however, that these Free Trade Zones offer few new advantages, and thus far investments in them have been modest.

Economists are divided on whether China will be able to maintain economic growth rates on the order of 7 percent a year, which are still rapid by global standards. Wages in Chinese factories are quickly increasing, convincing some foreign firms to relocate to poorer countries, such as Vietnam or Bangladesh. Corporate debt, especially in state-owned firms, has also increased rapidly, threatening long-term economic prospects. Yet despite the fact that critics have been predicting an economic collapse for years, the Chinese economy continues to perform strongly.

Another response to China's economic problems is increased emphasis on investments abroad. Infrastructure is often favored, as it facilitates both exports of Chinese goods and imports of raw materials into China. Prominent here is the New Silk Road project, part of China's **"One Belt, One Road"** initiative (see Chapter 10). The "road" part of this project is actually a maritime endeavor based on port development and linked transportation networks across the Indian Ocean basin. China is also planning ambitious rail projects in Latin America, including a "megarailway" that would span South America from Brazil's Atlantic coast to Peru's Pacific coast. Such projects, however, face pronounced local political and environmental opposition, and if China's economy continues to slip, it is uncertain whether adequate funds will be available.

Social and Regional Differentiation The Chinese economic surge unleashed by the reforms of the late 1970s and 1980s resulted in growing **social and regional differentiation**. In other words, certain groups of people—and certain portions of the country—prospered far more than others. Despite its official adherence to socialism, the Chinese state encouraged the formation of an economic elite, concluding that wealthy individuals are necessary to transform the economy. The least fortunate Chinese citizens were sometimes left without work, and many millions have moved to the booming coastal cities, generating one of the largest mass migrations the world has ever seen. The government has tried to control the transfer of population, but with only partial success. As China's economic boom accelerated, economic disparities mounted. The rapid growth of the elite population made China the world's fastest-growing market for luxury goods. At the same time, vulnerable state-owned enterprises have abandoned their provision of housing, medical care, and other social services.

Economic disparities in China, as in other countries, are geographically structured. Before the reform period, the communist government attempted to equalize the fortunes of the different regions. Such efforts were not wholly successful, and some provinces remain deprived (**Figure 11.46**). Urban areas are much better off than rural zones, with the average city dweller having three times as much disposable income as the average village resident.

The Booming Coastal Region The greatest benefits from China's economic transformation have flowed to the coastal region. The first beneficiaries were the southern provinces of Guangdong and Fujian. This region was predisposed to the new economy because the southern Chinese have long been noted for engaging in overseas trade. Guangdong and Fujian have also benefited from close connections with the overseas Chinese communities of Southeast Asia and North America. Proximity to Taiwan and especially Hong Kong also proved helpful.

By the early 2000s, the Yangtze Delta, centered on the city of Shanghai, reemerged as China's most dynamic region. The delta was the traditional economic (and intellectual) core of China, and, before the communist takeover, Shanghai was its premier industrial and financial center. The government, moreover, has encouraged the development of huge industrial, commercial, and residential complexes, hoping to take advantage of the region's dynamism.

▼ **Figure 11.45 Shenzhen** The city of Shenzhen, adjacent to Hong Kong, was one of China's first Special Economic Zones. It has recently emerged as a major city in its own right. **Q: Suggest reasons why the government of China selected Shenzhen as one of its first Special Economic Zones.**

▼ **Figure 11.46** **Development Issues in East Asia: Economic Differentiation in China** Although China has seen rapid economic expansion since the late 1970s, the benefits of growth have not been evenly distributed throughout the country. Economic prosperity and social development are concentrated on and near the coast. Most of the interior remains relatively poor.

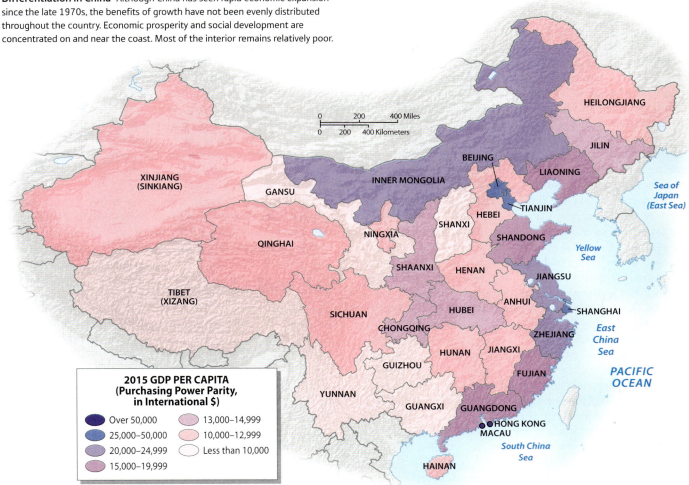

2015 GDP PER CAPITA
(Purchasing Power Parity, in International $)

- Over 50,000
- 25,000–50,000
- 20,000–24,999
- 15,000–19,999
- 13,000–14,999
- 10,000–12,999
- Less than 10,000

Interior and Northern China The interior regions of China have seen much less economic expansion. Central and northern Manchuria were formerly quite prosperous, owing to fertile soils and early industrialization, but have not participated much in the recent boom. Many of the state-owned heavy industries of the Manchurian **rust belt**, or zone of decaying factories, are not efficient. In 2012, six cities in Liaoning Province announced that they would convert several heavily polluted industrial sites to farmland. Manchuria's once-productive oil wells, moreover, are largely exhausted. China and Russia have, however, recently constructed an oil pipeline from Siberia to Daqing in Manchuria, hoping to revive the local economy.

Most of the interior provinces also largely missed the initial wave of growth in the 1980s and 1990s. By most measures, poverty increases with distance from the coast. As a result of such discrepancies, China is now encouraging development in the west, focusing on transportation improvement and natural resource extraction. As labor shortages begin to appear in the booming coastal areas,

industrialists build new factories in the interior, thus spreading the process of development.

Scholars debate the conditions found in the poorer parts of China. Some believe that official statistics are too positive, hiding significant poverty. Others argue that the country's economic boom has substantially raised living standards even in the poorest districts. According to the World Bank, China's poverty rate fell from 88 percent in 1981 to 6.5 percent in 2012 and perhaps as low as 4.1 percent in 2014. But in 2014, a senior government official stated that as many as 200 million Chinese people live on less than US$1.25 a day, indicating persistent destitution in some areas.

Whatever measurements are used, it is clear that China's coastal provinces are much more economically dynamic than its interior provinces. This difference has led to a massive migration stream. In 2000, inter-provincial migrants totaled 40 million, with 15 million of them living in Guangdong Province on the southeastern coast. By 2011, an estimated 158 million Chinese migrants lived in a province other than the one they were born in. Most of these migrants originated in the inland provinces of Anhui, Jiangxi, Hubei, Henan, Hunan, Guangxi, and Sichuan.

Rising Tensions China's explosive but uneven economic growth has generated its own problems. Corruption by state officials has been rampant; success often seems to depend on having the right connections. Not only must poorer people pay frequent bribes, but many have been evicted from their homes to make way for property developments. The government thus launched a massive anti-corruption campaign in 2012. By 2016, more than 100,000 people had been indicted on corruption charges, including 120 high-ranking officials. Critics complain, however, that the campaign has operated outside of normal legal channels, and some see it as a way for the key leaders to enhance their own powers.

An equally significant issue has been the struggle for free expression and political openness. In 1989, as the Cold War was ending, the state crushed a popular student-led protest movement for government accountability and democratic reform. After as many as 1 million citizens had gathered in Beijing's Tiananmen Square to demand change, the government sent in troops, declared martial law, and arrested hundreds of student leaders. Officials thought to be sympathetic to the protests were removed from office, and the opposition was forced to go underground.

In the years after the Tiananmen event, anti-government sentiments declined as most people concentrated on economic development. By the early 2000s, however, local protest movements gained strength, facilitated by tech-savvy activists using social media to organize. Experts now estimate that more than 100,000 organized protests occur each year. Although most such demonstrations focus on local issues, a few pro-democracy protests did break out in 2011. Strikes and other forms of labor unrest intensified in 2016, especially in coal-mining areas. Government plans to eliminate as many as 1.8 million jobs in the stressed coal and steel industries point to further labor conflicts in the coming years.

China's growing middle class, coupled with its slowing economic growth, presents a significant challenge to the country's political system. Although the middle class has not been demanding democracy, it is becoming discontented. People worry about income security as they grow old, fear that hospital bills could easily wipe out their savings, and agonize over their lack of property rights. Many are also angered by the luxurious habits and preferential treatment of the new elites. The government's renewed emphasis on Marxist ideology is unlikely to reduce such concerns.

China has responded to public discontent and protests in part by increasing its restrictions on free speech. Internet censorship is a particularly contentious issue. China's so-called Great Firewall is regarded as the world's most extensive and advanced Internet control mechanism. The government's "Internet police force" was reported to be 2 million strong in 2013. Chinese Internet users have learned to use sophisticated techniques to avoid official censorship, but they remain in a constant race with censors, who also employ cutting-edge technologies.

China's political and human rights policies have complicated its international relations. Tension with the United States and other wealthy countries also entails economic issues. China's large and growing trade surplus and its reluctance to enforce copyright and patent laws irritate many of its trading partners. Numerous U.S. firms have accused Chinese businesses of pirating music, software, and brand names. China passed a new trademark law in 2011, but enforcement has been spotty. Web restrictions and other obstacles placed on Western Internet-based firms operating in China also cause friction. This problem was highlighted in 2016 by the failure of the ride-sharing company Uber in the Chinese market, which was forced to sell out to a local rival.

Social Development in East Asia

Levels of social development in East Asia vary significantly across the region. Health and education standards are high in the more economically advanced parts of the region and lag behind in the poorer parts of China and in North Korea. But despite its poverty and political repression, even North Korea posts average life expectancy figures near the world average. The greatest advances in social development over the past few decades have occurred in China.

Social Conditions in China Since coming to power in 1949, China's communist government has made large investments in medical care and education, and today China boasts fairly impressive health and longevity figures. However, as China has moved to a market-based economy, its rural medical clinics have deteriorated. Some observers fear that levels of both health and education may be declining for the poorest segments of China's population.

Human well-being in China has a geographic dimension. The literacy rate, for example, remains lower in many of the poorer parts of China, including the uplands of Yunnan and Guizhou, than in the most prosperous coastal regions. Increasingly, the largest gap separates the booming cities from the languishing rural areas. People from poor, rural areas often move to the thriving cities, but the government strictly limits such movement under its *hukou* system of official residency, and as a result many urban migrants have no official residency status. They therefore live as "undocumented migrants" in their own country and face difficult access to governmental services, including education for their children. China's announcement in 2016 that it would gradually reform the hukou system by allowing more internal migration offered some hope for such people.

Gender Issues in East Asia Gender roles throughout East Asia have undergone significant changes over the last century as industrialization and changing social values have transformed local societies. Nevertheless, East Asian countries lag behind other industrialized countries in gender equality. According to the 2014 Global Gender Gap Report, China ranks 87th out of 142 countries, Japan ranks 104th, and South Korea ranks 117th (North Korea and Taiwan were not ranked).

In Japan, the first East Asian country to industrialize, women were denied most political and legal rights under the 1889 Meiji Constitution. A growing urban economy, however, opened up new opportunities for work outside the home. Since the end of World War II, women in Japan have enjoyed full political and legal rights, though their opportunities for career advancement lag behind. The traditional expectation that women devote themselves to caring for children forces many to choose between advancing their careers and starting families. Although this is a common dilemma in the industrialized world, the theme is particularly pronounced in Japan. The status of women in South Korea and Taiwan is similar to that in Japan. Though fully equal to males in the eyes of the law, few women reach positions of power in business or government.

The legacy of Confucianism, particularly in China, emphasizes obedience to a male patriarch within the home. The ideology both reflects and causes the difficulties women have faced historically in Chinese society. In the 1800s, the widespread practice of foot binding—involving the breaking and constriction of women's feet so as to achieve a dainty aesthetic—grew to involve nearly half the women in China before the custom's decline in the early 20th century. Economic growth and the relaxation of traditional cultural norms have clearly improved the position of women in China. Women now form over 42 percent of the Chinese workforce and made up 23 percent of the 2012 National Congress of the Communist Party of China. Nevertheless, the upper echelons of politics remain largely the domain of men. In 2015, women constituted only 17 percent of senior officials and managers in China.

Particularly troubling is China's gender imbalance (**Figure 11.47**). In 2004, 121 boys were born for every 100 girls, and in 2009 the province of Jiangxi posted an extraordinary figure of 143. This asymmetry partly reflects the practice of honoring one's ancestors; because family

▲ **Figure 11.47 China's Gender Imbalance** China has one of the world's most imbalanced gender ratios, with boys outnumbering girls in most parts of the country. This photograph shows a classroom in a primary school in Danzhou City on Hainan Island. Danzhou has the worst gender imbalance in China, with 170 males born for every 100 females, according to figures from the most recent Chinese national census.

lines are traced through male offspring, a family must produce a male heir to maintain its lineage. Many couples are, therefore, desperate to produce a son. One option is gender-selective abortion; if ultrasound reveals a female fetus, the pregnancy is sometimes terminated. Baby girls also are commonly abandoned, and rumors of female infanticide circulate. Baby boys, in contrast, are sometimes kidnapped and sold to young couples desperate to raise a son. Some evidence indicates, however, that attitudes are changing and the preference for sons is gradually decreasing. In 2015, China's sex ratio at birth had declined to 115 boys for every 100 girls.

REVIEW

11.13 How has the economic development of Japan, South Korea, Taiwan, and China been similar since the end of World War II, and how has it differed among these countries?

11.14 Why do levels of social and economic development vary so extensively from the coastal region of China to the interior portions of the country?

11.15 Do the geographical patterns of social development in East Asia differ from those of economic development? If so, how?

■ **KEY TERMS** *chaebol*, laissez-faire, Special Economic Zones (SEZs), "One Belt, One Road," social and regional differentiation, rust belt, *hukou*

Chapter 11 East Asia

Summary

- The economic success of East Asia has been accompanied by severe environmental degradation. Japan, South Korea, and Taiwan have responded by enacting strict environmental laws and by moving many of their most polluting industries overseas. The major environmental issues in the region today are tied to the rapid growth of China's economy.

- East Asia is a densely populated region, but its birth rates have plummeted in recent decades. Japan is experiencing population decline, which will put pressure on the Japanese economy. In China, the biggest demographic challenge results from the massive movement of people from the interior to the coast and from rural villages to the rapidly expanding cities.

- East Asia is unified by deep cultural and historical bonds. China has had the largest influence on East Asia because, at one time or another, it ruled most of the region. Although Japan has never been under Chinese rule, it still has profound historical connections to Chinese civilization. The more prosperous parts of East Asia have welcomed cultural globalization over the past several decades.

- East Asia has been characterized by much strife since the end of World War II. China and Korea are still suspicious of Japan, and they worry that it might rebuild a strong military force. Japan is concerned about the growing military power of China and about the nuclear arms and missiles of North Korea. Territorial disputes over islands complicate relations between China and Japan and also between Japan and South Korea.

- With the notable exception of North Korea, all East Asian countries have experienced rapid economic growth since the end of World War II. In the 2000s, the most important story was the rise of China. China's economic expansion has reduced poverty nationwide, but has also generated serious tensions between the wealthier, more globally oriented coastal regions and the less-prosperous interior provinces.

Review Questions

1. How have the natural hazards found in East Asia influenced settlement patterns and government policies in the region?

2. Describe how migration patterns, both with East Asia and between East Asia and the rest of the world, have influenced patterns of economic and cultural globalization.

3. How have particular features of religion and philosophy differentiated East Asia from the rest of the world, and how have the resulting differences influenced development in the region?

4. In what ways does the growing military power of China impact other regions of the world, and how have major countries outside of East Asia responded to this challenge?

5. Write a paragraph explaining how the economic growth of East Asia has transformed the global economy over the past half century.

Image Analysis

1. Examine the map of proposed and ongoing water diversion projects in China, which mostly entail transferring water from the Yangtze River basin to the Yellow River basin. Why is China investing in these massive engineering projects? What are the advantages and disadvantages of using the western routes, located on the Tibetan Plateau, as opposed to the other routes?

2. Search the Internet for information on the now-defunct North American Water and Power Alliance project for transferring water from Alaska and western Canada to southwestern and central North America. How would China's existing projects compare in scale to the NAWAPA project?

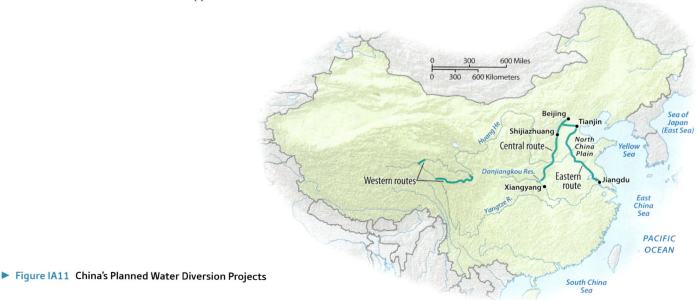

▶ Figure IA11 China's Planned Water Diversion Projects

JOIN THE DEBATE

Japan was radically transformed after WWII, changing almost overnight from a militaristic to a pacifist country. Japan's constitution places strict limits on military spending, and until recently Japan's policies prevented its armed forces from engaging abroad. Concerned about the rise of China, Japan is now reconsidering these policies, but any changes will be highly controversial.

Japan must rearm and enter alliances with foreign countries

- The United States is no longer able to singlehandedly guarantee Japan's security, and as a result resource-poor Japan has become too vulnerable to military disruptions.

- North Korea, with its nuclear weapons and advanced missiles, poses an existential threat to Japan, yet China has been unable or unwilling to restrain North Korea's leaders.

- Hostility toward Japan is pronounced in Chinese society, and China's leaders are increasingly assertive in demanding territory that Japan considers to be rightfully its own.

Remilitarization is a dangerous strategy

- Japan has been relatively secure under the U.S. military umbrella, and there is no need for it to provoke its powerful neighbor.

- China has no territorial ambitions in East Asia beyond the relatively insignificant Senkaku/Diaoyu islands, and therefore does not pose a threat to Japan.

- Remilitarization is expensive, and would therefore add to Japan's debt burden and pose its own threat to the Japanese economy.

▲ **Figure D11** Japan's Ground Self-Defense Forces (GSDF) take part in a live fire exercise.

■ KEY TERMS

anthropogenic landscapes *(p. 458)*
autonomous region *(p. 469)*
Burakumin *(p. 467)*
chaebol (p. 480)
China proper *(p. 443)*
Cold War *(p. 469)*
Confucianism *(p. 464)*
diaspora *(p. 468)*
dynastic succession *(p. 466)*
geomancy *(p. 465)*
hallyu (p. 470)
hukou (p. 486)
juche (p. 466)
ideographic writing *(p. 464)*
laissez-faire *(p. 482)*
loess *(p. 453)*
Mandarins *(p. 464)*

Marxism *(p. 466)*
Meiji Restoration *(p. 474)*
"One Belt, One Road" *(p. 483)*
"one country, two systems" *(p. 476)*
pollution exporting *(p. 451)*
regulatory lakes *(p. 452)*
rust belt *(p. 485)*
sediment load *(p. 452)*
Shogunate *(p. 474)*
social and regional differentiation *(p. 484)*
Special Economic Zones (SEZs) *(p. 483)*
spheres of influence *(p. 473)*
tonal *(p. 468)*
tsunamis *(p. 445)*
urban primacy *(p. 461)*

Mastering Geography™

Looking for additional review and test prep materials? Visit the Study Area in **MasteringGeography**™ to enhance your geographic literacy, spatial reasoning skills, and understanding of this chapter's content by accessing a variety of resources, including **MapMaster** interactive maps, videos, *In the News* RSS feeds, flashcards, self-study quizzes, and an eText version of *Diversity Amid Globalization*

DATA ANALYSIS

http://goo.gl/DXN3Yn

Rapidly developing China is known for its large regional economic disparities. Chinese officials seek to close the development gap. Are policies to balance economic development more evenly across China successful? We can compare the relative economic growth of China's provinces and province-level municipalities by examining gross regional product data on the National Bureau of Statistics of China website: http://data.stats.gov.cn/english/easyquery.htm?cn=E0103. A pull-down menu lets you select a province and view statistical measurements. At the bottom of the list, find the data for "Per Capita Gross Regional Product (yuan/person)."

1. Write down these figures for the last 10 years for these coastal regions: Tianjin, Shanghai, Zhejiang, Fujian, and Guangdong. Graph these five data sets, then do the same for the interior provinces of Gansu, Guizhou, Sichuan, Yunnan, and Tibet.

2. Compare the two sets of graphs, noting both current levels of economic disparity for the first and last years. Based on your graphs, are regional economic differences in China becoming more pronounced, less pronounced, or staying roughly the same?

3. Do your findings correlate with China's desire to create a geographically balanced national economy? Suggest reasons for the results you found.

Authors' Blogs

Scan to visit the
GeoCurrents blog
http://www.geocurrents.info/category/place/east-asia

Scan to visit the
Author's Blog
for field notes, media resources, and chapter updates
https://gad4blog.wordpress.com/category/east-asia/

Physical Geography and Environmental Issues

The arid parts of South Asia suffer from water shortages and salinization of the soil, whereas the humid areas often experience devastating floods. Pollution levels in most urban areas are extremely high.

Population and Settlement

South Asia has recently become the most populous world region. Birth rates in much of the region, however, have dropped sharply, falling below the replacement level in most of southern and western India. Fertility rates remain higher in northern India and Pakistan.

Cultural Coherence and Diversity

South Asia is one of the most culturally diverse regions of the world, with India alone having more than a dozen official languages, as well as numerous followers of most major religions. Religious and ethnic tensions have increased in many areas over the past several decades, especially in Pakistan.

Geopolitical Framework

The region is burdened by a large number of violent secession movements as well as by the struggle between the nuclear-armed countries of India and Pakistan. The relationship between Pakistan and the United States is tense, as is that between India and China.

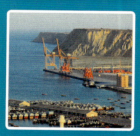

Economic and Social Development

Although South Asia is one of the poorest world regions, parts of India are experiencing rapid development based on the high-tech skills of its educated people. Despite extreme poverty, Bangladesh has also made significant social and economic progress, whereas Pakistan has lagged behind.

▶ Located in an area claimed by India but governed by Pakistan, Baltistan is one of the most rugged and remote parts of South Asia. In this Baltistan scene, a precarious bridge is suspended across the upper reaches of the Indus River as autumn begins to set in.

Gilgit-Baltistan, Pakistan

SOUTH ASIA

Pakistan, the world's sixth most populous country, faces severe water and energy problems. Most of the country is arid, and the rainfall received during the short summer monsoon is insufficient for most areas. Monsoon flooding can be devastating, but for the rest of the year water shortages are common. Energy shortages are equally serious. Pakistan's electricity generation meets only about half of the potential demand, resulting in long blackouts even in the major cities. Together, these two problems thwart economic development, imperil the food supply, and undermine public health.

To address the water and energy crises, Pakistan's government has initiated an ambitious dam-building program. Dams constructed in the rugged northern mountains could generate massive quantities of relatively clean energy, reducing Pakistan's carbon footprint while meeting its electricity demand. The reservoirs behind such dams, moreover, could store monsoon floodwaters for the dry season, thus reducing the flood risk while enhancing irrigation. The centerpiece of Pakistan's hydrological program is the huge Diamer-Bhasha Dam, now in an early stage of construction, in the northern region of Gilgit-Baltistan. If completed, the project would eliminate roughly half of the country's power deficit and irrigate millions of acres of farmland.

However, Diamer-Bhasha's completion is uncertain. The project will cost an estimated US$14 billion, a figure that Pakistan cannot easily meet. Earlier sponsors, including the World Bank and the Asian Development Bank, withdrew funding, and now China—Pakistan's main infrastructural backer—is showing reluctance. Environmental problems and the costs of relocating villages that will be flooded figure prominently, but the most serious obstacles are geopolitical. Diamer-Bhasha is located in territory that Pakistan controls but India claims, and India opposes construction. To understand contemporary South Asia, we must grasp the bitter and ongoing conflict between Pakistan and India, which together overshadow the rest of South Asia.

Defining South Asia

South Asia is easily defined in terms of physical geography. Most of the region forms a subcontinent—often called the **Indian Subcontinent**—separated from the rest of Asia by formidable mountain ranges (**Figure 12.1**). Located here are India, Pakistan,

and Bangladesh, as well as the mountainous states of Nepal and Bhutan. Also placed within South Asia are the Indian Ocean island countries of Sri Lanka and the Maldives, and the Indian territories of Lakshadweep and the Andaman and Nicobar islands.

India is by far the largest South Asian country, both in size and in population. Covering more than 1 million square miles (2,590,000 square km), India is the world's seventh largest country in area and, with just under 1.3 billion inhabitants, second only to China in population. Pakistan and Bangladesh are the next most populous countries, with more than 193 and 160 million inhabitants, respectively; compact Bangladesh is also one of the most densely populated parts of the world. Bangladesh shares a short border with Burma (Myanmar), but it is otherwise virtually surrounded by India.

South Asia is historically united by deep cultural commonalities. Religious ideas associated with Hinduism and Buddhism once dominated the region. For the past thousand years, however, Islam

> ## Religious tensions in South Asia have in turn influenced geopolitical conflicts, some spanning decades

has also played a major role, and South Asia has a larger Muslim population than any other world region. Religious tensions in the region have in turn influenced geopolitical conflicts, some spanning decades. For example, since independence from Britain in 1947, India and Pakistan have fought several wars and remain locked in a bitter dispute over the territory of Kashmir in the northern reaches of the region. Religious divisions are linked to this geopolitical turmoil, for India is primarily a Hindu country with a large Muslim minority, while Pakistan is almost entirely Muslim. Other forms of religious and ethnic tension are found elsewhere within the region.

Paralleling these geopolitical tensions are demographic and economic concerns. Although fertility levels have dropped dramatically across the region in recent years, they remain elevated in some areas, particularly north-central India and Pakistan. As a result, some experts worry about how South Asia will sustain its population. Agricultural production has kept pace with population growth over the past four decades, but many South Asian environments are experiencing pronounced stress. Poverty compounds such problems. Along with Sub-Saharan Africa, South Asia is the poorest part of the world.

Many observers are optimistic about South Asia's future, but others focus on the region's huge challenges. We first turn to the region's physical and environmental geography to understand these challenges.

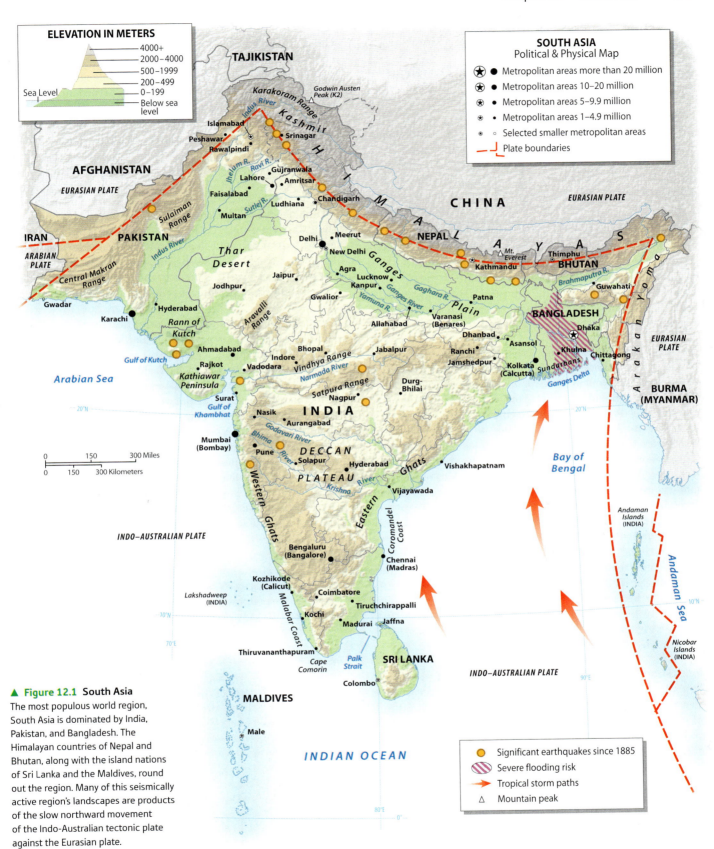

ELEVATION IN METERS

4000+
2000–4000
500–1999
200–499
0–199
Sea Level
Below sea level

SOUTH ASIA
Political & Physical Map

⊛ ● Metropolitan areas more than 20 million
⊛ ● Metropolitan areas 10–20 million
⊛ • Metropolitan areas 5–9.9 million
⊛ · Metropolitan areas 1–4.9 million
⊛ ○ Selected smaller metropolitan areas
— Plate boundaries

TAJIKISTAN

Godwin Austen Peak (K2)

Karakoram Range

Kashmir

Islamabad
Peshawar
Rawalpindi
Srinagar

AFGHANISTAN
EURASIAN PLATE

Gujranwala
Lahore
Amritsar
Faisalabad
Chandigarh
Ludhiana
Multan

CHINA   EURASIAN PLATE

H I M A L A Y A S

IRAN
ARABIAN PLATE

PAKISTAN

Central Makran Range
Sulaiman Range

Indus River

Jhelum R.
Ravi R.
Sutlej R.

Delhi   Meerut
New Delhi

NEPAL
Mt. Everest
Thimphu
Kathmandu   BHUTAN

Brahmaputra R.
Guwahati

EURASIAN PLATE

Gwadar

Thar Desert

Jaipur
Agra
Lucknow
Kanpur

Ganges
Gaghara R.
Ganges Plain
Patna

BANGLADESH

Dhaka

Arakan Yoma

Hyderabad
Karachi

Rann of Kutch

Aravalli Range

Gwalior
Yamuna R.
Ganges River

Allahabad
Varanasi (Benares)

Dhanbad
Asansol

Khulna
Chittagong

EURASIAN PLATE

Gulf of Kutch
Ahmadabad
Rajkot
Vadodara

Indore
Bhopal
Vindhya Range
Narmada River

Jabalpur

Ranchi
Jamshedpur

Kolkata (Calcutta)
Sundarbans

BURMA (MYANMAR)

Arabian Sea

Kathiawar Peninsula
Surat
Gulf of Khambhat

Nasik
Aurangabad

Satpura Range
Nagpur

Durg-Bhilai

Ganges Delta

0   150   300 Miles
0   150   300 Kilometers

Mumbai (Bombay)
Pune

Godavari River
Bhima River
DECCAN PLATEAU
Solapur
Hyderabad

INDIA

Krishna River

Vishakhapatnam

Bay of Bengal

20°N

Vijayawada

Andaman Islands (INDIA)

INDO–AUSTRALIAN PLATE

Western Ghats

Eastern Ghats

Bengaluru (Bangalore)

Coromandel Coast
Chennai (Madras)

Andaman Sea

Kozhikode (Calicut)

Coimbatore

Lakshadweep (INDIA)

Tiruchchirappalli
Kochi
Madurai   Jaffna

10°N

Nicobar Islands (INDIA)

70°E
Thiruvananthapuram

Malabar Coast

Cape Comorin
Palk Strait

SRI LANKA

INDO–AUSTRALIAN PLATE

90°E

MALDIVES

Colombo

INDIAN OCEAN

Male

80°E
0°

▲ Figure 12.1 South Asia
The most populous world region, South Asia is dominated by India, Pakistan, and Bangladesh. The Himalayan countries of Nepal and Bhutan, along with the island nations of Sri Lanka and the Maldives, round out the region. Many of this seismically active region's landscapes are products of the slow northward movement of the Indo-Australian tectonic plate against the Eurasian plate.

○ Significant earthquakes since 1885
▨ Severe flooding risk
➤ Tropical storm paths
△ Mountain peak

Physical Geography and Environmental Issues: From Tropical Islands to Mountain Rim

South Asia's environmental geography covers a wide spectrum, from the world's highest mountains to densely populated islands barely above sea level; from one of the wettest places on Earth to dry, scorching deserts; from tropical rainforests to degraded scrublands to coral reefs. Each ecological zone has its own distinct and complex environmental problems.

The Four Physical Subregions of South Asia

South Asia is separated from the rest of the Eurasian continent by a series of sweeping mountain ranges, including the Himalayas. Environmental conditions in this vast region can be described in terms of four physical subregions, starting with the lofty mountains of the northern fringe and extending to the tropical islands of the far south.

Mountains of the North South Asia's northern rim is dominated by the great Himalayan Range, forming the northern borders of India, Nepal, and Bhutan (see *Geographers at Work: Understanding the Urban Heat-Island Effect in a Changing Climate*). These mountains are linked to the equally high Karakoram Range to the west, extending through northern Pakistan. More than two dozen peaks exceed 25,000 feet (7600 meters), including the world's highest mountain, Everest, on the Nepal–China border at 29,028 feet (8848 meters). To the east are the lower Arakan Yoma Mountains, forming the border between India and Burma (Myanmar) and separating South Asia from Southeast Asia.

These formidable mountain ranges were produced by tectonic activity caused by peninsular India pushing northward into the larger Eurasia continental plate. The collision of these two tectonic plates folded and upthrust these great ranges. The entire region is seismically active, putting all of northern South Asia in serious earthquake danger (Figure 12.1). A massive earthquake in Nepal in April 2015 resulted in almost 9,000 deaths and 22,000 injuries, causing an estimated US$5 billion in damages.

Indus–Ganges–Brahmaputra Lowlands South of these highlands lie vast lowlands created by three major river systems that have carried sediments off the mountains over millions of years, building alluvial plains of fertile and easily farmed soils. These lowlands are densely settled and constitute the core population areas of Pakistan, India, and Bangladesh.

Of these three rivers, the Indus is the longest, covering more than 1800 miles (2880 km) as it flows southward from the Himalayas through Pakistan to the Arabian Sea, providing much-needed irrigation waters to Pakistan's deserts. The broad band of cultivated land in central and southern Pakistan that is watered by the Indus is highly fertile and densely populated. Pakistan is concerned about India's dam-building projects on several tributaries of the Indus, which could reduce the flow of this all-important river.

Even more heavily populated is the vast lowland of the Ganges, which, after flowing out of the Himalayas, travels southeasterly some 1500 miles (2400 km) to the Bay of Bengal. The Ganges not only provides the fertile alluvial soil that has long made northern India a densely settled area, but also serves as a vital transportation corridor. Given the central role of this river in Indian history, it is understandable that Hindus consider the Ganges sacred (**Figure 12.2**). But while the waters of the Ganges are reputed to have healing powers, the river is one of the world's most polluted major watercourses.

This lowland is often called the Indus–Ganges Plain, but the term neglects the Brahmaputra River. This river rises on the Tibetan Plateau and flows easterly, then south, and then westerly over 1700 miles (2700 km), joining the Ganges in central Bangladesh and spreading out over the world's largest delta. The Ganges–Brahmaputra Delta is also one of the most densely settled areas in the world, containing more than 3000 people per square mile (1200 per square km).

Peninsular India Jutting southward from the Indus–Ganges Plain is the familiar shape of peninsular India, made up in part by the Deccan Plateau. This plateau zone is bordered by narrow coastal plains backed by elongated north–south mountain ranges. The higher Western Ghats are generally about 5000 feet (1500 meters) in elevation, but exceed 8000 feet (2400 meters) near the peninsula's southern tip. The Eastern Ghats are lower and discontinuous, thus forming less of a transportation barrier to the broader eastern coastal plains. On both coastal plains, fertile soils and abundant water supplies support population densities comparable to those of the Ganges lowlands.

The Southern Islands At India's southern tip lies the island country of Sri Lanka, which is almost linked to India by a series

Explore the **Sights** of Ganges at Varanasi

http://goo.gl/Qya9jH

▼ **Figure 12.2 Ganges River** Most Hindus view the Ganges River of northern India as sacred, and therefore believe that they gain merit by bathing in its waters and praying along its banks. People here are dipping into the river in the city of Varanasi, widely regarded as the spiritual capital of India.

Geographers at Work

Understanding the Urban Heat-Island Effect in a Changing Climate

Chandana Mitra, a geographer at Auburn University, works extensively on two critical issues currently facing South Asia: climate change and the growth of huge cities (Figure 12.1.1). Her research shows that climate and urbanization must be investigated together, as land-use changes associated with urban growth significantly affect the local climate. In northern India's sweltering, pre-monsoon, summer season, average high temperatures already approach 104°F (40°C), a dangerously high level that is locally exacerbated in so-called urban heat islands. "Geographers are trained to see the different connections, the interdependency of phenomena," says Mitra about examining this interaction between the physical and human environments. Her GIS-based mapping, remote-sensing analysis, and weather and climate modeling allow her to delineate the extent and magnitude of such urban hot spots. The effect of megacities on climate is not limited to temperature; Mitra also found that increased rainfall in Kolkata just before the onset of the monsoon may be linked to the urban heat-island effect. As Kolkata often floods during heavy downpours, such work could help urban planners more effectively manage runoff.

Climate-Change Education Although most of Mitra's published work examines South Asia, she is more broadly interested in understanding the global ramifications of urbanization and climate change. "It's important to understand the spatial distribution of physical and human systems, as there's always a pattern," she notes. Mitra also seeks to improve climate education in U.S. high schools and middle schools, to "educate teachers and the young people about climate change." Working with the NASA-funded Innovations in Climate Education (NICE) Program in Alabama, Mitra and her colleagues have developed teaching modules that examine such issues as extreme climate events, the Earth's atmospheric composition, and the urban heat-island effect. "Geography is a special subject that connects different disciplines under a broad umbrella, and

▲ **Figure 12.1.1.** **Dr. Chandana Mitra** taking locational information, latitude, and longitude using a hand-held GPS instrument in Kolkata, West Bengal, India.

can lead to many types of jobs," she explains. "It's an all-encompassing subject, and the beautiful part is the blending."

1. Have you ever experienced the urban heat-island effect in the United States? What are some of the reasons why large cities tend to be warmer than the surrounding countryside?

2. What cities in the United States would tend to be most vulnerable to excessive temperatures from the urban heat-island effect, and why?

---

of small islands called Adam's Bridge. Sri Lanka is ringed by extensive coastal plains and low hills, but mountains reaching more than 8000 feet (2400 meters) cover the southern interior, providing a cool, moist climate. Because the main winds arrive from the southwest, that portion of the island is much wetter than the rain-shadow area of the north and east.

Forming a separate country are the Maldives, a chain of more than 1200 islands stretching south to the equator some 400 miles (640 km) off India's southwestern coast. The combined land area of these islands is only about 116 square miles (290 square km), and only a quarter of the islands are actually inhabited. The Maldives are flat, low coral atolls, with a maximum elevation just over 6 feet (2 meters) above sea level.

South Asia's Monsoon Climates

The dominant climatic factor for most of South Asia is the **monsoon**, the distinct seasonal change of wind direction that corresponds to wet and dry periods. Most of South Asia has three distinct seasons. The warm, rainy season of the southwest monsoon lasts from June through October, followed by a relatively cool and dry season from November until February, when winds from the northeast dominate. Only a few areas in far northwestern and southeastern South Asia get substantial rainfall during this winter monsoon.

Next comes the hot season from March to early June, during which heat and humidity build up until the summer monsoon's much anticipated and rather sudden "burst."

This monsoon pattern is caused by large-scale meteorological processes that affect much of Asia (Figure 12.3). During the Northern Hemisphere's winter, a large high-pressure system forms over the cold Asian landmass. Cold, dry winds flow outward from this high-pressure cell, over the Himalayas and down across South Asia. As winter turns to spring, these winds diminish, resulting in the hot, dry season. This buildup of heat over South and Southwest Asia eventually produces a large thermal low-pressure cell that becomes strong enough by June to draw in warm, moist air from the Indian Ocean.

Heavy **orographic rainfall** results from the uplifting and cooling of moist monsoon winds over the Western Ghats. As a result, some stations in the Am climate zone (Figure 12.4) receive more than 200 inches (508 cm) of rain during the four-month wet season. Inland, however, a strong rain-shadow effect dramatically reduces rainfall on the Deccan Plateau. Farther north, as the monsoon winds are forced up and over the mountains, copious amounts of rainfall are characteristic. Cherrapunji, India, at an elevation of 4000 feet (1200 meters), is a strong contender for the title of world's wettest place, with average annual rainfall of 450 inches (1150 cm).

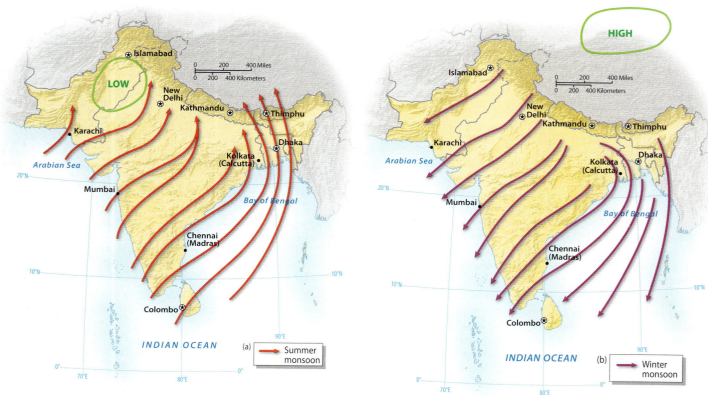

▲ **Figure 12.3 The Summer and Winter Monsoons** (a) Low pressure centered over South and Southwest Asia draws in warm, moist air masses during the summer, bringing heavy monsoon rains to most of the region. Usually, these rains begin in June and last for several months. (b) During the winter, high pressure forms over northern Asia. As a result, winds are reversed from those of the summer. Only a few coastal locations along India's east coast and in eastern Sri Lanka receive substantial rain during the winter monsoon.

Not all of South Asia receives substantial rainfall from the southwest monsoon. In much of Pakistan and the Indian state of Rajasthan, precipitation is low and variable, resulting in steppe and desert climates. Karachi's annual total is less than 10 inches (25 cm). But heavy rain can hit even in the drier parts of South Asia. In the summer of 2010, for example, prolonged downpours brought devastating flooding. With damages estimated at US$43 billion, the 2010 Pakistan flood is considered that country's worst natural disaster.

Whether rainfall is heavy or light, the monsoon rhythm affects all of South Asia in many different ways, from the delivery of much-needed water for crops and villages to the mood of millions of people as they eagerly welcome relief from oppressive heat (**Figure 12.5**). In some years the monsoon delivers its promise of abundant moisture, but in others it brings inadequate rainfall, resulting in crop failure and hardship. Both 2014 and 2015 saw below-average monsoon rainfall across most of western and northern India.

Climate Change in South Asia

Many areas of South Asia are highly vulnerable to the effects of global climate change. Even a minor rise in sea level will inundate large areas of the Ganges–Brahmaputra Delta in Bangladesh. Already, over 18,500 acres (7500 hectares) of swamplands have been submerged. If the most severe sea-level forecasts come to pass, the atoll nation of the Maldives will

vanish beneath the waves. The loss of glaciers in the Himalayas would severely imperil the water supplies of northern South Asia. Arid Pakistan is particularly exposed to this threat.

South Asian agriculture is likely to suffer from several problems linked to global climate change. Winter temperature increases of up to 6.4°F (3°C), along with decreased rainfall, could threaten the vital wheat crop of Pakistan and northwestern India, undermining the food supply of both countries. In parts of South Asia, however, climate change could result in increased rainfall due to an intensified summer monsoon. Unfortunately, the new precipitation regime will likely be characterized by more intense cloudbursts coupled with fewer episodes of gentle, prolonged rain. A 2013 World Bank report warned that "huge disruptive" monsoon events and subsequent flooding, which currently strike once a century on average, could occur once in every decade.

Responses to the Climate Challenge South Asia ranks low in per capita carbon output, trailed only by Sub-Saharan Africa among world regions, but its emissions are rapidly increasing. India is now the world's third largest carbon emitter, following China and the United States. The rapid industrialization of India, coupled with its reliance on coal for much of its electricity, means that its carbon dioxide output will continue to increase for some time.

But Indian leaders, like those of other South Asian countries, are concerned about climate change and are willing to

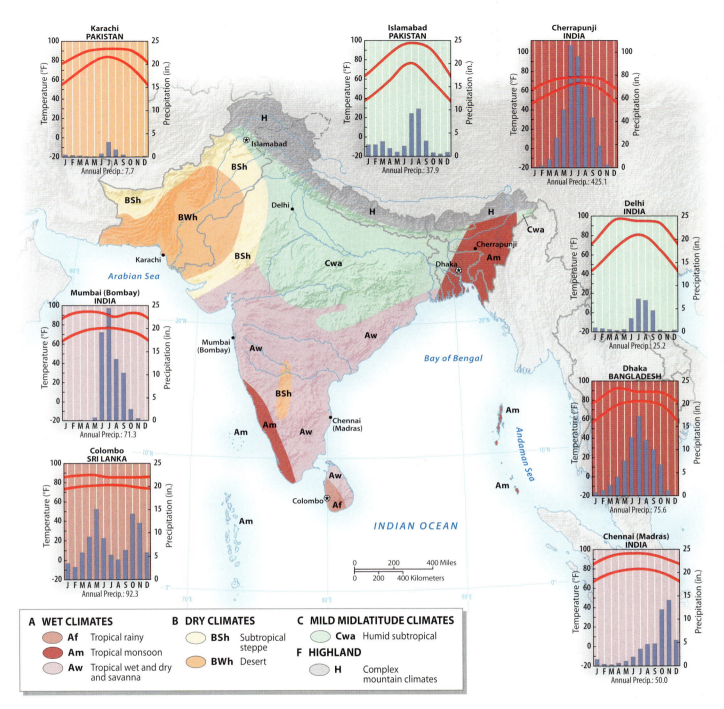

▲ **Figure 12.4 Climate of South Asia** Except for the extensive Himalayas, South Asia is dominated by tropical and subtropical climates. Many of these climates show a distinct summer rainfall season associated with the southwest monsoon. The climographs for Mumbai and Delhi are excellent illustrations. However, the climographs for locations on the east coast, such as Madras, India, and Colombo, Sri Lanka, show how some locations also receive rain from the northeast winter monsoon.

reduce their carbon emissions after achieving higher levels of economic development. They also insist that the greatest responsibility for action lies with the world's wealthy, industrialized countries. India signed the 2015 Paris Agreement on climate change, pledging that it would derive 40 percent of its power from renewables by 2030. But the Indian delegation complained that this agreement did not go far enough, expressing disappointment at the meager level of funding pledged by wealthy countries to finance climate initiatives.

Bangladesh's delegation even contemplated walking out of the Paris negotiations because money from the Green Climate Fund would come mostly in the form of loans rather than grants.

Energy in South Asia An adequate response to the challenge of climate change in South Asia will require major changes to the region's energy systems. The underdeveloped state of the electric grid in many parts of South Asia presents an enormous

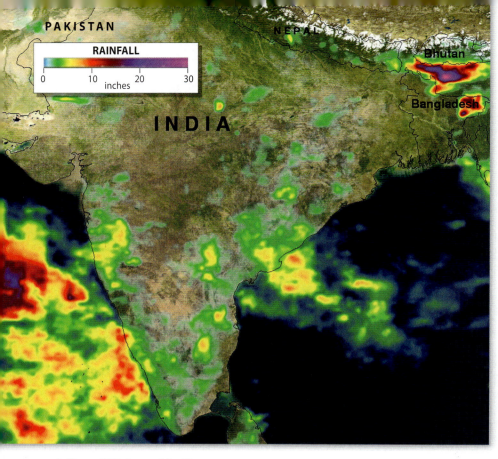

▲ **Figure 12.5 Start of the Monsoon** The rains of the southwest monsoon season typically reach southwestern South Asia in late May and then spread across the subcontinent by July. This map shows cumulative precipitation from June 1 to June 8, 2015, as the wet monsoon airmass approaches India.

vast coal reserves, mining more coal than any country except China and the United States, yet it still imports large quantities from Indonesia.

South Asia is, however, moving ahead on renewable energy, and here again India is the region's leader, launching a solar alliance of more than 120 countries in 2015, and pledging an initial US$30 million to set up the group's headquarters in the country. India is currently investing heavily in rooftop photovoltaic arrays, but is also moving ahead on wind power. According to current projections, more than two-thirds of India's new electricity installation by 2030 will come from a mix of solar and wind generation (**Figure 12.6**).

Other Environmental Issues in South Asia

As is true in other poor and densely settled world regions, other serious ecological issues plague South Asia, including the usual environmental problems of water and air pollution that accompany early industrialization and manufacturing (**Figure 12.7**). In all countries in the region, however, major efforts are under way to address the environmental crisis (see *Working Toward Sustainability: Community Development and Mangrove Conservation in Sri Lanka*).

The Precarious Situation of Bangladesh The link between population pressure and environmental degradation is nowhere clearer than in the delta area of Bangladesh, where the search for fertile land has driven people into hazardous areas, putting millions at risk from seasonal flooding and from the powerful tropical **cyclones** that form over the Bay of Bengal. With continual population

challenge for regional development, particularly in Bangladesh, where only 38 percent of the population had access to electricity in 2014. Even in areas with electricity access, power is often intermittent and unreliable, creating a significant barrier to investment. The lack of transmission lines in rural areas, coupled with reliably intense sunlight, makes South Asia an excellent candidate for small-scale solar electricity generation. Still, the lion's share of recent investment went toward gas-, coal-, and oil-based power plants. India in particular has

▶ **Figure 12.6 Future Energy Use in India** Coal is still an essential source of energy in India, but the country's government is planning to transition into renewables, as can been seen in this official projection. Skeptics doubt, however, whether India will really be able to rely so extensively on wind and solar power by 2030.

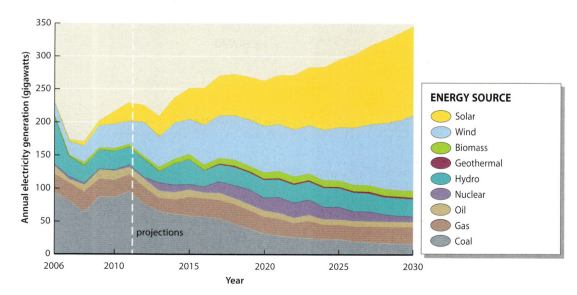

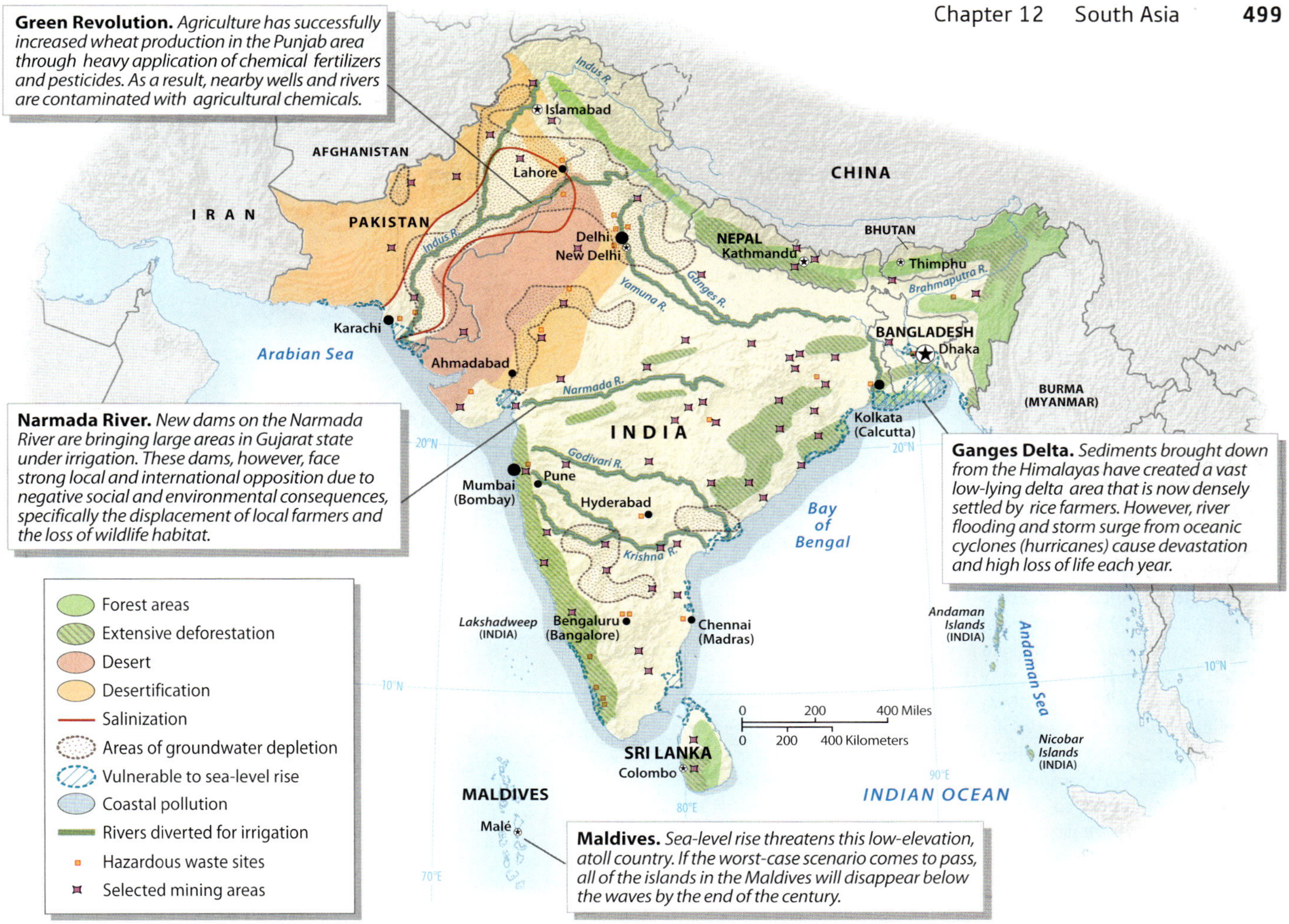

Green Revolution. *Agriculture has successfully increased wheat production in the Punjab area through heavy application of chemical fertilizers and pesticides. As a result, nearby wells and rivers are contaminated with agricultural chemicals.*

Narmada River. *New dams on the Narmada River are bringing large areas in Gujarat state under irrigation. These dams, however, face strong local and international opposition due to negative social and environmental consequences, specifically the displacement of local farmers and the loss of wildlife habitat.*

Ganges Delta. *Sediments brought down from the Himalayas have created a vast low-lying delta area that is now densely settled by rice farmers. However, river flooding and storm surge from oceanic cyclones (hurricanes) cause devastation and high loss of life each year.*

Maldives. *Sea-level rise threatens this low-elevation, atoll country. If the worst-case scenario comes to pass, all of the islands in the Maldives will disappear below the waves by the end of the century.*

Legend:
- Forest areas
- Extensive deforestation
- Desert
- Desertification
- Salinization
- Areas of groundwater depletion
- Vulnerable to sea-level rise
- Coastal pollution
- Rivers diverted for irrigation
- Hazardous waste sites
- Selected mining areas

▲ **Figure 12.7 Selected Environmental Issues in South Asia** As might be expected in a highly diverse and densely populated region, South Asia suffers from some major environmental problems. These range from salinization of irrigated lands in the dry areas of Pakistan and western India to groundwater pollution from Green Revolution fertilizers and pesticides. Deforestation and erosion are widespread in upland areas.

growth, people have transformed swamps into highly productive rice fields.

For millennia, drenching monsoon rains have eroded and transported huge quantities of sediment from the Himalayas to the Bay of Bengal, gradually building this low-lying, fertile delta environment. Deforestation of the Ganges and Brahmaputra headwaters magnifies the problem. Forest cover slows runoff to allow rainfall to soak into the ground and replenish groundwater supplies. Thus, deforestation in river headwaters increases flooding in the wet season, as well as lowers water level during the dry season, when groundwater supplements river flow.

Although periodic floods are natural and often beneficial events that enlarge deltas by depositing fertile river-borne sediment, flooding is a serious problem for people inhabiting these low-lying areas. In an average year, around 18 percent of Bangladesh floods, killing roughly 5,000 people and destroying several million homes. A particularly intense flood, like that of September 1998, can inundate as much as two-thirds of the country (**Figure 12.8**). Experts estimate that roughly 70 percent of the several hundred thousand people who move to Dhaka, Bangladesh's capital, each year do so to avoid flooding and

other environmental hazards. Bangladesh's water problems are further magnified by the fact that many of its aquifers are contaminated with arsenic from natural sources, threatening the health of as many as 80 million people.

Dam and Reservoirs Whereas Bangladesh and eastern India often suffer from excess water, Pakistan and much of northwestern and central India more often suffer from a lack of water. In many areas, farmers have traditionally stored runoff water from the monsoon rains in small reservoirs for use during the dry months. As these storage facilities have proved inadequate to meet the demands of a growing population, villagers have turned to deep wells and powerful pumps to provide groundwater for their crops. As a result, water tables are falling in many areas, requiring ever deeper wells and electricity expenditures.

As we have seen, the governments of India and Pakistan have been building large dams to generate electricity and supply additional water. These dams are almost always controversial, mostly because they displace hundreds of thousands of rural residents. The Sardar Sarovar Dam on the Narmada River in the state of Gujarat has been particularly

Working Toward Sustainability

Community Development and Mangrove Conservation in Sri Lanka

Mangrove swamp forests, which thrive in shallow coastal waters in many tropical regions, are endangered across southern Asia. In some areas they are being cleared to make room for export-oriented fish and shrimp ponds, and elsewhere the trees are cut for charcoal or firewood. But mangrove forests are vital components of coastal ecosystems, serving as nurseries for many fish and crustacean species. They also protect the shoreline from storm surges and can trap heavy metals and other pollutants under their roots, keeping them out of the water. As a result, mangrove conservation has become a high priority.

Sri Lanka has already lost many of its mangroves. The most extreme destruction occurred in the northwestern corner of the island, where swamp forests were converted into shrimp farms in the 1990s. That development, however, did not prove sustainable, as diseases later wiped out most of the shrimp. In 2015, Sri Lanka's government embarked on a new strategy, becoming the world's first country to officially protect all of its remaining mangrove forests.

A New Initiative Sri Lanka's new mangrove-protection scheme is attracting considerable global attention. Jointly run by the government, a local coastal-protection foundation, and a U.S.-based conservation group called Seacology, this US$3.4 million initiative combines environmental protection with rural development and women's empowerment (**Figure 12.2.1**). A central feature is the provision of small, low-interest loans of around US$100 to local residents, mostly women, who generally use the loans to start small businesses. Funds are also earmarked for planting fast-growing tree species on dry land to provide alternative sources of fuel.

In return for their loans, recipients are organized into 1500 groups of around 10 persons that are tasked with overseeing a specific area of mangroves. They are also responsible for teaching others about the importance of the mangrove ecosystem and for helping replant 9600 acres (3885 hectares) with mangrove seedlings. The groups must work with government rangers who patrol the 21,782 acres (8815 hectares) of mangroves now under official protection. In these areas, mangroves cannot be lawfully cut for commercial purposes.

Targeted Poverty Reduction Sri Lanka's new mangrove initiative seeks to reduce rural poverty as much as to protect nature. As a result, it focuses on the most vulnerable members of local communities. Half of all

▲ **Figure 12.2.1 New Mangrove Trees** Recently replanted mangroves thrive in Puttalam lagoon. Sri Lanka has made great efforts to preserve its remaining mangrove forests through community-oriented mangrove reforestation schemes.

loan recipients must be widows, and the other half must be school dropouts, whether male or female. The program appears to be quite successful in both ecological and economic terms: of the nearly 2000 loans made to local women thus far, the repayment rate has been over 96 percent.

Google Earth Virtual MG Tour Video

http://goo.gl/CNb2fe

1. Why have mangrove forests in particular been selected for conservation programs in Sri Lanka and elsewhere in the region? How do they differ from other forests in terms of their own vulnerability and their benefits to local communities?

2. Why has Sri Lanka decided to link a social-development program to its mangrove conservation initiative?

contentious (**Figure 12.9**). This dam is part of the much larger Narmada Project, which involves 30 major dams and extensive canal systems. In 2014, the Indian government approved plans to increase the height of the Sardar Sarovar Dam from its original 260 feet (80 meters) to 535 feet (163 meters). Proponents emphasize that this project will irrigate an additional 6,900 square miles (18,000 square kilometers) in drought-prone western Gujarat, but opponents argue that the human and environmental costs will outweigh such benefits.

Forests and Deforestation Forests and woodlands once covered most of South Asia outside of the northwestern deserts. In most places, however, tree cover has

▼ **Figure 12.8 Flooding in Bangladesh** Devastating floods are common in the low-lying delta lands of Bangladesh. Heavy rains come with the southwest monsoon, especially to the Himalayas, and powerful cyclones often develop over the Bay of Bengal.

▲ **Figure 12.9 Sardar Sarovar Dam** The Sardar Sarovar Dam on the Narmada River in the Indian state of Gujarat was completed in 2008.
Q: Why is this dam relatively popular in Gujarat but highly unpopular in the neighboring state of Madhya Pradesh?

disappeared as a result of human activities. The Ganges Valley and coastal plains of India were largely deforested thousands of years ago for agriculture. Elsewhere, forests were cleared more gradually for agricultural, urban, and industrial expansion. More recently, hill slopes in the Himalayas and in the remote lands of eastern India have been heavily logged for commercial purposes.

Beginning in the 1970s, India embarked on several reforestation projects. The central government claims that India's forest coverage actually increased by 2266 square miles (5871 square kilometers) between 2010 and 2012, but most Indian environmentalists are skeptical, pointing out that many existing forests are being degraded. Reforested areas in India, moreover, are often covered by nonnative trees, such as eucalyptus, that support little wildlife.

As a result of deforestation, most villages of South Asia suffer from a shortage of fuelwood for household cooking, forcing people to burn dung cakes from cattle. This low-grade fuel provides adequate heat, but diverts nutrients that could be used as fertilizers, and produces high levels of air pollution, both indoors and outside. Where wood is available, collecting it involves many hours of female labor, as the remaining sources of wood are often far from the villages (**Figure 12.10**).

Villagers in India's forested areas sometimes band together to protest deforestation. In the 1970s, the so-called Chipko movement mobilized women in northern India to engage in "tree-hugging" campaigns to stop logging. Such social pressure led to local residents becoming involved in forestry and conservation projects. Increasingly, villagers have been demanding their own rights to land and resources in wooded areas. The landmark 2006 Forest Rights Act granted extensive rights to forest-dwelling Indians. Some urban environmentalists, however, argue that this legislation could actually increase deforestation, as expanding villages seek to convert woodlands to agricultural fields.

Wildlife: Extinction and Protection Although the environmental situation in South Asia is worrisome, wildlife protection inspires some optimism. The region has managed to retain a diverse assemblage of wildlife despite population pressure and intense poverty. The only remaining Asiatic lions, for example, live in India's Gujarat state, and even Bangladesh retains a viable population of tigers in the Sundarbans, the mangrove forests of the southern Ganges Delta. Wild elephants still roam several large reserves in India, Sri Lanka, and Nepal.

Explore the **Sounds** of Wildlife in India

http://goo.gl/alunSY

Wildlife protection in India far exceeds that in most other parts of Asia. Project Tiger, which currently operates more than 53 preserves, is credited with increasing the country's tiger population from 1200 in the 1970s to roughly 3500 by the 1990s. Continued forest clearance and population increase, however, led to a drop in the tiger population to around 1700 by 2010 (**Figure 12.11**). As a result, the Tiger Protection Force was established to counter poaching and to relocate villagers from crucial habitat areas. In 2015, the Indian government announced that the tiger population increased by an astonishing 30 percent over the preceding four years. Wildlife experts, however, dispute this figure, given the high death rate of tigers in many areas.

Whatever the true figures, it is undeniable that India has done more to conserve tigers—and other wildlife species—than most other poor and crowded countries. Doing so, however, is never easy. In many areas, pressure is mounting to convert remaining wildlands to farmlands. The best remaining zone of extensive habitat is in India's far northeast, an area of rapid immigration and political unrest. Moreover, wild animals, particularly tigers and elephants, threaten crops, livestock, and even people living near the reserves. When a rogue elephant herd ruins a crop or a tiger kills livestock, government agents are usually forced to destroy the animal.

▼ **Figure 12.10 Indian Women Collecting Firewood** Across most rural areas of South Asia, firewood is still the main cooking fuel. Women are usually tasked with gathering wood, a chore that can take several hours a day. Here we see women returning home with bundles of wood in the Indian state of Rajasthan.

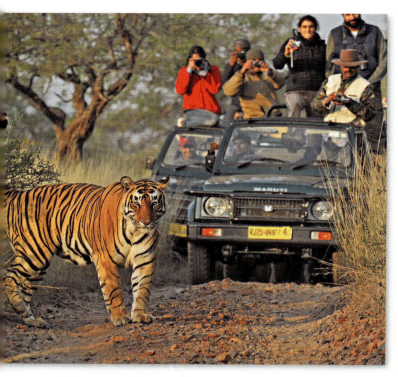

▲ **Figure 12.11 Human–Tiger Interactions** Tourists photograph a tiger in Ranthambhore National Park in the Indian state of Rajasthan. Ranthambhore is one of the best places in South Asia to see wild tigers.

REVIEW

12.1 Why is the monsoon so crucial to life in South Asia?

12.2 Why is flooding such an important environmental issue in Bangladesh and adjacent areas of northeastern India?

12.3 Describe the physical features of South Asia that make the region especially vulnerable to climate change.

■ **KEY TERMS** Indian subcontinent, monsoon, orographic rainfall, cyclone

---

Population and Settlement: The World's Emerging Demographic Core

South Asia recently surpassed East Asia to become the world's most populous region, and India will probably overtake China to become the world's most populous country in 2022 (Table 12.1). South Asia fertility levels have dropped markedly in recent years, but population growth continues. And although South Asia has made remarkable agricultural gains since the 1960s, there is still widespread concern about the region's ability to feed its expanding population.

The Geography of Population Growth

South Asia's recent decline in birthrates shows distinct geographical patterns. All the states of southern and western India, along with Sri Lanka, are now at or below replacement level and should soon see population stabilization. In Pakistan and the Ganges Valley of India, however, birth rates remain elevated. All South Asian countries have established family-planning programs, but commitment to these policies varies widely.

India Widespread concern over India's population growth began in the 1960s. Family-planning measures, along with general economic and social development, have been largely successful; the total fertility rate (TFR) dropped from 6 in the 1950s to the current rate of 2.3, and is expected to reach replacement level by 2020. Fertility rates vary widely within India, from 1.6 in West Bengal to a high 3.4 in Bihar. A strong relationship exists between women's education and family planning; where female literacy has increased most dramatically, fertility levels have rapidly declined.

A distinct cultural preference for male children is found in most of South Asia, which complicates family planning. Abortion rates of female fetuses are much higher than those of males, even though the practice of sex-selective abortion is illegal.

TABLE 12.1	Population Indicators							Mastering Geography™
Country	Population (millions) 2016	Population Density (per square kilometer)[1]	Rate of Natural Increase (RNI)	Total Fertility Rate	Percent Urban	Percent <15	Percent >65	Net Migration (Rate per 1000)
Bangladesh	162.9	1,222	1.5	2.3	34	33	6	−2
Bhutan	0.8	20	1.3	2.1	39	31	5	2
India	1,328.9	436	1.5	2.3	33	29	6	0
Maldives	0.4	1,337	1.5	2.5	46	27	5	1
Nepal	28.4	197	1.5	2.3	20	31	6	−1
Pakistan	203.4	240	2.3	3.7	39	36	4	−1
Sri Lanka	21.2	331	1.0	2.1	18	25	8	−4

Source: Population Reference Bureau, *World Population Data Sheet*, 2016.

[1]World Bank, *World Development Indicators*, 2016.

Login to Mastering Geography™ **& access** MapMaster **to explore these data!**

1) Does the urbanization rate among among South Asia countries show any correlation with fertility rates?

2) Is there any relationship between total fertility rates and population density in South Asia?

▶ **Figure 12.12 Population of South Asia** Except for the desert areas of the west and the high mountains of the north, South Asia is a densely populated region. Particularly high densities are found on the fertile plains along the Indus and Ganges rivers and in India's coastal lowlands. In rural areas, the population is typically clustered in villages, often located near water sources, such as streams, wells, canals, or small tanks that store water between monsoon rains.

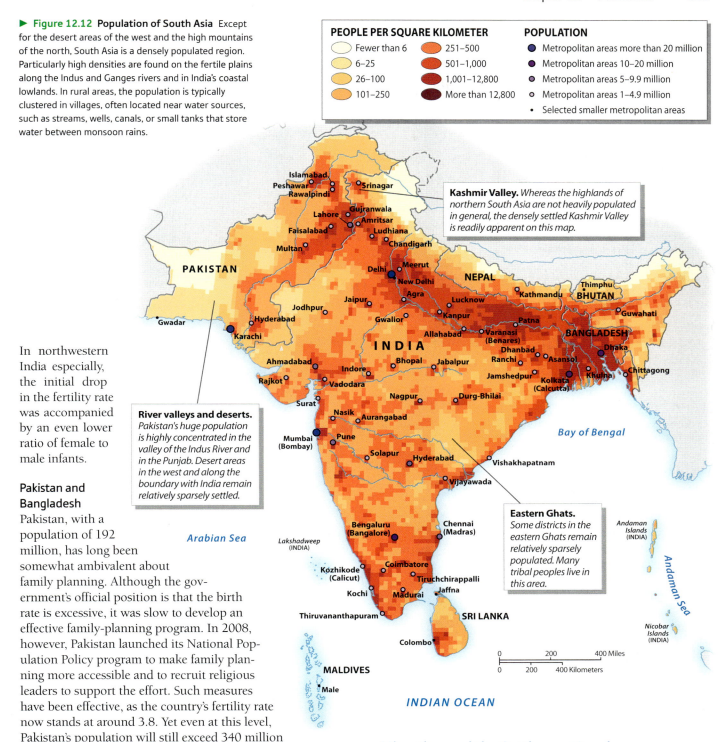

PEOPLE PER SQUARE KILOMETER

Fewer than 6	251–500
6–25	501–1,000
26–100	1,001–12,800
101–250	More than 12,800

POPULATION

- Metropolitan areas more than 20 million
- Metropolitan areas 10–20 million
- Metropolitan areas 5–9.9 million
- Metropolitan areas 1–4.9 million
- Selected smaller metropolitan areas

Kashmir Valley. *Whereas the highlands of northern South Asia are not heavily populated in general, the densely settled Kashmir Valley is readily apparent on this map.*

River valleys and deserts. *Pakistan's huge population is highly concentrated in the valley of the Indus River and in the Punjab. Desert areas in the west and along the boundary with India remain relatively sparsely settled.*

Eastern Ghats. *Some districts in the eastern Ghats remain relatively sparsely populated. Many tribal peoples live in this area.*

In northwestern India especially, the initial drop in the fertility rate was accompanied by an even lower ratio of female to male infants.

Pakistan and Bangladesh

Pakistan, with a population of 192 million, has long been somewhat ambivalent about family planning. Although the government's official position is that the birth rate is excessive, it was slow to develop an effective family-planning program. In 2008, however, Pakistan launched its National Population Policy program to make family planning more accessible and to recruit religious leaders to support the effort. Such measures have been effective, as the country's fertility rate now stands at around 3.8. Yet even at this level, Pakistan's population will still exceed 340 million by 2050.

Densely settled Bangladesh, with a population about half that of the United States packed into an area smaller than the state of Wisconsin, has made huge strides in population stabilization. As recently as 1975, its TFR was 6.3, but dropped to 2.2 by 2011 and is currently 2.3. Bangladesh's family-planning success can be attributed to strong governmental support through radio messages and billboards. Also important are more than 35,000 women fieldworkers who take information about family planning into every village in the country.

Migration and the Settlement Landscape

The most densely settled areas of South Asia still coincide with zones of fertile soils and dependable water supplies (**Figure 12.12**). The largest rural populations are found in the core area of the Ganges and Indus river valleys and on India's coastal plains. Settlement is less dense over much of the Deccan Plateau and is relatively sparse in the arid lands of the northwest. Although most of South Asia's northern mountains are too rugged and high to support heavy human settlement, major population clusters occur in the

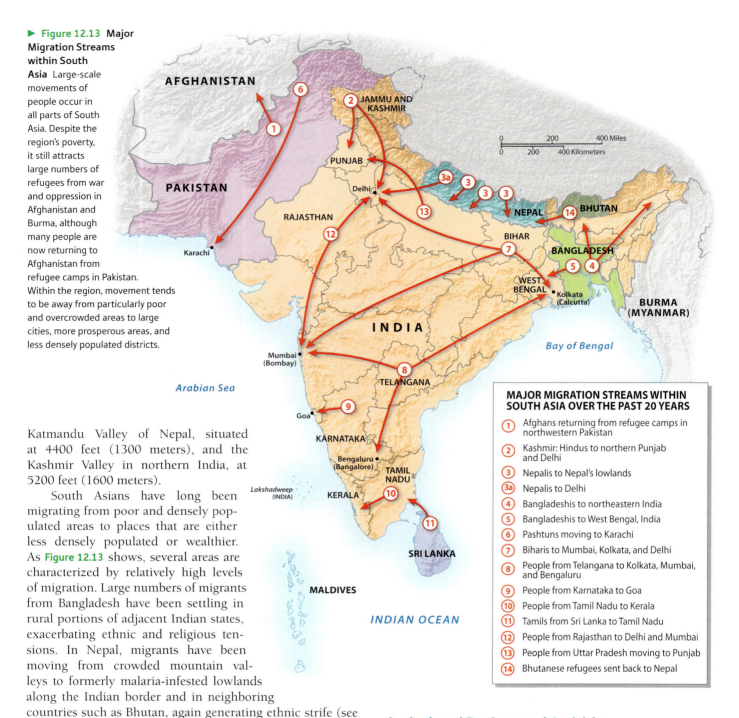

▶ **Figure 12.13 Major Migration Streams within South Asia** Large-scale movements of people occur in all parts of South Asia. Despite the region's poverty, it still attracts large numbers of refugees from war and oppression in Afghanistan and Burma, although many people are now returning to Afghanistan from refugee camps in Pakistan. Within the region, movement tends to be away from particularly poor and overcrowded areas to large cities, more prosperous areas, and less densely populated districts.

MAJOR MIGRATION STREAMS WITHIN SOUTH ASIA OVER THE PAST 20 YEARS

1. Afghans returning from refugee camps in northwestern Pakistan
2. Kashmir: Hindus to northern Punjab and Delhi
3. Nepalis to Nepal's lowlands
3a. Nepalis to Delhi
4. Bangladeshis to northeastern India
5. Bangladeshis to West Bengal, India
6. Pashtuns moving to Karachi
7. Biharis to Mumbai, Kolkata, and Delhi
8. People from Telangana to Kolkata, Mumbai, and Bengaluru
9. People from Karnataka to Goa
10. People from Tamil Nadu to Kerala
11. Tamils from Sri Lanka to Tamil Nadu
12. People from Rajasthan to Delhi and Mumbai
13. People from Uttar Pradesh moving to Punjab
14. Bhutanese refugees sent back to Nepal

Katmandu Valley of Nepal, situated at 4400 feet (1300 meters), and the Kashmir Valley in northern India, at 5200 feet (1600 meters).

South Asians have long been migrating from poor and densely populated areas to places that are either less densely populated or wealthier. As **Figure 12.13** shows, several areas are characterized by relatively high levels of migration. Large numbers of migrants from Bangladesh have been settling in rural portions of adjacent Indian states, exacerbating ethnic and religious tensions. In Nepal, migrants have been moving from crowded mountain valleys to formerly malaria-infested lowlands along the Indian border and in neighboring countries such as Bhutan, again generating ethnic strife (see *People on the Move: Bhutanese Refugees in Nepal and the United States*). Sometimes migrants are forced out by war; over the past 20 years, sizable streams of people from northern Sri Lanka and from Kashmir have sought security away from their battle-scarred homelands.

South Asia is one of the world's least urbanized regions, with only about one-third of its population living in cities, but large numbers are streaming into the region's rapidly growing metropolitan areas. As a result, India's urbanization rate is expected to exceed 40 percent by 2030. Such urbanization often stems as much from desperate conditions in the countryside as from the attractions of city life. As farms begin to mechanize, farm laborers often have no choice but to seek work in urban areas.

Agricultural Regions and Activities

South Asian agriculture has historically been less productive than that of East Asia. According to a 2012 UN report, the average rice yield in India is 2.3 metric tons per hectare as opposed to China's 6.5 metric tons. Although the reasons behind such poor yields are complex, many experts cite the relatively low social status of most cultivators and the fact that much farmland has long been controlled by absentee landlords. Some scholars also blame British colonialism, which emphasized export crops for European markets.

Whatever the causes, low agricultural yields in the context of a huge, hungry, and rapidly growing population constitute a pressing problem. Since the 1970s, however, agricultural production has grown faster than the population, primarily

People on the Move

Bhutanese Refugees in Nepal and the United States

Most Americans would be surprised to learn that a sizable percentage of the refugees admitted into their country come from tiny Bhutan. From 2008 to early 2014, some 75,000 Bhutanese refugees resettled in the United States. Settling primarily in New York, Pennsylvania, Texas, and several other states, many of these immigrants have faced profound difficulties in adjusting to their new lives, suffering from a suicide rate roughly twice the national average.

Nepalese People in Bhutan It might seem odd that Bhutan, a lightly populated country noted for its commitment to sustainable development, would generate such large numbers of refugees. The reason lies in ethnic conflict, as the refugees are almost all of Nepalese origin. Their ancestors left Nepal mainly in the late 1800s and early 1900s, settling in the sparsely inhabited foothills of southern Bhutan. At first, the Bhutanese authorities welcomed this influx, hoping to gain tax revenue from the development of the frontier zone. But attitudes changed by the second half of the 20th century. Bhutan's rulers then began to view the Nepalese as a foreign population that would never assimilate, thus threatening Bhutan's cultural integrity.

In the 1980s, the government of Bhutan began to impose harsh regulations on people of Nepalese origin. A "one nation, one people" policy sought cultural homogenization. Newcomers were forced to wear Bhutanese clothing and could no longer use the Nepalese language in their schools. Those who could not prove their residency in Bhutan since 1958 were deemed illegal immigrants. Conflicts flared as these restrictions were imposed, and by the 1990s tens of thousands of people were fleeing Bhutan. Many claim to have been forced out of the country by the Bhutanese army, which first made them sign "voluntary migration forms."

From Refugee Camps to Resettlement Most of these refugees ended up in seven camps in southeastern Nepal, run jointly by Nepal's government and the Office of the UN High Commissioner for Refugees (UNHCR) (**Figure 12.3.1**). Initially, camp conditions were grim, with high rates of malnutrition and widespread suffering from measles, scurvy, tuberculosis, and malaria. Conditions had improved greatly by the turn of the millennium, and today such camps as Beldangi, which houses almost 20,000 people, are widely viewed as models with relatively good schools, security, and sanitation. But a refugee camp is still a refugee camp, and most of their denizens want out.

But where can these refugees go? Bhutan and Nepal conducted 15 rounds of talks up to 2011 without agreeing on resettlement plans. Nepal claims to be too poor to accept the refugees, and Bhutan says that

▲ **Figure 12.3.1 Bhutanese Refugee Camp in Nepal** Large numbers of people of Nepalese origin have been expelled from Bhutan. Many are forced to live in refugee camps in Nepal, such as the Beldangi 2 Camp seen here. As of March 13, 2015, more than 18,000 people resided at this camp.

most of them would pose a national security threat. The one thing that both countries could agree on was a preference for settlement outside of the region. Several foreign countries then agreed to accept Bhutanese refugees, including the United States, Australia, Canada, Norway, Netherlands, Denmark, and New Zealand. The majority have settled in the United States, with most of the rest moving to Canada and Australia.

1. Why might the government of Bhutan think that their country is threatened by the presence of a few tens of thousands of people of Nepalese origin?

2. List some difficulties that many of these refugees likely face in the United States.

---

because of the **Green Revolution**, which refers to cultivation techniques based on hybrid crop strains and the heavy use of industrial fertilizers and pesticides. Because the Green Revolution also carries significant social and environmental costs, it remains controversial, as will be discussed shortly.

Crop Zones South Asia can be divided into several distinct agricultural regions, all with different problems and potentials. The most fundamental division is among the three primary subsistence crops of rice, wheat, and millet.

Rice is the main crop and foodstuff in the lower Ganges Valley, along the lowlands of India's eastern and western coasts, in the delta lands of Bangladesh, in Pakistan's lower Indus Valley, and in Sri Lanka. This distribution reflects the

large volume of irrigation water needed to grow rice. The amount of rice grown in South Asia is impressive: India ranks behind only China in rice production, and Bangladesh is the sixth largest producer.

Wheat is the principal crop in the northern Indus Valley and in the western half of India's Ganges Valley. South Asia's "breadbasket" lies in the northwestern Indian states of Punjab and Haryana, along with adjacent areas in Pakistan. Here the Green Revolution has been particularly successful in increasing grain yields. In the less-fertile areas of central India, millet and sorghum are the main crops, along with root crops such as manioc (**Figure 12.14**). Wheat and rice are the preferred staples throughout South Asia, and it is generally poorer people who consume "rough" grains such as the various millets.

▲ **Figure 12.14 Millet Growing** The various grain species known as millet form an essential part of the diet in the poorer and drier parts of South Asia. The people in this photograph are winnowing the finger-millet harvest in the Indian state of Karnataka.

Many other crops are widely cultivated in South Asia—some commercially, others for local subsistence. Oil seeds, such as sesame and peanuts, are grown in semiarid districts, while Sri Lanka and the Indian state of Kerala are noted for their coconut groves, spice gardens, and tea plantations. In both Pakistan and west central India, cotton is widely cultivated, and Bangladesh has long supplied most of the world's jute, a tough fiber used in rope manufacture.

Milk in the Northwest; Meat in the Northeast Many South Asians receive inadequate protein, and meat consumption is extremely low. Due primarily to religious restrictions, vegetarianism is widespread in India. But very few Indians follow a vegan diet; milk and other dairy products, derived from both cows and water buffalos, are avidly consumed in some areas, especially in the northwest (**Figure 12.15**). Indeed, India is the largest milk producer in the world, recently surpassing the entire European Union, and Pakistan ranks fourth. Milk is India's leading agricultural commodity, produced on some 75 million dairy farms. The Indian government has provided support for the dairy industry since 1970 through its "Operation Flood," which doubled per capita milk consumption.

But milk drinking and dairy consumption is less pronounced in southern South Asia, and is relatively rare in the northeast. The region's dairy disparity is partially based on genetics. In northeastern South Asia, the majority of people are lactose intolerant, and cannot digest milk as adults. In this area the lack of protein is partly made up by consuming meat, eggs, and fish. Vegetarianism is much weaker among Hindus in northeastern India than among those in the west. In Bengal, even Brahmins—whose dietary restrictions are pronounced—are allowed to eat fish, which can seem shocking to those from western India.

The Green Revolution Originating during the 1960s in agricultural research stations established by international development agencies, the Green Revolution allowed grain production in South Asia to keep up with population growth. Simply applying fertilizer to attain higher yields did not work with traditional crop varieties because the plants would grow too tall and then fall to the ground before maturation. The solution was to cross-breed new "dwarf" crop strains that respond to heavy fertilization by producing more grain rather than longer stems.

By the 1970s, it was clear that these efforts had achieved initial goals. The more prosperous farmers of the Punjab quickly adopted the new "miracle wheat" varieties, solidifying the state's position as the region's breadbasket. Green Revolution rice strains also were adopted in the more humid areas. As a result, India more than doubled its annual grain production between 1970 and the mid-1990s.

The Green Revolution was clearly an agricultural success, but many argue that its ecological and social effects have been disastrous. Frequent applications of harmful pesticides are often necessary because the new crop varieties lack natural resistance to local plant diseases and insects. They also need large quantities of industrial fertilizer, which is both expensive and polluting. A 2015 study concluded that toxic nitrates exceed permissible levels in wells sampled in 387 of India's 676 districts.

Social problems have also followed the Green Revolution. In many areas, only the more prosperous farmers can afford the new seed strains, along with the necessary irrigation equipment, farm machinery, fertilizers, and pesticides (**Figure 12.16**). As a result, poorer farmers often go into debt and can be forced off the land when they fail to repay their loans. This desperate agricultural situation has generated an epidemic of farmer suicides. Official statistics indicate that roughly 300,000 Indian famers committed suicide between 1995 and 2015. In 2016, India announced a new US$1.3 billion crop insurance scheme to reduce the suicide rate.

Future Food Supply The Green Revolution has fed South Asia's expanding population for nearly 50 years, but whether it can continue doing so remains unclear. Many of the crop improvements have seemingly exhausted their potential, and much of South Asia's agricultural economy is currently in a state of crisis.

Optimists believe, however, that South Asia's food production could be substantially increased. Further improvements in highways and railroads, for example, would reduce waste and increase profit margins for struggling farmers. Higher yields would also result if the techniques pioneered in northwestern India were applied throughout the country. In Punjab, for example, the average rice farm produces

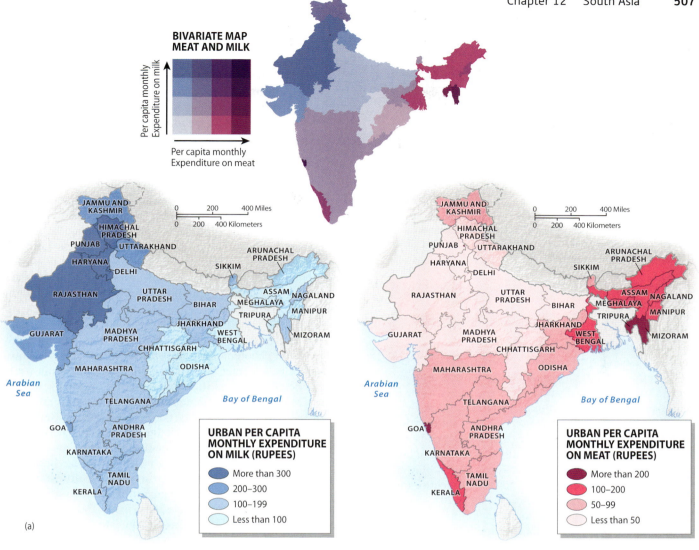

**BIVARIATE MAP
MEAT AND MILK**

Per capita monthly Expenditure on milk

Per capita monthly Expenditure on meat

URBAN PER CAPITA MONTHLY EXPENDITURE ON MILK (RUPEES)

- More than 300
- 200–300
- 100–199
- Less than 100

(a)

URBAN PER CAPITA MONTHLY EXPENDITURE ON MEAT (RUPEES)

- More than 200
- 100–200
- 50–99
- Less than 50

(b)

▲ **Figure 12.15 Milk vs. Meat in India** Dairy products are the most important source of animal protein in western South Asia. India's milk production has surged in recent decades as farmers embrace more productive techniques. Here we see an agricultural teacher demonstrating a new milking machine.

Explore the **Tastes** of North Indian Food

http://goo.gl/Yf8pzQ

▼ **Figure 12.16 Green Revolution Farming** "Miracle" wheat strains that have increased yields in the Punjab area have made this region the breadbasket of South Asia. Similarly, intensive methods are used in the Punjab for other crops, such as the tomatoes being grown under plastic covers. Increased production, however, has led to both social and environmental problems.

▲ **Figure 12.17 Mumbai Hutments** Hundreds of thousands of people in Mumbai live in crude hutments, with no sanitary facilities, built on formerly busy sidewalks. Hutment construction is forbidden in many areas, but wherever it is allowed, sidewalks quickly disappear.

3130 kilograms per acre, whereas the yield in Bihar is only 1370 kilograms per acre. Green Revolution techniques, moreover, might be profitably applied to secondary grain crops, and genetic engineering could usher in a second wave of increased crop yields. Most environmentalists, however, see serious dangers in this new technology.

Another option is expanded water delivery, as many fields remain unirrigated. Irrigation, however, brings its own problems, such as a drop in groundwater levels. In much of Pakistan and northwestern India, moreover, soil **salinization**—the buildup of salt in agricultural fields, is a major constraint (see Figure 12.7). An estimated 27 million acres (11 million hectares) in Pakistan are now too salty or alkaline for effective farming. Efforts are being made, however, to reduce the levels of salt and alkalinity in South Asian soils. A 2015 report from the Indian state of Telangana found that crop yields could be almost doubled by using established methods of reducing salt and improving the management of plant nutrients.

Urban South Asia

Although not heavily urbanized (see Table 12.1), South Asia does have some of the world's largest urban agglomerations. India alone lists some 53 cities with populations greater than a million, most of which are growing rapidly.

Because of this rapid growth, most South Asian cities have staggering problems with homelessness, poverty, congestion, water shortages, air pollution, and sewage disposal. Kolkata's homeless are legendary, with perhaps half a million people sleeping on the streets each night. In that city and others, sprawling squatter settlements, or **bustees**, mushroom in and around urban areas, providing temporary shelter for many urban migrants. A brief survey of the major cities illustrates the problems and prospects of the region's urban areas.

Mumbai Formerly Bombay, Mumbai vies with Delhi and Karachi for the position of South Asia's largest city. It is also India's financial, industrial, and commercial center. Mumbai generates most of India's foreign trade, has long been a manufacturing center, and is the focus of India's film industry—the world's largest. Mumbai's economic vitality draws people from all over India, resulting in ethnic tensions.

The city's restricted space has pushed most of Mumbai's growth to the north and east of the historic city. Building restrictions in the downtown area have resulted in skyrocketing commercial and residential rents—some of the highest in the world. Even members of the city's thriving middle class have difficulty finding adequate housing. Hundreds of thousands of less-fortunate immigrants live in "hutments," crude shelters built on formerly busy sidewalks (**Figure 12.17**). The least fortunate sleep on the street or in simple plastic tents, often placed along busy roadways.

Delhi India's sprawling capital, Delhi has roughly 25 million people in its greater metropolitan area. Delhi (or old Delhi), a former Muslim capital, is a congested town of tight neighborhoods; the landscape of New Delhi, in contrast, is a city of wide boulevards, monuments, parks, and expensive residential areas (**Figure 12.18**). New Delhi was born as a planned city when the British moved the colonial capital from Kolkata, (Calcutta) in 1911. Located here are embassies, luxury hotels, government office buildings, and airline offices necessary for a vibrant political capital. Rapid growth, along with the government's inability to control auto and industrial emissions, made Delhi one of the world's ten most polluted cities by the late 1990s.

▼ **Figure 12.18 New Delhi Ceremonial Landscape** India's capital of New Delhi, like Pakistan's Islamabad, is a planned city noted for its wide boulevards that are occasionally used for ceremonial purposes. Here we see several Indian Army marching contingents during a Republic Day parade in the city.

Explore the **Sights** of New Delhi's Rajpath

http://goo.gl/ObTVst

Kolkata To many, Kolkata (Calcutta) embodies the problems faced by rapidly growing cities in less developed countries. Not only is homelessness widespread, but this metropolitan area of some 15 million also falls far short of supplying its residents with water, power, or sewage treatment. Electricity is so inadequate that every hotel, restaurant, shop, and small business must have some sort of standby power system. During the wet season, many streets are routinely flooded.

Kolkata faces continued rapid growth as migrants pour in from the countryside, a mixed Hindu-Muslim population that generates ethnic rivalry, a troubled economic base, and an overloaded infrastructure. Clearly, Kolkata's future will have problems. Yet it remains a culturally vibrant city noted for its fine educational institutions, theaters, and publishing firms.

Dhaka The capital and major city of Bangladesh, Dhaka has experienced rapid growth due to migration from the surrounding countryside. In 1971, when Bangladesh gained independence from Pakistan, Dhaka had about 1 million inhabitants; today its greater metropolitan area numbers almost 17 million. Dhaka combines the administrative functions of government with the largest industrial concentration in Bangladesh. Cheap and abundant labor has made the city a global center for clothing, shoe, and sports equipment manufacturing.

Karachi Pakistan's largest city, Karachi, counts 24 million in its metropolitan area (**Figure 12.19**). Karachi was Pakistan's capital before the decision was made in 1960 to relocate the government to Islamabad. Karachi lies adjacent to the Arabian Sea, and its excellent harbors have long made it an important trading center. Today, 95 percent of Pakistan's foreign trade goes through Karachi's two main ports, and 90 percent of its financial and multinational companies are headquartered here. Karachi's economic centrality in modern Pakistan makes the city's increasing ethnic and religious tensions a major concern for Pakistan's leaders.

Islamabad Upon independence, Pakistan's leaders determined that Karachi was too far from the center of the country, requiring an entirely new capital. This planned city would make a statement through its name—Islamabad—about the religious foundation of Pakistan. Located close to the contested region of Kashmir, it would also make a geopolitical statement. Such a city is termed a **forward capital**, one that signals—both symbolically and geographically—a country's intentions. By building its new capital in the north, Pakistan sent a message that it would not abandon its claims to the portion of Kashmir controlled by India. Islamabad is closely linked to Rawalpindi, a major military center for centuries. These two cities form a single metropolitan region of about 4.5 million people, but are completely different in appearance and character. To avoid congestion, planners designed Islamabad around self-sufficient sectors, each with its own government buildings, residences, and shops.

The closest parallel in India to Islamabad is Chandigarh, the modern planned city that serves as the capital of two Indian states: Punjab and Haryana. All told, the cities of India and Pakistan have a similar feel, as might be expected considering the two countries' common historical and cultural backgrounds.

REVIEW

12.4 Why has the Green Revolution been so controversial, considering the fact that it has greatly increased South Asia's food supply?

12.5 Why is the Punjab region usually viewed as South Asia's "breadbasket"?

12.6 List the major advantages and disadvantages of the growth of South Asia's megacities.

■ **KEY TERMS** Green Revolution, salinization, bustees, forward capital

▼ **Figure 12.19 Karachi Cityscape** With a population of more than 24 million in its metropolitan area, Karachi is by far the largest and most cosmopolitan city in Pakistan. This strife-torn port city was also Pakistan's capital from 1947 until the early 1960s, when the seat of government was gradually transferred to Islamabad.

Cultural Coherence and Diversity: A Common Heritage Undermined by Religious Rivalries

In historical terms, South Asia forms a well-defined cultural region. A thousand years ago, virtually the entire area was united by ideas and social institutions associated with Hinduism. The subsequent arrival of Islam added a new religious dimension without undercutting the region's cultural unity. British imperialism subsequently imposed several cultural features over the entire region, from the widespread use of English to a common passion for cricket. Since the mid-20th century, however, religious and political strife has intensified, leading some to question whether South Asia can still be conceptualized as a culturally coherent world region.

India has been a secular country since its inception, with the Congress Party, its early guiding political organization, struggling to keep politics and religion separate. In doing so, it relied heavily on support from Muslims. Since the 1980s, this secular political tradition has come under increasing pressure from **Hindu nationalism**. Hindu nationalists promote the religious values of Hinduism as the essential fabric of Indian society.

Hindu nationalists gained considerable political power both at the federal level and in many Indian states through the Bharatiya Janata Party (BJP), leading to widespread pressure against the country's Muslim minority in the mid-1990s. In several high-profile instances, Hindu mobs demolished Muslim mosques that were allegedly built on the sites of ancient Hindu temples. In 2002, more than 2000 people—mostly Muslims—were killed during religious riots in the state of Gujarat. Over the past decade, however, the Hindu nationalist movement has moderated somewhat, and efforts are being made to promote religious understanding (**Figure 12.20**).

▼ **Figure 12.20 Religious Tension and Cooperation** Northern India has recently been the site of occasionally intense communal conflicts pitting Hindus against Muslims. Activists from both sides of the divide, however, are seeking to develop peaceful relations, and with some success. Here we see an Indian mother and her two-year-old son participating in a human chain to mark a "Communal Harmony Day" in New Delhi.

In predominantly Muslim Pakistan, rising Islamic fundamentalism has generated severe conflict. Radical religious leaders want to make Pakistan a fully religious state under Islamic law, a plan rejected by the country's secular intellectuals and international businesspeople. The government has attempted to intercede between the two groups, but is often viewed as biased toward the Islamists. Antiblasphemy laws, for example, have been used to persecute members of the country's small Hindu and Christian communities, as well as liberal Muslim writers. Large amounts of money from Saudi Arabia, moreover, have gone toward developing fundamentalist religious schools that are reputed to encourage extremism.

Religious and political activists in South Asia often argue that the current struggles reflect deeply rooted historical divisions. But most scholars regard religious conflict as a recent development. To weigh these contrasting claims, we must examine the cultural history of South Asian civilization.

Origins of South Asian Civilizations

The roots of South Asian culture extend back to the Indus Valley civilization, one of the world's first urban-oriented cultures, which flourished from 3300 to 1800 BCE in what is now Pakistan and northwestern India. This remarkable society is poorly understood, as its script has not been deciphered. Despite its long period of power, the Indus Valley civilization vanished almost entirely in the second millennium BCE, after which the record grows dim. By 800 BCE, however, a new urban focus had emerged in the middle Ganges Valley (**Figure 12.21**). The social, religious, and intellectual customs of this civilization eventually spread throughout the lowlands of South Asia.

Hindu Civilization The religious complex of this ancient South Asian civilization was an early form of **Hinduism**, a complicated faith that incorporates diverse forms of worship and lacks any standard system of beliefs (**Figure 12.22**). Certain deities are recognized, however, by all believers, as is the notion that these various gods are all manifestations of a single divine entity. All Hindus, moreover, share a set of epic stories, usually written in the sacred language of Sanskrit. One of its hallmarks is a belief in the transmigration of souls from being to being through reincarnation, wherein one's actions in the physical world influences the course of these future lives.

Scholars once confidently argued that Hinduism originated from the fusion of two distinct religious traditions: the mystical beliefs of the subcontinent's indigenous inhabitants and the sky-god religion of the Indo-European invaders who swept into the region from Central Asia in the second millennium BCE. Such a scenario also proved convenient for explaining India's **caste system**, the strict division of society into hierarchically ranked hereditary groups. The elite invaders, according to this theory, separated themselves from the people they had defeated through an elaborate

▶ **Figure 12.21 Early South Asian Civilizations** The roots of South Asian culture may extend back 5000 years to an Indus Valley civilization based on irrigated agriculture and vibrant urban centers. What happened to that civilization remains a topic of conjecture because the archaeological record grows dim by 1800 BCE. Later, a new urban focus emerged in the Ganges Valley, from which social, religious, and intellectual influences spread throughout lowland South Asia.

system of social division. Recent research, however, indicates that the caste system, like Hinduism itself, emerged through more gradual social and cultural evolution. Critics also note that the British imperial authorities tended to harden caste divisions in order to maintain power over the large Indian population.

Caste is actually an imprecise term for denoting the complex social order of the Hindu world. The word itself, of Portuguese origin, combines two distinct local concepts: *varna* and *jati*. *Varna* refers to the ancient fourfold social hierarchy of Hinduism, whereas *jati* refers to the hundreds of local endogamous ("marrying within") groups that exist at each varna level (jati groups are thus often called *subcastes*). Many argue that the essence of the caste system is the notion of social pollution: The lower one's position in the hierarchy, the more potentially polluting one's body supposedly is. Members of higher castes were traditionally not supposed to eat or drink with members of lower castes.

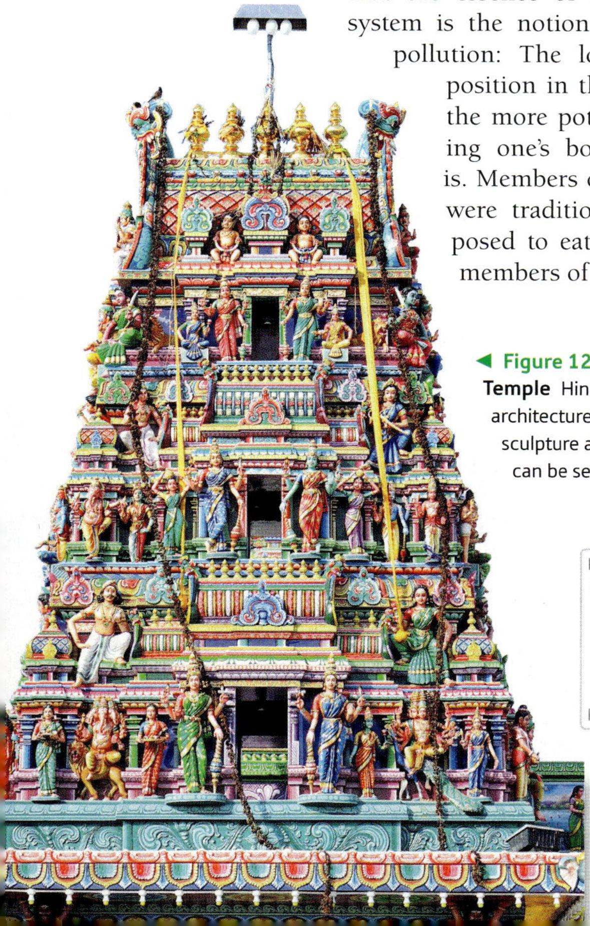

◀ **Figure 12.22 Hindu Temple** Hinduism's religious architecture often entails lavish sculpture and bright colors, as can be seen on this temple.

Explore the **Sights** of Sacred Pushkar Lake

http://goo.gl/EORY1w

Main Caste Groups Three varna groups constitute the traditional elite of Hindu society. At the apex sit the Brahmins, members of the traditional priestly caste. Brahmins perform the high rituals of Hinduism and form India's traditional intellectual elite. Below the Brahmins are the Kshatriyas, members of the warrior or princely caste. In premodern India, this group ruled the Hindu kingdoms. Next stand the Vaishyas, members of the traditional merchant caste. In earlier centuries, a near monopoly on long-distance trade and money lending in northern India gave many Vaishyas ample opportunities to accumulate wealth. The precepts of vegetarianism and nonviolence are particularly strong among certain merchant subcastes of western India. One prominent representative of this tradition was Mohandas Gandhi, the founder of modern India.

The majority of India's population fits into the fourth varna category, the Sudras. The Sudra caste is composed of an especially large array of subcastes (jati), most of which originally reflected occupational groupings. Most Sudra subcastes were traditionally associated with peasant farming, but others were based on craft occupations, including those of barbers, smiths, and potters.

The Brahmins, Kshatriyas, Vaishyas, and Sudras form the basic fourfold scheme of caste society, but another sizable group has long stood outside the varna system altogether—the so-called *untouchables,* or **dalits**, as they are now called. Dalits were not traditionally allowed to enter Hindu temples and their low status was derived historically from "unclean" occupations, such as leather working, scavenging, latrine cleaning, and swineherding.

Buddhism The early caste system that developed in the Ganges Valley civilization was challenged from within by **Buddhism**. Siddhartha Gautama, the Buddha, was born in 563 BCE to an elite caste. He rejected the life of wealth and power, however, and sought instead to attain enlightenment, or mystical union with the cosmos. He preached that the path to such "nirvana" was open to all, regardless of social position. His followers eventually established Buddhism as a new religion. Buddhism spread through most of South Asia, becoming something of an official faith under the Mauryan Empire, which ruled much of the subcontinent in the 3rd century BCE. Later centuries saw Buddhism expand through most of East, Southeast, and Central Asia.

But for all of its successes abroad, Buddhism never replaced Hinduism in India. It remained focused on monasteries rather than spreading throughout the wider society. Many Hindu priests, moreover, struggled against the new faith, although many of Buddhism's philosophical ideas were embraced and folded into Hinduism. By 500 CE, Buddhism was on the retreat throughout South Asia, and within another 500 years it had virtually disappeared. The only major exceptions were the island of Sri Lanka and the high Himalayas, both of which remain mostly Buddhist to this day.

Arrival of Islam The next major challenge to Hindu society—Islam—came from outside the region (see Chapter 7, Figure 7.27). Arab armies conquered the lower Indus Valley around 700 CE, but advanced no farther. Then around 1000 CE, Turkish-speaking Muslims began to move in from Central Asia. At first, they merely raided, but eventually settled and ruled on a permanent basis. By the 1300s, most of South Asia lay under Muslim power, although Hindu kingdoms persisted in southern India and in the arid lands of northwestern India. Later, during the 16th and 17th centuries, the **Mughal Empire** dominated much of the region from its power center in the upper Indus–Ganges Basin (**Figure 12.23**).

At first, Muslims formed a small ruling elite, but over time increasing numbers of Hindus converted to the new religion, particularly those from lower castes. Conversions were most pronounced in the northwest and northeast, and eventually the areas now known as Pakistan and Bangladesh became predominantly Muslim.

At first glance, Islam and Hinduism are strikingly divergent faiths. Islam is resolutely monotheistic, austere in its ceremonies, and spiritually egalitarian (all believers stand in the same relationship to God). Hinduism, by contrast, is polytheistic (at least on the surface), lavish in its rituals, and caste-structured. Because of these profound differences, Hindu and Muslim communities in South Asia are sometimes viewed as utterly distinct, sharing the same region, but not the same culture. Increasingly, such a view is expressed in South Asia itself, especially in Pakistan.

However, overemphasizing the separation of Hindu and Muslim communities risks missing much of what is historically distinctive about South Asia. Until the 20th century, Hindus and Muslims usually coexisted on friendly terms; the two faiths stood side by side for hundreds of years and influenced each other in many ways. Moreover, aspects of caste organization have persisted among South Asian Muslims, just as they have among India's Christians.

Contemporary Geographies of Religion

South Asia, as we have seen, has a predominantly Hindu heritage overlain by a substantial Muslim imprint. Such a picture fails, however, to capture the enormous diversity of modern religious expression in contemporary South Asia. The following discussion looks specifically at the geographical patterns of the region's main faiths (**Figure 12.24**).

Hinduism Fewer than 1 percent of the people of Pakistan are Hindu, and in Bangladesh and Sri Lanka, Hinduism is a distinctly minority religion. Almost everywhere in India, however—and in Nepal, as well—Hinduism is the faith of the majority. In east-central India, more than 95 percent of the population is Hindu. Hinduism is itself a geographically complicated religion, with aspects of faith varying from place to place and from caste to caste.

The caste aspect of Hinduism is undergoing significant changes in contemporary India. Its original occupational structure has long been undermined by the necessities of a modern economy, and various social reforms have chipped away at the discrimination that it embodies. The dalit community itself has produced several notable national leaders

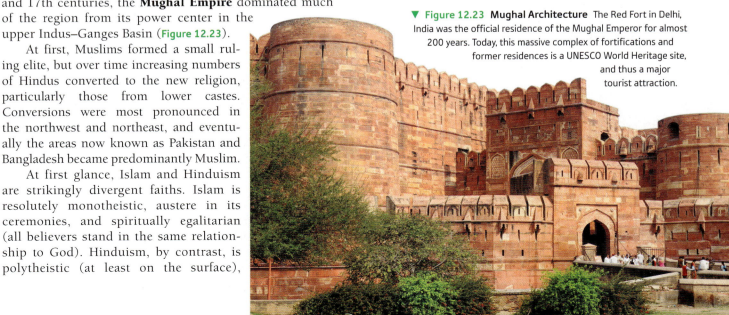

▼ **Figure 12.23 Mughal Architecture** The Red Fort in Delhi, India was the official residence of the Mughal Emperor for almost 200 years. Today, this massive complex of fortifications and former residences is a UNESCO World Heritage site, and thus a major tourist attraction.

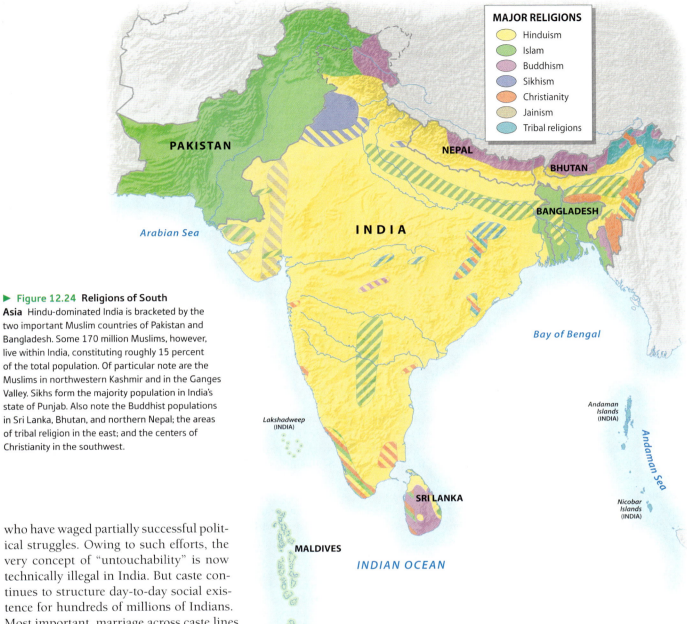

PAKISTAN

Arabian Sea

NEPAL

BHUTAN

BANGLADESH

I N D I A

Bay of Bengal

Lakshadweep
(INDIA)

*Andaman
Islands*
(INDIA)

Andaman Sea

*Nicobar
Islands*
(INDIA)

SRI LANKA

MALDIVES

INDIAN OCEAN

▶ **Figure 12.24** **Religions of South Asia** Hindu-dominated India is bracketed by the two important Muslim countries of Pakistan and Bangladesh. Some 170 million Muslims, however, live within India, constituting roughly 15 percent of the total population. Of particular note are the Muslims in northwestern Kashmir and in the Ganges Valley. Sikhs form the majority population in India's state of Punjab. Also note the Buddhist populations in Sri Lanka, Bhutan, and northern Nepal; the areas of tribal religion in the east; and the centers of Christianity in the southwest.

who have waged partially successful political struggles. Owing to such efforts, the very concept of "untouchability" is now technically illegal in India. But caste continues to structure day-to-day social existence for hundreds of millions of Indians. Most important, marriage across caste lines is still relatively rare.

To counter caste-based discrimination, the Indian government reserves a significant percentage of university seats and government jobs for students from low-caste backgrounds. A number of Indian states have set higher quotas and have even applied them to Muslims and Christians. Such "reservations," as they are called, are highly controversial, as many people think that they unfairly penalize people of higher-caste background, whereas others demand that their own subcastes be granted reservation status. Protests opposing national efforts to expand the quotas broke out in many parts of India in 2006. In 2015, heated demonstrations resulted in 11 deaths in Gujarat when the important Patidar community demanded "other backward caste" status that would include them in the reservation system.

Islam Islam may be a "minority" religion for the region as a whole, but such a designation obscures its tremendous importance in South Asia. With more than 400 million adherents, South Asia's Muslim community is the world's largest. Bangladesh and especially Pakistan are overwhelmingly Muslim. India's Islamic community, although constituting only some 14 percent of the country's population, is still roughly 170 million strong.

Although Muslims live in almost every part of India, they are concentrated in four main areas: in most of India's cities; in Kashmir, particularly in the densely populated Kashmir Valley; in the upper and central Ganges Plain; and in the southwestern state of Kerala. In northern India especially, Muslims tend to be poorer and less educated than their Hindu neighbors, leading some observers to call them the country's "new dalits." India's Muslim population is also growing faster than its Hindu population, due to a higher fertility rate. According to the Pew Research Center, India will have more Muslims than any other country by 2050.

Interestingly, Kerala is about 25 percent Muslim even though it was one of the few parts of India that never experienced prolonged Muslim rule. Islam in Kerala is historically connected to trade across the Arabian Sea. Kerala's Malabar Coast long supplied spices and other luxury products to Southwest Asia, and enticed many Arabian traders to settle there. Gradually, many of Kerala's native inhabitants converted to the new religion as well. Owing to a similar process, Sri Lanka is approximately 9 percent Muslim, and the Maldives is almost entirely Muslim.

As an overwhelmingly Muslim country, Pakistan officially calls itself an Islamic Republic. As such, Islamic law is technically supposed to override the country's secular laws, although in actuality the situation remains ambiguous. In Pakistan's two most conservative provinces, Balochistan and Khyber Pakhtunkhwa, Islamic law tends to be strictly enforced. Increasingly harsh interpretations of Islam, moreover, are spreading over most of the rest of the country.

Sikhism Tension between Hinduism and Islam in medieval northern South Asia helped give rise to a new religion called **Sikhism**. Sikhism originated in the late 1400s in Punjab, a site of religious fervor at the time; Islam was gaining converts, and Hinduism was increasingly on the defensive. The new faith combined elements of both religions and thus appealed to many who felt trapped between their competing claims. Many orthodox Muslims, however, viewed Sikhism as a heresy precisely because it incorporated elements of their own religion. Periodic bouts of persecution led the Sikhs to adopt a militantly defensive stance, and in the political chaos of the early 1800s they were able to carve out a powerful kingdom for themselves. Even today, Sikh men are noted for their work as soldiers and bodyguards.

At present, the Indian state of Punjab is approximately 60 percent Sikh. Small, but often influential groups of Sikhs are scattered across the rest of India. Devout Sikh men are immediately visible because they do not cut their hair or beards but wrap their hair in turbans and often tie their beards close to their faces.

Buddhism and Jainism Although Buddhism virtually disappeared from India in medieval times, it persisted in Sri Lanka. Among the island's dominant Sinhalese people, Theravada Buddhism developed into a virtual national religion, fostering close connections between Sri Lanka and mainland Southeast Asia. In the high valleys of the Himalayas, Buddhism in its Tibetan form survived as the majority religion. Tibetan Buddhism, with its esoteric beliefs and huge monasteries, has been better preserved in Bhutan and in the Ladakh region of northeastern Kashmir than in Chinese-controlled Tibet.

The small city of Dharamsala in the northern Indian state of Himachal Pradesh is the seat of Tibet's government-in-exile and of its spiritual leader, the Dalai Lama, who fled Tibet in 1959.

After being absent for centuries, Buddhism returned to central India in the 1950s. The crucial event here was the conversion of B. R. Ambedkar, the chief architect of India's constitution, from Hinduism to Buddhism, followed by that of several millions of his supporters. Ambedkar was from a dalit background, and he concluded that caste discrimination was too deeply entrenched in Hinduism to allow for meaningful reform of the religion.

At roughly the same time as the birth of Buddhism (circa 500 BCE), another religion emerged in northern India as a protest against orthodox Hinduism: **Jainism** (Figure 12.25). This faith took nonviolence to its ultimate extreme. Jains are forbidden to kill any living creatures, and so the most devout adherents wear gauze masks to prevent inhaling small insects. Agriculture is forbidden to Jains because plowing can kill small creatures. As a result, most members of the community have looked to trade for their livelihoods. Many have prospered, aided by the frugal lifestyles required by their religion. Today Jains are concentrated in northwestern India, particularly Gujarat.

Other Religious Groups Even more prosperous than the Jains are the Parsis, or Zoroastrians, concentrated in the Mumbai area. The Parsis arrived as refugees from Iran after the arrival of Islam in the 7th century. Zoroastrianism is an ancient religion that focuses on the cosmic struggle between good and evil. Although numbering only a few hundred thousand, the Parsi community has had a major impact on the Indian economy. Several of the country's largest industrial firms were founded by Parsi families. Intermarriage and low fertility, however, threaten the survival of the community.

▼ **Figure 12.25** **Jain Religious Statue** Bahubali, a man who is said to have reached a state of spiritual perfection, is a revered figure among India's Jain religious minority. Here we see a group of Jain pilgrims performing rituals at the base of a gigantic statue of Bahubali in the Indian state of Karnataka.

▲ **Figure 12.26 Syrian Christian Cathedral** The state of Kerala in southwestern India has had a substantial Christian presence for roughly 1700 years. St. Mary's Jacobite Syrian Cathedral, popularly known as Piravom Valiyapally, is one of the most prominent Syrian Christian churches in Kerala. Local legends claim that it is the oldest Christian church in the world.

Indian Christians are more numerous than either Parsis or Jains. Their religion arrived some 1700 years ago; early contact between the Malabar Coast and Southwest Asia brought Christian as well as Muslim traders. A Jewish population also established itself, but later declined and today numbers only a few hundred. Kerala's Christians, by contrast, are counted in the millions, constituting 20 percent of the state's population. Several Christian sects are represented, but the largest are historically affiliated with the Syrian Church of Southwest Asia (**Figure 12.26**). Another stronghold of Christianity is the small Indian state of Goa, a former Portuguese colony, where Roman Catholics make up a quarter of the population.

During the colonial period, British missionaries went to great efforts to convert South Asians to Christianity. They had little success in most Hindu, Muslim, and Buddhist communities, although in some areas many dalits converted to the new faith. The remote tribal districts of northeastern British India proved more receptive to missionary activity. Today the Indian state of Nagaland has a clear Baptist majority, whereas in Mizoram most people follow the Presbyterian faith. Christian missionaries are still active in many parts of India, but during the 1990s they began to experience pressure from Hindu nationalists. Pakistan's Christian community, some 2.8 million strong, has also been under pressure—in this case, from Muslim radicals, who sometimes use the country's strict antiblasphemy laws against them. In 2015, two bomb attacks on Christian churches in Lahore resulted in 15 deaths and 70 serious injuries.

Geographies of Language

South Asia's linguistic diversity matches its religious diversity. In fact, one of the world's most important linguistic boundaries runs directly across India (**Figure 12.27**). North of the line, almost all languages belong to the Indo-European group, the world's largest linguistic family. The languages of southern India belong to the **Dravidian family**, a linguistic group unique to South Asia. Along the mountainous northern rim of the region, a third linguistic family, Tibeto-Burman, prevails, but this area is marginal to the South Asian cultural sphere. Finally, scattered tribal groups in eastern India speak Austro-Asiatic languages related to those of mainland Southeast Asia.

How or when Indo-European languages came to South Asia is uncertain, but the traditional argument is that these were spoken by nomadic peoples from Central Asia who invaded the subcontinent in the second millennium BCE, pushing back the indigenous Dravidian peoples. According to this hypothesis, offshoots of the same original cattle-herding people also swept across both Iran and Europe, bringing their language to all three places. The ancestral Indo-European tongue introduced to India was similar to Sanskrit, the sacred language of Hinduism. This rather simplistic scenario, however, is now regarded with skepticism, as most scholars argue for a more gradual infiltration of Indo-European speakers from the northwest.

Any modern Indo-European language of India, such as Hindi or Bengali, is more closely related to English than to any Dravidian language of southern India, such as Tamil. But South Asian languages on both sides of this linguistic divide do share superficial features. Dravidian languages, for example, have borrowed many words from Sanskrit, particularly those associated with religion and scholarship, whereas the Indo-European languages have borrowed many sounds from the Dravidian languages.

The Indo-European North South Asia's Indo-European languages are divided into two subfamilies: Iranian and Indo-Aryan. Iranian languages, such as Baluchi and Pashto, are spoken in western Pakistan, near the border with Iran and Afghanistan. Languages of the strictly South Asian Indo-Aryan groups are closely related to each other, but are still distinctive, often written in different scripts. Each of the major languages of India is associated with one or more Indian states. Thus, Gujarati is spoken in Gujarat, Marathi in Maharashtra, Oriya in Odisha, and so on.

The most widely spoken language of South Asia is Hindi. With almost 500 million speakers, Hindi is the second most widely spoken language in the world. It occupies a prominent role in contemporary India, both because so many people speak it and because it is the main language of the Ganges Valley, the country's historical and demographic core. Hindi is the dominant tongue of several Indian states, including Uttar Pradesh, Madhya Pradesh, and Haryana. In addition, the main forms of speech found in Rajasthan and Bihar are often considered to be dialects of Hindi. Most students from other parts of the country learn at least some Hindi.

Bengali is South Asia's second most widely spoken language. It is the national language of Bangladesh and the main language of the Indian state of West Bengal. Spoken by more than 200 million people, Bengali is the world's eighth or ninth most widely spoken language. Its significance extends beyond its official status and numerical strength. Equally important

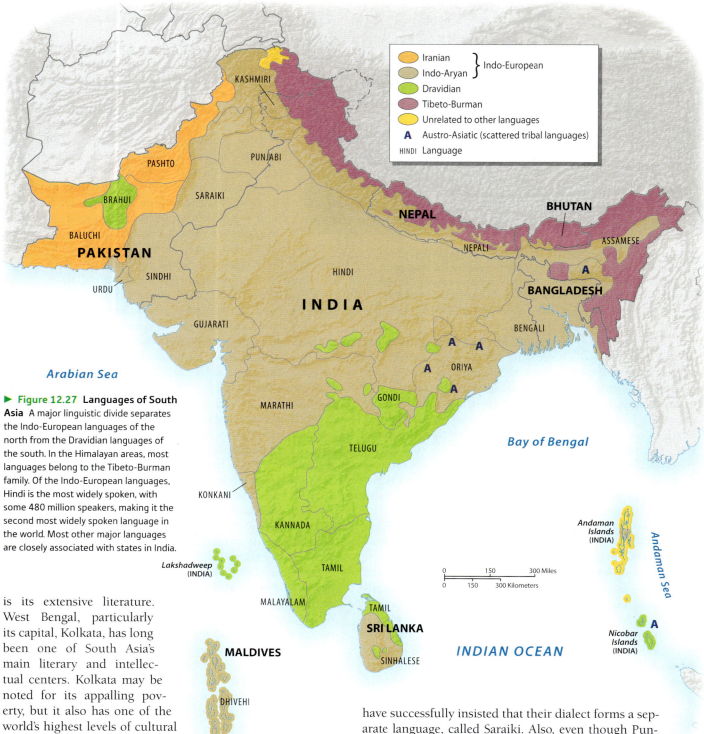

Legend:
- Iranian ⎫
- Indo-Aryan ⎬ Indo-European
- Dravidian
- Tibeto-Burman
- Unrelated to other languages
- **A** Austro-Asiatic (scattered tribal languages)
- HINDI Language

▶ **Figure 12.27 Languages of South Asia** A major linguistic divide separates the Indo-European languages of the north from the Dravidian languages of the south. In the Himalayan areas, most languages belong to the Tibeto-Burman family. Of the Indo-European languages, Hindi is the most widely spoken, with some 480 million speakers, making it the second most widely spoken language in the world. Most other major languages are closely associated with states in India.

is its extensive literature. West Bengal, particularly its capital, Kolkata, has long been one of South Asia's main literary and intellectual centers. Kolkata may be noted for its appalling poverty, but it also has one of the world's highest levels of cultural production as measured by the output of drama, poetry, novels, and film (**Figure 12.28**).

The Punjabi-speaking zone in the west was similarly split at the time of independence in 1947—in this case, between Pakistan and the Indian state of Punjab. An estimated 90 million people speak Punjabi, but it does not have the significance of Bengali. Although Punjabi is the main vehicle of Sikh religious writings, it lacks an extensive literary tradition. In recent years, moreover, the people of the southern half of Pakistan's province of Punjab

have successfully insisted that their dialect forms a separate language, called Saraiki. Also, even though Punjabi is Pakistan's most widely spoken language, it did not become the national language. Instead, that position was given to Urdu.

Urdu, like Hindi, originated on the plains of northern India. The difference between the two was largely one of religion: Hindi was the language of the Hindu majority, Urdu that of the Muslim minority, including the former ruling class. Owing to this distinction, they are written differently—Hindi in the Devanagari script (derived from Sanskrit) and Urdu in a modified version of the Arabic script. Although Urdu contains many words borrowed from Persian, its basic grammar and vocabulary are almost identical to those of Hindi.

▲ **Figure 12.28 Kolkata Bookstore** Although Kolkata is noted in the West mostly for its abject poverty, the city is also known in India for its vibrant cultural and intellectual life, illustrated by its large number of bookstores, theaters, and publishing firms.

With independence, millions of Urdu-speaking Muslims from the Ganges Valley fled to the new state of Pakistan. Because Urdu had a higher status than Pakistan's indigenous tongues, it was quickly established as the new country's official language. Karachi, Pakistan's largest city, is now mostly Urdu-speaking, but elsewhere other languages, such as Punjabi and Sindhi, remain primary. Most Pakistanis, however, do speak Urdu as their common language.

Languages of the South Four thousand years ago, Dravidian languages were probably spoken across most of South Asia. Indeed, a Dravidian tongue called Brahui is still found in the uplands of western Pakistan. The four main Dravidian languages, however, are confined to the south. As in the north, each language is closely associated with one or more Indian states: Kannada in Karnataka, Malayalam in Kerala, Tamil in Tamil Nadu, and Telugu in Andhra Pradesh and Telangana. Tamil is sometimes considered the most important member of the family because it has the longest history and the largest literature. Tamil poetry dates back to the first century CE, making it one of the world's oldest written languages.

Although Tamil is spoken in northern Sri Lanka, the country's majority population, the Sinhalese, speak an Indo-European language. Evidently, the Sinhalese migrated from northern South Asia several thousand years ago. Although this movement is lost to history, the migrants settled on the island's fertile and moist southwestern coastal and central highland areas, which formed the core of a succession of Sinhalese kingdoms. These same people also migrated to the Maldives, where the national language, Divehi, is essentially a Sinhalese dialect. The drier north and east coasts of Sri Lanka, in contrast, were settled long ago by Tamils from southern India. In the 1800s, British landowners imported Tamil peasants from the mainland to work on their tea plantations in the central highlands, giving rise to a second population of Tamil speakers in Sri Lanka.

Linguistic Dilemmas Multilingual Sri Lanka, Pakistan, and India have all experienced linguistic conflicts. Such problems are most complex in India, simply because India is so large and has so many different languages.

Indian nationalists have long dreamed of a national language that could help forge the different communities of the country into a more unified nation. But this **linguistic nationalism**, or the linking of a specific language with nationalistic goals, faces the stiff resistance of state-level loyalty, which is intertwined with local languages. Hindi was declared to be India's national language in 1947, a move that alienated speakers of other languages, especially Tamil. As a result of this cultural tension, India's 1950 constitution demoted Hindi to the position of one of 15 official languages. Additional languages were added over time, giving India today 23 official languages.

Regardless of such opposition, Hindi is expanding, especially in the north. Here local languages are closely related to Hindi, making Hindi relatively easy to learn. Hindi is spreading through education, but even more significantly through popular media, especially television and films. Movies and television programs are made in several Indian languages, but Hindi remains the most important vehicle. The most popular movies coming out of Mumbai's "Bollywood" film industry actually tend to be delivered in a neutral dialect that is as close to Urdu as to Hindi—and with a good deal of English thrown in as well.

The Role of English Even if Hindi is spreading, it still cannot be considered a common Indian language, as its role remains limited in the Dravidian south. National-level political, journalistic, and academic communication thus cannot be conducted in Hindi or in any other indigenous language. Only English, an "associate official language" of India, serves this function.

Before independence, many educated South Asians learned English for its political and economic benefits under British colonialism. It therefore became the common tongue of the upper and middle classes. Today, a few hard-core nationalists want to deemphasize English, but most Indians, and particularly those of the south, advocate it as a neutral national language that confers substantial international benefits. Roughly one-third of Indians can carry on a conversation in English, and English-medium schools abound in all parts of the region.

The English spoken in South Asia has forms and vocabulary elements that can make cross-cultural communication difficult. As a result, companies in India that run international call centers sometimes ask their employees to watch reruns of popular U.S. television shows in order to gain fluency in American pronunciation and slang.

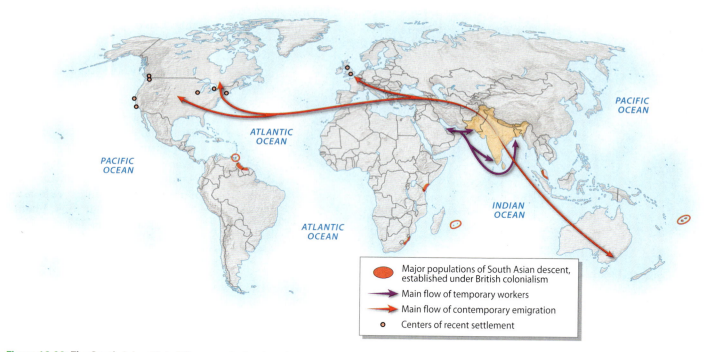

Major populations of South Asian descent, established under British colonialism

Main flow of temporary workers

Main flow of contemporary emigration

Centers of recent settlement

▲ **Figure 12.29 The South Asian Global Diaspora** During the British imperial period, large numbers of South Asian workers settled in other colonies. Today, roughly 50 percent of the population of such places as Fiji and Mauritius are of South Asian descent. More recently, large numbers have settled, and are still settling, in Europe (particularly the United Kingdom) and North America. Large numbers of temporary workers, both laborers and professionals, are employed in the wealthy oil-producing countries of the Persian Gulf.

South Asian in Global Cultural Context

The widespread use of English in South Asia not only facilitates the spread of global culture into the region, but also helps South Asian cultural production reach a global audience. The global spread of South Asian literature, however, is nothing new. As early as the turn of the 20th century, Rabindranath Tagore gained international acclaim for his poetry and fiction, earning the Nobel Prize for Literature in 1913. In the 1980s and 1990s, such Indian novelists as Salman Rushdie, Vikram Seth, and Arundhati Roy became major literary figures in Europe and North America.

South Asian films are also finding increasingly large audiences abroad. Bollywood's global reach became apparent to many observers in 2014 when its premier awards event was held in Tampa, Florida. Indian movies are now even being set in the United States, such as the recent blockbuster *Dhoom 3*, set largely in Chicago. Bollywood is the only foreign film industry that distributes its own movies in the United States. More recently, many Germans have taken to Indian films; 2006 saw the launch of *Ishq*, a glossy German-language Bollywood magazine. In 2015, a video of German girls doing a dubsmash of Bollywood dialogue became a viral sensation in India.

The spread of South Asian culture abroad has been accompanied by the spread of South Asians themselves. Migration from South Asia during the time of the British Empire led to the establishment of large communities in such far-flung places as eastern Africa, Fiji, and the southern Caribbean (**Figure 12.29**). Subsequent migration associated with this **Indian diaspora** mostly targeted the developed world. Many contemporary migrants to the United States are

doctors, software engineers, and other professionals, making Indian Americans one of the country's wealthiest ethnic groups. As of 2015, 15.5 million people who were born in India were living abroad, creating the world's largest diaspora.

As South Asian people have spread around the world, so has South Asian food. More than 15,000 South Asian eateries are now found in the United Kingdom, and London has more Indian restaurants than either Mumbai or Delhi. Chicken tikka masala has become so popular that it is widely considered Britain's "new national dish." Tellingly, while chicken tikka masala is classified as Indian food, it was probably invented in Britain by either a Bangladeshi or a Pakistani chef.

Globalization and Cultural Tensions In South Asia itself, the globalization of culture has brought tensions as severe as those felt anywhere in the world. Traditional Hindu and Muslim customs frown on any overt display of sexuality—a staple feature of global popular culture. Even though romance is a recurrent theme in the highly musical and often melodramatic Bollywood films, kissing is considered somewhat risqué (**Figure 12.30**).

Religious leaders thus often criticize Western films and television shows as immoral. Although India is a relatively open country, both the national and state governments periodically ban films and books that are considered too sexual. In 2012, for example, the film *The Girl with the Dragon Tattoo* was prohibited because of its adult scenes.

In the tourism-oriented Indian state of Goa, such cultural tensions are on full display. There, German and British

Explore the **Sounds** of Bollywood

http://goo.gl/ZVO065

sun-worshipers often wear nothing but thong bikini bottoms, whereas Indian women tourists go into the ocean fully clad. Young Indian men, for their part, often simply walk the beach and gawk at the outlandish foreigners.

Globalization and Sports The athletic culture of South Asia features an eclectic mix of immensely popular regional sports, increasingly popular global sports, and locally significant styles of martial arts. Rapid economic expansion and population growth provide an ever-larger audience for competition of many kinds. South Asia's most popular sports share their origins in the region's colonial past. Businesses associated with such sports have also expanded, including the manufacturing of soccer balls and other athletic equipment (see *Everyday Globalization: Soccer Balls from Sialkot*).

Cricket was introduced to India by sailors from the British East India Company during the early 18th century. The sport steadily grew in popularity, and well before 1900 India had become one of the world's centers of the sport. Currently, only ten countries are full members of the International Cricket Council, and four of those—India, Pakistan, Bangladesh, and Sri Lanka—lie in South Asia. The public popularity of top cricket stars has come to exceed that of Bollywood film stars. Over one billion people watched the televised India-Pakistan world cup cricket match in 2015.

Two other popular sports that link the region to British culture are field hockey and badminton. India's national field hockey team leads the world in Olympic gold medal victories with eight, earning its first in 1928. The first set of official badminton rules was devised in the Indian city of Pune in 1873, representing an evolution of the traditional English games of battledore and shuttlecock and the native Indian game now known as ball badminton. Although elite badminton competitions now tend to be dominated by Chinese, South Korean, and Southeast Asian

players, India remains competitive, and the sport continues to be one of the country's most popular pastimes.

Contemporary popular culture in South Asia thus reveals global linkages as well as divisions. The same tensions can be seen, and in much stronger form, in the region's geopolitical framework.

■ **REVIEW**

12.7 Why has religion become such a contentious issue in South Asia over the past several decades?

12.8 How have India and Pakistan tried to foster national unity in the face of ethnic and linguistic fragmentation?

12.9 What cultural features were spread by British imperialism in South Asia?

■ **KEY TERMS** Hindu nationalism, Hinduism, caste system, dalits, Buddhism, Mughal Empire, Sikhism, Jainism, Dravidian family, linguistic nationalism, Indian diaspora

Geopolitical Framework: A Deeply Divided Region

Before the 1800s, South Asia had never been politically united. A few empires covered most of the subcontinent at various times, but none spanned its entire extent. Whatever unity the region had was cultural, not political. The British, however, brought the region under a single political system by the middle of the 19th century. Independence in 1947 witnessed the traumatic separation of Pakistan from India; in 1971, Pakistan itself was divided when East Pakistan became independent Bangladesh. Serious internal tensions, moreover, increased in several South Asian countries in the late twentieth century (**Figure 12.31**). The region's most important geopolitical issue, however, continues to be the tension between Pakistan and India.

South Asia Before and After Independence in 1947

When Europeans began to arrive on the coasts of South Asia in the 1500s, most of the northern subcontinent came under the power of the Mughal Empire, a powerful Muslim state ruled by people of Central Asian descent (**Figure 12.32**). Southern India remained under a Hindu kingdom called Vijayanagara. European merchants, keen to obtain spices, textiles, and other products, established a series of coastal trading posts. The Mughals and other South Asian rulers were little concerned with the growing European naval power, as their own focus was the control of land. The Portuguese carved out an enclave in Goa on the west coast, while the Dutch gained control over much of Sri Lanka in the 1600s, but neither was a threat to the Mughals.

▼ **Figure 12.30 Bollywood Film Poster** India's Bollywood, one of the world's largest film production centers, extensively advertises its movies with large outdoor posters. Here we see famed actors Shahid Kapoor and Alia Bhatt posing for a photograph in front of one of their own movie posters, which advertises the Hindi film "Shaandaatr."

Everyday Globalization

Soccer Balls from Sialkot

Sports equipment manufacturing is a highly globalized business. All of the balls used in Major League Baseball, for example, are made by the Rawlings company in Costa Rica. Soccer balls, on the other hand, are closely associated with the city of Sialkot in northern Pakistan. Sialkot produces roughly 40 million soccer balls a year, or about 40 percent of the world's total supply (Figure 12.4.1). Its firms specialize in hand-stitched balls, used in World Cup competition and generally regarded as the best. Around 8 percent of Sialkot's soccer balls are exported to the United States.

Sialkot's soccer-ball industry faces a number of challenges. The city produces an array of other sporting goods, but its businesses have been losing customers to companies based in lower-cost countries, particularly China. As a result, Pakistani officials are keen to find new techniques that would allow them to maintain a competitive edge. In 2014, a team of researchers from Yale and Columbia Universities and Pakistan's Lahore School of Economics developed new procedures that would significantly reduce waste in the raw materials used to make high-quality soccer balls. They were disappointed, however, to learn that the factory workers resisted these new techniques, mostly because they did nothing to increase worker pay. In 2015, Pakistan's government announced its support for a new facility in Sialkot devoted to producing high-quality soccer balls using mechanized methods, hoping to eliminate the need for hand stitching.

1. How might Sialkot's sports-equipment firms respond to the challenge posed by Chinese exporters? What risks might they face in doing so?

▲ Figure 12.4.1 Soccer Balls from Sialkot The Pakistani city of Sialkot is noted for its export-oriented sporting-goods industry. Here we see workers sewing panels for an Adidas AG "Brazuca Replica Glider" soccer ball.

2. How many of your own sporting goods are imported? What countries have supplied them?

The Mughal Empire grew stronger in the 1600s, whereas Hindu power declined until it was limited to the peninsula's extreme south. In the early 1700s, however, the Mughal Empire weakened rapidly. Several contending states—some ruled by Muslims, others by Hindus, and one by Sikhs—emerged in former Mughal territories, creating political and military turmoil in South Asia.

The British Conquest These unsettled conditions provided an opening for European imperialism. The British and French, having largely displaced the Dutch and Portuguese, competed for trading posts. Because Indian cotton textiles were the best in the world prior to the Industrial Revolution, British and French merchants needed huge quantities for their global trading networks. Britain's overwhelming victory over France in the Seven Years' War (1756–1763), left France with a few marginal coastal possessions in southern India. Britain, or more specifically the **British East India Company**, the private firm that acted as an arm of the British government, now monopolized overseas trade in the area and staked out a South Asian empire of its own.

The company typically made strategic alliances with Indian states to defeat the latter's enemies, and then grabbed those enemy territories for itself. As time passed, its army, largely composed of South Asian mercenaries, grew ever more powerful. Several Indian states put up strong resistance, but none could ultimately resist the immense resources of the East India Company. Valuable local allies, as well as a few former enemies, were allowed to remain in power, provided that they no longer threatened British interests. The territories of these indigenous states, however, were gradually whittled back, and British advisors increasingly dictated their policies.

From Company Control to British Colony The continuing loss of Indian territory, coupled with the growing arrogance of British officials, led to a major rebellion in 1857. When this uprising (called the *Sepoy Mutiny* by the British) was finally crushed, a new political order was implemented. South Asia was now ruled by the British government, with the monarch of England serving as head of state.

Until 1947, the United Kingdom maintained direct control over South Asia's most productive and densely populated areas, including virtually the entire Indus–Ganges Valley and most of the coastal plains. The British also ruled Sri Lanka, having supplanted the Dutch in the 1700s. Major areas of indirect rule—where Indian rulers retained their princely states under British advisors—were in Rajasthan, the uplands of central India, southern Kerala, and along the northern frontiers (see Figure 12.32). The British administered this vast empire through three

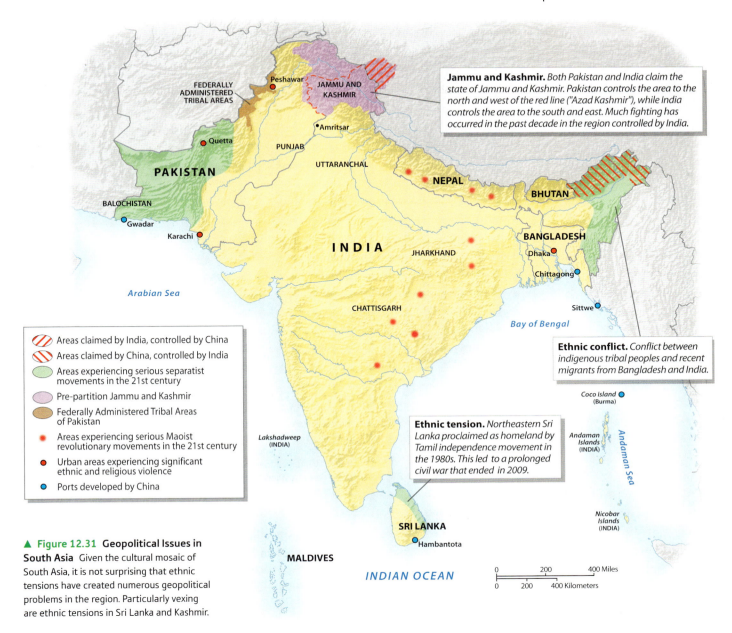

Jammu and Kashmir. *Both Pakistan and India claim the state of Jammu and Kashmir. Pakistan controls the area to the north and west of the red line ("Azad Kashmir"), while India controls the area to the south and east. Much fighting has occurred in the past decade in the region controlled by India.*

Ethnic conflict. *Conflict between indigenous tribal peoples and recent migrants from Bangladesh and India.*

Ethnic tension. *Northeastern Sri Lanka proclaimed as homeland by Tamil independence movement in the 1980s. This led to a prolonged civil war that ended in 2009.*

Legend:
- Areas claimed by India, controlled by China
- Areas claimed by China, controlled by India
- Areas experiencing serious separatist movements in the 21st century
- Pre-partition Jammu and Kashmir
- Federally Administered Tribal Areas of Pakistan
- Areas experiencing serious Maoist revolutionary movements in the 21st century
- Urban areas experiencing significant ethnic and religious violence
- Ports developed by China

▲ **Figure 12.31 Geopolitical Issues in South Asia** Given the cultural mosaic of South Asia, it is not surprising that ethnic tensions have created numerous geopolitical problems in the region. Particularly vexing are ethnic tensions in Sri Lanka and Kashmir.

coastal cities that were largely their own creation: Bombay (Mumbai), Madras (Chennai), and, above all, Calcutta (Kolkata). In 1911, they began building a new capital in New Delhi, near the strategic divide between the Indus and Ganges drainage systems.

While the political geography of British India stabilized after 1857, the empire's frontiers remained unsettled. British officials worried about threats to their colony, particularly from the Russians advancing across Central Asia, and attempted to expand as far to the north as possible. In some cases this merely entailed making alliances with local rulers. In such a manner, Nepal and Bhutan retained their independence. In the extreme northeast, some small states and tribal territories, most of which had never been part of the South Asian cultural sphere, were more directly

brought into British India. A similar policy was conducted on the vulnerable northwestern frontier, but local resistance was much more effective, and the British-Indian army suffered defeat at the hands of the Afghans. Afghanistan thus retained its independence, forming a buffer between the British and Russian empires. The British also allowed the tribal Pashto-speaking areas of what is now northwestern Pakistan to retain almost complete autonomy, thus forming a secondary buffer.

Independence and Partition British India began to unravel in the early 20th century as South Asians increasingly demanded independence. The British, however, were equally determined to stay, and by the 1920s the region was embroiled in massive political protests.

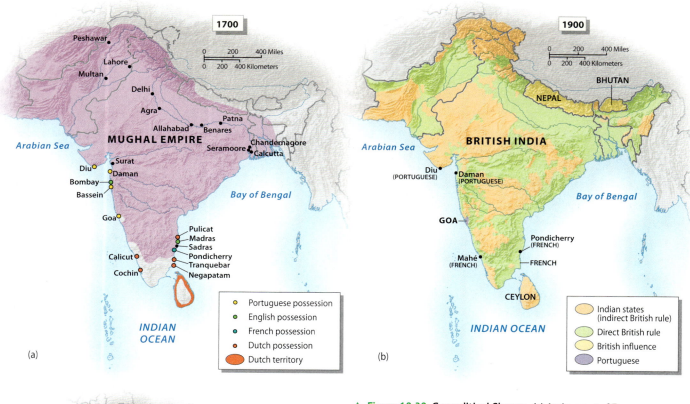

▲ **Figure 12.32 Geopolitical Change** (a) At the onset of European colonialism before 1700, much of South Asia was dominated by the powerful Mughal Empire. (b) Under Britain, the wealthiest parts of the region were ruled directly, but other lands remained under the partial authority of indigenous rulers. Independence for the region came after 1947, when the British abandoned their extensive colonial territory. (c) Bangladesh, formerly East Pakistan, gained its independence in 1971 after a short struggle against centralized Pakistani rule from the west.

Leaders of the rising nationalist movement faced a major dilemma in attempting to organize an independence movement. Many—including Mohandas Gandhi, the main figure of Indian independence—favored a unified state that would encompass all British territories in mainland South Asia (**Figure 12.33**). However, most Muslim leaders feared that a unified India would leave their people in a vulnerable position, and argued for dividing British India into two new countries: a Hindu-majority India and a Muslim-majority Pakistan. Yet parts of northern South Asia were settled by Muslims and Hindus in roughly equal proportions. A more significant obstacle was the fact that the areas of clear Muslim majority were located on opposite sides of the subcontinent, in present-day Pakistan and Bangladesh.

No longer able to maintain their world empire after World War II, the British withdrew from South Asia in 1947. As this occurred, the region was indeed divided into India and Pakistan. Partition itself was a horrific event, with fighting between Hindus and Muslims resulting in the deaths of hundreds of thousands of people. Roughly 7 million Hindus and Sikhs fled Pakistan, to be replaced by roughly 7 million Muslims fleeing India.

The Pakistan that emerged from partition was, for several decades, a clumsy two-part country, with its western section in the Indus Valley and its eastern portion in the Ganges Delta. The Bengalis, occupying the poorer eastern section, complained that they were treated as second-class citizens. In 1971, they launched a rebellion and, with the help of India, quickly prevailed. Bangladesh then emerged as a new country.

This second partition did not solve Pakistan's problems, however, as it remained politically unstable and prone to military rule. Pakistan retained the British policy of allowing

▲ **Figure 12.33 Gandhi Protesting British Rule** In 1930, Mohandas Gandhi, the key figure in the Indian independence movement, led his famous "salt march" to protest British policy and power. The British heavily taxed salt production, and as a result outlawed private salt gathering. After making their own salt at the sea coast, Gandhi and his followers were arrested. This event helped undermine British legitimacy, thus bolstering the Indian independence movement.

almost full autonomy to the Pashtun tribes of the Federally Administered Tribal Areas of the northwest, an area marked by clan fighting and vengeance feuds. The large and poor province of Balochistan in southwestern Pakistan posed another problem, as a long-simmering separatist movement continues to fight against Pakistan's military forces. Instability in neighboring Afghanistan presents yet another challenge. Registered Afghan refugees in Pakistan number around 1.5 million, and another 2.7 million unregistered Afghan refugees probably live there as well. These Afghans complain of persecution by the Pakistan military and police, and many claim to have been forcibly returned to Afghanistan.

Bangladesh has also struggled politically since achieving independence in 1971. Intense corruption, growing Islamic radicalism, and street-level fighting between members of its two major political parties have damaged its democratic institutions. In 2015, violent clashes stemming from a disputed national election in the previous year racked the country. Fifty-five people died and more than 10,000 were arrested, but no political solution was reached (**Figure 12.34**). Instead, the ruling Awami League party branded the opposing Bangladeshi Nationalist Party as a terrorist organization and confined its leader, Khaleda Zia, to her office.

Geopolitical Structure of India Following independence, Indian leaders, committed to democracy, faced a major challenge in organizing such a large and diverse

country. They decided to chart a middle ground between centralization and local autonomy and organized India as a **federal state**, with a significant amount of power given to individual states. The national government, however, retained full control over foreign affairs and a large degree of economic authority.

India's constituent states were reorganized to match the country's linguistic geography, so that each major language group should have its own state (with the massive Hindi-speaking population having several), and thus a degree of political and cultural autonomy. Yet only the largest groups received their own territories, which has led to recurring demands from smaller groups. Over time, several new states were added to the map. Goa, after the Portuguese were finally forced out in 1961, became a separate state in 1987 over the objection of its large neighbor, Maharashtra. In 2000, three new states were added (Jharkhand, Uttaranchal, and Chhattisgarh), and in 2014 Telangana was carved out of the northwestern districts of Andhra Pradesh. More than a dozen additional new states have been proposed elsewhere in India.

Ethnic Conflicts and Tensions in South Asia

The movement for new states in India has been largely rooted in ethnic tensions. Unfortunately, violent ethnic conflicts persist in many parts of South Asia. Of these conflicts, the most complex—and perilous—is in Kashmir.

Kashmir Relations between India and Pakistan were hostile from the start, and the situation in Kashmir has kept the conflict burning. During British rule, Kashmir was a large state with a primarily Muslim core joined to a Hindu district (Jammu)

▼ **Figure 12.34 Political Clashes in Bangladesh** Bangladesh's rival political parties have engaged in heated conflicts over the past several years. Here police barricades have been set up in front of the private offices of former Prime Minister Khaleda Zia in order to prevent access during a particularly intense period of political strife in 2015.

in the south and a Tibetan Buddhist district (Ladakh) in the far northeast. Kashmir was then ruled by a Hindu **maharaja**, a king subject to British advisors. During partition, Kashmir was severely pressured by both India and Pakistan. Troops linked to Pakistan gained control of western and much of northern Kashmir, at which point the maharaja opted for union with India. India thus retained the core areas of the state, but neither country would accept the other's control over any portion of Kashmir. As a result, India and Pakistan have fought several inconclusive wars.

Although the Indo-Pakistani boundary remained fixed, the struggle in Kashmir intensified, flaming into an open insurgency in 1989. Some Kashmiris would like to join their homeland to Pakistan, but polls indicate that most prefer independence. Opposition to Kashmir's independence is one of the few things on which India and Pakistan agree. India accuses Pakistan of supporting training camps for Islamist militants and of helping militants sneak across the border. As a result, India has been building a fortified fence along the line of control that divides Kashmir.

The situation in Kashmir began to improve in 2004 when India and Pakistan engaged in serious negotiations. India agreed that Pakistan must play a role in any Kashmir peace settlement, whereas Pakistan's government reduced its support of Muslim militants. By 2012, even tourism showed signs of revival in the Kashmir Valley, noted for lush gardens and orchards nestled among some of the world's most spectacular mountains. A successful state-level election in 2014 promised further improvements. But violence flared up later in the same year as Indian and Pakistani military contingents exchanged gunfire over the disputed border, killing 9 civilians in Pakistan and 7 in India. All told, the Kashmir conflict has claimed over 40,000 lives and displaced one out of every six inhabitants of the state.

India's Northeastern Fringe Another complicated ethnic conflict emerged late in the 20th century in the uplands of India's extreme northeast, in part due to demographic change and cultural collision. Much of this area is still relatively lightly populated, attracting many migrants from Bangladesh and adjacent Indian states. Tensions here have thus complicated India's relations with Bangladesh. India accuses Bangladesh of harboring separatists on its side of the border, and objects to continuing Bangladeshi emigration. As a result, India is building a 2500-mile (4000-km), US$1.2 billion fence along the border between the two countries (**Figure 12.35**).

Construction of this barrier has been complicated by the irregular boundary between the two countries, with numerous **exclaves** and **enclaves**. Until 2015, there were 102 pieces of Indian territory within the main body of Bangladesh and 71 pieces of Bangladeshi territory within the main body of India. Finally, the two countries agreed to swap out these territories, forming a clear, linear boundary for the first time. As a result, more than 50,000 people had their nationalities transformed overnight. It is widely hoped that this change will improve provision of education, welfare, and public health along the border zone.

Despite better relations between India and Bangladesh, far northeastern India remains a troubled area as insurgent groups continue to seek autonomy if not independence. After 2000, India's government began to invest more money in the region, hoping to reduce popular support for separatism. India is also eager to expand trade with Burma and has been working with the Burmese government to secure the border zone. As a result of these and other initiatives, several rebel movements have signed cease-fires with the Indian government. Other insurgent groups have even agreed to cooperate with the Indian army against those groups that continue to fight for independence. But as of early 2016, India was still actively conducting military operations in Assam, Manipur, Nagaland, and Tripura. According to official statistics, fighting in northeastern India resulted in roughly 40,000 fatalities between 1979 and 2015.

Sri Lanka Until recently, interethnic violence in Sri Lanka was also severe. Here the conflict has roots in both religious and linguistic differences. Northern Sri Lanka and parts of its eastern coast are dominated by Hindu Tamils, while the island's majority group is Buddhist in religion and Sinhalese

▼ **Figure 12.35 India–Bangladesh Border Barrier** India began building a fence between its territory and that of Bangladesh in 2003 in order to reduce illegal immigration and stop the influx of militants. Members of the Indian Border Security Force are patrolling a segment of the border barrier. **Q: Why is India so much more concerned about its border with Bangladesh than it is about its borders with Nepal, Bhutan, and Burma?**

Explore the **Sights** of India-Bangladesh Border

http://goo.gl/axOlmf

in language (**Figure 12.36**). Relations between the two communities have historically been fairly good, but tensions mounted soon after independence.

The basic problem is that Sinhalese nationalists favor a unitary government, some going so far as to argue that Sri Lanka should be defined as a Buddhist state. Most Tamils, on the other hand, support political and cultural autonomy, and have accused the government of discrimination. Overall, Tamils are better educated, but the government has favored the Sinhalese majority. In 1983, war erupted when the rebel force known as the **Tamil Tigers** attacked the Sri Lankan army. Both extreme Tamil and extreme Sinhalese nationalists remained unwilling to compromise, prolonging the war.

The conflict intensified in March 2007 when the Tamil Tigers cobbled together a rudimentary air force and bombed Sri Lanka's main airbase. The government subsequently abandoned negotiations, launching instead an all-out offensive. In May 2009, its military crushed the Tamil Tiger army, killing the organization's leaders. The defeat of the Tamil Tigers brought about a humanitarian disaster, as many civilians were killed and over 300,000 were displaced. Sri Lanka is now finally at peace, but ethnic tensions are still pronounced. In 2014, the UN Human Rights Council launched an inquiry into human rights abuses by the Sri Lankan military during the final period of the war.

As of 2015, an estimated 160,000 troops, almost entirely Sinhalese, were still stationed in the Tamil areas of northern Sri Lanka. The result, some say, is the highest ratio of civilian to security forces in the world. Local residents increasingly complain that the military has been grabbing land for itself, often to develop it for tourism. Several golf courses and resort hotels in the area are now reportedly run by the Sri Lankan army.

The Maoist Challenge

Not all of South Asia's current conflicts are rooted in ethnic or religious differences. Poverty and inequality in east-central India, for example, have generated a persistent revolutionary movement inspired by the former communist leader of China, Mao Zedong. Tribal people whose lands are being exploited for resources by outsiders form the bulk of these **Maoist** fighters. More than 13,000 people lost their lives in this struggle between 1996 and 2015. Such violence prevents investment in some of India's least-developed areas, intensifying the underlying economic and social problems that gave rise to the insurgency in the first place. But after peaking in 2010 at 1177, the annual death toll declined to 189 in 2015.

India's Maoist rebellion is too small to effectively challenge the state, but the same cannot be said in regard to Nepal. Nepalese Maoists, frustrated by the lack of development in rural areas, emerged as a significant force in the 1990s. In 2002, Nepal's king, citing this threat, dissolved parliament and took total control of the country's government. This move only intensified the struggle, and within a few years the rebels had gained control of most of Nepal. By 2005, Nepal's urban population also turned against the monarchy, and three years later the king was forced to step down. Nepal then became a republic, with the leader of the former Maoist rebels serving as prime minister.

The end of the monarchy, however, did not bring stability to Nepal, as several governments have been formed and then disbanded. After a series of political crises, a new constitution was finally announced in 2015 that seeks to reorganize the country into 7 new states that will be given enhanced powers. Many of the indigenous people of Nepal's southern lowlands were infuriated by this proposed change, as they were not granted an autonomous region of their own. These lowland people have been distressed by immigrants from the more densely populated Nepalese hill country. As a result, southern protesters blocked the delivery of fuel from India, plunging Nepal into an

Tamil Hindus
Sinhalese Buddhists
Hindus and Buddhists
Areas with significant Muslim populations
Areas with significant Christian populations
Formerly claimed as Tamil state (Tamil Eelam) by the Tamil Tigers

INDIA
Palk Strait
Jaffna
Mannar
Gulf of Mannar
Trincomalee
Anuradhapura
Puttalam
Polonnaruwa
SRI LANKA
Batticaloa
Matale
Kalmunai
Negombo
Kandy
Colombo
Badulla
Moratuwa
Ratnapura
Beruwala
Galle
INDIAN OCEAN

0   25   50 Miles
0   25   50 Kilometers

◀ **Figure 12.36 Ethnic Geography of Sri Lanka** The majority of Sri Lankans are Sinhalese Buddhist, many of whom maintain that their country should be a Buddhist state. A Tamil-speaking Hindu minority in the northeast strenuously resists this idea. Tamil militants, who waged war against the Sri Lankan government for several decades until defeat in 2009, hoped to create an independent country in their northern homeland. Separate Christian and Muslim populations make for a complex social environment in Sri Lanka.

economic crisis. Nepal's leaders accused India of organizing the protests and blockade, damaging relations between the two countries. The blockade was finally lifted in February 2016 after the Nepalese government offered a new constitutional amendment, but protest leaders were not fully satisfied and general unrest continues.

International Geopolitics

South Asia's major international geopolitical problem continues to be the struggle between India and Pakistan (**Figure 12.37**). The stakes are particularly high as both India and Pakistan have nuclear capabilities. As of 2015, Pakistan was estimated to have 120 warheads, whereas India is estimated to have between 110 and 120. Relations between the two countries sharply deteriorated in November 2008 when terrorists operating from Pakistan launched a series of attacks on tourist facilities in Mumbai, killing 173. Pakistan responded by investigating the event and arresting several alleged plotters, but many Indians suspect the attack had the support of certain elements of Pakistan's government and military. Talks between the two countries resumed, and by 2015 relations had improved considerably.

In early 2016, however, tensions intensified again when militants attacked an Indian airbase, killing 7 soldiers. Indian experts claimed that the heavily armed attackers could not have carried out the assault without the cooperation of Pakistani security forces. But Pakistan's government denounced the strike and offered to cooperate in finding those behind it. Many observers think that Pakistan's government is itself deeply divided over this issue, with some elements wanting rapprochement with India and others seeking to undermine any peace initiative.

Relations with China and the United States During the global Cold War, Pakistan allied itself with the United States and India remained neutral, leaning slightly toward the Soviet Union. Such entanglements fell apart with the end of the superpower conflict in the early 1990s. Subsequently, Pakistan forged an informal alliance with China, from which Pakistan obtained sophisticated military equipment. India, on the other hand, has strengthened its connections with the United States while forging increasingly close military ties with Israel.

China's military connection with Pakistan is rooted in its own animosity toward India. In 1962, China defeated India in a brief war, gaining control over the virtually uninhabited territory of Aksai Chin in northern Kashmir. Growing trade has brought the two countries closer together in some respects, but China's continued control of Aksai Chin and its claims to the entire northeastern Indian state of Arunachal Pradesh ensures that relations between the two countries remain tense. In 2013, India accused China of sending troops over the "line of control" established after the 1962 war in the Himalayas, resulting in a three-week standoff. Tensions eased in early 2016, however, when Indian and Chinese troops agreed to hold joint disaster-relief exercises near the disputed border.

Pakistan's Complex Geopolitics Pakistan's geopolitical situation became more complex following the September 11, 2001 attacks. Until that time, Pakistan had strongly supported Afghanistan's Taliban regime (see Chapter 10), but after the attacks on the World Trade Center and the Pentagon, the United States gave Pakistan a stark choice: either assist the United States in its fight against the Taliban and receive in return debt reductions and other forms of aid, or lose favor. Pakistan quickly agreed to help, offering both military bases and valuable intelligence to the U.S. military.

Pakistan's decision to help the United States came with large risks. The Afghan Taliban enjoy substantial support in Pakistan, particularly among the northwestern Pashtun people. After suffering several military reversals, Pakistan decided to negotiate with radical Islamists in this area, at one point giving them virtual control over sizable areas. From these bases, militants launched attacks on U.S. forces in Afghanistan and attempted to gain control over broader swaths of Pakistan's territory. The United States responded with drone aircraft attacks on insurgent leaders, resulting in large numbers of civilian casualties and generating pronounced anti-American sentiment throughout Pakistan.

U.S.-Pakistan relations further deteriorated in 2011 when U.S. forces launched a raid deep into Pakistan's territory to kill Osama bin Laden. An official Pakistani report released in July 2013 claimed that this raid was an "American act of war against Pakistan." Relations improved in 2015, however, after Pakistan's military attacked the Islamist militants who had previously enjoyed sanctuaries along the Afghanistan-Pakistan border. But the relationship between the United States and Pakistan remains difficult, and people in both countries increasingly object to the roughly US$2 billion in military and economic aid that Islamabad receives annually from Washington.

▼ **Figure 12.37 India–Pakistan Tensions** An Indian officer looks through binoculars in war-torn Kashmir. Relationships between India and Pakistan have remained extremely tense since independence in 1947. Because both countries are nuclear powers, the fear that border hostilities will escalate into wider warfare is a nightmarish possibility.

The security crisis in Pakistan has destabilized the country's internal politics as well. Radical Islamists consider the government of Pakistan an illegitimate ally of the United States, and are willing to attack in almost any way they can. In January 2016, for example, Islamists struck at a ceremony at a university in northwestern Pakistan, killing at least 22 people.

Pakistan is still a democratic state, though its democracy is troubled and its policies are somewhat repressive. Pakistan's 2013 election marked the first time in which one democratically elected government yielded power to another democratically elected government after its term in office was completed. But due to the Islamist challenge, ethnic tensions, and the mutual animosity of the country's main political parties, many observers fear a return to military rule. Skeptical outsiders also worry that Pakistan's military, particularly its extremely powerful Directorate for Inter-Services Intelligence (ISI), has been infiltrated by radical Islamist elements.

REVIEW

12.10 How have relations between India and Pakistan influenced South Asian geopolitical developments since independence?

12.11 Why has South Asia experienced numerous insurgencies and other political conflicts since the end of British rule in 1947?

12.12 How has the rise of China influenced geopolitical developments in South Asia over the past several decades?

■ **KEY TERMS** British East India Company, federal state, maharaja, exclaves, enclaves, Tamil Tigers, Maoist

Economic and Social Development: Rapid Growth and Rampant Poverty

South Asia is a land of developmental paradoxes. It is, along with Sub-Saharan Africa, the poorest world region, yet it is also the site of some immense fortunes. South Asia has achieved many world-class scientific and technological accomplishments, but it also has some of the world's highest illiteracy rates. Although South Asia's high-tech businesses are closely integrated with centers of the global information economy, the region's economy as a whole was long one of the world's most self-contained and inward looking.

It is difficult to exaggerate South Asia's poverty. Almost 200 million Indians live below their country's official poverty line, which is set at a very meager level (**Figure 12.38**). Approximately 20 percent of India's citizens are seriously undernourished, as are 30 percent of the people of Bangladesh. By measures such as infant mortality and average longevity, Nepal is in even worse condition. In urban slums throughout South Asia, rapidly growing populations have little chance of finding housing or basic social services. Observers estimate that up to half a million South Asian children work as virtual slaves in carpet-weaving workshops and other small-scale factories.

Despite such deep and widespread poverty, South Asia should not be regarded as a zone of uniform misery. India especially has a large and growing middle class as well as a

▲ Figure 12.38 **Poverty in India** Some of the world's highest rates of childhood poverty, as well as childhood labor, are found in South Asia. Here a 10-year old Bangladeshi boy collects waste materials from a trash-strewn waterfront to sell to a local retailer.

small, wealthy upper class. Roughly 270 million Indians are able to purchase such modern consumer items as televisions, motor scooters, and washing machines. Owing to strong economic growth, that figure is forecast to double by 2026. Falling energy prices helped India's economy expand by 7.5 percent in 2015, surpassing the growth rate of China. It is uncertain, however, whether such rapid growth is sustainable given India's deep environmental problems and economic inefficiencies. Prospects also vary significantly across the country, as several Indian states have shown real economic vitality whereas others have seen much less growth and development.

Geographies of Economic Development

After independence, South Asian governments attempted to create new economic systems that would benefit their own people rather than foreign countries or corporations. Planners focused on heavy industry and economic autonomy, and some major gains were realized, but the overall pace of development remained slow. Since the 1990s, the countries of South Asia have gradually opened their economies to the global economic system. This process has created core areas of economic development surrounded by peripheral areas that have lagged behind, creating landscapes of striking economic disparity (Table 12.2).

Himalayan Economies Both Nepal and Bhutan are disadvantaged by their rugged terrain and remote locations. Bhutan has purposely remained somewhat disconnected from the modern world economy, its small population living in a relatively pristine natural environment. Bhutan is so isolationist that it has only recently allowed tourists to enter—if the visitors agree to spend substantial amounts of money while in the country. Its government has made the unusual move of downplaying conventional measures of economic development, attempting to substitute "gross national happiness" for "gross national product."

Bhutan is not, however, cut off from the rest of the world. It exports substantial amounts of hydroelectric power to India, helping its economy expand at a rapid pace. Such growth

TABLE 12.2   Development Indicators

Mastering Geography™

Country	GNI per capita, PPP 2014	GDP Average Annual % Growth 2009–15	Human Development Index (2015)[1]	Percent Population Living Below $3.10 a Day	Life Expectancy (2016)[2]		Under Age 5 Mortality Rate (1990)	Under Age 5 Mortality Rate (2015)	Youth Literacy (% pop ages 15–24) (2005–2014)	Gender Inequality Index (2015)[3,1]
					Male	Female				
Bangladesh	3,330	6.2	0.570	81.5	71	73	139	38	81	0.503
Bhutan	7,280	6	0.605	28.9	69	70	138	33	–	0.457
India	5,630	6.9	0.609	67.9	67	70	114	48	86	0.563
Maldives	10,920	5.5	0.706	15.0	76	78	105	9	–	0.243
Nepal	2,410	4.4	0.548	74.4	66	69	135	36	85	0.489
Pakistan	5,090	3.4	0.538	53.7	66	67	122	81	73	0.536
Sri Lanka	10,300	6.8	0.757	16.8	72	78	29	10	98	0.370

Source: World Bank, *World Development Indicators*, 2016.

[1] United Nations, *Human Development Report*, 2015.

[2] Population Reference Bureau, *World Population Data Sheet*, 2016.

[3] Gender Inequality Index—A composite measure reflecting inequality in achievements between women and men in three dimensions: reproductive health, empowerment, and the labor market that ranges between 0 and 1. The higher the number, the greater the inequality

Login to Mastering Geography™ & access MapMaster to explore these data!

1) How closely does the Human Development index correlate with GNI per capita figures in South Asia?

2) Are there any other categories in this table that correlate more closely with GNI per capita than HDI?

has brought temporary migrants; approximately 100,000 Indian laborers work on roads and other infrastructure projects in Bhutan. Economic development has also led to the rapid growth of Bhutan's capital, Thimphu, which mushroomed from a small town of only a few thousand residents in the 1950s to a city of over 100,000 people at present. Such expansion has greatly strained Thimphu's infrastructure, especially its drainage system.

Nepal, more heavily populated and suffering more severe environmental degradation than Bhutan, is still rebuilding from the damage caused by the 2015 earthquake. Nepal is more closely integrated with India's economy and its own economy relies heavily on international tourism (**Figure 12.39**). Tourism has brought some prosperity to a few favored locations, but often at the cost of heightened ecological damage. Tourism in Nepal has suffered, moreover, since the country entered a period of political turmoil in 2002. Remittances from Nepalese workers living abroad currently help sustain its fragile economy.

Bangladesh Per capita economic figures for Bangladesh are higher than those of Nepal, but may be more indicative of widespread hardship because most Bangladeshis require cash to meet their basic needs. Partly because of the country's dense population, poverty is extreme and widespread, with 48 percent of its children estimated to be malnourished to some degree. More recently, however, Bangladesh's economy has grown at a relatively brisk pace, lifting millions of people out of dire poverty. The discovery of huge offshore natural gas deposits promises further economic gains, but thus far development of this resource has been slow.

Environmental degradation has contributed to Bangladesh's impoverishment, as did the partition of 1947. Prior to partition Bengal's businesses were located in the western area that went to India. The division of Bengal tore apart an integrated economic region, much to the detriment of the poorer, mainly rural eastern section. Bangladesh also has suffered because of its agricultural emphasis on jute, a plant that

Explore the **Sights** of Phewa Lake Resorts

http://goo.gl/HcyTSX

▼ **Figure 12.39** Tourism in Nepal Nepal has long been one of the world's main destinations for adventure tourism, although business has suffered in recent years because of the country's political instability. Tourists in Nepal often stay in rustic lodges, many of which post advertisements.

yields tough fibers for making ropes and burlap bags. When synthetic materials began to undercut the global jute market, Bangladesh failed to discover a major alternative export crop.

Over the past several decades, Bangladesh has pioneered several major programs to reduce poverty. The most important is **microfinance**, based on providing loans to small-scale business. Low-interest microcredit provided by Grameen Bank, which received the 2006 Nobel Peace Prize, has given hope to many poor women in Bangladesh, allowing the emergence of some vibrant small-scale enterprises. Critics, however, argue that these microloans are often used mainly to support consumption, that recipients can get trapped in a debt cycle, and the poorest of the poor have no collateral whatsoever and are thus ineligible for loans. Research is now being conducted on how to make microfinance more effective through such techniques as livelihood mapping, which helps determine what type of loan is most effective at a given location.

The economy of Bangladesh is most internationally competitive in clothing manufacture. Currently the world's second largest exporter of garments by some estimates, Bangladesh plans to double its apparel exports to US$50 billion by 2021. This massive industry has brought many benefits, but also negative consequences, as wages are low and working conditions are often brutal and extremely unsafe. Such problems were brought to global attention in 2013 when the eight-story Rana Plaza building, which contained several clothing factories, collapsed, killing 1127 people. Since that event, global clothing retailers have been working with the government of Bangladesh to improve working conditions.

Bangladesh, like India and Pakistan, has also found a lucrative niche in shipbreaking, the dismantling of large ships in order to recycle their steel and other materials. This activity, however, is also highly dangerous and environmentally damaging. Despite official efforts to improve safety conditions, from 2011 to 2015 60 Bangladeshi shipbreakers were killed and 125 seriously injured.

Pakistan Pakistan also suffered the effects of partition in 1947. But unlike Bangladesh, Pakistan inherited a reasonably well-developed urban infrastructure. It also has a productive agricultural sector, as it shares the fertile Punjab with India. In addition, Pakistan boasts a large textile industry, based on its huge cotton crop.

Yet in recent years, Pakistan's economy has been faltering, hampered by high inflation, slow growth, extremely high levels of military spending, and internal strife. A woefully inadequate power supply results in long brownouts that often force factories to shut down. Additionally, a small but powerful landlord class that pays virtually no taxes to the central government controls many of its best agricultural

▲ **Figure 12.40 Gwadar Port** Development of facilities at the excellent deep-water port of Gwadar, located in Pakistan's province of Balochistan, began in 2007. Pakistani and Chinese authorities hope that the port will enhance trade between the two countries. **Q: Why was Gwadar selected for the development of a major port, considering the fact that it is located in a sparsely populated part of Pakistan?**

lands. Unlike India, moreover, Pakistan has not been able to develop a successful IT industry. Falling oil prices in late 2014, however, gave Pakistan a boost, and in 2015 its economy grew by 4.4 percent, its best showing in eight years.

Like India, Pakistan has inadequate energy supplies, and so it looks to Central Asia and Southwest Asia for fossil fuels. One consequence of this policy has been the construction of a huge new deep-water port at Gwadar near its border with Iran, adjacent to some of the world's most important oil-tanker routes (**Figure 12.40**). Built with massive Chinese engineering and financial assistance, the port of Gwadar is managed by a state-owned Chinese firm. It must be heavily protected by Pakistan's military against Baluchi insurgents in the region.

Island Economies Sri Lanka's economy is more highly developed than those of India, Pakistan, and Bangladesh. Its exports are concentrated in textiles and agricultural products such as rubber and tea. By global standards, however, Sri Lanka is still a poor country, its progress long undercut by its civil war. Since the end of the war in 2009, its economy has bounced back, growing by 7.4 percent in 2014. Sri Lanka hopes to benefit from the prime location of its port of Colombo, its highly educated people, and its tourism potential. To do so, however, the country must attract large quantities of foreign investment, much of that from China.

The Maldives is South Asia's most prosperous country based on per capita Gross National Income (GNI), but its total economy, like its population, is tiny. Most of its revenues are gained from fishing and international tourism. The benefits from the tourist economy, however, flow mainly to the country's small elite population, resulting in large-scale public discontent and political repression. Tourism is also highly vulnerable to international recessions. The Maldives now looks to China to expand its tourism sector. In 2014, one-third of the country's tourists came from China, an increase of 9.6 percent over 2013. The Maldivian government is also working to establish special economic zones to help diversify its economic base.

India's Less-Developed Areas India's economy, like its population, dwarfs those of the other South Asian countries. India's per capita gross domestic product (GDP) is now well above that of Pakistan, and its total economy is many times larger. As the region's largest country, India also exhibits far more internal variation in economic development than its neighbors. The most basic economic division is between India's more prosperous west and south and its poorer districts in the north and east (**Figure 12.41**).

The remote states of northeastern India rank low on the economic ladder as measured by per capita GNI, but the prevalence of subsistence economies can make such statistics misleading. More extreme deprivation is found in the densely populated lower and middle Ganges Valley. Bihar is India's poorest state by most indicators, and neighboring Uttar Pradesh, India's most populous state, also is poverty-stricken. Other north-central states, such as Madhya Pradesh, Jharkhand, and Chhattisgarh, have likewise experienced relatively little economic development. Both Bihar and Uttar Pradesh have fertile soils, but have not profited as much from the Green Revolution as have Punjab and Haryana. In the Ganges Valley, the

▶ **Figure 12.41 Regional Differences in Indian Economic Development** India shows marked differences in regional levels of economic development. Its more prosperous areas are generally located in the west and south, while the north and east lag behind.

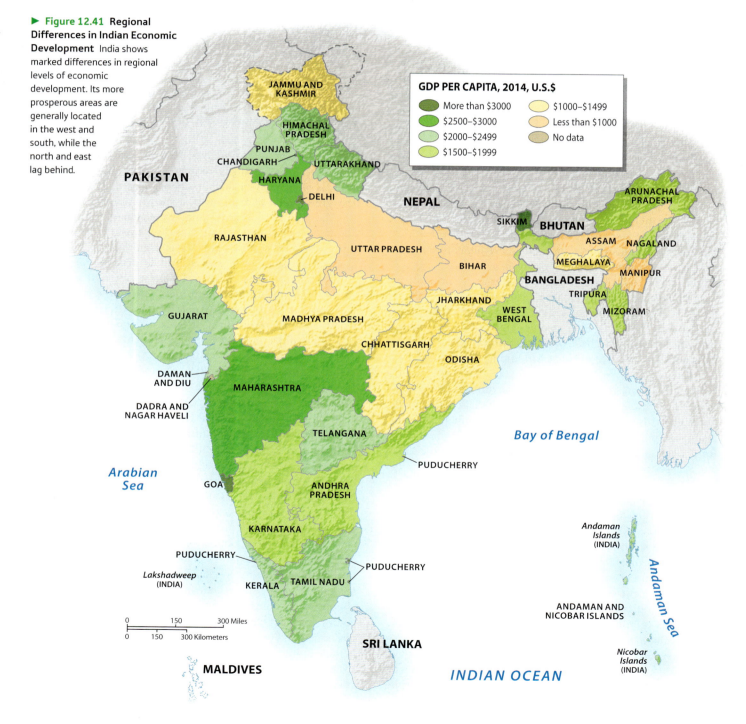

caste system is deeply entrenched, tension between Hindus and Muslims remains bitter, and opportunities for most peasants are limited. Ironically, South Asia's wealth was historically concentrated in these fertile lowlands, yet today the area ranks among the poorest parts of an impoverished world region.

Despite its deeply entrenched poverty, north-central India experienced a surprising resurgence beginning around 2008. Since then, Bihar has posted double-digit economic growth in most years, a turnaround partly attributable to reduced corruption. Several other poor states in the region have also done well of late, including Madhya Pradesh and Jharkhand. The economy of massive Uttar Pradesh exhibited strong growth for several years due mainly to a state-led highway-building boom, but growth has slowed since 2014.

Eastern Indian states such as Odisha (formerly Orissa) and Assam are also quite poor, but the large and important state of West Bengal ranks just below average for India as a whole. Some of the world's worst slums are located in West Bengal's Kolkata, yet Kolkata also supports a substantial well-educated middle class and a sizable industrial complex. For most of the time since independence, West Bengal has been governed by a leftist party that has fostered, with little success, state-led heavy industry. In a dramatic turnaround in the 1990s, West Bengal's Marxist leaders began to advocate internationalization, encouraging large multinational firms to build new factories in "special economic zones" with minimal taxation. Such programs have generated substantial opposition that occasionally becomes violent.

Western India is more prosperous than eastern India, but the large state of Rajasthan ranks below average. Rajasthan suffers from an arid climate; nowhere else in the world are deserts and semideserts so densely populated. It also is noted for its social conservatism. During the British period, almost all of Rajasthan remained outside the sphere of direct imperial power. Here, in the courts of maharajas, the military and political traditions of Hindu India persisted up until recent times. Rajasthan's rulers not only maintained elaborate courts and fortifications, but also supported many traditional Indian arts. This political and cultural legacy makes Rajasthan one of India's most important destinations for international tourists.

India's Centers of Economic Growth North of Rajasthan, the states of Punjab and Haryana showcase the Green Revolution. Their economies rely largely on agriculture, but investments in food processing and other industries have been substantial. Punjab has the lowest levels of malnutrition in India and the most highly developed infrastructure. But, in recent years, economic growth has lagged behind that of India as a whole, generating rural unrest. On Haryana's eastern border lies the capital district of New Delhi, where India's political power and much of its wealth are concentrated.

The west-central states of Gujarat and Maharashtra are noted for their industrial and financial clout as well as their agricultural productivity. Gujarat was one of the first parts of South Asia to experience substantial industrialization, and its textile mills remain highly productive (**Figure 12.42**). Gujaratis have long been famed as overseas traders, and are disproportionately represented in the Indian diaspora. As a result, cash remittances from these emigrants help to bolster the state's economy. Gujarat's government has recently invested in economic development, resulting in extremely rapid economic growth, yet critics contend that social development has lagged behind, due in part to Gujarat's focus on business.

The large state of Maharashtra is usually viewed as India's economic pacesetter. Huge Mumbai has long been India's financial center, media capital, and manufacturing powerhouse. According to official figures, Mumbai's metropolitan area accounts for 25 percent of India's industrial output, 70 percent of its major financial transactions, and as much as 70 percent of its maritime trade. Large industrial zones are located in several other cities of Maharashtra, especially Pune and Nagpur. In recent years, Maharashtra's economy has grown more quickly than those of most other Indian states, reinforcing its primacy.

The center of India's fast-growing high-technology sector lies farther to the south, especially in Karnataka's capital, Bengaluru (Bangalore). The Indian government selected the upland Bengaluru area for its fledgling aviation industry in the 1950s. Other technologically sophisticated ventures soon followed. In the 1980s and 1990s, a fast-growing computer software industry emerged, earning Bengaluru the label "Silicon Plateau." This growth was spurred by the investments of U.S. and other foreign corporations eager to hire relatively inexpensive Indian technical talent. Since the 1990s, these multinational companies have been joined by a rapidly expanding group of locally owned firms. Biotechnology is also thriving. Unfortunately, rapid growth has stretched Bengaluru's infrastructure to the breaking point. Roads are commonly jammed, electricity supplies are inadequate, and many parts of the city can count on only three hours of running water a day.

▼ **Figure 12.42 Textile Factory in Gujarat** The western Indian state of Gujarat is one of India's main manufacturing centers. This modern cotton mill is in the city of Ahmedabad.

Partly because of Bengaluru's problems, other cities in southern India have recently emerged as rival high-tech centers. Hyderabad in Telangana, often called "Cyberabad," is well known for its IT and pharmaceutical firms, as well as its film industry, India's second largest. Chennai (Madras) in Tamil Nadu, recently voted as having the highest quality of life among India's major cities, is noted for software production, financial services, and automobile industries. Tamil Nadu as a whole has exhibited consistently strong economic growth in recent years.

India has proved highly competitive in software because software development does not require a sophisticated infrastructure; computer code can be exported via wireless telecommunication systems instead of modern roads or port facilities. What is necessary, of course, is technical talent, and this India has in abundance. With the growth of the software industry and mobile telephony, India's brainpower is finally translating into economic gains, even for many of the poor. Drivers of auto rickshaws (essentially, expanded three-wheeled motorcycles) can now find passengers in real time through text messages, and farmers can find market prices and gain access to crop disaster insurance through mobile apps. Whether such developments can spread benefits throughout the country and to all sectors of society remains to be seen. What is certain, however, is that IT has tightly linked certain parts of India to the global economy.

Globalization and South Asia's Economic Future

South Asia is not one of the world's more globalized regions by conventional economic criteria. The volume of foreign trade is relatively small, foreign direct investment is modest, and international tourists are few (**Figure 12.43**). But some parts of South Asia have experienced more globalization than others, and almost everywhere global connections are intensifying.

From Protected Markets to Globalization To understand the region's low globalization figures, we must look at its recent economic history. India's postindependence economic policy, like those of other South Asian countries, was based on widespread private ownership combined with high-tariff barriers and government control of planning, resource allocation, and certain heavy industrial sectors. This mixed socialist-capitalist system initially led to fairly rapid development of heavy industry and allowed India to become virtually self-sufficient.

By the 1980s, however, problems with India's economic model were becoming apparent, and frustration was mounting among the business and political elite. Growth continued, but in most years at just a percentage point or two above the rate of population expansion. The proportion of Indians living below the poverty line, moreover, remained virtually constant. At the same time, countries such as

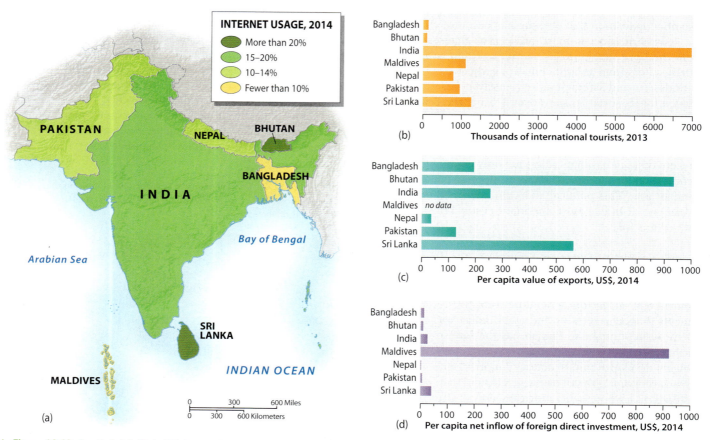

▲ **Figure 12.43** **South Asia's Global Linkages** Despite South Asia's growing global connections, the region as a whole is still relatively self-contained, especially in regard to finance. Internet use remains low, especially in Bangladesh. By some measure, the region's smallest nations, Sri Lanka, Bhutan, and the Maldives, are the most globally integrated.

China and Thailand were experiencing rapid development after opening their economies to global forces. Many Indian businesspeople chafed under governmental regulations that undermined their ability to expand, and foreign debt began to mushroom in the 1980s, further hampering growth.

In response, India's government began to liberalize its economy in 1991, modifying and eliminating many regulations and gradually opening the country to imports and multinational businesses. Firms in the advanced industrial countries increasingly turned to **outsourcing** many of their labor-intensive technical tasks to Indian companies. Other South Asian countries have followed a somewhat similar path, although generally with less success than India.

The gradual internationalization and deregulation of the Indian economy has generated substantial opposition. As early as 1984, opposition to foreign investment mounted when a gas leak from a poorly run Union Carbide pesticide factory in Bhopal killed almost 4000 people and injured as many as 500,000. More recently, foreign competitors began to challenge many domestic firms. Cheap manufactured goods from China are seen as a serious threat. India has a strong heritage of economic nationalism stemming from the colonial exploitation it long suffered. In 2006, India's government stalled further plans for privatization, but in 2014 it again reversed course and announced major sales of shares in state-owned and state-backed industries, particularly those in the energy sector.

South Asian Firms Expand Abroad Increasing technological sophistication and rising wages in India's high-tech centers have allowed Indian companies to acquire foreign firms and relocate some of their operations in the developed world. The famous British car brands Jaguar and Land Rover, for example, have been subsidiaries of Mumbai-based Tata Motors since 2008. Mumbai-based Aegis Communications recently opened a call center in New York, which focuses on providing medical information. Social media apps developed in India are also gaining a global market. Zomato, an Indian restaurant locator app, now operates in more than 150 countries, including the United States, through its Urbanspoon subsidiary (see *Exploring Global Connections: India's Emerging Computer Game Industry*).

Although India gets most of the media attention, other South Asian countries have also experienced significant economic globalization in recent years. Besides exporting textiles and other consumer goods, Bangladesh, Pakistan, and Sri Lanka send large numbers of their citizens to work abroad, particularly in the Persian Gulf countries, as does India. Remittances from foreign workers are Bangladesh's second largest source of income, reaching more than US$14 billion in 2014. On a per capita basis, remittances are much more important in Nepal, where they account for roughly one-quarter of the total economy, making it the world's third most remittance-dependent country. Remittances are also Sri Lanka's largest foreign exchange earner. Out of a total population of 21 million, some 1.5 million Sri Lankans work abroad, 90 percent of them in Southwest Asia.

The future of the South Asian remittance economy, however, is uncertain. In 2015 several governments in the region expressed concern as the growth in remittance earning dropped due to the economic slowdown in the oil-exporting countries of Southwest Asia. That same year, India placed restrictions on its citizens working in the Gulf region due to concerns about worker exploitation, particularly of women. More than 10,000 women from Kerala left for the Gulf region in 2014, but in 2015 that figure dropped to 837.

Social Development

South Asia's social indices show relatively low levels of health and education, which is hardly surprising considering the region's poverty. Levels of social well-being, however, vary greatly across the region: people in the more prosperous areas of western and southern India are healthier, live longer, and are better educated, on average, than those in poorer areas such as the lower Ganges Valley.

Several key measures of social welfare are now higher in India than in Pakistan. With a 58 percent literacy rate—as opposed to India's 72 percent—Pakistan has done a particularly poor job of educating its people. Still, it is important to recognize how much progress has been made. The province of Balochistan, for example, saw its literacy rate increase from 10 percent in 1981 to over 43 percent in 2015. Such gains have not occurred in the Federally Administered Tribal Areas (FATA), where the literacy rate is only 28 percent overall and under 8 percent for women. FATA's lack of formal education opened the door for radical Islamist organizations, which are often the only options for schooling.

In other areas of social development, however, Pakistan posts figures similar to those of India. Both countries, for example, have an average life expectancy of 66 years. Until the late twentieth century Pakistan suffered less malnutrition and hunger than India, but that situation has been reversed: the 2015 Global Hunger Index ranks Pakistan 11th from the bottom, whereas India is ranked in the 25th lowest position.

India has instituted several programs to reduce poverty and enhance health and education, especially among its most deprived communities. These initiatives have been plagued, however, by high levels of corruption. To reduce fraud in both welfare programs and elections, India has created an ambitious personal identification system in which every citizen will be given a unique ID number linked via a central database to such biometric data as photographs, fingerprints, and iris scans. As of January 2016, some 970 million Indians had been issued unique identification numbers. This program is expected to save the Indian government roughly US$1 billion a year, but many critics consider it an unwarranted intrusion on personal privacy.

Several discrepancies stand out when comparing South Asia's map of economic development with a map of social well-being. Parts of India's extreme northeast, for example, enjoy high literacy rates despite a general lack of economic development, due to the educational efforts of Christian missionaries. In Mizoram, for example, 94 percent of people over the age of seven can read and write. Greater Kolkata also stands out as relatively well-educated despite high levels of poverty in the lower Ganges Basin. The most pronounced discrepancies are found in the southern reaches of South Asia, which far outpaces the rest of the region in health, longevity, and education.

Exploring Global Connections

India's Emerging Computer Game Industry

India is well known for its information technology. Most Indian IT firms got their start by subcontracting for large western firms, often by providing specified forms software as well as a large array of back-office services. Companies such as Infosys, with revenue of US$8.7 billion in 2015, have been moving up the technology ladder, emerging as global leaders in certain niches of software engineering. But until recently, India lacked the "start-up" business culture that has been vital in developing cutting-edge technologies in Silicon Valley and other tech hubs. This is beginning to change, however, as a new generation of Indian entrepreneurs launch their own firms. Some of these companies are following in the steps of established firms, such as Amazon and Uber. Many of the smaller Indian start-ups, however, are focusing on developing video games for both Indian and global customers (**Figure 12.5.1**).

Developing an Indian Gaming Industry Video game-making in India got its start in the same way that many other tech industries did: by taking on certain highly specialized tasks, such as modeling the movements of racing cars, for American or European firms. Now companies like Mumbai-based Yellow Monkey Studios make their own games designed for a global audience, such as "Socioball," described by the company as a "stylish new isometric puzzle game." The rise of the Indian gaming industry was evident in the 2015 and 2016 meetings of Pocket Gamer Connect, one the world's leading mobile gaming events, in Bengaluru. Firms as large as Intel, Amazon, and Google sent representatives to the event, eager to establish a presence in this fast-growing sector.

A number of Indian gaming firms are connected with the film and television industries. Mumbai-based Reliance Games, for example, has racked up more than 70 million downloads of smartphone games linked to such movies as *Real Steel*, *Catching Fire*, and *Pacific Rim*. India's own massive film industry plays an increasingly important role as well. Ubisoft Pune, for example, has had marked success in smartphone music games that are endorsed by Bollywood stars. Game developers hope that such Indian-themed content will win over a global audience.

Obstacles and Opportunities India's gaming industry has been aided by the spectacular rise of smartphones in the country, now roughly as numerous in India as they are in the United States. But it has also been held back by India's poorly developed mobile communications infrastructure, which basically operates at 2G speeds. An equally serious problem is the lack of credit cards, held by only about 8 percent of Indians, which hinders purchase through online app stores. But, not surprisingly, Indian entrepreneurs are working on these problems through such measures as third-party wallet companies.

▲ **Figure 12.5.1 Indian Video Game Development** The Indian information technology industry has recently begun to take on video game development. Pictured here is Jaspreet Binda, head of the Entertainment Division of Microsoft India, with a game console at his office in Delhi, India.

However it is examined, the fast-growing Indian gaming industry is a highly globalized phenomenon. A prime example is UTV Software Communications, which aims to be the first Indian company to become a fully global player in the video game market—and which was acquired by Disney Enterprises in 2011. But globalization cuts both ways. A small Indian studio called On The Couch Entertainment recently created a game called Rooftop Mischief, but did not have the expertise to generate the accompanying music. For this feature the company turned to outsourcing, getting the necessary music from a firm in the United Kingdom.

Google Earth Virtual (MG) Tour Video

http://goo.gl/Pqu0xp

1. What cultural advantages might help India in its quest to develop video games for the global market?

2. What other obstacles might India face in trying to develop a start-up business culture?

The Educated South Southern South Asia's relatively high levels of social welfare are clearly visible in Sri Lanka. Despite its meager economic resources and long-lasting civil war, Sri Lanka must be considered one of the world's social-development success stories, with an average life span of 76 years and a literacy rate of 93 percent. It demonstrates that a country can achieve significant health and educational gains even in the context of an "undeveloped" economy. Sri Lanka's government achieved these results by funding universal primary education and inexpensive medical clinics.

Kerala in southwestern India has achieved even more impressive results. Kerala is not a particularly prosperous state. It is extremely crowded, has a high rate of unemployment, and

has long struggled to feed its population. Kerala's indices of social development, however, are the best in India, comparable to those of Sri Lanka: an estimated 94 percent of Kerala's adults are literate and average life expectancy stands at 74 years. Moreover, several diseases including malaria have been essentially eliminated from the state. Since Kerala is poorer than Sri Lanka, its social accomplishments are all the more impressive.

Kerala's social successes have been linked to its state policies. For most of the period since Indian independence, Kerala has been led by a socialist party that stressed mass education and community health care. Yet this may not be the only factor, as West Bengal also has a socialist political heritage, but its social programs have not been nearly as successful.

Kerala's neighboring state of Tamil Nadu, on the other hand, has made very rapid social progress in recent years despite different political leadership. Some researchers suggest that a key variable for explaining the far south's success is the relatively high social position of its women, whether Hindu, Muslim, or Christian. This has historical roots; among the Nairs—Kerala's traditional military and land-holding caste—all inheritance up to the 1920s passed through the female line.

Gender Relations in South Asia It is often said that South Asian women are accorded a very low social position in both the Hindu and the Muslim traditions. Higher-class families of both religions throughout the Indus–Ganges Basin traditionally secluded women to a large degree, and restricted their social relations with men outside the family. Throughout northern India, women traditionally leave their own families shortly after puberty to join their husbands' families. As outsiders, often in distant villages, young brides have few opportunities, and it is not uncommon for them to be bullied by parents-in-law. Higher-caste widows, moreover, are encouraged to go into permanent mourning rather than remarry.

Several social indices show that women in the Indus–Ganges Basin still suffer pronounced discrimination. In Pakistan, Bangladesh, and such Indian states as Rajasthan, Bihar, and Uttar Pradesh, female literacy rates are much lower than those of males. A disturbing statistic is the gender ratio—the relative proportion of males and females in the population. A 2014 UN report concluded that the declining child sex ratio in India—going from 976 girls to 1000 boys in 1961 to 918 girls in 2011—had reached "emergency proportions," thus requiring "urgent action." This problem is much more severe in north-central and northwestern India than elsewhere, but only in Kerala and Mizoram do we find biologically normal sex ratios. Efforts to address the issue include sting operations to catch doctors who perform illegal sex-selective abortions. In 2015, Punjab launched a program in which officials greet the parents of newborn girls in the most male-dominated districts and provide them with valuable gifts.

India's imbalance of males over females has several causes. Sex-selective abortion, though illegal, is not uncommon, resulting in the loss of an estimated 10 million girls in northern India over the past 20 years. In poor families, moreover, boys typically receive better nutrition and medical care than do girls, resulting in higher male survival rates. In rural households, boys are usually viewed as an economic asset because they typically remain with and work for the well-being of their families. In the poorest groups, elderly people (especially widows) subsist largely on what their sons provide. Girls, on the other hand, become an economic liability, marrying out of their families at an early age and requiring a dowry.

The social position of women appears to be improving, especially in the more prosperous parts of western India where employment opportunities outside the family context are emerging. But even in many middle-class households, women still suffer major disadvantages. Dowry demands seem to be increasing in many areas, and murders of young brides whose families failed to provide enough goods are not

▲ **Figure 12.44 Gulabi Gang** A grass-roots organization active across northern India, the Gulabi Gang is dedicated to women's empowerment and the struggle against oppression. Its members are noted for their pink saris and bamboo sticks that can be used for self-defense.

uncommon in north-central India, numbering over 8,000 in 2012. And even in Kerala the number of girls being born relative to the number of boys has declined, showing some evidence of sex-selective abortion.

Developments in Women's Rights Despite the striking social bias against South Asian women, women's rights organizations are gaining strength across India. One of their main concerns is sexual violence, a huge problem especially in the Ganges Valley. In 2013, more than 25,000 incidents of rape were reported in India, yet activists estimate that 90 percent of rapes go unreported. Women from low-caste backgrounds are the most common victims, but all social groups are vulnerable. After several horrific cases of sexual violence came to light, massive anti-rape protests erupted in many India cities in 2012 and again in 2014, several of which were broken up by the police with water cannons.

Women's empowerment movements, using both social media and traditional forms of organization, have led much of this resistance. One such organization is the Gulabi Gang, whose members wear pink saris and carry bamboo sticks for self-protection (**Figure 12.44**). The Gulabi Gang was recently the subject of a major Indian motion picture, and has gained the corporate sponsorship of several information technology firms. Other women's groups in both India and Pakistan are struggling for the mere freedom to enjoy public spaces. Currently, women are often harassed or accused of prostitution simply for relaxing or chatting with friends in the open. The "Why Loiter" campaign that emerged in early 2016 seeks both to change attitudes and to directly challenge men who bully women in streets and public squares.

REVIEW

12.13 How has India's economy been transformed since the reforms of 1991?

12.14 Explain why levels of social and economic development vary so much across South Asia.

12.15 How does the social position of women vary in different regions of South Asia?

■ **KEY TERMS** microfinance, outsourcing

Chapter 12 South Asia

Summary

- The region's monsoon climate generates both extensive floods and droughts. Rising sea level associated with global climate change may play havoc with the monsoon-dependent agricultural systems of India, Pakistan, and Bangladesh.

- Continuing population growth in densely populated South Asia demands attention. Although fertility rates have declined rapidly, Pakistan and northern India cannot easily meet the demands imposed by their expanding populations.

- South Asia's diverse cultural heritage, shaped by peoples speaking several dozen languages and following several major religions, creates a rich social environment. Unfortunately, cultural differences have often translated into political conflicts.

- The long-standing feud between Indian and Pakistan over the Kashmir issue generates huge problems for South Asia and for the world as a whole. Religious and ethnic conflicts plague several parts of the region.

- Much of South Asia has seen rapid economic expansion in recent years. India, in particular, has benefited from economic globalization, but hundreds of millions of Indians remain deeply impoverished.

Review Questions

1. Explain how South Asia's monsoon climate influences agricultural patterns and economic development in the region. How, and where, will climate change affect further development?

2. Describe the major migration flows in South Asia and explain the relationship between migration and geopolitical conflicts in the region.

3. Most scholars consider South Asia a culturally coherent region, but many South Asians increasingly stress the distinction between Muslim and Hindu cultures. Why? Relate this distinction to the creation of national identity and problems of ethnic insurgency in the region.

4. Explain how the conflict between India and Pakistan involved world powers found outside the region—China, Russia, and the United States. How have these dynamics changed over time?

5. Describe how economic globalization has impacted the countries of South Asia. Why has the process historically generated so much controversy in the region?

Image Analysis

1. This graph shows annual economic growth for India and Pakistan from 1960 to 2014. Was growth steady, or did it fluctuate, and why? Which country enjoyed higher growth prior to 1990, and which since 1990? What factors might account for the change?

2. Trade between India and the United States expanded considerably after 1991. Look at India's economic growth since that year. Describe the relationship between India's growth and its economic ties with the United States. What goods or services from India do you purchase or use?

▶ **Figure IA12 Annual Economic Growth for India and Pakistan, 1960–2014**

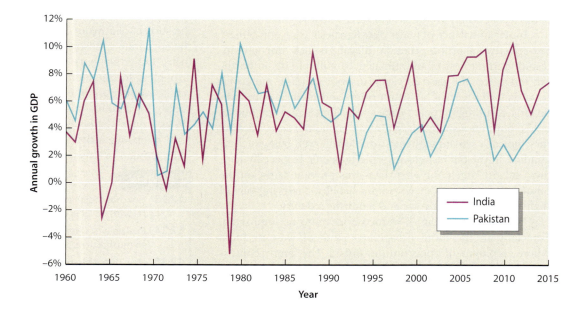

JOIN THE DEBATE

Pakistan's dam-building agenda has become controversial, as we saw at the beginning of this chapter in connection with the Diamer-Bhasha Dam project. Intriguingly, both opponents and proponents of dams often argue on ecological grounds, the former pointing to environmental degradation and the latter stressing the need for clean energy. Should dams be built?

Dam Building Should Proceed

- The region needs more energy, and hydroelectric power emits little carbon dioxide or conventional pollutants.
- Dams will reduce the flood threat in downstream areas, which could intensify due to climate change.
- Such dams will allow the expansion of irrigation, increasing the food supply of a hungry region.

Dam Building Should Come to an End

- The reservoirs behind the dams will displace large numbers of indigenous people, who will end up homeless, landless, and poor.
- Fish and other forms of wildlife will be disrupted by the dams and reservoirs, reducing biodiversity in the region.
- The reservoirs behind the dams will eventually fill with silt, and are therefore unsustainable.

▲ **Figure D12** **Protesters Rally Against the Construction of the Diamer-Bhasha Dam in Karachi**

■ KEY TERMS

British East India Company (p. 520)
Buddhism (p. 512)
bustees (p. 508)
caste system (p. 510)
cyclone (p. 498)
dalit (p. 511)
Dravidian family (p. 515)
enclaves (p. 524)
exclaves (p. 524)
federal state (p. 523)
forward capital (p. 509)
Green Revolution (p. 505)
Hindu nationalism (p. 510)
Hinduism (p. 510)

Indian diaspora (p. 518)
Indian subcontinent (p. 492)
Jainism (p. 514)
linguistic nationalism (p. 517)
maharaja (p. 524)
Maoist (p. 525)
microfinance (p. 529)
monsoon (p. 495)
Mughal Empire (also spelled *Mogul*) (p. 512)
orographic rainfall (p. 495)
outsourcing (p. 533)
salinization (p. 508)
Sikhism (p. 514)
Tamil Tigers (p. 525)

Mastering Geography™

Looking for additional review and test prep materials? Visit the Study Area in **MasteringGeography**™ to enhance your geographic literacy, spatial reasoning skills, and understanding of this chapter's content by accessing a variety of resources, including **MapMaster** interactive maps, videos, *In the News* RSS feeds, flashcards, web links, self-study quizzes, and an eText version of *Diversity Amid Globalization*.

DATA ANALYSIS

http://goo.gl/uzALmw

India, particularly northwestern India, is noted for its male-biased population, which generally indicates a low social standing for women. This bias is partly due to sex-selective abortion (illegal in India, but still widely practiced), but other factors are at work, such as different levels of nutrition and medical care provided for boys and girls. Go to the sex ratio page on the website for India's Planning Commission. http://planningcommission.nic.in/data/datatable/data_2312/DatabookDec2014%20215.pdf

1. Use these data to construct several graphs showing changes in sex ratios from 1901 to 2011 for India as a whole and for the following Indian states: Haryana, Kerala, Himachal Pradesh, and Bihar.

2. Sex-selective abortion only became widespread in the 1970s. What do the data suggest about the role of this procedure in maintaining unbalanced sex ratios in India? Describe the effects of campaigns to reduce sex-selective abortion and improve care for baby girls over the past several decades.

3. Compare the trend lines for the four states you graphed. Write a paragraph explaining the differences found across these four states, based on what you know about India's cultural geography and its economic and social development.

Authors' Blogs

Scan to visit the
GeoCurrents blog
http://geocurrents.info/category/place/south-asia

Scan to visit the
Author's Blog
field notes, media resources, and chapter updates
http://gad4blog.wordpress.com/category/south-asia/

Physical Geography and Environmental Issues
Southeast Asia is divided into a mainland area, and an island region. Southeast Asian rainforests are vital centers of biological diversity, but they are rapidly diminishing due to commercial logging and agricultural expansion.

Population and Settlement
Southeast Asia has a particularly uneven pattern of population distribution, with some areas experiencing serious crowding and others noted for their sparse settlement. Birthrates have recently dropped sharply across most of the region.

Cultural Coherence and Diversity
Culturally, Southeast Asia is characterized by more diversity than coherence, hosting significant areas of Muslim, Buddhist, and Christian religions, as well as a vast number of languages.

Geopolitical Framework
Southeast Asia is geopolitically unified, with all but one of its countries belonging to the Association of Southeast Asian Nations (ASEAN). Ethnic conflicts and maritime tensions with China, however, generate geopolitical problems in several Southeast Asian countries.

Economic and Social Development
Southeast Asia as a whole contains some of the world's most globalized economies, but until recently some of its countries were relatively isolated from the global system. Economic conditions in the region range from highly developed to deeply impoverished.

▶ Completed in 2011, the Third Thai-Lao Friendship Bridge spans the broad Mekong River, linking Thailand's Nakhon Phanom Province with Laos's Khammouane Province. The four Thai-Lao friendship bridges are important components of an emerging transportation system that connects the countries of Mainland Southeast Asia with each other and with the broader world economy.

Nakhon Phanom, Thailand

SOUTHEAST ASIA

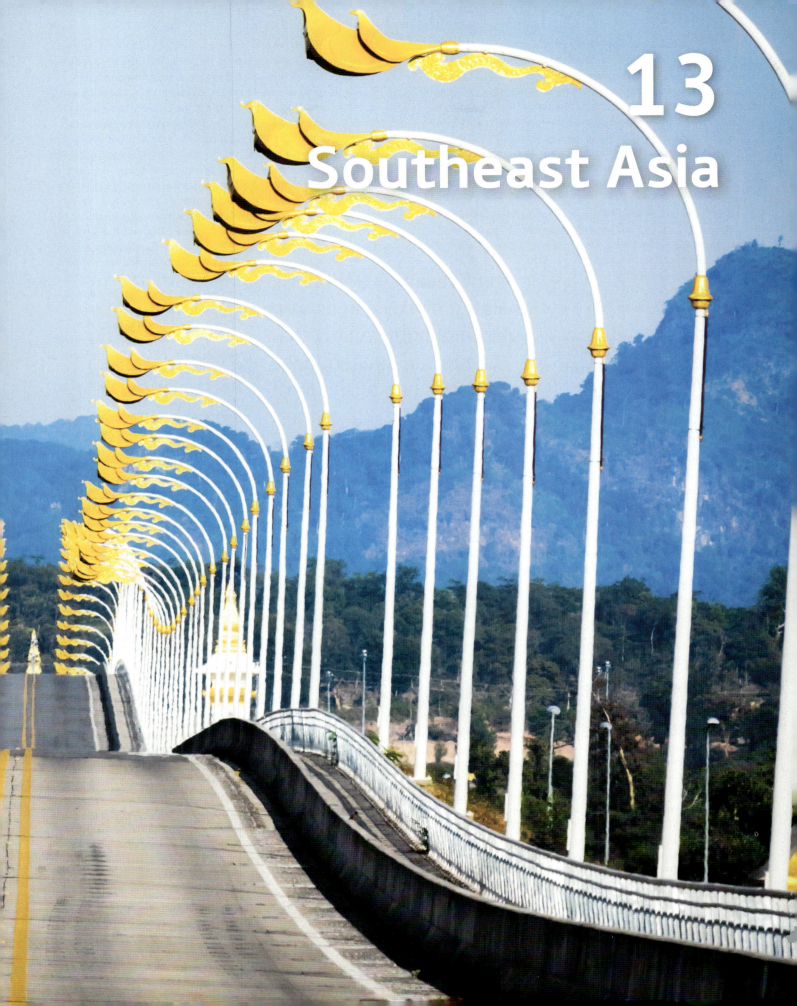

The Association of Southeast Asian Nations (ASEAN), has long been noted for its political success, which has greatly reduced tensions among the countries of the region. ASEAN's economic achievements have been more modest, as most Southeast Asian countries have focused more on their trading ties with East Asia, North America, and Europe than with each other. Recent signs, however, indicate that Southeast Asian economic integration is picking up. If successful, this development could enhance the global significance of both ASEAN and its major member states.

Infrastructural developments have been crucial to ASEAN's economic ambitions. In 2006 the area saw the initiation of the association's East-West Economic Corridor, a major road and port system that runs from Mawlamyine in Burma (Myanmar) through northern Thailand and southern Laos to terminate at the Vietnamese port of Da Nang. Thailand's government in particular has capitalized on this project by building a series of connecting roads and railroads, hoping to turn the country into Southeast Asia's "connectivity hub."

Thailand's economic position within the region was further enhanced by the 2015 creation of the ASEAN Economic Community, an organization that seeks to create a single Southeast Asian market, much like the European Union's single market. Hoping to benefit from improved economic integration fostered by both the transportation corridor and the economic community, Thailand announced in late 2015 that it would create special economic zones in six border provinces aimed at attracting massive foreign investment. Thailand's Commerce Ministry has conducted over 200 training sessions to teach businesspeople, farmers, educators, and government officials how to take advantage of Southeast Asian economic integration. Large firms have also stepped in. In 2014 alone, Siam Cement Group saw its annual sales in other ASEAN countries rise by 18 percent.

Defining Southeast Asia

Until recently, Southeast Asia's global linkages were growing more rapidly than its internal connections, making it one of the most globalized parts of the developing world. The region's involvement with the larger world is not new. Chinese and Indian connections date back many centuries. Later, commercial ties across the Indian Ocean opened the doors to Islam, making Indonesia the world's most populous Muslim-majority country. More recently came the impact of the West, as Britain, France, the Netherlands, and the United States gained control of Southeast Asian colonies. The region's resources and its strategic location made it a major battlefield during World War II. Long after peace was restored in 1945, Southeast Asia remained a Cold War battleground for world powers and their competing economic systems.

Southeast Asia is a clearly defined region consisting of 11 countries that vary widely in spatial extent, population, and cultural traits (Figure 13.1). Geographers divide these countries into those on the Asian continent—mainland Southeast Asia—and those on islands—insular Southeast Asia. The mainland includes Burma, Thailand, Cambodia, Laos, and Vietnam. Insular Southeast Asia includes the large countries of Indonesia, the Philippines, and Malaysia and the small countries of Singapore, Brunei, and East Timor. Although classified as part of the insular realm because of its cultural and historical background, Malaysia actually splits the difference between the continent and the islands. Part of its national territory is on the mainland's Malay Peninsula and part is on the huge island of Borneo, some 300 miles (480 km) distant. Borneo also includes Brunei, a small, oil-rich country of roughly 400,000 people. Singapore is essentially a city-state, occupying an island just to the south of the Malay Peninsula.

Controversies surround the names of two Southeast Asian countries. Since 1989, the government of Burma has insisted that the country's English name is *Myanmar*. Both terms are used in the country itself. Burma's democratic movement, as well as the governments of the United States, the United Kingdom, and Canada, have continued to use the term *Burma*, as does this book. If the country's government continues on its current path of reform, we may switch to *Myanmar* in future editions. Less controversial is the official name of East Timor: Timor-Leste. As *Leste* derives from the Portuguese word for "east," most English-language sources, including this book, continue to use the more familiar term *East Timor*.

> Until recently, Southeast Asia's global linkages were growing more rapidly than its internal connections, making it one of the most globalized parts of the developing world.

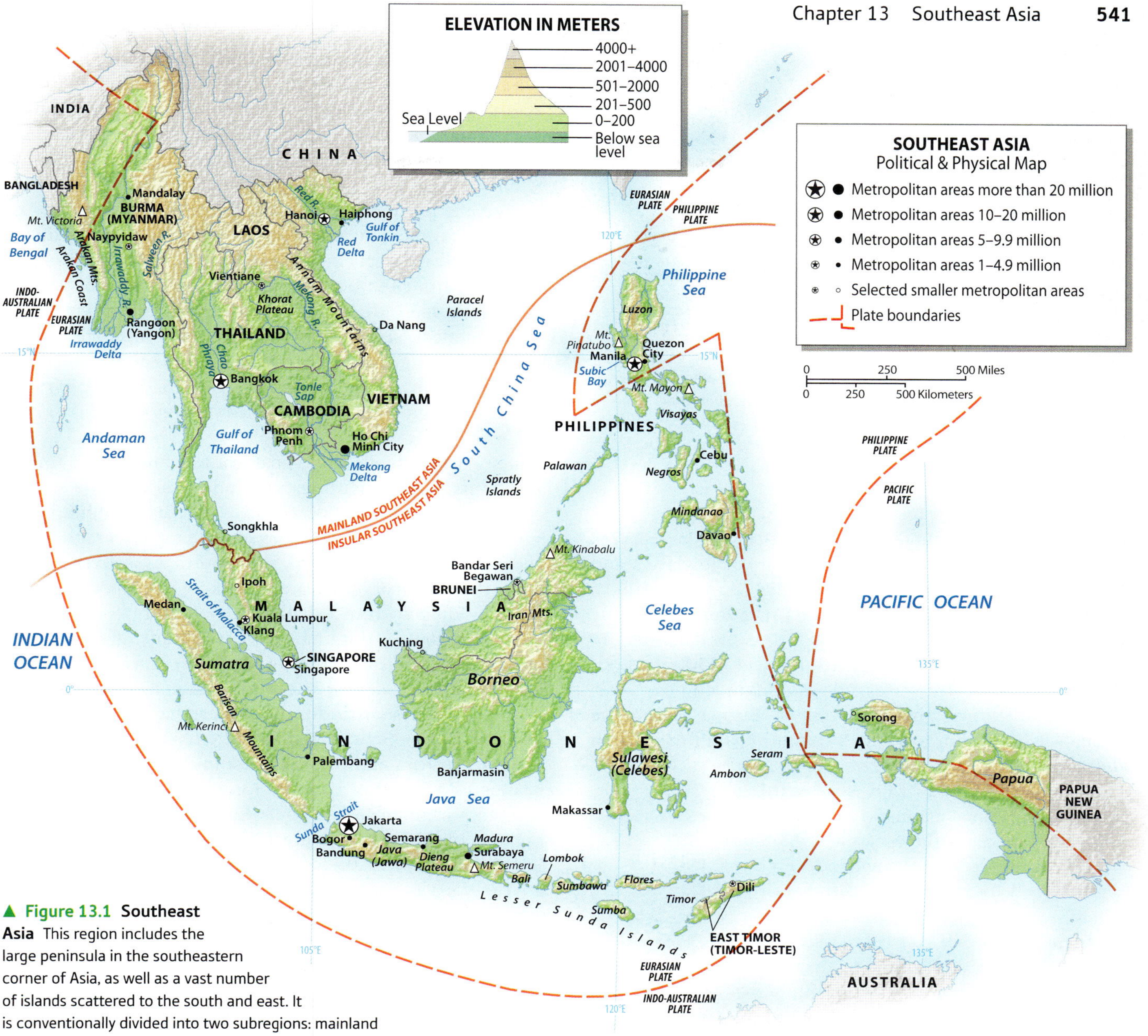

▲ **Figure 13.1 Southeast Asia** This region includes the large peninsula in the southeastern corner of Asia, as well as a vast number of islands scattered to the south and east. It is conventionally divided into two subregions: mainland Southeast Asia, which includes Burma (Myanmar), Thailand, Laos, Cambodia, and Vietnam; and insular (island) Southeast Asia, which includes Indonesia, the Philippines, Malaysia, Brunei, Singapore, and East Timor (Timor-Leste). Malaysia includes the tip of the mainland peninsula and most of the northern part of the island of Borneo (which Indonesians call Kalimantan).

Physical Geography and Environmental Issues: A Once-Forested Region

Before the 20th century, Southeast Asia was probably the most heavily forested world region. Although much of the region is still wooded, most areas have been cleared for agriculture and human settlement. The commercial logging of tropical forests, however, remains a controversial issue in several Southeast Asian countries. To understand regional patterns of both forest preservation and destruction, it is necessary to examine differences in landforms and climates across the region.

Patterns of Physical Geography

The physical environments of insular Southeast Asia differ significantly from those of the mainland part of the region. The island belt is mostly situated in the equatorial zone, constituting one of the world's three main zones of tropical rainforest. Mainland Southeast Asia, on the other hand, is mostly located in the tropical wet-and-dry zone, noted for its large seasonal differences in rainfall.

▲ **Figure 13.2 Delta Landscape** The Mekong Delta in southern Vietnam encompasses a complex maze of waterways. Extensive tracts of fertile farmland lie between the canals and river channels. The waterways are used for transportation and provide large quantities of fish and other aquatic resources.

Mainland Environments Mainland Southeast Asia is an area of rugged uplands interspersed with broad lowlands and deltas associated with large rivers. The region's northern boundary lies in a cluster of mountains connected to the highlands of eastern Tibet and south-central China. In Burma's far north, peaks reach 18,000 feet (5500 meters). From this point, a series of mountain ranges radiates out, extending through western Burma, along the Burma–Thailand border, and through Laos into southern Vietnam.

Several large rivers flow southward out of Tibet into mainland Southeast Asia. The longest is the Mekong, which courses through Laos, Thailand, and Cambodia before entering the South China Sea through an extensive delta in southern Vietnam (**Figure 13.2**). Second longest is the Irrawaddy, flowing through Burma's central plain before reaching the Andaman Sea, which also has a large delta. Also significant are two smaller rivers: the Red River, which forms a sizable and heavily settled delta in northern Vietnam, and the Chao Phraya, which created the fertile alluvial plain of central Thailand.

The centermost area of mainland Southeast Asia is Thailand's Khorat Plateau. Neither a rugged upland nor a fertile river valley, this low sandstone plateau averages about 500 feet (175 meters) in height and is noted for its thin soils. Water shortages and periodic droughts also plague the area.

Monsoon Climates Mainland Southeast Asia is affected by the seasonally shifting winds known as the monsoon (see Chapter 12). Its climate is characterized by a distinct hot and rainy season from May to October, followed by dry but still generally warm conditions from November to April (**Figure 13.3**). Only the central highlands of Vietnam and a few coastal areas receive significant rainfall during this period. In the far north, the winter months bring mild and sometimes rather cool weather.

Explore the **Sounds** of Mekong Delta Floating Market

http://goo.gl/Bcc0BV

Two tropical climate regions characterize mainland Southeast Asia. Although both are dominated by the monsoon, they differ in the amount of precipitation. Along the coasts and in the highlands, the tropical monsoon (Am) climate dominates, with annual rainfall usually exceeding 100 inches (250 cm) each year. The greater portion of mainland Southeast Asia falls into the drier tropical savanna (Aw) climate type. This reduced rainfall is largely explained by interior locations sheltered from the oceanic source of moisture by mountain ranges. Much of Burma's central Irrawaddy Valley is almost semiarid, with precipitation totals below 30 inches (75 cm). Forests in these areas are especially vulnerable, being easily converted by fire and agriculture into rough landscapes of brush and grass.

Insular Environments The signal feature of insular Southeast Asia is its island environments. Indonesia alone is said to contain more than 13,000 islands, whereas the Philippines supposedly contains 7000. Borneo and Sumatra are the third and sixth largest islands in the world, respectively, while many thousands of others are little more than specks of land rising at low tide from a shallow sea.

Indonesia is dominated by the four great islands of Sumatra, Borneo (the Indonesian portion is called Kalimantan), Java, and the oddly shaped Sulawesi. This island nation also includes the western half of New Guinea and the Lesser Sunda Islands, which extend to the east of Java. A prominent mountain spine runs through these islands as a result of tectonic forces. The two largest and most important islands of the Philippines are Luzon (about the size of Ohio) in the north and Mindanao (the size of South Carolina) in the south. Sandwiched between them are the Visayas Islands, which number roughly a dozen.

Closely related to this impressive collection of islands is one of the world's largest expanses of shallow seas. These waters cover the **Sunda Shelf**, which is an extension of the continental shelf stretching from the mainland through the Java Sea between Java and Borneo. Here waters are generally less than 200 feet (70 meters) deep. Some local peoples have adopted lifestyles that rely on the rich marine life of this region, essentially living on their boats and setting foot on land only as necessary.

Insular Southeast Asia is less geologically stable than the mainland. Four of Earth's tectonic plates converge here. As a result of tectonic processes, earthquakes occur frequently. Such seismic activity occasionally results in **tsunamis**, which can devastate coastal regions. A massive earthquake and tsunami in northern Sumatra on December 26, 2004, caused roughly 100,000 deaths in Indonesia alone. Insular Southeast Asia's tectonic activity has also generated a string of active volcanoes that extends the length of eastern Sumatra across Java and into the Lesser Sunda Islands. From late 2013 through March 2016, formerly inactive Mount Sinabung in northern Sumatra has erupted numerous times, forcing the government to relocate thousands of villagers (**Figure 13.4**).

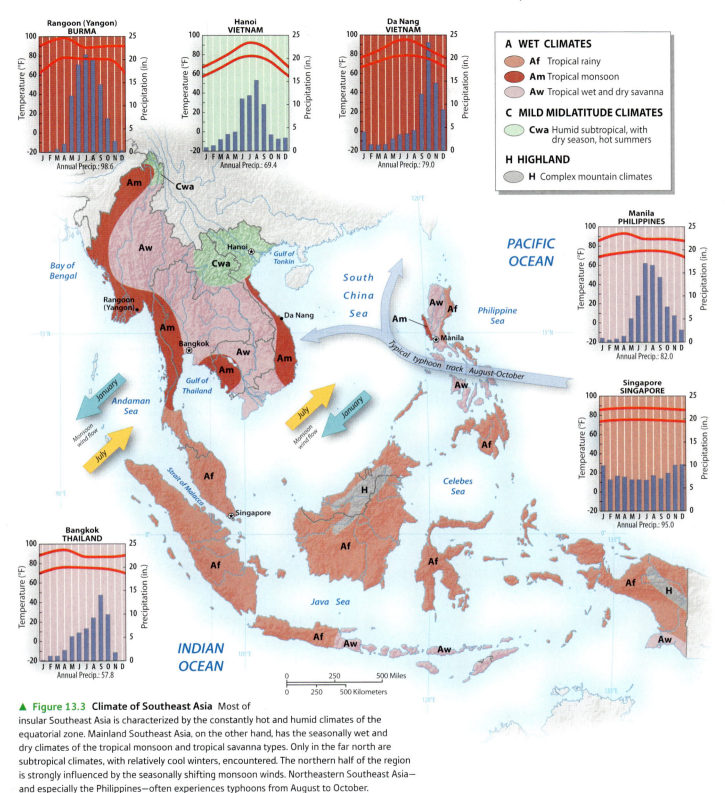

▲ Figure 13.3 Climate of Southeast Asia Most of insular Southeast Asia is characterized by the constantly hot and humid climates of the equatorial zone. Mainland Southeast Asia, on the other hand, has the seasonally wet and dry climates of the tropical monsoon and tropical savanna types. Only in the far north are subtropical climates, with relatively cool winters, encountered. The northern half of the region is strongly influenced by the seasonally shifting monsoon winds. Northeastern Southeast Asia—and especially the Philippines—often experiences typhoons from August to October.

Island Climates The climates of insular Southeast Asia are more varied than those of the mainland. Most of insular Southeast Asia, unlike the mainland, receives rain during the Northern Hemisphere's winter because the seasonal winds cross large areas of warm equatorial ocean, absorbing moisture. On Sumatra and Java, where north winds blow between November and March, heavy rains occur on the northern side of these east–west-running islands. But during the May–September period, the heaviest rains are found on the southern flanks of these same islands because of

indigenous peoples have long cleared small areas of forest for agricultural use, rampant deforestation came only in the last decades of the twentieth century. This surge in forest clearance was associated both with population growth and the resulting need for agricultural expansion and with large-scale commercial logging. The latter activity has largely been driven by the developed world's appetite for wood products such as tropical hardwood, plywood, and paper pulp. China plays a major role as well; by 2011, it was estimated that more than half of all timber shipped worldwide was destined for China.

Southeast Asian forests that have been commercially logged are sometimes opened for agricultural settlement, but more often they are replanted with trees or simply allowed to return to forest vegetation (**Figure 13.7**). Some of the logged-off lands are turned over to fast-growing, "weedy" tree species that support little wildlife. In Malaysia and Indonesia especially, vast areas of former rainforest have been planted with African oil palms, which yield large quantities of inexpensive edible oil.

▲ Figure 13.4 Mount Sinabung Erupting
Indonesia is famous for its numerous volcanic eruptions, which often cause extensive damage but also generate highly fertile soils. From 2014 through early 2015, Mount Sinabung in North Sumatra intermittently erupted, often violently, forcing many villagers to seek safety elsewhere.

Explore the **Sights** of Indonesia's Mt. Sinabung

http://goo.gl/cL5JrM

Local Patterns of Deforestation Malaysia has long been an exporter of tropical hardwoods. Most of the primary forests in western (or peninsular) Malaysia were cut in the late 20th century, which increased the logging pressure in the states of Sarawak and Sabah on the island of Borneo. In these areas, the granting of logging concessions to Malaysian and foreign firms has harmed local tribal people by disrupting their traditional resource base. Although Malaysia's conservation policies look good on paper, they are often not enforced. A 2013 study indicated that more than 80 percent of the forests on the Malaysian part of Borneo had been heavily logged.

the southwesterly winds associated with the Asian summer monsoon.

The climates of Indonesia, Singapore, Malaysia, and Brunei are heavily influenced by their equatorial locations, which almost eliminates seasonality. Here temperatures remain elevated throughout the year, with little variation, and rainfall is abundant and evenly distributed. As a result, large areas of island Southeast Asia are placed into the tropical rainforest (Af) climate category. The southeastern islands of Indonesia, however, experience a distinct dry season from May to October.

The southern part of the Philippines also has an equatorial climate, but most of the country experiences a dry period from December to April. The northern and central parts of the Philippines are frequently hit by tropical cyclones, or **typhoons**, especially from August to October. Each year several typhoons strike the Philippines with heavy damage through flooding and landslides. Typhoon Haiyan (Yolanda), which hit the central Philippines in November 2013, was the strongest storm ever to make landfall (**Figure 13.5**). Haiyan killed more than 6300 people, harmed roughly 14 million, and caused damage in excess of US$12 billion. After the storm passed, the Philippine government adopted a US$3.7 billion reconstruction plan aimed at fostering a sustainable, long-term recovery. As many as 1 million people may have to be moved away from the most vulnerable coastal areas.

The Deforestation of Southeast Asia

Deforestation has long been a major issue through most of Southeast Asia (**Figure 13.6**). Although colonial powers cut Southeast Asian forests for tropical hardwoods and naval supplies and

▼ Figure 13.5 Typhoon Haiyan The Philippines has suffered several devastating tropical storms in recent years, including Typhoon Bobha in 2012 and Typhoon Haiyan in 2013. **Q: What features of Philippines' physical and human geography make it particularly vulnerable to cyclone damage?**

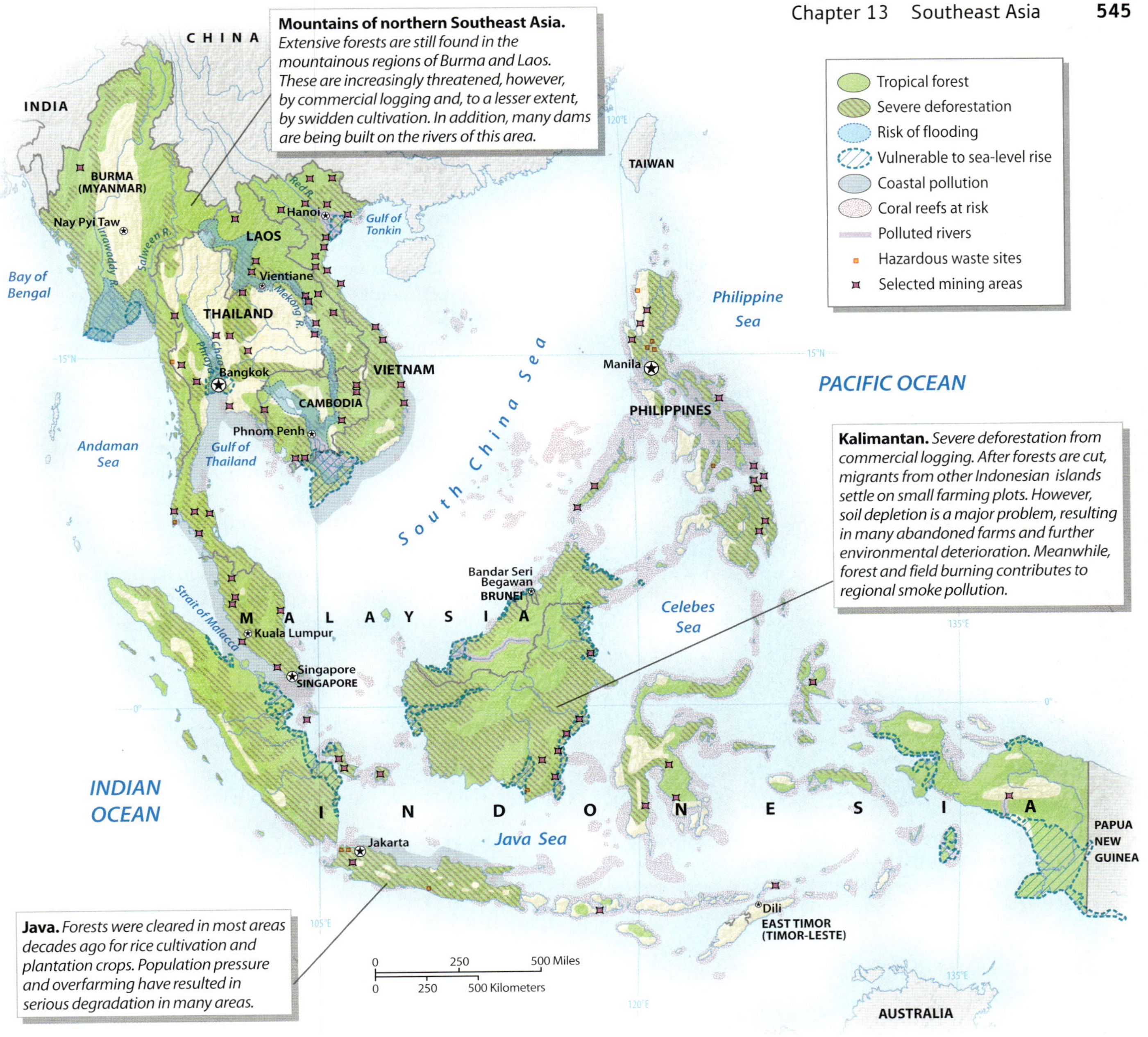

Mountains of northern Southeast Asia. *Extensive forests are still found in the mountainous regions of Burma and Laos. These are increasingly threatened, however, by commercial logging and, to a lesser extent, by swidden cultivation. In addition, many dams are being built on the rivers of this area.*

Legend:
- Tropical forest
- Severe deforestation
- Risk of flooding
- Vulnerable to sea-level rise
- Coastal pollution
- Coral reefs at risk
- Polluted rivers
- Hazardous waste sites
- Selected mining areas

Kalimantan. *Severe deforestation from commercial logging. After forests are cut, migrants from other Indonesian islands settle on small farming plots. However, soil depletion is a major problem, resulting in many abandoned farms and further environmental deterioration. Meanwhile, forest and field burning contributes to regional smoke pollution.*

Java. *Forests were cleared in most areas decades ago for rice cultivation and plantation crops. Population pressure and overfarming have resulted in serious degradation in many areas.*

▲ **Figure 13.6 Environmental Issues in Southeast Asia** This was once one of the most heavily forested regions of the world. Most of the tropical forests of Thailand, the Philippines, peninsular Malaysia, Sumatra, and Java have been destroyed by a combination of commercial logging and agricultural settlement. The forests of Borneo, Burma, Laos, and Vietnam, moreover, are now being rapidly cleared. Water and urban air pollution, as well as soil erosion, is also widespread.

Indonesia contains more than half of Southeast Asia's forest area, including about 10 percent of the world's true tropical rainforests. Its forest coverage, however, is highly threatened. Most of Sumatra's primary forests are gone, and those of Borneo (Kalimantan) are rapidly diminishing. Indonesia's last forestry frontier is on the island of New Guinea. One study found that Indonesia lost more than 6 million hectares of primary forest between 2000 and 2012. In response, Indonesia declared two-year logging moratoriums in 2011 and again in 2013. Unfortunately, these restrictions were often ignored, as an estimated 80 percent of the logging in the country occurs illegally.

Mainland Southeast Asia has also experienced extensive deforestation. Thailand cut more than 50 percent of its forests between 1960 and 1980. Damage to the landscape was severe; flooding increased in lowland areas, and erosion on hillslopes led to the accumulation of silt in irrigation works and hydroelectric facilities. As a result, a series of logging bans in the 1990s severely restricted commercial forestry. As the forests of Thailand disappeared, mainland Southeast Asia's logging frontier moved into Burma, Vietnam, Laos, and Cambodia. A 2013 study by the Worldwide Fund for Nature indicated that between 1973 and 2009, the countries of mainland Southeast Asia

▲ **Figure 13.7** **Forest Clearance** Insular Southeast Asia supports some of the world's most ecologically diverse tropical rainforests. Unfortunately, many of the region's forests have been cleared. After logging, forests are often replaced with oil palm plantations, as seen here in Indonesia's Papua province.

cleared almost a third of their forests for timber and agricultural expansion.

Forest clearance figures can be misleading, however, because in many parts of Southeast Asia, logged-over lands and other degraded areas are gradually returning to tree coverage. As a result, some experts think that Vietnam's total forest area has actually increased in recent years. Indonesia and other Southeast Asian countries also support reforestation projects, which often try to recruit local people to nurture tree growth. The resulting secondary forests are not as biologically rich as primary forests, but over time their biodiversity does increase.

Throughout the coastal areas of Southeast Asia, a more specific problem is the destruction of the mangrove forests that thrive in shallow and silty marine areas. Often mangrove forests are burned for charcoal, but many are converted to fish and shrimp ponds, as well as rice fields and oil palm plantations. Mangrove forests serve as nurseries for many fish species, so their destruction threatens to undercut several important Southeast Asian fisheries. Mangrove forests also protect coastal areas from storm surges associated

with the tropical cyclones that often strike the region. Such wetland forest zones are particularly widespread in coastal Borneo and on the east coast of Sumatra.

Protected Areas Despite rampant deforestation, most Southeast Asian countries have created large national parks and other protected areas (see *Working Toward Sustainability: Preserving the Pileated Gibbon in Cambodia*). Southeast Asian rainforests are among the most biologically diverse areas on the planet, containing a large number of species found nowhere else. Conservation officials hope that protected areas will allow animals such as the orangutan, which now lives only in a small portion of northern Sumatra and in a somewhat larger area of Borneo (Kalimantan), a chance to survive in the wild. Others are less optimistic, noting that Southeast Asian national parks often exist only on paper, receiving little real protection from loggers or immigrant farmers.

Southeast Asia is also one of the world's main centers in the trade in endangered species and animal products. Many animals of the region's tropical forests and seas are in high demand in neighboring China, both for food and for their alleged pharmaceutical properties. The pangolin, or scaly anteater, is particularly endangered by this trade.

Fires, Smoke, and Air Pollution

Logging operations typically leave large quantities of wood (small trees, broken logs, roots, branches, and so on) on the ground. Exposed to the sun, this remaining "slash" becomes highly flammable. Indeed, it is often burned on purpose in order to clear the ground for agriculture or tree replanting. Wildfires also are common in other Southeast Asian habitats. The most thoroughly deforested areas are often covered by rough grasses that are frequently burned on purpose so that cattle can graze on the tender shoots that emerge after a fire passes through. Grassland fires, in turn, prevent forest regeneration. In drained wetland areas, organic **peat** soils often burn, causing extensive air pollution (**Figure 13.8**).

In the late 1990s, wildfires associated with both logging and a severe drought raged so intensely across much of insular Southeast Asia that the region suffered from disastrous air pollution. As a result, Southeast Asian countries began to devote more attention to air quality. But as deforestation has continued and droughts recur, the fire threat remains. In September 2015, six Indonesian provinces declared states of emergency due to smoke and haze from uncontrolled fires associated with land clearing. Schools were forced to close, and more than 140,000 cases of respiratory illness were reported. The Indonesian government estimates that it would cost US$35 billion to solve this problem.

Efforts to protect Southeast Asia's air quality are also hampered by continuing industrial development, along with a major increase in vehicular traffic. The region's large metropolitan areas have extremely unhealthy levels of air and water pollution. Several cities, including Bangkok and Manila, have built rail-based public transportation systems in order to reduce traffic and vehicular emissions. Bangkok,

Preserving the Pileated Gibbon in Cambodia

Although the tropical rainforests of insular Southeast Asia receive far more attention, the seasonally dry forests of interior mainland Southeast Asia are also biologically rich and severely endangered. Some of the largest remaining lowland dry forests are found in the Northern Plains of Cambodia. Over the past several decades, sugar and rubber plantations have extended across much of this area, fragmenting the forest ecosystem. Rubber-cultivation concessions have even been granted in Kulen Promtep Wildlife Sanctuary, covering some 15 percent of this reserve's territory. As a result, wildlife has been diminishing rapidly. As roads are constructed to service the new farms, moreover, access to the forests becomes easier, intensifying pressure from hunters.

The Pileated Gibbon and Other Species of Concern Much of the concern for wildlife in this area centers on the pileated gibbon, a small arboreal ape that requires large expanses of intact forest (**Figure 13.1.1**). Classified as an endangered species, the pileated gibbon population has declined by more than 50 percent over the past 35 years. The largest remaining population, numbering about 35,000 individuals, is found in the forests of Cambodia's Northern Plains. These same forests also support populations of a number of other rare and diminishing species, including the Indochinese tiger, the Asian elephant, the Dhole (a wild dog), the Banteng (a wild cattle species), and the fishing cat. Another wild bovine, the kouprey, is also indigenous to this region but may already be extinct, as the last confirmed sighting was in 1988.

Until recently, the future of the remaining dry forests of northern Cambodia looked grim. But the Cambodian government has finally responded to the crisis, placing a moratorium on agricultural concessions. As a result, no new plantations are currently allowed in the Northern Plains. Local, national, and international wildlife organizations are also moving ahead with ambitious plans to preserve the pileated gibbon and the other wildlife species in the area, providing renewed environmental hope.

Patrolling the remaining forests to prevent poaching and unlawful land clearance is also increasingly emphasized. From 2013 to 2015, the wildlife patrol staff in this area nearly doubled, allowing a large area of land to be monitored more intensively. As a result, habitat protection has greatly improved, enhancing the prospects of the pileated gibbon and other species of concern.

New Conservation Tools Wildlife officials in northern Cambodia are also turning to technology for new conservation tools. Particularly useful has been the SMART (Spatial Monitoring And Reporting Tool) conservation software system. Now used in roughly 150 projects in 31 countries, SMART allows the easy conversion of data derived from patrols into visual information, including maps, for use by park managers. Perhaps most importantly, SMART facilitates the identification of poaching hotspots, which in turn helps generate rapid response measures.

Equally important for the conservation of Cambodia's dry forests has been community land-use planning, in which local villagers participate in—and benefit from—conservation efforts. Community land-use planning

▲ **Figure 13.1.1 Pileated Gibbon** The pileated gibbon is an endangered animal whose range is limited to the remaining forests of southeastern Thailand, western Cambodia, and far southern Laos. The Cambodian government is working with several environmental organizations to ensure the survival of the species.

also seeks to ensure that local people retain their land rights, and are thus not dispossessed by wildlife reserves. Through participatory planning, different areas in Cambodia's Northern Plains are being zoned for different uses, so that both people and animals can benefit from any proposed changes.

Google Earth Virtual (MG) Tour Video

http://goo.gl/XU263U

1. Suggest possible reasons why tropical rainforests have received much more international conservation attention than tropical dry forests.

2. What are some of the other ways in which mapping can be an effective conservation tool?

in particular, has substantially reduced its levels of ozone and sulfur dioxide, but particulate matter remains at unhealthy levels.

Controversies Over Dam Building

A number of aquatic species in Southeast Asia are more severely endangered by dam building, including freshwater dolphins and the Mekong giant catfish, the latter of which can weigh up to 660 pounds (300 kg). Dam construction is being most avidly pursued by Laos, a poor and rugged country that contains many sizable rivers, including the mighty Mekong. Although dams are built for several reasons, hydroelectric power is most significant. Demand for electricity in sparsely populated Laos is small, so much of the new power will be exported to neighboring countries,

▲ **Figure 13.8** **Burning Peatlands** The soil of most wetland areas in Southeast Asia is composed largely of peat, an organic substance that can burn when dry. Draining for agricultural expansion, as well as drought, often results in extensive peat fires.

low-lying cities, typhoons may be intensifying, and storm surges are becoming more common. Southeast Asian farmland is concentrated in delta environments and thus could suffer from saltwater intrusion, especially in dry years. In the 2015 drought, 22 percent of the vital rice crop in Vietnam's Mekong delta was harmed in this manner. Even if global warming remains below 2 degrees Celsius by 2100, Southeast Asia would probably experience a sea-level rise of nearly 30 inches (75 centimeters), resulting in widespread devastation.

Changes in precipitation across Southeast Asia brought about by global warming remain uncertain. Many experts foresee an intensification of the monsoon pattern, which could bring increased rainfall to much of the mainland. This would likely result in more destructive floods, but it could bring some agricultural benefits to dry areas, such as Burma's central Irrawaddy Valley. Complicating this scenario, however, is the prediction that climate change could intensify the **El Niño** effect, a complex weather pattern originating in the equatorial Pacific Ocean that would result in more extreme droughts in Southeast Asia. A strong El Niño in 2015 brought serious drought to much of the region. Overall, according to a recent report by the Asian Development Bank, climate change could cause the major countries of Southeast Asia to lose up to 11 percent of their total economic production by 2100.

Greenhouse Gas Emissions Southeast Asia's overall emissions from conventional sources remain low by global standards. Indonesia, Malaysia, and Vietnam, however,

particularly Thailand. Laos is currently planning 11 dams on the Mekong. In 2015, construction commenced on the final stage of the US$3.5 billion Xayaburi Dam, which will be the first to span the entire lower Mekong River (**Figure 13.9**).

Dam building in Laos, as elsewhere, is highly controversial, due both to local environmental impacts and to the fact that large numbers of people are usually displaced by the resulting reservoirs. Many fish species need to swim upstream to spawn, and massive dams on the main rivers will seriously disrupt fisheries. This problem is particularly acute in the lower Mekong River, which is estimated to have the world's most valuable inland fishery. As a result, both Vietnam and Cambodia have lodged protests against dam construction in Laos.

The Laotian government has responded to such criticism by redesigning some of its planned dams to allow some fish migration and to reduce sedimentation. Environmentalists, as well as the governments of the downstream countries, would like to see more serious reconsideration, arguing that hydroelectric facilities can be built by diverting water through electrical generators without having to dam the Mekong's entire flow. But Laos argues that it is more efficient simply to build dams, and has acquired adequate funding, mostly from China, to proceed with its ambitious plans.

Climate Change in Southeast Asia

Most of Southeast Asia's people live in coastal and delta environments, making the region particularly vulnerable to the rise in sea level associated with global warming. Periodic flooding is already a major problem in many of the region's

▼ **Figure 13.9** **Xayaburi Dam Site** The controversial Xayaburi Dam on the Mekong River in northern Laos is scheduled for completion in 2019. Designed primarily for hydroelectricity production, the dam will result in considerable damage to fisheries and wildlife in the lower Mekong.

have increased their reliance on coal-fired electricity generation, boosting the region's emissions. When greenhouse gas emissions associated with deforestation are factored in, moreover, Southeast Asia's role in global climate change becomes much larger. By some estimates, Indonesia emits more carbon than any country other than China and the United States. The largest problem here is the release of carbon from burning peat. Furthermore, Indonesia and Malaysia are actively draining coastal wetlands to make room for agricultural expansion, a process that results in the gradual oxidation of peat, which also releases carbon.

Southeast Asian countries are eager to reduce their greenhouse gas emissions, but argue that the developed countries need to provide more funding and enhanced technology transfer. Thailand has already committed to increasing its production of renewable energy to 25 percent of output by 2021, although this agreement is nonbinding. At the 2015 Paris climate negotiations, the Philippines was the most outspoken Southeast Asian country, owing in part to its extreme vulnerability to typhoons. Following the Paris Agreement, Southeast Asian leaders began to share more information on climate change and their planned responses to it, hoping that ASEAN can take a larger role in this crucial issue.

Southeast Asia's Energy Mix Southeast Asia was one of the world's first major oil-exporting regions, with production concentrated in Indonesia, Malaysia, and Burma. Production failed to keep up with demand in most areas, however, and now only tiny Brunei remains a significant oil exporter relative to the size of its economy. Indonesia, Malaysia, and Vietnam, however, still have significant oil reserves, and natural gas deposits in the region are larger than those of oil. Large offshore natural gas fields are also found between East Timor and Australia. Although the two countries have agreed to joint ownership, tussles persist over both revenue sharing and maritime border demarcation, slowing their development. Coal is an important source of energy in Vietnam, Thailand, and especially Indonesia. Indonesia is the world's fifth largest coal producer, and it exports much of its production, especially to India.

Most renewable energy in Southeast Asia comes from hydropower and geothermal plants. Laos gets 92 percent of its energy from hydro, one of the highest figures in the world. Hydropower is also important in Vietnam and Burma, and controversial new dam projects promise to supply increasing amounts of electricity. Geothermal power is concentrated in the Philippines and Indonesia, which are noted for their volcanoes and hot springs (**Figure 13.10**).

Southeast Asia does not produce nuclear power, although several countries in the region plan to do so in the future, in part because they hope to reduce carbon emissions.

▲ **Figure 13.10 Geothermal Plant in the Philippines** Owing to its geologically active location with numerous volcanoes, the Philippines is one of the world's leading countries in geothermal power production.

Here Vietnam is the leader, with 15 nuclear power plants in the planning stage. Malaysia and Thailand also have plans for nuclear plants, and Indonesia has recently completed its first experimental reactor. The Philippines, on the other hand, has already built a nuclear generator, but it never brought it online, due to environmental and safety concerns. Instead, the plant was turned into a tourist attraction.

REVIEW

13.1 Why do the mainland and insular regions of Southeast Asia have such distinctive climates and landforms, and how have these differences impacted the human communities of these two regions?

13.2 How have people changed the physical landscapes of Southeast Asia over the past 50 years?

13.3 What factors make climate change a particularly worrying matter in Southeast Asia?

■ **KEY TERMS:** Association of Southeast Asian Nations (ASEAN), Sunda Shelf, tsunamis, typhoons, peat, El Niño

Population and Settlement: Subsistence, Migration, and Cities

With a little over 600 million inhabitants, Southeast Asia is not heavily populated by Asian standards. One of the reasons for this relatively low density is the extensive tracts of rugged mountains, which generally remain thinly inhabited. In contrast, relatively dense populations are found in the region's deltas, coastal areas, and zones of fertile volcanic soil (**Figure 13.11**). Many of the favored lowlands of Southeast Asia experienced striking population growth in the second half of the twentieth century. As a result, demographic growth and family planning became important concerns

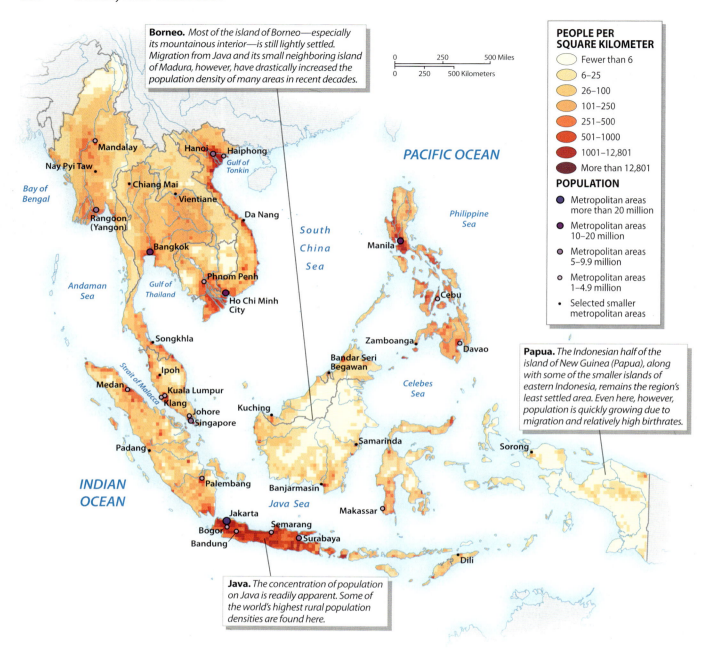

Borneo. *Most of the island of Borneo—especially its mountainous interior—is still lightly settled. Migration from Java and its small neighboring island of Madura, however, have drastically increased the population density of many areas in recent decades.*

PEOPLE PER SQUARE KILOMETER

- Fewer than 6
- 6–25
- 26–100
- 101–250
- 251–500
- 501–1000
- 1001–12,801
- More than 12,801

POPULATION

- Metropolitan areas more than 20 million
- Metropolitan areas 10–20 million
- Metropolitan areas 5–9.9 million
- Metropolitan areas 1–4.9 million
- Selected smaller metropolitan areas

Papua. *The Indonesian half of the island of New Guinea (Papua), along with some of the smaller islands of eastern Indonesia, remains the region's least settled area. Even here, however, population is quickly growing due to migration and relatively high birthrates.*

Java. *The concentration of population on Java is readily apparent. Some of the world's highest rural population densities are found here.*

▲ **Figure 13.11 Population of Southeast Asia** In mainland Southeast Asia, the population is concentrated in the valleys and deltas of the region's large rivers. Population density remains relatively low in the intervening uplands. In Indonesia, density is extremely high on Java, an island noted for its fertile soil and large cities. Some of Indonesia's outer islands, especially those of the east, remain lightly settled. Overall, population density is high in the Philippines, especially in central Luzon.
Q: Why is the population of Vietnam so unevenly distributed? What are some of the political and economic consequences of such uneven distribution?

through much of the region. Most Southeast Asian countries, and especially Singapore and Thailand, subsequently saw rapid reductions in their birth rates, but larger families are still common in Philippines, Laos, and particularly East Timor.

Settlement and Agriculture

Much of insular Southeast Asia has infertile soil, which cannot support intensive agriculture and high rural population densities. Island rainforests, though lush and biologically rich, typically grow on a poor base. Plant nutrients are locked up in the vegetation itself, rather than being stored in the soil where they would easily benefit agriculture. The incessant rain of the equatorial zone also tends to wash nutrients out of the soil. Agriculture must be carefully adapted to this limited fertility by constant field rotation or the heavy application of fertilizer.

Some notable exceptions to this generalization about soil fertility and settlement density can be found on these tropical islands. Unusually rich soils connected to volcanic activity are scattered through the region, particularly on the island of Java. With more than 50 volcanoes, Java is blessed with highly productive agriculture that supports a large array of tropical crops and a very high population density. Approximately 145 million people live on Java—more than half the total population of Indonesia—in an area smaller than the state of Iowa. Dense populations also are found in pockets of fertile alluvial soils along the coasts of insular Southeast Asia, where people have traditionally supplemented land-based farming with fishing and other commercial activities.

The demographic patterns in mainland Southeast Asia are less complicated than those of the island realm. In all the mainland countries, people are concentrated in the agriculturally intensive valleys and deltas of the large rivers. The population core of Thailand, for example, clusters around the Chao Phraya River, just as Burma's is focused on the Irrawaddy. Vietnam has two distinct foci: the Red River Delta in the far north and the Mekong Delta in the far south. In contrast to these densely settled areas, the middle reaches of the Mekong River provide only limited lowland areas in Laos, which reduces its population potential. In Cambodia, the population historically is centered around Tonle Sap, a large lake with an unusual seasonal flow reversal. During the rainy summer months, the lake receives water from the Mekong drainage, but during the drier winter months it contributes to the river's flow.

Agricultural practices and settlement forms vary widely across the complex environments of Southeast Asia. Generally speaking, however, three farming and settlement patterns are apparent.

Swidden in the Uplands Swidden, also known as shifting cultivation or "slash-and-burn" agriculture, is practiced throughout the uplands of both mainland and island Southeast Asia (**Figure 13.12**). In the **swidden** system, small plots of tropical forest or brush are periodically cut by hand. The fallen vegetation is then burned to transfer nutrients to the soil before subsistence crops are planted. Yields remain high for several years and then drop off dramatically as the soil nutrients are exhausted and insect pests and plant diseases multiply. These plots are abandoned after a few years and allowed to revert to woody vegetation. The cycle of cutting, burning, and planting is then moved to another small plot nearby. After a period of 10 to 75 years, farmers must return to the original plot, which once again has accumulated nutrients in the dense vegetation.

Swidden is sustainable when population densities remain low and when upland people control enough territory. Today, however, the swidden system is increasingly threatened. It cannot easily support the increased population resulting from relatively high birth rates and, in some cases, migration. With greater population density, the rotation period must be shortened, undercutting soil resources. The upland swidden system is also sometimes undermined by commercial logging. Road building presents another

▼ **Figure 13.12 Swidden Agriculture** In the uplands of Southeast Asia, swidden (or slash-and-burn) agriculture is widely practiced. When done by tribal peoples with low population densities, swidden is not environmentally harmful. When practiced by large numbers of immigrants from the lowlands, however, swidden can cause deforestation and extensive soil erosion.

threat. Vietnam, for example, has recently built a highway system through its mountainous spine, designed to aid economic development and more fully integrate its national economy. One result is the movement of lowlanders into the mountains, disrupting local ecosystems and indigenous societies.

When swidden can no longer support the local population, upland people sometimes adapt by switching to cash crops that allow them to participate in the commercial economy. In the mountains of northern Southeast Asia, often called the **Golden Triangle**, one of the main cash crops is opium, grown by local farmers for the global drug trade. Drug eradication programs between 1998 and 2006 proved relatively successful, but opium growing has since experienced resurgence (see *Exploring Global Connections: Opium Resurgence in Northern Southeast Asia*).

Plantation Agriculture With European colonization, Southeast Asia became a center of plantation agriculture, producing high-value specialty crops ranging from rice to rubber. Even in the 19th century, Southeast Asia was closely linked to a globalized economy through the plantation system. Forests were cleared and swamps drained to make room for these commercial farms. Labor was supplied, often unwillingly, by indigenous people or by contract laborers brought in from India or China.

Plantations are still an important part of Southeast Asia's geography and economy. Most of the world's natural rubber, for example, is grown in Malaysia, Indonesia, and Thailand, while sugarcane has long been a major plantation crop in the Philippines and Indonesia (**Figure 13.13**). More recently, pineapple plantations have spread in both the Philippines and Thailand, which are now the world's leading exporters of this fruit. Indonesia is the region's leading tea producer, while Vietnam has more recently become the world's second largest coffee grower. Coconut oil and **copra** (dried coconut meat) are widely produced in the Philippines, Indonesia, and elsewhere.

Over the past several decades, the African oil palm has emerged as the most important plantation crop in insular Southeast Asia. Indonesia and Malaysia together grow about 85 percent of this increasingly important crop that is used both for food and industrial products. Despite its popularity, palm oil is also highly criticized. Health advocates note that it is very high in saturated fats, which have been linked to heart disease and other health problems. The environmental problems associated with palm oil, such as deforestation and swamp draining, are equally serious. As a result of such criticisms, the Roundtable on Sustainable Palm Oil has sought to establish global standards for the industry. This nonprofit organization brings together palm oil producers, processors, and traders, as well as food and soap manufacturing companies and environmental and social organizations. Several prominent

▲ **Figure 13.13 Rubber Plantation** Most of the world's natural rubber is produced on plantations in Southeast Asia. This worker is tapping a rubber tree on the Indonesian island of Sumatra.

green activists, however, argue that the Roundtable merely seeks to create a smokescreen, hiding inherently unsustainable practices.

Rice in the Lowlands The lowland basins and deltas of mainland Southeast Asia are largely devoted to intensive rice cultivation. Throughout most of Southeast Asia, rice is the preferred staple food. Traditionally, rice was mainly cultivated on a subsistence basis by rural farmers. But as the number of wage laborers in Southeast Asia has grown as a result of economic development, so has the demand for commercial rice cultivation. Across most of the region, the use of agricultural chemicals and high-yield crop varieties, along with improved water control, has allowed rice production to keep pace with population growth. Both Thailand and Vietnam are among the world's top rice exporters.

In those areas without irrigation, yields remain relatively low. Rice growing on Thailand's Khorat Plateau, for example, depends largely on the uncertain rainfall, without the benefit of the more sophisticated water-control methods. In some lowland districts lacking irrigation, dry-field crops, especially sweet potatoes and manioc, form the staple foods of people too poor to buy market rice on a regular basis. But in most parts of Southeast Asia, economic growth and declining birth rates have significantly reduced the burden of poverty.

Recent Demographic Change

Most Southeast Asian countries have seen a sharp decrease in birth rates over the past several decades (Table 13.1). Because the region is not facing the same kind of population pressure as East or South Asia, governments have promoted a wide range of population policies. In countries with a highly uneven population distribution, internal relocation away from densely populated areas to outlying districts is a common response.

Exploring Global Connections

The Opium Resurgence in Northern Southeast Asia

Until late in the 20th century, the Golden Triangle of northern Southeast Asia, focused on Burma's Shan State, was the global center of opium growing and heroin production. In 1990, however, the Burmese army defeated several ethnic rebellions that were financed by narcotics production. These victories, along with antidrug campaigns by local governments and the United Nations, resulted in a massive opium decline (Figure 13.2.1). At the same time, the crop surged ahead in war-torn Afghanistan. By 2007, over 85 percent of the world's illicit opium originated in Afghanistan.

The Rise of Methamphetamine As opium growing declined in northern Southeast Asia, other drugs took its place, because both drug lords and ethnically-based militias were eager to retain their profits. Of particular importance was methamphetamine, a synthetic drug that is easy to make with the right chemicals. A lively trade thus developed, with "meth" flowing north into China and south into Thailand and with chemicals moving in the opposite directions. In 2014 alone, 36 tons of methamphetamine were seized by Southeast Asian officials. Even the United States found itself involved. In 2005, a New York grand jury indicted eight leaders of Burma's United Wa State Army on charges of trafficking amphetamines and other drugs.

Opium's Comeback Moreover, 2008 saw a resurgence in the growing of opium (Figure 13.2.2). Production in Afghanistan dropped, reducing competition, but more important was the rise of China as a new major market. According to the UN Office on Drugs and Crime (UNDODC), 55 percent of Asia's estimated 3.3 million heroin users currently reside in China. The opium fields of the Golden Triangle not only are much closer to this market than those of Afghanistan, but also produce a higher-quality

▼ **Figure 13.2.1** **The Fight Against Opium** The opium crop in northern mainland Southeast Asia results in widespread economic corruption and social damage. As a result, both governmental forces and ethnic militias have been involved in eradication programs. These soldiers from the Ta'ang (or Palaung) National Liberation Army in Burma are destroying an opium field in the ethnic Palaung area.

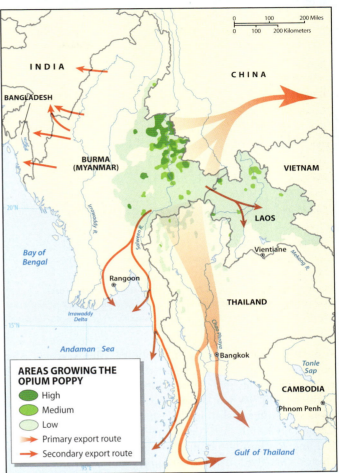

AREAS GROWING THE OPIUM POPPY
- High
- Medium
- Low
- → Primary export route
- → Secondary export route

▲ **Figure 13.2.2** **Opium Growing Areas** The opium poppy has long been commercially cultivated in the northern region of Mainland Southeast Asia. Although production declined significantly in the 1990s and early 2000s, it has recently made a dramatic comeback.

product. One survey found that the area devoted to opium growing in Burma and Laos increased by 6000 acres (2600 hectares) in 2014 alone. In that year, the two countries produced an estimated 762 metric tons of opium, generating 76 metric tons of heroin. Nearly 90 percent of this production comes from Burma's Shan State.

The resurgence of opium growing in the Golden Triangle has brought new eradication programs. Officials realize, however, that law enforcement efforts alone will not be adequate, as the value of opium is too high. A small field that would produce a rice crop worth US$30 could yield an opium crop worth US$585. Therefore, UNODC officials argue that broad-based economic and social development in the region's rugged uplands is needed. In particular, they emphasize the building of an effective transportation system to allow lower-value crops to be brought to market.

1. What are some of the political, cultural, and environmental factors that make the Golden Triangle a major source of illicit drugs?

2. Compare the possible advantages and disadvantages of an approach to drug production in the region based on economic and social development rather than law enforcement.

Google Earth Virtual (MG) Tour Video

http://goo.gl/Nr6XjO

553

TABLE 13.1   Population Indicators

MasteringGeography™

Country	Population (millions) 2016	Population Density (per square kilometer)[1]	Rate of Natural Increase (RNI)	Total Fertility Rate	Percent Urban	Percent < 15	Percent > 65	Net Migration (Rate per 1000)
Burma (Myanmar)	52.4	82	1.1	2.3	34	28	5	−1
Brunei	0.4	79	1.3	1.9	77	23	4	2
Cambodia	15.8	87	1.8	2.6	21	32	4	−2
East Timor	1.3	82	3.0	5.7	33	42	6	−9
Indonesia	259.4	140	1.3	2.5	54	28	5	−1
Laos	7.1	29	1.9	3.0	39	35	4	−3
Malaysia	30.8	91	1.2	2.0	75	25	6	2
Philippines	102.6	332	1.6	2.8	44	32	5	−1
Singapore	5.6	7,737	0.4	1.2	100	15	12	8
Thailand	65.3	133	0.4	1.6	50	18	11	−1
Vietnam	92.7	293	0.9	2.1	34	24	7	1

Source: Population Reference Bureau, *World Population Data Sheet*, 2016.

[1]World Bank, *World Development Indicators*, 2016.

Login to MasteringGeography™ **& access** MapMaster **to explore these data!**

1) In global terms, high levels of urbanization correlate with lower fertility patterns, but is this the case in Southeast Asia? Which Southeast Asia countries fit this patterns and which do not?

2) How do population density figures vary across the different countries of Southeast Asia, and why are the differences so extreme?

Population Contrasts The Philippines, the second most populous country in Southeast Asia, has a relatively high fertility rate, at just under three children per woman (Table 13.1). When a popular democratic government replaced a dictatorship in the 1980s, the Philippine Roman Catholic Church pressured the new government to cut funding for family-planning programs. As a result, many clinics that had dispensed family-planning information were closed. Although high birth rates are not associated with Catholicism in other parts of the world, the Church's outspoken stand on birth control in the Philippines has inhibited the dispersal of family-planning information. Still, fertility rates have been slowly declining. In Southeast Asia's other predominately Roman Catholic country, East Timor, the fertility rate remains extremely high, at around 5.7 children per woman.

Laos and Cambodia, countries of Buddhist religious tradition, also have fertility rates above the replacement rate, although these figures have dropped rapidly in recent years. Thailand, which shares cultural traditions with Laos yet is considerably more developed, saw a much earlier drop in its birth rate. Here the total fertility rate (TFR) fell from 5.7 in 1970 to 1.6 currently, a figure that will soon bring population decline if it persists. Thailand has promoted family planning for both population and health reasons, including the relatively high incidence of AIDS in the country. But the Thai government is now concerned about future labor shortages and as a result has contemplated eliminating its mandatory retirement age.

The city-state of Singapore stands out on the demographic charts with its extremely low fertility rate of 1.6, one of the lowest in the world. As a result, the population pyramid of Singapore has a completely different shape from that of the Philippines (**Figure 13.14**). If Singapore did not experience relatively high rates of immigration, its population would be steadily dropping. Concerned about its declining birth rate, Singapore announced a "Have Three or More" campaign in 1987, directed at the more highly educated segment of its population. Singaporeans can receive thousands of dollars of government subsidies for having a second or third child. Singapore also encourages the immigration of highly skilled workers.

Indonesia, with the region's largest population at 259 million, also has seen a dramatic decline in fertility in recent decades; its TFR is now slightly above the replacement level. As with Thailand, this drop in fertility seems to have resulted from a strong government family-planning effort, coupled with improvements in education.

Growth and Migration Until the end of the 20th century, Indonesia had an official policy of **transmigration**, with the government encouraging people to move from densely populated to lightly populated parts of the country. Primarily because of migration from Java and Madura, the population of the outer islands of Indonesia has grown rapidly since the 1970s. The province of East Kalimantan on Borneo, for example, experienced an astronomical growth rate of 30 percent per year during the last two decades of the 20th century. As a result of this population shift, many parts of Indonesia outside Java now have moderately high population densities, although some of the more remote districts remain lightly settled.

High social and environmental costs often accompany these relocation schemes. Javanese peasants, accustomed to working the highly fertile soils of their home island, often fail in their attempts to grow rice in the former rainforests

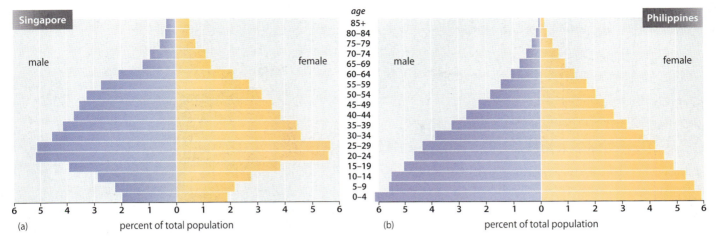

▲ **Figure 13.14** **Population Pyramids of Singapore and the Philippines** (a) Singapore has a very low birth rate and an aging population; as a result, its government encourages the immigration of skilled workers. (b) The Philippines, by contrast, has a high birth rate and a young population.

of Borneo. Abandoned fields are common after repeated crop failures. In some areas, farmers have little choice but to adopt a semi-swidden form of cultivation, moving to new sites once the old ones have been exhausted. The term **shifted cultivators** has been used for such displaced rural migrants.

▼ **Figure 13.15** **Indonesian Transmigration** Indonesia's population distribution shows a marked imbalance: Java, along with the neighboring island of Madura, is one of the world's most densely settled places, whereas most of the country's other islands remain rather lightly populated. The Indonesian government has encouraged resettlement to the outer islands, often paying the costs of relocation. The settlement visible in this satellite image is on the Indonesian side of the island of New Guinea.

Partly because of these problems, the Indonesian government restructured and substantially reduced its transmigration program in 2000. In 2014, however, it reversed course and pledged to relocate 700,000 families a year, but a year later fewer than 10,000 had actually been moved. Indonesia is currently creating a huge new agricultural project in south-central New Guinea, which could entail the transfer of over half a million people to the area (**Figure 13.15**).

The government of the Philippines has also used internal migration to reduce population pressure in the core areas of the country. Beginning in the late 19th century, people began streaming out of central and northwestern Luzon for the frontier zones. In the early 20th century, this settlement frontier still lay in Luzon; by the postwar years, it had moved south to the island of Mindanao. Today, however, internal migration is less appealing, both because population disparities between the islands have been reduced and because the south is economically weak and politically unstable. The emphasis is now on international migration, resulting in a Filipino overseas population of roughly 10 million (**Figure 13.16**).

Labor migration has recently become a controversial issue in both the Philippines and other Southeast Asian countries. Many young women from Indonesia and the Philippines find low-paying work as maids and nannies in wealthier countries. Exploitation, both sexual and financial, is common, particularly in the oil-rich countries of Southwest Asia. As a result, Indonesia announced in 2015 that it would gradually stop sending domestic workers to individual employers in the Middle East. The Philippines decreed in 2014 that it would no longer allow female workers to go to the United Arab Emirates after its government refused to guarantee a minimum salary of US$400 a month.

Population issues and policies thus vary greatly across Southeast Asia. Virtually every part of the region, however, has seen a rapid expansion of its urban population in recent decades.

▲ **Figure 13.16 Overseas Employment Fair in the Philippines** Due to a combination of uncertain economic conditions at home, rapid population growth, and good English-language skills, many Filipinos seek employment abroad. Here job seekers crowd an overseas employment fair.

▲ **Figure 13.17 Manila** The huge city of Manila is noted for both its elite business districts and its impoverished squatter settlements. Here we see shanty dwelling along a polluted waterway in the foreground of the gleaming Makati financial district.

Explore the **Sights** of Manila

http://goo.gl/jaZie2

Urban Settlement

Southeast Asia is not a heavily urbanized region. Even Thailand has only a 50 percent urbanization rate, which is unusual for a country that has experienced so much industrial development. Currently, however, the region's urban population is growing more quickly than the world average, and a number of extremely large cities are emerging across the region.

Several Southeast Asian countries have a **primate city**—a single, large urban settlement that overshadows all others. Thailand's urban system, for example, is dominated by Bangkok, just as Manila far surpasses all other cities in the Philippines. Both have recently grown into megacities with metropolitan areas of more than 15 million residents. More than half of all city dwellers in Thailand live in the Bangkok metropolitan area. In both Manila and Bangkok, as in other large Southeast Asian cities, explosive growth has led to housing shortages, congestion, and pollution (**Figure 13.17**). Manila and Bangkok also suffer from a lack of parks and other public spaces, enhancing the appeal of massive shopping malls.

Although Indonesia has historically had a more balanced urban system than the Philippines or Thailand, its capital city, Jakarta, has mushroomed over the past several decades. By some measures, metropolitan Jakarta is now the world's second largest urban agglomeration, with more than 30 million inhabitants. Unfortunately, Jakarta's infrastructure has not kept pace with its growth, generating horrific traffic jams. A rail-based mass transport system will be completed in 2017, but it will include only two lines and thus will be too limited to solve the transportation problem. Seasonal flooding is another serious issue, as 40 percent of Jakarta lies below sea level (**Figure 13.18**). To counter flooding, leaders have recently decided to build a US$40 billion seawall to protect the city.

Thailand, the Philippines, and Indonesia are all making efforts to encourage the growth of secondary cities by decentralizing economic functions. In the case of the Philippines, the city of Cebu has emerged in recent years as a relatively dynamic economic center, leading to

hopes that a more balanced urban system may be emerging. Several former colonial "hill stations" (resort cities) are also emerging as major cities in their own rights. One prime example is Baguio City, which sits at over 5,000 feet and thus has a perpetually spring-like climate. Baguio is now a major university center and is home to more than 300,000 people.

Urban primacy is less pronounced in other Southeast Asian countries. Vietnam, for example, has two main cities: Ho Chi Minh City (formerly Saigon) in the south, with a metropolitan population of roughly 9 million, and the capital city of Hanoi in the north, with 7.7 million residents in

▼ **Figure 13.18 Jakarta** The sprawling capital of Indonesia, Jakarta is a modern city plagued by severe flooding problems. On January 17, 2017, floodwaters covered the city's central business district, forcing businesspeople to wade to and from work.

its metro area (**Figure 13.19**). Rangoon (Yangon), until recently the capital of Burma, has doubled its population in the last two decades to roughly 6 million residents. Cambodia's capital, Phnom Penh, is still a relatively modest city of around 2.5 million. Vientiane, the capital and largest city of Laos, has only about 750,000 inhabitants.

Kuala Lumpur is Malaysia's largest city, with around 7.5 million inhabitants in its greater metro area. As a result of Malaysia's economic strength, it has received heavy investments from both the national government and the global business community. This has produced a modern city of grand ambitions that is free of most infrastructure problems plaguing other Southeast Asian cities. The Petronas Towers, owned by the country's national oil company, were the world's tallest buildings at almost 1500 feet (450 meters) when completed in 1996.

The independent republic of Singapore is essentially a city-state of 5.6 million people on an island of 240 square miles (600 square km) (**Figure 13.20**). Although space is at a premium, Singapore has successfully developed high-tech industries that

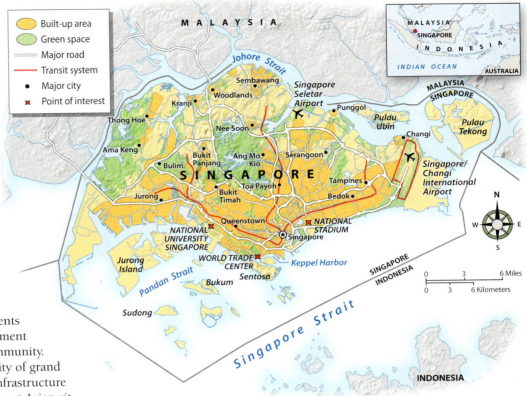

▲ **Figure 13.20 Singapore** The economic and technological hub of Southeast Asia, Singapore is famous for its clean, efficiently run, and very modern urban environment. Some residents complain, however, that Singapore lost much of its charm as it developed.

have brought it great prosperity. Unlike most other Southeast Asian cities, Singapore has no squatter settlements or slums. Only in the much-diminished Chinatown and historic colonial district does one find older buildings. Otherwise, Singapore is an extremely clean and orderly city of modern high-rise skyscrapers, apartment complexes, and space-intensive industry. Despite its crowded conditions, Singapore makes space available for large parks and other green areas.

Singapore is unique in Southeast Asia not only because it is a city-state, but also because most of its people have a Chinese cultural background. But as we shall see in the following section, strong Chinese influences are also found in most of the other large cities of the region.

▼ **Figure 13.19 The Old Quarter of Hanoi** Hanoi, the capital city of Vietnam, has retained much of its old commercial core, preserving most of its streets and buildings as they existed in the early 1900s. The Old Quarter is noted for its small shops and craft industries, particularly those related to silk. Local residents and tourists are attracted to the district's restaurants, bars, and clubs.

Explore the **Sounds** of Hanoi's Old Quarter

http://goo.gl/6C1lRt

REVIEW

13.4 Why do population density and population growth vary so much in different parts of Southeast Asia?

13.5 Explain why export-oriented plantation agriculture is so widespread in Southeast Asia, and describe the problems associated with its spread.

13.6 Why are some economists and planners concerned about the phenomenon of urban primacy countries such as Thailand and the Philippines?

■ **KEY TERMS** swidden, Golden Triangle, copra, transmigration, shifted cultivators, primate city

Cultural Coherence and Diversity: A Meeting Ground of World Cultures

Unlike many other world regions, Southeast Asia lacks the historical dominance of a single civilization. Instead, the region has been a meeting ground for cultural traditions from South Asia, China, the Middle East, Europe, and even North America. Abundant natural resources, along with the region's strategic location on oceanic trade routes, have long made Southeast Asia attractive to outsiders. As a result, the modern cultural geography of this diverse region reflects a long history of borrowing and combining external influences.

The Introduction and Spread of Major Cultural Traditions

In Southeast Asia, cultural diversity has been enhanced by the historical influences connected to the major religions of Hinduism, Buddhism, Islam, and Christianity, all of which originated in other world regions (**Figure 13.21**).

South Asian Influences The first major external influence arrived from India roughly 2000 years ago when small numbers of migrants from South Asia helped local rulers establish Hindu and Buddhist kingdoms in the western lowlands of Southeast Asia. Although Hinduism faded away in most locations, it persists on the Indonesian island of Bali, and vestiges of the faith remain in many other areas. The ancient Indian

◀ **Figure 13.21 Religions of Southeast Asia** Southeast Asia is one of the world's most religiously diverse regions. Buddhism dominates on the mainland, with Theravada Buddhism prevailing in Burma (Myanmar), Thailand, Laos, and Cambodia and Mahayana Buddhism (combined with other elements of the so-called Chinese religious complex) prevailing in Vietnam. The Philippines is primarily Christian (Roman Catholic), but the rest of insular Southeast Asia is primarily Muslim. Substantial Muslim minorities are found in the Philippines, Thailand, and Burma. Animist and Christian minorities can be found in remote areas throughout Southeast Asia, especially in Indonesia's portion of the large island of New Guinea.

▶ **Figure 13.22** **Chinese in Southeast Asia** People from the southern coastal region of China have been migrating to Southeast Asia for hundreds of years, a process that reached a peak in the late 1800s and early 1900s. Most Chinese migrants settled in the major urban areas, but in peninsular Malaysia large numbers were drawn to the countryside to work in the mining industry and in plantation agriculture. Today Malaysia has the largest number of people of Chinese ancestry in the region. Singapore, however, is the only Southeast Asian country with a Chinese majority.

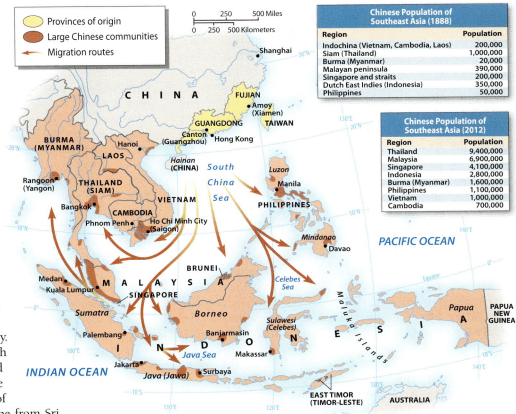

Legend:
- Provinces of origin
- Large Chinese communities
- Migration routes

Chinese Population of Southeast Asia (1888)	
Region	**Population**
Indochina (Vietnam, Cambodia, Laos)	200,000
Siam (Thailand)	1,000,000
Burma (Myanmar)	20,000
Malayan peninsula	390,000
Singapore and straits	200,000
Dutch East Indies (Indonesia)	350,000
Philippines	50,000

Chinese Population of Southeast Asia (2012)	
Region	**Population**
Thailand	9,400,000
Malaysia	6,900,000
Singapore	4,100,000
Indonesia	2,800,000
Burma (Myanmar)	1,600,000
Philippines	1,100,000
Vietnam	1,000,000
Cambodia	700,000

Brahmi script forms the basis for many Southeast Asian writing systems, and even in Muslim Java, the Hindu epic called the **Ramayana** remains a central cultural feature to this day.

A second wave of South Asian religious influence reached mainland Southeast Asia in the 13th century in the form of Theravada Buddhism, which came from Sri Lanka. Virtually all of the people in lowland Burma, Thailand, Laos, and Cambodia converted to Buddhism at that time. Today, this faith forms the foundation for many of the social institutions of mainland Southeast Asia. Saffron-robed monks, for example, are a common sight in Thailand and Burma, where Buddhist temples abound. In Burma especially, monks occasionally have been politically powerful.

Chinese Influences Unlike most other mainland peoples, the Vietnamese were not heavily influenced by South Asian civilization. Instead, their main early connections were with East Asia. Vietnam was actually a province of China until about a thousand years ago, when the Vietnamese rejected China's rule and established a kingdom of their own. But the Vietnamese retained many attributes of Chinese culture. The traditional religious and philosophical beliefs of Vietnam centered on Confucianism and Mahayana Buddhism, which is quite distinct from Theravada Buddhism.

East Asian cultural influences elsewhere in Southeast Asia are linked more directly to immigration from southern China (**Figure 13.22**). This migration stream reached a peak in the 19th and early 20th centuries. At first, most migrants were single men, many of whom married local women and established mixed communities. In the Philippines, the main elite population today, often called "Chinese Mestizo," is of mixed Chinese and Filipino descent. In the 19th century, Chinese women began to migrate in large numbers, allowing the creation of ethnically distinct Chinese settlements. Urban areas throughout most of Southeast Asia still have large and cohesive Chinese communities. In Malaysia, the

Chinese minority constitutes about a quarter of the population, whereas in the city-state of Singapore three-quarters of the people are of Chinese ancestry.

In many areas, relationships between the Chinese minority and the indigenous majority are strained. One source of tension is the fact that overseas Chinese communities tend to be wealthier than majority groups. Chinese emigrants frequently prospered by becoming merchants, a job that was often ignored by the local people. As a result, they have substantial economic influence, which is often resented. In 1998 and again in 2001, anti-Chinese rioting broke out in several Indonesian cities. Since then, the situation of the Chinese minority in Indonesia has improved considerably.

The Arrival of Islam Muslim merchants from India and Southwest Asia arrived in Southeast Asia more than a thousand years ago, and by the 1200s their religion began to spread. From an initial focus in northern Sumatra, Islam diffused into the Malay Peninsula, through the main population centers in the Indonesian islands, and east to the southern Philippines. By 1650, it had largely replaced Hinduism and Buddhism in Malaysia and Indonesia except on the small but fertile island of Bali. As Islam spread across Java, thousands of Hindu musicians and artists fled to Bali, giving the island an especially strong tradition of arts and crafts. Partly because of this artistic legacy, Bali is one of the premier destinations for international tourists.

Today, the world's most populous Muslim country is Indonesia, where 87 percent of the people follow Islam (**Figure 13.23**). This figure, however, hides significant internal religious diversity. In some parts of Indonesia, especially northern Sumatra (Aceh), orthodox forms of Islam took root. In others, such as central and eastern Java, a more lax form of worship emerged that retained certain Hindu and even animistic beliefs. Islamic reformers, however, have long been striving to instill more orthodox forms of worship among the Javanese, and they have recently found much success, especially among the young.

In insular Southeast Asia, Islamic fundamentalism has also gained ground in recent years. Whereas the current Malaysian government has generally supported the revitalization of Islam, it is wary of the growing power of the fundamentalist movement. Some Malaysian Muslims accuse fundamentalists of advocating social practices derived from the Arabian Peninsula that are not necessarily religious in origin. Adding to Malaysia's religious tension is the fact that almost all ethnic Malays, who form the country's majority population, follow Islam, whereas most members of the minority communities follow different religions. Religious tensions mounted in 2009, when the government banned Christians from using the term "Allah" to refer to God, and again in 2014, when the country's top Islamic council banned Halloween celebrations.

Brunei has moved further toward Islamic fundamentalism than any other Southeast Asian country. In 2014, the country's absolute monarch formally instituted *sharia*, or Islamic, law, including such harsh penalties as death by stoning and the cutting off of hands for theft. This move was widely criticized in both Brunei and abroad, and contributed to a decline in the country's tourism industry and its economy more generally.

Christianity in Southeast Asia Islam was still spreading eastward through insular Southeast Asia when the Europeans arrived in the 16th century, bringing Christianity. When Spain claimed the Philippine Islands in the 1570s, it found the southwestern portion of the archipelago to be thoroughly Islamic. The Muslims resisted the Roman Catholicism that the Spaniards insisted on spreading, and as a result parts of the southwestern Philippines remained an Islamic stronghold.

The Philippines as a whole is currently about 85 percent Roman Catholic, making it, along with East Timor, a former Portuguese colony, the only predominantly Christian country in Asia. Several Protestant sects have been spreading rapidly in the Philippines over the past several decades, however, creating a more complex religious environment.

Christian missions spread more widely through Southeast Asia in the late 19th and early 20th centuries, when European colonial powers controlled the region. French priests converted around 8 percent of the people of the lowlands of Vietnam to Catholicism, but they had little influence elsewhere. Missionaries were far more successful in Southeast Asia's highlands, where they found a wide array of tribal societies that had never accepted the major lowland religions. These peoples retained their indigenous belief systems, which generally focus on the worship of nature spirits and ancestors. Although some modern tribal groups retain such **animist** beliefs, many others converted to Christianity. In Indonesia, however, animist practices are technically illegal, as monotheism is one of the country's founding principles. As a result, most Indonesian tribal groups have been gradually converting to either Islam or Christianity.

Indonesia periodically experiences pronounced religious strife between its Muslim majority and its Christian minority. Despite its Muslim majority, Indonesia is a secular state that emphasizes tolerance among its officially recognized religions (Islam, Christianity, Buddhism, Hinduism, and Confucianism). With the Indonesian economic disaster of the late 1990s, however, relations deteriorated and violence erupted in the Maluku Islands of eastern Indonesia and in central Sulawesi. Religious tensions subsequently decreased in most parts of the country. In 2015, however, radical Muslim groups destroyed three churches and attacked a Christian village in Aceh in northern Sumatra, the only part of Indonesia that follows *sharia* law.

▼ **Figure 13.23 Indonesian Mosque** Indonesia is often said to be the world's largest Muslim country, as more Muslims reside here than in any other country in the world. Islam was first established in northern Sumatra, which is still the most devoutly Islamic part of Indonesia. Acehnese people are shown praying in front of the Baiturrahman Grand Mosque in the city of Banda Aceh. **Q: What are some of the political consequences of the high level of religious devotion found among most of the people of Aceh in northern Sumatra?**

Explore the **Sights** of Baiturrahman Grand Mosque

http://goo.gl/lcQEbl

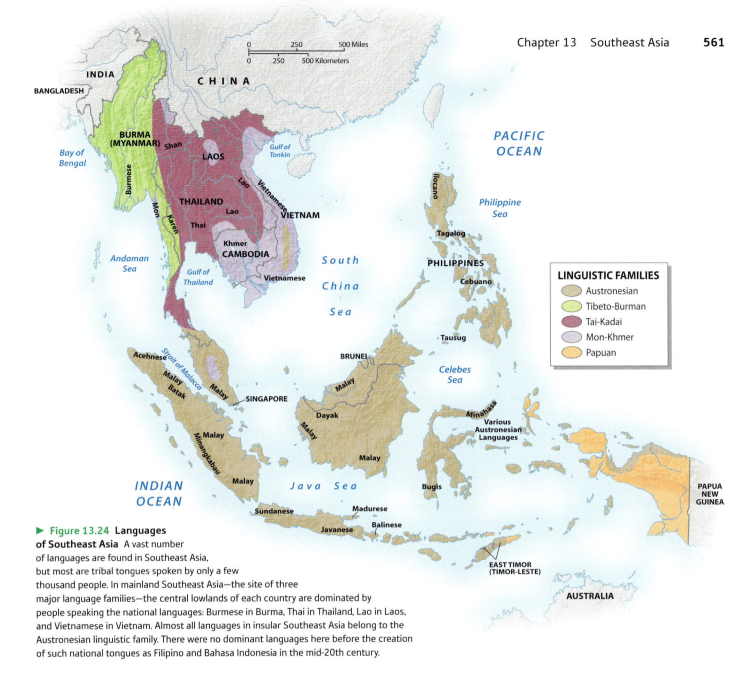

▶ **Figure 13.24 Languages of Southeast Asia** A vast number of languages are found in Southeast Asia, but most are tribal tongues spoken by only a few thousand people. In mainland Southeast Asia—the site of three major language families—the central lowlands of each country are dominated by people speaking the national languages: Burmese in Burma, Thai in Thailand, Lao in Laos, and Vietnamese in Vietnam. Almost all languages in insular Southeast Asia belong to the Austronesian linguistic family. There were no dominant languages here before the creation of such national tongues as Filipino and Bahasa Indonesia in the mid-20th century.

Religion and Communism By 1975, the political and economic system of communism had triumphed in Vietnam, Cambodia, and Laos, which resulted in the official discouragement of religious practices. At present, Vietnam's officially communist government is struggling against a revival of faith among the country's Buddhist majority and its 8 million Christians. Buddhist monks are frequently detained, and the government reserves for itself the right to appoint all religious leaders. Also expanding is Vietnam's indigenous religion of Cao Dai, a syncretic (or mixed) faith that venerates not only the Buddha, Confucius, and Jesus, but also the French novelist Victor Hugo.

Geography of Language and Ethnicity

As with religion, language in Southeast Asia expresses the long history of human movement. The several hundred distinct languages of the region can be placed into five major linguistic families, the first three of which certainly originated outside of the region. These are *Austronesian*, which covers most of the insular realm; *Tibeto-Burman,* which includes most of the languages of Burma; *Tai-Kadai*, centered on Thailand and Laos; *Mon-Khmer*, encompassing most of the languages of Vietnam and Cambodia; and *Papuan*, a Pacific language group found in eastern Indonesia (**Figure 13.24**).

Austronesian Languages Austronesian is one of the world's most widespread language families, extending from Madagascar to Easter Island in the eastern Pacific. Linguists believe that this family originated in Taiwan and adjacent areas of East Asia and then was spread widely across the Indian and Pacific oceans by seafaring people who migrated from island to island.

Today, almost all insular Southeast Asian languages belong to the Austronesian family. This means that elements of grammar and vocabulary are widely shared across the islands. But despite such linguistic commonality, more than 50 distinct Austronesian languages are spoken in Indonesia alone. And in far eastern Indonesia, many languages fall into the completely separate group called Papuan, closely associated with New Guinea.

One language, however, overshadows all others in insular Southeast Asia: Malay. Indigenous to the Malay Peninsula, eastern Sumatra, and coastal Borneo, Malay was spread historically by merchants and seafarers. As a result, it became a common trade language, or **lingua franca**, understood by peoples speaking different languages throughout much of the insular realm. When Indonesia became an independent country in 1949, its leaders decided to use the lingua franca version of Malay, written in the Roman alphabet, as the basis for a new national language that they called *Bahasa Indonesia* (or *Indonesian*). Although Indonesian differs from Malaysian, they essentially form a single language.

The goal of the new Indonesian government was to find a common language that would overcome ethnic differences throughout the far-flung state. This policy generally has been successful: Indonesian is now used in schools, the government, and the media, and most of the country's people understand it. But regional languages, such as Javanese, Balinese, and Sundanese, continue to be primary languages in most homes. More than 100 million people speak Javanese, making it one of the world's major tongues.

The eight major languages of the Philippines also belong in the Austronesian family. Despite more than 300 years of colonization by Spain, Spanish never became a unifying force for the islands. During the American period (1898–1946), English served as the language of administration and education. After independence following World War II, Philippine nationalists decided to adopt a national language to replace English and help unify the new country. They selected Tagalog, spoken in the Manila region. After Tagalog was standardized and modernized, it was renamed *Filipino* (alternatively, *Pilipino*). Today, mainly because of its use in education, television, and movies, Filipino has emerged as a generally successful national language.

Tibeto-Burman Languages Each mainland Southeast Asian country is closely identified with the national language spoken in its core territory. However, not all the inhabitants of these countries speak these official languages on a daily basis. In the mountains and other remote districts, minority languages remain primary. This linguistic diversity reinforces ethnic divisions despite national educational programs designed to foster unity.

Burma provides a good example of ethnic and linguistic diversity. Its national language is Burmese, which is closely related to Tibetan and written in its own script. People who speak Burmese as their first language, 68 percent of the population, are considered members of the nationally dominant **Burman** ethnic group. Although the nationalistic government of Burma attempted to force its language and version of unity on the entire population, a major schism developed with non-Burman tribal groups inhabiting the uplands. Most of these peoples speak languages in the Tibeto-Burman family, but their languages are quite distinct from Burmese.

Tai-Kadai Languages The Tai-Kadai linguistic family probably originated in southern China and then spread into Southeast Asia starting around 1200 CE. Today, closely related Tai languages are found through most of Thailand and Laos, in the uplands of northern Vietnam, in Burma's Shan Plateau, and in parts of southern China. Most of these languages are localized, spoken by small ethnic groups. However, two of them, Thai and Lao, are national languages, spoken in Thailand and Laos, respectively.

Linguistic terminology here is complicated. Historically, the main language of the Kingdom of Thailand (or Siam), called *Siamese*, was restricted to the national core in the lower Chao Phraya Valley. In the 1930s, however, the country changed its name from Siam to Thailand to emphasize the unity of all the peoples speaking closely related Tai languages within its territory. Siamese was similarly renamed *Thai*, and it has gradually become the unifying language for the country. Substantial variations in dialect persist, however, with those of the north often considered separate languages.

Lao, another Tai language, is the national tongue of Laos. More Lao speakers, however, reside in Thailand than in Laos, as the main language of the relatively poor Khorat Plateau of the northeast, called Isan, is essentially a dialect of Lao.

Mon-Khmer Languages The Mon-Khmer language family probably covered most of mainland Southeast Asia 1500 years ago. It contains two major languages—Vietnamese and Khmer (the national language of Cambodia)—as well as a host of minor languages spoken by hill peoples and a few lowland groups. Because of the historical Chinese influence in Vietnam, Vietnamese was usually written with Chinese characters until the French imposed the Roman alphabet, which remains in use. Khmer, like the other national languages of mainland Southeast Asia, is written in its own script, ultimately derived from India.

The most important aspect of mainland Southeast Asia's linguistic geography is the fact that in each country the national language is limited to the core area of densely populated lowlands, whereas the peripheral uplands are inhabited primarily by tribal peoples speaking separate languages. In Laos, up to 40 percent of the population is non-Lao. This linguistic contrast between the lowlands and the uplands poses an obvious problem for national integration. Such problems are most extreme in Burma, where the upland peoples are numerous and well organized—and have strenuously resisted the domination of the lowland Burmans.

Southeast Asian Culture in a Global Context

The imposition of European colonial rule ushered in a new era of globalization in Southeast Asia, bringing with it European languages, Christianity, and new educational systems. This period also deprived most Southeast Asians of their cultural autonomy. With decolonialization after World War II, several countries attempted to isolate themselves from the influences of the emerging global system. Burma retreated into its own form of Buddhist socialism, placing strict limits on foreign tourism and cultural globalization. Although the door has more recently opened, the government of Burma remains wary of foreign practices and influences. The Burmese government is relatively tolerant of music, however, which has led to a surge of spontaneous concerts and underground music festivals. At these events, hip-hop and techno-beat styles often mix with more traditional Burmese musical forms.

Other Southeast Asian countries have been more receptive to foreign cultural influences. This is particularly true of the Philippines, where American colonialism may have predisposed the country to many of the more popular forms of Western culture. As a result, Filipino musicians and other entertainers are in demand throughout East and Southeast Asia. Thailand, which was never subjected to colonial rule, has also been highly open to global culture.

Cultural globalization has also been challenged in Southeast Asia. The Malaysian government, for example, is critical of many American films and television programs. Islamic revivalism presents its own challenge to global culture in Malaysia and Indonesia. Risqué musicians such as Lady Gaga have had to cancel performances in Indonesia. But it is also true that several wildly popular Indonesian performers, most notably Jupe (Julia Perez, born Yuli Rachmawati), are equally sexually suggestive. Indonesia's popular president, Joko Widodo (called Jokowi), moreover, is a huge fan of hard-core heavy-metal music (**Figure 13.25**).

▼ **Figure 13.25 Joko Widodo: First Metalhead President** Joko Widodo ("Jokowi"), the president of Indonesia, is well-known as a fan of heavy-metal music. In this photograph, taken in 2013 when he was governor of Jakarta, Jokowi is holding a maroon bass guitar given to him by Robert Trujillo of the U.S. band Metallica. Jokowi often wears black T-shirts and leather jackets when attending heavy-metal concerts.

Some local firms are trying to capitalize on the tensions between traditional culture and cultural globalization. The online Southeast Asian retailer Zalora, for example, is pioneering the development of modest yet fashionable clothing that has a huge potential market both in the region and in Muslim countries elsewhere in the world.

The Globalization of Southeast Asian Food Southeast Asia also contributes heavily to global culture. Such influences are particularly notable in terms of cuisine, as Vietnamese, Indonesian, and especially Thai restaurants have spread across Europe, North America, and other parts of the world. In dispersing globally, Southeast Asian foods have undergone their own changes, as is evident in the saga of pho, the increasingly popular Vietnamese soup made from rice noodles, meat broth, sliced beef or chicken, and herbs. One study estimated that in 2005 alone, pho restaurants in the United States counted roughly US$500 million in revenue.

Despite its association with traditional Vietnamese culture, pho is actually a fairly recent innovation. Created in northern Vietnam in the early 1900s, pho was a simple breakfast food, widely considered too insubstantial to be lunch or dinner. Initially, it was sold by street vendors carrying large bowls of broth on shoulder poles. As pho gained popularity, it spread to central and southern Vietnam. In Saigon, now Ho Chih Minh City, the dish was transformed, becoming somewhat sweeter and spicier. It also became customizable, with customers requesting different sauces and herbs. In the process, pho transformed from a street food to a restaurant dish.

It was the southern Vietnamese version of pho that spread to the United States after the end of the Vietnam War, where it lost its association with breakfast and is instead often eaten for lunch or dinner (**Figure 13.26**). In Vietnam as well it is increasingly available at all hours of the day, but for most people it remains the quintessential breakfast dish.

Language and Globalization As the global language, English causes ambivalence in much of Southeast Asia. On the one hand, it is the language of questionable popular culture; yet on the other, citizens need proficiency in English if they are to participate in international business and politics.

Singapore's linguistic situation is particularly complex, as English, Mandarin Chinese, Malay, and Tamil all have official status, and the local languages of southern China remain common in home environments. Singapore's government strongly encourages Mandarin Chinese and discourages the southern Chinese dialects. It also supports English while struggling against *Singlish,* an English-based dialect containing many Malay and southern Chinese words and grammatical forms. Singaporean officials have occasionally slapped restrictive ratings on local films for "bad grammar" if they contain too much Singlish dialogue.

In the Philippines, nationalists have long criticized the use of English, even though widespread fluency has proved

(a)

(b)

▲ **Figure 13.26 Pho: Vietnamese Soup** (a) A customizable soup that was originally served for breakfast in northern Vietnam, pho has been transformed into a main meal as it has spread through Vietnam and across other parts of the world. (b) A popular pho restaurant in Ho Chi Minh City.

beneficial to the millions of Filipinos who work in call centers, as crew members in the globalized shipping industry, or have moved abroad. The Philippine government has favored Filipino in public schools, and as a result some fear that competence in English is slowly declining. Others counter that the spread of cable television, with its many U.S. shows, has enhanced Filipinos' English-language skills.

Sports and Globalization Southeast Asia is not particularly known for its athletic culture, although interest and participation have been growing, in part because of the Southeast Asian Games, an Olympic-style regional event. As is true for most of the world, soccer is highly popular. Basketball is also well-liked, and although a Southeast Asian championship event is held every other year, the Philippines remains the acknowledged leader. Indonesia and Malaysia dominate in badminton, another sport popular in the region.

An indigenous Southeast Asian game that is gaining popularity both at home and abroad is *sepak takraw,* perhaps best described as a cross between volleyball and soccer (you cannot use your hands), played with a rattan ball. Top players must have gymnastic skill, as they often do complete flips in order to kick the ball with a downward motion over the net (**Figure 13.27**).

Martial arts have widespread appeal across Southeast Asia. Most countries of the region have their own particular forms of fighting. In the Philippines, *arnis* (or *eskrima*) involves a well-developed set of techniques that generally involve the use of rattan sticks. *Pencak silat* is the umbrella term for martial arts in Indonesia and Malaysia. A wide array of weapons, both edged and blunt, are used in *Pencak silat* competitions. Similar martial arts are found in Thailand, but the country's best-known combat sport is *muay thai,* or Thai kickboxing. Competitors use a wide variety of punches, kicks, elbow jabs, and knee thrusts. Over the past several decades, *muay thai* has gained popularity across much of the world.

Although the diffusion of global cultural forms is beginning to merge cultures across Southeast Asia, religious revivalism operates in the opposite direction. This same combination of opposing forces is also found in the geopolitical realm.

REVIEW

13.7 Which major world religions have spread into Southeast Asia over the past 2000 years, and in what parts of the region did they become established?

13.8 How have such strongly multilingual countries as Indonesia, the Philippines, and Singapore dealt with the challenges posed by the lack of a dominant language?

13.9 How have different Southeast Asian countries reacted to the challenges of cultural globalization?

■ **KEY TERMS** Ramayana, animist, lingua franca, Burman

▼ **Figure 13.27 Sepak Takraw** A fast-paced game that might be considered something of a cross between volleyball and soccer, *sepak takraw* originated in Thailand and Malaysia. The sport is now played throughout Southeast Asia and, increasingly, in other parts of the world.

Geopolitical Framework: Ethnic Strife and Regional Cooperation

Southeast Asia is sometimes defined as a geopolitical entity composed of the 10 different countries that form the Association of Southeast Asian Nations, or ASEAN (**Figure 13.28**). But East Timor, which gained independence in 2002, is not a member. ASEAN is currently reviewing the readiness of East Timor to join the association, and in 2015 the East Timorese government announced that it had fulfilled all necessary obligations. Despite the general success of ASEAN in reducing international tensions in the region, several Southeast Asian counties are struggling with serious ethnic and regional conflicts.

Before European Colonialism

The modern countries of mainland Southeast Asia were all indigenous kingdoms at one time or another before the onset of European colonialism. Cambodia emerged earliest, reaching its height in the 12th century. By the 1300s, independent kingdoms had been established by the Burman, Siamese (Thai), Lao, and Vietnamese people in the major

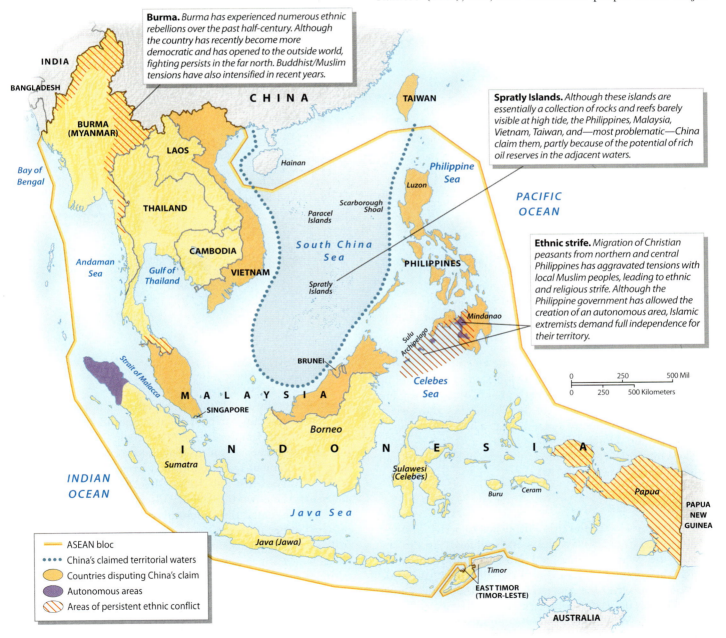

Burma. *Burma has experienced numerous ethnic rebellions over the past half-century. Although the country has recently become more democratic and has opened to the outside world, fighting persists in the far north. Buddhist/Muslim tensions have also intensified in recent years.*

Spratly Islands. *Although these islands are essentially a collection of rocks and reefs barely visible at high tide, the Philippines, Malaysia, Vietnam, Taiwan, and—most problematic—China claim them, partly because of the potential of rich oil reserves in the adjacent waters.*

Ethnic strife. *Migration of Christian peasants from northern and central Philippines has aggravated tensions with local Muslim peoples, leading to ethnic and religious strife. Although the Philippine government has allowed the creation of an autonomous area, Islamic extremists demand full independence for their territory.*

Legend:
- ASEAN bloc
- China's claimed territorial waters
- Countries disputing China's claim
- Autonomous areas
- Areas of persistent ethnic conflict

▲ **Figure 13.28 Geopolitical Issues in Southeast Asia** The countries of Southeast Asia have managed to solve most of their border disputes and other sources of potential conflicts through ASEAN (the Association of Southeast Asian Nations). Internal disputes, however, mostly focused on issues of religious and ethnic diversity, continue to plague several of the region's states, particularly Indonesia and Burma. ASEAN also experiences tension with China over islands in the South China Sea.

▶ **Figure 13.29 Colonial Southeast Asia** With the exception of Thailand, all of Southeast Asia was under Western colonial rule by the early 1900s. The Netherlands had the largest empire in the region, covering the territory that was later to become Indonesia. France maintained a substantial imperial realm in Vietnam, Laos, and Cambodia, as did Britain in Burma and Malaysia (including Singapore and Brunei). The Philippines was colonized by Spain, but passed the control to the United States in 1898.

Q: Note that Thailand was the only country in Southeast Asia that was not colonized by Western powers. How did Thailand avoid this fate, and what if any consequences did it have for the country's future development?

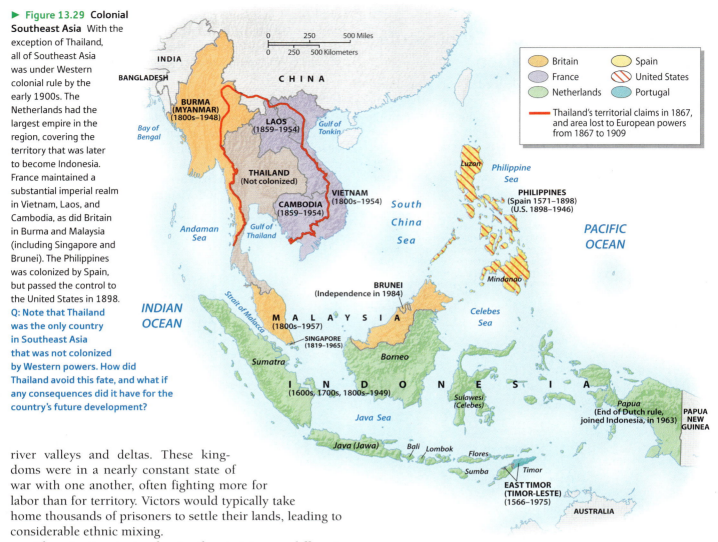

river valleys and deltas. These kingdoms were in a nearly constant state of war with one another, often fighting more for labor than for territory. Victors would typically take home thousands of prisoners to settle their lands, leading to considerable ethnic mixing.

The situation in insular Southeast Asia was different, with the premodern map bearing little resemblance to that of the modern nation-states. Many kingdoms existed on the Malay Peninsula and on the islands of Sumatra and Java, but few were territorially stable. In the Philippines and eastern Indonesia, societies were organized at a more local level. Indonesia, the Philippines, and Malaysia thus owe their modern territorial configurations almost wholly to European colonial influence (**Figure 13.29**).

The Colonial Era

The Portuguese were the first Europeans to arrive (around 1500), lured by the spices of eastern Indonesia. In the late 1500s, the Spanish conquered most of the Philippines, which they used as a base for their silver and silk trade between China and the Americas. By the 1600s, the Dutch had started staking out Southeast Asian territory, followed by the British. With superior naval weaponry, the Europeans quickly conquered key ports and strategic trading locales. Yet for the first 200 years of colonialism, except in the Philippines, the Europeans did not control large expanses of land.

Dutch Power By the 1700s, the Netherlands had become the most powerful force in insular Southeast Asia. As a result, a Dutch Empire in the "East Indies" began appearing on world maps. From its original base in northwestern Java, this colonial realm continued to grow into the early 20th century, when the Dutch defeated their last main adversary, the powerful Islamic sultanate of Aceh in northern Sumatra. Later, the Dutch invaded the western portion of New Guinea in response to German and British advances in the east. In a subsequent treaty, these imperial powers sliced New Guinea down the middle, with the Netherlands taking the west.

British, French, and U.S. Expansion The British initially focused their attention on the sea-lanes linking South Asia to China, establishing several fortified trading outposts along the vital Strait of Malacca. The most important of these was Singapore, founded in 1819. To avoid conflict, the British and Dutch agreed that the British would limit their attention to the Malay Peninsula and northern Borneo. Britain allowed Muslim sultans here to retain limited powers, and their descendants still enjoy token authority in several Malaysian states.

When the British left this area in the early 1960s, Malaysia emerged as an independent state. Two small portions of the former British sphere did not join the new country. In northern Borneo, the Sultanate of Brunei gained independence, backed by its large oil reserves. Singapore briefly joined Malaysia, but then withdrew and became fully independent in 1965. This divorce was carried out largely for ethnic reasons. Malaysia was to be a primarily Malay and Muslim state, but with Singapore included, Malaysia's population would have been almost half Chinese.

In the 1800s, European colonial power spread across most of mainland Southeast Asia. British forces in India fought several wars against the kingdom of Burma before annexing the entire extent in 1885, including considerable upland territories that had never been under Burmese rule. During the same period, the French moved into Vietnam's Mekong Delta, gradually expanding their territorial control westward into Cambodia and north to China's border. Thailand was the only country to avoid colonial rule, although it did lose substantial territories to the British in the Malay Peninsula and to the French in Laos.

The final colonial power to enter the area was the United States, which took the Philippines first from Spain and then, after a bitter war, from Filipino nationalists between 1898 and 1900. The U.S. army subsequently conquered the Muslim areas of the southwest, most of which had never been fully under Spanish authority.

Growing Nationalism Organized resistance to European rule began in the 1920s in mainland countries, but it took the Japanese occupation during World War II to shatter the myth of European invincibility. After Japan's surrender in 1945, agitation for independence was renewed throughout Southeast Asia. As Britain realized that it could no longer control its South Asian empire, it withdrew from adjacent Burma, which achieved independence in 1948. The Netherlands failed to reconquer Indonesia after World War II and was forced to acknowledge Indonesian independence in 1949. The United States granted long-promised independence to the Philippines in 1946, although it retained key military bases for several decades.

The Vietnam War and Its Aftermath

Unlike the United States, France was determined to maintain its Southeast Asian empire after World War II. Resistance to French rule was organized primarily by communist groups that were deeply rooted in northern Vietnam. Open warfare between French soldiers and the communist forces went on until France suffered a decisive defeat in 1954. An international peace council in Geneva then decided that Vietnam would be divided into two countries. As a result, communist leaders came to power in North Vietnam, allied with the Soviet Union and China. South Vietnam simultaneously emerged as an independent, capitalist-oriented state with close political ties to the United States.

The Geneva peace accord did not end the war. Communist guerrillas in South Vietnam fought to overthrow the new government and unite it with the north. North Vietnam sent troops and war materials across the border to aid the rebels. In Laos, the communist Pathet Lao forces challenged the government, while in Cambodia the **Khmer Rouge** guerrillas gained considerable power. In South Vietnam, the government gradually lost control of key areas, including much of the Mekong Delta, a region perilously close to the capital city of Saigon.

U.S. Intervention By 1962, the United States was sending large numbers of military advisors to South Vietnam. The **domino theory** became accepted U.S. foreign policy. According to this thesis, if Vietnam fell to the communists, then so would Laos, Cambodia, and eventually the rest of Southeast Asia. By 1968, over half a million U.S. troops were fighting a ferocious land war against the communist guerrillas. But despite superior arms and troops—and domination of the air—U.S. forces failed to gain control over much of the countryside (**Figure 13.30**). As casualties mounted and the antiwar movement back home strengthened, the United States began secret talks with North Vietnam toward a negotiated settlement. By the early 1970s, U.S. troop withdrawals began in earnest.

Communist Victory Withdrawal of U.S. forces and financial support led to the collapse of the noncommunist governments of the former French zone. After Saigon fell in 1975, Vietnam was officially reunited under the government of the north. Reunification was a traumatic event in southern Vietnam. Hundreds of thousands of people fled to other countries, especially the United States. The first wave of refugees were primarily wealthy professionals and businesspeople, but later migrants included many relatively poor ethnic Chinese. Most of these refugees fled on small, rickety boats; large numbers suffered shipwreck or pirate attack.

Vietnam proved fortunate compared with Cambodia, where the Khmer Rouge installed one of the most brutal regimes the world has ever seen. City dwellers were forced into the countryside to become peasants, and most wealthy and educated people were summarily executed. After several years of bloodshed that took an estimated 1.5 million lives, neighboring Vietnam invaded Cambodia and installed a far less brutal but still repressive regime. Fighting between different factions continued for more than a decade, but by the late 1980s the United Nations (UN) brokered a settlement and a parliamentary system of government subsequently brought peace to the shattered country. Although Cambodia is officially a constitutional monarchy with an elected government, corruption is

▲ **Figure 13.30 Vietnam War** The bloody and hard-fought Vietnam War involved the heavy use of aircraft by the U.S. military. This 1965 photograph shows U.S. Army helicopters covering the advance of ground troops by pouring machine gun fire into the nearby forest.

widespread, and democratic institutions remain weak and unstable.

Vietnam stationed significant numbers of troops in Laos after 1975. Large numbers of Hmong and other tribal peoples—many of whom had fought on behalf of the United States—fled to Thailand and the United States. The Communist Party still maintains political power in Laos, although much of the economy has been opened to private firms. Many Laotian Hmong refugees in the United States, however, continue to seek political change at home.

Geopolitical Tensions in Contemporary Southeast Asia

Most of the serious geopolitical problems in Southeast Asia today occur within countries rather than between them. In several areas, locally based ethnic groups struggle against national governments that inherited territory from the former colonial powers. Tension also results when tribal groups attempt to preserve their homeland from logging, mining, or interregional migration (see Figure 13.28). Such conflicts are especially pronounced in the large, multiethnic country of Indonesia.

Conflicts in Indonesia When Indonesia gained independence in 1949, it included all the former Dutch possessions in the region except western New Guinea. The Netherlands retained this territory, now called Papua, arguing that its cultural background distinguished it from Indonesia. Although Dutch authorities were preparing western New Guinea for independence, Indonesia demanded the entire territory. Bowing to pressure from the United States, the Netherlands relented and allowed Indonesia to take control in 1963.

Tensions in Papua increased in the following decades as Javanese immigrants, along with mining and lumber firms, arrived in the area. Faced with the loss of their land and the degradation of their environment, indigenous rebels formed the separatist organization, OPM (*Organisesi Papua Merdeka*). Rebel leaders demand independence, or at least autonomy, but they face a far stronger force in the Indonesian army. The struggle between the Indonesian government and the OPM is a sporadic, but occasionally bloody, guerrilla affair. Indonesia is determined to maintain control over Papua in part because it is home to one of the country's largest taxpayers, the highly polluting Grasberg copper and gold mine (**Figure 13.31**). In 2014, Indonesian's president, Joko Widodo, released Papuan political prisoners and promised a more accommodating strategy. But a 2016 Human Rights Watch report claimed that Indonesia was still suppressing the freedom of expression and association throughout Papua.

A more intensive war erupted in 1975 on the island of Timor. The eastern half of this poor island had been a Portuguese colony and had evolved into a mostly Christian society. The East Timorese expected independence when the Portuguese finally withdrew, but Indonesia viewed the area as its own territory and immediately invaded. A ferocious struggle ensued, which the Indonesian army won in part by starving the Timorese into submission.

But after an economic crisis in 1997, Indonesia's power in the region slipped. A new Indonesian government promised an election in 1999 to see if the East Timorese wanted independence. At the same time, however, Indonesia's army began to organize loyalist militias to intimidate the people of East Timor into voting to remain within the country. When it was clear that the vote would be for independence, the militias began to riot and loot. Under international pressure, Indonesia finally withdrew, UN forces arrived, and the East Timorese began to build a new country, with independence recognized in 2002.

Considering the devastation caused by the militias, the reconstruction of East Timor was not easy. Even selecting an official language proved difficult; after some deliberation, both Tetum (one of 16 indigenous languages) and Portuguese were selected. Ethnic rioting and coup attempts further weakened the country for a number of years. In 2012, East Timor was finally declared safe enough for the withdrawal of the UN peacekeeping force.

For many years, Indonesia's most serious regional conflict was that of Aceh in northern Sumatra. Many Acehnese—Indonesia's most devout Muslim group—have fought for the creation of an independent state. Ironically, the devastation caused by the December 2004 tsunami,

▲ **Figure 13.31 Grasberg Mine** Located in the Indonesian province of Papua, Grasberg is the world's largest gold mine and third largest copper mine. Employing more than 19,000 people, Grasberg is of great economic importance to Indonesia. It is also highly polluting, generating 253,000 tons (230,000 metric tons) of tailings per day and is thus opposed by many local inhabitants.

tensions by creating the Autonomous Region in Muslim Mindanao (ARMM), providing a degree of political and cultural autonomy to the main Muslim area. However, the more extreme Islamist factions continued to fight, demanding greater autonomy if not outright independence. A peace treaty signed in 2012 promised enhanced autonomy of the region, which would be renamed "Bangsamoro." But as splinter groups continue to fight the Philippine military, the treaty provisions have never been implemented.

The Philippines' political problems are not limited to the south. A long-simmering communist rebellion has affected almost the entire country. In the mid-1980s, the New People's Army (NPA) controlled one-quarter of the Philippines' villages scattered across all the major islands. Although the NPA's strength has declined since then, it remains a potent force in many areas, commanding an estimated 3,200 troops as of early 2015.

Due to the security threats posed by both Islamist fighters in the south and the NPA throughout the country, the Philippine government has turned to the United States for help. Several thousand U.S. military advisors and trainers are currently working with the Philippine military, mostly in the Mindanao region. As a result, many Filipino nationalists are concerned that their country is again falling under U.S. domination.

which left over 500,000 Acehnese homeless, allowed a peace settlement to be finally reached, as the needs of the area were so great that the separatist fighters agreed to lay down their weapons. In response, Indonesia allowed Aceh to become a special autonomous territory under Islamic (*sharia*) law, and subsequent elections brought former rebel leaders into the heart of the local government. Their policies have proved very controversial, both locally and in the rest of the country. In 2015, Aceh enacted laws criminalizing drinking alcohol, adultery, homosexuality, and even public displays of affection.

Indonesia has obviously had difficulties in building a unified nation over its vast extent. As a creation of the colonial period, Indonesia has weak historical foundations. Some critics have viewed it as something of a Javanese empire, even though many non-Javanese individuals have risen to high governmental positions. Indonesia, along with the Philippines, now has the highest degree of political freedom in Southeast Asia, but its influential military remains suspicious of any movements for regional autonomy.

Regional Tensions in the Philippines The Philippines has long struggled against a regional secession movement in its Islamic southwest. In the 1990s, it sought to defuse

The Opening of Burma Burma has long been a war-ravaged and relatively isolated country. Its military struggles of the 1970s, 1980s, and 1990s pitted the central government, dominated by the Burmans, against the country's varied non-Burman societies located in the peripheral areas of the country (**Figure 13.32**). Fighting gradually intensified after independence in 1948, until almost half of the country's territory had become a combat zone.

Burma's troubles were by no means limited to the country's ethnic minorities. From 1988 until 2011, it was ruled by a repressive military regime bitterly resented by much of the population. The Burmese state at the time was also extremely suspicious of the outside world. In 2005, it mandated the creation of a new capital city at Nay Pyi Taw, located in a remote, forested area 200 miles north of the old capital of Rangoon (Yangon). The new capital, which was supposedly picked on the basis of astrological calculations, was to be largely closed to the outside world. Such actions, however, did not generate stability. In 2007, Burma was convulsed by massive pro-democracy protests against the government, led by Buddhist monks.

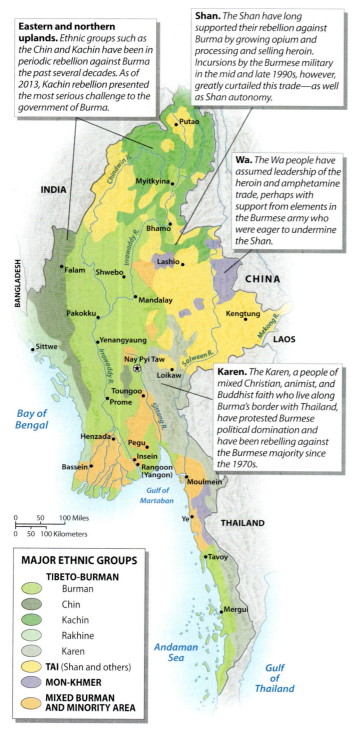

Eastern and northern uplands. *Ethnic groups such as the Chin and Kachin have been in periodic rebellion against Burma the past several decades. As of 2013, Kachin rebellion presented the most serious challenge to the government of Burma.*

Shan. *The Shan have long supported their rebellion against Burma by growing opium and processing and selling heroin. Incursions by the Burmese military in the mid and late 1990s, however, greatly curtailed this trade—as well as Shan autonomy.*

Wa. *The Wa people have assumed leadership of the heroin and amphetamine trade, perhaps with support from elements in the Burmese army who were eager to undermine the Shan.*

Karen. *The Karen, a people of mixed Christian, animist, and Buddhist faith who live along Burma's border with Thailand, have protested Burmese political domination and have been rebelling against the Burmese majority since the 1970s.*

INDIA

Putao

Myitkyina

Bhamo

BANGLADESH

Falam
Shwebo
Pakokku
Mandalay
Lashio
Kengtung

CHINA

LAOS

Sittwe
Yenangyaung
Nay Pyi Taw
Loikaw
Toungoo
Prome

Bay of Bengal

Henzada
Pegu
Insein
Bassein
Rangoon (Yangon)
Moulmein

Gulf of Martaban

Ye

THAILAND

Tavoy

Andaman Sea

Mergui

Gulf of Thailand

0 50 100 Miles
0 50 100 Kilometers

MAJOR ETHNIC GROUPS

TIBETO-BURMAN
- Burman
- Chin
- Kachin
- Rakhine
- Karen

TAI (Shan and others)

MON-KHMER

MIXED BURMAN AND MINORITY AREA

◀ **Figure 13.32 Ethnic Groups and Conflicts in Burma** Although the central lowlands of Burma are primarily populated by people of the dominant Burman ethnic group, the peripheral highlands and the southeastern lowlands are home to numerous non-Burman peoples. Most of these peoples have been in periodic rebellion against Burma since the 1970s, owing to their perception that the national government is attempting to impose Burman cultural norms. Economic stagnation and political repression by the central government have intensified these conflicts.

But despite these changes, Burma continues to be a troubled country, burdened by ethnic conflicts and the lack of effective governmental institutions. Despite her popularity and political victories, Aung San Suu Kyi is not legally allowed to become leader of the country due to her earlier marriage to a foreigner. Several non-Burman ethnic groups still deeply mistrust the government and continue to fight, particularly the mostly Christian Kachin people of the north and the Karen people of the east. They generally see the national government as a Burman institution seeking to impose its culture upon them, and they are concerned about proposed dams that could flood large portions of their lands. Such problems are exacerbated by ultra-nationalist monks, who stress the Buddhist nature of the country and oppose other religions. Most of their hostility has been directed against Burma's Muslim minority, who constitute about 4 percent of the country's population (see *People on the Move: Human Trafficking in Southeast Asia and the Plight of the Rohingya*).

Trouble in Thailand Compared to Burma's postwar history, that of Thailand has been relatively peaceful. Most observers thought that Thailand had become a stable democracy, but in 2006, Prime Minister Thaksin Shinawatra was overthrown by a military coup that apparently had the blessing of Thailand's revered king. Thaksin, a wealthy businessman, had gained the support of Thailand's poor by setting up national health-care and rice-subsidy systems, but he infuriated the middle and upper classes by taking too much power into his own hands. Huge protests and counterprotests followed for several years. In 2011, Thaksin's sister, Yingluck Shinawatra, became prime minister after winning a landslide victory. In May 2014, however, Thailand's military overthrew the elected government, dissolved the senate, repealed the constitution, and installed a military-dominated national legislature (**Figure 13.33**). Although martial law was finally lifted in 2015, Thailand has not yet returned to democratic rule.

Conflict persists, moreover, in Thailand's far south, a primarily Malay-speaking, Muslim region. Minor rebellions have flared up in southern Thailand for decades, but the violence sharply escalated in 2004. The Thai government initially responded with harsh military measures, a strategy that many Thai military leaders viewed as counterproductive. But the more conciliatory stance later taken by the Thai government proved no more effective, as a shadowy Islamist group continues to attack Thai-speaking government officials, as well as ordinary civilians. From 2004 to 2015, more than 6,500 Thai citizens lost their lives in this conflict. The war has also

More recently, Burma has undergone major changes, making progress toward democratization. In 2011, its government released the democratic opposition leader and Nobel Peace Prize recipient Aung San Suu Kyi from house arrest, and in 2015 her political party won a solid victory in a national election. During the election campaign, Suu Kyi pledged to end the hostility between the Burmese government and the country's rebellious ethnic groups. Burma has also begun to open itself to the rest of the world, seeking better relations with the European Union and the United States.

Human Trafficking in Southeast Asia and the Plight of the Rohingya

Southeast Asia is one of the world's major hubs of **human trafficking**—the forceful movement of people over long distances for the purposes of exploitation. Human trafficking in the region takes many forms, but usually involves transporting desperate people from the poorer countries of the region, such as Burma and Laos, to the region's more prosperous countries, such as Thailand and Malaysia. Thailand's large sex industry in particular has long been associated with trafficking. According to one report, more than 80,000 young women and girls were sold into prostitution in Thailand between 1990 and 2010.

Signs of Progress Some progress has been made in combatting Southeast Asian human trafficking. Since 2003, the Thai government has been actively cooperating with those of Laos and Burma to suppress the trade, but with uncertain success. More promising results have been obtained in the fight against the slave-labor practices found in some corners of the Thai seafood industry, a business that requires large amounts of labor for fish processing. In 2015, the Thai government arrested more than 100 suspected human traffickers after the European Union threatened to ban Thai seafood imports unless efforts were made to eliminate both labor abuses and illegal fishing. In early 2016, the United States banned imports of all Southeast Asian fish and shrimp produced through coerced labor.

The Plight of the Rohingya A more intractable human trafficking problem in Southeast Asia centers on the Rohingya people. A Muslim ethnic group, some 1.3 million strong, Rohingyas are of South Asian origin and live in eastern Burma, just south of the border with Bangladesh. The Rohingya have likely lived here for hundreds of years, but Burma's government regards them as recent immigrants from Bangladesh and thus denies them citizenship. After ethnic conflicts claiming more than 200 lives erupted in 2012, the Burmese government has been forcing the Rohingya out of the country. These displaced people end up in squalid refugee camps in Bangladesh, which also refuses to integrate the Rohingya into its own society.

With inadequate food, sanitation, and medical care, most Rohingyas are desperate to find a better life elsewhere. In 2014 alone, an estimated 53,000 Rohingya took to the sea, most hoping to reach Malaysia, a relatively prosperous Muslim-majority country. Rohingya refugees usually have to pay several hundred dollars to human traffickers, a huge sum for such impoverished people. They are then typically packed like cattle into rickety boats that are initially sent to makeshift camps in the forests of southern Thailand (**Figure 13.3.1**). Lucky refugees will eventually be transported to Malaysia, but the unfortunate may be captured by corrupt Thai officials and then sold to other human traffickers for work in fish processing or other exploitative industries. Death rates are high in the Thai camps and probably even higher at sea, as many boats flounder. In 2015, Thai officials found dozens of shallow graves in southern Thailand that are believed to contain the remains of Rohingya refugees.

The discovery of these graves prompted the Thai government to thoroughly investigate the area for signs of human smuggling. The government also issued warrants for local officials, politicians, and police officers believed to be involved in the trafficking rings. It remains to be seen whether such actions will be effective. Additional hope was provided, however, in early 2016 when Migrant Offshore Aid Station, a charitable organization based in Malta, sent a rescue ship to Southeast Asian waters to help Rohingya refugees at sea.

1. Why are the Rohingya sometimes referred to as "the world's most persecuted group of people"?

2. What techniques might prove useful for reducing if not eliminating human trafficking in Southeast Asia?

▼ **Figure 13.3.1 The Plight of the Rohingya** Tens of thousands of Rohingya people, sometimes described as the world's most persecuted ethnic group, have been forced to flee Burma in recent years. Many have died at sea in rickety boats. This overloaded boat was intercepted by a Thai naval patrol in 2013, and its occupants were taken to southern Thailand.

raised tensions between Thailand and Malaysia, as Thailand accuses its southern neighbor of not doing enough to prevent militants from slipping across the border.

International Dimensions of Southeast Asian Geopolitics

As events in southern Thailand show, geopolitical conflicts in Southeast Asia can be complex, involving several different countries as well as nonstate organizations.

Historically, some of the most serious tensions have emerged when two countries claimed the same territory. More recently, radical Islamist groups have posed the biggest challenge.

Territorial Conflicts In previous decades, several Southeast Asian countries quarreled over their boundaries. The Philippines, for example, still maintains a "dormant" claim to the Malaysian state of Sabah in northeastern Borneo, based on the fact that the Islamic Sultanate of the

▲ **Figure 13.33 Thai Military Coup** The government of Thailand has moved between military rule and democracy on several occasions in recent decades. Many civilians protested the 2014 military takeover of the country. Demonstrators hold signs as soldiers stand guard during a protest against Victory Monument in central Bangkok on May 26, 2014.

buildings, landing strips, and other facilities.

Although China claims that it is not planning to militarize its islands in the Spratly group, the Philippines and Vietnam are not convinced, nor is the United States. To demonstrate its opposition to the expanded Chinese presence, in 2015 the U.S. navy sent a guided-missile destroyer to sail within 12 nautical miles, the official limit of territorial waters, of a Chinese-held island. China quickly denounced this maneuver as a deliberate provocation. Although tensions between China and the United States subsequently receded, the South China Sea remains a global hotspot. Concerns about China's growing strength in the region has also led to closer military ties among the United States, the Philippines, and Vietnam, as well as Japan, South Korea, and Australia.

ASEAN, Global Geopolitics, and Maritime Threats Within Southeast Asia itself, the development and enlargement of ASEAN has generally reduced geopolitical tensions.

Sulu Archipelago in the southern Philippines formerly controlled much of that territory. With the rise of ASEAN, the Philippines government concluded that maintaining peaceful relations with neighbors was more important than pursuing claims to additional territory. It was, therefore, deeply embarrassed in early 2013 when a group of 200 armed Filipinos, led by a descendant of the former Sultan of Sulu, landed in a Sabah town to claim it as their own. When Malaysian troops retook the town, more than 50 Philippine citizens were killed.

A more difficult and complex conflict has emerged over the Spratly and Paracel islands in the South China Sea, two groups of rocks, reefs, and tiny islands that may well lie over substantial oil reserves (**Figure 13.34**). China controls the Paracel Islands but Vietnam claims them, while all or part of the Spratly Islands are claimed by the Philippines, Malaysia, Vietnam, Brunei, China, and Taiwan, all of which occupy individual islands. Tensions began to escalate in 2010 as China, which claims almost the entire South China Sea area, began to increase its military presence. By 2014 China was actively expanding its holdings in the region, dredging sand from the sea and depositing it around its islets until they were large enough to support

▼ **Figure 13.34 The Spratly Islands** The struggle over the Spratly Islands, some or all of which are claimed by China, Taiwan, Vietnam, Malaysia, Brunei, and the Philippines, has been heating up in recent years. In order to bolster its claims, China has been building facilities, as well as actual islands, in the area. This March 2015 photograph shows Chinese construction and dredging under way at Mischief Reef.

Explore the **Sights** of China's Airport on Fiery Cross Reef

http://goo.gl/yhiq5B

Although ASEAN is on friendly terms with the United States, one purpose of the organization is to prevent any outside country from gaining undue influence in the region. ASEAN's ultimate international policy is to encourage negotiation over confrontation and to enhance trade. As a result, the ASEAN Regional Forum (ARF) was established in 1994 as an annual conference in which Southeast Asian leaders could meet with the leaders of both East Asian and Western powers to discuss issues and ease tensions. Seeking to further improve relations with East Asia, ASEAN leaders also established an annual ASEAN+3 meeting, where its foreign ministers confer with those of China, Japan, and South Korea.

Thus far, ASEAN and its associated organizations have proven ineffective in addressing the conflict in the South China Sea. In early 2016, the Philippines urged ASEAN to agree to a binding code of conduct in the region to put pressure on China, but other member countries, particularly Laos and Cambodia, have strong ties with China and do not want to antagonize it. The Philippines has also turned to the International Court of Justice in the Netherlands, filing an arbitration case in 2013 against Chinese actions in the region. China has ignored this threat, however, and the legal case has moved forward very slowly.

One maritime issue that China, the United States, and ASEAN countries can agree on is the need to suppress **piracy**. In 2014, Southeast Asia surpassed Somalia as the most pirate-afflicted part of the world. Much global shipping passes through the Strait of Malacca between Peninsular Malaysia and Sumatra, a waterway characterized by numerous small channels where pirates can easily hide. Unlike the pirates of Somalia, those of Southeast Asia rarely hold and ransom crews, but instead concentrate instead on stealing cargo. In response, Malaysia, Indonesia, Singapore, and Thailand formed an anti-piracy effort called the Malacca Strait Patrol, but it has not been very effective. In 2015, Vietnam called on ASEAN to address the issue, but again the official response has been slow.

Global Terrorism and International Relations ASEAN has also sought to counter the rise of radical Islamism in the Muslim parts of the region, but again to uncertain effect. For many years, the strongest Islamist group in the region was Jemaah Islamiya (JI), which calls for the creation of a single Islamic state across insular Southeast Asia. JI agents are believed to have detonated several deadly bombs in Java and Bali in the early 2000s. Indonesia responded by creating an elite counterterrorism squad (Detachment 88) and by establishing a "deradicalization program." By 2008, most observers had concluded that these programs were so successful that JI was no longer a threat.

More recently, however, radical Islamism has again strengthened. Most worrisome is the presence of the so-called ISIL (Islamic State of Iraq and the Levant, also known as IS or ISIS). An ISIL-associated bombing of a Starbucks and a police post in Jakarta in early 2016 killed four civilians and generated a prolonged shootout with the local police. In response, Indonesia arrested 13 suspected terrorists, and both Malaysia and Singapore deported a number of additional suspects.

Despite the severity of these attacks, radical Islam has relatively few supporters in Southeast Asia. The region's people, however, have been generally wary of U.S. foreign policy initiatives designed to combat global terrorism. Resentment against the United States is especially pronounced in Indonesia and Malaysia, where most people see the conflicts in Iraq and Afghanistan as an assault on Islam.

▍REVIEW

13.10 How did European colonization influence the development of the modern countries of Southeast Asia, and how did that process differ in the insular and the mainland regions?

13.11 Why has Burma suffered from so many ethnic conflicts over the past 50 years?

13.12 How did the emergence and spread of ASEAN reduce geopolitical tensions in Southeast Asia, and why is the ASEAN process unable to resolve some conflicts in the region?

■ **KEY TERMS** Khmer Rouge, domino theory, piracy

Economic and Social Development: The Roller-Coaster Ride of Developing Economies

Over the past few decades, Southeast Asia has experienced major economic fluctuations. An economic boom between 1980 and 1997 was followed by a major recession. Southeast Asian economies once more grew quickly after 2000, but the global economic crisis of 2008–2009 resulted in another blow. Within a few years, however, the region had fully recovered. Overall, Southeast Asia accounted for some of the world's fastest-growing economies during the first 16 years of the new millennium. Some observers fear that a slowdown in China, along with the recent decline in global commodity prices, could hurt the region, but others think that the reduced price of oil will give it an economic boost.

Uneven Economic Development

Southeast Asia today is a region of strikingly uneven economic and social development (Table 13.2). Some countries, such as Indonesia, have experienced both booms and busts; others, such as Burma, Laos, and East Timor, missed the expansion of the 1980s and 1990s and

| TABLE 13.2 Development Indicators | | | | | | | | | | | MasteringGeography™ |

Country	GNI per capita, PPP 2014	GDP Average Annual %Growth 2009-15	Human Development Index (2015)[1]	Percent Population Living Below $3.10 a Day	Life Expectancy (2016)[2]		Under Age 5 Mortality Rate (1990)	Under Age 5 Mortality Rate (2015)	Youth Literacy (% pop ages 15-24) (2005-2014)	Gender Inequality Index (2015)[3,1]
					Male	Female				
Burma (Myanmar)	–	–	0.536	–	64	68	107	50	96	0.413
Brunei	72,190	0.6	0.856	–	77	81	12	10	99	–
Cambodia	3,080	7.1	0.555	8.8	61	66	117	29	87	0.477
East Timor	5,080	6.8	0.595	80.1	67	70	180	53	80	–
Indonesia	10,190	5.8	0.684	46.3	69	73	82	27	99	0.494
Laos	5,060	8.1	0.575	63.3	65	68	148	67	–	–
Malaysia	24,770	5.6	0.779	2.7	72	77	17	7	98	0.209
Philippines	8,450	6.1	0.668	37.6	65	72	57	28	98	0.420
Singapore	80,270	5.8	0.912	–	80	85	8	3	100	0.088
Thailand	14,870	3.9	0.726	<2.0	72	79	35	12	97	0.380
Vietnam	5,350	5.8	0.666	13.9	71	76	50	22	97	0.308

Source: World Bank, *World Development Indicators*, 2016.

[1]United Nations, *Human Development Report, 2015.*

[2]Population Reference Bureau, *World Population Data Sheet, 2016.*

[3]Gender Inequality Index—A composite measure reflecting inequality in achievements between women and men in three dimensions: reproductive health, empowerment and the labor market that ranges between 0 and 1. The higher the number, the greater the inequality.

Login to MasteringGeography™ **& access** MapMaster **to explore these data!**

1) How closely do the under-age-five mortality figures correlate with HDI figures in Southeast Asia? Do any Southeast Asian countries stand out in this regard?

2) What Southeast Asian countries have experienced the fastest and slowest economic growth in recent years? How do these same countries rank in terms of per capita economic output?

remain poor. Oil-rich Brunei and technologically sophisticated Singapore, on the other hand, are prosperous by any standards. Brunei's economic future, however, looks precarious due to the recent decline in the price of oil combined with its failure to develop non-petroleum-based industries (Table 13.2). In the Philippines, the situation is much more complicated.

The Philippine Decline Sixty years ago, the Philippines was the most highly developed and educated Southeast Asian country and was widely considered to have particularly bright prospects. Per capita gross national income (GNI) in 1960 was higher in the Philippines than in South Korea. By the late 1960s, however, Philippine development had been derailed. Through the 1980s and early 1990s, the country's economy failed to outpace its population growth, resulting in declining living standards. Filipinos are still well educated and reasonably healthy by world standards, but even the country's educational and health systems declined during the dismal decades of the 1980s and 1990s.

Why did the Philippines fail so spectacularly despite its earlier promise? Although there are no simple answers, it is clear that dictator Ferdinand Marcos (who ruled from 1968 to 1986) squandered billions of dollars, while failing to enact programs conducive to genuine development. The Marcos regime instituted a kind of **crony capitalism**

in which the president's many friends were granted huge sectors of the economy, while perceived enemies had their properties confiscated. After Marcos declared martial law in 1972 and suspended Philippine democracy, revolutionary activity intensified, and the country began to fall into a downward spiral. But although it is tempting to blame the failure of the Philippines on the Marcos regime, such an explanation is incomplete. Indonesia and Thailand also saw the development of crony capitalism, yet their economies proved much more competitive by the late 20th century.

More recently, the Philippine economy has showed signs of revival, expanding in 2015 at a brisk rate of 6 percent. Once the government finally turned its attention to infrastructure problems, such as electricity generation, foreign investments began to flow into the country. An especially vibrant local economy emerged in the former U.S. naval base of Subic Bay, which boasts both world-class shipping facilities and a highly competent local government (**Figure 13.35**). The Philippines as a whole has capitalized on language skills to develop a major "call center" industry, offering technical assistance and other services by telephone to customers in the United States and other English-speaking countries. It is estimated that by 2017, 1.3 million Filipinos will work in international call centers.

Despite its recent stability, the Philippine economy continues to be hampered by political and social problems.

▲ **Figure 13.35 Subic Bay, the Philippines** The former U.S. naval base in Subic Bay, the Philippines, is now a thriving industrial and export-processing center.

Explore the **Sights** of Subic Bay Freeport Zone

http://goo.gl/1VVukY

More recently, Singapore's economy has cooled down, growing by only 2 percent in 2015. Some observers worry that Singapore has become too expensive for manufacturing, which has been increasingly relocating to other countries. The Singaporean government has responded by encouraging high tech, particularly biotechnology development. Banking is also promoted, and by some measures Singapore is the world's fourth largest financial center, trailing only London, New York, and Hong Kong. Critics contend, however, that Singapore tolerates too much banking secrecy, allowing tainted money to be safely stored in its accounts.

The Philippine political system, modeled on that of the United States, entails elaborate checks and balances between the different branches of government. Critics contend that such "checks" are so effective that it is difficult to accomplish anything. Another obstacle is the Philippines' highly unequal distribution of wealth. Many members of the elite are fantastically wealthy, but almost 40 percent of the country's people subsist on lest than $3.10 a day. As a result, many Filipinos must seek employment abroad; women often work as nurses, maids, or nannies, whereas men often work in construction or on ships.

Singapore's political system has nurtured economic development, but it is somewhat repressive and by no means fully democratic. Although elections are held, the government manipulates the process, ensuring that the opposition never gains control of parliament. Freedom of speech, moreover, is limited by libel and slander laws that make it easy for the government to successfully sue its critics. Many Singaporeans object to such policies, but others counter that they have brought fast growth, as well as a clean, safe, and remarkably corruption-free society with a negligible unemployment rate.

Although the Singaporean government has thus far been able to repress dissent, its authoritarian form of capitalism confronts a challenge in the Internet. National leaders want online communication services, but worry about the free expression that this allows. Although Internet use in Singapore is currently rated as "largely unhindered," its Internet freedom score was downgraded in 2015 by the U.S.-based watchdog group Freedom House. It will be interesting to see how Singapore responds to the Internet challenge, in part because some experts claim that the governments of China and Vietnam are following the technocratic, authoritarian capitalism pioneered by Singapore.

The Regional Hub: Singapore Singapore presents a profound contrast to the Philippines—and indeed to all other Southeast Asian countries. Over the past half century, this small city-state has been Southeast Asia's great developmental success. It has transformed itself from an **entrepôt** port city—a place where goods are imported, stored, and then transshipped—to one of the world's most prosperous and modern states. A thriving high-tech manufacturing center, Singapore is also the communications and financial hub of Southeast Asia. Its government has played an active role in the development process, but has also relied heavily on market forces. Singapore has encouraged investment by multinational companies and has invested heavily in housing, education, and some social services (**Figure 13.36**). Roughly 85 percent of Singapore's residents live in public housing provided by the country's Housing and Development Board.

Malaysia's Economic Gains Although not nearly as prosperous as Singapore, Malaysia has experienced rapid economic growth, and as of 2015 it was still expanding at a healthy rate of 5 percent. Development was initially concentrated in agriculture and natural resource extraction, focused on tropical hardwoods, plantation products, and tin. More recently, manufacturing, especially in labor-intensive, high-tech sectors, has become the main engine of growth (**Figure 13.37**). As Singapore prospered, many of its enterprises moved their manufacturing operations into neighboring Malaysia. Increasingly, Malaysia's economy is multinational; many Western high-tech firms operate in the country, and several Malaysian companies have themselves

▲ **Figure 13.36 Housing in Singapore** Despite its free-market approach to economics, the government of Singapore has invested heavily in public housing. Many Singaporeans live in buildings similar to the ones depicted here. **Q: Singapore is, in general, a conservative country devoted to free-market economics. Considering this fact, why then has its government invested so heavily in public housing?**

The problem is particularly acute in Malaysia, however, simply because its Chinese minority accounts for 25 percent of its population. The government's response has been one of transferring economic clout to the numerically dominant Malay, or **Bumiputra** ("sons of the soil"), community. This policy has been somewhat successful, although it has not reached its main goal of placing 30 percent of the nation's wealth in the hands of the indigenous Malays. Because the Malaysian economy as a whole has grown rapidly, the Chinese community has thrived even as its relative share of the country's wealth declined. Considerable resentment, however, is still felt by the Chinese. Many of Malaysia's business leaders are also concerned about the policy, believing that it generates economic inefficiency.

These changing attitudes have made Malaysia's affirmative action program politically controversial. Several prominent leaders have promised to reform the system, but thus far no major actions have been taken. An associated problem is mounting political corruption, which in turn has led to heightened government restrictions on free speech. One consequence of these difficulties has been a marked **brain drain** of highly educated Malaysians, especially those of Indian and Chinese origin. In 2014, the World Bank reported that more than 300,000 high-skilled Malaysians had moved overseas during the previous year alone.

Thailand's Ups and Downs Thailand also climbed rapidly during the 1980s and 1990s into the ranks of the world's newly industrialized countries, although it remained less prosperous than Malaysia. Thailand experienced a major downturn in the late 1990s, and although recovery began in earnest after 2000, the military coup of 2006 and the global recession of 2008, coupled with the intensifying insurgency in the far south, resulted in another slowdown. Since then, the country's economic growth has lagged behind those of its neighbors, due in large part to continuing political instability.

Japanese companies were leading players in the Thai boom of the 1980s and early 1990s. As labor costs in Japan itself became too expensive for assembly and other manufacturing processes, Japanese firms began to relocate factories to such places as Thailand. They were particularly attracted by the country's low-wage, yet reasonably well-educated workforce. Thailand was also seen as politically stable, lacking the severe ethnic tensions found in many other parts of Asia. Although Thailand's Chinese population is large and economically powerful, relations between the Thai and the local Chinese have generally been good.

Thailand's economic expansion has by no means benefited the entire country to an equal extent. Most industrial

established subsidiaries in foreign lands. Malaysia's prosperity has attracted hundreds of thousands of illegal immigrants, mostly from Indonesia and the Philippines.

The economic geography of modern Malaysia shows large regional variations. Most industrial development has occurred on the west side of peninsular Malaysia, with most of the rest of the country remaining dependent on agriculture and resource extraction. More important, however, are disparities based on ethnicity. Malaysia's industrial wealth has been concentrated in the Chinese community. Ethnic Malays remain less prosperous and educated than Chinese Malaysians, and have recently been surpassed by Malaysians of Indian origin as well. Many of the country's tribal peoples have suffered as development proceeds and their lands are taken away.

The disproportionate prosperity of the local Chinese community is a feature of most Southeast Asian countries.

▲ **Figure 13.37 High-Tech Manufacturing in Malaysia** Many foreign companies have established high-tech manufacturing facilities in Malaysia. Most of their factories are located in the western part of Peninsular Malaysia.

development has occurred in the historical core, especially in the Bangkok metropolitan area and the eastern coastal zone. Yet even in Bangkok, the blessings of progress have been mixed. As the city begins to choke on its own growth, industry has begun to spread outward. The Chiang Mai area in northern Thailand has profited from its ability to attract large numbers of international tourists.

Other parts of the country have been less fortunate. Thailand's Lao-speaking northeast (the Khorat Plateau) is the country's poorest region. Soils here are too thin to support highly productive agriculture, yet the population is sizable. Local poverty has forced many Lao-speaking northeasterners to seek employment in Bangkok, where they often experience ethnic discrimination. Men typically find work in the construction industry; northeastern women more often make their living as prostitutes. Not coincidentally, northeastern Thailand has strongly supported the populist politicians who have struggled against the country's political establishment. The Muslim area of southern Thailand also remains quite poor, contributing to the region's insurgency.

Indonesian Economic Development At independence (1949), Indonesia was one of the poorest countries in the world. The Dutch had used their colony largely as an extraction zone for tropical crops and mineral resources, investing little in infrastructure or education. The population of Java mushroomed in the 19th and early 20th centuries, leading to serious land shortages.

The Indonesian economy finally began to expand in the 1970s. Oil exports fueled early growth, as did the logging of tropical forests. However, unlike most other petroleum exporters, Indonesia continued to grow even after prices plummeted in the 1980s. Production subsequently declined, and in 2004 Indonesia had to begin importing oil. But like Thailand and Malaysia, Indonesia proved attractive to multinational companies eager to export from a low-wage economy. Large Indonesian firms, some three-quarters of them owned by local Chinese families, also have capitalized on the country's low wages and abundant resources.

Despite its rapid growth, Indonesia is still burdened by poverty, and its economy still depends on the unsustainable exploitation of natural resources. Yet Indonesia was one of the few large countries to avoid the global economic crisis of 2008, and it has continued to expand at a healthy pace through 2016. The global fall in the price of oil allowed its government to reduce its expensive fuel subsidies, a move that could both boost its economic performance and provide more funds for social programs. In 2016, Indonesia announced ambitious plans for an economic "big bang," based on reducing restrictions on foreign investments and easing rules on e-commerce, health care, and entertainment.

As in Thailand, development in Indonesia exhibits pronounced geographical disparities. Northwest Java, close to the capital city of Jakarta, has boomed, and much of the resource-rich and moderately populated island of Sumatra has long been relatively prosperous. Eastern Borneo (Kalimantan), another resource area, also is relatively well-off. In the overcrowded rural districts of central and eastern Java, however, many peasants have inadequate land and remain near the margins of subsistence. Far eastern Indonesia has experienced little economic or social development, and throughout the remote areas of the "outer islands," tribal peoples have suffered as their resources have been taken by outsiders.

Vietnam's Uneven Progress The three countries of former French Indochina—Vietnam, Cambodia, and Laos—experienced only modest economic expansion during Southeast Asia's boom years of the 1980s and 1990s. This area endured almost continual warfare between 1941 and 1975, which persisted until the mid-1990s in Cambodia. Critics contend that the socialist economic system adopted by these countries prevented sustained economic growth. This debate is now moot, however, because a globalized capitalist model of development has largely been adopted in all three countries.

Vietnam's economy is much stronger than those of Cambodia and Laos. The country's per capita GNI, however, is still low by global standards. Reunification in 1975 did not initially bring the anticipated growth, and economic stagnation ensued. Conditions declined in the early 1990s after the fall of the Soviet Union, Vietnam's main supporter and trading partner. Until the mid-1990s, Vietnam remained under embargo by the United States. Frustrated with their country's economic performance, Vietnam's leaders began to embrace market economics while retaining the political forms of a communist state. In other words, Vietnam has followed the Chinese model. The country now welcomes multinational corporations, which are attracted by the low wages and relatively well-educated workforce.

Instant Coffee from Vietnam

Few people think of Vietnam as a major coffee-producing country, yet it now exports more coffee than any country other than Brazil. Vietnamese coffee is rarely advertised as such, as the country specializes in relatively low-value beans that are roasted elsewhere and then transformed into instant coffee. High-quality coffee generally comes from the seeds of the plant *Coffea arabica,* but the closely related *Coffea robusta,* favored by Vietnamese farmers, is more productive, easier to grow, and produces substantially more caffeine (**Figure 13.4.1**). The high caffeine level of *robusta* coffee gives it a bitter flavor but also makes it ideal for making instant coffee.

Large-scale coffee production began to revolutionize Vietnamese agriculture in 1986, when the government decided to prioritize the crop. An estimated 2.6 million small-scale Vietnamese farmers now cultivate coffee. Although most growers make a meager living, a few exporters have made fortunes in the hundreds of millions of dollars.

Though coffee growing has benefited the Vietnamese economy, it also generates many social and environmental problems. Most coffee is grown in the country's central highlands, the historical home of an array of non-Vietnamese minority groups. Ethnic tensions have mounted as Vietnamese farmers move into these uplands. Deforestation and land degradation are additional consequences of the coffee trade. Yet most signs indicate that Vietnam will continue to be a major coffee exporter: Globally, instant coffee production nearly tripled between 2000 and 2014, accounting for more than 34 percent of the world's retail coffee consumption.

▲ **Figure 13.4.1 Coffee Growing in Vietnam** Over the past 15 years, Vietnam has emerged as one of the world's main coffee producing countries. Here we see a coffee farmer harvesting her crop near the town of Buan in the central highlands of Vietnam.

1. Why has Vietnam emphasized coffee exports, favoring the production of relatively low-quality *robusta* beans?

2. What countries produce the coffee found in your town's coffee shops?

Such efforts seemingly began to pay off after 2000, with economic growth averaging more than 6 percent a year through 2015. Foreign investment poured into the country, and exports—especially of textiles—began to surge. Agricultural exports, particularly of coffee and rice, have also bolstered the Vietnamese economy (see *Everyday Globalization: Instant Coffee from Vietnam*). Vietnam joined the World Trade Organization in 2007, which both encourages exports and ensures the continuation of market-oriented reforms.

Yet Vietnam's recent economic development has generated a significant degree of social tension. Many lowland peasants and upland tribal peoples have been excluded from the boom and are growing increasingly discontented, and many small business owners complain about harassment by the government. Tensions between the north (the center of political authority) and the more capitalistic south (the center of economic power) may be increasing. Trade disputes have arisen with the United States, which has accused Vietnamese exporters of dumping such products as farm-raised catfish and shrimp on American markets.

Rapid Growth in Impoverished Laos and Cambodia Laos and Cambodia have long faced some of the most serious economic problems in the region. In Cambodia, the ravages of war, followed by continuing political instability, undermined

economic development for many years. Laos faces special difficulties owing to its rough terrain and relative isolation. Both countries also are hampered by a lack of infrastructure; outside the few cities, paved roads and reliable electricity are rare. As a result, both economies remain largely agricultural in orientation, and rely as well on environmentally destructive logging and mining operations. In Laos, subsistence farming still accounts for more than three-quarters of total employment.

The Laotian government is pinning its economic hopes on dam- and road-building projects. Laos is mountainous and has many large rivers, and could therefore generate large quantities of electricity, something that is in high demand in neighboring Thailand and Vietnam. Laos also benefits from the increasing volume of barge traffic going up the Mekong River to China and by Chinese-financed road and railroad building. In 2014, a Laotian government minister claimed that his country would secure its economic place by serving as a bridge between China and the ASEAN bloc. As a result of such investments, the Laotian economy has recently registered strong growth, averaging more than 7 percent a year from 2013 to 2015.

Like Laos, Cambodia is a poor country that has experienced a striking economic boom in recent years. Foreign investment has generated an expanding textile sector, tourism is thriving (Angkor Wat, the world's largest religious monument, receives well over half a

▲ **Figure 13.38 Cambodian Casino** The Cambodian border city of Poipet has recently emerged as a major gambling center. Most of the investments, and most of the tourists, come from Thailand, angering many Cambodians. Organized crime is also a major problem in the city.

million visitors annually), mining is taking off, and in 2005 important oil and natural gas fields were discovered in Cambodian territorial waters. Several Cambodian border towns, most notably Poipet, have set themselves up as highly profitable gambling centers, attracting investment—and criminal activities—from Thai underworld figures (**Figure 13.38**).

Despite its rapid development, Cambodia still faces major economic challenges. A lack of both skills and basic infrastructure, as well as problems of political instability and corruption, could easily undermine its recovery. The recent economic boom has also led to an epidemic of "land-grabbing," in which politically connected elites take over the properties of impoverished peasants in order to develop them. Many Cambodians, moreover, fear that their country could come under the economic domination of Thailand and Vietnam, and to some extent of China as well. The recent slowdown in China's economy presents a threat to the economies of both Cambodia and Laos, as well as that of Southeast Asia as a whole.

Despite their basic lack of development, Cambodia and Laos are not as impoverished as one might expect from the official economic statistics. Both countries have relatively low population densities and abundant resources. Their low per capita GNI figures, more importantly, partially reflect the fact that many of their people remain in a subsistence economy. Although Laotian highlanders make few contributions to GNI, most of them at least have adequate shelter and food.

Burma: An Emerging Economy Burma stands near the bottom of the scale of Southeast Asian economic development. For all of its many problems, however, Burma is a land of great potential. It has abundant resources—including natural gas, oil, minerals, water, and timber—as well as a large expanse of fertile farmland. Its population density is moderate, and its people are reasonably well educated. The country, however, experienced relatively little economic or social development during its first half-century of independence.

Burma's economic woes can be traced in part to the continual warfare it has experienced. Most observers also blame economic policy. Beginning in earnest in 1962, Burma attempted to isolate its economic system from global forces in order to achieve self-sufficiency under a system of Buddhist socialism. Its intentions were admirable, but the experiment was not successful. Instead of creating a self-contained economy, Burma found itself burdened by smuggling and black-market activities. The country's authoritarian rule and human rights abuses led the United States, the European Union, and other countries to impose economic sanctions, which caused additional havoc. In 2009, when the official exchange rate was 6.4 Burmese kyat to the U.S. dollar, up to 1090 kyat to the dollar could be obtained in street markets.

The dismal state of the Burmese economy was one of the main reasons why the country's government decided to open up to the rest of the world and begin to democratize in 2011. In doing so, it hoped to attract foreign investments, secure export markets, and end international sanctions. In 2012, the United States responded by easing investment and banking import restrictions. The Burmese economy subsequently began to attract large investments from other countries and to grow at a rapid pace, registering annual economic gains of over 7 percent from 2011 to 2015. But such growth has yet to translate into improved living conditions for the vast majority of the country's inhabitants.

Hope for East Timor The economic situation of East Timor is somewhat similar to that of Burma. The country is desperately poor, with 80 percent of its population living on less than $3.10 a day, and 80 percent surviving largely through subsistence farming. But discoveries of massive offshore oil and natural gas deposits significantly boosted East Timor's prospects, and from 2011 to 2014 it posted one of the world's fastest rates of economic expansion. The more recent drop in the price of oil, however, has clouded its economic prospects.

Globalization and the Southeast Asian Economy

Southeast Asia has undergone rapid but uneven integration into the global economy (**Figure 13.39**). Singapore has thoroughly staked its future to the success of multinational capitalism, and several neighboring countries are following suit. According to one measurement, Malaysia and Singapore

have, respectively, the fourth and fifth most trade-dependent economies in the world. Even communist Vietnam and once-isolationist Burma have opened their doors to international commerce, hoping to find advantages in economic globalization.

Southeast Asia's globalized economies depend heavily on exports to the international market. In the 1990s, many observers thought that the booming economies of Southeast Asia had become overly dependent on exports to the United States, but in recent years the rise of China has resulted in a more balanced trading regime. In 2015, for example, 11.2 percent of Thailand's exports went to the United States,

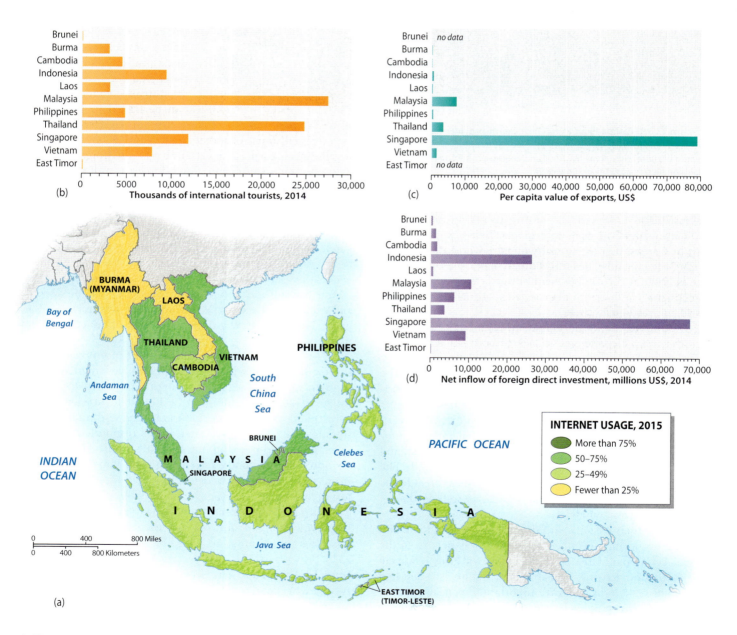

▲ **Figure 13.39 Globalization in Southeast Asia** Levels of globalization vary widely across Southeast Asia. Some countries, most notably Singapore, have high levels of Internet usage, international tourism, exports, and direct foreign investments. Other countries, most notably Burma, score low in all of these measurements. Burma has, however, seen a major increase in foreign investments in recent years.

▲ **Figure 13.40** **Anti-TPP Protest in Malaysia** The proposed Trans-Pacific Partnership Agreement, a free trade initiative, is highly controversial in Southeast Asia, just as it is in the United States. Here we see Malaysian protestors denouncing the TPP in a public protest.

not in their own factories, but rather in those of local companies that subcontract for them; activists, however, point out that multinationals have tremendous influence over their local subcontractors. Some Southeast Asian leaders object to the entire debate, accusing Westerners of wanting to prevent Southeast Asian development and protect their own home markets under the pretext of concern over worker rights.

Issues of Social Development

As might be expected, several key indicators of social development in Southeast Asia correlate well with levels of economic development. Singapore thus ranks among the world leaders in regard to health and education, as does the small yet oil-rich country of Brunei. East Timor, Laos, and Cambodia, not surprisingly, come out near the bottom for these measures. The people of Vietnam, however, are healthier and better educated than might be expected on the basis of their country's overall economic performance. Burma, on the other hand, has seen relative stagnation in some social indicators. Its per capita health budget is one of the lowest in the world, whereas its military budget, as a percentage of its overall economic output, is one of the highest.

Education and Health With the exceptions of Laos, Cambodia, Burma, and East Timor, Southeast Asia has achieved relatively high levels of social welfare. But even the poorest countries of the region have made substantial gains over the past several decades. In East Timor, for example, average life expectancy at birth rose from 33 years in 1978 to 68 years in 2015. Improvements in mortality under the age of five have also been striking, as is evident from the figures in Table 13.2. Overall, progress has been more pronounced in prosperous and stable countries such as Malaysia and Singapore, whereas war-torn Cambodia has made smaller gains.

Most Southeast Asian governments have placed a high priority on basic education. Literacy rates are relatively high in most countries of the region. Much less success, however, has been realized in university and technical education. As Southeast Asian economies continue to grow, this educational gap is beginning to have negative consequences, forcing many students to study abroad. High levels of basic education, along with general economic and social development, also have led to reduced birth rates throughout much of Southeast Asia. With population growing much more slowly than before,

while 16.6 percent went to China (including Hong Kong) and another 9.4 percent went to Japan. One prominent concern in Southeast Asia is competition from China in exporting inexpensive consumer goods to the rest of the world, and another is the possibility that a cooling Chinese economy will not need to import as many raw materials and other goods from the region.

Concerns about China's economic rise have encouraged the formation of trade agreements among Southeast Asian countries, the most important of which is the **ASEAN Economic Community (AEC)**. Some of these economic pacts have proved quite controversial, especially the signed but still un-ratified Trans-Pacific Partnership (TPP), which would link the economies of Vietnam, Malaysia, Singapore, and Brunei to those of Australia, New Zealand, Peru, Chile, Mexico, Canada, Japan, and the United States. (The Philippines, Thailand, Laos, and Indonesia have also expressed interest in joining this organization.) Critics have denounced the secrecy of the negotiations that led to the Trans-Pacific Partnership, and concerns about its environmental and labor standards have been mounting, generating large protests and placing ratification in serious doubt (**Figure 13.40**).

Globalized industrial production in Southeast Asia thus remains highly controversial. Consumers in the world's wealthy countries worry that many of their basic purchases, such as sneakers and clothing, are produced under exploitative conditions. Movements have thus emerged to pressure both multinational corporations and Southeast Asian governments to improve the working conditions of laborers in the export industries. The multinational firms in question counter that such exploitative conditions occur

Female Migrant Workers in Southeast Asia

"Geography provides a unique lens on contemporary world issues," notes **Rachel Silvey** of the University of Toronto. Silvey has studied the intersection of labor migration, gender roles, and economic development in Indonesia. Her research examines the movement of Indonesian workers to Persian Gulf countries, so in addition to Indonesia, Silvey has conducted fieldwork in Dubai and plans to travel to Saudi Arabia, allowing her to look at migration from different vantage points.

Connecting to Place and People Silvey "fell in love with Indonesia" well before discovering geography, spending her junior year in college there as part of a Volunteers in Asia program. She learned the language and lived in a rural community, studying gender and agriculture, and "felt like I had something to add to academic conversations when I came back." Her undergraduate thesis advisor asked her if she would consider geography for graduate studies. "I hadn't even heard of geography! But I found that geographers get to learn about the world firsthand through travel, and that it was really the disciplinary home for my diverse interests."

Silvey's research relies heavily on household surveys and in-depth interviews (**Figure 13.5.1**). As Silvey relates, her graduate advisor, geographer Victoria Lawson, "encouraged me to make the most of my fieldwork and the connection to place and the people that I'd gotten to know so deeply from my year abroad." Her interviews extend well beyond fact-finding: "To do a good interview is to have a good conversation with someone . . . when it works, you've not only collected data, you've developed a relationship with the person that then continues over the course of your life and theirs."

International Labor Migration Silvey's research on labor migration and gender looks beyond Southeast Asia to consider the position of Indonesian women working in the Gulf states, many of whom provide "care labor" for the elderly as well as children. She is part of a research team under the Centre for Global Social Policy (http://www.cgsp.ca/people), studying care pathways. The multidisciplinary team integrates

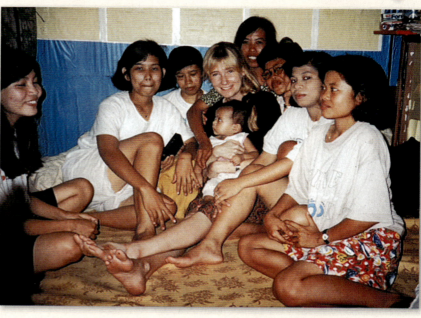

▲ **Figure 13.5.1 Geographer Rachel Silvey** Rachel Silvey of the University of Toronto is a leading expert in Indonesian labor migration and gender relations.

social work, gender and aging issues, migration paths, and public policy—an effort for which Silvey's geographic perspective and interview skills are invaluable. Says Silvey about her interviews: "There are histories . . . that need to be explored, along with the changing landscapes and politics of migration."

1. How might patterns of female labor migration in Indonesia differ from those of mainland Southeast Asian countries, such as Burma or Thailand?

2. Explain how interviewing immigrants could benefit a government or nongovernmental organization that tracks migration data.

---

economic gains are more easily translated into improved living standards.

Gender Relations The economic globalization that Southeast Asia has experienced has a pronounced gender dimension. Much of the labor in Southeast Asia's exporting companies comes from women, who are often paid much less than men doing the same work (see *Geographers at Work: Female Migrant Workers in Southeast Asia*). One study of factories in central Java found that most workers were young, single women from poor, landless families. A 2013 report claimed that even in relatively prosperous Malaysia, female factory workers earn 22 percent less than male factory workers. In Thailand, impoverished migrant women from Burma form a large, but mostly hidden labor force.

In historical terms, however, Southeast Asia has been characterized by relatively high levels of gender equity. Some scholars contend that women had a higher social position, on average, in Southeast Asia than in any other world region. In traditional kingdoms of the region, women often played important economic roles as marketers, merchants, and financiers, and many reached high political positions as diplomats, translators, and even royal bodyguards. When the Spaniards first reached the Philippines, they noted that almost all of the women, but not the men, of the Manila area were literate in the indigenous Tagalog script. Even in today's world, some anthropologists have gone so far as to describe the Muslim Minangkabau people of western Sumatra as living in a "modern matriarchy," as Minangkabau women still have a great deal of authority over their large households, which

▲ **Figure 13.41 Minangkabau House** The Minangkabau of Western Sumatra are known for their large traditional houses and for the high social position of women in their society.

are traditionally based on descent from female ancestors (**Figure 13.41**).

Historians have also argued that the position of women in Southeast Asia began to decline as religious and philosophical belief systems from other parts of the world spread through the region, as these tended to put men in the dominant position. But in some countries, women have been able to take advantage of modern education and changing ideas to reclaim positions of authority. According to the global Gender Gap Index of 2015, the Philippines has the seventh lowest gender gap in the world. Other Southeast Asia countries, however, do not rank nearly so high: Thailand and Vietnam received only average rankings (60th and 83rd, respectively, out of 145 countries), while Malaysia placed in the 111th position.

Despite some generally positive tendencies, Southeast Asia is also the site of some of the world's most intensive sexual exploitation. Commercial sex is a huge business in Thailand. According to one 2015 report, the sex trade forms a US$6.4 billion business that represents 10 percent of Thailand's GDP. But despite this massive scope, prostitution is technically illegal in Thailand, which means that it is a major source of corruption.

Other Southeast Asian countries, particularly the Philippines, Vietnam, and Cambodia, are also centers of a globally oriented commercial sex trade. Many workers in Southeast Asian

brothels are underage, often coerced into the activity, and some are reportedly held as virtual slaves. Young women, girls, and boys are frequently trafficked from the poorer parts of the region, often in connection with the drug trade. Sexually transmitted diseases, including HIV-AIDS, are associated with this activity, although the Thai government has engaged in a relatively successful public health campaign focused on condom use. Poor women from Indonesia, moreover, are frequently trafficked into Malaysia, where they are promised good jobs but often end up instead as underpaid maids or sex workers.

Two of the main centers of commercial sex in Southeast Asia developed around U.S. military bases during the Cold War: Pattaya in Thailand and Angeles City in the Philippines. Both cities lost their military bases, but subsequently expanded their economies by focusing on tourism, much of it sex-related. Pattaya now supports an estimated 27,000 sex workers. Here the high end of the business includes thousands of Russian and Ukrainian women. As a result, Pattaya now has a major Russian presence, attracting hundreds of thousands of Russian tourists every year, in addition to wealthy Russian investors. Russian organized crime now plays a major role in the city, illustrating one of the seamier aspects of globalization in modern Southeast Asia.

REVIEW

13.13 Why have some Southeast Asian countries experienced sustained economic growth and social development, whereas others have more generally experienced stagnation in the same period?

13.14 Why have the major Southeast Asian economies experienced such shared booms and busts over the past several decades?

13.15 Why does the position of women in Southeast Asia look favorable from some angles, but not from others?

■ **KEY TERMS** crony capitalism, entrepôt, Bumiputra, brain drain, ASEAN Economic Community (AEC)

Chapter 13 Southeast Asia

Summary

- Tectonic activity and tropical and monsoon climates have produced vast forests, river lowlands, and mountainous uplands. Commercial logging has caused extensive deforestation, swampland draining has led to massive forest and peat fires, while dam-building has generated clean electricity at the cost of habitat destruction.

- As people move from densely populated, fertile lowland areas into Southeast Asia's remote areas, both environmental damage and cultural conflicts have followed. Cities in the region have also seen explosive growth, placing Jakarta, Bangkok, and Manila among the world's largest urban aggregations.

- Most of the world's major religions are well represented in this culturally diverse region. Conflicts over language and religion have caused serious problems in several Southeast Asian countries, but the region as a whole has found a new sense of identity as expressed through the Association of Southeast Asian Nations (ASEAN).

- ASEAN has not solved all of Southeast Asia's political problems, as tensions persist between some countries of the region, insurgencies are still active in the Philippines, Burma, Indonesia, and Thailand, and the potential for conflict with China over disputed islands in the South China Sea grows.

- Most of Southeast Asia's trade is directed toward North America, Europe, and East Asia, and it is uncertain if efforts to develop an integrated regional economy will succeed. Whether social and economic development will be able to lift the entire region out of poverty instead of merely benefiting the more fortunate areas is an important question.

Review Questions

1. How have the natural hazards of Southeast Asia, including typhoons, tsunamis, and volcanoes, influenced the social and economic development of the region?

2. What is the relationship between human fertility patterns and both cultural and economic patterns across Southeast Asia? How are these patterns changing over time, and why?

3. How has the historical development of Southeast Asian cultures been influenced by relationships with other world regions, especially East Asia, South Asia, Southwest Asia, and Europe?

4. Why have Southeast Asian leaders been so keen to draw the region together through ASEAN, and what challenges do they face in deepening such ties?

5. Describe how the rise of China is influencing the economic and political development of Southeast Asia. How might such dynamics change the region's relationship with other important countries, such as the United States?

Image Analysis

1. Examine the changing fertility rates for selected Southeast Asian countries. Why do the fertility patterns for Singapore and the Philippines differ from those of the other countries? Describe the relatively sudden changes in the pattern for East Timor. What factors explain the pattern from 1978 to 2000? Since 2000?

2. Most Southeast Asian migrants to the United States come from the Philippines and Vietnam. Do you have a significant Southeast Asian community in your area, and if so, from what country? Explain how fertility patterns for their home country might be a factor in their migration decision. What other factors might be significant?

▶ **Figure IA13** Changing Fertility Rates in Selected Southeast Asian Countries, 1960–2015

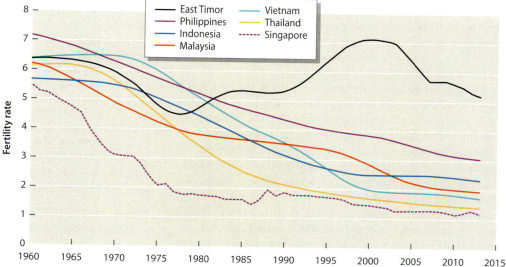

JOIN THE DEBATE

The Indonesian government has long supported the movement of people from densely populated Java to the country's so-called outer islands. Today, this controversial "transmigration" stream is focused mostly on Papua, the Indonesian portion of the island of New Guinea. Is transmigration an effective policy for Indonesia?

Transmigration is beneficial overall

- Java is one of the world's most densely populated places, and the availability of land for most of its inhabitants is simply inadequate.

- Papua and much of the rest of eastern Indonesia is poorly integrated into the Indonesian nation. Transmigration would enhance national unity by spreading Indonesia's core culture.

- Migrants generally have higher levels of education and economic productivity than the indigenous people, and as a result transmigration will in the long run enhance the Indonesian economy.

Transmigration is a detrimental policy

- Moving people out of Java over several decades has not had an appreciable effect on the island's population density, whereas the recent decline in the Javanese birthrate does point to eventual demographic stabilization.

- Transmigration poses a cultural and demographic threat to the people of Papua, and has provoked a good deal of violence in the region.

- The establishment of large settlements in areas that were previously lightly inhabited results in deforestation, the reduction of wildlife, and other forms of environmental degradation.

▲ **Figure D13** Papuan Students Rally in Surabaya, East Java Province, to Demand Independence for West Papua Province

KEY TERMS

animist *(p. 560)*
Association of Southeast Asian Nations (ASEAN) *(p. 540)*
ASEAN Economic Community (AEC) *(p. 581)*
brain drain *(p. 576)*
Bumiputra *(p. 576)*
Burman *(p. 562)*
copra *(p. 552)*
crony capitalism *(p. 574)*
domino theory *(p. 567)*
El Niño *(p. 548)*
entrepôt *(p. 575)*

Golden Triangle *(p. 552)*
human trafficking *(p. 571)*
Khmer Rouge *(p. 567)*
lingua franca *(p. 562)*
peat *(p. 546)*
piracy *(p. 573)*
primate city *(p. 556)*
Ramayana *(p. 559)*
shifted cultivators *(p. 555)*
Sunda Shelf *(p. 542)*
swidden *(p. 551)*
transmigration *(p. 554)*
tsunamis *(p. 542)*
typhoons *(p. 544)*

MasteringGeography™

Looking for additional review and test prep materials? Visit the Study Area in **MasteringGeography**™ to enhance your geographic literacy, spatial reasoning skills, and understanding of this chapter's content by accessing a variety of resources, including **MapMaster** interactive maps, videos, RSS feeds, flashcards, web links, self-study quizzes, and an eText version of *Diversity Amid Globalization*.

DATA ANALYSIS

http://goo.gl/2yUfKk

Southeast Asia is noted for its numerous indigenous languages. Many languages, however, are declining or endangered, and a number have died out in recent years, threatening cultural diversity. The *Ethnologue: Languages of the World* website (https://www.ethnologue.com) maintains a living language database. Access the site's "Browse Country" page and click on each Southeast Asian country. You will find a summary page listing the number of languages as well as the number of "extinct," "in trouble," and "dying" languages in that country.

1. Construct two graphs, one showing the total number of languages in each country and the other showing the number of "in trouble" and "dying" languages for each country.

2. Write a paragraph describing the patterns you see in the graphs. Does the total number of languages spoken in each country correlate with the number of endangered languages? Suggest reasons why the number of languages, and the number of threatened languages, vary so much from country to country.

3. Based on the graphs and what you know about the forces of globalization working in the region, how might Southeast Asia's linguistic geography change over the next 50 years?

Authors' Blogs

Scan to visit the
GeoCurrents blog
http://geocurrents.info/category/place/southeast-asia

Scan to visit the
Author's Blog
for field notes, media resources, and chapter updates
http://gad4blog.wordpress.com/category/southeast-asia/

Physical Geography and Environmental Issues

Diverse environments characterize this huge region, which includes a continent-sized landmass as well as thousands of small oceanic islands. Global climate change and rising sea levels threaten the very survival of some low-lying countries within the region.

Population and Settlement

Growing, dense cities punctuate the sparse rural settlement pattern of Oceania, with urban places as the magnets attracting migrants from both within and outside of the region.

Cultural Coherence and Diversity

Both Australia and New Zealand, originally products of European culture, are seeing new cultural geographies take shape because of immigrants from other parts of the world as well as their own native peoples, the Aborigines and Maori.

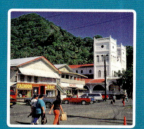

Geopolitical Framework

A heritage of colonial geographies overlaying native cultures is being replaced by contemporary power struggles among global powers, dominated by the tensions between China and the United States.

Economic and Social Development

While Australia and New Zealand are relatively wealthy because of world trade, most of island Oceania struggles economically. Even Hawaii has troubles during global downturns with its high cost of living and its boom-or-bust tourist economy.

▶ **The Island of Kosrae, Federated States of Micronesia** A local woman stands by her palm thatch house in the isolated village of Walung.

Kosrae, Federated States of Micronesia

AUSTRALIA AND OCEANIA

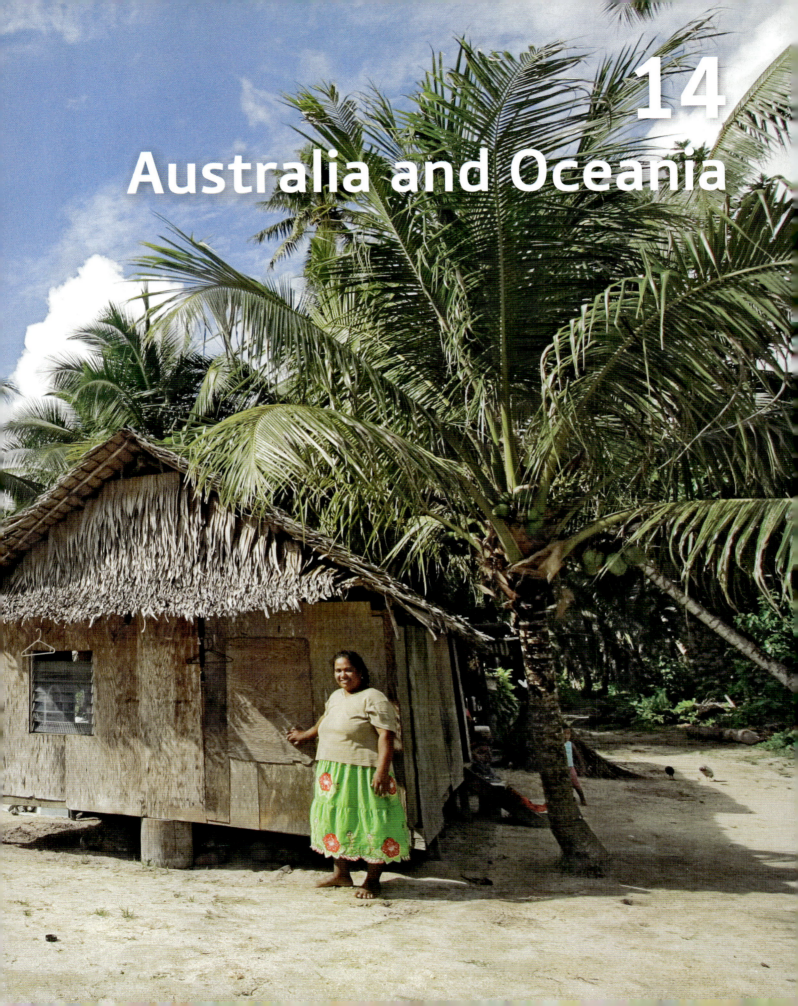

14

Australia and Oceania

The visual beauty and apparent tranquility of the Marshall Islands are deceiving. These tiny islands—scattered along 29 coral atolls in the South Pacific—are only a few feet above sea level, and global climate change has already irreversibly changed life here forever. On some islands, salt water has penetrated underground freshwater aquifers. Many islands have seen more frequent floods from passing storms. Changing trade winds have raised overall sea levels in this part of the Pacific about one foot since 1985.

This sprinkling of idyllic islands and atolls (covering 70 square miles, or 180 square kilometers, of land) historically consisted of many ethnic groups that made up small political units. In 1914, the Japanese moved into the islands, and the area remained under their control until 1944, when U.S. troops occupied the region. Following World War II, a UN trust territory (administered by the United States) was created across a wide swath of Micronesia, including the Marshall group. Demands for local self-government grew during the 1960s and 1970s, resulting in a new constitution and independence for the Marshall Islands by the early 1990s. Today,

Global climate change has already irreversibly changed life here forever.

still benefiting from U.S. aid, government officials in the modest capital city on Majuro Atoll struggle to unite island populations, protect large maritime sea claims, and resolve a generation of legal and medical problems that grew from U.S. nuclear bomb testing in the region. But the biggest challenge is the ocean itself and the Marshall Islands' population of more than 70,000 may eventually have to bid their homeland farewell. Other low-lying islands in this watery Pacific world face a similar plight.

Defining Australia and Oceania

This vast world region includes the island continent of Australia as well as **Oceania**, a collection of islands that reaches from New Guinea and New Zealand to the U.S. state of Hawaii in the mid-Pacific (**Figure 14.1**). Although native peoples settled the area

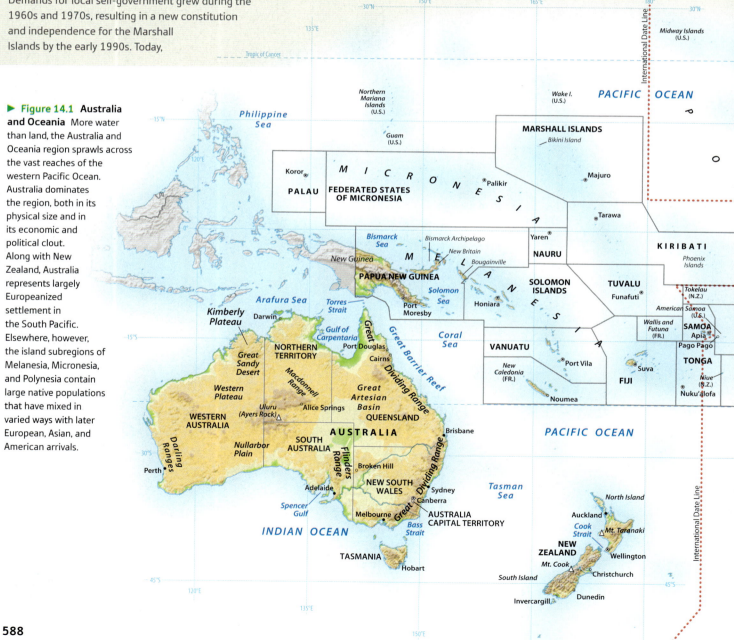

▶ **Figure 14.1 Australia and Oceania** More water than land, the Australia and Oceania region sprawls across the vast reaches of the western Pacific Ocean. Australia dominates the region, both in its physical size and in its economic and political clout. Along with New Zealand, Australia represents largely Europeanized settlement in the South Pacific. Elsewhere, however, the island subregions of Melanesia, Micronesia, and Polynesia contain large native populations that have mixed in varied ways with later European, Asian, and American arrivals.

long ago, more recent European and North American colonization began the process of globalization that is now producing new and sometimes unsettled environmental, cultural, and political geographies. The region's colonial legacy still dominates many agricultural and urban landscapes, and a number of political entities remain territories, still closely tied to distant colonial rule. In the past 30 years, however, growing linkages with Asia have produced major shifts in migration to the region as well as new economic ties. Today, Chinese tourists, South Asian immigrants, and Southeast Asian workers are all part of increasingly close connections between this region and its giant continental neighbor to the north and west.

The vast distances of the Pacific stretching from New Guinea to Hawaii help define the boundaries of this world region, but many of the national boundaries were born from political convenience during an earlier period of colonial globalization. Australia (or "southern land"), with its population of 24 million people, forms a coherent political unit and subregion. A three-hour flight to the east, New Zealand has a much smaller population of just over 4.5 million, and is linked to Australia by shared historical ties to Britain (**Figure 14.2**). However, New Zealand is considered part of **Polynesia** ("many islands") because of its native Maori people.

Hawaii, 4400 miles (7100 km) northeast of New Zealand, shares New Zealand's Polynesian heritage. Hawaii is thought of as the northeastern boundary of Oceania, while the region's southeastern boundary is usually delimited by the Polynesian islands of Tahiti, 3000 miles (4400 km) to the southeast.

Four thousand miles (6400 km) west of French Polynesia, well across the International Date Line, lies the island of New Guinea, the accepted, yet sometimes confusing boundary between Oceania and Asia. Today, an arbitrary boundary line bisects the island, dividing Papua New Guinea (the eastern half, usually considered part of Oceania) from neighboring Papua and West Papua (the western half, which, as part of Indonesia, is usually considered part of Southeast Asia). This western part of Oceania is sometimes called **Melanesia** (meaning "dark islands") because early explorers considered local peoples to be darker-skinned than those in Polynesia.

Finally, the more culturally diverse subregion of **Micronesia** (meaning "small islands") lies north of Melanesia and west of Polynesia. It includes microstates such as Nauru and the Marshall Islands as well as the U.S. territory of Guam.

ELEVATION IN METERS

	4000+
	2000–3999
	500–1999
	200–499
Sea Level	0–199
	Below sea level

AUSTRALIA & OCEANIA
Political & Physical Map

- ⭐● Metropolitan areas more than 20 million
- ✦● Metropolitan areas 10–19.9 million
- ✦• Metropolitan areas 5–9.9 million
- ✦• Metropolitan areas 1–4.9 million
- ✦○ Selected smaller metropolitan areas
- ⌐⌐ Plate boundaries

LEARNING Objectives

After reading this chapter you should be able to:

14.1 Describe the physical geographic characteristics of the region known as Oceania.

14.2 Identify major environmental issues in Australia and Oceania as well as pathways toward solving those problems.

14.3 Use a map to identify and describe major migration flows to (and within) the region.

14.4 Describe historical and modern interactions between native peoples and Anglo-European migrants in Australia and Oceania and their impacts on the region's cultures.

14.5 Identify and describe different pathways to independence taken by countries in Oceania.

14.6 List several geopolitical tensions that persist in Australia and Oceania.

14.7 Describe the diverse economic geographies of Oceania.

14.8 Explain the positive and negative interactions of Australia and Oceania with the global economy.

◀ **Figure 14.2 Mt. Taranaki, New Zealand** New Zealand's North Island contains several volcanic peaks, including Mt. Taranaki. The 8,000-foot (2,440-meter) peak offers everything from subtropical forests to challenging ski slopes, and attracts both local and international tourism.

Explore the **Sights** of Mt. Taranaki

http://goo.gl/BX3i7j

Regional Landforms and Topography

Three major topographic regions dominate Australia's physical geography. The vast, irregular Western Plateau, averaging only 1000 to 1800 feet (300 to 550 meters) in elevation, occupies more than half of the continent. To the plateau's east, the basins of the Interior Lowlands stretch for more than 1000 miles (1600 km), from the swampy coastlands of the Gulf of Carpentaria in the north to the valleys of the Murray and Darling rivers in the south, Australia's largest river system. Farthest east is the forested and mountainous country of the Great Dividing Range, extending over 2300 miles (3700 km) from the Cape York Peninsula in northern Queensland to southern Victoria (**Figure 14.4**).

As part of the Pacific Ring of Fire, New Zealand owes its geologic origins to volcanic mountain-building, which produced its two rugged and spectacular islands. The North Island's active volcanic peaks, reaching heights of more than 9100 feet (2800 meters), and its geothermal features reveal the country's fiery origins. Even higher and more rugged mountains comprise the western spine of the South Island. Mantled by high mountain glaciers and surrounded by steeply sloping valleys, the Southern Alps are some of the world's most visually spectacular mountains, complete with narrow, fjord-like valleys that indent much of the South Island's isolated western coast (**Figure 14.5**).

Physical Geography and Environmental Issues: Varied Landscapes and Habitats

The region's physical setting is amazingly diverse. Australia is made up of a vast semiarid interior—the **Outback**, a dry, sparsely settled land of scrubby vegetation—fringed by tropical environments in its far north and by hilly topography with summer-dry Mediterranean climates to the east, west, and south (**Figure 14.3**). In contrast, New Zealand is known for its green rolling foothills and rugged snow-capped mountains, landscapes that result from tectonic activity and more humid, cooler climates (see Figure 14.2). Surrounding Australia and New Zealand is the true island realm of Oceania, consisting of a varied array of both high, volcano-created islands and low-lying coral atolls.

▼ **Figure 14.3 The Australian Outback** Arid and generally treeless, the vast lands of the Australian Outback resemble some of the dry landscapes of the U.S. West. This view is in the interior of South Australia.

▼ **Figure 14.4 Grampian National Park** This park, located in Australia's Great Dividing Range west of Melbourne, was added to the national heritage list in 2006 because of its natural beauty and rich indigenous rock art sites.

In nearby Christchurch, major earthquakes, including a powerful 7.8 magnitude temblor late in 2016, were reminders of the region's vulnerability to seismic hazards.

Island Landforms Much of Melanesia and Polynesia is part of the seismically active Pacific Ring of Fire. As a result, volcanic eruptions, major earthquakes, and tsunamis are common across the region. For example, in 1994, volcanic eruptions and earthquakes on the island of New Britain (Papua New Guinea) forced more than 100,000 people from their homes. Four years later a massive tsunami triggered by an offshore earthquake swept across New Guinea's north coast, killing 3000 residents and destroying numerous villages. Such events are, unfortunately, a part of life in this geologically active part of the world.

Most of Oceania's islands were created by one of two distinct processes: either volcanic eruptions or, alternatively, coral reef-building. Those with a volcanic heritage are called **high islands** because most rise hundreds and even thousands of feet above sea level. The Hawaiian Islands are good examples, with volcanic peaks exceeding 13,000 feet (4000 meters) on the Big Island of Hawaii. Tonga, Samoa, Bora Bora, and Vanuatu provide other examples of high islands (**Figure 14.6**). Even larger and more geographically complex are the *continental high islands* of New Guinea, New Zealand, and the Solomon Islands.

Explore the **Sounds** of Kilauea Volcano

http://goo.gl/wpc9Fh

In contrast, **low islands**, as the name suggests, are formed from coral reefs, making the islands not just lower, but also flatter and usually smaller than high islands. Further, because the soil on these islands originated as coral, it

▲ **Figure 14.6 Bora Bora** The jewel of French Polynesia, Bora Bora displays many of the classic features of Pacific high islands. As the island's central volcanic core retreats, surrounding coral reefs produce a mix of wave-washed sandy shores and shallow lagoons.

is generally less fertile than the volcanic soil of high islands and supports less varied plant life. Low islands often begin as barrier reefs around or over sunken volcanic high islands, resulting in an **atoll** (**Figure 14.7**). The world's largest atoll, Kwajalein in Micronesia's Marshall Islands, is 75 miles (120 km) long and 15 miles (25 km) wide. Low islands dominate the countries of Tuvalu, Kiribati, and the Marshall Islands. Clearly, these low islands are the most vulnerable to rising sea levels associated with climate change.

Regional Climate Patterns

Zones of higher precipitation encircle Australia's arid heartland (**Figure 14.8**). In the tropical, low-latitude north, seasonal changes are dramatic and unpredictable. For example, Darwin can experience drenching monsoonal rains in the Southern Hemisphere summer, December to March, followed by bone-dry winters from June to September (see the climograph for Darwin in Figure 14.8).

Along the coast of Queensland, precipitation is high (60–100 in. or 150–250 cm), but rainfall diminishes rapidly in the state's western interior. Rainfall in central Australia, such as the Northern Territory's Alice Springs, averages less than 10 inches (25 cm) annually. In South Australia, summers are also hot and dry and these zones of Mediterranean climate produce scrubby eucalyptus woodlands known as **mallees**. South of Brisbane, more midlatitude influences dominate eastern Australia's climate. Coastal New South Wales, southeastern Victoria, and Tasmania experience the

▼ **Figure 14.5 The New Zealand Alps** The dominant topographic feature of the South Island, these picturesque mountains, referred to locally as the Southern Alps, rise to heights over 12,000 feet (3600 meters).

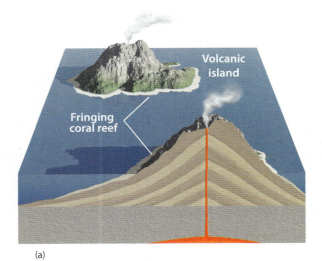

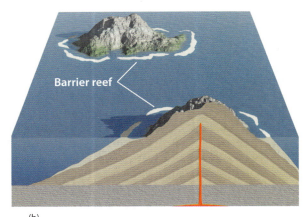

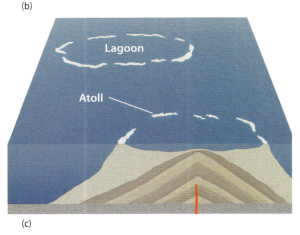

▲ **Figure 14.7 Evolution of an Atoll** (a) Many Pacific low islands begin as rugged volcanoes with fringing coral reefs. (b) However, as the extinct volcano subsides and erodes away, the coral reef expands, becoming a larger barrier reef. (The term *barrier reef* comes from the hazards these features pose to navigation when approaching the island from the sea.) (c) Finally, all that remains is a coral atoll surrounding a shallow lagoon.

country's most dependable year-round precipitation, averaging 40–60 inches (100–150 cm) per year; winter snow frequently covers the nearby mountains. Even here, however, extreme summer heat has added to recent wildfires during the dry season, threatening suburban settings in both Sydney and Melbourne. One series of fires scorched huge portions of interior Tasmania early in 2016 and destroyed plant species in a World Heritage Area that are found nowhere else on Earth (**Figure 14.9**). Extraordinarily dry conditions in that normally moist area were related to a strong El Niño Pacific circulation that some experts argue is a harbinger of more climate change and drier conditions ahead.

Climates in New Zealand are influenced by three factors: latitude, the moderating effects of the Pacific Ocean, and proximity to local mountain ranges. Most of the North Island is distinctly subtropical (see Figure 14.8); the coastal lowlands near Auckland, for example, are mild and wet year-round. On the South Island, conditions become distinctly cooler as you move closer to the South Pole. Indeed, the island's southern edge feels the seasonal breath of Antarctic chill, as it lies more than 46° south of the equator, at latitudes similar to Portland, Oregon. Mountain ranges on New Zealand's South Island also display incredible local variations in precipitation: West-facing slopes are drenched with more than 100 inches (250 cm) of precipitation annually, whereas lowlands to the east average only 25 inches (65 cm) per year. The Otago region, inland from Dunedin, sits partially in the rain shadow of the Southern Alps, and its rolling, open landscapes resemble the semiarid expanses of North America's West (**Figure 14.10**).

Island Climates The Pacific high islands usually receive abundant precipitation because of the orographic effect, resulting in dense tropical forests and vegetation. On the Hawaiian island of Kauai, Mt. Walaleale may be one of the wettest spots on Earth, receiving an average annual rainfall of 470 inches (1200 cm). In contrast to the high islands, low islands receive less precipitation, typically less than 100 inches (250 cm) annually. As a result, water shortages are common.

Unique Plants and Animals

Because of the Australian continent's long geologic history of separation and isolation from other landmasses, its bioregions contain an array of plants and animals found nowhere else in the world. More specifically, 83 percent of its mammals and 89 percent of its reptiles are unique to that country. Best known are the country's **marsupials** (mammals that raise their young in a pouch)—the kangaroo, koala, possum, wombat, and Tasmanian devil. Fully 70 percent of the world's known marsupials are found in Australia. Also unique is the platypus, an egg-laying mammal.

Exotic Species The introduction of **exotic species**—nonnative plants and animals—has caused problems for endemic (native) species throughout the Pacific region. In Australia, rabbits brought from Europe successfully multiplied in an environment that lacked the diseases and predators that elsewhere checked their numbers. Before long, rabbit populations had reached plague-like proportions, stripping large sections of land of its vegetation.

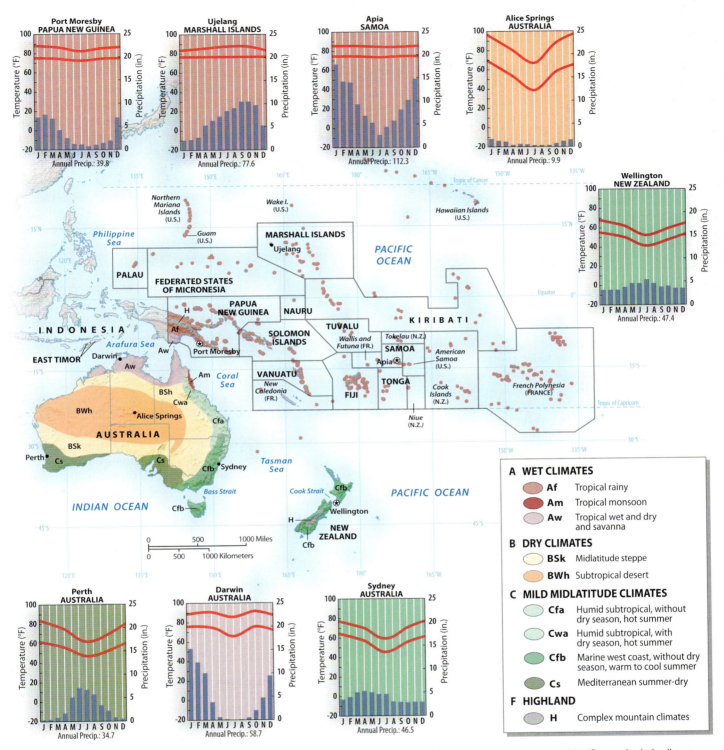

▲ **Figure 14.8 Climate Map of Australia and Oceania** Latitude and altitude shape regional climatic patterns. The equatorial Pacific zone basks in all-year warmth and humidity, while the Australian interior is dry, dominated by subtropical high pressure. Cool, moisture-laden storms of the southern Pacific Ocean provide midlatitude conditions across New Zealand and portions of Australia. Locally, mountain ranges dramatically raise precipitation totals in many highland zones. **Q: What part of the United States might have a climate that most resembles the climate of Perth? Alice Springs?**

▲ **Figure 14.9 Tasmanian Fires** Devastating fires scorched this unique World Heritage Area in Tasmania's interior in 2016.

The animals were brought under control only through the purposeful introduction of the rabbit disease myxomatosis. Feral cats have also multiplied. Probably arriving on ships bringing European convicts to the region, the cats have thrived and often weigh more than 33 pounds (15 kg). Officials are trying to establish fenced cat-free sanctuaries (to protect endangered native species) and impose a vigorous cat-culling initiative, but the adaptable felines are still wandering free in much of the Outback.

Exotic plants and animals in Oceania's island environments have had similar effects. For example, many small islands possessed no native land mammals, and their native bird and plant species proved vulnerable to the ravages of introduced rats, pigs, and other animals. The larger islands of the region, such as those of New Zealand, originally supported several species of large, flightless birds; the largest, called moas, were substantially larger than ostriches. During the first wave of human settlement in New Zealand some 1500 years ago, moa numbers fell rapidly, as they were hunted by humans and their eggs were consumed by invading rats. By 1800, moas had become extinct.

The spread of nonnative species continues today. In Guam, the brown tree snake, which arrived accidentally by cargo ship from the Solomon Islands in the 1950s, has taken over the landscape. The island's forest areas now contain more than 10,000 snakes per square mile. The snakes have wiped out nearly all the native bird species and also cause frequent power outages as they crawl along electrical wires. The brown tree snake has already done its damage to Guam, but it threatens other islands as well because it readily hides in cargo containers shipped elsewhere.

Complex Environmental Issues

Globalization has exacted an environmental toll on Australia and Oceania (**Figure 14.11**). Specifically, the region's considerable natural resource base has been opened to development, much of it by outside interests. Although benefiting from global investment, the region has also paid a considerable price for encouraging development, and the result is an increasingly threatened environment.

Historically, the region's peripheral economic and political status has often been environmentally costly. When the United States and France required atomic testing grounds for their nuclear weapons programs, the South Pacific was chosen as an ideal location (**Figure 14.12**). The environmental consequences have been long-lasting. Residents of Bikini Atoll in the Marshall Islands and across various parts of French Polynesia have been forced to evacuate their islands for decades. Elevated levels of toxic radioactive substances remain concentrated in soils, forever disrupting these island settings. Although the U.S.-backed Marshall Islands Nuclear Claims Tribunal has awarded more than $2 billion in health- and land-related claims to residents, many argue additional payments are necessary. In French Polynesia, more than 200 nuclear tests were conducted between 1966 and 1996. French government documents declassified in 2013 reported that islands such as Tahiti were exposed to 500 times more radiation than recommended, and increased cancers in the region have been traced to the testing. Now regional authorities are asking for more payments from the French government to address these persisting issues.

Major global mining operations also greatly affect Australia, Papua New Guinea, New Caledonia, and Nauru. Some of Australia's largest gold, silver, copper, and lead mines are located in sparsely settled portions of Queensland and New South Wales, polluting watersheds in these semi-arid regions. In Western Australia, huge open-pit iron mines dot the landscape, unearthing ore that is bound for global

▼ **Figure 14.10 Otago Valley on New Zealand's South Island** Many travelers have compared the scenery and open landscapes of New Zealand's Otago region to those of the American West.

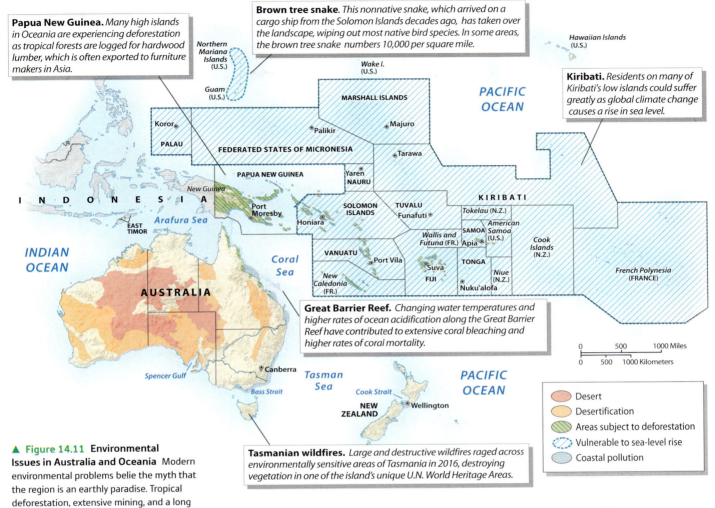

Papua New Guinea. *Many high islands in Oceania are experiencing deforestation as tropical forests are logged for hardwood lumber, which is often exported to furniture makers in Asia.*

Brown tree snake. *This nonnative snake, which arrived on a cargo ship from the Solomon Islands decades ago, has taken over the landscape, wiping out most native bird species. In some areas, the brown tree snake numbers 10,000 per square mile.*

Kiribati. *Residents on many of Kiribati's low islands could suffer greatly as global climate change causes a rise in sea level.*

Great Barrier Reef. *Changing water temperatures and higher rates of ocean acidification along the Great Barrier Reef have contributed to extensive coral bleaching and higher rates of coral mortality.*

Tasmanian wildfires. *Large and destructive wildfires raged across environmentally sensitive areas of Tasmania in 2016, destroying vegetation in one of the island's unique U.N. World Heritage Areas.*

Legend:
- Desert
- Desertification
- Areas subject to deforestation
- Vulnerable to sea-level rise
- Coastal pollution

▲ **Figure 14.11 Environmental Issues in Australia and Oceania** Modern environmental problems belie the myth that the region is an earthly paradise. Tropical deforestation, extensive mining, and a long record of nuclear testing by colonial powers have brought severe challenges to the region. Human settlements have also extensively modified the pattern of natural vegetation. Future environmental threats loom for low-lying Pacific islands as sea levels rise due to global climate change.

markets, particularly China and Japan. Off Australia's northeast coast, the Great Barrier Reef remains one of the region's most threatened ecosystems, but both local and global attention are now focused on preserving this unique marine resource (see *Working Toward Sustainability: Saving the Great Barrier Reef*). Further northeast, gold mining is transforming the Solomon Islands, while an even larger gold-mining venture has raised environmental concerns on the island of New Guinea. Elsewhere, Micronesia's tiny Nauru has been virtually turned inside out as much of the island's jungle cover has been removed to get at some of the world's richest phosphate deposits.

Other environmental threats are found across the region. Vast stretches of Australia's eucalyptus

▼ **Figure 14.12 Bikini Atoll Atomic Testing** The "Baker" atomic explosion, detonated by the United States in 1946, contributed to the long-term radioactive contamination of Bikini Atoll in the South Pacific.

Working Toward **Sustainability**

Saving the Great Barrier Reef

Scientists call it the world's single largest expression of a living organism. Stretching through the azure- and turquoise-tinted waters of the Coral Sea for more than 1400 miles (2300 km) off the coast of Queensland, the Great Barrier Reef (GBR) includes more than 900 small islands and a myriad of underwater coral reefs (**Figure 14.1.1**). This remarkable ecosystem is home to 1500 species of fish, 400 species of coral, whales, dolphins, sea turtles, sea eagles, terns, and plant species found nowhere else on Earth (**Figure 14.1.2**). Taking more than 10,000 years to form, the reef has been a UN World Heritage Site since 1981, and much of the area is protected by Australia's Great Barrier Reef Marine Park.

Fighting for Survival Today, however, the GBR is in the fight of its life. Thanks to global climate change, the reef has lost more than half its coral cover since 1985, much of it damaged by warmer ocean temperatures that have accelerated rates of seawater acidification and coral bleaching. Bleaching occurs when the organism experiences increased stress from changing temperature, light, or nutrient conditions. The stressed corals expel helpful algae, causing them to turn white. Bleached corals may survive, but are more susceptible to disease and death.

Coastal development also has added to its watery woes: More intensive agriculture in Queensland has produced ocean-bound sediment and increased runoff of toxic agricultural chemicals. Recent plans for expanding coal-loading depots at Abbot Point include potentially dumping waste rock onto the reef, a practice sure to disrupt the purity of local waters. On a hopeful note, the reef contributes to a huge tourist industry in Queensland, amounting to more than

▲ **Figure 14.1.2 Great Barrier Reef's Diverse Ecosystem** This undersea view of the Great Barrier Reef features yellow sea-fan corals, acroporas (hard and white corals), and purple anthias (fish).

$4.6 billion annually. That translates into powerful economic interests that are actually committed to preserving the reef's environmental health.

Adapting to Change Recently, the Australian government pledged $140 million toward a reef trust aimed at improving water quality. The government was also encouraged by a decision in 2015 by the United Nations *not* to include the GBR on its list of endangered Heritage Sites around the world. The UN panel cited positive efforts, both by the federal and Queensland governments, to invest in the GBR's environmental health and survival.

Still, the long-term story for the GBR will probably hinge upon responsible shoreline development along the Queensland coast as well as the further impacts of ocean warming and acidification that will no doubt affect the region in coming decades. In addition, better aerial surveillance (such as more satellite reconnaissance and use of high-tech drones) of the Marine Park may monitor rogue fishing vessels more effectively in the future. Longer term, scientists are also working on developing new genetic strains of coral that may be more resistant to global climate change and these could be used to repopulate damaged settings. For now, the reef's survival hangs in the balance, a giant poster child for a long list of damaging human impacts that threaten the environmental health of the entire South Pacific.

Google Earth Virtual MG Tour Video

http://goo.gl/clf2Rw

▼ **Figure 14.1.1 Australia's Great Barrier Reef** Off the eastern coast of Australia, the Great Barrier Reef stretches for more than 1400 miles (2300 km) through the Coral Sea.

1. How might economic development in Queensland proceed while at the same time preserving the environmental health of the GBR?

2. Cite a fragile and protected environmental area in your region and briefly outline future prospects for its survival.

woodlands have been destroyed to create pastures; in the Outback, overgrazing has increased desertification and salinization, resulting in saltier groundwater and less productive soils. In addition, coastal rainforests in Queensland cover only a fraction of their original area, although a growing environmental movement in the region is fighting to save the remaining forest tracts. Tasmania has also been an environmental battleground, particularly given the biodiversity of its midlatitude forest landscapes. The island's earlier European and Australian development featured many logging and pulp mill operations, but more than 20 percent of the land is now protected by national parks.

Many high islands in Oceania are also threatened by deforestation. With limited land areas, islands are subject to rapid tree loss, which, in turn, often leads to soil erosion. Although rainforests still cover 70 percent of Papua New Guinea, more than 37 million acres (15 million hectares) have been identified as suitable for logging (**Figure 14.13**). Some of the world's most biologically diverse environments are being threatened in these operations, but landowners see the quick cash sales to loggers as attractive, even though this nonsustainable practice is contrary to their traditional lifestyles.

Climate Change in Oceania

Even though Oceania contributes a relatively small amount of greenhouse gases to the atmosphere, the harbingers of climate change are widespread and problematic in this region.

▲ **Figure 14.14 Cyclone Winston Strikes Fiji** This structure was crushed by a tree when Cyclone Winston blew through Fiji in 2016. The strongest winds ever recorded in the Southern Hemisphere were associated with the monster storm which left thousands homeless in the island nation.

New Zealand mountain glaciers are melting. Australia has suffered from recent droughts, exceptionally destructive heat waves (which set records between 2013 and 2016), and devastating wildfires. Warmer ocean waters have caused widespread bleaching of Australia's Great Barrier Reef as microorganisms die, and rising sea levels threaten low-lying island nations. UN projections for the future are also highly disturbing: Sea levels may rise 4 feet (1.4 meters) by century's end, and island inhabitants will suffer from changed coastal resources as the ocean continues to warm. Additionally, this average sea-level rise does not take into account the tidal range for specific locations. In the South Pacific, seasonal extreme high tides can add another 10 feet (3.3 meters) to the average sea level. Furthermore, stronger tropical cyclones could devastate Pacific islands, with widespread damage to land and life. Tropical cyclone Pam, a category 5 storm, struck Vanuatu in 2015, wiping out crops and leaving more than 100,000 islanders homeless. A few months later Cyclone Winston hit Fiji in 2016, producing the strongest winds ever recorded in the Southern Hemisphere and leaving tens of thousands homeless (**Figure 14.14**).

Oceania's countries respond to these threats with a variety of actions and policies, depending on their susceptibility to climate change, the source and magnitude of their atmospheric emissions, and the state of their local economies. Australia, with its 24 million people, emits the most carbon in this large region, primarily because almost 80 percent of its electricity is generated from fossil fuels such as coal. Recent government proposals moved toward a carbon tax program

▼ **Figure 14.13 Logging in Oceania** Much of Oceania's tropical forest is being destroyed by logging. In this photo, hardwood tree trunks await loading for transfer to Asian mills where they will be made into furniture for the growing markets of China.

that would have accelerated emission reductions, but strong opposition to the plan, especially from mining and industrial interests, led the government to scrap it in 2014. As a result, greenhouse gas emissions from Australia's largest corporations rose again in 2015.

New Zealand has been more politically committed to addressing global climate change. More than half of the country's own energy comes from hydroelectric power, primarily generated in the high, wet mountains of the South Island. Wind and solar—particularly wind—supply fully 13 percent of the country's power. The government is now proposing a 10–20 percent reduction in greenhouse gas (GHG) emissions by 2020. To address New Zealand's livestock emissions of methane (a potent GHG), livestock specialists are experimenting with grass and grain fodder mixtures that could reduce these levels in the future. They may also impose a "belch tax" on New Zealand's livestock owners (there are 40 million sheep and cattle in the country) to encourage them to adopt methane-reducing management practices.

Oceania's low islands are already facing the most dramatic consequences of global climate change, especially as measured by rising sea levels. Low islands have very little topographic relief, because they are basically coral atolls that have grown from the sea as barrier reefs. Many Pacific nations—most notably Tuvalu, Kiribati, and the Marshall Islands—are already experiencing problems from sea-level rise, coral bleaching, and degraded fishery resources because of warming ocean waters related to climate change. These nations have collectively formed a strident political union lobbying for a global solution to climate change, demanding that developed nations such as the United States, Japan, and the countries of western Europe provide island nations with financial aid to mitigate damage from climate change.

Longer term, the questionable sustainability of Oceania's low islands has led Pacific islanders to plan for what many consider inevitable—abandoning their homeland. Aggravating the problems of sea-level rise due to climate change is the existing high population density that comes from decades of relatively rapid population growth and little vacant land. With growing island populations, there is no option but to leave. Tragically, that's what some countries are preparing for. Kiribati, for example, has adopted a "migration with dignity" program that emphasizes education and vocational training so its people can find jobs elsewhere. A new maritime training college has been created to help locals gain jobs with global shipping firms. Australian aid has created a nurse training program for islanders seeking jobs off-island in that understaffed field. Regionally, Kiribati is also one of five South Pacific nations (including Samoa, the Solomon Islands, Tuvalu, and Vanuatu) that qualify for UN assistance in developing National Adaptation Programs of Action (NAPAs). These adaptation schemes are designed to develop long-term options for island residents as sea levels rise.

REVIEW

14.1 How are high islands, coral atolls, and barrier reefs formed?

14.2 Describe the different climate regions found in Australia and New Zealand. What climate controls produce those regions?

14.3 List three key environmental issues for this region, and suggest why these have global importance.

■ KEY TERMS Oceania, Polynesia, Melanesia, Micronesia, Outback, high islands, low islands, atoll, mallees, marsupials, exotic species

Population and Settlement: Migration, Cities, and Empty Spaces

Modern population patterns reflect both indigenous and European settlement. In New Zealand, Australia, and the Hawaiian Islands, Anglo-European migration has structured the distribution and concentration of contemporary populations. Elsewhere in Oceania, population geographies are determined by the needs of native island peoples (Figure 14.15). Patterns of migration to both Australia and New Zealand also show more residents coming from Asia. This pattern is coupled with an increase in intraregional migration as people move about for complex reasons, including the push forces of unemployment, resource depletion, and even the threat of flooding associated with climate change.

Contemporary Population Patterns and Issues

Despite the stereotypes of life in the Outback, modern Australia has one of the most urbanized populations in the world (Table 14.1). Australia's eastern and southern coasts are home to the majority of its 24 million people. Inland, population densities decline as rapidly as the rainfall: Semiarid hills west of the Great Dividing Range still contain significant rural settlement, but Queensland's southwestern periphery remains sparsely populated.

New South Wales is the country's most heavily populated state; its sprawling capital city of Sydney (4.4 million people), focused around one of the world's most magnificent natural harbors, is the largest metropolitan area in the entire South Pacific (Figure 14.16). In the nearby state of Victoria, Melbourne (with 4.3 million residents) has long competed with Sydney for status as Australia's premiere city, claiming cultural and architectural supremacy over its slightly larger neighbor. Since 2010, employment growth in Melbourne has made it Australia's fastest-growing metropolitan area, and it may soon surpass Sydney in population. Both Sydney and Melbourne are dynamic, bustling metropolitan areas and like their North American counterparts, prone to sprawl and rapid suburban growth. Both cities have recently witnessed housing shortages and between 2012 and 2016 home prices in these settings have skyrocketed

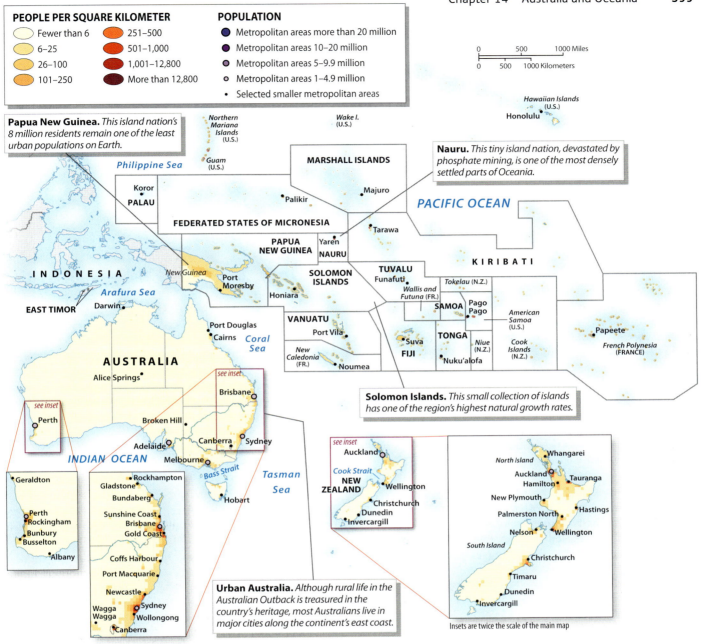

PEOPLE PER SQUARE KILOMETER

Fewer than 6
6–25
26–100
101–250
251–500
501–1,000
1,001–12,800
More than 12,800

POPULATION

Metropolitan areas more than 20 million
Metropolitan areas 10–20 million
Metropolitan areas 5–9.9 million
Metropolitan areas 1–4.9 million
Selected smaller metropolitan areas

Papua New Guinea. *This island nation's 8 million residents remain one of the least urban populations on Earth.*

Nauru. *This tiny island nation, devastated by phosphate mining, is one of the most densely settled parts of Oceania.*

Solomon Islands. *This small collection of islands has one of the region's highest natural growth rates.*

Urban Australia. *Although rural life in the Australian Outback is treasured in the country's heritage, most Australians live in major cities along the continent's east coast.*

Insets are twice the scale of the main map

▲ **Figure 14.15 Population of Australia and Oceania** Less than 40 million people occupy this world region. Although Papua New Guinea and many Pacific islands feature mainly rural settlements, most residents of the region live in the large urban areas of Australia and New Zealand. Sydney and Melbourne account for almost half of Australia's population, and most New Zealand residents live on the North Island, home to the cities of Auckland and Wellington.

by 25 percent (partly fueled by capital inflows from wealthy Chinese buyers).

Canberra (population 380,000), the much smaller federal capital, is located between these metropolitan giants. The capital represents a classic geopolitical compromise, in the same spirit that created Washington, DC, midway between the populous southern and northern portions of the eastern United States.

Outside of the Australian core, Aboriginal populations are widely but thinly scattered inland across Western Australia and Southern Australia as well as in the Northern Territory, creating smaller but regionally important centers of settlement. In the far southwest, sprawling Perth dominates the urban scene with its 1.9 million residents. Some planners fear the rapidly growing metropolitan area (projected 2050 population: 4.6 million) may become the "biggest low-density city on earth" because of its suburban fringe spreading into surrounding rural lands.

More than 70 percent of New Zealand's 4.5 million residents live on the North Island, with the Auckland metropolitan area (over 1.4 million) dominating the scene in the north and the capital city, Wellington (400,000), anchoring

TABLE 14.1   Population Indicators

MasteringGeography™

Country	Population (millions) 2016	Population Density (per square kilometer)	Rate of Natural Increase (RNI)	Total Fertility Rate	Percent Urban	Percent <15	Percent >65	Net Migration (Rate per 1000)
Australia	24.1	3	0.6	1.8	89	19	15	9
Fed. States of Micronesia	0.1	149	1.8	3.4	22	34	4	−14
Fiji	0.9	49	0.9	3.1	51	29	6	−6
French Polynesia	0.3	76	1.1	2.0	56	24	7	0
Guam	0.2	310	1.6	3.0	94	26	8	−6
Kiribati	0.1	136	2.2	3.9	54	35	4	−3
Marshall Islands	0.1	294	2.3	4.1	74	41	2	−17
Nauru	0.01	509	2.6	3.9	100	37	13	−8
New Caledonia	0.3	15	1.2	2.3	70	24	8	6
New Zealand	4.7	17	0.6	2.0	86	20	15	14
Palau	0.02	46	0.4	2.1	87	20	6	0
Papua New Guinea	8.2	16	2.0	3.7	13	39	3	0
Samoa	0.2	68	2.3	5.1	18	38	6	−28
Solomon Islands	0.7	20	2.5	4.0	20	39	3	0
Tonga	0.1	147	1.9	4.1	24	37	6	−19
Tuvalu	0.01	330	1.6	3.6	60	33	5	−3
Vanuatu	0.3	21	2.7	4.2	26	40	4	0

Source: Population Reference Bureau, *World Population Data Sheet*, 2016.

[1]World Bank, *World Development Indicators*, 2016.

Login to MasteringGeography™ **& access** MapMaster **to explore these data!**

1) List the region's six most densely settled nations. What do they have in common? What do their RNI data suggest about the future?

2) Examine data for the Percent of the population older than 65. Why are the data for Australia and New Zealand so much higher than for most of the rest of the region?

settlement along the Cook Strait in the south. Settlement on the South Island is mostly located in the somewhat drier lowlands and coastal districts east of the mountains, with Christchurch (375,000) serving as the largest urban center (**Figure 14.17**). Elsewhere, rugged and mountainous terrain on both the North and the South islands produces much lower densities.

In Papua New Guinea, less than 15 percent of the country's population is urban, with most people living in the isolated interior highlands. The nation's largest city is the capital, Port Moresby (400,000), located along the narrow coastal lowland in the far southeastern corner of the country. In stark contrast to Papua New Guinea, the largest urban area on the northern margin of Oceania is Honolulu (1 million), on the island of Oahu. Here rapid metropolitan growth since World War II is due to U.S. statehood and Hawaii's scenic attractions.

Various population-related issues face residents of the region today. Australian and New Zealand populations grew rapidly (mostly from natural increases) in the 20th century, but today's low birth rates parallel the pattern in North America, where population growth stems from immigration (**Figure 14.18**). Different demographic challenges grip many less-developed island nations of Oceania. High population growth rates of over 2 percent per year are not uncommon, as in the Solomon Islands, Vanuatu, and the Marshall Islands (see Table 14.1 and Figure 14.18). The larger islands of Melanesia contain some

Explore the **Sights** of Sydney Harbor

http://goo.gl/a4hxSK

▼ **Figure 14.16 Sydney Harbor, Australia** This aerial view, taken above North Sydney, shows Australia's most spectacular and famous harbor. Note the Opera House near the water on the opposite shore.

▲ **Figure 14.17 Christchurch, New Zealand** This image of Christchurch shows the downtown area and nearby Hagley Park. Note the sprawling Canterbury Plain bordered by the snow-clad Southern Alps in the distance.

remain attractive to migrants, as do portions of Oceania that have experienced resource-related booms (such as New Caledonia) or the growth of the tourist economy (such as Hawaii).

Historical Geography

The region's remoteness from the world's early population centers meant that it lay beyond the dominant migratory paths of earlier peoples. Even so, prehistoric settlers eventually found their way to the isolated Australian interior and even the far reaches of the Pacific. Later, once Europeans had explored the region and identified its resource potential, the pace of new in-migrations increased.

Peopling the Pacific The large islands of New Guinea and Australia, relatively near to the Asian landmass, were settled much earlier than the more distant islands of the Pacific. By 60,000 years ago, the ancestors of today's native Australians, or **Aborigines**, were making their way from Southeast Asia into Australia (**Figure 14.19**). The first Australians most likely arrived on some kind of watercraft, but because such boats were probably not capable of more lengthy voyages, the more distant islands remained inaccessible to settlement for tens of thousands of years. During the last glacial period, however, sea levels were much lower than they are now, allowing easier movement to Australia across relatively narrow spans of water from what is now called Southeast Asia. It is not known whether the original Australians arrived in one wave of people or in many, but the available evidence suggests that they soon occupied large portions of the continent, including Tasmania, which was once connected to the mainland by a land bridge because of the lower sea level.

Eastern Melanesia was settled much later than Australia and New Guinea. By 3500 years ago, some Pacific peoples had mastered long-distance sailing and navigation, which eventually opened the entire oceanic realm to human habitation.

room for settlement expansion, but competitive pressures from commercial mining and logging operations limit the availability of new agricultural land.

On some of the smaller island groups in Micronesia and Polynesia, population growth is a more pressing problem. Tuvalu (north of Fiji), for example, has just over 11,000 inhabitants crowded onto a land area of about 10 square miles (26 square kilometers), making it one of the world's most densely populated countries. Out-migration from several island nations is also very high. In Tonga and Samoa, for example, crowded conditions combine with high unemployment to encourage emigration. Elsewhere, as illustrated by Kiribati's "Migration with Dignity" initiative, climate change is already promoting the abandonment of low-lying Pacific island nations. In contrast, Australia and New Zealand

▼ **Figure 14.18 Population Pyramids of Australia and the Solomon Islands (2025)** (a) Like many developed countries, Australia has very low natural growth, as shown by the forecast for 2025. (b) In contrast, like many developing countries, the Solomon Islands forecast shows the classic pyramidal shape of a young and growing population.

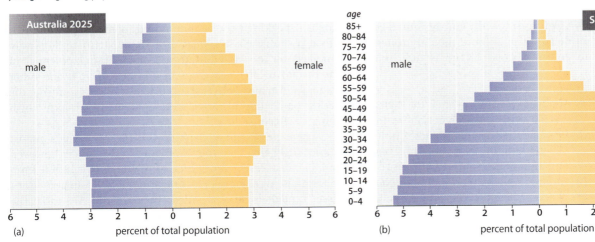

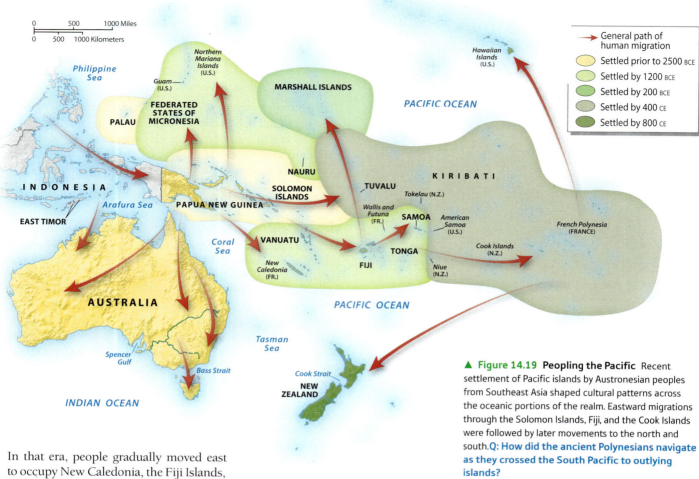

▲ **Figure 14.19 Peopling the Pacific** Recent settlement of Pacific islands by Austronesian peoples from Southeast Asia shaped cultural patterns across the oceanic portions of the realm. Eastward migrations through the Solomon Islands, Fiji, and the Cook Islands were followed by later movements to the north and south. **Q: How did the ancient Polynesians navigate as they crossed the South Pacific to outlying islands?**

In that era, people gradually moved east to occupy New Caledonia, the Fiji Islands, and Samoa. From there, later movements took seafaring folk north into Micronesia, with the Marshall Islands occupied around 2000 years ago.

Continuing movements from Asia further complicated the story of these migrating Melanesians. Some migrants mixed culturally and eventually reached western Polynesia, where they formed the core population of the Polynesian people. By 800 CE, they had reached such distant places as Tahiti, Hawaii, and Easter Island. Prehistorians hypothesize that population pressures may have quickly reached crisis stage on the relatively small islands, leading people to attempt dangerous voyages to colonize other Pacific islands. Equipped with sturdy outrigger sailing vessels and ample supplies of food, the Polynesians were quickly able to colonize most of the islands they discovered.

European Colonization About six centuries after the Maori people brought New Zealand into the Polynesian realm, Dutch navigator Abel Tasman spotted the islands on his global exploration of 1642, marking the beginning of a new chapter in the human occupation of the South Pacific. British sea captain James Cook surveyed the shorelines of both New Zealand and Australia between 1768 and 1780, with the belief that these distant lands might be worthy of European development. By 1800, other European expeditions were exploring the Pacific, placing most of Oceania's island groups on colonial maps.

European colonization of the region began in Australia when the British needed a remote penal colony to which it could exile convicts. Australia's southeastern coast was selected as an appropriate site, and in 1788 the First Fleet arrived with 750 prisoners in Botany Bay, near what is now Sydney. Other fleets and more convicts soon followed, as did boatloads of free settlers. Before long, free settlers outnumbered the convicts, who were gradually being released after serving their sentences. The growing population of English-speaking people soon moved inland and also settled other favorable coastal areas. British and Irish settlers were attracted by the agricultural and stock-raising potential of the distant colony as well as by the lure of gold and other minerals. A major gold rush occurred in Australia during the 1850s, paralleling historical developments in western North America (**Figure 14.20**).

These new settlers clashed with the Aborigines almost immediately after arriving. No treaties were signed, however, and in most cases Aborigines were simply displaced from their lands. In some places—most notably, Tasmania—they were hunted down and killed. In mainland Australia, the Aborigines were greatly reduced in numbers by disease, removal from their lands, and pure economic hardship.

▲ **Figure 14.20 Australian Gold Rush** This late-19th-century sketch of Victoria shows how the landscape was dramatically modified by miners as they searched for gold near Melbourne.

numerically dominant, but more recent colonization patterns have produced a scene mainly shaped by Europeans. The result includes everything from German-owned vineyards in South Australia to houses on New Zealand's South Island that appear to be plucked directly from the British Isles. In addition, processes of economic and cultural globalization have structured urbanization so that cities such as Perth or Auckland look strikingly similar to places such as San Diego or Seattle.

The Urban Transformation Both Australia and New Zealand are highly urbanized, Westernized societies, and thus the vast majority of their people live in cities and suburbs. As cities evolved, they took on many of the characteristics of their European counterparts, blended with a strong dose of North American influences. The result is an urban landscape in which many North Americans are quite comfortable, even though the varied accents heard on the street and many features of the metropolitan scene are reminders of the strong and lasting attachments to British traditions.

The affluent, Western-style urban environments of Australia and New Zealand contrast dramatically with the urban landscapes found in the region's less developed countries. The streets of Port Moresby, Papua New Guinea's capital and largest commercial center, reveal evidence of the large gap between rich and poor within Oceania (**Figure 14.21**). Rapid growth here has produced many of the classic problems of urban underdevelopment: a shortage of adequate housing, insufficient roads and schools, and rising street crime and alcoholism. Elsewhere, urban centers such

By the mid-19th century, Australia was primarily an English-speaking land, as the native peoples had been driven into submission.

The lush and fertile lands of New Zealand also attracted British settlers. European whalers and sealers arrived shortly before 1800, with more permanent agricultural settlement taking shape after 1840, when the British formally declared sovereignty over the region. As new arrivals grew in number and the scope of planned settlement colonies on the North and South islands expanded, tensions with the native Maori population increased. Organized in small kingdoms, or chiefdoms, the Maori were formidable fighters. Widespread Maori wars began in 1845 and engulfed New Zealand until 1870. The British eventually prevailed, however, and as in Australia, the native Maori lost most of their land.

Native Hawaiians also lost control of their lands to immigrants. Hawaii emerged as a united and powerful kingdom in the early 1800s, and for many years its native rulers limited U.S. and European claims to their islands. Increasing numbers of missionaries and settlers from the United States were allowed in, however, and by the late 19th century control of Hawaii's economy had largely passed to foreign plantation owners. In 1893, U.S. interests were strong enough to overthrow the Hawaiian monarchy, resulting in formal political annexation to the United States in 1898. Hawaii became a state in 1959.

Settlement Landscapes

Australia and Oceania's settlement geography is an interesting mix of local and global influences. The contemporary cultural landscape still reflects the imprint of indigenous peoples in those areas where native populations remain

▼ **Figure 14.21 Port Moresby, Papua New Guinea** These young Port Moresby residents live in the stilted village of Koki in one of the city's poorer urban neighborhoods.

as Suva (Fiji), Nouméa (New Caledonia), and Apia (Samoa) also reflect the economic and cultural tensions generated as local populations are exposed to Western influences. Rapid urban growth is a common problem in the smaller cities of Oceania because native people from rural areas and nearby islands gravitate toward the available job opportunities. In the past 50 years, the huge global growth of tourism in places such as Fiji and Samoa has also transformed the urban scene, replacing traditional village life with a landscape of souvenir shops, honking taxicabs, and crowded seaside resorts.

Rural Australia and New Zealand Rural landscapes across Australia and the Pacific region express a complex mosaic of cultural and economic influences. In some settings, Australian Aborigines can still be found in their familiar homelands, their traditional way of life and settlements barely changed from pre-European times. Yet such settlement landscapes are becoming rare. Global influences penetrate the scene as the cash economy, foreign tourism and investment, and the currents of popular culture flow from city to countryside.

Most of the rural Australian interior is too dry for farming and features range-fed livestock. Sheep and cattle dominate Australia's livestock economy, and rural landscapes in the interior of New South Wales, Western Australia, and Victoria are oriented around isolated sheep stations—ranch operations that move the flocks from one large pasture to the next. Cattle can sometimes be found in these same areas, although many of the more extensive range-fed cattle operations are concentrated farther north, in Queensland (**Figure 14.22**). Other isolated interior settings remain home to Aboriginal peoples who still pursue their traditional forms of hunting and gathering.

Croplands also vary across the region. A band of commercial wheat farming sometimes mingles with the sheep country across southern Queensland; the moister interiors

▼ **Figure 14.22 Cattle Ranch, Queensland Interior** This motorized cowboy is herding cattle at Longreach, deep in the interior Queensland Outback.

of New South Wales, Victoria, and South Australia; and a swath of favorable land east and north of Perth. Specialized sugarcane operations thrive along the narrow, warm, and humid coastal strip of Queensland. To the south and west, productive irrigated agriculture has developed in places such as the Murray River Basin, where orchard crops and vegetables are grown. **Viticulture**, or grape cultivation, dominates South Australia's Barossa Valley, the Riverina district in New South Wales, and Western Australia's Swan Valley.

New Zealand's rural settlement landscape includes a variety of agricultural activities. Ranching clearly dominates, with the vast majority of agricultural land devoted to livestock production, particularly sheep grazing and dairying. Commercial livestock outnumber people in New Zealand by a ratio of more than 20 to 1, and this is apparent throughout the countryside. Dairy operations are present mostly in the lowlands of the north, where they sometimes mingle with suburban landscapes in the vicinity of Auckland.

Rural Oceania The rural landscape varies considerably elsewhere in Oceania. On high islands with more water, denser populations take advantage of diverse agricultural opportunities; on the more barren low islands, fishing is often more important. Several types of rural settlement can be identified across the island region. In rural New Guinea, village-centered shifting cultivation dominates: Farmers clear a patch of forest and then, after a few years, shift to another patch, thus practicing the swidden agriculture common in Southeast Asia. Subsistence foods such as sweet potatoes, taro (another starchy root crop), coconut palms, bananas, and other garden crops are often found in the same field, an agricultural practice known as **intercropping**.

Commercial plantation agriculture has also made its mark in many of the more accessible rural settings. Unlike subsistence agriculture, these commercial operations generally feature a form of monoculture, where only one crop is grown in a field. In these places, settlements consist of worker housing near crops that is typically controlled by absentee landowners. For example, copra (coconut), cocoa, and coffee operations have transformed many agricultural settings in places such as the Solomon Islands and Vanuatu. Sugarcane and taro plantations have reshaped other island settings, particularly in Fiji and Hawaii (**Figure 14.23**).

REVIEW

14.4 Compare the populations of Australia and New Zealand in terms of size, density, and level of urbanization.

14.5 Describe the prehistoric peopling of the Pacific.

14.6 Compare the rural settlement patterns for Australia and the island countries, and explain why these differ.

■ **KEY TERMS** Aborigines, viticulture, intercropping

▲ **Figure 14.23 Sugar Cane Farming in Fiji** This commercial plantation operation in Fiji remains labor intensive. In this scene from Fiji's Nadroga Sigatoka region, workers harvest the crop in November.

Cultural Coherence and Diversity: A Global Crossroads

The Pacific world offers excellent examples of how culture is transformed as different groups migrate to a region, interact with one another, and evolve over time. As Europeans and other outsiders arrived in Oceania, colonization forced native peoples to resist or adjust. More recently, worldwide processes of globalization have redefined the region's cultural geography.

Multicultural Australia

Australia's cultural patterns illustrate globalization processes at work. Today, although the country is still dominated by European roots, its multicultural character is increasingly visible as native peoples assert their cultural identities and as immigrant populations play larger roles in society, particularly in metropolitan areas.

Aboriginal Imprints For thousands of years, Australia's indigenous Aborigines dominated the continent. They practiced hunting and gathering, a way of life that persisted up to the European conquest. Settlement densities remained low, tribal groups were often isolated from one another, and the overall population probably never exceeded 300,000 inhabitants. Cultural geographies were diverse and fragmented, producing many local languages. Even today about 50 indigenous languages can still be found.

Explore the **Sounds** of Australian Aboriginal Music

http://goo.gl/TE3jyq

Europeans brought radical demographic and cultural changes, decimating Aboriginal populations in the process. The geographic results of colonization were striking: Indigenous residents were relocated to the sparsely settled interior, particularly in northern and central Australia, where fewer Europeans competed for land. Historically, the European attitude toward the Aboriginal population was often more discriminatory than it was toward the native peoples of North America.

Today Aboriginal culture perseveres in Australia, and a native people's movement is growing, similar to what is happening in the Americas (**Figure 14.24**). Indigenous people account for approximately 3 percent (or 700,000) of Australia's population, but their geographic distribution has changed dramatically over the past century. Nearly a third of the Northern Territory's population is Aborigine (many located near Darwin), and other large native reserves are located in northern Queensland and Western Australia. Most native people, however, live in the same urban areas that dominate the country's overall population geography. Indeed, more than 70 percent of Aborigines live in cities, and very few of them still practice traditional hunting-and-gathering lifestyles. Processes of cultural assimilation are clearly at work: Urban Aborigines are frequently employed in service occupations, Christianity has often replaced traditional animist religions, and only 13 percent of Aborigines still speak a native language.

Still, there is a growing Aboriginal interest in preserving traditional cultural values, particularly in the Outback, where indigenous languages remain strong. Cultural leaders also are preserving Aboriginal spiritualism, and these

▼ **Figure 14.24 Australian Aborigines** The native people of Australia inhabited the continent before European colonization. Relocated to less desirable land by the European immigrants, they continue to struggle for equal rights and decent living conditions.

People on the Move

Recent Migrants to the Land Down Under

Australia has a long history of immigration, but recent policies fostering multiculturalism and a growing tide of global migrants have changed the "land down under" in important and enduring ways. Beginning about 1970, Australia made it easier for non-European, foreign-born immigrants to relocate there, especially if they possessed special skills. Today, more than one in four Australians is foreign-born, one of the highest totals in the developed world. The vast majority of immigrants have come for work, although an important and more contentious minority arrive as refugees asking for political asylum.

Source Regions and Destinations Where do Australia's migrants come from? A recent snapshot suggests the country's global appeal (**Figure 14.2.1**). Great Britain remains the largest contributing country: more than 1.2 million British-born citizens now live in Australia.

▼ **Figure 14.2.1. Origins of Australia's Foreign-Born Population, 2010 (50 top countries)** Great Britain is the source of an important part of Australia's immigrant population, but diverse Asian countries have contributed large numbers of migrants. Overall, about one-quarter of the country's population is foreign-born.

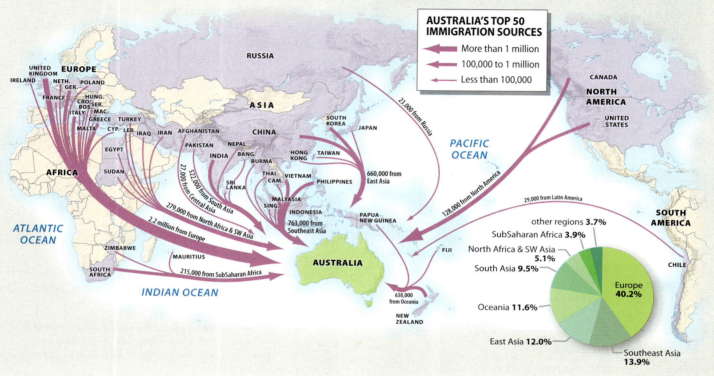

religious practices often link local populations to places and natural features considered sacred. In fact, a growing number of these sacred locations are at the center of land-use controversies (such as mining on sacred lands) between Aboriginal populations and Australia's European majority.

A Land of Immigrants Most Australians reflect the continent's more recent European-dominated migration history, but even these patterns have become more complex as more Asian immigrants arrive. Overall, more than 70 percent of Australians claim a British or Irish cultural heritage. These groups dominated 19th- and early-20th-century migrations into the country, and their close cultural ties to the British Isles remain strong.

European plantation owners along the Queensland coast imported inexpensive farm workers from the Solomons

and New Hebrides (now Vanuatu) in the late 19th century. These Pacific island laborers, known as **kanakas**, were spatially and socially segregated from their Anglo employers, but they further diversified the cultural mix of Queensland's "sugar coast" (**Figure 14.25**). Historically, however, non-white immigration was strictly limited by what is termed the **White Australia Policy**, in which the government after 1901 promoted European and North American immigration at the expense of other groups. The policy was not dismantled until 1973.

Recent migration trends feature more diverse inflows of new workers, adding to the country's ethnic diversity and multicultural character (see *People on the Move: Recent Migrants to the Land Down Under*). Indeed, today about 25 percent of Australia's people are foreign-born, reflecting the country's global popularity as a migration destination.

Italy, Germany, and Greece are also important European points of origin. But the big shift since the 1970s has been toward Asia. Today, about 10 percent of Australia's population has Asian roots. China, India, Vietnam, the Philippines, and Malaysia are key countries supplying mostly economic migrants. Where do immigrants settle? Resource-rich Western Australia has the country's highest percentage of immigrants, but the largest number of recent migrants venture to New South Wales and Victoria in search of jobs. About 20 percent of Sydney's and Melbourne's populations are Asian Australian (**Figure 14.2.2**). Fashionable suburbs such as Parramatta and Ryde on Sydney's northwest periphery are more than 30 percent Asian. While opportunities in Australian society have opened to these immigrants, some Asian Australians complain of a "bamboo ceiling" that has restricted access to the country's highest positions of government power and corporate influence.

Promoting Successful Immigrants The Australian government has created several important programs designed to foster immigrant success. The Adult Migrant English Program (AMEP) offers free English-language training to immigrants. AMEP provides over 500 hours of support during a migrant's first five years of residence. Additional hotlines and community service centers are also supported by the federal government. Cultural orientation classes are designed to facilitate smooth transitions into Australian life.

Political and Cultural Resistance Even though a recent survey suggested that 85 percent of Australia's population approved of the country's multiculturalism policies, migrants face ongoing resistance. Some environmental groups argue that the country is already overpopulated and that migrants simply add to the problem. Other critics point to high housing prices, especially in Sydney and Melbourne, arguing that migrants are driving affordable housing out of reach of

▲ **Figure 14.2.2. Asian-Australian Neighborhood, Melbourne** Little Bourke Street in Melbourne has long been home to a thriving community of Chinese Australian residents.

native-born Australians. Foreign-born migrants are also cited for taking jobs away from citizens, but there are few data that support this view. Finally, some residents, especially on the political right, reject multiculturalism altogether, and see migrants—especially those from Asia—as threatening the country's social cohesion, especially its immigrant European roots.

1. Cite some key similarities and differences between Australia's immigrant story and the situation for the United States.

2. Describe a setting near you that illustrates the strong relationship between a) areas of strong economic growth and b) higher rates of foreign-born population.

▼ **Figure 14.25 Kanaka Workers, Queensland Coast** This historical image, probably made in Queensland during the late 1870s, shows a group of young male Pacific Islander workers who labored in the area's sugar cane plantations.

Cultural Patterns in New Zealand

New Zealand's cultural geography broadly reflects the patterns seen in Australia, although the precise cultural mix differs slightly. Native **Maori** people are more numerically important and culturally visible in New Zealand than their Aboriginal counterparts are in Australia. While British colonization clearly mandated the dominance of Anglo cultural traditions by the late 19th century, the Maori survived, even though they lost most of their land in the process. After an initial decline, the native population rebounded in the 20th century, and today self-identified Maori account for about 15 percent of the country's 4.5 million residents (**Figure 14.26**). Geographically, the Maori remain most numerous on the North Island, including a sizable concentration in metropolitan Auckland. While urban living is on the rise, many Maori are committed to preserving their religion, traditional arts, and Polynesian culture.

▲ **Figure 14.26 Maori Artwork** Woodcarving is an active part of preserving Maori culture in New Zealand, as demonstrated by this carver at the Puia Cultural Center in Rotorua, North Island, New Zealand.

In addition, Maori joins English as an official language of New Zealand.

Although many New Zealanders still identify with their largely British heritage, the country's cultural identity has increasingly separated from its British roots. In many ways, popular culture ties the country ever more closely to Australia, the United States, and continental Europe, a function of increasingly global mass media. Several major movies have been filmed in New Zealand, including *The Adventures of Tintin, The Hobbit, Avatar,* the *Lord of the Rings* series, and *Whale Rider*, and the country is a favorite backdrop for India's Bollywood movies as well.

Paralleling the Australian pattern, many immigrants have also moved to New Zealand, especially to larger cities such as Auckland. Solid economic growth and construction-related jobs have contributed to growing employment opportunities for immigrants. In 2013, the New Zealand census reported that about one in four residents of New Zealand was foreign-born. Since 2010, the largest source nations for these migrants have been the United Kingdom, China, India, and the Philippines.

The Mosaic of Pacific Cultures

Native and colonial influences produce a range of cultures across the islands of the South Pacific. In more isolated places, traditional cultures are largely insulated from outside influences. In most cases, however, modern life in the islands revolves around an intricate cultural and economic interplay of local and Western influences.

Language Geography A modern language map reveals some significant cultural patterns that both unite and divide the region (**Figure 14.27**). Most native languages of Oceania belong to the **Austronesian** language family, which is found across wide expanses of the Pacific, much of insular Southeast Asia, and, somewhat surprisingly, Madagascar. Linguists hypothesize that the first prehistoric wave of oceanic mariners spoke Austronesian languages and thus spread them throughout this vast realm of islands and oceans. Within this broad language family, the Malayo-Polynesian subfamily includes most of the related languages of Micronesia and Polynesia, suggesting a common cultural and migratory history for these widespread peoples.

Melanesia's language geography is more complex. Although coastal peoples often speak languages brought to the region by the seafaring Austronesians, more isolated highland cultures, particularly on the island of New Guinea, speak varied Papuan languages. In fact, more than 1000 languages have been identified across the mountainous interior, creating such linguistic complexity that many experts question whether they even constitute a unified "Papuan family" of related languages. Some scholars estimate that half of New Guinea's languages are spoken by fewer than 500 people.

Given the frequency of contact between different island cultures, it is no surprise that people have generated new forms of intercultural communication. For example, several forms of **Pidgin English** (also known simply as *Pijin*) are found in the Solomons, Vanuatu, and New Guinea, where it is the major language used between ethnic groups. In Pijin, a largely English vocabulary is reworked and blended with Melanesian grammar. Pijin's origin is commonly traced to 19th-century Chinese sandalwood traders ("pijin" is the Chinese pronunciation of the word for "business"). While of historical origin, Pidgin English has become a globalized language of sorts in Oceania as trade and political ties have developed between different native island groups. About 300,000 people in Oceania speak Pidgin English on a regular basis, and it forms an important element of cultural identity among native Hawaiians, who use a version of Pidgin English in their everyday vernacular language.

Village Life Traditional patterns of social life are as complex and varied as the language map. Across much of Melanesia, including Papua New Guinea, most people live in small villages often occupied by a single clan or family group. Many of these traditional villages contain fewer than 500 residents,

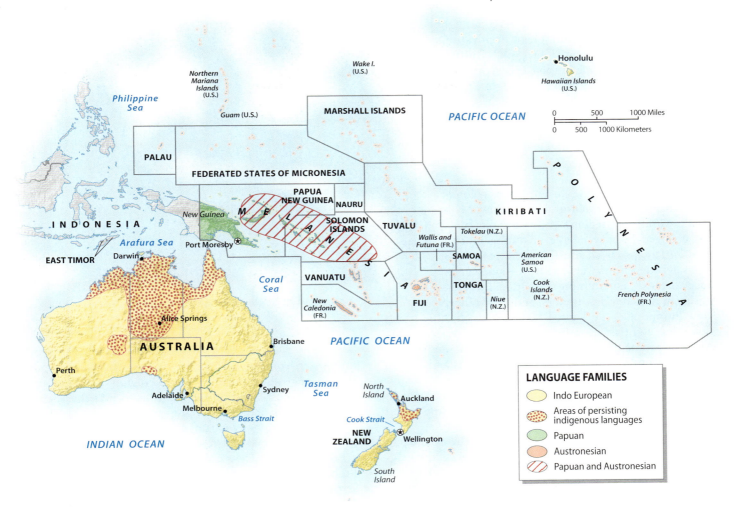

▲ **Figure 14.27 Languages of Australia and Oceania** While English is spoken by most residents, native peoples and their linguistic traditions remain an important cultural and political force in both Australia and New Zealand. Elsewhere, traditional Papuan and Austronesian languages dominate Oceania. The French colonial legacy also persists in some Pacific locations. Tremendous linguistic diversity has shaped the cultural geography of Melanesia, and more than 1000 languages have been identified in Papua New Guinea alone.

LANGUAGE FAMILIES

- Indo European
- Areas of persisting indigenous languages
- Papuan
- Austronesian
- Papuan and Austronesian

although some larger communities may house more than 1000 people. Life often revolves around the gathering and growing of food, annual rituals and festivals, and complex networks of kin-based social interactions (**Figure 14.28**).

Traditional Polynesian culture also focuses on village life, although strong class-based relationships often exist between local elites (who are often religious leaders) and ordinary residents. Polynesian villages are also more likely to be linked to other islands by wider cultural and political ties. Despite the Western stereotype of Polynesian communities as idyllic and peaceful, violent warfare was actually quite common across much of the region prior to European contact.

Interactions with the Larger World

While traditional cultures persist in some areas, most Pacific islands have witnessed tremendous cultural transformations in the past 150 years. Settlers from

Explore the **Sights** of Highlands of New Guinea

http://goo.gl/zmeVuJ

▼ **Figure 14.28 Highlands Village, Papua New Guinea** Residents of this village in the Papua New Guinea Highlands live in a loosely clustered array of homes created from local building materials.

Europe, the United States, and Asia brought new values and technological innovations that have forever changed Oceania's cultural geography and its place in the larger world. The result is a modern setting where Pidgin English has mostly replaced native languages, Hinduism is practiced on remote Pacific islands, and people from traditional fishing communities now work at resort hotels and golf course complexes.

Colonial Connections Anglo-European colonialism transformed the cultural geography of the Pacific world. The region's cultural makeup also was changed by new people migrating into the Pacific islands. Hawaii illustrates the pattern. By the mid-19th century, Hawaii's King Kamehameha was already entertaining assorted whalers, Christian missionaries, traders, and navy officers from Europe and the United States. A small, elite group of **haoles**, light-skinned Europeans and Americans, successfully profited from sugarcane plantations and Pacific shipping contracts. Labor shortages on the islands led to the importing of Chinese, Portuguese, and Japanese workers, who further complicated the region's cultural geography. By 1900, Japanese immigrants dominated the island workforce.

The United States formally annexed the islands in 1898, and the Hawaiian census of 1910 suggests the magnitude of culture change: More than 55 percent of the population was Asian (mostly Japanese and Chinese), native Hawaiians made up another 20 percent, and about 15 percent (mostly imported European workers) were white. By the end of the 20th century, however, the Asian population was less dominant, as about 40 percent of Hawaii's residents were white. In addition, native Hawaiians were joined by diverse migrants from other Pacific islands. Ethnic mixing among these groups has produced a rich mosaic of Hawaiian cultures with a unique blend of North American, Asian, Pacific, and European influences.

Hawaii's story has also played out on many other Pacific islands. In the Mariana Islands, Guam was absorbed into the United States' Pacific empire at the conclusion of the Spanish-American War in 1898. Thereafter, the native people were influenced not only by Americanization (the island remains a self-governing U.S. territory today), but also by the thousands of Filipinos who were moved in to supplement Guam's modest labor force. To the southeast, the British-controlled Fiji Islands offered similar opportunities for redefining Oceania's cultural mix. The same sugar-plantation economy that spurred changes in Hawaii prompted the British to import thousands of South Asian laborers to Fiji. The descendants of these Indians (most practicing Hinduism) now constitute almost half the island country's population and sometimes come

▲ **Figure 14.29 South Asians in Fiji** This retail clerk works in Nadi, Fiji's third largest city. **Q: What factors explain how migrants from South Asia settled in Fiji?**

into sharp conflict with native Fijians (**Figure 14.29**). In French-controlled portions of the region, small groups of traders and plantation owners filtered into the Society Islands (Tahiti), but a larger group of French colonial settlers (many originally part of a penal colony) had a major impact on the cultural makeup of New Caledonia. Still a French colony, New Caledonia is more than one-third French, and its capital city of Nouméa reveals a cultural setting forged from French and Melanesian traditions (see *Exploring Global Connections: Persisting French Influence in the South Seas*).

Sports and Globalization Like all aspects of culture, colonial influences have left their mark on Oceania's playing fields, particularly in those former British colonies where cricket, soccer, and rugby dominate the sporting scene. Netball, an English version of basketball played with seven players, is one of the most popular women's sports in both Australia and New Zealand, followed by field hockey. In New Zealand, where rugby is arguably the nation's favorite sport, the national team, the All Blacks (named for their uniform color, not their ethnicity), has integrated Polynesian warrior rituals and dances into its pre- and postgame activities, to international acclaim.

In recent years, many young men from the Pacific islands have looked to American football as their ticket to a better future, one that includes a U.S. college scholarship and perhaps even a chance to play in the National Football League (NFL). Many players come from American Samoa because they are American citizens and need no visas, although football players from Tonga and Fiji are also recruited. Coaches note that in addition to physical size and agility, these players often possess a strong work ethic. Carolina Panther defensive tackle Star Lotulelei,

Persisting French Influence in the South Seas

New Caledonia remains an unlikely combination: mix an oddly named tropical South Sea Island inhabited traditionally by Austronesian-speaking Melanesians with thousands of 19th-century French convicts and political prisoners. Add contemporary dashes of Asian immigrants, new infusions of French settlers, and a healthy dose of global capital investment (the island is home to about 25 percent of the world's nickel reserves). The result is a complex human mosaic and a fascinating landscape that reveals the island's diverse cultural origins.

A Complex Cultural Legacy Long before it was first sighted and named by Captain James Cook in the late eighteenth century (it reminded him of Scotland), New Caledonia was occupied by an assortment of Melanesian peoples, who still make up about 40 percent of the island's population. These *Kanaks* (native inhabitants) finally yielded to a more lasting European invasion after the island was annexed by France in 1853. For the next four decades, the French—taking a page from the British in Australia—deposited somewhere between 22,000 and 30,000 prisoners on the island, making it a huge penal colony. These older and involuntary settlers—largely male—formed the felonious foundations for the island's French culture: once they served their time, they often remained and were given land to settle. Amid the mix were about 2,000 Arab- and Berber-speaking deportees from French North Africa. This older French contingent, called

Google Earth Virtual **MG** Tour Video

http://goo.gl/bvCpXo

▼ **Figure 14.3.1.** **Cathédrale St. Joseph, Nouméa, New Caledonia** This view of New Caledonia's capital features the French-style Cathédrale St. Joseph in the foreground.

▲ **Figure 14.3.2.** **New Caledonia's Dual Flags** This woman holds two flags that represent both France and New Caledonia.

Caldoches, were later supplemented by subsequent (and more voluntary) migrants, the *Métros*, who arrived mainly from urban France. Today, these Europeans make up another third of the population. Additional migrants filtered onto the island from Polynesia (about 10 percent of the population) and East and Southeast Asia (another 5 percent of the population).

Persisting Ties to France Politically, New Caledonia remains French: it became a French overseas territory in 1946 and today is governed as a "special collectivity" in which territorial powers are gradually being transferred into local hands. While recent elections in 2014 slowed processes of independence, a referendum before 2020 will perhaps lead more clearly in that direction. Even so, it is hard to ignore the persisting patina of French influence that dominates the island, especially in the capital city of Nouméa (Figure 14.3.1). One can still stroll South Pacific streets where French colonial architecture pervades, the French language is spoken (over 95 percent of the island's population speaks French), and sidewalk cafes offer fresh-baked croissants, handmade chocolates, and tasty croque-monsieurs (baked or fried ham and cheese sandwiches). The French tricolor flag still flies, although since 2010 it is often accompanied by a second flag that acknowledges local, more indigenous influences (Figure 14.3.2). Ultimately, given the growing political weight of its large Melanesian minority, New Caledonia is likely to become another independent South Pacific nation, but it will wear the mantle of its colonial identity for a long time, much like its larger Australian and New Zealand neighbors to the west and south.

1. What similarities and differences could you cite between New Caledonia's story and that of Australia and New Zealand?

2. Select another small South Pacific island and trace out its colonial roots. What might remain today of these colonial ties?

for example, moved to Utah (as a Mormon) from Tonga (Figure 14.30). Lotulelei became part of that city's large Polynesian community and was recruited into the NFL. For many Polynesian athletes, however, success at the professional level remains elusive.

Fusion Diets in the Pacific Changing diets are another expression of globalization in the Pacific region. For example, some argue that globalization is partly responsible for Tonga being the most obese country in the world. Almost 40 percent of the population suffers with type 2 diabetes,

▲ **Figure 14.30 Star Lotulelei, Tongan NFL Player** Quite a few Pacific Islander men play in the National Football League. Lotulelei, born in Tonga, became a well-known defensive tackle for the Carolina Panthers.

▲ **Figure 14.31 Hawaiian Food Truck** A common sight across Hawaii, these mobile eateries offer both residents and visitors a quick, fresh meal. This Kahuku Shrimp Stand is a popular stop on Oahu's scenic north shore.

Explore the **Tastes** of Hawaiian Fusion Food

http://goo.gl/KsCmjm

and average lifespans are falling. Traditional healthy diets of fish, root vegetables, and coconuts have been replaced by high-fat meaty diets (imported turkey tails from the United States and mutton flaps [bellies] from New Zealand are popular) and a plethora of unhealthy fast-food options.

Elsewhere, changing diets represent a fusion of traditional and imported foods. Hawaii, home to a notable global mix of people from Polynesia, North America, and Asia, is also a melting pot for Pacific cuisine. Hawaiian "local food," often casually served from the side window of a food truck, is often a creole-style mix of fresh seafood, seaweed, rice, pineapple, beans, and yams. Toss in a little sushi, macaroni salad, or spam as a tasty accompaniment (**Figure 14.31**).

> **REVIEW**
>
> **14.7** Compare the Australian Aborigines and the New Zealand Maori in terms of their initial encounters with Europeans and the challenges they face today.
>
> **14.8** How does Australia's immigration story compare with that of New Zealand?
>
> **14.9** Describe Hawaiian cultural changes over the last century.
>
> ■ **KEY TERMS** kanakas, White Australia Policy, Maori, Austronesian, Pidgin English (Pijin), haoles

Geopolitical Framework: Diverse Paths to Independence

Pacific geopolitics reflects a complex interplay of forces (**Figure 14.32**). Begin with a diverse array of traditional indigenous place identities that vary greatly from island to island. These local political geographies were then overwhelmed by a European- and American-dominated colonial world that still shapes the region in important ways. Add to this mix other political players—mostly from Asia—that periodically have asserted their own presence in the region. Today, for example, China's influence is on the rise. Making matters even more interesting is the fact that even as the region enters the global stage, local native rights have been reasserted. No wonder that many observers suggest that the region's geopolitical framework is still very much in the making as we move farther into the 21st century.

Roads to Independence

The newness and fluidity of the region's political boundaries are remarkable. The oldest independent states are Australia and New Zealand, both 20th-century creations still considering whether to complete their formal political separation from the British Crown. Elsewhere, political ties between colony and mother country are closer and perhaps more enduring. Even many of the newly independent Pacific **microstates**, with their tiny overall land areas, keep special political and economic ties to countries such as the United States.

Independent Australia (1901) and New Zealand (1907) gradually created their own political identities, yet both still struggle with the final shape of these identities. Although Australia became a commonwealth in 1901 and New Zealand finally broke formal legislative links with the mother country in 1947, both nations still acknowledge

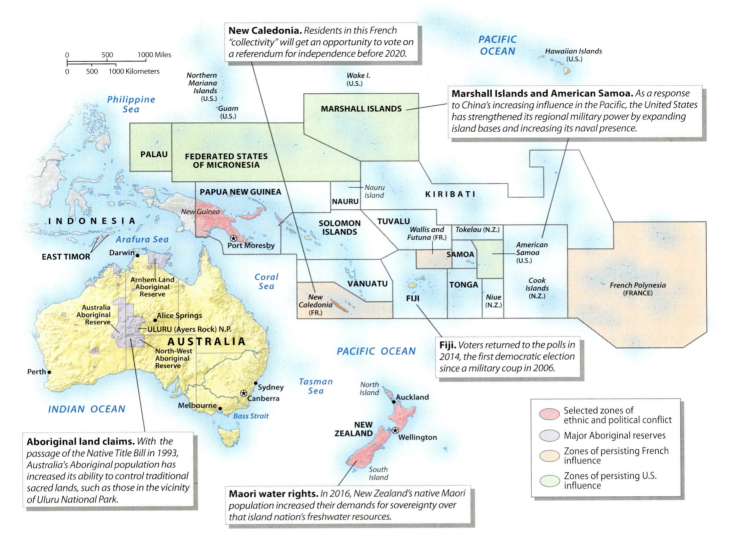

New Caledonia. *Residents in this French "collectivity" will get an opportunity to vote on a referendum for independence before 2020.*

Marshall Islands and American Samoa. *As a response to China's increasing influence in the Pacific, the United States has strengthened its regional military power by expanding island bases and increasing its naval presence.*

Fiji. *Voters returned to the polls in 2014, the first democratic election since a military coup in 2006.*

Aboriginal land claims. *With the passage of the Native Title Bill in 1993, Australia's Aboriginal population has increased its ability to control traditional sacred lands, such as those in the vicinity of Uluru National Park.*

Maori water rights. *In 2016, New Zealand's native Maori population increased their demands for sovereignty over that island nation's freshwater resources.*

- Selected zones of ethnic and political conflict
- Major Aboriginal reserves
- Zones of persisting French influence
- Zones of persisting U.S. influence

▲ **Figure 14.32 Geopolitical Issues in Australia and Oceania** Pacific geopolitics reflects a complex interplay of local, colonial-era, and current global-scale forces, resulting in a political geography that is still very much in the making today.

the British Crown as the symbolic head of government (**Figure 14.33**).

Colonial ties were cut even more slowly in other parts of the region, and the process has not yet been completed. New Caledonia remains affiliated (as a "special collectivity") with France after the island nation's citizens voted down a referendum for independence in 2014. In the 1970s, Britain and Australia began giving up their own colonial empires in the Pacific. Fiji (Great Britain) gained independence in 1970, followed by Papua New Guinea (Australia) in 1975 and the Solomon Islands (Great Britain) in 1978. The small island nations of Kiribati and Tuvalu (Great Britain) also became independent in the late 1970s. Independence, however, has not necessarily guaranteed political stability: Fiji experienced a military coup in 2006 and only recently returned to a representative and elected government, while Papua New Guinea suffered a constitutional crisis in 2012 as its judicial and legislative branches battled for power following a disputed election.

▼ **Figure 14.33 Prince William and the Duchess of Cambridge Visit Australia in 2014** Close ties between Australia and the British Crown are reflected on this visit to Sydney's Taronga Zoo by Prince William and the Duchess of Cambridge.

▲ **Figure 14.34 The Streets of Pago Pago, American Samoa** This view of downtown Pago Pago shows a small shopping area and a nearby church.

The United States has recently turned over most of its Micronesian territories to local governance, while still maintaining a large influence in the area. After gaining these islands from Japan in the 1940s, the U.S. government provided aid to the islanders and utilized a number of islands for military purposes. Bikini Atoll was destroyed by nuclear tests, and the large lagoon of Kwajalein Atoll was used as a giant missile target. A major naval base was established in Palau, the westernmost archipelago of Oceania.

By the early 1990s, the Marshall Islands and the Federated States of Micronesia (including the Caroline Islands) gained independence. Their ties to the United States remain close. Several other Pacific islands remain under U.S. administration. Palau is a U.S. "trust territory," which gives Palauans some local autonomy. The people of the Northern Marianas chose to become a "self-governing commonwealth in association with the United States," a rather vague political position that grants U.S. citizenship. Residents of self-governing Guam and American Samoa are also U.S. citizens (**Figure 14.34**). Hawaii became a full-fledged U.S. state in 1959, yet debate continues regarding native land claims and sovereignty.

Other colonial powers appear less inclined to give up their oceanic possessions. New Zealand still controls substantial territories in Polynesia, and France has even more extensive holdings in the region. In addition to New Caledonia in Melanesia, France's territories include French Polynesia, a large expanse of mid-Pacific territory, as well as the much smaller territory of Wallis and Futuna to the west.

Persistent Geopolitical Tensions

Cultural diversity, colonial legacy, youthful states, and a rapidly changing political map contribute to ongoing geopolitical tensions in the Pacific world. Indeed, some of these conflicts have consequences that extend far beyond the region's boundaries. Others are more local, but they are still reminders of the difficulties that occur as political space is redefined in varied natural and cultural settings.

Native Rights Issues Indigenous peoples in Australia and New Zealand have used the political process to gain more

control over land and resources in their two countries, paralleling similar efforts in North America and elsewhere. Australia's Aboriginal groups discovered newfound political power from effective lobbying during a time when a more sympathetic federal government moved to rectify historical discrimination that left native peoples with no legal land rights. The Australian government established several Aboriginal reserves, particularly in the Northern Territory, and expanded Aboriginal control over sacred national parklands such as Uluru (called Ayers Rock by the British settlers) (**Figure 14.35**). Further concessions to indigenous groups were made in 1993 when the government passed the **Native Title Bill**, which compensated Aborigines for lands already given up and gave them the right to gain title to unclaimed lands they still occupied. The bill also provided Aborigines with legal standing to deal with mining companies in native-settled areas.

Efforts to expand Aboriginal land rights, however, have met strong opposition. In 1996, an Australian court ruled that pastoral leases (held by ranchers who control most of the Outback) do not necessarily negate or replace Aboriginal land rights. Grazing interests were infuriated, which led the government to respond that Aboriginal claims allow them to visit sacred sites and do some hunting and gathering, but do not give them complete economic control over the land (**Figure 14.36**).

In New Zealand, native land claims are even more contentious because the Maori constitute a larger proportion of the overall population and the lands they claim are more valuable than rural Aboriginal lands in Australia. In 2016, the Maori also pressed their legal claim to all fresh water in the nation, noting their special economic and spiritual relationship with the valuable resource. The New Zealand prime minister, while acknowledging selective water rights belonging to particular Maori groups, has declined to cede full ownership to New Zealand's native population. Recent protests have included civil disobedience, ever-increasing Maori land claims over much of the North and South islands, and a call to return the

Explore the **Sights** of Uluru-Kata Tjuta National Park

http://goo.gl/Tw7vBf

▼ **Figure 14.35 Aborigines near Uluru Rock** An indigenous ranger patrols Uluru-Kata Tjuta National Park in Australia's Northern Territory. Uluru Rock is in the background.

▼ **Figure 14.36 Applications for Native Land Claims in Australia** This map shows the applications for native land claims in Australia filed by different Aboriginal groups as of 2013. Note that these are applications only, not government-approved claims. Nevertheless, the widespread extent of the claims shows why the topic is so contentious and controversial. **Q: What might account for the relative lack of native land claims in the center of Australia?**

country's name to the indigenous *Aotearoa*, "Land of the Long White Cloud." Overall, the government response has been to increasingly recognize Maori land and fishing rights.

Native Hawaiians, who call themselves *Kanaka maoli*, also have issues with the U.S. government concerning human rights, access to ancestral land, and the political standing of native people. Native Hawaiians are descendants of Polynesian people who arrived in the Hawaiian Islands about 1000 years ago. Until 1898, when the islands were annexed by the United States, Hawaiians lived in an independent and sovereign state that was recognized by major foreign powers. The legality of U.S. annexation is still contested today and underlies native Hawaiian demands for a return of their historical sovereignty. Some Hawaiian nationalists advocate a form of Polynesian sovereignty similar to that of the Navajo Nation in Arizona, Utah, and New Mexico.

Australia's Refugee Issue Many global refugees have headed for Australia, creating a major issue for that country. In 2013, for example, more than 20,000 asylum seekers applied for admission to the country. Afghanistan, Sri Lanka, Iraq, and Iran were major source areas for these arriving boat people. While many Australians are proud of permitting refugees to relocate there, critics note that in 2014 more than $1.2 billion was spent to operate detention centers where refugees are processed and housed while awaiting decisions regarding asylum. One of the most controversial

aspects of the refugee-asylum topic is Australia's so-called Pacific Solution, a government policy designed to dissuade refugees from targeting Australia as their goal. Originally instigated in 2001, the Pacific Solution had several components. The Australian navy is charged with intercepting refugee boats before they can make landfall on Australian territory. Australia also relinquished its ownership of small offshore islands where refugees often landed and where they could plead amnesty under Australian law.

In addition, detention centers created on Nauru, Christmas Island, and Manus (an island in Papua New Guinea) are designed to keep refugees off Australian territory and deprive them of Australian law and civil rights while their claims are processed. These centers have proven expensive and problematic: Through an agreement with Papua New Guinea, for example, the Manus Island facility has grown to almost 1000 detainees (**Figure 14.37**). While some Manus residents applaud the influx of people and cash, others point to potential environmental and social issues as asylum seekers try to resettle in nearby areas. The government of Papua New Guinea has also realized that the center is controversial and now is arguing that the center be closed in the near future.

The Strategic Pacific As was the case during World War II, Oceania is once again a strategic global region where numerous countries seek to expand their influence. The major players include the reigning superpowers, the United States and China; the Pacific Rim countries of Japan, South Korea, Russia, Taiwan, and Indonesia; France, as a former colonial master; and the two most powerful locals, Australia and New Zealand.

For several decades now, disputes between Taiwan and the giant People's Republic of China (PRC) have inflamed tensions in Oceania. Each country has vied for influence over island communities, particularly those countries with United Nations votes. Samoa, Tonga, and Fiji, for example, have received vast amounts of aid from the PRC and have often voted in the UN to support China's controversial policy toward Tibet. In late 2014, Chinese President Xi Jinping

▼ **Figure 14.37 Manus Island Detention Center, Papua New Guinea** Many people who seek asylum in Australia end up being processed and resettled in Papua New Guinea. This view shows some of the initial accommodations at the Manus Island Detention Center.

reinforced these connections by holding a summit of allied island nations in Fiji and by promising more economic aid to Pacific countries supportive of China's policies in the region. In contrast, Taiwan relies on the UN votes of Palau, Kiribati, and the Marshall Islands (all of them recent recipients of Taiwanese aid) to neutralize China's ambitions to reclaim Taiwan.

Because of the PRC's growing influence in Oceania, coupled with North Korea's continued hostility toward the West, the United States is increasing its diplomatic and military presence in the South Pacific. The term *Asia Pivot* is commonly used to describe the proposed shift in America's foreign and military policy away from Iraq and Afghanistan and toward the Asia–Pacific world. When China protested that this U.S. shift was a thinly veiled policy of containment, Washington's reply was that the U.S. policy was simply one of "Asia management."

Semantics aside, the Asia Pivot includes the establishment of numerous small military bases scattered around the Pacific (and, for that matter, around the world), commonly called "lily-pad" outposts (**Figure 14.38**). The recent—and controversial—agreement with Australia to post 2500 U.S. Marines near Darwin in northern Australia by 2016 is an example. Although the U.S. military understandably does not publicize its lily-pad bases, reportedly there are new or recently revitalized military bases in Australia's Cocos Islands, American Samoa, Tinian Island, the Marshall Islands, and the Marianas. Andersen Air Force Base in Guam has expanded its facilities, and similar expansion is taking place on existing U.S. military bases in Japan, the Philippines, and South Korea. The geopolitical goal behind this Asia Pivot seems clear enough: to reestablish the United States as a major political and military player in Oceania.

Since the policy emerged around 2011, the United States has successfully joined the East Asia Summit, an important regional discussion forum, but geopolitical distractions elsewhere in the world (the Middle East, Ukraine, etc.) have continued to dominate its daily geopolitical agenda. In addition, confronting China—politically and economically— in its own backyard in the western Pacific is proving to be daunting and expensive. Indeed, China's recent commitment to increased aid to sympathetic Pacific nations underscores the potentially high price tag of the Asia Pivot policy for the United States.

Not to be overlooked are Australia and New Zealand, which still play key political roles in the South Pacific. Although these two countries sometimes disagree on strategic and military matters, their size, wealth, and collective political influence in the region make them important forces for political stability. Also, special colonial relationships still connect these two nations with many Pacific islands. Australia maintains close political ties to its former colony of Papua New Guinea, and New Zealand's continuing control over Niue, Tokelau, and the Cook Islands in Polynesia confirms that its political influence extends well beyond its borders. How these two countries will respond to China's growing influence, however, is unclear.

▼ **Figure 14.38 United States Military Bases in the South Pacific** This U.S. base is on Kwajalein Atoll in the Marshall Islands. During the European colonial period, Germany used Kwajalein as a copra trading center, but after Germany's defeat in World War I, the Japanese took over the island. It was subsequently taken over by the U.S. Army in 1944 and served as a command center for postwar American nuclear testing at other Marshall Island atolls. Today the Kwajalein base is part of the Ronald Reagan Missile Defense Test Site. **Q: Describe how the landscape may have been changed by different uses since 1941.**

Australia's number one trade partner is China, yet it maintains unquestioned political ties to North America. Australia could be showing Oceania that a middle pathway is possible, that Pacific island countries do not necessarily have to choose between the United States and China (or between China and Taiwan). Small island states, however, may find such a conciliatory stance more problematic and less likely to produce supportive flows of economic aid and low-interest loans from their chosen allies.

REVIEW

14.10 Describe the colonial history of the Pacific islands, and contrast it to that of Australia and New Zealand.

14.11 Describe three settings in the region where native rights are being reasserted.

14.12 Explain the characteristics of, and the motivations behind, the United States' Asia Pivot policy.

■ **KEY TERMS**   microstates, Native Title Bill

Economic and Social Development: Increasing Ties to Asia

As with all other world regions, the Pacific region contains a diversity of economic situations creating both wealth and poverty. Even within affluent Australia and New Zealand, pockets of pronounced poverty occur, and large economic disparities exist as well between those Pacific countries with global trade ties and small island nations lacking resources and external trade. While tourism offers some relief from abject poverty, the tourist economy can be fickle, prone to unpredictable booms and busts. The global decline in commodity prices between 2013 and 2016 has also hit the region hard, because several major countries (especially Australia and Papua New Guinea) are important exporters of energy and mining resources. Overall, the economic future of the Pacific region remains uncertain and highly variable because of its small domestic markets, peripheral position in the global economy, and diminishing resource base (Table 14.2).

TABLE 14.2   Development Indicators — MasteringGeography™

Country	GNI per capita, PPP 2014	GDP Average Annual % Growth 2009–15	Human Development Index (2015)[1]	Percent Population Living Below $3.10 a Day	Life Expectancy (2016)[2] Male	Female	Under Age 5 Mortality Rate (1990)	Under Age 5 Mortality Rate (2015)	Youth Literacy (% pop ages 15–24) (2005–2014)	Gender Inequality Index (2015)[3,1]
Australia	44,700	2.7	0.935	–	80	84	9	4	–	0.110
Fed. States of Micronesia	3,590	–0.4	0.640	66.7	68	72	56	35	–	–
Fiji	8,410	2.3	0.727	17.0	67	73	30	22	–	0.418
French Polynesia	–	–	–	–	75	79	–	6	–	–
Guam	–	–	–	–	76	82	–	10	–	–
Kiribati	3,340	2.5	0.590	34.7	63	68	88	56	–	–
Marshall Islands	4,700	1.7	–	–	71	72	52	36	–	–
Nauru	–	–	–	–	63	70'	–	–	–	–
New Caledonia	–	–	–	–	74	80	–	13	100	–
New Zealand	36,200	2.3	0.914	–	80	83	11	6	–	0.157
Palau	14,280	3.9	0.780	–	70	76	32	16	100	–
Papua New Guinea	2,790	8.1	0.505	64.7	61	65	88	57	72	0.611
Samoa	5,610	1.2	0.702	8.4	73	76	30	18	99	0.457
Solomon Islands	2,020	6	0.506	69.3	66	73	42	28	–	–
Tonga	5,270	0.9	0.717	8.2	74	78	25	17	99	0.666
Tuvalu	5,410	2.1	–	–	67	72	58	27	–	–
Vanuatu	3,030	1.7	0.594	38.8	72	75	39	28	95	–

Source: World Bank, *World Development Indicators,* 2016.

[1] United Nations, *Human Development Report,* 2015.

[2] Population Reference Bureau, *World Population Data Sheet,* 2016.

[3] Gender Inequality Index—A composite measure reflecting inequality in achievements between women and men in three dimensions: reproductive health, empowerment, and the labor market that ranges between 0 and 1. The higher the number, the greater the inequality.

Login to MasteringGeography™ **& access** MapMaster **to explore these data!**

1) What might explain the fact that Papua New Guinea, the nation with the highest child mortality rate and the lowest Human Development Index, enjoys the region's highest economic growth rates between 2009 and 2015?

2) List the five nations with the lowest GNI per capita in 2014. What other variables in the table suggest similar patterns one would associate with a low-income population?

Australian and New Zealand Economies

Much of Australia's economic wealth has been built historically on the cheap extraction and export of abundant raw materials. Export-oriented agriculture, for example, has long been a key support of Australia's economy. Australian agriculture is highly productive in terms of labor input and produces varied temperate and tropical crops as well as huge quantities of beef and wool for world markets. Although farm exports are still important to the economy, the mining sector has grown more rapidly since 1970, making Australia one of the world's mining superpowers.

Mining's recent growth is primarily due to increased trade with China, which has made Australia the world's largest exporter of iron and coal. Besides these two resources, Australia produces an assortment of other materials—namely, bauxite (aluminum ore), copper, gold, nickel, lead, and zinc. As a result, the New South Wales-based Broken Hill Proprietary Company (BHP) is one of the world's largest mining corporations. China's appetite for natural resources is huge—equaled, apparently, only by Australia's willingness to sell its resources. In addition to energy and mineral resources, Chinese buyers are also eyeing the country's agricultural land, scooping up large cotton farms and cattle ranches in recent years. In return, Chinese consumer goods are eagerly purchased by Australian shoppers (**Figure 14.39**). Half a million Chinese tourists also visit Australia annually and add to the local economy by keeping Australian lifeguards, blackjack dealers, and real estate brokers busy. In addition, 88,000 Chinese students currently attend Australian universities.

With growing numbers of Asian immigrants and expanding economic links with Asian markets, Australia's

▲ **Figure 14.40 Queensland's Gold Coast** Many of these luxury hotels in the Surfer's Paradise section of the Gold Coast are owned by Japanese firms specializing in accommodations for Asian tourists. **Q: What are the environmental implications of this Gold Coast development?**

long-term economic future is promising, although a recent sharp slump in commodity prices (especially for coal and iron ore) hurt the resource-oriented economy of Western Australia in 2015 and 2016. On the positive side, an expanding tourism industry is helping to diversify the nation's economy. More than 7 percent of the nation's workforce is now devoted to serving the needs of more than 7.5 million tourists each year. Popular destinations include Melbourne and Sydney as well as Queensland's resort-filled Gold Coast, the Great Barrier Reef, and the vast, arid Outback. Along the Gold Coast, many luxury hotels are owned by Japanese firms and provide a bilingual resort experience for their Asian clientele (**Figure 14.40**).

New Zealand is also a wealthy country, although somewhat less well-off than Australia. Before the 1970s, New Zealand relied heavily on exports to the United Kingdom—primarily agricultural products such as wool and butter. These colonial trade linkages became problematic in 1973 when Britain joined the European Union, with its strict agricultural protection policies. Unlike Australia, New Zealand lacks a rich base of mineral resources to export to global markets. As a result, the country saw its economy stagnate in

◀ **Figure 14.39 Australian Trade with China** Containers from Asia testify to the recent explosion of two-way trade between Australia and China. Raw materials, mainly iron ore, are exported to China, and consumer goods flow from China into Australia, making that country a key beneficiary of China's recent economic growth.

the 1980s. Eventually, the New Zealand government enacted drastic neoliberal reforms. Tax rates fell, and many state-run industries were privatized. As a result, New Zealand has been transformed into one of the most market-oriented countries in the world.

Oceania's Divergent Development Paths

Varied economic activities characterize the Pacific island nations. One way of life is oriented around subsistence-based economies, such as shifting cultivation or fishing. In other places, a commercial extractive economy dominates, and large-scale plantations, mines, and timber activities compete with the traditional subsistence sector for both land and labor (**Figure 14.41**). Elsewhere, the huge growth in global tourism has transformed the economic geographies of many island settings, forever changing how people make a living. In addition, many island nations benefit from direct subsidies and economic aid from present and former colonial powers, assistance that is designed to promote development and stimulate employment.

Melanesia is the least developed and poorest part of Oceania because these countries have benefited the least from tourism and from subsidies from wealthy colonial and ex-colonial powers. Today many Melanesians still live in remote villages isolated from the modern economy. The Solomon Islands, for example, with few industries other than fish canning and coconut processing, has a per capita gross national income (GNI) below $2500 per year. In contrast, Fiji is a more affluent Melanesian country, largely because of its tourist economy, reflecting its popularity with Chinese and Japanese visitors. Recent investments in Papua New Guinea's energy sector may speed economic growth in that country. Beginning in 2014, a huge multibillion-dollar project to produce liquefied natural gas, mostly for consumption in East Asia, is raising hopes for more economic development there.

▼ **Figure 14.41 Ok Tedi Copper and Gold Mine, Papua New Guinea** A collapsed tailings dam at the giant mine resulted in damage downstream for miles along the Ok Tedi and Fly rivers, illustrating the environmental risks of commercial mining in such rugged, remote settings.

One key challenge for the government will be translating that increased energy-related wealth into an improved standard of living for the nation's 8 million residents.

Throughout Micronesia and Polynesia, economic conditions depend on both local subsistence economies and economic linkages to the wider world. Many islands export food products, but some native populations survive mainly on fish, coconuts, bananas, and yams. Some island groups enjoy large subsidies from either France or the United States, even though such support often comes with a political price. China is also increasingly involved with economic development plans that often imply political support.

Other Polynesian island groups have been completely transformed by tourism. In Hawaii, more than one-third of the state's income flows directly from tourist dollars. With more than 8 million visitors annually (including more than 1.2 million from Japan), Hawaii represents all the classic benefits and risks of the tourist economy. Job creation and economic growth have reshaped the islands, but congested highways, high prices, and the unpredictable spending habits of tourists have put the region at risk for future problems (see *Geographers at Work: Planning for the Future across the Pacific Basin*). Elsewhere, French Polynesia has long been a favored destination of the international jet set. More than 20 percent of French Polynesia's GNI is derived from tourism, making it one of the wealthiest areas of the Pacific (**Figure 14.42**). More recently, Guam has seen a resurgence of Japanese and Korean tourists, especially those on honeymoons.

The South Pacific Tuna Fishery The South Pacific is home to the world's largest tuna fishery, and this resource contributes significantly to local island economies, both directly and indirectly. Tuna fishing and processing account for about 10 percent of all wage employment in southern Oceania, although 90 percent of the working tuna boats come from far-flung Pacific nations—namely China, Japan, South Korea, and the United States. These foreign boats are charged access fees for fishing in each island's offshore territory.

International law allows the extension of sovereign rights 200 miles (320 km) beyond a country's coastline in an **Exclusive Economic Zone (EEZ)**. Each country has economic control over resources within its EEZ, such as fisheries and ocean-floor mineral rights. Because Oceania's countries are often composed of a series of islands and because each country's EEZ is delimited from its outermost points, the South Pacific tuna fishery is composed of a patchwork of intersecting EEZs, each with different fishing regulations and fees. Complicating matters, as tuna become more scarce (and more valuable as a resource), many island nations have pressured global fishing fleets for a larger share of tuna profits. Making matters more uncertain, the United States withdrew from the South Pacific Tuna Treaty (first signed in 1987) in 2015. The agreement had provided a framework for managing global fishing rights in the region, but growing disputes between Pacific Island nations and the United States over treaty details and payments have led to the recent impasse.

Planning for the Future across the Pacific Basin

" I don't know what it is about islands, but my whole life path has taken me from one to another," says **Laura Brewington**, a Research Fellow at the University of Hawaii's East–West Center (**Figure 14.4.1**). Brewington has worked in New Zealand, Thailand, Iceland, and the Galápagos Islands, and currently focuses on developing environmental policy and land-use guidelines that integrate traditional lifeways, economic development, and unfolding climate change scenarios across fragile Pacific islands.

GIS and Climate Change Brewington's earlier work was set in the eco-sensitive Galápagos Islands, and included mapping environmental vulnerabilities and developing ways to contain invasive species in the area as it became increasingly exposed to the effects of globalization. More recently, Brewington's attention shifted to climate change adaptations in the state of Hawaii and U.S.-affiliated Pacific Islands. She partners with climate change scientists, the U.S. Geological Survey, water conservation groups, farmers, developers, and local residents to develop a range of land- and water-use possibilities (all produced in a series of mapped layers) to inform policy decisions. Brewington says, "The Pacific region is huge and people very disparate, so our goal is to get those voices heard, but also to get the right climate science information into the right decision-makers' hands."

Brewington, a biostatistician in the public health sector, discovered geography when her work required her to learn spatial statistics and GIS. She is a passionate spokesperson for the discipline: "I love getting students engaged in hands-on experiences. I tell them to choose a career path in line with what they find exciting. I didn't do that until much later, and I wish I'd had that piece of advice."

▲ **Figure 14.4.1. Laura Brewington, East-West Center, Hawaii** Trained in conservation work, Brewington has turned her fieldwork skills and geographic tools toward meeting the challenges of climate change in island environments.

1. Explain how land use and resource mapping can help a government or nongovernment agency make policy decisions.

2. How might potential climate change scenarios shape local planning initiatives in your own community?

Mining in the South Pacific Mining dominates some regional settings. For New Caledonia, its nickel reserves, the world's second largest, are both a blessing and a curse. Although they currently sustain much of the island's export economy, income from nickel mining will decrease in the near future as reserves dwindle. Dramatic price fluctuations for this commodity also hamper economic planning for the French colony. To the north, the tiny island of Nauru has already seen its mining economy wither as phosphate deposits have dwindled, leaving its people with almost no job prospects. In Papua New Guinea, however, gold and copper mining has dramatically transformed the landscape and remains an important part of the nation's economy. Large open-pit mine operations have largely been financed by foreign corporations based in Australia, the United States, and elsewhere. Major projects are active in many districts on the main island as well as on the nearby island of Bougainville.

Deep-sea mining may be the region's next economic frontier, although environmentalists question its long-term effects. Many resources are a geological product of the Pacific Ring of Fire: Underwater volcanism and mineralization have produced vast deposits of commercially valuable mining resources. Since 2014, for example, several New Zealand mining interests have applied for permission to dredge deep-sea minerals

▼ **Figure 14.42 Tahiti, Outlier of Polynesia** Now a popular tourist destination, the islands of Tahiti were first settled between 300 and 800 CE and form the southeastern corner of Polynesia some 3000 miles (4800 kilometers) from New Zealand to the west and roughly the same distance from Hawaii to the north. This is a view of Cook's Bay on the island of Moorea.

Explore the **Sights** of Tahiti

http://goo.gl/FEOYWH

Everyday **Globalization**

Wine from Down Under Gains Global Appeal

Americans have enjoyed Australian wines since the 1990s, but the region's appeal is broadening: Australian wine bottles are found on dinner tables from China to India, and Americans are embracing additional offerings from New Zealand.

Thirsty Chinese consumers with a taste for European-style wines recently spurred an 8 percent increase in annual wine imports from Australia, now the world's fourth-largest wine exporter (**Figure 14.5.1**). A new bilateral trade agreement will end Chinese import tariffs on Australian wine by 2018, likely prompting more consumption. Australia is also the second-largest wine supplier to nearby India.

At the same time, U.S. wine imports from New Zealand have leaped since 2008, growing 13 percent in 2014. More than 80 percent of New Zealand's wine exports are its prize-winning sauvignon blanc varieties, but its pinot noir also appeals to Americans. The shift toward Kiwi wine is enhanced by the fact that savvy American growers, familiar with the challenges of importing alcoholic beverages into the United States (rules vary from state to state) and with marketing to North Americans, have bought up several major New Zealand vintners.

▲ **Figure 14.5.1. Barossa Valley Wine Country, South Australia** These vineyards in South Australia's beautiful Barossa Valley now produce wine for global export.

1. In addition to wine, what food products or other consumer goods from Australia or New Zealand do you commonly see for sale in your local community? Explain why.

2. Where do the beverages you consume come from? Keep a simple beverage journal for five days and summarize the patterns you find.

---

such as iron ore and phosphate nodules from the sea floor within that country's EEZ. Groups such as Kiwis Against Seabed Mining strongly oppose the applications, which still await approval. Off the Cook Islands, cobalt and manganese nodules the size of lettuce heads have also whetted the appetites of mining companies and government officials, although some in the area worry that mining could damage the region's pristine waters and large local tourist economy. However, Fiji, the Solomon Islands, Tonga, and Vanuatu have already moved forward with issuing exploration licenses. Similarly, a Canadian company hopes to initiate the world's first operational deep-sea mining project in the Bismarck Sea, off the coast of New Guinea, at the site of volcanic hot springs rich in gold and copper deposits. Time will tell whether there is a steep environmental price to be paid for these submarine riches.

Oceania in the Global Context

Many international trade flows link the region to the far reaches of the Pacific and beyond. Australia and New Zealand dominate global trade patterns in the region (see *Everyday Globalization: Wine from Down Under Gains Global Appeal*). In the past 30 years, ties to the United Kingdom, the British Commonwealth, and Europe have weakened

relative to growing trade links with Japan, East Asia, the Middle East, and the United States. Australia, for example, now imports more manufactured goods from China, Japan, and the United States than it does from Britain and Europe. New Zealand's trading pattern is similar: it exports dairy products, meat, wine, wool, and wood products to nations such as China and Australia and it imports consumer goods, machinery, and textiles from the EU, Australia, and China (**Figure 14.43**). Both Australia and New Zealand also participate in the **Asia–Pacific Economic Cooperation Group (APEC)**, an organization designed to encourage economic development in Southeast Asia and the Pacific Basin. In addition, both countries have signed on to the Trans-Pacific Partnership Agreement that eventually (once fully ratified by all participants) will facilitate more free trade between countries bordering the Pacific basin.

To promote more regional economic integration, Australia and New Zealand signed the **Closer Economic Relations (CER) Agreement** in 1982, slashing trade barriers between the two countries. New Zealand now benefits from larger Australian markets for its exports, and Australian corporate and financial interests have gained access to New Zealand business opportunities. Since the CER Agreement's signing, trade between the two countries has steadily grown. More than 20 percent of New Zealand's

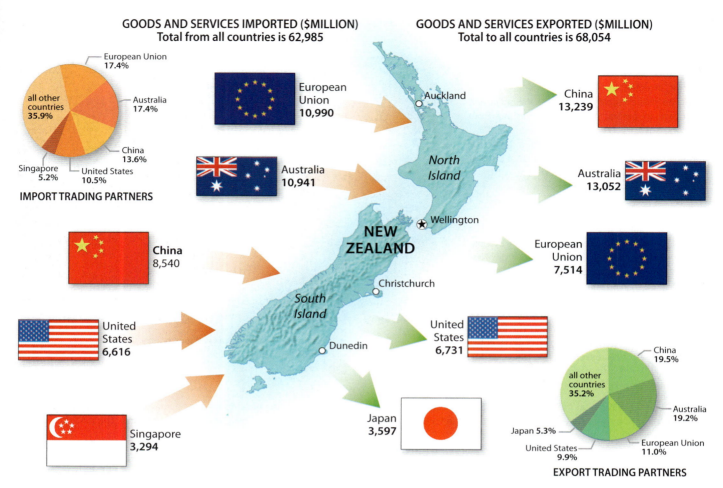

GOODS AND SERVICES IMPORTED ($MILLION)
Total from all countries is 62,985

GOODS AND SERVICES EXPORTED ($MILLION)
Total to all countries is 68,054

European Union
17.4%

Australia
17.4%

all other countries
35.9%

China
13.6%

Singapore
5.2%

United States
10.5%

IMPORT TRADING PARTNERS

European Union
10,990

Australia
10,941

China
8,540

United States
6,616

Singapore
3,294

Auckland

North Island

Wellington

NEW ZEALAND

Christchurch

South Island

Dunedin

China
13,239

Australia
13,052

European Union
7,514

United States
6,731

Japan
3,597

China
19.5%

all other countries
35.2%

Australia
19.2%

Japan 5.3%

United States
9.9%

European Union
11.0%

EXPORT TRADING PARTNERS

▲ **Figure 14.43 New Zealand's Trading Partners** Major exports from New Zealand target China and Australia, while the country imports products from a variety of sources including the European Union (especially the United Kingdom), Australia, and China.

imports and exports come from Australia, and this pattern of regional free trade is likely to strengthen in the future.

Smaller nations of Oceania, while often closely tied to countries such as China, Taiwan, Japan, the United States, and France, also benefit from their proximity to Australia and New Zealand. More than half of Fiji's imports come from those two nearby nations, and countries such as Papua New Guinea, Vanuatu, and the Solomon Islands enjoy a similarly close trading relationship with their more developed Pacific neighbors.

Continuing Social Challenges

Australians and New Zealanders enjoy high levels of social welfare but face some of the same challenges evident elsewhere in the developed world (see Table 14.2). Life spans average about 80 years in both countries, and rates of child mortality have fallen greatly since 1960. Like North America and Europe, however, the two countries count cancer and heart disease as the leading causes of death, and alcoholism is a continuing social problem, particularly in Australia. Furthermore, Australia's rate of skin cancer is among the

world's highest, the consequence of a largely fair-skinned, outdoors-oriented population from northwest Europe living in a sunny, low-latitude setting. Overall, high-quality medical care is offered by Australia's Medicare program (initiated in 1984) and by New Zealand's comprehensive system of social services.

Not surprisingly, the social conditions of the Aborigines and Maori are much less favorable than those of the population overall. Schooling is irregular for many native peoples, and levels of postsecondary education for Aborigines (12 percent) and Maori (14 percent) remain far below national averages (32–34 percent). Other social measures reflect the pattern. Less than one-third of Aboriginal households own their own homes, while more than 70 percent of white Australians are homeowners. Discrimination against native peoples continues in both countries, a situation aggravated and publicized since the assertion of indigenous political rights and land claims. As with African American, Hispanic, and Native American populations in North America, simple social policies have not yet solved these lasting problems.

Even in Hawaii, the proportion of native Hawaiians living below the poverty level is much higher than those of other ethnic groups. This group also has the shortest life expectancy and the highest infant mortality rate; rates of death from cancer and heart disease are almost 50 percent higher than for other U.S. groups, and native Hawaiian women have the highest rate of breast cancer in the world. Further, 55 percent of native Hawaiians do not complete high school, and only 7 percent have college degrees.

In other parts of Oceania, levels of social welfare are higher than might be expected based on the islands' economic situations. Sizable investments in health and education services have achieved considerable success. For example, the average life expectancy in the Solomon Islands, one of the world's poorer countries as measured by per capita GNI, is a respectable 67 years. By other social measures as well, the Solomon Islands and several other Oceania states have reached higher levels of human well-being than exist in most Asian and African countries with similar levels of economic output (**Figure 14.44**). This is partly a result of successful policies, but also reflects the relatively healthy natural environment of Oceania; many tropical diseases so troublesome in Africa simply do not exist in the Pacific region.

Gender, Culture, and Politics

Both Australia and New Zealand were early supporters of female suffrage, empowering women to vote in national elections in 1893 (New Zealand) and 1902 (Australia). In some cases, women were allowed to vote even earlier in local and state elections, although not until decades later could females actually hold office. Since that time, both countries have elected women as national leaders, with New Zealand's Helen Clark serving as prime minister between 1999 and 2008 and Australia's Julia Gillard serving

▼ **Figure 14.44 Solomon Islands** While not an economically wealthy part of the Pacific, the Solomon Islands enjoy relatively high levels of social development as measured by variables such as lifespan and under age 5 mortality.

in a similar office between 2010 and 2013. Yet both countries still have considerable gender gaps in terms of female representation in government, employment salaries, and social support services. As a result, in the global gender index New Zealand ranks 7th and Australia is 24th, both countries scoring lower than selected western European nations.

Indigenous gender roles were transformed by colonialism in New Zealand. Before European colonization, Maori women and men were equal in social status and power because of the overarching Maori principle of equity and balance in their nonhierarchical society. For example, the Maori language had no gender distinctions in personal pronouns (no "his" or "hers"). Further evidence for this gender equity comes from the prominent role women played in Maori proverbs and legends.

However, English colonial society was troubled by this gender-neutral native culture and sought to deprive Maori women of their status. In 1909, New Zealand law required Maori women to undergo legal marriage ceremonies that emphasized male ownership of all property—not just of land and livestock, but also of the wife. Missionary schools for Maori women had also long reinforced English notions of female domesticity where women were subject to male authority. As a result, gender roles in contemporary Maori society now reflect these colonial notions of male dominance. To further their European values, missionaries rewrote Maori proverbs and legends to emphasize male characters with heroic warrior attributes. Today Maori women never participate in the haka war dances; instead, female participation is limited to subservient songs and dances.

Across Australia, traditional Aboriginal society has distinct gender roles, with clear-cut distinctions between "women's business" and "men's business." These distinctions are played out not just in daily affairs, but also in the Dreamtime, an abstract parallel universe of central importance for Aboriginal people. In physical life as well as in Dreamtime, women are responsible for the vitality and resilience of family lives, while men's business centers on the larger group or tribe. These distinct gender roles also involve the landscape, with certain areas and locales linked closely to either men or women, but rarely to both. As a result, the Aboriginal territory is also highly gendered.

REVIEW

14.13 Define Exclusive Economic Zone (EEZ) and explain why it is important to Pacific island economies.

14.14 What is the Closer Economic Relations (CER) Agreement, and why is it important in understanding the economics of Oceania?

14.15 Describe the major challenges to social development in Oceania.

■ **KEY TERMS** Exclusive Economic Zone (EEZ), Asia–Pacific Economic Cooperation Group (APEC), Closer Economic Relations (CER) Agreement

Chapter 14 Australia and Oceania

Summary

- The region's natural environment has witnessed accelerated change in the past 50 years. Urbanization, tourism, resource extraction, exotic species, and global climate change are altering the landscape and increasing the vulnerability of island environments. On low-lying Pacific islands threatened by sea-level rise, residents increasingly face the grim reality of evacuating their homelands.

- Migration is a major theme in Oceania, beginning with the earliest Aboriginal peoples who came from the Asian mainland, to prehistoric island-hopping Polynesian people, to more recent European emigrants that populated Australia and New Zealand.

- Cultural geographies across the region include a mix of indigenous (Melanesian, Micronesian, and Polynesian) peoples, persisting colonial influences (especially British, French, and American), and a growing variety of Asian immigrants, primarily South Asian, Chinese, and Japanese.

- The region's contemporary cultural and political geography is changing as countries struggle to disentangle themselves from colonial ties by asserting their own identities. Also, native peoples are increasingly intent on regaining their land and civil rights. Oceania also has become entangled in the world's larger geopolitical tensions as China and the United States jockey for economic and political influence in the region.

- Australia is the region's economic powerhouse because of its rich mineral resources, with coal and iron dominating the list of trade commodities. In contrast, island nations, from Fiji to Hawaii, are largely at the mercy of global tourism, with boom-and-bust cycles linked to the vitality of the global economy.

Review Questions

1. How might global climate change shape the region in the future? Cite three settings that may already be impacted, describing their current human and physical geographies and the expected changes to both.

2. Describe parallels between the historical geography of settlement in North America and that of Australia and New Zealand.

3. Describe the types of farming you might observe in 1) South Australia, 2) the Papua New Guinea Highlands, and 3) Fiji. Explain the differences using both environmental and human variables in your discussion.

4. In a group of three, have one person each represent the issues facing Australian Aborigines, New Zealand Maori, and native Hawaiians in their different countries. Then, as a group, compare the similarities and differences.

Predict where each native group might be in 10 years.

5. Imagine yourself as the leader of a small Pacific Island nation. Discuss the advantages and disadvantages of building closer ties with China. How about the United States?

6. Write a short essay summarizing the economies of the region and explaining the wide economic gap separating affluent Australia and New Zealand from the rest of Oceania.

Image Analysis

1. Consult Table 14.1 (especially Rate of Natural Increase) and explain *why* the two population pyramids will continue to look quite different from one another in 2025. Given current regional growth rates, name another country that might look similar to the Solomon Islands in 2025.

2. What longer-term economic factors may explain why the two pyramids might look more similar by 2050?

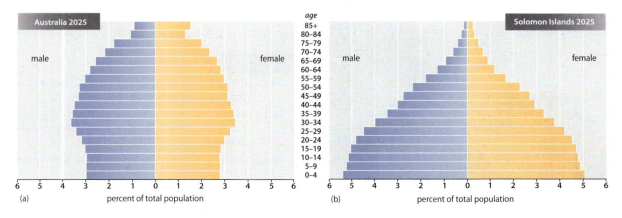

▲ Figure IA14 Population pyramids of Australia and the Solomon Islands

JOIN THE DEBATE

According to politicians, talk radio pundits, and activists of different stripes, Australia is facing a severe population crisis. However, the kind of crisis is not clear. The country's population geography is a striking mix of crowded, highly urbanized lands (mainly along the east and southwest coasts) and a vast, almost empty interior. Is Australia over- or underpopulated?

▲ Figure D14 Australian Population Map

Australia is already overpopulated!

- Activists and groups like Sustainable Population Australia (SPA) argue that Australia is already overpopulated by some 8 million people, and call for an environmentally responsible population policy. They say that because so much of the country is either arid or semiarid, it has a limited rural carrying capacity. While the interior looks empty, it cannot really support very many people.

- Global warming will make matters even worse, they argue, expanding Australia's arid interior and forcing even more people into the crowded coastal urban strips, where seasonal brushfires are already a serious problem.

- Some groups also see immigration as a growing issue, as migrants from Asia and rising refugee populations take jobs away from Australian citizens. Why should immigrants have that right?

Australia has plenty of room to grow!

- Australia's low birth rate will lead to a declining population unless immigration increases. Some observers say that the very low birth rates could halve the population by the end of the century.

- To combat this so-called baby bust, the Labor Party and business coalitions such as the Australian Population Institute are calling for maternity leave, tax breaks, and, most of all, increased immigration to shore up Australia's population. They argue that a larger market, labor force, and taxpayer base are essential to the national economy.

- Nearly as large as the mainland United States, yet with a population only about that of the Los Angeles metropolitan area, many find it difficult to believe Australia is overpopulated. This is particularly true for those living on densely populated Pacific islands, who watch rising sea levels and cast a longing eye at Australia's empty spaces, which might offer refuge if islanders have to leave their flooded homes.

KEY TERMS

Aborigines (p. 601)
Asia–Pacific Economic Cooperation Group (APEC) (p. 621)
atoll (p. 591)
Austronesian (p. 608)
Closer Economic Relations (CER) Agreement (p. 621)
Exclusive Economic Zone (EEZ) (p. 619)
exotic species (p. 592)
haoles (p. 610)
high islands (p. 591)
intercropping (p. 604)
kanakas (p. 606)
low islands (p. 591)
mallees (p. 591)
Maori (p. 607)
marsupial (p. 592)
Melanesia (p. 589)
Micronesia (p. 589)
microstates (p. 612)
Native Title Bill (p. 614)
Oceania (p. 588)
Outback (p. 590)
Pidgin English (Pijin) (p. 608)
Polynesia (p. 589)
viticulture (p. 604)
White Australia Policy (p. 606)

MasteringGeography™

Looking for additional review and test prep materials? Visit the Study Area in **MasteringGeography**™ to enhance your geographic literacy, spatial reasoning skills, and understanding of this chapter's content by accessing a variety of resources, including **MapMaster** interactive maps, videos, RSS feeds, flashcards, web links, self-study quizzes, and an eText version of *Diversity Amid Globalization*.

DATA ANALYSIS

At the broad scale of our regional climate maps, local detail for New Zealand's South Island (see Figure 14.8) is difficult to see. Yet a more detailed look at the island's diverse microclimates reveals an amazing and complex picture. Visit the website for Te Ara: The Encyclopedia of New Zealand (http://www.teara.govt.nz) and examine the map of annual precipitation for the South Island.

http://goo.gl/435lHN

1. Briefly describe and explain today's pattern of precipitation across the South Island.

2. Given your knowledge of New Zealand geography, write a paragraph about how this precipitation pattern may have shaped historical and contemporary geographies of settlement, agriculture, and economic development across the island.

3. Visit NIWA's website (https://www.niwa.co.nz) and access a map of projected changes for 2080–2099, based on global climate change models and predictions. See Figure 5 in the document. Briefly summarize what you see, and suggest possible social and economic implications of these changes for future residents and government planners.

Authors' Blogs

Scan to visit the **GeoCurrents blog**
www.geocurrents.info/category/place/australia-and-pacific

Scan to visit the **Author's Blog**
for field notes, media resources, and chapter updates
www.gad4blog.wordpress.com/category/australia-and-oceania

GLOSSARY

Aborigine   An indigenous inhabitant of Australia.

acid rain   Harmful form of precipitation high in sulfur and nitrogen oxides. Caused by industrial and auto emissions, acid rain damages aquatic and forest ecosystems in regions such as eastern North America and Europe.

adiabatic lapse rate   The rate an air mass cools or warms with changes in elevation, which is usually around 5.5°F per 1000 feet (1°C per 100 meters). Contrast with environmental lapse rate.

African diaspora   The forced removal of Africans from their native areas to localities around the globe, especially due to slavery. It can also refer to contemporary migration flows from Africa due to violence or poverty that result in African communities forming in other regions in the world.

African Union (AU)   Founded in 1963, the organization grew to include all the states of the continent except South Africa, which finally was asked to join in 1994. Morocco withdrew in 1984, but now wishes to rejoin. In 2004 the body changed its name from the Organization of African Unity (OAU) to the African Union. It is mostly a political body that has tried to resolve regional conflicts.

agrarian reform   A popular but controversial strategy to redistribute land to peasant farmers. Throughout the 20th century, various states redistributed land from large estates or granted land titles from vast public lands in order to reallocate resources to the poor and stimulate development. Agrarian reform occurred in various forms, from awarding individual plots or communally held land to creating state-run collective farms.

agribusiness   The practice of large-scale, often corporate farming in which business enterprises control closely integrated segments of food production, often extending from farm to grocery store.

agricultural density   The number of farmers per unit of arable land. This figure indicates the number of people who directly depend upon agriculture, and it is an important indicator of population pressure in places where rural subsistence dominates.

alluvial fan   A fan-shaped deposit of sediments dropped by a river or stream flowing out of a mountain range.

altitudinal zonation   The relationship between higher elevations, cooler temperatures, and changes in vegetation that result from the environmental lapse rate (averaging 3.5°F for every 1000 feet [6.5°C for every 1000 meters]). In Latin America, four general altitudinal zones exist: tierra caliente, tierra templada, tierra fria, and tierra helada.

anthropogenic   Caused by human activities.

anthropogenic landscapes   Landscapes that have been heavily transformed by humans.

apartheid   The policy of racial separateness that directed separate residential and work spaces for white, blacks, coloreds, and Indians in South Africa for nearly 50 years. It was abolished when the African National Congress came to power in 1994.

Arab League   A regional and economic organization of independent countries focused on the unity and development of the Arabic-speaking realm. Three non-Arabic-speaking countries, however, also belong to this organization: Somalia, Djibouti, and Comoros.

Arab Spring   A series of public protests, strikes, and rebellions in the Arab countries, often facilitated by social media, that have called for fundamental government and economic reforms.

areal differentiation   The geographic description and explanation of spatial differences on Earth's surface, including both physical as well as human patterns.

areal integration   The geographic description and explanation of how places, landscapes, and regions are connected, interact, and are integrated with each other.

Asia-Pacific Economic Cooperation Group (APEC)   An organization designed to encourage economic development in Southeast Asia and the Pacific Basin.

Association of Southeast Asian Nations (ASEAN)   An international organization linking together the 10 most important countries of Southeast Asia.

asylum laws   Protection for refugees who are victims of ethnic, religious, or political persecution in other parts of the world.

atoll   Low, sandy islands made from coral, often oriented around a central lagoon.

Austronesian   A language family that encompasses wide expanses of the Pacific, insular Southeast Asia, and Madagascar.

autonomous area   Minor political subunit created in the former Soviet Union and designed to recognize the special status of minority groups within existing republics. The term more generally can be used to designate any part of a country that has been given a certain degree of independence from the central government.

autonomous region   In China, province-level regions that have been given a degree of political and cultural autonomy due to the fact that they are inhabited in large part by minority groups.

Baikal–Amur Mainline (BAM) Railroad   Key central Siberian railroad connection completed in the Soviet era (1984), which links the Yenisey and Amur rivers and parallels the Trans-Siberian Railroad.

balkanization   Geopolitical process of fragmentation of larger states into smaller ones through independence of smaller regions and ethnic groups. The term takes its name from the geopolitical fabric of the Balkan region.

Berlin Conference   The 1884-1885 conference that set the framework for dividing Africa into European colonial territories. The boundaries created in Berlin satisfied European ambition but ignored indigenous cultural affiliations. Many of Africa's civil conflicts can be traced to ill-conceived territorial divisions crafted during this event.

biodiversity   The array of species, both flora and fauna, found in an ecosystem or bioregion.

biofuels   Energy sources derived from plants or animals. Throughout the developing world, wood, charcoal, and dung are primary energy sources for cooking and heating.

bioregion   A spatial unit or region of local plants and animals adapted to a specific environment such as a tropical savanna.

Bolsa Familia   A successful conditional cash transfer program developed in Brazil to address extreme poverty. Poor families who qualify receive a monthly check from the government provided they keep their children in school and vaccinated.

Bolsheviks   A faction within the Russian communist movement led by Lenin that successfully took control of the country in 1917.

boreal forest   Coniferous forest found in high-latitude or mountainous environments of the Northern Hemisphere.

brain drain   Migration of the best-educated people from developing countries to developed nations where economic opportunities are greater.

brain gain   The potential of return migrants to contribute to the social and economic development of their country of origin with the experiences they have gained abroad.

Brexit   The June 2016 referendum by United Kingdom voters to leave the European Union, which passed by a vote of 52 percent to 48 percent. Energizing the "Leave" vote were an array of anti-EU feelings ranging from unemployment allegedly caused by industry shifting to other EU countries, restrictive EU market regulations, "too many foreigners" because of EU free movement policies, and concerns about the EU's inability to deal with the increase of extra-legal immigration into Europe.

British East India Company   Private trade organization with its own army that acted as an arm of Britain in monopolizing trade in South Asia until 1857, when it was abolished and replaced by full governmental control.

Buddhism   A religion based on the teachings of Siddhartha Gautama, the Buddha, who was born in 563 BCE to an elite caste. He rejected the life of wealth and power and sought instead to attain enlightenment, or mystical union with the cosmos. He preached that the path to such "nirvana" was open to all, regardless of social position.

buffer state   A state located between two other states which provides a neutral zone. Mongolia, for example, emerged as an independent buffer state between China and Russia in 1911.

buffer zone   An array of nonaligned or friendly states that "buffer" a larger country from invasion. In Eurasia, maintaining a buffer zone has been a long-term policy of Russia (and also of the former Soviet Union) to protect its western borders from European invasion.

Bumiputra   The name given to native Malays (literally, "sons of the soil"), who are given preference for jobs and schooling by the Malaysian government.

Burakumin   A group of people in Japan who suffer from discrimination due to the employment of their ancestors in "polluting" jobs.

Burman   An ethnic group consisting of 68 percent of the population of Burma (Myanmar) who speak Burmese as their first language.

bustees   Sprawling squatter settlements in Kolkata and other Indian cities that appear in and around urban areas, providing temporary shelter for many urban migrants.

capital leakage   The gap between the gross receipts an industry (such as tourism) brings into a developing area and the amount of capital retained.

Caribbean Community and Common Market (CARICOM)   A regional trade organization established in 1972 that includes 20 member states mostly from the former English Caribbean colonies.

Caribbean diaspora   The recent large-scale movement of people from the Caribbean to major cities in North America and Europe.

caste system   Complex division of South Asian society into different hierarchically ranked hereditary groups. Most explicit in Hindu society, the caste system is also found in other South Asian cultures to a lesser degree.

centralized economic planning   An economic system in which the state sets production targets and controls the means of production.

chaebol   Huge Korean industrial conglomerates which moved from manufacturing inexpensive consumer goods to heavy industrial products and then to high-tech equipment.

chain migration   A pattern of migration in which people in a sending area become linked to a particular destination, such as Dominicans with New York City.

chernozem soils   A Russian term for dark, fertile soil, often associated with grassland settings in southern Russia and Ukraine.

China proper   The eastern half of the country of China where the Han Chinese form the dominant ethnic group. The vast majority of China's population is located in China.

choke point   A strategic setting where a narrow waterway is vulnerable to military blockade or disruption. A prominent example would be the Strait of Hormuz, which forms the entrance to the Persian Gulf.

choropleth map A thematic map in which areas are shaded or patterned in proportion to the measurement of the statistical variable being displayed on the map.

circular migration Temporary labor migration, in which an individual seeks short-term employment overseas, earns money, and then returns home.

clan A social unit that is typically smaller than a tribe or ethnic group but larger than a family, based on supposed descent from a common ancestor.

climate The average weather conditions for a place, usually based upon a minimum of 30 years of weather measurements.

climate change The measured change in climate from a previous state, contrasted with normal variability.

climographs Graphs of average annual temperature and precipitation data by month and season.

Closer Economic Relations (CER) An agreement signed in 1982 between Australia and New Zealand designed to eliminate all economic and trade barriers between the two countries.

Cold War The ideological struggle between the United States and the Soviet Union that occurred between 1946 and 1991.

Collective Security Treaty Organization (CSTO) A military alliance that links Russia to Kazakhstan, Kyrgyzstan, Tajikistan, Armenia, and Belarus. This organization was created in an attempt by Russia to enhance its power and provided that an attack on any CSTO member state be regarded as an attack on all others. The organization holds annual military exercises.

colonialism The formal and established rule over local peoples by a larger imperialist government.

coloured A racial category used throughout South Africa to define people of mixed European and African ancestry.

Columbian exchange An exchange of people, diseases, plants, and animals between the Americas (New World) and Europe/Africa (Old World) initiated by the arrival of Christopher Columbus in 1492.

communism A belief based on the writings of Karl Marx that promoted the overthrow of capitalism by the workers, the large-scale elimination of private property, state ownership and central planning of major sectors of the economy (both agricultural and industrial), and one-party authoritarian rule.

conflict diamonds Diamonds that are sold on the black market to fund armed conflict and civil war, especially in Sub-Saharan Africa.

Confucianism The philosophical system developed in China by Confucius in the 6th century BCE that stresses the creation of a proper social order.

connectivity The degree to which different locations are linked with one another through transportation and communication infrastructure.

continental climate The term describes inland climates with hot summers and cold winters. The interior portions of North America, Europe, and Asia have continental climates.

convergent plate boundaries Boundaries outlining plates that move toward one another, as opposed to divergent plate boundaries that surround plates that move apart.

copra Dried coconut meat.

core–periphery model A theoretical construct that describes how economic, political, and cultural power is spatially distributed between dominant core regions, and more marginal and dependent semi-peripheral and peripheral regions. According to this scheme, the United States, Canada, western Europe, and Japan constitute the global economic core, while other regions make up a less-developed economic periphery from which the core extracts resources.

Cossacks Highly mobile Slavic-speaking Christians of the southern Russian steppe who were pivotal in expanding Russian influence in 16th- and 17th-century Siberia.

counterinsurgency The suppression of a rebellion or insurgency by military and political means, which includes not just warfare but also winning the support of local peoples by improving community infrastructure (schools, roads, water supply, etc.).

creolization The blending of African, European, and even some Amerindian cultural elements into the unique sociocultural systems found in the Caribbean.

crony capitalism A system in which close friends of a political leader are either legally or illegally given business advantages in return for their political support.

cultural assimilation The process in which immigrants are culturally absorbed into the larger host society.

cultural homeland A culturally distinctive settlement in a well-defined geographic area, whose ethnicity has survived over time, stamping the landscape with an enduring personality.

cultural imperialism The active promotion of one cultural system over another, such as the implantation of a new language, school system, or bureaucracy. Historically, this has been primarily associated with European colonialism.

cultural landscape Primarily the visible and tangible expression of human settlement (house architecture, street patterns, field form, etc.), but also includes the intangible, value-laden aspects of a particular place and its association with a group of people.

cultural nationalism A process of protecting, either formally (with laws) or informally (with social values), the primacy of a specific cultural system against influences from another culture.

cultural syncretism The blending of two or more cultures, which then produces a synergistic third culture or specific behavior that exhibits traits from all cultural parents.

culture Learned and shared behavior by a group of people empowering them with a distinct "way of life"; it includes both material (technology, tools, etc.) and non-material (speech, religion, values, etc.) components.

culture hearth An area of historical cultural innovation that gave rise to a particular culture region.

cyclones Large storms, marked by well-defined air circulation around a low-pressure center. Tropical cyclones are typically called hurricanes in the Atlantic Ocean and typhoons in the western Pacific.

Cyrillic alphabet A writing system based on the Greek alphabet and used by Slavic languages heavily influenced by the Eastern Orthodox Church. Attributed to the missionary work of St. Cyril in the 9th century.

dacha A Russian country cottage used especially in the summer.

Dalai Lama The leader of the Tibetan Buddhist faith. The current Dalai Lama is an important advocate for Tibet rights.

dalits The so-called untouchable population of India; people often considered socially polluting because of their historical connections with occupations classified as unclean, such as leatherworking and latrine-cleaning.

decolonialization The process of a former colony's gaining (or regaining) independence over its territory and establishing (or reestablishing) an independent government.

demographic transition model A five-stage scheme that explains different rates of population growth over time through differing birth and death rates.

desertification The spread of desert conditions into semiarid areas owing to improper management of the land.

desiccation The process of extreme drying; in desert areas, such processes can lead to the disappearance of lakes when the rivers that previously fed them are diverted for agriculture or other purposes.

devolution The breaking apart or separation within a political unit such as a nation-state.

diaspora The scattering of a particular group of people over a vast geographical area. Originally, the term referred to the migration of Jews out of their original homeland, but now it has been generalized to refer to any ethnic dispersion.

divergent plate boundary A place where two tectonic plates move away from each other. Here, magma often flows from Earth's interior, producing volcanoes on the surface such as in Iceland. In other divergent zones large trenches or rift valleys are created.

diversity The differences between cultures, ethnicities, economies, landscapes, and regions.

dollarization An economic strategy in which a country adopts the U.S. dollar as its official currency. A country can be partially dollarized, using U.S. dollars alongside its national currency, or fully dollarized, when the U.S. dollar becomes the only medium of exchange and a country gives up its own national currency. Panama fully dollarized in 1904; more recently, Ecuador became fully dollarized in 2000, followed by El Salvador in 2001.

domestication The purposeful selection and breeding of wild plants and animals for cultural purposes.

domino theory A U.S. geopolitical policy of the 1970s that stemmed from the assumption that if Vietnam fell to the communists, the rest of Southeast Asia would soon follow.

dynastic succession A system of succession in which power flows from father to son. Although officially a communist republic, North Korea has embraced dynastic succession. When the leader Kim Jong-il died in late 2011, he was immediately replaced by his untested 28-year-old son, Kim Jong-un.

Eastern Orthodox Christianity A loose confederation of self-governing churches in eastern Europe and Russia that are historically linked to Byzantine traditions and to the primacy of the patriarch of Constantinople (Istanbul).

Economic and Monetary Union (EMU) An organization of 11 EU members states formed in 1999 that then agreed, in 2002, to use new euro coins and bills that replaced national currencies in the EMU countries. Today, 19 of the EU's 28 member states use the euro, with some refusing to join the Eurozone, while others await resolution of ongoing tensions between rich and poor euro countries.

Economic Community of West African States (ECOWAS) ECOWAS is an intergovernmental organization that promotes economic integration and security among its 15 member states in West Africa. It was founded in 1975.

edge city Suburban node of activity that features a mix of peripheral retailing, industrial parks, residential land uses, office complexes, and entertainment facilities.

El Niño An abnormally warm current that appears periodically in the eastern Pacific Ocean and usually influences storminess along the western coasts of the Americas. During an El Niño event, which can last several years, torrential rains often bring devastating floods to the Pacific coasts of North, Central, and South America. El Niño events are often associated with droughts in Australia and Southeast Asia.

enclaves Portions of a country or of groups of people completely surrounded by another country or a different group of people. For example, the Omani territory of Madha, is an enclave within the United Arab Emirates (UAE), as it is completely surrounded by other portions of the UAE.

entrepôt A city and port that specializes in the trans-shipment and warehousing of goods.

environmental lapse rate The decline in temperature as one ascends higher in the atmosphere. On average, the temperature declines 3.5°F for every 1000 feet ascended, or 6.5°C for every 1000 meters.

ethnic religions A religion closely identified with a specific ethnic or tribal group, often to the point of assuming the role of the major defining characteristic of that group. Normally, ethnic religions do not actively seek new converts.

ethnicity   A shared cultural identity held by a group of people with a common background or history, often as a minority group within a larger society.

Eurasian Economic Union (EEU)   A customs union (paralleling the European Union [EU]) designed to encourage trade as well as closer political ties between member states. Formed in 2015, the EEU contains five member states (Russia, Belarus, Kazakhstan, Armenia, and Kyrgyzstan).

European Union (EU)   The current association of 28 European countries that are joined together in an agenda of economic, political, and cultural integration.

Eurozone   The common monetary policy and currency of the European Union; those countries of Europe using the euro as its currency and who are members of the EU's common monetary system, contrasted to those countries having a national currency and monetary system. France is an example of the former, and the United Kingdom of the latter.

exclave   A portion of a country's territory that lies outside its contiguous land area.

Exclusive Economic Zone (EEZ)   A sea zone prescribed by the United Nations Convention on the Law of the Sea over which a state has special rights regarding the exploration and use of marine resources, including energy production from water and wind. It stretches from the baseline out to 200 nautical miles from its coast.

exotic river   A river that issues from a humid area and flows into a dry area otherwise lacking streams.

exotic species   Nonnative plants and animals, generally used for species that cause problems for native species. Exotic species are particularly troublesome in Australia and the Pacific.

family-friendly policies   Established in places such as Germany, France, and Scandinavia, these are policies that address concerns about population loss by offering full pay maternity and paternity leaves, guarantees of continued employment once these leaves conclude, extensive child-care facilities for working parents, outright cash subsidies for having children, and free or low-cost public education and job training for their offspring.

federal state   A country in which the major territorial subdivisions have a significant degree of political autonomy, such as the United States, India, or Mexico; federal states are contrasted to unitary states, in which the central governments set most policies.

Fertile Crescent   An ecologically diverse zone of lands in Southwest Asia that extends from Lebanon eastward to Iraq and that is often associated with early forms of agricultural domestication.

fjord   Flooded, glacially carved valley. In Europe, fjords are found primarily along Norway's western coast.

food deserts   Places where people do not have ready access to supermarkets and fresh, healthy, and affordable food.

food insecurity   When people lack physical or economic access to sufficient, safe, and nutritious food to meet their dietary needs for extended periods of time.

formal region   A geographic concept used to describe an area where a static and specific trait (such as a language or a climate) has been mapped and described. A geographic concept of areal or spatial similarity, large or small. A formal region contrasts with a functional region.

forward capital   A capital city deliberately positioned near a contested territory, signifying the state's interest and presence in this zone of conflict.

fossil water   Water supplies that were stored underground during wetter climatic periods.

fracking   See *hydraulic fracturing*.

free trade zone (FTZ)   A duty-free and tax-exempt industrial park created to attract foreign corporations and create industrial jobs.

functional region   A geographic concept used to describe the spatial extent dominated by a specific activity.

The circulation area of a newspaper is an example, as is the trade area of a large city.

gender   The social and cultural expressions of male- and femaleness, which contrasts with sex, which is the biological distinction between male and female.

gender gap   A term often used to describe gender differences in salary, working conditions, or political power.

gender inequality   Discrimination against women that takes many forms, from not allowing them to vote to discouraging school attendance.

gender roles   How female and male behavior differs in a specific cultural context.

gentrification   A process of urban revitalization in which higher income residents displace lower-income residents in central-city neighborhoods.

Geographic Information systems   (GIS) A computerized mapping and information system that is able to analyze vast amounts of data through both in its totality but also through layers of specific kinds of information, such as microclimates, hydrology, vegetation, or land-use zoning regulations.

geography   the spatial science that describes and explains physical and cultural phenomena on Earth's surface.

geomancy   The traditional Chinese and Korean practice of designing buildings in accordance with the principles of cosmic harmony and discord that supposedly course through the local topography.

geopolitics   The relationship between politics and space and territory.

geothermal   Subsurface heat that is tapped for electricity production by drilling wells to release the intense heat deep within the Earth, felt at the surface in geysers, hot springs, and volcanoes.

glasnost   A policy of greater political openness in the Soviet Union initiated during the 1980s by Soviet president Mikhail Gorbachev.

global positioning system (GPS)   Originally used to describe a very accurate satellite-based location system, but now also used in a general sense to describe smartphone location systems that may use cell phone towers as a substitute for satellites.

globalization   The increasing interconnectedness of people and places throughout the world through converging processes of economic, political, and cultural change.

glocalization   (which combines globalization with locale) is the process of modifying an introduced globalized product or service to accommodate local tastes or cultural practices.

Golden Triangle   An area of northern Thailand, Burma, and Laos that is known as a major source region for opium and heroin and is plugged into the global drug trade.

Gondwana   The name given for the southerly supercontinent that existed over 200 million years ago that included Africa, South America, Antarctica, Australia, the Arabian Peninsula, and the Indian subcontinent.

graphic (linear) scale   A ruler-like symbol on a map that translates the map's cartographic scale into visual terms.

grassification   The conversion of tropical forest into pasture for cattle ranching. Typically, this process involves introducing species of grasses and cattle, mostly from Africa.

Great Escarpment   A landform that rims southern Africa from Angola to South Africa. It forms where the narrow coastal plains meet the elevated plateaus in an abrupt break in elevation.

Great Rift Valley   Caused by the breakup of Gondwana, the ancient megacontinent that included Africa, South America, Antarctica, Australia, and the Arabian Peninsula, this term refers to the continental uplifts that left much of the area with vast plateaus—where the highest areas, found on the eastern edge of the African continent, forms a complex upland area of lakes, volcanoes, and deep valleys. In contrast, lowlands prevail in West Africa.

Greater Antilles   The four large Caribbean islands of Cuba, Jamaica, Hispaniola, and Puerto Rico.

Greater Arab Free Trade Area (GAFTA)   An organization, established in 2005 by 17 Arab League members, designed to eliminate all intraregional trade barriers and spur economic cooperation.

Green Revolution   A term applied to the development of new techniques, starting in the 1960s, that have transformed agriculture in India and many other developing countries; Green Revolution techniques usually involve the use of hybrid seeds that provide higher yields than native seeds when combined with high inputs of chemical fertilizer, irrigation, and pesticides.

greenhouse effect   The natural process of lower atmosphere heating that results from the trapping of incoming and reradiated solar energy by water moisture, clouds, and other atmospheric gases.

greenhouse gases (GHGs)   The gasses that create an envelope around the earth trapping the warmth from incoming and outgoing solar radiation. These include such natural constituents as water vapor, carbon dioxide CO_2, methane CH_4 and ozone O_3. Although the composition of these natural GHGs has varied somewhat over long periods of geologic time, it has been relatively stable since the last ice age ended 20,000 years ago.

gross domestic product (GDP)/gross national product (GNP)   GDP is the total value of goods and services produced within a given country (or other geographical unit) in a single year. GNP is a somewhat broader measure that includes the inflow of money from other countries in the form of the repatriation of profits and other returns on investments, as well as the outflow to other countries for the same purposes.

gross national income (GNI)   The value of all final goods and services produced within a country's borders (gross domestic product, or GDP), plus the net income from abroad (formerly referred to as gross national product, or GNP).

gross national income (GNI) per capita   The figure that results from dividing a country's GNI by the total population.

Group of Eight (G8)   A collection of powerful countries that confer regularly on key global economic and political issues. It includes the United States, Canada, Japan, Great Britain, Germany, France, Italy, and Russia.

Gulag Archipelago   A collection of Soviet-era labor camps for political prisoners, made famous by writer Aleksandr Solzhenitsyn.

Hajj   An Islamic religious pilgrimage to Makkah. One of the five essential pillars of the Muslim creed to be undertaken once in life, if an individual is physically and financially able to do it.

hallyu   South Korea's popular culture industry which continues to thrive, as Korean music, movies, and television shows have become popular around the world; a phenomenon also known as "Korean wave."

haoles   Light-skinned Europeans or U.S. citizens in the Hawaiian Islands.

high islands   Larger, more elevated islands, often focused around recent volcanic activity.

Hindu nationalism   A contemporary "fundamentalist" religious and political movement that promotes Hindu values as the essential—and exclusive—fabric of Indian society. As a political movement, it generally has less tolerance of India's large Muslim minority than other political movements.

Hinduism   A complicated faith that incorporates diverse forms of worship and lacks any standard system of beliefs. Certain deities are recognized by all believers, as is the notion that these various gods are all manifestations of a single divine entity. All Hindus, moreover, share a set of epic stories, usually written in the sacred language of Sanskrit. One of its hallmarks is a belief in the transmigration of souls from being to being through reincarnation, wherein one's actions in the physical world influences the course of these future lives.

homelands Nominally independent ethnic territories created for blacks under the grand apartheid scheme. Homelands were on marginal land, overcrowded, and poorly serviced. In the post-apartheid era, they were eliminated.

Horn of Africa The northeastern corner of Sub-Saharan Africa that includes the states of Somalia, Ethiopia, Eritrea, and Djibouti. Since the 1980s, drought, famine, ethnic conflict, and political turmoil have undermined development efforts in this area.

hukou In China, a hukou is a record in a government system of household registration that determines where citizens are allowed to live. Poor people moving from rural areas to cities are often unable to obtain a hukou that would officially allow them to do so. They, therefore, live as "undocumented migrants" in their own country and face difficult access to governmental services, including education for their children.

Human Development Index (HDI) For the past three decades, the United Nations has tracked social development in the world's countries through the Human Development Index (HDI), which combines data on life expectancy, literacy, educational attainment, gender equity, and income.

human geography The form of geography that concentrates on the spatial analysis of economic, political, social, and cultural systems.

human trafficking The illicit trade in human, usually for the purpose of sexual slavery, commercial sexual exploitation, or forced labor.

hurricanes Tropical storms in the Atlantic Basin with abnormally low-pressure centers and with sustaining winds of 74 mph or higher. Each year during hurricane season (July–October), a half dozen to a dozen hurricanes form in the warm waters of the Atlantic and Caribbean, bringing destructive winds and heavy rain.

hydraulic fracturing (fracking) A process for extracting oil from shale rock that involves injecting a fluid mixture into the ground that breaks apart (or fractures) the rock layer, thus making it easier to pump out the oil.

hydropolitics The interplay of water resource issues and politics. Countries often disagree about water resources, especially in regard to major rivers that cross international boundaries.

ideographic writing A writing system in which each symbol represents not a sound but rather a concept.

indentured labor A system of unfree labor in which a worker is bound by a contract to serve a particular employer for a fixed period of time. In the Caribbean, many immigrants came from South Asia as indentured laborers.

Indian diaspora The historical and contemporary movement of people from India to other countries in search of better opportunities. This process has led to large Indian populations in South Africa, the Caribbean, and the Pacific islands, along with Western Europe and North America.

Indian subcontinent The large Eurasian peninsula that extends south of the Himalayan Mountains and that encompasses most of South Asia.

Indo-European language family The world's largest group of languages that are descended from a common ancestral language. Most languages of Europe, northern and central South Asia, and Iran belong to this language family.

Industrial Revolution The period of time in the late 18th and early 19th centuries when European factories first changed from using animate power (human and animals) to inanimate power (water and coal) to power machines.

informal sector A much-debated concept that presupposes a dual economic system consisting of formal and informal sectors. The informal sector includes self-employed, low-wage jobs that are usually unregulated and untaxed. Street vending, shoe shining, artisan manufacturing, and even self-built housing are considered part of the informal sector. Some scholars include illegal activities such as drug smuggling and prostitution in the informal economy.

insolation Incoming solar energy that enters the atmosphere adjacent to Earth.

insurgency A violent political rebellion or uprising.

intercropping The agricultural practice in which subsistence foods such as sweet potatoes, taro, coconut palms, bananas, and other garden crops are often found in the same field.

internally displaced persons Groups and individuals who flee an area due to conflict or famine but still remain in their country of origin. These populations often live in refugee-like conditions but are harder to assist because they technically do not qualify as refugees.

Iron Curtain A term coined by British leader Winston Churchill during the Cold War that defined the western border of Soviet power in Europe. The notorious Berlin Wall was a concrete manifestation of the Iron Curtain.

irredentism A state or national policy of reclaiming lost lands or those inhabited by people of the same ethnicity in another nation-state.

ISIL (Islamic State of Iraq and the Levant) The Sunni extremist organization, also known as ISIS or as IS, which has brought about especially violent instability in the Syria and Iraq area in its attempts to create a new religious state (a *caliphate*).

Islamic fundamentalism A movement within both the Shiite and Sunni Muslim traditions to return to a more conservative, religious-based society and state. Often associated with a rejection of Western culture and with a political aim to merge civic and religious authority.

Islamism A political movement within the religion of Islam that challenges the encroachment of global popular culture and blames colonial, imperial, and Western elements for many of the region's problems. Adherents of Islamism advocate merging civil and religious authority.

Jainism A religious group in South Asia that emerged as a protest against orthodox Hinduism in the 6th century BCE. Jains are noted for their practice of nonviolence, which prohibits them from taking the life of any animal.

juche A unique official ideology also known as "self-reliance" with which North Korean leaders supplemented Marxism. It demands absolute loyalty to North Korea's political leaders. It also involves intense nationalism based on the idea of Korean racial purity.

kanakas Melanesian workers imported to Australia, historically concentrated along Queensland's "sugar coast." The term "kanak" is also used to refer to the indigenous Melanesian inhabitants of New Caledonia.

Khmer Rouge Literally, "Red (or communist) Cambodians." The left-wing insurgent group led by French-educated Marxists that ruled Cambodia from 1975 to 1979, during which time it engaged in genocidal acts against the Cambodian people.

kleptocracy A state where corruption is so institutionalized that politicians and bureaucrats siphon off a large percentage of a country's wealth for personal gain.

Kyoto Protocol Following the minimal success of the Rio Convention signers to reduce their atmospheric emissions, a more formal international agreement came from a 1997 meeting in Kyoto, Japan at which 30 Western industrialized countries agreed to cut their emissions to 1990 levels by 2012. The Kyoto Protocol had the force of international law, with penalties for those countries not reaching their emission reduction targets.

laissez-faire An economic system in which the state has minimal involvement and in which market forces largely guide economic activity.

latifundia A large estate or landholding in Latin America.

Latin alphabet The alphabet devised by the ancient Roman that is used today for writing most European languages, including English. It is also used for a number of non-European languages, such as Indonesian and Vietnamese. The Latin alphabet is one of two distinct alphabets which complicate the geography of Slavic languages. In countries with a strong Roman Catholic heritage, such as Poland and the Czech Republic, this language is used. In contrast, countries with close ties to the Orthodox Church—Bulgaria, Montenegro, Macedonia, parts of Bosnia–Herzegovina, and Serbia—use the Greek-derived Cyrillic alphabet.

latitude Lines, often called parallels, that run east–west around the globe and are used to locate places north and south of the equator (0 degrees latitude).

Lesser Antilles The arc of small Caribbean islands from St. Maarten to Trinidad.

Levant The eastern Mediterranean region.

life expectancy The statistical average for how long a person within a specific data category is expected to live.

lingua franca An agreed-upon common language to facilitate communication on specific topics such as international business, politics, sports, or entertainment.

linguistic nationalism The promotion of one language over others that is, in turn, linked to shared notions of political identity. In India, some nationalists promote Hindi as the unifying language of the country, yet this is resisted by many non-Hindi-speaking peoples.

lithosphere Earth's outer layer, consisting of large geologic platforms, or plates, that move very slowly across its surface. Driving the movement of these plates is a heat exchange deep within Earth.

location factors The various influences that explain why an economic activity takes place where it does.

locavore movement The trend which includes sustainable agriculture, where organic farming principles, limited use of chemicals, and an integrated plan of crop and livestock management combine to offer both producers and consumers environmentally friendly alternatives to large-scale commercial farming. It has encouraged more community gardens and farmers markets, making it easier for consumers to eat locally grown food.

loess A fine, wind-deposited sediment that makes fertile soil but is very vulnerable to water erosion.

longitude Lines, called meridians, that run from the North Pole (90 degrees north latitude) to the South Pole (90 degrees south latitude). Longitude values locate places east or west of the prime meridian, which is located at 0 degrees longitude at the Royal Naval Observatory in Greenwich, England (just east of London).

low island Flat, low-lying islands formed by coral reefs, and contrasting with high islands that were formed from volcanic eruptions.

Maghreb A region in northwestern Africa, including portions of Morocco, Algeria, and Tunisia.

maharaja Historical term for Hindu royalty, usually a king or prince, who ruled specific areas of South Asia before independence, but who was usually subject to overrule by British colonial advisers.

mallee A tough and scrubby eucalyptus woodland of limited economic value that is common across portions of interior Australia.

Mandarins A member of the high-level bureaucracy of Imperial China (before 1911). Mandarin Chinese based on the speech of these officials is the official spoken language of the country and is the native tongue of the vast majority of people living in north, central, and southwestern China.

Maoism The version of Marxism advocated by the Chinese revolutionary and political leader Mao Zedong in which the peasantry is regarded as the main revolutionary class. In east-central India, a Maoist insurgency has been fighting against the government for decades.

Maori Indigenous Polynesian people of New Zealand.

map projection The cartographic and mathematical solution to translating the surface of a rounded globe (usually Earth) to a flat surface (usually a piece of paper).

map scale The relationship between distances on a mapped object such as Earth and depiction of that space on a map. Large scale maps cover small areas in great detail, where small scale maps depict less detail but over large areas.

maquiladoras Assembly plants on or near the Mexican border built and owned by foreign companies. Most of their products are exported to the United States.

marine west coast climate A major climate type that is moderate and moist and modified by oceanic influences. Marine west coast climates are found along the west coast of Europe, in the Pacific Northwest in North America, in southern Chile, in Tasmania, and along the western coast of New Zealand's South Island.

maritime climate Climate moderated by proximity to oceans or large seas. It is usually cool, cloudy, and wet, and lacks the temperature extremes of continental climates.

maroons Runaway slaves who established communities rich in African traditions throughout the Caribbean and Brazil.

marsupial A class of mammals found primarily in the Southern Hemisphere with the distinctive characteristic of carrying their young in a pouch. Kangaroos are perhaps the best-known marsupial, with wallabies, koalas, wombats, and the Tasmanian Devil also found in Oceania.

Marxism A term referring to the political philosophy developed by Karl Marx in the 1800s and based on the ideas of communism.

medina The original urban core of a traditional Islamic city.

Mediterranean climate A mild, dry-summer climate found in the mid-latitude belt on the west sides of continents. Most of the Mediterranean Basin has a Mediterranean climate, as do California, central Chile, the Cape region of South Africa, and the southwestern corner of Australia.

megacities Urban conglomerations of more than 10 million people.

Megalopolis A large urban region formed as multiple cities grow and merge with one another. The term is often applied to the string of cities in eastern North America that includes Washington, DC; Baltimore; Philadelphia; New York City; and Boston.

Meiji Restoration The process that initiated the political transformation of Japan into a modern state in 1868. Although the Meiji Restoration supposedly restored the emperor of the country to power, it did not in actuality give him real governing authority.

Melanesia Pacific Ocean region that includes the culturally complex, generally darker-skinned peoples of New Guinea, the Solomon Islands, Vanuatu, New Caledonia, and Fiji.

mestizo A person of mixed European and Indian ancestry.

microfinance A program to reduce poverty based on providing loans to small-scale business. Low-interest microfinancing, for example, is provided by Grameen Bank to many poor women in Bangladesh, allowing the emergence of some vibrant small-scale enterprises.

Micronesia Pacific Ocean region that includes the culturally diverse, generally small islands north of Melanesia. Includes the Mariana Islands, Marshall Islands, and Federated States of Micronesia.

microstates Usually independent states that are small in both area and population, although no formal area and population limits have been accepted. Generally acknowledged microstates include Liechtenstein, Monaco, San Marino, Tuvalu, and Nauru.

mikrorayon Large, state-constructed urban housing project built during the Soviet period in the 1970s and 1980s.

minifundia Usually used in the context of Latin America, a small landholding farmed by peasants or tenants who produce food for subsistence and the market.

monocrop production Agriculture based upon a single crop, often for export.

monotheism A religious belief in a single God.

Monroe Doctrine A proclamation issued by U.S. President James Monroe in 1823 that the United States would not tolerate European military action in the Western Hemisphere. Focused on the Caribbean as a strategic area, the doctrine was repeatedly invoked to justify U.S. political and military intervention in the region.

monsoon The seasonal pattern of changes in winds, heat, and moisture in South Asia and other regions of the world that is a product of larger meteorological forces of land and water heating, the resultant pressure gradients, and jet-stream dynamics. The monsoon produces distinct wet and dry seasons.

monsoon winds Near Earth's surface, these are continent-scale winds that flow from high to low pressure areas in places such as Asia and North America; summer monsoons bring rainfall to the dry areas of interior India and the Southwest United States.

Mughal Empire (also spelled Mogul) The Muslim-dominated state that covered most of South Asia from the early 16th to late 17th centuries. The last vestiges of the Mughal dynasty were dissolved by the British following the rebellion of 1857.

nation-state A relatively homogeneous cultural group (a nation) with its own political territory (the state). While useful conceptually, the reality of today's globalized world is that there are very few countries that fit this simplistic definition because of the influx of migrants and/or the presence of minority ethnic groups.

Native Title Bill Australian legislation signed in 1993 that provides Aborigines with enhanced legal rights over land and resources within the country.

neocolonialism Economic and political strategies by which powerful states indirectly (and sometimes directly) extend their influence over other, weaker states.

neotropics Tropical ecosystems of the Americas that evolved in relative isolation and support diverse and unique flora and fauna.

net migration rate A statistic that depicts whether more people are entering or leaving a country through migration per year. It is usually expressed as a positive or negative number per 1,000 people in the population.

new urbanism An urban design movement stressing higher density, mixed-use, pedestrian-scaled neighborhoods where residents might be able to walk to work, school, and local entertainment.

nonrenewable energy Those energy sources such as oil and coal with finite reserves.

nonmetropolitan growth A pattern of migration in which people leave large cities and suburbs and move to smaller towns and rural areas.

North American Free Trade Agreement (NAFTA) An agreement made in 1994 among Canada, the United States, and Mexico that established a 15-year plan for reducing all barriers to trade among the three countries.

North Atlantic Treaty Organization (NATO) Initially NATO was a group of North Atlantic and European allies who came together in 1949 to counter the Soviet threat to western Europe. NATO today is a member alliance of 28 member states in North America, Europe, and Southwest Asia (Turkey) pledged to common defense in the event of war.

northern sea route An ice-free channel along Siberia's northern coast that will grow in importance given sustained global warming.

novel ecosystems Ecosystems completely new to Earth that are a result of human activity, urbanization, and industrialization.

Oceania A major world subregion that is usually considered to include New Zealand and the major island regions of Melanesia, Micronesia, and Polynesia.

offshore banking Financial services offered by certain European dependencies and microstates, often located on islands, that are typically confidential and tax exempt. As part of a global financial system, offshore banks have developed a unique niche, offering their services to individual and corporate clients for set fees. The Bahamas and the Cayman Islands are leaders in this sector.

oligarchs A small group of wealthy, very private businessmen who control (along with organized crime) important aspects of the Russian economy.

"One Belt, One Road" Also known as the "new Silk Road" initiative, this is a program that China is using to attempt to reorient global trade networks so that they focus on the Eurasian landmass rather than the Atlantic and Pacific ocean basins, a move that could have major geopolitical ramifications.

"one country, two systems" When Britain returned Hong Kong to China in the 1980s, this was the name of the model under which China promised that Hong Kong would retain its capitalist economic system and its partially democratic political system for 50 years. Civil liberties not enjoyed in China itself were to remain protected in Hong Kong.

Organization of American States (OAS) Founded in 1948 and headquartered in Washington, DC, the organization advocates hemispheric cooperation and dialogue. Most states in the Americas belong.

Organization of the Petroleum Exporting Countries (OPEC) An international organization, formed in 1960, that now includes 13 oil-producing countries that attempts to influence global prices and supplies of oil. Algeria, Angola, Ecuador, Gabon, Iran, Iraq, Kuwait, Libya, Nigeria, Qatar, Saudi Arabia, the United Arab Emirates, and Venezuela are members.

orographic effect The influence of mountains on weather and climate, usually referring to the increase of precipitation on the windward side of mountains, and a drier zone (or rain shadow) on the leeward or downwind side of the mountain.

orographic rainfall Enhanced precipitation over uplands that results from lifting (and cooling) of air masses as they are forced over mountains.

Ottoman Empire A large, Turkish-based empire (named for Osman, one of its founders) that dominated large portions of southeastern Europe, North Africa, and Southwest Asia between the 16th and 19th centuries.

Outback Australia's large, generally dry, and thinly settled interior.

outsourcing A business practice that transfers portions of a company's production and service activities to lower-cost settings, often located overseas.

Pacific Alliance A trade-oriented alliance formed in 2011 which includes Mexico, Colombia, Peru, and Chile with Costa Rica and Panama in the process of joining.

Palestinian Authority (PA) A quasi-governmental body that represents Palestinian interests in the West Bank and Gaza.

Pan-African Movement Founded in 1900 by U.S. intellectuals W. E. B. Du Bois and Marcus Garvey, this movement's slogan was "Africa for Africans," and its influence extended across the Atlantic.

Pangaea A supercontinent, which geologic evidence suggests existed some 250 million years ago, consisting of all the world's land masses tightly consolidated and centered on present-day Africa.

Paris Agreement A new international greenhouse gas reduction agreement approved at the Climate Change Conference held in Paris, France in December of 2015 which now serves as the world's blueprint for addressing the challenges of climate change.

pastoral nomadism A traditional subsistence agricultural system in which practitioners depend on the seasonal movements of livestock within marginal natural environments.

pastoralists Nomadic and sedentary peoples who rely upon livestock (especially cattle, camels, sheep, and goats) for their sustenance and livelihood.

peat Organic soils formed under wetland conditions of perennial saturation. When such areas are drained, organic peat soils can burn, often generating severe air pollution and adding large quantities of carbon dioxide into the atmosphere.

perestroika A program of partially implemented, planned economic reforms (or restructuring) undertaken during the Gorbachev years in the Soviet Union designed to make the Soviet economy more efficient and responsive to consumer needs.

permafrost A cold-climate condition in which the ground remains permanently frozen.

physical geography The form of geography that examines climate, landforms, soils, vegetation, and hydrology.

physiological density A population statistic that relates the number of people in a country to the amount of arable land.

Pidgin English (Pijin) A number of languages that are based on English but that also incorporate elements of other languages. Varieties of Pidgin English are often used to foster trade and basic communication between different culture groups.

piracy A practice of robbing and attacking ships at sea. Piracy is a maritime issue that China, the United States, and ASEAN countries agree needs to be suppressed, especially in Southeast Asia (which surpassed Somalia as the most pirate-afflicted part of the world).

place As a geographic concept, this is not just the characteristics of a location but also encompasses the meaning that people give to such areas, as in the sense of place. This diverse fabric of *placefulness* is of great interest to geographers because it tells us much about the human condition throughout the world. Places can tell us how humans interact with nature and how they interact among themselves; where there are tensions and where there is peace; where people are rich and where they are poor.

plantation America A cultural region that extends from midway up the coast of Brazil, through the Guianas and the Caribbean, and into the southeastern United States. In this coastal zone, European-owned plantations, worked by African laborers, produced agricultural products for export.

plate tectonics The theory that explains the gradual movement of large geological platforms (or plates) along Earth's surface.

podzol soils A Russian term for an acidic soil of limited fertility, typically found in northern forest environments.

polar jet stream Powerful atmospheric rivers of eastward-moving air that affect storms and pressure systems in both the Northern and the Southern hemispheres. These jets are products of Earth's rotation and global temperature differences. The two polar jets (north and south) are the strongest and the most variable, flowing 23,000–39,000 feet (7–12 km) above the surface at speeds reaching 200 miles per hour (322 km/h).

pollution exporting The process of exporting industrial pollution and other waste material to other countries. Pollution exporting can be direct, as when waste is simply shipped abroad for disposal, or indirect, as when highly polluting factories are constructed abroad.

Polynesia Pacific Ocean region, broadly unified by language and cultural traditions, that includes the Hawaiian Islands, Marquesas Islands, Society Islands, Tuamotu Archipelago, Cook Islands, American Samoa, Samoa, and Tonga.

population density The number of people per areal unit, usually measured in people per square kilometer or per square mile.

population pyramid A graph representing the structure of a population, including the percentage of young and old. The percentages of all different age groups are plotted along a vertical axis that divides the population into male and female. In a fast growing population with a large percentage of young people and a small percentage of elderly, the graph will have a wide base and a narrow tip, giving it a pyramidal shape.

postindustrial economy An economy in which the tertiary and quaternary sectors dominate employment and expansion.

prairie An extensive area of grassland in North America. In the more humid eastern portions, grasses are usually taller than in the drier western areas, which are in the rain shadow of the Rocky Mountain range.

primate cities Massive urban settlements that dominate all other cities in a given country. A primate city is usually the capital of the country in which it is located.

prime meridian Zero degrees longitude, from which locations east and west are measured in a system of latitude and longitude. Currently, the most used Prime Meridian is that established in 1851 at the Naval Observatory in Greenwich, England (in southeastern London). Before that other countries and cultures established their own prime meridians upon which to base their maps and navigation systems.

pristine myth The idea put forward by geographer William Denevan that the Americas were more ecologically disturbed in 1492 than they were in the mid-1700s due to the demographic collapse of Amerindian peoples.

protectorate During the period of Western global imperialism, a state or other political entity that remained autonomous but sacrificed its foreign affairs to an imperial power in exchange for "protection" from other imperial powers.

proven reserves The amount of a non-renewable energy source (oil, coal, and gas) still in the ground that is feasible to exploit under current market conditions.

purchasing power parity (PPP) A method of comparing the economic output of different countries or regions that that takes into account the strength or weakness of local currencies. Poorer countries typically have higher levels of economic production under PPP terms than they do under nominal terms because a single dollar can buy more goods and services there than the same dollar could buy in a wealthy country.

Quran (also spelled Koran) The holy book of Islam, which Muslims believe to be the divine revelations received by the prophet Muhammad.

rain shadow A drier area of precipitation, usually on the leeward or downwind side of a mountain range, that receives less rain and snowfall than the windward or upwind side. A rain shadow is caused by the warming of air as it descends down a mountain range; this warming increases the ability of an air mass to hold moisture.

Ramayana One of the two main epic poems of the Hindu religion, the Ramayana is also commonly performed in the shadow puppet theaters of the predominately Muslim island of Java.

rate of natural increase (RNI) The standard statistic used to express natural population growth per year for a country, region, or the world based upon the difference between birth and death rates. RNI does not consider population change from migration. Though most often a positive figure (such as 1.7 percent), RNI can also be expressed either as zero or even as a negative number for no-growth countries.

refugee A person who flees his or her country because of a well-founded fear of persecution based on race, ethnicity, religion, ideology, or political affiliation.

region A geographic concept of areal or spatial similarity, large or small.

regional geography A basic division in geography that focuses on a specific topic or theme as opposed to analyzing a specific place or a region.

regulatory lakes A term applied to a series of lakes in the middle Yangtze Valley of China. Regulatory lakes take excess water from the river during flood periods and supply water to the river during dry periods.

remittances Monies sent by immigrants working abroad to family members and communities in countries of origin. For many countries in the developing world, remittances often amount to billions of dollars each year. For small countries, remittances can equal 5 to 10 percent of a country's gross domestic product.

remote sensing A method of digitally photographing Earth's surface from satellites or high altitude aircraft so that the information captured can be manipulated by computers to translate information into certain electromagnetic bandwidths, which, in turn, emphasizes certain features and patterns on Earth's surface.

renewable energy Energy sources, such as solar, wind, and hydro, that are replenished by nature at a faster rate than they are used or consumed and generally have a low environmental impact.

replacement rate The total fertility rate at which women give birth to enough babies to sustain population levels. In countries with low levels of childhood mortality, the replacement rate is around 2.1 children per woman.

reradiate As land and water surfaces are warmed by the incoming short-wave solar energy, which has passed through the atmosphere and been absorbed, the heat is sent back into the lower atmosphere as infrared, long-wave energy. This reradiating energy, in turn, is absorbed by water vapor and other atmospheric gases such as carbon dioxide CO_2 creating the envelope of warmth that makes life possible on our planet.

rift valleys The deep depression formed where tectonic plates diverge. An example is the Red Sea between northern Africa and Saudi Arabia; movement of the African and Arabian plates away from one another created this body of water.

rimland The mainland coastal zone of the Caribbean, beginning with Belize and extending along the coast of Central America to northern South America.

rural-to-urban migration The flow of internal migrants from rural areas to cities.

Russification A policy of the Soviet Union designed to spread Russian settlers and influences to non-Russian areas of the country.

rust belt Regions of heavy industry that experienced marked economic decline after their factories ceased to be competitive.

Sahel The semidesert region at the southern fringe of the Sahara, and the countries that fall within this region, which extends from Senegal to Sudan. Droughts in the 1970s and early 1980s caused widespread famine and dislocation of population.

salinization The accumulation of salts in the upper layers of soil, often causing a reduction in crop yields, resulting from irrigation with water of high natural salt content and/or irrigation of soils that contain a high level of mineral salts.

Schengen Agreement The 1985 agreement between some—but not all—European Union member countries to reduce border formalities in order to facilitate free movement of citizens between member countries of this new "Schengenland." For example, today there are reduced border controls between France and Germany, or between France and Italy.

sectarian violence Conflicts that divide people along ethnic, religious, and sectarian lines.

sectoral transformation The evolution of a labor force from being highly dependent on the primary sector to being oriented around more employment in the secondary, tertiary, and quaternary sectors.

secularism The term describes both the separation of politics and religion, as well as the non-religious segment of a population. An example of the first usage is the secularism of the United States constitution, which clearly separates State from Church; whereas in the second usage it is common to refer to the growing secularism of Europe, referring to the disinterest in religion by a large part of the population.

sediment load The amount of sand, silt, and clay carried by a river.

Shanghai Cooperation Organization (SCO) An international organization composed of China, Russia, Kazakhstan, Kyrgyzstan, Tajikistan, and Uzbekistan that aims to enhance security and economic cooperation in Central Asia. India and Pakistan are expected to join the organization in 2017.

shields Large upland areas of very old exposed rocks that range in elevation from 600 to 5000 feet (200 to 1500 meters). The three major shields in South America are the Guiana, Brazilian, and Patagonian.

shifted cultivators Migrant farmers who are either transplanted by government relocation schemes or forced to move on their own when their lands are expropriated.

Shiites Muslims who practice one of the two main branches of Islam; especially dominant in Iran and nearby southern Iraq.

shogunate The political order of Japan before 1868, in which power was held by the military leader known as the shogun, rather than by the emperor, whose authority was merely symbolic.

Sikhism A South Asian religion, concentrated in the Indian state of Punjab, that shows some similarities with both Islam and Hinduism.

Silk Road A historical trade route that extended across Central Asia, linking China with Europe and Southwest Asia.

siloviki Members of elite Russian military and security forces.

Slavic peoples A group of peoples in eastern Europe and Russia who speak Slavic languages, a distinctive branch of the Indo-European language family.

social and regional differentiation "Social differentiation" refers to a process by which certain classes of people grow richer when others grow poorer; "regional differentiation" refers to a process by which places grow more prosperous while others become less prosperous.

socialist realism An artistic style once popular in the Soviet Union that was associated with realistic depictions of workers in their patriotic struggles against capitalism.

Southern African Development Community (SADC) A trade bloc formed in the 1970s to facilitate intraregional exchange and development. It became even more important in the 1990s and is anchored by South Africa, one of the region's largest economies.

Soviet Union A sprawling communist state of the Russian domain, also known as the Union of Soviet Socialist Republics (USSR), which dominated the region from its formation in 1922 to its collapse in 1991.

sovereignty A term geopolitically defined as the ability (or the inability) of a government to control activities within its borders. A sovereign state is essentially the same as an independent country.

space An idea in the field of geography that represents a more abstract, quantitative, and model-driven approach to understanding how objects and practices are connected and impact each other.

Spanglish A hybrid combination of English and Spanish spoken by Hispanic Americans.

Special Economic Zones (SEZs) Relatively small districts in China that were fully opened to global capitalism after China began to reform its economy in the 1980s.

spheres of influence In countries not formally colonized in the 19th and early 20th centuries (particularly China and Iran), areas called "spheres of influence" were gained by particular European countries for trade purposes and more generally for economic exploitation and political manipulation.

steppe Semiarid grasslands found in many parts of the world. Steppe grasses are usually shorter and less dense than those found in prairies.

structural adjustment programs Controversial yet widely implemented programs used to reduce government spending, encourage the private sector, and refinance foreign debt. Typically, these IMF and World Bank policies trigger drastic cutbacks in government-supported services and food subsidies, which disproportionately affect the poor.

subduction zone Areas where two tectonic plates are converging or colliding. In these areas, one plate usually sinks below another. They are characterized by earthquakes, volcanoes, and deep oceanic trenches.

subnational organizations Groups that form along ethnic, ideological, or regional lines that can induce serious internal divisions within a state.

subtropical jet stream Powerful atmospheric rivers of eastward-moving air that affect storms and pressure systems in both the Northern and the Southern hemispheres. These jets are products of Earth's rotation and global temperature differences. The subtropical jets are usually higher and somewhat weaker than the polar jet streams.

Suez Canal Pivotal waterway connecting the Red Sea and the Mediterranean, opened in 1869.

Sunda Shelf An extension of the continental shelf from the Southeast Asia mainland to the large islands of Indonesia. Because of the shelf, the overlying sea is generally shallow (less than 200 feet [61 meters] deep).

Sunnis Muslims who practice the dominant branch of Islam.

supranational organization Governing bodies that include several states, such as trade organizations, and often involve a loss of some state powers to achieve organizational goals.

sustainable agriculture A system of agriculture where organic farming principles, a limited use of chemicals, and an integrated plan of crop and livestock management combine to offer both producers and consumers environmentally friendly alternatives.

Sustainable Development Goals (SDGs) Goals resulting from a UN–led effort to end extreme poverty by focusing on 17 key indicators, the top five of which are no poverty, zero hunger, good health, quality education, and gender equality, with key benchmarks for 2030.

swidden A form of cultivation in which forested or brushy plots are cleared of vegetation, burned, and then planted in crops, only to be abandoned a few years later as soil fertility declines. Also called slash-and-burn agriculture or shifting cultivation.

Sykes-Picot Agreement A secret agreement in 1916 between Great Britain and France, and agreed to by the Russian Empire, that defined imperial spheres of influence in the disintegrating Ottoman Empire. According to the pact, France was to have a dominant position after WWI in what is now Syria and Lebanon, whereas Britain was to receive the area extending from what is now Israel and Palestine through Iraq. As Britain had previously promised independence to the Arabs of the region who were rebelling against the Ottoman Empire, the Sykes-Picot Agreement is widely seen in Southwest Asia as a prime example of European dishonesty and double-dealing.

syncretic religions The blending of different belief systems. In Latin America, many animist practices were folded into Christian worship.

taiga The vast coniferous forest of Russia that stretches from the Urals to the Pacific Ocean. The main forest species are fir, spruce, and larch.

Taliban A harsh, Islamic fundamentalist political group that ruled most of Afghanistan in the late 1990s. In 2001, the Taliban lost power, but it later regrouped in Pakistan and continues to fight against the Afghan government in southern and eastern Afghanistan.

Tamil Tigers The common name of the rebel forces in Sri Lanka (officially known as the Liberation Tigers of Tamil Eelam, or LTTE) that fought the Sri Lankan army from 1983 until their defeat in 2009.

territory A concept integral to the practice of sovereignty delimiting the area over which a state exercises control and which is recognized by other states. A sovereign state must have a territory that is recognized by other states.

terrorism The systematic use of terror to achieve political or cultural goals.

thematic geography A basic division in geography which focuses on a specific topic or theme as opposed to analyzing a specific place or a region.

theocracy A government run by religious leaders.

theocratic state A political state led by religious authorities.

tonal language A language in which the meaning of a syllable varies in accordance with the tone in which it is uttered.

total fertility rate (TFR) The average number of children who will be borne by women of a hypothetical, yet statistically valid, population, such as that of a specific cultural group or within a particular country. Demographers consider TFR a more reliable indicator of population change than the crude birthrate.

townships Racially segregated neighborhoods created for nonwhite groups under apartheid in South Africa. They are usually found on the outskirts of cities and classified as black, coloured, or South Asian.

Trans-Pacific Partnership (TPP) Signed in 2016, this was a comprehensive agreement among 12 countries including Canada and the United States, designed to stimulate Pacific trade and lower tariff barriers between member states. Under the Trump administration, the US may withdraw from the TPP in 2017.

Trans-Siberian Railroad Key southern Siberian railroad connection completed during the Russian empire (1904) that links European Russia with the Russian Far East terminus of Vladivostok.

transform fault An earthquake fault where the ground on each side of the fault moves in opposite directions because of tectonic forces.

transhumance A form of pastoralism in which animals are taken to high-altitude pastures during the summer months and returned to low-altitude pastures during the winter.

transmigration The planned, government-sponsored relocation of people from one area to another within a state territory; the term is usually associated with Indonesia.

transnational migration Complex social and economic linkages that form between home and host countries through international migration. Unlike earlier generations of migrants, 21st century immigrants can maintain more enduring and complex ties to their home countries as a result of technological advances.

Treaty of Tordesillas A treaty signed in 1494 between Spain and Portugal that drew a north–south line some 300 leagues west of the Azores and Cape Verde islands. Spain received the land to the west of the line and Portugal the land to the east.

tribalism Allegiance to a particular tribe or ethnic group rather than to the nation-state. Tribalism is often blamed for internal conflict within Sub-Saharan states.

tribes A group of families or clans with a common kinship, language, and definable territory but not an organized state.

tsar A Russian term (also spelled czar) for "Caesar," or ruler; the authoritarian rulers of the Russian empire before its collapse in the 1917 revolution.

tsunamis Very large sea waves induced by earthquakes.

tundra Arctic biome with a short growing season in which vegetation is limited to low shrubs, grasses, and flowering herbs.

typhoons Large tropical storms, synonymous with hurricanes, that form in the western Pacific Ocean in tropical latitudes and cause widespread damage to the Philippines, Vietnam, and coastal regions in East Asia.

Union of South American Nations (UNASUR)   An intergovernmental organization designed to improve the economic integration of South American countries. UNA-SUR currently includes every independent country in South America.

unitary state   A political system in which power is centralized at the national level.

universalizing religions   A religion, usually with an active missionary program, that appeals to a large group of people regardless of local culture and conditions. Christianity, Buddhism, and Islam all have strong universalizing components.

urban decentralization   The process in which cities spread out over a larger geographical area.

urban heat island   An effect in built-up areas in which development associated with cities often produces nighttime temperatures some 9 to 14°F (5 to 8°C) warmer than nearby rural areas.

urban primacy   The situation found in a country in which a disproportionately large city, such as London, Seoul, or Bangkok, dominates the urban system and is the center of economic, political, and cultural life.

urbanized population   That percentage of a country's population living in settlements characterized as cities. Usually, high rates of urbanization are associated with higher levels of industrialization and economic development because these activities are usually found in and around cities. Conversely, lower urbanized populations (less than 50 percent) are characteristic of developing countries.

viticulture   Grape cultivation, usually for the purpose of making wine.

Warsaw Pact   The Cold War military pact between the Soviet Union and its eastern European satellite countries, formed to counter NATO.

Washington Consensus   This is a term referring to a set of policies for economic development in which the key players include the U.S. Treasury and Federal Reserve, the World Bank, the International Monetary Fund (IMF), and the World Trade Organization.

water stress   An environmental planning tool used to predict areas that have—or will have—serious water problems based upon the per capita demand and supply of freshwater.

White Australia Policy   Before 1975, a set of stringent Australian limitations on nonwhite immigration to the country. Largely replaced by a more flexible policy today.

World Trade Organization (WTO)   Formed as an outgrowth of the General Agreement on Tariffs and Trade (GATT) in 1995, the WTO is a large collection of member states dedicated to reducing global barriers to trade. The WTO currently includes 164 countries.

PHOTO CREDITS

CHAPTER 1 **Opening Photo** Antonio Masiello/ZUMA Press/Newscom **Figure 1.1** Jason Langley/Alamy **Figure 1.2** NASA **Figure 1.3** Robert Harding World Imagery **Figure 1.6** Rob Crandall **Figure 1.7** Rob Crandall **Figure 1.9** Ben Curtis/AP Images **Figure 1.11** Noor Khamis/Reuters **Figure 1.12** Rob Crandall **Figure 1.13** Edgar Su/Reuters **Figure 1.14** ZUMA Press, Inc./Alamy **Figure 1.15** Jay Directo/AFP/Getty Images **Figure 1.20** NASA Earth Observatory image created by Jesse Allen **Figure 1.2.1** NielsDK/imageBROKER/AGE Fotostock **Figure 1.23** Jake Lyell/Alamy **Figure 1.24** Moodboard/Alamy **Figure 1.27** John Zada/Alamy **Figure 1.3.1** Rob Crandall **Figure 1.28** P. Batchelder/Alamy **Figure 1.4.1** Marie Price **Figure 1.29** Rob Crandall **Figure 1.30** Rob Crandall **Figure 1.32** RosaIreneBetancourt 7/Alamy **Figure 1.34** ThavornC/Shutterstock **Figure 1.36** EPA european pressphoto agency b.v./Alamy **Figure 1.37** Roz Gaizkaroz/AFP/Getty Images **Figure 1.40** Handout/Alamy **Figure 1.5.1** Courtesy of Susan Wolfinbarger **Figure 1.5.2** DigitalGlobe/ScapeWare3d/AAAS/Getty Images **Figure 1.41** Jake Lyell/Alamy **Figure 1.43** Manish Lakhani/Alamy **Figure 1.46** Bill Mullen/KRT/Newscom **Figure 1.47** B.O'Kane/Alamy **Figure D1** Dbimages/Alamy

CHAPTER 2 **Opening Photo** Russ Bishop/AGE Fotostock **O2.1** Ivan Alvarado/Thomson Reuters (Markets) LLC **O2.2** Andrew Biraj/Thomson Reuters (Markets) LLC **O2.3** Chris Harris/Glow Images **O2.4** Paul Strawson/Alamy **O2.5** Otmar Smit/Shutterstock **Figure 2.3** Ivan Alvarado/Thomson Reuters (Markets) LLC **Figure 2.4** Ragnar Th. Sigurdsson/Arctic Images/Alamy **Figure 2.5** USGS, U.S. Geological Survey Library **Figure 2.7** Greg Vaughn/VW Pics/AGE Fotostock **Figure 2.1.1** M. Jackson **Figure 2.10** NASA Goddard MODIS Rapid Response Team/Jeff Schmaltz **Figure 2.2.2** Blickwinkel/Hummel/Alamy **Figure 2.13** NOAA **Figure 2.19** Pallava Bagla/Corbis Historical/Getty Images **Figure 2.21a** Rob Crandall **Figure 2.21b** Karen Desjardin/Getty Images **Figure 2.22a** 06photo/Fotolia **Figure 2.22b** Tuul and Bruno Morandi/The Image Bank/Getty Images **Figure 2.23a** Mshch/Fotolia **Figure 2.23b** Shsphotography/Fotolia **Figure 2.24a** John E Marriott/Glow images **Figure 2.24b** Ryan DeBerardinis/Shutterstock **Figure 2.25** Kevin Foy/Alamy **Figure 2.26** Chris Harris/Glow Images **Figure 2.4.1** MediaWorldImages/Alamy **Figure 2.3.1** Paul Strawson/Alamy **Figure 2.3.2** Koenig/Wello/Splash/Newscom **Figure 2.28** McClatchy Tribune Content Agency, LLC/Alamy **Figure 2.30** Otmar Smit/Shutterstock **Figure 2.31** Jake Lyell/Alamy **Figure IA2** Chris Harris/Glow Images **Figure D2** Andrew Biraj/Thomson Reuters (Markets) LLC

CHAPTER 3 **Opening Photo** Bas Vermolen/Getty Images **O3.1** Caleb Foster/Fotolia **O3.2** Klaus Lang/AGE Fotostock **O3.3** John Mitchell/Alamy **O3.4** NASA **O3.5** Lynne Sladky/AP Images **Figure 3.2** Hans Blossey/Alamy **Figure 3.3** George Rose/Contributor/Getty Images **Figure 3.4** NASA **Figure 3.5** Caleb Foster/Fotolia **Figure 3.8** NASA **Figure 3.1.2** Pete Mcbride/National Geographic Creative **Figure 3.11** Steve Proehl/Corbis Documentary/Getty Images **Figure 3.14** J.Castro/Getty Images **Figure 3.16** Peter Titmuss/Alamy **Figure 3.18** Alex Garcia/Los Angeles Times/Getty Images **Figure 3.19** Jerry Jackson/TNS/Newscom **Figure 3.20** Klaus Lang/AGE Fotostock **Figure 3.21** William Wyckoff **Figure 3.22** NASA **Figure 3.2.2** Rob Crandall **Figure 3.24** William Wyckoff **Figure 3.27** John Mitchell/Alamy **Figure 3.29** James Nord/AP Images **Figure 3.30** James Nord/AP Images **Figure 3.31b** William Wyckoff **Figure 3.3.1** Courtesy of Diane Papineau **Figure 3.33** EPA european pressphoto agency b.v./Alamy **Figure 3.34** Michael Interisano/Newscom **Figure 3.36** NASA **Figure 3.37** Michael Dwyer/Alamy **Figure 3.5.1** Mark E. Groth: St. Louis City Talk **Figure 3.39** Can Balcioglu/Shutterstock **Figure 3.40** Michael Forsberg/National Geographic/Getty Images **Figure 3.41** Advanced Aerial Photography/DCVB **Figure 3.42** Raquel Lonas/Getty Images **Figure 3.43** Lynne Sladky/AP Images **Figure 3.45** Lucy Nicholson/Reuters **Figure D3** Jason Cohn/MCT/Getty Images

CHAPTER 4 **Opening Photo** Rodrigo Arangua/AFP/Getty Images **O4.1** Toni Sanchez Poy/Alamy **O4.2** Brazil Photo Press/LatinContent/Getty Images **O4.3** Mario Tama/Staff/Getty Images **O4.4** Orlando Sierra/AFP/Getty Images **O4.5** Keith Dannemiller/Alamy **Figure 4.2** Rob Crandall **Figure 4.3** Rob Crandall **Figure 4.4** Toni Sanchez Poy/Alamy **Figure 4.5** Danny Lehman/Corbis/VCG/Getty Images **Figure 4.7** Phil Clarke Hill/Contributor/Getty Images **Figure 4.9** Rob Crandall **Figure 4.11** Salesian Museum/ZUMA Press/Newscom **Figure 4.13** NASA **Figure 4.14** Rob Crandall **Figure 4.15** EPA european pressphoto agency b.v./Alamy **Figure 4.1.1** Rob Crandall **Figure 4.1.2** Rob Crandall **Figure 4.16** Matt Mawson/Getty Images **Figure 4.18** ImageBroker/Alamy **Figure 4.20** Aurora Photos/Alamy **Figure 4.21** Rob Crandall **Figure 4.22** Brazil Photo Press/LatinContent/Getty Images **Figure 4.24** Daniel Guimaraes/AFP/Getty Images **Figure 4.2.1** Orlando Sierra/AFP/Getty Images **Figure 4.25** Rob Crandall **Figure 4.26** Sergi Reboredo/Alamy **Figure 4.28** Alexandro Auler/STR/LatinContent WO/Getty Images **Figure 4.3.1** Noel Celis/AFP/Getty Images **Figure 4.30** Rob Crandall **Figure 4.4.1** Courtesy of Corrie Drummond Garcia **Figure 4.33** Paulo Fridman/Corbis Historical/Getty Images **Figure 4.34** Kaveh Kazemi/Getty Images **Figure 4.35** Philippe Psaila/Science Source **Figure 4.5.1** Mario Tama/Staff/Getty Images **Figure 4.37** Keith Dannemiller/Alamy **Figure 4.39** Rob Crandall **Figure 4.40** Rob Crandall **Figure D4** Phil Clarke Hill/Contributor/Getty Images

CHAPTER 5 **Opening Photo** travelstock44/Alamy **O5.1** Arco Images GmbH/G.A. Rossi/Alamy **O5.2** Margaret S/Alamy **O5.3** Rob Crandall **O5.4** Ana Martinez/Reuters **O5.5** Frank Fell/Robert Harding World Imagery **Figure 5.2** Rob Crandall **Figure 5.3** Arco Images GmbH/G.A. Rossi/Alamy **Figure 5.4** Hufton and Crow/View Pictures/AGE Fotostock **Figure 5.5** Jody Amiet/AFP/Getty Images **Figure 5.7** Desmond Boylan/Reuters **Figure 5.9** NASA **Figure 5.10** Rob Crandall **Figure 5.11** Rob Crandall **Figure 5.1.2** Helene Valenzuela/Stringer/AFP/Getty Images **Figure 5.2.1** Javier Teniente/Contributor/Cover/Getty Images **Figure 5.2.2** Roger Bacon/REUTERS/Alamy **Figure 5.15** De Agostini Picture Library/Contributor/Getty Images **Figure 5.16** Orlando barria/EPA/Newscom **Figure 5.17** Rob Crandall **Figure 5.18** Margaret S/Alamy **Figure 5.20** Ranu Abhelakh/Reuters **Figure 5.22** Randy Brooks/Contributor/Sportsfile/AFP/Getty Images **Figure 5.3.1** Shi Rong Xinhua News Agency/Newscom **Figure 5.24** Rob Crandall **Figure 5.25** Alexander Hassenstein/Staff/Getty Images **Figure 5.4.1** Rob Crandall **Figure 5.4.2** The Washington Post/Contributor/Getty Images **Figure 5.27** Ana Martinez/Reuters **Figure 5.28** Rob Crandall **Figure 5.30** David Evans/Contributor/Bloomberg/Getty Images **Figure 5.5.1** Courtesy of Sarah Blue **Figure 5.32** Frank Fell/Robert Harding World Imagery **Figure 5.34** Huntstock/Disability Images/Alamy **Figure D5** Yamil Lage/Getty Images

CHAPTER 6 **Opening Photo** Joerg Boethling/Alamy **O6.1** Flowerphotos/Alamy **O6.2** Rob Crandall **O6.3** Charles O. Cecil/Alamy **O6.4** Roger Bacon/Reuters/Alamy **O6.5** Jake Lyell/Alamy **Figure 6.3** Rob Crandall **Figure 6.4** AfriPics.com/Alamy **Figure 6.5** Robert Caputo/Aurora/Getty Images **Figure 6.6** Yann Arthus-Bertran/Getty Images **Figure 6.8** Rob Crandall **Figure 6.9** Heeb Christian/Prisma Bildagentur AG/Alamy **Figure 6.11** Daniel Berehulak/Staff/Getty Images **Figure 6.1.1** Flowerphotos/Alamy **Figure 6.13** Akintunde Akinleye/Thomson Reuters (Markets) LLC **Figure 6.14** Rob Crandall **Figure 6.18** Tommy Trenchard/Alamy **Figure 6.19** Rob Crandall **Figure 6.21** Yannick Tylle/Corbis Documentary/Getty Images **Figure 6.22** Melba Photo Agency/Alamy **Figure 6.2.1** Jeremy Graham/dbimages/Alamy **Figure 6.23** EPA european pressphoto agency b.v./Alamy **Figure 6.24** Max Milligan/AWL Images/Getty Images **Figure 6.25** NASA **Figure 6.26** Henrique NDR Martins/E+/Getty Images **Figure 6.29** Neil Cooper/Alamy **Figure 6.30** Charles O. Cecil/Alamy **Figure 6.31** Shashank Bengali/MCT/Newscom **Figure 6.3.1** Seyllou/Stringer/AFP/Getty Images **Figure 6.32** Yann Latronche/Gamma-Rapho/Contributor/Getty Images **Figure 6.33** Ian Walton/Staff/Getty Images **Figure 6.36** Ulrich Doering/Alamy **Figure 6.4.1** Stuart price/Stringer/AFP/Getty Images **Figure 6.38** Roger Bacon/Reuters/Alamy **Figure 6.39** AFP/Stringer/Getty Images **Figure 6.41** Jake Lyell/Alamy **Figure 6.42** Pan Siwei/Xinhua News Agency/Newscom **Figure 6.43** Tom Gilks/Alamy **Figure 6.5.1** Courtesy of Fenda Akiwumi **Figure 6.46** Robert Harding/Godong/Alamy **Figure D6** United Nations

CHAPTER 7 **Opening Photo** Sam Tarlin/Corbis News/Getty Images **O7.1** Rieger Bertrand/Hemis/Alamy **O7.2** Bill Bachmann/Alamy **O7.3** Raga Jose Fuste/Prisma Bildagentur AG/Alamy **O7.4** Umit Bektas/Reuters **O7.5** Tom Hanley/Alamy **Figure 7.2** Wigbert Roth/ImageBroker/Alamy **Figure 7.3** NASA **Figure 7.4** A.A.M. Van der Heyden/Independent Picture Service/Alamy **Figure 7.5** Duby Tal/Albatross/Alamy **Figure 7.7** Rieger Bertrand/Hemis/Alamy **Figure 7.8** Michael Runkel/ImageBroker/Alamy **Figure 7.10** Jochen Tack/ArabianEye/Getty Images **Figure 7.11** Muratart/Shutterstock **Figure 7.13** NASA **Figure 7.15** Travelstock44/Look Die Bildagentur der Fotografen GmbH/Alamy **Figure 7.18** Bill Bachmann/Alamy **Figure 7.19** Peter Horree/Alamy **Figure 7.20** Raga Jose Fuste/Prisma Bildagentur AG/Alamy **Figure 7.22** Wael hamdan/Alamy **Figure 7.23** Iain Masterton/Alamy **Figure 7.24** Abaca/Newscom **Figure 7.1.2** Fabrizio Villa/Polaris/Newscom **Figure 7.28** Al Jazeera English/Alamy **Figure 7.30** Duby Tal/Albatross/Superstock **Figure 7.32** K.M. Westermann/Corbis Documentary/Getty Images **Figure 7.33** EPA european pressphoto agency b.v./Alamy

Figure 7.2.1 Stepanek Photography/Shutterstock **Figure 7.34** Action Plus Sports Images/Alamy **Figure 7.37b** Jelle vanderwolf/Alamy **Figure 7.38** Debbie Hill/UPI/Newscom **Figure 7.39** Umit Bektas/Reuters **Figure 7.3.1** Courtesy of Karen Culcasi **Figure 7.3.2** PixelPro/Alamy **Figure 7.4.2** Jalal Morchidi/Anadolu Agency/Getty Images **Figure 7.5.2** Gareth Dewar/Alamy **Figure 7.43** Tom Hanley/Alamy **Figure 7.44** Xi Li, Wuhan University/University of Maryland/Courtesy of #withSyria **Figure 7.45** van der Meer Marica/Arterra Picture Library/Alamy **Figure D7** K.M. Westermann/Corbis Documentary/Getty Images

CHAPTER 8 **Opening Photo** Michael Spring/Alamy **O8.1** Tom Till/Alamy **O8.2** Newzulu/Alamy **O8.3** Clynt Garnham Education/Alamy **O8.4** Petras Malukas/AFP/Getty Images **O8.5** Sakis Mitrolidis/AFP/Getty Images **Figure 8.1** Nicolas Maeterlinck/AFP/Stringer/Getty Images **Figure 8.4** Hemis/Alamy **Figure 8.5** Tom Till/Alamy **Figure 8.6** ImageBroker/Alamy **Figure 8.9** Magnus Qodarion/Alamy **Figure 8.10** Lars Ruecker/Getty Images **Figure 8.1.1** Courtesy of Weronika Kusek **Figure 8.1.2** Courtesy of Weronika Kusek **Figure 8.13** Newzulu/Alamy **Figure 8.14** Dpa picture alliance/Alamy **Figure 8.15** Lester Rowntree **Figure 8.2.1** NielsDK/imageBROKER/AGE Fotostock **Figure 8.2.2** Bruce yuanyue Bi/Alamy **Figure 8.16** John Kellerman/Alamy **Figure 8.17** Clynt Garnham Education/Alamy **Figure 8.3.1** C3744 Lars Halbauer Deutsch Presse Agentur/Newscom **Figure 8.19a** Albert Gea/Reuters **Figure 8.19b** Studioimagen73/Shutterstock **Figure 8.20** Peter Forsberg/EU/Alamy **Figure 8.22** KorayErsin/Fotolia **Figure 8.23** Anadolu Agency/Getty Images **Figure 8.24** Vyacheslav Prokofyev/TASS/Getty Images **Figure 8.27a** Bettmann/Getty Images **Figure 8.27b** Picture Alliance/Dpa/Newscom **Figure 8.4.1** Petras Malukas/AFP/Getty Images **Figure 8.29** Dylan Martinez/Thomson Reuters (Markets) LLC **Figure 8.30** Robert Morris/Alamy **Figure 8.32a** Mariola S/Shutterstock **Figure 8.32b** Photocreo Bednarek/Fotolia **Figure 8.5.1** PA Images/Alamy **Figure 8.35** Sakis Mitrolidis/AFP/Getty Images **Figure 8.36** Kerkla/Getty Images **Figure D8** Dpa picture alliance/Alamy

CHAPTER 9 **Opening Photo** Wojtek Buss/AGE Fotostock **O9.1** Ilya Naymushin/Reuters **O9.2** Andrey Rudakov/Bloomberg/Getty Images **O9.3** Alexander Utkin/AFP/Getty Images **O9.4** Mikhail Svetlov/Getty Images **O9.5** James P. Blair/National Geographic/Getty Images **Figure 9.2** Hemis/Alamy **Figure 9.5** Mykola Mazuryk/Shutterstock **Figure 9.6** Derek watt/Alamy **Figure 9.7** Yuliya/AGE Fotostock America Inc. **Figure 9.8** Paul Doyle/Alamy **Figure 9.10** Ilya Naymushin/Reuters **Figure 9.11** Katvic/Fotolia **Figure 9.12** Tom Brakefield/Stockbyte/Getty Images **Figure 9.1.2** Volodymyr Shuvayev/AFP/Getty Images **Figure 9.2.2** Allan White/Getty Images **Figure 9.14** NASA **Figure 9.15** ITAR-TASS Photo Agency/Alamy **Figure 9.16** Yulenochekk/Fotolia **Figure 9.3.1** Courtesy of Kelsey Nyland **Figure 9.18** Misha Japaridze/AP Images **Figure 9.19** Alex's Pictures Moscow/Alamy **Figure 9.20** Andrey Rudakov/Bloomberg/Getty Images **Figure 9.21** ITAR TASS Photo Agency/Alamy **Figure 9.25** Vova Pomortzeff/Alamy **Figure 9.27** ITAR TASS Photo Agency/Alamy **Figure 9.28** Alexander Utkin/AFP/Getty Images **Figure 9.4.2** Karen Minasyan/AFP/Getty Images **Figure 9.29** Mara Zemgaliete/Fotolia **Figure 9.30** Dpa picture alliance/Alamy **Figure 9.31** Mikhail Svetlov/Getty Images **Figure 9.35** STR/AFP/Getty Images **Figure 9.36** Kay Nietfeld/AFP/Getty Images **Figure 9.37a** AFP/Getty Images **Figure 9.39** James P. Blair/National Geographic/Getty Images **Figure 9.5.1** ITAR TASS Photo Agency/Alamy **Figure 9.40** Kommersant/Getty Images **Figure D9** Mikhail Svetlov/Getty Images

CHAPTER 10 **Opening Photo** Bill Ingalls/NASA **O10.1** UncorneredMarket.com/Getty Images **O10.2** Aurora Photos/Alamy **O10.3** Paula Bronstein/Getty Images **O10.4** Bashir Safai/AFP/Getty Images **O10.5** Ton Koene/AGE Fotostock **Figure 10.2** UncorneredMarket.com/Getty Images **Figure 10.4** Vladimir Pirogov/Reuters **Figure 10.6** NASA **Figure 10.7** Aurora Photos/Alamy **Figure 10.1.1** China Photos/Getty Images **Figure 10.8** Theodore Kaye/Alamy **Figure 10.9** Michal Cerny/Alamy **Figure 10.12** Jesse Allen/Robert Simmon/NASA Earth Observatory **Figure 10.13** Aurora Photos/Alamy **Figure 10.14** Alenvl/Shutterstock **Figure 10.15** GM Photo Images/Alamy **Figure 10.2.1** REUTERS/Ivan Stolpnikov **Figure 10.16** Ilya Postnikov/Fotolia **Figure 10.3.1** Courtesy of Holly Barcus **Figure 10.20** Robert Nickelsberg/The LIFE Images Collection/Getty Images **Figure 10.21** J Marshall/Tribaleye Images/Alamy **Figure 10.22** Ryumin Alexander TASS/Newscom **Figure 10.23** Paula Bronstein/Getty Images **Figure 10.26** Karen Minasyan/AFP/Getty Images **Figure 10.27** Bashir Safai/AFP/Getty Images **Figure 10.29** Oliviero Olivieri/robertharding/Newscom **Figure 10.4.1** Jesse Allen/Robert Simmon/NASA Earth Observatory **Figure 10.30** View Stock/Alamy **Figure 10.31** Ton Koene/AGE Fotostock **Figure 10.5.1** Shamil Zhumatov/Reuters **Figure 10.33** AFP/Getty Images **Figure IA10** NASA **Figure D10** Vladimir Pirogov/Reuters

CHAPTER 11 **Opening Photo** JTB Photo/Contributor/Universal Images Group/Getty Images **O11.1** AFP Photography, LLC/Contributor/Getty Images **O11.2** Lintao Zhang/Staff/Getty Images **O11.3** Iain Masterton/Alamy **O11.4** Lucas Vallecillos/AGE Fotostock **O11.5** Bloomberg/Getty Images **Figure 11.4** George F. Mobley/National Geographic Stock **Figure 11.5** AFP Photography, LLC/Contributor/Getty Images **Figure 11.7** NASA/GSFC/MODIS Land Rapid Response Team **Figure 11.8** Tuomas Lehtinen/Alamy **Figure 11.9** Tessa Bunney/Contributor/Getty Images **Figure 11.11** EPA european pressphoto agency b.v./Alamy **Figure 11.12** Marco Brivio/Photographer's Choice/Getty Images **Figure 11.13a** Panorama Images/The Image Works **Figure 11.13b** Liu Xiaoyang/China Images/Alamy **Figure 11.14** Liu Xiaoyang/China Images/Alamy **Figure 11.1.1** Hufton and Crow/VIEW Pictures Ltd/Alamy **Figure 11.15** China Color Photo/AP Images **Figure 11.16** Jim Richardson/National Geographic/Getty Images **Figure 11.17** Gamma/Gamma-Rapho/Contributor/Getty Images **Figure 11.21** Japan Stock Photography/Alamy **Figure 11.22** Liu Xiaoyang/ China Images/Alamy **Figure 11.23** Roger Bacon/REUTERS/Alamy **Figure 11.2.1** Lintao Zhang/Staff/Getty Images **Figure 11.24** Brian Lawrence/A1 images/Alamy **Figure 11.25** Lou Linwei/Alamy **Figure 11.26** Im ChanKyung/Topic Photo Agency/Getty Image **Figure 11.3.1** Courtesy of Cary Karacas **Figure 11.3.2** US National Archives, Cartographic and Architectural Section, Record Group 226: 330/20/8 **Figure 11.28** Iain Masterton/Alamy **Figure 11.29** Rob Crandall **Figure 11.30** Silvio Vicente, Igreja Bela Vista, São Paulo, Brasil/Wikipedia **Figure 11.32** Lester Ledesma/Contributor/ASAblanca/Getty Images **Figure 11.33** Olaf Schubert/ImageBROKER/Alamy **Figure 11.35** Xinhua/Alamy **Figure 11.36** Victor Paul Borg/Alamy **Figure 11.38** Universal Images Group/Superstock **Figure 11.40** Yonhap/AP Images **Figure 11.41** Lucas Vallecillos/AGE Fotostock **Figure 11.42** Kyodo/Newscom **Figure 11.4.1** Chung Sung-Jun/Getty Images **Figure 11.43** Chris McGrath/Getty Images **Figure 11.5.1** Bloomberg/Getty Images **Figure 11.44** Roger Bacon/Reuters/Alamy **Figure 11.45** Yuan shuiling/Imaginechina/AP Images **Figure 11.47** Rex/Newscom **Figure D11** Asia File/Alamy

CHAPTER 12 **Opening Photo** Amir Mukhtar/Moment Open/Getty Images **O12.1** Sam Panthaky/Stringer/AFP/Getty Images **O12.2** Rob Crandall **O12.3** Rafal Cichawa/Fotolia **O12.4** Danish Ismail FK/TW/Reuters **O12.5** Iqbal Khatri/Moment Open/Getty Images **Figure 12.2** David Pearson/Alamy **Figure 12.1.1** Courtesy of Dr. Chandana Mitra **Figure 12.5** NASA **Figure 12.2.1** Mohammed Abidally/Alamy **Figure 12.8** Barcroft Media/Getty Images **Figure 12.9** Sam Panthaky/Stringer/AFP/Getty Images **Figure 12.10** David Crossland/Alamy **Figure 12.11** Aditya "Dicky" Singh/Alamy **Figure 12.3.1** Omar Havana/Getty Images News/Getty Images **Figure 12.14** Sean Sprague/Alamy **Figure 12.15b** The India Today Group/Getty Images **Figure 12.16** Earl & Nazima Kowall/Corbis Documentary/Getty Images **Figure 12.17** Rob Crandall **Figure 12.18** Blaine Harrington III/Corbis Documentary/Getty Images **Figure 12.19** Aamir Qureshi/AFP/Getty Images **Figure 12.20** EPA european pressphoto agency b.v./Alamy **Figure 12.22** Shahril KHMD/Shutterstock **Figure 12.23** Rafal Cichawa/Fotolia **Figure 12.25** Frank Bienewald/ImageBroker/AGE Fotostock **Figure 12.26** Captain of Hope **Figure 12.28** Olaf Kruger/ImageBroker/Newscom **Figure 12.30** Strdel/Stringer/AFP/Getty Images **Figure 12.4.1** Bloomberg/Getty Images **Figure 12.33** Sueddeutsche Zeitung Photo/Alamy **Figure 12.34** Monirul Alam/ZUMA Press/Newscom **Figure 12.35** Anupam Nath/AP Images **Figure 12.37** Danish Ismail FK/TW/Reuters **Figure 12.38** NurPhoto/Getty Images **Figure 12.39** Hemis/Alamy **Figure 12.40** Iqbal Khatri/Moment Open/Getty Images **Figure 12.42** John Henry Claude Wilson/Glow Images **Figure 12.5.1** The India Today Group/Getty Images **Figure 12.44** Joerg Boethling/Alamy **Figure D12** Athar Hussain/Reuters

CHAPTER 13 **Opening Photo** Adison pangchai/Shutterstock **O13.1** Y. T. Haryono/Anadolu Agency/Getty Images **O13.2** EPA european pressphoto agency b.v./Alamy **O13.3** Zainal Abd Halim/Thomson Reuters (Markets) LLC **O13.4** Athit Perawongmetha/Reuters **O13.5** Jonathan Drake/Bloomberg/Getty Images **Figure 13.2** Julia Rogers/Alamy **Figure 13.4** Y. T. Haryono/Anadolu Agency/Getty Images **Figure 13.5** Kevin Frayer/Getty Images News/Getty Images **Figure 13.7** ROMEO GACAD/Staff/AFP/Getty Images **Figure 13.1.1** Terry Whittaker/Alamy **Figure 13.8** STR/EPA/Newscom **Figure 13.9** Ulet Ifansasti/Stringer/Getty Images **Figure 13.10** Darren Whiteside/Reuters **Figure 13.12** The Asahi Shimbun Premium/The Asahi Shimbun Premium Archive/Getty Images **Figure 13.13** Peter & Georgina Bowater Stock Connection Worldwide/Newscom **Figure 13.2.1** Thierry Falise/LightRocket/Getty Images **Figure 13.15** NASA **Figure 13.16** Cheryl Ravelo/Reuters **Figure 13.17** EPA european pressphoto agency b.v./Alamy **Figure 13.18** Ed Wray/Getty Images News/Getty Images **Figure 13.19** Michael Brooks/Alamy **Figure 13.23** EPA european pressphoto agency b.v./Alamy **Figure 13.25** STR/AFP PHOTO/Getty Images **Figure 13.26a** Joshua Resnick/Shutterstock **Figure 13.26b** Dan Herrick/Lonely Planet Images/Getty Images

Figure 13.27 Zainal Abd Halim/Thomson Reuters (Markets) LLC **Figure 13.30** Horst Faas/AP Images **Figure 13.31** Stewart Cohen/Tetra Images/Alamy **Figure 13.3.1** YONGYOT PRUKSARAK/EPA/Newscom **Figure 13.33** Athit Perawongmetha/Reuters **Figure 13.34** DigitalGlobe/ScapeWare3d/Getty Images **Figure 13.35** National Archives **Figure 13.36** Roslan Rahman/Newscom **Figure 13.37** Jonathan Drake/Bloomberg/Getty Images **Figure 13.4.1** Nguyen Huy Kham/Thomson Reuters (Markets) LLC **Figure 13.38** Sukree Sukplang/Reuters **Figure 13.40** Pacific Press/Sipa USA/Newscom **Figure 13.5.1** Rachel Silvey **Figure 13.41** Robin Laurance/Look Die Bildagentur der Fotografen GmbH/Alamy **Figure D13a** Adrian Arbib/Alamy **Figure D13b** AFP/Staff/Getty Images

CHAPTER 14   **Opening Photo** Yvette Cardozo/Alamy **O14.1** Chad Ehlers/Alamy **O14.2** Don Fuchs/Look/AGE Fotostock **O14.3** Penny Tweedie/Alamy **O14.4** Education Images/Universal Images Group/Getty Images **O14.5** Atiger/Shutterstock **Figure 14.2** Daniel Bosma/Moment/Getty Images **Figure 14.3** Outback Australia/Alamy **Figure 14.4** Jochen Schlenker/Robert Harding World Imagery/ WireImage/Getty Images **Figure 14.5** David Wall/Alamy **Figure 14.6** Chad Ehlers/Alamy **Figure 14.9** Daniel Broun **Figure 14.10** 42pix/Alamy **Figure 14.12** US Department of Defense **Figure 14.1.1** William Chopart/Mond Image/ AGE Fotostock **Figure 14.1.2** Hemis/Alamy **Figure 14.13** Friedrich Stark/Alamy **Figure 14.14** Handout/Getty Images **Figure 14.16** Don Fuchs/Look/AGE Fotostock **Figure 14.17** Colin Monteath/AGE Fotostock **Figure 14.20** Historical Image Collection by Bildagentur-Online/Alamy **Figure 14.21** Dave G. Houser/Alamy **Figure 14.22** Egmont Strigl/Image Broker/Alamy **Figure 14.23** The Sydney Morning Herald/Fairfax Media/Getty Images **Figure 14.24** Penny Tweedie/Alamy **Figure 14.2.2** Van der Meer Marica/ ArTerra Picture Library/AGE Fotostock **Figure 14.25** John Oxley Library, State Library of Queensland Neg: raw00108 **Figure 14.26** Emily Riddell/AGE Fotostock **Figure 14.28** Blickwinkel/Lohmann/Alamy **Figure 14.29** Thomas Cockrem/Alamy **Figure 14.3.1** Hemis/Alamy **Figure 14.3.2** Marc Le Chelard/Stringer/AFP/Getty Images **Figure 14.30** Kent Smith/AP Images **Figure 14.31** Andrew Woodley/Alamy **Figure 14.33** Danny Martindale/WireImage/Getty Images **Figure 14.34** Education Images/Universal Images Group/Getty Images **Figure 14.35** Angela Prati/AGE Fotostock **Figure 14.37** Australian Department of Immigration and Citizenship/Handout/Getty Images **Figure 14.38** Andre Seale/Alamy **Figure 14.39** Bloomberg/Getty Images **Figure 14.40** Atiger/Shutterstock **Figure 14.41** Wayne G. Lawler/Science Source **Figure 14.4.1** Courtesy of Laura Brewington **Figure 14.42** Raga Jose Fuste/Prisma Bildagentur AG/Alamy **Figure 14.5.1** Stephen Andrews/LatitudeStock Images/AGE Fotostock **Figure 14.44** Philip Game/Alamy

TEXT AND ILLUSTRATION CREDITS

CHAPTER 1 **Table 1.1 page 22** Population Reference Bureau, World Population Data Sheet, 2016; World Bank, World Development Indicators, 2016 **Table 1.2 page 44** World Bank, World Development Indicators, 2016; United Nations, Human Development Report, 2015; Population Reference Bureau, World Population Data Sheet, 2016 **Text 1.1 page 20** UN World Commission on Environment and Development

CHAPTER 2 **Figure 2.16 page 60** From Greenland Ice Sheet Today (2015) 2015 melt season in review. Copyright © 2016, National Snow and Ice Data Center **Figure 2.18 page 61** Copyright © by Carbon Dioxide Information Analysis Center **Figure 2.20 page 62** Based on Clawson and Fisher, 2004, World Regional Geography, 8th ed., Upper Saddle River, NJ: Prentice Hall **Table 2.1 page 70** BP Statistics, 2015 (June 2016)

CHAPTER 3 **Table 3.1 page 87** Population Reference Bureau, World Population Data Sheet, 2016; World Bank, World Development Indicators, 2016 **Figure 3.25 page 98** American Community Survey Reports, U S Census Bureau **Figure 3.28 page 100** Modified from Jordan, Domosh,and Rowntree, 1998, The HumanMosaic, Upper Saddle River, NJ: Prentice Hall **Figure 3.31a page 102** Rubenstein, James M., The Cultural Landscape: an Introduction to Human Geography, 9th Ed., © 2008. Reprinted and Electronically reproduced by permission of Pearson Education, Inc. Upper Saddle River, New Jersey **Figure 3.32 page 103** Adapted from Jerome Fellman, Arthur Getis, and Judith Getis, Human Geography (Dubuque, Iowa: Brown and Benchmark, 1997), p.164. Used by permission of The McGraw-Hill Companies, Inc. **Text 3.1 page 106** President John F. Kennedy summarized the links in a speech to the Canadian parliament in 1962 **Table 3.2 page 112** World Bank, World Development Indicators, 2016; United Nations, Human Development Report, 2015; Population Reference Bureau, World Population Data Sheet, 2016. **Figure 3.38 page 113** Modified from Clawson and Fisher, 2004,World Regional Geography, 8th ed., Upper Saddle River, NJ: Prentice Hall, and Howard Veregin, ed., 2010, Goode's World Atlas, 22nd ed., Upper Saddle River, NJ: Prentice Hall

CHAPTER 4 **Text 4.1 page 168** Felipe Calderón **Table 4.1 page 140** Population Reference Bureau, World Population Data Sheet, 2016; World Bank, World Development Indicators, 2016 **Figure 4.19 page 141** Based on Clawson, 2000, Latin America and the Caribbean: Lands and People, 2nd ed., Boston: McGraw-Hill **Table 4.2 page 160** World Bank, World Development Indicators, 2016; United Nations, Human Development Report, 2015; Population Reference Bureau, World Population Data Sheet, 2016 **Figure 4.41 page 169** World Bank Development Indicators, 2015

CHAPTER 5 **Figure 5.6 page 179** Copyright © by Pearson Education, Upper Saddle River, NJ **Figure 5.8 page 180** Based on DK World Atlas, London: DK Publishing, 1997, pp. 7, 55 **Table 5.1 page 185** Population Reference Bureau, World Population Data Sheet, 2016; CIA World Factbook; World Bank, World Development Indicators, 2016 **Figure 5.14 page 187** Copyright © by Pearson Education, Upper Saddle River, NJ **Figure 5.19 page 192** Data based on Philip Curtin, The Atlantic Slave Trade, A Census, Madison: University of Wisconsin Press, 1969, p. 268 **Figure 5.21 page 193** Robert Voeks, "African Medicine and Magic in the Americas," Geographical Review 83, no. 1, (1993): 66–78. Used by permission of the American Geographical Society **Figure 5.26 page 199** Data from Barbara Tenenbaum, ed., Encyclopedia of Latin American History and Culture, 1996, vol. 5, p. 296, with permission of Charles Scribner's Sons; and John Allcock, Border and Territorial Disputes, 3rd ed., Harlow, Essex, UK: Longman Group, 1992 **Table 5.2 page 203** World Bank, World Development Indicators, 2016; United Nations, Human Development Report, 2015; Population Reference Bureau, World Population Data Sheet, 2016; Gender Inequality Index; Additional data from the CIA World Factbook **Figure 5.33 page 208** United Nations, Human Development Report, 2009

CHAPTER 6 **Text 6.1 page 225** Michael Watts **Table 6.1 page 229** Population Reference Bureau, World Population Data Sheet, 2016; World Bank, World Development Indicators, 2016 **Text 6.2 page 233** Excerpt from Things Fall Apart by Chinua Achebe. Published by Heinemann, © 1986 **Box 6.5 page 250** Based on Africa in their Words: A Study of Chinese Traders in South Africa, Lesotho, Botswana, Zambia and Angola (2012) by Terence McNamee. Discussion Paper 2012/2013 for the Brenthurst Foundation, South Africa **Figure 6.37 page 251** Data from UNHCR 2012 **Table 6.2 page 254** World Bank, World Development Indicators, 2016; United Nations, Human Development Report, 2015; Population Reference Bureau, World Population Data Sheet, 2016; Gender Inequality Index

CHAPTER 7 **Figure 7.12 page 275** Modified from Soffer, Arnon, Rivers of Fire: The Conflict over Water in the Middle East, 1999, p. 180. Reprinted by permission of Rowman and Littlefield Publishers, Inc. **Figure 7.17 page 280** Modified from Clawson and Fisher, 2004, World Regional Geography, 8th ed., Upper Saddle River, NJ: Prentice Hall, and Bergman and Renwick, 1999, Introduction to Geography, Upper Saddle River, NJ: Prentice Hall **Table 7.1 page 278** Population Reference Bureau, World Population Data Sheet, 2016; World Bank, World Development Indicators, 2016 **Text 7.1 page 281** In Palace of Desire, by Naguib Mahfouz **Figure 7.27 page 288** Modified from Rubenstein, 2005, An Introduction to Human Geography, 8th ed., Upper Saddle River, NJ: Prentice Hall **Figure 7.29 page 289** Modified from Rubenstein, 2011, An Introduction to Human Geography, 10th ed., Upper Saddle River, NJ: Prentice Hall, and National Geographic Society, 2003, Atlas of the Middle East, Washington, DC **Figure 7.31 page 291** Modified from Rubenstein, 2011, An Introduction to Human Geography, 10th ed., Upper Saddle River, NJ: Prentice Hall and National Geographic Society, 2003, Atlas of the Middle East, Washington, DC **Figure 7.36 page 297** Modified from Rubenstein, 2011, An Introduction to Human Geography, 10th ed., Upper Saddle River, NJ: Prentice Hall **Figure 7.37 page 298** Modified from Rubenstein, 2011, An Introduction to Human Geography, 10th ed., Upper Saddle River, NJ: Prentice Hall **Table 7.2 page 301** World Bank, World Development Indicators, 2016; United Nations, Human Development Report, 2015; Population Reference Bureau, World Population Data Sheet, 2016; Gender Inequality Index; CIA World Factbook **Figure 7.41 page 302** Modified from Rubenstein, 2011, An Introduction to Human Geography, 10th ed., Upper Saddle River, NJ: Prentice Hall

CHAPTER 8 **Table 8.1 page 325** Population Reference Bureau, World Population Data Sheet, 2016; World Bank, World Development Indicators, 2016 **Figure 8.28 page 342** Data from CIA World Factbook, 2010 **Table 8.2 page 344** World Bank, World Development Indicators, 2016; United Nations, Human Development Report, 2015; Population Reference Bureau, World Population Data Sheet, 2016; Gender Inequality Index **Figure 8.33 page 348** From Eurostat Statistics Explained. Copyright © 1995 by Eurostat - European Commission. Reproduced or Used by permission of the Eurostat - European Commission.

CHAPTER 9 **Table 9.1 page 370** Population Reference Bureau, World Population Data Sheet, 2016; World Bank, World Development Indicators, 2016 **Table 9.2 page 440** World Bank, World Development Indicators, 2016; United Nations, Human Development Report, 2015; Population Reference Bureau, World Population Data Sheet, 2016; Gender Inequality Index

CHAPTER 10 **Table 10.1 page 416** Population Reference Bureau, World Population Data Sheet, 2016; World Bank, World Development Indicators, 2016 **Figure 10.17 page 419** ©2012 Pearson Education, Inc. **Table 10.2 page 431** World Bank, World Development Indicators, 2016; United Nations, Human Development Report, 2015; Population Reference Bureau, World Population Data Sheet, 2016; Gender Inequality Index

CHAPTER 11 **Table 11.1 page 456** Population Reference Bureau, World Population Data Sheet, 2016; World Bank, World Development Indicators, 2016 **Figure 11.2 page 479** World Bank, World Development Indicators, 2016; United Nations, Human Development Report, 2015; Population Reference Bureau, World Population Data Sheet, 2016; Gender Inequality Index

CHAPTER 12 **Table 12.1 page 502** Population Reference Bureau, World Population Data Sheet, 2016; World Bank, World Development Indicators, 2016 **Text 12.1 page 504** Zillur Rahman **Table 12.2 page 528** World Bank, World Development Indicators, 2016; United Nations, Human Development Report, 2015; Population Reference Bureau, World Population Data Sheet, 2016; Gender Inequality Index

CHAPTER 13 **Table 13.1 page 554** Population Reference Bureau, World Population Data Sheet, 2016; World Bank, World Development Indicators, 2016 **Table 13.2 page 574** World Bank, World Development Indicators, 2016; United Nations, Human Development Report, 2015; Population Reference Bureau, World Population Data Sheet, 2016; Gender Inequality Index

CHAPTER 14 **Table 14.1 page 600** Population Reference Bureau, World Population Data Sheet, 2016; World Bank, World Development Indicators, 2016 **Table 14.2 page 617** World Bank, World Development Indicators, 2016; United Nations, Human Development Report, 2015; Population Reference Bureau, World Population Data Sheet, 2016; Gender Inequality Index

INDEX

Note to the Reader: Throughout this index, page numbers followed by *f* indicate illustrations; page numbers followed by *t* indicate tables.

World – Physical

Great Basin	Land features
Caribbean Sea	Water bodies
Aleutian Trench	Underwater features

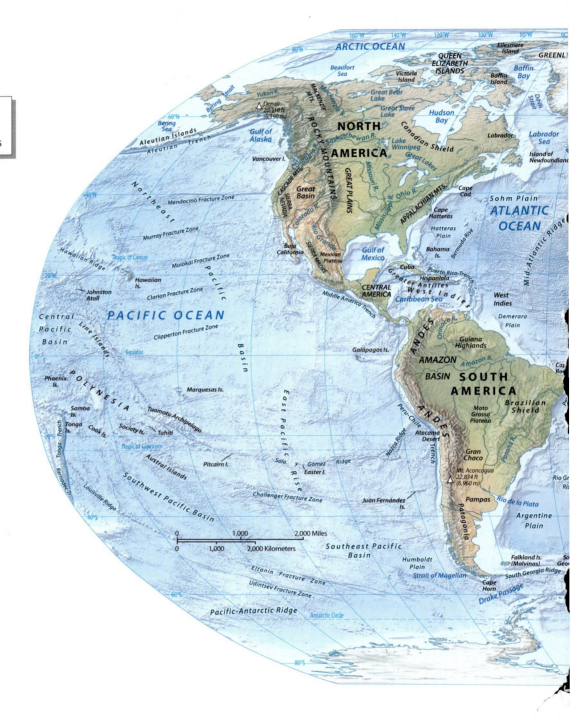

ARCTIC OCEAN

QUEEN ELIZABETH ISLANDS

Ellesmere Island

GREENL

Beaufort Sea

Victoria Island

Baffin Island

Baffin Bay

Davis Strait

Bering Strait

Yukon R.

MACKENZIE MTS.

Mackenzie R.

Great Bear Lake

Great Slave Lake

Hudson Bay

Denali 20,310 ft (6,190 m)

NORTH AMERICA

Canadian Shield

Labrador

Labrador Sea

Bering Sea

Gulf of Alaska

ROCKY MOUNTAINS

Saskatchewan R.

Lake Winnipeg

Great Lakes

Island of Newfoundland

Aleutian Islands

Aleutian Trench

Vancouver I.

Missouri R.

GREAT PLAINS

Mississippi R.

Ohio R.

APPALACHIAN MTS.

Cape Cod

Sohm Plain

Northeast

Mendocino Fracture Zone

CASCADE MTS.

SIERRA NEVADA

Great Basin

Colorado R.

Cape Hatteras

Hatteras Plain

Bermuda Rise

ATLANTIC OCEAN

Mid-Atlantic Ridge

Murray Fracture Zone

Hawaiian Ridge

Tropic of Cancer

Molokai Fracture Zone

Baja California

SIERRA MADRE

Rio Grande

Mexican Plateau

Gulf of Mexico

Bahama Is.

Cuba

Puerto Rico Trench

Hispaniola

Greater Antilles

Hawaiian Is.

Johnston Atoll

Clarion Fracture Zone

Central Pacific Basin

PACIFIC OCEAN

Line Islands

Middle America Trench

CENTRAL AMERICA

West Indies

Caribbean Sea

West Indies

Demerara Plain

Pacific

Clipperton Fracture Zone

Orinoco R.

Guiana Highlands

Equator

Galápagos Is.

AMAZON

Amazon R.

Cap Ro

Phoenix Is.

POLYNESIA

Basin

BASIN

SOUTH AMERICA

Marquesas Is.

ANDES

Brazilian Shield

Samoa Is.

Tonga Is.

Cook Is.

Society Is.

Tahiti

Tuamotu Archipelago

East Pacific Rise

Mato Grosso Plateau

Tropic of Capricorn

Peru-Chile

ANDES

Nazca Ridge

Atacama Desert

Austral Islands

Pitcairn I.

Sala y Gómez Ridge

Easter I.

Gran Chaco

Paraná R.

Tonga Trench

Kermadec Tr.

Southwest Pacific Basin

Challenger Fracture Zone

Juan Fernández Is.

Mt. Aconcagua 22,834 ft (6,960 m)

Pampas

Patagonia

Rio de la Plata

Rio Gr Ris

Louisville Ridge

0	1,000	2,000 Miles
0	1,000	2,000 Kilometers

Argentine Plain

Southeast Pacific Basin

Humboldt Plain

Falkland Is. (Malvinas)

So Geor

Eltanin Fracture Zone

Strait of Magellan

South Georgia Ridge

Udintsev Fracture Zone

Cape Horn

Drake Passage

Pacific-Antarctic Ridge

Antarctic Circle